KIRK-OTHMER

ENCYCLOPEDIA OF CHEMICAL TECHNOLOGY

FOURTH EDITION

VOLUME **11**

FLAVOR CHARACTERIZATION
TO
FUEL CELLS

EXECUTIVE EDITOR
Jacqueline I. Kroschwitz

EDITOR
Mary Howe-Grant

KIRK-OTHMER

ENCYCLOPEDIA OF CHEMICAL TECHNOLOGY

FOURTH EDITION

VOLUME 11

FLAVOR CHARACTERIZATION
TO
FUEL CELLS

A Wiley-Interscience Publication
JOHN WILEY & SONS

New York • Chichester • Brisbane • Toronto • Singapore

Copyright © 1994 by John Wiley & Sons, Inc.

Library of Congress Cataloging-in-Publication Data

Encyclopedia of chemical technology / executive editor, Jacqueline
    I. Kroschwitz; editor, Mary Howe-Grant.—4th ed.
        p. cm.
    At head of title: Kirk-Othmer.
    "A Wiley-Interscience publication."
    Includes index.
    Contents: v. 11, Flavor characterization to fuel cells.
    ISBN 0-471-52680-0 (v. 11)
    1. Chemistry, Technical—Encyclopedias. I. Kirk, Raymond E.
    (Raymond Eller), 1890–1957. II. Othmer, Donald F. (Donald
    Frederick), 1904– . III. Kroschwitz, Jacqueline I., 1942– .
    IV. Howe-Grant, Mary, 1943– . V. Title: Kirk-Othmer encyclopedia
    of chemical technology.
    TP9.E685   1992   91-16789
    660'.03—dc20

Printed in the United States of America

10 9 8 7 6 5 4 3

CONTENTS

EDITORIAL STAFF FOR VOLUME 11

Executive Editor: **Jacqueline I. Kroschwitz**
Editor: **Mary Howe-Grant**
Assistant Editor: **Cathleen A. Treacy**
Editorial Supervisor: **Lindy J. Humphreys**
Copy Editors: **Christine Punzo**
            Lawrence Altieri

CONTRIBUTORS TO VOLUME 11

Terry E. Acree, *Cornell University, Geneva, New York,* Flavor characterization

Amos A. Avidan, *Mobil Research and Development Corporation, Paulsboro, New Jersey,* Fluidization

Webb I. Bailey, *Air Products and Chemicals, Inc., Allentown, Pennsylvania,* Halogens (under Fluorine compounds, organic)

William X. Bajzer, *Dow Corning Corporation, Midland, Michigan,* Introduction; Poly(fluorosilicones) (both under Fluorine compounds, organic)

vii

Max M. Boudakian, *Chemical Consultant, Pittsford, New York,* Fluorinated aromatic compounds (under Fluorine compounds, organic)

Aaron L. Brody, *Rubbright-Brody, Inc., Devon, Pennsylvania,* Food packaging

Thomas H. Burgess, *Fischer & Porter Company, Warminster, Pennsylvania,* Flow measurement

Elton J. Cairns, *University of California, Berkeley,* Fuel cells

Barry A. Crouch, *E. I. du Pont de Nemours & Co., Inc., Wilmington, Delaware,* Fracture mechanics

Julius E. Dohany, *Consultant, Berwyn, Pennsylvania,* Poly(vinylidene fluoride) (under Fluorine compounds, organic)

David J. Drury, *BP Chemicals, Ltd., Middlesex, United Kingdom,* Formic acid (under Formic acid and derivatives)

Douglas J. Durian, *University of California, Los Angeles,* Foams

S. Ebnesajjad, *E. I. du Pont de Nemours & Co., Inc., Wilmington, Delaware,* Poly(vinyl fluoride) (under Fluorine compounds, organic)

Arthur J. Elliott, *Halocarbon Products Corporation, North Augusta, South Carolina,* Bromotrifluoroethylene; Fluorinated acetic acids; Fluoroethanols (all under Fluorine compounds, organic)

Francis E. Evans, *Consultant, Hamburg, New York,* Boron, boron trifluoride; Sulfur (both under Fluorine compounds, inorganic)

Daniel F. Farkas, *Oregon State University, Corvallis,* Food processing

Richard E. Fernandez, *E. I. du Pont de Nemours & Co., Inc., Wilmington, Delaware,* Fluorinated aliphatic compounds (under Fluorine compounds, organic)

Frank Fischetti, Jr., *Craftmaster Flavor Technology Inc., Amityville, New York,* Flavors (under Flavors and spices)

Barry A. J. Fisher, *Scientific Services Bureau, Los Angeles, California,* Forensic chemistry

Richard M. Flynn, *3M Company, St. Paul, Minnesota,* Fluoroethers and fluoroamines (under Fluorine compounds, organic)

Peter R. Foster, *Scottish National Blood Transfusion Service, Edinburgh, United Kingdom,* Plasma fractionation (under Fractionation, blood)

Leslie J. Friedman, *Arthur D. Little, Inc., Cambridge, Massachusetts,* Food additives

Subhash V. Gangal, *E. I. du Pont de Nemours & Co., Inc., Wilmington, Delaware,* Perfluorinated ethylene–propylene copolymers; Polytetrafluoroethylene; Tetrafluoroethylene–ethylene copolymers; Tetrafluoroethylene–perfluorovinyl ether copolymers (all under Fluorine compounds, organic)

H. Robert Gerberich, *Hoechst-Celanese, Corpus Christi, Texas,* Formaldehyde

C. Gail Greenwald, *Arthur D. Little, Inc., Cambridge, Massachusetts,* Food additives

Howard I. Heitner, *Cytec Industries, Stamford, Connecticut,* Flocculating agents

Philip B. Henderson, *Air Products and Chemicals, Inc., Allentown, Pennsylvania,* Nitrogen; Tungsten (both under Fluorine compounds, inorganic)

A. Höhn, *BASF AG, Lugwigshafen, Germany,* Formamide (under Formic acid and derivatives)

Fred H. Hoskins, *Washington State University, Pullman,* Food toxicants, naturally occurring

Yung K. Kim, *Dow Corning Corporation, Midland, Michigan,* Introduction; Poly(fluorosilicones) (both under Fluorine compounds, organic)

Desmond F. King, *Chevron Research and Technology Company, Richmond, California,* Fluidization

Kimio Kinoshita, *University of California, Berkeley,* Fuel cells

Ted M. Knowlton, *Institute of Gas Technology, Chicago, Illinois,* Fluidization

Jack L. Kosmala, *3M Company, St. Paul, Minnesota,* Polychlorotrifluoroethylene (under Fluorine compounds, organic)

Richard J. Lagow, *University of Texas at Austin,* Direct fluorination (under Fluorine compounds, organic)

Charles B. Lindahl, *Elf Atochem North America, Inc., Tulsa, Oklahoma,* Introduction; Antimony; Arsenic; Barium; Calcium; Germanium; Phosphorus; Tantalum; Tin; Zinc (all under Fluorine compounds, inorganic)

John H. Litchfield, *Battelle Memorial Institute, Columbus, Ohio,* Foods, nonconventional

Tariq Mahmood, *Elf Atochem North America, Inc., Tulsa, Oklahoma,* Introduction; Antimony; Arsenic; Calcium; Germanium; Phosphorus; Tantalum; Tin; Zinc (all under Fluorine compounds, inorganic)

Ganpat Mani, *AlliedSignal Inc., Morristown, New Jersey,* Boron, boron trifluoride; Sulfur (both under Fluorine compounds, inorganic)

John A. Marsella, *Air Products and Chemicals, Inc., Allentown, Pennsylvania,* Dimethylformamide (under Formic acid and derivatives)

Harold J. McElhone, Jr., *Ciba-Geigy Corporation, Greensboro, North Carolina,* Fluorescent whitening agents

Dayal T. Meshri, *Advance Research Chemicals, Inc., Catoosa, Oklahoma,* Aluminum; Cobalt; Copper; Iron; Lead; Mercury; Molybdenum; Nickel; Rhenium; Silver; Titanium; Zirconium (all under Fluorine compounds, inorganic)

George H. Millet, *3M Company, St. Paul, Minnesota,* Polychlorotrifluoroethylene (under Fluorine compounds, organic)

Werner H. Mueller, *Hoechst-Celanese Corporation, Charlotte, North Carolina,* Sodium (under Fluorine compounds, inorganic)

George A. Olah, *University of Southern California, Los Angeles,* Friedel-Crafts reactions

John R. Papcun, *Atotech, Cleveland, Ohio,* Ammonium; Boron, fluoroboric acid and fluoroborates; Lithium; Magnesium; Potassium (all under Fluorine compounds, inorganic)

Mel Pell, *E. I. du Pont de Nemours & Co., Inc., Wilmington, Delaware,* Fluidization

G. K. Surya Prakash, *University of Southern California, Los Angeles,* Friedel-Crafts reactions

V. Prakash Reddy, *University of Southern California, Los Angeles,* Friedel-Crafts reactions

Paul R. Resnick, *E. I. du Pont de Nemours & Co., Inc., Wilmington, Delaware,* Perfluoroepoxides (under Fluorine compounds, organic)

James A. Rogers, *Consultant, Ramsey, New Jersey,* Spices (under Flavors and spices)

Patricia Savu, *3M Company, St. Paul, Minnesota,* Fluorinated higher carboxylic acids; Perfluoroalkanesulfonic acids (both under Fluorine compounds, organic)

George C. Seaman, *Hoechst-Celanese, Corpus Christi, Texas,* Formaldehyde

Philip E. Shaw, *U.S. Department of Agriculture, Winter Haven, Florida,* Fruit juices

George Shia, *AlliedSignal, Buffalo, New York,* Fluorine

Jean'ne M. Shreeve, *University of Idaho, Moscow,* Oxygen (under Fluorine compounds, inorganic)

Theodore Hein Smit Sibinga, *Haemonetics, Braintree, Massachusetts,* Cell separation (under Fractionation, blood)

Bruce E. Smart, *E. I. du Pont de Nemours & Co., Inc., Wilmington, Delaware,* Fluorinated aliphatic compounds (under Fluorine compounds, organic)

Robert A. Smith, *AlliedSignal, Morristown, New Jersey,* Hydrogen (under Fluorine compounds, inorganic)

L. G. Snow, *E. I. du Pont de Nemours & Co., Inc., Wilmington, Delaware,* Poly(vinyl fluoride) (under Fluorine compounds, organic)

I. J. Solomon, *IIT Research Institute, Chicago, Illinois,* Oxygen (under Fluorine compounds, inorganic)

J. S. Son, *Shell Development Company, Houston, Texas,* Fluid mechanics

Kyung W. Suh, *The Dow Chemical Company, Granville, Ohio,* Foamed plastics

David A. Weitz, *Exxon Research & Engineering Company, Annandale, New Jersey,* Foams

Andrew J. Woytek, *Air Products and Chemicals, Inc., Allentown, Pennsylvania,* Halogens; Nitrogen; Tungsten (all under Fluorine compounds, inorganic)

Baki Yarar, *Colorado School of Mines, Golden, Colorado,* Flotation

NOTE ON CHEMICAL ABSTRACTS SERVICE REGISTRY NUMBERS AND NOMENCLATURE

Chemical Abstracts Service (CAS) Registry Numbers are unique numerical identifiers assigned to substances recorded in the CAS Registry System. They appear in brackets in the *Chemical Abstracts* (CA) substance and formula indexes following the names of compounds. A single compound may have synonyms in the chemical literature. A simple compound like phenethylamine can be named β-phenylethylamine or, as in *Chemical Abstracts*, benzeneethanamine. The usefulness of the *Encyclopedia* depends on accessibility through the most common correct name of a substance. Because of this diversity in nomenclature careful attention has been given to the problem in order to assist the reader as much as possible, especially in locating the systematic CA index name by means of the Registry Number. For this purpose, the reader may refer to the CAS Registry Handbook—Number Section which lists in numerical order the Registry Number with the *Chemical Abstracts* index name and the molecular formula; eg, **458-88-8**, Piperidine, 2-propyl-, (*S*)-, $C_8H_{17}N$; in the *Encyclopedia* this compound would be found under its common name, coniine [*458-88-8*]. Alternatively, this information can be retrieved electronically from CAS Online. In many cases molecular formulas have also been provided in the *Encyclopedia* text to facilitate electronic searching. The Registry Number is a valuable link for the reader in retrieving additional published information on substances and also as a point of access for on-line data bases.

In all cases, the CAS Registry Numbers have been given for title compounds in articles and for all compounds in the index. All specific substances indexed in *Chemical Abstracts* since 1965 are included in the CAS Registry System as are a large number of substances derived from a variety of reference works. The CAS Registry System identifies a substance on the basis of an unambiguous computer-language description of its molecular structure including stereochemical detail. The Registry Number is a machine-checkable number (like a Social Security number) assigned in sequential order to each substance as it enters the registry system. The value of the number lies in the fact that it is a concise and unique means of substance identification, which is independent of, and therefore bridges, many systems of chemical nomenclature. For polymers, one Registry Number may be used for the entire family; eg, polyoxyethylene (20) sorbitan monolaurate has the same number as all of its polyoxyethylene homologues.

Cross-references are inserted in the index for many common names and for some systematic names. Trademark names appear in the index. Names that are incorrect, misleading, or ambiguous are avoided. Formulas are given very frequently in the text to help in identifying compounds. The spelling and form used, even for industrial names, follow American chemical usage, but not always the usage of *Chemical Abstracts* (eg, *coniine* is used instead of *(S)-2-propylpiperidine*, *aniline* instead of *benzenamine*, and *acrylic acid* instead of *2-propenoic acid*).

There are variations in representation of rings in different disciplines. The dye industry does not designate aromaticity or double bonds in rings. All double bonds and aromaticity are shown in the *Encyclopedia* as a matter of course. For example, tetralin has an aromatic ring and a saturated ring and its structure

appears in the *Encyclopedia* with its common name, Registry Number enclosed in brackets, and parenthetical CA index name, ie, tetralin [*119-64-2*] (1,2,3,4-tetrahydronaphthalene). With names and structural formulas, and especially with CAS Registry Numbers, the aim is to help the reader have a concise means of substance identification.

CONVERSION FACTORS, ABBREVIATIONS, AND UNIT SYMBOLS

SI Units (Adopted 1960)

The International System of Units (abbreviated SI), is being implemented throughout the world. This measurement system is a modernized version of the MKSA (meter, kilogram, second, ampere) system, and its details are published and controlled by an international treaty organization (The International Bureau of Weights and Measures) (1).

SI units are divided into three classes:

BASE UNITS

length	meter[†] (m)
mass	kilogram (kg)
time	second (s)
electric current	ampere (A)
thermodynamic temperature[‡]	kelvin (K)
amount of substance	mole (mol)
luminous intensity	candela (cd)

SUPPLEMENTARY UNITS

plane angle	radian (rad)
solid angle	steradian (sr)

[†]The spellings "metre" and "litre" are preferred by ASTM; however, "-er" is used in the *Encyclopedia*.

[‡]Wide use is made of Celsius temperature (t) defined by

$$t = T - T_0$$

where T is the thermodynamic temperature, expressed in kelvin, and $T_0 = 273.15$ K by definition. A temperature interval may be expressed in degrees Celsius as well as in kelvin.

DERIVED UNITS AND OTHER ACCEPTABLE UNITS

These units are formed by combining base units, supplementary units, and other derived units (2–4). Those derived units having special names and symbols are marked with an asterisk in the list below.

Quantity	Unit	Symbol	Acceptable equivalent
*absorbed dose	gray	Gy	J/kg
acceleration	meter per second squared	m/s^2	
*activity (of a radionuclide)	becquerel	Bq	1/s
area	square kilometer	km^2	
	square hectometer	hm^2	ha (hectare)
	square meter	m^2	
concentration (of amount of substance)	mole per cubic meter	mol/m^3	
current density	ampere per square meter	A//m^2	
density, mass density	kilogram per cubic meter	kg/m^3	g/L; mg/cm^3
dipole moment (quantity)	coulomb meter	C·m	
*dose equivalent	sievert	Sv	J/kg
*electric capacitance	farad	F	C/V
*electric charge, quantity of electricity	coulomb	C	A·s
electric charge density	coulomb per cubic meter	C/m^3	
*electric conductance	siemens	S	A/V
electric field strength	volt per meter	V/m	
electric flux density	coulomb per square meter	C/m^2	
*electric potential, potential difference, electromotive force	volt	V	W/A
*electric resistance	ohm	Ω	V/A
*energy, work, quantity of heat	megajoule	MJ	
	kilojoule	kJ	
	joule	J	N·m
	electronvolt[†]	eV[†]	
	kilowatt-hour[†]	kW·h[†]	
energy density	joule per cubic meter	J/m^3	
*force	kilonewton	kN	
	newton	N	kg·m/s^2

[†]This non-SI unit is recognized by the CIPM as having to be retained because of practical importance or use in specialized fields (1).

Quantity	Unit	Symbol	Acceptable equivalent
*frequency	megahertz	MHz	
	hertz	Hz	1/s
heat capacity, entropy	joule per kelvin	J/K	
heat capacity (specific), specific entropy	joule per kilogram kelvin	J/(kg·K)	
heat transfer coefficient	watt per square meter kelvin	W/(m²·K)	
*illuminance	lux	lx	lm/m²
*inductance	henry	H	Wb/A
linear density	kilogram per meter	kg/m	
luminance	candela per square meter	cd/m²	
*luminous flux	lumen	lm	cd·sr
magnetic field strength	ampere per meter	A/m	
*magnetic flux	weber	Wb	V·s
*magnetic flux density	tesla	T	Wb/m²
molar energy	joule per mole	J/mol	
molar entropy, molar heat capacity	joule per mole kelvin	J/(mol·K)	
moment of force, torque	newton meter	N·m	
momentum	kilogram meter per second	kg·m/s	
permeability	henry per meter	H/m	
permittivity	farad per meter	F/m	
*power, heat flow rate, radiant flux	kilowatt	kW	
	watt	W	J/s
power density, heat flux density, irradiance	watt per square meter	W/m²	
*pressure, stress	megapascal	MPa	
	kilopascal	kPa	
	pascal	Pa	N/m²
sound level	decibel	dB	
specific energy	joule per kilogram	J/kg	
specific volume	cubic meter per kilogram	m³/kg	
surface tension	newton per meter	N/m	
thermal conductivity	watt per meter kelvin	W/(m·K)	
velocity	meter per second	m/s	
	kilometer per hour	km/h	
viscosity, dynamic	pascal second	Pa·s	
	millipascal second	mPa·s	
viscosity, kinematic	square meter per second	m²/s	
	square millimeter per second	mm²/s	

Quantity	Unit	Symbol	Acceptable equivalent
volume	cubic meter	m^3	
	cubic decimeter	dm^3	L (liter) (5)
	cubic centimeter	cm^3	mL
wave number	1 per meter	m^{-1}	
	1 per centimeter	cm^{-1}	

In addition, there are 16 prefixes used to indicate order of magnitude, as follows:

Multiplication factor	Prefix	Symbol	Note
10^{18}	exa	E	
10^{15}	peta	P	
10^{12}	tera	T	
10^{9}	giga	G	
10^{6}	mega	M	
10^{3}	kilo	k	
10^{2}	hecto	h^a	[a]Although hecto, deka, deci, and centi
10	deka	da^a	are SI prefixes, their use should be
10^{-1}	deci	d^a	avoided except for SI unit-multiples
10^{-2}	centi	c^a	for area and volume and nontech-
10^{-3}	milli	m	nical use of centimeter, as for body
10^{-6}	micro	μ	and clothing measurement.
10^{-9}	nano	n	
10^{-12}	pico	p	
10^{-15}	femto	f	
10^{-18}	atto	a	

For a complete description of SI and its use the reader is referred to ASTM E 380 (4) and the article UNITS AND CONVERSION FACTORS which appears in Vol. 24.

A representative list of conversion factors from non-SI to SI units is presented herewith. Factors are given to four significant figures. Exact relationships are followed by a dagger. A more complete list is given in the latest editions of ASTM E 380 (4) and ANSI Z210.1 (6).

Conversion Factors to SI Units

To convert from	To	Multiply by
acre	square meter (m^2)	4.047×10^3
angstrom	meter (m)	$1.0 \times 10^{-10\dagger}$
are	square meter (m^2)	$1.0 \times 10^{2\dagger}$

[†]Exact.

To convert from	To	Multiply by
astronomical unit	meter (m)	1.496×10^{11}
atmosphere, standard	pascal (Pa)	1.013×10^{5}
bar	pascal (Pa)	$1.0 \times 10^{5\dagger}$
barn	square meter (m²)	$1.0 \times 10^{-28\dagger}$
barrel (42 U.S. liquid gallons)	cubic meter (m³)	0.1590
Bohr magneton (μ_B)	J/T	9.274×10^{-24}
Btu (International Table)	joule (J)	1.055×10^{3}
Btu (mean)	joule (J)	1.056×10^{3}
Btu (thermochemical)	joule (J)	1.054×10^{3}
bushel	cubic meter (m³)	3.524×10^{-2}
calorie (International Table)	joule (J)	4.187
calorie (mean)	joule (J)	4.190
calorie (thermochemical)	joule (J)	$4.184^{\dagger}$
centipoise	pascal second (Pa·s)	$1.0 \times 10^{-3\dagger}$
centistokes	square millimeter per second (mm²/s)	$1.0^{\dagger}$
cfm (cubic foot per minute)	cubic meter per second (m³/s)	4.72×10^{-4}
cubic inch	cubic meter (m³)	1.639×10^{-5}
cubic foot	cubic meter (m³)	2.832×10^{-2}
cubic yard	cubic meter (m³)	0.7646
curie	becquerel (Bq)	$3.70 \times 10^{10\dagger}$
debye	coulomb meter (C·m)	3.336×10^{-30}
degree (angle)	radian (rad)	1.745×10^{-2}
denier (international)	kilogram per meter (kg/m)	1.111×10^{-7}
	tex‡	0.1111
dram (apothecaries')	kilogram (kg)	3.888×10^{-3}
dram (avoirdupois)	kilogram (kg)	1.772×10^{-3}
dram (U.S. fluid)	cubic meter (m³)	3.697×10^{-6}
dyne	newton (N)	$1.0 \times 10^{-5\dagger}$
dyne/cm	newton per meter (N/m)	$1.0 \times 10^{-3\dagger}$
electronvolt	joule (J)	1.602×10^{-19}
erg	joule (J)	$1.0 \times 10^{-7\dagger}$
fathom	meter (m)	1.829
fluid ounce (U.S.)	cubic meter (m³)	2.957×10^{-5}
foot	meter (m)	$0.3048^{\dagger}$
footcandle	lux (lx)	10.76
furlong	meter (m)	2.012×10^{-2}
gal	meter per second squared (m/s²)	$1.0 \times 10^{-2\dagger}$
gallon (U.S. dry)	cubic meter (m³)	4.405×10^{-3}
gallon (U.S. liquid)	cubic meter (m³)	3.785×10^{-3}
gallon per minute (gpm)	cubic meter per second (m³/s)	6.309×10^{-5}
	cubic meter per hour (m³/h)	0.2271

†Exact.
‡See footnote on p. xiii.

To convert from	To	Multiply by
gauss	tesla (T)	1.0×10^{-4}
gilbert	ampere (A)	0.7958
gill (U.S.)	cubic meter (m³)	1.183×10^{-4}
grade	radian	1.571×10^{-2}
grain	kilogram (kg)	6.480×10^{-5}
gram force per denier	newton per tex (N/tex)	8.826×10^{-2}
hectare	square meter (m²)	$1.0 \times 10^{4\dagger}$
horsepower (550 ft·lbf/s)	watt (W)	7.457×10^{2}
horespower (boiler)	watt (W)	9.810×10^{3}
horsepower (electric)	watt (W)	$7.46 \times 10^{2\dagger}$
hundredweight (long)	kilogram (kg)	50.80
hundredweight (short)	kilogram (kg)	45.36
inch	meter (m)	$2.54 \times 10^{-2\dagger}$
inch of mercury (32°F)	pascal (Pa)	3.386×10^{3}
inch of water (39.2°F)	pascal (Pa)	2.491×10^{2}
kilogram-force	newton (N)	9.807
kilowatt hour	megajoule (MJ)	$3.6^{\dagger}$
kip	newton(N)	4.448×10^{3}
knot (international)	meter per second (m/S)	0.5144
lambert	candela per square meter (cd/m³)	3.183×10^{3}
league (British nautical)	meter (m)	5.559×10^{3}
league (statute)	meter (m)	4.828×10^{3}
light year	meter (m)	9.461×10^{15}
liter (for fluids only)	cubic meter (m³)	$1.0 \times 10^{-3\dagger}$
maxwell	weber (Wb)	$1.0 \times 10^{-8\dagger}$
micron	meter (m)	$1.0 \times 10^{-6\dagger}$
mil	meter (m)	$2.54 \times 10^{-5\dagger}$
mile (statute)	meter (m)	1.609×10^{3}
mile (U.S. nautical)	meter (m)	$1.852 \times 10^{3\dagger}$
mile per hour	meter per second (m/s)	0.4470
millibar	pascal (Pa)	1.0×10^{2}
millimeter of mercury (0°C)	pascal (Pa)	$1.333 \times 10^{2\dagger}$
minute (angular)	radian	2.909×10^{-4}
myriagram	kilogram (kg)	10
myriameter	kilometer (km)	10
oersted	ampere per meter (A/m)	79.58
ounce (avoirdupois)	kilogram (kg)	2.835×10^{-2}
ounce (troy)	kilogram (kg)	3.110×10^{-2}
ounce (U.S. fluid)	cubic meter (m³)	2.957×10^{-5}
ounce-force	newton (N)	0.2780
peck (U.S.)	cubic meter (m³)	8.810×10^{-3}
pennyweight	kilogram (kg)	1.555×10^{-3}
pint (U.S. dry)	cubic meter (m³)	5.506×10^{-4}
pint (U.S. liquid)	cubic meter (m³)	4.732×10^{-4}

†Exact.

To convert from	To	Multiply by
poise (absolute viscosity)	pascal second (Pa·s)	$0.10^{\dagger}$
pound (avoirdupois)	kilogram (kg)	0.4536
pound (troy)	kilogram (kg)	0.3732
poundal	newton (N)	0.1383
pound-force	newton (N)	4.448
pound force per square inch (psi)	pascal (Pa)	6.895×10^3
quart (U.S. dry)	cubic meter (m³)	1.101×10^{-3}
quart (U.S. liquid)	cubic meter (m³)	9.464×10^{-4}
quintal	kilogram (kg)	$1.0 \times 10^{2\dagger}$
rad	gray (Gy)	$1.0 \times 10^{-2\dagger}$
rod	meter (m)	5.029
roentgen	coulomb per kilogram (C/kg)	2.58×10^{-4}
second (angle)	radian (rad)	$4.848 \times 10^{-6\dagger}$
section	square meter (m²)	2.590×10^6
slug	kilogram (kg)	14.59
spherical candle power	lumen (lm)	12.57
square inch	square meter (m²)	6.452×10^{-4}
square foot	square meter (m²)	9.290×10^{-2}
square mile	square meter (m²)	2.590×10^6
square yard	square meter (m²)	0.8361
stere	cubic meter (m³)	$1.0^{\dagger}$
stokes (kinematic viscosity)	square meter per second (m²/s)	$1.0 \times 10^{-4\dagger}$
tex	kilogram per meter (kg/m)	$1.0 \times 10^{-6\dagger}$
ton (long, 2240 pounds)	kilogram (kg)	1.016×10^3
ton (metric) (tonne)	kilogram (kg)	$1.0 \times 10^{3\dagger}$
ton (short, 2000 pounds)	kilogram (kg)	9.072×10^2
torr	pascal (Pa)	1.333×10^2
unit pole	weber (Wb)	1.257×10^{-7}
yard	meter (m)	$0.9144^{\dagger}$

†Exact.

Abbreviations and Unit Symbols

Following is a list of common abbreviations and unit symbols used in the *Encyclopedia*. In general they agree with those listed in *American National Standard Abbreviations for Use on Drawings and in Text (ANSI Y1.1)* (6) and *American National Standard Letter Symbols for Units in Science and Technology (ANSI Y10)* (6). Also included is a list of acronyms for a number of private and government organizations as well as common industrial solvents, polymers, and other chemicals.

Rules for Writing Unit Symbols (4):

1. Unit symbols are printed in upright letters (roman) regardless of the type style used in the surrounding text.
2. Unit symbols are unaltered in the plural.
3. Unit symbols are not followed by a period except when used at the end of a sentence.
4. Letter unit symbols are generally printed lower-case (for example, cd for candela) unless the unit name has been derived from a proper name, in which case the first letter of the symbol is capitalized (W, Pa). Prefixes and unit symbols retain their prescribed form regardless of the surrounding typography.
5. In the complete expression for a quantity, a space should be left between the numerical value and the unit symbol. For example, write 2.37 lm, *not* 2.37lm, and 35 mm, *not* 35mm. When the quantity is used in an adjectival sense, a hyphen is often used, for example, 35-mm film. *Exception:* No space is left between the numerical value and the symbols for degree, minute, and second of plane angle, degree Celsius, and the percent sign.
6. No space is used between the prefix and unit symbol (for example, kg).
7. Symbols, not abbreviations, should be used for units. For example, use "A," not "amp," for ampere.
8. When multiplying unit symbols, use a raised dot:

$$N \cdot m \quad \text{for} \quad \text{newton meter}$$

In the case of W·h, the dot may be omitted, thus:

$$Wh$$

An exception to this practice is made for computer printouts, automatic typewriter work, etc, where the raised dot is not possible, and a dot on the line may be used.
9. When dividing unit symbols, use one of the following forms:

$$m/s \quad or \quad m \cdot s^{-1} \quad or \quad \frac{m}{s}$$

In no case should more than one slash be used in the same expression unless parentheses are inserted to avoid ambiguity. For example, write:

$$J/(mol \cdot K) \quad or \quad J \cdot mol^{-1} \cdot K^{-1} \quad or \quad (J/mol)/K$$

but *not*

$$J/mol/K$$

10. Do not mix symbols and unit names in the same expression. Write:

$$\text{joules per kilogram} \quad or \quad \text{J/kg} \quad or \quad \text{J·kg}^{-1}$$

but *not*

$$\text{joules/kilogram} \quad nor \quad \text{joules/kg} \quad nor \quad \text{joules·kg}^{-1}$$

ABBREVIATIONS AND UNITS

A	ampere	AOAC	Association of Official Analytical Chemists
A	anion (eg, HA)		
A	mass number	AOCS	Americal Oil Chemists' Society
a	atto (prefix for 10^{-18})		
AATCC	American Association of Textile Chemists and Colorists	APHA	American Public Health Association
		API	American Petroleum Institute
ABS	acrylonitrile–butadiene–styrene		
		aq	aqueous
abs	absolute	Ar	aryl
ac	alternating current, *n.*	ar-	aromatic
a-c	alternating current, *adj.*	as-	asymmetric(al)
ac-	alicyclic	ASHRAE	American Society of Heating, Refrigerating, and Air Conditioning Engineers
acac	acetylacetonate		
ACGIH	American Conference of Governmental Industrial Hygienists		
		ASM	American Society for Metals
ACS	American Chemical Society	ASME	American Society of Mechanical Engineers
AGA	American Gas Association		
Ah	ampere hour	ASTM	American Society for Testing and Materials
AIChE	American Institute of Chemical Engineers		
		at no.	atomic number
AIME	American Institute of Mining, Metallurgical, and Petroleum Engineers	at wt	atomic weight
		av(g)	average
		AWS	American Welding Society
		b	bonding orbital
AIP	American Institute of Physics	bbl	barrel
		bcc	body-centered cubic
AISI	American Iron and Steel Institute	BCT	body-centered tetragonal
		Bé	Baumé
alc	alcohol(ic)	BET	Brunauer-Emmett-Teller (adsorption equation)
Alk	alkyl		
alk	alkaline (not alkali)	bid	twice daily
amt	amount	Boc	*t*-butyloxycarbonyl
amu	atomic mass unit	BOD	biochemical (biological) oxygen demand
ANSI	American National Standards Institute		
		bp	boiling point
AO	atomic orbital	Bq	becquerel

C	coulomb	DIN	Deutsche Industrie Normen
°C	degree Celsius		
C-	denoting attachment to carbon	*dl*-; DL-	racemic
		DMA	dimethylacetamide
c	centi (prefix for 10^{-2})	DMF	dimethylformamide
c	critical	DMG	dimethyl glyoxime
ca	circa (approximately)	DMSO	dimethyl sulfoxide
cd	candela; current density; circular dichroism	DOD	Department of Defense
		DOE	Department of Energy
CFR	Code of Federal Regulations	DOT	Department of Transportation
cgs	centimeter-gram-second	DP	degree of polymerization
CI	Color Index	dp	dew point
cis-	isomer in which substituted groups are on same side of double bond between C atoms	DPH	diamond pyramid hardness
		dstl(d)	distill(ed)
		dta	differential thermal analysis
cl	carload		
cm	centimeter	(*E*)-	entgegen; opposed
cmil	circular mil	ϵ	dielectric constant (unitless number)
cmpd	compound		
CNS	central nervous system	*e*	electron
CoA	coenzyme A	ECU	electrochemical unit
COD	chemical oxygen demand	ed.	edited, edition, editor
coml	commercial(ly)	ED	effective dose
cp	chemically pure	EDTA	ethylenediaminetetra-acetic acid
cph	close-packed hexagonal		
CPSC	Consumer Product Safety Commission	emf	electromotive force
		emu	electromagnetic unit
cryst	crystalline	en	ethylene diamine
cub	cubic	eng	engineering
D	debye	EPA	Environmental Protection Agency
D-	denoting configurational relationship		
		epr	electron paramagnetic resonance
d	differential operator		
d	day; deci (prefix for 10^{-1})	eq.	equation
d-	*dextro*-, dextrorotatory	esca	electron spectroscopy for chemical analysis
da	deka (prefix for 10^1)		
dB	decibel	esp	especially
dc	direct current, *n.*	esr	electron-spin resonance
d-c	direct current, *adj.*	est(d)	estimate(d)
dec	decompose	estn	estimation
detd	determined	esu	electrostatic unit
detn	determination	exp	experiment, experimental
Di	didymium, a mixture of all lanthanons	ext(d)	extract(ed)
		F	farad (capacitance)
dia	diameter	*F*	faraday (96,487 C)
dil	dilute	f	femto (prefix for 10^{-15})

FAO	Food and Agriculture Organization (United Nations)	hyd	hydrated, hydrous
		hyg	hygroscopic
fcc	face-centered cubic	Hz	hertz
FDA	Food and Drug Administration	i (eg, Pr^i)	iso (eg, isopropyl)
		i-	inactive (eg, i-methionine)
FEA	Federal Energy Administration	IACS	International Annealed Copper Standard
FHSA	Federal Hazardous Substances Act	ibp	initial boiling point
		IC	integrated circuit
fob	free on board	ICC	Interstate Commerce Commission
fp	freezing point		
FPC	Federal Power Commission	ICT	International Critical Table
		ID	inside diameter; infective dose
FRB	Federal Reserve Board		
frz	freezing	ip	intraperitoneal
G	giga (prefix for 10^9)	IPS	iron pipe size
G	gravitational constant = 6.67×10^{11} N·m^2/kg^2	ir	infrared
		IRLG	Interagency Regulatory Liaison Group
g	gram		
(g)	gas, only as in $H_2O(g)$	ISO	International Organization Standardization
g	gravitational acceleration		
gc	gas chromatography	ITS-90	International Temperature Scale (NIST)
gem-	geminal		
glc	gas–liquid chromatography	IU	International Unit
g-mol wt; gmw	gram-molecular weight	IUPAC	International Union of Pure and Applied Chemistry
GNP	gross national product	IV	iodine value
gpc	gel-permeation chromatography	iv	intravenous
		J	joule
GRAS	Generally Recognized as Safe	K	kelvin
		k	kilo (prefix for 10^3)
grd	ground	kg	kilogram
Gy	gray	L	denoting configurational relationship
H	henry		
h	hour; hecto (prefix for 10^2)	L	liter (for fluids only) (5)
ha	hectare	l-	$levo$-, levorotatory
HB	Brinell hardness number	(l)	liquid, only as in NH_3(l)
Hb	hemoglobin	LC$_{50}$	conc lethal to 50% of the animals tests
hcp	hexagonal close-packed		
hex	hexagonal	LCAO	linear combination of atomic orbitals
HK	Knoop hardness number		
hplc	high performance liquid chromatography	lc	liquid chromatography
		LCD	liquid crystal display
HRC	Rockwell hardness (C scale)	lcl	less than carload lots
		LD$_{50}$	dose lethal to 50% of the animals tested
HV	Vickers hardness number		

LED	light-emitting diode	N-	denoting attachment to nitrogen
liq	liquid		
lm	lumen	n (as n_D^{20})	index of refraction (for 20°C and sodium light)
ln	logarithm (natural)		
LNG	liquefied natural gas	^{n} (as Bu^n),	
log	logarithm (common)	n-	normal (straight-chain structure)
LPG	liquefied petroleum gas		
ltl	less than truckload lots	n	neutron
lx	lux	n	nano (prefix for 10^9)
M	mega (prefix for 10^6); metal (as in MA)	na	not available
		NAS	National Academy of Sciences
M	molar; actual mass		
$\overline{M}_w$	weight-average mol wt	NASA	National Aeronautics and Space Administration
$\overline{M}_n$	number-average mol wt		
m	meter; milli (prefix for 10^{-3})	nat	natural
		ndt	nondestructive testing
m	molal	neg	negative
m-	meta	NF	*National Formulary*
max	maximum	NIH	National Institutes of Health
MCA	Chemical Manufacturers' Association (was Manufacturing Chemists Association)	NIOSH	National Institute of Occupational Safety and Health
MEK	methyl ethyl ketone	NIST	National Institute of Standards and Technology (formerly National Bureau of Standards)
meq	milliequivalent		
mfd	manufactured		
mfg	manufacturing		
mfr	manufacturer		
MIBC	methyl isobutyl carbinol	nmr	nuclear magnetic resonance
MIBK	methyl isobutyl ketone		
MIC	minimum inhibiting concentration	NND	New and Nonofficial Drugs (AMA)
min	minute; minimum	no.	number
mL	milliliter	NOI-(BN)	not otherwise indexed (by name)
MLD	minimum lethal dose		
MO	molecular orbital	NOS	not otherwise specified
mo	month	nqr	nuclear quadruple resonance
mol	mole		
mol wt	molecular weight	NRC	Nuclear Regulatory Commission; National Research Council
mp	melting point		
MR	molar refraction		
ms	mass spectrometry	NRI	New Ring Index
MSDS	material safety data sheet	NSF	National Science Foundation
mxt	mixture		
μ	micro (prefix for 10^{-6})	NTA	nitrilotriacetic acid
N	newton (force)	NTP	normal temperature and pressure (25°C and 101.3 kPa or 1 atm)
N	normal (concentration); neutron number		

NTSB	National Transportation Safety Board	qv	quod vide (which see)
O-	denoting attachment to oxygen	R	univalent hydrocarbon radical
o-	ortho	(R)-	rectus (clockwise configuration)
OD	outside diameter	r	precision of data
OPEC	Organization of Petroleum Exporting Countries	rad	radian; radius
o-phen	o-phenanthridine	RCRA	Resource Conservation and Recovery Act
OSHA	Occupational Safety and Health Administration	rds	rate-determining step
owf	on weight of fiber	ref.	reference
Ω	ohm	rf	radio frequency, *n.*
P	peta (prefix for 10^{15})	r-f	radio frequency, *adj.*
p	pico (prefix for 10^{-12})	rh	relative humidity
p-	para	RI	Ring Index
p	proton	rms	root-mean square
p.	page	rpm	rotations per minute
Pa	pascal (pressure)	rps	revolutions per second
PEL	personal exposure limit based on an 8-h exposure	RT	room temperature
		RTECS	Registry of Toxic Effects of Chemical Substances
pd	potential difference	^{s} (eg, Bu^{s});	
pH	negative logarithm of the effective hydrogen ion concentration	*sec*-	secondary (eg, secondary butyl)
		S	siemens
phr	parts per hundred of resin (rubber)	(S)-	sinister (counterclockwise configuration)
p-i-n	positive-intrinsic-negative	S-	denoting attachment to sulfur
pmr	proton magnetic resonance	s-	symmetric(al)
p-n	positive-negative	s	second
po	per os (oral)	(s)	solid, only as in H_2O(s)
POP	polyoxypropylene	SAE	Society of Automotive Engineers
pos	positive	SAN	styrene-acrylonitrile
pp.	pages	sat(d)	saturate(d)
ppb	parts per billion (10^9)	satn	saturation
ppm	parts per million (10^6)	SBS	styrene–butadiene–styrene
ppmv	parts per million by volume		
ppmwt	parts per million by weight	sc	subcutaneous
PPO	poly(phenyl oxide)	SCF	self-consistent field; standard cubic feet
ppt(d)	precipitate(d)		
pptn	precipitation	Sch	Schultz number
Pr (no.)	foreign prototype (number)	sem	scanning electron microscope(y)
pt	point; part		
PVC	poly(vinyl chloride)	SFs	Saybolt Furol seconds
pwd	powder	sl sol	slightly soluble
py	pyridine	sol	soluble

soln	solution	*trans-*	isomer in which
soly	solubility		substituted groups are
sp	specific; species		on opposite sides of
sp gr	specific gravity		double bond between C
sr	steradian		atoms
std	standard	TSCA	Toxic Substances Control
STP	standard temperature and		Act
	pressure (0°C and 101.3	TWA	time-weighted average
	kPa)	Twad	Twaddell
sub	sublime(s)	UL	Underwriters' Laboratory
SUs	Saybolt Universal seconds	USDA	United States Department
syn	synthetic		of Agriculture
^{t} (eg, Bu^t),		USP	*United States*
t-, tert-	tertiary (eg, tertiary		*Pharmacopeia*
	butyl)	uv	ultraviolet
T	tera (prefix for 10^{12}); tesla	V	volt (emf)
	(magnetic flux density)	var	variable
t	metric ton (tonne)	*vic-*	vicinal
t	temperature	vol	volume (not volatile)
TAPPI	Technical Association of	vs	versus
	the Pulp and Paper	v sol	very soluble
	Industry	W	watt
TCC	Tagliabue closed cup	Wb	weber
tex	tex (linear density)	Wh	watt hour
T_g	glass-transition	WHO	World Health
	temperature		Organization (United
tga	thermogravimetric		Nations)
	analysis	wk	week
THF	tetrahydrofuran	yr	year
tlc	thin layer chromatography	(*Z*)-	zusammen; together;
TLV	threshold limit value		atomic number

Non-SI (Unacceptable and Obsolete) Units		Use
Å	angstrom	nm
at	atmosphere, technical	Pa
atm	atmosphere, standard	Pa
b	barn	cm^2
bar†	bar	Pa
bbl	barrel	m^3
bhp	brake horsepower	W
Btu	British thermal unit	J
bu	bushel	m^3; L
cal	calorie	J
cfm	cubic foot per minute	m^3/s
Ci	curie	Bq
cSt	centistokes	mm^2/s
c/s	cycle per second	Hz

†Do not use bar (10^5 Pa) or millibar (10^2 Pa) because they are not SI units, and are accepted internationally only for a limited time in special fields because of existing usage.

Non-SI (Unacceptable and Obsolete) Units		Use
cu	cubic	exponential form
D	debye	$C \cdot m$
den	denier	tex
dr	dram	kg
dyn	dyne	N
dyn/cm	dyne per centimeter	mN/m
erg	erg	J
eu	entropy unit	J/K
°F	degree Fahrenheit	°C; K
fc	footcandle	lx
fl	footlambert	lx
fl oz	fluid ounce	m^3; L
ft	foot	m
ft·lbf	foot pound-force	J
gf den	gram-force per denier	N/tex
G	gauss	T
Gal	gal	m/s^2
gal	gallon	m^3; L
Gb	gilbert	A
gpm	gallon per minute	(m^3/s); (m^3/h)
gr	grain	kg
hp	horsepower	W
ihp	indicated horsepower	W
in.	inch	m
in. Hg	inch of mercury	Pa
in. H_2O	inch of water	Pa
in.-lbf	inch pound-force	J
kcal	kilo-calorie	J
kgf	kilogram-force	N
kilo	for kilogram	kg
L	lambert	lx
lb	pound	kg
lbf	pound-force	N
mho	mho	S
mi	mile	m
MM	million	M
mm Hg	millimeter of mercury	Pa
mμ	millimicron	nm
mph	miles per hour	km/h
μ	micron	μm
Oe	oersted	A/m
oz	ounce	kg
ozf	ounce-force	N
η	poise	Pa·s
P	poise	Pa·s
ph	phot	lx
psi	pounds-force per square inch	Pa
psia	pounds-force per square inch absolute	Pa
psig	pounds-force per square inch gage	Pa
qt	quart	m^3; L
°R	degree Rankine	K
rd	rad	Gy
sb	stilb	lx
SCF	standard cubic foot	m^3
sq	square	exponential form
thm	therm	J
yd	yard	m

BIBLIOGRAPHY

1. The International Bureau of Weights and Measures, BIPM (Parc de Saint-Cloud, France) is described in Appendix X2 of Ref. 4. This bureau operates under the exclusive supervision of the International Committee for Weights and Measures (CIPM).
2. *Metric Editorial Guide (ANMC-78-1)*, latest ed., American National Metric Council, 5410 Grosvenor Lane, Bethesda, Md. 20814, 1981.
3. *SI Units and Recommendations for the Use of Their Multiples and of Certain Other Units (ISO 1000-1981)*, American National Standards Institute, 1430 Broadway, New York, N.Y. 10018, 1981.
4. Based on *ASTM E 380-89a (Standard Practice for Use of the International System of Units (SI))*, American Society for Testing and Materials, 1916 Race Street, Philadelphia, Pa. 19103, 1989.
5. *Fed. Regist.*, Dec. 10, 1976 (41 FR 36414).
6. For ANSI address, see Ref. 3.

R. P. LUKENS
ASTM Committee E-43 on SI Practice

F

Continued

FLAVOR CHARACTERIZATION

The aroma of fruit, the taste of candy, and the texture of bread are examples of flavor perception. In each case, physical and chemical structures in these foods stimulate receptors in the nose and mouth. Impulses from these receptors are then processed into perceptions of flavor by the brain. Attention, emotion, memory, cognition, and other brain functions combine with these perceptions to cause behavior, eg, a sense of pleasure, a memory, an idea, a fantasy, a purchase. These are psychological processes and as such have all the complexities of the human mind. Flavor characterization attempts to define what causes flavor and to determine if human response to flavor can be predicted. The ways in which simple flavor active substances, flavorants, produce perceptions are described both in terms of the physiology, ie, transduction, and psychophysics, ie, dose-response relationships, of flavor (1,2). Progress has been made in understanding how perceptions of simple flavorants are processed into hedonic behavior, ie, degree of liking, or concept formation, eg, crispy or umami (savory) (3,4). However, it is unclear how complex mixtures of flavorants are perceived or what behavior they cause. Flavor characterization involves the chemical measurement of individual flavorants and the use of sensory tests to determine their impact on behavior.

Human perception creates difficulty in the characterization of flavor; people often, if not always, perceive flavors differently due to both psychological and physiological factors. For example, certain aryl thiocarbamates, eg, phenylthiocarbamide, taste exceedingly bitter to some people and are almost tasteless to others (5). This difference is genetically determined, and the frequency of its occurrence differs from one population to another; 40% of U.S. caucasians are nontasters, whereas only 3% of the Korean population cannot perceive the strong

1

bitter taste of the aryl thiocarbamates (6). Similar differences were found in the sense of smell for compounds such as menthol, carvone, and ethyl butyrate (7).

Sensory Analysis. Sensory analysis is concerned with the similarities in human flavor perception using methods that are designed to average out certain differences and to detect others. A collection of people (a panel) tastes or smells the same material and reports their perceptions according to previously explained guidelines. Using statistical methods, the similarities, if any, in the panelists' perceptions can be isolated (8,9). Sensory analysis requires a large amount of time to design, execute, and analyze, and is therefore expensive. Consequently, manufacturers concerned with flavor are motivated to find less labor intensive instrumental procedures to predict the flavor perceptions of people (10). Although instrumental methods used for routine quality control are often less expensive than sensory tests, they are indirect and their accuracy must be established using direct sensory methods.

Flavor Perception

Flavor, ie, the human perceptions of flavorants, is generally defined in terms of odor and taste (2); a third component, texture, also may be included (4). Odor is a result of stimuli interacting with specialized receptors in the nose. Taste results from the interactions between stimuli and receptor organs on the tongue and in the mouth (11,12). There are, however, no single identifiable organs involved with the perception of texture. The presence of pain, the sense of touch, and the detection of sound all contribute to the perception of texture; thus the texture of a food includes all perceptions detected in the mouth that are owing to neither odor nor taste. In the broadest definition, flavor is the combined perception of odor, taste, and texture. Flavor is also used to denote a collection of flavorants that might be added to a food (see FLAVORS AND SPICES).

The idea that a stimulus applied to a sense organ produces a response is easily understood since this phenomenon is observed as an external process. For example, the application of an electric current (a stimulus) to a frog leg results in an observable muscle contraction (a response). However, the perception of a stimulus by an organism is not directly observable but is inferred from behavior; with humans this involves the use of language. The data collected in a sensory test may not be quantifications of perceptions alone. Nevertheless, the analyst characterizing flavor operates under the belief that perception is being measured. The resolution of this issue of what is being measured will not necessarily decrease the value of the measurements, eg, adequate accuracy and precision for quality control measurements, but it may change the validity of underlying assumptions, eg, in a complex mixture of stimuli, subjects may be relating to different but correlated sensations. Confidence in sensory methods comes from the accumulated success of sensory testing by food manufacturers and research scientists since the 1940s.

Sensory perception is both qualitative and quantitative. The taste of sucrose and the smell of linalool are two different kinds of sensory perceptions and each of these sensations can have different intensities. Sweet, bitter, salty, fruity, floral, etc, are different flavor qualities produced by different chemical compounds;

the intensity of a particular sensory quality is determined by the amount of the stimulus present. The saltiness of a sodium chloride solution becomes more intense if more of the salt is added, but its quality does not change. However, if hydrochloric acid is substituted for sodium chloride, the flavor quality is sour not salty. For this reason, quality is substitutive, and quantity, intensity, or magnitude is additive (13). The sensory properties of food are generally complicated, consisting of many different flavor qualities at different intensities. The first task of sensory analysis is to identify the component qualities and then to determine their various intensities.

Psychophysics. Psychophysics, ie, the study of the relationship between sensory perceptions and the stimuli that produce them, has produced some very useful concepts for flavor characterization. In the early 1800s, E. H. Weber observed a mathematical relationship between the perceived intensity of a sensation and the physical intensity of the stimulus that produced it. He observed that the minimum detectable difference between the intensity of two stimuli, called the just noticeable difference (JND), is proportional to the total intensity of the stimulus. This relationship has been verified for a large number of sensory qualities, eg, the taste and odor of many different chemicals, the loudness of sounds, the brightness of lights, the pain produced by electric shock, etc (14). This relationship between sensory intensity and stimulus intensity is called Weber's law. For determination of JND, a discriminability test is used in which people are asked to discriminate between a stimulus at two different intensities. The JND is the value of this minimally detectable difference. Data from such experiments contain only physical variables, ie, stimulus intensities. The gradual development, by Fechner, Plateau, and Stevens, of psychological variables corresponding to the perceived intensity of sensations led to a more refined form of Weber's law (13), generally referred to as Stevens' law:

$$\Psi = k\Phi^n \tag{1}$$

Ψ is a measure of the perceived intensity of the sensation and Φ is a measure of the physical intensity of the stimulus which produced the response; k and n are constants. The constant k depends only on the units chosen for the variables Ψ and Φ, ie, the scales. However, the exponent n is characteristic of what is being measured and is independent of stimulus intensity. Two stimuli used for the same function, eg, sweetness, may have different (15) Stevens' law exponents, as is the case with sucrose, $n = 1.5$, and saccharin, $n = 0.8$. Furthermore, saccharin is partially bitter to some whereas sucrose is not. Since the taste of saccharin is both quantatively and qualitatively different from that of sucrose, it must be in a different but similar taste class.

Table 1 lists several different sensory qualities and their corresponding Stevens' law exponents. These values were determined by a process called magnitude estimation, ie, people associate numbers proportional to their perception of the intensity of a sensation (13). A plot of the log of these numbers vs the log of the stimulus magnitude yields a straight line with a slope equal to the exponent n. Those sensory qualities having an exponent less than one, eg, the odor of n-heptane, $n = 0.6$, are said to be compressive. A large range of stimulus magnitudes are compressed into a small sensory scale. This allows detection and quan-

Table 1. Stevens' Law Exponents for Different Sensory Stimuli[a]

Stimulus	Exponent, n
electric shock	3.5
temperature	1.6
loudness of sound	0.6
brightness of light	0.3
sweetness of sucrose	1.5
bitterness of quinine	0.6
saltiness	1.0
sourness	1.0
odor of n-heptane	0.6

[a]Stevens' law, $\Psi = k\Phi^n$, where k and n are constants.

tification of a large range of odor concentrations but limits the ability to detect small changes or differences in odor intensities. In contrast, an expansive sensory quality, eg, the taste of sucrose, $n = 1.5$, has a small total range of detectable concentrations; minute changes in concentration are easily detected (14). Practical implications for the manufacture of food are profound; the components that cause taste must be controlled within very narrow limits, whereas the components that cause odor can be allowed to vary much more. The fact that taste and odor have very different Stevens' law exponents explains why adding 50% water to a glass of wine changes its acid, sweet, and astringent tastes greatly whereas the aroma is only slightly, if at all, changed.

The existence of many different sensory exponents may seem unnecessarily complicated, but compelling arguments may justify these complicated perceptual processes. Consider those female insects that use volatile chemicals, pheromones, to attract males from great distances. During the male's flight toward the female (see INSECT CONTROL TECHNOLOGY) the concentration of the pheromone that must be detected varies over many orders of magnitude. The insect can deal with this variation by compressing a large range of stimulus concentrations into a narrower range of neural responses (16). On the other hand, it is extremely useful to be able to detect small changes in skin temperature over the narrow range that most organisms can tolerate and, indeed, the exponent for the sensation of warmth is 1.5 in humans (14). The collection of methods used to determine the quality and quantity of sensory perceptions are called sensory tests and their application is termed sensory analysis. A knowledge of psychophysics plays an essential role in their design and execution (2).

Flavor Intensity.   In most sensory tests, a person is asked to associate a name or a number with his perceptions of a substance he sniffed or tasted. The set from which these names or numbers are chosen is called a scale. The four general types of scales are nominal, ordinal, interval, and ratio (17). Each has different properties and allowable statistics (4,14). The measurement of flavor intensity, unlike the evaluation of quality, requires an ordered scale, the simplest of which is an ordinal scale.

Nominal Scale. A nominal scale is always used to determine the quality or qualities of a flavor. It has no order, distance, or origin and is just a collection of names, eg, sweet, sour, bitter, salty, umami, etc. It has been demonstrated that the nominal scaling of odors requires only that the person has frequently encountered the odor, has a long-standing connection between the odor and the name, and has aid in recalling the name (18,19). The first two requirements are satisfied by training and the third is satisfied by presenting the person with a familiar list of names from which to choose. The nominal scaling of the flavor qualities of a real material is complicated by the fact that a number of different flavor qualities are perceived simultaneously; tests to demonstrate how well people discriminate components of mixtures have been very disappointing (20,21). However, the study of simple mixtures has revealed the existence of quantitative relationships between odor qualities.

Ordinal Scale. An ordinal scale has both name and order but no distance or origin; eg, weak, moderate, strong, or the numbers 1, 2, 3 can form an ordinal scale with three points. There are no points between weak and moderate or between 1 and 2 because there is no concept of distance between these points when they are labels on an ordinal scale. All that is known is that one is smaller than the other, eg, $1 < 2 < 3$ and weak < moderate < strong. Although ordinal scales are sometimes used to measure flavor intensity, they are more often used to determine attitudes toward a sensation, eg, like or dislike of a sensation. This aspect of flavor, called hedonics, is an attempt to determine what is the psychological impact of flavor perceptions. It is distinct from flavor characterization, which deals with the perceptual impact of a stimulus. In flavor characterization, ordinal scales are used to arrange a set of samples in their order of increasing intensity (ranking) and to estimate the magnitude of these intensities.

Interval Scale. The interval scale has meaningful distances as well as name and order but no meaningful origin. Like elevation with an arbitrary origin at sea level, all interval scales associate meaning to differences between measurements but not to their absolute value. The Quartermaster Nine-point Acceptance Scale uses a scale of 1 (like extremely) to 9 (dislike extremely) with a midpoint of 5 (neither like nor dislike); other expressions include like very much, moderately, and slightly, and dislike in reverse degree of intensity. This has been shown to be an interval scale (22) and seems to have an optimum design in terms of both performance and consistency with psychological theory (23). Data collected on an interval scale can be analyzed with parametric statistical procedures. Ordinal data, in which the differences between points are not known or highly variable, require nonparametric statistical procedures. An excellent guide to the proper application of statistics to the most common types of sensory data is available (9). Although the nine-point scale is most frequently used to record degree of liking or similar hedonic properties, similar nine-point scales interval properties make it a useful tool for measuring perceived intensity.

The problems posed by psychological scaling have led to an approach called cross-modal matching or magnitude production (14,24). Panelists associate their perception of flavor intensity with the perception of some other stimulus, eg, loudness of a sound, which they produce by turning a dial. The association of the intensity of one kind of stimulus with that of another can be achieved with remarkable ease, probably because it requires no conception of numbers. Fortuitously, the perceived length of a line has been shown to follow Stevens' law with an exponent of one (14). Thus, asking people to indicate a line using a pencil, a lever, a computer mouse, etc, the length of which represents their perception of a flavor intensity, yields interval data in one-to-one correspondence with their perceptions. Graphical line marking scales, like the one in which a taster simply marks a position on a line to indicate perceived intensity, produce data that are parametric.

Ratio Scale. The ratio scale has name, order, distance, and a meaningful origin. A zero value on the scale means the absence of any of the property, eg, zero Kelvin means the absence of motion and gives meaning to the gas law, $PV = nRT$, whereas zero Celsius is arbitrary and meaningless in terms of the gas law. The mathematical form of Stevens' law has been used to argue that a ratio scale could be developed to measure flavor intensity (14). The magnitude estimation method yields a ratio scale when the data follow certain rules. In this method a panelist is instructed to associate the flavor intensity of a second flavor with a number, Y, that is perceived to be a multiple of the flavor intensity, X, of the first sample. If the ratio of X to Y is always the same no matter what the value of X given for the first sample, then the panelist is estimating the flavor intensity on a ratio scale. However, if the difference between X and Y is constant for different values for X the flavor is being estimated on an interval scale. Zero on a ratio scale means the absence of the perception being measured and this is a controversial conclusion for some psychologists (25). Nevertheless, magnitude estimation is frequently used and often defended as an appropriate scaling method for sensory data (24). There has been considerable discussion (26–29) of the many psychological scales used to quantitate sensory perceptions, motivated by the desire to devise analytical methods that are consistent with the demands of Weber's and Stevens' laws and appropriate for the use of parametric statistical methods. Although the method of magnitude estimation has some theoretical appeal, data produced with nine-point interval and graphical line marking scales are much easier to obtain and are statistically similar. An important consideration is that interval data should be used in models without a fixed intercept or defined origin whereas ratio scaled data requires the inclusion of a zero intercept in most models (29).

Flavor Description. Typically, a sensory analyst determines if two samples differ, and attempts to explain their differences so that changes can be made. The Arthur D. Little flavor profile (FP), quantitative descriptive analysis (QDA), and spectrum method are three of the most popular methods designed to answer these and more complicated questions (30–33). All three methods involve the training of people in the nominal scaling of the flavor qualities present in the food being studied, but they differ in their method for quantitation.

The FP procedure uses an ordinal scale to quantify each flavor quality. Participants are asked to indicate flavor qualities, eg, sweetness, fruitiness, mercaptan, crispness, and dryness by marking space indicated as threshold, slight, moderate, and strong. The type of scaling process employed in the QDA procedure is not obvious. Participants are asked to quantify flavor qualities by marking a point on a given line. The authors of the QDA system consider it an interval scale (33), but it has been described as a ratio scale because there are no intervals marked off on the line (34). Three words, ie, weak, moderate, and strong, are indicated on the line and these form an ordinal scale above the line. This points out one of the most serious, practical problems in perceptual measurement, ie, the responses that a panelist gives in a sensory analysis depend on his or her perception of the questionnaire as well as perception of the stimulus. However, rigorous training with most ballot designs can minimize these problems.

Threshold, Saturation, and Adaptation. Several aspects of flavor perception are not accounted for in the Weber-Stevens' laws, eg, threshold, saturation, and adaptation. For every sense there is a minimum detectable stimulus intensity called the threshold. When the concentration of the stimulus is below the threshold value there is little or no perception. The threshold value is not absolute but is greatly affected by the presence of other stimuli, eg, the threshold for geosmin in both beet juice and fish flesh is ca 50 times higher than it is in water (35,36).

Saturation is the concentration of a stimulus above which no increase in perception can be detected. It is true that Weber-Stevens' laws can predict the relationship between stimulus intensity and sensory response with some precision; however, they do not describe the very common situation of stimuli at or near the threshold or point of saturation.

Exposure to a flavor over time always results in a decrease in the perceived intensity. This dynamic effect of flavorants, called adaptation, is a central part of the process by which people experience flavors in foods as well as in sensory tests. Measuring the dynamics of flavor perception is an emerging technology made possible by inexpensive computing. Called time-intensity analysis, these methods are finding wide applications in taste analysis.

Discriminant Sensory Analysis. Discriminant sensory analysis, ie, difference testing, is used to determine if a difference can be detected in the flavor of two or more samples by a panel of subjects. These differences may be quantitative, ie, a magnitude can be assigned to the differences but the nature of the difference is not revealed. These procedures yield much less information about the flavor of a food than descriptive analyses, yet are extremely useful; eg, a manufacturer might want to substitute one component of a food product with another safer or less expensive one without changing the flavor in any way. Several formulations can be attempted until one is found with flavor characteristics that cannot be discriminated from the original or standard sample.

In the most popular discriminant sensory analysis, called the triangle test, panelists are presented with three samples, two of which are the same. The panelists examine the three samples and try to identify the odd sample. Their probability of choosing the correct sample by chance is one in three, or 33%. Invariably, the samples analyzed in these tests are very similar and statistical procedures must be used to determine if the panelists detected a difference, usually expressed as probability of randomness or level of significance. The one-third chance of guessing correctly on a triangle test is a decided advantage over other methods, eg, the duo–trio and the paired comparison. In the duo–trio method, a standard sample is identified and the panelist must determine which of two other samples is the same as the standard; the probability of guessing correctly is one in two, or 0.5. The paired comparison is a discrimination test since it contains an element of direction, eg, a person is asked to choose which of two samples is weaker or stronger in some particular flavor quality. Nevertheless, the chance of choosing any one of them is 50%. The higher probability of guessing correctly in the duo–

trio and paired tests requires more trials to establish significance, thus increasing the cost of analysis. For both descriptive and discriminant sensory analyses, proper experimental design requires a clear idea of the question being asked and the factors that must be controlled. It requires appropriate panel selection, choice of test environment, method of sample preparation, type of statistical analysis, etc (9,37).

The development of precise and reproducible methods of sensory analysis is prerequisite to the determination of what causes flavor, or the study of flavor chemistry. Knowing what chemical compounds are responsible for flavor allows the development of analytical techniques using chemistry rather than human subjects to characterize flavor (38,39). Routine analysis in most food production for the quality control of flavor is rare (40). Once standards for each flavor quality have been synthesized or isolated, they can also be used to train people to do more rigorous descriptive analyses.

Flavor Chemistry

Chemical compounds having odor and taste number in the thousands. In 1969 a description of the odor characteristics of more than three thousand chemical compounds used in the flavor and perfume industries were described (41). The list of volatile compounds found in food that may contribute to odor and taste is even larger (42), and the list of all possible flavor compounds, including those that have yet to be synthesized, is greater than a thousand. Many different compounds have the same flavor character or quality, differing perhaps in their relative intensity but indistinguishable in the type of flavor they elicit. The exact number of different flavor qualities is not known, but it appears to be much less than the total number of compounds with flavor.

A persistent idea is that there is a very small number of flavor qualities or characteristics, called primaries, each detected by a different kind of receptor site in the sensory organ. It is thought that each of these primary sites can be excited independently but that some chemicals can react with more than one site producing the perception of several flavor qualities simultaneously (12). Sweet, sour, salty, bitter, and umami qualities are generally accepted as five of the primaries for taste; sucrose, hydrochloric acid, sodium chloride, quinine, and glutamate, respectively, are compounds that have these primary tastes. Sucrose is only sweet, quinine is only bitter, etc; saccharin, however, is slightly bitter as well as sweet and its Stevens' law exponent is 0.8, between that for purely sweet (1.5) and purely bitter (0.6) compounds (34). There is evidence that all compounds with the same primary taste characteristic have the same psychophysical exponent even though they may have different threshold values (24). The flavor of a complex food can be described as a combination of a smaller number of flavor primaries, each with an associated intensity. A flavor may be described as a vector in which the primaries make up the coordinates of the flavor space.

Table 2 lists examples of compounds with taste and their associated sensory qualities. Sour taste is primarily produced by the presence of hydrogen ion slightly modified by the types of anions present in the solution, eg, acetic acid is more sour than citric acid at the same pH or molar concentration (43). Saltiness is due to

Table 2. Compounds with Taste

Name	CAS Registry Number	Structure	Taste
hydrogen ion	[12408-02-5]	H^+	sour
sodium ion	[17341-25-2]	Na^+	salty
potassium ion	[24203-36-9]	K^+	salty
potassium iodide	[7681-11-0]	KI	bitter
quinine	[130-95-0]		bitter
β-D-mannose	[27710-21-0]		bitter
α-D-mannose	[27710-21-9]		sweet
β-D-glucose	[492-61-5]		sweet
glycine	[56-40-6]		sweet
L-glutamate	[142-47-2][a]		umami
inosine 5′-monophosphate	[131-99-7][b]		umami

[a] Monosodium L-glutamate (MSG) (see AMINO ACIDS, (MSG)).
[b] Inosine 5′-(dihydrogen phosphate).

9

the salts of alkali metals, the most common of which is sodium chloride. However, salts such as cesium chloride and potassium iodide are bitter; potassium bromide has a mixed taste, ie, salty and bitter (44). Thus saltiness, like sourness, is modified by the presence of different anions but is a direct result of a small number of cations.

Sweetness and bitterness are complicated because of the great variety of chemical compounds having these taste qualities. Very different compounds can have the same flavor quality, such as quinine and β-D-mannose, both of which are bitter. Conversely, two very similar compounds can have different flavor qualities, eg, α-D-mannose is sweet and β-D-mannose is bitter. Examination of the structures in Table 2 with sweet, bitter, and umami taste exemplifies the problem of identifying those molecular features that distinguish sweet from bitter tasting compounds. Many chemical and physical properties have been examined to determine the requirements for a chemical to elicit a particular sensory response (15). The acid hydrogen–base (AH–B) theory of sweetness postulates that a molecule must have a proton attached to an electronegative center located 0.3 nm away from another electronegative center in order to taste sweet (45). This theory has been extended to include a hydrophobic site near the AH–B system, but it is still unable to predict the taste of all structures (46). All sweet compounds have an AH–B system as demonstrated in Figure 1, but some bitter and some tasteless compounds also have such a system in their structures.

Odor Compounds

The relationship between molecular structure and sensory properties is very unclear for compounds with odor. It seems likely that there is a set of odors that

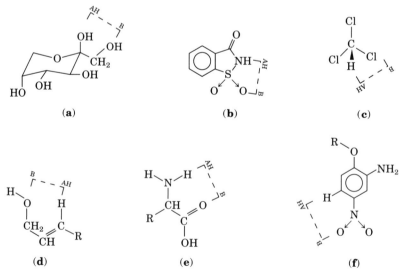

Fig. 1. Sweet-tasting compounds of various chemical classes and their common (AH-B) unit. (**a**) β-D-fructose; (**b**) saccharin; (**c**) chloroform; (**d**) unsaturated alcohols; (**e**) amino acids; and (**f**) 2-alkoxy-5-nitroanilines.

could be called primaries, but a widely accepted list of such primary odor qualities has not been devised. Molecular size and shape have been used to describe the features that distinguish different odor qualities (47). There are seven odor primaries: ethereal, camphoraceous, musty, floral, minty, pungent, and putrid. Like all theories attempting to relate chemical structure with sensory properties, these ideas have little predictive value. The compounds shown in Table 3 (see also AIR POLLUTION CONTROL METHODS) are not meant to represent odor primaries but are examples of naturally occurring substances that have very low odor thresholds and diverse structures; eg, geosmin, the compound responsible for the characteristic odor of beets and of wet soil, can be detected in water at a concentration of 10^{-12}g/g (50,51,57,58). The isolation and identification of trace components, such as geosmin, responsible for the odor of natural products is a significant challenge in flavor chemistry.

Identification of Odor Components. The methods used to isolate, concentrate, and identify odor components are not unique to flavor chemistry; the use of sensory analyses to monitor their progress is. Once a particular compound has been identified and a standard unambiguously synthesized, its odor characteristic must be verified by sensory analysis. The complexity of natural products requires the use of separation techniques having the highest available resolution, such as gas and liquid chromatography. It is difficult to detect a compound occurring at a concentration of 10^{-9} to 10^{-12}g/g without concentrating it at least several thousandfold. Even after concentration there are often only a few nanograms available for structure analysis. The extraction of 100,000 kg of beets would be required to obtain 20 mg of geosmin for a standard proton magnetic resonance (pmr) spectrum (59). Therefore, in addition to high resolution, the most sensitive analytical techniques such as mass spectroscopy must be used to detect the highly potent trace components.

Since the early 1980s gas chromatography–olfactometry (gco) has emerged from a long history as a simple bioassay for odor in gas chromatography effluents

Table 3. Naturally Occuring Compounds with Characteristic Odors

Name	CAS Registry Number	Structure[a]	Odor	Source	Reference
isoamyl acetate	[123-92-2]	(1)	fruity	banana	41
linalool	[78-70-6]	(2)	citrus	orange	48
β-damascenone	[23726-93-4]	(3)	floral	rose, apple	49
geosmin	[19700-21-1]	(4)	earthy	beet, soil	50, 51
furaneol	[3658-77-3]	(5)	strawberry	strawberry	49
eremophilone	[562-23-2]	(6)	woody	sandalwood	52
o-aminoaceto-phenone	[537-93-9]	(7)	foxy	grape, weasel	53
muscone	[541-91-3]	(8)	musk	musk deer	54
5α-androst-16-en-3-one	[18339-16-7]	(9)	boar urine	boar urine	55
2-isobutyl-3-methoxypyrazine	[24683-00-9]	(10)	bell peppers	bell pepper	49
cadaverine	[462-94-2]	(11)	putrid	carrion	56

[a]See Figure 2.

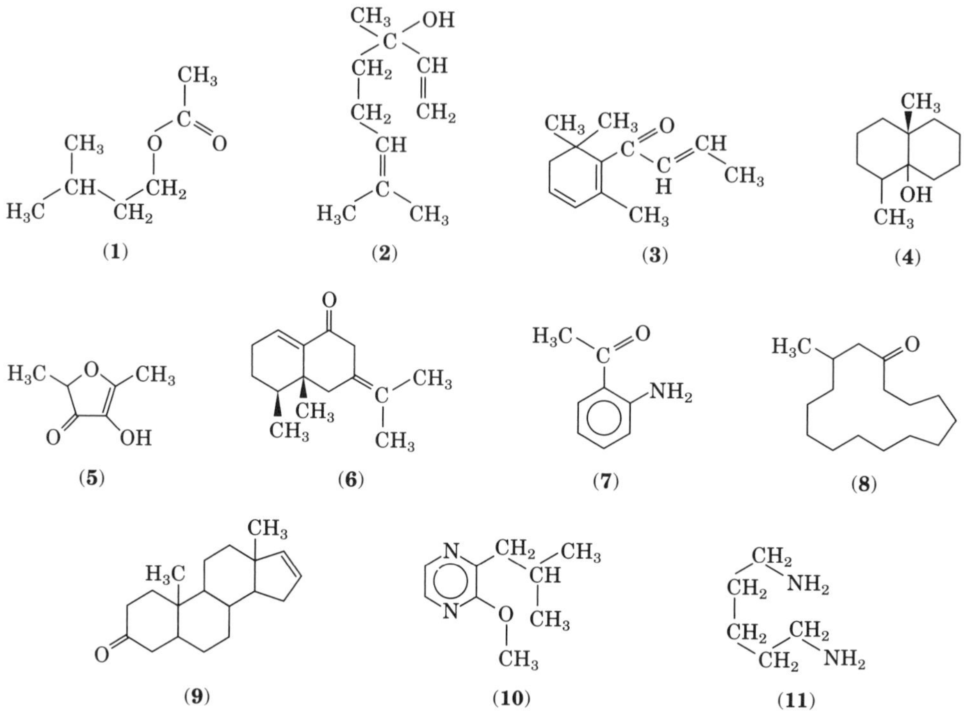

Fig. 2. Naturally occurring compounds with characteristic odors. See Table 3.

(39,60,61) to become a quantitative method for the characterization of odor and aroma. Quantitative gco is exemplified by CharmAnalysis (62) and Aroma Extraction Dilution Analysis (63) in which a series of dilutions are chromatographed separately. As each diluted extract is sniffed eluting from the gas chromatograph, the weaker odors drop below the threshold and cannot be detected. Combination of the data produced in these sessions produces a chromatogram like the lower chromatogram shown in Figure 3 of an extract of Valencia orange juice (64). The taller peaks indicate odors that are well above their thresholds whereas the smaller peaks represent compounds that may be below their thresholds. The upper chromatogram shows the Flame Ionization Detector (fid) response to the same sample. The most abundant volatile shown by the fid chromatogram at index 1021 is limonene. In contrast, the gco chromatogram shows no response at 1021, indicating that limonene does not contribute directly to the flavor of orange juice. The dominant odor is linalool (index 1083) a compound from the peel of the orange. Between 1200 and 1400 indexes are several extremely potent odors that produce little response in the fid chromatogram, the causes of these odors are still unknown. The quantitative bioassay data produced in dilution based gco indicate the relative potency of odorants in a complex mixture; however, there are gco methods based on the perception of intensity instead of the measurement of potency. Most of the earlier methods involved the scaling of perceived intensity. A more developed method of this type, called Osme, is based on the computerized recording of lever position matched with perception of intensity (65). Although

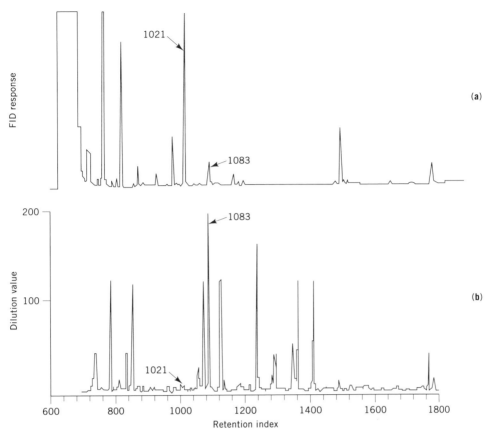

Fig. 3. (a) Flame ionization detector (fid) response to an extract of commercially processed Valencia orange juice. (b) Gas chromatography–olfactometry (gco) chromatogram of the same extract. The abscissa in both chromatograms is a normal paraffin retention index scale ranging between hexane and octadecane (Kovats index). Dilution value in the gco is the -fold that the extract had to be diluted until odor was no longer detectable at each index.

the meaning of the data produced by the several different gco techniques are slightly different, they all can be used to focus analytical chemistry on the most odor-potent components of complex foods (39,61).

Gco has been used to find a subtle but distinctive odor note common in the grapes grown in North America, ie, *Vitis rotundifolia* (southern Fox Grape) and *Vitis labruscanna* (northern Fox Grape). Grapes of these species have been characterized as having a sweet fox-like smell; methyl anthranilate traditionally has been the only component identified in grapes that could explain this character. However, many cultivars, eg, Catawaba, have a sweet foxy smell but have subthreshold levels of methyl anthranilate [*134-20-3*]. A serendipitous discovery in Japan led to the recognition that the waxy contents of the anal sac of the Japanese and Korean weasels contain an intense odor similar to that of the native American grapes (52,66). The odor was found to be due to the presence of *o*-aminoacetophenone in both the weasel and the grape. The most odor-potent compound in the grape was phenylethyl alcohol, a characteristic odor of all grapes and yeast fermented juices; however, the o-aminoacetophenone was the odorant that characterized these grapes as *labruscanna*. The high level in the weasel allowed easy identification and once this standard was available verification of the presence at much lower levels in the grape was easy.

This example demonstrates the most challenging problem of flavor chemistry, ie, each flavor problem may require its own analytical approach; however, a sensory analysis is always required. The remaining unknown odorants demand the most sensitive and selective techniques, and methods of concentration and isolation that preserve the sensory properties of complex and often delicate flavors. Furthermore, some of the subtle odors in one system will be first identified in very different systems, like o-aminoacetophenone in weasels and fox grapes.

BIBLIOGRAPHY

"Organoleptic Testing" in *ECT* 2nd ed., Vol. 14, pp. 336–344, by L. B. Sjöström, Arthur D. Little, Inc.; "Flavor Characterization" in *ECT* 3rd ed., Vol. 10, pp. 444–455, by T. E. Acree, Cornell University.

1. D. Lancet, *Society Of General Physiologists Series* **47**, 73–91 (1991).
2. H. T. Lawless and B. P. Klein, *Sensory Science Theory and Application in Foods, ift Basic Symposium Series*, Marcel Dekker, Inc., New York, 1991, p. 441.
3. Z. Vickers and M. C. Bourne, *Food Sci.* **41**(5), 1158–1164 (1976).
4. M. O'Mahony, in Ref. 2, pp. 223–268.
5. A. L. Fox, *Proc. Nat. Acad. Sci.* **18**, 115 (1932).
6. J. Cohen and D. P. Ogdon, *Psychol. Bull.* **46**, 490 (1949).
7. A. B. Marin, T. E. Acree, and J. Barnard, *Chem. Senses* **13**(3), 435–444 (1988).
8. M. A. Amerine, R. M. Pangborn, and E. B. Roessler, *Principles of Sensory Evaluation of Food*, Academic Press, Inc., New York, 1965.
9. M. O'Mahony, *Sensory Evaluation of Foods*, Marcel Dekker, Inc., New York, 1986, p. 487.
10. T. E. Acree, in T. E. Acree and R. Teranishi, eds., *Flavor Science: Sensible Principles and Techniques*, ACS Books, Washington, D.C., 1993.
11. J. Brozek, *Am. J. Clin. Nutr.* **6**, 332 (1957).
12. R. W. Moncrieff, *The Chemical Senses*, 3rd ed., CRC Press, Cleveland, Ohio, 1967.
13. S. S. Stevens, *Science* **127**, 383 (1958).
14. S. S. Stevens, *Am. Sci.* **48**, 226 (1960).
15. M. G. J. Beets, *SAR: Structure–Activity Relationships in Human Chemoreception*, Applied Science Publishers, London, 1978.
16. H. Arn, *Z. Naturforsch. Teil B.* **30**, 722 (1975).
17. S. N. Deming, *CHEMTECH*, (Feb. 19–21, 1993).
18. E. Gibson, *Principles of Perceptual Learning and Development*, Appleton Century, New York, 1969.
19. W. S. Cain, *Science* **203**, 467 (1979).
20. D. G. Laing and A. Glemarec, *Physiol. Behav.* **52**(6), 1047–1053 (1992).
21. D. G. Laing, B. A. Livermore, and G. W. Francis, *Twelfth Annual Meeting of the Association For Chemoreception Sciences*, Sarasota, Fla., Apr. 1990.
22. N. L. Jones and L. L. Thurstone, *J. Appl. Psychol.* **39**, 31–36 (1955).
23. H. T. Lawless, in Ref. 2, pp. 1–36.
24. H. R. Moskowitz, *Percept. Psychophys.* **9**, 51 (1971).

25. J. C. Baird, *Fundamentals of Scaling and Psychophysics*, John Wiley & Sons, Inc., New York, 1978, pp. 126–155.
26. S. S. Stevens, *Psychol. Rev.* **70**, 153 (1963).
27. H. Eisler, *Psychol. Rev.* **70**, 243 (1963).
28. H. R. Moskowitz and J. L. Sidel, *J. Food Sci.* **36**, 677 (1971).
29. J. H. Pearce, B. Korth, and C. B. Warren, *J. Sensory Studies*, 27 (1986).
30. S. E. Cairncross and L. B. Sjöström, *Food Technol.* **4**, 308 (1950).
31. H. Stone and co-workers, *Food Technol.* **29**, 24 (1974).
32. G. V. Civille and C. A. Dus, *Fourth Chemical Congress Of North America*, Vol. 202, New York, Aug., 1991, pp. 1 and 2.
33. H. Stone and co-workers, *Food Technol.* **28**(11) 24–34 (1974).
34. J. J. Powers and H. R. Moskowitz, *Am. Soc. Test. Mater. Spec. Tech. Publ.* **594**, 35 (1974); R. A. Scanlan, ed., *Flavor Quality: Objective Measurement*, ACS Symposium Series No. 51, Washington, D.C., 1977.
35. L. D. Tyler, T. E. Acree, and N. L. Smith, *J. Food Sci.* **44**, 79 (1979).
36. M. Yurkowski and J. L. Tabachek, *J. Fish Res. Board Can.* **31**, 1851 (1974).
37. J. L. Sidel and H. Stone, *Food Technol.* **30**, 32 (1976).
38. R. Teranishi and co-workers, *Flavor Research, Principles and Techniques*, Marcel Dekker, Inc., New York, 1971.
39. T. E. Acree and R. Teranishi, in Ref. 10, pp. 1–21.
40. M. Gillette, *Perfume Flavor* **15**(3), 33–40 (1990).
41. S. Arctander, *Perfume and Flavor Chemicals*, 2 Vols., S. Arctander, Montclair, N.J., 1969.
42. H. Maarse and C. A. Visscher, eds. *Volatile Compounds in Food, Qualitative Data*, 5th ed., Zeist, the Netherlands, 1983.
43. F. W. Fabian and H. B. Blum, *Food Res.* **8**, 179 (1943).
44. L. M. Beidler, *Handbook of Sensory Physiology*, Vol. IV, Springer-Verlag, Berlin, 1971, p. 256.
45. R. S. Shallenberger and T. E. Acree, *Nature* **216**, 480 (1967).
46. L. B. Kier, *J. Pharm. Sci.* **61**, 1394 (1972).
47. J. E. Amoore, *Proc. Sci. Sect. Toilet Coods Assoc. Special. Suppl.* **7**, 1 (1962).
48. T. E. Furia and N. Bellanca, *Fenaroli's Handbook of Flavor Ingredients*, Chemical Rubber Co., Cleveland, Ohio, 1971.
49. H.-D. Belitz and W. Grosch, *Food Chemistry*, trans. by D. Hadziyev, Springer-Verlag, New York, 1986.
50. T. E. Acree and co-workers, *J. Agric. Food Chem.* **24**, 430 (1976).
51. K. E. Murray, P. A. Bannister, and R. G. Buttery, *Chem. Ind.* **26**, 973 (1975); C. Charalambous and I. Katz, eds., *Phenolics, Sulfur and Nitrogen Compounds in Food Flavors*, ACS Symposium Series No. 26, Washington, D.C., 1976.
52. J. L. Willis, in G. Ohloff and A. F. Thomas, eds., *Gustation and Olfaction*, Academic Press, Inc., New York, 1971, p. 169.
52. T. E. Acree and co-workers, *Flavour Sci. Technol.*, 49–52 (1990).
54. L. Ruzicka, *Helv. Chim. Acta* **9**, 1008 (1926).
54. R. P. Pomeroy and H. Cruse, *J. Am. Water Works Assoc.* **61**, 21 (1969).
55. R. L. S. Patterson, *J. Agric. Food Chem.* **19**, 31 (1968).
56. D. Ackerman, *Z. Physiol. Chem.* **54**, 16 (1907).
57. R. G. Buttery and J. Garibaldi, *J. Agric. Food Chem.* **24**, 1247 (1976).
58. L. D. Tyler, T. E. Acree, and R. M. Butts, *J. Agric. Food Chem.* **26**, 1415 (1978).
59. L. D. Tyler and co-workers, *J. Agric. Food Chem.* **26**, 775 (1978).
60. T. E. Acree and co-workers, *Anal. Chem.* **48**, 1821 (1976).
61. T. E. Acree, in C. T. Ho and C. H. Manley, eds., *Flavor Measurement: ift Basic Symposium Series*, Marcel Dekker, Inc., New York, 1993.

62. T. E. Acree, J. Barnard, and D. Cunningham, *Food Chem.* **14**, 273–286 (1984).
63. F. Ullrich and W. Grosch, *Z. Lebensm. Unters. Forsch.* **184**(4), 277–282 (1987).
64. A. B. Marin and co-workers, *J. Agric. Food Chem.* **40**(4), 650–654 (1992).
65. N. B. Sanchez and co-workers, in G. Charalambous, ed., *Proceedings of the 6th International Flavour Conference*, Rethymnon, Crete, Greece, 1992, pp. 403–426.
66. R. R. Nelson and co-workers, *J. Food Sci.* **42**, 57 (1977).

TERRY E. ACREE
Cornell University

FLAVORS AND SPICES

FLAVORS

Flavor has been defined as a memory and an experience (1). These definitions have always included as part of the explanation at least two phenomena, ie, taste and smell (2). It is suggested that in defining flavor too much emphasis is put on the olfactory (smell) and gustatory (taste) aspects (3), and that vision, hearing, and tactile senses also contribute to the total flavor impression. Flavor is viewed as a division between physical sense, eg, appearance, texture, and consistency, and chemical sense, ie, smell, taste, and feeling (4). The Society of Flavor Chemists, Inc. defines flavor as "the sum total of those characteristics of any material taken in the mouth, perceived principally by the senses of taste and smell and also the general senses of pain and tactile receptors in the mouth, as perceived by the brain" (5).

The acceptability of food is determined by its flavor, and a large variety of industrial flavorings are used for the commercial preparation of foods. Most of the daily food intake, even in industrialized countries, contains flavor naturally or flavor formed during cooking and preparation for human consumption. Only a minor part of the daily food intake is covered by foods containing added flavorings.

Function.   Flavors do several things in food systems. Foremost among these functions is their ability to render food more acceptable and enjoyable.

Flavors are often used to create the impression of flavor where little or none exist, and they impart food products with a recognizable character. Some food products would not exist without the addition of flavorings, eg, soft drinks, water ices, confectionery, milk desserts, etc. Many food products need a specific flavor note to characterize them among other similar products of the same food category, eg, citrus soft drinks, mint candies, gingerbread, yogurt, and cottage cheese.

Flavors can be used to alter the flavor of a product, eg, the flavor of dairy products; to modify, supplement, or enhance an existing flavor, eg, the butter flavor in margarine and the meat or chicken flavor in bouillon; and to compensate for the loss of flavor during food processing, eg, pasteurized foods, concentrated citrus fruit juices, syrups, alcoholic beverages, freezing, filtration, pasteurization, and long-term storage. Flavorings also transform nutritionally valuable materials of bland taste, eg, grasses, weeds, seeds, and roots, into well-accepted foods.

Food Acceptance. Four features of food are recognized to determine acceptance, ie, flavor, nutritive value, appearance, and mouthfeel. When all four aspects are in proper quantitative proportions, a food finds general acceptance. When all four are interdependent, appearance takes precedence over the others. However, a report by the Food Marketing Institute has shown that consumers placed nutrition second to flavor in importance (6).

A food must have the expected or proper appearance and color before it will be readily consumed (7). There are many prepared foods in which artificial flavors and colors are used whose flavor is sufficiently bland to make color essential for flavor identification, eg, margarine. The preservation of color in natural food during processing or the development of color by processing are aspects of primary importance in food acceptance.

Mouthfeel (8), ie, a texture or kinetic feature evaluated by the skin or muscles in the mouth, includes smoothness, roughness, stickiness, slickness, brittleness, and viscosity. Texture is not only detected during mastication, but also by the ease of cutting or dividing before consumption. A particular flavor observed without the proper mouthfeel may not only be unappreciated, but go unrecognized, ie, nutty flavor is associated with brittleness, and butter flavor with slickness and smoothness. In low fat products, when one removes or lowers the fat the mouthfeel component of the product and the perception of the flavor changes. The flavor generally has to be revised or its amount changed to accommodate this problem.

The acceptability of certain foods may be because of availability and habit, which in turn affects certain customs, eg, the sugar maple, indigenous to northern United States, provides a syrup generally enjoyed and familiar in this region but generally unacceptable in the southern states. The greater mobility of people in the late twentieth century demands variety in flavorings to produce these regional tastes.

Taste. Certain basic principles are involved in the physiology of flavor perception. Researchers studying taste generally agree that there are at least five tastes, ie, salty (9), sour (10), bitter (11), sweet (12), and umami (13). Umami can be defined to the Japanese as the taste of three broths, ie, Kombu, Shiitake, and Katsuobushi. In English, the narrower definition of the taste of monosodium L-glutamate [142-47-2] (MSG), or the broad but vague concept of savory, meaty, or brothy are used (13) (see FLAVOR CHARACTERIZATION). Tastes are perceived by certain sensory cells or taste buds, contained in the approximately 10,000 papillae located on the tongue. On the human tongue the taste buds are localized in three general areas. The fungiform papillae, located on the front two-thirds of the tongue, are innervated by the chorda tympani branch of the facial nerve. The foliate papillae, situated in the rear and on the sides of the tongue, are innervated by the glossopharyngeal nerve. The larger circumvallate papillae are situated principally on the back third of the tongue. Taste buds are also found on the palate, pharynx, epiglottis, larynx, and esophagus (14).

Taste-active chemicals react with receptors on the surface of sensory cells in the papillae causing electrical depolarization, ie, drop in the voltage across the sensory cell membrane. The collection of biochemical events that are involved in this process is called transduction (15,16). Not all the chemical steps involved in transduction are known; however, it is clear that different transduction mechanisms are involved in different taste qualities; different transduction mechanisms exist for the same chemical in different species (15). Thus the specificity of chemosensory processes, ie, taste and smell, to different chemicals is caused by differences in the sensory cell membrane, the transduction mechanisms, and the central nervous system (14).

Several aspects affect the extent and character of taste and smell. People differ considerably in sensitivity and appreciation of smell and taste, and there is lack of a common language to describe smell and taste experiences. A hereditary or genetic factor may cause a variation between individual reactions, eg, phenylthiourea causes a bitter taste sensation which may not be perceptible to certain people whose general ability to distinguish other tastes is not noticeably impaired (17). The variation of pH in saliva, which acts as a buffer and the charge carrier for the depolarization of the taste cell, may influence the perception of acidity differently in people (15,18). Enzymes in saliva can cause rapid chemical changes in basic food ingredients, such as proteins and carbohydrates, with variable effects on the individual.

When food contains both sweet and bitter substances, the temporal pattern of reception, ie, the order in which sweet and bitter tastes are perceived, affects the total qualitative evaluation. This temporal effect is caused by the physical location of taste buds. The buds responding to sweet are located on the surface and the tip of the tongue, the bitter in grooves toward the rear. Therefore, the two types of taste buds can be activated sequentially.

Texture also influences the evaluation of taste. Sweetness in a liquid is associated with body or viscosity. An artificially sweetened beverage that lacks body, therefore, may be rated qualitatively lower than one equally sweet but containing sucrose.

The metallic taste (12,19,20) is not ascribed to any special taste buds or mouth area. Along with pungency (the hot taste of peppers), astringency (the puckering taste of alum), and cold taste (the cool effect of menthol), the metallic taste is called a common chemical sense (21).

Although the values cannot be considered absolute, approximate magnitude of taste sensitivity has been measured (Table 1). Certain taste interrelationships should be considered in the evaluation of taste magnitude. The apparent sourness of citric acid is depressed by both sucrose and sodium chloride. Although the saltiness of sodium chloride is reduced by sucrose, it can be significantly enhanced by acid (23).

Simultaneous stimulation of the tongue with the application of different taste stimuli produces an interaction, modification, or blending of the stimuli in some instances but not in others. Warm and cold sensations are reported to act similarly on the tongue in two groups: bitter, warm, and sweet; and sour, cold,

Table 1. Approximate Taste Thresholds[a]

Basic taste	Test standard	CAS Registry Number	Wt % in water
sour	hydrochloric acid	[7647-01-0]	0.007
salty	sodium chloride	[7647-14-5]	0.25
sweet	sucrose	[57-50-1]	0.50
bitter	quinine	[130-95-0]	0.00005

[a]Ref. 22.

and salty (24). The theory of the specificity of the taste buds may be subject to modification (25).

Generalizations.   Several generalizations can be made regarding taste (16,26). A substance must be in water solution, eg, the liquid bathing the tongue (saliva), to have taste. Water solubility is the first requirement of the taste stimulus (12). The typical stimuli are concentrated aqueous solution in contrast with the lipid-soluble substances which act as stimuli for olfaction (22). Many taste substances are hydrophilic, nonvolatile molecules (15). Taste detection thresholds for lipophilic molecules tend to be lower than those of their hydrophilic counterparts (16).

Only acids are sour. Sourness is not identical to chemical acidity or pH, which is a function of the hydrogen ion concentration, but also appears to be a function of the entire acid molecule. A combination of pH and acid concentration determines the actual degree of the sour taste. At the same pH, any organic acid, eg, citric acid, exhibits a far greater sourness than a mineral acid, eg, hydrochloric acid (27,28).

Only salts are salty; however, not all salts are salty. Some are sweet, bitter, or tasteless. The salty taste is exhibited by ionized salts, and the greatest contribution to salty taste comes from the cations (29). The salt taste is produced by monovalent cations (15).

Organic aromatic molecules are usually sweet, bitter, a combination of these, or tasteless, probably owing to lack of water solubility. Most characteristic taste substances, especially salty and sweet, are nonvolatile compounds. Many different types of molecules produce the bitter taste, eg, divalent cations, alkaloids, some amino acids, and denatonium (14,15).

Odor.   The physiology of odor, which is the determining characteristic of flavor, is more complex and less understood than that of taste. It has been claimed that odor is 80% of flavor. A large number of odors are distinguishable, but it is not known how this is accomplished. Olfactory response is only observed when the substance contacts the olfactory membrane (30), called the olfactory mucosa or olfactory epithelium, which occupies an area of about 2.5 cm^2 in each nostril. Above the nasal passages, the two olfactory clefts are separated by the nasal septum. For a substance to have an odor, it must be capable of reaching the olfactory epithelium high up in the nose, and must come in contact with the olfactory cilia membrane. The actual olfactory receptors are specialized protein molecules located in the cilia. These proteins are responsible for detecting odor-producing molecules. It must have sufficient volatility and low molecular weight so that it may make contact in the nasal passages by inspiration. It also appears that solubility in lipids is essential. However, a normally nonodorous or nonvolatile substance placed in contact by a spray or mist may become odorous. Many materials now regarded as useless (30) may have practical odorant applications because of this phenomenon.

The odor of a substance is most logically attributed to its molecular structure. As in taste, its perception is preceded by the process called transduction in which a chemical reaction with a receptor cell excites a nerve center, giving a sensation. Brain functions such as emotion, attention, cognition, etc, mediate these sensations into perceptions. Other theories that have been ascribed to nonchemical reactions, such as molecular vibration or infrared absorption, seem to be without solid foundation (31,32). Although odor quality appears to be associated with chemical structure, it has not been possible to predict odor types accurately on this basis (33,34). Generally, compounds of carbon and hydrogen, ie, terpenes, are odorless or of uncharacteristic odor until such atoms as oxygen, nitrogen, or sulfur are introduced. These atoms, combining to form certain functional groups such as alcohols (RC–OH), aldehydes (RCHO), amines (RC–NH$_2$), esters (RCOOR), ketones (RCOR), etc, make up what has been called an osmophore, or odor-bearing group. The feature that affects odor is the structural form of the molecule, especially its isomeric and spatial variations which includes its enantiomeric properties. An excellent example of this is found in the difference in taste and odor between L-carvone [6485-40-10] and D-carvone [2244-16-8]. L-carvone has a spearmint character and D-carvone has a dillweed, caraway, or ryebread odor and taste.

Whatever the physiology of odor perception may be, the sense of smell is keener than that of taste (22). If flavors are classed into odors and tastes as is common practice in science, it can be calculated that there are probably more than 10^4 possible sensations of odor and only a few, perhaps five, sensations of taste (13,21,35–37). Just as a hereditary or genetic factor may cause taste vari-

Table 2. Odor Detection Threshold Levels in Water and Mineral Oil,[a] ppm

Compound	CAS Registry Number	Water	Mineral oil
2-methylpyrazine	[109-08-8]	105	27
2,5-dimethylpyrazine	[123-32-0]	35	17
2,6-dimethylpyrazine	[108-50-9]	54	8
2,3,5-trimethylpyrazine	[14667-55-1]	9	27
2,3,5,6-tetramethylpyrazine	[1124-11-4]	10	38
2,5-dimethyl-3-ethylpyrazine	[13360-65-1]	43	24
2,6-dimethyl-3-ethylpyrazine	[13925-07-0]	15	24
5-isopentyl-2,3-dimethylpyrazine	[75492-01-2]	6	
6-isobutyl-2,3,-dimethylpyrazine		4.1	
5-isobutyl-2,3-dimethylpyrazine		0.77	
5-isobutyl-2-methoxy-3-methylpyrazine	[78246-20-5]	0.00018	

[a]Ref. 42.

Table 3. Odor Detection Thresholds of Organic Compounds[a]

Compound	CAS Registry Number	Detection threshold[b]	
		mg/m^3 Air	ppm/H$_2$O
acetaldehyde	[75-07-0]	0.066[c]	0.000688
acetic acid	[64-19-7]	0.025–76	24.3
acetophenone	[98-86-2]	1.5	65.0
amyl acetate	[628-63-7]	0.04–31.0	0.08
benzaldehyde	[100-51-6]	0.042[c]	0.00425
butanol	[71-36-3]	0.158–1000	2.5
butyric acid	[107-92-6]	0.001	
1,8-cineole	[470-82-6]	0.055–0.19	0.012
ethanol	[64-17-5]	4000.0[c]	100.0
ethyl acetate	[141-78-6]	3.6–1.120	5.0
ethyl acrylate	[140-88-5]	0.0010	0.067
ethyl benzoate	[93-89-0]	0.62[c]	
ethyl lactate	[97-64-3]	8.0	1.4
ethyl 2-methylbutyrate	[7452-79-1]	4.6	0.0001
ethyl vanillin	[121-32-4]	0.000007	0.1
hexanol	[111-27-3]	0.04–1.5	5.2
2-hexenal	[505-57-7]	0.034	0.017
isobutyl alcohol	[78-83-1]	1–500	7.0
isobutyric acid	[79-31-2]		8.1
isovaleric acid	[503-74-2]	0.005	0.7
α-ionone	[127-41-3]	0.0001–0.008	0.5
DL-menthone	[1074-95-9]	42.0	0.17
methyl anthranilate	[134-20-3]	0.0094[c]	
methyl salicylate	[119-36-8]	0.002–112	0.10
phenol	[108-95-2]	0.001[c]	
6-isobutyl-2,3-dimethyl pyrazine[d]			4.1
pyridine	[110-86-1]	0.03[c]	
skatole	[83-34-1]	0.0000004	
vanillin	[121-33-5]	0.000001–0.005	

[a]Ref. 39.
[b]The minimum physical intensity detected by a subject (40) who is not required to identify but just detect the existence of stimuli.
[c]Value is ppm/air.
[d]Ref. 41.

ations between individuals toward phenylthiourea, a similar factor may be in operation with odor. The odor of the steroid androsterone, found in many foods and human sweat, may elicit different responses from different individuals. Some are very sensitive to it and find it unpleasant. To others, who are less sensitive to it, it has a musk or sandalwood-like smell. Approximately 50% of the adults tested cannot detect any odor even at extremely high concentrations. It is believed that this ability is genetically determined (38).

The odor detection-threshold values of organic compounds, water, and mineral oil have been determined by different investigators (Table 2 and 3) and may vary by as much as 1000, depending on the test methods, because human senses are not invariable in their sensitivity. Human senses are subject to adaption, ie, reduced sensitivity after prolonged response to a stimulus, and habituation, ie, reduced attention to monotonous stimulation. The values give approximate magnitudes and are significant when the same techniques for evaluation are used. Since 1952, the chemistry of odorous materials has been the subject of intense research (43). Many new compounds have been identified in natural products (37–40,42,44–50) and find use in flavors.

Flavor Materials

Materials for flavoring may be divided into several groups. The most common groupings are either natural or artificial flavorings. Natural materials include spices and herbs; essential oils and their extracts, concentrates, and isolates; fruit, fruit juices, and fruit essence; animal and vegetable materials and their extracts; and aromatic chemicals isolated by physical means from natural products, eg, citral from lemongrass and linalool from bois de rose.

Artificial materials include aliphatic, aromatic, and terpene compounds that are made synthetically as opposed to those isolated from natural sources. As an example, benzaldehyde may be made synthetically or obtained from oil of bitter almond (51); and L-menthol may be made synthetically or isolated from oil of *Mentha arvensis* var. to give Brazilian mint oil or corn mint oil.

Natural and artificial flavors are defined as a combination of natural flavors and artificial flavors. It is assumed that whichever portion is in greater amount becomes the first portion of the name. For example, if the natural portion is in greater amount the flavor name is natural and artificial; if the artificial part predominates the name of the flavor is artificial and natural.

In 1992 a large number of flavor materials were allowable on the Flavor and Extract Manufacturers Association (FEMA) and FDA lists (Tables 4 and 5).

Natural Flavorings. U.S. regulations define natural flavorings as the essential oil, oleoresin, essence, extractive, protein hydrolysate, distillate, or product of roasting, heating, or enzymolysis, which contain the flavor constituents derived from a spice, fruit or fruit juice, vegetable or vegetable juice, edible yeast, herb, bark, bud, root, leaf or similar plant material, meat, seafood, eggs, dairy products, or fermentation products thereof, whose significant function in food is flavoring rather than nutrition. Natural flavoring agents include those items listed in the *Code of Federal Regulations* (3,53).

Many essential oils are used for flavoring and perfumery, eg, neroli, geranium, and ylang (see OILS, ESSENTIAL). The whole fruit, crushed fruit, and puree may be used directly in foods, ice cream, cakes, and confections. Fruit juices, concentrates, and essences are more commonly employed (see FRUIT JUICES).

Another group of natural flavoring ingredients comprises those obtained by extraction from certain plant products such as vanilla beans, licorice root, St. John's bread, orange and lemon peel, coffee, tea, kola nuts, catechu, cherry, elm bark, cocoa nibs, and gentian root. These products are used in the form of alcoholic infusions or tinctures, as concentrations in alcohol, or alcohol–water extractions termed fluid or solid extracts. Official methods for their preparation and specifi-

Table 4. Chemical Classes Approved for Use in Flavors

	Compounds		
Chemical class	1992	1965[a]	Example
sulfur	152	13	thioester, thiol, mercaptan
nitrogen	99	21	amino acids, pyrazine, ester
acids	67	42	
esters	546	372	methyl, ethyl, allyl, terpene
acetals	28	21	
aldehydes	122	21	terpene
ketones	144	64	terpene, ionone, pyrone
alcohols	143	80	terpene, phenols
ethers	52	51	dioxane, furan, oxide
hydrocarbon	8		terpene
miscellaneous	11		
Total	*1415*	*730*	

[a]Ref. 52.

Table 5. Natural and Food Ingredients Used in Flavors[a]

Compound	Number of items
Natural flavors[b]	
absolutes	19
botanical extracts	114
botanicals	249
concretes	3
essential oils	144
oleoresins	20
miscellaneous	9
Total	*558*
Food ingredients[c]	
emulsifiers	51
preservatives	49
anticaking agents	12
multipurpose	180
flavor	18
Total	*310*

[a]Information courtesy of Flavor Knowledge Systems, Glenview, Illinois.
[b]FEMA and FDA listings.
[c]FDA listing.

cations for all products used in pharmaceuticals are described (54,55). There are many flavor extracts for food use for which no official standards exist; the properties of these are solely based on suitability for commercial applications (56).

Artificial Flavorings. Artificial flavorings are defined in the *Code of Federal Regulations* (CFR) as any substance or substances, the function of which is to impart flavor, which are not derived from natural sources. These items include the list of substances found in CFR 21 parts 172.515, 182.60, and FEMA GRAS (Generally Recognized as Safe) lists. The largest group of flavoring materials is composed of hundreds of isolates from natural materials derived by chemical means and synthetic materials, eg, aldehydes, ketones, acids, esters, phenols, phenol ethers, lactones, organic derivatives of sulfur, and aliphatic, aromatic, and terpene alcohols. More than one of these chemically active or osmophoric groups may be contained in a single chemical compound, eg, vanillin (4-hydroxy-3-methoxybenzaldehyde) with phenolic, aldehydic, and ether groups; methyl anthranilate (methyl-2-aminobenzoate) with ester and amine groups; and ethyl maltol [*4940-11-8*] (2-ethyl-3-hydroxy-4*H*-pyran-4-one) with hydroxy and ketone groups, etc.

Several manuals devoted, at least in part, to flavor formulation have been published (52–63), eg, literature from the Fragrance Materials Association of the United States, Washington, D.C. The increasing number of materials available has resulted in the improvement of flavor characteristics and has permitted a closer rendition of natural flavors. Often such materials bear a scant sensory relationship to the true natural flavor character. When used as a component and judiciously applied, these materials serve a useful purpose in a properly compounded flavor.

Character Impact Items. The character impact item is a chemical or blend of chemicals that provide the principal portion of a flavor's sensory identity, ie, when tasted and/or smelled, the item is reminiscent of the named character, eg, vanillin is the character impact item for vanilla flavors (Table 6). A character item for one flavor can contribute to another flavor in a different way, for example, ethyl oenanthate is a character item for the grape flavor of the *Vinus vinifera* type and is a contributor to the flavor of the concord grape, ie, the labruska-type grape.

Applications for synthetic character impact items include cough drops, toothpaste, chewing gum, candies, soft drinks, baked goods, gelatin deserts, ice cream, margarine, and cheese.

Classification. In commerce, several classifications of flavoring and compounded flavorings are listed according to composition to allow the user to conform to state and federal food regulations and labeling requirements, as well as to show their proper application. Both supplier and purchaser are subject to the control of the FDA, USDA, and the Bureau of Alcohol, Tobacco, and Firearms (BATF). The latter regulates the alcoholic content of flavors and the tax drawbacks on alcohol, ie, return of a portion of the tax paid on ethyl alcohol used in flavoring.

One class of flavorings, known as true fruit, is composed of fruit juices, their concentrates, and their essences. A second group, fruit flavor with other natural flavors (WONF), contains fruit concentrates or extracts that may be fortified with natural essential oils or extractives (isolates), or other naturally occurring plants (64,65). This class of flavor is employed when the manufacturer is compelled by regulation to use only natural products, as in wines and cordials in the United States.

Table 6. Selected Character Impact Items

Flavor character	Materials	CAS Registry Number
almond	5-methylthiophen-2-carboxaldehyde	[13679-70-4]
	benzaldehyde	[100-51-6]
anise	anethole	[4180-23-8]
	methyl chavicol (estragol)	[140-67-0]
apple	isoamyl acetate	[123-92-2]
	ethyl 2-methylbutyrate	[7452-79-1]
	damaseneone	[23696-85-7]
	n-hexanal	[66-25-1]
	trans-2-hexenal	[6728-26-3]
asparagus	dimethyl sulfide	[75-18-3]
banana	isoamyl acetate	[123-92-2]
bergamot	linalyl acetate	[115-95-7]
blueberry	isobutyl 2-buteneoate	[589-66-2]
blue cheese	2-heptanone	[110-43-0]
	2-nonanone	[821-55-6]
bramble (artic)	2,5-dimethyl-4-methoxy-3(2*H*)furanone	[4077-47-8]
butter	diacetyl	[431-03-8]
caramel	2,5-dimethyl-4-hydroxy-3(2*H*)furanone	[3658-77-3]
caraway	D-carvone	[2244-16-8]
cassia	cinnamic aldehyde	[104-55-2]
	methoxycinnamic aldehyde	[1504-74-1]
celery	3-propylidene-1(3*H*)-isobenzofuranone	[17369-59-4]
	cis-3-hexenyl pyruvate	[68133-76-6]
cherry	benzaldehyde	[100-52-7]
	tolyl aldehyde	[1334-78-7]
	benzyl acetate	[140-11-4]
chocolate	5-methyl-2-phenyl-2-hexenal	[21834-92-4]
	isoamyl butyrate	[106-27-4]
	vanillin	[121-33-5]
	ethyl vanillin	[121-32-4]
	isoamyl phenylacetate	[102-19-2]
	2-methoxy-5-methylpyrazine	[2882-22-6]
cinnamon	cinnamic aldehyde	[104-55-2]
coconut	γ-nonalactone	[104-61-0]
coffee	furfuryl mercaptan	[98-02-2]
	furfuryl thiopropionate	[59020-85-8]
cognac	ethyl oenanthate	[106-30-9]
clove	eugenol	[97-53-0]
coriander	linalool	[78-70-6]
cream	*cis*-4-heptenal	[6728-31-0]
cucumber	nona-*trans*-2-*cis*-6-dienal	[557-48-2]
	2-nonenal	[2463-53-8]
fresh fruit	2-methyl-2-pentenoic acid	[3142-72-1]
fruity	ethyl butyrate	[105-54-4]
garlic	diallyl sulfide	[592-88-1]
	di-2-propenyl disulfide	[2179-57-9]
grape	methyl anthranilate	[134-20-3]
	ethyl 3-hydroxybutyrate	[5405-41-4]
grapefruit	nootkatone	[4674-50-4]
green bell pepper	2-methoxy-3-isobutylpyrazine	[24683-00-9]

Table 6. (*Continued*)

Flavor character	Materials	CAS Registry Number
green leafy	*cis*-3-hexenol	[928-96-1]
guava	cinnamyl acetate	[103-54-8]
	β-caryophyllene	[87-44-5]
hazelnut (filbert)	methyl(methylthio) pyrazine	[21948-70-9]
	5-methyl-2-hepten-4-one	[81925-81-7]
honey (clover)	methyl anthranilate	[134-20-3]
horseradish	1-penten-3-one	[1629-58-9]
jasmine	benzyl acetate	[140-11-4]
	indole	[120-72-9]
lamb	4-methylnonanoic acid	[45019-28-1]
lemon	citral	[5392-40-5]
lime (distilled)	α-terpineol	[98-55-5]
	citral	[5392-40-5]
mandarin	β-sinensal	[8028-48-6]
	dimethyl anthranilate	[85-91-6]
	thymol	[89-83-8]
maple	2-hydroxy-3-methyl-2-cyclopenten-1-one	[80-71-7]
meat	methyl-5-(β-hydroxyethyl) thiazole	[137-00-8]
melon	2-methyl-3-*p*-tolylpropionaldehyde	[16251-78-8]
	hydroxycitronellal dimethylacetal	[141-92-4]
	2,6-dimethyl-5-heptenal	[106-72-9]
	2-phenylpropionaldehyde	[93-53-8]
	2-methyl-3-(4-isopropylphenyl) propionaldehyde	[103-95-7]
mint	menthol	[89-78-1]
mushroom	1-octen-3-ol	[3391-86-4]
	2-octanol	[123-96-6]
	1-octen-3-one	[4312-99-6]
	2-octene-1-ol	[10849-17-1]
	1-octanol	[111-87-5]
	3-octanol	[589-98-0]
	3-octanone	[106-68-3]
mustard	allyl isothiocyanate	[57-06-7]
okra	2-methoxy-4-vinylphenol	[7786-61-0]
onion	dipropyl disulfide	[629-19-6]
orange	β-sinensal	[8028-48-6]
	octyl aldehyde	[124-13-0]
	decyl aldehyde	[112-31-2]
paprika	2-isobutyl-3-methoxypyrazine	[24683-00-9]
	2-nonene-4-one (cooked)	[14309-57-0]
passion fruit	3-methythio-1-hexanol	[5155-66-9]
peach	γ-undecalactone	[104-67-6]
	6-amyl-α-pyrone	[27593-23-3]
peanut	2,5-dimethylpyrazine	[123-32-0]
	2-methoxy-5-methylpyrazine	[68358-13-5]
pear (bartlett)	ethyl decane-*cis*-4-*trans*-2-dienoate	[3025-30-7]
peppermint	menthol	[89-78-1]
pineapple	allyl caproate	[123-68-2]
	methyl β-methylthiopropionate	[13532-18-8]
	ethyl butyrate	[105-54-4]
	allyl cyclohexanpropionate	[2705-87-5]

Table 6. (*Continued*)

Flavor character	Materials	CAS Registry Number
popcorn	methyl-2-pyridyl ketone	[5910-89-4]
potato	methional	[3268-49-3]
	2-ethyl-3-methoxypyrazine	[25680-58-4]
	2,3-dimethylpyrazine	[5910-89-4]
prune	benzyl-4-heptanone	[7497-37-7]
	dimethylbenzylcarbinyl isobutyrate	[59354-71-1]
quince	ethyl tiglate	[5837-78-5]
raspberry	6-methyl-α-ionone	[79-69-6]
	trans-α-ionone	[127-41-3]
	p-hydroxypheny-1-2-butanone	[5471-51-2]
	damasceneone	[23696-85-7]
red currant	*trans*-2-hexenol	[928-95-0]
seafood	pyridine	[110-86-1]
	piperidine	[110-89-4]
	trimethylamine	[75-50-3]
scallop	dimethyl sulfide	[75-18-3]
soursop	methyl caproate	[106-70-7]
	methyl 2-hexenoate	[32585-08-3]
soy sauce	4-ethylguaicol	[2785-89-9]
	maltol	[118-71-8]
,	phenylethyl alcohol	[60-12-8]
	methionol	[505-10-20]
	2-acetylpyrrole	[1072-83-9]
smoke	guaicol	[90-05-1]
	2,6-dimethoxyphenol	[91-10-1]
	p-vinylguaicol	[7786-61-0]
spearmint	L-carvone	[6485-40-10]
strawberry	ethylmethylphenylglycidate	[77-83-8]
	ethyl maltol	[4940-11-8]
	methyl cinnamate	[103-26-4]
	4-hydroxy-2,5-dimethyl-3(2*H*)-furanone	[3658-77-3]
sugar	4,5-dimethyl-3-hydroxy-2,5-dihydrofuran-2-one	[28664-35-9]
tangerine	dimethyl anthranilate	[85-91-6]
	thymol	[89-83-8]
	β-sinensal	[8028-48-6]
tomato	2-isobutylthiazole	[18640-74-9]
	cis-4-heptenal	[6728-31-0]
tomato (cooked)	dimethyl sulfide	[75-18-3]
	β-damaseneone	[23696-85-7]
	isovaleraldehyde	[590-86-3]
	2-methylbutyric acid	[116-53-0]
vanilla	vanillin	[121-33-5]
	ethyl vanillin	[121-32-4]
	propenyl quaethol	[94-86-0]
vinegar	ethyl acetate	[141-78-6]
	acetic acid	[64-19-7]
watercress	2-phenylethyl isothiocyanate	[2257-09-2]
wintergreen	methyl salicylate	[119-36-8]

A third class, artificial fruit flavors, includes fruit concentrates fortified with synthetic materials. These may be subdivided into two or more groups according to price, use of the proportionate strengths of the natural fruit, and synthetic fortification. Flavors other than fruit flavor can also be fortified with synthetic materials, ie, the making of an artificial maple flavor as well as an artificial meat flavor.

Specifications. Specifications for many of the essential oils and artificial flavorings are available (66). Physical specifications encourage standardization and uniformity in basic flavor and perfume materials. Although compliance with specifications does not guarantee that flavor quality standards will be acceptable, the specifications fill a need and provide a valuable reference for the flavor industry.

The *Food Chemicals Codex* defines food-grade quality for the identity and purity of chemicals used in food products. In the United States, the FDA adopts many of the *Food Chemicals Codex* specifications as the legal basis for food-grade quality of flavor and food chemicals.

Specifications also appear in other publications, including publications of the Fragrance Materials Association (FMA) of the United States (53,57) (see also FINE CHEMICALS). The FMA specifications include essential oils, natural flavor and fragrance materials, aromatic chemicals, isolates, general tests, spectra, suggested apparatus, and revisions adopted by the FMA.

In 1993, the FDA lists 199 flavorings and food adjuncts accepted, proposed, and under investigation (64). Day-to-day status of materials can be found in the *Federal Register*. The Flavor Extract Manufacturers' Association (FEMA) lists approximately 1774 GRAS chemicals for food and flavor use, including nonflavoring agents (58,67). The *Food Chemical Codex* lists approximately 845 flavoring materials (66); 116 additional specifications will be added in the fourth edition scheduled to be published in 1994.

The flavor chemist is responsible for the basic knowledge of sensory and application properties of each of this large number of raw materials; the large number of possible combinations of these items to produce specifically flavored finished compounds is readily apparent. It is not uncommon to develop a flavor that combines essential oils, plant extractive, fruit juices, and synthetics. The choice of materials depends on type of product, conditions of manufacture, labeling, and intended use.

The terms synthetic, artificial, and chemical have aroused the doubts and suspicions of consumers in some instances (68,69). However, many such chemical components also occur in nature, ie, nature identical (37,68) (see FOOD ADDITIVES). It has been noted by the FDA that an artificial flavor is no less safe, nutritious, or desirable than a natural flavor, and that the purpose for distinguishing between a natural and artificial flavor is for economic reasons, ie, the natural flavor is often more expensive than the artificial flavor (70). Since it is generally economically impractical to isolate many of the components now used in flavorings from natural product, synthesis creates effective replacements.

Compounding

In the compounding technique, constituents are selected or rejected because of their odor, taste, and physical chemical properties, eg, boiling point, solubility, and chemical reactivity, as well as the results of flavor tests in water, syrup, milk, or an appropriate medium. A compound considered to be characteristic is then combined with other ingredients into a flavor and tested as a finished flavor in the final product by an applications laboratory.

A flavor is tried at several different levels and in different mediums until the most characteristic one is selected. This is important because the character of a material is known to change quality with concentration and environment. For example, anethole, benzaldehyde, and citral taste different with and without acid. Gamma-decalactone has different characters at different levels of use. *p-tert*-Butyl phenylacetate with acid is strawberry or fruity; without acid it is creamy milk chocolate. 2,5-Dimethyl-4-hydroxy-3-(2*H*)-furanone with acid is strawberry; without acid it is caramel or meat.

Once the characteristic level is determined, the flavor is put into panel tests. After it passes these panel tests it is then subjected to storage stability.

Flavoring proportions, often referred to as dosage, used in high fat foods must be greater than in low fat foods because flavors are absorbed or suppressed by the fat, thereby lowering flavor impact (71). The distribution coefficient between the fat (oil) and water also changes, causing different flavor perception. Often the flavor must be revised to accommodate this phenomenon. A hard candy with a high glucose content compared to the cane sugar content requires either more flavoring or an increase in the more potent ingredients of the flavoring than a hard candy with a low glucose content.

Flavor and Diluent Portions. *Flavor Portion.* A flavor compound consists of a flavor portion, ie, flavor character impact item(s), as well as flavor contributory and flavor differential items, and a diluent portion.

A flavor contributory item is an additive that when smelled and/or tasted helps to create, enhance, or potentiate the named flavor. It is not characteristic of the flavor, but essential in that it acts with other substances to produce a definite character.

A flavor differential item is an additive or combination of additives that when smelled or tasted has little, if any, character reminiscent of the named flavor. It gives roundness and fixation to the flavor. It may be added by the flavor chemist to confuse simulation of the flavor, and it is neither characteristic of, nor essential to, the intended flavor. The greatest examples of creativity are found in this area.

The flavor portion of a flavor compound gives it its name, acceptability, and palatability, and provides character fixation of the flavor, ie, relatively high boiling point solids, usually in combination, are used at concentrations above their threshold values at use level so that upon dilution the levels remain above threshold value and the perception of the flavor does not change.

Diluent Portion. The diluent portion of a flavor compound is the carrier for the color and the flavor, ie, the solvent for the flavor portion. It keeps the flavor homogenous, ie, keeps solids in solution; retards chemical reactions from occurring; and regulates flavor strength, ie, the greater the amount of solvent, the weaker the flavor.

The diluent gives the flavor a physical fixation. Relatively high boiling point materials are used in the diluent to make the flavor less heat labile. They are included when a flavor is to be used at temperatures above the boiling point of water; examples include vegetable oils and isopropyl myristate.

The diluent portion also determines the form, or physical appearance, of the flavor, ie, liquid, powder, or paste. Liquid flavor forms include water-soluble, oil-soluble, and emulsion forms; powder flavor forms include plated (including dry solubles), extended, occluded, inclusion complexes, and other encapsulated forms; and paste flavor forms include fat, protein, and carbohydrate-based paste.

Flavor Formulas. Tables 7 and 8 give examples of modern flavor formulas. In Table 7 formula A is composed of fruit juice concentrate and essence distilled

Table 7. Modern Flavor Formulas, % by Weight

Ingredient	A	B	C
fruit juice, 72 brix[a]	80.0	40.0	60.0[b]
fruit essence[c]	5.0	20.0	
fortifier			
natural[d]		1.0	
artificial[e]			1.0
ethyl alcohol, 95%	15.0	15.0	15.0
water		24.0	24.0

[a]Brix = g of sugar per 100 g liquid. [b]60 brix.[a] [c]150-fold, ie, one gallon (3.785 L) of concentrated distillate is obtained from 150 gallons of single-fold juice. [d]A blend of botanical extracts; natural chemicals, ie, isolates and those derived via natural processes, eg, ethyl acetate, absolute tagette; and oil petitigrain mandarin. [e]Composed of some, if not all, artificial chemicals, botanical extractive, essential oil, etc.

Table 8. Pineapple Flavor, Artificial

Ingredient	Wt %
Characterizing flavor items	
allyl cyclohexane propionate	1.4
allyl caproate	13.0
methyl β-methylthiolpropionate	0.2
Contributory flavor items	
geranyl propionate	0.5
ethyl isovalerate	1.0
ethyl butyrate	1.0
γ-nonalactone	0.1
Differential flavor items	
maltol	1.0
vanillin	0.5
2,5-dimethyl-3(OH)-4-(2H)-furanone	0.2
oil orange	1.0
Diluent portion	
ethyl alcohol, 95%	46.0
propylene glycol	34.1

or extracted from the fruit juice. It is all natural and all from the named fruit, and is therefore termed a "natural flavor." It has a characterizing natural flavor. In Formula B the flavor is all natural, but is not all from the named fruit, ie, the fortifier is all natural but is not totally derived from the named fruit. Since the fortifier simulates, resembles, or reinforces the named flavor, eg, apple or pine-

Table 9. Essential Oil, Herb, and Spice Equivalents

Herb or spice	Essential oil equivalent,[a] kg
allspice (pimento berries)	2.5
almond, bitter	0.5
angelica root	0.75
angelica seed	1.0
anise seed	2.5
basil, sweet reunion	0.15
caraway seed	2.5
cardamom seed	3.0
cassia cinnamon	1.5
celery seed	2.0
cinnamon Sri Lanka	0.5
clove	15.0
coriander seed	0.75
cumin seed	3.0
dill seed	4.0
dill weed	1.0
tarragon	0.13
fennel seed	5.0
garlic	0.5
ginger	0.25
horseradish	1.0
laurel leaves (bay)	1.0
lovage root	0.5
mace	5.0
marjoram, sweet	0.5
mustard seed	0.25
nutmeg	5.0
onion	0.05
origanum	0.75
parsley seed	3.0
pepper, black	1.0
rosemary	0.5
sage	1.25
savory	0.25
thyme	2.0
valerian root	1.0

[a]To 100 kg of corresponding best quality dry herb or spice.

apple, the flavor must be called "flavor with other natural flavors." It has a natural flavor with characterizing naturals added. Formula C is composed of both natural and artificial components with the natural usage outweighing the artificial. Therefore, it is a "flavor natural and artificial." It has a characterizing natural and artificial flavor.

The formula of an artificial pineapple flavor is given in Table 8. The flavor contains no natural pineapple components, ie, juice and essence, and the artificial portion far outweighs the natural portion of the flavor; this flavor is a "flavor artificial." It has a characterizing artificial flavor.

The characterizing, contributory, and differential flavor items are listed. The diluent portion of the flavor makes the flavor applicable and is generally the largest part of every flavor formula.

Essential Oil Equivalents.   Essential oil equivalents (Table 9) are now commonly used to replace, in whole or in part, the dry spices from which they are derived. The latter serve primarily to enhance the attractiveness or appearance of the products, which are usually packed in glass containers. For example, a blend of 12.7 grams of the essential oils of black pepper, ginger, allspice, cinnamon, nutmeg, and clove is equivalent to the flavoring strength of one kilogram of the dry seasoning blend.

Enzyme Flavor Precursors

The characteristic flavors of foods, such as in fruits and vegetables, are considered to result from enzymatic action upon certain more complex components during the normal developmental or ripening process (72). The formation of volatile flavor components has been studied *in vitro* (73–75). Fruits and vegetables were deflavored by heat, usually by boiling, and their flavor was partially restored by the addition of an enzyme. Although many different enzymes develop from a precursor only certain ones produce the characteristic flavor (73,74). Thus a true raspberry flavor is produced in a deflavored raspberry puree only by enzymes obtained from raspberry fruit (76). A vegetable product contains several different enzymes in varying proportions that cause the formation of volatile components in ratios that may be different from the actual ratio of precursors present in that particular product.

It has been found that the flavor of fruit can be increased by a process called precursor atmosphere (PA) (77). When apples were stored in a controlled atmosphere containing butyl alcohol [71-36-3], the butyl alcohol levels increase by a factor of two, and the polar products, butyl ester, and some sesquiterpene products increase significantly. The process offers the possibility of compensating for loss of flavor in fruit handling and processing due to improper transportation conditions or excessive heat.

Another process employed to increase the formation of volatile compounds in fruit is that of bioregulators. When a bioregulator is applied to lemon trees an increase in both the aldehyde and alcohol fractions of the lemon oil extracted from the fruit of the treated lemon trees was observed (78).

Enzymes not only produce characteristic and desirable flavor (79) but also cause flavor deterioration (80,81) (see ENZYMES, INDUSTRIAL). The latter enzyme types must be inactivated in order to stabilize and preserve a food. Freezing depresses enzymatic action. A more complete elimination of enzymatic action is accomplished by pasteurization.

The creation of flavor by enzymes is used in the fermentation process to prepare products such as alcoholic beverages, cheese, pickles, vinegar, bread, and sauerkraut. In some vegetables, such as Cruciferae (mustard) and Alliacae (onion and garlic), the flavor components are released enzymatically when the tissue is crushed or broken. In the case of Cruciferae, the enzyme myrosinase [9025-38-1] degrades the naturally occurring glucosinates to isothiocyanates. In the Alliacae, the enzyme alliinase [9031-77-0] acts on the substrate, ie, alkyl cysteine-S-oxide, to release the flavor and lachrymator found in the onion. Several essential oils are also created enzymatically. In the case of oil sweet birch, the oil is released from the bark enzymatically; in the case of oil bitter almond, the enzyme emulsin [9001-22-3] attacks the glucoside (amygdalin) and releases the oil of bitter almond (82,83) (see FERMENTATION).

A more complex flavor development occurs in the production of chocolate. The chocolate beans are first fermented to develop fewer complex flavor precursors; upon roasting, these give the chocolate aroma. The beans from unfermented cocoa do not develop the chocolate notes (84–88) (see CHOCOLATE AND COCOA). The flavor development process with vanilla beans also allows for the formation of flavor precursors. The green vanilla beans, which have little aroma or flavor, are scalded, removed, and allowed to perspire, which lowers the moisture content and retards the enzymatic activity. This process results in the formation of the vanilla aroma and flavor, and the dark-colored beans that after drying are the product of commerce.

The use of dry heat, as in roasting (89,90), baking, and frying, develops flavor characteristics not found in the unheated product (91). The roasting of meats (47,92), coffee, (93), peanuts (94), and sesame seeds (95); the baking of bread (96,97); and the baking (98) or frying of potatoes (99) are methods by which desirable flavor is created. The distinctive flavor development in breakfast foods, ie, the crust of baked bread, the aroma of roasted coffee, etc, can be directly attributed to the chemical combinations brought about during the heat-treatment operation. These types of flavor are generally characterized by the presence of pyrazines in the product. The reactions of sugars and amino acids (or proteins), ie, the Maillard reaction (100), is accomplished by a loss of free amino groups, increased acidity (101), the evolution of carbon dioxide, and the development of flavors and a brown color. Glucose combines with α-aminobutyric acid [80-60-4] to give a product with a maple flavor, and glucose and/or fructose reacts with proline [609-36-9] to produce a bread-like flavor (102) (Table 10). A honey-flavored syrup can be obtained by heating glucose, β-phenylalanine [63-91-2], invert sugar, and glutamic acid [56-86-0]. Many amino acids generate a typical aldehyde odor when heated with glucose, probably due to the Strecker degradation portion of the Maillard reaction; ie, the decarboxylation and deamination of an α-amino acid to an aldehyde or ketone of one carbon less than the starting amino acid (eq. 1 and 2) (104,105).

$$(CH_3)_2-CH_2-CH_2-CH(NH_2)-COOH \xrightarrow{H_2O} (CH_3)_2-CH_2-CH_2-CHO + CO_2 + NH_3 \quad (1)$$

$$2\ HSCH_2-\underset{NH_2}{\overset{|}{CH}}-COOH + 2\ H_2O \rightarrow HSCH_2-\underset{O}{\overset{\|}{C}}-H + 2\ NH_3 + H_2S + H_3C-\underset{O}{\overset{\|}{C}}-H + 2\ CO_2 \quad (2)$$

Other examples are glycine → formaldehyde, alanine → acetaldehyde, valine → isobutyraldehyde, phenylalanine → phenylacetaldehyde, and methionine → methional (106). Products such as dried skim milk, dried eggs, and dehydrated vegetables and fruits are particularly susceptible to deteriorative flavor changes ascribed to this reaction (Table 10).

Table 10. Amino Acid Sugar Reactions[a]

Amino acid	Sugar	Temperature, °C	Aroma
cysteine	glucose	100	meat
	ribose	180	sulfur, spicy meat
	arabinose	100	beef
cystine	glucose	100	meat, burnt turkey skin
	ribose	180	meat with H_2S note
glutamic acid	ribose	180	roasted meat
methionine	ribose	180	crust of roasted meat

[a]Ref. 103.

Economic Aspects

The flavoring extract and syrups industry employs approximately 8200 employees (107) in approximately 280 companies; 99 companies employ more then 20 workers (Table 11). The total number of flavor companies has diminished since 1972, but the percentage of larger flavor companies has remained fairly constant (108).

Table 11. Estimated Employment of Flavoring Extract and Syrups Industry

Year	Companies	Companies with > 20 employees, (%)
1972	400	115 (28.8)
1977	368	132 (35.9)
1982	343	131 (38.2)
1987	280	99 (35.4)

Flavor Regulations

The Pure Food and Drug Act of 1906 introduced federal regulations to combat food and drug alteration and fraud (109). The law was superseded by the Food, Drug, and Cosmetic Act of 1938 and the Amendments of 1954, 1958, 1960, and 1962 (110) (see FOOD ADDITIVES). Flavor regulations differ from country to country but progress is being made to harmonize regulations in the interest of international trade (111). There are at least five different ways countries can control and administer flavor regulations: countries may have a positive list of flavor materials, a negative list of flavor materials, a mixed-system having both a positive and negative list, no flavor regulations, or countries may demand prior government approval before a material can be used.

Countries that use a positive list, eg, Japan, Switzerland, the former Soviet Union, and the United States, list all substances allowable in flavor and foods; any material not listed is not allowable. Any material included on a positive list is considered safe for its intended use. This system works well only if there is a specific procedure to allow for the addition of new materials. In the United States there is the Food Additive Petition and the GRAS route. No new material can be used in flavor or foods unless it undergoes one of these two procedures. This assures that any new technology can be added and used after it is reviewed for safety.

Countries with a negative list system, eg, Australia, Brazil, Canada, Chile, India, New Zealand, and Singapore, define flavoring substances that cannot be used or may only be used in very limited and strictly defined ways. All materials not on such lists may be used without limitation. This system works well with all natural and nature identical flavor materials, but it is not good for controlling the use of new artificial materials. Any new flavor material created will not be specifically listed, and can theoretically be used.

Countries that use a mixed-system, eg, Argentina, Germany, Italy, Spain, and the Netherlands, have a positive list for artificial flavor materials, and a restricted list of natural and nature identical flavor materials. This system addresses some of the shortcomings of the positive and negative list systems.

Several countries have no specific flavor regulations, eg, Austria, Belgium, Denmark, Ireland, France, Greece, Jamaica, Luxembourg, Norway, Portugal, South Africa, Sweden, and the United Kingdom. This can cause difficulty when trading with these countries. Although these countries have no codified legislation, they have definite ideas as to what can be used. It is best to consult with the specific agencies in these countries before selling flavors or foods to these countries.

In some countries, eg, Finland, Hungary, Peru, Poland, and Bulgaria, one must petition the appropriate government agency and receive permission to manufacture or sell a flavor or a flavored product. Newly emerging countries, although they have no specific regulations, often accept other countries' legislation. This is often the case when dealing with the countries of Africa.

Sensory Evaluation

The type of food and its processing affect flavoring efficiency; therefore, flavor materials must be taste-tested in the food itself. Because there has been a lack of standardization of testing techniques, a committee on sensory evaluation of the Institute of Food Technologists has offered a guide (112) which is designed to help in developing standard procedures.

For each type of problem, appropriate taste tests are suggested together with the type of panel, number of samples per test, and analysis of data. This useful summary eliminates a large amount of difficult to interpret data and contains a list of original references. Specific tests are outlined for new product development, product and process improvement, cost reduction, selection of new source of supply, quality or rating selection of the best sample, market testing of new or improved product, consumer preference tests, and selection of trained panelists (113–116). Ten types of test, designed for trained, untrained, semitrained, and highly trained panels, are explained (117,118). The methods of sample presentation include single-sample, paired-comparison, and triangle test, ie, three samples are employed, two of them alike. Other methods are fitted to the attainment of specific objectives; for these, the type of panel and number of panelists are also suggested. Flavor evaluation is adequate only when the origin of flavor is known and differences can be measured by a scientific method based on units or significant qualities of flavor.

The order of sample presentation has been shown to lead to errors in judgment (119), eg, the later samples in a series were rated lower; serving good samples first lowered the rating of poor samples and serving poor samples first lowered the rating of good samples. The literature on sensory testing includes References 120–123 (see FLAVOR CHARACTERIZATION).

Glossary

Aftertaste. The experience that, under certain conditions, follows removal of the taste stimulus; it may be continuous with the primary experience or may follow as a different quality after a period during which swallowing, saliva, dilution, and other influences may have affected the stimulus substance. The result of the persistence of a flavor note, particularly after swallowing.

Ageusia. Lack or impairment of sensitivity to taste stimuli.

Agnosia. Inability to recognize sensations; may be primarily in one sense, eg, olfactory agnosia.

Anosmia. Inability to smell, either totally or a particular substance or group of substances.

Antetaste. A prior taste, or foretaste, usually of short duration, preceding the main taste or flavor characteristic.

Aroma. The fragrance or odor of food, perceived by the nose by sniffing. In wines, the aroma refers to odors derived from the variety of grape, eg, muscat aroma. It is the overall odor impression as perceived by the nasal cavity.

Autosmia. Disorder of the sense of smell in which odors are perceived when none are present.

Bland. Having no distinctive taste or odor property.

Cacogeusia. Persistent or intermittent unpleasant taste in the mouth.

Cacosmia. Perception of persistent or intermittent unpleasant odor.

Cloying. A taste sensation that stimulates beyond the point of satiation; frequently used to describe overly sweet products.

Compatibility. In flavor terminology the ability of one substance to enhance the flavor characteristic of another.

Compensation. The result of interaction of the components in a mixture of stimuli, each component of which is perceived as less intense than it would be alone.

Convergence. The tendency of a test sample, regardless of quality, to be perceived as similar to prior sample(s); sometimes called the halo effect.

Cryptosmia. Impairment of olfaction by obstruction of the nasal passages.

Dysosmia. Difficulty in ability to smell.

Essential Oil. The volatile material, derived by a physical process, usually distillation, from odorous plant material of a single botanical form and spices with which it agrees in name and odor.

Flavor. The sensation produced by a material taken into the mouth, perceived principally by the senses of taste and smell, but also by the common chemical sense produced by pain, tactile, and temperature receptors in the mouth.

Flavor or Flavorant. A substance, added to food, whose significant function is to affect odor, imparting a characteristic flavor to that food.

Flavors. Those mixtures of ingredients whose exact composition is usually known only to their suppliers, sold in bulk to food and beverage manufacturers. They are to be labeled as flavors per CFR 21 part 101 and may contain adjuncts that are nonflavor ingredients.

Flavoring Ingredient. Any single chemical entity or natural mixture added to food, drugs, or other products taken in the mouth, the clearly predominant purpose and effect of which is to provide all or part of the particular flavor of the final product.

Flavor Adjunct. A substance used in or with a flavor but not essentially a part of it. These include solvents, antioxidants, enzymes, adjusting agents, emulsifiers, and acidulants.

Flavor Enhancer. A substance added to supplement, modify, or enhance the original taste and/or aroma of a food without imparting a characteristic taste or odor of its own.

Fold. Strength of concentrated flavoring materials. The concentration is expressed as a multiple of a standard, eg, citrus oil is compared to cold pressed oil. In the case of vanilla, folded flavors are compared to a standard extract with minimum bean content.

Gustation. A taste sense, the receptors of which lie in the mucous membrane covering the tongue, and the stimuli for which consist of certain soluble chemicals, eg, salts, acid, and sugar.

Hyperosmia. Unusually keen olfactory sensitivity.

Hypogeusia. Diminished sense of taste.

Hyposmia. Diminished sense of smell.

Insipid. Tasteless, flat, vapid.

Isolate. A relatively pure chemical produced from natural raw materials by physical means, eg, distillation, extraction, crystallization, etc, and therefore natural; or by chemical means, ie, via hydrolysis, bisulfite addition products, and regeneration, etc, and therefore artificial by 1993 U.S. labeling regulations.

Macrosmatic. Abnormally keen olfactory sense.

Merosmia. A condition analogous to color blindness, in which certain odors are not perceived.

Microsmatic. Having a poorly developed sense of smell.

Nature Identical Flavor Material. A flavor ingredient obtained by synthesis, or isolated from natural products through chemical processes, chemically identical to the substance present in a natural product and intended for human consumption either processed or not; eg, citral obtained by chemical synthesis or from oil of lemongrass through a bisulfite addition compound.

Odor and Odorant. That which is smelled. Odor may refer to the odorant or to the sensation resulting from the stimulation of olfactory receptors in the nasal cavity by gaseous material.

Odoriphore (Osmophore). Odor-producing group.

Osmics. The science of smell.

Osmyl. An odorant.

Parageusia. Gustatory disturbance resulting in erroneous identification of taste stimuli.

Parosmia. A disturbance to the sense of smell resulting in smelling the wrong odors, usually perceived as repulsive.

Sapid. Having the power of affecting the taste receptor.

Savory. Appetizing; having an agreeable flavor.

Sensory Analysis. The science of measuring and evaluating the properties of food products by one or more human senses.

Stevens Power Law, $S = I^n$. The increase in perceived intensity, S, is equal to the concentration, I, to the nth power.

Volatile Oil. That portion of a botanical that codistills with water during steam distillation and is generally flavorful.

BIBLIOGRAPHY

"Flavors and Spices" in *ECT* 1st ed., Vol. 6, pp. 581–594, by E. C. Crocker, Arthur D. Little, Inc.; in *ECT* 2nd ed., Vol. 9, pp. 347–380, by E. H. Hamann and E. Guenther, Fritzsche Brothers Inc.; in *ECT* 3rd ed., Vol. 10, pp. 456–488, by J. A. Rogers, Jr., and F. Frischetti, Jr., Fritzsche Dodge & Olcott, Inc.

1. *Food Tech.* **23**, 11 (1969).
2. E. von Sydow, *Food Tech.* **25**,40 (1971).
3. F. J. Hammerschhmid, *Dragoco Report.*, 244 (Oct. 1977).
4. J. F. Caul, *Cereal Sci. Today* **12**(1),275 (1967).
5. *Soc. Flavor Chemists Candy Ind.* **5**, 10 (Aug 6, 1968).
6. *Food Eng.*, 26 (Feb. 1992).
7. G. Mackinney and A. C. Little, *Colors of Foods*, Avi Publications Co., Westport, Conn., 1962.
8. S. A. Matz, *Food Texture*, Avi Publications Co., Westport, Conn., 1962.
9. R. W. Moncrieff, *Flavour Ind.*, 828 (Dec. 1971).
10. *Ibid.*, 84 (Feb. 1971).
11. *Ibid.*, 583 (Sept. 1970).
12. R. W. Moncrieff, in H. W. Schultz, H. Day, and L. M. Libby, eds., *Symposium of Foods: The Chemistry and Physiology of Flavor*, Avi Publications Co., Westport, Conn., 1967.

13. M. O'Mahony and R. Ishii, in Y. Kawamure and M. R. M. Kare, eds., *Umami: A Basic Taste*, Marcel Dekker, New York, 1987.
14. M. H. Akabas, *Intern. Review of Neurobiology*, **32**, 255 (1990).
15. S. E. Kinnamon and T. A. Cummings, *Ann. Rev. Physiol.* **54**, 715–731 (1992).
16. R. J. Gardner, *Chem. Senses Flavour* **4**, 275 (1987).
17. *Flavor Research and Food Acceptance*, Reinhold Publishing Corp., New York, 1958.
18. J. J. Brennan and U. Leffler, *Proceedings Canadian Food Technologists Conference*, Montreal, May 1964.
19. R. W. Moncrieff, *Perfumery Res. Oil Record* **55**, 205 (1964).
20. S. S. Schiffman and R. P. Erickson, *Physiology* **7**, 617 (1971).
21. B. G. Green, in H. T. Lawless and B. P. Klein, eds. *Sensory Science Theory and Applications in Foods*, Marcel Dekker, Inc., New York, 1991.
22. R. W. Moncrieff, *The Chemical Senses*, CRC Press, Cleveland, Ohio, 1967.
23. R. M. Pangborn and R. B. Chrisp, *J. Food Sci.* **29**(4), 490 (1964).
24. G. von Bekesy, *Science* **145**, 834 (1964).
25. H. L. Meiselman, *Crit. Rev. Food Technol.*, 89 (Apr. 1972).
26. R. W. Bragg and co-workers, *J. Chem. Ed.* **55**, 2812 (1978).
27. K. Kulka, *Agric. Food Chem.* **15**, 48 (Jan/Feb 1967).
28. S. Price and J. A. Desiomone, *Chem. Senses and Flavor* **2**, 448 (1977).
29. *Ibid.*, p. 432.
30. D. Ottoson, "Odor and Taste," in *Proceedings First International Symposium Wenner Gren Center*, Stockholm, 1962, New York, 1963.
31. F. Tomel, *Dragoco Rep.*, 191 (Sept. 1977).
32. *Chem. Ind.*, 8 (Mar. 25, 1967); R. H. Adrian in Ref. 30.
33. Y. R. Hayes, *Soc. Chem. Ind. (London)*, 38 (1957).
34. M. Stoll, *Soc. Chem. Ind. (London)*, 1 (1957).
35. H. R. Schiffman, *Sensations and Perception: An Integrated Approach*, John Wiley & Sons, Inc., New York, 1976, p. 123.
36. F. Drawert, *Proceedings of the International Symposium on Aroma Research*, Zeist, the Netherlands, 1975, p. 13.
37. S. Van Straten, ed., *Volatile Compounds in Food*, 6th ed., Central Institute for Nutrition and Food Research, Zeist, the Netherlands, 1983.
38. C. J. Wysocki and G. K. Beauchamp, *Proc. Nat. Acad. of Sci. U.S.A.*, **81**, 4899 (1984).
39. L. J. van Gemert and A. H. Nettenbreiger, *Compilation of Odor Threshold Values in Air and Water*, National Institute for Water Supply, Voolburg, the Netherlands, Central Institute for Nutrition and Food Science, Zeist, the Netherlands.
40. W. H. Stahl, ed., *Compilation of Odor and Taste Threshold Values Data, DS 48*, American Society for Testing Materials, Philadelphia, Pa., 1973.
41. *J. Food Sci.* **51**(4), (1986).
42. P. E. Koehler, M. E. Mason, and G. V. Odel, *J. Food Sci.* **36**, 817 (1971).
43. E. C. Bate-Smith and T. N. Morris, *Food Science*, Cambridge University Press, U.K., 1952.
44. J. A. Maga, *Crit. Rev. Food Sci. Nutr.* **5** (July 1975).
45. *Ibid.*, (Sept. 1975).
46. *Ibid.*, (Jan. 1975).
47. B. M. Dwivedi, *Crit. Rev. Food Sci. Nutr.* **5** (Apr. 1975).
48. L. Schutte, *Crit. Rev. Food Sci. Nutr.* **4** (Mar. 1974).
49. J. A. Maga and C. E. Sizer, in T. F. Furia and N. Bellanca, eds., *Fenaroli's Handbook of Flavor Ingredients*, Vol. 1, 2nd ed., CRC Press, Boca Raton, Fla., 1975, p. 47.
50. H. E. Nursten and A. A. Williams, *Chem. Ind.*, 487 (Mar. 25, 1967).
51. N. A. Shaath and B. Benveniste, *Natural Oil of Bitter Almond, Per. Flav.* **16** (Nov./Dec., 1991).

52. "Multifaceted Nature of the Flavorists," papers presented by the *Society of the Flavor Chemists*, Rutgers University, New Brunswick, N.J., 1974.

53. *Code of Federal Regulations,* CFR 21, Part 172.510, 182.10–182.50, and 101.22, Washington, D.C., 1988.

54. *U.S. Pharmacopeia XX, (USP XX-NF XV)*, The United States Pharmacopeial Convention Inc., Rockville, Md., 1980; *National Formulary XIVV*, American Pharmaceutical Association, Washington, D.C., 1975.

55. "Food Flavoring" in *Code of Federal Regulations*, Part 22, Sections 22.1–22.7, 1986.

56. A. V. Saldarini, *Flavour Ind.*, 247 (Sept./Oct. 1974).

57. T. E. Furia and N. Bellanca, *Fenaroli's Handbook of Flavor Ingredients*, Vol. 2, 2nd ed., CRC Press, Boca Raton, Fla., 1975, pp. 3–564.

58. R. A. Ford and G. M. Cramer, *Perf. Flav.* **2**(1) (1977).

59. M. B. Jacobs, *Synthetic Food Adjuncts*, D. Van Nostrand Co., Inc., New York, 1947.

60. J. Merory, *Food Flavoring (Composition, Manufacture and Use)*, Avi Publications Co. Inc., Westport, Conn., 1968.

61. H. Heath, *Source Book of Flavors*, Avi Publications Co. Inc., Westport, Conn., 1981.

62. H. B. Heath, *Flavor Technology: Profiles, Products, Applications*, Avi Publications Co. Inc., Westport, Conn., 1978.

63. H. B. Heath, *Flavor Chemistry and Technology*, Avi Publications Co. Inc., Westport, Conn., 1986.

64. *Fed. Reg.* **42**, 14,315 (Mar. 15, 1977).

65. *Code of Federal Regulations 21*, paragraph 101.22, 3(a) iii,i, 1988.

66. *Food Chemical Codex*, 3rd ed., National Academy of Sciences–National Research Council, Washington, D.C., 1980.

67. F. Grundschhober and co-workers, *Flavours*, (July/Aug. 1975).

68. N. J. Leinen, *Food Proc.*, 28 (Mar. 1978).

69. A. E. Sloan, *Food Eng.*, 72 (Sept. 1985).

70. *Fed. Reg.* **38**, 33,285 (Dec. 3, 1973).

71. W. Grab, E. Beck, and T. Bernegger, *Intern. Flav. Food Add.*, 63 (Mar./Apr. 1977).

72. M. S. Konigsbacker and E. J. Hewitt, *Ann. N.Y. Acad. Sci.* **116**, 705 (1964).

73. E. J. Herwitt and co-workers, *Food Tech.* **487**, 10910 (1956).

74. K. S. Konigsbacher, E. J. Hewitt, and R. L. Evans *Food Tech.* **13**(2),128(1959).

75. D. A. M. Mackay and E. J. Hewitt, *Food Res.* **24**, 253 (1959).

76. C. Weurman, *Food Tech.* **15**, 531 (1961).

77. R. G. Berger, *Perf. Flav.* **15**, 34 (Mar./Apr. 1990).

78. H. Yokoyama and co-workers, *Food Tech.*, 111–113 (Nov. 1986).

79. B. C. deLumen and co-workers, *J. Food Sci.* **43**, 698 (1969).

80. G. Reed, ed., *Enzymes in Food Processing*, Academic Press, Inc., New York, 1975.

81. T. Godfrey, *Int. Flav. Food Add.*, 163, (July/Aug. 1978).

82. G. G. Freeman, in D. G. Land and H. E. Nursten, eds., *Progress in Flavour Research*, Applied Science Publishers, London, 1979.

83. E. Guenther, *The Essential Oils*, Vols. 5 and 6, D. Van Nostrand Co. Inc., New York, 1952.

84. T. A. Rohan, *J. Food Sci.* **29**, 457 (1964).

85. T. A. Rohan and M. Connell, *J. Food Sci.* **29**, 460 (1964).

86. T. A. Rohan, *Flavour Ind.*, 147 (1971).

87. G. A. Reineccius, P. G. Kenney, and W. Weisberger, *J. Agr. Food Chem.* **20**, 202 (1972).

88. G. C. Harlee and J. C. Leffingwey, *Tobacco International* **46**, 407 (Mar. 9, 1979).

89. L. C. Maillard, *Compt. Rend.* **154**, 66 (1912).

90. J. Adrian, *Labo. Pharma. Prob. Tech.*, (244) (June 1975).

91. J. E. Hodge, *Chemistry and Physiology of Flavors*, Avi Publications Co., Westport, Conn., 1967.

92. R. A. Lawrew, *Flavour Ind.*, 591 (1970).
93. C. G. Tassan and G. F. Russell, *J. Food Sci.* **39**, 64 (1974).
94. J. A. Newell, M. E. Mason, and R. S. Matio, *J. Agric. Food Chem.* **15**, 767 (1967).
95. C. H. Manley, P. P. Vallon, and R. E. Erickson, *J. Food Sci.* **39**, 73 (1974).
96. I. R. Hunter and co-workers, *Cereal Sci. Today* **11**, 493,496 (1965).
97. C. E. Sizer, J. A. Maga, and K. Lorenz, *Lebensm. Wiss. Technol.* **8**, 267 (1975).
98. S. R. Parales and S. S. Chang, "Identification of Compounds Responsible for Baked Potato Flavor," in *Paper of the Journal Series*, New Jersey Agricultural Experimental Station, Rutgers University, New Brunswick, N.J., Oct. 1, 1973.
99. S. S. Chang, R. J. Peterson, and C. T. Ho, in M. K. Supran, ed., *Chemistry of Deep Fat Fried Flavor in Lipids as a Source of Flavor*, ACS Symposium Series 75, American Chemical Society, Washington, D.C., 1978.
100. G. R. Waller and M. S. Feather, eds., *The Maillard Reaction in Foods and Nutrition*, ACS Symposium Series No. 215, American Chemical Society, Washington, D.C., 1983.
101. T. F. Stewart, *A Survey of the Chemistry of Amino Acids-Reducing Sugar Reaction in Relation to Aroma Production*, Scientific and Technical Surveys No. 61, British Food Manufacturing Industries Research Association, London, Dec. 1969.
102. Brit. Pat. 1,223,796 (Mar. 3, 1971), E. Hitoshi and S. Okumura (to Ajinomoto Co, Inc.).
103. R. A. Wilson and I. Katz, *Flavor Ind.* 5,30 (1974).
104. R. H. Walter and J. S. Fagerson, *J. Food Sci.*, 33 (1968).
105. W. J. Herz and R. G. Schallenberger, *Food Res.* **25**, 491 (1960).
106. *Die Nahrung* **24**(2), 115–127 (1980).
107. *1989 Annual Survey of Manufacturers*, Statistics for Industry Group and Industries, M 89(AS-1), U.S. Department of Commerce, Washington, D.C., 1989.
108. *1987 Census of Manufacturers*, Industries Series, Beverages, MC 87-1-20H, U.S. Department of Commerce, Washington, D.C., 1990.
109. W. W. Lawrence, *Of Acceptable Risk, Science and the Determination of Safety*, William Kaufman Inc., Los Altos, Calif., 1976.
110. R. L. Hall and B. L. Oser, *Food Tech.* **19**(2) (Feb. 1965).
111. *Food Tech.* **47**(3), 106, 118, 125 (1993).
112. *Inst. Food Technologists* **18**(8), 25 (1964).
113. E. Larmond, *Laboratory Methods for Sensory Evaluation of Food*, publication No. 1637, Canada Department of Agriculture, Ottawa, 1977.
114. W. H. Stahl and M. A. Einstein, in F. D. Snell and L. S. Ettre, eds., *Encyclopedia of Industrial Chemical Analysis*, Vol. 17, Wiley-Interscience, New York, 1973, pp. 608–644.
115. *A General Guide to Sensory Evaluation*, McCormick and Co., Hunt Valley, Md., 1978.
116. *Manual of Sensory Testing Methods*, ASTM Special Technical Publication 434, Philadelphia, Pa., 1968.
117. J. F. Caul, *Advan. Food Res.* **7**, 1 (1957).
118. S. E. Cairncross and L. B. Sjostron, *Food Tech.* **4**, 308 (1950).
119. J. Eindoven and co-workers, *J. Food Sci.* **29**, 520 (1964).
120. R. M. Pangborn, *Food Technol.* 9(11), 63 (1964).
121. M. A. Amerine, R. M. Pangborn, and E. B. Roessler, *Principles of Sensory Evaluation of Food*, Academic Press, Inc., New York, 1965.
122. A. W. Williams, *Intern. Flav. and Food Add.*, 80 (Mar./Apr. 1978);131,133,135 (May/June 1978); (July/Aug. 1978).
123. H. W. Spencer, *Flavour Ind.*, 293 (May 1971).

FRANK FISCHETTI, JR.
Craftmaster Flavor Technology, Inc.

SPICES

A spice is any aromatic, pungent, or colored vegetable product used in preparation or cooking to give zest and a piquant or pleasing flavor or color to foods; it is not a food of itself. The vegetable product may be fresh, dried, or ground. Examples of spices include ground dried cinnamon, whole dried cloves, ground fresh basil leaves, pressed garlic, and powdered paprika. Traditionally, what distinguished a spice was that it was used in the cooking process. This distinction was particularly applicable to the tropical or ancient spices. While modern cooking methods have made this difference somewhat less specific, it remains a reasonable distinction to maintain in the definition of spice.

A condiment is a vegetable or food product that has aroma, pungency, or color and is added to food after it is cooked and when it is served. By this definition, a spice or group of spices may be considered a condiment, depending on use. A condiment may be a single spice or composed of several spices, and often contains other food products such as oil or vinegar. Examples of condiments include ground black pepper, salad dressings, chutney, and ketchup. Tomatoes, since they are a food of themselves, are not spices but may be condiments. Salt, since it is neither a vegetable nor a food, is neither a spice nor a condiment, but is classified as a seasoning.

A seasoning is anything that enhances in flavor or appearance, or gives relish to foods. Spices, condiments, and salt or monosodium glutamate (MSG) are all seasonings (see AMINO ACIDS (MSG)).

In the 1990s, flavorists, traders, and culinary experts have expanded the term spice to include many vegetable products. Spices can be divided into four general categories (1). The traditional or so-called true or tropical spices, eg, black and white pepper, cloves, cinnamon, ginger, nutmeg, and mace, may be buds, fruit, bark, roots, or other parts of tropical plants. Herbs, eg, sage, rosemary, marjoram, and oregano, are usually the leafy parts of plants that grow in the temperate zones. Spice seeds, eg, mustard, celery, caraway, dill, fennel, and anise, grow in both tropical and temperate zones. Lastly, dehydrated aromatic vegetables, eg, onion, garlic, parsley, and sweet pepper, are often dried to prevent spoilage.

The reduction of water limits mold formation, but only slightly affects the aroma or pungency. The dried product maintains its character and pound for pound is stronger in aroma and flavor than the fresh spice, since a nonessential component has been substantially removed. In areas where a spice is grown, the same product that is dried for storage and shipment is often used fresh for flavoring.

Many spices are processed (2) to produce essential oils, oleoresins, essences, tinctures, extracts, resinoids, etc. These processes separate nonflavor components and further concentrate the aromatic or pungent principles of the spices. Such products allow a wider variety of uses and applications of the vital spice components.

A spice oleoresin is a concentrated form of the dried spice, processed by extraction, usually with a volatile nonaqueous solvent. The pharmacological herbs are usually processed by extraction with water or dilute alcohol, yielding powdered, fluid, or solid extracts. Oleoresins contain the odor and flavor of the spice including nonvolatile principles and, unlike spice essential oils, are usually viscous liquids, semisolid, or solid materials. Natural botanicals contain many components that make no contribution to the odor, flavor, or spice value of the item. Selection of the proper solvent removes such things as resins and resin acids, water, cellulose and pentosans, lignin, carbohydrates, sugars, starches, fiber, tannins, and minerals, while retaining such things as fixed oils (glycerides), essential oils, di- and triterpenoids, proteins, amides, phenols, and some waxes.

The oleoresin extraction process requires advanced technology which was accomplished prior to World War II only in industrialized countries that imported the dried spice. Modern factories now exist (ca 1993) in countries that grow the spice as well as in many developing nations. Advantages to manufacturing the finished goods locally include reduced volume of the product with no loss of total flavor intensity, and reduced bulk and shipping costs. The product also is more stable to temperature and time, is bacteria-free and bacteria-stable, and is more uniform and easier to specify and control. Additionally, the raw spice need not pass the rigid scrutiny of international customs prior to extraction; thus broken spices, even perhaps some dustings, which do contain flavor but otherwise have little commercial value, can be extracted and salvaged.

Labeling of Spices for Foods and Beverages. For the labeling of spices used in foods and beverages the United States Food and Drug Administration (FDA) does not permit every item contained in the four general categories to be labeled simply as spice. The FDA does not differentiate between culinary herbs and spices, but it does require that those substances that traditionally have been regarded as foods, eg, bell peppers, onion, garlic, and celery, be labeled separately. Also, spices used to impart color, eg, turmeric, saffron, paprika, or annatto, must be listed separately by name or as spice and coloring (3) (see COLORANTS FOR FOOD, DRUGS, COSMETICS, AND MEDICAL DEVICES).

For the products under its jurisdiction, eg, meat and meat products, the United States Department of Agriculture (USDA) has requirements similar to those of the FDA. However, mustard and spices that impart color must always be listed separately; onion and garlic powder may be listed simply as flavors.

The American Spice Trade Association (ASTA) (4) accepts spice as any dried plant product used primarily for seasoning purposes. This broad definition was designed so that items labeled only as spice could give adequate protection to proprietary formulas for spice mixtures. However, ASTA recommends that the dehydrated vegetables and the color spices be listed separately by name on all labels. ASTA also has recommended that the capsicums, no matter the species, be delisted as spices and labeled separately.

History

Detailed information on the history of spices is available (4–6). Until the eighteenth century, the country or area that controlled the spice traffic was economically and politically the most powerful nation of the era.

The most ancient uses of spices appear to be therapeutic in nature. The use of spices was common in China but little, if any, authentic Chinese records exist to confirm this. According to Chinese myths and legends, Shen Nung, the Divine Cultivator, founded Chinese medicine and discovered the curative powers of many herbs. He is said to have described more than 100 plants in a treatise reportedly written in 2700 BC. It has been shown, however, that no written language was available in China at that time. Although some of the herbal uses in the treatise go back several centuries BC, the work seems to have been produced by unknown authors in the first century AD. Other records on the use of cassia and ginger are known to have been written in the fifth and fourth centuries BC, in the latter case by Confucius.

The first authentic records of the applications of spices and herbs date from Egypt earlier than the twentieth century BC. Burial chambers and ancient temples show written proof of such products. An Egyptian medical scroll, dating from about 1515 BC, describes almost 800 medical drugs, herbs, and spices used not only in medicine, but also in embalming, purifications, cooking, and for fragrant ointments and perfumes. Cinnamon and cassia, available only from China and other parts of Asia, are among the spices recorded; this indicates a spice trade between the edge of the Mediterranean Sea and the Far East long before recorded history. Before the eighteenth century, control of the spice trade was one of the motivating economic forces of entire nations. From the ninth century BC to the fifteenth century AD, the spice trade was completely controlled by the Arabs in principal cities along the land and sea trade routes. The Greek, Roman, and later European civilizations increased the demand for spices, but despite European expansion eastward, the Arabs controlled the flow of products from the Orient.

In 1096 AD the Crusades opened up the northern Italian ports of Genoa and Venice to Oriental trade through Alexandria. When the Portuguese found the sea route around the coast of Africa by way of the Cape of Good Hope, bypassing Alexandria, the Western spice centers were shifted from Alexandria and the ports of northern Italy to Portugal and Spain. Destruction of the Spanish Armada in 1588 removed Spain from competitive commercial shipping. By the 1620s the Dutch had driven the Portuguese from the Spice Islands and regulated the spice trade by creating artificial shortages and high prices.

By 1799, the British held northern Borneo, India, Ceylon, Singapore, and the mainland of Malay to the borders of Siam, and allowed the Dutch to retain the islands of the Malay Archipelago. The center of the spice trade in the West shifted to London.

From about 1815, fast vessels gave Salem, Massachusetts a virtual monopoly of the pepper trade with Sumatra. In the nineteenth and twentieth centuries, the spice trade began to play a secondary role in economic and political influence. The demand for spices in the United States directed the shift of the occidental spice center to New York while the oriental center was at Singapore.

Japanese expansion throughout the East Indian islands in the 1930s and 1940s, then anticolonialism and the emergence of the Third World nations, finally wrested control of the Spice Islands from the Dutch.

In the 1990s the United States is the leading factor in the spice trade. The western hemisphere has become a large producer of important spices, among them aromatic seeds and herbs, eg, mustard, capsicum, dehydrated onion and garlic, basil, tarragon, coriander, cardamom, ginger, and sesame seed. Brazil has become a factor in black pepper production. New York remains the main port of entry for spices (ca 1993) and the United States is the largest user.

Production and Economics of Spices

Spices have become commercial products in over 70 countries of the world and may be produced in almost every country that can grow crops. However, many species of botanicals can be grown only in particular climates or have particular soil requirements. The warm, moist, tropical climates foster the growth of more species than any other areas; the traditional or tropical spices originated in these areas.

The purpose of the flavor function of a botanical to the plant is not well understood, eg, whether it assists in propagation or repels infestation. Different parts of botanicals have use as spices and more than one part of a botanical may be useful, eg, the bark and leaves of the cinnamon tree; the leaves and seeds of the cumin bush or celery plant; and the buds, leaves, and stems of the clove tree. The parts of plants that have spice value include leaves (mint and parsley), twigs (cassia and cloves), bark (cinnamon), roots (ginger), tubers (orris and lemongrass), grasses (lemongrass), buds (cloves), stigmas (saffron), berries (black pepper and juniper), seeds (cardamom and coriander), and fruit (nutmeg and vanilla).

The growing, cultivation, and harvesting of spices began as a primitive or cottage industry run by single families in isolated areas of the world who grew and picked indigenous crops to sell to an agent or collector. Since the beginning of the twentieth century more modern methods have been used. Botanical selection is scientific and sophisticated, analyses of soil and other growing conditions are carefully determined and maintained, and entire farms and cultivated plantings are set aside. However, modern agricultural science has not been universally applied throughout the spice industry.

Crops have been introduced to different parts of the world for specific advantages, eg, better yield, closeness to markets, and more abundant labor. Attempts have been made to reduce the labor required for planting, growing, and harvesting even in areas where labor is cheap.

Attempts to grow a botanical in other areas of the world where land and labor are more available, but where soil and climate may differ slightly, does not always result in greater availability of a comparable product. Vanilla beans grown in Tahiti and Indonesia do not compare in flavor strength or quality, or command the price, of beans grown in Reunion or Madagascar. Brazil grows large amounts of black pepper but the price and quality are less favorable when compared to that grown in India. Indonesia has replaced India as the prime supplier to the United States of black pepper with the price equivalent to that of Indian Malabar type, but Tellicherry Extra black pepper from India still commands a premium price, ie, about 50% higher than Brazilian, Indonesian, or Malabar. Ginger is now grown in Hawaii, commanding a prime price, and increasing in volume tenfold since 1977.

The most important considerations in marketing and establishing a crop from a new source are constancy of supply and quality. For some spices, it is difficult to reduce labor costs, as some crops demand individual manual treatment even if grown on dedicated plantations. Only the individual stigmas of the saffron flower must be picked; cinnamon bark must be cut, peeled, and rolled in strips; mature unopened clove buds must be picked by hand; and orchid blossoms must be hand pollinated to produce the vanilla bean.

Economic Market. The spice trade is controlled by many direct elements and responds slowly to supply and demand fluctuations. Resupply depends on growth to plant maturity, which for certain items, such as black pepper or nutmeg, can be several years. The raw material is directly affected by climate, adverse weather conditions, and control of plant diseases and insect and animal pests. Limited agricultural scientific advances are applied to the cultivation of the botanicals, and there are many grades of product and degrees of quality caused by different growing or processing conditions, sometimes by unknown factors as well.

Local government control, nationalization, and political unrest also can cause shortages, and business manipulations to alter prices and supply are common.

United States Imports of Spices and Oleoresins. The consumption of spices has continued to increase in the United States into 1993 (7). The demand for ethnic foods, and the trend toward less salt, glycerides, and fat, has stimulated more spice and condiment use. The United States consumes approximately 25% of the spices produced in the world. In 1993, imports accounted for about 65% of U.S. seasoning needs compared to 80% in the early 1980s. In 1991 approximately 50% by value of U.S. imported spices entered New York, the principal port of entry; around 1983 more than 75% was imported through this port. The volume of spices and oleoresins (spice extracts) into the United States has been increasing steadily, but the value of imports has varied because of specific shortages and large price variations (Table 1).

The value of imports rose to record levels in 1986 due to a twofold increase in the price of pepper, the second highest imported spice in volume, averaging about $4.00/kg. The effect on import values was pronounced, ie, approximately one-third higher, than 1985 imports. The upward price trend on black pepper continued in 1987 because the vines had been exhausted and the volume was lower.

Table 2 compares United States spice imports, according to the U.S. Department of Commerce, for 1992 and 1978.

Table 1. United States Spice and Oleoresin Imports[a]

	Spices		Oleoresins	
Year	Amount, t	Value, 10^6 \$	Amount, t	Value, 10^6 \$
1984	195,855	271.3	451	8.5
1985	190,409	288.5	640	12.7
1986	202,942	383.6	680	17.0
1989	226,892	374.0	939	22.9
1990	238,933	355.0	1027	30.3
1991	241,633	362.3	1086	32.6
1992	253,206	368.7	1139	30.1

[a]Refs. 7 and 8.

Table 2. Imports of Individual Spices, 1978 and 1992[a]

Spice	1978		1992	
	Amount, t	Value, 10^3 $	Amount, t	Value, 10^3 $
allspice (pimenta)	830	1,301	861.3	1,707.9
anise seed	489	636	1,027.7	2,234.5
basil	479	447	2,152.8	3,038.5
capers			1,146.2	5,891.3
capsicum (red pepper)	4,394	5,266	26,906.4[b]	54,336.1
caraway seed	3,089	4,137	3,269.0	3,007.1
cardamom seed	169	1,531	168.9	777.8
cassia	7,711	5,538	14,578.7[b,c]	24,226.3
celery seed	2,160	1,824	2,666.0	1,599.4
cinnamon (Sri Lanka)	886	1,540	467.2[b,c]	2,382.9
cloves	1,145	6,609	1,155.7	1,503.0
coriander seed	4,279	2,262	2,313.8	1,496.7
cumin seed	3,339	5,803	6,435.2	12,596.2
curry and curry powder			466.5[b]	1,511.9
dill seed	557	462	722.4	764.4
fennel seed	906	703	3,155.1	3,413.6
garlic (dehydrated)			2,657.2	2,649.1
ginger	3,556	4,703	8,970.8[b,d]	9,771.6
laurel (bay) leaves	393	497	1,812.4[e]	4,554.8
mace	253	681	219.9	456.3
marjoram	380	351		
mint leaves			290.6	444.2
mustard seed	32,040	9,835	63,911.6[b,e]	25,240.4
nutmeg	2,110	4,066	1,685.2	2,357.6
onions (dehydrated)			1,203.5	1,619.0
origanum leaves	2,385	3,741	5,543.2[e]	11,702.1
paprika	5,005	7,569	3,077.3	6,054.0
parsley	132	181	227.0	731.2
pepper, black	26,142	53,053	40,590.2	41,705.2
pepper, white	2,382	6,522	5,543.5	7,186.0
poppy seed			4,881.7	3,126.7
rosemary	259	147		
saffron			3.2	3,173.4
sage	1,309	2,168	2,414.7	5,567.7
savory	73	104		
sesame seed	32,000	27,256	34,992.9	42,729.7
tarragon	22	127		
thyme	603	785	54.4	132.1
turmeric	1,839	1,986	2,605.6	4,075.5
vanilla bean	1,185	25,281	1,261.6	65,700.0
Total	*142,501*	*187,112*	*249,998.2[f]*	*360,901.9[f]*

[a]Ref. 7.　[b]Including ground spice.
[c]Estimated, products combined in USDA report (7).　[d]Including sweet and candied.
[e]Including other than crude.　[f]Excluding ground pepper and mixed spices.

47

In 1991, vanilla beans were the highest valued spice import, with shipments totaling $69.0 million, followed by black and white pepper at $60.6 million, capsicum peppers and paprika at $42.6 million, sesame seed at $40.6 million, and cassia and cinnamon at $27.8 million. The most expensive spices, on a unit value basis, include saffron, $1116/kg average New York spot; vanilla beans, $80.50/kg for Bourbon beans from Madagascar, Comoros, and Reunion, and $22.05/kg for Java beans; and cardamom, $38.54/kg for grade AA bleached Indian and $3.88/kg for Guatemalan mixed greens.

The annual United States import of vanilla beans jumped 34% in 1991, as importers and users replenished inventories (7). Indonesia was the largest supplier, accounting for 47% of the total; the lower price for Indonesian beans is responsible for the increase in shipments. Users have been replacing Bourbon beans with Indonesian types, or blending in Indonesian beans, because they may still label their product as natural and meet other FDA specifications.

Vanilla flavoring in bakery goods, confectionery, and many frozen desserts need not be natural vanilla. The artificial and synthetic vanilla flavors that are used include vanillin [121-33-5] from lignin (wood pulp), ethyl vanillin [121-32-4], and vanitrope [94-86-0]; the latter two are synthetics. Over 90% of the U.S. market for vanilla flavor contains vanillin. These synthetics continue to dominate the market because of availability, quality, and relatively low and stable prices.

U.S. Oleoresin Imports.   Reports of the USDA list only paprika and black pepper oleoresins by name, and label all others, eg, capsicum, celery seed, and turmeric, as "other" (Table 3). The steady increase in the value of oleoresin imports is expected to continue.

Table 3. United States Import of Oleoresins[a]

	1980		1985		1992	
Oleoresin[a]	t	10^6 $	t	10^6 $	t	10^6 $
paprika	128.3	4.36	334.4	6.83	449.0	18.65
black pepper	92.0	1.67	158.0	3.14	225.1	3.57
other	115.6	1.51	147.7	2.68	465.2	7.88
Total		*7.54*		*12.65*		*30.10*

[a]Amount in t, value in 10^6 $.

United States Exports of Spices and Oleoresins.   The United States (ca 1993) is the foremost grower of peppermint, spearmint, orange, lemon, lime, and grapefruit products. The mints are processed to essential oils, and the citrus fruit are sold as fresh fruit or processed to frozen concentrates and essential oils; thus they do not qualify as spices.

United States exports of spices in 1992 totaled $87.7 million, up from $85.2 million in 1991 (Table 4). Dehydrated onion was the most important export spice product at $36.1 million; followed by dehydrated garlic, $7.5 million; capsicum peppers, $8.9 million; black and white pepper, $5.8 million; and prepared mustard products, $4.8 million. Canada, Japan, and Germany are the principal markets for United States spice exports. The production in the United States of temperate zone spices is increasing every year; some of this output is exported as raw spice.

Table 4. United States Exports of Spices[a]

Spice	1990		1992	
	Amount, t	Value, 10^3 \$	Amount, t	Value, 10^3 \$
anise (badian)	37.6	60.9	37.8	64.9
capsicum (red pepper)	2,138.3	4,922.1	3,275.5	8,864.6
caraway seed	38.6	37.5	24.3	40.2
cardamom seed	72.0	394.0	27.1	81.9
cassia and cinnamon	455.8	1,405.7	578.8	2,193.8
cloves	30.1	120.0	27.8	85.5
coriander seed	65.0	154.2	38.8	73.9
cumin seed	70.0	165.7	119.8	339.3
curry	72.5	244.2	71.4	186.0
fennel (juniper)	32.2	76.9	6.8	14.6
garlic (dehydrated)	3,595.0	8,090.0	2,726.0	7,297.0
ginger	1,043.5	2,254.3	1,278.5	2,243.1
mace	57.5	195.6	57.7	216.9
mustard seed	1,649.5	610.3	758.2	341.5
mustard, prepared	2,868.8	3,383.9	4,481.7	4,822.7
nutmeg	186.9	850.2	45.7	172.6
onions (dehydrated)	14,471.0	33,215.3	14,718.0	36,089.0
pepper, black and white	1,621.5	4,725.5	2,003.6	5,794.2
poppy seed	228.8	101.1	48.0	45.3
saffron	26.3	83.5	5.8	29.4
sesame seed	2,112.4	2,523.9	1,252.8	1,709.7
thyme leaves	68.1	198.0	92.1	314.7
turmeric	36.8	164.1	43.4	282.7
vanilla bean	386.4	2,367.2	[b]	[b]
other	3,397.6	14,121.0	4,552.7	16,349.8
Total	*35,062.2*	*80,465.1*	*36,338.3*	*87,653.3*

[a]Ref. 7. [b]Vanilla beans misclassified-(USD Commerce 1991 audit).

Many imported raw spices are processed and packaged in the United States and then exported. Because of strict United States food laws, many foreign purchasers prefer spice available from the United States, whether domestic or imported. Thus imported products may end up as a significant part of the export list.

Information regarding U.S. production of oleoresins is not available. It is estimated that there is a decline in domestic production of oleoresins of those spices imported in large volume, such as black pepper, capsicums of all types, and turmeric, since these oleoresins are more frequently produced in the growing areas. However, the manufacture of specialty oleoresins produced from selected imports will continue, and oleoresin production from domestically grown spices is expected to increase (7,8).

Market Trends of the 1990s. The United States spice market can be divided into three sectors based on application: industrial, ie, food processing and manufacture; institutional, ie, restaurants, hospitals, schools, and military; and retail. The food manufacturers and institutions account for almost 65% of U.S. spice usage, an increase from about 40% in the 1980s. Retail food outlets make up most of the remainder.

Several factors contribute to the general increase in spice usage in the United States. The spices used in pizza continue to expand, and the demand among the general population for other ethnic foods has increased. This applies mostly to the popular hot Oriental foods, other than Chinese and Japanese which have been around for a while, such as Thai and East Indian. Jamaica Jerk and American Southwestern diets and cultures also have been responsible for the increase in capsicum and red pepper usage. Cajun seasoning and ginger are being added to Mexican foods. Salsa, containing hot pepper, onions, tomatoes, etc, is now the foremost condiment in retail sales.

In Europe there has been an increased demand for green peppercorns especially for use in sauces, meats, and cheeses.

The health-conscious trend toward development of fat replacements (see FAT REPLACERS), reduced salt and MSG intake, and use of artificial sweeteners (qv) to reduce caloric intake has influenced the increase in spice usage. All of these trends require spices to compensate for flavor loss or to overcome a perceived or actual difference in flavor.

Interest continues in the industrial sector in expanding the use of processed spices, ie, essential oils, oleoresins, and extracts. Except for the visual effect of ground or chopped spice on salads, sauces, and in extruded products such as chips, or ground nutmeg on eggnog, the processed products offer many advantages to the industrial food processer, eg, uniformity, cleanliness and sterility, constancy, and stability. The products can be solubilized, clarified, and colored. Processed products include oleoresins; solubilized or standardized oleoresins, ie, oleoresins to which solubilizing agents such as propylene glycol or corn oil have been added to maintain consistency of vital components to specifications; essential oils, ie, the steam distilled volatile active flavor principles of the spice; dry-soluble spices, ie, dispersions of extracts or essential oils on free-flowing carriers such as sugars, dextrins, salt, and modified starches; and encapsulated spices, ie, emulsification of extracts or essential oils with gums, starches, waxes, or maltodextrins, with subsequent spray drying, spray cooling, or coacervation.

The amount of vital components removed from the whole spice by processing to the oleoresin or essential oil varies according to the particular spice, eg, from 0.1% in oil of garlic to 20% in oleoresin clove. Based on the yield, the concentrate can proportionally replace the dried spice. Since 1985, spices have been extracted using a cryogenic solvent, such as carbon dioxide, to commercially produce oleoresins. The lack of heat treatment in this process presumably offers flavor advantages. Countercurrent extraction, using two immiscible solvents, has also been used. Neither of these new techniques have been used to any large commercial degree.

More and more raw spices are converted to finished products near the growing sites. This saves shipping costs of bulk vs concentrate. Rapid processing also assures less loss of flavor volatiles resulting from evaporation, reduction of colored components due to oxidation or isomerization, and reduction of losses due to insect and rodent infestation.

A few spices, particularly rosemary and sage, are known to act as antioxidants which prevent rancidity due to oxidation in fats and fatty foods.

Synthetics. The lack of spice products to satisfy demand and the wide variation in price and availability have caused the manufacture of selected synthetics, chemically identical to the component in the natural spice, to replace the vital components of some spices. However, synthetic organic chemistry is not yet able to manufacture economically the many homologous piperine [94-62-2] components in black pepper or those capsaicin [404-86-4] amides in red pepper.

Some spice characters can be synthesized. Cinnamic aldehyde [14371-10-9], the principal flavor component of cinnamon and cassia, is produced synthetically from petrochemical raw materials to the extent that it would require the production of at least 10 times the present amount of spice to satisfy demand. Natural cinnamon bark contains less than 0.5% of cinnamic aldehyde and costs \$2.00– 3.00/kg, and synthetic cinnamic aldehyde contains 99% aldehyde and costs approximately the same price per kilogram; thus over 200-fold savings and usage. Vanillin, from lignin, has significant use in many baking flavor extracts. The cost ratio of natural vs synthetic in this case is several hundredfold depending on the ever fluctuating cost of vanilla beans. The vanilla bean and its products perhaps have the strictest FDA controls and the most extensive analytical specifications of all spices.

Other synthetics with cost advantages and large volume productions are L-carvone [6485-40-1], the primary component in natural spearmint essence; D-carvone [2244-16-8], the primary component in natural dill and caraway; anethol [4180-23-8], in place of anise and fennel spices; and smaller amounts of thymol [89-83-8] replacing thyme and disulfide synthetics for onion and garlic. All of these synthetics must be labeled as artificial which may limit their use among consumers.

Specifications, Analysis, and Quality

Spices are natural agricultural products and exhibit a range of variations of many specific characteristics. The most important quality assessment is the subjective physical observation of the whole or ground spice by an expert. The macroscopic and microscopic examination of spice is the criterion for the continued analysis of the product to determine adherance to specifications.

Physical appearance, as well as flavor quality and strength, can be influenced by soil conditions, rainfall, storms, blights, insects, growing and harvesting methods, storage, etc. All of these must be considered to evaluate a particular lot and to harvest, sell, or buy the lot and use it in a food product.

The FDA applies the Federal Food, Drug and Cosmetic Act (3) to the spice industry and its products. The FDA has established a definition for spice which is somewhat general. It states, however, that vegetables such as onions, garlic, and celery are regarded as foods, not spices, even if dried.

The USDA considers most spices generally recognized as safe (GRAS). There are no standards of identity or legal definitions of spices. Spices used in drugs must meet the official standards of the *U.S. Pharmacoepia* in force. Advisory specifications may also be applied in commercial spice trading.

The American Spice Trade Association (ASTA), established by the industry, works with U.S. government organizations to set up specifications (9) for spice products that conform specifically to the laws, however generalized they are.

Initially, there was some overlap on proposed analytical methods to accomplish a particular analysis. The Association of Official Analytical Chemists (AOAC) methods and Bacteriological Analytical Manual (BAM) methods in some cases duplicated ASTA methods, but the procedures differed. Most spice companies, particularly those who are members of ASTA, use ASTA recommended methods. In an attempt to ensure that equivalent specifications are reported, the Technical Group of ASTA develops specifications and in some cases recommends that a BAM or AOAC method be used.

ASTA has developed a series of Official Analytical Methods (9) which includes sections on general methods and methods for specific spices. General methods discussed include preparation of sample, moisture by distillation and drying, total ash and acid-insoluble ash, steam-volatile oil by modified Clevenger or Lee and Ogg methods, crude fiber, direct acid hydrolysis of starch, alcohol extract, sieve analysis, and nonvolatile methylene chloride extract. Methods presented for specific spices include piperine content of black and white pepper, light berries and extraneous matter in pepper, volatile oil in mustard seed and flour, steam-volatile oil in cassia, cinnamic aldehyde in cassia oils, color power of turmeric, phenols in nutmeg and mace, extractable color in capsicums, pungency of capsicums and their oleoresins by Scoville and hplc, microanalytical analyses of paprika and ground capsicums other than paprika, aflatoxin determination, and ethylene oxide and ethylene chlorohydrin residue in black pepper.

ASTA has also made available a manual of microbiological methods (10) which contains 22 different procedures for the detection and enumeration of microorganisms occuring in spices. A handbook on clean spices for ASTA members (11) is available.

The FDA has published methods for the determination of residual solvents in spice extracts such as oleoresins and has limited the concentrations of those specific solvents that are permitted. Chlorinated hydrocarbons and benzene have been almost completely removed from use as extracting solvents in the United States; their use continues overseas where toxicity regulations are less stringent. The presence of pesticides or herbicides in spices is rigidly controlled by the FDA.

Spices and herbs can play an important indirect role in good nutrition. They are not high in nutrient values, but they help to increase the appeal and satisfaction of foods that are highly nutritious. Spices do contain fat, protein and carbohydrates, electrolytic minerals, iron and B vitamins, and others, but even the highest calorie spice, poppy seeds, contains only two to three calories per serving in normal use (12).

Individual Spices

Allspice (*Pimenta*). Allspice is the dried, nearly ripe berry of the tree *Pimenta dioica* L., formerly called *Pimenta officinalis-Lindley* (Myrtaceae), which is a tropical evergreen growing semiwild in Jamaica and other Latin American countries. Allspice should not be confused with pimento or pimiento, a variety of red pepper. Allspice is the only significant commercial spice exclusive to the Western Hemisphere. *Pimenta* berries resemble black pepper corns; they are round and up to 0.65 cm in diameter. The berries are hand-picked while green, then slightly fermented and sun-dried, turning them reddish brown. The aroma of *pimenta* berries resembles that of cloves, cinnamon, black pepper, and nutmeg, hence the name allspice or *quatre épices*. Allspice is used to flavor vegetables, fruits, pickles, and spicy table sauces.

Anise Seed. Anise seed is the dried, ripe fruit of *Pimpinella anisum* L. (Umbelliferea), a member of the parsley family indigenous to Asia Minor, Greece, and Egypt. The bulk of supply comes from Turkey, China, Spain, and Egypt, but it can be cultivated in most temperate climates. Anise seed resembles caraway seed in appearance, but has a sweet, licorice flavor popular in Mediterranean and Latin countries, particularly in beverages, anisette, confectionery, salad dressings, and sausage flavors.

The star anise (*Illicium verum Hook* F.) is from a small evergreen tree, native to southwest China. When ripe, the hard brown fruits of this tree open up into an eight-pointed star, hence the name. The flavor and aroma of this spice is similar to that of *P. anisum*; the essential oils are of similar composition. The uses are similar but more localized.

Basil (Sweet Basil). Basil consists of the brown, dried leaves and tender stems of *Ocimum basilicum* L. (Labiatae), an annual native to India, Africa, and Asia, and cultivated in Egypt, southern France, Morocco, the Mediteranean countries, and the United States. Basil is one of the oldest known herbs, and it is reported that there are perhaps 50–60 poorly defined *Ocimum* species which can only be identified according to their chemical components. The flavor of the *basilicum* type is warm, sweet, somewhat pungent, and peculiar, ie, methyl chavicol and linalool. It is used with meats, fish, certain cheeses, and tomato-based salads. The fresh leaves are ground and known as pesto with pastas. It is the main component of the liqueur Chartreuse.

Bay (Laurel) Leaves. These are the dried leaves of *Laurus nobilis* L. (Lauraceae), also called sweet bay or laurel tree, an evergreen with shiny green leaves up to 7.6 cm in length. It is not to be confused with the bay rum tree (*Pimenta racimosa Mill* (Myrtaceae)) from Puerto Rico and neighboring islands, or California bay laurel (*Umbellularia californica Nutt.* (Lauraceae)). *L. nobilis* has been cultivated since antiquity in Mediterranean countries and is now grown extensively in Turkey, the former Yugoslavia, France, and Central America. The odor of the leaves is delicate but distinctly aromatic; the flavor is slightly bitter and burning. It is used extensively for meats, sauces, meat dishes, bouillabaisse, stews, and for pickling spice.

Capsicum. Several important condiments are derived from the dozens of spices and varieties of the genus *Capsicum* (Solanaceae), the nightshade family, particularly from *C. annum* L., ie, paprika, red pepper, and cayenne pepper. These plants are indigenous to Mexico, Central America, the West Indies, and much of South America. Capsicums are sensitive to climatic and soil conditions. Many horticultural varieties have been developed, differing in size, shape, taste, pungency, and color. Capsicum can be divided into two general groups, the sweet and the pungent.

Paprika, also called sweet pepper or pimento (pimiento, in Spanish), is the dried ripe fruit (stemless pod) of chiefly *Capsicum annum* L. The varieties of this annual herb, grown primarily for its bright red color, not pungency, are mild and slightly sweet. The odor is pleasant and somewhat aromatic. This herb is grown primarily in the United States, Spain, Mexico, Hungary, Bulgaria, and Morocco. It is a rich source of vitamin C and serves as a garnish and flavorant in many foods, such as meats, soups, poultry, salad dressings, etc. It is an important constituent of ketchup and chili sauces.

Red pepper, cayenne pepper, and chilis belong to the pungent group of capsicums derived from the varieties of *C. annum* L. and *C. frutescens* L. They are usually small in size and exhibit a very hot, pungent flavor. They are cultivated primarily in tropical and subtropical latitudes, eg, southern United States, Mexico, China, Pakistan, India, Africa, and Japan, and occur in dozens of types, ie, African chilis, Birdseye, tabasco, Jalapeno peppers, Louisiana Sports, Japanese Santaka, etc, that vary in degree of pungency and color. Often ground peppers are blended to arrive at a required or specified strength of flavor, pungency, and color. Red pepper serves where pungency is essential, eg, in barbecue and spicy table sauces, chili powder, hot curries, and in medicinal preparations such as hair lotions and liniments. Capsicum is the main component of salsa, the best selling table sauce.

Caraway Seed.   This spice is the dried ripe fruit of *Carum carvi* L. (Umbelliferae). It is a biennial plant cultivated extensively in the Netherlands and Hungary, Denmark, Egypt, and North Africa. The seed is brown and hard, about 0.48 cm long, and is curved and tapered at the ends. It is perhaps the oldest condiment cultivated in Europe. The odor is pleasant and the flavor is aromatic, warm, and somewhat sharp (carvone). Caraway is used in dark bread, potatoes, sauerkraut, kuemmel liqueurs, cheese, applesauce, and cookies.

Cardamom Seed.   The dried, ripe seed of *Elettaria cardamomum* M. (Zingiberaceae) is from this large perennial herb, indigenous to and cultivated in southeastern India and Sri Lanka, and also cultivated extensively in Guatemala. The seeds are removed from the capsules (pods) and when dried are irregularly round and about 0.23 cm long. The aroma and flavor are pleasant, aromatic, and somewhat camphoraceous. It is a tropical spice, known since the fourth century BC. Throughout Arab countries, cardamom is the most popular spice, used in a coffee-like beverage, and is a symbol of hospitality. The spice is also used in baking pastries and pies in Nordic countries, and is used as well in ground meats. It is an ingredient of curry powder and pickle flavor.

Celery Seed.   Celery spice is the dried ripe fruit of *Apium graveolens* L. (Umbelliferae) a biennial, sometimes annual, herb native to southern Europe and grown extensively in India, China, Mexico, and the United States. The seed is 0.42 cm long and brown. The odor of the seed is characteristic and warm and the taste somewhat bitter. It is used in tomato ketchup, sauces, soups, pickles, pastries, salads, and certain cheeses.

Chili Powder.   Chili powder is a commercial blend of several spices, ie, chili peppers, oregano, cumin seed, onion and garlic powders, allspice, perhaps cloves, and others. Chili powder is the basic flavor for many highly spiced dishes, among them chili con carne, and is used in cocktail sauces and ground meats.

Cinnamon.   There are two distinct types of cinnamon, each with its own varieties.

Cassia.   Cinnamomum cassia Blum (Lauraceae) is the so-called cassia. It is native to southeastern China and has not been grown successfully outside of this area. The dried bark of this evergreen tree is stripped, ground, and sold almost exclusively in China. The leaves and twigs of the tree contain the same flavor components as the bark and are steam distilled to yield the cassia oil of commerce. Infrequently, small amounts of bark are bundled and exported as cassia lignea but cannot compete with the other varieties as bark spice. The Chinese prefer to sell the essential oil.

Other botanical varieties are called cassia, but the leaves of these varieties differ in flavor components from those of the bark. Saigon cinnamon, *C. loureirii* Nees, from Viet Nam, closely resembles Chinese cassia in appearance but is grown on the other side of the mountains and has an entirely different flavor character, containing no orthomethoxy cinnamic aldehyde. *C. burmani Blume*, ie, Korintje or Kerintje cinnamon and Padang or Batavia cinnamon, is from Sumatra and Indonesia. *C. sintok Blume* is native to Malaysia and of minor commercial importance.

The spice from Saigon and Indonesia is sold in the form of quills somewhat the same as the true Ceylon cinnamon but rougher in appearance. The Saigon and Indonesian barks have a higher oil content with a different flavor character than the Ceylon type.

Zeylanicum.   The second type of cinnamon is the dried inner bark of the shoots of the tree *Cinnamonium zeylanicum* Nees, a moderately sized coppiced evergreen bush of the laurel family cultivated in Sri Lanka. The bark is stripped, rolled into quills, dried, and then shipped in large bundles. The aroma and flavor of the *zeylanicum* type, which contains eugenol as well as cinnamic aldehyde, is much milder than the *cassia* type.

Owing to the more pungent flavor, higher oil yield, and cheaper price, grinders in the United States generally prefer the Saigon to the Sri Lankan cinnamon. These spice varieties are used widely in baked goods and confectionery, in pickling, curry powders, and spicy table sauces.

Cloves.   The clove spice is the dried unopened buds of the evergreen tree, *Eugenia caryophyllus* Thumb (Myrtaceae). This tree is also called *Syzygium aromaticum* L. Other botanical names are used, but some discrepancies exist as to the proper nomenclature. The tree is indigenous to the Molucca Islands. In 1770, shoots were smuggled to the islands of Mauritius and Reunion, and eventually were cultivated on extremely large plantings in Zanzibar and the Pemba Islands in Tanzania, and Madagascar in Malagasy. The term clove is derived from the French *clou* for nail; the dried clove buds resemble round-headed nails, about 1.3 to 1.9 cm long. The color is dark reddish brown, the aroma is powerful, and the taste is hot and spicy. Cloves, both whole and ground, are widely used in baked goods, confectioneries, puddings, desserts, sweet syrups, pickling and pickled fruit, and spicy table sauces. About half of the world's clove production is mixed with tobacco to supply kretek cigarettes to Indonesia.

Coriander Seed.   This spice is the dried ripe fruit of *Coriandrum sativum* L. (Umbelliferae) and is indigenous to southern Europe and the Mediterranean. It is now cultivated in Morocco, Canada, Bulgaria, and Romania. The seed is about

0.5 cm in diameter, yellowish to brown, and was one of the first herbs grown in the United States, ie, Massachusetts, in 1670. The foliage and unripe fruit have a fetid, buggy odor, described by the Greeks as *koris* or bedbuggy. As the fruit ripens, the disagreeable odor fades away and is replaced by a warm, fragrant, and spicy odor. It is used in ground meats (sausage and frankfurters), cheeses, curry powders, and pickling spices; it has many medicinal uses as well.

Cumin Seed.   Cumin spice is the dried ripe fruit of *Cuminum cyminum* L. (Umbelliferae). The seed-like fruit is elongated about 0.31 to 0.63 cm and is yellowish brown. It is native to upper Egypt and the eastern Mediterranean, but is now cultivated in Pakistan, Turkey, India, China, and Syria. It has a strong peculiar aroma and flavor, pleasing to some and offensive to others. Cumin seed is used in chutney and curry, chili powders, chili con carne, cheeses, and the pickling of cabbage.

Curry Powder.   This powder is a commercial blend of many ground spices. The composition of curry varies in the East according to the region, eg, Bombay curry, Madras curry, etc. It is used in the flavoring of exotic dishes, especially Indian, Indonesian, and Chinese specialties. Typical constituents of curry powder are coriander, ginger, nutmeg, clove, cinnamon, fenugreek, red pepper, onion salt, turmeric, etc.

Dill Seed.   Dill spice is the dried ripe fruit of *Anethum graveolens* L. (Umbelliferae), a plant indigenous to the Mediterranean region and to the southern former Soviet Union, but also cultivated commercially in India, Egypt, Sweden, and Pakistan. The seed is oval shaped, light brown, and up to 0.16 cm in length. The aroma resembles that of caraway seed, but is thinner. The herb is grown extensively in the United States, principally for the leaves which must be harvested before the plant flowers. The leaves, known as dillweed, are used mainly for essential oil production. Dill seed is used as a condiment, in soups, processed meats, sausages, spicy sauces, salads, and sauerkraut, and particularly in pickling.

Fennel Seed.   Fennel is the dried ripe fruit of *Foeniculum vulgare* Mill. (Umbelliferae), a biennial or perennial aromatic herb native to India, southern Europe, and the Mediterranean, particularly in the vicinity of the sea. The seed is oblong–oval, straight or slightly curved, up to 0.80 cm in length, and yellowish brown. The flavor is aromatic and sweet and resembles that of anise seed. It is popular in Mediterranean countries for the flavoring of baked goods, soups, fish dishes, and sweet pickles. It is used in cordials and pharmaceutical preparations.

Fenugreek Seed (Foenugreek).   This spice is the dried ripe fruit of the plant *Trigonella foenum graecum* L. (Leguminosae), an annual of the pea family native to western Asia and southern Europe and now cultivated commercially in India, Turkey, and Morocco. The seed is irregularly oval, up to 0.48 cm in length, and is a brownish yellow. It has a slightly bitter, peculiar flavor and odor, somewhat pleasant, resembling maple or burnt sugar. It is used in curry powders, chutney, and imitation maple flavors. Medicinal uses continue from ancient times.

Garlic Powder.   This powdered spice is the ground, selected dehydrated clove of *Allium sativum* (Liliaceae), a widely cultivated perennial herb with an underground compound segmented bulb. The aroma is powerful and offensive, developing by enzymatic action when the clove is crushed or moistened. In diluted form, it is attractive and indispensable in many Italian and French dishes, salad dressings, and sauces. Garlic is widely used in the United States for meat packing, sausages, frankfurters, and luncheon meats. One of the oldest of spices, it is indigenous to Egypt, China, and India and is grown throughout the world.

Ginger. This tuberous spice is the washed or peeled sundried rhizome (underground stem) of a perennial, *Zingiber officinale* Rose (Zingiberaceae). The color is an even, light buff and the tubers are referred to as hands because of their size and shape. Ginger was cultivated by the ancient Chinese and Hindus and is now grown throughout the world, particularly in China, Fiji, India, Nigeria, Sierra Leone, Nicaragua, and Jamaica. The aroma and flavor is distinctively pleasant, aromatic, sweet, and pungent with a slight bite. African ginger is camphoraceous, Cochin ginger is lemon-like, and Jamaican ginger is perhaps the finest combination of both characteristics. Ginger is used in soft drinks, baked goods (gingerbread), confectionery, curry powders, meats, pickling, and certain cordials.

Mace. This unique spice is the dried aril or skin surrounding the kernel of *Myristica fragrans* Houtt (Myristicaceae), or nutmeg. It is grown and produced in Indonesia. The color is yellow orange; the flavor is similar to nutmeg, but softer, somewhat less pungent but fuller. Mace is used in a wide range of products, particularly baked goods, meats, and vegetables.

Marjoram (Sweet Marjoram). Marjoram spices are the dried leaves, with or without small portions of the flowering tops, of *Majorana hortensis* Moench (Labiatae). It is a perennial herb cultivated in Egypt, France, Turkey, Greece, Mexico, and the United States. It may be confused, sometimes purposely, with oregano. Marjoram is a gray-green color and the flavor is highly aromatic, warm, and somewhat bitter. It is used in salad dressings, vegetables, meat stews, sausages, soups, and poultry stuffings.

Mint. The dried leaves of *Mentha spicata* L. (spearmint) and *Mentha piperita* L. (peppermint), hardy perennial herbs of the family Labiatae, are included in this spice category. These are indigenous to Europe and the Mediterranean but also are grown extensively in regions of the United States, ie, in the far west (Washington and Oregon) and the midwest (Michigan and Indiana). Although it has many medicinal uses, it is used particularly in toothpaste, mouth washes, chewing gum, and confectionery as well as sauces, desserts, salads, jellies, vinegars, and teas.

Mustard. Mustard is composed of the dried, ripe seeds of several closely related genera, species, and varieties of herbaceous annuals of the family Cruciferae, eg, *Sinapsis alba* L. (syn. *Brassica hirta* Moench) known as white or yellow mustard; *Brassica nigra* L. Koch, black mustard; and *Brassica juncea* Coss and Czern, which is brown mustard. The first two are probably indigenous to southern Europe and the Mediterranean region, and *B. juncea* to the northern Himalayan area. They are all of the mustard family and are cultivated commercially in Canada, France, the United Kingdom, China, South America, and the United States. The seed is extremely small, globular, and hard; at one time it was considered the smallest known seed. Within one growing season the plant may reach a height of 3 m or more.

When ground and mixed with water, the seeds of black mustard ferment and yield a sharp, irritating odor and a pungent taste; white mustard does not react in this manner.

Mustard is marketed in three forms, ie, as a seed; as a prepared blend of ground mustard seed, vinegar, salt, sugar, and other spices; and as powdered dry mustard, also known as ground mustard or mustard flour. The seeds of the white or yellow mustard add pungency to any preparation, and that of the black mustard is required for aroma. Mustard is the largest volume spice imported into the United States and its use covers almost every flavor category except dessert items.

Nutmeg. Derived from an apricot-like fruit of *Myristica fragrans* Houtt (Myristicaceae), nutmeg is an evergreen indigenous to the Moluccas and other East Indian islands. It is also cultivated in Indonesia and in Grenada in the West Indies. The fleshy fruit is globose, lemon-yellow to light brown, and when ripened splits in half exposing a beautiful scarlet reticulated areola (mace) investing the thin, hard, brown, and shiny shell which contains the seed (nutmeg). The oval nutmegs range in size up to 4 cm in diameter.

The flavor of nutmeg is pleasant, warm, spicy, and aromatic. Nutmeg is used to flavor baked goods, processed meats, curries, and beverages.

Onion Powder. Dehydrated onions, *Allium cepa* L. and other *Allium* species (Liliaceae), are ground to produce this spice. These botanicals are cultivated throughout the world. They generate their odor and flavor by enzymatic action in the presence of moisture. A sharp, pungent odor and flavor develops which is desirable on dilution. It is widely used in meats, sausages, vegetables, and many dishes.

Origanum (Oregano). Origanum is the dried leaves of *Lippia graveolens* (Verbenaceae) and *Origanum*, mostly *vulgare* species (Labiatae). The *lippia* species are grown in Mexico and the *origanum* in the Mediteranean countries, Greece, and Turkey. The light green leaf is about 1.59 cm long having a strong and aromatic odor. The flavor is pungent, warm, and slightly bitter. The aroma resembles marjoram. Oregano is used in chili powders, Italian specialties, gravies, salad dressings, vegetables, and various meats.

Parsley (Parsley Flakes). The dried leaves of *Petroselinum hortense*, syn. *P. crispum* (Mill) (Umbelliferae), are from a hardy biennial, native to the Mediterranean region and now cultivated commercially in the United States and southern Europe. The aroma is green and the flavor is pleasant, characteristic, and mild. Parsley is used for the seasoning of fish, meats, soups, salads, etc. Parsley seed, the dried ripe fruit of the parsley herb, has an aroma and flavor less pronounced than the leaves.

Pepper, Black. Black pepper spice consists of the dried immature berries (fruit) of *Piper nigrum* L. (Piperaceae). This vine-like perennial is native to the Malabar coast of southern India and is now commercially cultivated in the tropics of both hemispheres. There are many varieties of pepper, differing in appearance, size, and flavor. Dried black pepper consists of a light-colored kernel surrounded by a dark-colored, shriveled outer hull. There are about 3926 kernels to the kilogram. It has a strong characteristic fragrant odor, a hot, biting flavor, and is the most extensively used seasoning known.

Pepper, White. This spice is derived from the same plant as black pepper except that the berries are allowed to ripen fully. The dark outer hulls are removed by rubbing and washing after soaking and fermentation. White pepper has generally the same flavor and aroma as black pepper; however, it is milder and has a slight fecal odor.

Poppy Seed. The tiny dried seed of *Papaver somniferum* L. (Papaveraceae) is the poppy seed spice. The plant is an annual cultivated widely in Turkey, Iran, India, and China mostly for the preparation of opium. It is grown also in Australia, the Netherlands, and the former Yugoslavia. Poppy seed is one of the most ancient of spices, found among the ruins of the prehistoric lake dwellings in Switzerland. The black, kidney-shaped seeds, about 0.12 cm long, are not narcotic and have an agreeable nutty flavor when roasted. Poppy seeds are used in whole form as a topping for bread, rolls, and pastries, and also in pasta, salads, and some vegetables.

Rosemary. The dried leaves of *Rosmarinus officinalis* L. (Labiatae) are from an evergreen shrub growing wild in the Mediterranean countries. It is grown commercially in the former Yugoslavia, Albania, France, Portugal, Spain, Morocco, and the United States. The leaves are brownish green and resemble curved pine needles in shape, seldom exceeding 2.5 cm in length. The odor is strong, fragrant, and aromatic, and the taste is somewhat bitter and camphoraceous. Rosemary is used to flavor stuffings, poultry, beef, lamb, fish, stews, and soups.

Saffron. Saffron spice is the dried stigmas of *Crocus sativus* L. (Iridaceae), a bulbous perennial native to southern Europe and Asia Minor and cultivated in the Mediterranean countries, particularly Spain. True saffron should not be confused with either meadow saffron, ie, *Colchicum autumnale* L. (Liliaceae) also called safflower, or bastard saffron, ie, *Carthamus tinctorius* L. (Compositae), both of which are occasionally used to adulterate true saffron.

Saffron is the world's most expensive spice. As the flower blooms the stigmas from 75,000 flowers (three per flower) must be hand picked to obtain about 500 grams of spice after drying. The color is bright yellow-red and the aroma is powerful, somewhat bitter, peculiar, and exotic. It is used in exotic dishes, particularly the coloring and flavoring of rice in Spanish specialties, eg, arroz con pollo, paella, and bacalao vixcaino. It is important in bouillabaisse, a shellfish and fish mixture from the French Mediterranean coast, and is also used in cheese, pastry, and confectionery.

Sage. This spice consists of the dried leaves of *Salvia officinalis* L. (Labiatae), a perennial, low shrub which grows spontaneously in Mediterranean countries. Of the numerous sage species, the Dalmatian variety is considered the finest for flavoring and is widely cultivated in Albania, the former Yugoslavia, France, Germany, and Morocco. The color of the dried spice is grayish green or silver-gray and the aroma is strong, fragrant, warm, and slightly bitter with no camphoraceous or piney off-notes. Sage is used in the flavoring of meat, poultry, fish, certain cheeses, salad dressings, and sausages as well as poultry seasonings.

Savory (Summer Savory). This spice is the dried leaves and flowering tops of *Satureia hortensis* L. (Labiatae), an annual plant cultivated in southern Europe, particularly the former Yugoslavia, France, and Spain. The leaves are brown-green and up to 0.95 cm in length. The spice odor is strong, warm, and highly aromatic; the flavor is sharp and somewhat resinous. Savory is used in poultry seasoning, meats, soups, eggs, salads, and sauces and is sometimes substituted for parsley.

Sesame Seed (Benne, Benni, Bene Seed). Sesame seed is the whole dried seed of *Sesamum indicum* L. (Pedaliaceae), an annual plant now cultivated in Mexico and Central America, although indigenous to Indonesia and tropical Africa. It may be the oldest condiment known. The seed is small, shiny, and oval shaped, about 0.32 cm long. The unhulled seeds are dark and the hulled seeds are pearly white. Sesame seeds, when baked, have a pleasant, roasted, nutty flavor. They are used in baked goods and in confections, eg, halvah.

Tarragon (Estragon). Tarragon spice is the dried leaves and flowering tops of *Artemisia dracunculus* L. (Compositae), a perennial herb indigenous to the southern former Soviet Union and western Asia. The herb is also cultivated throughout southern Europe, particularly France and Spain, and in the United States. The aroma and flavor of the spice are warm, aromatic, and reminiscent of anise, although bittersweet. It is used in the flavoring of vinegar and in salads, soups, stews, and sauces such as bearnaise, tartare, etc.

Thyme. This spice is the dried leaves and flowering tops of *Thymus vulgaris* L. and *T. zygis* L. (Labiatae), small, perennial shrubs native to the Mediterranean area and Asia Minor, and grown commercially in Jamaica, Syria, Jordan, France, Spain, and the United States. The tiny brown-gray-green, narrow, curled leaves rarely exceed 0.63 cm in length. The flavor of the spice is aromatic, warm, and pungent. Thyme is used to flavor fish chowders, sausages, meats, poultry dressings, fish sauces, and in the liqueur Benedictine.

Turmeric (Curcuma). Turmeric is the cleaned, sound, dried rhizome of *Curcuma longa* L. (Zingiberaceae), a herbaceous perennial indigenous to southern Asia and now cultivated in India, China, Central America, and Indonesia. Turmeric has a thick round rhizome or underground stem with blunt fingers. The flavor is distinctly aromatic, earthy, and somewhat bitter. Turmeric is used as much for its orange-yellow color as it is for flavor. It is used in curry powder, for coloring pickles, and for flavoring poultry, seafood, and rice dishes.

Vanilla. Vanilla is the dried, cured, full-sized, but not fully ripe fruit pods (beans) of *Vanilla planifolia* And. and *V. tahitensis*, J. W. Moore (Orchidaceae). The vine is native to the tropical rain forests of southern Mexico, Central America, the West Indies, and northern South America. Plantings were started in Madagascar, Reunion, Java, Mauritius, and Zanzibar in 1840. The Madagascar-type bean is still the most important, but Indonesia produces more than Malagasy. The structure of the flower prevents self-pollination and therefore, where insects are not prevalent, hand pollination is necessary.

A large portion of beans go to make extracts and tinctures, which are then used for ice cream, bakery goods, liqueurs, and many other items.

BIBLIOGRAPHY

"Flavors and Spices" in *ECT* 1st ed., Vol. 6, pp. 581–594, by E. C. Crocker, Arthur D. Little, Inc.; in *ECT* 2nd ed., Vol. 9, pp. 347–380, by E. H. Hamann and E. Guenther, Fritzsche Brothers Inc.; in *ECT* 3rd ed., Vol. 10, pp. 456–488, by J. A. Rogers, Jr. and F. Frischetti, Jr., Fritzsche Dodge & Olcott, Inc.

1. M. W. Neale, *Canner/Packer* **32**(3), 41 (1963).
2. W. E. Dorland and J. A. Rogers, Jr., *The Fragrance and Flavor Industry*, Dorland Publishing Co., Mendham, N.J., 1977, p. 55.
3. *The Federal Food, Drug, and Cosmetic Act*, (21 U.S.C. 301–392), USFDA, Washington, D.C., 1984.
4. *History of Spices*, American Spice Trade Association Inc., New York, 1960.
5. F. Rosengarten, Jr., *The Book of Spices*, Pyramid Books, New York, 1973.
6. J. W. Parry, *The Story of Spices*, Chemical Publishing Co., New York, 1962.
7. *U.S. Spice Trade, FTEA 1-92*, United States Department of Agriculture, Foreign Agricultural Service, Washington, D.C., Apr. 1992.
8. *U.S. Spice Trade, FTEA 1-93,* United States Department of Agriculture, Foreign Agricultural Service, Washington, D.C., Apr. 1993.
9. *Official Analytical Methods of The American Spice Trade Association*, 3rd ed., American Spice Trade Association, Englewood Cliffs, N.J., 1985.
10. *Official Microbiological Methods of The American Spice Trade Association*, 1st ed., American Spice Trade Association, Englewood Cliffs, N.J., 1976.
11. *Cleanliness Specifications for Unprocessed Spices, Seeds, and Herbs*, American Spice Trade Association, Englewood Cliffs, N.J., Aug. 1, 1991.
12. *The Nutritional Composition of Spices*, Research Committee, ASTA, Englewood Cliffs, N.J., Feb. 1977.

JAMES A. ROGERS, JR.
Consultant

FLAX.   See FIBERS, VEGETABLE.

FLAXSEED OIL.   See DRYING OILS; FATS AND FATTY OILS.

FLOCCULATING AGENTS

Flocculation is defined as the process by which fine particles, suspended in a liquid medium, form stable aggregates called flocs. The degree of flocculation can be defined mathematically as the number of particles in a system before flocculation divided by the number of particles (flocs) after flocculation. Flocculation makes the suspension nonhomogeneous on a macroscopic scale. A complete or partial separation of the solid from the liquid phase can then be made by using a number of different mechanical devices. Flocculating agents are chemical additives which, at relatively low levels compared to the weight of the solid phase, increase the degree of flocculation of a suspension. They act on a molecular level on the surfaces of the particles to reduce repulsive forces and increase attractive forces.

Applications

The principal use of flocculating agents is to aid in making solid–liquid separations. These applications include:

(1) Removing small amounts of suspended inorganic or organic particles from surface water prior to its use as drinking water or industrial process water.

(2) Concentrating the organic solids in municipal or industrial wastewater to produce a sludge with a minimum volume and water content for incineration or other means of disposal, and a clarified (very low suspended solids) water that can be discharged or recycled. This operation is often called dewatering (qv).

(3) Removing suspended inorganic material from waste streams generated in the beneficiation of ores or nonmetallic minerals, to form a concentrated slurry that can be used for reclamation of mined out areas or other uses and a clarified water that can be discharged or recycled.

(4) Separating the solid and liquid phases in leaching operations, where a valuable material is contained in the liquid phase, so its recovery is to be maximized.

(5) Binding fine cellulose fibers and solid inorganic additives to long cellulose fibers as the paper pulp is being formed into sheets on a paper machine (see PAPERMAKING MATERIALS AND ADDITIVES).

The environmental legislation in the United States, such as the Clean Water Act and the Safe Drinking Water Act, as well as similar regulations in Europe and Japan have focused considerable attention and a large research effort on flocculation in the first three areas because the way in which both industry and local government agencies use water has changed considerably. The fourth area in extractive metallurgy has been driven primarily by the economic necessity to increase recovery. The fifth application is specific to the paper industry, and the materials used are referred to as retention aids. They are similar in chemistry and mechanism to the flocculants used for other applications and are included here for that reason. Research in this area is driven by the goal of improving the quality of the paper and the economics of its production, as well as the necessity to reduce the amount of waste.

Flocculants are also used for solid–liquid separations in other industries such as the sugar industry (1). Recent experiments have shown flocculants to be effective soil conditioners for agricultural land, which effectively reduce run-off and erosion (2). However, they have not been used on a large scale for this application.

Chemical Composition

Flocculants can be classified as inorganic or organic. The inorganic group as well as some highly charged cationic organic flocculants are sometimes referred to as coagulants; however, no such distinction is made in this article.

Inorganic Flocculating Agents. The inorganic flocculating agents are water-soluble salts of divalent or trivalent metals. For all practical purposes these metals are aluminum, iron, and calcium. Sodium silicate is also used in some applications. The principal materials currently in use are described in the following.

Aluminum sulfate (hydrate) [17927-65-0, 57292-32-7] is commonly known as alum [10043-01-3, 10043-67-1] (see ALUMINUM COMPOUNDS, ALUMINUM SULFATE AND ALUMS). The use of this material has been known for centuries (3). It is made by the leaching of aluminous ores, such as bauxite, with sulfuric acid. It is now well known that alum forms various polymeric species when used (4). Their rate of formation and the species formed is controlled by the pH and the presence of other ions. Although its use has not grown rapidly over the past few years, it is still widely used in municipal wastewater treatment, drinking water treatment, and in the paper industry (5). The principal disadvantages of alum are that it lowers the pH of the system, which often necessitates addition of base, and it leaves soluble aluminum in the effluent. It is sold both as a solution or as a dry chemical. The former is easier to dispense but must be kept warm to prevent crystallization. Grades lower in iron are used in papermaking and command a higher price than high iron grades used in waste treatment.

Aluminum chloride hydroxide [1327-41-9] also called polyaluminum chloride or PAC, is made by partial hydrolysis of aluminum chloride to form a mixture of polymeric species. It is more expensive than alum on a weight basis, but has advantages over alum such as not lowering the pH as much and better cost-effectiveness in some applications. Residual aluminum in the water is said to be lower and performance in cold water is better (6,7). It is sold as a solution (see ALUMINUM COMPOUNDS, POLYALUMINUM CHLORIDES).

Sodium aluminate [1302-42-7] is another source of soluble aluminum made by leaching bauxite with caustic soda. As with alum, the active species are really its hydrolysis products which depend on the chemistry of the system to which it is added. It tends to raise the pH. It is available both as a solid and as a solution (see ALUMINUM COMPOUNDS, ALUMINATES).

A more recently introduced product is polyaluminum–silicate–sulfate or PASS, made by Handy Chemical of Quebec. It is said to be more effective than alum in cold water, have only a slight effect on pH, and leave less residual aluminum in the water. It is sold as a solution (8).

Iron compounds (qv) include ferric chloride [7705-08-0], ferric sulfate [10028-22-5], and ferrous sulfate [7720-78-7]. For the most part they are by-products of other industries, mainly the steel industry (pickle liquor), and the processing of titanium ore (ilmenite). This has the advantage of reducing their price; however, the shift of the source industries between countries and changes in production levels have produced some supply problems (9). In recent years iron salts have replaced aluminum salts particularly in treatment of drinking water. Concern about metallic impurities in by-product iron salts has led at least one company to produce ferric sulfate from iron ore and virgin sulfuric acid (3). As with the aluminum salts, the active agents are the hydrolysis products of the additive. Ferrous salts are rapidly oxidized in solution by air and the active species are ferric compounds. Ferric chloride is sold as either a solution or as a solid. Other iron salts are usually sold as solids.

The principal calcium salt used as a flocculant is calcium hydroxide [1305-62-0] or lime. It has been used in water treatment for centuries (see CALCIUM COMPOUNDS). Newer products are more effective, and its use in water and effluent treatment is declining (10). It is still used as a pH modifier and to precipitate metals as insoluble hydroxides. Lime is also sometimes used in combination with polymeric flocculants.

Iron, calcium, and aluminum compounds precipitate phosphate from aqueous solutions, whereas organic flocculants do not. This is an advantage if a low phosphate effluent is desired, but it is a disadvantage if phosphate is required as a nutrient for bacteria used for biological sludge treatment. Iron and aluminum salts also precipitate organic color bodies, and iron salts also remove hydrogen sulfide by forming an insoluble precipitate.

Sodium silicate is usually added to slurries as a dispersant (see DISPERSANTS). Small amounts of sodium silicate are used as flocculants. The active species are polymeric silicates formed by hydrolysis.

Organic Flocculants.   The organic flocculants are all water-soluble natural or synthetic polymers.

Natural Products.   The use of natural polymers has been known for a long time; for example, the soluble protein albumin is used for clarifying wine. Since the 1950s the use of natural products as flocculating agents has steadily declined as more effective synthetics have taken their place. The only natural polymers used to a significant degree as flocculants are starch (qv) and guar gum (see GUMS).

Starch is a polysaccharide found in many plant species. Corn and potatoes are two common sources of industrial starch. The composition of starch varies somewhat in terms of the amount of branching of the polymer chains (11). Its principal use as a flocculant is in the Bayer process for extracting aluminum from bauxite ore. The digestion of bauxite in sodium hydroxide solution produces a suspension of finely divided iron minerals and silicates, called red mud, in a highly alkaline liquor. Starch is used to settle the red mud so that relatively pure alumina can be produced from the clarified liquor. It has been largely replaced by acrylic acid and acrylamide-based (11,12) polymers, although a number of plants still add some starch in addition to synthetic polymers to reduce the level of residual suspended solids in the liquor. Starch [9005-25-8] can be modified with various reagents to produce semisynthetic polymers. The principal one of these is cationic starch, which is used as a retention aid in paper production as a component of a dual system (13,14) or a microparticle system (15).

Guar gum [9000-30-0], derived from the seed of a legume (11,16), is used as a flocculant in the filtration of mineral pulps leached with acid or cyanide for the recovery of uranium and gold (16). It is also used as a retention aid, usually in a chemically modified form (14,17). Starch and guar gum are subject to biological degradation in solution, so they are usually sold as dry powders that are dissolved immediately before use. Starch requires heating in most cases to be fully dissolved.

Synthetic Polymers.   Examples of polymers in this class include acrylamide–acrylic polymers and their derivatives, polyamines and their derivatives, poly-(ethylene oxide), and allylamine polymers.

Acrylamide–acrylic polymers are made by free-radical polymerization of monomers containing the acrylic structure, where R is —H or —CH$_3$ and R' is —NH$_2$ or a substituted amide or the alkoxy group of an ester.

$$CH_2=CR$$
$$|$$
$$C=O$$
$$|$$
$$R'$$

The principal monomer is acrylamide [*79-06-1*], where R = H and R' = NH$_2$, made by the hydrolysis of acrylonitrile. The homopolymer [*9003-05-8*] of acrylamide, which in theory has no electrical charge, has some use as a flocculant; however, the majority of acrylamide-based flocculants are copolymers with acrylic monomers containing charged functional groups, such as those shown in Figure 1, or polymers containing functional groups formed by modification of acrylamide homopolymers or copolymers (Fig. 2). The chemistry of polyacrylamides has been reviewed by several authors (18–20) (see ACRYLAMIDE POLYMERS).

There are two main advantages of acrylamide–acrylic-based flocculants which have allowed them to dominate the market for polymeric flocculants in many application areas. The first is that these polymers can be made on a commercial scale with molecular weights up to 10–15 million which is much higher

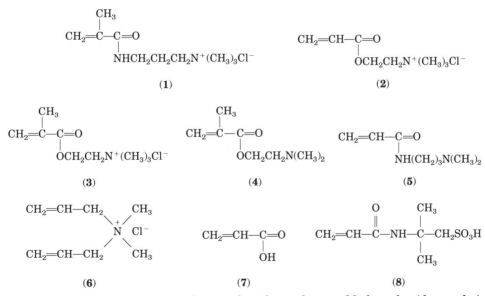

Fig. 1. Functional monomers used in acrylamide copolymers. Methacrylamidopropyltrimethylammonium chloride [*51410-72-1*] (**1**), acryloyloxyethyltrimethylammonium chloride [*44992-01-0*] (**2**), methacryloyloxyethyltrimethylammonium chloride [*50339-78-1*] (**3**), *N,N*-dimethylaminoethyl methacrylate [*2867-47-2*] (**4**), *N,N*-dimethylaminopropylacrylamide [*3845-76-9*] (**5**), diallyldimethylammonium chloride [*7398-69-8*] (**6**), acrylic acid (and its salts) [*79-10-7*] (**7**), and 2-acrylamido-2-methylpropanesulfonic acid (and its salts) [*15214-89-8*]) (**8**).

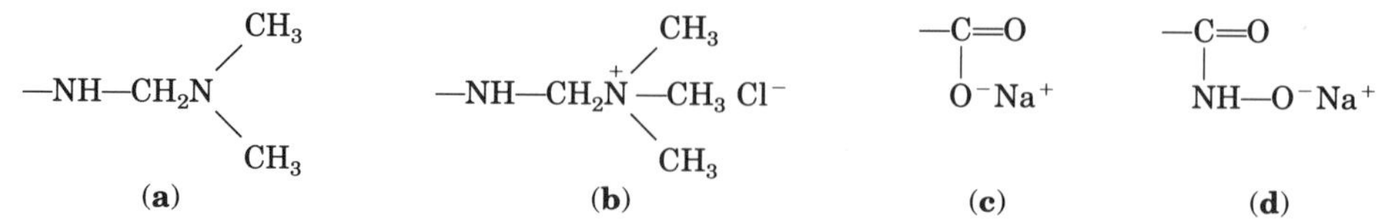

Fig. 2. Functional groups on modified polyacrylamides: (**a**) formed by reaction with dimethylamine and formaldehyde (Mannich reaction); (**b**), quaternized Mannich amine; (**c**), carboxylate formed by acid or base-catalyzed hydrolysis or copolymerization with sodium acrylate; and (**d**), hydroxamate formed by transamination with hydroxylamine.

than any natural product. The second is that their electrical charge in solution and the charge density can be varied over a wide range by copolymerizing acrylamide with a variety of functional monomers or by chemical modification.

The high molecular weight of these polymers makes their solutions very viscous, which presents a problem on an industrial scale with regard to shipping, handling, and dissolving. The two principal forms in which these polymers are sold, dry powders and inverse emulsions, represent two different solutions to this problem. The dry powder form can be made by two different routes. In the first, a concentrated monomer solution is polymerized, producing a gel which is cut up, granulated or extruded, and dried in a fluidized-bed drier to produce a free-flowing powder. In the second method, the concentrated monomer solution is suspended as droplets in an immiscible organic liquid. After polymerization the polymer is in the form of spherical beads containing polymer and water. The beads can be removed and dried to give a dry powder made up of spherical particles. Drying of nonionic polyacrylamides may introduce a slight anionic charge because of hydrolysis.

The inverse emulsion form is made by emulsifying an aqueous monomer solution in a light hydrocarbon oil to form an oil-continuous emulsion stabilized by a surfactant system (21). This is polymerized to form an emulsion of aqueous polymer particle ranging in size from 1.0 to about 10 μm dispersed in oil. By addition of appropriate surfactants, the emulsion is made self-inverting, which means that when it is added to water with agitation, the oil is emulsified and the polymer goes into solution in a few minutes. Alternatively, a surfactant can be added to the water before addition of the inverse polymer emulsion (see EMULSIONS).

If either dry powders or inverse emulsions are not properly mixed with water, large lumps of polymer form that do not dissolve. This not only wastes material, but can also cause downstream problems. This is especially true for paper where visible defects may be formed. Specialized equipment for dissolving both dry polymers and inverse emulsions on a continuous basis is available (22,23). Some care must be taken with regard to water quality when dissolving polyacrylamides. Anionic polymers can degrade rapidly in the presence of ferrous ion sometimes present in well water (24). Some cationic polymers can lose charge by hydrolysis at high pH (25).

Polyamines are condensation polymers containing nitrogen; they are made by a variety of synthetic routes. Most of the commercial polyamines are made by reaction of epichlorohydrin with amines such as methylamine [25988-97-0] or dimethylamine [*39660-17-8*] (18,19). Branching can be increased by adding small amounts of diamines such as ethylenediamine [*42751-79-1*]. A typical structure of this type of polyamine is structure (**9**).

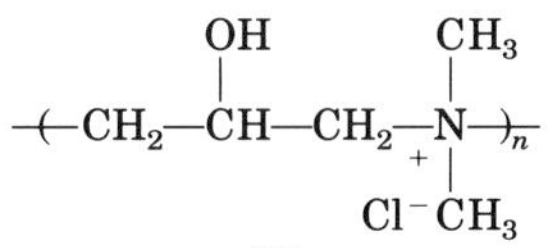

$$-\!\!\left(\!-CH_2-\underset{\underset{OH}{|}}{CH}-CH_2-\underset{\underset{Cl^-}{\overset{+}{|}}}{\overset{\overset{CH_3}{|}}{N}}-\!\right)_{\!\!n}^{\!CH_3}$$

(**9**)

Polyamines can also be made by reaction of ethylene dichloride with amines (18). Products of this type are sometimes formed as by-products in the manufacture of amines. A third type of polyamine is polyethyleneimine [*9002-98-6*] which can be made by several routes; the most frequently used method is the polymerization of aziridine [*151-56-4*] (18,26). The process can be adjusted to vary the amount of branching (see IMINES, CYCLIC). Polyamines are considerably lower in molecular weight compared to acrylamide polymers, and therefore their solution viscosities are much lower. They are sold commercially as viscous solutions containing 1–20% polymer, and also any by-product salts from the polymerization reaction. The charge on polyamines depends on the pH of the medium. They can be quaternized to make their charge independent of pH (18).

Poly(ethylene oxide)s [*25372-68-3*] are made by condensation of ethylene oxide with a basic catalyst. In order to achieve a very high molecular weight, water and other compounds that can act as chain terminators must be rigorously excluded. Polymers up to a molecular weight of 8 million are available commercially in the form of dry powders (27). These must be dissolved carefully using similar techniques to those used for dry polyacrylamides. Poly(ethylene oxide)s precipitate from water solutions just below the boiling point (see POLYETHERS, ETHYLENE OXIDE POLYMERS).

Allyl polymers are made by free-radical polymerization of diallyl compounds, most frequently diallyldimethylammonium chloride (DADMAC) [*7398-69-8*] forming a chain containing a five-membered ring (28) poly(DADMAC) [*26062-79-3*].

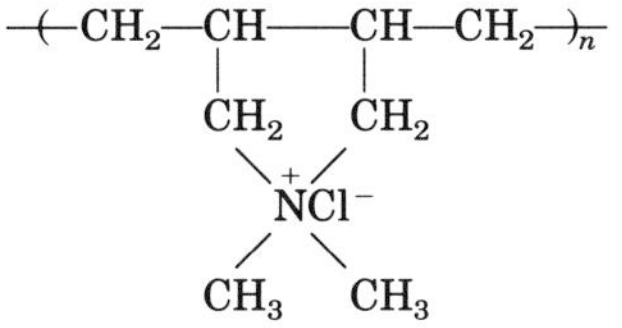

$$-\!\!\left(\!-CH_2-CH\!-\!\!-\!CH-CH_2-\!\right)_{\!\!n}$$

$$\begin{array}{cc} | & | \\ CH_2 & CH_2 \\ \diagdown \overset{+}{N}Cl^- \diagup \\ \diagup \quad \diagdown \\ CH_3 \quad CH_3 \end{array}$$

(**10**)

The monomer can also be copolymerized with acrylamide. Because of the high chain-transfer rate of allylic radicals, the molecular weights tend to be lower than for acrylic polymers. These polymers are sold either as a viscous solution or a dry powder made by suspension polymerization (see ALLYL MONOMERS AND POLYMERS).

Mechanism of Flocculation

In order to form flocs the individual particles must move and collide. Flocculation can be classified as either orthokinetic or perikinetic. In the first case particle motion results from turbulence in the suspension, and in the latter from Brownian motion (29). Orthokinetic motion is almost always the case in industrial applications. At very close distances, polar materials are attracted by dipole-induced dipole interactions commonly called van der Waals forces. In most aqueous suspensions, ionization of surface groups gives the particle an overall negative charge. The charged particles in suspension are surrounded by a group of positive ions referred to as the double layer. As particles approach each other the resulting electrostatic repulsion of the double layers prevents flocculation. Increasing the ionic strength of the liquid medium reduces the repulsion until the particles start to aggregate at the critical flocculation concentration. As the charge of these positive ions forming the double layer is increased by adding higher charged ions to the system, the double layer gets nearer to the surface allowing the particles to become closer and be attracted by the van der Waals forces. This is the explanation for the empirically derived Schulze-Hardy rule that the critical flocculation concentration of positive ions for a particular system decreases proportionally with the sixth power of the charge (30). This mechanism is called double-layer compression and is often cited for the inorganic flocculating agents, such as alum and ferric salts, which add trivalent ions to the system. However, this explanation of the action of aluminum and ferric salts does not take into account the fact that they are present at least partially as polymeric species when added to many systems (4), and that polymeric precipitates may be formed (3) at the usual concentrations and the pH range that they are used.

In some systems, such as lake and river waters, the suspended inorganic particles may be coated by biological polymers, termed humic substances, which prevent flocculation by either steric or electrostatic mechanisms. These can also interact with added inorganic salts (31) that can neutralize charged functional groups on these polymers.

The second flocculation mechanism is referred to as the charge patch or electrostatic mechanism (32). A highly cationic polymer is adsorbed on a negative particle surface in a flat conformation. That is to say most of the charged groups are close to the surface of the particle, as illustrated in Figure 3. This promotes flocculation by first reducing the overall negative charge on the particle thus reducing interparticle repulsion. This effect is called charge neutralization and is associated with reduced electrophoretic mobility. In addition, the areas of polymer adsorption can actually have a net positive charge because of the high charge density of the polymer. The positive regions are also attracted to negative regions on other particles, which is called heterocoagulation. Polymeric inorganic materials may also adsorb on surfaces and cause flocculation by a similar mechanism. A third mechanism is called bridging. Some individual segments of a very high molecular weight polymer, usually a high molecular weight anionic polyacrylamide, adsorb on a surface. As shown in Figure 4a, large segments of the polymer extend into the liquid phase where other segments are adsorbed on other particles, effectively linking the particles together with polymer bridges. In con-

Fig. 3. Adsorbed cationic polymer forming charge patch on particle surface (32).

trast to the first two mechanisms, bridging is strongly affected by molecular weight and the ionic content of the solution. Only large molecules (33) can bridge between particles. Low molecular weight anionic polymers actually act as dispersants in the same systems. The partial adsorption of the anionic polymer on a negatively charged particle is promoted by the presence of divalent and trivalent ions (34). The charge density of the polymer is also critical. As the negative charge on the polymer increases, the mutual repulsion of negatively charged groups along the chain causes the molecule to have a more extended conformation in solution that favors bridging. The higher charge, however, works against adsorption on negatively charged particles. Increasing the ionic strength of the medium pro-

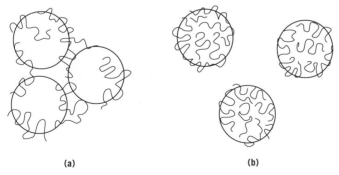

(a)   (b)

Fig. 4. (**a**) Polymer bridging between particles; and (**b**), particle stabilization by adsorbed polymer (32).

motes adsorption; however, the ions shield the negatively charged groups along the chain, which favors a less extended conformation. For this reason, for each combination of aqueous and solid phases there is an optimal charge (35). This effect was first reported in 1954 (36). This principle is well illustrated in the Bayer process, where the residue from bauxite leaching is alternately flocculated and repulped in solutions with decreasing ionic content. As the ionic content goes down, the optimal charge, in terms of settling rate, of the anionic polymer used as a flocculant decreases (37).

Cationic polymers can also bridge between particles, if the molecular weight is high enough. Bridging is cited as the mechanism for cationic retention aids (38). If the substrate has a high negative charge, the cationic polymer tends to adsorb in a flatter conformation than an anionic polymer, with fewer loops extending out to bridge with other particles.

In most applications, the flocs formed by these mechanisms are composed of chemically similar particles. However, in the case of retention aids the substrate is a heterogeneous mixture of cellulose fibers and inorganic fillers and pigments such as TiO_2, $CaCO_3$, and clay. The flocculant must have the ability to hold all of these together. The process may be complicated by the fact that some of this material, ie, clay, may have been treated with dispersants which can block some of the available adsorption sites on their surface (39). In some cases, however, having a heterogeneous substrate may promote flocculation, if the different components such as titanium oxide pigment and paper fibers have opposite electrical charge at the same pH. This is another example of heterocoagulation.

A fourth mechanism is called sweep flocculation. It is used primarily in very low solids systems such as raw water clarification. Addition of an inorganic salt produces a metal hydroxide precipitate which entrains fine particles of other suspended solids as it settles. A variation of this mechanism is sometimes employed for suspensions that do not respond to polymeric flocculants. A solid material such as clay is deliberately added to the suspension and then flocculated with a high molecular weight polymer. The original suspended matter is entrained in the clay flocs formed by the bridging mechanism and is removed with the clay.

Small particles of silica or clay can also be used in combination with polymers as retention aids. These are called microparticle systems (15). Low molecular weight cationic polymers on the surface of the inorganic particle bind to the fine cellulose fibers by bridging. The solid particles extend the effective length of the flocculant molecules and give the flocs rigidity. These small, rigid flocs are bound tightly to the larger fibers. In many systems, more than one of these mechanisms may be operative at the same time. Cationic–anionic combinations are often used in mineral processing and retention aid applications. The cationic polymer is usually added first to neutralize the charge on the particles and form charge patches. Alum or ferric salts can also be used for this purpose. These can serve as adsorption sites for higher molecular weight anionic flocculants. For retention aids, a cationic polymer with a moderately high charge density is usually preferred (39). Very small flocs are formed which are then flocculated by a higher molecular weight anionic polymer. This often results in an efficient removal of all suspended particles and good retention of fine fibers. Small hydroxide flocs produced by hydrolysis of inorganic salts can be flocculated by organic polymeric flocculants.

Flocculant Performance and Selection

There is no comprehensive quantitative theory for predicting flocculation behavior that can be used for flocculant selection. This must ultimately be determined experimentally. There are three variables that affect the results obtained in any particular flocculation system. These are the type of flocculant, type of substrate, and type of mechanical treatment of the flocculated substrate. The size and physical properties of the flocs that form, rather than the degree of flocculation, are the key elements in determining the practical effectiveness of a flocculant in any specific application. The effect of mechanical treatment can be viewed in terms of the type of force applied to the flocs. In thickeners and settling basins the flocs are acted on by gravity and by the weight of material added on top of them. In vacuum filters the flocs are subjected to atmospheric pressure. In belt presses and plate-and-frame filters the flocs are subjected to mechanical pressure and in centrifuges they are subject to centrifugal forces. In a flowing system, such as a continuous paper machine, they are subjected to shear and elongational forces on the same scale as the particle size. In addition to the type of force that is applied to the flocs, the kinetics of floc formation also plays an important role in the results obtained in their application.

　　The effect of mechanical treatment on floc behavior is illustrated in Figure 5. In one work (40), identical slurries were treated with varying doses of the same polymer. At each dosage, it can be assumed that the same type of floc formed at the same rate. However, the dosage response was completely different depending on which parameter of the flocculated slurry was measured. Thus the term opti-

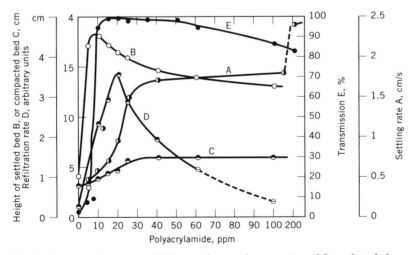

Fig. 5. Effect of polymer dosage on different observed properties of flocculated slurry (40). Comparison of five parameters in a flocculation system (8% fluorite suspension + polyacrylamide Cyanamer P250). A, Rate of settling of floc boundary, in cm/s; B, height of settled bed, cm; C, height of consolidated filter-cake, cm; D, refiltration rate, arbitrary units; and E, clarification, % optical transmission of 1 cm of supernatant liquid after 3 min settling time.

mal flocculation cannot be applied to any flocculant–substrate combination if the solid–liquid separation process or process parameter is not specified.

There are some general principles that can serve as guidelines for initial screening in terms of both flocculant chemistry and molecular weight. In general, the large flocs formed by high molecular weight polymers tend to settle faster than smaller ones. In the upper part of a thickener or settling basin, the settling rate of an individual floc is governed by Stokes law. Using a spherical model to approximate an individual floc, at the terminal velocity the viscous drag is equal to the gravitational force on the floc. The downward velocity is given by equation 1 where g is the gravitational constant, a is the radius of the particle, d_1 is the density of the particle, d_2 is the density of the liquid, and η is the viscosity of the liquid.

$$V = 2\, ga^2(d_1 - d_2)/9\, \eta \qquad (1)$$

Although floc density has an effect on the settling rate, it is overshadowed by the effect of size. For most substrates, high molecular weight polymeric flocculants give the largest flocs, when the charge density is optimized for the particular system. In one case, the settling rate was found to be proportional to the sixth power of the polymer molecular weight (33). The rate of floc formation and the initial settling rate is very high with these high molecular weight flocculants. This may remove the flocculant adsorbed on the flocs from the system so fast that some unflocculated material is left in suspension. If this occurs in a thickener, solids may appear in the overflow, which may have an adverse effect on the process. This can often be prevented by adding the polymer in two stages (12). The initial dose gives small flocs which form and settle slowly and therefore give a more complete removal of solids. The second dose forms larger flocs which give the desired high settling rate. Alternatively, a lower molecular weight polymer of the same or opposite charge can be added first to form the small flocs, which are then flocculated with a higher molecular weight polymer.

In the case of thickeners, the process of compaction of the flocculated material is important. The flocs settle to the bottom and gradually coalesce under the weight of the material on top of them. As the bed of flocculated material compacts, water is released. Usually the bed is slowly stirred with a rotating rake to release trapped water. The concentrated slurry, called the underflow, is pumped out the bottom. Compaction can often be promoted by mixing coarse material with the substrate because it creates channels for the upward flow of water as it falls through the bed of flocculated material. The amount of compaction is critical in terms of calculating the size of the thickener needed for a particular operation. The process of compaction has been extensively reviewed in the literature (41,42).

For most substrates the operating dosage of flocculant necessary to give the settling rate necessary to operate a thickener is well below the maximum amount that can be adsorbed on the substrate. As more and more polymer is added above this operating dosage the flocs can become larger and somewhat sticky. The bed of flocculated material then becomes very viscous. The rake mechanism may become overloaded and the flocculated material may not flow into the underflow pump. The dosage response and the sensitivity to overdosing may affect the selection of flocculating agent.

For filter belt presses and centrifuges, resistance to shear and mechanical pressure is the most important parameter. In general, flocs produced by charge patch neutralization are stronger than those produced by inorganic salts alone. If these flocs are broken, the cationic polymer remains strongly bound to the surface and the flocs can reform. Very strong flocs can be made with high molecular weight polymers that bridge between particles. However, these may not reform if broken because the bridging segments have been broken. The residual polymer fragments on the surface may even act as a dispersant by covering the particle surface, as shown in Figure 4**b**.

For vacuum filters, both the rate of filtration and the dryness of the cake may be important. The filter cake can be modeled as a porous solid, and the best flocculants are the ones that can keep the pores open. The large, low density flocs produced by high molecular weight polymers often collapse and cause blinding of the filter. Low molecular weight synthetic polymers and natural products that give small but rigid flocs are often found to be the best.

Retention aid polymers are used in a very high shear environment, so floc strength and the ability for flocs to reform after being sheared (43) is important. The optimum floc size is a compromise. Larger flocs give better free drainage, but tend to produce an uneven sheet due to air breakthrough in the suction portions of the paper machine (38). In some cases the type of floc needed for retention can be seen as similar to that needed for vacuum filtration. The substrate materials are the inorganic fillers and fine fibers, whereas the filter is the mat of long fibers formed on the paper machine. Floc size can be controlled by both the type of flocculant and the addition point.

General guidelines concerning the initial selection of flocculant chemistry are (1) suspensions of organic materials, such as municipal waste, are usually treated with a cationic flocculant, either inorganic or organic; and (2) suspensions of inorganic materials such as clay are usually treated with an anionic polymer or a combination of an anionic polymer with a cationic flocculating agent. There are also some special cases where particular flocculants are applied.

Acidic suspensions such as those produced by acid leaching often respond to natural products such as guar as well as nonionic polyacrylamides and anionic polyacrylamides containing sulfonic acid groups.

Poly(ethylene oxide) (PEO) forms a unique type of floc with a number of substrates in the mining industry (44–46) such as phosphate slimes and coal refuse. These large, strong flocs release water readily when subjected to mechanical force. Extensive testing has been done by the U.S. Bureau of Mines to dewater mining wastes by flocculating with PEO and dewatering (qv) with a rotating trommel screen. However, this process has not been widely adopted on a large scale in mineral processing (see MINERAL RECOVERY AND PROCESSING). Poly(ethylene oxide)s are also used as components of dual retention aid systems (13).

A similar technique to the Bureau of Mines trommel process called pellet flocculation has been used in Japan on a number of substrates on an industrial scale (47) using equipment made by the Ebara-Infilco Co. Combinations of inorganic salts such as lime with polyacrylamides are used as flocculants.

Laboratory Flocculant Testing. The objective of laboratory testing of flocculants is to determine which chemical composition and molecular weight will
give the best cost performance. The usual method is to simulate on a laboratory
scale the formation of flocs and then subject them to the same or similar types of
forces as would be encountered in a full-scale dewatering device. For applications
in thickeners and settling basins, the substrate is usually mixed with the flocculant in a graduated cylinder using a plunger or inverting the cylinder. As the flocs
form and settle there is usually a sharp boundary between clear liquid and the
suspension of flocs. The settling rate is determined from the downward velocity
of the interface. Initially this velocity is constant, but as the concentration of flocs
increases in the lower portion of the cylinder the rate slows down because of the
interaction between the flocs. The compaction of the flocculated material can be
measured by measuring the height of the interface over a longer time period (42).
Laboratory-scale thickeners are also available to evaluate flocculant performance
(48,49). Small pumps are used to add substrate and flocculant and remove the
compacted material. Transparent sides permit direct observation of floc formation. A graduated cylinder equipped with a rake is recommended for experiments
to determine the required thickener size (50). The density of flocculated red muds
and other substrates can be measured using a γ ray densitometer and the results
used to evaluate flocculants and to calculate the required thickener size (51).

For evaluation of flocculants for pressure belt filters, both laboratory-scale
filters and filter simulators are available (52,53) in many cases from the manufacturers of the full-scale equipment. The former can be run either batchwise or
continuously; the simulators require less substrate and are run batchwise. The
observed parameters include cake moisture, free drainage, release of the cake
from the filter cloth, filter blinding, and retention of the flocculated material during application of pressure.

Vacuum filters are usually simulated with a Buchner funnel test or filter
leaf test (54). The measured parameters are cake weight, cake moisture, and filtration rate. Retention aids are usually evaluated using the Britt jar test, also
called the Dynamic Drainage Jar, which simulates the shear conditions found on
the paper machine and predicts performance (55).

Operating Parameters and Control

Flocculating agents differ from other materials used in the chemical process industries in that their effect not only depends on the amount added, but also on
the concentration of the solution and the point at which it is added. The process
streams to which flocculants are added often vary in composition over relatively
short time periods. This presents special problems in process control.

Dilution. In many applications, dilution of the flocculant solution before it
is mixed with the substrate stream can improve performance (12). The mechanism
probably involves getting a more uniform distribution of the polymer molecules.
Since the dosage needed to form flocs is usually well below the adsorption maximum, a high local concentration is effectively removed from the system at that
point, leaving no flocculant for the rest of the particles. A portion of the clarified
overflow can be used for dilution so no extra water is added to the process.

Addition Point.　The flocculant addition point in a continuous system can also have a significant effect on flocculant performance. The turbulence as the flocculant is mixed in and the flocs travel toward the point where they enter the thickener or filter causes both the formation and breakup of flocs. Usually there is an optimal addition point or points which have to be determined empirically. In cases where the same polymer is being added at two or more points, the relative amounts added at each point may also affect performance. Thus providing multiple addition points in the design of new installations is recommended (56).

Automatic Control.　In some industries, the waste streams can vary in composition over a relatively short time period. When the solids level of a slurry changes, the entire dosage response may change (12,57). Automatic systems are available for thickeners that adjust the dosage according to the incoming solids level, overflow turbidity (58), and streaming current potential (59). Appropriate control software is used, which takes into consideration response times and other factors. These systems can improve operation and reduce flocculant usage (60). On-stream optical sensors that measure floc size are available (61,62). Self-cleaning cells are used to overcome the problem of fouling of optical surfaces. Sensors of this type are useful in continuous processes such as papermaking, where a sudden shift in floc size caused by a dosage change could be a serious problem (63).

Analysis

Inorganic flocculants are analyzed by the usual methods for compounds of this type. Residual metal ions in the effluent are measured by spectroscopic techniques such as atomic absorption. Polymeric aluminum species formed in solution have been characterized by ^{27}Al-nmr (64).

The detection of organic polymers in solution represents a more difficult problem, especially in industrial water and wastewater. In theory, charged polymers react with polymers of the opposite charge in solution and such reactions can be used to titrate the concentration of polymer present. There are a number of techniques using this method (65).

Polymers can also be labeled with fluorescent chromophores (66) or radioactive monomers. Other methods have been reviewed (65). Polyacrylamide, whether charged or not, can be detected by reactions of the amide group (67,68); however, a number of substances can interfere with the determination. If the molecular weight is high enough, flocculation of a standard slurry of clay or other substrate is a sensitive method for detecting low levels of polyacrylamide (69). Once polymers are adsorbed on a surface, many of these methods cannot be used. One exception is the use of a labeled polymer.

The molecular weights and molecular weight distributions of lower molecular weight polymeric flocculants are determined by viscosity measurements, such as the intrinsic viscosity, and by size exclusion chromatography. High molecular weight acrylamide-based polymers are characterized by light scattering techniques (21). ^{13}C-nmr can be used to determine sequence distribution (70) and the composition of polyacrylamides and their reaction products (25).

Toxicology and Environmental Issues

All materials used to treat public water supplies have come under increasing scrutiny in recent years. One controversy is based on a possible link between dietary aluminum and Alzheimer's disease. This stems from the finding of high aluminum levels in the brains of deceased Alzheimer's patients and epidemiological evidence (71,72) linking the aluminum level in drinking water with the incidence of Alzheimer's disease. Both PAC and PASS are claimed to leave less residual aluminum in solution. All of the epidemiological studies on Alzheimer's disease and ingestion of aluminum (from water, food, medications, etc) has been critically reviewed (73). The review concludes that none of the studies show an epidemiological linkage between Alzheimer's disease and the amount of aluminum ingested. Aluminum from drinking water is a relatively minor source of dietary aluminum. Exposure to very high levels of aluminum by dialysis treatment or accidental overdose of alum has been shown to cause a variety of medical problems (74,75).

Based on animal studies and mutagenicity studies, trace amounts of organic polymers do not appear to present a toxicity problem in drinking water (76). The reaction products with both chlorine and ozone also appear to have low toxicity (77). The principal concern is the presence of unreacted monomer and other toxic and potentially carcinogenic nonpolymeric organic compounds in commercial polymeric flocculants. The principal compounds are acrylamide in acrylamide-based polymers, dimethyldiallyammonium chloride in allylic polymers, and epichlorohydrin and chlorinated propanols in polyamines, as well as the reaction products of these compounds with ozone and chlorine (77). In most cases, the concentrations of unreacted monomer and organic contaminants can be kept very low by controlling the manufacturing process. Careful analysis and quality control in production are therefore essential to the safe use of polymeric flocculants. There has also been concern over the presence of trace metals in by-product inorganic flocculating agents, especially ferric salts.

Until 1990 the EPA maintained a list of chemicals suitable for potable water treatment in the United States. Since then the entire question of certification and standards has been turned over to a group of organizations headed by the National Sanitation Foundation, which has issued voluntary standards. As of January 1992, standards had been issued for most of the principal inorganic products, but only for two polymers, poly(DADMAC) and Epi-DMA (epichlorohydrin-dimethylamine) polymers (78). Certifications for commercial products meeting specified standards are issued by the National Sanitation Foundation, Underwriter Laboratories, and Risk Focus/Versar (79).

The same questions about the safety of organic flocculants have been raised in other countries. The most drastic response has occurred in Japan (7,77) and Switzerland (77) where the use of any synthetic polymers for drinking water treatment is not permitted. Alum and PAC are the principal chemicals used in Japan (7). Chitin, a biopolymer derived from marine animals, has been used in Japan (80,81). Maximum allowed polymer doses have been set in France and Germany (77).

In the area of municipal and industrial wastewater treatment, the principal environmental issue is the toxicity of residual flocculating agents in the effluent.

Laboratory studies have shown that cationic polymers are toxic to fish because of the interaction of these polymers with gill membranes. Nonionic and anionic polymers show no toxicity (82,83). Other studies have shown that in natural systems the suspended inorganic matter and humic substances substantially reduce the toxicity of added cationic polymer, and the polymers have been used successfully in fish hatcheries (84–86). Based on these results, the EPA has added a protocol for testing these polymers for toxicity toward fish in the presence of humic acids (87). The addition of anionic polymers to effluent streams containing cationic polymers to reduce their toxicity has been mentioned in the patent literature (83).

Economic Aspects

The principal trend in the flocculants market is the gradual replacement of low price inorganics, especially alum, with higher priced polymers. In one report (88), the total value of flocculating agents used for drinking water and wastewater treatment in the United States in 1990 was $427.2 million with $115.2 million for alum, $40.0 million for iron salts, $51.0 million for other inorganics, and $221.0 million for polymers. The strongest growth is expected for the polymers, which have increased in value in the United States at an annualized rate of 12.5% (89).

In terms of value, the alum market share is expected to decline. Alum is facing strong competition from polyaluminum chloride both in water treatment and paper (8), and from iron salts (9) in water treatment. Alum is being replaced in papermaking by the introduction of dual retention and microparticle retention systems which use synthetic polymers as well as modified starches (90). The changeover from acid to alkaline papermaking will also decrease alum usage. Also, to some extent, the change in papermaking pH may increase the usage of anionic retention aids (39,90). Because alum is a high volume/low unit price commodity, alum pricing is affected by increases in shipping and raw materials costs, which have pushed prices up recently (91).

The polymer market in the United States is dominated by synthetics with natural polymers constituting about one-eighth in monetary terms (88). Of the synthetic polymers, most are based on acrylamide. A list of producers is as follows; producers in the left-hand column also produce polyamines and polyquaternaries.

Producers of Synthetic Flocculants Based on Acrylamide

BASF	Allied Colloids
Betz Laboratories	Diafloc
Calgon	Dow
Cytec Industries	Chemische Fabrik Stockhausen
Floerger	KEMIRA Oy
Nalco	Kyoritzu Yuki
Polypure	Mitsubishi Chemical Industries
Rohm GmbH	Mitsui-Cyanamid
Sankyo Chemical Industries	Toagosei Chemical Industries
Sanyo Chemical Industries	
Sumitomo Chemical Co.	

Total U.S. demand for acrylamide in 1991 was put at 45,500 t with a domestic capacity of 54,500 t (92). The overcapacity can be attributed to the virtual disappearance of the enhanced oil recovery polymer market in the 1980s. The principal monomer producers throughout the world are also principal polymer producers. Product prices cover a wide range depending on the amount of more expensive comonomers used. In general, anionics are less expensive than cationics on a weight basis. However, because of the need to maximize the dewatering of wastewater sludges in many areas, cost performance rather than unit cost is the overriding factor, and cationic polyacrylamide is displacing alum and other polymers.

Prices of natural products such as starch, which is produced in many countries, and guar, which is produced mainly in India and Pakistan, are affected by unpredictable factors such as the weather. Toward the end of 1991 prices were rising (93); however, in the future an oversupply might cause a large drop in prices. Because of the amounts used, starch is usually purchased locally, and pricing fluctuates with local farm prices and conditions.

BIBLIOGRAPHY

"Flocculating Agents" in *ECT* 3rd ed., Vol. 10, pp. 489–523, by F. Halverson and H. P. Palver, American Cyanamid Co.

1. J. C. P. Chen, *Cane Sugar Handbook*, John Wiley & Sons, Inc., New York, 1985, pp. 149–155.
2. H. J. C. Smith, G. J. Levy, and I. Shainberg, *Soil Sci. Soc. Am. J.* **54**, 1084–1087 (1990).
3. K. Dentel and J. M. Gosset, *J. Am. Water Works Assoc.*, 187–198 (Apr. 1988).
4. J. E. Van Benschoten and J. K. Edzwald, *Water Res.* **24**, 1519–1526 (1990).
5. *Chem. Mark. Rep.*, 6 (Oct. 29, 1991).
6. A. M. Simpson, W. Hatton, and M. Brockbank, *Environ. Tech. Lett.* **9**, 907–916 (1988).
7. S. Kawamura and R. R. Trussel, *J. Am. Water Works Assoc.*, 56–62 (June 1991).
8. G. Busch, *Chem. Mark. Rep.*, 9, 31 (Sept. 9, 1991).
9. *Chem. Mark. Rep.*, 5, 21 (July 29, 1991).
10. *Chem. Mark. Rep.*, 3 (Dec. 24, 1990).
11. N. M. Levine, in W. L. K. Schwoyer, ed., *Polyelectrolytes for Water and Wastewater Treatment*, CRC Press, Boca Raton, Fla., 1981, pp. 47–60.
12. L. J. Connelly, D. O. Owen, and P. F. Richardson, *Light Metals* **2**, 61–68 (1986).
13. D. F. Honig, *Advanced Topics in Wet End Chemistry*, TAPPI Press, Atlanta, Ga., 1987, pp. 1–5.
14. F. Halverson, in K. J. Hipolit, ed., *Chemical Processing Aids in Papermaking: A Practical Guide*, TAPPI Press, Atlanta, Ga., 1992, pp. 103–127.
15. K. Moberg, *TAPPI 1989 Retention and Drainage Short Course Notes*, TAPPI Press, Atlanta, Ga., 1989, pp. 65–86.
16. J. M. W. Mackenzie, *Eng. Mining J.*, 80–87 (Oct. 1980).
17. J. Farewell, ed., *Commercially Available Chemical Agents for Paper and Paperboard Manufacture*, TAPPI Press, Atlanta, Ga., 1990, pp. 17, 58–59.
18. N. Vorchheimer, in Ref. 11, pp. 1–46.
19. D. A. Mortimer, *Polym. Int.*, **25**, 29–41 (1991).
20. V. A. Myagchenkov and V. F. Kurenkov, *Polym.-Plast. Technol. Eng*, **30** 109–135 (1991).
21. D. Hunkeler, *Polym. Int.* **27**, 23–33 (1992).

22. P. A. Rey and R. G. Varsanik, *Adv. Chem.* **213**, 113–143 (1986).
23. R. J. Chamberlain, in Ref. 11, pp. 243–266.
24. H. I. Heitner, in B. M. Moudgil and B. J. Scheiner, eds., *Flocculation and Dewatering*, Engineering Foundation, New York, 1989, pp. 215–220.
25. D. R. Draney and co-workers, *ACS Polym. Prepr.* **32**, 500–501 (1990).
26. D. N. Roark and B. C. McKusick, in *Ullmanns Encyclopedia of Industrial Chemistry*, Vol. A3, VCH Verlagsgesellschaft, Weinheim, Germany, 1985, pp. 239–244.
27. *Polyox Water Soluble Resins*, Union Carbide Corp., Danbury, Conn., 1988.
28. J. E. Lancaster, L. Baccei, and H. P. Panzer, *J. Polym. Sci., Polymer Lett. Ed.* **14**, 549 (1976).
29. J. G. Janssens, *Aqua* **1987**, 91–97 (1987).
30. J. T. G. Overbeek, *Pure Appl. Chem.* **52**, 1151–1161 (1980).
31. M. R. Jekel, *Water Res.* **20**, 1543–1554 (1986).
32. J. Gregory, in B. M. Moudgil and P. Somasundaran, eds., *Flocculation, Sedimentation and Consolidation*, American Institute of Chemical Engineers, New York, 1985, pp. 125–138.
33. W. E. Walles, *J. Colloid Interface Sci.* **27**, 797–803 (1968).
34. A. Sommerauer, D. L. Sussman, and W. Stumm, *Kolloid Zeit.* **225**, 147–154 (1968).
35. F. Halverson, *Proceedings of the 10th Annual Meeting Canadian Mineral Processors*, Jan. 1978, Ottawa, Canada, pp. 404–450.
36. A. S. Michaels, *Ind. Eng. Chem.* **46**, 1485 (1954).
37. U.S. Pat. 4,678,585 (July 7, 1987), N. J. Brownrigg (to American Cyanamid Co.).
38. R. A. Stratton in Ref. 15, pp. 1–4.
39. D. F. Honig, *1989 Papermakers Conference*, TAPPI Press, Atlanta, Ga., 1989, pp. 161–168.
40. R. W. Slater and J. A. Kitchener, *Disc. Faraday Soc.* **42**, 267–275 (1966).
41. B. Fitch, *Am. Inst. Chem. Eng. Jr.* **25**, 913–929 (1979).
42. R. M. Schlauch, in Ref. 11, pp. 91–144.
43. M. A. Hubbe, *TAPPI J.*, 116–117 (Aug. 1986).
44. A. G. Smelley and B. J. Scheiner, in *Proceedings of the Progress in the Dewatering of Fine Particles Conference*, University of Alabama, Tuscaloosa, 1981.
45. B. J. Scheiner and P. M. Brown, RI 8824, U. S. Bureau of Mines, Washington, D.C., 1983.
46. B. J. Scheiner and G. M. Wilemon, in Y. A. Attia ed., *Flocculation in Biotechnology and Separation Systems*, Elsevier Science Publishing Co., Inc., New York, 1987, pp. 175–186.
47. M. Yusa, in Ref. 46, pp. 755–764.
48. N. P. Chironis, *Coal Age*, 140–145 (Jan. 1976).
49. A. A. Terchick, D. T. King, and J. C. Anderson, *Trans. SME AIME* **258**, 148–151 (1975).
50. D. A. Dahlstrom, *Eng. Mining J.*, 120–133 (1980).
51. R. D. Brassinga and G. D. Fulford, in *Light Metals 1986* **2**, 51–59 (1986).
52. R. W. Kaesler, L. J. Connelly, and P. F. Richardson, in Ref. 24, pp. 473–490.
53. T. M. Camus, in Ref. 24, pp. 449–459.
54. W. L. K. Schwoyer, in Ref. 11, pp. 159–210.
55. D. S. Honig, in Ref. 15, pp. 29–33.
56. *EPA Process Design Manual for Suspended Solids Removal*, U.S. Environmental Protection Agency Technology Transfer, Washington, D.C., 1975, pp. 5–6.
57. H. I. Heitner, in Ref. 46, pp. 793–801.
58. J. Zhang, U. Wiesmann, and A. Grohmann, in H. H. Hahn and R. Klute, eds., *Chemical Water and Wastewater Treatment*, Springer-Verlag, New York, 1990, pp. 257–269.
59. C. W. Converse, J. F. Foley, and J. B. Carling, *Proc. Water Qual. Technol. Conf.* **15**, 513–522 (1987).

60. G. Schrank, *Wat. Sci. Tech.* **22**, 233–243 (1990).
61. J. Gregory and D. W. Nelson, in J. Gregory, ed., *Solid–Liquid Separation*, Ellis Horwood, Chichester, UK, 1984, pp. 172–182.
62. J. Eisenlauer and D. Horn, *Colloids Surfaces* **14**, 121–134 (1985).
63. H. P. Pendse, in Ref. 24, pp. 657–662.
64. P. M. Bertsch, in G. Sposito, ed., *The Environmental Chemistry of Aluminum*, CRC Press, Boca Raton, Fla., 1989, pp. 87–114.
65. G. B. Wickramanayake, B. W. Vigon, and R. Clark, in Ref. 46, pp. 125–148.
66. P. Somasundaran and R. Ramachandran, in Ref. 24, pp. 21–41.
67. M. W. Scoggins and J. W. Miller, *Soc. Petroleum Eng. J.*, **19**, 151–154 (1979).
68. J. L. Hoyt, *Proceedings of Technicon International Congress*, Chicago, 1969, pp. 69–72.
69. H. Burket, *Gas Wasserfach, Wasser-Abwasser* **11**, 282–286 (1970).
70. H. P. Panzer and F. Halverson, in Ref. 24, pp. 239–249.
71. C. N. Martyn an co-workers, *Lancet* **1**, 59–62 (1989).
72. C. N. Martyn, *Lancet* **336**, 430–431 (1990).
73. L. F. Smith, *Am. Water Works Assoc., Ontario Section*, Apr. 1992 meeting, in press, 1993.
74. J. B. V. Eastwood and co-workers, *Lancet* **336**, 462 (1990).
75. D. R. McLachan, and co-workers, *Can. Med. Assoc. J.* **145**, 793–805 (1991).
76. J. Mallevialle, A. Bruchet, and F. Fiessinger, *J. Am. Water Works Assoc.*, 87–93 (June 1984).
77. R. D. Letterman and R. W. Pero, *J. Am. Water Works Assoc.*, 87–97 (Nov. 1990).
78. *A.N.S.I./A.W.W.A. Standards*, American Water Works Association, Denver, Colo., 1987 and 1988, B451-87 and B452-90.
79. E. Baruth, A.W.W.A., personal communication, Apr. 1992.
80. *Chem. Mark. Rep.*, 16, 171 (Mar. 30, 1987).
81. *Chem. Week*, 40–42 (Sept. 19, 1984).
82. K. E. Biesinger and co-workers, *J. Water Pollution Control Fed.* **48**, 183–187 (1976).
83. K. E. Biessinger and G. M. Stokes, *J. Water Pollution Control Fed.* **58**, 207–213 (1986).
84. G. A. Cary, J. A. McMahon, and W. J. Kuc, *Environ. Toxicol. Chem.* **6**, 469–474 (1987).
85. W. S. Hall and R. J. Mirenda, *Res. J. Water Pollution Control Fed.* **63**, 895–899 (1991).
86. M. S. Goodrich and co-workers, *Environ. Toxicol. Chem.* **10**, 509–515 (1991).
87. *Code of Federal Regulations*, section 850.1075, *Fish Acute Toxicity Mitigated by Humic Acid, Title 40*. Washington, D.C., 1992.
88. *Water Treatment Chemicals and Equipment Outlook for the 1990's*, Leading Edge Reports, Cleveland Heights, Ohio, 1991, pp. 113–128.
89. *Chem. Mark. Rep.*, 21 (Oct. 14, 1991).
90. H. Tilton, *Chem. Mark. Rep.*, SR18 (Sept. 23, 1991).
91. *Chem. Mark. Rep.*, 31–32 (Oct. 15, 1990).
92. *Chem. Mark. Rep.*, 36 (Jan. 21, 1991).
93. *Chem. Mark. Rep.* **5**, 14 (Sept. 30, 1991).

HOWARD I. HEITNER
Cytec Industries

FLOTATION

Flotation or froth flotation is a physicochemical property-based separation process. It is widely utilized in the area of mineral processing also known as ore dressing and mineral beneficiation for mineral concentration. In addition to the mining and metallurgical industries, flotation also finds applications in sewage treatment, water purification, bitumen recovery from tar sands, and coal desulfurization. Nearly one billion tons of ore are treated by this process annually in the world. Phosphate rock, precious metals, lead, zinc, copper, molybdenum, and tin-containing ores as well as coal are treated routinely by this process; some flotation plants treat 200,000 tons of ore per day (see MINERAL RECOVERY AND PROCESSING). Various aspects of flotation theory and practice have been treated in books and reviews (1–9).

Technology

The flotation process is based on the exploitation of wettability differences of particles to be separated. Differences of wettability among solid (mineral) particles can be natural, or can be induced by the use of chemical adsorbates. Because the largest segment of industrial applications is conducted in water, with air, the following discussion is confined mainly to these fluids.

The flotation process applies to a particle size range of about 500 μm (eg, coal cleaning) to 2–10 μm (eg, copper ore concentration); however, 65 mesh (230 μm) to 270 mesh (53 μm) is typical. Figure 1 shows the relationship between flotation recovery and particle size in a sulfide ore processing operation and illustrates the optimum range (10). Figure 2 summarizes the main steps in mineral processing using froth flotation.

The raw ROM (run of mine) ore is reduced in size from boulders of up to 100 cm in diameter to about 0.5 cm using jaw crushers as well as cone, gyratory, or roll-type equipment. The crushed product is further pulverized using rod mills and ball mills, bringing particle sizes to finer than about 65 mesh (230 μm). These size reduction (qv) procedures are collectively known as comminution processes. Their primary objective is to generate mineral grains that are discrete and liberated from one another (11). Liberation is essential for the exploitation of individual mineral properties in the separation process. At the same time, particles at such fine sizes can be more readily buoyed to the top of the flotation cell by air bubbles that adhere to them.

The flotation step is accomplished by the preparation of a pulp, consisting of a solid–liquid slurry that may contain up to 40% solids, to which chemical reagents known as collectors are added in a conditioning tank. The reagents are added to render some minerals hydrophobic so that they selectively adhere to air bubbles introduced into the pulp in a flotation cell. On the other hand, some reagents enhance selectivity through activation and depression phenomena. Frothers are also used to generate a mineral-laden froth layer and enhance particle-bubble adhesion. The products from the flotation cell are a concentrate and a tailings stream. The concentrate proceeds to the next step for further cleaning or

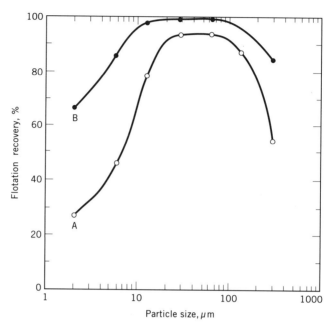

Fig. 1. Effect of particle size on the flotation recovery of a sulfide mineral. Mineral: chalcocite [2112-20-9], Cu_2S; reagent: potassium ethyl xanthate, C_2H_5OCSSK, 3.7 mg/L; flotation time: A, 0.5 min. and B, 2.0 min. Ref. 10.

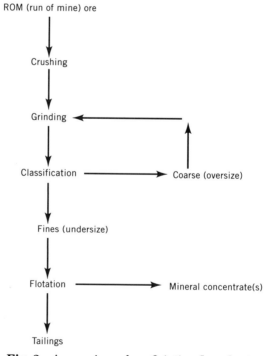

Fig. 2. A generic ore beneficiation flow sheet.

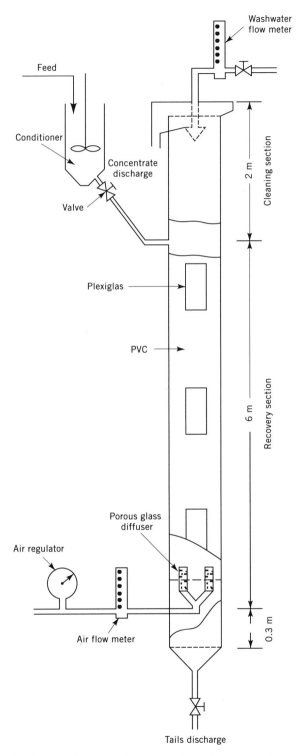

Fig. 3. A typical column flotation cell and peripherals.

treatment by hydro- or pyrometallurgical methods for the extraction of metals and other valuable compounds, while the tailings, which are ore components stripped of their valuable mineral content, are collected in lagoons known as tailings ponds. A typical froth flotation process can treat a ROM ore that assays 0.5% to a few percent copper to give a mineral concentrate analyzing 35% copper with a recovery of more than 85% of the copper content of the original ore.

The actual flotation phenomenon occurs in flotation cells usually arranged in batteries (12) and in industrial plants and individual cells can be any size from a few to 30 m^3 in volume. Column cells have become popular, particularly in the separation of very fine particles in the minerals industry and colloidal precipitates in environmental applications. Such cells can vary from 3 to 9 m in height and have circular or rectangular cross sections of 0.3 to 1.5 m wide. They essentially simulate a number of conventional cells stacked up on top of one another (Fig. 3). Microbubble flotation is a variant of column flotation, where gas bubbles are consistently in the range of 10–50 μm.

Process Design and Machinery.   Following the field work of geologists and mining engineers and analyses (assays) to establish the grades (concentrations) of values in ores, a mineral concentration flow sheet is established on the basis of a number of preliminary tests. These include studies of comminution properties of the ore, liberation properties of the minerals, and optimization of conditions at which they occur. Reagent testing, choice of flotation conditions, pH, collectors, frothers, and auxiliary reagents follow. The locked cycle test is a design aid that allows the simulation of a full-scale flotation procedure prior to pilot-plant testing (13).

Flotation cells, also called flotation machines, exist in numerous designs differing in mode of agitation and method of gas introduction and dispersion. Flotation machine designs are largely empirical although once designed their specification and performance data can be expressed by mathematical relations (14). Figure 4 outlines one classification scheme of flotation machine types (15).

The processes that occur in a typical flotation cell are schematically shown in Figure 5 and consist of agitation, particle–bubble collision and attachment, flotation of particle–bubble aggregates, collection of aggregates in a froth layer at the top of the cell, removal of mineral-laden froth as concentrate, and flow of the nonfloating fraction as tailings slurry.

Figure 6 shows air dispersion and pulp agitation mechanisms in commercial open flow machines (16).

Interfacial Phenomena

Flotation is a surface chemistry-based process, where numerous phenomena that simultaneously occur at the solid–liquid–air interfacial region determine its outcome (17). In this context, the variable known as contact angle θ illustrated in Figure 7, is an important correlative parameter. At θ = 0°, the liquid spreads on the solid; in aqueous media in contact with air such a solid is said to be hydrophilic and is wetted by water. Air bubbles do not adhere to hydrophilic solids in water. Conversely, hydrophobic solids are not wetted by water; air bubbles do adhere to them and the value of the contact angle is larger than zero degrees, ie, θ > 0°.

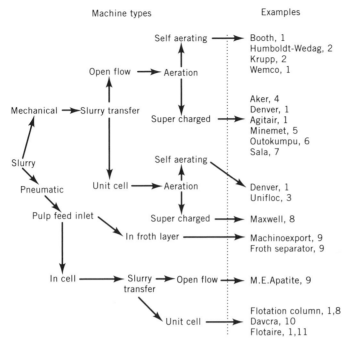

Fig. 4. Classification of flotation machine types and examples of brand names. Numbers indicate countries of origin of machines. 1, United States; 2, Germany; 3, United Kingdom; 4, Norway; 5, France; 6, Finland; 7, Sweden; 8, Canada; 9, former USSR; 10, Australia; and 11, South Africa. Ref. 15.

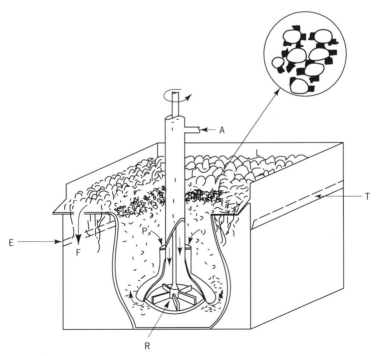

Fig. 5. Processes that occur in a flotation cell: A, air supply; E, slurry inlet; F, froth overflow; L, froth layer; inset, mineralized bubbles; P, flotation pulp; R, pulp agitation assembly (see Fig. 6); and T, tailings exit port.

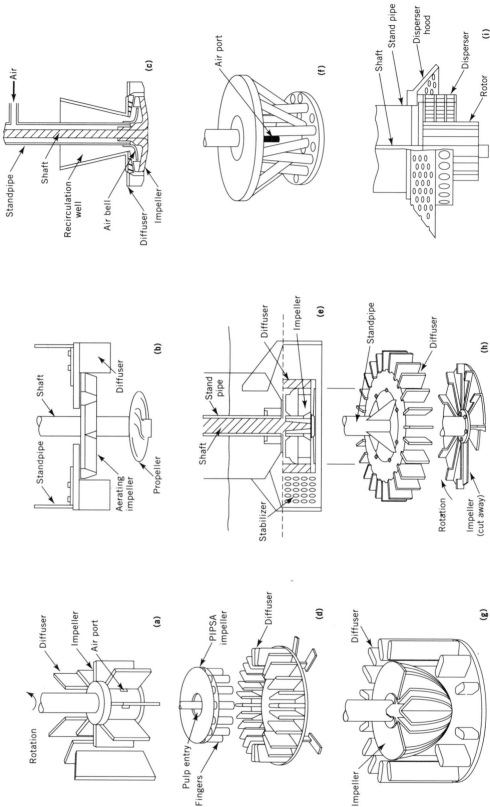

Fig. 6. Air dispersion and pulp agitation mechanisms in commercial flotation machines (16): (**a**) Aker, (**b**) Booth, (**c**) Denver D-R, (**d**) Agitair, (**e**) Wedag, (**f**) BCS, (**g**) OK, (**h**) Sala, and (**i**) Warmo.

86

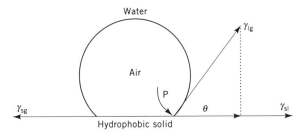

Fig. 7. The concept of contact angle with a captive bubble in an aqueous medium, adhering to a hydrophobic solid: P is the three-phase contact point. Here, the vector γ_{lg} passes through P and forms a tangent to the curved surface of the air bubble. The contact angle θ is drawn into the liquid.

The three interfacial tensions at equilibrium (Fig. 7) conform to Young's equation (eq. 1): where γ represents solid–gas, solid–liquid, and liquid–gas interfacial tension as indicated by subscripts.

$$\gamma_{sg} - \gamma_{sl} = \gamma_{lg} \cos \theta \tag{1}$$

Strictly speaking, equation 1 represents a special case that does not take into consideration the effects of gravity or external forces such as electric and magnetic fields (18). It also needs to be modified for rough (nonflat) and heterogenous (impure) surfaces as well as corner and edge effects. However, it has a thermodynamic basis and is widely utilized to account for wetting and spreading phenomena that occur between the three phases when they are in contact. Soldering, welding, joining, and detergency (qv), are but a few examples of systems besides flotation where wetting and spreading phenomena play significant roles. Some contact angle values for solids in contact with aqueous media are given in Table 1.

Electrical Phenomena at the Solid–Liquid Interface. Solid particles, such as minerals, in contact with water, as in a flotation pulp, undergo an electrical charge rearrangement at the water–solid interface because of hydration and ion dissolution from the lattice, ion adsorption from the aqueous environment, as well as lattice defects and substitutions. Thus an electrical double layer surrounding the particles is established. The double layer can consist of wall charges, then a layer of chemically bonded ions followed by, moving further into the liquid, a layer that consists of compactly packed, solvated ions (Stern layer) after which the diffuse part of the double layer starts. In Figure 8, only the wall, the Stern layer, and diffuse part of the double layer are shown. Specific adsorptions occur within the first few atomic diameters (0.1–0.6 nm), whereas the diffuse part of the double layer can extend to tens of nanometers depending on the ionic strength (salt concentration) of the medium. According to the DLVO theory of colloid stability, the thickness of the double layer is highest at the lowest ionic strength (see COLLOIDS).

The chemical composition, stoichiometry, and crystal structures of the solids in contact with water also play important roles in the degree of hydration that

Table 1. Contact Angle Values for Solids in Contact with Aqueous Media

Solid	CAS Registry Number	Conditions	Contact angle, θ^a
colemanite	[12291-65-5]	$5 \times 10^{-3}\,M$ sodium oleate	43
copper (metal)	[7440-50-8]	$1.5 \times 10^{-4}\,M$ sodium oleate	93
fluorite	[7789-75-5]	$10^{-5}\,M$ sodium oleate, pH = 8.1	91
galena	[12179-39-4]	$10^{-3}\,M$ potassium ethyl xanthate	60
graphite	[7782-42-5]	water	86–96
ilmenite	[12168-52-4]	sodium oleate solution, $T = 75°C$, pH = 8	80
coal (high rank)		water	45–60
oil shale^b		water	59.5
paraffin wax		water	108–111
silica	[7631-86-9]	$1.1 \times 10^{-5}\,M$ dodecylammonium chloride, pH = 10	81
stibnite	[1317-86-8]	water	84
Teflon	[9002-84-0]	water	160
Teflon		methanol in water solutions^c	0

^{a}Degrees.
^{b}From Colorado, organic carbon = 28%.
^{c}All concentrations with $\gamma_{lv} \leq 20$ mN/m(=dyn/cm).

occurs at the solid–liquid interface and adsorption phenomena that affect the flotation process. The manner of cleavage or breakage of crystals is significant. A higher number of broken bonds at the cleavage (or breakage) surface indicates the possibility of a solid wall being more likely to take up water molecules.

The zeta potential (Fig. 8) is essentially the potential that can be measured at the surface of shear that forms if the solid was to be moved relative to the surrounding ionic medium. Techniques for the measurement of the zeta potentials of particles of various sizes are collectively known as electrokinetic potential measurement methods and include microelectrophoresis, streaming potential, sedimentation potential, and electroosmosis (19). A numerical value for zeta potential from microelectrophoresis can be obtained to a first approximation from equation 2, where η = viscosity of the liquid, ϵ = dielectric constant of the medium within the electrical double layer, V_e = electrophoretic velocity, and E = electric field.

$$\zeta = (4\pi\eta/\epsilon)\,(V_e/E) \tag{2}$$

The zeta potential and contact angle as well as flotation recovery correlate well in some flotation systems as shown in Figure 9 (20).

In principle, zeta potential allows the definition of an isoelectric point (IEP) for each mineral (or suspended solid) that defines the conditions at which the measured zeta potential is equal to zero. IEP is also known as point of zero charge (PZC) when the surface carries no net charge. This is the condition at which the net charge on the wall of the particle is electrostatically compensated by an equal and oppositely signed quantity of charge in the diffuse part of the electrical double layer. Most IEP values of solids are reported to reflect the pH at which this con-

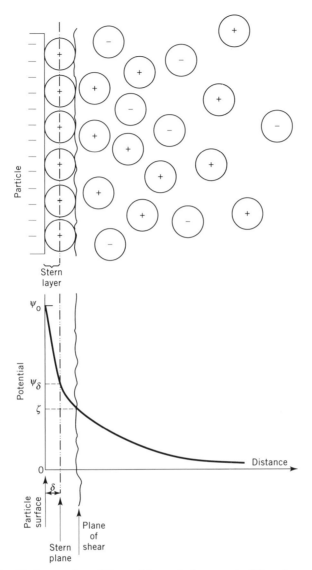

Fig. 8. Electrical double layer of a solid particle and placement of the plane of shear and zeta potential. ψ_o = Wall potential, ψ_δ = Stern potential (potential at the plane formed by joining the centers of ions of closest approach to the solid wall), ζ = zeta potential (potential at the shearing surface or plane when the particle and surrounding liquid move against one another). The particle and surrounding ionic medium satisfy the principle of electroneutrality.

dition is satisfied. However, the condition of ($\zeta = 0\,V$) can also be readily expressed in terms of the concentration of ions other than H^+. Examples of IEP values for some solids are tabulated in Table 2. Figure 10 shows the relationship between the IEP and the floatability of a typical oxide mineral (21). When pH < IEP and $\zeta > 0$ anion-type surfactants adsorb at the solid–liquid interface, whereas at pH > IEP and $\zeta < 0$, cation-type surfactants act as flotation collectors.

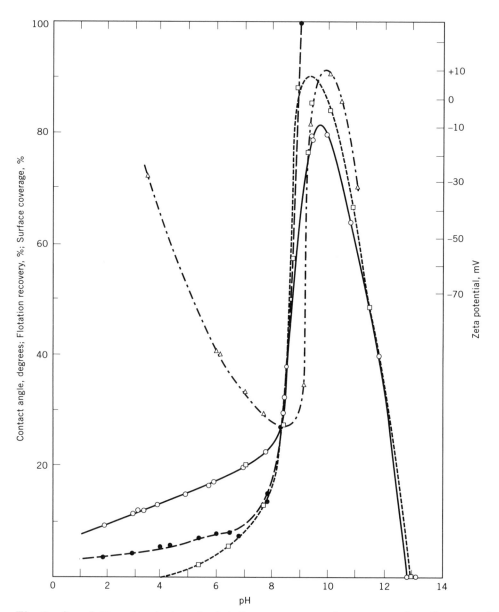

Fig. 9. Correlation of contact angle, flotation recovery, surface coverage by collector, and zeta potential. Solid, quartz, collector reagent, 4×10^{-4} M dodecylammonium acetate. □ = recovery, %; △ = zeta potential, mV; ○ = contact angle, degrees; and ● = surface coverage, % of one monolayer. Ref. 20.

Table 2. Isoelectric Points of Some Solids in Aqueous Media

Solid	pH^a
AgCl	$pAg^+ = 4$
AgBr	$pAg^+ = 5.4$
Ag_2S	$pAg^+ = 10.2$
$CuSiO_3 \cdot 2H_2O$	$pCu^{2+} = 4$, at pH $= 7$
$CaCO_3$	5–12
CaF_2	$pCa^{2+} = 3$
Fe_2O_3	6.5 and 8.5
Fe_3O_4	6.5
FeS_2	7
$MgCO_3$	2–11.5
PbS	3.5
SiO_2	2–3.7
SiO_2	7^b

^{a}Unless otherwise noted. p indicates $-\log_{10}$ of concentration.
^{b}In the presence of 3×10^{-3} M dodecylammonium acetate.

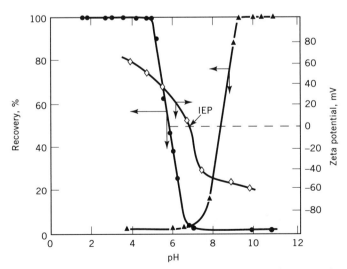

Fig. 10. Relationship between the IEP and the floatability of a typical oxide mineral (Goethite: $Fe_2O_3 \cdot H_2O$). At pH $>$ pH_{IEP} the zeta potential has a negative sign and the collector is cationic, whereas at pH $<$ pH_{IEP} the zeta potential carries a positive sign and the effective collector is anionic. ●, recovery with sodium dodecyl sulfate or sodium dodecyl sulfonate; ▲, recovery with dodecyl ammonium chloride; and ◇, zeta potential.

Chemicals in the Flotation Process

Flotation reagents are used in the froth flotation process to (1) enhance hydrophobicity, (2) control selectivity, (3) enhance recovery and grade, and (4) affect the velocity (kinetics) of the separation process. These chemicals are classified based on utilization: collector, frother, auxiliary reagent, or based on reagent chemistry: polar, nonpolar, and anionic, cationic, nonionic, and amphoteric. The active groups of the reagent molecules are typically carboxylates, xanthates, sulfates or sulfonates, and ammonium salts.

According to the first classification, collectors (also sometimes known as promoters) are the reagents that impart hydrophobicity to the solid to be floated (Table 3). Frothers lead to the creation of a froth layer at the top of the flotation cell and also enhance collector action by reducing the induction time, the time it takes for air bubbles to adhere to the hydrophobic solid grains with which they collide (Table 4). Auxiliary reagents comprise pH regulators, oxidizing-reducing agents, and colloidal and polymeric additives (Tables 5 and 6). The flotation of naturally hydrophobic solids using chemicals, not conventionally known as flotation collectors, is called collectorless flotation.

An inherent drawback of classifying flotation reagents according to their function in the flotation process is that what acts as a frother in one flotation system might play both collecting and frothing roles in another. For example, long-chain alcohols (see Table 4) are recognized as frothers in most flotation systems, but they act both as collectors and frothers in coal, talc, or graphite-containing flotation pulps.

Interaction of Solids With Flotation Reagents. For flotation to occur with the aid of reagents, such compounds must adsorb at the solid–liquid interface unless the solid to be floated is naturally hydrophobic. In this latter case only depression can be attempted by the use of additional ions or depressants that hinder bubble–particle adhesion. Frothers (typically long-chain alcohols) and/or modifying agents such as hydrocarbon oils can, however, be used to enhance the collection of naturally hydrophobic solids such as MoS_2, talc, or plastics.

Adsorption Mechanisms. The following mechanisms of adsorption are responsible for the formation of mineral–reagent bonds.

Electrostatic Interactions. This is the mechanism that operates when adsorption sites and reagents carry opposite electrical charge signs.

Hydrogen Bond Formation. This facilitates adsorption if the mineral and the adsorbate have any of the highly electronegative elements S,O,N,F, and hydrogen. A weak (physical) bond is established between the solid wall and the reagent through the alignment of the cited elements.

Collectors Fitting into Lattice Cavities. Lattice site fitting of collectors at solid walls has been invoked as a means of explaining the selective behavior of amines (cationic collectors) as reagents in the flotation-separation of soluble salt minerals such as KCl and NaCl (22).

Chemical Bond Formation (Chemisorption). This is the mechanism that leads to the formation of the strongest bonds between collectors and mineral surfaces. Chemically adsorbed reagents usually form surface compounds at the active wall sites. The flotation of calcite ($CaCO_3$) and apatite ($Ca_3(PO_4)_2$) by fatty acids

Table 3. Flotation Collectors Used in the Minerals Industry and Their Areas of Application

Compound	Active ingredient	Common application
alkyl thiocarbonates (xanthates)	$R-O-\underset{\underset{S}{\|\|}}{C}-SNa(K)^{a}$	sulfides, metallic minerals
alkyl morpholines	(morpholine ring structure with R on N)	potash
alkyl sulfonates	$R-SO_3^-M^+$	carbonates, hematite, borates, magnetite
alkyl sulfates	$R-O-SO_3^-M^+$	similar to sulfonates
dialkyl dithiophosphates (aerofloats)	R_1-O, R_2-O bonded to P with $=S$ and S^-M^+	native gold, copper, and sulfides
dixanthogens	$S=CSSC=S$ with OR OR	sulfides
hydrocarbon oils^{b}	C_nH_{2n+2}	coal, molybdenite, borates
mercaptobenzothiazole	(benzothiazole ring) $C-SH$	pyrite
naphthenic acids	R_2C ... CR_2 ... $R_2C-CR(CH_2)_nCOOH$	fluorite, borates
O-ethyl isopropyl-thiocarbamate	$\underset{H}{\overset{C_3H_7}{>}}N-C\underset{OC_2H_5}{\overset{S}{<}}$	copper sulfides
oximes	$R-\underset{HO}{\overset{\|}{C}H}-\underset{NOH}{\overset{\|\|}{C}}-R$	chrysocolla, cassiterite
p-tolyl arsonic acid	H_3C-(ring)$-AsO(OH)_2$	cassiterite
primary amine salts	$RNH_3^+Cl^-$	silica, silicates, sylvite
quaternary ammonium salts	$RN(CH_3)_3^+Cl^-$, $R = C_{10}-C_{16}$	silica, silicates, oxides

Table 3. *(continued)*

Compound	Active ingredient	Common application
sodium alkyl hydroxamates	R—C=N—O⁻Na⁺ with OH below C	wolframite, cassiterite, hematite
sodium carboxylates	$RCOO^-Na^+$	carbonates, phosphate rock, fluorite, hematite
thiocarbanilide[c]		sulfides
xanthogen formates (minerec)		sulfides

[a] R is an alkyl group with one to five carbons.
[b] Eg, vapor oil, kerosene, fuel oils.
[c] N,N-Diphenylthiourea [102-08-9].

Table 4. Examples of Flotation Frothers Widely Utilized in the Minerals Industry

Common name and composition	Chemical structure of active component
aliphatic alcohols (long-chain)	$CH_3—(CH_2)_n—CH_2OH$
eucalyptus oil[a]	
methylisobutyl carbinol[b] MIBC	$CH_3CHCH_2CHCH_3$ with CH_3 and OH above
pine oil	mixture of terpineols
poly(propylene glycol) monoalkyl ethers[c]	$R(OC_3H_6)_nOH$ $n = 2–5$, $R = CH_3, C_4H_9$
poly(ethylene glycol)s	$R(OC_2H_4)_nOC_2H_4OH$ $n = 2–5$

[a] Cineole or eucalyptol [470-82-6] is the principal component.
[b] 4-Methyl-2-pentanol [108-11-2].
[c] Trade name = Dowfroth-250.

Table 5. Inorganic Chemicals Used as Auxiliary Reagents

Name	CAS Registry Number	Composition	Area of use
cupric sulfate	[7758-98-7]	$CuSO_4$	sphalerite and arsenopyrite activator
lead acetate	[301-04-2]	$Pb(CH_3COO)_2$	stibnite activator
lime	[1305-78-8]	CaO	pH regulator, depressant, activator
Nokes reagent		complex mixture[a]	selective depressant in molybdenite circuits; improves molybdenite grade
sodium silicate	[1344-09-8]	Na_2SiO_3	dispersant–depressant for siliceous gangue, clays
sodium dichromate	[10588-01-9]	$Na_2Cr_2O_7$	galena depressant
sodium cyanide	[143-33-9]	$NaCN$	metal sulfide depressant
sodium hydroxide	[1310-73-2]	$NaOH$	pH regulator, dispersant
sodium sulfide	[1313-82-2]	Na_2S	sulfide depressant, sulfidizer
sodium carbonate	[497-19-8]	Na_2CO_3	pH regulator, dispersant
sulfur dioxide	[7446-09-5]	SO_2	depressant for sphalerite
sulfuric acid	[7664-93-9]	H_2SO_4	pH regulator

[a]Includes P_2S_5, As_2O_3, Sb_2O_3, and $NaOH$.

such as oleic acid ($C_{17}H_{35}COOH$) is a typical example of where strong mineral–collector bonds of this nature form.

Crystal Field Adsorption and Hydrophobic Bonding. These are two other mechanisms that operate in reagent adsorption processes (23,24). The former has been suggested to explain the adsorption of polyacrylamide (a potential flotation depressant) onto fluorite (CaF_2) and the latter is a mechanism that is more frequently invoked where the hydrocarbon ends of surfactant molecules used as flotation reagents aggregate to form micelles in the solution or hemimicelles near the wall of the solid (25). The hydrocarbon ends of amphipatic molecules (surfactants) adsorb onto hydrophobic surfaces by the same mechanism. The action of vapor oil in molybdenite (MoS_2) flotation and the use of Vaseline-type collectors in the grease table collection of diamonds are further examples of systems where hydrophobic bonding can be suggested as the operational bonding mechanism.

Electrochemical processes at some sulfide mineral surfaces lead to the formation of oxidation products as in the case of the hydrophobization of galena (PbS) by xanthates ($ROCSS^-$) or the oxidation of pyrite (FeS_2) or chalcopyrite ($CuFeS_2$) surfaces, which generate hydrophobic layers. Arguably these represent mechanisms of mineral surface hydrophobization and not necessarily flotation reagent adsorption mechanisms.

Other Interaction Processes. The selectivity of flotation reagents in a pulp and their functions depend on their interactions with the mineral phases to be separated, but other physicochemical and hydrodynamic processes also play roles. All adsorption–desorption phenomena occur at the solid–liquid interfacial region.

Table 6. Organic Auxiliary Reagents Used in Froth Flotation Technology

Name	Active component	Area of application
poly(ethylene oxide) [25322-68-3]	$-(CH_2-CH_2-O)_n-$	flocculant, filtration aid
quebracho[a]	mixture of polyhydroxy-cyclic carboxylic acids and other cyclic components	carbonate and fluorite depressant, defoaming agent
sodium isopropyl naphthalene sulfonate[b] [28348-64-3]	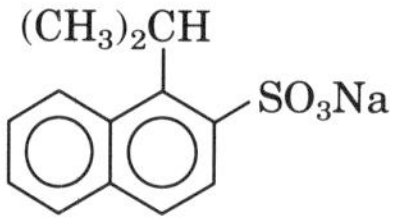 $(CH_3)_2CH$ ... SO_3Na	wetting agent, defoaming agent, emulsifier
sodium dioctyl-sulfosuccinate[c]	$CH_2COOC_8H_{17}$ $NaO_3SCHCOOC_8H_{17}$	similar to aerosol OS
sodium polyacrylate	$-(CH-CH_2)_n-$ $C=O$ O^-Na^+	flocculant, thickening and filtration aid
starch [9005-25-8]	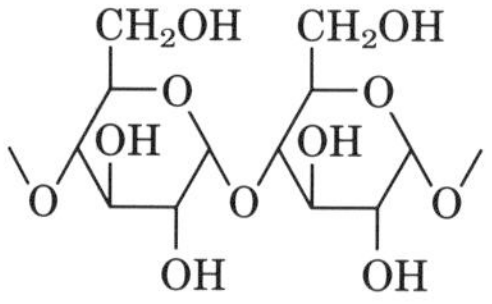 CH_2OH CH_2OH	slime depressant, hematite depressant
tannic acid [1401-55-4]	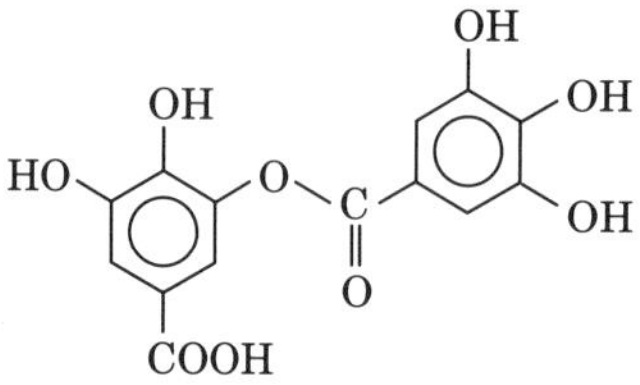	fluorite and carbonate depressant, defoamer

[a]Shinopsis tree extract.
[b]Aerosol OS.
[c]Aerosol OT.

Surface processes that influence such adsorptions include activation and depression. Activators and depressants are auxiliary reagents.

Activators enhance the adsorption of collectors, eg, Ca^{2+} in the fatty acid flotation of silicates at high pH or Cu^{2+} in the flotation of sphalerite, ZnS, by sulfohydryl collectors. Depressants, on the other hand, have the opposite effect; they hinder the flotation of certain minerals, thus improving selectivity. For example, high pH as well as high sulfide ion concentrations can hinder the flotation of sulfide minerals such as galena (PbS) in the presence of xanthates (ROCSS$^-$). Hence, for a given fixed collector concentration there is a fixed critical pH that defines the transition between flotation and no flotation. This is the basis of the

Barsky relationship which can be expressed as $[X^-]/[OH^-]$ = constant, where $[X^-]$ is the xanthate ion concentration in the pulp and $[OH^-]$ is the hydroxyl ion concentration indicated by the pH. Similar relationships can be written for sulfide ion, cyanide, or thiocyanate, which act as typical depressants in sulfide flotation systems.

Flotation Kinetics

It is possible to analyze froth flotation as a probability process, ie,

$$P_f = P_c \cdot P_a \cdot P_s \tag{3}$$

where P_f, P_c, P_a, and P_s are probability of flotation, particle–bubble collision, particle–bubble adhesion, and formation of a stable particle–bubble aggregate, respectively. Numerous variables affect these probability functions, eg: (1) particle and bubble sizes, pulp density, and number of bubbles in the pulp, as well as the intensity of agitation are primary variables that affect the probability of particle––bubble collisions; (2) the probability of particle–bubble adhesion, that is, the formation of particle–bubble bond(s) depends on the value of the contact angle, directional angle of impact and velocity of collision, the presence of frother in the medium and the resistance of the disjoining film at the interface; (3) the probability of formation of particle–bubble aggregates that can sustain the particle––bubble bond depends on the strength of this bond at the hydrodynamic conditions which prevail in the agitated flotation pulp.

Disjoining pressure measures the strength of the liquid film between the bubble and the particle when the two collide (26). Thus a hydrophilic solid has a film that exhibits a high disjoining pressure whereas the disjoining pressure on a hydrophobic solid is very small. When a particle and bubble collide the disjoining film deforms, thins, ruptures, and recedes, upon which particle–bubble attachment is completed as illustrated in Figure 11. The time it takes for this process to occur is known as induction time and varies from 10 ms to one millisecond.

Flotation process kinetics determine the residence time, the average time a given particle stays in the flotation pulp from the instant it enters the cell until it exits. One way to study flotation kinetics is to record flotation recoveries as a function of time under a given set of conditions such as pulp pH, collector concen-

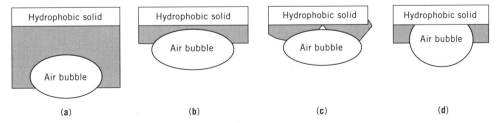

Fig. 11. Stages of disjoining film behavior in flotation upon particle–bubble collision: (**a,b**) film thinning; (**c**) film rupture; and (**d**) film recession and particle–bubble adhesion.

tration, particle size, etc. The data allow the derivation of an expression that describes the rate of the process.

$$(dC/dt) = -kC^n \tag{4}$$

where C = concentration of solids left in the flotation cell, t = time, k = rate constant, and n = order of the process.

First-order kinetics (ie, $n = 1$) is frequently assumed and seems adequate to describe the kinetics of most flotation processes. However, highly hydrophobic particles float faster and very fine particles or coarse ones outside the optimal flotation size range (see Fig. 1) take longer to collect in the froth layer. Excellent reviews of the subject are available in the literature (27).

Two technologically significant concepts in mineral concentration processes including froth flotation are recovery and grade. Recovery quantifies the percentage of value mineral collected in the froth layer whereas grade represents the chemical analyses of starting materials and products. The grade of an ore or a concentrate with respect to a valuable component indicates the percentage of this component in it. For example, a copper ore concentrate that assays 35% Cu has a grade of 35% Cu. Similarly, recovery indicates the percentage of total valuable material initially available in the feed which ends up in the concentrate fraction. Thus 80% recovery of copper indicates that 80% of the copper in the ore fed to the flotation circuit has been recovered in the concentrate, the remaining 20% having been lost into the tailings stream. Recovery and grade tend to vary inversely to one another.

Applications

Sulfide Ore Flotation.   Sulfide minerals, frequently cited as metallic minerals, occur mostly in complex ore bodies that bear a multitude of sulfides together with gangue. Thus it is quite common to find a complex sulfide ore body that contains galena [12179-39-4], PbS; sphalerite [12169-28-7], ZnS; chalopyrite [1308-56-1], $CuFeS_2$; and pyrite [1309-36-0], FeS_2, as primary sulfide minerals, and gangue usually as silica (SiO_2), silicates, or carbonate minerals. The treatment of such an ore aims at the recovery of individual concentrates of lead, zinc, and copper-bearing minerals plus a pyrite concentrate whereas gangue minerals end up in the tailings stream. Figure 12, for example, is a flow sheet suggested after a laboratory study of an ore that contains copper and zinc as the valuable minerals.

A number of generalizations can be made regarding the use of froth flotation technology in sulfide mineral concentrations: (1) xanthates and dithiophosphates are suitable sulfide mineral flotation collectors; (2) the action of xanthates in sulfide mineral systems require the presence of oxygen, from air, for flotation to occur; (3) longer chain homologues of a straight-chain collector series are more effective collectors per unit weight. This is also true for collector systems other than xanthates; (4) auxiliary reagents, as with nonsulfide systems, can be used to enhance selectivity.

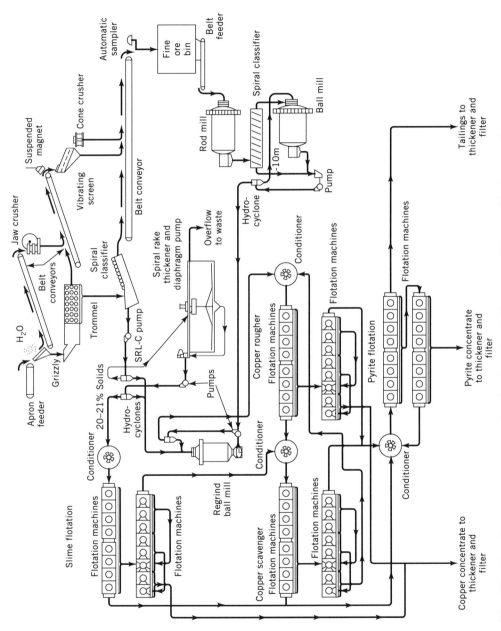

Fig. 12. Complex copper ore concentration flow sheet. Courtesy of Denver Equipment Co.

99

Nonsulfide Ore Flotation.   Nonsulfide minerals recovered by flotation include native elements such as graphite, diamonds, copper, gold, and numerous oxides as well as salts such as carbonates, phosphates, tungstates, and the like. Examples of value-bearing nonsulfide, noncoal minerals include

apatite [1306-05-4] $Ca_3(PO_4)_2$ kaolinite [1318-74-7]
calcite [471-34-1] $CaCO_3$ $Al_4Si_4O_{10}(OH)_8$
cassiterite [1317-45-9] SnO_2 magnesite [13717-00-5] $MgCO_3$
chrysocolla $CuSiO_3 \cdot nH_2O$ [26318-99-0]   rutile [13463-67-7] TiO_2
colemanite [12291-65-5] $Ca_2B_4O_{11} \cdot 5H_2O$ scheelite [14913-80-5] $CaWO_4$
hematite [1309-37-1] Fe_2O_3 silica [7631-86-9] SiO_2
ilmenite [12168-52-4] $FeO \cdot TiO_2$ smithsonite [14476-25-6] $ZnCO_3$

The basic flow sheet for the flotation-concentration of nonsulfide minerals is essentially the same as that for treating sulfides but the family of reagents used is different. The reagents utilized for nonsulfide mineral concentrations by flotation are usually fatty acids or their salts (RCOOH, RCOOM), sulfonates (RSO_3M), sulfates (RSO_4M), where M is usually Na or K, and R represents a linear, branched, or cyclic hydrocarbon chain; and amines [$R_1N(R)_3$]A where R and R_1 are hydrocarbon chains and A is an anion such as Cl^- or Br^-. Collectors for most nonsulfides can be selected on the basis of their isoelectric points. Thus at pH > pH_{IEP} cationic surfactants are suitable collectors whereas at lower pH values anion-type collectors are selected as illustrated in Figure 10 (28). Figure 13 shows an iron ore flotation flow sheet as a representative of high volume oxide flotation practice.

Soluble Salt Flotation.   KCl separation from NaCl and media containing other soluble salts such as $MgCl_2$ (eg, The Dead Sea works in Israel and Jordan) or insoluble materials such as clays is accomplished by the flotation of crystals using amines as collectors. The mechanism of adsorption of amines on soluble salts such as KCl has been shown to be due to the matching of collector ion size and lattice vacancies (in KCl flotation) as well as surface charges carried by the solids floated (22). Although cation-type collectors (eg, amines) are commonly used, the utility of sulfonates and carboxylates has also been demonstrated in laboratory experiments.

Coal Flotation.   Coal is a conglomerate of minerals, some of which are combustible. The primary objective of coal processing practice is to remove two types of components from the ROM ore: noncombustibles which are the ash-forming materials and sulfur-bearing compounds. The ash-forming components are shale, clays, calcite, and a small quantity of oxides such as SiO_2 or Fe_2O_3; the sulfur-bearing components are mostly pyrite (FeS_2). Sulfurous minerals are responsible for SO_x emissions during combustion. There is also a sulfur-bearing component in most coals which is not removable by physical means such as flotation. Methods for the elimination of this small percentage component known as structurally bound sulfur are available in the literature (29).

The reagents used in coal flotation are either those that collect coal, leaving the ash-forming compounds in the tailing, or those that float pyrite leaving a low sulfur coal concentrate. For coal flotation, nonionic reagents such as long-chain alcohols alone or together with nonpolar hydrocarbon liquids such as kerosene or

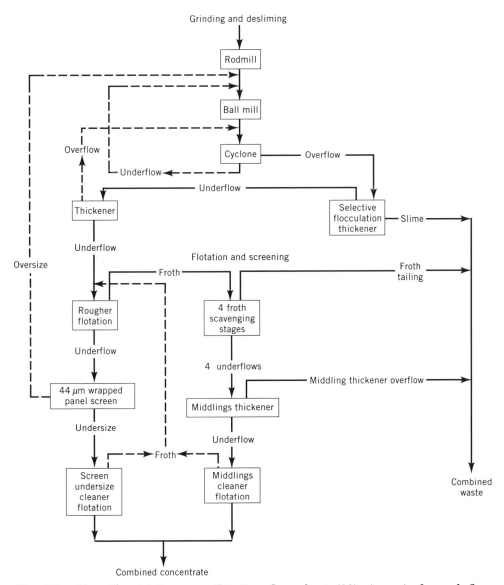

Fig. 13. Experimental iron ore flotation flow sheet (28): (———), forward flow; (— — —), return flow. Courtesy of USBM. Note the combination of flotation and flocculation in the same process. A plant based on this type of study has been built at the Tilden Mine (Michigan).

fuel oil are commonly used. For pyrite flotation xanthates are widely applied whereas for shales and oxides cationic collectors such as amines together with a frother such as a long-chain alcohol or pine oil are used. A coal flotation flow sheet is given in Figure 14.

Water Treatment. Flotation in water treatment is used both for the removal of dissolved ions such as Cu^{2+}, Cr^{3+}, or $(PO_4)^{3-}$ or surfactants and suspended

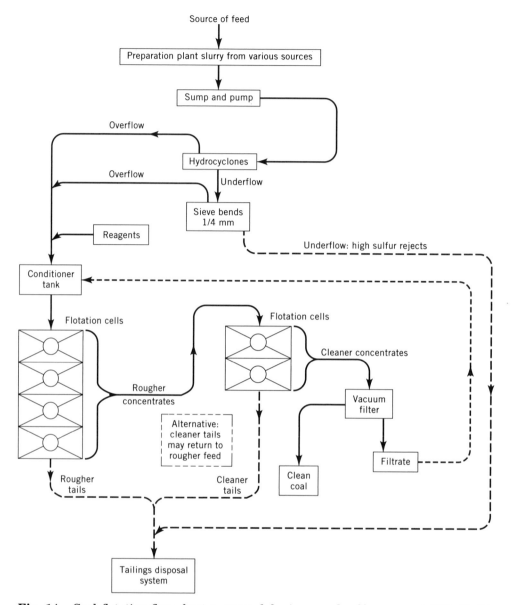

Fig. 14. Coal flotation flow sheet suggested for increased sulfur removal (29). Pyritic sulfur removal from coal makes it imperative to closely control pulp densities and reagent regimes.

solids as in the case of sludge treatment. The final product in this case is purified water rather than a mineral concentrate. Furthermore, water is treated either for drinking purposes (potable water preparation) or safe disposal to the environment.

In the removal of contaminating ions such as $(PO_4)^{3-}$ or Fe^{2+} a precipitate such as $Ca_3(PO_4)_2$ or $Fe(OH)_3$, after oxidizing ferrous ion to ferric, is formed and the solid is removed. The addition of surfactants is usually not essential (nor desirable) since most waters contain natural surfactants that would render the solids sufficiently hydrophobic for flotation to occur. Such surfactants derive from the degradation of organic matter, and humic substances abundantly available in nature (30).

Two main operational variables that differentiate the flotation of finely dispersed colloids and precipitates in water treatment from the flotation of minerals is the need for quiescent pulp conditions (low turbulence) and the need for very fine bubble sizes in the former. This is accomplished by the use of electroflotation and dissolved air flotation instead of mechanically generated bubbles which is common in mineral flotation practice. Electroflotation is a technique where fine gas bubbles (hydrogen and oxygen) are generated in the pulp by the application of electricity to electrodes. These very fine bubbles are more suited to the flotation of very fine particles encountered in water treatment. Its industrial usage is not widespread. Dissolved air flotation is similar to vacuum flotation. Air-saturated slurries are subjected to vacuum for the generation of bubbles. The process finds limited application in water treatment and in paper pulp effluent purification. The need to run it batchwise renders it less versatile.

Two air-saturation systems suited for use in water treatment are shown in Figure 15 (31). Such mechanisms facilitate the release of air that generates much finer bubbles than mechanical air dispersion methods used in mineral flotation practice.

Ion Flotation and Foam Separation. Ions and dissolved surfactant molecules can be removed from solutions by the agency of foam. In this case ions are sandwiched in foam films. The scientific basis of these processes is well understood and successes of metal ion recovery from solutions including U, Pt, Au, as well as different surfactants (detergents) have been reported in the literature.

Gamma Flotation Process. From consideration of the variables that influence the value of the contact angle on hydrophobic solids and Young's equation it

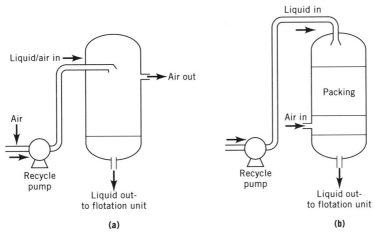

Fig. 15. Air-saturator systems used in the practice of water treatment by flotation (31). (**a**), Injection of air into suction line of recycle pump; (**b**), packed saturator.

can be seen that it should be possible to obtain $\theta = 0°$ by manipulation of the surface tension of the liquid. Thus a plot of cos θ against γ_{lg} gives a straight line that extrapolates to cos $\theta = 1$, $(\theta = 0°)$ which allows the definition of a critical surface tension of wetting (γ_c). The critical surface tension of wetting of a solid can also be determined using flotation instead of contact angle measurements.

The concept of critical surface tension of wetting has been introduced (32) in connection with wetting phenomena in general and its application to froth flotation demonstrated (33,34) using artificial mixtures of hydrophobic materials. The term gamma flotation has been coined to describe the use of surface tension control as a means of achieving selective flotation (35). The gamma flotation process relies on the exploitation of differences between the critical surface tension of wetting values of two hydrophobic solids in the same flotation pulp. By arranging the surface tension of the solution (pulp) to a value between the (γ_c) values of solids 1 and 2, solid 1 should float whereas solid 2 should remain hydrophilic and not adhere to air bubbles. The thermodynamical validity (36) of this approach and its applicability in flotation–separation has been demonstrated for various minerals (33,34,36–39). It has also been recognized that the process called salt flotation can be explained using the principles of gamma flotation (35).

Two-Liquid Flotation.   There are certain advantages to the substitution of oil droplets for air bubbles in a flotation pulp. Among these is the fact that oil droplets can be produced at much finer sizes than mechanically dispersed air bubbles and thus finer particles collected. Attractive forces operating among molecules in air bubbles and in saturated hydrocarbon oils are mainly van der Waals forces which lead to hydrophobic bonding. The three-phase contact among oil, water, and solid can be represented as in Figure 7 with oil replacing air so that $\gamma_{gl} = \gamma_{ow}$, $\gamma_{sg} = \gamma_{so}$, and $\gamma_{sl} = \gamma_{sw}$. The spreading coefficients for oil on solid in the presence of water $(S_{o(w/s)})$ and water on solid in the presence of oil $(S_{w(o/s)})$ correlate as follows:

$$S_{o(w/s)} = \gamma_{sw} - \gamma_{ow} - \gamma_{so} \tag{5}$$

$$S_{w(o/s)} = \gamma_{so} - \gamma_{ow} - \gamma_{sw} \tag{6}$$

γ_{so}, γ_{ow}, and γ_{sw} are the interfacial tensions between solid–oil, oil–water, and solid–water, respectively. The numerical values of these could be modified using surface-active agents (detergents) so as to affect the equilibrium location of the solid at interfaces, ie,

$$\gamma_{so} \geq \gamma_{sw} + \gamma_{ow} \tag{7}$$

$$\gamma_{sw} \geq \gamma_{sw} + \gamma_{so} \tag{8}$$

$$\gamma_{ow} > \gamma_{sw} + \gamma_{so} \tag{9}$$

If equation 7 holds, then the solid is exclusively in the aqueous phase; equation 8 defines the condition at which the solid resides in the oil phase whereas if equation 9 is satisfied then the solid collects at the water–oil interfacial region.

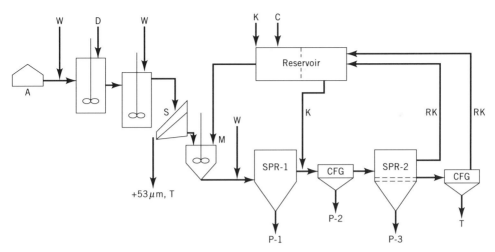

Fig. 16. Two-liquid flotation flow sheet (39). The original ROM is kaolin (white clay) that contains 11% impurity in the form of mica, anatase, and silica. Treatment produces high purity kaolin and a TiO_2-rich fraction. A, Kaolin stockpile; D, dispersant (sodium silicate plus alkali); W, water; K, kerosene; C, collector (sodium oleate); RK, recycled kerosene; S, screen; M, inline mixer; SPR, separator; CFG, centrifuge; P, product; and T, to waste.

Figure 16 is the flow sheet of a bench-scale study that demonstrates the concept of two-liquid flotation (40).

The term flotoflocculation is used to describe the process of aggregating dispersed oil droplets by the aid of polymeric flocculants (flocculation) then subjecting them to conventional flotation. It is also used, generically, to describe situations where particles are first aggregated then floated.

Other Applications. There are a variety of flotation processes employing the principles described.

Skin Flotation. Hydrophobic particles can be removed in the form of a thin, usually one particle thick layer on top of a trough, giving rise to the skin flotation process.

Piggyback Flotation. This process has also been called carrier flotation. The principle is based on the flotation of fine particles adhering to others by mutual coagulation. Thus when one is floated the other, which is usually more difficult to float, is also collected.

Environment, Safety, and Future Developments

New technologies based on the process of froth flotation in areas outside mineral technology are being developed. These include plastics recycling, glass recycling, and recovery of radioactive contaminants or heavy-metals removal from soil; newsprint deinking is an established flotation technology. Similarly, ion and precipitate flotation as well as foam fractionation (41,42) are areas poised for increased activity due to their potential usefulness in environmental site cleanup operations. In fact, the utilization of the unit operations and unit processes of mineral processing is fundamental to the separations technology on which recycling relies.

Ore flotation processes treat millions of tons of minerals per year, and since these are associated with mining activity they appear to be associated with physical damage to the environment. However, these technologies have a long history of practice and associated environmental control procedures. Specially lined tailings ponds, turbidity, and toxic chemical abatement approaches designed to eliminate environmental damage, as well as revegetation of old mining sites tailings ponds, dams, and dikes, are widespread practices. Furthermore, the flotation process is a technology well established in sewage treatment and water purification, and it can be used for the removal of harmful ions from effluents. Therefore it is safe and the flotation industry is self-regulating.

BIBLIOGRAPHY

"Flotation" in *ECT* 1st ed., Vol. 6, pp. 595–614, by H. R. Spedden, Massachusetts Institute of Technology; in *ECT* 2nd ed., Vol. 9, pp. 380–398, by F. F. Aplan, Union Carbide Corp., Mining and Metals Division; in *ECT* 3rd ed., Vol. 10, pp. 523–547, by F. F. Aplan, Pennsylvania State University.

1. K. L. Sutherland and I. W. Wark, *Principles of Flotation*, Australian Institute of Mining and Metallurgy Inc., Melbourne, Australia, 1955.
2. A. M. Gaudin, *Flotation*, McGraw-Hill Book Co., Inc., New York, 1957.
3. V. I. Klassen and V. A. Mokrausov, *An Introduction to the Theory of Flotation*, transl. J. Leja and G. W. Poling, Butterworths, London, 1963.
4. M. C. Fuerstenau, ed., *Flotation: A.M. Gaudin Memorial Volume*, 2 vols., American Institution of Mining Metallurgical and Petroleum Engineers, Inc., New York, 1976.
5. J. Leja, *Surface Chemistry of Froth Flotation*, Plenum Press, New York, 1982.
6. K. J. Ives, ed., *The Scientific Basis of Flotation*, Martinus-Nijhof, The Hague, the Netherlands, 1984.
7. R. P. King, ed., *Principles of Flotation*, South African Institute of Mining and Metallurgy, Johannesburg, South Africa, 1982.
8. P. Somasundaran, ed., *Advances in Mineral Processing*, Society Mining Engineers (SME), New York, 1986.
9. N. L. Weiss, ed., *SME Mineral Processing Handbook*, Vol. 1, AIME Inc., New York, 1985, pp. 5.1–5.110.
10. T. W. Healy and W. J. Trahar, in K. V. Sastry and M. C. Fuerstenau, eds., *Challenges in Mineral Processing*, SME, Littleton, Colo., 1989, p. 3.
11. G. Barbery, *Mineral Liberation*, Editions GB, Montreal, Canada, 1991.
12. S. G. Malghan, in A. L. Mular and M. A. Anderson, eds., *Design and Installation of Concentration and Dewatering Circuits*, Society of Mining Engineers, Littleton, Colo, 1986, p. 76.
13. M. R. Smith and R. J. Gochin, in B. Yarar and Z. Dogan, eds., *Mineral Processing Design*, Martinus-Nijhof, Boston, 1987, p. 166.
14. C. C. Harris, in Ref. 4, p. 753.
15. G. Barbery, in Ref. 6, p. 289.
16. P. Young, *Mining Mag.* **146**, 35 (1982).

17. B. Yarar and D. J. Spottiswood, eds., *Interfacial Phenomena in Mineral Processing*, Engineering Foundation Publishers, New York, 1982.
18. Y. Zimmels and B. Yarar, *J. Colloid Interface Sci.* **99**(1), 59 (1984).
19. R. J. Hunter, *Zeta Potential in Colloid Science*, Academic Press, Inc., London, 1981.
20. D. W. Fuerstenau, *Trans. AIME* **208**, 1365 (1973).
21. I. Iwasaki, S. R. B. Cooke, and A. F. Colombo, *USBM, Report of Investigation*, (5593) (1960).
22. D. W. Fuerstenau and M. C. Fuerstenau, *Trans. AIME* **205**, 302 (1956); R. J. Roman, M. C. Fuerstenau, and D. C. Seidel, *Trans. AIME* **241**, 56 (1968).
23. O. Griot and J. A. Kitchener, *Trans. Faraday Soc.* **61**(509), 1026 (1965).
24. H. E. Garrett, *Surface Active Chemicals*, Pergamon Press, London, 1972.
25. D. W. Fuerstenau, T. W. Healy, and P. Somasundaran, *Trans. AIME* **229**, 321 (1964).
26. B. V. Derjaguin and N. V. Churaev, *J. Colloid Interface Sci.* **66**(3), 389 (1978).
27. C. Ek, in P. Mavros and K. A. Matis, eds., *Innovations in flotation Technology*, Kluwer Academic Press, London, 1992; T. Inoue, M. Nonaka, and T. Imaizumi, in Ref. 8, p. 209.
28. A. F. Colombo and H. D. Jacobs, *USBM Report of Investigation*, (8180) (1976).
29. J. W. Leonard, ed., *Coal Preparation*, 4th ed. AIME, New York, 1979.
30. G. A. Aiken and co-workers, *Humic Substances in Soil, Sediment and Water*, John Wiley & Sons, Inc., New York, 1985.
31. Th. F. Zabel, in Mavros and Matis, Ref. 27, p. 431.
32. W. A. Zisman, in K. J. Mysels and co-workers, eds., *20 Years of Surface and Colloid Chemistry, The Kendall Award Addresses*, American Chemical Society, Washington, D.C., 1973, p. 109.
33. J. A. Finch and G. W. Smith, *Can. Metall. Q.* **14**(1), 44 (1975).
34. D. T. Hornsby and J. Leja, *Colloids and Surfaces* **7**, 339 (1983).
35. B. Yarar, in S. H. Castro and J. Alvarez, eds., *Froth Flotation*, Elsevier, Amsterdam, the Netherlands, 1988, p. 41.
36. J. Laskowski, in Ref. 8, p. 189.
37. B. Yarar and J. Kaoma, *Colloids and Surfaces* **11**(3), 429 (1984).
38. B. Yarar and J. Kaoma, *Trans. AIME* **276**, 1878 (1985).
39. G. P. Hemphill and B. Yarar, in F. A. Curtis, ed., *Proc. Energy-84*, Pergamon Press, Toronto, Canada, 1984, p. 111.
40. H. L. Shergold, in Ref. 17, p. 303.
41. D. J. Wilson and A. N. Clarke, *Topics in Foam Flotation*, Marcel Dekker, Inc., New York, 1983.
42. W. Walkowiak, in Mavros and Matis, Ref. 27, p. 455.

BAKI YARAR
Colorado School of Mines

FLOUR. See BAKERY PROCESSES AND LEAVENING AGENTS.

FLOW MEASUREMENT

Flow measurement is a broad field covering a spectrum ranging from the min-
uscule flow rates associated with the pharmaceutical industry to the immense
volumes involved in rivers. This measurement is an essential part of the production, distribution, consumption, and disposal of all liquids and gases including
fuels, chemicals, foods, and wastes. Fluids to be measured may be pure compounds
or contaminated mixtures, under vacuum or at high pressure, and at temperatures ranging from cryogenic to molten metal (see also FLUIDIZATION; FLUID ME-
CHANICS). This breadth and diversity have led to the development of a multitude
of flow measurement devices. All meet the requirements of certain applications
and some achieve broad utility. None as of this writing, however, comes close to
being universal in scope. The field of flow measurement is thus application oriented. The specific requirements of a particular measurement must be analyzed
in detail before proper equipment selection can be made.

Flow Meter Selection

A number of considerations should be evaluated before a flow measurement
method can be selected for any application. These considerations can be divided
into four general classifications: fluid properties; ambient environment; measurement requirements; and economics.

Fluid Properties.   A great variety of equipment exists for measuring clean,
low viscosity, single-phase fluids at moderate temperatures and pressures. Fluid-
related factors that are normally considered are operating pressure, temperature,
viscosity, density, corrosive or erosive characteristics, flashing or cavitation tendencies, and fluid compressibility. Any extreme fluid characteristic or condition,
such as a corrosive nature or high operating temperature, greatly reduces the
range of available equipment and should be given first consideration in any selection procedure. Other fluid properties important for use of certain meter types
are heat capacity, an important consideration for thermal meters, and fluid electrical conductivity, required for magnetic flow meter operation. In some cases
particular fluid requirements may limit the metering choices. An example of this
is the requirement for the sanitary design of meters used in food processing (qv).

Reynolds Number.   One important fluid consideration in meter selection is
whether the flow is laminar or turbulent in nature. This can be determined by
calculating the pipe Reynolds number, Re, a dimensionless number which represents the ratio of inertial to viscous forces within the flow. Because

$$q = \frac{\pi D^2}{4} V, Re = \frac{4eq}{\pi \mu D}$$

where ρ is the fluid density; μ, the fluid absolute viscosity; V, the average fluid
velocity; D, the pipe or meter inlet diameter; and q the volumetric flow rate. When
q is in units of m^3/h, ρ in kg/m^3, D in m, and μ in Pa·s, the equation becomes

$$Re = \frac{(m^3/h)(kg/m^3)(m)}{2827(Pa \cdot s)}$$

When q is in gal/min, D in in., ρ = specific gravity, and μ in cP,

$$Re = \frac{3160 \; (gal/min)(in.)}{cP}$$

A low Reynolds number indicates laminar flow and a parabolic velocity profile of the type shown in Figure 1a. In this case, the velocity of flow in the center of the conduit is much greater than that near the wall. If the operating Reynolds number is increased, a transition point is reached (somewhere over $Re = 2000$) where the flow becomes turbulent and the velocity profile more evenly distributed over the interior of the conduit as shown in Figure 1b. This tendency to a uniform fluid velocity profile continues as the pipe Reynolds number is increased further into the turbulent region.

Most flow meters are designed and calibrated for use on turbulent flow, by far the more common fluid condition. Measurements of laminar flow rates may be seriously in error unless the meter selected is insensitive to velocity profile or is specifically calibrated for the condition of use.

Ambient Environment. The environment around the flow conduit must be considered in meter selection. Such factors as the ambient temperature and humidity, the pipe shock and vibration levels, the availability of electric power, and the corrosive and explosive characteristics of the environment may all influence flow meter selection. Special factors such as possible accidental flooding, the need for hosedown or steam cleaning, and the possibility of lightning or power transients may also need to be evaluated.

Enough space must be available to properly service the flow meter and to install any straight lengths of upstream and downstream pipe recommended by the manufacturer for use with the meter. Close-coupled fittings such as elbows or reducers tend to distort the velocity profile and can cause errors in a manner similar to those introduced by laminar flow. The amount of straight pipe required depends on the flow meter type. For the typical case of an orifice plate, piping

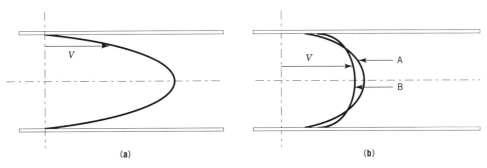

(a)   (b)

Fig. 1. Flow profiles, where V is velocity: (**a**) laminar, and (**b**) turbulent for fluids having Reynolds numbers of A, 2×10^5, and B, 2×10^6.

requirements are normally listed in terms of the β or orifice/pipe bore ratio as shown in Table 1 (1) (see PIPING SYSTEMS).

Measurement Requirements.   Any analysis of measurement requirements must begin with consideration of the particular accuracy, repeatability, and range needed. Depending on the application, other measurement considerations might be the speed of system response and the pressure drop across the flow meter. For control applications repeatability may be the principal criterion; conversely for critical measurements, the total installed system accuracy should be considered. This latter includes the accuracy of the flow meter and associated readout devices as well as the effects of piping, temperature, pressure, and fluid density. The accuracy of the system may also relate to the required measurement range.

Accuracies of the flow meters discussed herein are specified as either a percentage of the full-scale flow or as a percentage of the actual flow rate. It may be convenient in some applications to compare the potential inaccuracies in actual volumetric flow rates. For example, in reading two liters per minute (LPM) on a flow meter rated for five LPM, the maximum error for a $\pm 1\%$ of full-scale accuracy specification would be $0.01 \times 5 = \pm 0.05$ LPM. If another flow meter of similar range, but having $\pm 1\%$ of actual flow rate specification, were used, the maximum error would be $0.01 \times 2 = \pm 0.02$ LPM. To minimize errors, meters having full-scale accuracy specifications are normally not used at the lower end of their range. Whenever possible, performance parameters should be assessed for the expected installation conditions, not the reference conditions that are the basis of nominal product performance specifications.

Economic Considerations.   The principal economic consideration is, of course, total installed system cost, including the initial cost of the flow primary, flow secondary, and related ancillary equipment as well as material and labor required for installation. Other typical considerations are operating costs and the requirements for scheduled maintenance. An economic factor of increasing importance is the cost of disposal at the end of normal flow meter service life. This may involve meter decontamination if hazardous fluids have been measured.

Flow Calibration Standards

Flow measuring equipment must generally be wet calibrated to attain maximum accuracy, and principal flow meter manufacturers maintain extensive facilities for this purpose. In addition, a number of governments, universities, and large flow meter users maintain flow laboratories.

Calibrations are generally performed with water or air using one or more of four basic standards: weigh tanks, volumetric tanks, pipe provers, or master flow meters. Most standards can be used statically, ie, where the flow rate is quickly started and stopped at the beginning and end of the test; dynamically, ie, where readings are taken at the instant the test is initiated and again at the instant it is completed; or in a hybrid dynamic start-and-stop static reading mode. Static systems operate best for flow meters that have good accuracy at low flow rates and fast dynamic response. These methods do not give optimum results for vortex or turbine meters because of errors obtained during the short periods of low flow at the beginning and end of the test. Completely dynamic systems are limited by

Table 1. Required Straight Pipe Lengths for Orifice Plates, Nozzles, and Venturis[a]

β	On upstream side of the primary device							On downstream side[f]
	Single 90° bend or tee[b]	Two or more 90° bends in same plane	In different planes	Reducer[c]	Expander[d]	Globe[e] valve	Gate[e] valve	
0.20	10 (6)	14 (7)	34 (17)	5	16 (8)	18 (9)	12 (6)	4 (2)
0.25	10 (6)	14 (7)	34 (17)	5	16 (8)	18 (9)	12 (6)	4 (2)
0.30	10 (6)	16 (8)	34 (17)	5	16 (8)	18 (9)	12 (6)	5 (2.5)
0.35	12 (6)	16 (8)	36 (18)	5	16 (8)	18 (9)	12 (6)	5 (2.5)
0.40	14 (7)	18 (9)	36 (18)	5	16 (8)	20 (10)	12 (6)	6 (3)
0.45	14 (7)	18 (9)	38 (19)	5	17 (9)	20 (10)	12 (6)	6 (3)
0.50	14 (7)	20 (10)	40 (20)	6 (5)	18 (9)	22 (11)	12 (6)	6 (3)
0.55	16 (8)	22 (11)	44 (22)	8 (5)	20 (10)	24 (12)	14 (7)	6 (3)
0.60	18 (9)	26 (13)	48 (24)	9 (5)	22 (11)	26 (13)	14 (7)	7 (3.5)
0.65	22 (11)	32 (16)	54 (27)	11 (6)	25 (13)	28 (14)	16 (8)	7 (3.5)
0.70	28 (14)	36 (18)	62 (31)	14 (6)	30 (15)	32 (16)	20 (10)	7 (3.5)
0.75	36 (18)	42 (21)	70 (35)	22 (11)	38 (19)	36 (18)	24 (12)	8 (4)
0.80	46 (23)	50 (25)	80 (40)	30 (15)	54 (27)	44 (22)	30 (15)	8 (4)

[a]Nonparenthetical values are zero additional uncertainty. Parenthetical values are ± 0.5% additional uncertainty. All straight lengths are expressed as multiples of the pipe diameter D. They are measured from the upstream face of the primary device.
[b]Flow from one branch only.
[c]$2D$ to D over a length of $1.5D$ to $3D$.
[d]$0.5D$ to D over a length of $1D$ to $2D$.
[e]Fully open.
[f]All fittings are included in this table.

speed of response considerations and the general difficulties encountered using on-the-fly readings. Because of these limitations hybrid dynamic start-and-stop static reading weight and volume systems have been developed to provide more accurate liquid calibrations than purely static or dynamic systems. In such systems, the desired test flow rate is first obtained but diverted around the weight or volume standard. The test run is initiated by diverting the flow into the standard and completed by diverting it out of the standard. The weight or volume is then read after an appropriate settling time.

The key to the performance of a dynamic start-and-stop static reading system is the design of the flow diverter valve that switches the flow in and out of the standard. In a well-designed system the actual diversion time is much shorter than the collection time and the flow profile exiting the diverter relatively independent of flow rate. Under these circumstances, the limiting factor in system accuracy is the basic accuracy and resolution of the weight or volume standard. Errors can be reduced to less than 0.1% of actual flow rate. Calibration systems of this type are generally restricted to large test facilities because of the high cost involved and the lack of portability.

Liquid Displacement Gas Meter Provers. The liquid displacement prover is the most prevalent standard for the calibration of flow meters at low to moderate gas flow rates. The method consists of displacing a known volume of liquid with gas (Fig. 2). Gas entering the inverted bell causes it to rise and a volume increment can be timed. Typical prover capacities are 1 m^3 or less although capacities as large as 20 m^3 are available. Accuracies can be on the order of 0.5% of actual flow rate.

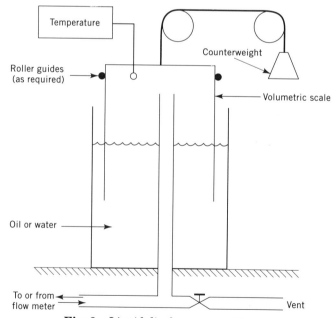

Fig. 2. Liquid displacement calibrator.

Pipe Provers. In pipe proving systems a sealed piston or elastic sphere is driven through a pipe section of precise bore using either the fluid energy or external activation. This measuring section may be coupled into the process pipe or part of a portable closed loop also containing the meter under test. Piston travel is detected to establish calibration volume. On closed-loop systems valves normally permit bidirectional actuation. Figure 3 shows one form of a commercial closed-loop prover system. In this system temperature and pressure inputs are used to compute fluid density changes and provide an overall accuracy rating of up to 0.05% of actual flow rate.

Liquid provers are available for use with flow rates from approximately 0.005 up to rates of thousands of liters per minute. They are suitable for clean liquids of almost any viscosity.

Prover systems are relatively compact and rugged. These features, combined with accuracy, result in wide usage in the custody transfer of petroleum (qv) products. Many open-loop-type pipe provers are permanently installed in pipeline metering stations where these provers are used to check operating meters (see PIPELINES). The readings have a high economic impact, affecting the transfer of large sums of money. Self-contained proving systems are also commonly used as flow laboratory standards.

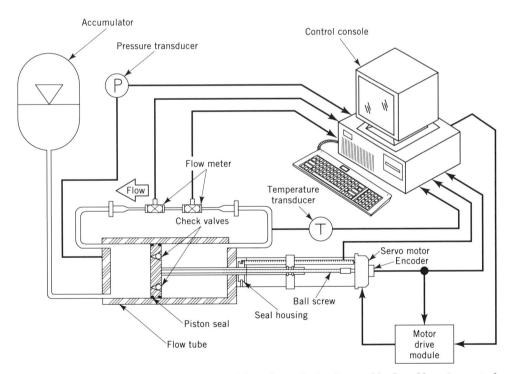

Fig. 3. Comtrack 921 pipe prover. Liquid flow through the Comtrak's closed loop is created by the movement of a sealed piston. Flow meters being tested are installed in the loop upstream from the piston. As the piston advances, the calibration fluid travels through the meters and returns to the back side of the piston.

Master Flow Meters. Perhaps the most common method of flow meter calibration is to compare the output of the meter under test with one or more meters of high resolution and proven accuracy, called master flow meters. This method is both convenient and quick. To obtain optimum accuracy, a system can be set up using multiple masters of overlapping range where the masters are regularly compared against each other and periodically calibrated using weight or volume standards. This combines the convenience of the master meter method and the accuracy capability of the basic standard (2).

Flow Meter Classifications

Flow meters have traditionally been classified as either electrical or mechanical depending on the nature of the output signal, power requirements, or both. However, improvement in electrical transducer technology has blurred the distinction between these categories. Many flow meters previously classified as mechanical are now used with electrical transducers. Some common examples are the electrical shaft encoders on positive displacement meters, the electrical (strain) sensing of differential pressure, and the ultrasonic sensing of weir or flume levels.

The flow meters discussed herein are divided into two groups based on the method by which the basic flow signal is generated. The first group consists of meters in which the signal is generated from the energy of the flowing fluid. For example, a differential-pressure meter generates a signal from the flow itself. The second group comprises those flow meters that derive their basic signal from the interaction of the flow and an external stimulus. The manner in which the flow signal is transduced, conditioned, or transmitted does not determine the classification.

Meters can be further divided into three subgroups depending on whether fluid velocity, the volumetric flow rate, or the mass flow rate is measured. The emphasis herein is on common flow meters. Devices of a highly specialized nature, such as biomedical flow meters, are beyond the scope of this article.

Fluid Energy Activated Flow Meters

Positive-Displacement Flow Meters. Positive-displacement flow meters separate the incoming fluid into chambers of known volume which, using the energy of the fluid, advance through the meter and discharge into the downstream pipe. The total volume of fluid passing through the meter is the product of the internal-meter swept volume and the number of fillings. Meter sizing is based on the relationship between flow meter capacity, pressure drop across the meter, and fluid viscosity (3).

Positive-displacement meter chambers are housed in a pressure-containing vessel which may be of single- or double-wall construction. In single-wall construction the housing forms part of the measuring chamber walls. In double-wall construction the housing is separate from the measuring portion and serves only as a pressure vessel. Double-wall construction has two distinct advantages: first, chamber walls can be thinner and more precisely formed as these do not need to withstand the full fluid static pressure; second, this construction allows piping stress to be confined to the external housing leaving the chambers free of potentially distorting forces.

The output signal from positive-displacement meters may be mechanical, where the motion is transmitted by an output shaft through a housing seal, or it may be magnetically or inductively coupled.

All positive-displacement meters depend on very close clearance dimensions between rotating and moving parts and thus are not suitable for fluids containing abrasive particles. These meters have broad application in the distribution of natural gas for two reasons: the completely mechanical nature and the ability to maintain good accuracy over long periods of time. Wear in positive-displacement meters tends to increase leakage so that errors are in the direction of under-registration, the most acceptable mode of error for commercial billing meters within the gas industry. Meters are normally periodically recalibrated and adjusted to read within 1% of the actual volumetric flow.

Positive-displacement meters also find broad application in the measurement of viscous liquids because high viscosities provide lubrication and minimize seal leakage. Positive-displacement designs are inherently insensitive to incoming velocity profile and thus to piping configuration and Reynolds number. They normally do not require specific upstream or downstream piping. Good accuracy can be obtained at conditions of transitional Reynolds numbers where many other meters exhibit nonlinearity.

Positive-displacement meters are normally rated for a limited temperature range. Meters can be constructed for high or low temperature use by adjusting the design clearance to allow for differences in the coefficient of thermal expansion of the parts. Owing to small operating clearances, filters are commonly installed before these meters to minimize seal wear and resulting loss of accuracy.

There are at least five types of positive-displacement meters commercially available.

Reciprocating Piston Meters. In positive-displacement meters of the reciprocating piston type, one or more pistons similar to those in an internal-combustion engine are used to convey the fluid. Capacity per cycle can be adjusted by changing the piston stroke.

Bellows or Diaphragm Meters. Bellows meters use flexible diaphragms as the metering chambers. A series of valves and linkages control the filling and emptying of the chambers. Movement of the flexible walls is regulated for a constant displacement per stroke. Meters of this type are widely used in the gas industry as residential meters (see GAS, NATURAL).

Nutating Disk Meters. In positive-displacement meters of nutating disk design, the chambers are formed by a disk mounted on a central ball. The disk is held in an inclined position so that it is in contact with the chamber bottom along a radial section on one side of the ball and in contact with the top at a section 180° away. A radial partition prevents the disk from rotating about its own axis. Inlet and outlet ports are located on each side of the partition. Liquid alternately enters above and below the disk and flows around the conical chamber toward the outlet port. This movement causes the disk to nutate, ie, to undergo a circular nodding motion. This disk motion is coupled to a mechanical meter register which integrates volumetric flow rate. Nutating disk meters are mechanically simple and rugged. They are widely used as commercial water meters.

Rotary Impeller Vane and Gear Meters. One group of positive-displacement meters depends on shaped impellers or gears to form the measuring chambers. Figure 4 illustrates a two-lobed rotary meter of the type used to measure gas flow. The impellers are designed to maintain a continuous seal during rotation. Close tolerances and the use of precision bearings permit these meters to have minimal leakage while keeping overall pressure loss low. Rotating-vane meters are somewhat similar in design but include a timed gate to isolate the inlet and outlet ports. Figure 5 shows one cycle of a rotating-vane gas meter. The pressure of the entering gas rotates the vane assembly counterclockwise and, through timing gears, rotates the gate. In successive positions, the annular segment of gas is isolated by vanes 1 and 2, rotated through the housing, and discharged by the action of the gate.

Differential-Pressure Flow Meters. Differential-pressure or variable head flow meters are the oldest, most common group of flow measurement devices. This general category includes orifice plates, venturi tubes, flow nozzles, elbow meters, wedge meters, pitot tubes, and laminar flow elements. All are based on the

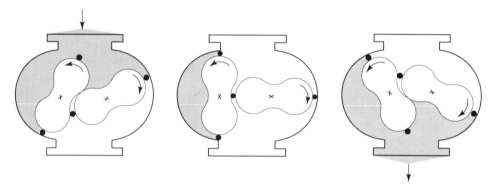

Fig. 4. Operating sequence for a two-lobed rotary gas flow meter where the shaded area represents the flowing fluid.

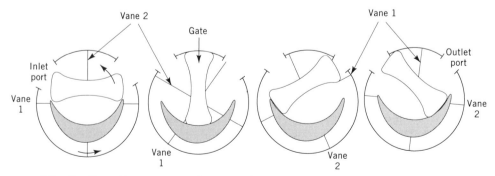

Fig. 5. Operating sequence for a rotating-vane positive-displacement meter.

Bernoulli principle that, in a flowing stream, the total energy, ie, the sum of the pressure head, velocity head, and elevation, remains constant. Differential-pressure devices all create some restriction in the fluid conduit causing a temporary increase in fluid velocity and a corresponding decrease in local head or pressure. For these conditions, the Bernoulli principle can be applied to give a general equation for head meters:

$$q = kA(2gh)^{1/2}$$

where q is the volumetric rate of flow; k, the dimensionless experimentally determined flow coefficient; A, the inside cross-sectional area of the pipe; h, the differential produced by the restriction measured in height of the flowing fluid; and g, the gravitational constant.

The basic form of the equation is normally modified so that the differential is expressed in pressure units and the flow coefficient is divided into the product of an experimentally determined discharge coefficient, K, and a series of calculated coefficients. In this form, for concentric restrictions:

$$q = K\beta^2 AYFa2g \left(\frac{P_1 - P_2}{\rho}\right)^{1/2}$$

where β is the ratio of the restriction to the pipe diameter, d/D, known as the beta ratio; Y, the gas expansion factor for an adiabatic charge from P_1 to P_2; P_1 is the upstream pressure; P_2, the restriction pressure; Fa, the thermal expansion factor for the restriction; ρ, the fluid density; and K, the discharge coefficient.

An outstanding advantage of common differential pressure meters is the existence of extensive tables of discharge coefficients in terms of beta ratio and Reynolds numbers (1,4). These tables, based on historic data, are generally regarded as accurate to within 1–5% depending on the meter type, the beta ratio, the Reynolds number, and the care taken in manufacture. Improved accuracy can be obtained by running an actual flow calibration on the device.

Improvements in low differential readout devices and the desire to minimize permanent pressure losses have resulted in a trend toward higher beta devices. This is generally at the price of increased sensitivity to upstream piping configuration (see Table 1). To obtain accurate differential-pressure measurements, the installation should conform as closely to reference conditions as possible. The pipe should be inspected with respect to diameter, roundness, smoothness, and tap location. The fluid restriction should be mounted concentric to the pipe internal diameter and any gaskets required cut so as not to protrude into the flow stream. Both pressure taps should be of the same diameter with no roughness or burrs. The pressure lines must be carefully installed in accordance with the guidelines in Reference 5. Improperly installed lines are probably the most common cause of orifice measurement errors. The differential-pressure transmitter should be as close to the taps as possible and the coupling lead lines sloped to permit condensate or gas bubble removal.

In liquid service the lines must constantly slope downward toward the transmitter from the taps to prevent possible gas pockets. In gas service the lines should drain to prevent condensate accumulation or, if the condensate is used to transmit the pressure from the taps to transmitter, the condensate legs must be of equal height.

Orifice Plates. The various types of orifice plates are shown in Figure 6. A square-edge orifice, the most commonly applied type, is a thin flat plate set perpendicular to the flow with a clean, sharp edged circular opening. This opening is normally concentric with respect to the pipe centerline, although eccentric plates having the opening tangent to the pipe axis are often used for steam (qv), gas with entrained liquids, or sediment bearing fluids. Segmented orifice plates having the opening in the shape of a circular segment are applied to viscous flows and slurries. At pipe Reynolds numbers below 10,000 the coefficient of discharge for a square edge orifice tends to become nonlinear and quadrant edge orifices are commonly used. These are thicker plates where the inlet edge is rounded making them less sensitive to fluid viscosity effects. For example, the change in discharge coefficient for a quadrant-edge orifice between Reynolds numbers of 5,000 and 10,000 is approximately 2% compared with a change several times greater for a sharp edged orifice over the same *Re* range.

There are five locations in use for the taps used to couple the differential to the measurement device. These locations are depicted in Figure 7. Corner taps, drilled in the orifice mounting flanges on either side and as close to the plate as possible, are commonly used in Europe for pipe sizes under 50 mm. Flange taps, each located 25 mm from the respective faces of the plate, are easily constructed and in greatest general use. For larger size pipes, radius taps, or *D* and *D*/2 taps,

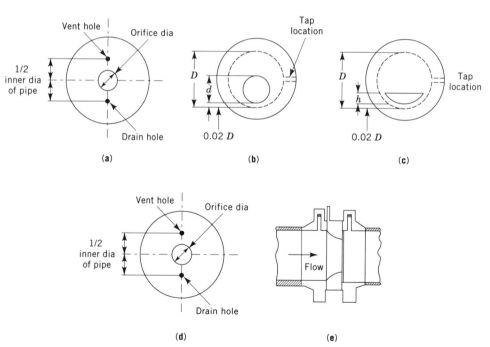

Fig. 6. Orifice plates: (**a**) concentric, (**b**) eccentric, (**c**) segmental, (**d**) universal, and (**e**) quadrant-edge.

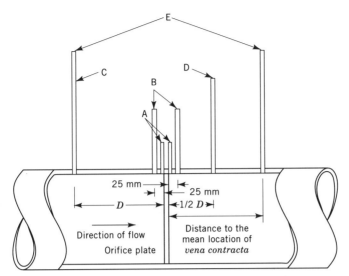

Fig. 7. Orifice plate pressure tap locations. A, corner taps; B, flange taps; C, D taps; D, 1/2 D taps; and E, *vena contracta* taps. See text.

are located 1 pipe diameter upstream and 0.5 pipe diameters downstream from the upstream face of the plate. These taps are normally used in North America for small pipe sizes. Pipe taps (not shown) are located 2.5 pipe diameters upstream and 8 pipe diameters downstream from each plate face. This downstream connection increases orifice fitting length and limits usage of this type of tap.

The fifth type of tap is unique in that the downstream tap location varies depending of the orifice β ratio. This tap is located at the *vena contracta*, the location where the stream issuing from the orifice attains its minimum cross section. The location of this tap is defined from the upstream face of the orifice as is the $D/2$ tap. The downstream tap for corner, flange, and pipe taps is measured from the downstream face of the orifice. *Vena contracta* taps maximize the measured differential pressure. For modern transmitters this is not an important consideration and this type of tap is no longer widely used.

For very low flow rates the orifice plate is often incorporated into a manifold, an integral part of the differential-pressure transmitter. This provides a convenient compact installation.

Two equations for the calculation of square-edge orifice coefficients are in general use: one discharge (4) is based on extensive calibrations performed in the 1930s and is widely used in natural gas measurement; the second is accepted by the American Society of Mechanical Engineers (ASME) and the International Organization for Standardization (ISO) (1). In larger pipes the difference between the coefficients calculated by these two methods is generally small (6).

Orifice plates should be periodically inspected for signs of edge wear, warping, scratches, and deposits, any of which may result in a loss of accuracy from a change in discharge coefficient. Orifice plates have the advantages of being simple, hydraulically predictable, readily interchangeable, and reliable. These meters find wide use in both liquid and gas service where moderate accuracy and limited range meet the needs of many applications.

Venturi Tubes.   A venturi tube consists of two hollow truncated cones, the smaller diameters of which are connected by a short circular section known as the throat. Pressure differential is measured between the upstream and throat sections and can be related to flow from equations or tables in a manner similar to orifice plates. The sole purpose of the downstream cone is to recover part of the differential pressure. For an exit cone angle of 7%, the permanent pressure loss is only about 10% of the differential. A number of short-tube, larger-exit angle venturi tubes are commercially available; these require less space for installation but have a higher permanent pressure loss. Advantages of venturi tubes relative to orifice plates are the lower permanent pressure drop and reduced sensitivity to dirty flow conditions of the former. The smooth contours of a venturi allow entrained particles to flow past instead of building up as these do when an orifice is used. Disadvantages of the venturi are the greater cost, longer installation length, and lack of easy interchangeability.

Flow Nozzles.   A flow nozzle is a constriction having an elliptical or nearly elliptical inlet section that blends into a cylindrical throat section as shown in Figure 8. Nozzle pressure differential is normally measured between taps located 1 pipe diameter upstream and 0.5 pipe diameters downstream of the nozzle inlet face. A nozzle has the approximate discharge coefficient of an equivalent venturi and the pressure drop of an equivalent orifice plate although venturi nozzles, which add a diffuser cone to proprietary nozzle shapes, are available to provide better pressure recovery.

Flow nozzles are commonly used in the measurement of steam and other high velocity fluids where erosion can occur. Nozzle flow coefficients are insensitive to small contour changes and reasonable accuracy can be maintained for long periods under difficult measurement conditions that would create unacceptable errors using an orifice installation.

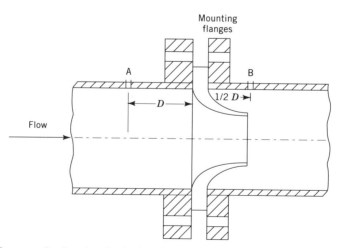

Fig. 8.   Flow nozzle showing A, the high pressure tap, and B, the low pressure tap.

Critical Nozzles.   As the pressure differential across a nozzle or venturi is increased the rate of discharge of a gas also increases until the linear velocity in the throat reaches the velocity of sound. Any further increase in pressure differential does not cause an increase in velocity. In this condition the nozzle is referred to as choked or at critical flow. The ratio of downstream/upstream pressure where critical flow is first obtained is called the critical pressure ratio, R_c. For nozzles of the shape shown in Figure 8, R_c is approximately equal to 0.5 but the addition of a venturi-type outlet can provide pressure recovery so that values of R_c as high as 0.96 can be obtained. As long as critical conditions are maintained only the upstream pressure and temperature are needed to determine flow rate. Nozzles operated at critical conditions are rugged and provide repeatable measurements; they make a good standard for the calibration of other gas flow meters. One principal application area is in the low flow testing of automotive carburetor and emission-control systems (see EXHAUST CONTROL, AUTOMOTIVE). Critical nozzles are also widely used at high gas flow rates where actual nozzle calibration is impractical. For these applications a theoretical discharge coefficient is used.

Elbow Meters.   Fluid passing through a common pipe elbow generates a differential pressure between the inside and outside of the elbow resulting from centrifugal force. This differential can be measured to provide an estimate of flow, on an uncalibrated elbow, to approximately 4% uncertainty. Experimental tests indicate the elbow flow coefficient to be insensitive to changes in relative elbow roughness. Elbow differentials are generally used for the balancing of flow rates in multiple manifold systems or in efficiency testing.

Wedge Meters.   The wedge flow meter consists of a flanged or wafer-style body having a triangular cross section dam across the top of the fluid conduit. Pressure taps are on either side of this restriction. Overall meter sizes range from 10 to 600 mm. Within each size several restrictions are available to provide the range of differential pressure desired for the application.

The wedge design maintains a square root relationship between flow rate and differential pressure for pipe Reynolds numbers as low as approximately 500. The meter can be flow calibrated to accuracies of approximately 1% of actual flow rate. Accuracy without flow calibration is about 5%.

The wedge restriction has no critical surface dimensions or sharp edges and tends to retain accuracy despite visible corrosive or erosive wear. It is commonly applied to high viscosity liquids, slurries, and hot multiphase mixtures. A similar device is also available using a cone, positioned so that its large diameter is upstream, mounted on the meter centerline.

Pitot Tubes.   The fundamental design of a pitot tube is shown in Figure 9**a**. The opening into the flow stream measures the total or stagnation pressure of the stream whereas a wall tap senses static pressure. The velocity at the tip opening, V, can be obtained by the Bernoulli equation:

$$V = C \, (2g/(P_1 - P_2))^{1/2}$$

This equation is applicable for gases at velocities under 50 m/s. Above this velocity, gas compressibility must be considered. The pitot flow coefficient, C, for some designs in gas service, is close to 1.0; for liquids the flow coefficient is dependent on the velocity profile and Reynolds number at the probe tip. The coefficient drops

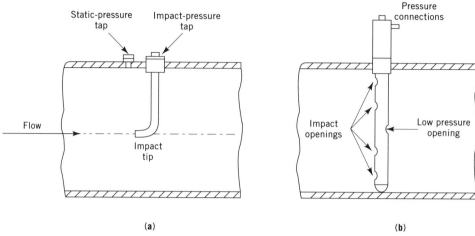

Fig. 9. Pitot tube designs: (**a**) basic and (**b**) averaging.

appreciably below 1.0 at Reynolds numbers (based on the tube diameter) below 500.

Standard pitot tubes provide a measurement of point velocity only; any attempt at determining total flow involves the assumption of a velocity profile. To overcome this disadvantage averaging pitot tubes have been developed. As shown in Figure 9**b**, these tubes extend across the pipe and use a series of holes at specific annular spacing to obtain an average velocity along the probe length. A single downstream-facing opening senses the downstream pressure.

Advantages of the pitot method of measurement are low pressure loss and easy installation. In some cases, installations in existing lines are made without process shutdown by hot tapping the line (7).

Laminar Flow Elements.   Each of the previously discussed differential-pressure meters exhibits a square root relationship between differential pressure and flow; there is one type that does not. Laminar flow meters use a series of capillary tubes, rolled metal, or sintered elements to divide the flow conduit into innumerable small passages. These passages are made small enough that the Reynolds number in each is kept below 2000 for all operating conditions. Under these conditions, the pressure drop is a measure of the viscous drag and is linear with flow rate as shown by the Poiseuille equation for capilary flow:

$$\text{capillary flow} = \frac{\pi \ CD^4(P_1 - P_2)}{128 \ \mu L}$$

where D is the diameter of the capillary tube; L, the tube length; C, the experimentally determined flow coefficient; μ, the fluid viscosity; and $P_1 - P_2$ is the differential pressure over the tube length L. Because of the small passage sizes and dependence on D^4, laminar flow meters are suitable for use only with very clean fluids.

Target Flow Meters.   Target flow meters use a drag-producing body in the flow stream to generate a force proportional to velocity. This force is sensed using strain gauges or a force balance system. The basic equation governing operation is that for the drag of a body in a flow stream and is similar to the equation for differential pressure flow meters:

$$F = C_D A \frac{\rho V^2}{2g}$$

where F is the drag force; C_D, the drag coefficient; A, the frontal area of the body; ρ, the fluid density; V, the upstream fluid velocity; and g, the gravitational constant. In general the target, commonly a circular disk, is mounted on the pipe centerline to form an annular orifice. For pipe Reynolds numbers above 4000, the drag coefficient of such a design is essentially constant. At lower Reynolds numbers the drag and meter coefficient depend on the d/D ratio and operating Reynolds numbers. The target meter has greatest application in the measurement of hot, viscous, or sediment-bearing fluids which would plug or congeal in the pressure taps of a differential-pressure meter. Accuracy is typically 2% of full-scale flow and turndown 5:1.

Variable-Area Flow Meters. In variable-head flow meters, the pressure differential varies with flow rate across a constant restriction. In variable-area meters, the differential is maintained constant and the restriction area allowed to change in proportion to the flow rate. A variable-area meter is thus essentially a form of variable orifice. In its most common form, a variable-area meter consists of a tapered tube mounted vertically and containing a float that is free to move in the tube. When flow is introduced into the small diameter bottom end, the float rises to a point of dynamic equilibrium at which the pressure differential across the float balances the weight of the float less its buoyancy. The shape and weight of the float, the relative diameters of tube and float, and the variation of the tube diameter with elevation all determine the performance characteristics of the meter for a specific set of fluid conditions. A ball float in a conical constant-taper glass tube is the most common design; it is widely used in the measurement of low flow rates at essentially constant viscosity. The flow rate is normally determined visually by float position relative to an etched scale on the side of the tube. Such a meter is simple and inexpensive but, with care in manufacture and calibration, can provide readings accurate to within several percent of full-scale flow for either liquid or gas.

A variety of other float shapes are available, some of which are designed to be insensitive to fluid viscosity changes. Tubes having various tapers are made to give linear or logarithmic scales and long slow taper tubes are available for higher resolution. Tubes may contain flutes, triangular flats, or guide rods to center the float and prevent chatter. Metal tubes are available for high pressure service. In these, a magnetic coupling typically detects the float position and an external indicator or transmitter is used to provide the flow reading. Other somewhat less common forms of the variable-area meter use a tapered plug riding vertically within the bore of an orifice. The area of the restriction is controlled to maintain the differential-pressure constant. Flow rate is proportional to the effective restriction area and is derived from the motion of the plug.

Because of the design, variable-area meters are relatively insensitive to the effects of upstream piping and have a pressure loss which is essentially constant over the whole flow range. These meters have greatest application where direct visual indication of relatively low flow rates of clean liquids or gases are required. Common applications are laboratory measurements, process purge flows, chemical analyzers, and medical gas dispensing. Variable-area meters can be readily fitted with high or low flow alarms. A common use for these meters is to provide a relatively low cost method of protecting critical equipment from lubrication failure.

Head-Area Meters.   The Bernoulli principle, the basis of closed-pipe differential-pressure flow measurement, can also be applied to open-channel liquid flows. When an obstruction is placed in an open channel, the flowing liquid backs up and, by means of the Bernoulli equation, the flow rate can be shown to be proportional to the head, the exact relationship being a function of the obstruction shape.

Weirs.   Weirs are dams or obstructions built across open channels that have, along their top edge, an opening of fixed dimensions and shape through which the stream can flow. This opening is called the weir notch and its bottom edge is designated the crest. Weirs are commonly used in irrigation, water works, wastewater discharge lines, electrical generating facilities (see POWER GENERATION), and pollution monitoring.

Predictable forms of weirs have been developed that are classified according to the shape of the notch. The discharge of each type can be determined from tables (8) or by actual flow calibration. Selection of weir type is dependent on the nature of the application. A broad-crested rectangular notch (Fig. 10**a**) allows streamline development to pass most floating debris and works at lower heads than a sharp-crested weir. Triangular, sharp-crested weirs (Fig. 10**d**) provide maximum flow range but do not transport floating material. The trapezoidal notch (Fig. 10**c**) is a combination of the rectangular and triangular forms. In the Cip-

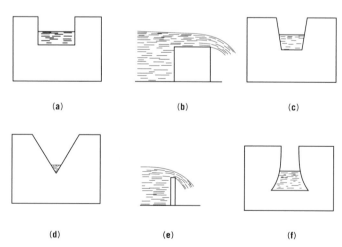

Fig. 10.   Stream flow over (**a**) a broad-crested, rectangular weir; (**b**) a cross-current view of the rectangular and Cipolletti weirs; (**c**) a trapezoidal-notch or Cipolletti weir; (**d**) a sharp-crested, triangular, or V-notch weir; (**e**) a cross-current view of the V-notch and hyperbolic-notch weirs; and (**f**) a hyperbolic-notch weir.

poletti design, the slope of the ends has a value so that the additional discharge through the triangular portions of the notch exactly compensates for the effects of end contractions. Special forms of notches can be constructed to simplify the flow head relationship. Figure 10**f** shows one such form.

For accurate flow measurement the channel area upstream of the weir should be large enough to allow the flow to develop a smooth flow pattern and a velocity of 0.1 m/s or less. The downstream channel must be large enough to prevent high flow rates from submerging the weir and the flow over the notch must be sufficient for it to clear the downstream face as shown in Figures 10**b** and 10**e**. This free-discharge mode is the basis of weir capacity tables.

Flumes. Flumes, open channels that have gradual rather than sharp restrictions, are closely analogous to venturi meters for closed pipes. Weirs are analogous to orifice plates. The flume restriction may be produced by a contraction of the sidewalls, by a raised portion of the channel bed (a low broad-crested weir), or by both. One common design is the Parshall flume shown in Figure 11. Dimensions and capacity tables for Parshall flumes are available (9). Flumes, widely used in measuring irrigation water, are alternatives to weirs where lower head requirements, higher capacity, and reduced sensitivity to silting are advantageous. Flumes are generally considered more expensive and less accurate than sharp-crested weirs.

A number of flume designs have been created specifically for use in partially filled circular conduits such as sewers. These are available in molded fiber glass and can be lowered through a manhole if required. As with all open-channel head-area meters, flumes must be sized to prevent submergence of the restriction.

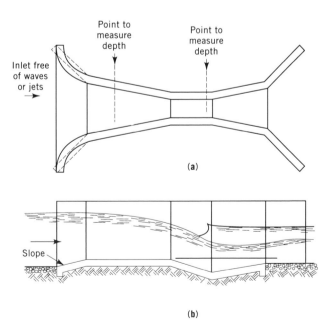

Fig. 11. The Parshall flume showing (**a**) flume construction and (**b**) stream flow (9).

Cup and Vane Anemometers. A number of flow meter designs use a rotating element kept in motion by the kinetic energy of the flowing stream such that the speed is a measure of fluid velocity. In general, these meters, if used to measure wind velocity, are called anemometers; if used for open-channel liquids, current meters; and if used for closed pipes, turbine flow meters.

Cup anemometers have shaped cups mounted on the spokes of a wheel. The cups, under the action of the fluid forces, spin in a horizontal plane about a vertical shaft mounted in bearings. Vane or propeller types use a multibladed rotor, the axis of which is parallel to the flow direction as the rotating member. Both designs are commonly used for wind speed measurement or similar applications such as the velocity in ventilation ducts. Because of inertia, anemometers are most accurate under steady conditions. Velocity fluctuations cause readings that are too high.

Current Meters. Various vane designs have been adapted for open-channel flow measurement. The rotating element is partially immersed and rotates rather like a water wheel. Operation is similar to that of vane anemometers.

Turbine Meters. The turbine meter represents a refinement of the anemometer or current-meter design for use in a closed conduit. A typical turbine cross section is shown in Figure 12. Flow entering the meter is directed through a flow-straightening section that shapes the flow and acts as a rotor support. The flow then turns the helically bladed rotor and passes through a rear-support section. A magnetic pick-off coil, or other externally mounted proximity detector, senses the passage of the rotor blade. At a steady velocity the rotor comes to a speed at which the angle of the fluid striking the blade produces a driving force that is just sufficient to balance the drag forces resisting rotation. This angle of attack is a measure of the total forces on the rotor resisting rotation. For maximum range, it is essential that these drag forces and the attack angle are as small as possible.

The output of turbine meters is inherently digital, provides a high information rate, and has excellent accuracy and repeatability over a wide range. Turbines are available in sizes from 6 to 600 mm. Typical accuracy is $\pm 0.5\%$ of reading over flow ranges of 10:1 or greater in large sizes. In smaller sizes friction and

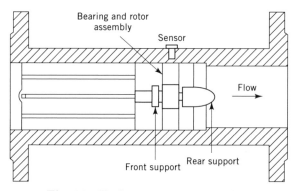

Fig. 12. Turbine meter cross section.

viscosity effects reduce the linear range. Lower cost meters having reduced accuracy specifications are also available.

A typical universal viscosity performance curve is shown in Figure 13 where meter output is plotted against operating frequency divided by kinematic viscosity, which is analogous to Reynolds number. Calibration at four different viscosities all overlay down to some low flow rate where either mechanical or magnetic drag forces, or both, become significant. This point defines the lower limit of the predictable-performance range. Because of excellent curve repeatability, the output can be made linear to low Reynolds numbers using microprocessor-based programmable secondaries.

Turbine meters designed for clean liquids use ball bearing mounted rotors for low friction and wide range. Typical uses include aerospace and aircraft testing, cryogenic liquid measurement (see CRYOGENICS), and the digital blending of petroleum products. Models are also available using self-cleaning journal bearings which provide a shorter measurement range but superior service life in the presence of fluid contamination. These meters are used in a broad range of industrial environments including liquids, gases, and steam.

Oscillatory Flow Meters. Three different oscillatory fluid phenomena are used in flow measurement.

Fluid Oscillation. Fluidic flow meters are based on wall attachment, ie, the Coanda effect, and the technology of bistable fluid oscillators such as those used in fluid logic. A fluidic meter (Fig. 14a) consists of an entrance nozzle section and a diverging section where the walls are designed to permit the flow to attach to one side or the other, but not to both. Downstream feedback passages connect with control ports upstream of the point of attachment. In operation, natural turbulence and the Coanda effect cause the flow jet to attach to one side wall. As the flow is biased to this side, a portion of it is directed through the feedback passage to the control port causing the jet to switch to the opposite side where

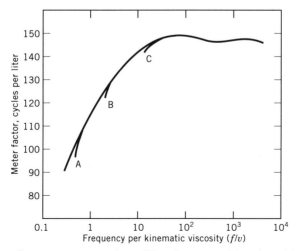

Fig. 13. A turbine flow meter composite-calibration curve. For viscosities of A, 50 mm^2/s; B, 10 mm^2/s; C, 1 mm^2/s, where 1 mm^2/s = 1 cSt.

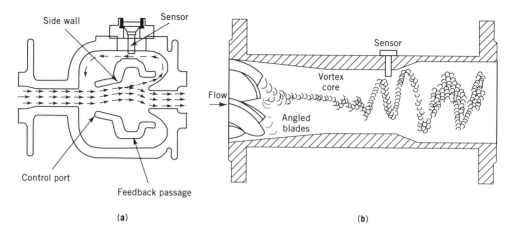

Fig. 14. Cross sections of oscillatory flow meters: (**a**) fluidic, where (➤) represents the main flow and (➤—➤) the feedback, and (**b**) vortex precession.

the same feedback action is repeated. The result is a continuous oscillation having a frequency linearly related to the fluid velocity. This frequency is detected by means of a sensor in one of the feedback passages (see SENSORS). Fluidic flow meters have been applied to the measurement of clean, low viscosity liquids in pipe sizes 80 mm and smaller.

Vortex Precession. When a swirling body of fluid enters a divergent section of the center of rotation precesses, ie, it leaves a straight line and takes up a helical path. Using a meter based on this principle (Fig. 14**b**), entering fluid is given a rotational component of velocity by a set of fixed blades creating a rotational fluid profile, the centerline of which coincides with the meter centerline. This flow profile is stabilized and accelerated by a convergent section before it enters an enlarged section that causes precession to take place. The frequency of this precession, detected by a dynamic pressure sensor, is linear with volumetric flow rate over a wide range. Typical accuracy is ±1% of actual flow.

Vortex precession meters feature no moving parts and a relatively high frequency digital output. They are used in the measurement of gas and liquid flows, generally in pipe sizes 80 mm and smaller.

Vortex Shedding. When a streamlined body is placed in a flowing stream, the fluid follows the contours of the body without separating from its surface. If, however, the body is bluff or nonstreamlined, the fluid separates at some point from the surface and rolls into a vortex. For two-dimensional symmetrical bodies, the changes in local velocity and pressure associated with the separation on one side interact with vortex formation on the opposite side. This feedback quickly causes a stable pattern of alternate vortex shedding so that the downstream wake becomes a staggered pattern of vortices commonly referred to as a Karmen-vortex street. This pattern is shown for a cylindrical obstruction in Figure 15. It is this phenomena that causes the flapping of flags and the clear turbulence behind jet aircraft.

The frequency of vortex formation is a linear function of the fluid velocity and the width of the obstruction at the point where shedding occurs. Vortex-

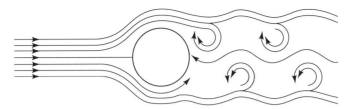

Fig. 15. Karmen-vortex pattern behind a circular cylinder.

shedding flow meters use various forms of well-defined symmetrical obstructions to optimize vortex formation and detect the vortices using sensors which respond to local velocity or pressure changes. Since these meters originally became available in the early 1970s, improved sensors have broadened the application range of the vortex-shedding meter. Vortex-sensing techniques include differential-pressure-sensing diaphragms (having capacitive or inductive pick-off), strain gauges, piezoelectric (see PIEZOELECTRICS) crystals, and thermistors.

Vortex-shedding flow meters typically provide 1% of flow rate accuracy over wide ranges on liquid, gas, and steam service. Sizes are available from 25 to 200 mm. The advantages of no moving parts and linear digital output have resulted in wide usage in the measurement of steam, water, and other low viscosity liquids.

External Stimulus Flow Meters

External stimulus flow meters are generally electrical in nature. These devices derive their signal from the interaction of the fluid motion with some external stimulus such as a magnetic field, laser energy, an ultrasonic beam, or a radioactive tracer.

Electromagnetic Flow Meters. Faraday's law of electromagnetic induction states that relative motion, at right angles, between a conductor and a magnetic field induces a voltage in the conductor. The magnitude of the induced voltage is proportional to the relative velocity of the conductor and the magnitude of the magnetic field. This principle is used to measure the flow of conducting liquids using meter designs similar to that of Figure 16. A pair of coils produces an electromagnetic field through an insulating tube carrying the liquid. Electrodes at a right angle to both the flow and field sense the induced voltage, E:

$$E = CBdV$$

where C is the meter calibration factor; B, the average magnetic flux density; d, the distance between electrodes; and V, the average fluid velocity.

Electromagnetic flow meters are available using either alternating current (a-c) or pulsed direct current (d-c) coil drives. The a-c-actuated meters require zero adjustment at full pipe and no flow conditions. These meters provide a high accuracy and wide turndown and are the preferred meter in certain applications. The main limitation on performance is a tendency to zero shift in coating appli-

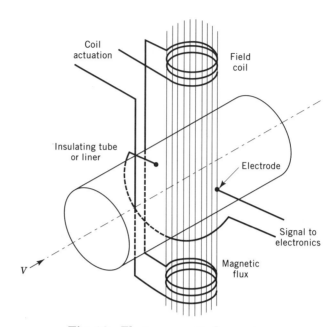

Fig. 16. Electromagnetic flow meter.

cations. Electromagnetic flow meters that use a pulsed d-c voltage coil excitation, eliminating the zero shift problem, have also become available. This meter type uses a duty cycle where the signal M_1 is measured during steady-state conditions $(d\theta/dt = 0)$ with the field coils on, and M_2 is similarly measured with the coil off. The on–off period is synchronized at a multiple of the line frequency so any a-c power noise averages out. The difference $M_1 - M_2$ is thus directly proportional to the flow. Meters of this design function accurately under conditions where sinusoidal excitation meters do not provide acceptable results. Because of design, pulsed d-c meters have a slower speed of response than a-c excitation types.

The exact magnitude of the generated voltage of an electromagnetic flow meter is an integration of the individual velocity and field vectors along the three-dimensional path between the electrodes. Modern designs use characterized fields to weigh all velocities equally. In this manner, the meter is made less sensitive to changes in velocity profile than other common meters.

Electromagnetic flow meters are available with various liner and electrode materials. Liner and electrode selection is governed by the corrosion character-istics of the liquid. For corrosive chemicals, fluoropolymer or ceramic liners and noble metal electrodes are commonly used; polyurethane or rubber and stainless steel electrodes are often used for abrasive slurries. Some fluids tend to form an insulating coating on the electrodes introducing errors or loss of signal. To over-come this problem, specially shaped electrodes are available that extend into the flow stream and tend to self-clean. In another approach, the electrodes are peri-odically vibrated at ultrasonic frequencies.

A more recent approach to the problem of electrode coating is the electrode-less magnetic flow meter. In this design there are no electrodes in contact with the process. Large plates placed on the outside of a ceramic spool perform the same function and are capacitively coupled to the transmitter through a high impedance amplifier. This meter has been found to provide satisfactory service at very low fluid conductivities and under coating conditions where other magnetic flow meters required cleaning after short periods of operation.

Additional magnetic flow meter innovations have been the addition of low flow and empty pipe cutoffs. The low flow cutoff is normally set to a value of from 1 to 5% of range setting and drops the output signal to zero when the flow rate reaches this level. This feature is useful in preventing erroneous signals owing to sloshing when there is no net flow. Many piping systems allow the magnetic flow meter to drain empty when flow stops. When this occurs the meter has an unstable output. To prevent errors under this condition some magnetic flow meters contain a pair of contacts that drive the signal to zero when actuated by a dry contact closure. Contact closure is provided from a pump or valve as appropriate.

Electromagnetic flow meters are available in essentially all pipe sizes, ie, 1 mm to 3 m, and provide measurement accuracy of 1% of rate or better over wide ranges. The meters are obstructionless, have no moving parts, and are extremely rugged. Pressure loss is that of an equivalent section of pipe. The meters are insensitive to viscosity, density, and temperature changes. Fluid conductivity has no effect on meter performance provided it is above a minimum level which is a function of meter design. Models having thresholds as low as 2 μS/cm (tap water typically has a value of 500 μS/cm) are available. Electromagnetic flow meters have the additional advantage of sensing flows containing entrained gas or solids on a flowing volume basis provided the flow is well mixed and traveling at a common overall velocity.

Because of these characteristics electromagnetic flow meters have been widely applied to the measurement of difficult liquids such as raw sewage and wastewater flows, paper pulp slurries, viscous polymer solutions, mining slurries, milk, and pharmaceuticals. They are also used in less demanding applications such as the measurement of large domestic water volumes.

Several special forms of electromagnetic flow meters have been developed. A d-c field version is used for liquid metals such as sodium or mercury. Pitot and probe versions provide low cost measurements within large conduits. Another design combines a level sensor and an electromagnetic meter to provide an indication of flow within partially full conduits such as sewer lines.

Momentum Flow Meters. Momentum flow meters operate by superimposing on a normal fluid motion a perpendicular velocity vector of known magnitude thus changing the fluid momentum. The force required to balance this change in momentum can be shown to be proportional to the fluid density and velocity, the mass-flow rate.

Coriolis-Type Flow Meters. In Coriolis-type flow meters the fluid passes through a flow tube being electromechanically vibrated at its natural frequency. The fluid is first accelerated as it moves toward the point of peak vibration amplitude and is then decelerated as it moves from the point of peak amplitude. This creates a force on the inlet side of the tube in resistance to the acceleration and an opposite force on the outlet side resisting the deceleration. The result of these forces is an angular deflection or twisting of the flow tube that is directly proportional to the mass flow rate through the tube.

A number of meter designs have been developed based on this principle. Some are shown in Figure 17. Certain advantages are claimed for each, but all share a number of characteristics. Perhaps the most important property is a full-scale deflection on the order of 0.001 mm. The sensors for these meters are extremely sensitive, stable, and capable of being temperature compensated.

Coriolis flow meters are available in line sizes from 0.1 to 200 mm and are sensitive enough to measure liquid flow rates of several grams per hour. Typical rangeability is 25:1 with accuracies reaching ±0.25% of actual flow rate. Installation procedures vary although rigid pipe supports just beyond the meter ends are normally recommended. A downstream shutoff valve is also desirable to enable in-place zero adjustment. In some systems, such as those using positive displacement pumps, a bypass line may be required to relieve shutoff pressure during zeroing.

Coriolis meters were first developed in the 1970s and continue to be improved. These meters are being applied in many areas that were commonly metered by other flow meter technologies. The greatest application has been in food and chemical processing.

Axial-Flow Angular-Momentum Flow Meter.   In this design an impeller is rotated at constant speed in the flow field imparting a constant angular momentum to it. A downstream rotor absorbs this momentum but is restrained from rotating. Using Newton's second law, the torque generated by the change in momentum is

$$T = WR^2M$$

where T is the fluid torque; W, the fluid angular velocity; R, the radius of gyration of the fluid; and M, the mass-flow rate. If a balance torque is supplied by a closed-loop servo system, the torque of which may be expressed by

$$T = KWV$$

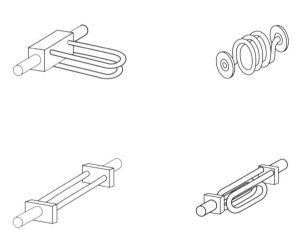

Fig. 17. Coriolis flow meter tube configurations.

where K is a controlled constant and V is a voltage signal to the servo, then

$$V = \frac{R^2 M}{K}$$

The servo voltage is a function of mass-flow rate. Axial-flow angular-momentum meters are sometimes used in measuring jet engine fuel flow as the fuel energy content correlates much more closely with mass than volume.

Ultrasonic Flow Meters.   Ultrasonic flow meters can be divided into three broad groups: passive or turbulent noise flow meters, Doppler or frequency-shift flow meters, and transit time flow meters.

Passive Detectors.   Passive or turbulent noise detectors are ultrasonic microphones clamped on to the flow conduit (see ULTRASONICS). These microphones respond to some portion of the frequency spectrum of turbulent noise within the pipe. This noise increases with increasing velocity although the exact relationship is dependent on the particular installation. Passive detectors can be used for liquids, gases, or slurries to activate flow switches or to provide a low cost general indication of relative flow. Because this signal is generated by the flow itself, passive detectors are actually self-generating flow meters. Other types, however, form the great majority of ultrasonic flow meters and are not self-generating.

Doppler Flow Meters.   Doppler flow meters sense the shift in apparent frequency of an ultrasonic beam as it is reflected from air bubbles or other acoustically reflective particles that are moving in a liquid flow. It is essential for operation that at least some particles are present, but the concentration can be low and the particles as small as ca 40 μm. Calibration tends to be influenced by particle concentration because higher concentrations result in more reflections taking place near the wall, in the low velocity portion of the flow profile. One method used to minimize this effect is to have separate transmitting and receiving transducers focused to receive reflections from an intercept zone near the center of the pipe.

Both wetted-sensor and clamp-on Doppler meters are available for liquid service. A straight run of piping upstream of the meter and a Reynolds number of greater than 10,000 are generally recommended to ensure a well-developed flow profile. Doppler meters are primarily used where stringent accuracy and repeatability are not required. Slurry service is an important application area.

Transit Time Flow Meters.   This type of ultrasonic meter depends on measuring the transit time of an ultrasonic beam through the flow. In most designs, a pair of ultrasonic transducers are mounted diagonally on opposite sides of a pipe section. These transducers may be wetted, ie, built into the flow meter, or clamped onto the outside of an existing pipe. In one design these transducers simultaneously transmit upstream and downstream ultrasonic pulses and measure the transit time until the leading edge of the pulse is received at the opposite transducer. The transit time for the pulse moving in the direction of the flow is less than for the pulse moving against the flow. Transit times are given by the expressions:

$$t_u = \frac{L}{C + V_1} \qquad t_d = \frac{L}{C - V_1}$$

where t_u is the transit time in the upstream direction; t_d, the transit time in the downstream direction; L, the acoustic path length; C, the speed of sound in the fluid; and V_1, the average component of liquid velocity V along the acoustic path. The difference in transit times, $t_d - t_u = \Delta t$, is

$$\Delta t = \frac{2LV_1}{C^2 - V_1^2}$$

Because $C^2 >> V^2$ and $V_1 = V \cos \theta$,

$$\Delta t = \frac{2LV \cos \theta}{C^2}$$

or

$$V = C^2 \frac{\Delta t}{2L \cos \theta}$$

The flow velocity is thus proportional to the difference in transit time between the upstream and downstream directions and to the square of the speed of sound in the fluid. Because sonic velocity varies with fluid properties, some designs derive compensation signals from the sum of the transit times which can also be shown to be proportional to C.

The angle θ can also change owing to changes in the refraction angle in accordance with Snell's law. This is primarily a problem in clamp-on designs where the pipe wall material is not controlled. Thermal gradients in the fluid may also cause problems by distorting the acoustic path. Designs that use a single acoustical path are inherently sensitive to velocity profile and swirl. These require good upstream piping and fully turbulent flow to provide accurate measurement. Multiple path designs are also available. These provide greater precision where well-controlled conditions cannot be maintained. The relatively long transit path of large pipes also permits operation at lower velocities than is possible in small pipes. Greatest application has been on liquid service in large pipes where flow profile and fluid properties are relatively constant, although gas designs are now available.

A variation on the transit time method is the frequency-difference or sing-around method. In this technique, pulses are transmitted between two pairs of diagonally mounted transducers. The receipt of a pulse is used to trigger the next pulse. Alternatively this can be done using one pair of transducers where each acts alternately as transmitter and receiver. The frequency of pulses in each loop is given by

$$f_u = \frac{1}{t_u} \qquad f_a = \frac{1}{t_d}$$

where f_u is the frequency of pulses in the upstream loop and f_d is the frequency of pulses in the downstream loop.

$$\Delta f = f_u - f_d = \frac{(C + V_1)}{L} - \frac{(C - V_1)}{L} = \frac{2V_1}{L} = \frac{2V \cos \theta}{L}$$

so that

$$V = \frac{L \, \Delta f}{\cos \theta}$$

The flow velocity in this design is therefore proportional to the difference between the frequencies but independent of sonic speed within the fluid.

In practice Δf is a small number and the sing-around frequencies are scaled up for display. In one example, for a pipe 1 m in diameter and water flowing at 2 m/s, the frequency difference is 1.4 Hz (10). Frequency difference transit time meters provide greater resolution than normal transit time ultrasonic meters. The greatest application is in sizes from 100 mm to 1 m diameter.

Transit time ultrasonic flow meters require a homogeneous fluid without a high density of reflective particles; in this sense fluid requirements are opposite those of a Doppler-type although there is an area of overlap. These flow meters have the advantages of being obstructionless with low pressure drop, having a wide range, and being bidirectional. The accuracy capability depends on the technique selected and the control of flow profile and fluid properties. Typical specifications range from $\pm 5\%$ of full scale to $\pm 1\%$ of actual flow rate.

Smart designs using microprocessor-based electronics are available using clamp on nonwetted transducers in both Doppler and transit time designs. These meters are convenient for making flow measurements on existing pipes. The pipe material and thickness, as well as any liner material and thickness, are programmed in the unit to permit flow area and sonic velocity corrections. These meters can be repeatable to 0.5% of full-scale flow. Absolute accuracy is dependent on application and installation parameters. Sound absorbing liners such as cement or glass may affect the measurement.

Laser Doppler Velocimeters. Laser Doppler flow meters have been developed to measure liquid or gas velocities in both open and closed conduits. Velocity is measured by detecting the frequency shift in the light scattered by natural or added contaminant particles in the flow. Operation is conceptually analogous to the Doppler ultrasonic meters. Laser Doppler meters can be applied to very low flows and have the advantage of sensing at a distance, without mechanical contact or interaction. The technique has greatest application in open-flow studies such as the determination of engine exhaust velocities and ship wake characteristics.

Correlation Flow Meters. *Tracer Type.* A discrete quantity of a foreign substance is injected momentarily into the flow stream and the time interval for this substance to reach a detection point, or pass between detection points, is measured. From this time, the average velocity can be computed. Among the tracers that have historically been used are salt, anhydrous ammonia, nitrous oxide, dyes, and radioactive isotopes. The most common application area for tracer methods is in gas pipelines where tracers are used to check existing metered sections and to spot-check unmetered sections.

Cross Correlation. Considerable research has been devoted to correlation techniques where a tracer is not used. In these methods, some characteristic pattern in the flow, either natural or induced, is computer-identified at some point or plane in the flow. It is detected again at a measurable time later at a position slightly downstream. The correlation signal can be electrical, optical, or acoustical. This technique is used commercially to measure paper pulp flow and pneumatically conveyed solids.

Thermal Flow Meters. *Hot-Wire and Hot-Film Anemometers.* Hot-wire devices depend on the removal of heat from a heated wire or film sensor exposed to the fluid velocity. The sensor is typically connected in a bridge circuit with a similar sensor that is not exposed to the velocity in the opposite leg of the bridge. This provides compensation for fluid temperature changes. Hot-wire anemometers are normally operated in a constant temperature mode. The resistance of the sensor, and therefore its temperature, is maintained constant at a value slightly over the fluid temperature by a servo-amplifier. In this mode, the current to the sensor becomes the flow-dependent variable. Constant temperature operation minimizes thermal inertia and makes the system capable of sensing rapid changes in velocity. Hot-wire signals are dependent on the heat transfer from the sensor and thus on both the fluid velocity and density, ie, the mass-flow rate. These signals are also dependent on the thermal conductivity and specific heat of the fluid and are susceptible to any contamination that changes the heat transfer. For these reasons hot-wire and hot-film anemometers are primarily used in clean liquids and gases where they can be calibrated for the exact condition of use. Applications are in the measurement of low air velocities both in the atmosphere and in building ventilation studies (see AIR POLLUTION; AIR POLLUTION CONTROL METHODS).

Differential-Temperature Thermal Flow Meters. Meters of this type inject heat into the fluid and measure the resulting temperature rise or, alternatively, the amount of power required to maintain a constant temperature differential. The power required to raise the temperature of a flowing stream by an amount ΔT is given by the relation:

$$P = MC_p\Delta T$$

where M is the mass-flow rate; P, the required power; C_p, the specific heat at constant pressure, and ΔT, the temperature rise. The thermal meter can therefore measure the mass-flow rate of a particular gas independent of pressure provided the specific heat is constant, a condition that is approximately true for most changes in temperature or pressure. The original differential temperature design heated the entire stream via a grid network and measured the temperature upstream and downstream with resistance grids. Because of high power consumption this design has been supplanted by several forms that retain the essential features but provide lower power consumption and better corrosion protection. In one form the outside of the meter tube is symmetrically heated by an external coil. Thermocouples are located on the tube wall equidistant from ends of the tube (Fig. 18). Heat sinks placed at the ends of the tube cause a symmetrical temper-

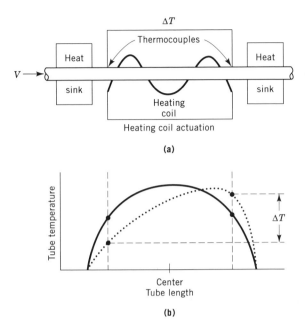

Fig. 18. (**a**) Differential-temperature thermal flow meter. (**b**) Tube length-temperature profiles for (———) zero flow and ($\cdots$) flow. ΔT is the temperature differential measured by the thermocouples.

ature pattern at zero flow and no temperature differential is measured between thermocouples. When fluid flows through the tube, the temperature distribution becomes skewed in the downstream direction and a differential is generated between the thermocouples which is dependent on the mass flow. Small differential temperature thermal meters are used to meter corrosive gases such as chlorine.

BIBLIOGRAPHY

"Flow Measurement" in *ECT* 3rd ed., Suppl. Vol., pp. 466–494, by T. H. Burgess, Fischer & Porter Co.

1. *Measurement of Fluid Flow by Means of Orifice Plates, Nozzles and Venturi Tubes Inserted in Circular Cross Section Conduits Running Full*, ISO 5167-1980(e), International Organization for Standardization, Geneva, Switzerland, 1980.
2. T. H. Burgess and co-workers, *Flow 2*, Instrument Society Of America, Research Triangle Park, N.C., 1981.
3. D. W. Spitzer, *Industrial Flow Measurement*, Instrument Society Of America, Research Triangle Park, N.C., 1984.
4. *Orifice Metering of Natural Gas and Other Related Hydrocarbons*, ANSI/API 2530, AGA report No. 3, American Gas Association, Arlington, Va., 1985.
5. *Fluid Meters—Their Theory and Application*, 6th ed., American Society of Mechanical Engineers, New York, 1971.
6. R. W. Miller, *J. Fluid Eng.* **101**, 483 (1979).
7. D. W. Spitzer, ed., *Flow Measurement*, Instrument Society of America, Research Triangle Park, N.C., 1991.

8. *Liquid Flow Measurement in Open Channels Using Thin Plate Weirs and Venturi Flumes*, ISO 1438-1975(E), International Organization for Standardization, Geneva, Switzerland, 1975.
9. P. Ackers and co-workers, *Weirs and Flumes*, John Wiley & Sons, Inc., New York, 1978.
10. W. K. Genthe and M. Yomamoto, *Flow 1*, Instrument Society Of America, Research Triangle Park, N.C., 1974.

THOMAS H. BURGESS
Fischer & Porter Company

FLOW METERS.   See FLOW MEASUREMENT; FLUID MECHANICS.

FLUIDIZATION

Gas–solids fluidization is the levitation of a bed of solid particles by a gas. Intense solids mixing and good gas–solids contact create an isothermal system having good mass transfer (qv). The gas-fluidized bed is ideal for many chemical reactions, drying (qv), mixing, and heat-transfer applications. Solids can also be fluidized by a liquid or by gas and liquid combined. Liquid and gas–liquid fluidization applications are growing in number, but gas–solids fluidization applications dominate the fluidization field. This article discusses gas–solids fluidization.

The basic concepts of a gas-fluidized bed are illustrated in Figure 1. Gas velocity in fluidized beds is normally expressed as a superficial velocity, U, the gas velocity through the vessel assuming that the vessel is empty. At a low gas velocity, the solids do not move. This constitutes a packed bed. As the gas velocity is increased, the pressure drop increases until the drag plus the buoyancy forces on the particle overcome its weight and any interparticle forces. At this point, the

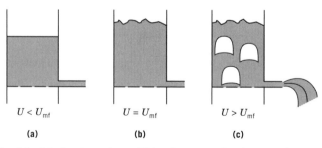

$$U < U_{mf} \qquad\qquad U = U_{mf} \qquad\qquad U > U_{mf}$$

(a)   (b)   (c)

Fig. 1.   Fluidized-bed behavior where U is the superficial gas velocity and U_{mf} is the minimum fluidization velocity: (**a**) packed bed, no flow; (**b**) fluid bed, uniform expansion; and (**c**) bubbling fluid bed, flow.

bed is said to be minimally fluidized, and this gas velocity is termed the minimum fluidization velocity, U_{mf}. The bed expands slightly at this condition, and the particles are free to move about (Fig. **1b**). As the velocity is increased further, bubbles can form. The solids movement is more turbulent, and the bed expands to accommodate the volume of the bubbles.

Once fluidized, the bed behaves as if it were a fluid. A level is maintained and a static pressure head is generated. No flow of solids through a side outlet occurs in a packed bed; however, flow through the opening does occur after a fluidized state has been achieved (Fig. **1c**).

A generic multipurpose fluidized bed is illustrated in Figure 2 (1). The solids are contained in a vessel and gas is introduced into the system via a distributor, which is typically a drilled plate at the bottom of the vessel. A plenum chamber is provided below the distributor plate. The height of the solids level above the distributor is called the bed height, and the vertical space above the bed height is called the freeboard. A splash zone may exist as a transition between the bed and freeboard. Cyclones, located either in the freeboard or external to the vessel, are used to remove solids from the gas stream. Diplegs can return entrained solids directly to the bed.

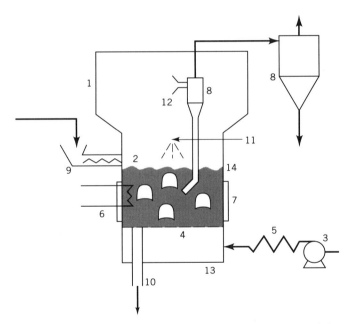

Fig. 2. Multipurpose fluidized bed where 1 represents the shell; 2, solid particles; 3, the blower; 4, the gas distributor; 5, the heat exchanger for fluidizing gas; 6, internal heating or cooling; 7, external heating or cooling; 8, cyclones; 9, the solids feeder; 10, solids offtake; 11, liquid feed; 12, the freeboard; 13, the plenum; and 14, the solids level. Adapted from Reference 1.

History of Fluidization and Examples of Applications

Although earlier references dating back to the nineteenth century exist, industrial fluidized-bed applications began with a large-scale Winkler gasifier in 1926. This was the first application of coarse-powder fluidization. Fluidized-bed catalytic cracking (FCC) of crude oil to gasoline was commercialized in 1942, and as of this writing is still the principal application of fine-powder fluidization (2) (see GASOLINE AND OTHER MOTOR FUELS; PETROLEUM). Several catalytic applications, such as acrylonitrile (qv) synthesis, phthalic anhydride manufacture, and Fischer-Tropsch synthesis of fuels from coal-based gas extended the range of catalytic applications following FCC. Spouted beds and moving beds were developed during this same period. Spouted beds process very large particles and are used, for example, in wheat drying and ore roasting (see MINERAL RECOVERY AND PROCESSING; WHEAT AND OTHER CEREAL GRAINS;).

Three-phase biochemical processes were developed in the 1960s as were bubbling bed combustors. In the 1970s, Lurgi commercialized the circulating fluidized bed (CFB) for coarse powders, which operates above the terminal velocity of all the bed particles. The bed inventory in a CFB is continually entrained out of the vessel, recovered, and recirculated. Polyethylene began to be produced in fluidized beds, and this is a principal application of these beds. The 1980s saw commercialization of circulating bed combustion, and production of polypropylene in fluidized beds. Newer areas of application were the production of semiconductor and ceramic materials via chemical vapor deposition in a fluidized bed, and the use of liquid fluidized beds for biological applications (see CERAMICS; ELECTRONIC MATERIALS; ENZYME APPLICATIONS; SEMICONDUCTORS).

Fluidized-bed applications in the 1990s may be separated into catalytic reactions, noncatalytic reactions, and physical processes. Examples of fluidized-bed applications include the following:

Chemical Catalytic Processes

Catalytic cracking of heavy petroleum fractions (FCC)

Phthalic anhydride

Acrylonitrile

Aniline (hydrogenation of nitrobenzene)

Synthesis of polyethylene and polypropylene

Fischer-Tropsch synthesis

Oxidation of SO_2 to SO_3

Chlorination or bromination of methane, ethylene, etc

Maleic anhydride (from butane)

Pyridine

Chemical Noncatalytic Processes

Roasting of sulfide and sulfate ores (ZnS, pyrites, Cu_2S, $CuCoS_4$, nickel sulfides)

Calcination (limestones, phosphates, aluminum hydroxide)

Incineration of waste liquids and solids refuse

Coking (thermal cracking)

Combustion of coal and other fuels
Gasification of coal, peat, wood wastes
Carbonization of coal (decomposition without oxygen)
Fluoridation of UO_2 pellets
Catalyst regeneration
Hydrogen reduction of ores
Titanium dioxide

Physical Processes

Drying (eg, phosphates, coal, PVC, polypropylene, foods)
Granulation (eg, pharmaceuticals, fertilizers)
Classification
Blending
Coating (eg, polymer coat on metal object)
High temperature baths
Airslide conveying
Absorption (eg, CS_2)
Filtering of aerosols
Medical beds
Quenching, annealing, tempering

Particle Properties

Fluidized-bed design procedures require an understanding of particle properties. The most important properties for fluidization are particle size distribution, particle density, and sphericity.

Particle Size.   The solids in a fluidized bed are never identical in size and follow a particle size distribution. An average particle diameter, d_p, is generally used for design. It is necessary to give relatively more emphasis to the low end of the particle size distribution (fines), which is done by using the surface mean diameter, d_{sv}, to calculate an average particle size:

$$d_{sv} = 1/\Sigma(x_i/d_{pi}) \tag{1}$$

The surface mean diameter is the diameter of a sphere of the same surface area-to-volume ratio as the actual particle, which is usually not a perfect sphere. The surface mean diameter, which is sometimes referred to as the Sauter mean diameter, is the most useful particle size correlation, because hydrodynamic forces in the fluid bed act on the outside surface of the particle. The surface mean diameter is directly obtained from automated laser light diffraction devices, which are commonly used to measure particle sizes from 0.5 to 600 μm. X-ray diffraction is commonly used to measure smaller particles (see SIZE MEASUREMENT OF PARTICLES).

Before the advent of light diffraction devices, screen sieving devices or sedimentation methods were commonly used, resulting in the weight-average particle size, d_{pw}. Screen sieving is still used for particles greater than 500 μm in size. A third particle size indicator is the volume average diameter, d_v, which is defined as the diameter of a sphere having the same volume as the particle. For spherical particles, all three particle diameters are nearly equal.

Particle size distribution is usually plotted on a log-probability scale, which allows for quick evaluation of statistical parameters. Many naturally occurring and synthetic powders follow a normal distribution, which gives a straight line when the log of the diameter is plotted against the percent occurrence. However, bimodal or other nonnormal distributions are also encountered in practice.

Solid Density.   Solids can be characterized by three densities: bulk, skeletal, and particle. Bulk density is a measure of the weight of an assemblage of particles divided by the volume the particles occupy. This measurement includes the voids between the particles and the voids within porous particles. The skeletal, or true solid density, is the density of the solid material if it had zero porosity. Fluid-bed calculations generally use the particle density, ρ_p, which is the weight of a single particle divided by its volume, including the pores. If no value for particle density is available, an approximation of the particle density can be obtained by multiplying the bulk density by two.

Sphericity.   Sphericity, ψ, is a shape factor defined as the ratio of the surface area of a sphere the volume of which is equal to that of the particle, divided by the actual surface area of the particle.

$$\psi = d_{sv}/d_v \qquad (2)$$

Rounded materials such as catalyst and round sand have sphericities on the order of 0.9 or higher. Examples of sphericity of various powders are shown in Table 1.

Angles of Repose and Internal Friction.   The angle of repose is the angle that a pile of solids forms with the horizontal plane. The angle of internal friction is the angle with the horizontal that the flow, no-flow boundary forms when solids are flowing over themselves. This angle is a slight function of the solids flow rate. However, a typical angle of internal friction for a nonsticky material without sharp corners generally exceeds 65°. When designing fluidized-bed internal baffles, the baffles generally are angled at greater than 65° to the horizontal to prevent a zone of stagnant solids forming on top.

Table 1. Properties of Powders

Geldart group[a]	Powder	Average particle size, d_p, μm	Particle density, ρ_p, kg/m³	Angles of		Sphericity, ψ
				Internal friction, deg	Repose, deg	
A	FCC catalyst	60	1400	79	32	0.99
B	sand	500	2000	64	36	0.92
C	ion-exchange resin	30	800	82	29	0.86
D	TCC beads[b]	3000	1000	72	35	1.0

[a]See text.
[b]TCC = Thermofor catalytic cracking.

Terminal Velocity.   The single-particle terminal velocity, U_t, is the gas velocity required to maintain a single particle suspended in an upwardly flowing gas stream. A knowledge of terminal velocity is important in fluidized beds because it relates to how long particles are retained in the system. If the operating superficial gas velocity in the fluidized bed far exceeds the terminal velocity of the bed particles, the particles are quickly removed.

Equations 3 to 7 indicate the method by which terminal velocity may be calculated. From a hydrodynamic force balance that considers gravity, buoyancy, and drag, but neglects interparticle forces, the single particle terminal velocity is

$$U_t = \left[\frac{4g \, d_p(\rho_p - \rho_g)}{3\rho_g \, C_d} \right]^{1/2} \tag{3}$$

Assuming spherical particles, the drag coefficient, C_d, in the laminar, the Stokes flow regime is

$$C_d = \frac{24}{Re_p} \tag{4}$$

where the particle Reynolds number, Re_p, is defined as

$$Re_p = d_p U \rho_g / \mu \tag{5}$$

where μ is the fluid viscosity. The single spherical particle terminal velocity is then

$$U_t = \frac{g(\rho_p - \rho_g)d_p^2}{18 \, \mu} \quad \text{for } Re_p < 0.4 \tag{6}$$

For large particles, C_d is 0.43 and

$$U_t = [3.1 \, (\rho_p - \rho_g) \, gd_p/\rho_g]^{1/2} \quad \text{for } Re_p > 500 \tag{7}$$

This equation indicates that, for small particles, viscosity is the dominant gas property and that for large particles density is more important. Both equations neglect interparticle forces.

The single-particle terminal velocity is only a mathematical limit, because most gas–solids operations operate at a high concentration of solids. Particles interact strongly with each other hydrodynamically, such as by drag reduction owing to shielding and via interparticle forces. The actual slip velocity between particles and gas is then much higher than the single-particle terminal velocity, in many cases tens, or even hundreds of times higher.

In general, the slip velocity, U_{slip}, or the effective terminal velocity for a particle in suspension, U_t^* is

$$U_{\text{slip}} = U_t^* = U_t \cdot f(\epsilon) \tag{8}$$

where $f(\epsilon)$ is a correction for voidage which accounts for interparticle interactions. The voidage, ϵ, of a fluidized bed is the volume fraction occupied by gas. Many empirical proposals for $f(\epsilon)$ have been made. One of the earliest is the Kozeny-Carman approximation:

$$f(\epsilon) = 0.1 \frac{\epsilon^2}{(1 - \epsilon)} \tag{9}$$

Another approximation, one of the most enduring empirical correlations in multiphase systems, is the Richardson-Zaki correlation for a single particle in a suspension [3]:

$$\frac{U}{U_t} = \epsilon^n \tag{10}$$

where n is a function of d_p/D and the Reynolds number and varies from 2.4 to 4.7.

The dimensionless plot of particle and gas properties shown in Figure 3 was constructed [4] to cover the entire particle size range. It shows particle properties

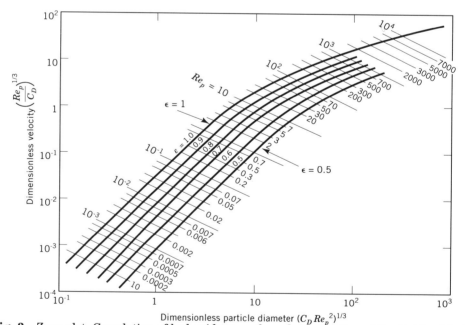

Fig. 3. Zenz plot. Correlation of bed voidage, ϵ, the volume fraction of the fluidized bed that is occupied by gas, for values of ϵ from 1.0 to 0.5, and dimensionless velocity and particle properties, where $C_D = 4 g d_p (\rho_p - \rho_g)/(3 \rho_g U^2)$. The horizontal lines represent the different values of Re_p; $\epsilon = 0.5$ corresponds to the minimum fluidization velocity; $\epsilon = 1$, the terminal velocity. Adapted from Reference 4.

plotted against gas properties. Voidage, ϵ, is a parameter. The single-particle terminal velocity is determined at $\epsilon = 1.0$, and the minimum fluidization velocity is found at the voidage of loosely packed solids, usually 0.4 to 0.5.

Minimum Fluidization Velocity.   There is a minimum superficial gas velocity required to just fluidize a bed of solids. The minimum fluidization velocity can be estimated from the Zenz plot (Fig. 3) assuming the voidage at minimum fluidization is 0.5. Alternatively, it can be estimated via a correlation that gives a result equivalent to the plot using a voidage of 0.4. Using both methods defines the range within which a measured value for minimum fluidization velocity falls. The Wen and Yu equation (5), suitable for solids that have a voidage near 0.4, is

$$U_{\text{mf}} = \mu[(1135.7 + 0.0408\,Ar)^{0.5} - 33.7]/(\rho_g d_p) \tag{11}$$

where the Archimedes number, Ar, is defined as

$$Ar = \rho_g d_p^3 (\rho_p - \rho_g) g/\mu^2 \tag{12}$$

Particle Regimes.   In 1973, particles were classified with respect to how they fluidize in air at ambient conditions into Geldart groups (6) (Fig. 4). Particles that formed bubbles immediately after the gas superficial velocity exceeded U_{mf} were designated as Group B particles. For these particles, the gas velocity at which bubbles first appear in the bed is $U_{\text{mb}} = U_{\text{mf}}$. Group B particles typically have an average particle size from 100 to 700 μm. Dense particles, eg, glass, sand, and ore, are likely to be in Group B. Bubbles in a bed of Group B particles can grow to sizes on the order of a meter in tall, large-diameter beds fluidized at relatively low gas velocities.

Group A particles are smaller or lighter than Group B particles. Most manufactured fluidized-bed catalysts are in the Group A category. Particle sizes range from about 10 to 130 μm. These particles are cohesive owing to interparticle forces and when the gas velocity is increased beyond U_{mf}, the bed continues to expand smoothly without forming bubbles. A velocity is eventually reached when bubbles

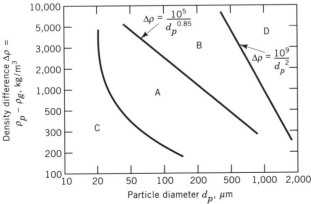

Fig. 4.   Geldart group particle classification diagram for air at ambient conditions (6). Group A consists of fine particles; B, coarse particles; C, cohesive, very fine particles; and D, moving and spouted beds.

start to form. Because U_{mb} is greater than U_{mf}, these particles were designated as Group A for aeratable. The ability to hold aeration enables Group A particles to flow well in transfer pipes in and out of fluidized beds. The small average particle size and the presence of very fine particles promote bubble splitting, and a small maximum stable bubble size can be shown to exist for Group A particles (7).

The maximum bubble size for Group A powders is of great significance for design. The single most important parameter controlling bubble size is particle size distribution, and in particular the <44-μm fines fraction. About 25% fines are optimal for minimal bubble size, and hence for best conversion and highest heat transfer. Industrial processes are normally operated under conditions which suppress bubble formation and growth. The effect of particle size distribution on bubble size for Group A is shown in Figure 5**a**. Increasing the fines fraction decreases bubble size until the bubbles are smaller than 25 mm in diameter.

Classical bubbles do not exist in the vigorously bubbling, or turbulent fluidization regimes. Rather, bubbles coalesce constantly, and the bed can be treated as a pseudohomogenous reactor. Small bubble size improves heat transfer and conversion, as shown in Figure 5**b**. Increasing fines levels beyond 30–40% tends to lower heat transfer and conversion as the powder moves into Group C.

Group C particles, smaller and lighter than Group A particles, are designated Group C because they are cohesive. They are usually less than 30 μm in average particle diameter. The large external surface area and low mass of these particles produce large attractive forces. The particles do not flow well in pipes and are difficult to fluidize. Thus gas flows through the bed in channels called ratholes. When fluid-bed measurements are performed on Group C systems at low gas velocities, a low pressure drop is observed, ie, the gas is flowing in a channel without encountering most of the particles. Often, Group C particles can be fluidized by using a high gas velocity to overcome the cohesive forces between the particles, or using fluidization aids such as larger particles, fibrous carbon particles, etc.

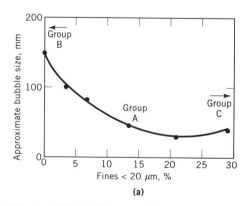
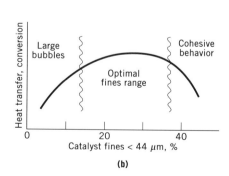

Fig. 5. Effect of fines particle size on (**a**) bubble size for FCC catalyst, of $\rho_p = 1250$ kg/m^3, d_{sv} decreases with increasing fines content, $U = 0.1$ m/s and $D = 0.15$ m (8), and (**b**) on heat transfer and conversion (9,10).

Group D particles are large, on the order of 1 or more millimeters (1000 μm) in average particle size. In a fluidized bed, they behave similarly to Group B particles. Because of the high gas velocities required to fluidize Group D particles, it is often more economical to process these particles in spouted or in moving beds, where lower gas rates suffice.

Whereas Geldart's classification relates fluidized-bed behavior to the average particle size in a bed, particle feed sizes may be quite different. For example, in fluidized-bed coal (qv) combustion, large coal particles are fed to a bed made up mostly of smaller limestone particles (see COAL CONVERSION PROCESSES).

Interparticle Forces. Interparticle forces are often neglected in the fluidization literature, although in many cases these forces are stronger than the hydrodynamic ones used in most correlations. The most common interparticle forces encountered in gas fluidized beds are van der Waals, electrostatic, and capillary.

Capillary forces, caused by liquid bridges between the particles, can predominate in applications such as drying, flow out of bins, etc. Electrostatic forces are repelling forces caused by exchange of electrons between two nonconducting materials, such as sand and a plastic wall. Electrostatic forces are especially significant in cold-flow models and when dry fluidizing gas is used. Capillary forces and electrostatic forces can act on all sizes of particles, large or small.

van der Waals' forces operate between molecules at a very short distance, and hence are significant only for the small, ie, Group C and A, particles. However, these forces can be as important as hydrodynamic forces in many fluidization applications of practical interest. Unlike bubbles, interparticle forces have received little attention, and are usually not included in the basic equations describing fluid bed flow. A noted exception may be found in Reference 11.

The effect of interparticle forces on the entrainment of Group A powders has been studied (12). When a bed is fluidized at a superficial gas velocity equal to the terminal velocity of the average particle, it takes many hours to entrain the bed because of the interparticle forces holding the particles within. As the gas velocity reaches the terminal velocity of the largest 1% of the particles, it can still take nearly 20 minutes to empty the bed. These thermodynamic interparticle forces combine with hydrodynamic drag reduction forces so that Group A particles behave as larger clusters of particles. These clusters vary in size from the single particle at low solids concentration to as high as 25 mm for a dense bed (13).

At higher gas velocities, appreciable entrainment of solids starts, but still large clumps of particles fall back along the walls (small tubes) or along the walls and elsewhere (large tubes). This results from clustering of fine powders even in the dilute concentration in the freeboard which is caused by interparticle forces and by drag reduction owing to shielding effects (see POWDERS, HANDLING).

Fluidization Regimes

The different fluidized-bed regimes are a function of gas velocity. At a low gas velocity, the solids are in a packed-bed or fixed-bed state. As the gas velocity is increased, the drag and buoyancy forces eventually overcome the weight of the particles and interparticle forces, and the particles are completely supported by the gas. This is the particulate regime. At minimum fluidization, particles display minimal motion, and the bed is slightly expanded. The bubbling regime, the beginning of aggregative fluidization, occurs when the gas velocity is increased. Bubbling occurs immediately after minimum fluidization for Group B particles, but there is a gas-velocity range of bubble-free expansion for Group A particles which results from interparticle forces. Poor contacting of gas with solids is a potential problem in the bubbling regime unless bubble size is kept small.

For small columns, growing bubbles can expand to more than half the vessel diameter. When bubbles reach this size, the bed is said to be in the slug flow regime. Bubble size is then limited by the column diameter. Scale-up problems can occur when the ratio of bubble size to bed diameter in pilot-scale vessels is greatly exceeded by that ratio in a commercial unit. However, bubble size is kept small in most commercial units by proper choice of particle size distribution and high gas velocities. Gas velocities in the turbulent fluidization regime are high enough for the gas and solids to occupy similar volumes, so that there are no distinct bubbles, just random gas voids. For Group A particles the gas velocity in this regime is above the single-particle terminal velocity of all the particles, and turbulent beds require a system to return the large volumes of entrained solids back to the bed.

To escape aggregative fluidization and move to a circulating bed, the gas velocity is increased further. The fast-fluidization regime is reached where the solids occupy only 5 to 20% of the bed volume. Gas velocities can easily be 100 times the terminal velocity of the bed particles. Increasing the gas velocity further results in a system so dilute that pneumatic conveying (qv), or dilute-phase transport, occurs. In this regime there is no actual bed in the column.

Figure 6 shows how three reactors might appear when operating in the three most common commercial fluidization regimes (12). The vigorously bubbling bed unit (shown with large bubbles, as with Group B particles, or Group A particles which are devoid of <44 μm fines) has a dense bed containing a solids volume

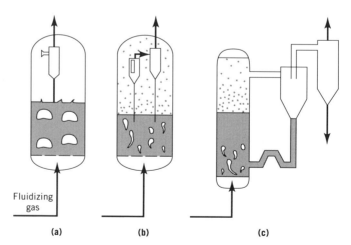

Fig. 6. Schematics of commercially used beds, where the shaded area represents the solids: (**a**) vigorously bubbling, (**b**) turbulent, and (**c**) fast fluidized. Adapted from Reference 12.

fraction typically ranging from 0.5 to 0.35. There is a distinct bed level, and a relatively small amount of solids are entrained with the gas as it leaves the bed. A cyclone is generally used to separate the entrained solids from the gas stream, and this may be either internal, as illustrated in Figure 6**a**, or external to the reactor vessel.

The turbulent fluidized bed has a similar or slightly lower solids volume fraction than the vigorously bubbling bed. There is considerable transport of solids out of the turbulent bed and the bed level is not very distinct. Large-scale cyclones are needed to return solids to the bed. On average, the bed inventory passes through the cyclones several times per hour.

In the fast-fluidized bed, a distinct bed level does not exist, but there is a long transition zone from the dense turbulent section at the bottom to a dilute regime at the top of the reactor. Large cyclones are needed to collect the large amounts of entrained solids. Fast-fluidized beds are more properly called circulating fluidized beds (CFBs), because the solids are constantly recirculated at high rates around the system. Fast-fluidized beds have the advantages of high throughputs, good mixing, and good gas–solids contact. On the other hand, they are generally more expensive to construct and operate than vigorously bubbling or turbulent fluidized beds. Typical axial density profiles for these fluidization regimes are illustrated in Figure 7.

Many phase diagrams have been proposed to describe fluidization regimes. One such diagram for FCC catalyst, which is the quintessential Group A powder, is shown in Figure 8 (10). Solids volume fraction decreases with increasing gas velocity in the bubbling and turbulent regimes. The density of the dilute transport regime above the turbulent bed is shown via a tie-line B–b. A similar line shows the approximate decrease in axial density between the bottom and the top of a fast bed (C–c) and a more dilute, transported bed, as in FCC riser flow (D–d). As can be seen from the schematic diagram of an FCC riser and regenerator system, all of these regimes can exist simultaneously in a complex circulating fluidized-bed system.

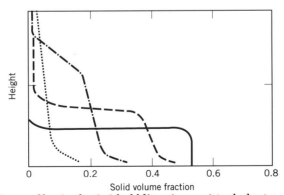

Fig. 7. Axial density profiles in the (—) bubbling, (— — —) turbulent, and (—·—) fast and (·····) riser circulating fluidization regimes. Typical gas velocities for fine particles are 0.1 and 0.5 m/s for the bubbling and turbulent regimes, respectively, and 3 and 10 m/s for the fast and riser circulating beds, respectively.

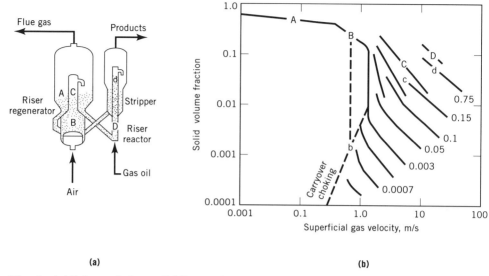

(a) **(b)**

Fig. 8. (**a**) Schematic for an FCC unit showing where the various fluidization regimes are found and (**b**) a corresponding phase diagram for Group A powder (FCC catalyst) where the numbers on the curves represent the superficial solid velocity in m/s. A represents the bubbling regime; B, the turbulent; C, the fast; and D, the riser flow. The lowercase b, c, and d represent solids concentrations near the top of the bed.

A more general phase diagram, based on the dimensionless groups of Figure 3, is shown in Figure 9 (14). In addition to showing the minimum fluidization velocities and the single-particle terminal velocity, regimes of commercial operation are shown for conventional fluidized, ie, vigorously bubbling and turbulent, beds; circulating beds; transport reactors; spouted beds; and moving packed beds. The transition to turbulent fluidization occurs at lower gas velocities for commercial-scale beds than for small-scale beds.

Pressure Drop. The pressure drop across a two-phase suspension is composed of various terms, such as static head, acceleration, and friction losses for both gas and solids. For most dense fluid-bed applications, outside of entrance or exit regimes where the acceleration pressure drop is appreciable, the pressure drop simply results from the static head of solids. Therefore, the weight of solids in the bed divided by the height of solids gives the apparent density of the fluidized bed, ie

$$\Delta P/L = \text{suspension density} = \rho_p(1 - \epsilon)\left(\frac{g}{g_c}\right) \qquad (13)$$

where ϵ is the voidage, the volume fraction of the bed occupied by interparticle gas. The measurement of pressure drop across the bed is the most common and useful diagnostic technique employed for control of fluidized beds.

Effects of Temperature and Pressure on Minimum Fluidization Velocity. Many basic fluid-bed properties are affected by temperature and pressure. Pres-

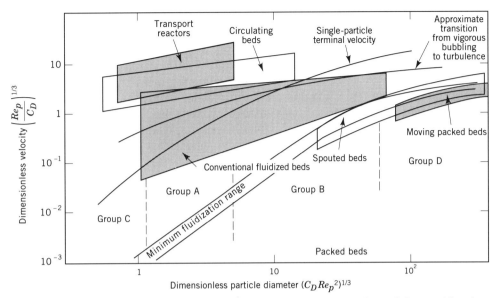

Fig. 9. Nondimensional phase diagram showing Group A, B, C, and D particles (see Fig. 3) (14).

sure has little effect on the minimum fluidization velocity, U_{mf}, of fine Group A particles because frictional resistance is mostly a function of viscosity and inter-particle forces. However, the larger Group B and D particles show a decrease in minimum fluidization velocity with increasing pressure. The interparticle forces in these groups are usually negligible. The Wen and Yu equation (eq. 11) and the Zenz plot (Fig. 3) both predict this trend. Increasing temperature increases viscosity and, therefore, reduces U_{mf} for Group A and most Group B particles. U_{mf} for larger particles is relatively insensitive to temperature, as the effects of increasing viscosity and decreasing density tend to cancel. However, for particle sizes of about 2500 μm and larger, U_{mf} increases with increasing temperature.

Bubbles and Fluidized Beds. Bubbles, or gas voids, exist in most fluidized beds and their role can be important because of the impact on the rate of exchange of mass or energy between the gas and solids in the bed. Bubbles are formed in fluidized beds from the inherent instability of two-phase systems. They are formed for Group A powders when the gas velocity is sufficient to start breaking inter-particle forces at U_{mb}. For Group B powders, where interparticle forces are usually negligible, $U_{mb} = U_{mf}$ and bubbles form immediately upon fluidization. Bubbles, which are inherently undesirable, can grow to a large size and cause contact inefficiencies brought on by significant gas bypassing.

Bubble size control is achieved by controlling particle size distribution or by increasing gas velocity. The data as to whether internal baffles also lower bubble size are contradictory. (Internals are commonly used in fluidized beds for heat exchange, control of solids backmixing, and other purposes.) In some cases it seems that internals can increase bubble size.

Control of bubble size via particle-size control is easy using Group A powders. A typical Group A powder, having an average particle size of 70 μm and a normal particle size distribution, has nearly 25% of the particles as fines, ie, below 44 μm. If these particles are spherical and if interparticle forces are absent, bubble size is typically smaller than 25 mm (see Fig. 5a). Well-defined bubbles are not present in the typical industrial turbulent fluidized-bed reactor. The turbulent fluid bed can be viewed as a single-phase system for most practical design considerations. Bubble size control is more difficult for Group B powders, especially in applications where there is little control of particle properties, eg, in processing of naturally occurring materials. In some cases, increasing the gas velocity and moving into turbulent or fast-fluidization regimes has been the preferred solution.

Two-Phase Theory. According to the bubble two-phase theory of fluidization, all gas above that required for minimum fluidization passes through the bed in the form of bubbles. This is nearly true for Group B particles, but is only an approximation for Group A particles fluidized at low velocities. Because fluid beds containing Group A solids are usually operated at high multiples of U_{mf}, this two-phase theory has little application in the instance of Group A particles.

Bubbles rise through the bed in two different regimes. Slow bubbles rise at a gas velocity less than V_{mf} (equal to U_{mf}/ϵ_{mf}), and present an opportunity for gas to bypass the bed material and short-circuit through the bubble on the way to the bed surface. Most commercial bubbling fluidized-bed applications have fast-moving bubbles of small size which rise at a gas velocity greater than V_{mf}.

A single bubble rises through a fluid bed at a velocity, U_{br}, proportional to the square root of its diameter, D_b, or more accurately, the diameter of a sphere of equivalent volume:

$$U_{br} = 0.71(gD_b)^{0.5} \tag{14}$$

If the bed is slugging, bubble motion is retarded by the bed wall, and the bed or tube diameter, D, rather than the actual bubble diameter, determines the bubble rise velocity, ie

$$U_{br} = U_{slug} = 0.35(gD)^{0.5} \tag{15}$$

The velocity of a bubble in a bubbling bed has been observed to be higher than equation 14 predicts, and it has been suggested that the actual bubble rise velocity in a bubbling bed (15) is

$$U_b = (U - U_{mf}) + U_{br} \tag{16}$$

This empirical equation attempts to account for complex bubble coalescence, splitting, irregular shapes, etc. Apparent bubble rise velocity in vigorously bubbling beds of Group A particles is lower than equation 16 predicts.

As bubbles rise through the bed, they coalesce into larger bubbles. The actual bubble size at any height above the distributor, H_{ad}, in the bed is a function of the initial bubble size as it emerges from the gas distributor and the gas flow rate (16):

$$D_b = 0.54(U - U_{mf})^{0.4}[H_{ad} + 4(A_d/N)^{0.5}]^{0.8} \tag{17}$$

where A_d/N is the cross-sectional area of the distributor per distributor hole. For a porous distributor, the distributor area per hole is assumed to be equal to zero. This correlation does not predict the effect of particle-size distribution on bubble diameter.

Bubbles can grow to on the order of a meter in diameter in Group B powders in large beds. The maximum stable bubble size is limited by the size of the vessel or the stability of the bubble itself. In large fluidized beds, the limit to bubble growth occurs when the roof of the bubble becomes unstable and the bubble splits. Empirically, it has been found that the maximum stable bubble size (D_{bmax}) may be calculated for Group A particles from

$$D_{bmax} = 2U_t^2/g \tag{18}$$

where U_t is the terminal velocity of a particle 2.7 times the average particle size in the bed. Again, this correlation is only approximate because it contains no information on the most significant factor affecting bubble size, namely particle-size distribution.

Bed Expansion and Bed Density. Bed density can readily be determined for an operating unit by measuring the pressure differential between two elevations within the bed. This is a highly useful measurement for control and monitoring purposes.

Because bubbles occupy space in a bubbling fluid bed, the expansion of the bed becomes a function of both the bubble velocity and the volume of the gas entering the bed:

$$(H_{max} - H_{mf})/H_{mf} = (U - U_{mf})\frac{(U - U_{mf})}{U_{br}} \tag{19}$$

where U_{br} is the single-bubble rise velocity, H_{max} is the operating level of the bed, and H_{mf} is the bed height at minimum fluidization. Extending this volume balance to predict bed densities for bubbling beds gives

$$\rho_{bed} = \rho_{bulk}U_{br}/U_b \tag{20}$$

For turbulent and fast-fluidized beds, bubbles are not present as distinct entities. The following expression for bed voidage, ϵ_{bed}, the volume fraction of the bed occupied by gas, where U is in m/s, has been suggested (17):

$$\epsilon_{bed} = (U + 1)/(U + 2) \tag{21}$$

Bed density can then be predicted from $\rho_{bed} = \rho_p (1 - \epsilon_{bed})$. The expression for ρ_{bed} is applicable for turbulent fluidized beds and for the turbulent bottom of fast-fluidized beds fluidized at gas velocities up to 5 m/s.

A more general expression for the expansion of two-phase systems was given in equation 10. It holds for a surprisingly wide range of systems at low and high velocities. This expression can also be used to describe the expansion of gas–solids systems if interparticle forces are negligible. When interparticle forces cannot be neglected, as for Group C and A particles, U_t and n have been shown to depend on the stable cluster size at a given gas velocity (18). The effective U_t^* is then much larger than the single-particle terminal velocity. Bed expansion of a typical Group A powder (FCC catalyst) is shown in Figure 10.

Solids and Gas Mixing.   Solids in an unrestricted fluidized bed can be almost completely backmixed, giving the bed uniform solids properties and a constant temperature throughout. The engine driving the solids mixing and circulation is the drag exerted by the gas on the particles. Mixing increases with bed diameter if the bed is not restricted with internals. This increase is linear at first but slows down as the turbulent eddy size is restricted by bed diameter. Beyond a diameter of approximately 2 m, the eddies no longer grow to the bed diameter, and axial solids (and gas) mixing starts tapering off (19). At small bed diameters the ratio of effective axial dispersion to bed diameter is 1:1; for large diameter beds, this ratio drops to 1:10.

Mass Transfer.   Mass transfer in a fluidized bed can occur in several ways. Bed-to-surface mass transfer is important in plating applications. Transfer from the solid surface to the gas phase is important in drying, sublimation, and desorption processes. Mass transfer can be the limiting step in a chemical reaction system. In most instances, gas from bubbles, gas voids, or the conveying gas reacts with a solid reactant or catalyst. In catalytic systems, the surface area of a catalyst can be enormous. For Group A particles, surface areas of 5 to over 1000 m^2/g are possible.

Because particles close to each other are shielded from free gas flow around them, the mass-transfer coefficient for a fluidized-bed system is always less than that for an individual particle in a freely flowing gas. Also, gas entering the bed in a bubble must leave the bubble to react. Interchange with the solid-rich phase occurs mostly by bubble breakage, and more slowly owing to molecular diffusion and gas recirculation from the bubble to the dense phase. The high degree of gas–void splitting in turbulent beds results in a low gas–solids mass-transfer resistance. The section of the fluidized bed near the distributor also shows enhanced mass transfer.

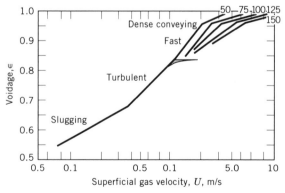

Fig. 10.   Expansion curve for FCC catalyst in a 0.15-m inner diameter column showing the fluidization regimes where the numbers on the lines correspond to the solids rate in kg/(m²·s) (18).

Correlating of mass-transfer data in industrial fluidized beds is most successful when neglecting bubble properties. An example is using an analogy to a staged process where the mass-transfer resistance is related to the height of a diffusion stage. The overall resistance to mass transfer is then the number of diffusion stages or the bed height divided by the individual stage height, ie

$$N_\alpha = H/H_\alpha \tag{22}$$

where N_α is the number and H_α the height of a diffusion stage. For Group A powders in pilot- and commercial-sized units, the height of a stage (20) is

$$H_\alpha = (1.8 - 1.06/D^{0.25})(3.5 - 2.5/H^{0.25}) \tag{23}$$

This equation predicts that the height of a theoretical diffusion stage increases, ie, mass-transfer resistance increases, both with bed height and bed diameter. The diffusion resistance for Group B particles where the maximum stable bubble size and the bed height are critical parameters may also be calculated (21).

Heat Transfer. One of the reasons fluidized beds have wide application is the excellent heat-transfer characteristics. Particles entering a fluidized bed rapidly reach the bed temperature, and particles within the bed are isothermal in almost all commercial situations. Gas entering the bed reaches the bed temperature quickly. In addition, heat transfer to surfaces for heating and cooling is excellent.

Gas-to-Particle Heat Transfer. Heat transfer between gas and particles is rapid because of the enormous particle surface area available. A Group A particle in a fluidized bed can be considered to have a uniform internal temperature. For Group B particles, particle temperature gradients occur in processes where rapid heat transfer occurs, such as in coal combustion.

In a quiescent fluid, the dimensionless mass-transfer coefficient, or the Nusselt number, $h_{gs} d_p/k_f$, for a sphere is two. In fluidized beds the Nusselt number is much less than two for the same reasons given for mass transfer, ie, the particles shield each other from the gas flow.

Bed-to-Surface Heat Transfer. Bed-to-surface heat-transfer coefficients in fluidized beds are high. In a fast-fluidized bed combustor containing mostly Group B limestone particles, the dense bed-to-boiling water heat-transfer coefficient is on the order of 250 W/(m²·K). For an FCC catalyst cooler (Group A particles), this heat-transfer coefficient is around 600 W/(m²·K).

The heat-transfer coefficient of most interest is that between the bed and a wall or tube. This heat-transfer coefficient, h, is made up of three components. To obtain the overall dense bed-to-boiling water heat-transfer coefficient, the additional resistances of the tube wall and inside-tube-wall-to-boiling-water must be added. Generally, the conductive heat transfer from particles to the surface, h_{cond}; the convective heat transfer from interstitial gas to the surface, h_{conv}; and the radiative heat transfer, h_{rad}, to the surface are added to give

$$h = h_{cond} + h_{conv} + h_{rad} \tag{24}$$

Radiative heat transfer is negligible if the bed or the heat-transfer surface is below 600°C.

The heat-transfer coefficient depends on particle size distribution, bed void-age, tube size, etc. Thus a universal correlation to predict heat-transfer coefficients is not available. However, the correlation of Andeen and Glicksman (22) is adequate for approximate predictions:

$$h = (900k_f(1 - \epsilon)/D)(GD\rho_s/[\rho_f\mu])(\mu^2/(d_p^3\rho_s^2 g))^{0.326}Pr^{0.3} \tag{25}$$

where Pr is the Prandtl number, $C_p\mu/k_f$.

Fundamental models correctly predict that for Group A particles, the conductive heat transfer is much greater than the convective heat transfer. For Group B and D particles, the gas convective heat transfer predominates as the particle surface area decreases. Figure 11 demonstrates how heat transfer varies with pressure and velocity for the different types of particles (23). As superficial velocity increases, there is a sudden jump in the heat-transfer coefficient as gas velocity exceeds U_{mf} and the bed becomes fluidized.

For Group A solids (Fig. 11**a**) a heat-transfer coefficient of about 900 W/(m²·K) is typical for beds at a fluidized density of 720 kg/m³ operating in the vigorously bubbling or turbulent fluidization regimes. This coefficient decreases if fines are absent, and decreases as bed density decreases. For Group B solids (Fig. 11**b**) an overall heat-transfer coefficient of 300 W/(m²·K) is typical for beds at a voidage of 0.5 (bubbling fluidized bed). The overall heat-transfer coefficient from the bed to heat-transfer fluid, such as boiling water, inside the tubes is about half that value.

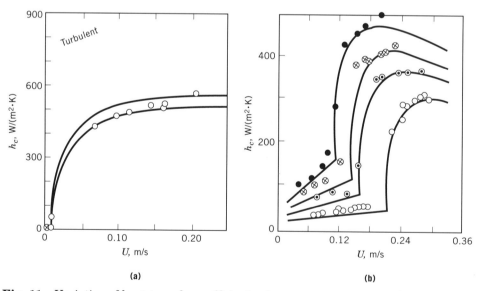

(a)   **(b)**

Fig. 11. Variation of heat-transfer coefficient, where ○ represents experimental results at 100 kPa; ⊙, 500 kPa; ⊗, 1000 kPa; and ●, 2000 kPa, of pressure (23) for (**a**) a 0.061-mm glass–CO₂ system (Group A particles) and (**b**) a 0.475-mm glass–N₂ system (Group B and D particles). To convert kPa to psi, multiply by 0.145.

Distributor Design

Good gas distribution is necessary for the bed to operate properly, and this requires that the pressure drop over the distributor be sufficient to prevent maldistribution arising from pressure fluctuations in the bed. Because gas issues from the distributor at a high velocity, care must also be taken to minimize particle attrition. Many distributor designs are used in fluidized beds. The most common ones are perforated plates, plates with caps, and pipe distributors.

Perforated Plates. A perforated plate can be flat, concave, convex, or double-dished. The main advantages of the perforated plate are that it is simple, inexpensive, easy to modify, and easy to clean. The disadvantages of a perforated plate are the possiblity of solids leaking, ie, weeping through it into the plenum; lower turndown capability than other distributors; the requirement of a peripheral seal; and a relatively high pressure drop required for good distribution.

Plates with Caps. Several cap-type distributors are shown in Figure 12. These minimize weeping and have good turndown, but are difficult to clean and modify, and are more expensive than perforated plates. A peripheral seal is also required as for a perforated plate.

Pipe Distributors. Figure 13 shows two pipe distributors, one in a branched and one in a ring configuration. These distributors minimize weeping, have good turndown, may require the lowest pressure drop, and avoid the need for a plenum chamber. They are also well suited to multiple-level fluid injection. The disadvantages of these distributors are that there are defluidized solids beneath the distributor and the mechanical design is more complex.

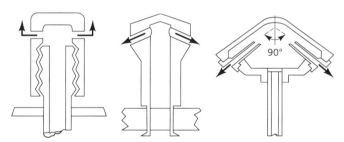

Fig. 12. Cap-type gas distributors where the fluid is directed laterally.

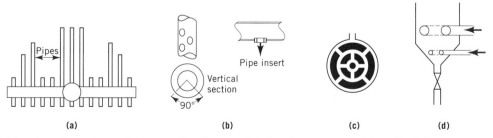

Fig. 13. Examples of pipe gas distributors: (**a**) simple sparger, (**b**) details of the pipe, (**c**) wagon wheel, and (**d**) multilevel distributor.

Design Considerations.   For a perforated plate, the pressure drop across the distributor should be at least 30% of the bed pressure drop when operating at the lowest expected gas velocity. The number of holes in the distributor should exceed 10 per square meter. The pressure drop, ΔP, across the distributor is given by

$$\Delta P = \frac{\rho_g}{2} \left(\frac{U_{\text{or}}}{C_d} \right)^2 \tag{26}$$

where the drag coefficient C_d is usually 0.8, and gas conditions are taken as they exist in the hole and plenum. The space between the holes L_h for a triangular pitch is given as

$$L_h = 1/(N_{\text{or}} \sin 60°)^{0.5} \tag{27}$$

where N_{or} is the hole density (holes per square meter).

In pipe distributors, the pressure drop required for good gas distribution is 30% of the bed pressure drop for upward facing holes, but only 10% for downward facing ones. The pressure drop calculation and the recommended hole density are the same as for a perforated plate. To maintain good gas distribution within the header system, it is recommended the relation

$$[D_{\text{header}}^2/(ND_{\text{or}}^2)]^2 > 5 \tag{28}$$

be maintained for good gas distribution, where D_{header} is the pipe header inner diameter and N is the total number of holes in the pipe. The velocity in each header pipe should not exceed 25 m/s, and the holes should be located at least one pipe diameter from a tee or a sharp bend.

Jet Penetration.   At the high gas velocities used in commercial practice, there are jets of gas issuing from distributor holes. It is essential that jets not impinge on any internals, otherwise the internals may be quickly eroded. Figure 14 is a graphical correlation used to determine the jet penetration length as a function of gas velocity and gas density. Jets from horizontal and downflow holes are considerably shorter than those that are pointed upward.

Particle Attrition.   Distributor jets are a potential source of particle attrition. Particles are swept into the jet, accelerated to a high velocity, and smash into other particles as they leave. To reduce attrition at distributors, a shroud or larger-diameter pipe is often added concentric to the jet hole, as shown in Figure 15. The required length of the concentric shroud is given by the relation

$$L_{\text{shroud}} > \cot 5.5° \, (D_{\text{noz}} - D_{\text{or}})/2 \tag{29}$$

This shroud length allows the jet issuing from the orifice to expand and fill the shroud. The gas velocity leaving the shroud should not exceed 70 m/s, to minimize attrition.

There are many other potential sources of attrition in a fluidized bed other than high velocity impacts at the distributor. Cyclone inlets and elbows in con-

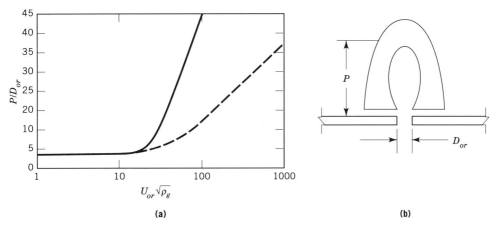

(a)   **(b)**

Fig. 14. (**a**) Correlation of jet penetration, P, from distributors into fluidized beds where (——) represents upwardly directed jets and (— — —) downwardly and horizontally directed jets, and U_{or} = throat velocity at point of entry into bed in m/s, and ρ_g = density of gas at nozzle throat in, kg/m³. (**b**) Schematic defining jet penetration, P, and nozzle or grid hole diameter, D_{or}.

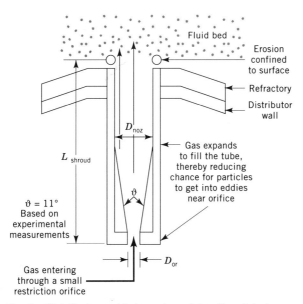

Fig. 15. Schematic of a distributor grid shroud used to allow jets to expand and enter the fluid bed at lower velocity.

veying lines are sources of attrition. Additionally, changes in temperature and chemical reactions can also cause the particles to attrit. A particle's resistance to attrition can be ranked according to its hardgrove grindability index (HGI), a standard ASTM test (24) originally developed for coal grinding. Special test procedures in ball mills and jet attritors are also common (25).

Bed Internals.   Various types of internals that may be found in commercial fluidized beds include solids and gas distributors; cyclones and cyclone diplegs; solids return and withdrawal lines; heat-transfer tubes; supports, hangers, and guides for heat-transfer tubes; baffles; secondary gas-injection nozzles; and pressure, temperature, and sample probes. Fluidized beds may be completely empty, having only instrumentation tubes penetrating through the vessel walls, or in some cases, they may have significant internals. Heat exchangers can fill a vessel completely and reduce the space between tubes to the minimum required for solids circulation and good maintenance practices.

Entrainment

Entrainment, or elutriation, is the carryover of particles from a fluidized bed with the exiting gas. When the gas velocity exceeds the terminal velocity of a Group B particle, the particle is usually removed from the bed. For Group A and C powders, the gas drag needs to overcome interparticle forces as well, and gas velocities of many times the single-particle velocity are needed to entrain particles from the bed at high rates. Knowledge of the entrainment rate is important in order to estimate cyclone inlet loading, solids loss rates, and to predict bed particle size changes resulting from the selective loss of fines.

Transport Disengaging Height.   When the drag and buoyancy forces exerted by the gas on a particle exceed the gravitational and interparticle forces at the surface of the bed, particles are thrown into the freeboard. The ejected particles can be coarser and more numerous than the saturation carrying capacity of the gas, and some coarse particles and clusters of fines particles fall back into the bed. Some particles also collect near the wall and fall back into the fluidized bed.

The height above the bed at which entrainment becomes essentially constant with height is termed the transport disengaging height (TDH). This is a somewhat arbitrary definition, but it is useful for design. It is desirable to locate cyclones and vessel outlets above TDH so as to minimize solids loading to the cyclones. A schematic drawing illustrating the concept of the TDH is shown in Figure 16. An empirical correlation (4) for estimating the TDH is given in Figure 17 as a function of gas velocity and column size. The estimates are good for TDH at low pressures. However, TDH increases with system pressure (26) and the correlation underestimates TDH at high pressures.

Entrainment Above TDH.   A relatively simple procedure to calculate the amount of entrainment above the TDH is given herein. A more detailed treatment can be found in Reference 21.

Fine particles in a fluidized bed are analogous to volatile molecules in a boiling solution. Therefore, the concentration of particles in the gas above a fluidized bed is a function of the saturation capacity of the gas. To calculate the entrainment rate, it is first necessary to determine what particle sizes in the bed can be entrained. These particles are the ones which have a terminal velocity less than the superficial gas velocity, assuming that interparticle forces in a dilute zone of the freeboard are negligible. An average particle size of the entrainable particles is then calculated. If all particles in the bed are entrainable, the entrained material has the same size distribution as the bed material.

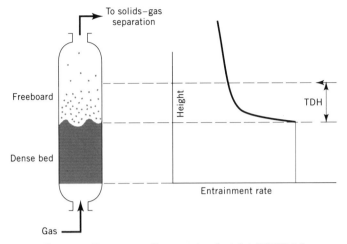

Fig. 16. Transport disengaging height (TDH) (4).

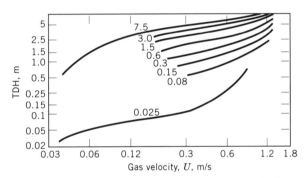

Fig. 17. TDH above vigorously bubbling or turbulent fluidized beds as a function of bed diameter from 0.025 to 7.5 m (27).

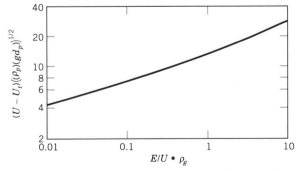

Fig. 18. Correlation of entrainment rate and composition (27).

Figure 18 is an entrainment or gas-carrying capacity chart (25). The operating conditions and particle properties determine the vertical axis; the entrainment is read off the dimensionless horizontal axis. For entrainment purposes, the particle density effect is considered through the ratio of the particle density to the density of water. When the entrainable particle-size distribution is smaller than the particle-size distribution of the bed, the entrainment is reduced by the fraction entrainable, ie, the calculated entrainment rate from Figure 18 is multiplied by the weight fraction entrainable.

In practice, the entrained material is enriched in fines even when the entire bed is entrainable. However, as the gas velocity is increased to many multiples of the terminal velocity, the composition of the entrainable material approaches the bed composition.

Normally vessels are designed with the gas outlet location well above TDH. If circumstances force operation with a bed height so that the outlet is below TDH, an equivalent velocity, an effective velocity higher than the actual superficial gas velocity, is used in the above calculation. The effective gas velocity can be determined from Figure 19 (27).

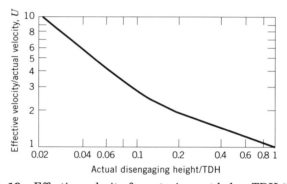

Fig. 19. Effective velocity for entrainment below TDH (27).

Cyclones.   Cyclones are an integral part of most fluidized-bed systems. A cyclone is an inexpensive device having no moving parts that separates solids and gases using centrifugal force. A basic cyclone is shown in Figure 20. The solid–gas mixture enters the cyclone tangentially through a rectangular duct. The gas velocity and the curvature of the cyclone body combine to subject the particles to a centrifugal force up to 100 times gravity. The particles are forced to the wall, spiral to the bottom, and the solids-free gas exits through the center outlet tube at the top of the cyclone. Cyclones are used commercially to remove solids with high efficiency down to about 15 μm. Other aspects of cyclone design may be found in Reference 27 (see also AIR POLLUTION CONTROL METHODS).

Circulating Fluidized Beds

Circulating fluidized beds (CFBs) are high velocity fluidized beds operating well above the terminal velocity of all the particles or clusters of particles. A very large

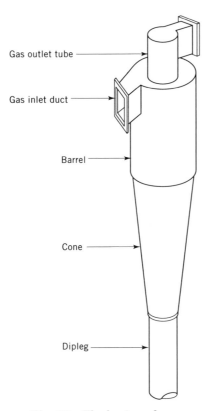

Gas outlet tube

Gas inlet duct

Barrel

Cone

Dipleg

Fig. 20. The basic cyclone.

cyclone and seal leg return system are needed to recycle solids in order to maintain a bed inventory. There is a gradual transition from turbulent fluidization to a truly circulating, or fast-fluidized bed, as the gas velocity is increased (Fig. 6), and the exact transition point is rather arbitrary. The solids are returned to the bed through a conduit called a standpipe. The return of the solids can be controlled by either a mechanical or a nonmechanical valve.

The bed level is not well defined in a circulating fluidized bed, and bed density usually declines with height. Axial density profiles for different CFB operating regimes show that the vessel does not necessarily contain clearly defined bed and freeboard regimes. The solids may occupy only between 5 and 20% of the total bed volume.

Pressure Balance and Standpipes. The pressure balance around the loop of a circulating fluidized bed is illustrated in Figure 21. The driving force of a high gas velocity entering the bed expands the bed volume and carries solids out of the bed. The pressure drop is lower than in a dense fluidized-bed system, because the solids occupy a smaller fraction of the volume. Another pressure drop exists across the cyclone system. The solids flow from the cyclone into a standpipe which returns them to the bottom of the bed. Pressure builds up in the standpipe to allow the solids to flow back into the high pressure zone at the bottom of the CFB. The maximal pressure buildup in the standpipe is obtained by keeping the solids in

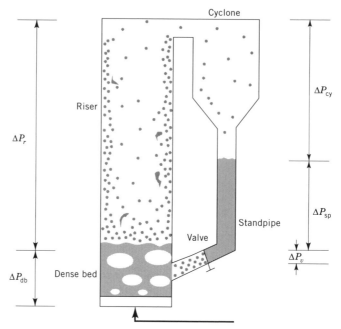

Fig. 21. CFB pressure balance, where $\Delta P_{sp} = \Delta P_v + \Delta P_{db} + \Delta P_r + \Delta P_{cy}$. A high gas velocity in a fast bed results in a high solids entrainment rate. Head buildup in the standpipe overcomes head buildup in the dense bed, riser, cyclone, and sealing device (valve), and allows solids to recirculate.

the standpipe fluidized. The fluidized solids in the standpipe develop a pressure head like any fluid. Some of this pressure buildup can then be dissipated across a control valve to regulate solids flow and form a seal against gas flow back into the standpipe.

Whereas standpipes have been in operation for many years, many aspects of their operation are not well understood. The purpose of a standpipe is to transfer solids from a region of low pressure to a region of higher pressure. In a properly operating standpipe of Group A solids, the solids are moving downward at a relative gas–solids velocity greater than U_{mf}/ϵ. The relative gas solids velocity, V_r, is defined as

$$V_r = V_s - U/\epsilon \tag{30}$$

This causes the solids to be fluidized. In practice, the solids are moving so fast that interstitial gas and gas bubbles are dragged downward with them.

The gas volume dragged down the standpipe is compressed owing to the pressure buildup, and may cause the solids to become defluidized. Figure 22 shows that the pressure buildup in a standpipe can be reduced, or even become negative. Aeration can be added to the standpipe to prevent the gas volume reduction causing defluidization because of the pressure increase. The suggested amount of aeration to add to the standpipe to overcome the effects of gas compression may be calculated from the following expression:

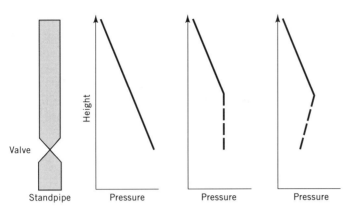

Fig. 22. Standpipe and standpipe pressure profiles showing (——) fluidized flow and (— — —) packed bed or defluidized flow.

$$Q_{\min} = \left[\frac{P_b}{P_t}\left(\frac{1}{\rho_{\mathrm{mf}}} - \frac{1}{\rho_s} \right) - \left(\frac{1}{\rho_t} - \frac{1}{\rho_s} \right) \right] \tag{31}$$

Note that the solids density used in this equation should be the true solids, ie, skeletal, density, because the gas in the pores is also compressed. For Group A solids the aeration gas should also be added evenly along the standpipe.

Group B solids have higher minimum fluidization velocities than Group A solids. For best results for Group B solids flowing in standpipes, standpipe aeration should be added at the bottom of the standpipe, not uniformly along the standpipe.

Nonmechanical Valves. Nonmechanical valves, which have no moving parts in the solids flow path, are often used to control the flow of Group B solids. Examples of nonmechanical valves are an L-valve, J-valve, loop seal, and reverse seal. A nonmechanical L-valve joins a pneumatic conveying line or a fluidized bed in the shape of a capital L. The elbow of the L-valve does not allow solids to flow unless aeration is added in the vertical part of the L-valve. The optimum location for aeration is approximately one pipe diameter above the horizontal centerline. The amount of solids flow is a function of the amount of aeration added to the L-valve (28). This system can operate under hot, corrosive, and erosive environments which would be too difficult for mechanical valves. L-valves also cost significantly less than mechanical valves. L-valves work best with Group B powders. Nonmechanical valves have also been used for many years in moving beds of Group D particles, such as in the Thermofor catalytic cracker (TCC).

Scale Up of Fluidized Beds

The greatest problems in scaling up fluidized beds have been encountered using Group B solids where bubbles can grow to very large sizes and bubble size control is more difficult than with Group A solids. A pilot-plant or laboratory-scale fluidized bed can operate where bubbles are no more than several centimeters in diameter because of the limiting small diameter of the bed (see PILOT PLANTS

AND MICROPLANTS). However, a larger commercial unit can have bubbles up to one meter in diameter. The larger bubbles flow through the bed at a high velocity, and the gas inside the bubble has little interaction with the solids.

Several attempts have been made to determine proper scaling relationships for Group B fluidized beds. These early attempts, and a rigorous derivation by nondimensionalizing the basic differential equations governing fluid and solids flow, have been summarized (29). The basic scaling factors are Reynolds number (eq. 5); Froude number where

$$\text{Froude number} = \frac{U^2}{gd_p} \tag{32}$$

density ratio, ρ_p/ρ_g; length ratios, L/d_p and D/d_p; and other factors such as ψ, sphericity, particle size distribution, and bed geometry.

There are some data to suggest that hydrodynamic similarity improves scale up for two-phase systems such as fluidized beds, even though it is not as convincing as single-phase evidence. To use these scaling factors to simulate large-scale hot systems, small-scale cold flow testing needs to hold the ratios of several variables constant. For example, to scale from a cold (15°C) to a large, hot (800°C) system, the following ratios are recommended (29):

Variable	Ratio
velocity, U	0.5
particle size, d_p	0.25
particle density, ρ_p	3.5
bed height, H	0.25
bed diameter, D	0.25

This scaling method has several limitations. There has been concern about changing particle properties for complex, wide distributions of naturally occurring materials. This analysis ignores interparticle forces, and hence does not apply to Group C and A powders. In addition, electrostatic forces can become significant in cold-flow models of large particles. Moreover, this analysis is too simplistic to take into consideration rapidly changing conditions, such as occur near a distributor region in a fluidized bed.

Group A particles cause fewer scale-up problems because fluidized beds of Group A particles generally are operated in the vigorously bubbling or turbulent fluidization regimes. Also, it is not unusual for a maximum stable bubble–gas void to be on the order of 25 mm or less for these particles. Thus a pilot-plant facility can generally be operated using the same gas void size that a commercial unit would experience. An example of scale-up concerns and ways to avoid them for Group A powders is shown in Figure 23 (30). In this case, the efficiency was maintained at 80% for a turbulent fluidized bed, ie, there was no scale-up loss, but efficiency decreased with scale up for a bubbling bed. Adding fines and operating at a higher gas velocity in a bubbling bed, ie, moving it toward turbulence, can offset scale-up loss.

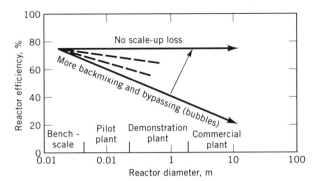

Fig. 23. Turbulent and bubbling beds scale-up comparison where increasing gas velocity, fines content, and *H/D* staging can help maintain reactor efficiency as the reactor diameter increases. A 100% efficiency is equivalent to plug flow.

Bubble size is controlled by maintaining the content of fines in a catalyst blend at a concentration of between 15 and 30%. This helps to keep gas voids unstable and small, and minimizes gas bypassing via the gas voids in the bed. Using a high gas velocity also helps to approximate plug flow of the gas. To minimize scale-up problems using Group A solids, the bed should be operated in the turbulent regime, which requires a gas velocity of 0.5 to 1 m/s. Bed internals can be beneficial in reducing solids backmixing, which can lower the efficiency of most reactions. Because of inherent backmixing, the best fluidized-bed reactor has an efficiency equivalent to approximately 80% of a plug-flow reactor.

Some backmixing is important to ensure temperature uniformity, but uncontrolled mixing can lead to high axial dispersions and hence rapid decrease in performance (Fig. 23). Bed internals, a combination of vertical heat-exchange tubes and horizontal baffles and hangers, are used in many Group A catalytic reactors such as those for acrylonitrile, ethylene dichloride, etc, to eliminate scale-up losses. Scale up for such reactions using Group A powders is simple as long as basic sound design principles are used. Successful scale ups of such reactions from 25 mm inner diameter pilot plants to 10 m dia commercial units have been

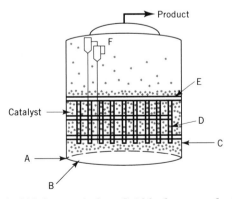

Fig. 24. Elements of a bubbleless turbulent fluid-bed reactor design where the internals create four stages. A represents the shrouded grid; B, the first feed; C, the second feed; D, heat exchange; E, staging; and F, the cyclones and/or filters.

achieved in one step, as was the case for acrylonitrile in the 1950s. The design principles of a commercial turbulent fluidized-bed reactor of Group A powders are illustrated schematically in Figure 24. The key variables which ensure successful and simple scale up are particle properties (sufficient fines); gas velocity in the turbulent fluidization regime; proper distributor design; internals, with appropriate open area; and primary cyclone diplegs which return the entrained fines to the region where they are needed most for bubble size control, ie, near the distributor. Using these principles, scale up of Group A catalytic fluidized-bed reactors can be easily accomplished.

NOMENCLATURE

A	area of bed
A_d	cross-sectional area of distributor
Ar	Archimedes number, $\rho_g d_p^3 (\rho_p - \rho_g) g / \mu^2$
C_d	drag coefficient
D	bed diameter
D_b	diameter of bubble
D_{bmax}	maximum stable bubble diameter
D_{noz}	diameter of nozzle
D_{or}	diameter of orifice
D_{header}	diameter of pipe header
d_p	"average" particle diameter
d_{pw}	weight average particle diameter
d_{pi}	particle diameter of an individual cut
d_{sv}	surface mean particle diameter
d_v	volume average diameter
E	entrainment rate (kg/s·m²)
G	solids mass flux
g	gravitational acceleration
g_c	conversion factor (in SI units $g_c = 1$)
H	height of fluidized bed
H_{ad}	height above distributor
H_α	height of a diffusion stage
H_{max}	maximum bed height
H_{mf}	bed height at minimum fluidization
h	heat-transfer coefficient from a fluidized bed to a heat-transfer surface
h_{cond}	conductive heat-transfer coefficient
h_{conv}	convective heat-transfer coefficient
h_{gs}	gas–solids heat-transfer coefficient
h_{rad}	radiative heat-transfer coefficient
k_f	thermal conductivity of gas
L	length
L_h	distributor pitch
L_{shroud}	length of shroud
n	Richardson-Zaki coefficient
N	number of distributor holes
N_a	number of diffusion stages
N_{or}	distributor hole density
P	jet penetration length
P_b	absolute pressure at bottom of standpipe
Pr	Prandtl number, $C_p \mu / K$

P_t absolute pressure at top of standpipe
Q gas volumetric flow rate
Q_{bubble} volume flow rate of bubbles
Q_{mf} gas volumetric flow rate at minimum fluidization
Q_{min} minimum aeration required to maintain fluidized conditions in a standpipe,
            actual cubic meters per second of aeration required (at standpipe conditions)
            per kg of solids flowing down the standpipe
Re_p particle Reynolds number, $d_p U \rho_g / \mu$
TDH   transport disengaging height
U superficial gas velocity
U_b bubble rise velocity in a bubbling bed
U_{br} single bubble rise velocity
U_{mb} minimum bubbling velocity
U_{mf} minimum fluidization velocity
U_{or} gas velocity through an orifice in a distributor
U_{slip} gas–solids velocity
U_{slug} riser velocity of slug
U_t single-particle terminal velocity
U_t^* effective or "cluster" terminal velocity
V_{mf} interstitial minimum fluidization velocity, U_{mf}/ϵ_{mf}
V_r relative gas–solids velocity
V_s solids velocity
X_i weight fraction
Δ characteristic system length
ΔP pressure drop
ΔP_{cy} cyclone pressure drop
ΔP_{db} dense bed pressure drop
ΔP_r riser pressure drop
ΔP_{sp} standpipe head buildup
ΔP_v valve pressure drop
ϵ gas voidage (volume fraction occupied by gas)
ϵ_{bed} voidage in bed
ϵ_{mf} voidage at minimum fluidization
μ gas viscosity
ρ_{bulk} bulk density of solids
ρ_{mf} bed density at minimum fluidization
ρ_g gas density
ρ_p solids density
ρ_s true solids (skeletal) density
ρ_t fluidized density at top of standpipe
ψ particle sphericity

BIBLIOGRAPHY

"Fluidization" in *ECT* 1st ed., Suppl. 1, pp. 365–400, by F. A. Zenz, Consultant; in *ECT* 2nd ed., Vol. 9, pp. 398–445, by F. A. Zenz, Consultant; in *ECT* 3rd ed., Vol. 10, pp. 548–581, by F. A. Zenz, Consultant.

1. D. Geldart, ed., *Gas Fluidization Technology*, John Wiley & Sons, Inc., Chichester, U.K., 1986.
2. A. A. Avidan, M. Edwards, and H. Owen, *O & G J.*, 33 (Jan. 8, 1990).
3. J. F. Richardson and W. N. Zaki, *Trans. Inst. Chem. Eng.* **32**, 35 (1954).

4. F. A. Zenz and D. F. Othmer, *Fluidization and Fluid Particle Systems*, Reinhold Publishing Corp., New York, 1960.
5. C. Y. Wen and Y. H. Yu, *AIChE J.* **12** 610–612, (1966).
6. D. Geldart, *Powder Technol.* **7**, 285–292 (1973).
7. J. M. Matsen, *AIChE Symp. Ser.* **69**, 30 (1973).
8. D. Geldart and H. Y. Xie, in O. E. Potter and D. J. Wicklin, eds., *Fluidization VII Proceedings*, Engineering Foundation, New York, 1992, pp. 749–756.
9. R. J. de Vries, W. P. M. van Swaaij, C. Mantovan, and A. Heijkoop, *Chem. React. Eng. Proc. Eur. Symp.* **5**, B9–59 (1972).
10. A. M. Squires, M. Kwauk, and A. A. Avidan, *Science* **230**(4732), 1329–1337 (1988).
11. K. Rietema, *The Dynamics of Fine Powders*, Elsevier Applied Science Publishers Ltd., London, 1991.
12. J. Yerushalmi and A. A. Avidan, in J. F. Davidson, R. Clift, and D. Harrison, eds., *Fluidization*, 2nd ed., Academic Press, London, 1985.
13. J. Yerushalmi, N. T. Cankurt, D. Geldart, and B. Liss, *AIChE Symp. Ser.* **74**(176), 1–13 (1978).
14. J. R. Grace, *Can. J. of Chem. Eng.* **64**, 353–363 (June 1986).
15. J. F. Davidson and D. Harrison, *Fluidised Particles*, Cambridge University Press, Cambridge, U.K., 1963.
16. R. C. Darton, R. D. LaNauze, J. F. Davidson, and D. Harrison, *Trans. Inst. Chem. Eng.* **55**, 274–280 (1977).
17. D. F. King, in *Fluidization VI Conference*, Engineering Foundation, New York, 1989, pp. 1–8.
18. A. A. Avidan, *Bed Expansion and Solids Mining in High-Velocity Fluidized Beds*, Ph.D. dissertation, City University of New York, 1980.
19. A. A. Avidan and J. Yerushalmi, *AIChE J.* **31**(5), 835–841 (1985).
20. W. P. M. van Swaaij and F. J. Zuiderweg, in H. Angelino, ed., *Proceedings of the International Symposium on Fluidization and its Applications*, Cepadues-edit, Toulouse, France, 1973, pp. 454–467.
21. M. Pell, *Gas Fluidization*, Elsevier, Amsterdam, the Netherlands, 1990.
22. B. R. Andeen and L. R. Glicksman, ASME/AIChE Heat Transfer Conference, St. Louis, Mo., 1976, paper 76HT-67.
23. D. F. King, *Fluidization Under Pressure*, Ph.D. dissertation, University of Cambridge, U.K., 1979.
24. ASTM D409-85 *1988 Annual Book of ASTM Standards*, 05.05, American Society for Testing and Materials, Philadelphia, Pa., 1988, pp. 186–190.
25. J. E. Gwyn, *AIChE J.* **15**, 35–39 (1969).
26. I. Chan and T. M. Knowlton, *AIChE Symp. Ser.* No. 241, Vol. 80, 24–33 (1984).
27. *Manual of Disposal of Refinery Wastes*, Volume on Atmospheric Emissions, API Publication 931, American Petroleum Institute, Washington, D.C., May 1975, Chapt. 11.
28. T. M. Knowlton and I. Hirsan, *Hydrocarbon Process.* **57**, 149 (1978).
29. L. R. Glicksman, *Chem. Eng. Sci.* **39**(9), 1373–1379 (1984).
30. A. A. Avidan and M. Edwards, in K. Østergaard and A. Sørensen, eds., *Fluidization V Conference*, Engineering Foundation, New York, 1986, pp. 457–464.

AMOS A. AVIDAN
Mobil Research and Development Corporation

DESMOND F. KING
Chevron Research and Technology Company

TED M. KNOWLTON
Institute of Gas Technology/Particulate
Solids Research Inc.

MEL PELL
E. I. du Pont de Nemours & Co., Inc.

FLUIDIZED-BED COMBUSTION. See FLUIDIZATION; REACTOR TECHNOLOGY.

FLUID MECHANICS

Fluid mechanics is both a descriptive science of the phenomena that occur when fluids flow and a quantitative science showing how these phenomena may be described in mathematical terms. To a practicing chemical technologist, fluid mechanics is an entire body of knowledge, theoretical and empirical, qualitative and quantitative, allowing analysis of the performance of complex plant equipment handling moving fluids. Calculation of the details of the flow is secondary to understanding the phenomena well enough to accomplish the process task. At times the technologist's needs are best satisfied by an empirical correlation; at other times the necessary skills consist largely of knowing how to fit the idealized solutions of the mathematician into a practical situation.

This article is intended to provide a useful first understanding of flow phenomena and techniques and to provide an entry to more precise and detailed methods where these are required. Although the main concern is the proper design and operation of plant equipment, the importance of preservation of the environment is recognized. Thus data from the fields of meteorology and oceanography are occasionally needed by the technologist (see also FLOW MEASUREMENT; FLUIDIZATION).

General Principles

Fluid mechanics became an exact science with the application of the conservation principles. These include the law of the conservation of mass, including the law of conservation of individual species, electric charge, etc; the law of conservation of energy; and the law of conservation of momentum, ie, Newton's second law. Fluid mechanics differs from conventional mechanics in that the former mainly treats bodies that are capable of unlimited deformation. Flow is regarded as primary and elasticity as secondary; in solid mechanics elasticity is primary and flow is regarded as secondary. Thus instabilities in flow, eg, turbulence, play a large role in fluid mechanics, whereas these are of minor concern in solid mechanics. In some instances the two converge, as in the treatment of extrusion of very viscous or plastic materials in which fluid-like and solid-like behavior are of equal importance.

The most useful mathematical formulation of a fluid flow problem is as a boundary value problem. This consists of two main parts: a set of differential equations to be satisfied within a region of interest and a set of boundary conditions to be satisfied on the surfaces of that region. Sometimes additional conditions are also of interest, eg, when one is investigating the stability of a flow.

The choice of the differential equation to be used within the region depends on what approximations can legitimately be made. A central assumption to most situations is that the fluid can be treated as a continuum, ie, mass, momentum, and energy are strictly conserved for any subdivision of space, no matter how small, and the material properties of the fluid are not affected by the process of subdivision. This idealization allows one to define the velocity at a point V as the time rate of change of the center of mass of a material body. Similarly, the rates of deformation or strain of the body constitute the components of a second-order tensor. Clearly, if the subdivision were extended to molecular or mean-free-path dimensions, the velocity would become highly fluctuating and random. Thus the results of fluid-mechanical calculations are limited to situations in which the changes of interest occur over length scales much larger than these dimensions. Where the continuum assumption breaks down, eg, in highly rarefied gases or in the behavior of submicrometer aerosol particles, recourse must be had to more general methods of calculation (see DIMENSIONAL ANALYSIS).

The boundary conditions for a situation are usually easy to state, at least where the boundaries are fixed in position. These usually correspond to the physical requirement that there is a specified flow through (or stress at) the boundary. Real fluids, which exhibit viscosity (a resistance to continuous deformation), also obey the principle of continuity of velocity, which asserts that the velocity of a flowing fluid does not suffer discontinuous changes at interfaces with other fluids or solids (see also RHEOLOGICAL MEASUREMENT). As applied to flow past a stationary solid, it asserts that the fluid velocity decreases to zero as the wall is approached. In classical inviscid hydrodynamics, solid surfaces serve to confine or deflect the flow but offer no retardation to the velocity, which is assumed to maintain its mainstream value vanishingly close to the surface. Although classical hydrodynamics is suitable for describing some features of flow, it fails completely in other key areas; its most notable failure is the incorrect prediction that a cylinder placed with its axis normal to the flow experiences no net force from the fluid (D'Alembert's Paradox). The assumption of a velocity decreasing continuously to zero at the surface and the introduction of shear stresses related to this gradient provide a more accurate prediction of flow phenomena consistent with experimental observations. The statement serves to emphasize that fluid mechanics still depends significantly on experiment to verify that a given mathematical solution does describe reality. This is particularly true when multiple solutions are possible.

Equations of Motion

The starting point for obtaining quantitative descriptions of flow phenomena is Newton's second law, which states that the vector sum of forces acting on a body equals the rate of change of momentum of the body. This force balance can be made in many different ways. It may be applied over a body of finite size or over each infinitesimal portion of the body. It may be utilized in a coordinate system moving with the body (the so-called Lagrangian viewpoint) or in a fixed coordinate system (the Eulerian viewpoint). Described herein is derivation of the equations of motion from the Eulerian viewpoint using the Cartesian coordinate system. The equations in other coordinate systems are described in standard references (1,2).

General Equation of Motion.   Neglecting relativistic effects, the rate of accumulation of mass within a Cartesian volume element $dx \cdot dy \cdot dz$ must equal the sum of the rates of inflow minus outflow. This is expressed by the equation of continuity:

$$\frac{\partial \rho}{\partial t} + \frac{\partial}{\partial x}\rho u + \frac{\partial}{\partial y}\rho v + \frac{\partial}{\partial z}\rho w = 0 \tag{1}$$

where ρ is the fluid density; u, v, and w are the x-, y-, and z-components of velocity; and t is the time.

Similarly, a momentum balance can be made on the same infinitesimal volume element. First it is necessary to have a firm idea of the nature of the forces that might be exerted on the element. There may be body forces which, in Newtonian mechanics, are pictured as acting at a distance and include gravitational, electrostatic, and magnetic forces and, if a noninertial frame of reference is used, those forces resulting from accelerations of the frame, eg, centrifugal force. In addition there are forces arising from the continuum stresses exerted on the faces or edges of the element. These stresses arise from intermolecular forces and motions of molecules that express themselves macroscopically as pressure, viscosity, yield stress, etc. These stresses behave mathematically as the components of a symmetrical, second-order tensor. At boundaries between phases, additional forces, which are generalizations of surface tension, must be included. These forces are discussed in standard texts.

In Figure 1, the force balance in Cartesian coordinates for a body not intersected by phase boundaries is

$$\rho \frac{Du}{Dt} = X + \frac{\partial \sigma_x}{\partial x} + \frac{\partial \tau_{yx}}{\partial y} + \frac{\partial \tau_{zx}}{\partial z}$$

$$\rho \frac{Dv}{Dt} = Y + \frac{\partial \sigma_y}{\partial y} + \frac{\partial \tau_{zy}}{\partial z} + \frac{\partial \tau_{xy}}{\partial x} \tag{2}$$

$$\rho \frac{Dw}{Dt} = Z + \frac{\partial \sigma_z}{\partial z} + \frac{\partial \tau_{xz}}{\partial x} + \frac{\partial \tau_{yz}}{\partial y}$$

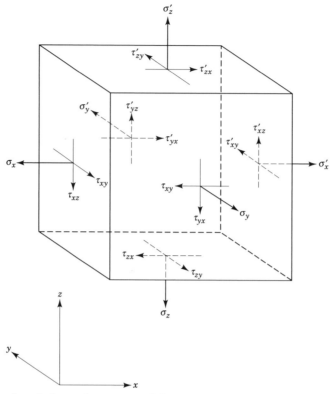

Fig. 1. Normal and shear forces on a differential volume $dx \cdot dy \cdot dz$, where $\sigma_i' = \sigma_i + \left(\dfrac{\partial \sigma_i}{\partial i}\right) di$ and $\tau_{ij}' = \tau_{ij} + \left(\dfrac{\partial \tau_{ij}}{\partial i}\right) di$.

where Du/Dt, Dv/Dt, and Dw/Dt are the components of acceleration; X, Y, and Z are body forces per unit volume; σ_i is the normal stress on the i-face; and τ_{ij} is the shear stress on the i-face exerted in the j-direction. The operator D/Dt is called the material, total, or substantial derivative and can be expressed as:

$$\frac{D}{Dt} = \frac{\partial}{\partial t} + u\frac{\partial}{\partial x} + v\frac{\partial}{\partial y} + w\frac{\partial}{\partial z} \tag{3}$$

Equation 2 gives the relation between stresses and accelerations obtained from momentum balances. To proceed further requires use of the constitutive equations which codify the material properties through additional relations between the stresses (τ_{xy}, etc) and the rates of strain ($\partial u/\partial x$, etc). The constitutive equation for a given fluid is found empirically or theoretically by use of some theory of material properties. The simplest model is one in which the various stresses are expressed as linear combinations of the rates of strain. When the fluid is homogeneous and isotropic, this relation leads to the Navier-Stokes equations. Fluids that obey these equations are by definition Newtonian. The conditions of homogeneity and isotropy ensure that only two material constants, the shear viscosity, μ, and the dilational viscosity, λ, are needed to describe the fluid. By defining pressure as the negative of the average of the three normal stresses, σ_i, one finds that $\lambda = 2/3\,\mu$, hence only a single material viscosity μ is required. As defined, the pressure is usually identified with the thermodynamic pressure for purposes such as determining physical properties. Equations 4 and 5 show the constitutive equation so derived, and equation 6 shows the form taken by the equation of motion in the X-direction when these are inserted into the force balance.

$$\sigma_x = -P + 2\mu\frac{\partial u}{\partial x} - \frac{2}{3}\mu\left(\frac{\partial u}{\partial x} + \frac{\partial v}{\partial y} + \frac{\partial w}{\partial z}\right)$$

$$\sigma_y = -P + 2\mu\frac{\partial v}{\partial y} - \frac{2}{3}\mu\left(\frac{\partial u}{\partial x} + \frac{\partial v}{\partial y} + \frac{\partial w}{\partial z}\right) \tag{4}$$

$$\sigma_z = -P + 2\mu\frac{\partial w}{\partial z} - \frac{2}{3}\mu\left(\frac{\partial u}{\partial x} + \frac{\partial v}{\partial y} + \frac{\partial w}{\partial z}\right)$$

$$\tau_{xy} = \tau_{yx} = \mu\left(\frac{\partial u}{\partial y} + \frac{\partial v}{\partial x}\right)$$

$$\tau_{yz} = \tau_{zy} = \mu\left(\frac{\partial v}{\partial z} + \frac{\partial w}{\partial y}\right) \tag{5}$$

$$\tau_{zx} = \tau_{xz} = \mu\left(\frac{\partial w}{\partial x} + \frac{\partial u}{\partial z}\right)$$

$$\rho\frac{Du}{Dt} = X - \frac{\partial P}{\partial x} + \frac{\partial}{\partial x}\left\{\mu\left[2\frac{\partial u}{\partial x} - \frac{2}{3}\left(\frac{\partial u}{\partial x} + \frac{\partial v}{\partial y} + \frac{\partial w}{\partial z}\right)\right]\right\}$$

$$+ \frac{\partial}{\partial y}\left[\mu\left(\frac{\partial u}{\partial y} + \frac{\partial v}{\partial x}\right)\right] + \frac{\partial}{\partial z}\left[\mu\left(\frac{\partial w}{\partial x} + \frac{\partial u}{\partial z}\right)\right] \tag{6}$$

The shear stresses are proportional to the viscosity, in accordance with experience and intuition. However, the normal stresses also have viscosity-dependent components, not an intuitively obvious result. For flow problems in which the viscosity is vanishingly small, the normal stress component is negligible, but for fluid of high viscosity, eg, polymer melts, it can be significant and even dominant.

For some materials the linear constitutive relation of Newtonian fluids is not accurate. Either stress depends on strain in a more complex way, or variables other than the instantaneous rate of strain must be taken into account. Such fluids are known collectively as non-Newtonian. Many different types of behavior have been observed, ranging from fluids for which the viscosity in the Navier-Stokes equation is a simple function of the shear rate to the so-called viscoelastic fluids, for which the constitutive equation is so different that the normal stresses can cause the fluid to flow in a manner opposite to that predicted for a Newtonian fluid.

Frictionless Flow. There is actually a constitutive equation that is even simpler than that for the Newtonian fluid. This arises when internal fluid friction is neglected. Although this is not precisely true of a real fluid, with the possible exception of liquid helium-3 at temperatures below 2 K, in many situations the flows calculated under this assumption are very close to those actually observed, at least in a significant portion of the flow field. Examples are high speed flows around obstacles in regions well away from solid boundaries or from wake regions. The importance of the frictionless flow theory lies in the wide variety of available solutions, and in the powerful techniques of calculation available. Notable is Bernoulli's theorem which, to take a useful special case, states that for irrotational, time-independent, barotropic (ie, density is a function of pressure), flow in a gravitational field, the sum of the three terms given in equation 7 is constant along any streamline:

$$\int \frac{dP}{\rho} + \frac{V^2}{2} + gh = \text{constant} \tag{7}$$

Equation 7 should not be construed to mean that the constant is the same for all streamlines. Once the velocities are known, this immediately gives the pressure. For a constant density, irrotational flow, the velocity components themselves are derivable from a velocity potential, ϕ:

$$u = \frac{\partial \phi}{\partial x} \qquad v = \frac{\partial \phi}{\partial y} \qquad w = \frac{\partial \phi}{\partial z} \tag{8}$$

where

$$\frac{\partial^2 \phi}{\partial x^2} + \frac{\partial^2 \phi}{\partial y^2} + \frac{\partial^2 \phi}{\partial z^2} = 0 \tag{9}$$

Equation 9 is Laplace's equation which also occurs in several other fields of mathematical physics. Where the flow problem is two-dimensional, the velocities are also derivable from a stream function, ψ.

$$u = \frac{\partial \psi}{\partial y} \qquad v = -\frac{\partial \psi}{\partial x} \tag{10}$$

where

$$\frac{\partial^2 \psi}{\partial x^2} + \frac{\partial^2 \psi}{\partial y^2} = 0 \tag{11}$$

Both equation 11 and the two-dimensional counterpart of equation 9 can be solved by several standard mathematical techniques, one of the more useful being that of conformal mapping. A numerical solution is often more practical for complicated configurations.

Flow Phenomena

Flow Past Bodies. A fluid moving past a surface of a solid exerts a drag force on the solid. This force is usually manifested as a drop in pressure in the fluid. Locally, at the surface, the pressure loss stems from the stresses exerted by the fluid on the surface and the equal and opposite stresses exerted by the surface on the fluid. Both shear stresses and normal stresses can contribute; their relative importance depends on the shape of the body and the relationship of fluid inertia to the viscous stresses, commonly expressed as a dimensionless number called the Reynolds number (Re), $LV\rho/\mu$. The character of the flow affects the drag as well as the heat and mass transfer to the surface. Flows around bodies and their associated pressure changes are important.

Flow Along Smooth Surfaces.   When the flow is entirely parallel to a smooth surface, eg, in a pipe far from the entrance, only the shear stresses contribute to the drag; the normal stresses are directed perpendicular to the flow (see PIPING SYSTEMS). The shear stress is usually expressed in terms of a dimensionless friction factor:

$$f = \frac{2\tau_w}{\rho \overline{V}^2} \tag{12}$$

Two distinct regimes of flow are observed in a pipe. At low velocities the flow is laminar, so-called because the fluid flows in concentric cylindrical sheaths or laminae that do not mix with each other. A stream of dye introduced into the fluid proceeds down the pipe as a thread, spreading only slightly by molecular diffusion. The velocity profile is parabolic, with the velocity decreasing from twice the average velocity at the axis to zero at the wall. This is easily demonstrated by solving the Navier-Stokes equations, which take on a rather simple form in this configuration (see eqs. 31 and 32). The shear stress is found to be proportional to the velocity and to the viscosity of the fluid, ie, as fluid velocity rises, the friction factor drops in inverse proportion. At some high Reynolds number $(D\overline{V}\rho/\mu)$, typically about 2100 in commercial pipe, an abrupt change occurs. Random instabilities, which decay at lower velocities, grow, destroying the laminar flow pattern. At a Reynolds number of ca 10,000 the transformation is nearly complete. The velocity over most of the pipe becomes fairly uniform, about equal to the average velocity. Only in a thin region near the wall are significant radial velocity gradients observed.

A more quantitative description can be obtained by dividing the field conceptually into three regions: a turbulent core in which momentum, heat, and mass are readily transported radially by strong eddying motions; a thin, nearly laminar, sublayer hugging the wall; and a somewhat thicker transition or buffer layer bridging the two. Within the laminar sublayer, viscous stresses are important and the steady-state Navier-Stokes equations are valid. In the other regions they are inadequate. The fluctuating velocities in these regions can be described by the time-dependent Navier-Stokes equations, but the solution is enormously complex. It has been estimated that modern computers would require a time equal to the age of the universe to adequately describe turbulent pipe flow at a Reynolds number of 10^7. To avoid this complexity, the time-dependent equations can be averaged in the manner originally suggested by Reynolds. This yields equations that are similar to those for laminar flow, but which contain additional terms involving time-averaged products of the velocity fluctuations. Mathematically these terms can be treated as additional stresses and are termed the Reynolds stresses. Physically the terms can be thought of as arising from the transport of momentum by the fluctuating eddies in a manner analogous to that by which viscous stresses arise from the transport of momentum by diffusing molecules. Much research in fluid mechanics has been devoted to developing physical models and mathematical descriptions for these stresses. Although some useful concepts have resulted, the solution of practical problems still rests heavily on empirical correlations developed for the configurations of interest (see ENGINEERING, CHEMICAL DATA CORRELATION).

As velocity continues to rise, the thicknesses of the laminar sublayer and buffer layers decrease, almost in inverse proportion to the velocity. The shear stress becomes almost proportional to the momentum flux (ρV^2) and is only a modest function of fluid viscosity. Heat and mass transfer (qv) to the wall, which formerly were limited by diffusion throughout the pipe, now are limited mostly by the thin layers at the wall. Both the heat- and mass-transfer rates are increased by the onset of turbulence and continue to rise almost in proportion to the velocity.

Figure 2 shows the friction factor both for smooth and roughened pipes. The higher frictional losses for rough pipes arise when the surface irregularities extend past the buffer zone and exert the higher drag characteristic of bluff bodies.

Flow Past Bluff Bodies.   When a fluid passes around a bluff body, both the shear and normal stresses can contribute to the drag force exerted on the body. The force produced by the shear stresses is commonly referred to as skin friction drag to distinguish it from the form drag arising from an imbalance in normal stresses. Skin friction drag is dominant at low velocities, and form drag is dominant at high velocities. The combined drag force is usually expressed in terms of a dimensionless drag coefficient:

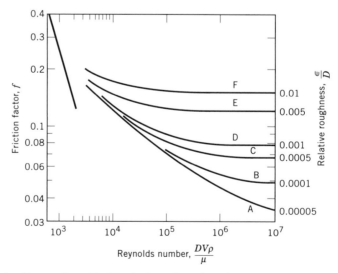

Fig. 2. Friction factors for cylindrical pipe where line A represents drawn tubing; line B, commercial steel; C, galvanized iron; and lines D, E, and F, concrete.

$$C_D = \frac{2F}{A_p \rho V_\infty^2} \tag{13}$$

Flow around a cylinder with axis normal to the flow provides a good example of the many unusual features of such flows. These are illustrated in Figure 3.

At very low velocities the flow around the cylinder is symmetrical front-to-back as well as side-to-side. Fluid passing near the cylinder accelerates over the upstream face and decelerates over the downstream face. Normal stresses on these faces largely cancel each other and drag results almost entirely from skin friction. Flow is laminar and the drag coefficient behaves like the friction factor in pipe flow, varying inversely with velocity. The drag force is directly proportional to the velocity and to fluid viscosity. As the velocity rises the flow becomes increasingly unsymmetrical. Under the influence of fluid inertia, the fluid streamlines crowd together on the upstream face of the cylinder and spread further apart on the downstream side. This imbalance is reflected in an imbalance in normal stresses and a growing importance of form drag. At a Reynolds number, $DV_\infty \rho/\mu$, of about 5, flow separation occurs. The fluid flowing next to the surface of the cylinder departs abruptly from the surface, the gap formed being filled by a pair of recirculating, laminar eddies in a bubble or wake behind the cylinder.

Separation is a result of the inability of the slowly moving fluid near the surface to move against an increasing pressure gradient. As the bulk of the fluid passes over the forward face of the cylinder and accelerates, it decreases in pressure. The fluid close to the surface, the boundary layer, also accelerates, although by a much smaller amount because of the retarding effects of viscous stresses. Over the rear face of the cylinder, the bulk fluid decelerates and pressure rises. This pressure gradient decelerates the already slowly moving fluid near the

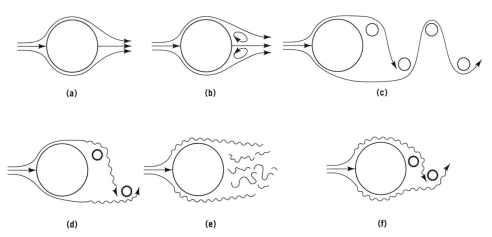

(a)   (b)   (c)

(d)   (e)   (f)

Fig. 3. Flow past a circular cylinder for (**a**), $Re < 5$ where no separation is evident; (**b**) $5 < Re < 40$ and fixed vortices exist in a separation bubble or wake; (**c**) $40 < Re < 100$ and laminar vortices are shed from the eddies in the wake; (**d**) $100 < Re < 4 \times 10^5$ and the vortex street becomes turbulent but the boundary layer remains laminar; (**e**) $4 \times 10^5 < Re < 4 \times 10^6$, the boundary layer becomes turbulent before separation and the vortex street disappears; and (**f**) $4 \times 10^6 < Re$ and a narrow vortex street reappears (3).

boundary, brings it to rest, and, as pressure continues to rise, reverses the flow at the surface and lifts the boundary layer (Fig. 4). The position at which the velocity becomes zero is the point of separation.

As the Reynolds number rises above about 40, the wake begins to display periodic instabilities, and the standing eddies themselves begin to oscillate laterally and to shed some rotating fluid every half cycle. These still laminar vortices are convected downstream as a vortex street. The frequency at which they are shed is normally expressed as a dimensionless Strouhal number which, for Reynolds numbers in excess of 300, is roughly constant:

$$Sr = \frac{n_s D}{V_\infty} \sim 0.2 \tag{14}$$

The periodic shedding produces lateral forces of the same period on the cylinder. Should the cylinder be weakly supported and have a natural frequency close to the shedding frequency, it oscillates strongly in concert with the vortex street. Such behavior is responsible for the singing of power lines, the oscillation of tall smokestacks, and, most spectacularly, for the collapse in 1940 of the newly built Tacoma Narrows suspension bridge, in Washington state, under the influence of a steady 65 km/h wind.

Above a Reynolds number of about 100, the standing eddies are no longer apparent. The vortex street undergoes a laminar-to-turbulent transition, becoming fully turbulent at a Reynolds number of about 300. The drag coefficient has become nearly independent of Reynolds number. The pressure on the downstream side of the cylinder is almost constant, roughly equal to the low value at the point of separation. This results in a considerable imbalance of pressure between the upstream and downstream surfaces. Skin friction now provides only a minor part of the total drag. No further changes occur until the Reynolds number reaches about 4×10^5, at which point the drag coefficient undergoes a sudden downward step change so large that the drag force actually decreases. This sudden change is caused by the development of turbulence within the boundary layer. Although

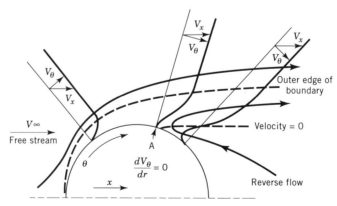

Fig. 4. Boundary layer development around a circular cylinder where A represents the point of separation.

the wake has been turbulent for some time, flow around the cylinder has remained laminar up to the point of separation. The onset of turbulence, by allowing a freer exchange of momentum between faster and slower moving regions at the surface, stabilizes the boundary layer against decelerations and postpones its separation. The separation point, which hitherto has been about 82° from the forward stagnation point of the cylinder, moves to about 110° and the partial recovery of pressure over the downstream surface reduces the net drag force. The wake is narrower and more disorganized and no vortex street is evident. The vortex street does not reappear until the Reynolds number exceeds 4×10^6. The laminar-to-turbulent transition can be initiated at somewhat lower Reynolds numbers by roughening the surface, eg, with sand or a trip wire. Sucking fluid through the surface can also postpone separation and reduce drag.

The behavior of the boundary layer also affects local heat transfer to the surface. At low Reynolds numbers, the thin boundary layer on the upstream face of the cylinder exhibits a higher heat-transfer coefficient than does the thickened layer on the downstream face. Despite the onset of separated flow and the large energy losses in the wake, the upstream coefficient remains larger until the Reynolds number reaches 5×10^4. Above this point the washing of the surface by the shedding eddies makes the downstream coefficient larger. The point of maximum heat transfer occurs at the forward stagnation point (0°) for Reynolds numbers up to 5×10^4, at the rear stagnation point (180°) for Reynolds numbers between 5×10^4 and 2×10^5, and at about the 115° position for Reynolds numbers above 2×10^5. Minimum heat transfer occurs near the point of separation.

Although the above description has concentrated on separation from a smooth surface, separation also occurs at sharp edges. Where separation is undesirable for a process reason, it can often be eliminated by redirecting the flow using turning vanes, ie, forcing it to hug the surface.

The Boundary Layer. At several points in the previous descriptions passing reference was made to the boundary layer, the fluid flowing close to the surface. The concept of the boundary layer was developed principally by Prandtl to facilitate the solution of the equations of motion in the presence of large velocity gradients perpendicular to the flow (4). It expresses the observation that the flow field often can be divided into two reasonably well-defined regions: a thin region close to the surface, the boundary layer, in which the gradient of tangential velocity is large and shear stresses are important; and a region outside the boundary in which velocity gradients are small and the potential flow solutions of inviscid flow theory, eg, Bernoulli's equation, are very nearly valid.

The transition between the two regions is gradual, not abrupt. The edge of the boundary layer is taken to be the point at which the velocity differs from that for inviscid flow by some small amount, say one percent. This approximation allows the main flow field and the normal stresses on the surface to be calculated by the many techniques applicable to Laplace's equation, with the boundary layer calculation then appended to yield the shear stresses at the surface. Extensive treatises have been devoted to analyzing flows in terms of boundary layer phenomena, particularly the growing boundary layers of entrance flows, in which a previously undisturbed fluid impinges on an obstruction (5). Although boundary layer analysis is widely applicable, it can fail badly in regions of strong deceleration where the boundary can separate under the influence of the unfavorable pressure gradient imposed by the main flow.

Entrance and Exit Effects. When a flowing fluid encounters an obstruction, the flow is disturbed and a new, steady-flow pattern is not reestablished until some distance downstream of the obstruction. The effects most commonly encountered are associated with the flow of fluids through ducts in which sudden changes in cross section occur. In Figure 5**a**, flow through a sudden contraction is normally accompanied by separation, except possibly at very low Reynolds numbers, $D\overline{V}\rho/\mu$. The inertia of the converging flow causes it to contract to a smaller area than that of the duct. The area at the *vena contracta* (point of minimum area) depends on the geometry, but for turbulent flows it is about 60% of the duct's area. Flow down to the *vena contracta* is substantially frictionless, and Bernoulli's equation can be used to describe the pressure change. Expansion beyond the *vena contracta* is not frictionless. The overall irreversible loss of energy depends on the geometry but is about $\overline{V}_2^2/4$ for turbulent flow through a large, abrupt change in area. Where separation is undesirable it can be eliminated by rounding the inlet to the downstream duct.

Entrance flow is also accompanied by the growth of a boundary layer (Fig. 5**b**). As the boundary layer grows to fill the duct, the initially flat velocity profile

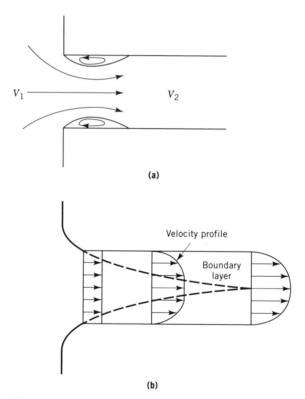

Fig. 5. Entrance flows in a tube or duct: (**a**) separation at sharp edge; (**b**) growth of a boundary layer (illustrated for laminar flow).

is altered to yield the profile characteristic of steady-state flow in the downstream duct. For laminar flow in a tube, the distance required for the velocity at the center line to reach 99% of its asymptotic value is given by

$$\frac{L}{D} = 0.057 \left(\frac{D\overline{V}\rho}{\mu} \right) \tag{15}$$

For turbulent flow the entrance length is 50–100 pipe diameters. Throughout the entrance region, the velocity gradient at the wall is larger than for the fully developed profile, producing a higher pressure drop and greater heat transfer per unit length than for fully developed flow. Although entrance phenomena are often ignored, these can be of significance, especially in capillary viscometers and compact (short tube) heat exchangers (see HEAT-EXCHANGE TECHNOLOGY).

When a flow expands abruptly, separation occurs producing a jet of fluid flanked by recirculating eddies. For turbulent flows, separation occurs in a diverging duct when the angle of divergence exceeds about 7°. When separation occurs, the ability of the fluid to recover pressure upon deceleration is seriously impaired because the kinetic energy is lost to friction. When the ratio of duct areas is large, all of the kinetic energy is lost and no pressure is recovered. Under these circumstances, the loss of energy per mass of fluid is $\overline{V}_1^2$ for laminar flow and $\overline{V}_1^2/2$ for turbulent flow, where $\overline{V}_1$ is the average velocity in the upstream section of the duct. For the limiting case, proper laminar flow exit loss is obtained by doubling the turbulent flow exit loss.

When the entrance flow is bounded on only one side, the boundary layer can, in theory, grow indefinitely (albeit slowly) into the surrounding fluid. Flow past a flat plate illustrates the behavior. As the fluid passes the leading edge of the plate, a boundary layer begins to grow. Initially this layer is laminar and its growth can be calculated easily by solving the Navier-Stokes equations using the concept of boundary layer approximation (2). At a length defined by a Reynolds number, $LV\rho/\mu$, of about 4×10^5, the laminar boundary has grown so thick that it is unstable to fluctuations. The oscillations caused by disturbances from the outer flow are not damped but are transformed into vortices that break down into the random eddies of turbulence. This process is almost complete when the Reynolds number reaches ca 4×10^6. Turbulence allows a higher rate of transfer of momentum within the fluid; consequently, transition is accompanied by a considerable increase in the thickness of the boundary layer. The turbulent boundary layer is similar to that noted for turbulent flow in pipes, consisting of a laminar sublayer, a transition zone, and a much thicker turbulent zone. In both cases the velocity profiles and drag forces can be described by the so-called universal velocity distribution (5). Table 1 illustrates the boundary layer properties for a 100 m/s flow of air past a thin wing.

Flow Past Deformable Bodies. The flow of fluids past deformable surfaces is often important, eg, contact of liquids with gas bubbles or with drops of another liquid. Proper description of the flow must allow for both the deformation of these bodies from their shapes in the absence of flow and for the internal circulations that may be set up within the drops or bubbles in response to the external flow. Deformability is related to the interfacial tension and density difference between

Table 1. Boundary Layer Properties for 100 m/s Flow of Air

Property	Value
distance to laminar turbulent transition, cm	6.0
thickness of laminar boundary layer at transition, cm	0.05
distance to complete transition, cm	60.0
thickness of turbulent boundary layer, cm	1.1
laminar sublayer, cm	0.002
buffer layer, cm	0.010
velocity at edge of laminar sublayer, m/s	19
buffer layer, m/s	53

the phases; internal circulation is related to the drop viscosity. A proper description of the flow involves not only the Reynolds number, $dV\rho/\mu$, but also other dimensionless groups, eg, the viscosity ratio, μ_d/μ_c; the Eötvos number ($E\ddot{o}$), $g\Delta\rho d^2/\sigma$; and the Morton number (Mo), $g\mu_c^4\Delta\rho/\rho_c^2\sigma^3$ (6).

Where surface-active agents are present, the notion of surface tension and the description of the phenomena become more complex. As fluid flows past a circulating drop (bubble), fresh surface is created continuously at the nose of the drop. This fresh surface can have a different concentration of agent, hence a different surface tension, from the surface further downstream that was created earlier. Neither of these values need equal the surface tension developed in a static, equilibrium situation. A proper description of the flow under these circumstances involves additional dimensionless groups related to the concentrations and diffusivities of the surface-active agents.

Because of high buoyancy and frequently large size, gas bubbles rising in a liquid can deviate greatly from spherical shapes. Figure 6 illustrates the behavior observed. In pure liquids small bubbles rise faster than would be predicted from the drag correlations developed for solid spheres because internal circulation permits a higher fluid velocity at the surface and less drag than for the corresponding solid. Very large bubbles rise much more slowly than do undeformed spheres of the same volume. This is caused by deformation to a shape of large frontal area but small thickness. Very small bubbles obey Stokes law for solid (immobile) surfaces. This behavior is reflected also in the mass-transfer coefficients for such bubbles, which are lower than would be expected from the correlations developed for larger, circulating bubbles.

Free Flows: Jets, Wakes, and Plumes.   When a smoothly flowing stream is disturbed by inserting a solid body or by injecting additional fluid, the disturbance is confined to a narrow region downstream of the initial point. When the disturbance is caused by the injection of a fluid at high velocity, the disturbed region is called a jet. When the velocity is lower than that in the main stream, the disturbance is termed a wake. The term plume is usually applied to a case in which an additional effect, such as buoyancy, plays a significant role and for which the turbulence characteristics of the main stream play the deciding role in its spread.

Jets are used industrially to perform many mixing operations ranging from gradual mixing of tanks to the rapid mixing needed in chemical reactors or flames

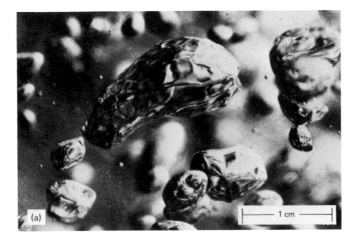

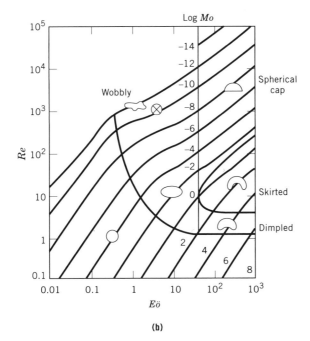

Fig. 6. Shape of drops and bubbles. (**a**) Bubble rising in sparged tower system: air–water. Courtesy of Shell Development Co. (**b**) Bubble and droplet regimes where $\otimes$ represents a 0.5-cm gas bubble rising in water at 20°C (6).

(see MIXING AND BLENDING). Mixing in turbulent wakes is important in the disposal of wastes at sea and in the dispersion of potential atmospheric pollutants downwind of structures. Most interest in plumes is centered around the behavior of gases emitted from smokestacks into the atmosphere.

    The development of the flow field around a jet or wake may be regarded as a boundary layer phenomenon in which the jet grows by entraining fluid from its surroundings. The boundary in this case is the envelope over which the velocity is some small (about 1%) fraction of the velocity of the jet along its center line. Jets may be either laminar or turbulent, although most important applications involve turbulent jets, which begin to form when the Reynolds number, $D\bar{V}\rho/\mu$, of the inlet stream exceeds a few hundred. The flow field around such a jet can be divided into three main regions (Fig. 7) (7). In the entry region the jet displays a diminishing core in which the inlet velocity profile is preserved. This is analogous to the entry region of a pipe but there is no pressure drop and no acceleration of flow. The entry region disappears about five nozzle diameters past the inlet. A transition region extends an additional two nozzle diameters, after which the jet enters the similarity region. In this region the velocity profile follows a similarity relationship:

$$\frac{u}{u_C} \simeq \exp\left[-4.6\left(\frac{y}{\delta}\right)^2\right] \tag{16}$$

It has been experimentally verified that initially jets spread roughly linearly in the similarity region, ie, $\delta \alpha (x - x_0)$ where x_0 is the virtual point source of the jet. For both axially symmetric and plane parallel (slot) jets, the jet width parameter, δ, is given by

$$\delta = 0.25x \tag{17}$$

In a free jet the absence of a pressure gradient makes the momentum flux at any cross section equal to the momentum flux at the inlet, ie, equations 16 and 17 define jet velocity at all points. For a cylindrical jet this leads to a center-line velocity that varies inversely with $(x - x_0)$, whereas for slot jets it varies inversely with the square root of $(x - x_0)$. As the jet proceeds still further downstream the turbulent entrainment initiated by the jet is gradually subordinated to the turbulence level in the surrounding stream and the jet, as such, disappears.

Mass and heat transport within the jet follow the same general pattern as does momentum transport. For gases it is found experimentally that the thermal

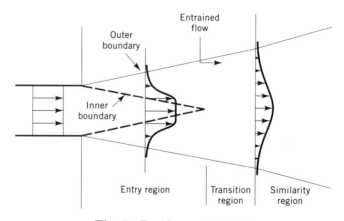

Fig. 7. Development of a jet.

or concentration jet spreads somewhat faster and decays more rapidly than does the momentum jet. All can be described by the turbulent equations of motion and useful results can be obtained by using eddy diffusivities, which are nearly constant across the jet but decay with time (distance downstream). All of the jets are gradually swallowed up in the surrounding fluid. The concentration jet, because of its association with a particular chemical species, remains identifiable longest, even though its behavior becomes completely dominated by the surrounding fluid. An important example of such a concentration jet is the plume, as might be released from a smokestack or vent. Except right at the source, the behavior of the plume is dominated by its buoyancy and/or the turbulence of the fluid into which it is injected. Figure 8 describes several of the plume patterns that can be obtained as they are affected by the temperature gradient in the surrounding air. The quantitative description of the dispersion of such plumes, especially their ground-level concentration, is still imperfectly developed and one must usually resort to empirical correlations. Where the effects of buoyancy are subordinate to the effects of wind velocity, useful results for dispersion over complex terrain or around structures can be obtained by wind tunnel experiments using scale models.

Most of the remarks above refer to unconfined or free flows. Many industrial applications involve the use of confined jets. It is customary to consider a jet

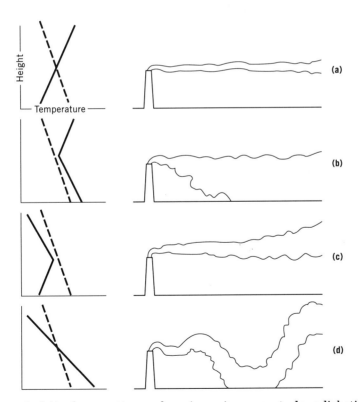

Fig. 8. Characteristic plume patterns where (– – –) represents dry-adiabatic lapse rate and (——), air: (**a**) fanning; (**b**) fumigation; (**c**) lofting; and (**d**) looping. Courtesy of *Scientific American* (8).

confined when the ratio of the confinement radius to the source radius lies in the range 4–100. Below a ratio of 2, the jet does not develop its similarity profile before striking the wall, whereas above a ratio of 100 the jet itself may usually be considered free. Under certain conditions, flow in confined jets is accompanied by the existence of a recirculation zone which significantly affects the jet behavior by returning material upstream (9). This recirculation can be particularly important in combustion processes.

Other Flow Phenomena. *Compressible Flow.* The flow of easily compressible fluids, ie, gases, exhibits features not evident in the flow of substantially incompressible fluid, ie, liquids. These differences arise because of the ease with which gas velocities can be brought to or beyond the speed of sound and the substantial reversible exchange possible between kinetic energy and internal energy. The Mach number, the ratio of the gas velocity to the local speed of sound, plays a central role in describing such flows.

Consider the converging–diverging nozzle described in Figure 9, and for simplicity, consider only frictionless, adiabatic, one-dimensional flow. When the pressure in the downstream reservoir is reduced, flow commences. In the converging section, velocity increases with distance, and both pressure and temperature fall in accordance with the overall energy balance and the equation of state. In the diverging section, the fluid slows and some pressure is recovered, B. As the downstream pressure is reduced further, the behavior remains qualitatively the

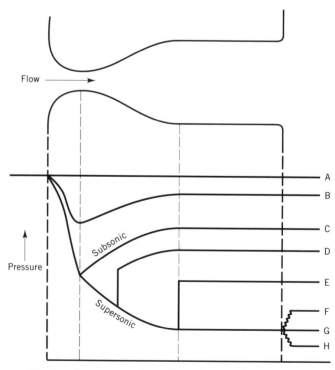

Fig. 9. Compressible flow in a converging–diverging nozzle where A represents no flow, C subsonic flow, and F through H supersonic flow. See text.

same until a point is reached at which the velocity in the nozzle throat equals the local speed of sound. At this point it is found that there are two possible paths that the downstream flow can follow and remain thermodynamically reversible. One corresponds to recompression and deceleration to subsonic flow C, the other to further expansion and acceleration to supersonic flow G. Operation at an intermediate pressure in the reservoir D is possible only by sacrificing thermodynamic reversibility. Under these circumstances, a standing normal shock front is set up at some point in the diverging section. At this front, pressure, temperature, and entropy rise, and velocity drops abruptly over a distance of a few molecular mean-free paths. Flow changes from supersonic upstream of the shock to subsonic downstream. The relationship between the two conditions is established by conservation of energy and by conservation of momentum across the shock front.

Further reductions in reservoir pressure move the shock front downstream until it reaches the outlet of the nozzle E. If the reservoir pressure is reduced further, the shock front is displaced to the end of the tube, and is replaced by an oblique shock, F, no pressure change, G, or an expansion fan, H, at the tube exit. Flow is now thermodynamically reversible all the way to the tube exit and is supersonic in the tube. In practice, frictional losses limit the length of the tube in which supersonic flow can be obtained to no more than 100 pipe diameters.

While the behavior downstream of the throat is displaying these complexities, nothing is changing upstream. Once the velocity at the throat becomes sonic, no downstream perturbation, eg, a sound wave, can be propagated upstream. Throughout the change from C to H the mass rate of flow remains constant at the value first established at C. An increase in flow can be accomplished only by raising the upstream pressure. In the absence of any temperature change in the upstream reservoir, the increase in flow results solely from a change in fluid density, not from a higher velocity in the throat. This isolation of flow rate from downstream effects can be employed to meter gases through orifices or nozzles at a constant rate, irrespective of downstream fluctuations. For ideal gases and where the upstream velocity is low, sonic flow through the throat is established whenever the ratio of upstream to downstream pressure across the constriction exceeds the quantity

$$\left(\frac{\gamma + 1}{2}\right)^{\frac{\gamma}{\gamma - 1}} \tag{18}$$

where $\gamma = C_p/C_v$. For simple gases, γ is about 2.

Phenomena analogous to shock waves in gases can occur in open-channel flow of liquids. The Froude number, $(V^2/gh)^{1/2}$, is the ratio of fluid velocity to the velocity of a small surface wave, and plays the same role in open-channel flow as the Mach number does in compressible flow. Thus liquid flowing under a sluice gate is often discharged at a shallow depth at high velocity, corresponding to a Froude number greater than unity. At a distance downstream, the liquid is observed to undergo a hydraulic jump to a greater depth at a slower velocity, for which the Froude number is less than unity. The decreased liquid momentum appears as greater pressure (depth). In contrast to the normal shock, a great deal of turbulence is generated at the jump because of the sudden lateral expansion, and energy is dissipated through friction.

Flow in Porous Media.   Flow of fluids through fixed beds of solids occurs in situations as diverse as oil-field reservoirs, catalyst beds and filters, and absorption (qv) towers. The complex interconnected pore structure of such systems makes it necessary to use simplified models to make practical quantitative predictions. One of the more successful treatments of single-phase pressure drop through such systems employs the results for flow through tubes, using average velocities and tube diameters. The average velocity through the pores is related to the superficial velocity, ie, the apparent velocity based on the entire cross section, via

$$\overline{V} = \frac{V_s}{\epsilon} \tag{19}$$

For spherical particles the average diameter of the pores, defined as four times the pore volume divided by the surface area, can be shown to be

$$\overline{D} = \frac{6\epsilon}{1 - \epsilon} d_p \tag{20}$$

The overall pressure drop is expressed as the sum of a laminar term proportional to $\overline{V}/\overline{D}^2$ and a turbulent term proportional to $\overline{V}^2/\overline{D}$ to yield the Ergun equation (1):

$$\frac{\Delta P}{L} = A\frac{\mu V_s}{d_p^2}\frac{(1 - \epsilon)^2}{\epsilon^3} + B\frac{\rho V_s^2}{d_p}\frac{(1 - \epsilon)}{\epsilon^3} \tag{21}$$

For randomly packed spherical particles, the constants A and B have been determined experimentally to be 150 and 1.75, respectively. For nonspherical particles, equivalent spherical diameters are employed and additional corrections for shape are introduced.

When two phases are present the situation is quite complex, especially in beds of fine solids where interfacial forces can be significant. In coarse beds, eg, packed towers, the effects are often correlated empirically in terms of pressure drops for the single phases taken individually.

Non-Newtonian Fluids: Die Swell and Melt Fracture.   For many fluids the Newtonian constitutive relation involving only a single, constant viscosity is inapplicable. Either stress depends in a more complex way on strain, or variables other than the instantaneous rate of strain must be taken into account. Such fluids are known collectively as non-Newtonian and are usually subdivided further on the basis of behavior in simple shear flow, ie, flow between sliding planes or, to a good approximation, between two mutually rotating cylinders. Figure 10 illustrates the behavior of several of these. The types illustrated are examples of the generalized Newtonian fluid. This simple generalization uses the Navier-Stokes equations but takes viscosity to depend in some way on a single quantity: the rate of strain.

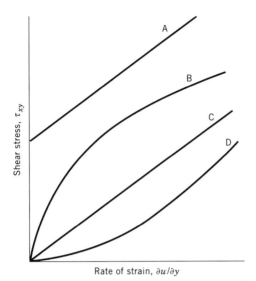

Fig. 10. Fluid behavior in simple shear flow where A is Bingham; B, pseudoplastic; C, Newtonian; and D, dilatant.

Pseudoplastic fluids are the most commonly encountered non-Newtonian fluids. Examples are polymeric solutions, some polymer melts, and suspensions of paper pulps. In simple shear flow, the constitutive relation for such fluids is

$$\tau_{yx} = K\left(\frac{\partial u}{\partial y}\right)^n \tag{22}$$

The apparent viscosity, defined as $\tau_{yx}/(\partial u/\partial y)$, drops with increased rate of strain. Dilatant fluids follow a constitutive relation similar to that for pseudoplastics except that the viscosities increase with increased rate of strain, ie, $n > 1$ in equation 22. Dilatancy is observed in highly concentrated suspensions of very small particles such as titanium oxide in a sucrose solution. Bingham fluids display a linear stress–strain curve similar to Newtonian fluids, but have a nonzero intercept termed the yield stress (eq. 23):

$$\eta\left(\frac{\partial u}{\partial y}\right) = \tau_{yx} - \tau_0, \text{ when } \tau_{yx} \geq \tau_0$$

$$\frac{\partial u}{\partial y} = 0, \text{ when } \tau_{yx} < \tau_0 \tag{23}$$

The coefficient η is termed the modulus of rigidity. The viscosities of thixotropic fluids fall with time when subjected to a constant rate of strain, but recover upon standing. This behavior is associated with the reversible breakdown of structures within the fluid which are gradually reestablished upon cessation of shear. The smooth spreading of paint following the intense shear of a brush or spray is an example of thixotropic behavior. When viscosity rises with time at constant rate of strain, the fluid is termed rheopectic. This behavior is much less common but is found in some clay suspensions, gypsum suspensions, and certain sols.

When these relatively simple fluids are conveyed in laminar flow in pipes, the behaviors can be deduced directly and quantitatively from constitutive relations and the linear variation of shear stress with radius (see eq. 55). Pseudoplastic fluids display lower viscosities in the high shear region on the wall than at the pipe axis. Their velocity profiles are steeper at the wall and flatter at the axis than the parabolic shape developed by Newtonian fluids. Dilatant fluids show opposite behavior, tending to freeze at the wall and finger along the axis. Bingham fluids show no velocity gradient up to the radius at which the yield stress is reached, giving the impression of a solid core of material moving along the axis. Pipelines (qv) conveying thixotropic and rheopectic fluids are usually designed on a most viscous basis. For thixotropic fluids this corresponds to viscosity at time zero, but for rheopectic fluids the viscosity after extended shear is used.

The pressure drop accompanying pipe flow of such fluids can be described in terms of a generalized Reynolds number, which for pseudoplastic or dilatant fluids takes the form:

$$Re = \frac{D^n \overline{V}^{2-n} \rho}{8^{n-1} K \left(\dfrac{3n+1}{4} \right)^n} \tag{24}$$

The transition from laminar to turbulent flow occurs at Reynolds numbers varying from ca 2000 for $n \geq 1$ to ca 5000 for $n = 0.2$. In the laminar region the Fanning friction factor (Fig. 2) is identical to that for Newtonian fluids. In the turbulent region the friction factor drops significantly with decreasing values of n, producing a family of curves.

In configurations more complex than pipes, eg, flow around bodies or through nozzles, additional shearing stresses and velocity gradients must be accounted for. More general equations for some simple fluids in laminar flow are described in Reference 1.

Many industrially important fluids cannot be described in simple terms. Viscoelastic fluids are prominent offenders. These fluids exhibit memory, flowing when subjected to a stress, but recovering part of their deformation when the stress is removed. Polymer melts and flour dough are typical examples. Both the shear stresses and the normal stresses depend on the history of the fluid. Even the simplest constitutive equations are complex, as exemplified by the Oldroyd expression for shear stress at low shear rates:

$$\tau + \lambda_1 \frac{d\tau}{dt} = \mu \left(\frac{d\dot{\gamma}}{dt} + \lambda_2 \frac{d^2 \dot{\gamma}}{dt^2} \right) \tag{25}$$

The relaxation times, λ_1 and λ_2, describe the times required to relieve stress on the cessation of strain and to relieve strain on the cessation of stress, respectively. The full Oldroyd tensor requires knowledge of eight material properties.

The development of significant normal stresses in such fluids can lead to bizarre behaviors when the fluids are deformed. Stirred viscoelastic fluids can move radially inward and climb the stirring shaft (the Weissenberg effect) if the inward-directed normal stresses accompanying shear are greater than the centrifugal forces developed by the stirring. When viscoelastic polymer melts are extruded through dies, they swell, recovering in part the lateral dimensions they possessed upstream of the die. At some critical rate the polymer undergoes melt fracture. Stress is relieved in an unsymmetrical manner and the extruded rod takes on a screw-like form. As the flow increases, additional modes of instability appear and the rod takes on progressively more complex shapes.

Gas–Liquid Flow. When two or more fluids flow together, a much greater range of phenomena occurs as compared to flow of a single phase. In a conduit many of the technically significant phenomena have to do with the positions assumed by the phase boundaries, and these are governed by the flow conditions rather than by the walls of the conduit. In addition to the densities and viscosities, surface properties can be important. Methods for quantitative calculation for gas–liquid flow are poorly developed in comparison to those for single-phase flow. Consider, for example, pressure drop for fully developed, steady-state, incompressible flow in smooth conduits. For single-phase flow, dimensional analysis (qv) shows that only two dimensionless numbers are involved. These are the Fanning friction factor $\tau_w/(\rho\bar{V}^2/2)$ and the Reynolds number. A set of experiments in which only two measurable quantities are varied suffices to establish the desired empirical correlation. For fully developed gas–liquid flow in horizontal pipes alone, there are at least six such groups and a six-dimensional space must be mapped. This has not yet been fully accomplished although thousands of experimental measurements have been made. The methods available run the gamut from empirical correlations only weakly based, if at all, on an understanding of the phenomena, to correlations based on conceptual models that use simplifying relations connecting the variables of the multidimensional problem. Even with the best correlations, however, errors of a factor of two are not uncommon. Much of the data on which available correlations are based are obtained in small (dia <10 cm) pipes using air and water. There is still much uncertainty as to how or even whether such data can be extrapolated to large pipes conveying fluids having significantly different properties.

Despite these difficulties, many useful calculations can be made, particularly when the effects of changes in an already existing operation are of interest. All of these calculations depend on an accurate picture of the flow regime, used to denote the significant configurational characteristics of the flow, eg, unidirectional, recirculating, steady, unsteady, liquid-dispersed, gas-dispersed, etc. For illustration, consider the simultaneous flow of air and water at various velocities in a 2.5-cm horizontal pipe. It is desirable to know such things as which phase is continuous, the degree of dispersion or atomization, and the presence or absence of flow surges. At low velocities, both phases are continuous and flow smoothly; the flow is stratified as sketched in the lower left-hand corner of Figure 11. As either the liquid-flow rate or gas-flow rate is raised, the flow becomes unstable by a Kelvin-Helmholtz instability. This causes waves to rise at the phase interface, producing wave-stratified flow. With further increase in liquid-flow rate, the waves grow in amplitude, and at some flow rate touch the upper wall of the pipe,

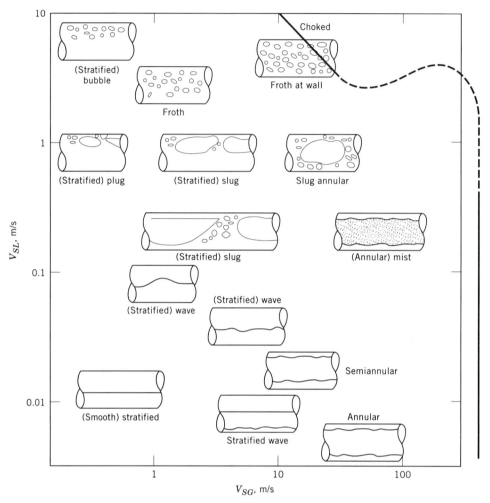

Fig. 11. Flow regimes for air–water in a 2.5-cm horizontal pipe where V_{SL} is superficial liquid velocity and V_{SG} is superficial gas velocity. Courtesy of Shell Development Co.

destroying the continuity of the gas phase. This is plug flow or slug flow, the former term connoting the more gentle action where the gas bubble moves forward without intense agitation, and the latter expressing the more violent action observed at higher gas-flow rates. As liquid-flow rate is raised still further, the gas plugs break up into rather small bubbles. The flow may then be described as bubble flow or froth flow, according to whether the gas volume fraction is small or large, respectively.

If the gas-flow rate is increased, one eventually observes a phase transition for the abovementioned regimes. Coalescence of the gas bubbles becomes important and a regime with both continuous gas and liquid phases is reestablished, this time as a gas-filled core surrounded by a predominantly liquid annular film. Under these conditions there is usually some gas dispersed as bubbles in the liquid and some liquid dispersed as droplets in the gas. The flow is then annular. Various qualifying adjectives may be added to further characterize this regime. Thus there are semiannular, pulsing annular, and annular mist regimes. Over a wide variety of flow rates, the annular liquid film covers the entire pipe wall. For very low liquid-flow rates, however, there may be insufficient liquid to wet the entire surface, giving rise to rivulet flow.

At very high flow rates, phenomena occur that are equivalent to the sonic flow observed with gases. This is illustrated in Figure 11 by the envelope labeled choked flow. Without special effort, it is impossible to produce velocities in a pipe at rates higher than indicated by the envelope. It is especially noteworthy that the limiting gas velocity can be very low provided the liquid rate is high. (Realization of this fact has often come as a surprise to a designer who has calculated higher velocities in some pipes.)

In inclined or vertical pipes, the flow regimes are similar to those described for horizontal pipes when both gas- and liquid-flow rates are high. At lower flow rates, the effects of gravity are important and the regimes of flow are quite different. For liquid velocities near 30 cm/s and gas velocities near 150 cm/s, piston or vertical slug flow is observed (Fig. 12). Large gas bubbles, having approximately spherical top surfaces, bridge the pipe. The calculation of the rate of rise of such

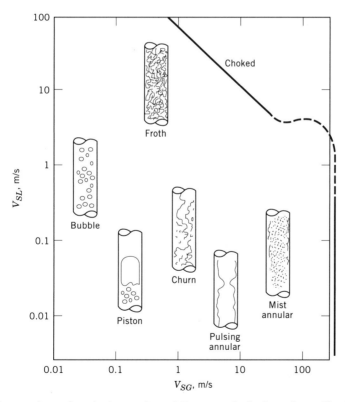

Fig. 12. Flow regimes for air–water in a 2.5-cm vertical pipe where V_{SL} is superficial liquid velocity and V_{SG} is superficial gas velocity. Courtesy of Shell Development Co.

bubbles can be made fairly accurately by assuming frictionless flow near the bubble apex. The success of this rather simple calculation illustrates one of the advantages of having a firm knowledge of flow regimes.

The upward flow of gas and liquid in a pipe is subject to an interesting and potentially important instability. As gas flow increases, liquid holdup decreases and frictional losses rise. At low gas velocity the decrease in liquid holdup and gravity head more than compensates for the increase in frictional losses. Thus an increase in gas velocity is accompanied by a decrease in pressure drop along the pipe, a potentially unstable situation if the flows of gas and liquid are sensitive to the pressure drop in the pipe. Such a situation can arise in a thermosyphon reboiler, which depends on the difference in density between the liquid and a liquid–vapor mixture to produce circulation. The instability is manifested as cyclic surging of the liquid flow entering the boiler and of the vapor flow leaving it.

Flow Instability. In many flow situations it is found that a mathematically valid solution to the Navier-Stokes equations is closely verified by experiment over some ranges of the variables, but when the variables are changed new flow patterns that are not in keeping with that solution are observed. The change is often rather sudden, eg, the laminar–turbulent transition. Such behavior occurs because solutions to the Navier-Stokes equations are generally not unique. When more than one solution exists, it is possible to observe one flow pattern under one set of circumstances and a different pattern under another. At the present stage of development of fluid mechanics, an experiment must be performed to determine whether a given solution applies. In some cases, however, it is possible, by analyzing the equations of motion, to determine the criteria by which one flow pattern becomes unstable in favor of another. The mathematical technique used most often is linearized stability analysis, which starts from a known solution to the equations and then determines whether a small perturbation superimposed on this solution grows or decays as time passes.

Fluids in Motion. Many of the instabilities associated with fluids in motion are of the shear-flow type. In shear flow the velocity varies principally in a direction perpendicular to the flow direction, eg, pipe flow, boundary layer flow, jet flow, and wake flow. Such flows change from laminar to turbulent when the Reynolds number based on some characteristic length scale exceeds a critical value, which may be different for each flow configuration. In pipe flows and some types of boundary layer flows, the instability is usually referred to as Tollmien-Schlichting instability. This instability occurs in essentially four steps: (*1*) small two-dimensional waves form and are linearly amplified; (*2*) the two-dimensional waves develop into finite three-dimensional waves and are amplified by nonlinear interactions; (*3*) a turbulent spot forms at some localized point in the flow; and (*4*) the turbulent spot propagates until the spot fills the entire flow field with turbulence (2). At low Reynolds numbers, viscosity damps out the instability. At high Reynolds numbers, however, viscosity provides the destabilizing mechanisms.

The phenomena are quite complex even for pipe flow. Efforts to predict the onset of instability have been made using linear stability theory. The analysis predicts that laminar flow in pipes is stable at all values of the Reynolds number. In practice, the laminar–turbulent transition is found to occur at a Reynolds number of about 2000, although by careful design of the pipe inlet it can be postponed to as high as 40,000. It appears that linear stability analysis is not applicable in this situation.

In jet flow and wake flow there occurs another type of instability called Kelvin-Helmholtz instability. This instability may be illustrated by considering the development of a small disturbance in the flow situation given in Figure 13. Suppose a small disturbance causes a slight waviness of the boundary between the two flows. The fluid on the convex sides of each flow moves slightly faster and that on the concave sides moves slightly slower. According to the Bernoulli equation, this disturbance decreases the pressure on the convex sides of each flow, and thus the initial disturbance is amplified. Kelvin-Helmholtz instability is also observed in horizonal concurrent flows of stratified immiscible fluids (Fig. 14). Raising of a wave at the interface forces the upper fluid to increase in velocity and drop in pressure as it passes the wave. If the pressure decrease is large enough, it overcomes the increase in potential associated with the raising of the wave and the disturbance grows.

An instability involving transition from one laminar flow pattern to another occurs in the flow between concentric, mutually rotating cylinders. When the inner cylinder is rotated at an angular velocity below a critical value, motion is purely circumferential (Couette flow). Above this value, however, centrifugal force

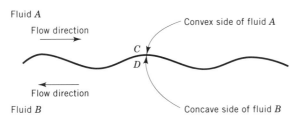

Fig. 13. Sketch of Kelvin-Helmholtz instability, where C, the convex side of fluid A, is at a lower pressure than D, the concave side of fluid B.

Fig. 14. Shear instability in stably stratified fluid; lower denser fluid is dyed. Courtesy of Cambridge University Press (10).

destabilizes the flow and a series of laminar, cellular vortices known as Taylor cells is superimposed on the main flow (Fig. 15). The pattern of cells depends on the geometry and whether the outer cylinder is stationary or also rotating. As the angular velocity is increased, the cells become wavy and then irregular. At still higher velocities the cells break up and the flow becomes turbulent. If only the outer cylinder is rotated, Taylor cells do not form, and this arrangement is customarily used in Couette-type viscometers so as to maximize their useful ranges of shear rates.

When a fluid of low viscosity is used to displace a fluid of higher viscosity from a porous medium or a pipe, the displacement can become unstable by a process known as fingering. Small fingers of the low viscosity fluid, once formed, become regions of lower pressure drop. The displacing fluid flows preferentially into these fingers which grow and ultimately reach the outlet, leaving a portion of the viscous fluid undisplaced. Such displacements are widely practiced to increase the productivities of oil fields, and much effort is devoted to minimizing fingering by proper adjustment of the physical and chemical properties of the displacing medium, as well as of the rate of displacement.

Fluids at Rest.   Fluids at rest may be set into motion by impressing upon them gradients in body or surface forces. Benard instability refers to the formation of convection cells within a fluid as a result of the action of a gravitational field on density differences induced by a temperature gradient in the fluid. Such behavior is observed, for example, when fluid is confined between two horizontal plates of which the lower is heated. The heated fluid on the bottom expands, resulting in an unstable density gradient and providing a driving force to redistribute the fluid. Redistribution is inhibited by viscosity, which slows the flow,

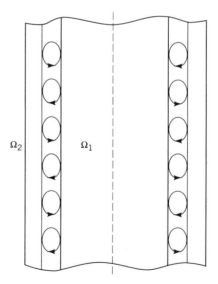

Fig. 15.   Flow pattern in rotating Couette flow where Ω_1 and Ω_2 represent the outer and inner rotational speeds. Courtesy of Cambridge University Press (11).

and by thermal conductivity, which tends to even out the temperature difference. The onset of instability is described by a critical Rayleigh number (Ra):

$$Ra = \frac{g\left(\dfrac{\partial \rho}{\partial T}\right)\Delta T l_b^3}{\mu \alpha} \tag{26}$$

Hot fluid rises on one side and cold fluid falls on the other side of the cell. The rising fluid cools as it nears the top, and the falling fluid warms as it nears the bottom, thus maintaining a steady flow.

Cellular motions may also arise from gradients in surface tension caused by variations in temperature or concentration. This behavior is commonly referred to as Marangoni instability. To illustrate this, consider a solution from which a volatile solute is evaporating and suppose the surface tension of the solution decreases as the concentration of solute increases. Because the solute is evaporating, the surface has a lower concentration of solute than the bulk. If a small disturbance occurs and causes a spot of high solute concentration (low surface tension) to appear, this spot spreads. Spreading is a result of the attempt by the interface to minimize its free energy by expansion of regions of low surface tension and contraction of those of high surface tension. The local spreading is accompanied by movement of fluid from the bulk to fill the spreading spot. Because the new fluid has lower surface tension too, the original disturbance is amplified and convection cells are established. In some cases the driving force is so strong that the surface undergoes violent twitching with a large increase in the overall mass-transfer rate. Such motion is commonly termed interfacial turbulence. If the direction of mass transfer is reversed, disturbances usually are not amplified and the surface remains quiescent. This effect is responsible for the occasionally observed large differences in mass-transfer rates for absorption and desorption in otherwise the same system. The transfer of acetic acid between water and carbon tetrachloride is a classic example.

Circulation of fluid is promoted by surface tension gradients but inhibited by viscosity, which slows the flow, and by molecular diffusion, which tends to even out the concentration differences. The onset of instability is described by a critical Marangoni number (Ma), an analogue of the Rayleigh number:

$$Ma = \frac{\left(\dfrac{\partial \sigma}{\partial c}\right)\left(\dfrac{\partial c}{\partial x}\right)l_M^2}{\mu \mathscr{D}} \tag{27}$$

Linear stability analysis has been successfully applied to derive the critical Marangoni number for several situations.

Surface tension is also responsible for the varicose or Rayleigh breakup of liquid strands into droplets. By virtue of surface tension the pressure within a strand is slightly higher than that in the ambient gas by the amount:

$$\Delta P = \frac{2\sigma}{D} \tag{28}$$

A small perturbation (reduction) in the radius results in a locally higher pressure than in the rest of the strand, causing liquid to flow away from that region, reducing the radius still further and making the thread unstable. Rayleigh analyzed this phenomenon for a low velocity, inviscid, cylindrical jet. He determined that one wavelength of disturbance grows more rapidly than all others and thus, in the absence of forced oscillations, determines the drop diameter. For an inviscid fluid this is about twice the strand diameter. At higher flows a jet undergoes sinuous breakup in which the entire strand oscillates like a vibrating string. At still higher rates, as in commercial swirl atomizing nozzles, the liquid exits as a sheet that undergoes Kelvin-Helmholtz breakup into thread-like rings which in turn disintegrate further, via Rayleigh instability, into fine droplets. The drop size produced is a function primarily of nozzle diameter and fluid velocity. One widely used empirical correlation (12) yields the relationship

$$\overline{X} \propto D^{0.65} V^{-0.55} \rho_L^{-0.35} \sigma^{0.2} \mu_L^{0.15} \tag{29}$$

Atomization. A gas or liquid may be dispersed into another liquid by the action of shearing or turbulent impact forces that are present in the flow field. The steady-state drop size represents a balance between the fluid forces tending to disrupt the drop and the forces of interfacial tension tending to oppose distortion and breakup. When the flow field is laminar the ability to disperse is strongly affected by the ratio of viscosities of the two phases. Dispersion, in the sense of droplet formation, does not occur when the viscosity of the dispersed phase significantly exceeds that of the dispersing medium (13).

More commonly, atomization occurs under turbulent conditions. The mechanism of atomization and its quantitative description are still incompletely understood. It is possible that breakup occurs because of the impact forces exerted by the small, usually isotropic turbulent eddies in the manner proposed in Reference 14. The effects of these eddies are measured by the local rate of energy dissipation, but since this quantity is seldom known, correlations are often expressed in terms of average energy dissipation per unit mass or volume. In fact, in most important situations, eg, stirred tanks, breakup occurs in highly anisotropic regions, eg, the vortices shed from the turbine blades, where the local energy dissipation bears little relationship to the average and where nonturbulent mean shearing stresses may be large. In pipe flow it appears that, at least initially, breakup occurs quite close to the wall, quite likely as a result of mean flow shear. Drop size is related to velocity (gradient) and not to average energy dissipation. Under these circumstances the most practical approach is to employ empirical correlations derived for the geometries and residence times of interest.

Drop or bubble sizes are also strongly affected by the tendency of the drops or bubbles to coalesce once they enter regions of low turbulence or low mean shear. When coalescence is rapid, as in dispersion of gas in pure liquids, average bubble size seems to be determined by the average rate of energy dissipation and is insensitive to how the energy is applied. Correlations obtained for one geometry can be applied with reasonable success to other geometries. In many liquid–liquid systems where coalescence rates are lower, the drops made in the regions of intense shear can persist throughout. In a stirred vessel, for example, these drops are characteristic of conditions near the impeller and the same impeller operating at the same speed in two vessels of different sizes gives about the same size drops, despite the lower average energy dissipation in the larger vessel. In the extreme case of no coalescence, eg, emulsions stabilized by surface-active agents, drop sizes are determined by the highest shear rates available. In a stirred vessel this results in a sensitivity of drop size only to tip speed and not to impeller shape or energy input. In a practical situation the time needed to emulsify is as important as ultimate drop size, and impellers providing reasonable circulation times are required.

Secondary Flows. In many cases a cursory examination of the flow pattern might indicate a rather simple type of flow with a high degree of symmetry, whereas in fact the flow realized is more complex. In the flow around a bend, for example, one might imagine that the individual streamlines simply follow the general course of the curvature of the pipe; in fact they do not. Instead, a pattern of secondary flow develops that is superimposed on the main flow so that the streamlines are actually helical. The choice is somewhat arbitrary as to which flow to call main and which to call secondary. Adopting the convention that the main flow is parallel to the tube axis, then at the axis the secondary flow is directed outward toward the section of pipe having the weakest curvature, returning inward along the pipe wall. This type of secondary flow is a consequence of the inertia of the fluid. It is most obvious at low Reynolds numbers but is also significant in turbulent flow. By improving the lateral transport of momentum it postpones the transition from laminar to turbulent flow. Even relatively small curvatures can have a significant effect as is evidenced by the empirical correlation for helices:

$$\left(\frac{D\overline{V}\rho}{\mu}\right)_{\text{transition}} = 2100(1 + 12\,(D/D_H)^{1/2}) \qquad (30)$$

Secondary flows also occur in channels that have a noncircular cross section, but are otherwise straight. In these cases the secondary flows are directed outward into the corners and return along the walls.

Solutions to Equations of Motion

Three basic approaches have been used to solve the equations of motion. For relatively simple configurations, direct solution is possible. For complex configurations, numerical methods can be employed. For many practical situations, particularly three-dimensional or one-of-a-kind configurations, scale modeling is employed and the results are interpreted in terms of dimensionless groups. This section outlines the procedures employed and the limitations of these approaches (see COMPUTER-AIDED ENGINEERING (CAE)).

Exact Solutions to the Navier-Stokes Equations. As was true for the inviscid flow equations, exact solutions to the Navier-Stokes equations are limited to fairly simple configurations that allow for considerable simplification both in the equation and in the boundary conditions. For the important situation of steady, fully developed, laminar, Newtonian flow in a circular tube, for example, the Navier-Stokes equations reduce to

$$\frac{dP}{dX} = \frac{\mu}{r} \frac{d}{dr} \left(r \frac{du}{dr} \right) \qquad (31)$$

This equation may be integrated and the constant of integration evaluated using the boundary conditions: $(du/dr)_{r=0}$ and $u(R) = 0$. The solution is the well-known Hagen-Poiseuille relationship given by

$$\frac{\Delta P}{L} = \frac{32\mu\overline{V}}{D^2} \qquad (32)$$

Solutions for other simple flow situations are also available (1).

Numerical Solution of the Equation of Motion. Usually an analytical solution to the equation of motion cannot be found; recourse must be had to numerical (computational) methods (see ENGINEERING, CHEMICAL DATA CORRELATIONS). The field of computational fluid dynamics (CFD) is moderately mature. A vast repertoire of techniques exists (15) and solutions to many complex flows directly relevant to industrial situations can be obtained. Also, a number of CFD flow codes are now commercially available. However, in spite of the significant progress, many problems remain in the treatment of turbulent flows. Furthermore, three-dimensional turbulent flows are computationally expensive and their solution, for most practical problems, requires the use of supercomputers (see COMPUTER TECHNOLOGY).

In general, a computational method includes the task of dividing the flow domain of interest into a network of elements with adequate grid resolution, obtaining a set of algebraic equations for the unknown dependent values at the grid intersections or nodes by discretization of the set of pertinent partial differential equations (PDE), and then prescribing an efficient algorithm for solving the set of algebraic equations. The computational results are values of the dependent variables, such as velocity, pressure, temperature, etc, at the grid points which are studied with the help of a graphics display package.

Discretization of the Governing Equations. The basic equations of fluid mechanics are the time-dependent Navier-Stokes equations. These include the equations for conservation of mass and momentum and are given by equations 1–6. Other PDE such as the turbulent transport, energy, and other scalar equations may be included depending on whether or not the flow is turbulent and whether heat transfer or other scalar transport is involved. Many alternative techniques are available for the discretization of the partial differential equation set. The commonly used techniques are finite-difference, finite-volume, finite-element, and spectral methods. The finite-volume method (FVM) is the most widely used method in the CFD codes used by the process industries. In the FVM, each partial differential equation is replaced by a set of algebraic finite-difference equations obtained by integration of the equation over a fictitious control volume surrounding the point of location of the relevant variable in the grid structure. In the so-called staggered grid system, the pressure and the other scalar quantities are calculated at the nodes and the velocities are computed at locations midway between the nodes (Fig. 16). The staggered arrangement avoids checkerboard oscillations in pressure and velocity. Techniques have been developed to avoid checkerboard oscillations using the so-called nonstaggered system where the velocity components and the scalar quantities are calculated using the same set of nodes. For simulations using a boundary fitted coordinate system, the nonstaggered arrangement requires significantly less computer time and memory to execute. Second-ordered centered differencing schemes are used in the discretization with the exception of the advection terms where special differencing schemes are used to improve numerical stability by properly allocating the variables between the nodes based on the ratio of the advective to diffusive flux coefficients. Upwind differencing, hybrid differencing, high order upwind differencing, and quadratic upwind differencing are but a few examples. The resulting algebraic equations have the general form

$$(a_P - S_P)\phi_P = \sum_n a_n \phi_n + S_0 \tag{33}$$

where a contains the advection and diffusion effects, S represents the effects of sources and sinks, and P and n denote central and neighboring nodes. Summation occurs over the six neighboring nodes. Further details may be found in Reference 16.

Algorithm and Solution of the Linearized Equations. The procedure is to solve the finite difference equation sets for each dependent variable (inner iteration)

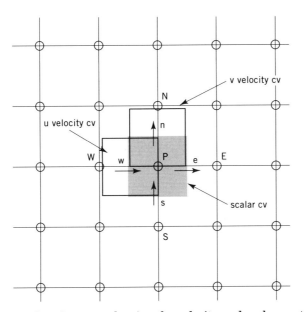

Fig. 16. Nonstaggered grid system showing the velocity and scalar control volumes (cv). The scalar quantities are calculated at the nodes (eg, P, N, S, E, W) and the velocities are calculated where the small arrows (eg, n, s, e, w) are shown. See text.

sequentially and repeatedly. Each cycle of calculation (outer iteration) consists of computing the velocity components from the momentum equations using the most recently calculated pressure field (a guessed pressure field is used during the first cycle of iteration). The velocities and the pressure fields are then corrected to satisfy the continuity equation. This is followed by the calculation of the other field variables, eg, turbulence quantities, temperature, and concentration, if these influence the flow field. The procedure is repeated until the solution converges, ie, until the so-called residual source (the sum of the absolute values of the residuals of all the equations) is less than a predetermined value. This procedure is followed in the semi-implicit method for pressure-linked equation (SIMPLE) algorithm (17). Other variants of the SIMPLE algorithms, SIMPLER, SIMPLEC, PISO, PISOC, etc, differ primarily in the procedure used in correcting for continuity. A large proportion of the computer time is used to solve the individual transport equations. Each of the transport equations consists of a single diagonally dominant equation. Good solvers for handling such equations are available. The alternating direction method, line relaxation method, the Stone algorithm, and the incomplete Cholesky conjugate gradient method are but a few examples. To avoid numerical instabilities, relaxation parameters are used in that only a fraction of the change of the dependent variables calculated at the current iteration step are applied to the next iteration step. The relaxation parameters can be different for each of the dependent variables.

Turbulence Modeling. The time-dependent Navier-Stokes equations are generally considered adequate to represent turbulent flows. Direct numerical solution (DNS) of these equations is limited to low Reynolds numbers. At higher Reynolds numbers, the number of grid points required to resolve small eddies and the small time step size needed to obtain meaningful results make the computation of turbulent flows encountered in engineering practice by DNS outside the capability of present computers. A less computationally intensive technique called large eddy simulation (LES) calculates the three-dimensional time-dependent details of the largest scales of motion using a simple subgrid scale model for the smaller eddies. However, the method is still very computationally intensive. Both DNS and LES are used primarily for studying the physics of turbulence, but LES has the potential of becoming an engineering tool in the near future

As of this writing, the only practical approach to solving turbulent flow problems is to use statistically averaged equations governing mean flow quantities. These equations, which are usually referred to as the Reynolds equations of motion, are derived by Reynolds' decomposition of the Navier-Stokes equations (18). The randomly changing variables are represented by a time mean and a fluctuating part:

$$\phi = \overline{\phi} + \phi' \tag{34}$$

Following Reynolds, a time-averaged quantity $\overline{\phi}$ is defined:

$$\overline{\phi} = \frac{1}{t_1} \int_{t_0}^{t_0 + t_1} \phi(t)\, dt \tag{35}$$

where t_1 is long compared to the period of the random fluctuations. By definition, the time average of the fluctuating quantity is zero, thus:

$$\overline{\phi'} = 0 \tag{36}$$

Equations 34 and 36 are applied to equations 1–6. After time-averaging, the following equations are obtained for incompressible flows:

continuity
$$\frac{\partial}{\partial x_j} \overline{u}_j = 0 \tag{37}$$

momentum
$$\frac{\partial}{\partial t} (\rho\, \overline{u}_i) + \frac{\partial}{\partial x_j} (\rho\, \overline{u}_i\, \overline{u}_j) = -\frac{\partial \overline{P}}{\partial x_i} + \frac{\partial}{\partial x_j} \left\{ \mu \left(\frac{\partial}{\partial x_j} \overline{u}_i + \frac{\partial}{\partial x_i} \overline{u}_j \right) - \rho \overline{u_i' u_j'} \right\} \tag{38}$$

Each of the three time-averaged momentum equations contains three unknown turbulent stresses, $\rho \overline{u_i' u_j'}$, commonly termed Reynolds stresses, only six of which are independent. The Reynolds stress $\rho \overline{u_1' u_2'}$, for example, is the rate at which x-momentum, $\rho u_1'$, is being transported in the y-direction by the velocity fluctuation u_2'. A hierarchy of equations for velocity correlation functions, ie, averages of products of u_i', can be obtained, but each equation so derived involves an unknown higher order correlation function and hence the set of equations is not closed. A turbulence model is needed to determine the turbulent transport terms before the set of equations can be solved. Turbulence modeling is concerned with the development and testing of closure assumptions for the Reynolds stresses. A large number of closure models are available. They are usually divided into two groups, eddy viscosity models and Reynolds stress models, according to whether or not the Boussinesq assumption is applied.

Eddy Viscosity Models. A large number of closure models are based on the Boussinesq concept of eddy viscosity:

$$-\overline{u'v'} = \nu_t \frac{\partial u}{\partial y} \tag{39}$$

These models are usually categorized according to the number of supplementary partial differential transport equations which must be solved to supply the modeling parameters. The so-called zero-equation models do not use any differential equation to describe the turbulent quantities. The best known example is the Prandtl (19) mixing length hypothesis:

$$\nu_t = l_m^2 \left| \frac{\partial \overline{u}}{\partial y} \right| \tag{40}$$

where l_m, the mixing length of turbulent motion, must be obtained from experimental data. The model gives good predictions for many boundary layer flows. Equation 40 assumes that turbulence production is equal to the dissipation at each point in the flow field; thus it is not affected by turbulence in the neighborhood of that point. This deficiency leads to unrealistic simulations in many cases.

One-equation models relax the assumption that production and dissipation of turbulence are equal at all points of the flow field. Some effects of the upstream turbulence are incorporated by introducing a transport equation for the turbulence kinetic energy k (20) given by

$$\frac{\partial k}{\partial t} + \overline{u}_i \frac{\partial k}{\partial x_i} = \frac{\partial}{\partial x_i}\left(\frac{\nu_t \partial k}{\sigma_k \partial x_i}\right) - \overline{u_i' u_j'}\frac{\partial \overline{u}_i}{\partial x_j} - \epsilon \tag{41}$$

The quantity k is related to the intensity of the turbulent fluctuations in the three directions, $k = 0.5\,\overline{u_i' u_i'}$. Equation 41 is derived from the Navier-Stokes equations and relates the rate of change of k to the advective transport by the mean motion, turbulent transport by diffusion, generation by interaction of turbulent stresses and mean velocity gradients, and destruction by the dissipation ϵ. One-equation models retain an algebraic length scale, which is dependent only on local parameters. The Kolmogorov-Prandtl model (21) is a one-dimensional model in which the eddy viscosity is given by

$$\nu_t = c_\mu (k l_t)^{1/2} \tag{42}$$

where l_t is the length scale of turbulence and c_μ is an empirical constant. This is a much better model than Prandtl's mixing length model for nonequilibrium shear layers if the distribution of length scale can be estimated with confidence. Its main shortcoming is the assumption that the length scale is dependent only on local flow parameters and therefore not affected by the upstream processes. This difficulty is addressed by use of a two-equation model which adds a transport equation for the length scale or a length scale parameter. Of the many two-equation models proposed in the literature, the k-ϵ model has been the most widely used and has proven successful in many calculations having practical relevance. It is one of the most tested turbulence models and is used in almost all of the commercial CFD packages used in the process industries.

The two-equation model of Launder and co-workers (22,23) and the one most often used employs the ϵ equation given by

$$\frac{\partial \epsilon}{\partial t} + \overline{u}_i \frac{\partial \epsilon}{\partial x_i} = \frac{\partial}{\partial x_i}\left(\frac{\nu_t \partial \epsilon}{\sigma_\epsilon \partial x_i}\right) - c_{\epsilon 1}\frac{\epsilon}{k}\overline{u_i' u_j'}\frac{\partial \overline{u}_i}{\partial x_j} - c_{\epsilon 2}\frac{\epsilon^2}{k} \tag{43}$$

The turbulent kinetic energy is calculated from equation 41. Equation 43 defines the rate of energy dissipation, ϵ, which is related to the length scale via

$$l_t = c_D \frac{k^{2/3}}{\epsilon} \tag{44}$$

where c_D is an empirical constant. The k-ϵ model contains five empirical constants. In addition, turbulent Prandtl and Schmidt numbers must be included when turbulent heat and mass transfer are present. The values ($c_\mu = 0.09$, $c_{\epsilon 1} = 1.44$,

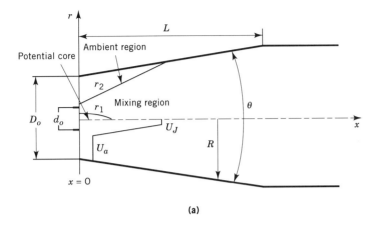

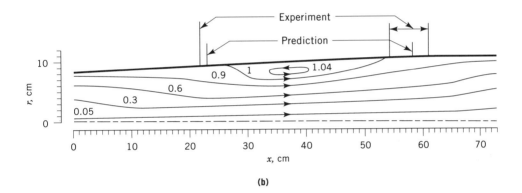

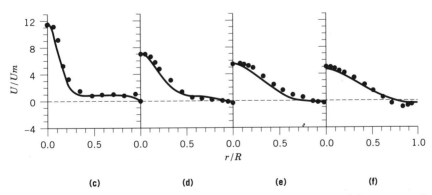

Fig. 17. Comparison of the predictions of k-ϵ model with experimental data for a turbulent jet inside a 5° conical duct. (**a**) Flow geometry and inlet conditions, where geometry d_o = 1.6 cm, D_o = 16 cm, L = 64 cm, θ = 5°; flow conditions, ρ = 0.998 g/cm³, μ = 0.01 g/cm·s, U_j = 40 cm/s, U_a = 2.33 cm/s. (**b**) Predicted streamlines (5). (**c**–**f**) Axial velocity profiles, where x = 10, 20, 30, and 40 cm, respectively. Courtesy of John Wiley & Sons, Inc. (25).

$c_{\epsilon 2} = 1.92, \sigma_k = 1.0, \sigma_\epsilon = 1.3$) have been recommended (24). Using these empirical constants, the model has done well in predicting quite a variety of flows, including confined flows with separation and complex three-dimensional flows with weak swirl. In general, the k-ϵ predictive capability is poor for flows subject to and influenced by anisotropic stress such as curved shear layers. The k-ϵ model is not appropriate in the immediate vicinity of the wall where the local Reynolds number of turbulence is low. Empirical relations called wall functions are used to bridge the viscous sublayer (24). These functions are derived from one-dimensional analysis of Couette flow in the near wall region.

Examples of flow simulations using the k-ϵ model are shown in Figures 17–19. The first example is a numerical simulation of a turbulent jet inside a 5°

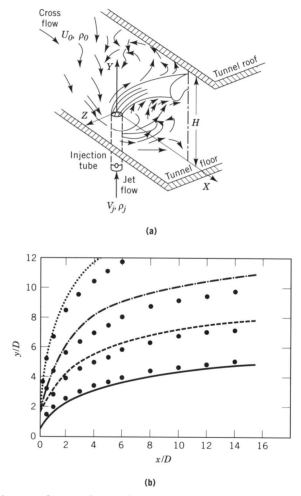

Fig. 18. Jet trajectory of a round jet in bounded cross flow where $J = \rho_j V_j^2 / \rho_0 U_0^2$ (**a**) flow geometry, ratio of height of tunnel to diameter of injection tube $(H/D) = 12$; and (**b**) flow streamlines where the data points are experimental determinations and the lines correspond to calculated predictions for (——) $J = 8$, (– – –) $J = 18$, (—·—) $J = 32$, and (····) $J = 72$. Courtesy of American Institute of Aeronautics and Astronautics (20).

conical duct (25). The flow geometry and inlet conditions are shown in Figure 17**a** and the calculated streamlines in Figure 17**b**. The flow exhibits a recirculation bubble at the duct wall whose separation and reattachment locations are in good agreement with experimental data. Figures 17**c**–17**f** show a comparison of the predicted and experimentally measured axial velocity profiles at four different locations downstream of the duct inlet. The predictions are in good agreement with experimental data except for a slight underprediction of the extent of the reverse flow region at a distance of 40 cm from the duct inlet. The turbulent shear stress components and the turbulent kinetic energy profiles (not shown) are only qualitatively predicted. Figure 18 contains the development of a jet in a bounded cross flow for various jet-to-cross flow momentum ratios, J. The predictions are in good agreement with experimental data except at high values of the parameter J (20).

The third example illustrates the capability of the k-ϵ model in simulating confined swirling flows having internal circulation zones, which are encountered in many process applications where intense mixing is needed. Figure 19**a** shows

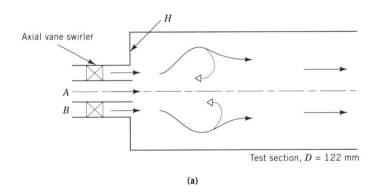

(a)

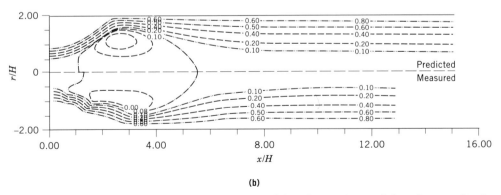

(b)

Fig. 19. Comparison of the predictions of k-ϵ model and experimental data for a confined swirling flow. (**a**) Flow configuration where A is the primary inlet, $D = 25$ mm, and B is the secondary inlet, $D_i = 31$ mm, $D_0 = 59$ mm; and the step height, $H = 31.5$ mm. (**b**) Predicted and measured streamline values where r/H is the ratio of the radial distance from the centerline to the step height. Courtesy of John Wiley & Sons, Inc. (26).

the configuration of a confined double concentric jet which expands suddenly into a cylindrical chamber (26). The simulation swirl number (dimensionless ratio of angular momentum to linear momentum) and Reynolds number are 0.38 and 47,000, respectively. The calculated and experimentally measured streamlines are shown in Figure 19**b**. The shape and size of the recirculating zone is well predicted although the penetration of the primary jet in the early part of the cylindrical chamber is somewhat slightly underpredicted. Axial velocity profiles are also well predicted but not so the predicted tangential velocity profiles which at large distances from the inlet to the cylindrical chamber show large discrepancies from measured values (26). The turbulence intensity profiles predicted are much higher than the measurements although the shapes are approximately correct. Discrepancies very likely result from the moderately high swirl number of the flow. In general, however, the mean flow properties are predicted with sufficient accuracy for process applications. At higher swirl numbers, more accurate turbulence models such as the so-called Reynold-stress models would have to be used.

Reynolds Stress Models. Eddy viscosity is a useful concept from a computational perspective, but it has questionable physical basis. Models employing eddy viscosity assume that the turbulence is isotropic, ie, $\overline{u'_1 u'_1} = \overline{u'_2 u'_2} = \overline{u'_3 u'_3}$ and $\overline{u'_1 u'_2} = \overline{u'_2 u'_3} = \overline{u'_1 u'_3} = 0$. Another limitation is that the turbulent stresses are related only to one velocity scale, $k^{1/2}$. In the so-called Reynolds stress models, transport equations for $\overline{u'_i u'_j}$ are introduced to allow for the individual stresses to develop quite differently. Transport equations for the Reynolds stresses (27) are

$$\frac{\partial}{\partial t}\overline{u'_i u'_j} + \overline{u}_l \frac{\partial}{\partial x_l}\overline{u'_i u'_j} = c_s \frac{\partial}{\partial x_l}\left(\frac{k}{\epsilon}\overline{u'_k u'_l}\frac{\partial}{\partial x_k}\overline{u'_i u'_j}\right) - \overline{u'_i u'_l}\frac{\partial}{\partial x_l}\overline{u'_j} - \overline{u'_j u'_l}\frac{\partial}{\partial x_l}\overline{u'_i}$$

$$- c_1\frac{\epsilon}{k}\left(\overline{u'_i u'_j} - 2/3\delta_{ij}k\right) + c_2\left(\overline{u'_i u'_l}\frac{\partial}{\partial x_l}\overline{u'_j} + \overline{u'_j u'_l}\frac{\partial}{\partial x_l}\overline{u'_i}\, 2/3\delta_{ij}\overline{u'_i u'_j}\frac{\partial}{\partial x_j}\overline{u'_i}\right) \quad (45)$$

$$- 2/3\;\epsilon\delta_{ij}$$

These transport equations are often referred to as second-order closure and are solved simultaneously with the dissipation equation (eq. 43) and the Reynolds averaged equations (37,38). Thus RSM models require more computer time and memory because of the larger number of PDE involved and the nonlinearity and strong coupling of the set of equations. This extra effort, however, yields more realistic results for flows having rotation, curvature, and strong swirl. Simplified relations in the form of eddy viscosity equations have been used to model the convective and diffusive terms in the differential stress transport equations, thus reducing them to algebraic expressions. These simplified models, which retain many of the features of the RSM models, are called algebraic stress models (ASM).

Figure 20 compares the predictions of the k-ϵ, RSM, and ASM models and experimental data for the growth of the layer width δ and the variation of the maximum turbulent kinetic energy k and turbulent shear stress $\overline{uv}$ normalized with respect to the friction velocity $(\overline{\tau}_w)^{1/2}/\rho$ for a curved mixing layer (20). The

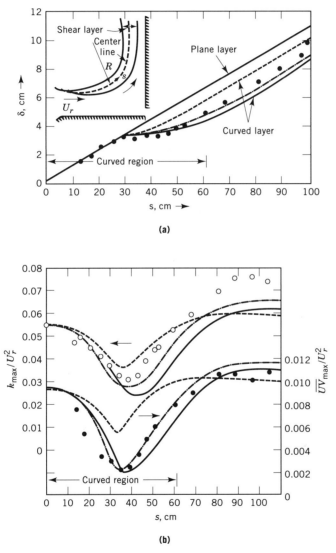

Fig. 20. Comparison of the predictions of the (- - -)k-ϵ, (——) ASM, and (—) RSM models with experimental data for a curved mixing layer. (**a**) Values of the shear-layer thickness, δ, plotted against distance, s, measured along the centerline. (**b**) Calculated ($\circ$) and measured ($\bullet$) streamwise variation of maximum turbulence intensity, k, and maximum shear stress, $\overline{UV}$. Courtesy of American Institute of Aeronautics and Astronautics (21).

experimental data are predicted well by the ASM and RSM models but the k-ϵ model fails to account for most of the curvature effects.

Convergence and Accuracy.   Converged numerical solutions can be quite tedious to obtain for complex flows because the nonlinearity of the finite difference equations usually makes the numerical procedure susceptible to numerical instabilities. These instabilities are often caused by poor estimate of the initial (guessed) field variables and/or rather poor choice of the relaxation parameters. In the SIMPLE algorithm, insufficiently converged solutions of the pressure correction equation at each cycle of the outer iteration often also cause numerical instabilities. The accuracy of the converged numerical solution depends on the values of the residual sources when the iterative calculations are terminated and the extent by which the finite-difference equations approximate the PDE. In general, smaller grid sizes are needed, particularly in areas where the gradients are steep, to get accurate solutions. A good check for the accuracy is to reduce the grid size until a grid independent solution is obtained. This practice can be computationally expensive for complex flows in a large domain. Most of the CFD work reported in the literature does not provide adequate assessment of the accuracy of the simulations.

Complex Geometries.   To extend the capabilities of finite-difference and finite-volume methods to handle complex geometries, a boundary fitted coordinate system (BFCS) is now being used and is available in many commercial flow codes. The underlying concept of the BFCS is to map (28) the complex flow domain in physical space to a regular, eg, rectangular, flow domain in the computational space. The boundaries of the physical flow domain coincide with the curvilinear coordinate lines. The boundary conditions can then be conveniently implemented in the rectangular computational domain, albeit at the expense of making the PDE more complicated.

Modeling.   The majority of technological flow problems are not solved by integrating the equations of motion. Instead, most are solved by carrying out laboratory experiments which are then correlated, ie, interpreted, so as to yield useful information about systems that may differ greatly in size and in fluid properties. For the behavior of the experimental model to duplicate that of a system of interest, two criteria must, in principle, be met: the experimental apparatus must be geometrically similar to the system of interest; and certain dimensionless groupings of variables must be duplicated on the two scales. There are two basic methods available for determining the dimensionless groups appropriate to a given situation: dimensional analysis (qv), which can be applied when the equations governing the process are not known; and similarity analysis, which proceeds from the governing equations and offers physical insights into the meanings of the groups.

Dimensional Analysis.   Dimensional analysis is a mathematical technique that proceeds from the general principle that physical laws must be independent of the units of measurement used to express them. If one quantity is related to a group of other quantities, the quantities comprising the group must be related in such a manner that the net units or dimensions of the group are the same as those of the dependent quantity. A dimensionless group can then be formed immediately by division. A useful tool, the Pi theorem, asserts that the number of dimensionless groups needed to describe a situation is equal to the total number of variables less the number of fundamental dimensions needed to express them. Fundamental dimensions are generally taken to be length, time, mass, temperature, and heat content. The Pi theorem is a powerful tool because it limits the amount of experimental work needed to establish a general relationship. For example, consider the problem of determining the drag force on a smooth sphere around which a Newtonian fluid is flowing. The diameter of the sphere and the fluid viscosity, density, and velocity may vary. Each of the five variables can be expressed using various combinations of the three dimensions mass, length, and time (force = mass times length per the square of time). The Pi theorem leads immediately to the conclusion that only two dimensionless numbers are needed to describe the relationship. These may be taken to be a drag coefficient $2F/A_p\rho V_\infty^2$ and a Reynolds number $DV_\infty\rho/\mu$. An accurate set of measurements for one sphere in one fluid provides a universal relationship between these numbers that is applicable to all spheres in all Newtonian fluids.

The strength of dimensional analysis lies in its ability to limit the number of studies that need to be made and to handle situations in which the governing equations are not known. It can even handle lack of geometrical symmetry by using ratios of important dimensions as additional dimensionless groups. Its weakness lies in the need to know, a priori, which variables must be included, which can be ignored, and in its awkwardness when handling several variables that have the same dimensions, eg, densities in multiphase mixtures. This last difficulty reflects the fact that dimensional analysis is not capable of describing the functional relationship among the dimensionless groups, nor is it capable of describing how an empirical relationship might be extrapolated outside the range covered by the original data, nor is it even capable of selecting the best set of dimensionless groups, since products, sums, quotients, and differences of dimensionless groups are also dimensionless. To achieve the simplest, most meaningful relationship and to judge how it might best be extrapolated, it is necessary for the investigator to use previous experience, similarity analysis, or physical insight.

Similarity Analysis. Similarity analysis starts from the equation describing a system and proceeds by expressing all of the dimensional variables and boundary conditions in the equation in reduced or normalized form. Velocities, for example, are expressed in terms of some reference velocity in the system, eg, the average velocity. When the equation is rewritten in this manner certain dimensionless groupings of the reference variables appear as coefficients, and the dimensional variables are replaced by their normalized relatives. If another physical system can be described by the same equation with the same numerical values of the coefficients, then the solutions to the two equations (normalized variables) are identical and either system is an accurate model of the other.

The principle can be illustrated by examining the Navier-Stokes equation for two-dimensional incompressible flow. The x-component of the equation is

$$\rho\frac{Du}{Dt} = \rho g_x - \frac{\partial P}{\partial x} + \mu\left(2\frac{\partial^2 u}{\partial x^2} + \frac{\partial^2 u}{\partial y^2} + \frac{\partial^2 v}{\partial x\partial y}\right) \tag{46}$$

All variables in the system can be expressed in reduced form. Velocity can be expressed as $u = V_0 u'$ where V_0 is a fixed reference velocity and u' is the dimensionless reduced velocity. Because time, t_0, is the quotient of length, L_0, and velocity, V_0, the equation can be manipulated to yield

$$\rho' \frac{Du'}{Dt'} = \left[\frac{g_0 L_0}{V_0^2}\right] \rho' g_x' - \left[\frac{P_0}{\rho_0 V_0^2}\right] \frac{\partial P'}{\partial x'} + \left[\frac{\mu_0}{\rho_0 L_0 V_0}\right]\left(2\mu' \frac{\partial^2 u'}{\partial x'^2} + \ldots\right) \quad (47)$$

The dimensionless quantities in brackets are, respectively, the reciprocal of the Froude number, the Euler number, and the reciprocal of the Reynolds number for the system.

Because the Navier-Stokes equation is basically a force balance stemming from Newton's law, the dimensionless groups that arise from it are often related conceptually to the forces acting upon a unit volume of fluid. It is correct to say that when all the dimensionless groups are matched between the two systems, then the ratio of viscous stresses to body forces or to fluid inertia in one system is the same as that ratio at the corresponding point of the second system; the flows are kinematically similar. It is often said, however, that the Reynolds number measures the ratio of fluid inertia to viscous forces. This is not correct, except perhaps in an average sense. The Reynolds number of a system is a single number, and the ratio of fluid inertia to viscous stresses may vary widely throughout the system. Perhaps the most useful general statement that can be made is that when a system's behavior is significantly affected by the interaction of fluid inertia and viscous drag, then the Reynolds number is an important parameter. When the system is affected by the interaction of gravity and inertia, eg, by a falling ball, the Froude number is important. It is the strength of similarity analysis that such groupings arise automatically from careful attention to the properties of the defining equation(s).

Dimensionless numbers are not the exclusive property of fluid mechanics but arise out of any situation describable by a mathematical equation. Some of the other important dimensionless groups used in engineering are listed in Table 2.

The principal difficulty faced by the experimenter in applying the results of dimensional or similarity analysis is that often there is insufficient freedom with respect to the physical properties of the modeling fluids to match all of the potentially important dimensionless groups. Construction and interpretation of the model rests largely on the experience and judgment of the experimenter as to which groups can safely be ignored within the accuracy desired. In some cases

Table 2. Dimensionless Numbers Used in Engineering

Dimensionless number	Contributing effects	Area of use
Euler	inertia–pressure	compressible flow
Prandtl	conduction–convection	heat transfer
Nusselt	surface transfer–convection	heat transfer at boundary
Fanning	surface drag–inertia	drag at boundary
Weber	surface tension–inertia	multiphase contacting

useful results can be obtained even when none of the dimensionless groups can be duplicated. An example is the use of scale models in wind tunnels to study the turbulent dispersion of materials released to the atmosphere in the vicinity of complex structures such as refineries or orchards (29). It is not practical to maintain the Reynolds number by increasing wind velocity in inverse proportion to the reduction in scale. Instead, use is made of the observation that when the Reynolds number based on the characteristic dimension of the structure exceeds about 10^4, one can reasonably expect flows to be similar on all scales, ie, turbulent fluctuations will be proportional to the mean velocity and the wakes of the structures will be geometrically similar. Although flow fields around small elements of the structure are not duplicated, the overall error from this approximation is less than would result from an attempt to calculate the dispersion by combining the correlations for individual structures.

Integral Forms of Equations of Motion

The solutions of some problems in fluid mechanics require the detailed integration of the differential equations of motion, whereas many others can be solved to a sufficient degree of accuracy by examining only the overall balances of mass, momentum, and energy. In applying these balances, an appropriate control volume is established first and the rates of accumulation of the quantities within the volume are balanced against their rates of generation within the volume and their rates of transport through the control surface. In most cases the volume of interest is a structure fixed in space, eg, a pipe or tank. In some cases, eg, shock waves, a moving coordinate system is more appropriate.

Conservation of Mass. The general equations for the conservation of mass are the scalar equations (Fig. 21a):

$$\frac{\partial}{\partial t} \iiint c_i \, dB + \iint c_i V \cos \theta \, dS + \iint J_i \, dS - \iiint R_i \, dB = 0 \tag{48}$$

$$\underbrace{\qquad\qquad}_{\text{accumulation}} \qquad \underbrace{\qquad\qquad}_{\text{convection}} \qquad \underbrace{\qquad}_{\text{diffusion}} \qquad \underbrace{\qquad}_{\text{chemical reaction}}$$

A separate equation is obtained for each component i. Consider the important case of steady-state flow of a single fluid through the pipe section of Figure 21b, where the flow is taken to be perpendicular to the cross sections S_1 and S_2. Making use of the concept of average velocity:

$$S_1 \overline{V}_1 \rho_1 = S_2 \overline{V}_2 \rho_2 \tag{49}$$

where

$$\overline{V}_j = \frac{1}{\pi R_j^2} \int_0^{R_j} V_j 2\pi r \, dr \tag{50}$$

The conservation of mass gives comparatively little useful information until it is combined with the results of the momentum and energy balances.

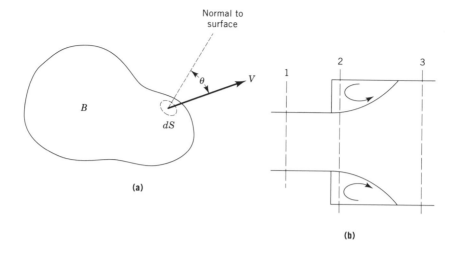

(a)

(b)

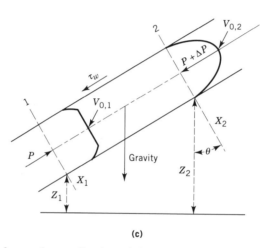

(c)

Fig. 21. Control volumes for application of the integral equations of motion where 1, 2, and 3 are the location of control surfaces S_1, S_2, and S_3: (**a**) general, (**b**) sudden expansion, and (**c**) pipe flow. See text.

Conservation of Momentum.   The general equation for the conservation of momentum is

$$\frac{\partial}{\partial t} \iiint \rho \vec{V} dB + \iint \rho \vec{V} V \cos \theta \, dS - \iiint \vec{F}_B dB - \iint \vec{F}_s dS = 0$$

$$\underbrace{\hspace{2cm}}_{\text{accumulation}} \quad \underbrace{\hspace{2cm}}_{\text{convection}} \quad \underbrace{\hspace{2cm}}_{\substack{\text{creation by} \\ \text{body forces}}} \quad \underbrace{\hspace{2cm}}_{\substack{\text{creation by} \\ \text{surface forces}}} \tag{51}$$

The creation terms embody the changes in momentum arising from external forces in accordance with Newton's second law ($F = ma$). The body forces arise from gravitational, electrostatic, and magnetic fields. The surface forces are the shear and normal forces acting on the fluid; diffusion of momentum, as manifested in viscosity, is included in these terms. In practice the vector equation is usually resolved into its Cartesian components and the normal stresses are set equal to the pressures over those surfaces through which fluid is flowing.

To illustrate the use of the momentum balance, consider the situation shown in Figure 21c in which the control volume is bounded by the pipe wall and the cross sections 1 and 2. The forces acting on the fluid in the x-direction are the pressure forces acting on cross sections 1 and 2, the shear forces acting along the walls, and the body force arising from gravity. The overall momentum balance is

$$\left[-\Delta P + g \int_1^2 \rho dz \right] \pi R^2 - 2\pi R \int_1^2 \tau_w dx = \int_0^R 2\pi r (\rho_2 V_2^2 - \rho_1 V_1^2) \, dr \quad (52)$$

The momentum terms can be rewritten using the mass balance to yield:

$$\int_0^R 2\pi r (\rho_2 V_2^2 - \rho_1 V_1^2) \, dr = (\pi R^2 W) \left[\frac{\overline{V_2^2}}{\overline{V}_2} - \frac{\overline{V_1^2}}{\overline{V}_1} \right] = (\pi R^2 W) \left[\frac{\overline{V}_2}{\beta_2} - \frac{\overline{V}_1}{\beta_1} \right] \quad (53)$$

In general, $\overline{V^2} \neq (\overline{V})^2$. For laminar Newtonian flow the radial velocity profile is parabolic and $\beta = 3/4$. For fully developed turbulent flow the radial velocity profile is almost flat, $\beta = 0.95–0.99$; hence it can be taken as unity with little error.

When the fluid is incompressible and the velocity profiles are identical, $\rho_1 = \rho_2$, $\beta_1 = \beta_2$, the velocity terms disappear and τ_w is independent of x. Equation 52 reduces to:

$$\frac{1}{\Delta x} (-P - \rho g \Delta z) = -\frac{\Delta P^*}{\Delta x} = \frac{2\tau_w}{R} \quad (54)$$

A similar analysis can be carried out for any axisymmetric tube to yield:

$$-\frac{\Delta P^*}{\Delta x} = \frac{2\pi}{r} \quad \text{or} \quad \frac{\tau}{\tau_w} = \frac{r}{R} \quad (55)$$

The shear stress is linear with radius. This result is quite general, applying to any axisymmetric fully developed flow, laminar or turbulent. If the relationship between the shear stress and the velocity gradient is known, equation 50 can be used to obtain the relationship between velocity and pressure drop. Thus, for laminar flow of a Newtonian fluid, one obtains:

$$-\mu \frac{dV}{dr} = \tau = -\frac{r}{2} \left(\frac{\Delta P^*}{\Delta x} \right) \quad (56)$$

This integrates directly to yield a parabolic velocity profile:

$$V(o) - V(r) = \frac{r^2}{4\mu}\left(\frac{\Delta P^*}{\Delta x}\right) \tag{57}$$

Averaging the velocity using equation 50 yields the well-known Hagen-Poiseuille equation (see eq. 32) for laminar flow of Newtonian fluids in tubes. The momentum balance can also be used to describe the pressure changes at a sudden expansion in turbulent flow (Fig. 21**b**). The control surface 2 is taken to be sufficiently far downstream that the flow is uniform but sufficiently close to surface 3 that wall shear is negligible. The additional important assumption is made that the pressure is uniform on surface 3. The conservation equations are then applied as follows:

momentum $$(P_3 - P_2)S_2 = \rho(S_2V_2^2 - S_1V_3^2) \tag{58}$$

mass $$V_1S_1 = V_2S_2 \tag{59}$$

At the inlet the velocity and pressure are presumed continuous, ie, $P_1 = P_3$, $V_1 = V_3$. Simple algebraic manipulations yield the result

$$\frac{P_2 - P_1}{\rho} = V_1^2\left(\frac{S_1}{S_2}\right)\left[1 - \left(\frac{S_1}{S_2}\right)\right] \tag{60}$$

Conservation of Energy. The energy associated with a unit mass of the flowing fluid may be considered as the sum of its potential, kinetic, and internal energies:

$$E = gZ + \frac{V^2}{2} + U \tag{61}$$

The equation describing the conservation of energy is the scalar equation:

$$\frac{\partial}{\partial t}\iiint \rho E dB + \iint \rho EV \cos\theta \, dS + \iint J_E dS = 0 \tag{62}$$

$$\underbrace{\qquad\qquad}_{\text{accumulation}} \qquad \underbrace{\qquad\qquad}_{\text{convection}} \qquad \underbrace{\qquad\qquad}_{\text{nonconvective flux}}$$

The nonconvective energy flux across the boundary is composed of two terms: a heat flux and a work term. The work term in turn is composed of two terms: useful work delivered outside the fluid, and work done by the fluid inside the control volume B on fluid outside the control volume B, the so-called flow work. The latter may be evaluated by imagining a differential surface moving with the fluid which at time zero coincides with a differential element of the surface, S. During the time dt the differential surface sweeps out a volume $VcosdSdt$ and does work on the fluid outside B at a rate of $PVcos\theta dS$. The total flow work done on the fluid outside B by the fluid inside B is

$$\iint PV \cos\theta \, dS \tag{63}$$

This is substituted into equation 62 to yield

$$\frac{\partial}{\partial t} \iiint \rho E \, dB + \iint \left(E + \frac{P}{\rho} \right) \rho V \cos\theta \, dS + Q - W_E = 0 \tag{64}$$

where Q and W_E are the rates at which heat is added to the system and work, both mechanical and electrochemical, is done by the system on its surroundings. For the steady-state system of Figure 21**b**, the overall energy balance is

$$Q - W_E + (U_1 - U_2) + \left(\frac{P_1}{\rho_1} - \frac{P_2}{\rho_2} \right) + g(Z_1 - Z_2) + \frac{1}{2}\left(\frac{\overline{V}_1^3}{\overline{V}_1} - \frac{\overline{V}_2^3}{\overline{V}_2} \right) = 0 \tag{65}$$

In terms of enthalpy and average velocities, this becomes

$$Q - W_E + (H_1 - H_2) + g(Z_1 - Z_2) + \frac{1}{2}\left(\frac{\overline{V}_1^2}{\alpha_1} - \frac{\overline{V}_2^2}{\alpha_2} \right) = 0 \tag{66}$$

For laminar flow $\alpha = 0.5$. For fully developed turbulent flow $0.88 < \alpha < 0.98$, but can be taken to be unity with little error.

The total energy balance may be regarded as the first law of thermodynamics for flowing fluid. It is an essential tool in analyzing the performance of machines that interconvert heat, work, and internal energy. It contains, however, no explicit terms relating to such matters as dissipation of energy within the system by friction, and there must be recourse to more specific equations such as the mechanical energy balance.

The Mechanical Energy Balance. The mechanical energy, often termed the engineering Bernoulli equation, is, strictly speaking, neither. Mechanical energy is not always a conserved quantity and the Bernoulli equation refers to a single streamline in an inviscid fluid. Instead, the mechanical energy equations can be viewed as a definition of a dissipation term ΔF which describes the irreversible conversion into heat of energy available for mechanical work.

Two approaches to this equation have been employed. (*1*) The scalar product is formed between the differential vector equation of motion and the vector velocity and the resulting equation is integrated (1). This is the most rigorous approach and for laminar flow yields an explicit equation for ΔF in terms of the velocity gradients within the system. (*2*) The overall energy balance is manipulated by asserting that the local irreversible dissipation of energy is measured by the difference:

$$-\delta(\Delta F) = \delta Q - T dS = \delta Q - dU - P d\left(\frac{1}{P} \right) - \sum \overline{F}_i \, dn_i \tag{67}$$

Both approaches yield the same general result, which is

$$\Delta F = -W_M + \int_2^1 \frac{dP}{\rho} + g(Z_1 - Z_2) + \frac{1}{2}\left(\frac{\overline{V}_1^2}{\alpha_1} - \frac{\overline{V}_2^2}{\alpha_2} \right) \tag{68}$$

The work term W_M is restricted to the mechanical work delivered to the outside via normal and shear forces acting on the boundary. Electrochemical work, ie, by electrolysis of the fluid, is excluded. Evaluation of the integral requires knowledge of the equation of state and the thermodynamic history of the fluid as it passes from S_1 to S_2, neither of which is needed in the overall energy balance.

In practice, the loss term ΔF is usually not determined by detailed examination of the flow field. Instead, the momentum and mass balances are employed to determine the pressure and velocity changes; these are substituted into the mechanical energy equation and ΔF is determined by difference. For the sudden expansion of a turbulent fluid depicted in Figure 21**b**, which delivers no work to the surroundings, application of equations 49, 60, and 68 yields

$$\Delta F = \frac{(V_1 - V_2)^2}{2} = \frac{V_1^2}{2}\left(1 - \frac{S_1}{S_2}\right)^2 \geq 0 \tag{69}$$

Flow Measurement

There are dozens of flow meters available for the measurement of fluid flow (30). The primary measurements used to determine flow include differential pressure, variable area, liquid level, electromagnetic effects, thermal effects, and light scattering. Most of the devices discussed herein are those used commonly in the process industries; a few for the measurement of turbulence are also described.

Measurement by Differential Pressure.   The most widely used devices for the flow of fluids are those that utilize a fixed constriction in the path of flow to produce a difference in pressure between the upstream and downstream measuring points. The devices discussed here are orifice meters, venturi meters, and target meters. The metering orifice (Fig. 22**a**) has been standardized in most countries as a centered, circular, square-edged opening in a thin plate. It is simple and convenient to use, but has the disadvantage that a large fraction of the pressure drop generated for the measurement of flow is dissipated as frictional energy.

The venturi meter (Fig. 22**b**) consists of a standardized conical nozzle followed by a gradually expanding downstream cone. The pressure change is measured between the upstream fluid and the narrow throat. The gradual expansion downstream of the throat allows the fluid to recover as much as 90% of the pressure difference that would otherwise be lost with a sudden expansion. The venturi meter is a rugged, precision meter but suffers from the considerable disadvantages of high cost as well as large space requirements. In order that one installation may reproduce the results of another within a tolerable error, orifice and venturi installations have been rigorously standardized, largely as the result of work by the ASME Special Research Committee on Fluid Meters and the Joint AGA-ASME Committee on Orifice Coefficients. Their publications should be consulted before making an installation. To a reasonably good approximation, the pressure change measured is proportional to the quantity W^2/ρ_{fluid} for a turbulent fluid.

The target meter is a circular disk or target supported concentrically in the pipe thus giving an annular orifice (Fig. 22**c**). The force exerted on the target is

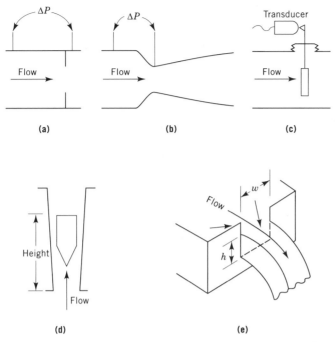

Fig. 22. Common flow meters: (**a**) orifice, (**b**) venturi, (**c**) target, (**d**) rotameter, and (**e**) weir. See text.

proportional to the square of the flow rate. This force is transmitted to an external electrical or pneumatic transducer by the support rod that passes through a diaphragm seal. The open spaces at the top and bottom of the pipe make the device particularly well suited to measuring fluids containing small amounts of gas or solids that tend to settle.

Measurement by Variable Area.   Meters that operate on the principle of variable area incorporate an adjustable constriction in the path of the flow that may be varied so as to maintain a constant differential pressure. The most commonly used device of this type is the rotameter. This is a tapered vertical tube containing a free plummet or float (Fig. 22**d**). Fluid entering the lower, small end of the tube lifts the plummet until the velocity in the annulus between plummet and tube provides a dynamic force balance around the plummet. The level of the plummet indicates the rate of flow. Rotameters are supplied with glass and metal tubes to meet a variety of service conditions in metering clean liquids and gases. They can be used for manual or automatic control of flow, and can be adapted to indicating and recording instruments. For precise metering, a rotameter calibrated with the fluid that is to be metered is unsurpassed. With very large flow rates, especially for gases, the cost and bulk of the rotameter, as compared to an orifice meter, are important disadvantages. For a given position of the plummet, the mass rate of turbulent flow through the meter is proportional to the quantity: $[\rho_{\text{fluid}} (\rho_{\text{plummet}} - \rho_{\text{fluid}})]^{1/2}$. Corrections for nonturbulent flow are generally available from the meter manufacturer.

Measurement by Liquid Level.   The flow rate of liquids flowing in open channels is often measured by the use of weirs (see LIQUID-LEVEL MEASUREMENT). The most common type is the rectangular weir shown in Figure 22e. The flow rate across such a weir varies approximately with the quantity $wh^{3/2}$. Other shapes of weirs are also employed. Standard civil engineering handbooks describe the precautions necessary for constructing and interpreting data from weirs.

Measurement by Electromagnetic Effects.   The magnetic flow meter is a device that measures the potential developed when an electrically conductive flow moves through an imposed magnetic field. The voltage developed is proportional to the volumetric flow rate of the fluid and the magnetic field strength. The process fluid sees only an empty pipe so that the device has a very low pressure drop. The device is useful for the measurement of slurries and other fluid systems where an accumulation of another phase could interfere with flow measurement by other devices. The meter must be installed in a section of pipe that is much less conductive than the fluid. This limits its applicability in many industrial situations.

Measurement by Thermal Effects.   When a fine wire heated electrically is exposed to a flowing gas, it is cooled and its resistance is changed. The hot-wire anemometer makes use of this principle to measure both the average velocity and the turbulent fluctuations in the flowing stream. The fluid velocity, V, is related to the current, i, and the resistances, R, of the wire at wire, w, and gas, g, temperatures via

$$\frac{I^2 R_w}{R_w - R_g} = A' + B' (V)^{1/2} \tag{70}$$

The constants A' and B' can be calculated theoretically, but in practice are usually determined experimentally.

Two modes of operation are available. In the simpler mode, a fixed current is applied to the wire. In the other mode, the sensor temperature (hence R_w) is maintained at a fixed value above that of the gas by means of a feedback amplifier. In both cases, velocity is measured by the voltage drop.

Because of its small size and portability, the hot-wire anemometer is ideally suited to measure gas velocities either continuously or on a troubleshooting basis in systems where excess pressure drop cannot be tolerated. Furnaces, smokestacks, electrostatic precipitators, and air ducts are typical areas of application. Its fast response to velocity or temperature fluctuations in the surrounding gas makes it particularly useful in studying the turbulence characteristics and rapidity of mixing in gas streams. The constant current mode of operation has a wide frequency response and relatively lower noise level, provided a sufficiently small wire can be used. Where a more rugged wire is required, the constant temperature mode is employed because of its insensitivity to sensor heat capacity. In liquids, hot-film sensors are employed instead of wires. The sensor consists of a thin metallic film mounted on the surface of a thermally and electrically insulated probe.

Laser Anemometry.   When light is scattered from a particle moving at a velocity, V, its frequency is altered by the fraction $(\cos\alpha' - \cos\beta')V/C$, where C is the velocity of light in the fluid and α' and β' are the angles between the particle track and the incident and scattered beams, respectively. This principle is employed in the laser anemometer, which uses a laser beam as the source of monochromatic light, particles in the fluid as scattering centers, and appropriate electronic circuitry to measure the frequency shift (see LASERS). This method has been successfully employed to study turbulent fluctuations and to measure local velocities in stirred vessels. It has an advantage over hot-wire/hot-film anemometers in that it does not disturb the flow, but the latter can measure over a smaller volume, with lower background noise and probably better frequency response. Complete hardware systems are commercially available. Because of its cost and the need for an optical path into the fluid, laser anemometry is largely confined to experimental work.

NOMENCLATURE

A_p	area of a body projected in a plane perpendicular to the flow
B	control volume
C, C_i	concentration, concentration of component i
C_D	drag coefficient
C_p, C_v	specific heat at constant pressure, constant volume
c_D	empirical constant in length scale equation
$c_{\epsilon1}, c_{\epsilon2}$	empirical constant in k-ϵ model
c_1, c_2, c_3	empirical constant in RSM model
c_μ	empirical constant
D	diameter, eg, of pipe, nozzle, cylinder, strand
D	average pore diameter in a porous bed
D_H	diameter of helix
d	diameter of drop or bubble
d_p	diameter of particle in a porous bed
$\mathscr{D}$	molecular diffusivity
E	total energy of a unit mass of fluid
$E\ddot{o}$	Eötvos number, $g\Delta\rho d^2 /\sigma$
f	Fanning friction factor
F	force exerted on a body by a fluid
F_B, F_S	body force, surface force acting on fluid in the control volume
F_i	partial molal free energy of component i
ΔF	irreversible energy dissipation per unit mass of fluid
g_i	component of gravitational acceleration in the i direction
h	distance above datum plane, height of crest above weir, depth of liquid in channel
H	enthalpy per unit mass of fluid
J_E	nonconvective flux of energy through surface of control volume
J_i	nonconvective flux of component i through surface of control volume
K	constant in rheological equation for pseudoplastic or dilatant fluids
k	turbulent kinetic energy, $0.5 \overline{u_i' u_i'}$
L	characteristic length in a system, length to achieve a specified approach to fully developed flow, thickness of porous bed
l_b, l_M	distance between plates (Benard instability), characteristic length (Marangoni instability)

l_m, l_t	Prandtl mixing length, length scale of turbulence
Ma	Marangoni number, $(\partial\sigma/\partial c)\,(\partial c/\partial x)l_M^2/\mu\mathscr{D}$
Mo	Morton number, $g\mu_c^4\Delta\rho/\rho_c^2\sigma^3$
n	exponent or index in power-law rheological equation
n_i	moles of component i per unit mass of fluid
n_s	frequency of vortex shedding
P	thermodynamic pressure
P	mean pressure
$P*$	P $+ \rho gz$
Q	heat added to a unit mass of fluid
r, r_o	radial distance from axis, inner tube radius
Ra	Rayleigh number, $g(\partial\rho/\partial T)\Delta Tl_b^3/\mu\alpha$
Re	Reynolds number, $DV\rho/\mu$, $dV\rho/\mu$, $LV\rho/\mu$, etc
R, R_j	radius of tube, radius at location j
R_i	rate of generation of component i per unit volume
s	distance along the jet centerline
S	entropy, surface area
S_j	cross-sectional area at location j
Sr	Strouhal number, n_sD/V_∞
$t, \Delta t$	time, time increment
$T, \Delta T$	temperature, temperature difference
u, v, w	components of velocity in the x, y, z directions
u', u_m	dimensionless reduced velocity, mean flow velocity
u_c	velocity at the center line of a jet at distance x from the nozzle
u_i	instantaneous fluid velocity in the i direction
$\bar{u}_i$	mean fluid velocity in the i direction
u_i'	fluctuating fluid velocity in the i direction
$\overline{u_i'v_j'}$	Reynolds stresses
U, U_m	internal energy, axial velocity, mean velocity
V, V_0	velocity, characteristic system velocity, fixed reference velocity
$\overline{V}$	average velocity, eg, in a pipe
V_j	velocity at location j: 1 = upstream, 2 = downstream
$V_{o,j}$	velocity at centerline of pipe at location j
V_s	superficial velocity in a porous bed
V_∞	uniform velocity at a great distance from a bluff body
w	width of weir
W_E	work done by a fluid on its surroundings
W	weight rate of flow
W_M	mechanical work done by a fluid on its surroundings
$\bar{x}$	Sauter mean drop size
x, y, z	space coordinates used to describe flow; z also indicates height above datum plane
x_i, x_j, x_l, x_k	coordinate in tensor notation
X, Y, Z	body forces acting in x, y, z directions
α	thermal diffusivity (Rayleigh number)
α_j	factor expressing flux of kinetic energy in terms of average velocity $= \overline{V_j^3}/\overline{V}_j^3$
β_j	factor expressing flux of momentum in terms of average velocity $= \overline{V_j^2}/\overline{V}_j^2$
γ	shear rate
δ	characteristic width of a freely expanding jet at a distance x from the nozzle

ϵ	porosity, size of roughness elements on pipe wall
ϵ	energy dissipation
θ	angle from normal to surface or datum plan
η	modulus of rigidity
$\mu,\ \mu_c,\ \mu_d$	shear viscosity, viscosity of continuous phase, viscosity of dispersed phase
λ	dilational viscosity, relaxation time
ν	kinematic viscosity $= \mu/\rho$
ν_t	eddy (or turbulent) viscosity
$\rho,\ \rho_i$	density; density of fluid i
$\Delta\rho$	density difference
$\rho_c,\ \rho_d$	density of continuous phase, dispersed phase
$\bar{\rho}$	mean density
σ	interfacial tension
σ_i	normal stress exerted in the i direction on a plane perpendicular to the i axis
$\sigma_k,\ \sigma_\epsilon$	constants in the k-ϵ model
$\tau,\ \tau_w$	shear stress, shear stress at the wall
τ_{ij}	shear stress exerted in the j direction on a plane perpendicular to the i axis
ϕ	velocity potential function
ϕ_P,ϕ_n	dependent variable at central, P, and neighboring nodes n
ψ	stream function
ω_1,ω_2	density ratios $\rho 1/(\rho 1 + \rho 2)$, $\rho 2/(\rho 1 + \rho 2)$

BIBLIOGRAPHY

"Principles" under "Fluid Mechanics" in *ECT* 1st ed., Vol. 6, pp. 614–640, by T. Baron, University of Illinois, and M. Souders, Jr., Shell Development Co.; "Flow Measurement" under "Fluid Mechanics" in *ECT* 1st ed., Vol. 6, pp. 640–648, by M. Souders, Jr., Shell Development Co.; "Principles" under "Fluid Mechanics" in *ECT* 2nd ed., Vol. 9, pp. 445–473, by M. Souders, Chemical Engineer; "Flow Measurement" under "Fluid Mechanics" in *ECT* 2nd ed., Vol. 9, pp. 473–483, by M. Souders, Chemical Engineer; "Fluid Mechanics" in *ECT* 3rd ed., Vol. 10, pp. 582–629, by A. M. Benson, G. Q. Martin, J. S. Son, and C. V. Sternling, Shell Development Co.

1. R. B. Bird, W. E. Stewart, and E. H. Lightfoot, *Transport Phenomena*, John Wiley & Sons, Inc., New York, 1960.
2. R. S. Brodkey, *The Phenomena of Fluid Motions*, Addison-Wesley Publishing Co., Inc., Reading, Mass., 1967.
3. J. H. Lienhard, *Bulletin 300*, College of Engineering, Washington State University, Pullman, Wash., 1966.
4. L. Prandtl, *Uber Flussigkeitsbewegung bei sehr kleiner Reibung.*, Proceedings of the *Third International Mathematics Congress*, Heidelberg, Germany, 1904.
5. H. Schlichting, *Boundary Layer Theory*, McGraw-Hill Book Co., Inc., New York, 1960.
6. J. R. Grace, T. Wairegi, and T. H. Nguyen, *Trans. Inst. Chem. Eng.* **54**, 167 (1976).
7. G. N. Abramovich, *The Theory of Turbulent Jets*, The M.I.T. Press, Cambridge, Mass., 1963.
8. J. Walker, *Sci. Am.* **238**, 168 (1978).
9. J. M. Beer and N. A. Chigier, *Combustion Aerodynamics*, John Wiley & Sons, Inc., New York, 1972.

10. S. A. Thorpe, *J. Fluid Mech.* **46**, 299 (1971).
11. D. Coles, J. *Fluid Mech.* **21**, 385 (1965).
12. R. A. Mugele and H. D. Evans, *Ind. Eng. Chem.* **43**, 1317 (1951).
13. H. J. Karam and J. C. Bellinger, *Ind. Eng. Chem. Fundam.* **7**, 576 (1967).
14. V. G. Levich, *Physicochemical Hydrodynamics*, Prentice-Hall, Inc., Englewood Cliffs, N.J., 1962.
15. P. J. Roache, *Computational Fluid Dynamics*, Hermosa Publishers, Albuquerque, N.M., 1982.
16. A. D. Gosman, E. E. Khalil, and J. H. Whitlaw, in F. Durst and co-workers, eds., *Turbulent Shear Flow I*, Springer-Verlag, Berlin, 1979.
17. D. A. Anderson, J. C. Tannehill, and R. H. Pletcher, *Computational Fluid Mechanics and Heat Transfer*, McGraw-Hill Co., Inc., New York, 1984.
18. J. O. Hinzi, *Turbulence*, McGraw-Hill Book Co., Inc., New York, 1959.
19. L. Prandtl, *ZAMM* **5**, 136 (1925).
20. W. Rodi, *AIAA J.* **20**, 872 (1981).
21. A. N. Kolmogorov, *Izo Akad Nauk SSR Seria Fizicheska Vi* **1–2**, 56 (1942); English trans., Rep ON/6, M.E. Dept., Imperial College, London, 1968.
22. W. A. Jones and B. E. Launder, *Int. J. Heat Mass Trans.* **16**, 1119 (1973).
23. B. E. Launder and D. B. Spalding, *Mathematical Models of Turbulence*, Academic Press, Inc., New York, 1972.
24. B. E. Launder and D. B. Spalding, *Comp. Meth. Appl. Mech. Eng.* **3**, 269 (1974).
25. J. Zhu and W. Rodi, *Int. J. Num. Meth. Fluids* **14**, 241 (1992).
26. F. Durst and D. Wennerberg, *Int. J. Num. Meth. Fluids* **12**, 203 (1991).
27. B. E. Launder, G. J. Reese, and W. Rodi, *J. Fluid Mech.* **68**, 537 (1975).
28. J. F. Thompson, in W. Kolmann, ed., *Computational Fluid Dynamics*, Hemisphere, Washington, D.C., 1980.
29. J. E. Cermak, *J. Fluids Eng.* **97**, 9 (1975).
30. L. K. Spink, *Principles and Practice of Flow Meter Engineering*, The Foxboro Co., Foxboro, Mass., 1976.

J. S. SON
Shell Development Co.

FLUORAPATITE. See FERTILIZERS.

FLUORESCEIN. See XANTHENE DYES.

FLUORESCENT PIGMENTS (DAYLIGHT). See LUMINESCENT
MATERIALS.

FLUORESCENT WHITENING AGENTS

The operation of whitening, ie, bleaching or brightening, is concerned with the preparation of fabrics whose commercial value is dependent on the highest possible whiteness. In bleaching, textile converters and paper manufacturers are concerned with the removal of colored impurities or their conversion into colorless substances. In chemical bleaching, impurities are oxidized or reduced to colorless products. Physical bleaching involves the introduction of a complementary color whereby the undesired color is made invisible to the eye in an optical manner, eg, in blueing the yellow cast of substrates such as textiles, paper, sugar, etc, is eliminated by means of blue or blue-violet dyes. Through color compensation the treated product appears whiter to the eye; however, it is actually grayer than the untreated material.

With the aid of fluorescent whitening agents (FWAs), also referred to as optical brighteners or fluorescent brightening agents, optical compensation of the yellow cast may be obtained. The yellow cast is produced by the absorption of short-wavelength light (violet-to-blue). With FWAs this lost light is in part replaced; thus a complete white is attained without loss of light. This additional light is produced by the whitener by means of fluorescence. Fluorescent whitening agents absorb the invisible uv portion of the daylight spectrum and convert this energy into the longer-wavelength visible portion of the spectrum, ie, into blue to blue-violet light (Fig. 1). Fluorescent whitening, therefore, is based on the addition of light, whereas the blueing method achieves its white effect through the removal of light.

A fluorescent whitener should be optically colorless on the substrate, and should not absorb in the visible part of the spectrum. In the application of FWAs it is possible to replace the light lost through absorption, thereby attaining a neutral, complete white. Further, through the use of excess whitener, still more uv radiation can be converted into visible light, so that the whitest white is made more sparkling. Since the fluorescent light of a fluorescent whitener is itself col-

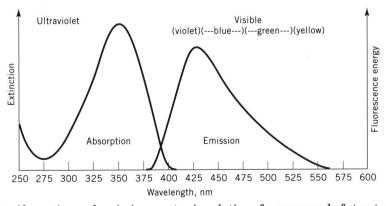

Fig. 1. Absorption and emission spectra in solution of a compound of structure (**1**).

ored, ie, blue-to-violet, the use of excess whitener always gives either a blue-to-violet or a bluish green cast.

The principle of fluorescent whitening was described in 1929 (1), but the industrial use of FWAs began about 10 years later. Since that time FWAs have found increasing use in the most diverse fields (2–5). The toxicological properties of fluorescent whiteners have been summarized (6). Commercial products investigated thus far have been found to be completely harmless. More than 2000 patents for FWAs exist, there are several hundred commercial products, and approximately one hundred producers and distributors.

In 1992, world consumption of FWAs was estimated at 60,000 metric tons. Fifty percent was consumed by the detergent industry, 33% by the paper industry, and 17% by the textile industry. Whitener levels in detergents have stabilized and growth rates track population growth. The paper industry has shown higher growth rates due to a trend to higher whites. The rate of growth has been 4% a year (ca 1992). The textile industry has shown moderate growth (2–4%) largely due to greater usage of cotton fabric. The usage of FWAs in plastics is less than 1% of the total consumption.

Many chemical compounds have been described in the literature as fluorescent, and since the 1950s intensive research has yielded many fluorescent compounds that provide a suitable whitening effect; however, only a small number of these compounds have found practical uses. Collectively these materials are aromatic or heterocyclic compounds; many of them contain condensed ring systems. An important feature of these compounds is the presence of an uninterrupted chain of conjugated double bonds, the number of which is dependent on substituents as well as the planarity of the fluorescent part of the molecule. Almost all of these compounds are derivatives of stilbene [588-59-0] or 4,4'-diaminostilbene; biphenyl; 5-membered heterocycles such as triazoles, oxazoles, imidazoles, etc; or 6-membered heterocycles, eg, coumarins, naphthalimide, s-triazine, etc.

Types of Whitening Agents

Stilbene Derivatives. Most commercial brighteners are bistriazinyl derivatives (**1**) of 4,4'-diaminostilbene-2,2'-disulfonic acid (Table 1). The usual compounds are symmetric; preparation begins with reaction of 2 moles of cyanuric chloride derivatives with 1 mole of 4,4'-diaminostilbene-2,2'-disulfonic acid [81-11-8]. Asymmetric derivatives can be synthesized via 4-amino-4'-nitrostilbene-2,2'-disulfonic acid; however, their preparation is more expensive, and they show little advantage over the symmetrical compounds (see STILBENE DYES). The principal effects of structural variations are changes in solubility, substrate affinity, acid fastness, etc. The bistriazinyl compounds are not stable toward hypochlorite; however, some compounds show some fastness after application to the fiber. Mono(azol-2-yl)stilbenes arose from efforts to find hypochlorite-stable products with neutral fluorescence. The bistriazinyl brighteners are employed principally on cellulosics, such as cotton or paper. Some products also show affinity for nylon at the weakly alkaline pH of most commercial detergents.

2-(Stilben-4-yl)naphthotriazoles (**2**) are prepared by diazotization of 4-amino-stilbene-2-sulfonic acid or 4-amino-2-cyano-4'-chlorostilbene, coupling

Table 1. Symmetrical Stilbene Derivatives[a] used as Fluorescent Whitening Agents

(1)

CAS Registry Number	Geometric isomer	R	R'	References
[17118-48-8]	E	—NHC$_6$H$_5$	—OCH$_3$	7
[31900-04-6], disodium salt		—NHC$_6$H$_5$	—NHCH$_3$	8, 9
[17118-46-6]	E	—NHC$_6$H$_5$	—N(CH$_3$)(CH$_2$CH$_2$OH)	10, 11
[17118-44-4]	E	—NHC$_6$H$_5$	—N(CH$_2$CH$_2$OH)$_2$	12–14
[32466-46-9]	E	—NHC$_6$H$_5$	—N(morpholino)	15
[17863-51-3]	E	—NHC$_6$H$_5$	—NHC$_6$H$_5$	16
[16470-24-9], tetrasodium salt		—NH—C$_6$H$_4$—SO$_3$H	—N(CH$_2$CH$_2$OH)$_2$	17
[17118-40-0]	E	—NH—C$_6$H$_4$(SO$_3$H)	—N(CH$_2$CH$_2$OH)$_2$	18, 19
[41098-56-0], hexasodium salt		—NH—C$_6$H$_3$(SO$_3$H)$_2$	—N(CH$_2$CH$_3$)$_2$	20, 21

[a]R and R' are substituted or unsubstituted amino groups, substituted hydroxyl groups, etc.

with an ortho-coupling naphthylamine derivative, and finally, oxidation to the triazole.

where R = —SO$_3$H, R' = —H (**2a**) and R = —C≡N, R' = —Cl (**2b**)

With water-solubilizing groups, eg, —SO$_3$H as in (**2a**) [*4434-38-2*] (22), these types of compounds are suitable for whitening cellulosic materials or nylon from soap and detergent baths. In solution and on the substrate these compounds show good fastness to hypochlorite and to light. Water-insoluble derivatives of this family, eg, compounds having the nitrile group as in (**2b**) [*5516-20-1*] (23), are suitable for brightening synthetic fibers and resins.

2-(4-Phenylstilben-4-yl)benzoxazoles are prepared by means of the anil synthesis from 2-(4-methylphenyl)benzoxazoles and 4-biphenylcarboxaldehyde anil, and used for brightening polyester fibers (24,25). An example is (**3**) [*16143-18-3*].

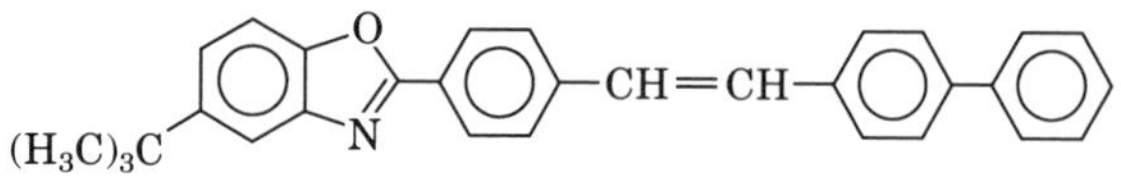

(**3**)

Bis(azol-2-yl)stilbenes (26,27) such as (**4**) have been prepared. 4,4′-Dihydrazinostilbene-2,2′-disulfonic acid, obtained from the diamino compound, on treatment with 2 moles of oximinoacetophenone and subsequent ring closure, leads to the formation of (**4**) [*23743-28-4*]. Such compounds are used chiefly as washing powder additives for the brightening of cotton fabrics, and exhibit excellent light- and hypochlorite-stability.

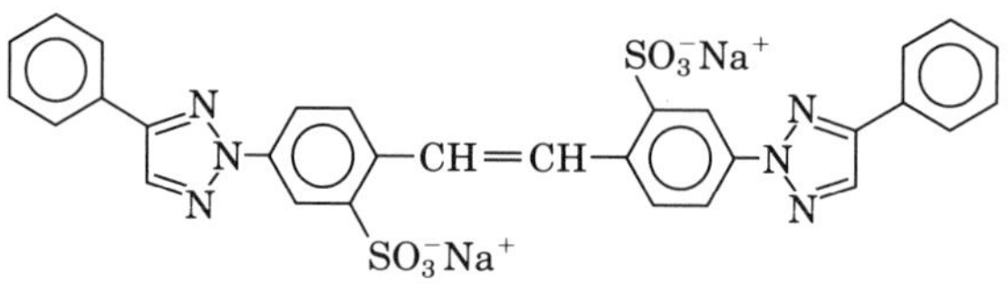

(**4**)

Styryl Derivatives of Benzene and Biphenyl.   Other compounds based on the styryl group were prepared to lengthen the conjugated system of stilbene.

1,4-Bis(styryl)benzenes are obtained by the Horner modification of the Wittig reaction, eg, 1,4-bis(chloromethyl)benzene is treated with 2 moles of triethyl phosphite and the resulting phosphonate reacts with 2 moles of *o*-cyanobenzaldehyde to yield (**5**) [*13001-39-3*] (28). A strong brightening effect with a reddish cast is obtained on polyester fibers.

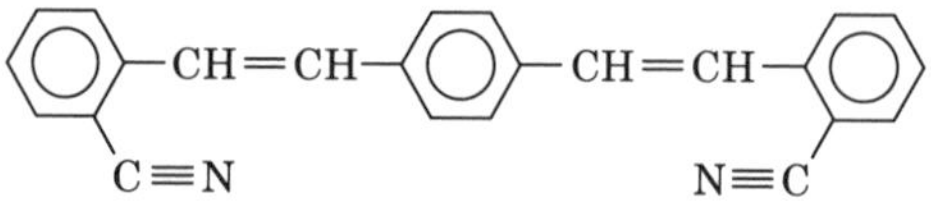

(**5**)

4,4′-Bis(styryl)biphenyls are also obtained by the Horner-Wittig reaction of the phosphonate derived from 1,4-bis(chloromethyl)biphenyl and triethyl phosphite with benzaldehyde-*o*-sulfonic acid, giving the corresponding bisstyrylbiphenyl disodium salt (**6**) [*27344-41-8*] (29). They are used in washing powders for brightening cotton to a very high degree of whiteness with improved hypochlorite stability.

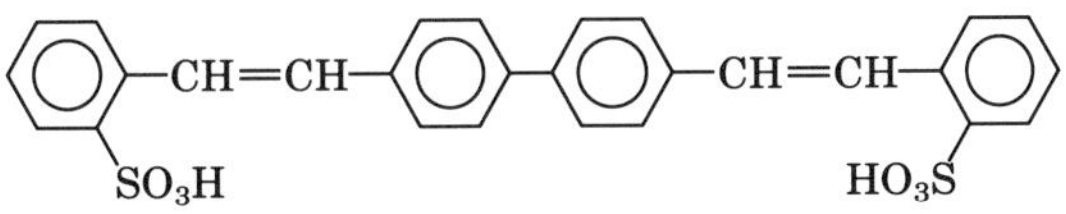

(6)

Pyrazolines. 1,3-Diphenyl-2-pyrazolines (**7**) (Table 2) are obtainable from appropriately substituted phenylhydrazines by the Knorr reaction with either β-chloro- or β-dimethylaminopropiophenones (30,31). They are employed for brightening synthetic fibers such as polyamides, cellulose acetates, and polyacrylonitriles.

Table 2. Pyrazoline Fluorescent Whiteners[a]

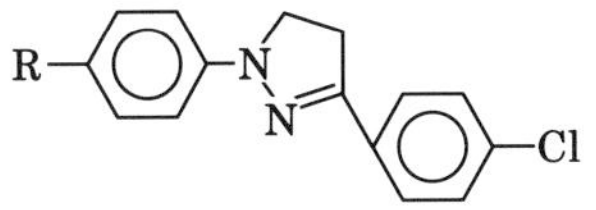

(7)

CAS Registry Number	R	References
[*2697-84-9*]	—SO_3H	32
[*2744-49-2*]	—SO_2NH_2	33, 34
[*38848-70-3*]	—$SO_2NHCH_2CH_2CH_2N^+(CH_3)_3\ ^-SO_3OCH_3$	35
[*27441-70-9*]	—$SO_2CH_2CH_2SO_3H$, sodium salt	36
[*6608-82-8*]	—$SO_2CH_2CH_2O$—CH—CH_2—$N(CH_3)_2$ CH_3	37

[a]Examples wherein the chlorinated phenyl ring is substituted have also been made (38,39).

Bis(benzoxazol-2-yl) Derivatives. Bis(benzoxazol-2-yl) derivatives (**8**) (Table 3) are prepared in most cases by treatment of dicarboxylic acid derivatives of the central nucleus, eg, stilbene-4,4′-dicarboxylic acid, naphthalene-1,4-dicarboxylic acid, thiophene-2,5-dicarboxylic acid, etc, with 2 moles of an appropriately substituted *o*-aminophenol, followed by a ring-closure reaction. These compounds are suitable for the brightening of plastics and synthetic fibers.

Table 3. Bis(benzoxazol-2-yl) Derivatives

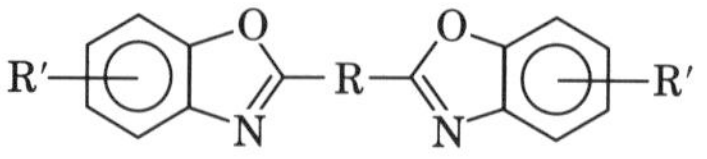

(8)

R^a	R'	CAS Registry Number	References
—CH=CH—	alkyl, 5-CH$_3$	[1041-00-5]	40, 41
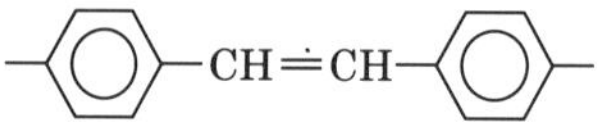	H, alkyl	[1533-45-5]	42, 43
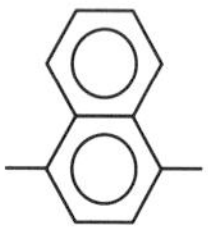	H	[5089-22-5]	44–46
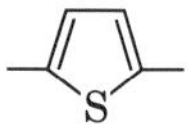	COO-alkyl, SO$_2$-alkyl H, alkyl	[2866-43-5]	47

^{a}R represents the conjugated system of the central nucleus.

A large number of patents cover bis(benzimidazol-2-yl) derivatives (**9**) (48–50). Besides being effective on cotton, compounds of this type show good affinity for nylon.

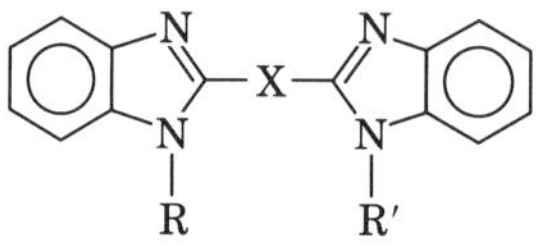

(**9**)

where X =   —CH=CH—   or

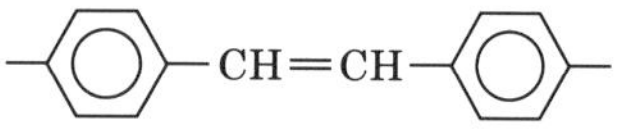

2-(Benzofuran-2-yl)benzimidazoles may be synthesized by reaction of substituted benzofuran-2-carboxylic acid chlorides with substituted o-phenylene-diamines and ring closure of the resulting o-aminoamide, followed by quaternization (51–54). Such products are brighteners for synthetic fibers, in particular those of polyacrylonitrile.

Coumarins. By treatment of flax with esculin, a glucoside of esculetin [305-01-1] (**10**), a brightening effect is achieved; however, this effect is not fast to washing and light. The use of β-methylumbelliferone [90-33-5] (**11**) and similar compounds as brighteners for textiles and soap has been patented.

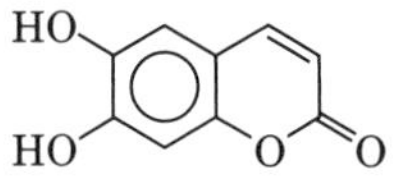

(10) 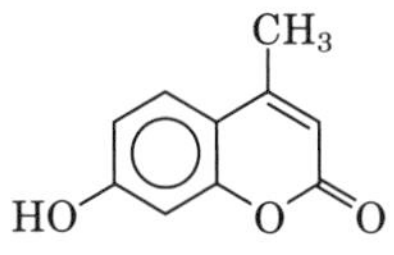 (11) 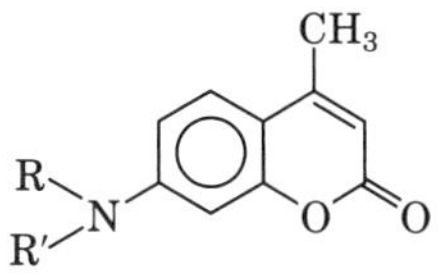 (12)

R = R' = H (12a)
R = R' = CH₃ (12b)

As improvements over β-methylumbelliferone (55–57), 4-methyl-7-amino-coumarin [26093-31-2] (12a) and 7-dimethylamino-4-methylcoumarin [87-01-4] (12b) (58–61) were proposed. These compounds are used for brightening wool and nylon either in soap powders or detergents, or as salts under acid dyeing conditions. They are obtained by the Pechmann synthesis from appropriately substituted phenols and β-ketocarboxylic acid esters or nitriles in the presence of Lewis acid catalysts (see COUMARIN).

A further development in the coumarin series is the use of derivatives of 3-phenyl-7-aminocoumarin ((13) where R, R′ = Cl or substituted amines) as building blocks for a series of light-stable brighteners for various plastics and synthetic fibers, and, as the quaternized compounds, for brightening polyacrylonitrile (62).

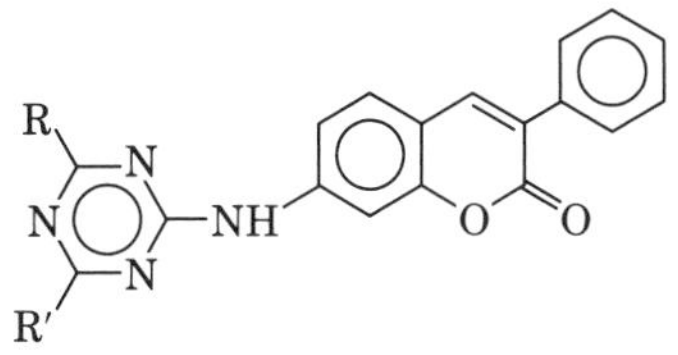

(13)

3-Phenyl-7-aminocoumarin is obtained by a Knoevenagel reaction of substituted salicylaldehydes with phenylacetic acid or benzyl cyanide. Further synthesis of the individual end products is carried out by usual procedures.

Other related substances are 3-phenyl-7-(azol-2-yl)coumarins (63–65) and 3,7-bis(azolyl)coumarins (66,67).

Carbostyrils. Carbostyrils such as (14) [33934-60-0] are prepared by the reaction of 2-alkylamino-4-nitrotoluene with ethyl glyoxalate in the presence of piperidine, reduction of the resulting 3-phenyl-7-nitrocarbostyril derivatives, and finally methylation of the corresponding amino compounds (68). They are whiteners for polyamides, wool, and cellulose acetates (see FIBERS, CELLULOSE ESTERS).

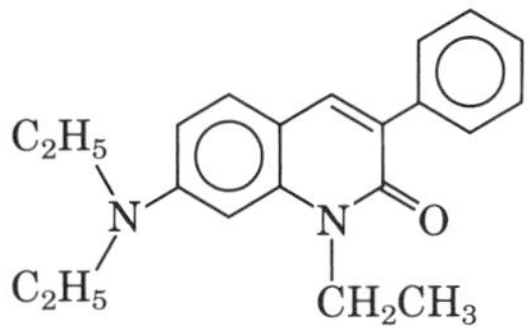

(14)

Other Heterocyclic Systems.   Naphthalimides (**15**), where $R' = -NHCOR$, are derivatives of 4-aminonaphthalimide (69) and are used in plastics. The alkoxynaphthalimides (**15**, $R' = OCH_3$) (70,71) are of particular interest as fluorescent whitening agents.

Naphthalimides are prepared from naphthalic anhydride obtained from naphthalene-1,8-dicarboxylic acid, ie, the oxidation product of acenaphthene or its derivatives, by reaction with amines. They are utilized for synthetic fibers such as polyesters.

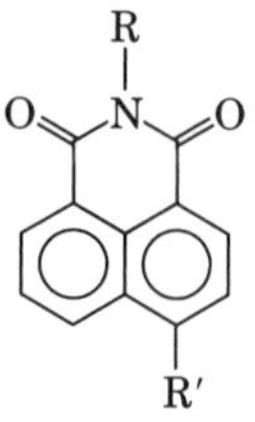

(**15**)

A further group of whiteners was found in the acylamino (R,R') derivatives (**16**) of 3,7-diaminodibenzothiophene-2,8-disulfonic acid-5,5-dioxide. The preferred acyl groups are alkoxybenzoyls (72–74). These compounds give a greenish fluorescence and are relatively weak in comparison with stilbene derivatives on cotton; however, they show good stability to hypochlorite.

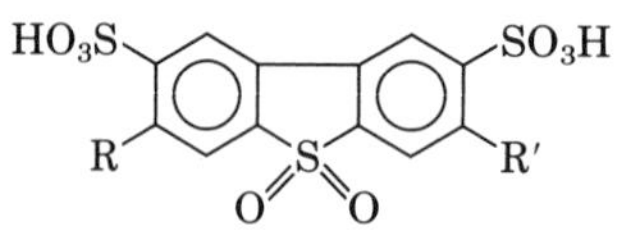

(**16**) 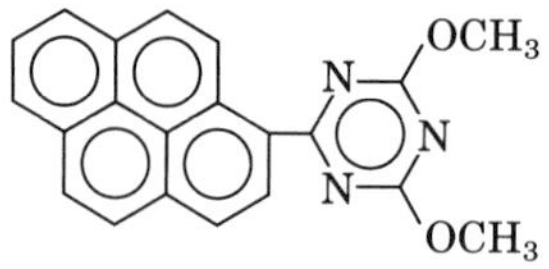   (**17**)

The pyrene derivative (**17**) [3271-22-5] is obtainable by the Friedel-Crafts reaction of pyrene with 2,4-dimethoxy-6-chloro-s-triazine, and is used for brightening polyester fibers (75).

Quaternized pyridotriazoles can be used for brightening acrylic fibers (76).

Uses

Initially, fluorescent whitening agents (FWAs) were used exclusively in textile finishing; the detergent and paper industries followed thereafter. These products are also used in fiber spinning masses, plastics, and paints.

There are more than a thousand known products derived from 200 compounds based on ca 40 fundamental structures. The approximate use distribution of whiteners is shown in Table 4.

Textile Applications.   A 1971 estimate of world textile fiber consumption (2) showed that approximately 60% of textile goods are dyed and about 30% are

Table 4. Use of Fluorescent Whitening Agents

Industry	Proportion,[a] %	Number of fundamental structures
textile	20	>30
detergents	45	7[a]
paper	35	
synthetic fibers, plastics	1	9[a]

[a]Values are approximate.

whites. The proportion of white goods (>40%) is highest for cotton. These percentages also hold true in the 1990s.

Textile substrates of natural or synthetic fibers are contaminated in the raw state by substances of varying degrees of yellowness. Bleaching is required to remove the yellowish cast. Chemical bleaching agents destroy the yellow coloring matter in fibers. However, even if bleaching processes are carried to the technically acceptable limits of damage to the fibers, they never succeed in completely removing this intrinsic color (see BLEACHING AGENTS).

To produce the color white it is necessary to dye with a fluorescent whitener. FWAs used in textiles can be roughly divided into products containing sulfonic acid groups, corresponding to acid dyes, for cotton, wool, and polyamides; cationic whiteners that behave in the same way as basic dyes, for polyacrylonitrile fibers; and whiteners containing no solubilizing groups, corresponding to disperse dyes, for polyester and secondary acetate fibers. This is not a strict division since nonionic FWAs can whiten polyacrylonitrile and polyamide, and certain cationic FWAs produce effects on polyester (77). The second generation of synthetic fibers includes types that have been acid- or base-modified and consequently display different dyeing characteristics (see DYES, APPLICATION AND EVALUATION). For dyeing fiber blends such as viscose–polyamide, polyamide–Spandex, or polyester–cotton, only compatible FWAs may be used that do not interfere with one another or have any detrimental effect on fastness properties.

In conjunction with the increased use of synthetic fibers and blends of synthetic and natural fibers, and the modernization of application processes which has taken place simultaneously, the technique of textile whitening has been improved considerably.

Fixing the dye to a fiber can be performed by an exhaust procedure or a padding procedure. In an exhaust procedure the fluorescent whitener is exhausted from a long liquor onto the substrate until an approximate equilibrium is reached between whitener in the bath and whitener on the substrate. In this procedure the equilibrium is biased primarily toward FWA on the substrate, ie, the highest possible degree of exhaustion is desired. Exhaust procedures are used for loose stock, yarns, woven fabrics, and knit goods which give poor or unsatisfactory results in padding processes, and for garments and garment parts.

Padding methods, ie, application from short liquors, are increasingly important for whitening piece-goods. Woven fabrics or knit goods are passed in an unfolded, open-width state through a small trough charged with treatment liquor containing FWA and subsequently between squeeze rollers to express the liquor to a precisely defined liquor pick-up.

During the drying and, if required, the heat treatment that follows, the fluorescent whitener is fixed on the substrate. FWAs and dyes used in padding procedures must have low substantivity during the padding operation. This is an important prerequisite for level whitening with no tailing.

In contrast to dyes, fluorescent whiteners are not applied exclusively in special processes, but often in combination with bleaching and finishing steps. Fluorescent whiteners used in such processes must be stable and should not interfere with the operation.

The most common chemical bleaching procedures are hypochlorite bleach for cotton; hydrogen peroxide bleach for wool and cotton; sodium chlorite bleach for cotton, polyamide, polyester, and polyacrylonitrile; and reductive bleaching with dithionite for wool and polyamide.

Whitening in combination with the finishing process is used primarily for woven fabrics of cellulosic fibers and their blends with synthetic fibers.

Detergent Applications.   The primary function of FWAs in the laundry process is to whiten fabric load and maintain the original appearance of the white, laundered articles. Laundering is characterized by repeated application to the same item. Fluorescent whiteners used in this repetitive process have to compensate for the reduction in whiteness and contribute toward prolongation of the useful life of the textile material.

In textile whitening there is an unlimited degree of freedom in application, and the processing method can be adapted to suit the optimum color behavior of the substrate. In laundering, the condition of the application bath is predetermined by the composition of the detergent, and the soil loading of the laundry varies from batch to batch. The pH of the washbath is always neutral to alkaline. The bath temperature does not correspond to an optimum dyeing temperature; laundering temperatures vary from cold to boiling water (2). Detergent fluorescent whiteners must remain stable under conditions that prevail during the washing, rinsing, drying, and ironing cycle (2), and they should possess lightfastness. Detergent whiteners must not cause a shade change in the original, generally neutral white of the goods. Fluorescent whiteners that tend to accumulate in excessive amounts on the substrate, thus lending it a greenish cast, are not suitable for use in detergents.

Where laundry is done at boiling point, sodium perborate is added to the detergent as a bleaching agent; in the United States, South Africa, and Australia, sodium hypochlorite is generally added to the washbath; and where laundry is done cold the goods are treated in rinsebaths containing hypochlorite. Only FWAs stable or fast to hypochlorite can be used successfully in these countries (2). The types of FWAs used in the detergent industry are listed in Table 5.

A secondary benefit of FWAs is obtained in improving the appearance of the detergent powder itself. Many significant detergent ingredients are off-white to light yellow in color, resulting in a dull white detergent granule base powder. Through the proper selection and incorporation of a fluorescent whitening agent, the powder aspect can be improved to provide a brilliant white appearance. The bis(styryl)diphenyls (**6**) are especially effective product whiteners.

Table 5. Fluorescent Whitening Agents in the
Detergent Industry

Substrate	FWAs used[a]
wool	(**12**)
cellulosics	(**1**) (**2**) (**4**) (**6**) (**9**)
secondary acetate	(**7**) (**8**) (**12**)
polyamide	(**1**) (**2**) (**4**) (**7**) (**8**) (**9**) (**12**)
polyester[b]	(**8**)

[a]Structure numbers.
[b]Good effects can be obtained only in special processes with
high washing and/or drying temperatures.

Paper Industry Applications.   Derivatives of bistriazinylstilbene (**1**) are used in the paper industry.

Most papers are whitened by addition of FWA to both the pulp and surface coating. Roughly one-third of the FWA is added at the pulp stage and the remaining two-thirds are added to the preformed sheet to give surface whiteness. In the first operation, the FWA in aqueous solution is added to the stock at the Hollander or other beater. For this application the products must be inexpensive and readily soluble in water. Besides satisfactory exhaustion at low temperatures, good paper FWAs also require good acid and alum stability as well as compatibility with fillers (qv). Good affinity for pulp is also required, since any unabsorbed FWA is lost in the effluent (see PAPER; PAPERMAKING ADDITIVES; PULP).

The surface of various types of paper are treated with size coatings and pigment coatings to improve their writability and printability. In some cases, the coating itself may have a low degree of whiteness; in others, a higher degree of whiteness can be obtained with high white pigments. FWAs can be incorporated into size press coatings or off-machine coating operations to enhance the appearance of the finished sheet.

Synthetic Fiber and Plastics Industries.   In the synthetic fibers and plastics industries, the substrate itself serves as the solvent, and the whitener is not applied from solutions as in textiles. Table 6 lists the types of FWAs used in the synthetic fibers and plastic industries. In the case of synthetic fibers, such as polyamide and polyester produced by the melt-spinning process, FWAs can be added at the start or during the course of polymerization or polycondensation. However, FWAs can also be powdered onto the polymer chips prior to spinning. The above types of application place severe thermal and chemical demands on FWAs. They must not interfere with the polymerization reaction and must remain stable under spinning conditions.

In the case of solvent spinning, ie, secondary acetate, polyacrylonitrile, and poly(vinyl chloride), the FWA is added to the polymer solution. An exception is gel-whitening of polyacrylonitrile, where the wet tow is treated after spinning in a washbath containing FWA.

In the case of poly(vinyl chloride) plastics, the FWA is mixed dry with the PVC powder before processing or dissolved in the plasticizing agent (see VINYL POLYMERS). Polystyrene, acrylonitrile–butadiene–styrene (ABS), and polyolefin

Table 6. Fluorescent Whitening Agents Used in the
Synthetic Fibers and Plastics Industries

Substrate	FWAs used[a]
polyamide	(**1**) (**3**) (**5**)
polyester	(**2**) (**8**)
secondary acetate	(**8**)
polyacrylonitrile	(**2**) (**7**) (**9**)
PAN gel whitening	(**7**) (**12**)
poly(vinyl chloride)	(**2**) (**8**) (**9**)
polystyrene	(**2**) (**8**) (**9**)
acrylonitrile–butadiene–styrene	(**8**) (**9**)

[a]Structure numbers.

granulates are powdered with FWA prior to extrusion (2,78) (see STYRENE PLAS-
TICS; ACRYLONITRILE POLYMERS; OLEFIN POLYMERS).

Measurement of Whiteness. The Ciba-Geigy Plastic White Scale is effective in the visual assessment of white effects (79), but the availability of this scale
is limited. Most evaluations are carried out (ca 1993) by instrumental measurements, utilizing the CIE chromaticity coordinates or the Hunter Uniform Color
System (see COLOR). Spectrophotometers and colorimeters designed to measure
fluorescent samples must have reversed optics, ie, the sample is illuminated by a
polychromatic source and the reflected light passes through the analyzer to the
detector.

In order to facilitate rating the preference for white substrates, numerous
whiteness equations have been developed based on comparisons of instrumental readings vs visual evaluations. Among the most commonly used equations
are those developed by Ganz, ie, $W = Y - 800x - 1700y + 813.7$, which corresponds to the Ciba-Geigy White Scale (79,80); Hunter, ie, $W = L - 3b$; and
Stensby, ie, $W = L + 3a - 3b$ (81). In the paper industry, the TAPPI Brightness
value often is used as a measure of pulp or paper whiteness. However, the methodology for measuring TAPPI Brightness was developed on nonfluorescent substrates, ie, bleached pulp, and does not fully assess the effects of fluorescent whitening agents. In instrumental comparisons of fluorescently whitened samples, full
colorimetric measurements are recommended.

Safety and Environmental Aspects

Several studies on FWAs have concluded that diaminostilbenedisulfonic acid/
cyanuric chloride (DAS/CC) and distyrylbiphenyl (DSBP) type whiteners are of a
low order of toxicity. Their safety has been extensively reviewed by governmental
agencies; there is no evidence of human health hazards. FWA producers and users
consider these products to be both safe and beneficial to the ultimate consumer.
This view is supported by appropriate trade associations. A comprehensive review
of available safety and environmental data has been published (82). In addition,

principal suppliers are conducting life cycle analyses on the primary whiteners in use (ca 1993).

BIBLIOGRAPHY

"Whitening Agents" in *ECT* 1st ed., Vol. 15, pp. 45–48, by H. B. Freyermuty, General Aniline & Film Corp.; "Brighteners, Optical" in *ECT* 2nd ed., Vol. 3, pp. 737–750, by R. Zweidler and H. Haüsermann, J. R. Geigy S. A.; "Brighteners, Fluorescent" in *ECT* 3rd ed., Vol. 4, pp. 213–226, by R. Zweidler and H. Hofti, Ciba-Geigy Ltd.

1. P. Krais, *Melliand Textilber.* **10**, 468 (1929).
2. "Fluorescent Whitening Agents," in F. Coulston and F. Korte, eds., *Environmental Quality and Safety*, Suppl. Vol. 4, Georg Thieme Verlag, Stuttgart, and Academic Press, Inc., New York, 1975.
3. H. Gold, "Fluorescent Brightening Agents," in K. Venkataraman, ed., *The Chemistry of Synthetic Dyes*, Vol. 5, Academic Press, Inc., New York, 1971, pp. 535–679.
4. A. Dorlars, C. W. Schellhammer, and J. Schroeder, *Angew. Chem.* **87**, 693 (1975).
5. R. Zweidler, *Textilveredlung* **4**, 75 (1969).
6. R. Anliker and co-workers, *Fluorescent Whitening Agents, MVC-Report 2, Proceedings of a Symposium Held at the Royal Institute of Technology*, Miljövardscentrum, Stockholm, Sweden, Apr. 11, 1973.
7. U.S. Pat. 2,713,046 (Oct. 9, 1951), W. W. Wilson and H. B. Freyermuth (to General Aniline & Film Corp.).
8. U.S. Pat. 2,612,501 (Sept. 30, 1952), R. H. Wilson (to Imperial Chemical Industries).
9. U.S. Pat. 2,763,650 (Sept. 18, 1956), F. Ackermann (to Ciba AG).
10. U.S. Pat. 2,762,801 (Sept. 11, 1956), H. Häusermann (to J. R. Geigy AG).
11. Brit. Pat. 1,116,619 (June 6, 1968), H. Häusermann (to J. R. Geigy AG).
12. Fr. Pat. 874,939 (Aug. 3, 1942), B. Wendt (to I.G. Farbenindustrie).
13. Ger. Pat. 752,677, B. Wendt (to I. G. Farbenindustrie).
14. U.S. Pat. 3,211,665 (Oct. 12, 1965), W. Allen and R. F. Gerard (to American Cyanamid).
15. U.S. Pat. 2,618,636 (Nov. 18, 1952), W. W. Williams and W. E. Wallace (to General Aniline & Film Corp.).
16. U.S. Pat. 2,171,427 (Aug. 29, 1939), J. Eggert and B. Wendt (to I. G. Farbenindustrie).
17. U.S. Pat. 3,025,242 (Mar. 13, 1962), R. C. Seyler (to E. I. du Pont de Nemours & Co., Inc.).
18. Ger. Pat. 1,090,168 (Oct. 6, 1960), J. Hegemann, A. Mitrowsky, and H. Roos (to Bayer AG).
19. Ger. Pat. 1,250,830 (Sept. 28, 1967), E. A. Kleinheidt and H. Gold (to Bayer AG).
20. Ger. Pat. 55,668 (May 5, 1967), B. Noll and co-workers (to VEB Farbenfabrik).
21. U.S. Pat. 3,479,349 (Nov. 18, 1969), R. C. Allison, F. Fischer, and H. Häusermann (to Geigy Chemical Corp. Ardsley).
22. U.S. Pat. 2,784,183 (Mar. 5, 1957), E. Keller, R. Zweidler, and H. Häusermann (to J. R. Geigy AG).
23. U.S. Pat. 2,972,611 (Feb. 21, 1961), R. Zweidler and E. Keller (to J. R. Geigy AG).
24. U.S. Pat. 3,725,395 (Apr. 3, 1973), A. E. Siegrist and co-workers (to Ciba AG).
25. U.S. Pat. 3,781,278 (Dec. 25, 1973), A. E. Siegrist and co-workers (to Ciba AG).
26. U.S. Pat. 3,485,831 (Dec. 23, 1969), A. Dorlars, O. Neuner, and R. Putter (to Bayer AG).
27. U.S. Pat. 3,666,758 (May 30, 1972), A. Dorlars and O. Neuner (to Bayer AG).
28. U.S. Pat. 3,177,208 (Apr. 6, 1965), W. Stilz and H. Pommer (to Badische Anilin- und Soda-Fabrik).

29. U.S. Pat. 3,984,399 (Oct. 5, 1976), K. Weber and co-workers (to Ciba AG).
30. A. Wagner, C. W. Schellhammer, and S. Petersen, *Angew. Chem. Int. Ed. Engl.* **5**, 699 (1966).
31. U.S. Pat. 2,610,969 (Sept. 16, 1952), J. D. Kendall and G. F. Duffin (to Ilford Ltd.).
32. U.S. Pat. 2,640,056 (May 26, 1953), J. D. Kendall and G. F. Duffin (to Ilford Ltd.).
33. U.S. Pat. 2,639,990 (May 26, 1953), J. D. Kendall and G. F. Duffin (to Ilford Ltd.).
34. U.S. Pat. 3,135,742 (June 2, 1964), A. Wagner, A. Schlachter, and H. Marzolph (to Bayer AG).
35. U.S. Pat. 3,131,079 (Apr. 28, 1964), A. Wagner and S. Petersen (to Bayer AG).
36. U.S. Pat. 3,255,203 (June 7, 1966), E. Schinzel and K. H. Lebkücher (to Hoechst AG).
37. U.S. Pat. 3,560,485 (Feb. 2, 1971), E. Schinzel, S. Bildstein, and K. H. Lebkücher (to Hoechst AG).
38. Brit. Pat. 1,360,490 (Oct. 14, 1971), H. Mengler (to Hoechst AG).
39. U.S. Pat. 3,865,816 (Feb. 11, 1975), H. Mengler (to Hoechst AG).
40. U.S. Pat. 2,488,289 (Nov. 15, 1949), J. Meyer, Ch. Gränacher, and F. Ackermann (to Ciba AG).
41. U.S. Pat. 3,649,623 (Mar. 14, 1972), F. Ackermann, M. Dünnenberger, and A. E. Siegrist (to Ciba AG).
42. U.S. Pat. 3,260,715 (July 12, 1966), D. G. Saunders (to Eastman Kodak).
43. U.S. Pat. 3,322,680 (May 30, 1967), D. G. Hedberg, M. S. Bloom, and M. V. Otis (to Eastman Kodak).
44. Brit. Pat. 1,059,687 (Feb. 22, 1967), (to Hoechst AG).
45. U.S. Pat. 3,709,896 (Jan. 9, 1973), H. Frischkorn, U. Pinschovius, and H. Behrenbruch (to Hoechst AG).
46. U.S. Pat. 3,993,659 (Nov. 23, 1976), H. R. Meyer (to Ciba-Geigy AG).
47. U.S. Pat. 2,995,564 (Aug. 8, 1961), M. Dünnenberger, A. E. Siegrist, and E. Maeder (to Ciba AG).
48. U.S. Pat. 2,488,094 (Nov. 15, 1949), Ch. Gränacher and F. Ackermann (to Ciba AG).
49. U.S. Pat. 2,488,289 (Nov. 15, 1949), J. Meyer, Ch. Gränacher, and F. Ackermann (to Ciba AG).
50. U.S. Pat. 2,838,504 (June 10, 1958), N. C. Crounse (to Sterling Drug Inc.).
51. U.S. Pat. 3,900,419 (Aug. 19, 1975), H. Schläpfer and G. Kabas (to Ciba-Geigy AG).
52. U.S. Pat. 3,772,323 (Nov. 13, 1973), H. Schläpfer and G. Kabas (to Ciba-Geigy AG).
53. U.S. Pat. 3,940,417 (Feb. 24, 1976), H. Schläpfer (to Ciba-Geigy AG).
54. Swiss Pat. 560,277 (Mar. 27, 1955), H. Schläpfer (to Ciba-Geigy AG).
55. Ger. Pat. 765,901, (to Hoffmanns Stärkefabriken AG).
56. U.S. Pat. 2,590,485 (Mar. 25, 1952), H. Meyer (to Lever Brothers Co.).
57. U.S. Pat. 2,424,778 (July 29, 1947), P. W. Tainish (to Lever Brothers Co.).
58. U.S. Pat. 2,610,152 (Sept. 9, 1952), F. Ackermann (to Ciba AG).
59. U.S. Pat. 2,600,375 (June 17, 1952), F. Ackermann (to Ciba AG).
60. U.S. Pat. 2,654,713 (Oct. 6, 1953), F. Fleck (to Sandoz AG).
61. U.S. Pat. 2,791,564 (May 7, 1957), F. Fleck (to Saul & Co.).
62. U.S. Pat. 2,945,033 (July 12, 1960), H. Häusermann (to J. R. Geigy AG).
63. U.S. Pat. 3,123,617 (Mar. 3, 1964), H. Häusermann (J. R. Geigy AG).
64. U.S. Pat. 3,646,052 (Feb. 29, 1972), O. Neuner and A. Dorlars (to Bayer AG).
65. U.S. Pat. 3,288,801 (Nov. 29, 1966), F. Fleck, H. Balzer, and H. Aebli (to Sandoz AG).
66. Brit. Pat. 1,201,759 (Aug. 12, 1970), (to Bayer AG).
67. U.S. Pat. 3,839,333 (Oct. 1, 1974), A. Dorlars and W. D. Wirth (to Bayer AG).
68. U.S. Pat. 3,420,835 (Jan. 7, 1969), W. D. Wirth and co-workers (to Bayer AG).
69. Brit. Pat. 741,798 (Dec. 14, 1955), E. Nold (to Badische Anilin- und Soda-Fabrik AG).
70. U.S. Pat. 3,310,564 (Mar. 21, 1967), T. Kasai.
71. Jpn. Pat. 71 13,953 (Apr. 14, 1971), T. Kasai.

72. U.S. Pat. 2,563,795 (Aug. 7, 1951), M. Scalera and D. R. Eberhart (to American Cyanamid Co.).

73. U.S. Pat. 2,573,652 (Oct. 30, 1951), M. Scalera and W. S. Forster (to American Cyanamid Co.).

74. U.S. Pat. 2,702,759 (Feb. 22, 1955), M. Scalera and D. E. Eberhart (to American Cyanamid Co.).

75. U.S. Pat. 3,157,651 (Nov. 17, 1964), J. R. Atkinson and S. Hartley (to Imperial Chemical Industries Ltd.).

76. U.S. Pats. 3,058,989 (Oct. 16, 1962), and 3,049,438 (Aug. 14, 1962), B. G. Buell and R. S. Long (to American Cyanamid).

77. R. Anliker and co-workers, *Textilveredlung* **11**, 369 (1976).

78. G. W. Broadhurst and A. Wieber, *Plast. Eng.*, 36 (1973).

79. E. Ganz, *J. Color Appearance* **1**, 33 (1972).

80. E. Ganz, *Textilveredlung* **9**, 10 (1974).

81. P. Stensby, *Soap Chem. Special.* **43**(I–V), 41 (Apr.; 84, May; 80, July; 94, Aug.; 96, Sept. (1967)).

82. J. B. Kramer, in O. Hutzinger, ed., *The Handbook of Environmental Chemistry*, Vol. 3, Part F, Springer-Verlag, Berlin, 1992.

HAROLD J. MCELHONE, JR.
CIBA-GEIGY Corporation

FLUORINE

Fluorine [7782-42-4], F_2, is a diatomic molecule existing as a pale yellow gas at ordinary temperatures. Its name is derived from the Latin word *fleure*, meaning to flow, alluding to the well-known fluxing power of the mineral fluorite [7789-75-5], CaF_2, which is the most abundant naturally occurring compound of the element. Although radioactive isotopes between atomic weight 17 and 22 have been artificially prepared and have half-lives between 4 s for ^{22}F and 110 min for ^{18}F, fluorine has a single naturally occurring isotope, ^{19}F, and has an atomic weight of 18.9984 (1). Fluorine, the most electronegative element and the most reactive nonmetal, is located in the upper right corner of the Periodic Table. Its electron configuration is $1s^2 2s^2 2p^5$.

The only commercially feasible method of preparing elemental fluorine is by the electrolysis of molten fluoride-containing salts. Fluorine was first isolated in 1886 by the French chemist Moissan (2) who applied a method originally suggested and unsuccessfully tried by Davey and Ampere in 1810–1812. Moissan used potassium fluoride in anhydrous hydrogen fluoride resulting in an electrically conductive electrolyte. The only chemical route, which does not rely on compounds derived from F_2, has more recently been discovered (3). Both starting materials are easily prepared from HF, and react at 150°C evolving fluorine gas.

$$K_2MnF_6 \ + \ 2\ SbF_5 \to 2\ KSbF_6 \ + \ MnF_3 \ + \ 1/2\ F_2$$

Fluorine was first produced commercially ca 50 years after its discovery. In the intervening period, fluorine chemistry was restricted to the development of various types of electrolytic cells on a laboratory scale. In World War II, the demand for uranium hexafluoride [7783-81-5], UF_6, in the United States and United Kingdom, and chlorine trifluoride [7790-91-2], ClF_3, in Germany, led to the development of commercial fluorine-generating cells. The main use of fluorine in the 1990s is in the production of UF_6 for the nuclear power industry (see NUCLEAR REACTORS). However, its use in the preparation of some specialty products and in the surface treatment of polymers is growing.

Fluorine, which does not occur freely in nature except for trace amounts in radioactive materials, is widely found in combination with other elements, accounting for ca 0.065 wt % of the earth's crust (4). The most important natural source of fluorine for industrial purposes is the mineral fluorspar [14542-23-5], CaF_2, which contains about 49% fluorine. Detailed annual reports regarding the worldwide production and reserves of this mineral are available (5). A more complete discussion of the various sources of fluorine-containing minerals is given elsewhere (see FLUORINE COMPOUNDS, INORGANIC).

Physical Properties

Fluorine is a pale yellow gas that condenses to a yellowish orange liquid at $-188°C$, solidifies to a yellow solid at $-220°C$, and turns white in a phase transition at $-228°C$. Fluorine has a strong odor that is easily detectable at concentrations as low as 20 ppb. The odor resembles that of the other halogens and is comparable to strong ozone (qv).

Because of the extreme difficulty in handling fluorine, reported physical properties (Table 1) show greater than normal variations among investigators. A detailed summary and correlation of the physical, thermodynamic, transport, and electromagnetic properties of fluorine is given in Reference 20.

Chemical Properties

Fluorine is the most reactive element, combining readily with most organic and inorganic materials at or below room temperature. Many organic and hydrogen-containing compounds, in particular, can burn or explode when exposed to pure fluorine. With all elements except helium, neon, and argon, fluorine forms compounds in which it shows a valence of -1. Fluorine reacts directly with the heavier helium-group gases xenon, radon, and krypton to form fluorides (see HELIUM-GROUP GASES, COMPOUNDS).

Fluorine is the most electronegative element and thus can oxidize many other elements to their highest oxidation state. The small size of the fluorine atom facilitates the arrangement of a large number of fluorines around an atom of another element. These properties of high oxidation potential and small size allow

Table 1. Physical Properties of Fluorine

Property	Value	References
melting point, °C	−219.61[a]	6
	−217.9	7
	−223	8
boiling point, °C	−188.13[a]	6
	−187.7	7
	−187.0	8
	−188.22	9
	−188.03	10
solid transition temperature, °C	−227.60	6
critical temperature, °C	−129.2[a]	11
	−129.00	9
critical pressure, kPa[b]	5571	9,11
heat of vaporization, ΔH_{vap}, at −188.44°C and 98.4 kPa, J/mol[c]	6544	6
heat of fusion, ΔH_{fus}, J/mol[c]	510	6
heat of transition, J/mol[c]	727.6	6
heat capacities, J/(mol·K)[c]		
solid at −223°C	49.338	6
at −238°C	31.074	6
	23.267	7
at −253°C	12.987	6
	9.372	7
liquid	57.312[a]	6
	45.35	7
gas, C_p	31.46[a]	12
	31.456	13
	31.380	14
	31.325	15
density of liquid at bp, kg/m³	1516[a]	16
	1514	17
density of solid, kg/m³	1900[d]	18
refractive index		
liquid at bp	1.2	19
gas at 0°C and 101.3 kPa[b]	1.000214	20
surface tension, liquid, mN/m(= dyn/cm)		
at −193.26°C	14.81	21
at −192.16°C	14.60	16
viscosity, mPa·s(= cP)		
liquid at −187.96°C	0.257	16
at −203.96°C	0.414	16
gas at 0°C and 101.3 kPa[b]	0.0218	22
	0.0209	7
thermal conductivity, gas at 0°C and 101.3 kPa,W/(m·K)	0.02477	23
dielectric constant, ϵ		
at −189.95°C	1.517	7
at −215.76°C	1.567	7
vapor pressure, kPa[b]		
at 53.56 K	0.22	6
at 63.49 K	2.79	6
at 72.56 K	18.62	6
at 83.06 K	80.52	6
at 89.40 K	162.11	6

[a]Generally accepted value.   [b]To convert kPa to mm Hg, multiply by 7.5.
[c]To convert J to cal, divide by 4.184.   [d]Mean estimate value.

the formation of many simple and complex fluorides in which the other elements are at their highest oxidation states.

The reactivity of fluorine compounds varies from extremely stable, eg, compounds such as sulfur hexafluoride [2551-62-4], nitrogen trifluoride [7783-54-2], and the perfluorocarbons (see FLUORINE COMPOUNDS, ORGANIC); to extremely reactive, eg, the halogen fluorides. Another unique property of nonionic metal fluorides is great volatility. Volatile compounds such as tungsten hexafluoride [7783-82-6], WF_6, and molybdenum hexafluoride [7783-77-9], MoF_6, are produced by the reaction of the particular metal with elemental fluorine.

Fluorine is the first member of the halogen family. However, many of its properties are not typical of the other halogens. Fluorine has only one valence state, -1, whereas the other halogens also form compounds in which their valences are $+1$, $+3$, $+5$, or $+7$. Fluorine also has the lowest enthalpy of dissociation relative to the other halogens, which is in part responsible for its greater reactivity. Furthermore, the strength of the bond fluorine forms with other atoms is greater than those formed by the other halogens.

Table 2 shows bond energies for the four diatomic molecular halogens, as well as for the halides of hydrogen, carbon, boron, and aluminum. Examination of these data indicates that the enthalpies of fluorination are much greater than those of other halogenations. Less energy is required to form fluorine molecules than that needed for chlorine or bromine molecules, and much more energy is evolved in the formation of the fluorides. Therefore, fluorination reactions occur more readily, generating intense heat, and these frequently occur in situations where other halogenations do not.

Reactions. *Metals.* At ordinary temperatures, fluorine reacts vigorously with most metals to form fluorides. A number of metals, including aluminum, copper, iron, and nickel, form an adherent and protective surface film of the metal fluoride salt thus allowing the metal's use in the storage and handling of the gas. A metal's susceptibility to reaction with fluorine depends, to a great extent, on its physical state. For example, powdered iron of 0.84-mm size (20 mesh) is not attacked by liquid fluorine, whereas in the 0.14-mm size (100 mesh) it ignites and burns violently. There is no apparent reaction between liquid fluorine and powdered nickel as fine as 0.14 mm. Massive copper burns at 692°C, whereas copper

Table 2. Average Bond Energies, kJ[a,b]

Halogen	XX[c]	HX[d]	BX$_3$[d]	AlX$_3$[e]	CX$_4$[d]
F	157.8	569	645	582	456
Cl	243.6	431	444	427	327
Br	193.0	368	368	360	272
I	151.1	297	272	285	239

[a]To convert J to cal, divide by 4.184.
[b]X = halogen.
[c]Ref. 24.
[d]Ref. 25.
[e]Ref. 26.

wool ignites at a much lower temperature. Nickel burns in fluorine at 1147°C, and aluminum burns above its melting point (27).

Tin reacts completely with fluorine above 190°C to form tin tetrafluoride [7783-62-2], SnF_4. Titanium reacts appreciably above 150°C at a rate dependent on the size of the particles; the conversion to titanium tetrafluoride [7783-63-3], TiF_4, is complete above 200°C. Fluorine reacts with zirconium metal above 190°C. However, the formation of a coating of zirconium tetrafluoride [7783-64-4], ZrF_4, prevents complete conversion, the reaction reaching only 90% completion even at 420°C (28). Tungsten powder reacts with fluorine at a temperature above 250°C to produce the volatile tungsten hexafluoride, WF_6.

Nonmetals. Sulfur reacts with fluorine to yield the remarkably stable sulfur hexafluoride, SF_6. Operating conditions must be controlled because a mixture of the lower fluorides such as disulfur difluoride [13709-35-8], S_2F_2, disulfur deca-fluoride [5714-22-7], S_2F_{10}, and sulfur tetrafluoride [7783-60-0], SF_4, may also be formed. When this reaction is carried out between 310 and 340°C, SF_4 is primarily obtained and essentially no SF_6 and only trace amounts of lower fluorides. Below 300°C, and preferably at ca 275°C, SF_6 is the primary product. At 450–500°C, a mixture comprising ca 50% SF_4 and the lower sulfur fluorides is formed (see FLUORINE COMPOUNDS, INORGANIC–SULFUR).

Silicon and boron burn in fluorine forming silicon tetrafluoride [7783-61-1], SiF_4, and boron trifluoride [7637-07-2], BF_3, respectively. Selenium and tellurium form hexafluorides, whereas phosphorus forms tri- or pentafluorides. Fluorine reacts with the other halogens to form eight interhalogen compounds (see FLUOR-INE COMPOUNDS, INORGANIC–HALOGENS).

Water. Fluorine reacts with water to form hydrofluoric acid [7664-39-3], HF, and oxygen difluoride [7783-41-7], OF_2. In dilute (<5%) caustic solutions, the reaction proceeds as follows:

$$2\,F_2 + 2\,NaOH \rightarrow OF_2 + 2\,NaF + H_2O$$

In the presence of excess caustic, the oxygen difluoride is gradually reduced to oxygen and fluoride:

$$OF_2 + 2\,NaOH \rightarrow 2\,NaF + O_2 + H_2O$$

The overall reaction under controlled conditions provides a method for the disposal of fluorine by conversion to a salt:

$$2\,F_2 + 4\,NaOH \rightarrow 4\,NaF + O_2 + 2\,H_2O$$

Oxygen. Oxygen does not react directly with fluorine under ordinary conditions, although in addition to oxygen difluoride, three other oxygen fluorides are known (29). Dioxygen difluoride [7783-44-0], O_2F_2, trioxygen difluoride [16829-28-0], O_3F_2, and tetraoxygen difluoride [12020-93-8], O_4F_2, are produced in an electric discharge at cryogenic temperatures by controlling the ratio of fluorine to oxygen.

Nitrogen. Nitrogen usually does not react with fluorine under ordinary conditions and is often used as a diluent to moderate fluorinations. However, nitrogen can be made to produce nitrogen trifluoride, NF_3, by radiochemistry (30), glow discharge (31), or plasma (32) synthesis (see PLASMA TECHNOLOGY).

Noble Gases. Fluorine has the unique ability to react with the heavier noble gases to form binary fluorides. Xenon reacts at room temperature under uv radiation forming xenon difluoride [*13709-36-9*], XeF_2 (16–18,21,33,34). Xenon tetrafluoride [*13709-61-0*], XeF_4, is obtained by mixing an excess of fluorine with xenon and heating the mixture to 400°C (33,35). Fluorine and xenon at 300°C under 6–6.8 MPa (60–67 atm) yield the xenon hexafluoride [*13693-09-9*], XeF_6 (36,37). Fluorine reacts with radon at 400°C to yield a compound of low volatility, probably the difluoride (38). Krypton reacts with fluorine in an electric discharge at liquid air temperatures to yield krypton difluoride [*13773-81-4*], KrF_2 (39).

Hydrogen. The reaction between fluorine and hydrogen is self-igniting and extremely energetic. It occurs spontaneously at ambient temperatures as evidenced by minor explosions which sometimes occur in fluorine-generating cells from the mixing of the H_2 and F_2 streams. The controlled high temperature reaction of fluorine atoms, whether generated thermally or photolytically from fluorine gas, with hydrogen or deuterium is an energy source for high power chemical lasers (qv) (40). However, NF_3 has become the preferred fluorine source because it is easier to handle (41).

Ammonia. Ammonia (qv) reacts with excess fluorine in the vapor phase to produce N_2, NF_3, N_2F_2, HF, and NH_4F. This reaction is difficult to control in the vapor phase because of the intense heat of reaction, and in some cases only N_2 and HF are produced. Nitrogen trifluoride was obtained in 6% yields in a gas-phase reaction over copper (42). Yields of ca 60% are achieved by the reaction of fluorine and ammonia in a molten ammonium acid fluoride solution (43,44).

Organic Compounds. The reaction of pure or undiluted fluorine and organic compounds is usually accompanied by either ignition or a violent explosion of the mixture because of the very high heat of reaction. However, useful commercial-scale syntheses using fluorine are undertaken. Volatile compounds may be fluorinated in the gas phase by moderating the reaction using an inert gas such as nitrogen, by reducing reaction temperatures ($\leq -78°C$), and/or by the presence of finely divided packing materials. Solutions or dispersions of higher boiling materials may be fluorinated in inert solvents such as 1,1,2-trichloro-1,2,2-trifluoroethane [*76-13-1*] or some perfluorocarbon fluids, eg, Fluorinert FC-27 or FC-75 (3M) (45,46). Efficient removal of the very high reaction heat, which leads to molecular fragmentation and runaway reactions, is the underlying principle in any of the aforementioned approaches.

Saturated hydrocarbons (qv) under controlled conditions react with elemental fluorine to produce perfluorocarbons; the reaction is usually accompanied by some fragmentation and polymerization (47,48). The fluorination of aromatic compounds gives degradation products, polymers, unstable unsaturated compounds, or highly fluorinated cyclohexane derivatives, but no aromatic compounds. Methanol (qv) and acetone (qv) produce a variety of fluorinated carbonyl compounds (49–51). Various polyethers have been successfully fluorinated to give perfluoropolyethers in high yields (45,46) (see FLUORINE COMPOUNDS, ORGANIC–DIRECT FLUORINATION). Fluorine is also used in the preparation of cobalt trifluoride [*10026-18-3*], CoF_3, and other higher oxidation state metal fluorides, which can be used in high temperature fluorinations of aliphatic compounds (52).

Selective fluorination, where only one or two fluorines are introduced into a molecule, is becoming more prominent. One well-known example is the direct fluorination of uracil (2,4-pyrimidinedione), $C_4N_2H_4O_2$, in aqueous solution to produce 5-fluorouracil [51-21-8] (5-fluoro-2,4-pyrimidinedione), $C_4N_2H_3FO_2$ (53). Certain nitrogen- and oxygen-containing organics react with fluorine to yield a group of compounds known as electrophilic fluorinating agents (54–57). These reagents are used in regiospecific fluorinations of other organic substrates, particularly those used in pharmaceuticals (qv). For more information regarding selective fluorination, see References 58 and 59.

Polymers. The dilution of fluorine using an inert gas significantly reduces the reactivity, thus allowing controlled reactions to take place with hydrocarbon polymers, even at elevated temperatures. High density polyethylene containers can be blow-molded using 1–10% fluorine in nitrogen mixtures to produce barrier layers on the inside of the containers (see BARRIER POLYMERS; OLEFIN POLYMERS) (60). The permeation rate of nonpolar solvents such as *n*-pentane can be reduced by a factor of almost 500 when containers are blow-molded with a 1% fluorine-in-nitrogen mixture (61,62). Mixtures of 1–10% fluorine in nitrogen have been used to improve the surface properties, ie, moisture transport, soil release, and soil redeposition, of polyester, polyamide, polyolefin, and polyacrylonitrile fibers (63,64) (see FIBERS, POLYESTER; POLYAMIDES, FIBERS).

Fluorine may also be used in conjunction with other reactive gases, eg, oxygen and water vapor, to activate polymer surfaces in order to improve chemical bonding and adhesion (65). For example, ethylene–propylene–diene monomer (EPDM) rubber moldings exposed to a very dilute fluorine mixture, eg, 1 to 5% F_2 in N_2, and subsequently to atmospheric oxygen and moisture have higher surface energy and can be bonded with adhesives (qv) (66). Adhesives do not wet and thus do not adhere to the untreated rubber. Fluorine activates the polymer surface by initiating reactions which form reactive C—OH, C—OOH, or C—OF surface groups. Also, contaminates such as plasticizers (qv) and polymer processing aids are oxidized and removed from the surface. Similarly, other materials including polyolefins (67), polyethylene–vinyl acetate foams (68), and rubber tire scrap (69), can be treated with fluorine.

Carbon and Graphite. Fluorine reacts with amorphous forms of carbon, such as wood charcoal, to form carbon tetrafluoride [75-73-0], CF_4, and small amounts of other perfluorocarbons. The reaction initiates at ambient conditions, but proceeds to elevated temperatures as the charcoal burns in fluorine.

Fluorine reacts with high purity carbon or graphite at elevated temperatures under controlled conditions to produce fluorinated carbon, $(CF_x)_n$. Compounds having colors ranging from black to white have been prepared with fluorine contents ranging from $x = 0.1$ to $x = 1.3$ (70–74). The material was first obtained (70) in 1934, when graphite was heated to 420–460°C in a stream of fluorine to produce a gray product of composition $CF_{0.92}$. Subsequently, a white material of composition $CF_{1.12}$ was obtained (71) by accurate control of the reaction temperature to 627 ± 3°C. Applications utilizing the unique properties of these materials began to be developed in the late 1960s. Fluorinated carbon is a specialty product used in lithium batteries (qv) (75) and lubricants (76). For a complete review of these materials see References 77 and 78.

Manufacture

Fluorine is produced by the electrolysis of anhydrous potassium bifluoride [*7789-29-9*], KHF_2 or KF·HF, which contains various concentrations of free HF. The fluoride ion is oxidized at the anode to liberate fluorine gas, and the hydrogen ion is reduced at the cathode to liberate hydrogen. Anhydrous HF cannot be used alone because of its low electrical conductivity (see ELECTROCHEMICAL PROCESSING, INORGANIC).

Fluorine-generating cells are classified into three distinct types, based on operating temperatures: low (−80 to 20°C) temperature cells, medium (60–110°C) temperature cells, and high (220–300°C) temperature cells. Figure 1 is a melting point diagram for the KF–HF system showing the three distinct areas of temperature operation and the corresponding HF concentrations in the electrolyte. Reference 79 presents a complete description of various laboratory and industrial fluorine cells worldwide.

Cells operating at low (2,80,81) and high (79,82) temperatures were developed first, but discontinued because of corrosion and other problems. The first medium temperature cell had an electrolyte composition corresponding to KF·3HF, and operated at 65–75°C using a copper cathode and nickel anodes. A later cell operated at 75°C and used KF·2.2HF or KF·2HF as electrolyte (83,84), and nickel and graphite as anode materials.

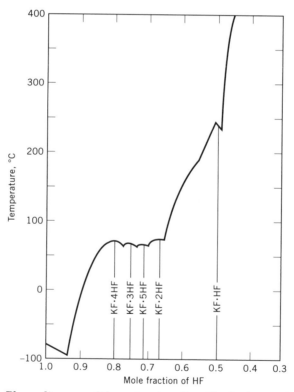

Fig. 1. Phase diagram of the potassium fluoride–hydrogen fluoride system.

Commercial Cells. All commercial fluorine installations employ medium temperature cells having operating currents of ≥ 5000 A. The medium temperature cell offers the following advantages over low and high temperature cells: (*1*) the vapor pressure of HF over the electrolyte is less; (*2*) the composition of the electrolyte can vary over a relatively wide range for only a small variation in the operation of the cell; (*3*) less corrosion or deterioration of the anode occurs; (*4*) tempered water can be used as cell coolant; and (*5*) the formation of a highly resistant film on the anode surface is considerably reduced compared to the high temperature cell.

The C and E type of the Atomic Energy Commission (AEC) (now the Department of Energy) cell designs (85–88) predominate in the United States and Canada. These were developed by Union Carbide Nuclear Co., under the auspices of the AEC, as part of the overall manufacturing process for uranium hexafluoride, UF_6. Large fluorine-generating plants using these designs were installed at the Paducah, Oak Ridge, and Portsmouth gaseous diffusion plants (which are no longer in operation) (see DIFFUSION SEPARATION METHODS). The AEC designs have been made available to industry and are used by several commercial producers. The other cell type used in the United States is a proprietary design developed by Allied Chemical, Corp. (now AlliedSignal, Inc.). This latter cell has a capacity of 5000 A and is used by AlliedSignal Inc. at its Metropolis, Illinois, plant. Table 3 gives the operating characteristics of a typical commercial size cell (AEC E-type).

AEC Cell. A diagram of the AEC cell is shown in Figure 2. The main components are the cell tank, cell head, anode assembly, cathode assembly, screen diaphragm, and packing gland (86,88,89). The E-type and C-type are similar in design and are both rated at 6000 A maximum capacity. The E-type incorporated design changes to improve heat removal efficiency and to prolong life of the anode-contact connection. The cell tank is constructed of 95-mm Monel plate surrounded by a thin-gauge Monel jacket to overcome the corrosion problems encountered by using a steel jacket. A recessed bolt carbon-plug fastens the anodes to a copper support bar. This design gives improved cell life over the C-type copper pressure-plate design by using a steel anode support bar.

The cell head is fabricated from a 2.54-cm steel plate and has separate compartments for fluorine and hydrogen. The outlet-gas manifolds, hydrogen fluoride feed and purge lines, and electrical connections are on top of the head. The gas

Table 3. AEC E-Type Cell Operating Characteristics

Characteristic	Value
current, A	6000
operating voltage, V	9–12
cell operating temperature, °C	90–105
hydrogen fluoride in electrolyte, %	40–42
effective anode area, m^2	3.9
anode current density, A/m^2	1500
anodes	32
anode life, A·h	$40–80 \times 10^6$

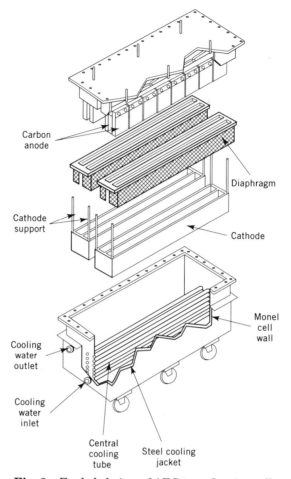

Carbon anode

Diaphragm

Cathode support

Cathode

Cooling water outlet

Monel cell wall

Cooling water inlet

Central cooling tube

Steel cooling jacket

Fig. 2. Exploded view of AEC-type fluorine cell.

separation skirt is made of Monel. An insulating gasket maintains the seal between the tank and the head. The anode assembly consists of 32 carbon blades bolted onto a copper bar, each of which contains three copper conductor posts. The cathode assembly consists of three vertical, 0.6-cm parallel steel plates. The plates surround the anode assembly and are supported by three steel posts which also serve as conductors.

AlliedSignal Cell.   A cross section of AlliedSignal's cell is shown in Figure 3. The cell body is a rectangular steel box on wheels with an outside water-cooling jacket. A central partition divides the box lengthwise. The negative side of the d-c bus is connected directly to the cell container. The interior of the box and the central partition act as the cathodic area of the cell.

Other Cell Designs.   Although not used in the United States, another important cell is based on designs developed by ICI (90). Cells of this type are used by British Nuclear Fuels plc and differ from the cells shown in Figures 2 and 3

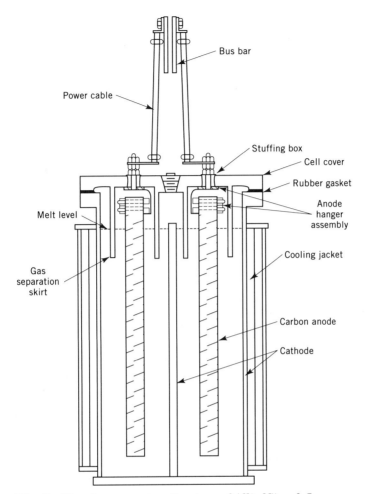

Fig. 3. Fluorine generator. Courtesy of AlliedSignal, Inc.

in two ways: (*1*) the anodes used are made of the same hard, nongraphitized carbon, but are more porous; and (*2*) the cathodes are formed from coiled tubes and provide additional cooling (91).

Anodes. Fluorine cell anodes are the most important cell component, and their design and materials of construction are key factors in determining productivity and cell life. Nickel and graphite, used in early cells, were abandoned when it was discovered that hard, nongraphitized carbon gave longer life. In the 1990s, anodes are made from petroleum coke and a pitch binder which is calcined at temperatures below that needed to convert the material to graphite. The anode carbon has low electrical resistance, high physical strength, and is resistant to reaction with fluorine. Historically, Union Carbide's YBD-grade carbon made the best anodes. More recently other carbon manufacturers have begun to offer improved anodes for fluorine service, eg, Carbone of America Ind. Corp.; Société Des

Electrodes et Refractaires Savoié, part of the Pechiney group; and Toyo Tanso USA, Inc. The nature and quality of the starting materials (petroleum coke and pitch) and careful control of the calcining process are generally responsible for the improvements (92).

About 30% of the cell's operating voltage (9–12 V) is consumed by the anode overvoltage. Although some electrode overvoltage is common in all electrochemical processes, 3 to 4 V is exceptional. There are three generally accepted reasons for the high anode overvoltage: ohmic overvoltage, bubble overvoltage, and inhibition of charge transfer. Ohmic losses are generally small and result from the resistivity of the anode carbon and contact resistance between the anode and the metal current carriers. However, severe losses can arise if the contact between the anode and the current carrier becomes corroded. All commercial fluorine producers have proprietary designs for these connections, which mitigate this problem.

Bubble overvoltage and inhibition of charge transfer are related problems that are caused by the formation of a layer of fluorinated carbon, $(CF_x)_n$, on the anode surface (93,94). Because this material has very low electrical conductivity and is strongly nonwetting, electronic charge transfer is impeded and contact with the electrolyte is lost. As a result of the nonwetting character of the $(CF_x)_n$, bubbles of fluorine gas cling to the anode and grow in size, thereby reducing its effective surface area. Localized hot spots within the cell usually form under these conditions as the remaining working portions of the anode are subjected to higher than normal current densities. The locally excessive temperature also leads to the deterioration or burning of the anode. Signs of this problem are higher than normal cell voltage and higher levels of CF_4 in the fluorine product stream. In extreme cases the working surfaces of the anodes may become so restricted that fluorine production ceases.

Studies of anode electrochemistry have shown that water (>500 ppm) in the electrolyte can increase the formation of $(CF_x)_n$ on the anode surface (95). Under normal operating voltages, carbon can be electrochemically oxidized in the presence of water to form a graphite oxide, C_xO, film on the electrode surface, which then readily reacts with fluorine to form $(CF_x)_n$. Highly graphitic forms of carbon are much more susceptible to this problem. By starting out new cells at lower than normal operating voltages (6 V or less) trace amounts of water may be electrolyzed to H_2 and O_2 and removed from the cell, while avoiding the formation of C_xO.

Further improvements in anode performance have been achieved through the inclusion of certain metal salts in the electrolyte, and more recently by direct incorporation into the anode (92,96,97). Good anode performance has been shown to depend on the formation of carbon–fluorine intercalation compounds at the electrode surface (98). These intercalation compounds resist further oxidation by fluorine to form $(CF_x)_n$, have good electrical conductivity, and are wet by the electrolyte. The presence of certain metals enhance the formation of the intercalation compounds. Lithium, aluminum, or nickel fluoride appear to be the best salts for this purpose (92,98).

Other Cell Components.   American fluorine manufacturers use Monel or steel cathodes. The early German investigators used magnesium cathodes without excessive corrosion in the high temperature cells. Welded steel or Monel construction is used for the cell body. Skirts are used to separate the hydrogen and fluorine above the electrolyte. The solid metal skirt is welded to the cell cover plate and extends vertically downward 10–15 cm into the electrolyte. In the AEC cells, the skirt extends further into the electrolyte as a wire mesh. This extension is called the diaphragm and acts to direct the flow of gases as they are liberated. The AEC cells use Monel skirts and diaphragms. The AlliedSignal cell uses a magnesium alloy skirt and does not employ a diaphragm.

Polytetrafluoroethylene (PTFE) provides the most satisfactory electrical insulation. Concentric rings of PTFE and PTFE impregnated with calcium fluoride are used for the packing glands which support the anode and cathode posts. Rubber is used as the gasket material to form a seal between the cover and the cell body.

Cells must be fitted with mild steel jackets and/or coils to remove heat during cell operation and to provide heat to maintain the electrolyte molten during shutdown. All commercial cells are totally jacketed. However, the accumulation of corrosive products can cause flow restrictions, resulting in decreased heat-transfer capacity. This problem was overcome in the AEC E-type cell by using a water jacket constructed of thin-gauge Monel with vertical corrugations to provide strength. The AEC cells also use internal Monel tubes, manifolded to the external jacket, for additional heat-transfer area. Because one of the operating limits is heat removal, improvements in jacket and tube design were a key factor in increasing the current-handling capacity of commercial cells.

Heat Transfer. A large portion of cell operating voltage is consumed in ohmic processes which generate heat and are a result of the large separation between anode and cathode and the resistivity of the electrolyte. Approximately 34.8 MJ (33,000 Btu) must be removed per kilogram of fluorine produced from any fluorine cell. This is accomplished by jacketing the cell and/or by using cooling tubes. The temperature of the cooling water should not drop below 58°C in order to avoid crystallization of bifluoride on the cell wall. The overall heat-transfer coefficient, U (thermal conductance), depends on operating current and inlet cooling water temperature (86). At 3000 A and an inlet water temperature of 40°C, an average U value is 109.5 W/(m²·K) (19.3 Btu/(h·ft²·°F)); for an operating current of 4000 A and a water temperature of 57°C, an average U value is 177.6 W/(m²·K) (31.3 Btu/(h·ft²·°F)).

Raw Material. The principal raw material for fluorine production is high purity anhydrous hydrofluoric acid. Each kilogram of fluorine generated requires ca 1.1 kg HF. Only a small portion of the hydrofluoric acid produced in the United States is consumed in fluorine production. The commercial grade is acceptable for use as received, provided water content is less than 0.02%. Typical specifications for hydrofluoric acid are

Assay	Wt %
HF, min	99.95
SO_2, max	0.005
H_2SiF_6, max	0.001
H_2O, max	0.02
nonvolatile acid (as H_2SO_4), max	0.01

Potassium bifluoride, KF·HF, is used as a raw material to charge the cells initially and for makeup when cells are rebuilt. A newly charged cell requires about 1400 kg KF·HF. Overall consumption of KF·HF per kilogram of fluorine generated is small. Commercial-grade flake potassium bifluoride is acceptable. Its specifications are

Assay	Wt %
KF·HF, min	99.3
Cl, max	0.01
K_2SiF_6	0.50
SO_4, max	0.01
Fe, max	0.02
Pb, max	0.005
H_2O, max	0.10

Process. The generation of fluorine on an industrial scale is a complex operation (89,99). The basic raw material, anhydrous hydrogen fluoride, is stored in bulk and charged to a holding tank from which it is continuously fed to the cells. Electrolyte for the cells is prepared by mixing KF·HF with HF to form KF·2HF. The newly charged cells are started up at a low current, which is gradually increased at a conditioning station separate from the cell operating position until full current is obtained at normal voltages. After conditioning, cells are connected in series using ca 12 V provided for each cell by a low voltage, 6000 A d-c rectifier. Hydrogen fluoride content is maintained between 40 and 42% by continuous additions. The electrolyte level must be set and controlled at a certain level below the cell head in order to maintain a seal between the fluorine and hydrogen compartments. The cells are operated at 95–105°C and cooled with water at 75°C.

Approximately 142.3 MJ/h (135,000 Btu/h) must be removed at an operating current of 6000 A. The hydrogen and fluorine gas leaving the cell contains ca 10 vol % HF. The individual gas streams from each cell are joined into separate hydrogen and fluorine headers for further processing in the plant. Demisters and filters are provided in the product gas streams to remove entrained electrolyte. The gas streams are then cooled to − 110°C in refrigerant-cooled condensers to reduce the HF concentration to approximately 3 mol %. The condensed HF is recycled, and hydrogen stream is scrubbed with a caustic solution and vented or burned. Several possibilities are available for utilizing the fluorine stream. If lower than 3% HF levels are required, sodium fluoride towers or further cooling are employed to freeze out the HF and reduce the concentration to less than 0.2 mol %. Compressors or exhausters are normally required in both fluorine and hydrogen streams at some point in the system beyond the HF condensers.

Figure 4 presents the equipment flow sheet for a 9-t/d fluorine plant.

Equipment

Fluorine can be handled using a variety of materials (100–103). Table 4 shows the corrosion rates of some of these as a function of temperature. System clean-

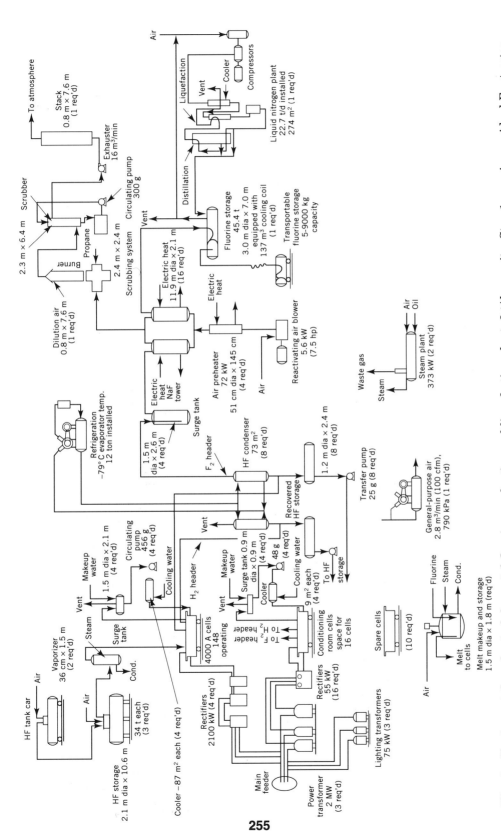

Fig. 4. Equipment flow sheet of elemental fluorine production and liquefaction plant, 9 t/d capacity. Step 1: purging residual F_2 at rates indicated; all but a trace of residual F_2 is removed in 15min; N_2 purge is maintained for 1 h to remove last traces. Step 2: HF removal at rates indicated; all but a trace of HF is removed for 10h; air purge is maintained for 10 h. To convert kPa to psi, multiply by 0.145.

255

Table 4. Corrosion of Metals at Various Temperatures, mm/mo[a]

Material	Temperature, °C					
	200	300	400	500	600	700
nickel			0.018	0.129	0.74	0.86
Monel			0.013	0.051	1.5	3.8
Inconel			0.96	1.6	4.3	13
copper			4.1	3.0	25	74
aluminum			0	0.33	0.46	
magnesium	0	0				
iron	0	0.23	0.61	295		
steel						
0.27% carbon	0.051	0.23	0.38	503		
stainless, 310	0	0.79	14.2			

[a] Table abstracted from Reference 104.

liness and passivation are critical to success. Materials such as nickel, Monel, aluminum, magnesium, copper, brass, stainless steel, and carbon steel are commonly used. More information is available in the literature (20,104).

Copper, brass, and steel are generally used for gaseous service at temperatures below 200°C; Monel and nickel are used at elevated temperatures. For critical applications and where there is any danger of temperature buildup, only nickel or Monel should be used because of the stability of the nickel fluoride film. Only highly fluorinated polymers, such as polytetrafluoroethylene, are resistant to the gas under nominally static conditions. The PTFE must be free of any impurities, dirt, or foreign materials that could initiate ignition with fluorine. However, fluoropolymers are more susceptible to ignition than metals, and thus should be used as little as possible in fluorine service, especially in contact with flowing streams at high velocities or pressures.

All equipment, lines, and fittings intended for fluorine service must be leak-tight, dry, and thoroughly cleansed of all foreign matter before use. The system should be checked for leaks, at least to its working pressure. It should be flushed with a nonaqueous degreasing solvent, such as methylene chloride, thoroughly purged with a stream of dry nitrogen, and evacuated to make certain no volatiles are present. Any foreign matter, particularly organics, not removed could burn with fluorine and initiate the burning of the metal equipment. After cleaning, the system should be filled with dry nitrogen.

The corrosion resistance of all materials used with fluorine depends on the passivation of the system. This is a pickling operation intended to remove the last traces of foreign matter, and to form a passive fluoride film on the metal surface. The dry nitrogen in the system is slowly replaced with gaseous fluorine in small increments until the concentration and pressure approach working conditions. Carbon steel is most commonly used for pipe and fittings (2.5 cm or greater) in gaseous fluorine service at ambient temperature conditions and pressures up to 2.86 MPa (415 psi).

Carbon steel or bronze-body gate valves are commonly used in gaseous fluorine service at low pressure. Plug valves, having Monel bodies and plugs, are recommended for moderate pressure service below 500 kPa (<5 atm). For valve-stem packing PTFE polymer is recommended and it must be maintained leak-tight. Valves lubricated or packed with grease or other organics should never be used. Bellows-type valves having Monel or stainless steel bellows are recommended for high pressure service, but not ball valves.

Compressors and blowers for gaseous fluorine service vary in design from multistage centrifugal compressors to diaphragm and piston types. Standard commercial instrumentation and control devices are used in fluorine systems. Pressure is measured using Bourdon-type gauges or pressure transducers. Stainless steel or Monel construction is recommended for parts in contact with fluorine. Standard thermocouples are used for all fluorine temperature-measuring equipment, such as the stainless-steel shielded type, inserted through a threaded compression fitting welded into the line. For high temperature service, nickel-shielded thermocouples should be used.

Dilute mixtures (eg, 10 or 20% F_2 in N_2) are generally less hazardous than pure fluorine, but the same precautions and procedures should be employed.

Economic Aspects

Availability and Shipping.   Fluorine gas is packaged and shipped in steel cylinders conforming to Department of Transportation (DOT) specifications 3A1000 and 3AA1000 under a pressure of 2.86 MPa (415 psi). Table 5 lists world fluorine producers. Cylinders containing 2.2 and 0.7 kg are available from Air Products and Chemicals, Inc. All cylinders are equipped with special fluorine valves, the outlets of which have a left-hand thread conforming to the Compressed Gas Association (CGA) Specification 679 or 670. DOT regulations stipulate that cylinders must be shipped without safety relief devices and be equipped with valve protection caps. The total quantity of pure, gaseous fluorine in any size container is limited to 2.7 kg and the pressure must not exceed 2.86 MPa (415 psi) at 21°C.

Mixtures of 10 and 20% fluorine in nitrogen or other inert gases are commercially available in cylinders and tube trailers from Air Products and Chemicals, Inc. Blends can be safely packaged and stored at high, eg, 13.8 MPa (2000 psi), pressure. Filled with a 20% fluorine blend, tube trailers can contain up to 500 kg of fluorine. Such high pressure mixtures permit larger quantities of fluorine to be safely shipped.

Price.   The 1993 U.S. price for fluorine in cylinders was $109/kg for 2.2 kg and $260/kg for 0.7 kg cylinders. The price in large volumes is determined by (1) the price of hydrofluoric acid; (2) power costs, ca 4.5 kW·h electricity is required for each kilogram of fluorine produced; (3) labor costs; (4) costs to maintain and rebuild cells; and (5) amortization of fixed capital. Fluorine production is highly capital intense. In addition, purification, compression, packaging, and distribution in cylinders increase the cost significantly.

Manufacturers.   Besides manufacturers in the United States, commercial fluorine plants are operating in Canada, France, Germany, Italy, Japan, and the United Kingdom (see Table 5). Fluorine is also produced in the Commonwealth of Independent States (former Soviet Union); however, details regarding its manufacture, production volumes, etc, are regarded as secret information. The total

Table 5. Fluorine Producers

Name	Location	Products[a]	Capacity[b]
Air Products & Chemicals	Allentown, Pa.	SF_6, NF_3, WF_6, SF_4, ClF_3, BrF_3, IF_5, perfluorinated hydrocarbons, F_2 gas	
AlliedSignal, Inc.	Morristown, N.J.	UF_6, SF_6, IF_5, SbF_5, $(CF_x)_n$	large
Asahi Glass Co., Ltd.	Tokyo	SF_6	
Ausimont SpA	Milan, Italy	SF_6, fluoropolymer fluids, fluoromonomers, F_2 gas	large
British Nuclear Fuels plc	Preston, U.K.	UF_6, F_2 gas	large
Cameco Corp.	Port Hope, Ontario, Canada	UF_6	large
Central Glass Co., Ltd.	Tokyo	WF_6, NF_3, ClF_3, F_2 gas	moderate
Comurhex (Pechiney group)	Paris	UF_6, ClF_3, N_2F_2, WF_6, F_2 gas	large
Daikin Industries	Osaka, Japan	$(CF_x)_n$, IF_5, perfluorinated hydrocarbons	moderate
Kanto Denka Kogyo Co., Ltd.	Tokyo	SF_6, CF_4, CHF_3, WF_6, C_2F_6, NF_3, F_2 gas	moderate
Sequoyah Fuels Corp.	Gore, Okla.	UF_6	large
Solvay Fluor und Derivate, GmbH	Hannover, Germany	SF_6, IF_5, perfluorinated hydrocarbons, CF_4, WF_6, F_2 gas	large

[a]Not all companies that produce fluorine sell F_2 gas.
[b]Large, >1000 t/yr; moderate, >100 but < 1000 t/yr.

commercial production capacity of fluorine in the United States and Canada is estimated at over 5000 t/yr, of which 70–80% is devoted to uranium hexafluoride production. Most of the gas is used in captive uranium-processing operations.

Analysis

Direct analysis of gaseous fluorine is not possible by conventional methods because of its reactivity, therefore fluorine is converted (105) quantitatively to chlorine and the effluent stream analyzed. First, the fluorine sample is passed through a bed of sodium fluoride which retains the hydrogen fluoride. This step can be eliminated if the fluorine is free of HF as received. The sample is then passed over granular sodium chloride which reacts quantitatively with fluorine to release chlorine. This latter is reduced to chloride, which is determined by the Volhard method. Impurities such as oxygen, nitrogen, carbon tetrafluoride, sulfur hexafluoride, and carbon dioxide are determined by conventional gas chromatography techniques. Hydrogen fluoride is determined by infrared analysis of a separate sample. For processes using fluorine, the concentration may be monitored using uv analyzers such as those manufactured by Du Pont Instruments.

Fluorine in the atmosphere can be detected by chemical methods involving the displacement of halogens from halides. Dilute fluorine leaks are easily detected by passing a damp piece of starch iodide paper around the suspected area. The paper should be held with metal tongs or forceps to avoid contact with the gas stream and immediately darkens when fluorine is present.

Continuous monitoring for the presence of fluorine gas in the workplace may be accomplished using detectors available from Mine Safety Appliances (Pittsburgh, Pa.) or EIT (Exton, Pa.).

Specifications. Fluorine, having a dewpoint of 40°C and containing a maximum of 1.0 ppm water, is sold in cylinders according to the following specifications:

Assay	Mol %
fluorine, min	98
oxygen, max	0.5
nitrogen, max	1.0
carbon tetrafluoride, max	0.1
sulfur hexafluoride, max	0.1
hydrogen fluoride, max	0.2
carbon dioxide, max	0.1

Health and Safety

Fluorine, the most reactive element known, is a dangerous material but may be handled safely using proper precautions. In any situation where an operator may come into contact with low pressure fluorine, safety glasses, a neoprene coat, boots, and clean neoprene gloves should be worn to afford overall body protection. This protection is effective against both fluorine and the hydrofluoric acid which may form from reaction of moisture in the air.

In addition, face shields made of conventional materials or, preferably, transparent, highly fluorinated polymers, should be worn whenever operators approach equipment containing fluorine under pressure. A mask having a self-contained air supply or an air helmet with fresh air supply should always be available. Leaks in high pressure systems usually result in a flame from the reaction of fluorine with the metal. Shields should be provided for valves, pressure-reducing stations, and gauges. Valves are a particularly susceptible area for fluorine fires that can be initiated by foreign material accumulated at the valve seat. High pressure cylinders and valves should be remotely located with proper personnel protection and the latter should be operated using handle extensions. An excellent guide for the safe handling of fluorine in the laboratory is available (106).

Toxicity. Fluorine is extremely corrosive and irritating to the skin. Inhalation at even low concentrations irritates the respiratory tract; at high concentrations fluorine inhalation may result in severe lung congestion.

The American Conference of Governmental Industrial Hygienists (ACGIH) has established the 8-hour time-weighted average TLV as 1 ppm or 1.6 mg/m^3, and the short-term exposure limit TLV as 2 ppm or 3.1 mg/m^3. Fluorine has a sharp, penetrating odor detectable at levels well below the TLV. Manifestations of overexposure to fluorine include irritation or burns of the eyes, skin, and respiratory tract. The following emergency exposure limits (EEL) for humans have been suggested (107): 15.0 ppm for 10 min; 10 ppm for 30 min; and 7.5 ppm for 60 min.

Toxicity studies (108–110) established tolerance levels and degrees of irritations, indicating that the eye is the area most sensitive to fluorine. Comprehensive animal studies (111–113) determined a rat LC_{50} value of 3500 ppm·min for a single 5-min exposure and of 5850 ppm·min for a 15-min exposure. A no-effect concentration corresponded to a concentration-time value of ca 15% of the LC_{50} levels.

Because of the corrosive effects and discomfort associated with inhalation of fluorine, chronic toxicity does not occur. Although the metabolic fate of fluorine is not clear, it does not seem that much is converted to fluoride ion in the body (107). Therefore comparisons to effects of fluoride ion poisoning, known as fluorosis, are probably incorrect.

Burns. Skin burns resulting from contact with pure fluorine gas are comparable to thermal burns and differ considerably from those produced by hydrogen fluoride (114). Fluorine burns heal much more rapidly than hydrofluoric acid burns.

Disposal. Fluorine can be disposed of by conversion to gaseous perfluorocarbons or fluoride salts. Because of the long atmospheric lifetimes of gaseous perfluorocarbons (see ATMOSPHERIC MODELS), disposal by conversion to fluoride salts is preferred. The following methods are recommended: scrubbing with caustic solutions (115,116); reaction with solid disposal agents such as alumina, limestone, lime, and soda lime (117,118); and reaction with superheated steam (119). Scrubbing with caustic solution and, for dilute streams, reaction with limestone, are practiced on an industrial scale.

In a caustic scrubbing system, caustic potash, KOH, is preferred to caustic soda, NaOH, because of the higher solubility of the resulting potassium fluoride. Adequate solution contact and residence time must be provided in the scrub tower to ensure complete neutralization of the intermediate oxygen difluoride, OF_2. Gas residence times of at least one minute and caustic concentrations in excess of 5% are recommended to prevent OF_2 emission from the scrub tower.

Uses

Elemental fluorine is used captively by most manufacturers for the production of various inorganic fluorides (Table 5). The market for gaseous fluorine is small, but growing. The main use of fluorine is in the manufacture of uranium hexafluoride, UF_6, by

$$UF_4 + F_2 \rightarrow UF_6$$

Uranium hexafluoride is used in the gaseous diffusion process for the separation and enrichment of uranium-235, which exists in low concentration in natural uranium. The enriched UF_6 is converted back into an oxide and used as fuel for the nuclear power industry.

Another large use for elemental fluorine is in production of sulfur hexafluoride, SF_6, a gaseous dielectric for electrical and electronic equipment (see ELECTRONIC MATERIALS; FLUORINE COMPOUNDS, INORGANIC–SULFUR). Its high dielectric strength, inertness, thermal stability, and ease of handling have led to increased use as an electrically insulating medium, permitting reductions in size, weight, and cost of high voltage electrical switch gear, breakers, and substations. Elemental fluorine is also used to produce sulfur tetrafluoride, SF_4, by the reaction of sulfur and fluorine under controlled conditions. Sulfur tetrafluoride is a selective fluorinating agent used to produce fluorochemical intermediates in the pharmaceutical and herbicide industry (see HERBICIDES).

Fluorine reacts with the halogens and antimony to produce several compounds of commercial importance: antimony pentafluoride [7783-70-2], bromine trifluoride [7787-71-5], chlorine trifluoride [7790-91-2], and iodine pentafluoride [7783-66-6]. Chlorine trifluoride is used in the processing of UF_6 (see URANIUM AND URANIUM COMPOUNDS). Bromine trifluoride is used in chemical cutting by the oil well industry (see PETROLEUM). Antimony and iodine pentafluorides are used as selective fluorinating agents to produce fluorochemical intermediates (see FLUORINE COMPOUNDS, INORGANIC).

Fluorination of tungsten and rhenium produces tungsten hexafluoride, WF_6, and rhenium hexafluoride [10049-17-9], ReF_6, respectively. These volatile metal fluorides are used in the chemical vapor deposition industry to produce metal coatings and intricately shaped components (see THIN FILMS, FILM FORMATION TECHNIQUES).

Fluorine reacts with ammonia in the presence of ammonium acid fluoride to give nitrogen trifluoride, NF_3. This compound can be used as a fluorine source in the high power hydrogen fluoride–deuterium fluoride (HF/DF) chemical lasers and in the production of microelectronic silicon-based components.

Fluorine is used by a number of manufacturers to produce polyolefin containers that are resistant to permeation by organic liquids. In one application, the air which is normally used to blow-mold containers is replaced by a low concentration of fluorine in a mixture with nitrogen. In another approach, the containers are placed in a large enclosure subsequently flooded with very dilute fluorine–nitrogen or fluorine–air mixtures. Containers may vary in size from small bottles to automotive fuel tanks and show an outstanding resistance to nonpolar solvents and fuels. However, fuels containing polar additives, eg, alcohols, have been more difficult to contain and tank manufacturers are modifying the polyolefin and the fluorination process in an effort to meet Environmental Protection Agency (EPA) mandated fuel loss guidelines.

An important newer use of fluorine is in the preparation of a polymer surface for adhesives (qv) or coatings (qv). In this application the surfaces of a variety of polymers, eg, EPDM rubber, polyethylene–vinyl acetate foams, and rubber tire scrap, that are difficult or impossible to prepare by other methods are easily and quickly treated. Fluorine surface preparation, unlike wet-chemical surface treatment, does not generate large amounts of hazardous wastes and has been demonstrated to be much more effective than plasma or corona surface treatments. Figure 5 details the commercially available equipment for surface treating plastic components. Equipment to continuously treat fabrics, films, sheet foams, and other web materials is also available.

Fig. 5. Equipment for surface treating plastic components. Parts are loaded into one of the two lower chambers which is then evacuated to remove most of the air. This chamber is then flooded with a dilute mixture of fluorine and nitrogen which is made and stored in the upper chamber. After the treatment is completed, the fluorine mixture is pumped back up to the upper chamber for storage and the lower chamber repeatedly flooded with air and evacuated to remove any traces of fluorine gas. Two treatment chambers are cycled between the loading/unloading operation and the treatment step to increase equipment output. The fluorine–nitrogen blend may be used several times before by-products from the treatment process begin to interfere. All waste gases are purged through the scrubber shown to the right. Courtesy of FluorTec GmbH, Marksuhl, Germany.

Fluorine is used in the production of fluorinated organics both for the manufacture of perfluorinated materials and for the selective and regiospecific introduction of fluorine. Perfluorinated aliphatics, cycloaliphatics, and polyethers are made by fluorination of the hydrogen-containing analogue using F_2 or CoF_3. The superior chemical and thermal stability of perfluorocarbons has led to uses in high temperature lubrication, thermal testing of electronic components, and as specialty fluids for vacuum pumps, liquid seals, and hydraulic applications. Because of the high solubility of oxygen in these materials, perfluorinated aliphatics and cycloaliphatics have also been used as synthetic blood substitutes (see BLOOD, ARTIFICIAL) (120).

Although the selective introduction of fluorine into biologically important molecules has been demonstrated to provide dramatic improvements in efficacy and toxicity as compared to unfluorinated analogues, commercial methods using fluorine gas are uncommon. A notable exception is the production of 5-fluorouracil, made by the direct reaction of fluorine and uracil, which is used in cancer chemotherapy (see CHEMOTHERAPEUTICS, ANTICANCER). To overcome the difficulties encountered in direct fluorinations, electrophilic fluorinating agents have been commercialized by AlliedSignal, Inc.; Air Products & Chemicals, Inc.; and the Onoda Cement Co. These reagents are made from nitrogen-containing compounds, which when reacted with F_2 yield compounds that act as positive fluorine, F^+, sources. The selective introduction of fluorine into many types of organic molecules, including steroids, nucleosides, heterocycles, and aromatic compounds, can be effected using these reagents (54–57).

BIBLIOGRAPHY

"Fluorine" in *ECT* 1st ed., Vol. 6, pp. 656–667, by H. C. Miller and F. D. Lommis, Pennsylvania Salt Manufacturing Co.; in *ECT* 2nd ed., Vol. 9, pp. 506–525, by H. R. Neumark and J. M. Siegmund, Allied Chemical Corp.; in *ECT* 3rd ed., Vol. 10, pp. 630–654, by A. J. Woytek, Air Products & Chemicals, Inc.

1. D. N. Lapedes, ed., *Encyclopedia of Science and Technology*, Vol. 5, McGraw-Hill Book Co., Inc., New York, 1977, pp. 389–393.
2. H. Moissan, *Comp. Rend.* **102**, 1534 (1886); **103**, 202, 256 (1886); *Gmelins, Hanbuch der Anorganischen Chemie, System 5*, 8th ed., Deutsche Chemische Gesellschaft, Verlag Chemie, Berlin, 1926, pp. 4–16.
3. K. O. Christe, *Inorg. Chem.* **25**, 3721 (1986).
4. G. C. Finers, in M. Stacey, J. C. Tatlow, and A. G. Sharpe, eds., *Advances in Fluorine Chemistry*, Vol. 2, Butterworths, London, 1962.
5. M. M. Miller, *Fluorspar, Annual Report*, U.S. Dept. of Interior, Bureau of Mines, Washington, D.C., Sept. 1991.
6. J. H. Hu, D. White, and H. Johnson, *J. Am. Chem. Soc.* **75**, 5642 (1953).
7. E. Kanda, *Bull. Chem. Soc. Jpn.* **12**, 473 (1937).
8. J. D. Collins, L. S. Stone, and P. A. Juvner, *Background Chemistry for Development of Liquid Rocket Oxidizers, AD-18-283*, Callery Chemical Co., Callery, Pa., 1953.
9. G. H. Cady and J. H. Hildenbrand, *J. Am. Chem. Soc.* **52**, 3829 (1930).
10. W. H. Claussen, *J. Am. Chem. Soc.* **56**, 614 (1934).
11. D. Horovitz, *A Review of the Physical and Chemical Properties of Fluorine and Certain of Its Compounds, Report No. RMI-293-85*, Reaction Motors, Inc., Rockaway, N.J., 1950.
12. G. M. Murphy and J. E. Vance, *J. Chem. Phys.* **7**, 806 (1939).
13. V. N. Huff and S. Gordon, *Tables of Thermodynamics Functions for Analysis of Aircraft-Propulsion Systems, Tech. No. 2161*, National Advisory Committee for Aeronautics, Washington, D.C., Aug. 1950.
14. K. V. Butkov and R. B. Bozenbaum, *J. Phys. Chem. USSR* **24**, 706 (1950).
15. L. G. Cole, M. Farber, and G. W. Eluerum, Jr., *J. Chem. Phys.* **20**, 586 (1952).
16. G. W. Elverum, Jr., and R. N. Doeschev, *J. Chem. Phys.* **20**, 1834 (1952).
17. R. L. Jarry and H. C. Miller, *J. Am. Chem. Soc.* **78**, 1553 (1956).
18. W. T. Ziegler and J. C. Mullins, *Calculation of the Vapor Pressure and Heats of Vaporization and Sublimation of Liquids and Solids, Especially Below One Atmosphere, IV, Nitrogen and Fluorine. Technical Report no. 1*, Engineering Experiment Station, Georgia Institute of Technology, to Cryogenic Engineering Laboratory, National Bureau of Standards, Boulder, Colo., under NBS contract CST-7404, Apr. 1963.
19. E. U. Franck, *Naturwissenschaften* **41**, 37 (1954).
20. J. F. Tompkins and co-workers, *The Properties and Handling of Fluorine, Technical Report no. ASD-TDR-62-273*, Air Products & Chemicals, Inc., Allentown, Pa., 1963.
21. J. L. Weeks, C. L. Chernick, and M. S. Matheson, *J. Am. Chem. Soc.* **84**, 4612 (1962).
22. E. U. Franck and W. Stober, *Z. Naturforsch.* **7a**, 822 (1952).
23. E. U. Franck and E. Wicke, *Z. Elektrochem.* **55**, 643 (1951).
24. I. G. Stamper and R. F. Barrow, *Trans. Faraday Soc.* **54**, 1592 (1958).
25. D. A. Johnson, *Some Thermodynamic Aspects of Inorganic Chemistry*, Cambridge University Press, New York, 1968, p. 158.
26. E. L. Meutterties and C. W. Tullock, in W. L. Jolly, ed., *Preparative Inorganic Reactions*, Vol. 2, Interscience Publishers, New York, 1965, p. 243.
27. T. W. Godwin and C. F. Lorenzo, "Ignition of Seven Metals in Fluorine," paper no. 740, *American Rocket Society, 13th Annual Meeting*, New York, Nov. 17–21, 1958.
28. H. M. Haendler and co-workers, *J. Am. Chem. Soc.* **76**, 2177 (1954).

29. A. G. Streng, *Chem. Rev.* **63**, 607 (1963).

30. V. A. Dmitrievskii, V. N. Cherednikow, and E. K. Illin, *Khim Vys. Energ.* **7**(3), 206 (1973).

31. I. V. Nikitin and V. Ya. Rosolovski, *Izv. Adad, Nauk USSR Ser. Khim* **7**, 1464 (1970).

32. U.S. Pat. 3,304,248 (Feb. 14, 1967), H. T. Fullan and H. V. Scklemain (to Stauffer Chemical Co.).

33. C. L. Chernick and co-workers, *Science* **138**, 136 (1962).

34. J. H. Holloway, *Chem. Commun.*, 22 (1966).

35. H. H. Claassen, H. Selig, and J. G. Malm, *J. Am. Chem. Soc.* **84**, 3593 (1962).

36. J. G. Malm, I. Sheft, and C. L. Chernick, *J. Am. Chem. Soc.* **85**, 110 (1963).

37. E. E. Weaver, B. Weinstock, and C. P. Knop, *J. Am. Chem. Soc.* **85**, 111 (1963).

38. P. R. Fields, L. Stein, and M. H. Zirin, *J. Am. Chem. Soc.* **84**, 4164 (1962).

39. A. V. Grosse and co-workers, *Chem. Eng. News* **41**, 47 (Jan. 7, 1963).

40. G. C. Pimentel and J. H. Packer, *J. Chem. Phys.* **51**, 91 (1961).

41. M. C. Lin, M. E. Umstead, and N. Djeu, *Ann. Rev. Phys. Chem.* **34**, 557 (1983).

42. S. I. Morrow and co-workers, *J. Am. Chem. Soc.* **82**, 5301 (1960).

43. U.S. Pat. 4,091,081 (May 23, 1978), A. J. Woytek and J. T. Likeck (to Air Products and Chemicals, Inc.).

44. *Chem. Eng.* **84**(26), 116 (1977).

45. WO Pat. 90/06296 (June 14, 1990), M. G. Costello and G. I. Moore (to Minnesota Mining and Manufacturing Co.).

46. Eur. Pat. 0,332,601 (Sept. 13, 1989), F. R. Feher, and co-workers (to Monsanto Co.).

47. E. H. Hadley and L. A. Bigelow, *J. Am. Chem. Soc.* **62**, 3302 (1940).

48. E. A. Tyczkowski and L. A. Bigelow, *J. Am. Chem. Soc.* **77**, 3007 (1955).

49. K. B. Kellogg and G. H. Cady, *J. Am. Chem. Soc.* **70**, 3968 (1948).

50. N. Fukuhara and L. A. Bigelow, *J. Am. Chem. Soc.* **63**, 778 (1941).

51. W. D. Clark and R. J. Lagow, *J. Fluor. Chem.* **52**, 37 (1991).

52. M. Stacey and J. C. Tatlow, in M. Stacey, J. C. Tatlow, and A. G. Sharpe, eds., *Advances in Fluorine Chemistry*, Butterworths Publications, London, 1960, pp. 166–198.

53. Can. Pat. 3,954,749 (Mar. 16, 1976), P. D. Schuman and co-workers (to PCR, Inc.).

54. L. German and S. Zemskov, eds., *New Fluorinating Agents in Organic Synthesis*, Springer-Verlag, New York, 1989.

55. A. J. Poss and co-workers, *J. Org. Chem.* **56**, 5962 (1991).

56. E. Differding and co-workers, *Synlett*, 187 (1991); *Synlett*, 395 (1991).

57. T. Umemoto and co-workers, *J. Am. Chem. Soc.* **112**, 8563 (1990).

58. S. T. Purrington and B. S. Kagen, *Chem. Rev.* **86**, 997 (1986).

59. V. Grakauskas, *Intra-Science Chem. Rep.* **5**, 85 (1971).

60. U.S. Pat, 3, 862,284 (Jan. 21, 1975), D. D. Dixson, D. G. Manly, and G. W. Recktenwald (to Air Products and Chemicals, Inc.).

61. A. J. Woytek and J. F. Gentilecore, "A New Blow Molding Process to Reduce Solvent Permeation of Polyolefin Containers," paper no. 13 presented at *Advances in Blow Molding Conference*, Rubber and Plastics Institute, London, Dec. 6, 1977.

62. J. F. Gentilecore, M. A. Triolo, and A. J. Woytek, *Plast. Eng.* **34**(9), 40 (1978).

63. U.S. Pat. 4,020,223 (Apr. 26, 1977), D. D. Dixson and L. J. Hayes (to Air Products & Chemicals, Inc.).

64. U.S. Pat. 3,988,491 (Oct. 26 1976), D. D. Dixson and L. J. Hayes (to Air Products & Chemicals, Inc.).

65. R. Milker and A. Koch, in D. Satas, ed., *Coatings Technology Handbook*, Marcel Dekker, Inc., New York, 1990.

66. R. Milker and A. Koch, *Coating* **1**, 8 (1988).

67. B. D. Bauman, "Novel Polyurethane Composites with Surface-Modified Polymer Particles," paper presented at *SPI 32nd Annual Technical/Marketing Conference*, 1989.
68. R. Milker and A. Koch, *Kunststoffberater* **7/8**, 56 (1989).
69. B. D. Bauman, "Scrap Tire Reuse Through Surface-Modification Technology," paper presented at *International Symposium on Research and Development for Improving Solid Waste Management*, Cincinnati, Ohio, Feb. 7, 1991.
70. O. Ruff, D. Bretschneider, and F. Elert, *Z. Anorg. Chem.* **217**, 1 (1934).
71. U.S. Pat, 3,674,432 (July 4, 1972), R. J. Lagow and co-workers (to R. I. Patents, Inc.).
72. W. Rudorff, *Adv. Inorg. Chem. Radiochem.* **1**, 230 (1959).
73. G. R. Hennig, *Prog. Inorg. Chem.* **1**, 125 (1959).
74. N. Watanabe and K. Kumon, *Denki Kagaku* **35**, 19 (1967).
75. R. L. Fusaro and H. E. Sliney, *NASA Tech. Note D-5097*, National Aeronautics & Space Administration, Washington, D.C., 1969; *ALSE Trans.* **13**, 56 (1970).
76. M. Fukuda and T. Iijima, in J. P. Gabano, ed., *Lithium Batteries*, Academic Press, Inc., New York, 1983.
77. N. Watanabe, T. Nakajima, and H. Touhara, *Studies in Inorganic Chemistry 8, Graphite Fluorides*, Elsevier Science Publishers, New York, 1988.
78. G. A. Shia and G. Mani, in R. E. Banks, B. Smart, and J. C. Tatlow, eds., *Organofluorine Chemistry: Principles and Commercial Applications*, Plenum Publishing Corp., New York, in press.
79. R. J. Ring and D. Royston, *A Review of Fluorine Cells and Fluorine Production Facilities*, Australian Atomic Energy Commission, AAEC/E 281/, Sept. 1973.
80. A. J. Rudge, in A. Kuhn, ed., *Industrial Electrochemical Processes*, Elsevier Publishing Co., Amsterdam, the Netherlands, 1971, Chapt. 1.
81. A. J. Rudge, *Chem. Ind.* **22**, 504 (1956).
82. R. D. Fowler and co-workers, *Ind. Eng. Chem.* **39**, 3, 266 (1947).
83. G. H. Cady and H. S. Booth, eds., *Inorganic Synthesis*, Vol. 1, McGraw-Hill Book Co., Inc., New York, 1939, pp. 136–137.
84. C. H. Cady, D. A. Rogers, and C. A. Carlson, *Ind. Eng. Chem.* **34**, 4, 443 (1942).
85. S. H. Smiley and D. C. Brater, *USAEC Report TID-5295*, U.S. Atomic Energy Commission (USAEC), Washington, D.C., 1956.
86. J. Dykstra and co-workers, *Ind. Eng. Chem.* **47**, 5, 883 (1955).
87. B. W. Clark, *USAEC Report KY-326*, USAEC, Washington, D.C., 1960.
88. R. C. Kelley and W. E. Clark, eds., *USAEC Report TID4100, Suppl. 42, CAPE-55*, 1967, and *Suppl. 45, CAPE-486*, 1968, USAEC, Washington, D.C.
89. J. Dykstrra, A. P. Huber, and B. H. Thompson, "Multi-Ton Production of Fluorine for Manufacture of Uranium Hexafluoride," paper presented at *Second United Nations International Conference on the Peaceful Use of Atomic Energy*, A/CONF.15/P/524, June 1958.
90. Brit. Pat. 861,978 (1961), A. J. Rudge and A. Davies (to ICI).
91. J. F. Ellis and G. F. May, *J. Fluor. Chem.* **33** 133 (1986).
92. U.S. Pat. 4,312,718 (Jan. 26, 1982), N. Watanabe, M. Aramaki, and Y. Kita (to N. Watanabe, Central Glass., Ltd., and Toyo Tanso Co.).
93. L. Bai and B. E. Conway, *J. Appl. Electrochem.* **18**, 839 (1988).
94. L. Bai and B. E. Conway, *J. Appl. Electrochem.* **20**, 916 (1990).
95. T. Nakajima, T. Ogawa, and N. Watanabe, *J. Electrochem. Soc.* **134**, 8 (1987).
96. N. Watanabe, *Proc. Int. Symp. Molten Salt Chem. Technol., Molten Salt Comm. of the Electrochem. Soc. of Japan*, 21 (1983).
97. U.S. Pat. 4,915,809 (Apr. 10, 1990), O. Brown and M. Wilmott (to British Nuclear Fuels plc).
98. T. Nakajima and M. Touma, *J. Fluor. Chem.* **57**, 83 (1992).
99. J. Jacobson and co-workers, *Ind. Eng. Chem.* **47**, 5, 878 (1955).

100. J. R. McGuffey, R. Paluzelle, and W. E. Muldrew, *Ind. Eng. Chem.* **54**, 5, 46 (1962).
101. J. M. Siegmund, *Chem. Eng. Prog.* **63**, 6, 88 (1967).
102. W. C. Robinson, ed., *Fluorine Systerms Handbook*, Douglas Aircraft Co., Inc., Long Beach, Calif., 1967.
103. D. L. Endicott and L. H. Donahue, *Development and Demonstration of Criteria for Liquid Fluorine Feed System Components*, Report AFRPL-TR-65-133, McDonnell Douglas Astronautics Co., St. Louis, Mo., 1965.
104. W. R. Meyers and W. B. DeLong, *Chem. Eng. Progress* **44**, 359 (1948).
105. N. S. Nikolaev and co-workers, *Analytical Chemistry of Fluorine*, trans. by J. Schmorak, Halsted Press, New York, 1972, Chapt. 8.
106. E. A. Ranken and C. V. Borzileri, "The Safe Handling of Fluorine," *Health and Safety Manual, Supplement 21.12*, University of California, Lawrence Livermore National Laboratory, Berkeley, Apr. 1987.
107. Board on Toxicology and Environmental Health Hazards, National Research Council, *Emergency and Continuous Exposure Limits for Selected Airborne Contaminants*, Vol. 1, National Academy Press, Washington, D.C., Apr. 1984.
108. M. L. Keplinger and L. W. Suissa, *Am. Ind. Hyg. Assoc. J.* **29**, 10 (1968).
109. P. M. Ricca, *Am. Ind. Hyg. Assoc. J.* **31**, 22 (1970).
110. J. S. Lyon, *J. Occup. Med.* **4**, 199 (1962).
111. H. E. Stockinger, in C. Voegtlin and H. C. Hodge, eds., *The Pharmacology and Toxicology of Uranium*, McGraw-Hill Book Co., New York, 1949, Chapt. 17.
112. N. Ericksen and co-workers, *A Study of the Toxicological Effects of the Inhalation of Gaseous Fluorine at Concentrations of Approximately 25, 8, 3 and 0.7 mg/m³*, United States Atomic Energy Report 397, 407, 427, and 429, University of Rochester, New York, 1945.
113. N. Ericksen and co-workers, *A Study of the Lethal Effect of the Inhalation of Gaseous Fluorine at Concentrations from 100 ppm to 10,000 ppm*, United States Atomic Energy Report 435, University of Rochester, New York, 1945.
114. R. Y. Eagers, *Toxic Properties of Inorganic Fluorine Compounds*, Elsevier Publishing Co., Ltd., Amsterdam, the Netherlands, 1969, p. 43.
115. R. Landau and R. Rosen, *Ind. Eng. Chem.* **40**, 1239 (1948).
116. J. B. Ruch, *USAEC Report CF-60-4-38*, USAEC, Washington, D.C., 1960.
117. R. C. Liimatainer and W. L. Merchan, *Report ANL-5429*, Argonne National Laboratory, Northbrook, Ill., 1955.
118. J. D. Davratel, *USAEC Report RFP-1200*, USAEC, Washington, D.C., 1968.
119. S. H. Smiley and C. R. Schmitt, *Ind. Eng. Chem.* **46**, 244 (1954).
120. K. C. Lowe, *Adv. Mater.* **3**, 87 (1991).

General References

F. A. Cotton, ed., *Progress in Inorganic Chemistry*, Vol. 2, Interscience Publishers, New York, 1960.
R. Y. Eagers, *Toxic Properties of Inorganic Fluorine Compounds*, Elsevier Publishing Co. Ltd., London, 1969.
H. J. Emeleus, *J. Chem. Soc.*, 441 (1942).
Gmelins Handbuch der Anorganishcen Chemie, 8th ed., Suppl., Verlag Chemie, Weinheim, Germany, 1959, pp. 66–79.
M. Hudlicky, *Chemistry of Organic Fluorine Compounds*, Ellis Harwood Limited, Sussex, UK, 1976.
A. T. Kuhn, ed., *Industrial Electrochemical Processes*, Elsevier Publishing Co. Ltd., Amsterdam, the Netherlands, 1971.
Mellor's Comprehensive Treatise on Inorganic and Theoretical Chemistry, Vol. 2, Suppl., I. Longmans Green, New York, 1956, pp. 15–45.

N. S. Nikolaev and co-workers, *Analytical Chemistry of Fluorine*, trans. by J. Schmorak, Halsted Press, New York, 1972.

A. J. Rudge, *The Manufacture and Use of Fluorine and Its Compounds*, Oxford University Press, Inc., New York, 1962.

O. Ruff, *Chem. Ber.* **69A**, 181 (1936).

J. H. Simons, ed., *Fluorine Chemistry*, Vol. 1, Academic Press, Inc., New York, 1950.

C. Slesser and S. R. Schram, *Preparation Properties and Technology of Fluorine and Organic Fluoro-Compounds*, National Nuclear Energy Series, Div. VII, Vol. 1, McGraw-Hill Book Co., Inc., New York, 1951.

M. Stacey, J. C. Tatlow, and A. G. Sharpe, eds., *Advances in Fluorine Chemistry*, Vol. 2, Butterworths Inc., Washington, D.C., 1961.

Handbook of Compressed Gases, 3rd ed., Van Nostrand Reinhold, New York, 1990.

GEORGE SHIA
AlliedSignal, Inc.

FLUORINE COMPOUNDS, INORGANIC

INTRODUCTION

Fluorine (qv), the most electronegative element, is much more reactive than the other elements. On the Pauling scale of electronegativities, fluorine (value 4.0) lies well above oxygen (3.5), chlorine (3.0), and nitrogen (3.0) (1). Indeed, fluorine reacts with virtually every other element, including the helium group elements. These last were commonly called inert gases until 1962 when xenon, radon, and krypton were shown to react with fluorine (see HELIUM-GROUP GASES). Because of unique properties, fluorine has been called a superhalogen (1) and several of its compounds called superacids. The term superacid (2) is used for systems having higher acidities than anhydrous sulfuric or fluorosulfuric acid. A number of fluorine species exhibit superacid properties in HSO_3F or HSO_3F-SO_3 solutions. The SbF_5-HSO_3F system is an example.

The basic fluorine-containing minerals are fluorite [14542-23-5], commonly called fluorspar, CaF_2; and fluorapatite [1306-05-4], commonly called phosphate rock. The reaction of calcium fluoride and sulfuric acid produces hydrogen fluoride. Fluorosilicic acid is produced from fluorapatite as a by-product in the production of phosphoric acid. The boiling point of hydrogen fluoride, 19.54°C, is much higher than that of HCl, -84.9°C, owing to extensive molecular association via hydrogen bonding in the former. Hydrogen fluoride is the most common reagent for production of fluorine compounds. The first pure sample of anhydrous hydrogen fluoride was produced from thermal decomposition of KF·HF (3,4). Elemental fluorine, a pale greenish yellow gas, is produced by electrolysis of anhydrous potassium fluoride–hydrogen fluoride melts (see also ELECTROCHEMICAL PROCESSING, INORGANIC). IUPAC has recommended the prefix *fluoro* rather than the frequently used *fluo* for inorganic fluorine compounds. Terms such as fluoborate, fluosilicate, and silicofluoride are frequently used, however, rather than the preferred nomenclature, fluoroborate and fluorosilicate.

The fluoride ion is the least polarizable anion. It is small, having a diameter of 0.136 nm, 0.045 nm smaller than the chloride ion. The isoelectronic F^- and O^{2-} ions are the only anions of comparable size to many cations. These anions are about the same size as K^+ and Ba^{2+} and smaller than Rb^+ and Cs^+. The small size of F^- allows for high coordination numbers and leads to different crystal forms and solubilities, and higher bond energies than are evidenced by the other halides. Bonds between fluorine and other elements are strong whereas the fluorine–fluorine bond is much weaker, 158.8 kJ/mol (37.95 kcal/mol), than the chlorine–chlorine bond which is 242.58 kJ/mol (57.98 kcal/mol). This bond weakness relative to the second-row elements is also seen in O–O and N–N single bonds and results from electronic repulsion.

A number of elements exhibit the highest oxidation state only because fluorides and oxidation states of $+6$ and $+7$ are not uncommon. Examples of volatile fluorides of high oxidation state include VF_5, CrF_5, TaF_5, WF_6, MoF_6, PtF_6, ReF_6, ReF_7, and IF_7. Many complex fluorides demonstrate coordination not found for the other larger halides, eg, NiF_6^{2-}, SiF_6^{2-}, TaF_7^{2-}, TaF_8^{3-}, PF_6^-, AsF_6^-, ZrF_7^{3-}, and PbF_7^{3-}.

Fluorine forms very reactive halogen fluorides. Reaction of Cl_2 and F_2 at elevated temperatures can produce ClF, ClF_3, or ClF_5; BrF_3 and BrF_5 can be obtained from the reaction of Br_2 and F_2. These halogen fluorides react with all nonmetals, except for the noble gases, N_2, and O_2 (5). Fluorine also forms a class of compounds known as hypofluorites, eg, CF_3OF (6). Fluorine peroxide [7783-44-0], O_2F_2, has also been reported (6).

Fluorine's special properties lead to many applications. Its complexing properties account for its use as a flux in steelmaking and as an intermediate in aluminum manufacture. The reaction of fluorides with hydroxyapatite, $Ca_5(PO_4)_3OH$, which is found in tooth enamel, to form less soluble and/or more acid-resistant compounds, led to the incorporation of fluorides in drinking water and dentifrices (qv) to reduce dental caries. Many fluorides are volatile and in many cases are the most volatile compounds of an element. This property led to the use of UF_6 for uranium isotope enrichment, critical to the nuclear industry (see NUCLEAR REACTORS), and the use of metal fluorides in chemical vapor deposition (WF_6, MoF_6, ReF_6), in ion implantation (qv) for semiconductors (qv) (BF_3, PF_3, AsF_5, etc), and as unreactive dielectrics (SF_6). Because fluorine forms stable bonds, its compounds can be both extremely reactive (F_2, HF, interhalogens, hypofluorides, fluorinated peroxides, ionic MF_x) and extremely stable (CF_4, SF_6, covalent MF_x). Fluorinated steroids, other fluorinated drugs, and anesthetics have medical applications. The stability, lack of reactivity and, therefore, lack of toxicity of some fluorine compounds are also demonstrated by studies reporting survival of animals in an atmosphere of 80% SF_6 and 20% oxygen, and use of perfluorochemicals as short-term blood substitutes because of the ability to efficiently transport oxygen and carbon dioxide (see BLOOD, ARTIFICIAL; FLUORINE COMPOUNDS, ORGANIC). Fluorides including HF, BF_3, SbF_5, PF_5, and several complexes, eg, BF_4^-, PF_6^-, SbF_6^-, and AsF_6^-, are used in many applications in catalysis (qv).

History

The names fluorine and fluorospar are derived from the Latin *fluere* meaning flow or flux. In 1529 the use of fluorspar as a flux was described. In 1670 the etching of glass by acid-treated fluorspar was reported. Elemental fluorine was isolated by Moissan in 1886 (7).

Fluorspar has been used as a flux in the steel industry since the introduction of the open-hearth process. Historically, hydrogen fluoride was used in limited quantities for glass etching, polishing, scale removal, and small-volume production of fluorides. In the 1930s the first significant commercial HF production was applied toward the production of aluminum (see ALUMINUM AND ALUMINUM ALLOYS) and chlorofluorohydrocarbons for refrigerants (see REFRIGERATION AND REFRIGERANTS). During World War II HF was used in alkylation catalysis to produce aviation gasoline and in the manufacture of fluorine to produce volatile UF_6 for isotopic enrichment of the uranium essential for nuclear devices. The use of the very corrosive UF_6 also stimulated development of fluorinated organic compounds for lubricants and seals that are resistant to UF_6 (see LUBRICATION AND LUBRICANTS). Anhydrous HF is used both as a reactant and as a solvent (4) in the manufacture of inorganic fluorides.

Sources and Applications

The earth's crust consists of 0.09% fluorine. Among the elements fluorine ranks about thirteenth in terrestrial abundance.

The ores of most importance are fluorspar, CaF_2; fluorapatite, $Ca_5(PO_4)_3F$; and cryolite [*15096-52-3*], Na_3AlF_6. Fluorspar is the primary commercial source of fluorine. Twenty-six percent of the world's high quality deposits of fluorspar are in North America. Most of that is in Mexico. United States production in 1987–1991 was 314,500 metric tons, most of which occurred in the Illinois–Kentucky area. Imported fluorspar in 1990–1991 represented about 82% of U.S. consumption; 31% of U.S. fluorspar imports were from Mexico and 29% from China compared to 66% from Mexico in the 1973–1978 period. The majority of the fluorine in the earth's crust is in phosphate rock in the form of fluorapatite which has an average fluorine concentration of 3.5%. Recovery of these fluorine values as by-product fluorosilicic acid from phosphate production has grown steadily, partially because of environmental requirements (see PHOSPHORIC ACID AND THE PHOSPHATES).

Production of hydrogen fluoride from reaction of CaF_2 with sulfuric acid is the largest user of fluorspar and accounts for approximately 60–65% of total U.S. consumption. The principal uses of hydrogen fluoride are in the manufacture of aluminum fluoride and synthetic cryolite for the Hall aluminum process and fluoropolymers and chlorofluorocarbons that are used as refrigerants, solvents, aerosols (qv), and in plastics. Because of the concern that chlorofluorocarbons cause upper atmosphere ozone depletion, these compounds are being replaced by hydrochlorofluorocarbons and hydrofluorocarbons. The balance of hydrogen fluoride is used in applications such as stainless steel pickling, inorganic fluoride production, alkylation (qv), uranium enrichment, and fluorine production. Hydrogen fluoride is used to convert uranium oxide to UF_4 which then reacts with elemental fluorine to produce volatile UF_6. The UF_6 is then isotopically enriched by gaseous diffusion or gas centrifuge processes for nuclear applications.

The steel (qv) industry is also an extremely large user of fluorspar which is added to slag to make it more reactive. Smaller amounts are also used in the aluminum, ceramic, brick, cement, glass fiber, and foundry industries.

Synthesis

Most inorganic fluorides are prepared by the reaction of hydrofluoric acid with oxides, carbonates, hydroxides, chlorides, or metals. Routes starting with carbonate, hydroxide, or oxide are the most common and the choice is determined by the most economical starting material. In many cases, the water produced by the reaction cannot be removed without at least partial hydrolysis of the metal fluoride. This hydrolysis frequently can be reduced by dehydrating in a stream of hydrogen fluoride. If hydrolysis is unavoidable, reaction of anhydrous HF and the metal or the metal chloride may be required. The reaction of the metal and HF can be the most desirable if the metal is inexpensive relative to its salts and if the metal has an oxidation potential higher than hydrogen. If the metal is not finely divided, formation of a fluoride coating on the metal surface may occur. This may slow the reaction. Another route to metal fluorides is by reaction of the metal or its salts with elemental fluorine or with interhalogen fluorides. These reactions occur rapidly and frequently are violent. Because hydrofluoric acid is much less expensive than fluorine, it normally is used whenever possible. However, many of the compounds containing elements in the higher oxidation states can be achieved only by use of elemental fluorine.

Analysis and Characterization

The most popular device for fluoride analysis is the ion-selective electrode (see ELECTROANALYTICAL TECHNIQUES). Analysis using the electrode is rapid and this is especially useful for dilute solutions and water analysis. Because the electrode responds only to free fluoride ion, care must be taken to convert complexed fluoride ions to free fluoride to obtain the total fluoride value (8). The fluoride electrode also can be used as an end point detector in titration of fluoride using lanthanum nitrate [10099-59-9]. Often volumetric analysis by titration with thorium nitrate [13823-29-5] or lanthanum nitrate is the method of choice. The fluoride is preferably steam distilled from perchloric or sulfuric acid to prevent interference (9,10). Fusion with a sodium carbonate–sodium hydroxide mixture or sodium may be required if the samples are covalent or insoluble.

Because fluorine has a nuclear spin of one-half, a strong signal, and a large coupling constant, nmr spectroscopic analysis is an invaluable tool (see MAGNETIC SPIN RESONANCE). The use of gas chromatography on the many volatile fluorine compounds also has greatly enhanced separation and identification. However, nuclear magnetic resonance and infrared spectroscopy remain key techniques in the study of fluorine compounds (see INFRARED AND RAMAN SPECTROSCOPY). Raman spectroscopy and mass spectrometry (qv) are also useful tools.

Safety, Toxicity, and Handling

Hazards associated with fluorides are severe. Anhydrous or aqueous hydrogen fluoride is extremely corrosive to skin, eyes, mucous membranes, and lungs; it can cause permanent damage and even death. Detailed information about safety, toxicity, and handling can be obtained from the producers of hydrogen fluoride, eg, Elf Atochem North America, Inc., Du Pont, and AlliedSignal. Fluorides susceptible to hydrolysis can generate aqueous hydrogen fluoride. Ingestion of excess fluorides may cause poisoning or damage to bones and/or teeth. Fluorine-containing oxidizers can react with the body in addition to causing burns.

Hydrogen fluoride or compounds that can produce it and fluorine-containing oxidizers should be handled with adequate safety equipment and extreme care by well-trained personnel. Often the effect of skin exposure is not immediately evident, especially when dilute solutions are handled. Pain may develop several hours later.

Fluorides in small (1 ppm in water, 0.1% in dentifrices) quantities have been shown to provide dramatic reduction in dental decay. Fluorides also show promise for bone treatment and in pharmaceuticals (qv) (see also CHEMOTHERAPEUTICS, ANTICANCER; STEROIDS). However, larger quantities of fluorides can lead to dental fluorosis, bone fracture, and even death. The oral LD_{50} for free fluoride ion in rats appears to be 50 to 100 mg/kg body weight based on LD_{50} values for several fluorides.

Because hydrogen fluoride is extremely reactive, special materials are necessary for its handling and storage. Glass reacts with HF to produce SiF_4 which leads to pressure buildup and potential ruptures. Anhydrous hydrogen fluoride is produced and stored in mild steel equipment. Teflon or polyethylene are frequently used for aqueous solutions.

The OSHA permissible exposure limit (11) and the American Conference of Governmental Industrial Hygienists (ACGIH) established threshold limit value (TLV) (12) for fluorides is 2.5 mg of fluoride per cubic meter of air. This is the TLV–TWA concentration for a normal 8-h work day and a 40-h work week.

BIBLIOGRAPHY

"Fluorine Compounds, Inorganic," in *ECT* 1st ed., Vol. 6, pp. 667–668, by C. R. Hough, Polytechnic Institute of Brooklyn; in *ECT* 2nd ed., Vol. 9, pp. 527–529, by J. B. Beal, Jr., Ozark-Mahoning Co.; in *ECT* 3rd ed., Vol. 10, pp. 655–659, by C. B. Lindahl and D. T. Meshri, Ozark-Mahoning Co.

1. L. Pauling, *The Nature of the Chemical Bond*, 3rd ed., Cornell University Press, Ithaca, N.Y., 1960, pp. 82 and 90.
2. R. J. Gillespie, *Acc. Chem. Res.* **1**, 202 (1968).
3. E. Frémy, *Ann. Chim. Phys.* **47**, 5 (1856).
4. R. E. Banks, *J. Fluorine Chem.* **33**, 3–26 (1986).
5. O. Glemser, *J. Fluorine Chem.* **33**, 45–69 (1986).
6. J. M. Shreeve, *J. Fluorine Chem.* **33**, 179–193 (1986).
7. H. Moissan, *Compt. Rend.* **12**, 1543 (1886).
8. *Analytical Procedures for Fluoride Analyses*, Orion Research Inc., Boston, Mass., 1990–1991.
9. G. H. Cady, *Anal. Chem.* **48**, 655–660 (1976).
10. H. H. Willard and O. B. Winter, *Ind. Eng. Chem. Analyt. Edn.* **5**, 7–10 (1933).
11. *Code of Federal Regulations*, Title 29, Part 1910.1000, Washington, D.C.
12. *Threshold Limit Values for Chemical Substances and Physical Agents, 1992–1993*, The American Conference of Governmental Industrial Hygienists, Cincinnati, Ohio.

General References

Chemistry and characteristics

P. Tarrant, ed., *Fluorine Chemistry Reviews*, Vol. 6, Marcel Dekker Inc., New York, 1971; Vol. 1, 1967; Vol. 2, 1968; Vol. 3, 1969; Vol. 4, 1969; Vol. 5, 1973; Vol. 7, 1974; Vol. 8, 1977.

J. H. Simons, ed., *Fluorine Chemistry*, Academic Press, Inc., New York, Vol. 1, 1950; Vol. 2, 1954; Vol. 3, 1963; Vol. 4, 1965; Vol. 5, 1964.

M. Stacey and co-eds., *Advances in Fluorine Chemistry*, Butterworth Inc., Washington, D.C., Vol. 1, 1960; Vol. 2, 1961; Vol. 3, 1963; Vol. 4, 1965; Vol. 5, 1965; Vol. 6, 1970; Vol. 7, 1973.

G. Brauer, *Handbuch der Praparativen Anorganischen Chemie*, Ferdinand Enke, Stuttgart, Band 1, 1960; Band 2, 1962.

I. G. Ryss, *The Chemistry of Fluorine and Its Inorganic Compounds*, State Publishing House of Scientific, Technical, and Chemical Literature, Moscow, 1956; Eng. trans., AEC-tr-3927, Office of Technical Services, U.S. Dept. of Commerce, Washington D.C., 1960.

F. A. Cotton, ed., *Progress in Inorganic Chemistry*, Vol. 2, Interscience Publishers, Inc., New York, 1960.

Mellor's Comprehensive Treatise of Inorganic and Theoretical Chemistry, Vol. 2, Suppl. 1, Longmans Green & Co. Ltd., London, 1962.

Gmelin, *Handbuch der Anorganischen Chemie*, Fluorination System-Number 5, Verlag Chemie, GmbH, Weinheim, Germany, 1959.

M. F. A. Dove and A. F. Clifford, *Inorganic Chemistry in Liquid Hydrogen Fluoride*, Pergamon Press Ltd., Oxford, U.K., 1971.

E. Newbrun, ed., *Fluorides and Dental Caries*, Charles C Thomas Publishers, Springfield, Ill., 1975.

W. E. Jones and E. G. Skolnik, *Chem. Rev.* **76**, 563 (1976).

R. E. Banks, D. W. A. Sharp, and J. C. Tatlow, eds., *Fluorine, The First Hundred Years*, Elsevier, Science Publishing Co., Inc., New York, 1986, and *Journal of Fluorine Chemistry*, Vol. 33.

Production, consumption, uses, prices, and imports

Minerals Yearbooks Chapter on Fluorspar (annual comprehensive reports but delayed approximately two years).

Mineral Industry Surveys on Fluorspar, quarterly, and annual (less comprehensive but current).

U.S. Dept. of the Interior, Bureau of Mines, Washington, D.C. (periodic).

CHARLES B. LINDAHL
TARIQ MAHMOOD
Elf Atochem North America, Inc.

ALUMINUM

Both the binary and complex fluorides of aluminum have played a significant role in the aluminum industry. Aluminum trifluoride [*7784-18-1*], AlF_3, and its trihydrate [*15098-87-0*], $AlF_3 \cdot 3H_2O$, have thus far remained to be the only binary fluorides of industrial interest. The nonahydrate [*15098-89-2*], $AlF_3 \cdot 9H_2O$, and the monohydrate [*12252-28-7, 15621-55-3*], $AlF_3 \cdot H_2O$, are of only academic curiosity. The monofluoride [*13595-82-9*], AlF, and the difluoride [*13569-23-8*], AlF_2, have been observed as transient species at high temperatures.

Of the fluoroaluminates known, cryolite, ie, sodium hexafluoroaluminate [*15096-52-2*], Na_3AlF_6, has been an integral part of the process for production of aluminum. Recently, the mixtures of potassium tetrafluoroaluminate [*14484-69-6*], $KAlF_4$, and potassium hexafluoroaluminate [*13575-52-5*], K_3AlF_6, have been employed as brazing fluxes in the manufacture of aluminum parts.

Aluminum Monofluoride and Aluminum Difluoride

Significant vapor pressure of aluminum monofluoride [*13595-82-9*], AlF, has been observed when aluminum trifluoride [*7784-18-1*] is heated in the presence of reducing agents such as aluminum or magnesium metal, or is in contact with the cathode in the electrolysis of fused salt mixtures. AlF disproportionates into AlF_3 and aluminum at lower temperatures. The heat of formation at 25°C is -264 kJ/mol (-63.1 kcal/mol) and the free energy of formation is -290 kJ/mol (-69.3 kcal/mol) (1). Aluminum difluoride [*13569-23-8*] has been detected in the high temperature equilibrium between aluminum and its fluorides (2).

Aluminum Trifluoride

Aluminum trifluoride trihydrate [*15098-87-0*], $AlF_3 \cdot 3H_2O$, appears to exist in a soluble metastable α-form as well as a less soluble β-form (3). The α-form can be obtained only when the heat of the reaction between alumina and hydrofluoric acid is controlled and the temperature of the reaction is kept below 25°C. Upon warming the α-form changes into a irreversible β-form which is insoluble in water and is much more stable. The β-form is commercially available.

Aluminum trifluoride trihydrate is prepared by reacting alumina trihydrate and aqueous hydrofluoric acid. The concentration of acid can vary between 15 to 60% (4). In the beginning of the reaction, addition of $Al(OH)_3$ to hydrofluoric acid produces a clear solution which results from the formation of the soluble α-form of $AlF_3 \cdot 3H_2O$. As the addition of $Al(OH)_3$ is continued and the reaction temperature increases, irreversible change takes place and the α-form of $AlF_3 \cdot 3H_2O$ gets converted to the β-form and precipitation is observed. After all the alumina is added, the reaction mixture is continuously agitated for several hours at 90–95°C. After the precipitate settles down, the supernatant liquid is removed using rotary or table vacuum filters and the slurry is centrifuged. The cake is washed with cold water, dried, and calcined in rotating horizontal kilns (5), flash dryers, or fluid-bed calciners to produce anhydrous AlF_3 for aluminum reduction cells. This process is known as a wet process.

Aluminum trifluoride can also be advantageously made by a dry process in which dried $Al(OH)_3$ is treated at elevated temperatures with gaseous hydrogen fluoride. High temperature corrosion-resistant alloys, such as Monel, Inconel, and titanium are used in the construction of fluidized-bed reactors. In one instance, an Inconel reactor is divided into three superimposed compartments by two horizontal fluidizing grid sieve plates. Aluminum hydroxide is fed into the top zone where it is dried by the existing gases. The gases such as HF and SiF_4 are scrubbed from stack gases with water. These gases are recycled or used in the manufacture of cryolite [*15096-52-3*]. Solids are transported from top to bottom by downcomers while HF enters at the bottom zone getting preheated by heat exchange from the departing AlF_3. The bulk of the reaction occurs in the middle compartment which is maintained at 590°C.

The third process involves careful addition of aluminum hydroxide to fluorosilicic acid (6) which is generated by fertilizer and phosphoric acid-producing plants. The addition of $Al(OH)_3$ is critical. It must be added gradually and slowly so that the silica produced as by-product remains filterable and the $AlF_3 \cdot 3H_2O$ formed is in the soluble α-form. If the addition of $Al(OH)_3 \cdot 3H_2O$ is too slow, the α-form after some time changes into the insoluble β-form. Then separation of silica from insoluble β-$AlF_3 \cdot 3H_2O$ becomes difficult.

$$H_2SiF_6 + 2\,Al(OH)_3 \rightarrow 2\,AlF_3 \cdot 3H_2O + SiO_2 + H_2O$$

Environmentally sound phosphate fertilizer plants recover as much of the fluoride value as H_2SiF_6 as possible. Sales for production of $AlF_3 \cdot 3H_2O$ is one of the most important markets (see FERTILIZERS; PHOSPHORIC ACID AND THE PHOSPHATES).

Dehydration of $AlF_3 \cdot 3H_2O$ above 300°C leads to a partial pyrohydrolysis forming HF and Al_2O_3 which can be avoided by heating the trihydrate gradually to 200°C to remove 2.5 moles of water and then rapidly removing the remainder at 700°C. This latter procedure yields a product having less than 3.5% water content and Al_2O_3 content below 8% (7). This product is a typical material used in aluminum reduction cells. The presence of alumina does not interfere in the process of aluminum reduction because it replaces part of the alumina that is fed to the cells.

The principal producers of aluminum trifluoride in North America are Alcan, Alcoa, and AlliedSignal. It is also produced in other countries, eg, France, Mexico, Norway, Italy, Tunisia, and Japan. Total worldwide production of aluminum trifluoride in 1990 was 400,000 metric tons and the price was $1100/t. In 1993, because of excess recovery of fluorine values, use of energy efficient smelters, and the worldwide economic climate, the price was down to $750/t.

The principal use of AlF_3 is as a makeup ingredient in the molten cryolite, $Na_3AlF_6 \cdot Al_2O_3$, bath used in aluminum reduction cells in the Hall-Haroult process and in the electrolytic process for refining of aluminum metal in the Hoopes cell. A typical composition of the molten salt bath is 80–85% Na_3AlF_6, 5–7% AlF_3, 5–7% CaF_2, 2–6% Al_2O_3, and 0–7% LiF with an operating temperature of 950°C. Ideally fluorine is not consumed in the process, but substantial quantities of fluorine are absorbed by the cell lining and fluorine is lost to the atmosphere. Modern aluminum industry plants efficiently recycle the fluorine values.

Minor uses of aluminum fluoride include flux compositions for casting, welding (qv), brazing, and soldering (see SOLDERS AND BRAZING ALLOYS) (8,9); passivation of stainless steel (qv) surfaces (10); low melting glazes and enamels (see ENAMELS, PORCELAIN OR VITREOUS); and catalyst compositions as inhibitors in fermentation (qv) processes. Table 1 gives typical specifications for a commercial sample of AlF_3.

Other hydrates of aluminum trifluoride are the nonahydrate [15098-89-2], $AlF_3 \cdot 9H_2O$, which is stable only below 8°C, and aluminum trifluoride monohydrate [12252-28-7], [15621-55-3], $AlF_3 \cdot H_2O$, which occurs naturally as a rare mineral, fluellite found in Stenna-Gwyn Cornwall, U.K. (11).

Table 1. Specification for Commercial Aluminum Trifluoride

Parameter	Specification
assay as AlF_3, %	90–92
Al_2O_3, typical, %	8–9
SiO_2, max, %	0.1
iron as Fe_2O_3, %	0.1
sulfur as SO_2, %	0.32
bulk density, g/cm^3	
loose	1.3
packed	1.6
screen analysis, % retained	
105 μm (140 mesh)	20
74 μm (200 mesh)	60
44 μm (325 mesh)	90

High Purity Aluminum Trifluoride.   High purity anhydrous aluminum trifluoride that is free from oxide impurities can be prepared by reaction of gaseous anhydrous HF and $AlCl_3$ at 100°C, gradually raising the temperature to 400°C. It can also be prepared by the action of elemental fluorine on metal/metal oxide and subsequent sublimation (12) or the decomposition of ammonium fluoroaluminate at 700°C.

Relatively smaller amounts of very high purity AlF_3 are used in ultra low loss optical fiber–fluoride glass compositions, the most common of which is ZBLAN containing zirconium, barium, lanthanum, aluminum, and sodium (see FIBER OPTICS). High purity AlF_3 is also used in the manufacture of aluminum silicate fiber and in ceramics for electrical resistors (see CERAMICS AS ELECTRICAL MATERIALS; REFRACTORY FIBERS).

Anhydrous aluminum trifluoride, AlF_3, is a white crystalline solid. Physical properties are listed in Table 2. Aluminum fluoride is sparingly soluble in water (0.4%) and insoluble in dilute mineral acids as well as organic acids at ambient temperatures, but when heated with concentrated sulfuric acid, HF is liberated, and with strong alkali solutions, aluminates are formed. AlF_3 is slowly attacked by fused alkalies with the formation of soluble metal fluorides and aluminate. A series of double salts with the fluorides of many metals and with ammonium ion can be made by precipitation or by solid-state reactions.

Health and Safety.   Owing to very low solubility in water and body fluids, AlF_3 is relatively less toxic than many inorganic fluorides. The toxicity values are oral LD_{LO}, 600 mg/kg; subcutaneous, 3000 mg/kg. The ACGIH adopted (1992–1993) TLV for fluorides as F^- is TWA 2.5 mg/m^3. Pyrohydrolysis and strong acidic conditions can be a source of toxicity owing to liberated HF.

Table 2. Physical Properties of Anhydrous Aluminum Trifluoride

Property	Value
mol wt	83.977
mp, °C	1278[a]
transition point, °C	455
density, g/cm^3	3.10
dielectric constant	6
heat of transition at 455°C, kJ/mol[b]	0.677
heat of sublimation for crystals at 25°C, kJ/mol[b]	300
ΔH_f at 25°C, kJ/mol[b]	−1505
ΔG_f at 25°C, kJ/mol[b]	−1426
S at 25°C, J/(mol·K)[b]	66.23
C_p at 25°C J/(mol·K)[b]	
α-crystals	74.85
β-crystals	100.5

[a]Sublimes.

[b]To convert J to cal, divide by 4.184.

Fluoroaluminates

Several fluoroaluminates are known to exist but sodium hexafluoroaluminate [15096-52-2], Na_3AlF_6, has dominated industrial applications. More recently potassium tetrafluoroaluminate [14484-69-6], $KAlF_4$, has provided a noncorrosive and inexpensive flux in the manufacture of aluminum parts for various applications. The naturally occurring fluoroaluminates are listed in Table 3.

Table 3. Naturally Occurring Fluoroaluminates

Name	Cas Registry Number	Molecular formula
cryolite	[15096-52-2]	Na_3AlF_6
chiolite	[1302-84-7]	$Na_5Al_3F_{14}$
cryolithionate	[15491-07-3]	$Na_3Li_3(AlF_6)_2$
thomsenolite, hagemannite	[16970-11-9]	$NaCaAlF_6 \cdot H_2O$
ralstonite	[12199-10-9]	$Na_{2x}(Al_{2x},Na_x)(F,OH)_6 \cdot yH_2O$
prosopite	[12420-95-0]	$CaAl_2(F,OH)_8$
jarlite, *meta*-jarlite	[12004-61-4]	$NaSr_3Al_3F_{16}$
weberite	[12423-93-7]	Na_2MgAlF_7
gearksutite	[12415-96-2]	$CaAl(F,OH)_5 \cdot H_2O$
pachnolite	[15489-46-0]	$NaCaAlF_6 \cdot H_2O$

The common structural element in the crystal lattice of fluoroaluminates is the hexafluoroaluminate octahedron, AlF_6^3. The differing structural features of the fluoroaluminates confer distinct physical properties to the species as compared to aluminum trifluoride. For example, in AlF_3 all corners are shared and the crystal becomes a giant molecule of very high melting point (13). In $KAlF_4$, all four equatorial atoms of each octahedron are shared and a layer lattice results. When the ratio of fluorine to aluminum is 6, as in cryolite, Na_3AlF_6, the AlF_6^{3-} ions are separate and bound in position by the balancing metal ions. Fluorine atoms may be shared between octahedrons. When opposite corners of each octahedron are shared with a corner of each neighboring octahedron, an infinite chain is formed as, for example, in Tl_2AlF_5 [33897-68-6]. More complex relations exist in chiolite, wherein one-third of the hexafluoroaluminate octahedra share four corners each and two-thirds share only two corners (14).

Cryolite. Cryolite constitutes an important raw material for aluminum manufacturing. The natural mineral is accurately depicted as $3NaF \cdot AlF_3$, but synthetic cryolite is often deficient in sodium fluoride. Physical properties are given in Table 4.

Cryolite derives its name from its resemblance to ice when immersed in water as a result of the closely matched refractive indexes. The only commercially viable source of cryolite deposits has been found in the south of Greenland at Ivigtut (15). Minor localities, not all authenticated, are in the Ilmen Mountains in the former USSR; Sallent, in the Pyrenees, Spain; and Pikes Peak, Colorado (16). For the most part the ore from Ivigtut is a coarse-grained aggregate carrying 10–30% of admixtures, including siderite, quartz, sphalerite, galena, chalcopyrite, and pyrite, in descending order of frequency.

Table 4. Physical Properties of Cryolite

Property	Value
mol wt	209.94
mp, °C	1012
transition temperature, °C	
monoclinic-to-rhombic	565
second-order	880
dimensions of unit cell, nm	
a	0.546
b	0.561
c	0.780
vapor pressure of liquid at 1012°C, Pa[a]	253
heat of fusion at 1012°C, kJ/mol[b]	107
heat of vaporization at 1012°C, kJ/mol[b]	225
heat of transition, kJ/mol[b]	
monoclinic-to-rhombic at 565°C	8.21
second-order at 880°C	0.4
heat capacity, J/(mol·K)[b]	
monoclinic crystal at 25°C	215
cubic crystal at 560°C	281
liquid at 1012°C	395
S, J(mol·K)[b,c]	238
ΔH_f^0 at 25°C,[c] kJ/mol[b]	-3297
ΔG_f^0 at 25°C,[c] kJ/mol[b]	-3133
density, g/cm^3	
monoclinic crystal at 25°C	2.97
cubic crystal from x-ray	2.77
solid at 1012°C	2.62
liquid at 1012°C	2.087
hardness, Mohs'	2.5
refractive index	
α-fom	1.3385
β-fom	1.3389
τ-fom	1.3396
electrical conductivity, $(\Omega \cdot cm)^{-1}$	
solid at 400°C	4.0×10^{-6}
liquid at 1012°C	2.82
viscosity, liquid at 1012°C, mPa·s(=cP)[a]	6.7
surface tension, liquid in air, mN/m(=dyn/cm)	125
activity product constant in water at 25°C	1.46×10^{-34}
solubility in water, g/100 g	
at 25°C	0.0042
at 100°C	0.0135

[a]To convert Pa to mm Hg, multiply by 7.
[b]To convert J to cal, divide by 4.184.
[c]Monoclinic crystal.

The mineral cryolite is usually white, but may also be black, purple, or violet, and occasionally brownish or reddish. The lustre is vitreous to greasy, sometimes pearly, and the streak is white. The crystals are monoclinic, differing only slightly from orthorhombic symmetry, and have an axial angle of 90°11'. The space group is P2$_1$/m. The [001] and [110] axes are usually dominant, giving the crystals a cubic appearance. Twinning is ubiquitous, and because the lamellae tend to be perpendicular, cleavage appears to be cubic. The fracture of individual crystals, however, is uneven. Because its refractive indexes are close to that of water, powdered cryolite becomes nearly invisible when immersed in water, but because the optical dispersion is different for the two materials the suspension shows Christiansen colors.

Upon heating the crystallographic angles approach 90° and the transition to the cubic form at 565°C is accompanied by a small heat change. The transition also involves a substantial change in density as evidenced by a characteristic decrepitation (17). The second transformation occurs at 880°C as indicated by the slope of the heating curve. It is also accompanied by a sharp rise in electrical conductivity. The heat change is very small and the transitions with rising temperatures probably mark the onset of a lattice disorder. The more plastic character of the solid near the melting point seems to corroborate this view (18).

Liquid cryolite is an equilibrium mixture of the products of the dissociation:

$$Na_3AlF_6 \rightarrow 2\,NaF + NaAlF_4$$

The composition to the melting point is estimated to be 65% Na_3AlF_6, 14% NaF, and 21% $NaAlF_4$ [1382-15-3]. The ions Na^+ and F^- are the principal current carrying species in molten cryolite whereas the AlF^{4-} is less mobile. The structural evidences are provided by electrical conductivity, density, thermodynamic data, cryoscopic behavior, and the presence of $NaAlF_4$ in the equilibrium vapor (19,20).

Molten cryolite dissolves many salts and oxides, forming solutions of melting point lower than the components. Figure 1 combines the melting point diagrams for cryolite–AlF$_3$ and for cryolite–NaF. Cryolite systems are of great importance

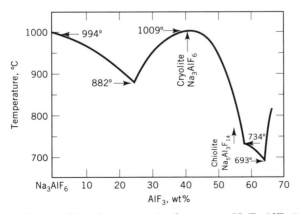

Fig. 1. Liquidus curves in the system NaF–AlF$_3$ (21).

in the Hall-Heroult electrolysis process for the manufacture of aluminum (see ALUMINUM AND ALUMINUM ALLOYS). Table 5 lists the additional examples of cryolite as a component in minimum melting compositions.

The vapor from molten cryolite is largely $NaAlF_4$, the vapor pressures of Na_3AlF_6, NaF, and $NaAlF_4$ near the melting point are about in the ratios 5:1:30. Therefore, the liquid tends to become depleted in AlF_3, and the composition of the aluminum cell electrolyte has to be regularly adjusted by the addition of AlF_3 (20,22).

In contact with moist air, molten cryolite loses HF and is depleted in AlF_3.

$$2\ Na_3AlF_6\ +\ 3\ H_2O \rightarrow 6\ NaF\ +\ 6\ HF +\ Al_2O_3$$

The more electropositive metals react with cryolite, liberating aluminum or aluminum monofluoride (22,23). The reduction of cryolite by magnesium is a current method for removal of magnesium in the refining of aluminum. Upon contact with strong acids cryolite liberates hydrogen fluoride.

Synthetic Cryolite. As of this writing, the supply of cryolite is almost entirely met by synthetic material which possesses the same properties and composition with a minor difference in that it is deficient in NaF. Synthetic cryolite also commonly contains oxygen, hydroxyl group, and/or sulfate groups. The NaF

Table 5. Minimum Melting Compositions Containing Cryolite

Material composition	Added component Wt %	Melting point, °C
NaF	24.5	882
AlF_3	64	693
Al_2O_3	10.5	962
Li_3AlF_6	62	710
CaF_2	25.8	945
ZrO_2	14	969
MgO	7.5	902
CaO	11.3	896
ZnO	2.4	974
CdO	6.0	971
TiO_2	4.0	970
BaF_2	62.5	835
PbF_2	40	730
feldspar	70	830
NaF	34.0	870
Al_2O_3	12.0	
CaF_2	23.0	867
Al_2O_3	17.7	
CaF_2	37.8	675
AlF_3	6.2	
SiO_2	17	ca 800
Al_2O_3	50	

deficiency does not interfere for most applications but the presence of moisture leads to the fluorine losses as HF on heating. Because synthetic cryolite is lighter than the natural mineral, losses by dusting are also higher.

There are several processes available for the manufacture of cryolite. The choice is mainly dictated by the cost and quality of the available sources of soda, alumina, and fluorine. Starting materials include sodium aluminate from Bayer's alumina process; hydrogen fluoride from kiln gases or aqueous hydrofluoric acid; sodium fluoride; ammonium bifluoride, fluorosilicic acid, fluoroboric acid, sodium fluosilicate, and aluminum fluorosilicate; aluminum oxide, aluminum sulfate, aluminum chloride, alumina hydrate; and sodium hydroxide, sodium carbonate, sodium chloride, and sodium aluminate.

The manufacture of cryolite is commonly integrated with the production of alumina hydrate and aluminum trifluoride. The intermediate stream of sodium aluminate from the Bayer alumina hydrate process can be used along with aqueous hydrofluoric acid, hydrogen fluoride kiln gases, or hydrogen fluoride-rich effluent from dry-process aluminum trifluoride manufacture.

$$NaAlO_2 + Na_2CO_3 + 6\,HF \rightarrow Na_3AlF_6 + 3\,H_2O + CO_2$$

The HF and Na_2CO_3 give a sodium fluoride solution. Bayer sodium aluminate solution is added in the stoichiometric ratio. Cryolite is precipitated at 30–70°C by bubbling CO_2, until the pH reaches 8.5–10.0. Seed crystals are desirable. The slurry is thickened and filtered, or settled and decanted, or centrifuged. The resulting product is calcined at 500–700°C. The weight ratio of fluorine to aluminum in the product should exceed 3.9. The calculated value is 4.2 (24). Cryolite can also be made by passing gaseous HF over briquettes of alumina hydrate, sodium chloride, and sodium carbonate at 400–700°C, followed by sintering at 720°C (25).

In addition, there are other methods of manufacture of cryolite from low fluorine value sources, eg, the effluent gases from phosphate plants or from low grade fluorspar. In the former case, making use of the fluorosilicic acid, the silica is separated by precipitation with ammonia, and the ammonium fluoride solution is added to a solution of sodium sulfate and aluminum sulfate at 60–90°C to precipitate cryolite (26,27):

$$12\,NH_4F + 3\,Na_2SO_4 + Al_2(SO_4)_3 \rightarrow 2\,Na_3AlF_6 + 6\,(NH_4)_2SO_4$$

The ammonia values can be recycled or sold for fertilizer use. The most important consideration in this process is the efficient elimination of the phosphorus from the product, because as little as 0.01% P_2O_5 in the electrolyte causes a 1–1.5% reduction in current efficiency for aluminum production (28).

Significant amounts of cryolite are also recovered from waste material in the manufacture of aluminum. The carbon lining of the electrolysis cells, which may contain 10–30% by weight of cryolite, is extracted with sodium hydroxide or sodium carbonate solution and the cryolite precipitated with carbon dioxide (28). Gases from operating cells containing HF, CO_2, and fluorine-containing dusts may be used for the carbonation (29).

The specifications for natural cryolite include 95% content of sodium aluminum fluorides as Na_3AlF_6, 4% of other fluorides calculated as CaF_2, and 88%

of the product passing through 44 μm sieve (325 mesh). Product for the ceramic industry contains a small amount of selected lump especially low in iron. The following is a typical analysis for commercial-grade cryolite: cryolite as Na_3AlF_6, 91%; fluorine, 48–52%; sodium, 31–34%; aluminum, 13–15%; alumina, 6.0%; silica (max), 0.70%; calcium fluoride, 0.04–0.06%; iron as Fe_2O_3, 0.10%; with moisture at 0.05–0.15%, bulk density at 1.4–1.5 g/cm^3, and screen analysis passing through 74 μm (200 mesh) at 65–75%.

In spite of the fact that cryolite is relatively less soluble, its fluoride toxicity by oral routes are reported to be about the same as for soluble fluorides: LD_{50} = 200 mg/kg; for NaF, 180 mg/kg; KF, 245 mg/kg (30). Apparently, stomach fluids are acid enough to bring the solubility of cryolite up to values comparable with other fluorides. Chronic exposure may eventually lead to symptoms of fluorosis. The toxicity to insects is in many cases high enough for control. Because of its variable composition, synthetic cryolite may show physiological activity greater than the natural mineral (31).

The effective dissolution of Al_2O_3 by molten cryolite to provide a conducting bath has spurred the need for its use in manufacture of aluminum. Additives enhance the physical and electrical properties of the electrolyte, for example the lowering of melting point by AlF_3 (Fig. 1). Figure 2 illustrates the effect of various additives on the electrical conductivity of liquid cryolite. AlF_3 has the adverse effect of decreasing the electrical conductivity. Calcium fluoride is better in this regard but again too much of it can lead to rise in density of the melt close to that of aluminum (ca 2.28 g/cm^3), inhibiting the separation of metal and electrolyte as indicated in Figure 3. Sodium fluoride has the disadvantage of reducing the current efficiency while increasing density and conductivity. Small amounts of lithium fluoride may also improve electrical conductivity. Compromises on all of these factors have led to the following composition of the electrolyte: 80–85% cryolite, 5–7% AlF_3, 5–7% CaF_2, 0–7% LiF, and 2–8% Al_2O_3.

Another use for cryolite is in the production of pure metal by electrolytic refining. A high density electrolyte capable of floating liquid aluminum is needed, and compositions are used containing cryolite with barium fluoride to raise the density, and aluminum fluoride to raise the current efficiency.

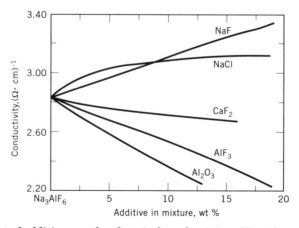

Fig. 2. Effect of additives on the electrical conductivity of liquid cryolite at 1009°C (32).

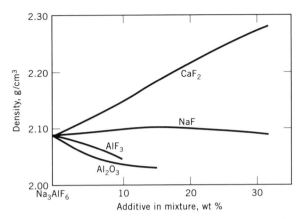

Fig. 3. Effect of additives on the density of liquid cryolite at 1009°C.

Other applications of cryolite include use in reworking of scrap aluminum as flux component to remove magnesium by electrochemical displacement; as a flux in aluminizing steel as well as in processing a variety of metals; in the compounding of welding-rod coatings; as a flux in glass manufacture owing to its ability to dissolve the oxides of aluminum, silicon, and calcium, and also because of the low melting compositions formed with the components; for lowering the surface tension in enamels and thereby improving spreading (33); as a filler for resin-bonded grinding wheels for longer wheel life, reducing metal buildup on the wheel, and faster and cooler grinding action; and in insecticide preparations making use of the fines residue from the refining operation of the cryolite.

Canada, the United States, and South America are the principal exporters of cryolite and Russia and Europe import cryolite. Primary producers in North America are Alcan, Alcoa, and Reynolds Aluminum. The 1993 price of recovered-grade cryolite, which has SO_4^{2-} as impurity, was $400/t, and of high purity cryolite, $800/t. There was a surplus of cryolite in 1992 in the United States and Canada.

Potassium Tetrafluoroaluminate. Potassium tetrafluoroaluminate, $KAlF_4$, is a more recent addition to the industrially important fluoroaluminates, mainly because of developments in the automotive industry involving attempts to replace the copper and solder employed in the manufacture of heat exchangers. The source mineral for aluminum radiator manufacture, bauxite, is highly abundant and also available in steady supply. Research and developmental work on the aluminum radiators started in the 1960s using chloride salt mixtures for brazing. The resulting products and the process itself could not compete with conventional radiators because these processes were comparatively uneconomical. This led to the development of an all fluoride-based flux which confers corrosion-resistant features to the product as well as to the process. Potassium tetrafluoroaluminate in mixtures with other fluoroaluminates, potassium hexafluoroaluminate [13775-52-5], K_3AlF_6, and potassium pentafluoroaluminate monohydrate [41627-26-3], $K_2AlF_5 \cdot H_2O$, has emerged as a highly efficient, noncorrosive, and nonhazardous flux for brazing aluminum parts of heat exchangers. Nocolok 100 Flux (Alcan Aluminum Corp.) developed by Alcan (Aluminum Co. of Canada) has been the first commercial product. Its use and mechanistic aspects of the associated brazing process have been well documented (33–37).

The important task performed by all brazing processes is the removal of oxide films lying on the surfaces of metals to be joined. The process should also permit wetting and flow of the molten filler metal at the brazing temperature (38). The fluxes employed should melt and become active for a successful brazing action. Thus if the flux melts at a temperature higher than that of the filler metal, it leads to the development of thick oxide films on the liquid filler metal inhibiting the flux action. The system $KF \cdot AlF_3$ (Fig. 4) (39) provides the most suitable flux for this application. The system presents a eutectic mixture of $KAlF_4$ and K_3AlF_6 which melts at $559 \pm 2°C$ (40). This is just below the eutectic temperature of the Al–Si filler metal, which is 577°C. The melting point of pure $KAlF_4$ is $574 \pm 1°C$ and that of K_3AlF_6 is 990°C (40).

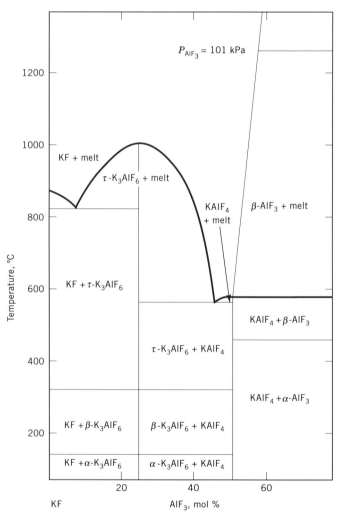

Fig. 4. KF–AlF_3 phase diagram.

Both $KAlF_4$ and K_3AlF_6 are white solids. The former is less soluble (0.22%) in water than the latter (1.4%). The generally cubic form of $KAlF_4$ inverts to the orthorhombic modification between -23 and $50°C$. On heating the cubic form is stable to its congruent melting temperature. The materials are generally inert and infinitely stable under ambient conditions. At melting temperatures and more significantly at temperatures above $730°C$ they react with water releasing hydrogen fluoride (41). Dissolution in strong acids is also slow but is enhanced at higher temperatures leading to the evolution of HF. Several possible interactions of $KAlF_4$ and the metal oxides in the brazing processes have been proposed as part of the mechanism for the latter (34).

An early method of preparation of $KAlF_4$ (42) involved combining aqueous solutions of HF, AlF_3, and KHF_2 in stoichiometric proportions and evaporating the suspension to a dry mixture. The product was subsequently melted and recrystallized. Some of the other conventional technical methods comprise reacting hydrated alumina, hydrofluoric acid, and potassium hydroxide followed by separation of the product from the mother liquor; concentrating by evaporation, a suspension obtained by combining stoichiometric amounts of components; and melting together comminuted potassium fluoride and aluminum fluoride at $600°C$ and grinding the resulting solidified melt.

Several other proprietary methods have been reported, which in general have the aim of producing lower melting products thereby aiming more at the preparation of a eutectic mixtures of the fluoroaluminates as discussed in the beginning of this section. One process (42) describes the making of $KAlF_4$, melting below $575°C$, by addition of potassium hydroxide to the aqueous solution of fluoroaluminum acid. The fluoroaluminum acid is prepared from a reaction of hydrofluoric acid and hydrated alumina. A fairly similar method has been reported in making a flux mixture comprising of K_2AlF_5 or $K_2AlF_5 \cdot H_2O$ and $KAlF_4$, wherein a potassium compound is added to the mixed aqueous fluoroaluminic acid ($HAlF_4$, H_2AlF_5, and H_2AlF_6) solution (43).

The toxicity of these fluoroaluminates is mainly as inorganic fluorides. The ACGIH adopted (1992–1993) values for fluorides as F^- is TLV 2.5 mg/m^3. The oral toxicity in laboratory animal tests is reported to be LD_{50} rat 2.15 mg/kg (41). Because of the fine nature of the products they can also be sources of chronic toxicity effects as dusts.

BIBLIOGRAPHY

"Aluminum Fluoride" under "Fluorine Compounds, Inorganic," in *ECT* 1st ed., Vol. 6, pp. 668–671 by R. G. Danehower, Pennsylvania Salt Manufacturing Co.; "Aluminum Fluorides" under "Fluorine Compounds, Inorganic," in *ECT* 2nd ed., Vol. 9, pp. 529–533, by J. F. Gall, Pennsalt Chemicals Corp.; "Fluoroaluminates" under "Fluorine Compounds, Inorganic," in *ECT* 1st ed., Vol. 6. pp. 671–675, by I. Mockrin, Pennsylvania Salt Manufacturing Co.; "Fluoroaluminates" under "Fluorine Compounds, Inorganic," in *ECT* 2nd ed., Vol. 9, pp. 534–548, by J. F. Gall, Pennsalt Chemicals Corp.; "Aluminum" under "Fluorine Compounds, Inorganic," in *ECT* 3rd ed., Vol. 10, pp. 660–675, by J. F. Gall, Philadelphia College of Textiles and Science.

1. *JANAF Thermochemical Tables*, 2nd ed., NSR DS-NBS 37, National Bureau of Standards, Washington, D.C., 1985.

2. T. C. Ehlert and J. L. Margrave, *J. Am. Chem. Soc.* **86**, 3901 (1964).
3. W. F. Ehret and F. J. Frere, *J. Am. Chem. Soc.* **67**, 64 (1945).
4. U.S. Pat. 2,958,575 (Nov. 1, 1960), D. R. Allen (to The Dow Chemical Co.).
5. J. K. Callaham, *Chem. Met. End.* **52**(3), 94 (1945).
6. F. Weinratter, *Chem. Eng.* **71**, 132 (Apr. 27, 1964).
7. J. K. Bradley, *Chem. Ind., London*, 1027 (1960).
8. Jpn. Kokai Tokkyo Koho 04 04,991 [92 04,991] (Jan. 9, 1992), T. Usui and S. Kagoshige (to Showa Aluminum Corp.).
9. Jpn. Kokai Tokkyo Koho 04 09,274 [92 09,274] (Jan. 14 1992), K. Toma and co-workers (to Mitsubishi Aluminum Co. Ltd.).
10. Jpn. Kokai Tokkyo Koho 03 215,656 [91 215,656] (Sept. 20, 1991), T. Omi and co-workers (to Hashimoto Industries Co. Ltd.).
11. J. D. Dana and co-workers, *The System of Mineralogy*, 7th ed., Vol. 2, John Wiley & Sons, Inc., New York, 1951, pp. 124–125.
12. U.S. Pat. 4,983,373 (Jan. 8, 1991), H. P. Withers Jr. and co-workers (to Air Products & Chemicals, Inc.).
13. P. J. Durrant and B. Durrant, *Introduction to Advanced Inorganic Chemistry*, John Wiley & Sons, Inc., New York, 1970, p. 570.
14. A. F. Wells, *Structural Inorganic Chemistry*, 4th ed., Clarendon Press, Oxford, U.K., 1975, pp. 388–390.
15. H. Pauly, *Met. Assoc. Acid. Magmat.* **I**, 393 (1974).
16. C. Palache and co-workers, in Ref. 4, pp. 110–113.
17. P. P. Fedotiev and V. Hyinskii, *Z. Anorg. Chem.* **80**, 113 (1913).
18. G. G. Landon and A. R. Ubbelohde, *Trans. Faraday Soc.* **52**, 647 (1955).
19. P. A. Foster, Jr. and W. B. Frank, *J. Electrochem. Soc.* **107**, 997 (1960).
20. L. M. Foster, *Ann. N.Y. Acad. Sci.* **79**, 919 (1960).
21. N. W. F. Philips and co-workers, *J. Electrochem. Soc.* **102**, 648–690 (1955).
22. K. Grjotheim and co-workers, *Light. Met.* **1**, 125 (1975).
23. M. Feinleib and B. Porter, *J. Electrochem. Soc.* **103**, 231 (1956); W. E. Haupin, *J. Electrochem. Soc.* **107**, 232 (1960).
24. U.S. Pat. 3,061,411 (Oct. 30, 1962), D.C. Gernes (to Kaiser Aluminum & Chemicals Corp.).
25. U.S. Pat. 3,104,156 (Sept. 17, 1963), P. Saccardo and F. Gozzo (to Sicedoison SpA).
26. G. Tarbutton and co-workers *Ind. Eng. Chem.* **50**, 1525 (1958).
27. U.S. Pat. 2,687,341 (Aug. 24, 1954), I. Mockrin (to Pennsylvania Salt Manufacturing Co.).
28. E. Elchardus, *Compt. Rend.* **206**, 1460 (1938).
29. U.S. Pat. 3,065,051 (Nov. 20, 1962), H. Mader (to Vereinigte Metallwerke Ranshofen-Berndorf A.G.).
30. *The Toxic Substances List*, 1974 ed., U.S. Dept. of Health, Education & Welfare, National Institute for Occupational Safety & Health, Rockville, Md., June 1974.
31. E. J. Largent, *J. Ind. Hyg. Toxicol.* **30**, 92 (1948).
32. J. D. Edwards and co-workers, *J. Electrochem. Soc.* **100**, 508 (1953); K. Matiasovsky and co-workers, *J. Electrochem. Soc.* **111**, 973 (1964).
33. R. Marker, *Glas Email Keramo Tech.* **4**, 117 (1957); **5**, 178 (1957).
34. Y. Ando and co-workers, *SAE Technical Paper Series, International Congress and Exposition*, paper no. 870180, Detroit, Mich., Feb. 23–27, 1987; D. J. Field and N. I. Steward, *ibid.*, paper no. 870186.
35. D. G. W. Claydon and A. Sugihara, in Ref. 34, paper no. 830021.
36. W. E. Cooke and H. Bowman, *Welding J.* (Oct. 1980).
37. W. E. Cooke and co-workers, *SAE Technical Paper Series, Congress and Exposition* paper no. 780300, Detroit, Mich., Feb. 27–Mar. 3, 1978.
38. J. R. Terril and co-workers, *Welding J.* **50**(12), 833–839 (1971).

39. B. Jensen, *Phase and Structure Determination of a New Complex Alkali Aluminum Fluoride*, Institute of Inorganic Chemistry, Norwegian Technical University, Trandheim, 1969.
40. B. Philips and co-workers, *J. Am. Ceram. Soc.* **49**(2), 631–634 (1966).
41. *Nocolok 100 Flux, Material Safety Data Sheet*, Alcan Aluminum Corp., Apr. 1986.
42. U.S. Pat. 4,428,920 (Jan. 31, 1984), H. Willenberg and co-workers (to Kali-Chemie Aktiengesellachaft).
43. U.S. Pat. 4,579,605 (Apr. 1, 1986), H. Kawase and co-workers (to Furukawa Aluminum Co., Ltd.).

DAYAL T. MESHRI
Advance Research Chemicals, Inc.

AMMONIUM

Two well-known salts of ammonia (qv) are the normal ammonium fluoride [*12125-01-8*], NH_4F, and ammonium bifluoride [*1341-49-7*], NH_4HF_2; the latter is sometimes named ammonium acid, or hydrogen, difluoride. Much of the commercial interest in the ammonium fluorides stems from their chemical reactivity as less hazardous substitutes for hydrofluoric acid.

Ammonium Fluoride

Ammonium fluoride is a white, deliquescent, crystalline salt. It tends to lose ammonia gas to revert to the more stable ammonium bifluoride. Its solubility in water is 45.3 g/100 g of H_2O at 25°C and its heat of formation is -466.9 kJ/mol (-116 kcal/mol). Ammonium fluoride is available principally as a laboratory reagent. If it is needed in large quantities, one mole of aqueous ammonia can be mixed with one mole of the more readily available ammonium bifluoride (1).

Ammonium Bifluoride

Properties. Ammonium bifluoride, NH_4HF_2, is a colorless, orthorhombic crystal (2). The compound is odorless; however, less than 1% excess HF can cause an acid odor. The salt has no tendency to form hydrates yet is hygroscopic if the ambient humidity is over 50%. A number of chemical and physical properties are listed in Table 1.

Corrosion. Ammonium bifluoride dissolves in aqueous solutions to yield the acidic bifluoride ion; the pH of a 5% solution is 3.5. In most cases, NH_4HF_2 solutions react readily with surface oxide coatings on metals; thus NH_4HF_2 is used in pickling solutions (see METAL SURFACE TREATMENTS). Many plastics, such as polyethylene, polypropylene, unplasticized PVC, and carbon brick, are resistant to attack by ammonium bifluoride.

Manufacture. Anhydrous ammonium bifluoride containing 0.1% H_2O and 93% NH_4HF_2 can be made by dehydrating ammonium fluoride solutions and

Table 1. Properties of Ammonium Bifluoride, NH_4HF_2

Property	Value	Reference
melting point, °C	126.1	3
boiling point, °C	239.5	3
index of refraction, n_D	1.390	3
solubility at 25°C, wt %		
water	41.5	3
90% ethanol	1.73	3
specific gravity	1.50	3
standard heat of formation, kJ/mol[a]	−798.3	4
heat of fusion, kJ/mol[a]	19.1	4
heat of vaporization, kJ/mol[a]	65.3	4
heat of solution, kJ/mol[a]	20.3	4
heat of dissociation,[b] kJ/mol[a]	141.4	4
heat capacity, C_p, J/(mol·K)[a] at 25°C	106.7	4
vapor pressure,[c] $\log P_{Pa} = a - bT^{-1}$		
153−207°C	$a = 11.72, b = 3370$	5
207−245°C	$a = 9.38, b = 2245$	5
oral LD (guinea pig), mg/kg	150	6

[a] To convert kJ to kcal, divide by 4.184.
[b] $NH_4HF_2 \rightarrow NH_3 + 2\,HF$.
[c] To convert Pa to mm Hg, multiply by 7.5×10^{-3}.

by thermally decomposing the dry crystals (7). Commercial ammonium bifluoride, which usually contains 1% NH_4F, is made by gas-phase reaction of one mole of anhydrous ammonia and two moles of anhydrous hydrogen fluoride (8); the melt that forms is flaked on a cooled drum. The cost of the material in 1992 was $1.48/kg.

Production of bifluoride from fluoride by-products from the phosphate industry (9) has had little if any commercial significance.

Precautions in Handling. Ammonium bifluoride, like all soluble fluorides, is toxic if taken internally. Hydrofluoric acid burns may occur if the material comes in contact with moist skin. Ammonium bifluoride solutions should be thoroughly washed from the skin with mildly alkaline soap as soon as possible; however, if contact has been prolonged, the affected areas should be soaked with 0.13% solution of Zephiran chloride, or 0.2% Hyamine 1622 (Lonza, Inc.) or calcium gluconate, the treatment recommended for hydrofluoric acid burns. If any of these solutions come in contact with the eyes, they should be washed with water for at least 10 min and a physician should be consulted.

Applications. Ammonium bifluoride solubilizes silica and silicates by forming ammonium fluorosilicate [16919-19-0], $(NH_4)_2SiF_6$. Inhibited 15% hydrochloric acid containing about 2% ammonium bifluoride has been used to acidize oil wells in siliceous rocks to regenerate oil flow (10) (see PETROLEUM). Ammonium fluoride solution is made on-site near the well bore from ammonium bifluoride and ammonia and mixed with methyl formate to prevent rapid consumption of most of the HF (11). The use of ammonium bifluoride is important in locations where dissolved silicates foul boiler tubes with scale that cannot be removed using usual cleaning aids (12). Ammonium bifluoride is also used as an etching agent for silicon wafers.

Rapid frosting of glass is accomplished in a concentrated solution of ammonium bifluoride and hydrofluoric acid with nucleating agents that assure uniform frosts (13). A single dip in an aqueous solution of NH_4HF_2, HF, and sorbitol at $<20°C$ for less than 60 s produces the low specular-reflecting finish on television face plates and on glass (qv) for picture framing (14). Treating glass, eg, often badly weathered window panes, with 2–5% solutions of ammonium bifluoride results in a polishing effect. Glass ampuls for parenteral solutions (15) and optical lenses (16) are best cleaned of adhering particulate matter in dilute ammonium bifluoride solutions.

Ammonium bifluoride is used as a sour or neutralizer for alkalies in commercial laundries and textile plants. Treatment also removes iron stain by forming colorless ammonium iron fluorides that are readily rinsed from the fabric (17).

Ammonium fluorides react with many metal oxides or carbonates at elevated temperatures to form double fluorides; eg,

$$2\ NH_4HF_2\ +\ BeO\ \xrightarrow{-H_2O}\ (NH_4)_2BeF_4\ \xrightarrow{\Delta}\ BeF_2\ +\ 2\ NH_3\ +\ 2\ HF$$

The double fluorides decompose at even higher temperatures to form the metal fluoride and volatile NH_3 and HF. This reaction produces pure salts less likely to be contaminated with oxyfluorides. Beryllium fluoride 7787-49-7], from which beryllium metal is made, is produced this way (18) (see BERYLLIUM AND BERYLLIUM ALLOYS). In pickling of stainless steel and titanium, NH_4HF_2 is used with high concentrations of nitric acid to avoid hydrogen embrittlement. Ammonium bifluoride is used in acid dips for steel (qv) prior to phosphating and galvanizing, and for activation of metals before nickel plating (19,20). Ammonium bifluoride also is used in aluminum anodizing formulations. Ammonium bifluoride is used in treatments to provide corrosion resistance on magnesium and its alloys (21). Such treatment provides an excellent base for painting and good abrasion resistance, heat resistance, and protection from atmospheric corrosion. A minor use for ammonium bifluoride is in the preservation of wood (qv) (22).

BIBLIOGRAPHY

"Ammonium" under "Fluorine Compounds, Inorganic," in *ECT* 1st ed., Vol. 6, p. 676, by G. C. Whitaker, The Harshaw Chemical Co.; "Ammonium Fluoride" under "Fluorine Compounds, Inorganic," in *ECT* 2nd ed., Vol. 9, pp. 548–549, by G. C. Whitaker, The Harshaw Chemical Co.; "Ammonium" under "Fluorine Compounds, Inorganic," in *ECT* 3rd ed., Vol. 10, pp. 675–678, by H. S. Halbedel and T. E. Nappier, The Harshaw Chemical Co.

1. Fr. Pat. 1,546,234 (Nov. 15, 1968), (to Farbenfabriken Bayer A.-G.).
2. O. Hassel and H. Luzanski, *Z. Kristallogr.* **83**, 448 (1932).
3. R. C. Weast, ed., *Handbook of Chemistry and Physics*, 59th ed., The Chemical Rubber Co., Cleveland, Ohio, 1978.
4. H. Schutza, M. Eucken, and W. Namesh, *Z. An. All. Chem.* **292**, 293 (1957).
5. L. N. Lazarev and B. V. Andronov, *J. Appl. Chem. USSR* **46**, 2087 (1973).

6. H. C. Hodge and F. A. Smith, in J. H. Simon, ed., *Fluorine Chemistry*, Vol. 4, Academic Press, Inc., New York, 1965, p. 192.
7. U.S. Pat. 3,310,369 (Mar. 21, 1967), J. A. Peterson (to Hooker Chemical Corp.).
8. U.S. Pat. 2,156,273 (Apr. 28, 1939), A. R. Bozarth (to Harshaw Chemical Co.).
9. U.S. Pat. 3,501,268 (Mar. 17, 1970), R. J. Laran, A. P. Giraitix, and P. Kobetz (to Ethyl Corp.).
10. H. K. van Poolen, *Oil Gas J.* **65**, 93 (Sept. 11, 1967).
11. U.S. Pat. 3,953,340 (Apr. 27, 1976), C. C. Templeton, E. H. Street, Jr., and E. A. Richardson (to Shell Oil Co.).
12. W. S. Midkiff and H. P. Foyt, *Mater. Perform.* **17**(2), 17 (1978).
13. *Glass Frosting and Polishing Technical Service Bulletin 667*, Harshaw Chemical Co., Solon, Ohio.
14. U.S. Pat. 3,373,130 (Mar. 19, 1968), E. E. Junge and J. Chabal (to PPG Industries).
15. A. L. Hinson, *Bull. Parenter. Drug. Assoc.* **25**, 266 (1971).
16. R. L. Parkes and M. R. Browne, *Appl. Opt.* **17**, 1845 (1978).
17. *Control of Souring Operations, Special Report #7*, American Institute of Laundering, Joliet, Ill.
18. Brit. Pat. 833,808 (Apr. 27, 1960), A. R. S. Gough and E. W. Bennet (to the United Kingdom Atomic Energy Commission).
19. U.S. Pat. 3,767,582 (Oct. 23, 1973), G. A. Miller (to Texas Instruments, Inc.).
20. U.S. Pat. 3,296,141 (Jan. 3, 1967), W. A. Lieb and E. Billow (to R. O. Hull Co.).
21. L. F. Spencer, *Met. Finish.* **68**(10), 52 (1970); H. K. DeLong, *Met. Prog.* **97**, 105 (June 1970); *Met Prog.* **98**, 43 (Mar. 1971); W. F. Higgins, *Light Met. Age* **17**(12), 8 (1959); A. E. Yaniv and H. Schick, *Plating* **55**, 1295 (1968).
22. E. Panck, *Am. Wood Preservers Assoc.* **59**, 189 (1963).

JOHN R. PAPCUN
Atotech

ANTIMONY

Antimony forms both a trifluoride and a pentafluoride. It also forms the very stable hexafluoroantimonate ion [*17111-95-4*], SbF_6^-, present in solution and a number of salts.

Antimony Trifluoride

Properties. Antimony trifluoride [*7783-56-4*], SbF_3, is a very hygroscopic, white, crystalline solid, mp = 292°C. It can be sublimed under vacuum. It is very soluble in water, hydrofluoric acid, and polar organic solvents such as alcohols and ketones. Its solubility in water is 384.7 g/100 g at 0°C, 492.4 g/100 g at 25°C, and 563.6 g/100 g at 30°C (1). The solubility at 25°C is 154 g/100 mL CH_3OH, 33 g/100 mL C_3H_7OH, and 55.3 g/100 mL acetone. It is practically insoluble in benzene, chlorobenzene, and heptane. The density of SbF_3 at 25°C is 4.385 g/cm^3. It does hydrolyze in water, but the rate of hydrolysis is very slow, much slower than $SbCl_3$.

Antimony trifluoride is a mild fluorinating reagent. However, it is much more effective in the Swarts reactions where its effectiveness as a fluorinating reagent is dramatically increased by addition of Cl_2, Br_2, or $SbCl_5$ to the reaction mixture (2). Antimony trifluoride can be used for the replacement of chlorine or bromine in halocarbons, hydrohalocarbons, and nonmetal and metal halides. Typical reactions can be summarized as follows:

$$3\ RCCl_3 + 2\ SbF_3 \rightarrow RCCl_2F + RCClF_2 + RCF_3 + 2\ SbCl_3$$

In aliphatic compounds, the ease of fluorination is of the order of $-CCl_3 >$ $-CCl_2F > -CClF_2$. Other groups, eg, $-C{=}CCl_2$ and $-CHCl_2$, react, but not readily. Antimony trifluoride is not a suitable reagent for the replacement of hydrogen in organic compounds.

Inorganic compounds also can be fluorinated using SbF_3, eg,

$$PCl_3 + SbF_3 \xrightarrow{SbCl_5} PF_3 + SbCl_3 \cdot SbCl_5$$

$$SiHCl_3 + SbF_3 \xrightarrow{SbCl_5} SiHF_3 + SbCl_3 \cdot SbCl_5$$

In aqueous solutions SbF_3 reacts with many metal fluorides to form compounds such as $MSbF_4$ where M = Li [72121-39-2], Na [34109-83-6], K [15273-81-1], Cs [36195-09-0], and Tl [54189-44-5], and M_2SbF_5 where M = K [20645-41-4], Cs [40902-54-3], and NH_4 [32516-50-0]. In addition, triantimonate [65176-04-7], Na_3SbF_6, and MSb_4F_{13} where M = Tl [60719-48-4], Na [56094-73-6], K [56094-72-5], Rb [12776-50-0], Cs [12775-92-7], and NH_4 [52015-24-4] have been reported (3).

Preparation. Antimony trifluoride can be readily prepared by dissolving Sb_2O_3 in an excess of anhydrous hydrogen fluoride or in aqueous acid of 40% or higher strength hydrofluoric acid, followed by evaporation of the solution to dryness (4). It can also be prepared by thermal decomposition of the graphite intercalation compound with SbF_3Cl_2 (5), by heating ammonium hexafluorantimonate (6), and by the reaction of metal with anhydrous hydrogen fluoride in the presence of nitrile (7).

Uses. Early manufacturing processes for fluorocarbons and chlorofluorocarbons used SbF_3 on a large scale, but development of alternative routes to the Swarts reactions have greatly reduced usage. Its main use is in the manufacture of antimony pentafluoride. The market for SbF_3 in the United States is less than 5 t/yr. More recent uses of SbF_3 have been in the manufacture of fluoride glass and fluoride glass optical fiber preform (8), and fluoride optical fiber (9) in the preparation of transparent conductive films (10) (see FIBER OPTICS).

Antimony Pentafluoride

Properties. Antimony pentafluoride [7783-70-2], SbF_5, is a colorless, hygroscopic, very viscous liquid that fumes in air. Its viscosity at 20°C is 460 mPa·s($=$cP) which is very close to the value for glycerol. The polymerization of high purity SbF_5 at ambient temperature can be prevented by addition of 1% anhydrous hydrogen fluoride, which can be removed by distillation prior to the use of SbF_5. The pure product melts at 7°C (11), boils at 142.7°C, and has a specific gravity (12) of 3.145 g/cm^3 at 15.5°C. The viscous, pure liquid can be handled briefly in glass if moisture and air are carefully excluded. However, it must never be stored in glass because any HF or moisture present leads to a dangerous reaction. Any moisture reacts with SbF_5 to produce HF which reacts with glass to produce SiF_4 and water which, in turn, reacts with SbF_5 to again produce HF. The reaction continues until the SiF_4 pressure ruptures the container. Commercial antimony pentafluoride is shipped in steel cylinders or polytetrafluoroethylene bottles (generally 1-kg or less). Nickel is rapidly attacked by a mixture of SbF_5 and HF, although there is little effect on mild steel or aluminum.

Preparation.   Antimony pentafluoride can be prepared by direct fluorination of SbF_3 or antimony or by reaction of $SbCl_5$ with HF (13). The reaction of $SbCl_5$ with anhydrous hydrogen fluoride proceeds with the formation of intermediate products. These chlorofluoroantimonates can be prepared separately (14). Reaction of $SbCl_5$ with anhydrous hydrogen fluoride at -60°C produces $SbCl_4F$ [14913-58-7], mp = 83°C, which polymerizes on sublimation to $(SbCl_4F)_4$. $SbCl_4F$ can also be prepared by the reaction of $SbCl_5$ and anhydrous hydrogen fluoride in chlorofluorocarbons (15). Pure $SbCl_3F_2$ [24626-20-6], mp ~55°C, can be crystallized from a mixture of $SbCl_3F_2$ and $SbCl_2F_3$ [7791-16-4]. $SbCl_2F_3$, which is a thick liquid, can be prepared from the reaction of SbF_3 and Cl_2 at 135°C (16). $SbClF_4$ [15588-48-4] is obtained at 100°C by the reaction of $SbCl_5$ with a large excess of hydrogen fluoride. Other methods for preparation of antimony pentafluoride include reacting Sb with HF/F_2 (17) and fluorination of Sb using F_2 in a quartz tube (18).

Uses.   Antimony pentafluoride is a moderate fluorinating reagent and a powerful oxidizer. It spontaneously inflames phosphorus and sodium but it is practically inert toward arsenic. Powdered antimony reduces SbF_5 to solid $SbF_5 \cdot 2SbF_3$. SbF_5 reacts with water to form the solid antimony pentafluoride dihydrate [65277-49-8], $SbF_5 \cdot 2H_2O$, which reacts violently with an additional amount of water to form a clear solution. Antimony pentafluoride undergoes very slow hydrolysis in the presence of a dilute NaOH solution to form $Sb(OH)_6^-$. Sulfur dioxide and nitrogen dioxide react with SbF_5 to form the adducts $SbF_5 \cdot SO_2$ [19344-14-0] and $SbF_5 \cdot NO_2$ [72121-47-2], respectively. These adducts decompose in water.

Antimony pentafluoride is used to saturate double bonds in straight-chain olefins, cycloolefins, aromatic rings (19–21), and in the fluorination of halocarbons and CrO_2Cl_2, $MoCl_5$, WCl_6, PCl_3, P_4O_{10}, $SiCl_4$, $TiCl_4$, and SiO_2.

Antimony pentafluoride forms intercalation compounds with graphite (22,23) and fluorinated graphite (24), CF_x, where $x = 1.06$, which have much higher conductivity than graphite and fluorinated graphite, respectively (25). These nonstoichiometric substances may have potential use as superconducting materials. When a mixture of O_2, F_2, and SbF_5 or NF_3, F_2, and SbF_5 is subjected to elevated temperature and pressure, it gives the dioxygenyl salt O_2SbF_6 [51681-88-0] (26) and the perfluoroammonium salt NF_4SbF_6 [16871-76-4] (27), respectively. The dioxygenyl salt is a solid that can oxidize xenon (28) and has been used for removal of xenon, radon, and radon daughter elements from contaminated atmospheres. SbF_5 has also been used in the conversion of methane to gasoline range hydrocarbons (29), in the syntheses of fluorocarboranes (30), in superacids (31), and in the preparation of stable carbocations (32).

Hexafluoroantimonates

Hexafluoroantimonic acid [*72121-43-8*], $HSbF_6 \cdot 6H_2O$, is prepared by dissolving freshly prepared hydrous antimony pentoxide in hydrofluoric acid or adding the stoichiometric amount of 70% HF to SbF_5. Both of these reactions are exothermic and must be carried out carefully.

The superacid systems $HSO_3F \cdot SbF_5$ [*33843-68-4*] and $HF \cdot SbF_5$ [*16950-06-4*] (fluoroantimonic acid) are used in radical polymerization (33) and in carbocation chemistry (34). Addition of SbF_5 drastically increases the acidities of HSO_3F and HF (35,36).

Anhydrous salts, $MSbF_6$, where M = H, NH_4, and alkali metal, and $M(SbF_6)_2$, where M is an alkaline-earth metal, can be prepared by the action of F_2 on MF or MF_2 and SbF_3 (37) by the oxidation of Sb(III) with H_2O_2 or alkali metal peroxide in HF (38), by the action of HF on a mixture of $SbCl_5$ and MF where M = NH_4, Li, Na, K, Ru, Cs, Ag, and Tl (39). These compounds can be used as photoinitiators for the production of polymers (40).

Environmental and Safety Aspects

OSHA has a TWA standard on a weight of Sb basis of 0.5 mg/m^3 for antimony in addition to a standard TWA of 2.5 mg/m^3 for fluoride. NIOSH has issued a criteria document on occupational exposure to inorganic fluorides. Antimony pentafluoride is considered by the EPA to be an extremely hazardous substance and releases of 0.45 kg or more reportable quantity (RQ) must be reported. Antimony trifluoride is on the CERCLA list and releasing of 450 kg or more RQ must be reported.

BIBLIOGRAPHY

"Antimony Compounds" under "Fluorine Compounds, Inorganic" in *ECT* 1st ed., Vol. 6, pp. 676–677, by F. D. Loomis and C. E. Inman, Pennsylvania Salt Manufacturing Co.; "Antimony" under "Fluorine Compounds, Inorganic" in *ECT* 2nd ed., Vol. 9, pp. 549–551, by W. E. White, Ozark-Mahoning Co.; in *ECT* 3rd ed., Vol. 10., pp. 679–681, by D. T. Meshri and C. B. Lindahl, Ozark-Mahoning Co., a subsidiary of the Pennwalt Corp.

1. A. Rosenheim and H. Grünbaum, *Z. Anorg. Chem.* **61**, 187 (1909).
2. F. Swarts, *Bull. Acad. Roy. Belg.* **24**, 309 (1892).
3. I. G. Ryss, *The Chemistry of Fluorine and Its Inorganic Compounds*, State Publishing House of Scientific, Technical, and Chemical Literature, Moscow, USSR, 1956; English trans., AEC-tr-3927, Office of Technical Services, U.S. Department of Commerce, Washington, D.C., 1960, pp. 283–295 (Part I).
4. Z. Xie, *Huaxue Shijie* **26**(5), 165–166 (1985).
5. H. Preiss, E. Alsdorf, and A. Lehman, *Carbon* **25**(6), 727–733 (1987).
6. Eur. Pat. 156,617 A2 (Oct. 2, 1985), M. Watanabe and S. Nishimura.

7. U.S. Pat. 4,0340,780 (July 5, 1977), J. A. Wojtowicz and D. F. Gavin.
8. Eur. Pat. 331,483 A2 (Sept. 6, 1989), K. Fujiura, Y. Ohishi, M. Fujiki, T. Kanamori, and S. Takahashi.
9. Jpn. Pat. 6011239 A2 (Jan. 21, 1983), (to Nippon Telegraph & Telephone Public Co.).
10. Jpn. Pat. 63314713 A2 (Dec. 22, 1988), N. Sonoda and N. Sato.
11. O. Ruff and co-workers, *Chem. Ber.* **42**, 4021 (1909).
12. R. C. Shair and W. F. Shurig, *Ind. Eng. Chem.* **43**, 1624 (1951).
13. O. Ruff and W. Plato, *Ber.* **37**, 673 (1904).
14. M. F. A. Dove and Md. O. Ali, *J. Inorg. Nucl. Chem.* 77. (1976 Suppl.).
15. E. Santacesaria and M. DiSerio, Jr., *J. Fluorine Chem.* **44**, 87–111 (1989).
16. A. L. Henne and P. Trott, *J. Am. Chem. Soc.* **69**, 1820 (1947).
17. Jpn. Pat. 03242326 A2 (Oct. 29, 1991), K. Kuge, S. Saito, A. Chuma, and S. Takenuki.
18. D. Ganter, A. Boles, and B. Erlec, *Vestn. Slov. Kem. Drus.* **30**(3), 289–93 (1983).
19. E. T. McBee, P. A. Wisemen, and G. B. Bachman, *Ind. Eng. Chem.* **39**, 415 (1947).
20. U.S. Pat. 2,488,216 (Nov. 15, 1949), E. T. McBee, V. V. Lindgren, and W. B. Ligett (to Purdue Research Foundation).
21. U.S. Pat. 2,533,217 (May 15, 1951), F. B. Stilmar (to the United States of America as represented by the Atomic Energy Commission).
22. J. M. Lalancette and J. Lafontaine, *J. Chem. Soc., Chem. Commun.*, 815 (1973).
23. A. A. Opalvskii, A. S. Nazarov, and A. A. Uminskii, *Zh. Neorg. Khim.* **19**, 1518 (1974); English ed., *Russ. J. Inorg. Chem.* **19**, 827 (1974).
24. L. B. Ebert, R. A. Muggins, and J. I. Brauman, *Mater. Res. Bull.* **2**, 615 (1976).
25. L. Vogel, *J. Mater. Sci.* **12**, 982 (1977).
26. J. B. Beal, Jr., C. Pupp, and W. E. White, *Inorg. Chem.* **8**, 828 (1969).
27. U.S. Pat. 3,708,570 (Jan. 2, 1973), W. E. Tolberg, R. S. Stringham, and R. T. Rewich (to Stanford Research Institute).
28. U.S. Pat. 3,829,551 (Aug. 13, 1974), L. Stein (to the United States Atomic Energy Commission).
29. U.S. Pat. 4,973,776A (Nov. 27, 1990), V. M. Allenger and R. N. Pandey.
30. V. N. Lebedev and co-workers *J. Organomet. Chem.* **385**(3), 307–308 (1990).
31. U.S. Pat. 4,369,107A (Jan. 18, 1983), S. C. Amendola.
32. G. A. Olah and co-workers, *J. Am. Chem. Soc.* **97**(19), 5477–5481 (1975).
33. K. K. Laali, E. Geleginter, and R. Filler, *J. Fluorine Chem.* **53**(1), 107–126 (1991).
34. G. A. Olah, A. Germain, and H. C. Lin, *J. Am. Chem. Soc.* **97**(19), 5481–5488 (1975).
35. R. J. Gillespie and T. E. Peel, *J. Am. Chem. Soc.* **95**, 5173 (1973).
36. R. J. Gillespie, in V. Gold, ed., *Proton Transfer Reactions*, Chapman and Hall, London, 1975, p. 27.
37. Jpn. Pat. 62027306 A2 (Feb. 5, 1987), Y. Mochida and co-workers.
38. Jpn. Pat. 62108730 A2 (May 20, 1987), Y. Mochida and co-workers.
39. Ger. Offen. DE 3432221 A1 (Mar. 13, 1986), A. Guenther.
40. U.S. Pat. 4,136,102 (Jan. 23, 1979), J. V. Crivello (to General Electric Co.).

TARIQ MAHMOOD
CHARLES B. LINDAHL
Elf Atochem North America, Inc.

ARSENIC

Arsenic forms the binary compounds arsenous trifluoride and arsenic pentafluoride, as well as a series of compounds and the acid of the very stable hexafluoroarsenate ion.

Great care should be exercised in the handling and use of all arsenic compounds (qv) because NIOSH has determined inorganic arsenic to be a carcinogen and OSHA considers inorganic arsenic to be a cancer hazard. The OSHA permissible exposure limit is 10 $\mu g/m^3$, averaged over any 8-h period. The OSHA action level is 5 $\mu g/m^3$, averaged over any 8-h period. The OSHA limits have the force of law and are much lower than the 0.2 mg/m^3 of ACGIH.

Arsenous Fluoride

Arsenous fluoride [7784-35-2], AsF_3, is a colorless liquid, mp = $-5.95°C$, bp = $57.13°C$ at 99 kPa (742.5 mm Hg) (1), and sp gr = 2.67, having a standard enthalpy of formation of -858.1 kJ/mol (-205.1 kcal/mol) (2). Arsenic(III) fluoride can be prepared by fluorination of arsenous oxide using sulfuric acid and calcium fluoride (3), or using hydrofluoric acid or fluorosulfuric acid; from thermal decomposition of $AsBr_4AsF_6$ (4); from the fluorination of gallium arsenide using F_2 or NF_3 (5); from As_2O_3, CaF_2, and concentrated H_2SO_4 (6); from disproportionation of graphite intercalated compounds of AsF_5 (7); from the reaction of arsenous trichloride with NaF at 300°C in the presence of $ZnCl_2$ or KCl (8), and from the fluorination of arsenous trichloride with antimony trifluoride or zinc fluoride (9).

It is used as a fluorinating reagent in semiconductor doping, to synthesize some hexafluoroarsenate compounds, and in the manufacture of graphite intercalated compounds (10) (see SEMICONDUCTORS). AsF_3 has been used to achieve >8% total area simulated air-mass 1 power conversion efficiencies in Si *p-n* junction solar cells (11) (see SOLAR ENERGY). It is commercially produced, but usage is estimated to be less than 100 kg/yr.

Arsenic Trifluoride Oxide

Arsenic trifluoride oxide [15120-14-6], $AsOF_3$, has been reported to be produced by the uv photolysis of O_3 or HOF in the presence of AsF_3 (12,13).

Arsenic Pentafluoride

Arsenic pentafluoride [7784-36-3], AsF_5, melts at $-79.8°C$ and boils at $-52.8°C$ (14). At the boiling point the liquid has a density of 2.33 g/mL. The standard enthalpy of formation is -1237 kJ/mol (-295.6 kcal/mol), and the average bond strength is 387 kJ/mol (92.4 kcal/mol), compared to 484.1 kJ/mol (115.7 kcal/mol) for AsF_3 (15).

Arsenic pentafluoride can be prepared by reaction of fluorine and arsenic trifluoride or arsenic; from the reaction of NF_3O and As (16); from the reaction of

$Ca(FSO_3)_2$ and H_3AsO_4 (17); or by reaction of alkali metal or alkaline-earth metal fluorides or fluorosulfonates with H_3AsO_4 or H_2AsO_3F (18).

It is used as a fluorinating reagent and in syntheses of some hexafluoroarsenate compounds. Arsenic pentafluoride is also used to dope semiconductors (19); to produce conductive polymers (20,21); and in conducting-oriented fibers (22). Arsenic pentafluoride has been found to react with graphite to form AsF_5- graphite intercalation compounds (23) having electrical conductivity as high as that of silver (24,25). Arsenic pentafluoride is produced commercially and usage is estimated to be less than 100 kg/yr.

Hexafluoroarsenic Acid and the Hexafluoroarsenates

The AsF_6^- ion is very stable toward hydrolysis in aqueous solution. It is not hydrolyzed by boiling a strongly basic solution almost to dryness (26), although it is hydrolyzed in sulfuric acid (27) or in boiling perchloric acid (26). The hydrolysis of AsF_6^- in concentrated sulfuric acid (27) and in base (28) at 193–222°C is first order in AsF_6^-. The hydrolysis of AsF_6^- in alkaline solution is slower than either PF_6^- or SbF_6^-.

Hexafluoroarsenic acid [17068-85-8] can be prepared by the reaction of arsenic acid with hydrofluoric acid or calcium fluorosulfate (29) and with alkali or alkaline-earth metal fluorides or fluorosulfonates (18). The hexafluoroarsenates can be prepared directly from arsenates and hydrofluoric acid, or by neutralization of $HAsF_6$. The reaction of 48% HF with potassium dihydrogen arsenate(V), KH_2AsO_4, gives potassium hydroxypentafluoroarsenate(V) [17068-84-7], $KAsF_5OH$, which hydrolyzes rapidly in water solution (26). Anhydrous HF reacts with KH_2AsO_4 or $KAsF_5OH$ to produce $KAsF_6$ [17029-22-0]. O_2AsF_6 [12370-43-3] can be prepared from the reaction of OF_2 and AsF_5 or a mixture of O_2, F_2, and AsF_5 (30). Reactions of $XeF(AsF_6)$ and water give $H_2OF(AsF_6)$ which reacts with SF_4 to produce $OSF_3(AsF_6)$ and with ClF_3 to produce $OClF_2AsF_6$ (31). The compound $SCl_3 \cdot AsF_6$ has also been reported (32).

Because of the special stability of the hexafluoroarsenate ion, there are a number of applications of hexafluoroarsenates. For example, onium hexafluoroarsenates (33) have been described as photoinitiators in the hardening of epoxy resins (qv). Lithium hexafluoroarsenate [29935-35-1] has been used as an electrolyte in lithium batteries (qv). Hexafluoroarsenates, especially alkali and alkaline-earth metal salts or substituted ammonium salts, have been reported (34) to be effective as herbicides (qv). Potassium hexafluoroarsenate [17029-22-0] has been reported (35) to be particularly effective against prickly pear. However, environmental and regulatory concerns have severely limited these applications.

BIBLIOGRAPHY

"Arsenic Compounds" in *ECT* 1st ed., Vol. 2, pp. 119–123, by I. E. Campbell, Battelle Memorial Institute; in *ECT* 2nd ed., Vol. 2, pp. 718–733, by G. O. Doak, L. D. Freedman, and G. G. Long, North Carolina State of the University of North Carolina at Raleigh;

"Arsenic" under "Fluorine Compounds, Inorganic" in *ECT* 3rd ed., Vol. 10, pp. 682–683, by C. B. Lindahl, Ozark-Mahoning Co., a subsidiary of the Pennwalt Corp.

1. H. Russell, Jr., R. E. Rundle, and D. M. Yost, *J. Am. Chem. Soc.* **63**, 2825 (1941).
2. A. A. Woolf, *J. Fluorine Chem.* **5**, 172 (1975).
3. O. Ruff, *Die Chemie des Fluors*, Springer-Verlag, Berlin, 1920, p. 27.
4. B. Ponsold, and H. Kath, *Z. Gesamte Hyg. Ihre Grenzgeb.* **37**(2), 58–63 (1991).
5. Eur. Pat. 333084 A2 (Sept. 20, 1989), I. Harada, Y. Yoda, N. Iwanaga, T. Nishitsuji, and A. Kikkawa.
6. Ger. Pat. DD248249 A3 (Aug. 5, 1987), P. Wolter, M. Schoenherr, D. Hass.
7. J. G. Hooley, *Ext. Abstr. Program*, 16th, Biennial Conference on Carbon, 1983, pp. 240–241.
8. U.S. Pat. 4,034,069 (July 5, 1977), D. M. Curtis.
9. F. Kober, *J. Fluorine Chem.* **2**(3), 247–256 (1973).
10. Y. Yacoby, *Synth. Met.*, **34**(1-3), 437–438 (1989).
11. E. J. Caine and E. J. Charlson, *J. Electron. Mater.* **13**(2), 341–372 (1984).
12. E. A. Evans, A. J. Downs, and C. J. Gardner, *J. Phys. Chem.* **93**(2), 598–608 (1989).
13. A. J. Downs, G. P. Gaskill, and S. B. Saville, *Inorg. Chem.* **21**(9), 3385–3393 (1982).
14. O. Ruff, A. Braida, O. Bretschneider, W. Menzel, and H. Plaut, *Z. Anorg. Allgem. Chem.* **206**, 59 (1932).
15. P. A. G. O'Hare and W. N. Hubbard, *J. Phys. Chem.* **69**, 4358 (1965).
16. O. D. Gupta, R. L. Kirchmeier, and J. M. Shreeve, *Inorg. Chem.* **29**(3), 573–574 (1990).
17. U.S. Pat. 3,875,292 (Apr. 1, 1975), R. A. Wiesboeck and J. D. Nickerson.
18. U.S. Pat. 3,769,387 (Oct. 30, 1973), R. A. Wiesboeck and J. D. Nickerson.
19. D. G. H. Ballard, A. Courtis, I. M. Shirley, and S. C. Taylor, *Air Force Off. Sci. Res.* (Technical Report), AFOSR-TR (U.S.), AFSOR-TR-87 1884, Biotechnol. Aided Synth. Aerosp. Compos. Resins 53-92, CA110(24):213428x.
20. M. Aldissi, *Polymer Prepr. (Am. Chem. Soc., Div. Polym. Chem.)* **26**(2), 269–270 (1985).
21. Jpn. Pat. 59 133,216 A2 (July 31, 1984) Showa (to Orient Watch Co.).
22. M. Stamm, *Mol. Cryst. Liquid Cryst.* **105**(1-4), 259–271 (1984).
23. L. Chun-Hsu, H. Selig. M. Rabinovitz, I. Agranat, and S. Sarig, *Inorg. Nucl. Chem. Lett.* **11**, 601 (1975).
24. E. R. Falardeau, G. M. T. Foley, C. Zeller, and F. L. Vogel, *Chem. Commun.*, 389 (1977).
25. G. M. T. Foley, C. Zeller, E. R. Falardeau, and F. L. Vogel, *Solid State Commun.* **24**, 371 (1977).
26. H. M. Dess and R. W. Parry, *J. Am. Chem. Soc.* **79**, 1589 (1957).
27. W. L. Lockhart, Jr., M. M. Jones, and D. O. Johnston, *J. Inorg. Nucl. Chem.* **31**, 407 (1969).
28. I. G. Ryss, V. B. Tul'chinskii and Y. A. Mazurov, *Izv. Sib. Otd. Akad. Nauk. SSSR Ser., Khim. Nauk*, 81 (1968).
29. U.S. Pat 3,875,292 (Apr. 1, 1975), R. A. Wiesboeck and J. D. Nickerson (to U.S. Steel Corp.).
30. J. B. Beal, Jr., P. Christian, and W. E. White, *Inorg. Chem.* **8**(4), 828–830 (1969).
31. R. Minkwitz and G. Nowicki, *Angew. Chem.* **102**(6), 692–693 (1990).
32. F. Claus and R. Minkwitz, *J. Fluorine Chem.* **19**(3-6), 243–252 (1982).
33. Ger. Offen. 2,618,871 (Nov. 11, 1976) and 2,518,652 (May 2, 1974), J. V. Crivello (to General Electric Co.).
34. Belg. Pat. 659,342 (Aug. 5, 1965), T. N. Russell (to Pennsalt Chemicals Corp.).
35. P. E. Buckley, J. D. Dodd, and W. H. Culver, *Proc. West. Soc. Weed Sci.* **22**, 17 (1968).

CHARLES B. LINDAHL
TARIQ MAHMOOD
Elf Atochem North America, Inc.

BARIUM

Barium Fluoride

Barium fluoride [7782-32-8], BaF_2, is a white crystal or powder. Under the microscope crystals may be clear and colorless. Reported melting points vary from 1290 (1) to 1355°C (2), including values of 1301 (3) and 1353°C (4). Differences may result from impurities, reaction with containers, or inaccurate temperature measurements. The heat of fusion is 28 kJ/mol (6.8 kcal/mol) (5), the boiling point 2260°C (6), and the density 4.9 g/cm^3. The solubility in water is about 1.6 g/L at 25°C and 5.6 g/100 g (7) in anhydrous hydrogen fluoride. Several preparations for barium fluoride have been reported (8–10).

High purity BaF_2 can be prepared from the reaction of barium acetate and aqueous HF (11), by dissolving the impure material in 2-12N HCl and recrystallizing at −40°C (12), by vacuum distillation of the metal fluoride impurities from a BaF_2 melt (13), by purification of the aqueous acetate solution by ion exchange followed by fluorination (14), by solvent extraction using dithiocarbamate and CCl_4 (15–17), and by solvent extraction using acetonitrile (18).

A typical analysis of the commercial product is 99% with a loss on ignition of 0.9%; sulfates as SO_4, 0.2%; hexafluorosilicate as SiF_4, 0.02%; heavy metals as lead, 0.02%; and iron, 0.005%.

Barium fluoride is used commercially in combination with other fluorides for arc welding (qv) electrode fluxes. However, this usage is limited because of the availability of the much less expensive naturally occurring calcium fluoride.

Other reported uses of barium fluoride include the manufacture of fluorophosphate glass (19); stable fluoride glass (20); fluoroaluminate glass (21); fluorozirconate glass (22); infrared transmitting glass (23); in oxidation-resistant ceramic coatings (24); in the manufacture of electric resistors (25,26); as a superconductor with copper oxide (27); and as a fluoride optical fiber (28) (see FIBER OPTICS; GLASS; SUPERCONDUCTING MATERIALS).

The toxicity of barium fluoride has received only little attention. A value for oral LD_{LO} of 350 mg/kg in guinea pigs has been reported (29). OSHA has a TWA standard on the basis of Ba of 0.5 mg/m^3 for barium fluoride (29) in addition to a standard TWA on the basis of F of 2.5 mg/m^3 (30). NIOSH has issued a criteria document (30) on occupational exposure to inorganic fluorides.

BIBLIOGRAPHY

"Barium Fluoride" under "Fluorine Compounds, Inorganic," in *ECT* 1st ed., Vol. 6, p. 677, by F. D. Loomis, Pennsylvania Salt Manufacturing Co.; in *ECT* 2nd ed., Vol. 9, p. 551, by W. E. White, Ozark-Mahoning Co.; "Barium" under "Fluorine Compounds, Inorganic," in *ECT* 3rd ed., Vol. 10, p. 684, by C. B. Lindahl, Ozark-Mahoning Co.

1. I. Barin and O. Knache, *Thermochemical Properties of Inorganic Substances*, Springer Verlag, Berlin, 1973.
2. H. Kojima, S. G. Whiteway, and C. R. Masson, *Can. J. Chem.* **46**, 2698 (1968).
3. I. Jackson, *Phys. Earth Planet. Inter.* **14**, 143 (1977).
4. B. Porter and E. A. Brown, *J. Am. Ceram. Soc.* **45**, 49 (1962).

5. G. Petit and A. Cremieo, *C. R. Acad. Sci.* **243**, 360 (1956).
6. O. Ruff and L. LeBoucher, *Z. Anorg. Chem.* **219**, 376 (1934).
7. A. W. Jache and G. H. Cady, *J. Phys. Chem.* **56**, 1106 (1952).
8. SU 1325018 Al (July 23, 1985), V. A. Bogomolov and co-workers.
9. SU 998352 A1 (Feb. 23, 1983), A. A. Luginina and co-workers.
10. A. A. Lugina and co-workers *Zh. Neorg. Khim 1981*, **26**(2), 332–336.
11. Jpn. Pat. 90-144378 (June 4, 1990), K. Kobayashi, K. Fujiura, and S. Takahashi.
12. EP 90-312689 (Nov. 21, 1990), J. A. Sommers, R. Ginther, and K. Ewing.
13. A. M. Garbar, A. N. Gulyaikin, G. L. Murskii, I. V. Filimonov, and M. F. Churbanov, *Vysokochist, Veshchestva* (6), 84–85 (1990).
14. A. M. Garbar, A. V. Loginov, G. L. Murskii, V. I. Rodchenkov, and V. G. Pimenov, *Vysokochist, Veshchestva* (3), 212–213 (1989).
15. Jpn. Pat. 01028203 A2 (Jan. 30, 1989), K. Kobayashi (to Heisei).
16. K. Kobayashi, *Mater. Sci. Forum*, **32–33**(5), 75–80 (1988).
17. DE 3813454 A1 (Nov. 3, 1988); Jpn. Pat. 87-100025 (Apr. 24, 1987), H. Yamashita and H. Kawamoto.
18. J. Guery and C. Jacoboni, in Ref. 16, pp. 31–35.
19. V. D. Khalilev, V. G. Cheichovskii, M. A. Amanikov, and Kh. V. Sabirov, *Fiz. Khim. Stekla* **17**(5), 740–743 (1991).
20. Y. Wang, *J. Non. Cryst. Solids* **142**(1–2), 185–188 (1992).
21. H. Hu, F. Lin, and J. Feng, *Guisudnyan Xuebao* **18**(6), 501–505 (1990).
22. M. N. Brekhovskikh, V. A. Fedorov, V. S. Shiryaev, and M. F. Churbanov, *Vysokochist Veshchestva* (1), 219–223 (1991).
23. A. Jha and J. M. Parker, *Phys. Chem. Glasses* **32**(1), 1–12 (1991).
24. EP 392822 A2 17 (Oct. 17, 1990), L. M. Niebylski.
25. Jpn. Pat. 63215556 A2 (Sept. 8, 1988), T. Honda, T. Yamada, K. Onigata, and S. Tosaka (to Showa).
26. Jpn. Pat. 63215553 A (Sept. 28, 1988), T. Honda, T. Yamada, K. Onigata, and S. Tosaka (to Showa).
27. S. R. Ovshinsky, R. T. Young, B. S. Chao, G. Fournier, and D. A. Pawlik, *Rev. Solid State Sci.* **1**(2), 207–219 (1987).
28. J. Chen and co-workers, *J. Non-Cryst. Solids* **140**(1–3), 293–296 (1992).
29. *Registry of Toxic Effects of Chemical Substances*, Vol. II, NIOSH, Washington, D.C., 1977, p. 141.
30. *Criteria for a Recommended Standard-Occupational Exposure to Inorganic Fluorides*, PB 246 692, NIOSH 76-103, U.S. Department of Health, Education, and Welfare, Washington, D.C., 1975.

TARIQ MAHMOOD
CHARLES B. LINDAHL
Elf Atochem North America, Inc.

BORON

BORON TRIFLUORIDE

Boron trifluoride [*7637-07-2*] (trifluoroborane), BF_3, was first reported in 1809 by Gay-Lussac and Thenard (1) who prepared it by the reaction of boric acid and fluorspar at dull red heat. It is a colorless gas when dry, but fumes in the presence of moisture yielding a dense white smoke of irritating, pungent odor. It is widely used as an acid catalyst (2) for many types of organic reactions, especially for the production of polymer and petroleum (qv) products. The gas was first produced commercially in 1936 by the Harshaw Chemical Co. (see also BORON COMPOUNDS).

The boron atom in boron trifluoride is hybridized to the sp^2 planar configuration and consequently is coordinatively unsaturated, ie, a Lewis acid. Its chemistry centers around satisfying this unsaturation by the formation with Lewis bases of adducts that are nearly tetrahedral (sp^3). The electrophilic properties

Table 1. Physical Properties of Boron Trifluoride

Property	Value	Reference
molecular weight	67.8062	5
melting point, °C	-128.37	6
boiling point, °C	-99.9	6
vapor pressure of liquid, kPa^a		
at 145 K	8.43	
at 170 K	80.19	
at 220 K	1156	
at 260 K	4842	7
triple point at 8.34 kPa,^{a} K	144.78	5
critical temperature, T_c, °C	-12.25 ± 0.03	7
critical pressure, P_c, kPa^a	4984	7
density		
critical, d_c, g/cm^3	ca 0.591	8
gas at STP, g/L	3.07666	9
gas limiting, L_N, g/L	3.02662	9
liquid, for 148.9 to 170.8 K, g/cm^3	$1.699 - 0.00445\,(t + 125.0)$	10
enthalpy of fusion, $\Delta H_{144.45}$, kJ/mol^b	4.2417	11
enthalpy of vaporization, $\Delta H_{154.5}$, kJ/mol^b	18.46	12
entropy, $S_{298.15}$, J/(mol·K)^b	254.3	
Gibbs free energy of formation, $\Delta G_{f298.15}$, kJ/mol^b	-1119.0	13
enthalpy of formation, $\Delta H_{f298.15}$, kJ/mol^b	-1135.6	13
infrared absorption frequencies, cm^{-1}		
v_1	888	
v_2	696.7	
v_3	1463.3	
v_4	480.7	

^{a}To convert kPa to mm Hg, multiply by 7.5. ^{b}To convert J to cal, divide by 4.184.

(acid strengths) of the trihaloboranes have been found to increase in the order $BF_3 < BCl_3 < BBr_3 < BI_3$ (3,4).

Physical Properties. The physical properties are listed in Table 1. The molecule has a trigonal planar structure in which the F—B—F angle is 120° and the B—F bond distance is 0.1307 ± 0.0002 nm (13).

Nuclear magnetic resonance ^{11}B spectral studies of BF_3 have given a value of 9.4 ± 1.0 ppm for the chemical shift relative to $BF_3 \cdot O(C_2H_5)_2$ as the zero reference (14). Using methylcyclohexane as a solvent at 33.5°C and $BF_3 \cdot O(CH_2CH_3)_2$ as the internal standard, a value of 10.0 ± 0.1 ppm was obtained for the chemical shift (15). A value for the ^{19}F chemical shift of BF_3 in CCl_3F relative to CCl_3F is reported to be 127 ppm (16). The coupling constant $J_{11_B - 19_F}$ is reported to be 15 ± 2 Hz for BF_3 (17). Additional constants are available (3,18). See Table 2 for solubilities.

Aqueous mineral acids react with BF_3 to yield the hydrates of BF_3 or the hydroxyfluoroboric acids, fluoroboric acid, or boric acid. Solution in aqueous alkali gives the soluble salts of the hydroxyfluoroboric acids, fluoroboric acids, or boric acid. Boron trifluoride, slightly soluble in many organic solvents including saturated hydrocarbons (qv), halogenated hydrocarbons, and aromatic compounds, easily polymerizes unsaturated compounds such as butylenes (qv), styrene (qv), or vinyl esters, as well as easily cleaved cyclic molecules such as tetrahydrofuran (see FURAN DERIVATIVES). Other molecules containing electron-donating atoms such as O, S, N, P, etc, eg, alcohols, acids, amines, phosphines, and ethers, may dissolve BF_3 to produce soluble adducts.

Chemical Properties. In addition to the reactions listed in Table 3, boron trifluoride reacts with alkali or alkaline-earth metal oxides, as well as other inorganic alkaline materials, at 450°C to yield the trimer trifluoroboroxine [13703-95-2], $(BOF)_3$, MBF_4, and MF (29) where M is a univalent metal ion. The trimer is stable below -135°C but disproportionates to B_2O_3 and BF_3 at higher temperatures (30).

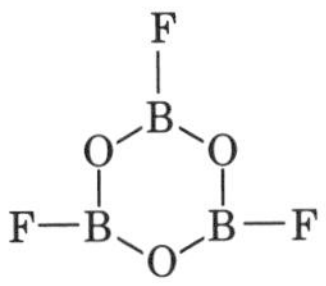

The reaction of metal hydrides and BF_3 depends on the stoichiometry as well as the nature of the metal hydride. For example, LiH and $BF_3 \cdot O(C_2H_5)_2$ may form diborane (6) or lithium borohydride (31,32):

$$6 \text{ LiH} + 8 \text{ } BF_3 \cdot O(C_2H_5)_2 \rightarrow B_2H_6 + 6 \text{ } LiBF_4 + 8 \text{ } (C_2H_5)_2O$$

$$4 \text{ LiH} + 4 \text{ } BF_3 \cdot O(C_2H_5)_2 \rightarrow LiBH_4 + 3 \text{ } LiBF_4 + 4 \text{ } (C_2H_5)_2O$$

The first method is commonly used for preparing diborane.

Metal halides react with BF_3 (33) when heated to form BX_3 and the metal fluoride. For example,

$$AlBr_3 + BF_3 \rightarrow BBr_3 + AlF_3$$

Table 2. Solubilities of Boron Trifluoride

BF_3, g	Solvent, g	Temperature, °C	Product	CAS Registry Number	Reference
369.4	water,[a] 100[b]	6	$BF_3 \cdot H_2O$ $HBF_3(OH)$	[15799-89-0] [16903-52-9]	19
2.06	sulfuric acid, conc, 100%	25			20
	nitric acid[a]	20	$HNO_3 \cdot 2BF_3$	[20660-63-3]	21
	orthophosphoric acid[a]	25	$H_3PO_4 \cdot BF_3$	[13699-76-6]	22
2.18	hydrofluoric acid,[c]	4.4			24
	hydrochloric acid, anhydrous (l)	24	miscible		25

[a]Dissolves with reaction to form complexes and other species.
[b]A higher dilution results in a mixture of $H[BF_2(OH)_2]$, HBF_4, and H_3BO_3.
[c]Equations for the solubility of BF_3 in liquid HF at 24, 49, and 90°C and up to 6.8 kPa (51 mm Hg) may be found in Reference 23.

Table 3. Reactions of Boron Trifluoride

Reactant	Temperature, °C	Products	Formula	Reference
sodium[a]		boron, amorphous, sodium fluoride	NaF	26
magnesium, molten alloys	no reaction			
calcium	1600	calcium hexaboride	CaB_6	
aluminum	1200	aluminum boride (1:12), tetragonal boron	AlB_{12}	
	1650[b]	β-rhombohedral boron		
titanium	1600	titanium boride	TiB_2	27
copper, mercury, chromium, iron	RT or below	no reaction[c]		
sodium nitrate, sodium nitrite	180	sodium fluoroborate, boric oxide	$NaBF_4$	28

[a]With incandescence.
[b]Further reaction.
[c]Even when subjected to pressure for a considerable length of time; also no reaction with red-hot iron.

The reaction of BF_3 with alkali halides yields the respective alkali fluoroborates (34):

$$3\ KCl\ +\ 4\ BF_3 \rightarrow 3\ KBF_4\ +\ BCl_3$$

Alkyl and arylboranes are obtained (35) from BF_3 using the appropriate Grignard reagent, alkylaluminum halide, or zinc alkyl, using diethyl ether as the solvent (see also ORGANOMETALLICS):

$$BF_3\ +\ 3\ RMgX \rightarrow BR_3\ +\ 3\ MgXF$$

Tetraorganylborate complexes may be produced when tetrahydrofuran is the solvent (36).

Alkylfluoroboranes result from the reaction of the appropriate alkylborane and BF_3 under suitable conditions (37):

$$BR_3 + 2(C_2H_5)_2O \cdot BF_3 \rightarrow 3\ RBF_2 + 2\ (C_2H_5)_2O$$

Adducts of BF_3 and some organic compounds having labile hydrogen atoms in the vicinity of the atom bonding to the boron atom of BF_3 may form a derivative of BF_3 by splitting out HF. For example, β-diketones such as acetylacetone or benzoylacetone react with BF_3 in benzene (38):

$$BF_3 + CH_3COCH_2COCH_3 \rightarrow CH_3COCH{=}C(CH_3)OBF_2 + HF$$

In Group 14 (IV), carbon serves as a Lewis base in a few of its compounds. In general, saturated aliphatic and aromatic hydrocarbons are stable in the presence of BF_3, whereas unsaturated aliphatic hydrocarbons, such as propylene or acetylene, are polymerized. However, some hydrocarbons and their derivatives have been reported to form adducts with BF_3. Typical examples of adducts with unsaturated hydrocarbons are 1:1 adducts with tetracene and 3,4-benzopyrene (39), and 1:2 BF_3 adducts with α-carotene and lycopene (40).

In Group 15 (V), nitrogen compounds readily form molecular compounds with BF_3. Phosphorus compounds also form adducts with BF_3. Inorganic or organic compounds containing oxygen form many adducts with boron trifluoride, whereas sulfur and selenium have been reported to form only a few (41–43).

Boron trifluoride forms two hydrates, $BF_3 \cdot H_2O$ and boron trifluoride dihydrate [*13319-75-0*], $BF_3 \cdot 2H_2O$, (also $BF_3 \cdot D_2O$ [*33598-66-2*] and $BF_3 \cdot 2D_2O$ [*33598-66-2*]). According to reported nmr data (43,44), the dihydrate is ionic, $H_3O^+F_3BOH^-$. The trihydrate has also been reported (45). Acidities of BF_3–water systems have been determined (46). Equilibrium and hydrolysis of BF_3 in water have been studied (47–49).

Most of the coordination compounds formed by trifluoroborane are with oxygen-containing organic compounds (Table 4). Although the other boron halides frequently react to split out hydrogen halide, boron trifluoride usually forms stable molecular compounds. The reason is attributed to the back coordination of electrons from fluorine to boron forming a strong B—F bond which is 28% ionic (50).

It has been reported (51) that some adducts of alkyl ethers and/or alcohols are unstable and decompose at −80°C to yield BF_3, H_2O, and the polyalkene. Adducts of BF_3 have been reported with hydrogen sulfide, sulfur dioxide, thionyl fluoride, and the sulfur analogues of many of the kind of oxygen-containing organic molecules cited in Table 4. The carbonyl oxygen or the carbonyl sulfur is the donor to BF_3 in 1:1 adducts such as $CH_3COOCH_3 \cdot BF_3$ [*7611-14-5*], $CH_3COSCH_3 \cdot BF_3$ [*52913-04-9*], and $CH_3CSOCH_3 \cdot BF_3$ [*52912-98-8*] (52).

Compounds containing fluorine and chlorine are also donors to BF_3. Aqueous fluoroboric acid and the tetrafluoroborates of metals, nonmetals, and organic radicals represent a large class of compounds in which the fluoride ion is coordinating with trifluoroborane. Representative examples of these compounds are given in Table 5. Coordination compounds of boron trifluoride with the chlorides of sodium,

Table 4. Boron Trifluoride Adducts with Oxygen-Containing Compounds

Donor	Adduct name	CAS Registry Number	Molecular formula
alcohols	ethanol trifluoroborane	[353-41-3]	$C_2H_5OH \cdot BF_3$
	bis(ethanol) trifluoroborane	[373-59-1]	$2C_2H_5OH \cdot BF_3$
	bis(2-chloroethanol) trifluoroborane	[72985-81-0]	$2ClCH_2CH_2OH \cdot BF_3$
	benzyl alcohol trifluoroborane	[456-31-5]	$C_6H_5CH_2OH \cdot BF_3$
acids	acetic acid trifluoroborane	[753-53-7]	$CH_3COOH \cdot BF_3$
	bis(acetic acid) trifluoroborane	[373-61-5]	$2CH_3COOH \cdot BF_3$
	stearic acid trifluoroborane	[60274-92-2]	$CH_3(CH_2)_{16}COOH \cdot BF_3$
	bis(phenol) trifluoroborane	[462-05-5]	$2C_6H_5OH \cdot BF_3$
ethers	diethyl ether trifluoroborane	[109-63-7]	$(C_2H_5)_2O \cdot BF_3$
	tetrahydrofuran trifluoroborane	[462-34-0]	$(CH_2)_4O \cdot BF_3$
	anisole trifluoroborane	[456-31-5]	$CH_3OC_6H_5 \cdot BF_3$
acid anhydride	acetic anhydride trifluoroborane	[591-00-4]	$(CH_3CO)_2O \cdot BF_3$
esters	ethyl formate trifluoroborane	[462-33-9]	$HCOOC_2H_5 \cdot BF_3$
	phenyl acetate trifluoroborane	[30884-81-6]	$CH_3COOC_6H_5 \cdot BF_3$
ketones	acetone trifluoroborane	[661-27-8]	$(CH_3)_2CO \cdot BF_3$
	benzophenone trifluoroborane	[322-21-4]	$(C_6H_5)_2CO \cdot BF_3$
	acetophenone trifluoroborane	[329-25-9]	$C_6H_5COCH_3 \cdot BF_3$
aldehydes	acetaldehyde trifluoroborane	[306-73-0]	$CH_3CHO \cdot BF_3$
	neopentanal trifluoroborane	[306-78-5]	$(CH_3)_3CCHO \cdot BF_3$
	benzaldehyde trifluoroborane	[456-30-4]	$C_6H_5CHO \cdot BF_3$

Table 5. Boron Trifluoride Adducts with Compounds Containing Chlorine and Fluorine

Name	CAS Registry Number	Molecular formula
potassium tetrafluoroborate	[14075-53-7]	KBF_4
hexaamminenickel(II) tetrafluoroborate	[13877-20-8]	$[Ni(NH_3)_6](BF_4)_2$
nitrosyl tetrafluoroborate	[14635-75-7]	$NOBF_4$
acetylium tetrafluoroborate	[2261-02-1]	CH_3COBF_4
tetramethylammonium tetrafluoroborate	[661-36-9]	$(CH_3)_4NBF_4$
difluorobromine tetrafluoroborate	[14282-83-8]	BrF_2BF_4
anilinium tetrafluoroborate	[15603-97-1]	$C_6H_5NH_2HBF_4$

aluminum, iron, copper, zinc, tin, and lead have been indicated (53); they are probably chlorotrifluoroborates.

Trifluoroborane may form adducts with some of the transition elements. See Reference 54 for a detailed discussion of complexes of trifluoroborane with various Group 6–10 (VI, VII, and VIII) species.

Manufacture. Boron trifluoride is prepared by the reaction of a boron-containing material and a fluorine-containing substance in the presence of an acid. The traditional method used borax, fluorspar, and sulfuric acid.

In another process fluorosulfonic acid is treated with boric acid:

$$3\ HSO_3F + H_3BO_3 \rightarrow BF_3 + 3\ H_2SO_4$$

Numerous other reactions are available for the preparation of small quantities of boron trifluoride, some of which are of high purity (55).

Shipment and Handling. The gas is nonflammable and is shipped in DOT 3A and 3AA steel cylinders at a pressure of approximately 12,410 kPa (1800 psi). Boron trifluoride is classified as a poison gas, both domestically and internationally. Cylinders must have a poison gas diamond and an inhalation hazard warning label. Tube trailers carry both a poison gas placard and an inhalation hazard warning. Cylinders containing 27.2 kg and tube trailers containing 4.5–10 metric tons are available. If boron trifluoride is compressed using oil as a compressor lubricant, it must not be used with oxygen under pressure nor with gauges, valves, or lines that are to be used with oxygen.

Inasmuch as the gas hydrolyzes readily, all equipment should be purged repeatedly using inert dry gas before admitting boron trifluoride. Under anhydrous conditions, carbon steel equipment is satisfactory. Stainless steel and aluminum silicon bronze may also be used. Stainless steel tubing is recommended for both temporary and permanent connections.

In the presence of moisture, boron trifluoride may be handled in polytetrafluoroethylene (PTFE), polyethylene, Pyrex glass (limit to atmospheric pressure), or Hastelloy C containers. At 600°C, stainless steel (304 L) and Hastelloy N are attacked by BF_3; Hastelloy C is more resistant (56). Kel F and PTFE serve as satisfactory gasket and packing materials, whereas rubber, fiber, polymerizable materials, or organic oxygen- and nitrogen-containing compounds must be avoided. Because boron trifluoride is soluble in, and reacts with, many liquids, the gas must not be introduced into any liquid unless a vacuum break or similar safety device is employed.

Economic Aspects, Standards, and Analyses. The sole United States producer of boron trifluoride is AlliedSignal, Inc. The 1992 price of boron trifluoride was $9.59–12.46/kg, depending on purity and the quantity purchased.

Commercial boron trifluoride is usually approximately 99.5% pure. The common impurities are air, silicon tetrafluoride, and sulfur dioxide. An excellent procedure for sampling and making a complete analysis of gaseous boron trifluoride has been developed (57).

Health and Safety Factors. Boron trifluoride is primarily a pulmonary irritant. The toxicity of the gas to humans has not been reported (58), but laboratory tests on animals gave results ranging from an increased pneumonitis to death. The TLV is 1 ppm (59,60). Inhalation toxicity studies in rats have shown that exposure to BF_3 at 17 mg/m^3 resulted in renal toxicity, whereas exposure at 6 mg/m^3 did not result in a toxic response (61). Prolonged inhalation produced dental fluorosis (62). High concentrations burn the skin similarly to acids such as HBF_4 and, if the skin is subject to prolonged exposure, the treatment should be the same as for fluoride exposure and hypocalcemia. No chronic effects have been observed in workers exposed to small quantities of the gas at frequent intervals over a period of years.

Uses. Boron trifluoride is an excellent Lewis acid catalyst for numerous types of organic reactions. Its advantages are ease of handling as a gas and the absence of undesirable tarry by-products. As an electrophilic molecule, it is an excellent catalyst for Friedel-Crafts and many other types of reactions (63–65) (see FRIEDEL-CRAFTS REACTIONS).

$BF_3 \cdot HF$ compositions have been reported to act as super acids in catalyzing condensation reactions (66). BF_3-catalyzed preparation of 1- or 2-naphthol is reported to be regioselective (67). Dehydration reactions may also be regioselective (68). Selected fluorinations may be catalyzed by BF_3 using HF as the fluoride source (69). BF_3 is widely used for the preparation of hydrocarbon resins (70), tall oil (qv) resins (71), and tackifier resins (72). Alpha olefin-based synthetic lubricants are commonly made using BF_3-based catalysts (73–75). BF_3 is widely used as a polymerization catalyst (76–78). A developing use for BF_3 is as an ion implant medium for semiconductor materials (79). BF_3 may be used as a chemical reagent for the manufacture of fluoroboro complexes (80), boron nitride [10043-11-5] (81), and boron trichloride [10294-34-5] (82). Carboxylic acids and esters may be prepared by reacting CO with olefins in the presence of BF_3-containing catalysts (83).

In addition, boron trifluoride and some of its adducts have widespread application as curing agents for epoxy resins (qv), and in preparing alcohol-soluble phenolic resins (qv) (41).

Boron trifluoride catalyst is used under a great variety of conditions either alone in the gas phase or in the presence of many types of promoters. Many boron trifluoride coordination compounds are also used.

Boron trifluoride catalyst may be recovered by distillation, chemical reactions, or a combination of these methods. Ammonia or amines are frequently added to the spent catalyst to form stable coordination compounds that can be separated from the reaction products. Subsequent treatment with sulfuric acid releases boron trifluoride. An organic compound may be added that forms an adduct more stable than that formed by the desired product and boron trifluoride. In another procedure, a fluoride is added to the reaction products to precipitate the boron trifluoride which is then released by heating. Selective solvents may also be employed in recovery procedures (see CATALYSTS, REGENERATION).

Boron trifluoride is also employed in nuclear technology by utilizing several nuclear characteristics of the boron atom. Of the two isotopes, ^{10}B and ^{11}B, only ^{10}B has a significant absorption cross section for thermal neutrons. It is used in $^{10}BF_3$ as a neutron-absorbing medium in proportional neutron counters and for controlling nuclear reactors (qv). Some of the complexes of trifluoroborane have been used for the separation of the boron isotopes and the enrichment of ^{10}B as $^{10}BF_3$ (84).

Boron trifluoride is used for the preparation of boranes (see BORON COMPOUNDS). Diborane is obtained from reaction with alkali metal hydrides; organoboranes are obtained with a suitable Grignard reagent.

Boron trifluoride has been used in mixtures to prepare boride surfaces on steel (qv) and other metals, and as a lubricant for casting steel (see LUBRICATION AND LUBRICANTS).

BIBLIOGRAPHY

"Boron Trifluroide" under "Fluroine Compounds, Inorganic" in *ECT* 1st ed., Vol. 6, pp. 678–684, by D. R. Martin, University of Illinois; "Boron Trifluoride" under "Boron" under "Fluorine Compounds, Inorganic" in *ECT* 2nd ed., Vol. 9, pp. 554–562, by D. R. Martin, The

Harshaw Chemical Co.; in *ECT* 3rd ed., Vol. 10, pp. 685–693 by D. R. Martin, University of Texas at Arlington.

1. J. L. Gay-Lussac and J. L. Thénard, *Rech. Phys.* **2**, 38 (1811); *Ann. Chim Phys.* **69**, 204 (1809).
2. J. A. Nieuwland, R. R. Vogt, and W. L. Foohey, *J. Am. Chem. Soc.* **52**, 1018 (1930).
3. H. C. Brown and R. R. Holmes, *J. Am. Chem. Soc.* **78**, 2173 (1956).
4. A. Oliva, *THEOCHEM 1991*, **82**(1–2), 75–84, 1991.
5. J. C. G. Calado and L. A. K. Staveley, *Trans. Faraday Soc.* **67**, 1261 (1971).
6. E. Pohland and W. Harlos, *Z. Anorg. Allgem. Chem.* **207**, 242 (1932).
7. H. S. Booth and J. M. Carter, *J. Phys. Chem.* **36**, 1359 (1932).
8. R. F. Smith, U.S. Atomic Energy Commission, *NAA-SR-5286*, 1960.
9. C. F. Rumold, PhD. dissertation, Case Western Reserve University, Cleveland, Ohio, 1931.
10. E. Wiberg and W. Mäthing, *Ber. Dtsch. Chem. Ges. B.* **70B**, 690 (1937).
11. A. Eucken and E. Schröder, *Z. Physik. Chem.* **341**, 307 (1938).
12. H. M. Spencer, *J. Chem. Phys.* **14**, 729 (1946).
13. D. R. Stull and H. Prophet, *Natl. Stand. Ref. Data Ser. Natl. Bur. Stand.* **37** (1971).
14. T. P. Onak and co-workers, *J. Phys. Chem.*, 63 (1959).
15. M. F. Lappert and co-workers, *J. Chem. Soc. A.*, 2426 (1971).
16. T. D. Coyle, S. L. Stafford, and F. G. Stone, *J. Chem. Soc.*, 3103 (1961).
17. T. D. Coyle and F. G. A. Stone, *J. Chem. Phys.* **32**, 1892 (1960); I. S. Jaworiwsky and co-workers, *Inorg. Chem.* **18**, 56 (1979).
18. *Gmelins Handbuch der Anorganischen Chemie*, Vol. 13, 8th ed., Verlag Chemie, GmbH, Weinheim/Bergstrasse, Germany, 1954, pp. 167–196.
19. S. Pawlenko, *Z. Anorg. Allegem. Chem.* **300**, 152 (1959).
20. N. N. Greenwood and A. Thompson, *J. Chem. Soc.*, 3643 (1959).
21. H. Gerding and co-workers, *Rec. Trav. Chim.* **71**, 501 (1952).
22. N. N. Greenwood and A. Thompson, *J. Chem. Soc.*, 3493 (1959).
23. R. J. Mikovsky, S. D. Levy, and A. L. Hensley, Jr., *J. Chem. Eng. Data* **6**, 603 (1961).
24. E. C. Hughes and S. M. Darling, *Ind. Eng. Chem.* **43**, 746 (1951).
25. H. S. Booth and D. R. Martin, *J. Am. Chem. Soc.* **64**, 2198 (1942).
26. K. L. Khachishvile and co-workers, *Zh. Neorg. Khim.* **6**, 1493 (1961).
27. P. Pichat, *C. R. Acad. Sci. Paris Ser. C* **265**, 385 (1967).
28. R. N. Scott and D. F. Shriver, *Inorg. Chem.* 5, 158 (1966).
29. P. Baumgarten and W. Bruns, *Ber. Dtsch. Chem. Ges. B.* **B72**, 1753 (1939); *Ibid.* **B74**, 1232 (1941).
30. H. D. Fishcher, W. J. Lehmann, and I. Shapiro, *J. Phys. Chem.* **65**, 1166 (1961).
31. H. I. Schlesinger and co-workers, *J. Am. Chem. Soc.* **75**, 195 (1953).
32. *Ibid.*, p. 199.
33. E. L. Gamble, *Inorg. Synth.* **3**, 27 (1950).
34. Brit. Pat. 226,490 (Dec. 20, 1923), A. F. Meyerhofer.
35. E. Krause and R. Nitsche, *Chem. Ber.* **54B**, 2784 (1921).
36. H. C. Brown and U. S. Racherla, *Organometallics 1986* **5**(2), 391–393 (1986).
37. B. M. Mikhailov and T. A. Schhegoleva, *J. Gen. Chem. U.S.S.R.* **29**, 3404 (1959).
38. G. T. Morgan and R. B. Tunstall, *J. Chem. Soc.* **125**, 1963 (1924).
39. W. I. Aalbersberg and co-workers, *J. Chem. Soc.*, 3055 (1959).
40. W. V. Bush and L. Zechmeister, *J. Am. Chem. Soc.* **80**, 2991 (1958).
41. H. S. Booth and D. R. Martin, *Boron Trifluoride and Its Derivatives*, John Wiley & Sons, Inc., New York, 1949.
42. P. Baumgarten and H. Henning, *Chem. Ber.* **72B**, 1743 (1939).
43. C. Gascard and G. Mascherpa, *J. Chim. Phys. Phys. Chim. Biol.* **70**, 1040 (1973).
44. R. J. Gillespie and J. L. Hartman, *Can. J. Chem.* **45**, 859 (1967).

45. H. S. Booth and D. R. Martin, *Boron Trifluoride and its Derivatives*, John Wiley & Sons, Inc., New York, 1948.
46. D. Farcasiu and A. Ghenciu, *J. Catal.* **134**(1), 126–133 (1992).
47. J. S. McGrath and co-workers, J.A.C.S., *66*, 126 (1944)
48. C. A. Wamser, *J. Am. Chem. Soc.* **73**, 409 (1951).
49. C. A. Wamser, *J. Am. Chem. Soc.* **70**, 1209 (1948)
50. V. I. Durkov and S. S. Batsanov, *Zh. Strukt. Khim.* **2**, 456 (1961).
51. E. F. Mooney and M. A. Qaseem, *Chem. Commun.*, 230 (1967).
52. M. J. Bula, J. S. Hartman, and C. V. Raman, *J. Chem. Soc. Dalton Trans.*, 725 (1974).
53. Brit. Pat 486,887 (June 13, 1938), (to E. I. du Pont de Nemours & Co., Inc.).
54. D. R. Martin and J. M. Canon, in G. A. Olah, ed., *Friedel-Crafts and Related Reactions*, Vol. 1, Wiley-Interscience, New York, 1963, pp. 399–567.
55. H. S. Booth and K. S. Wilson, *Inorg. Synth.* **1**, 21 (1939).
56. J. W. Koger, Oak Ridge National Laboratory, TM-4172, 1972; *Nucl. Sci. Abstr.* **28**, 11,211 (1973).
57. C. F. Swinehart, A. R. Bumblish, and H. F. Flisik, *Anal. Chem.* **19**, 28 (1947); *Ann. Proc.* 35-0049, internal document, Harshaw Chemical Co., Mar. 23, 1964.
58. K. H. Jacobson, R. A. Rhoden, and R. L. Roudabush, *HEW Pub. (NIOSH) Publ. 77*, (1976).
59. Code of Fed. Reg. 29, part. 1901, U.S. Govt. Printing Office, Washington, D.C., 1988.
60. A.C.G.I.H., *Threshold Limit Values for Chemical Substances*, 1989–1990.
61. G. M. Rusch and co-workers, *Toxicology and Applied Pharmacology* **83**, 69–78 (1986).
62. C. J. Spiegl, *Natl. Nucl. Energy Ser. Div. VI 1 (Book 4)*, 2291 (1953).
63. G. A. Olah, ed., in Ref. 54, pp. 228–235.
64. Ref. 41, Chapt. 6.
65. A V. Topchiev, S. V. Zavgorodnii, and Y. M. Paushkin, *Boron Fluoride and Its Compounds as Catalysts in Organic Chemistry*, Pergamon Press, New York, 1959.
66. Fr. Pat. 2,647,108 (Nov. 23, 1990), L. Gilbert and co-workers (to Rhône-Poulenc).
67. U.S. Pat. 4,419,528 (Dec. 6, 1983), G. A. Olah (to PCUK Ugine Kuhlman).
68. G. H. Posner and co-workers, *Tetrahedran Lett.* **32**(45) 6489–6492 (1991).
69. U.S. Pat. 4,962,244 (Oct. 9, 1990), M. Y. Elsheikh (to Atochem, N. Amer. Inc.).
70. U.S. Pat. 4,657,773 (Apr. 14, 1987), S. C. Durkee (to Hercules Inc.)
71. U.S. Pat. 4,657,706 (Apr. 14, 1987), S. C. Durkee (to Hercules Inc.)
72. U.S. Pat. 5,051,485 (Sept. 24, 1991), J. J. Schmid and J. W. Booth (to Arizona Chem.)
73. U.S. Pat. 4,434,309 (Feb. 28, 1984), J. M. Larkin and W. H. Brader (to Texaco Inc.)
74. U.S. Pat. 4,484,014 (Nov. 20, 1984), W. I. Nelson and co-workers (to Phillips Pet. Co.)
75. U.S. Pat. 4,935,570 (June. 19, 1990), M. B. Nelson and co-workers (to Ethyl Corp.)
76. U.S. Pat. 5,068,490 (Feb. 29, 1988), B. E. Eaton (to Amoco Corp.)
77. U.S. Pat. 5,071,812 (Mar. 31, 1989), D. R. Kelsey (to Shell Oil Co.)
78. M. C. Throckmorton, *J. Appl. Polym. Sci.* **42**(11), 3019–3024 (1991).
79. M. H. Juang and H. C. Cheng, *J. Appl. Phys.* **71**(3), 1265–1270 (1992).
80. B. K. Mohapatra and co-workers, *Indian. J. Chem., Sect. A* **30A**(11), 944–947 (1991).
81. W. Ahmed and co-workers, *J. Phys. IV*, **1**(C2) 119–126 (1991).
82. Jpn. Pat. 03,218,917[91,218,917] (Sept. 26, 1991), (to Hashimoto Chem. Ind. Co. Ltd.).
83. U.S. Pat. 5,034,368 (July 23, 1991), E. Drent (to Shell Int. Res. MIJ BV).
84. A. A. Palko and J. S. Drury, *J. Chem. Phys.* **47**, 2561 (1967).

FRANCIS EVANS
GANPAT MANI
AlliedSignal, Inc.

FLUOROBORIC ACID AND FLUOROBORATES

Fluoroboric Acid and the Fluoroborate Ion

Fluoroboric acid [16872-11-0], generally formulated as HBF_4, does not exist as a free, pure substance. The acid is stable only as a solvated ion pair, such as $H_3O^+BF_4^-$; the commercially available 48% HBF_4 solution approximates H_3O^+ $BF_4^- \cdot 4H_2O$. Other names used infrequently are hydrofluoroboric acid, hydroborofluoric acid, and tetrafluoroboric acid. Salts of the acid are named as fluoroborates or occasionally borofluorides. Fluoroboric acid and its salts were investigated as early as 1809 (1,2). The acid and many transition-metal salts are used in the electroplating (qv) and metal finishing industries. Some of the alkali metal fluoroborates are used in fluxes.

   Properties. Fluoroboric acid is stable in concentrated solutions, and hydrolyzes slowly in aqueous solution to hydroxyfluoroborates. For the stability of the fluoroborate species, see Reference 3. The equilibrium quotients Q (4,5) in 1 molal NaCl at 25°C show the strong affinity of boron for fluoride:

$$B(OH)_3 + F^- \rightleftharpoons BF(OH)_3^- \qquad\qquad \log Q = -0.36 \pm 0.19$$

$$B(OH)_3 + 2\ F^- + H^+ \rightleftharpoons BF_2(OH)_2^- + H_2O \quad \log Q = 7.06 \pm 0.02$$

$$B(OH)_3 + 3\ F^- + 2\ H^+ \rightleftharpoons BF_3OH^- + 2\ H_2O \quad \log Q = 13.689 \pm 0.003$$

$$B(OH)_3 + 4\ F^- + 3\ H^+ \rightleftharpoons BF_4^- + 3\ H_2O \qquad \log Q = 19.0 \pm 0.1$$

The hydrolysis of BF_4^- occurs stepwise to BF_3OH^-, $BF_2(OH)_2^-$, and $BF(OH)_3^-$. By conductivity measurements the reaction of boric acid and HF was found to form $H[BF_3(OH)]$ [15433-40-6] rapidly; subsequently HBF_4 formed much more slowly from HBF_3OH. These studies demonstrate that BF_4^- is quite stable to hydrolysis yet is slow to form from BF_3OH^- and HF:

$$BF_4^- + H_2O \rightleftharpoons BF_3OH^- + HF$$

Kinetic results (5) and ^{19}F nmr experiments (6) illustrate clearly that the hydroxyfluoroborates are in rapid equilibrium and easily exchange fluoride.

   Table 1 lists some of the physical properties of fluoroboric acid. It is a strong acid in water, equal to most mineral acids in strength and has a pK_{H_2O} of -4.9 as compared to -4.3 for nitric acid (9). The fluoroborate ion contains a nearly tetrahedral boron atom with almost equidistant B–F bonds in the solid state. Although lattice effects and hydrogen bonding distort the ion, the average B–F distance is 0.138 nm; the F–B–F angles are nearly the theoretical 109° (10,11). Raman spectra on molten, ie, liquid $NaBF_4$ agree with the symmetrical tetrahedral structure (12).

   The fluoroborate ion has traditionally been referred to as a noncoordinating anion. It has shown little tendency to form a coordinate–covalent bond with transition metals as do nitrates and sulfates. A few exceptional cases have been reported (13) in which a coordinated BF_4^- was detected by infrared or visible spectroscopy.

Table 1. Physical Properties of Fluoroboric Acid

Property	Value	Reference
heat of formation, kJ/mol[a]		
aqueous, 1 molal, at 25°C	−1527	
from boric oxide and HF (aq)	−123.34	2
BF_4^-, gas	−1765 ± 42	7
entropy of the BF_4^- ion, J/(mol·K)[a]	167	
specific gravity		
48% soln	1.37	
42% soln	1.32	
30% soln	1.20	
surface tension, 48% soln at 25°C, mN/m(=dyn/cm)	65.3	
ir absorptions,[b] cm^{-1}	ca 1100	8
	ca 530	

[a]To convert J to cal, divide by 4.184.
[b]Generally observed as strong absorptions.

Hydroxyfluoroborates are products of the reaction of BF_3 with water; $BF_3 \cdot 2H_2O$ [13319-75-0] is actually $H_3O^+BF_3OH^-$. Salts such as sodium hydroxyfluoroborate [13876-97-6], $NaBF_3OH$, are made by neutralizing the acid. The BF_3OH^- anions are distorted tetrahedra (14). In the HBO_2–HF system, $HBO_2 \cdot 2HF$ was found to be $HBF_2(OH)_2$, dihydroxyfluoroboric acid [17068-89-2] (15).

Manufacture, Shipping, and Waste Treatment. Fluoroboric acid (48%) is made commercially by direct reaction of 70% hydrofluoric acid and boric acid, H_3BO_3 (see BORON COMPOUNDS). The reaction is exothermic and must be controlled by cooling.

The commercial product is usually a 48–50% solution which contains up to a few percent excess boric acid to eliminate any HF fumes and to avoid HF burns. Reagent-grade solutions are usually 40%. A 61% solution can be made from metaboric acid, HBO_2, and 70% HF, and a lower grade by direct combination of fluorospar, CaF_2, sulfuric acid, and boric acid (16). The product contains a small amount of dissolved calcium sulfate. A silica-containing (0.11% SiO_2) fluoroboric acid is produced from inexpensive fluorosilicic acid (17). Boric acid is added to a 10% H_2SiF_6 solution and then concentrated in several steps to 45% HBF_4. Granular silicon dioxide must be filtered from the product.

Vessels and equipment must withstand the corrosive action of hydrofluoric acid. For a high quality product the preferred materials for handling HBF_4 solutions are polyethylene, polypropylene, or a resistant rubber such as neoprene (see ELASTOMERS, SYNTHETIC). Where metal must be used, ferrous alloys having high nickel and chromium content show good resistance to corrosion. Impregnated carbon (Carbate) or Teflon can be used in heat exchangers. Teflon-lined pumps and auxilliary equipment are also good choices. Working in glass equipment is not recommended for fluoroboric acid or any fluoroborate.

Fluoroboric acid and some fluoroborate solutions are shipped as corrosive material, generally in polyethylene-lined steel pails and drums or in rigid nonreturnable polyethylene containers. Acid spills should be neutralized with lime or soda ash.

Waste treatment of fluoroborate solutions includes a pretreatment with aluminum sulfate to facilitate hydrolysis, and final precipitation of fluoride with lime (18). The aluminum sulfate treatment can be avoided by hydrolyzing the fluoroborates at pH 2 in the presence of calcium chloride; at this pH, hydrolysis is most rapid at elevated temperature (19).

Economic Aspects. In the United States fluoroboric acid is manufactured by Atotech USA, Inc., General Chemical, C.P. Chemical Co., Fidelity Chemical Products, and Chemtech Harstan. Research quantities of reagent grade are made by Advance Research Chemical Co., Johnson-Mathey, and Ozark-Mahoning Co. The price for 48% fluoroboric acid in truckload quantities in 1993 was $2.13–2.25/kg (20).

Many specialty fluoroborates are available in research quantities from Advance Research Chemicals.

Analysis. Fluoroboric acid solutions and fluoroborates are analyzed gravimetrically using nitron or tetraphenylarsonsium chloride. A fluoroborate ion-selective electrode has been developed (21).

Toxicity. Fluoroborates are excreted mostly in the urine (22). Sodium fluoroborate is absorbed almost completely into the human bloodstream and over a 14-d experiment all of the $NaBF_4$ ingested was found in the urine. Although the fluoride ion is covalently bound to boron, the rate of absorption of the physiologically inert BF_4^- from the gastrointestinal tract of rats exceeds that of the physiologically active simple fluorides (23).

Uses. Printed circuit tin–lead plating is the main use of fluoroboric acid (24). However, the Alcoa Alzak process for electropolishing aluminum requires substantial quantities of fluoroboric acid. A 2.5% HBF_4 solution is used to produce a highly reflective surface (25). The high solubility of many metal oxides in HBF_4 is a decided advantage in metal finishing operations (see METAL SURFACE TREATMENTS). Before plating or other surface treatment, many metals are cleaned and pickled in fluoroboric acid solution; eg, continuous strip pickling of hot-rolled low carbon steel is feasible in HBF_4 solutions (26). Nontempered rolled steel requires 80°C for 60 s in HBF_4 130 g/L, whereas tempered rolled steel requires only 65°C for 60 s in 65 g/L. The spent pickling solution is recovered by electrodialysis.

Fluoroboric acid is used as a stripping solution for the removal of solder and plated metals from less active substrates. A number of fluoroborate plating baths (27) require pH adjustment with fluoroboric acid (see ELECTROPLATING).

A low grade fluoroboric acid (16) is used in the manufacture of cryolite (28) for the electrolytic production of aluminum:

$$4 \ Na_2SO_4 \cdot NaF + 5 \ HBF_4 + 2 \ Al_2O_3 + 9 \ H_2O \rightarrow 4 \ Na_3AlF_6 + 5 \ H_3BO_3 + 4 \ H_2SO_4$$

The boric and sulfuric acids are recycled to a HBF_4 solution by reaction with CaF_2. As a strong acid, fluoroboric acid is frequently used as an acid catalyst, eg, in synthesizing mixed polyol esters (29). This process provides an inexpensive route to confectioner's hard-butter compositions which are substitutes for cocoa butter in chocolate candies (see CHOCOLATE AND COCOA). Epichlorohydrin is polymerized in the presence of HBF_4 for eventual conversion to polyglycidyl ethers (30) (see CHLOROHYDRINS). A more concentrated solution, 61–71% HBF_4, catalyzes the addition of CO and water to olefins under pressure to form neo acids (31) (see CARBOXYLIC ACIDS).

Main Group

Properties.   A summary of the chemical and physical properties of alkali-metal and ammonium fluoroborates is given in Tables 2 and 3. Chemically these compounds differ from the transition-metal fluoroborates usually separating in anhydrous form. This group is very soluble in water, except for the K, Rb, and Cs salts which are only slightly soluble. Many of the soluble salts crystallize as hydrates.

Lithium fluoroborate crystallizes from aqueous solutions as $LiBF_4 \cdot 3H_2O$ [39963-05-8] and $LiBF_4 \cdot H_2O$ [39963-03-6]. The heat of dehydration of the monohydrate at 91°C is 70.9 kJ/mol (16.95 kcal/mol); the melting point is 117°C (45). Magnesium, calcium, strontium, and barium fluoroborates crystallize as hydrates: $Mg(BF_4)_2 \cdot 6H_2O$ [19585-07-0], $Ca(BF_4)_2 \cdot 2H_2O$ [27860-81-7], $Sr(BF_4)_2 \cdot 4H_2O$ [27902-05-2], and $Ba(BF_4)_2 \cdot 2H_2O$ [72259-09-7], respectively. These hydrated fluoroborates can be dehydrated completely to the anhydrous salts, which show decreasing stabilities: Ba > Sr > Ca > Mg.

The anhydrous magnesium salt is least stable thermally. It forms MgF_2, which has the highest lattice energy. This has been confirmed by differential thermal analysis (dta) of the crystalline hydrates (46). Aluminum fluoroborate [14403-54-4], $Al(BF_4)_3 \cdot (H_2O)_n$, is soluble in strongly acid solutions and displays a tendency for fluoride exchange with BF_4^- to form aluminum fluorides. The aluminyl compound, $AlO^+BF_4^-$, is extremely hygroscopic and is prepared by the reaction of AlOCl, BF_3, and HF (47). Differential thermal analysis experiments show thermal decomposition beginning at 85°C, corresponding to removal of BF_3 and formation of AlOF.

Differential thermal analysis studies of ammonium fluoroborate showed the orthorhombic to cubic transition at 189 ± 5°C and BF_3 generation from 389 to 420°C (48). Sodium hydroxide reacts with NH_4BF_4 liberating ammonia and forming $NaBF_4$. When sodium fluoroborate was studied by infrared spectroscopy, sodium hydroxyfluoroborate, $NaBF_3OH$, was found to be present (49). Although pure sodium hydroxyfluoroborate is thermally unstable, decomposing to $Na_2B_2F_6O$ [18953-03-2] and H_2O, in a melt of $NaBF_4^-NaF$ no instability of the small amount of $NaBF_3OH$ present was detected. Fusion of $NaBF_4$ or KBF_4 with boric oxide generates BF_3 and complex borates such as KFB_4O_6 (50). Most fluoroborates decompose readily to give BF_3 when treated with sulfuric acid or when calcined (see Table 3 for dissociation pressure). Under strongly basic conditions the chemical equilibrium is shifted away from BF_4^- to borates and fluorides.

Manufacture.   Fluoroborate salts are prepared commercially by several different combinations of boric acid and 70% hydrofluoric acid with oxides, hydroxides, carbonates, bicarbonates, fluorides, and bifluorides. Fluoroborate salts are substantially less corrosive than fluoroboric acid but the possible presence of HF or free fluorides cannot be overlooked. Glass vessels and equipment should not be used.

Table 2. General Properties of Metal Fluoroborates

Compound	CAS Registry Number	Molecular weight	Color	Physical form	Mp, °C	Density,[a] g/cm³	Solubility H$_2$O g/100 mL[b]	Solubility Other	References
LiBF$_4$	[14283-07-9]	93.74	white				very soluble		2,32
NaBF$_4$	[13755-29-8]	109.79	white	orthorhombic <240°C $a = 0.68358, b = 0.62619, c = 0.67916$ nm noncubic >240°C	406 dec	2.47 210[c]	108 (26°C)	sl alcohol	13,32,33
KBF$_4$	[14075-53-7]	125.92	colorless	rhombic <283°C $a = 0.7032, b = 0.8674, c = 0.5496$ nm cubic >283°C	530 dec	2.498	0.45 (20°C) 6.27 (100°C)	sl ethanol insol alkali	32–34
RbBF$_4$	[18909-68-7]	172.27		orthorhombic <245°C $a = 0.7296, b = 0.9108, c = 0.5636$ nm cubic >245°C	612 dec	2.820 10[c]	0.6 (17°C)		32–34
CsBF$_4$	[18909-69-8]	219.71	white	orthorhombic <140°C $a = 0.7647, b = 0.9675, c = 0.5585$ nm cubic >140°C	555 dec	3.20 30[c]	1.6 (17°C)		32–34
NH$_4$BF$_4$	[13826-83-0]	104.84	white	orthorhombic <205°C $a = 0.7278, b = 0.9072, c = 0.5678$ nm	487 dec	1.871[d]	3.09 (−1.0°C) 5.26 (−1.5°C) 10.85 (−2.7°C) 12.20 (0°C) 25 (16°C) 25.83 (25°C) 44.09 (50°C) 67.50 (75°C) 98.93 (100°C) 113.7 (108.5°C)	HF[e]	32,34,35 36
NaBF$_3$OH	[13876-97-6]			hexagonal $a = 0.8084, c = 0.7958$ nm		2.46			10

[a] Unless otherwise stated, at 20°C. [b] Temperature given in parentheses. [c] At 100°C. [d] At 15°C. [e] Value at 0°C is 19.89%.

Table 3. Thermodynamic Data^a for Metal Fluoroborates, kJ/mol^b

Compound	ΔH_{diss}	Lattice energy, $-U$	ΔH_{fus}	ΔH_f	Other	$\log P_{Pa} = -aT^{-1} + b$			References
						a	b^c	T, °C	
LiBF$_4$	15.9	699		−1838.4	$\Delta H^d = -89.54$	833	6.40	210–320	37–39
NaBF$_4$	69.83	657.3	13.6	−1843.5	$\Delta H^e = -134.1$	3650	8.75	400–700	33, 38–40
KBF$_4$	121	598	18.0	−1881.5	$\Delta H^f = -180.5$	6317	8.15	510–830	33, 38, 39, 41, 42
					$\Delta H_{sub} = 330$				
					$S = 130^g$				
					$C_p = 112.1^g$				
RbBF$_4$	112.8	577	19.6			5960	9.57	600–1000	33, 38, 41
CsBF$_4$	112.5	556	19.2			5880	9.47	610–1040	33, 38, 41
NaBF$_3$OH	77.0			−1754		4024	9.11	400–700	40
NH$_4$BF$_4$		607^h			$\Delta H_{sub} = 47.3$	2469	8.94		43, 44

$^a \Delta H_{diss}$ = heat of dissociation, ΔH_{fus} = heat of fusion, ΔH_f = heat of formation, ΔH_{sub} = sublimation. All thermodynamic data at 25°C, unless otherwise stated.

^{b} To convert J to cal, divide by 4.184.

^{c} To convert $\log P_{Pa}$ to $\log P_{mm\ Hg}$, subtract 2.12 from b.

^{d} LiF(s) + BF$_3$(g) $\rightarrow$ LiBF$_4$(s).

^{e} NaF(s) + BF$_3$(g) $\rightarrow$ NaBF$_4$(s).

^{f} KF(s) + BF$_3$(g) $\rightarrow$ KBF$_4$(s).

^{g} Units are in J/(mol·K).

^{h} At 260°C.

Sodium Fluoroborate. Sodium fluoroborate is prepared by the reaction of NaOH or Na_2CO_3 with fluoroboric acid (51), or by treatment of disodium hexafluorosilicate with boric acid.

Potassium Fluoroborate. Potassium fluoroborate is produced as a gelatinous precipitate by mixing fluoroboric acid and KOH or K_2CO_3. Alternatively, fluorosilicic acid is treated with H_3BO_3 in a 2:1 molar ratio to give HBF_3OH, which reacts with HF and KCl to yield 98% of KBF_4 in 98.5% purity (52). Commercial KBF_4 normally contains less than 1% KBF_3OH.

Ammonium and Lithium Fluoroborates. Ammonia reacts with fluoroboric acid to produce ammonium fluoroborate (53). An alternative method is the fusion of ammonium bifluoride and boric acid (54):

$$2\ NH_4HF_2 + H_3BO_3 \rightarrow NH_4BF_4 + 3\ H_2O + NH_3$$

The water and ammonia must be removed from the melt. Lithium hydroxide or carbonate react with HBF_4 to form $LiBF_4$.

Magnesium Fluoroborate. Treatment of magnesium metal, magnesium oxide, or magnesium carbonate with HBF_4 gives magnesium fluoroborate [14708-13-5]. The MgF_2 is filtered and the product is sold as a 30% solution.

Economic Aspects. In the United States the sodium, potassium, ammonium, and magnesium fluoroborates are sold by Advance Research Chemicals, Atotech USA, Inc., and General Chemical. The lithium compound is available from Advance Research Chemicals, Cyprus Foote Mineral, and FMC Lithium Corp. of America. Small amounts of other fluoroborates are sold by Alfa Inorganics, Inc. and Ozark-Mahoning Co. Prices in 1993 for truckload quantities were $NaBF_4$, \$4.95–6.25/kg; KBF_4, \$3.55/kg; and NH_4BF_4 \$5.03–6.35/kg.

Uses. Alkali metal and ammonium fluoroborates are used mainly for the high temperature fluxing action required by the metals processing industries (see METAL SURFACE TREATMENTS; WELDING). The tendency toward BF_3 dissociation at elevated temperatures inhibits oxidation in magnesium casting and aluminum alloy heat treatment.

The molten salts quickly dissolve the metal oxides at high temperatures to form a clean metal surface. Other uses are as catalysts and in fire-retardant formulations (see FLAME RETARDANTS).

Potassium Fluoroborate. The addition of potassium fluoroborate to grinding wheel and disk formulations permits lower operating temperatures (55). Cooler action is desirable to reduce the burning of refractory materials such as titanium and stainless steels. Excellent results in grinding wheels are also obtained with $NaBF_4$ (56). A process for boriding steel surfaces using B_4C and KBF_4 as an activator improves the hardness of the base steel (57). Fluxes for aluminum bronze and silver soldering and brazing contain KBF_4 (58) (see SOLDERS). Fire retardance is imparted to acrylonitrile polymers by precipitating KBF_4 within the filaments during coagulation (59). In polyurethanes, KBF_4 and NH_4BF_4 reduce smoke and increase flame resistance (60). Both the potassium and ammonium salts improve insulating efficiency of intumescent coatings (61). The endothermic characteristics of these fillers (qv) (release of BF_3) counteract the exothermic nature of the intumescent agents (nitroaromatic amines) in the coating. The sodium and potassium salts are claimed to have a synergistic effect with polyhalogenated aromatics that improve flame-retardant properties of polyesters (62). Elemental boron is prepared by the Cooper electrolysis of a KBF_4 melt with B_2O_3 and KCl (63). The boron may be up to 99.5% purity and, if KBF_4 containing the ^{10}B isotope is used, the product is ^{10}B which is used in the nuclear energy field as a neutron absorber (see NUCLEAR REACTORS).

Sodium Fluoroborate.   Sodium fluoroborate can be used in the transfer of boron to aluminum alloys but the efficiency is lower than for KBF_4 (64). Sodium fluoroborate in an etching solution with sulfamic acid and H_2O_2 aids in removing exposed lead in printed circuit manufacture (65). During the annealing of galvanized iron (galvannealing), the surface becomes oxidized. The resulting oxide coating, which causes difficulty in soldering, can be removed by aqueous $NaBF_4$ or NH_4BF_4 (66). Work at Oak Ridge National Lab (Tennessee) has shown that a $NaBF_4$, with 8 mol % NaF, salt mixture could be used as the coolant in the molten breeder reactor (67); in this molten salt at nearly 600°C the corrosion rate of Hastelloy N is about 8 μm/yr. Sodium fluoroborate acts as a catalyst for cross-linking cotton cellulose with formaldehyde (68); transesterification in the preparation of polycarbonates (69); and preparation of cyclic oligoethers from ethylene oxide (70). Sodium and lithium fluoroborates are effective flame retardants for cotton and rayon (71).

Ammonium Fluoroborate.   Ammonium fluoroborate blends with antimony oxide give good results in flame-retarding polypropylene (72). The complete thermal vaporization makes ammonium fluoroborate an excellent gaseous flux for inert-atmosphere soldering (73). A soldering flux of zinc chloride and ammonium fluoroborate is used in joining dissimilar metals such as Al and Cu (74). Ammonium fluoroborate acts as a solid lubricant in cutting-oil emulsions for aluminum rolling and forming.

Lithium Fluoroborate.   Lithium fluoroborate is used in a number of batteries (qv) as an electrolyte, for example in the lithium–sulfur battery (75).

Miscellaneous.   Flame-resistant cross-linked polyethylene can be made with a number of fluoroborates and antimony oxide. This self-extinguishing material may contain the fluoroborates of NH_4^+, Na^+, K^+, Ca^{2+}, Mg^{2+}, Sr^{2+}, or Ba^{2+} in amounts of 4–20% (76). Magnesium fluoroborate catalyzes the epoxy treatment of cotton fabrics for permanent-press finishes (77) (see TEXTILES).

Transition-Metal and Other Heavy-Metal Fluoroborates

The physical and chemical properties are less well known for transition metals than for the alkali metal fluoroborates (Table 4). Most transition-metal fluoroborates are strongly hydrated coordination compounds and are difficult to dry without decomposition. Decomposition frequently occurs during the concentration of solutions for crystallization. The stability of the metal fluorides accentuates this problem. Loss of HF because of hydrolysis makes the reaction proceed even more rapidly. Even with low temperature vacuum drying to partially solve the decomposition, the dry salt readily absorbs water. The crystalline solids are generally soluble in water, alcohols, and ketones but only poorly soluble in hydrocarbons and halocarbons.

Table 4. Properties of Metal Fluoroborates[a]

Compound	CAS Registry Number	Color	Specific gravity	Solubility	Miscellaneous
$Mn(BF_4)_2 \cdot 6H_2O$	[26044-57-5]	pale pink	1.982	water, ethanol	
$Fe(BF_4)_2 \cdot 6H_2O$	[13877-16-2]	pale green	2.038	water, ethanol	
$Co(BF_4)_2 \cdot 6H_2O$	[15684-35-2]	red	2.081	water, alcohol	
$Ni(BF_4)_2 \cdot 6H_2O$	[14708-14-6]	green	2.136	water, alcohol	
$Cu(BF_4)_2 \cdot 6H_2O$	[72259-10-0]	blue	2.175	water, alcohol	
$AgBF_4 \cdot H_2O$	[72259-11-1]	colorless		water, less sol in alcohol, sol benzene, sol ether	dec 200°C, light sensitive
$Zn(BF_4)_2 \cdot 6H_2O$	[27860-83-9]	white	2.120	water, alcohol	dehydrates at 60°C
$Cd(BF_4)_2 \cdot 6H_2O$	[27860-84-0]	white	2.292	water, alcohol	
$In(BF_4)_3 \cdot xH_2O$	[27765-48-6]	colorless		water	
$TlBF_4 \cdot H_2O$	[72259-12-2]	colorless		water	orthorhombic, $a = 0.947$, $b = 0.581$, $c = 0.740$ nm, light sensitive
$Sn(BF_4)_2 \cdot xH_2O$	[72259-13-3]	white		water	$Sn(BF_4)_2 \cdot SnF_2 \cdot 5H_2O$ crystallizes from soln
$Pb(BF_4)_2 \cdot H_2O$	[26916-34-7]	colorless			

[a]Crystalline solids (2,78).

Differential thermal analysis in air on the crystalline hexahydrates of Zn, Cd, Fe, Co, and Ni fluoroborates show the loss of BF_3 and H_2O simultaneously at 195, 215, 180, 185, and 205°C, respectively (46,79). The dta curves also indicate initial melting at 107, 117, and 150°C for Zn, Cd, and Fe fluoroborates, respectively. The anhydrous metal fluoride and/or oxide is usually isolated. The copper salt also decomposes with liberation of BF_3 and H_2O (80).

The water of hydration of these complexes can be replaced with other coordinating solvents. For example, the ethanol and methanol solvates were made by dissolving the hydrates in triethyl and trimethyl orthoformate, respectively (81,82). The acetic acid solvates are made by treating the hydrates with acetic anhydride (83). Conductivity and visible spectra, where applicable, of the Co, Ni, Zn, and Cu fluoroborates in N,N-dimethylacetamide (L) showed that all metal ions were present as the ML_6^{2+} cations (84). Solvated fluoroborate complexes of Cr^{3+}, Fe^{2+}, Co^{2+}, Ni^{2+}, Cu^{2+}, Cu^+, and Zn^{2+} in diethyl ether, nitromethane, and benzene solutions have been prepared. Solutions of $Ti(BF_4)_3$, $V(BF_4)_3$, and $Fe(BF_4)_3$ could not be prepared probably because of formation of BF_3 and the metal fluoride (85). Ammonia easily replaces the coordinated water; the products are usually tetrammine or hexammine complexes (2) (see COORDINATION COMPOUNDS). The hexahydrate of $Ni(BF_4)_2$ was found to be stable from 25 to 100°C; solubility also was determined to 95°C (86). At 120°C the solid decomposed slowly to NiF_2 with loss of HF, H_3BO_3, and H_2O.

Manufacture. The transition- and heavy-metal fluoroborates can be made from the metal, metal oxide, hydroxide, or carbonate with fluoroboric acid. Because of the difficulty in isolating pure crystalline solids, these fluoroborates are usually available as 40–50% solutions, $M(BF_4)_x$. Most of the solutions contain about 1–2% excess fluoroboric acid to prevent precipitation of basic metal complexes. The solutions are usually sold in 19 and 57 L polyethylene containers.

In some cases, particularly with inactive metals, electrolytic cells are the primary method of manufacture of the fluoroborate solution. The manufacture of Sn, Pb, Cu, and Ni fluoroborates by electrolytic dissolution (87,88) is patented. A typical cell for continous production consists of a polyethylene-lined tank with tin anodes at the bottom and a mercury pool (in a porous basket) cathode near the top (88). Fluoroboric acid is added to the cell and electrolysis is begun. As tin fluoroborate is generated, differences in specific gravity cause the product to layer at the bottom of the cell. When the desired concentration is reached in this layer, the heavy solution is drawn from the bottom and fresh HBF_4 is added to the top of the cell continuously. The direct reaction of tin with HBF_4 is slow but can be accelerated by passing air or oxygen through the solution (89). The stannic fluoroborate is reduced by reaction with mossy tin under an inert atmosphere. In earlier procedures, HBF_4 reacted with hydrated stannous oxide.

Anhydrous silver fluoroborate [1404-20-2] is made by the addition of BF_3 gas to a suspension of AgF in ethylbenzene (90). An $AgBF_4 \cdot C_8H_{10}$ complex is precipitated with pentane and the complex is washed with pentane to give anhydrous $AgBF_4$.

Economic Aspects.   Most fluoroborate solutions listed in Table 5 are manufactured by Atotech USA, Inc., General Chemical, Chemtec/Harstan, C.P. Chemical Co., and Fidelity Chemical Products. Prices are shown in Table 5.

Uses.   Metal fluoroborate solutions are used primarily as plating solutions and as catalysts. The Sn, Cu, Zn, Ni, Pb, and Ag fluoroborates cure a wide range of epoxy resins at elevated or ambient room temperature (91,92). In the textile industry zinc fluoroborate is used extensively as the curing agent in applying resins for crease-resistant finishes (93). Emulsions of epoxy resins (94), polyoxymethylene compounds (95), or aziridinyl compounds (96) with $Zn(BF_4)_2$ and other additives are applied to the cloth. After the excess is removed, the cloth is dried and later cured at a higher temperature. Similarly treated acrylic textiles using epoxy resins take on an antistatic finish (97), or the acrylic textiles can be coated

Table 5. Commercial Metal Fluoroborate Solutions

Metal cation	CAS Registry Number	Formula	% Metal	Specific gravity, g/cm^3	1992 price, $/kg
antimony(II)	[14486-20-5]	Sb(BF$_4$)$_3$	12.8	1.42	
cadmium	[14886-19-2]	Cd(BF$_4$)$_2$	19.7	1.60	8.15
cobalt(II)	[26490-63-1]	Co(BF$_4$)$_2$	11.8	1.42	
copper(II)	[38465-60-0]	Cu(BF$_4$)$_2$	12.2	1.48	3.40–4.0
indium	[27765-48-6]	In(BF$_4$)$_3$	15.3	1.55	
iron(II)	[13877-16-2]	Fe(BF$_4$)$_2$	10.3	1.47	
lead(II)	[13814-96-5]	Pb(BF$_4$)$_2$	28.9	1.75	1.85–2.16
nickel(II)	[14708-14-6]	Ni(BF$_4$)$_2$	11.2	1.47	10.25–11.8
tin(II)	[13814-96-5]	Sn(BF$_4$)$_2$	20.2	1.61	6.38–6.6
zinc	[13826-88-5]	Zn(BF$_4$)$_2$	11.0	1.39	1.75–2.1

with 20% $Zn(BF_4)_2$ which results in up to 5.5% added solids for a fire-resistant finish (98).

The use of silver fluoroborate as a catalyst or reagent often depends on the precipitation of a silver halide. Thus the silver ion abstracts a Cl^- from a rhodium chloride complex, $((C_6H_5)_3As)_2(CO)RhCl$, yielding the cationic rhodium fluoroborate [30935-54-7] hydrogenation catalyst (99). The complexing tendency of olefins for $AgBF_4$ has led to the development of chemisorption methods for ethylene separation (100,101). Copper(I) fluoroborate [14708-11-3] also forms complexes with olefins; hydrocarbon separations are effected by similar means (102).

The manufacture of linear polyester is catalyzed by Cd, Sn (103), Pb, Zn, or Mn (104) fluoroborates. The Beckmann rearrangement of cyclohexanone oxime to caprolactam is catalyzed by $Ba(BF_4)_2$ [13862-62-9] or $Zn(BF_4)_2$ [13826-88-5] (105). The caprolactam is polymerized to polyamide fibers using $Mn(BF_4)_2$ [30744-82-2] catalyst (106). Nickel and cobalt fluoroborates appear to be good catalysts for the polymerization of conjugated dienes to cis-1,4-polydienes; the cis configuration is formed in up to 96% yields (107–109).

Electroplating. Metal fluoroborate electroplating (qv) baths (27,110,111) are employed where speed and quality of deposition are important. High current densities can be used for fast deposition and near 100% anode and cathode efficiencies can be expected. Because the salts are very soluble, highly concentrated solutions can be used without any crystallization. The high conductivity of these solutions reduces the power costs. The metal content of the bath is also easily maintained and the pH is adjusted with HBF_4 or aqueous ammonia. The disadvantages of using fluoroborate baths are treeing, lack of throwing power, and high initial cost. Treeing and throwing power can be controlled by additives; grain size of the deposits can also be changed. As of this writing, metals being plated from fluoroborate baths are Cd, Co, Cu, Fe, In, Ni, Pb, Sb, and Zn. Studies on Fe (112,113), Ni (113), and Co (113) fluoroborate baths describe the compositions and conditions of operation as well as the properties of the coatings. Iron foils electrodeposited from fluoroborate baths and properly annealed have exceptionally high tensile strength (113).

The Fe, Co, and Ni deposits are extremely fine grained at high current density and pH. Electroless nickel, cobalt, and nickel–cobalt alloy plating from fluoroborate-containing baths yields a deposit of superior corrosion resistance, low stress, and excellent hardenability (114). Lead is plated alone or in combination with tin, indium, and antimony (115). Sound insulators are made as lead–plastic laminates by electrolytically coating Pb from a fluoroborate bath to 0.5 mm on a copper-coated nylon or polypropylene film (116) (see INSULATION, ACOUSTIC). Steel plates can be simultaneously electrocoated with lead and poly(tetrafluoroethylene) (117). Solder is plated in solutions containing $Pb(BF_4)_2$ and $Sn(BF_4)_2$; thus the lustrous solder-plated object is coated with a Pb–Sn alloy (118).

BIBLIOGRAPHY

"Fluoroboric Acid" under "Fluorine Compounds, Inorganic," in *ECT* 1st ed., Vol. 6, pp. 684–688, by F. D. Loomis, Pennsylvania Salt Manufacturing Co.; "Fluoroboric Acid and Fluoroborates" under "Fluorine Compounds, Inorganic" in *ECT* 2nd ed., Vol. 9, pp. 562–572, by H. S. Halbedel, The Harshaw Chemical Co.; in *ECT* 3rd ed., Vol. 10, pp. 693–706, by H. S. Halbedel and T. E. Nappier, The Harshaw Chemical Co.

1. J. W. Mellor, *Comprehensive Treatise on Inorganic and Theoretical Chemistry*, Vol. 5, Longman, Green and Co., New York, 1929, pp. 123–129.
2. H. S. Booth and D. R. Martin, *Boron Trifluoride and Its Derivatives*, John Wiley & Sons, Inc., New York, 1949, pp. 87–165.
3. R. E. Mesmer, K. M. Palen, and C. F. Baes, *Inorg. Chem.* **12**(1), 89 (1973).
4. I. G. Ryss, *The Chemistry of Fluorine and Its Inorganic Compounds*, State Publishing House for Scientific, Technical, and Chemical Literature, Moscow, USSR, 1956; F. Haimson, English trans., *AEC-tr-3927*, U.S. Atomic Energy Commission, Washington, D.C., 1960, pp. 505–579.
5. C. H. Wamser, *J. Am. Chem. Soc.* **70**, 1209 (1948); **73**, 409 (1951).
6. R. E. Mesmer and A. C. Rutenberg, *Inorg. Chem.* **12**(3), 699 (1973).
7. R. D. Srinastava, M. O. Uy, and M. Faber, *J. Chem. Soc. Farad. Trans. 1* **70**, 1033 (1970).
8. H. Bonadeo and E. Silberman, *J. Mol. Spect.* **32**, 214 (1969).
9. J. Bessiere, *Anal. Chim. Acta* **52**(1), 55 (1970).
10. M. J. R. Clark, *Can. J. Chem.* **47**, 2579 (1969).
11. G. Brunton, *Acta Crystallogr. Sect. B* **24**, 1703 (1968).
12. A. S. Quist and co-workers, *J. Chem. Phys.* **54**, 4896 (1971); **55**, 2836 (1971).
13. M. R. Rosenthal, *J. Chem. Ed.* **50**(5), 331 (1973).
14. M. J. R. Clark and H. Linton, *Can. J. Chem.* **48**, 405 (1970).
15. I. Pawlenko, *Z. Anorg. Allgem. Chem.* **340**(3–4), 201 (1965).
16. H. W. Heiser, *Chem. Eng. Prog.* **45**(3), 169 (1949); U.S. Pats. 2,182,509–11 (Dec. 5, 1939), (to Alcoa).
17. U.S. Pat. 2,799,559 (July 16, 1957), T. J. Sullivan, C. H. Milligan, and J. A. Grady.
18. U.S. Pat. 3,959,132 (May 25, 1976), J. Singh (to Gilson Technical Services, Inc.).
19. U.S. Pat. 4,045,339 (Aug. 30, 1977), T. F. Korenowski, J. L. Penland, and C. J. Ritzert (to Dart Industries Inc.).
20. *Chemical Economics Handbook*, Stanford Research Institute, Menlo Park, Calif., 1975, p. 739.5030H.
21. D. C. Cornish and R. J. Simpson, *Meas. Contr.* **4**(11), 308 (1971).
22. E. J. Largent, "Metabolism of Inorganic Fluoride" in *Fluoridation as a Public Health Measure*, American Association for the Advancement of Science, Washington, D.C., 1954, pp. 49–78.
23. I. Zipkin and R. C. Likens, *Am. J. Physiol.* **191**, 549 (1957).
24. U.S. Pat. 3,888,778 (Mar. 13, 1973), M. Beckwith and G. F. Hau.
25. J. F. Jumer, *Met. Finish.* **56**(8), 44 (1958); **56**(9), 60 (1958).
26. R. M. Hudson, T. J. Butler, and C. J. Warning, *Met. Finish.* **74**(10), 37 (1976); U.S. Pat. 3,933,605 (Jan. 20, 1976), T. J. Butler, R. M. Hudson, and C. J. Warning (to U.S. Steel Corp.).
27. R. D. Mawiya and K. P. Joshi, *Indian Chem. J.* **6**(2), 19 (1971).
28. U.S. Pat. 2,925,325 (Feb. 16, 1960), J. Kamlet (to Reynolds Metals Co.).
29. U.S. Pat. 3,808,245 (Apr. 30, 1974), D. E. O'Connor and G. R. Wyness (to Procter & Gamble Co.).
30. U.S. Pat. 3,305,565 (Feb. 21, 1967), A. C. Mueller (to Shell Oil Co.).
31. U.S. Pat. 3,349,107 (Oct. 24, 1967), S. Pawlenko (to Schering Akliengessellshaft).
32. R. C. Weast, ed., *Handbook of Chemistry and Physics*, Vol. 59, The Chemical Rubber Co., Cleveland, Ohio, 1978.
33. A. S. Dworkin and M. A. Bredig, *J. Chem. Eng. Data* **15**, 505 (1970).
34. M. J. R. Clark and H. Lynton, *Can. J. Chem.* **47**, 2579 (1969).
35. V. S. Yatlov and E. N. Pinaevskays, *Zh. Obshch. Khim.* **15**, 269 (1945).

36. H. Boch, *Z. Naturforsch.* **17b**, 426 (1962).
37. L. J. Klinkenberg, doctoral thesis, Leiden, Germany, 1937.
38. T. C. Waddington, *Adv. Inorg. Chem. Radiochem.* **1**, 158 (1959).
39. P. Gross, C. Hayman, and H. A. Joel, *Trans. Faraday Soc.* **64**, 317 (1968).
40. L. J. Klinkenberg, *Rec. Trav. Chim.* **56**, 36 (1937).
41. J. H. de Boer and J. A. H. Van Liempt, *Rec. Trav. Chim.* **46**, 24 (1927).
42. *JANAF Thermochemical Tables*, Clearinghouse for Federal Scientific and Technical Information, U.S. Dept. of Commerce, Springfield, Va., Dec. 1963.
43. A. W. Laubengayer and G. F. Condike, *J. Am. Chem. Soc.* **70**, 2274 (1948).
44. A. P. Altschuller, *J. Am. Chem. Soc.* **77**, 6515 (1955).
45. V. N. Plakhotnik, V. B. Tul'chinski, and V. K. Steba, *Russ. J. Inorg. Chem.* **22**, 1398 (1977).
46. T. V. Ostrovskaya and S. A. Amirova, *Russ. J. Inorg. Chem.* **15**, 338 (1970).
47. A. V. Pankratov and co-workers, *Russ. J. Inorg. Chem.* **17**, 47 (1972).
48. R. T. Marano and J. L. McAtee, *Thermochimica Acta* **4**, 421 (1972).
49. J. B. Bates and co-workers, *J. Inorg. Nucl. Chem.* **34**, 2721 (1972).
50. L. Maya, *J. Am. Ceram. Soc.* **60**(7–8), 323 (1977).
51. V. Pecak, *Chem. Prum.* **23**(2), 71 (1973).
52. Ger. Pat. 2,320,360 (Nov. 7, 1974), H. K. Hellberg, J. Massonne, and O. Gaertner (to Kali-Chemie Fluor GmbH).
53. U.S. Pat. 2,799,556 (Feb. 1, 1954), T. J. Sullivan and C. G. Milligan (to American Agriculture Chemical Co.).
54. H. S. Booth and S. Rhemar, *Inorganic Synthesis*, Vol. 2, McGraw-Hill Book Co., New York, 1946, p. 23.
55. U.S. Pat. 3,541,739 (Nov. 24, 1970), J. P. Bryon and A. G. Rolfe (to English Abrasives Limited).
56. U.S. Pat. 3,963,458 (June 15, 1976), M. T. Gladstone and S. J. Supkis (to Norton Co.).
57. G. von Matuschka, *Kunstofftechnik* **11**(11), 304 (1972).
58. USSR Pat. 495,178 (Dec. 15, 1975), V. Boiko.
59. U.S. Pat. 3,376,253 (Apr. 2, 1968), E. V. Burnthall and J. J. Hirshfeld (to Monsanto Co.).
60. Ger. Pat. 2,121,821 (Dec. 2, 1971), K. C. Frisch (to Owens Corning Fiberglass Co.).
61. P. M. Sawko and S. R. Riccitiello, *Tech. Brief ARC-11043*, NASA-Ames Research Center, Moffett Field, Calif., July 1977.
62. U.S. Pat. 3,909,489 (Sept. 30, 1975), D. D. Callander (to Goodyear Tire and Rubber Co.).
63. U.S. Pats. 2,572,248-9 (Oct. 23, 1951), H. S. Cooper (to Walter M. Weil).
64. J. D. Donaldson, C. P. Squire, and F. E. Stokes, *J. Mater. Sci.* **13**, 421 (1978).
65. U.S. Pat. 3,305,416 (Feb. 21, 1967), G. J. Kahan and J. L. Mees (to International Business Machines Corp.).
66. U.S. Pat. 3,540,943 (Nov. 17, 1970), E. M. Grogan (to U.S. Steel Corp.).
67. W. R. Huntley and P. A. Gnadt, *Report ORNL-TM-3863*, Oak Ridge National Laboratory, Oak Ridge, Tenn., 1973.
68. L. Kravetz and G. R. Ferrante, *Text. Res. J.* **40**, 362 (1970).
69. Fr. Pat. 1,578,918 (Aug. 22, 1968), J. Borkowski.
70. J. Dale and K. Daasvet'n, *J. Chem. Soc. Chem. Commun.*, (8), 295 (1976).
71. M. A. Kasem and H. R. Richard, *Ind. Eng. Chem. Prod. Res. Dev.* **11**(2), 114 (1972).
72. *Technical Bulletin FR175*, Harshaw Chemical Co., Cleveland, Ohio, 1975.
73. U.S. Pat. 2,561,565 (July 24, 1951), A. P. Edson and I. L. Newell (to United Aircraft Corp.).
74. Br. Pat. 1,181,753 (Feb. 18, 1970), (to Aluminum Co. of America).
75. Ger. Pat. 2,334,660 (Jan. 23, 1975), H. Lauck.

76. U.S. Pat. 3,287,312 (Nov. 22, 1966), T. H. Ling (to Anaconda Wire and Cable Co.).

77. T. Hongu, *S. Gakkaishi* **26**(1), 38 (1970).

78. D. W. A. Sharp, in M. Stacy, J. C. Tatlow, and A. G. Sharpe, eds., *Advances in Fluorine Chemistry*, Vol. 1, Academic Press, Inc., New York, 1960, pp. 68–128.

79. T. V. Ostrovskaya, S. A. Amirova, and N. V. Startieva, *Russ. J. Inorg. Chem.* **12**, 1228 (1967).

80. R. T. Marano and J. L. McAtee, *Therm. Anal. Proc. Int. Conf. 3rd, 1971* **2**, 335 (1972).

81. A. D. Van Ingen Schenau, W. L. Groenveld, and J. Reedijk, *Recl. Trav. Chim. Pays-Bas* **9**, 88 (1972).

82. P. W. N. M. Van Leeuwen, *Recl. Trav. Chim. Pays-Bas* **86**, 247 (1967).

83. U.S. Pat. 3,672,759 (July 4, 1972), T. Yamawaki and co-workers.

84. E. Kamienska and I. Uruska, *Bull. Akad. Pol. Sci. Ser. Sci. Chim.* **21**, 587 (1973).

85. D. W. A. Sharp and co-workers, *Proc. Int. Conf. Coord. Chem. 8th, Vienna*, 322 (1964).

86. V. N. Plakhotnik and V. V. Varekh, *Izv. Vyssh. Uchebn. Zaved. Khim. Khim. Tekhnol.* **16**, 1619 (1973).

87. U.S. Pat. 3,795,595 (Mar. 5, 1974), H. P. Wilson (to Vulcan Materials Co.).

88. U.S. Pat. 3,300,397 (Jan. 24, 1967), G. Baltakmens and J. P. Tourish (to Allied Chemical Corp.).

89. U.S. Pat. 3,432,256 (Mar. 11, 1969), H. P. Wilson (to Vulcan Materials Co.).

90. S. Buffagni and I. M. Vezzosi, *Gazz. Chim. Ital.* **97**, 1258 (1967).

91. U.S. Pat. 4,092,296 (May 30, 1978), R. A. Skiff.

92. U.S. Pat. 3,432,440 (Mar. 11, 1969), D. A. Shimp, W. F. McWhorter, and N. G. Wolfe (to Celanese Coatings Co.).

93. *Technical Bulletin ZBF873*, Harshaw Chemical Co., Solon, Ohio.

94. A. Zemaitaitis and J. Zdanavicius, *Cellul. Chem. Technol.* **4**, 621 (1970).

95. U.S. Pat. 3,854,869 (Dec. 17, 1974), Y. Yanai (to Nisshin Spinning Co., Ltd.).

96. C. E. Morris and G. L. Drake, Jr., *Am. Dyestuff Rep.* **58**(4), 31 (1969).

97. Jpn. Pat. 71 11,080 (Mar. 20, 1971), S. Hiroaka and K. Mitsumura (to Mitsubishi Rayon Co., Ltd.).

98. U.S. Pat. 3,577,342 (May 4, 1971), L. I. Fidell (to American Cyanamid Co.).

99. U.S. Pat. 3,697,615 (Oct. 10, 1972), W. B. Hughes (to Phillips Petroleum Co.).

100. E. Rausz and S. Hulisz, *Chemik* **28**(7), 256 (1975).

101. H. W. Quinn and R. L. Van Gilder, *Can. J. Chem.* **48**, 2435 (1970).

102. U.S. Pat. 3,514,488 (May 26, 1970), C. E. Uebele, R. K. Grasselli, and W. C. Nixon (to Standard Oil Co. of Ohio).

103. Jpn. Pat. 714,030 (Dec. 3, 1971), Y. Fujita and T. Morimoto (to Mitsui Petrochemical Industries, Ltd.).

104. Jpn. Pat. 70 19,514 (July 3, 1970), I. Hiroi (to Toho Rayon Co., Ltd.).

105. Jpn. Pat. 76 04,163 (Jan. 14, 1976), J. Takeuchi, F. Iwata, and K. Kubo (to Ube Industries, Ltd.).

106. Jpn. Pat. 72 18,227 (May 26, 1972), S. Sugiura and co-workers (to Ube Industries, Ltd.).

107. Jpn. Pat. 73 06,185 (Feb. 23, 1973), T. Yamawaki, T. Suzuki, and S. Hino (to Mitsubishi Chemical Industries Co. Ltd.).

108. Fr. Pat. 2,039,808 (Jan. 15, 1971), (to Mitsubishi Chemical Industries, Co., Ltd.).

109. Jpn. Pat. 72 06,411 (May 4, 1972), T. Yamawaki and co-workers (to Mitsubishi Chemical Industries Co., Ltd.).

110. *Plating Processes*, Harshaw Chemical Co., Solon, Ohio, Mar. 1977.

111. Y. M. Faruq Marikan and K. I. Vasu, *Met. Finish.* **67**(8), 59 (1969).

112. F. Wild, *Electroplat. Met. Finish.* **13**, 331 (Sept. 1960).

113. E. M. Levy and G. J. Hutton, *Plating* **55**(2), 138 (1968).

114. U.S. Pat. 3,432,338 (Mar. 11, 1969), R. E. Sickles (to Diamond Shamrock Corp.).

115. N. J. Spiliotis, *Galvanotech. Oberflaechenschutz* **7**(8), 192 (1966).

116. Jpn. Pat. 76 02,633 (Jan. 10, 1976), J. Hara, R. Miyashata, and Y. Fukuoka (to Nippon Kayaku Co., Ltd.).

117. Ger. Pat. 2,146,908 (Mar. 23, 1972), K. Ishiguro and H. Shinohara (to Toyota Motor Co., Ltd.).

118. *Plating Processes, Tin-Lead Solder Alloy Fluoborate Plating Process for Printed Circuit Applications, HTPB5N 0272*, Harshaw Chemical Co., Solon, Ohio.

JOHN R. PAPCUN
Atotech

CALCIUM

Fluorine chemistry began with observations by Georgius Agricola as early as 1529 that fluorspar lowers the melting point of minerals and reduces the viscosity of slags. This property of fluxing (Latin *fluoere*, to flow) is the origin of the name fluorine. The term fluorspar correctly describes ores containing substantial amounts of the mineral fluorite [*14542-23-5*], CaF_2, but the word fluorspar is often used interchangeably with fluorite and calcium fluoride (see also CALCIUM COMPOUNDS).

Calcium Fluoride

Significant mining of fluorspar began in England about 1775 and in the United States after 1820. Substantial use of fluorspar began about 1880 in the basic open-hearth process for making steel (qv). Large increases in demand came with the need for fluorides in the aluminum industry, starting about 1900. A large fluorine chemicals industry based on hydrogen fluoride made from fluorspar followed in production of refrigerants (see REFRIGERATION AND REFRIGERANTS) (1930), alkylation (qv) catalysts for gasoline (1942), materials for nuclear energy (ca 1942), aerosol propellants (see AEROSOLS) (ca 1942), fluoroplastics (ca 1942), and fluorocarbons for soil-repellant surface treatments (early 1950s). Fluorspar is used directly in the manufacture and finishing of glass (qv), in ceramics (qv) and welding (qv) fluxes, and in the extraction and processing of nonferrous metals (see METALLURGY, EXTRACTIVE).

In the geochemistry of fluorine, the close match in the ionic radii of fluoride (0.136 nm), hydroxide (0.140 nm), and oxide ion (0.140 nm) allows a sequential replacement of oxygen by fluorine in a wide variety of minerals. This accounts for the wide dissemination of the element in nature. The ready formation of volatile silicon tetrafluoride, the pyrohydrolysis of fluorides to hydrogen fluoride, and the low solubility of calcium fluoride and of calcium fluorophosphates, have provided a geochemical cycle in which fluorine may be stripped from solution by limestone and by apatite to form the deposits of fluorspar and of phosphate rock (fluoroapatite [*1306-01-0*]), approximately $CaF_2 \cdot 3Ca_3(PO_4)_2$ which are the world's main resources of fluorine (1).

On average, fluorine is about as abundant as chlorine in the accessible surface of the earth including oceans. The continental crust averages about 650 ppm fluorine. Igneous, metamorphic, and sedimentary rocks all show abundances in the range of 200 to 1000 ppm. As of 1993, fluorspar was still the principal source of fluorine for industry.

Fluorspar deposits are commonly epigenetic, ie, the elements moved from elsewhere into the country rock. For this reason, fluorine mineral deposits are closely associated with fault zones. In the United States, significant fluorspar deposits occur in the Appalachian Mountains and in the mountainous regions of the West, but the only reported commercial production in 1993 was from the faulted carbonate rocks of Illinois.

Worldwide, large deposits of fluorspar are found in China, Mongolia, France, Morocco, Mexico, Spain, South Africa, and countries of the former Soviet Union. The United States imports fluorspar from most of these countries (Table 1).

Properties. Some of the important physical properties of calcium fluoride are listed in Table 2. Pure calcium fluoride is without color. However, natural fluorite can vary from transparent and colorless to translucent and white, wine-yellow, green, greenish blue, violet-blue, and sometimes blue, deep purple, bluish black, and brown. These color variations are produced by impurities and by radiation damage (color centers). The color of fluorite is often lost upon heating, sometimes with luminescence. Mineral specimens are usually strongly fluorescent, and the mineral thus gives its name to this phenomenon. Specimens vary from well-formed crystals (optical grade) to massive or granular forms.

The crystal structure of fluorite gives its name to the fluorite crystal type. The lattice is face-centered cubic (fcc), where each calcium ion is surrounded by eight fluoride ions situated at the corners of a cube, and each fluoride ion lies within a tetrahedron defined by four calcium ions (3). The bonding is ionic. The unit cell (space group O_h^5) can be pictured as made up of eight small cubes, each containing a fluoride ion, and the eight forming a cube with a calcium ion on each corner and one in the center of each face (Fig. 1). The lattice constant is 0.54626 nm at 25°C (4). The habit is usually cubic, less frequently octahedral, rarely dodecahedral. Cleavage on the [111] planes is perfect. The crystals are brittle with flat-conchoidal or splintery fracture. Luster is vitreous, becoming dull in massive varieties.

Systems of metal oxides with calcium fluoride usually have a simple freezing point composition diagram, commonly exhibiting a eutectic point and no abnormal lowering of the melting point (5). When silicates are present, the systems become more complicated, and a striking decrease in the viscosity of the glassy melts is observed. The viscosity most likely decreases because of depolymerization of chains or networks of SiO_4 tetrahedra via the replacement of oxide ion by the singly charged fluoride ion which is close in both size and electronegativity to oxide ion (1). The benefits of calcium fluoride as a metallurgical flux result from both the freezing point depression and the decrease in slag viscosity.

Although stable at ambient temperature, calcium fluoride is slowly hydrolyzed by moist air at about 1200°C, presumably to CaO and HF. Calcium fluoride is not attacked by alkalies or by reactive fluorine compounds, but is decomposed by hot, high boiling acids, as in the reaction with concentrated sulfuric acid which is the process used to produce hydrogen fluoride. Calcium fluoride is slightly sol-

Table 1. U.S. Imports for Consumption of Fluorospar[a,b]

Country	1991[c] Quantity, t	1991[c] Value, $\times 10^6$	1992[d] Quantity, t	1992[d] Value, $\times 10^6$	1993[e] Quantity, t	1993[e] Value, $\times 10^6$	1993[e] Quantity, t	1993[e] Value, $\times 10^6$	1993[e] Quantity, t	1993[e] Value, $\times 10^6$
					$CaF_2 > 97\%$					
Canada	18	9								
China	56,311	6,352	128,960	13,127	73,942	7,175	34,817	3,077	40,173	4,142
France	36	15	52	26						
Japan	4,627	416								
Mexico	52,475	6,570	34,058	4,228	8,917	1,096	7,289	884	6,010	744
Morocco	9,828	1,382								
Republic of South Africa	79,495	11,107	106,066	13,147	5,932	702	25,519	2,990	32,413	3,754
Spain	11,278	1,545								
Total	*214,350*	*27,387*	*267,328*	*30,528*	*88,791*	*8,973*	*77,791*	*7,895*	*78,611*	*8,642*
					$CaF_2 \leq 97\%$					
Canada	285	22					3,937	335	5,181	369
China	11,051	909	59,130	4,619	17,968	1,270	9,117	712		
Japan	4,674	387								
Mexico	33,970	3,449	54,405	5,069	822	55			9,353	754
Republic of South Africa	15,860	2,108	11,152	1,370						
Total	*65,840*	*6,875*	*124,687*	*11,058*	*18,790*	*1,325*	*13,054*	*1,047*	*14,534*	*1,123*

[a]Imports for consumption include imports of immediate entry plus warehouse withdrawals (2).
[b]Cost, insurance, freight (cif) at U.S. ports.   [c]1991 Numbers represent the total of 2nd, 3rd, and 4th quarters only.
[d]1992 Numbers represent the total of 1st, 2nd, 3rd, and 4th quarters.   [e]1993 Numbers represent 1st, 2nd, and 3rd quarters, respectively.

325

Table 2. Physical Properties of Calcium Fluoride

Property	Value	Reference
formula weight	78.08	
composition, wt %		
Ca	51.33	
F	48.67	
melting point, °C	1402	6
boiling point, °C	2513	7
heat of fusion, kJ/mol[a]	23.0	8
heat of vaporization at bp, kJ/mol[a]	335	9
vapor pressure at 2100°C, Pa[b]	1013	9
heat capacity, C_p, kJ/(mol·K)[a]		
solid at 25°C	67.03	10
solid at mp	126	11
liquid at mp	100	11
entropy at 25°C, kJ/(mol·K)[a]	68.87	10
heat of formation, solid at 25°C, kJ/mol[a]	-1220	10
free energy of formation, solid at 25°C, kJ/mol[a]	-1167	10
thermal conductivity, crystal at 25°C, W/(m·K)	10.96	12
density, g/mL		
solid at 25°C	3.181	13
liquid at mp	2.52	14
thermal expansion, average 25 to 300°C, K^{-1}	22.3×10^{-6}	15
compressibility, at 25°C and 101.3 kPa (=1 atm)	1.22×10^{-8}	16
hardness		
Mohs' scale	4	
Knoop, 500-g load	158	17
solubility in water, g/L at 25°C	0.146	18
refractive index at 24°C, 589.3 nm	1.43382	19
dielectric constant at 30°C	6.64	20
electrical conductivity of solid, (Ω·cm)$^{-1}$		
at 20°C	1.3×10^{-18}	21
at 650°C	6×10^{-5}	22
at mp	3.45	23
optical transmission range, nm	150 to 8000	24

[a]To convert J to cal, divide by 4.184.
[b]To convert Pa to mm Hg, multiply by 7.5×10^{-3}.

uble in cold dilute acids, and somewhat more soluble in solutions of aluminum halides.

Preparation. CaF_2 is manufactured by the interⸯction of H_2SiF_6 with an aqueous carbonate suspension (25–28); by the reaction ⸯ$CaSO_4$ with NH_4F (29); by the reaction of HF with $CaCO_3$ in the presence of ⸯH_4F (30); by reaction of $CaCO_3$ and NH_4F at 300–350°C followed by calcining at 700–800°C (31); by reaction of NH_4F and $CaCO_3$ (32–37); and from the thermal decomposition of calcium trifluoroacetate (38).

High purity CaF_2 is obtained from micro- and ultrafiltration (qv) of raw materials and then crystallization of CaF_2 (39) from the reaction of $CaCO_3$ and the

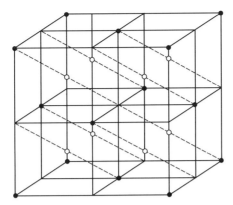

Fig. 1.   Structure of fluorite where ● is Ca and ○ is F (3).

product of Li_2SiF_6 and NH_3 (40). High purity CaF_2 having particles of 0.0005–0.5 mm is produced from the reaction of NaF, KF, or NH_4F and $CaCO_3$ with a particle size distribution of 0.01–0.05 nm (41). High purity CaF_2 is also prepared from the reaction of $Ca(NO_3)_2 \cdot 4H_2O$ and NH_4F or a mixture of NH_4HF_2 and NH_4F (42), and obtained by heating impure CaF_2 with 10–15% HCl at 95–100°C (43). Very pure calcium fluoride for the manufacture of special glasses is made by the reaction of hydrofluoric acid with precipitated calcium carbonate. Acicular (whisker-form) CaF_2 particles have been manufactured by continuous feeding of an aqueous $Ca(OH)_2$ solution into water containing CO_2 and subsequent reaction with HF (44). Coarse grain CaF_2 crystals can be prepared by several routes (45–58). CaF_2 can be is crystallized from a wastewater containing fluoride by adding $CaCl_2$ (59). Calcium fluoride can be produced from waste H_2SiF_6 from phosphate product operations and from treating fluoride solution from industrial wastewater with KOH and then with lime (60).

The large amount of fluorine values released from phosphate rock in the manufacture of fertilizers (qv) gives a strong impetus to develop fluorine chemicals production from this source (see PHOSPHORIC ACID AND THE PHOSPHATES). Additional incentive comes from the need to control the emission of fluorine-containing gases. Most of the fluorine values are scrubbed out as fluorosilicic acid, H_2SiF_6, which has limited usefulness. A procedure to convert fluorosilicic acid to calcium fluoride is available (61).

Mining.   Underground mining procedures are used for deep fluorspar deposits, and open-pit mines are used for shallow deposits or where conditions do not support underground mining techniques (see MINERAL RECOVERY AND PROCESSING).

Fluorspar occurs in two distinct types of formation in the fluorspar district of southern Illinois and Kentucky; in vertical fissure veins and in horizontal bedded replacement deposits. A 61-m bed of sandstone and shale serves as a cap rock for ascending fluorine-containing solutions and gases. Mineralizing solutions come up the faults and form vein ore bodies where the larger faults are plugged by shale. Bedded deposits occur under the thick sandstone and shale roofs. Other elements of value associated with fluorspar ore bodies are zinc, lead, cadmium, silver, germanium, iron, and thorium. Ore has been mined as deep as 300 m in this district.

In 1993, Illinois was the only state reporting production of acid-grade fluorspar, typically 96.5–97.5% CaF_2, and accounted for 100% of all reported shipments. Ozark-Mahoning Co., a wholly owned subsidiary of Elf Atochem North America, Inc., operated three deep mines and a flotation mill in Hardin County, Illinois. A limited amount of metallurgical-grade gravel was produced by Hastie Mining, also located in Hardin County, Illinois (1).

Outside of the United States, there are six primary producers in China, France, Mexico, Morocco, South Africa, and Spain. Mines in Newfoundland, Canada, were closed in 1990. Both Mexico and South Africa have lost market share to China which has high grade, low cost fluorspar. China is expected to dominate world markets because reserves are vast and production cost is low. Table 3 (2) shows a list of world producers by country of fluorspar in the early 1990s.

Beneficiation. Most fluorspar ores as mined must be concentrated or beneficiated to remove waste. Metallurgical-grade fluorspar is sometimes produced by

Table 3. World Production of Fluorspar,[a] t

Country	1990	1991	1992
Argentina	20,000[b]	20,000	19,000
Brazil	70,383	90,000	81,000
Canada	25,000		
China	1,700,000	1,600,000	1,600,000
Czechoslovakia	46,966	40,000	40,000
France	201,000	200,000	160,000
Germany	155,300	135,000	60,000
India	21,700	22,000	25,000
Iran	4,767	6,000	12,000
Italy	122,503	100,000	80,000
Kenya	112,295	90,000	80,630
North Korea	40,000	41,000	41,000
Mexico	634,000	352,000	364,000
Mongolia	614,000	520,000	277,000
Morocco	86,500	86,000	80,000
Namibia	25,980	29,246	40,480
Pakistan	5,312	5,300	5,000
Romania	15,000	14,000	15,000
Republic of South Africa	311,032	270,340	258,105
Spain	100,000[b]	90,000	90,000
Thailand	94,757	100,000	52,000
Tunisia	40,974	40,000	15,000
Turkey	13,000	13,000	13,000
former USSR	380,000	350,000	300,000
United Kingdom	118,498	80,000	85,000
United States (shipments)[b]	63,500	58,000	51,000
Total	*5,024,626*	*4,354,036*	*3,844,315*

[a]Only countries producing more than 2000 metric tons are listed (2).
[b]Values are estimated.

hand sorting lumps of high grade ore. In most cases the ore is beneficiated by gravity concentration with fluorspar and the waste minerals, having specific gravity values of > 3 and < 2.8, respectively. In preconcentration of fluorspar, barite and valuable sulfide minerals are separated from waste as the higher density valuable minerals sink while the waste floats and is discarded. This preconcentration can enrich ores as low as 14% to a concentration of 40%. Multistage froth flotation (qv) is used to take this preconcentrate and produce acid and ceramic grades of fluorspar as well as zinc and lead sulfides. In this process air bubbles are forced through a suspension of pulverized ore which float the ore into a froth that is continuously skimmed off. After flotation the fluorspar products are filtered and dried in rotary kilns.

In steelmaking, the preferred form of fluorspar flux is a washed gravel, 0.6–5 cm in diameter, containing less than 3% water, and assaying 60 to 80% effective CaF_2 units. The higher CaF_2 ranges are hard to supply in large amounts from some sources of fluorspar. The use of fluorspar briquettes and pellets in the steel (qv) industry has declined but these are still preferred by some producers. The briquettes contain 25–90% CaF_2, are frequently made to customer specifications, and may include fluxing agents and recycled steel mill wastes. Binders used include molasses, lime, and sodium silicate (see METALLURGY, EXTRACTIVE).

Economic Aspects. Pertinent statistics on the U.S. production and consumption of fluorspar are given in Table 4. For many years the United States has relied on imports for more than 80% of fluorspar needs. The principal sources are Mexico, China, and the Republic of South Africa. Imports from Mexico have declined in part because Mexican export regulations favor domestic conversion of fluorspar to hydrogen fluoride for export to the United States.

There were very high U.S. exports in 1991. However, this number reflects material imported into U.S. foreign trade zones and then transhipped to other countries.

Consumer stocks (2) at the end of 1993 were 65,000 t. The National Defense Stockpile of fluorspar inventory, at year end 1992, contained 809,000 t of acid-grade material, 281,000 t of metallurgical-grade material, 816 t of nonstockpile, acid-grade material, and 105,938 t of nonstockpile, metallurgical-grade material.

Table 4. United States Fluorspar Production and Consumption, t

Parameter	1987	1988	1989	1990	1991	1992
production[a]	63,500	63,500	66,000	63,500	58,000	51,000
exports	2,595	3,136	5,134	14,921	73,943	13,646
imports for consumption	531,530	689,139	655,590	513,921	376,081[b]	407,169
consumption						
reported	542,830	651,055	641,882	564,545	483,589	48,544
apparent[c]	643,659	723,804	693,121	566,885	365,831	480,376
world production	4,600,441	5,086,376	5,529,184	5,024,626	4,354,036	3,846,443

[a]Values are estimated (2).
[b]At a value of $50,780,000 cif.
[c]U.S. primary and secondary production plus imports minus exports plus adjustments for government and industry stock changes.

The prices for acid-grade fluorspar dry basis from Mexico and the Republic of South Africa for 1993 were Mexican spar, fob Tampico, $100–$112/t; South Africa acid spar, fob Durban, $95–$105/t. Mid-1992 quotes for Chinese acid-grade fluorspar delivered to Louisiana Gulfport were $89–$97/t and to Northern Europe, $105–$110/t. The price reductions generated by the Chinese caused producers to reduce production as market share was lost. Table 5 (2) contains U.S. imports of fluorspar by country.

Table 5. U.S. Imports for Consumption of Fluorspar by Country, t[a]

Country	1990	1991	1992	1993[b]
	$CaF_2 > 97\%$			
Canada	22,137			
China	125,190	74,853	128,960	148,932
Mexico	142,964	70,322	34,248	22,216
Morocco	10,950	9,828		
Republic of South Africa	86,660	105,031	106,866	63,864
	$CaF_2 \leq 97\%$			
China	28,088	28,228	59,130	27,086
Mexico	71,006	49,541	54,405	19,272
Republic of South Africa	8,280	17,360		

[a]Countries not having 10,000 metric tons in any one year were omitted (2).
[b]Through September 1993.

Grades, Quality Control.   Fluorspar is marketed in several grades: metallurgical fluorspar (metspar) is sold as gravel, lump, or briquettes. The minimum acceptable assay is 60% effective calcium fluoride. The effective value is determined by subtracting from the contained calcium fluoride 2.5% for every percent of SiO_2 found in the complete analysis apparently based on the following stoichiometry (1):

$$2\ CaF_2 + SiO_2 \rightarrow 2\ CaO + SiF_4$$

Ceramic-grade fluorspar and acid-grade fluorspar have the typical analyses shown in Table 6. Both types are usually finely ground, the bulk of the powder passing a 0.23 mm (65 mesh) screen, and 22 to 81% held on a 44 μm (325 mesh) screen. Optical-grade calcium fluoride, for special glasses and for growing single crystals, is supplied in purities up to 99.99% CaF_2. This grade is especially low in transition elements. For process control (qv) and product specification, fluorspar is commonly analyzed for fluorine, calcium, silica, carbonate, sulfide, iron, barium, and where significant, for metal values.

Analytical Methods.   Fluorite is readily identified by its crystal shape, usually simple cubes or interpenetrating twins, by its prominent octahedral cleavage, its relative softness, and the production of hydrogen fluoride when treated with sulfuric acid, evidenced by etching of glass. The presence of fluorite in ore speci-

Table 6. Analyses of Ceramic- and Acid-Grade Fluorspar, wt %[a]

Assay	Ceramic	Acid
CaF_2	90.0–95.5	96.5–97.5
SiO_2	1.2–3.0	1.0
$CaCO_3$	1.5–3.4	1.0–1.5
MgO		0.15
B		0.02
Zn		0.02
Fe_2O_3	0.10	0.10
P_2O_5		0.03
$BaSO_4$		0.2–1.3
R_2O_3[b]	0.15–0.25	0.1–0.3

[a]Refs. 62 and 63.
[b]R_2O_3 is any trivalent metal oxide, eg, Al_2O_3.

mens, or when associated with other fluorine-containing minerals, may be determined by x-ray diffraction.

For many years fluorine has been determined by the Willard-Winters method in which finely ground ore, after removal of organic matter, is distilled with 72% perchloric acid in glass apparatus. The distillate, a dilute solution of fluorosilicic acid, is made alkaline to release fluoride ion, adjusted with monochloroacetic acid at pH 3.4, and titrated with thorium nitrate, using sodium alizarine sulfonate as indicator.

The direct determination of fluoride using ion-selective electrodes has allowed analysis of fluorspar without the tedious distillation step (see ELECTRO-ANALYTICAL TECHNIQUES). The fluoride electrode uses a single crystal of lanthanum fluoride doped with europium in contact with the solution, and a high impedance detector similar to that used for glass electrode pH measurements. The finely ground fluorspar sample is dissolved in boiling aluminum chloride and hydrochloric acid solution, and the solution is buffered using ammonium sulfosalicylate. Ethylenediaminetetraacetic acid (EDTA) is added to displace the fluoride ion from complexes with positive ions (64–66). X-ray fluorescence is also used to analyze for fluoride content.

Fluorspar assay may be completed by fluoride determination alone, because the mineralogical grouping rarely includes fluorine minerals other than fluorite. Calcium can be determined as oxalate or by ion-selective electrodes (67). Silica can be determined in the residue from solution in perchloric acid–boric acid mixture by measuring the loss in weight on fuming off with hydrofluoric acid. Another method for determining silica in fluorspar is the ASTM Standard Test Method E463-72.

Carbonate is measured by evolution of carbon dioxide on treating the sample with sulfuric acid. The gas train should include a silver acetate absorber to remove hydrogen sulfide, a magnesium perchlorate drying unit, and a CO_2-absorption bulb. Sulfide is determined by distilling hydrogen sulfide from an acidified slurry of the sample into an ammoniacal cadmium chloride solution, and titrating the precipitated cadmium sulfide iodimetrically.

Health and Safety Factors. The low solubility of calcium fluoride reduces the potential problem of fluoride-related toxicity. Water saturated with calcium fluoride has a fluoride concentration of 8.1 ppm as compared to the recommended water fluoridation level of 1 ppm fluoride ion. However, because the solubility of calcium fluoride in stomach acid is higher, continued oral ingestion of calcium fluoride could produce symptoms of fluorosis. The adopted TWA limit for fluorides as F is 2.5 mg/m^3 (68,69).

A significant hazard results from contact of calcium fluoride with high concentrations of strong acids because of evolution of toxic concentrations of hydrogen fluoride. A less recognized danger is the production of hydrogen fluoride by pyrohydrolysis when calcium fluoride is exposed to water vapor at high temperature, as in a direct-fired kiln.

Beneficiation facilities require air and water pollution control systems, including efficient control of dust emissions, treatment of process water, and proper disposal of tailings (see AIR POLLUTION CONTROL METHODS). In handling finished fluorspar, operators must avoid breathing fluorspar dust and contacting fluorspar with acids. Proper disposal of spills and the use of respirators and other personnel protective equipment must be observed. Contact with fluorspar may irritate the skin and eyes.

Consumption and Uses. Acid-grade fluorspar, which is > 97% calcium fluoride, is used primarily in the production of hydrogen fluoride. Ceramic-grade fluorspar, containing 85 to 95% CaF$_2$ content, is used in the production of glass and enamel, to make welding rod coatings, and as a flux in the steel industry. Metallurgical-grade fluorspar, containing 60 to 85% or more CaF$_2$, is used primarily as a fluxing agent by the steel industry. Fluorspar is added to the slag to decrease its melting point and to increase its fluidity, thus increasing the chemical reactivity of the slag. Reducing the melting point of the slag brings lime and other fluxes into solution to allow the absorption of impurities.

Reported domestic consumption (2) by the HF industry in 1992 was 347,367 t. Reported consumption by the steel industry decreased in 1991 by about 52% from 1990 partly because of a 13% decrease in the production of steel.

In the ceramic industry, fluorspar is used as a flux and as an opacifier in the production of flint glass, white or opal glass, and enamels (see ENAMELS, PORCELAIN OR VITREOUS). Fluorspar is used in the manufacture of aluminum, brick, cement, and glass fibers, and is used by the foundry industry.

A small but artistically interesting use of fluorspar is in the production of vases, cups, and other ornamental objects popularly known as Blue John, after the Blue John Mine, Derbyshire, U.K. Optical quality fluorite, sometimes from natural crystals, but more often artificially grown, is important in use as infrared transmission windows and lenses (70) and optical components of high energy laser systems (see INFRARED AND RAMAN SPECTROSCOPY; LASERS; OPTICAL FILTERS) (71).

Calcium fluoride is also used in the thermal plasma process for purification of silicon (72); in the manufacture of fluoroaluminate glass (73,74); and fluorophosphate glasses (75); in the removal of lead in flotation waste (76); in the manufacture of clean steel without deoxidation with aluminum (77); in the electrochemical preparation of lanthanum (78); in the leaching process to treat domestic Mn-bearing silicate resources (79); as pigments for paper (qv) with titania (80);

in ir-transmitting window material (81); in the formation of passivated layers on superconductor oxides (82); in the manufacture of electrically resistive pastes (83) and electric resistors (84); in the refining of molten pig steel (85); in the recovery and purification of uranium (86); and as a solid lubricant for hot rolling (87). It is also used to strengthen cement (88,89) and for microcrystalline fluoride fibers useful for making seals for liquid propellant engines (90).

BIBLIOGRAPHY

"Calcium Fluoride" under "Fluorine Compounds, Inorganic," in *ECT* 1st ed., Vol. 6, pp. 689–692, by H. C. Miller, Pennsylvania Salt Manufacturing Co.; "Calcium Fluoride" under "Fluorine Compounds, Inorganic," in *ECT* 2nd ed., Vol. 9, pp. 573–582, by J. F. Gall, Pennsalt Chemicals Corp.; "Calcium" under "Fluorine Compounds, Inorganic," in *ECT* 3rd ed., Vol. 10, pp. 707–717, by J. F. Gall, Philadelphia College of Textiles and Science.

1. D. R. Shawe, ed., *Geology and Resources of Fluorine in the United States*, U.S. Geological Survey Professional Paper 933, Washington, D.C., 1976, pp. 1–5, 18, 19, 82–87.
2. M. M. Miller, *Fluorspar 1991 Annual Report*, U.S. Department of Interior, Bureau of Mines, Washington, D.C.
3. A. F. Wells, *Structural Inorganic Chemistry*, 3rd ed., Clardon Press, Oxford, U.K., 1962, p. 77.
4. H. E. Swanson and E. Tatge, *Natl. Bur. Stand. U.S. Circ. 539*, **1**, 69 (1953).
5. H. Krainer, *Radex Rundsch.*, 19 (1949).
6. B. Porter and E. A. Brown, *J. Am. Ceram. Soc.* **45**, 49 (1962).
7. D. A. Schulz and A. W. Searcy, *J. Phys. Chem.* **67**, 103 (1963).
8. G. Petit and A. Cremieu, *Compt. Rend.* **243**, 360 (1956).
9. O. Ruff and L. Leboucher, *Z. Anorg, Allg. Chem.* **219**, 376 (1934).
10. *National Bureau of Standards Technical Notes*, Washington, D.C., 1971, pp. 270–276.
11. B. F. Naylor, *J. Am. Chem. Soc.* **67**, 150 (1945).
12. K. A. McCarthy and S. S. Ballard, *J. Appl. Phys.* **36**, 1410 (1960).
13. Ref. 4, p. 69.
14. A. V. Grosse and C. S. Stokes, U.S. Department of Commerce, Office of Technical Service, PB Report 161460, Washington, D.C., 1960.
15. O. J. Whittemore, Jr. and N. N. Ault, *J. Ceram. Soc.* **39**, 443 (1956).
16. E. W. Washburn, ed., *International Critical Tables*, Vol. 3, McGraw-Hill Book Co., Inc., New York, 1929, p. 50.
17. S. S. Ballard, L. S. Combes, and K. A. McCarthy, *J. Opt. Soc. Am.* **42**, 684 (1952).
18. D. W. Brown and C. E. Roberson, *J. Res. U.S. Geol. Surv.* **5**, 509 (1977).
19. *Natl. Bur. Stand. U.S., Tech. News Bull.* **47**, 91 (1963).
20. J. L. Pauley and H. Chessin, *J. Am. Chem. Soc.* **76**, 3888 (1954).
21. E. W. Washburn, ed., *International Critical Tables*, Vol. 6, McGraw-Hill Book Co., Inc., New York, 1929, p. 154.
22. R. W. Ure, Jr., *J. Chem. Phys.* **26**, 1365 (1957).
23. T. Baak, *J. Chem. Phys.* **29**, 1195 (1958).
24. Data sheet, *IR Transmission Materials*, Barnes Engineering Co., Instrument Division, Stamford, Conn., 1992.
25. V. V. Babkin, V. V. Koryakov, T. A. Sokolova, and N. K. Petrova, *Khim. Prom-St (Moscow)* **3**, 1963–164.25, (1992).
26. H. Gabryel, L. Kacalski, and U. Glabisz, *Chem. Stosow* **33**(4), 673–678 (1989).
27. SU 1286520 A1, (Jan. 30, 1987), I. A. Elizarov and co-workers.
28. U.S. Pat. 4,264,563 (Apr. 28, 1981) S. K. Sikdar.

29. SU 1708762 A1 (Jan. 30, 1992), I. G. Saiko, A. A. Perebeinos, L. M. Pupyshevea, N. A. Orel.

30. SU 1699922 A1 (Dec. 23, 1991), M. E. Rakhimov, D. D. Ikrami, L. F. Mansurhodzhaeva, and Sh. A. Khalimov.

31. SU 998352 A1 (Feb. 23, 1983), A. A. Luginina, L. A. Ol'Khovaya, V. A. Reiterov, and D. D. Ikrami.

32. SU 802185 (Feb. 7, 1981), V. I. Rodin and co-workers.

33. A. A. Luginina and co-workers, *Zh. Neorg. Khim* **26**(2) 332–336 (1981).

34. V. V. Tumanov and co-workers, *Khim. Prom-St (Moscow)* (9), 668–671 (1989).

35. SU 83-3558912 (Mar. 2, 1983), M. I. Lyapunov, V. V. Tumonov, L. P. Belova, and G. H. Alekseeva.

36. V. V. Tumanov, L. P. Belova, and G. N. Alekseeva, *Prom-St (Moscow)* **9**, 551–553 (1983).

37. PL 104419 (Nov. 30, 1979), W. Augustyn, M. Dziegielewska, and A. Kossuth.

38. C. Russell, *J. Mater. Sci. Lett.* **11**(3), 152–154 (1992).

39. T. N. Naumova and co-workers, *Zh. Priki. Khim (Leningrad)* **64**(3), 480–484 (1991).

40. W. Augustyn, and co-workers, *Prezm. Chem.* **68**(4), 153–155 (1989).

41. PL 106787 (Jan. 31, 1980), W. Augustyn, M. Dziegielewska, and A. Kossuth.

42. RO 88593 B1 (Mar. 31, 1986), H. Glieb, E. Apostol, and C. Dan.

43. SU 983052 A1 (Dec. 23, 1982), V. K. Fomin, N. I. Varlamova, O. V. Leleedev, and A. P. Krasnov.

44. Jpn. Pat. 01083514 A2 (Mar. 29, 1989), Y. Oata, N. Goto, I. Motoyama, T. Iwashita, and K. Nomura.

45. PL 85616 (Sept. 15, 1976), W. Augustyn and co-workers.

46. U.S. Pat. 77,810,047 (June 27, 1977), W. C. Warneke.

47. U. Glabisz and co-workers, *Prezm. Chem.* **68**(1), 20, 29–30 (1989).

48. U.S. Pat. 685,100, AO (Aug. 1, 1986), A. B. Kreuzmann and D. A. Palmer.

49. V. S. Sakharov and co-workers, *Khim Prom-St (Moscow)* (1), 257 (1982).

50. SU 79-2829664 (July 30, 1978), G. A. Loptkina, V. I. Chernykh, and O. D. Fedorova.

51. SU 709537 (Jan. 15, 1980), G. A. Lopatking and V. I. Chernykh.

52. EP 210937 A1 (Feb. 4, 1987), L. Siegneurin.

53. M. S. Nesterova and T. Yu Magda, *Tekhnal. Obogashch. Polezn. Iskop. Sredni.* **A3** 3, 96–99 (1981).

54. R. V. Chernov and D. L. Dyubova, *Zh. Priki Khim. (Leningrad)* **56**(5), 1133–1135 (1983).

55. U. Glabisz, H. Gabryel, L. Kacalski, and B. Kic, *Pr. Nauk. Akad. Ekon. Im. Oskara Langego Wroclawiu* **338**, 165–169 (1986).

56. SU 1224263 A1 (Apr. 15, 1986), V. K. Fomin, N. I. Varlamova, V. P. Kozma, and M. N. Esin.

57. V. V. Pechkovskii, E. D. Dzyuba, and L. P. Valyu, *Zh. Priki (Leningrad)* **53**(5), 961–965 (1980).

58. A. A. Opalovskii and co-workers, *Zh. Neorg. Khim.* **20**(5), 1179–1183 (1975).

59. EP 476773 A1 (Mar. 25, 1992), J. Dijkhorst.

60. U.S. Pat. 82,406,420 A (Nov. 8, 1982), J. P. Harrison.

61. R. C. Kirby and A. S. Prokopovitsh, *Science* **191**, 717 (Feb. 1976).

62. P. L. Braekner, Allied Chemical, Industrial Chemicals Division, Morristown, N.J., private communication, Nov. 1978.

63. Data sheet, *Fluorspar*, Reynolds Chemicals, Richmond, Va., Mar. 1978.

64. *Analytical Methods Guide*, 6th ed., Orion Research, Inc., Cambridge, Mass., Aug. 1973, p. 18.

65. R. T. Oliver and A. G. Clayton, *Anal. Chim. Acta* **51**, 409 (1970).

66. R. B. Fisher, *J. Chem. Educ.* **51**, 387 (1974). (Note: reference to lithium fluoride on p. 389 is in error. It should read lanthanum fluoride.)

67. A. Julaicki and M. Trojanowicz, *Anal. Chim. Acta* **68**, 155 (1974).

68. *Criteria for a Recommended Standard—Occupational Exposure to Inorganic Fluorides*, NTIS Document PB-246692, National Institute for Occupational Safety and Health, Rockville, Md., 1975.

69. *1992–1993 Threshold Limit Values for Chemical Substances and Physical Agents*, American Conference of Governmental Industrial Hygienists, Cincinnati, Ohio, 1992.

70. Data sheet, *IR Transmission Materials*, Barnes Engineering Co., Instrument Division, Stamford, Conn., 1992.

71. C. B. Willingham and R. T. Newberg, *Exploratory Development of Fusion Cost Calcium Fluoride for 1.06 Micrometer Pulsed LaserOptics*, Progress Report, Sept. 25 to Dec. 25, 1976, NTIS Document C00-4029-2, Raytheon Co., for U.S. Energy Research and Development Administration, Jan. 1977.

72. P. Humbert and co-workers, *E. C. Photovoltair Sol. Energy Conference, Proc. Int. Conf. 1991*, 10th, pp. 261–266.

73. H. Hu, F. Lin, and Y. Yhan, *J. Feng. Mater. Scien. Forum*, 67–68; (Halide Glasses VI) 239–243 (1991).

74. H. Hu, and F. Lin, *J. Feng. Guisuanyan Xuebau* **18**(6), 501–505 (1990).

75. V. D. Khalilev and co-workers, *Fiz. Khim. Stekla* **17**(5), 740–743 (1991).

76. A. G. Nimchik and Kh. L. Usmanov, *N.A. Sirazhiddinov Uzb, Khim. Zh.* (6), 68–70 (1991).

77. Jpn. Pat. 03291324 A2 (Dec. 20, 1991), K. Masame and T. Matsuo (to Heisei).

78. FR 2661425 A1 (Oct. 31, 1991), Y. Bertaud and co-workers.

79. P. Comba, K. P. V. Lei, and T. G. Carnahan, *Bur. Mines Rep. Invest.*, RI **9372** 7 pp. (1991).

80. F1 83664 B (Apr. 30, 1991), T. Helttula and O. Jokinen.

81. Jpn. Pat. 03023251 A2 (Jan. 31, 1991), K. Shibata.

82. WO 9003265 A1 (Apr. 5, 1990), J. H. Weaver and co-workers; PCT Int. Appl. (1990), J. H. Weaver and co-workers.

83. Jpn. Pat. 63215553 A2 (Sept. 8, 1988), T. Honda, T. Yamada, K. Onigata, and S. Tosaka (to Showa).

84. Jpn. Pat. 63215547 A2 (Sept. 8, 1988) T. Honda, T. Yamada, K. Omigata, and S. Tosaka (to Showa); Jpn. Pat. 63215555 A2 (Sept. 8, 1988) (to Showa); Jpn. Pat. 63215557 A2 (Sept. 8, 1988) (to Showa); Jpn. Pat. 63215554 A2 (Sept. 8, 1988) (to Showa).

85. Belg. Pat. 905858 A1 (Apr. 1, 1987), M. Palchetti, S. Palella, and A. Crisafull.

86. U.S. Pat. 4,591,382 A (May 27, 1986), G. R. B. Elliott.

87. Czech. Pat. 221212 B (Jan. 15, 1986), J. Kotrbaty, J. Bar, J. Gocal, and T. Pazdiora.

88. I. Vulkova and co-workers, *Stroit. Mater. Silik, Prom-St.* **24**(12), 18–20 (1983).

89. Jpn. Pat. 53091932 (Aug. 12, 1978), H. Kitagawa (to Showa).

90. U.S. Pat. 3,832,451 (Aug. 27, 1974), E. F. Abrams and R. G. Shaver.

TARIQ MAHMOOD
CHARLES B. LINDAHL
Elf Atochem North America, Inc.

COBALT

Cobalt Difluoride

Cobalt difluoride [*10026-17-2*], CoF_2, is a pink solid having a magnetic moment of 4.266×10^{-23} J/T (4.6 Bohr magneton) (1) and closely resembling the ferrous (FeF_2) compounds. Physical properties are listed in Table 1. Cobalt(II) fluoride is highly stable. No decomposition or hydrolysis has been observed in samples stored in plastic containers for over three years.

CoF_2 is manufactured commercially by the action of aqueous or anhydrous hydrogen fluoride (see FLUORINE COMPOUNDS, INORGANIC–HYDROGEN) on cobalt carbonate (see COBALT COMPOUNDS) in a plastic, ie, polyethylene/polypropylene, Teflon, Kynar, rubber, or graphite-lined container to avoid metallic impurities. The partially hydrated mass is lavender pink in color. It is dried at 150–200°C and then pulverized to obtain the anhydrous salt. A very high (99.9%) purity CoF_2 having less than 0.05% moisture content has also been prepared by reaction of $CoCO_3$ and liquid hydrogen fluoride. This is a convenient synthetic route giving quantitative yields of the pure product. The reaction of $CoCl_2$ and anhydrous HF is no longer commercially practical because of environmental considerations. The various hydrates, eg, the cobalt(II) fluoride dihydrate [*13455-27-1*], $CoF_2 \cdot 2H_2O$, cobalt(II) fluoride trihydrate [*13762-15-7*], $CoF_2 \cdot 3H_2O$, and cobalt(II) fluoride tetrahydrate [*13817-37-3*], $CoF_2 \cdot 4H_2O$, have been obtained by the reaction of freshly prepared oxide, hydroxide, or carbonate of cobalt(II) and aqueous hydrogen fluoride (2).

Cobalt difluoride, used primarily for the manufacture of cobalt trifluoride, CoF_3, is available from Advance Research Chemicals, Inc., Aldrich Chemicals, and PCR in the United States, Fluorochem in the UK, and Schuhardt in Germany. The 1993 price varied from \$60 to \$200/kg depending on the quantity and the price of cobalt metal. CoF_2 is shipped as a corrosive and toxic material in DOT-approved containers.

Table 1. Physical Properties of the Cobalt Fluorides

Parameter	Cobalt difluoride[a]	Cobalt trifluoride
molecular weight	96.93	115.93
melting point, °C	1127	926
solubility, g/100 g[b]		
water	1.36	dec
anhydrous HF	0.036	
density, g/cm^3	4.43	3.88
ΔH_f, kJ/mol[c]	−672	−790
ΔG_f, kJ/mol[c]	−627	−719
S, J/(mol·K)[c]	82.4	95
C_p, J/(mol·K)[c]	68.9	92

[a]The bp of CoF_2 is 1739°C.
[b]CoF_2 is also soluble in mineral acids.
[c]To convert J to cal, divide by 4.184.

Cobalt Trifluoride

Cobalt(III) fluoride [10026-18-3] or cobalt trifluoride, CoF_3, is one of the most important fluorinating reagents. Physical properties may be found in Table 1. It is classified as a hard fluorinating reagent (3) and has been employed in a wide variety of organic and inorganic fluorination reactions. CoF_3, a light brown, very hygroscopic compound, is a powerful oxidizing agent and reacts violently with water evolving oxygen. It should be handled in a dry box or in a chemical hood and stored away from combustibles, moisture, and heat. The material should not be stored in plastic containers for more than two years. The crystals possess a hexagonal structure.

Cobalt trifluoride is readily prepared by reaction of fluorine (qv) and $CoCl_2$ at 250°C or CoF_2 at 150–180°C. Direct fluorination of CoF_2 leads to quantitative yields of 99.9% pure CoF_3 (4).

CoF_3 is used for the replacement of hydrogen with fluorine in halocarbons (5); for fluorination of xylylalkanes, used in vapor-phase soldering fluxes (6); formation of dibutyl decalins (7); fluorination of alkynes (8); synthesis of unsaturated or partially fluorinated compounds (9–11); and conversion of aromatic compounds to perfluorocyclic compounds (see FLUORINE COMPOUNDS, ORGANIC). CoF_3 rarely causes polymerization of hydrocarbons. CoF_3 is also used for the conversion of metal oxides to higher valency metal fluorides, eg, in the assay of uranium ore (12). It is also used in the manufacture of nitrogen fluoride, NF_3, from ammonia (13).

CoF_3 is available from Advance Research Chemicals, Inc., Aldrich Chemicals, Aesar, Johnson/Matthey, PCR, Pfaltz & Bauer, Noah Chemicals, and Strem Chemicals of the United States, Fluorochem of the UK, and Schuhardt of Germany. Demand for cobalt trifluoride varies from 100 to 1500 kg/yr and the 1993 price for smaller quantities ranged from $300 to $350/kg.

The ACGIH adopted TLV/TWA for 1992–1993 for fluorides as F^- is TWA 2.5 mg/m^3, and for cobalt as Co metal dust TWA 0.05 mg/m^3. Dust masks should be used while handling both the cobalt fluorides and all other cobalt compounds. CoF_3 is shipped as an oxidizer and a corrosive material.

BIBLIOGRAPHY

"Cobalt Compounds" under "Fluorine Compounds, Inorganic" in *ECT* 1st ed., Vol. 6, p. 693, by F. D. Loomis; "Cobalt" under "Fluorine Compounds, Inorganic" in *ECT* 2nd ed., Vol. 9, pp. 582–583, by W. E. White; in *ECT* 3rd ed., Vol. 10, pp. 717–718, by D. T. Meshri, Advance Research Chemicals Inc.

1. A. G. Sharp, *Quart. Rev. Chem. Soc.* **11**, 49 (1957).
2. I. G. Ryss, *The Chemistry of Fluorine and its Inorganic Compounds*, State Publishing House for Scientific and Chemical Literature, Moscow, 1956, Eng. Trans. ACE-Tr-3927, Vol. II, Office of Technical Services, U.S. Department of Commerce, Washington, D.C., 1960, pp. 659–665.
3. D. T. Meshri and W. E. White, "Fluorinating Reagents in Inorganic and Organic Chemistry," in the *Proceedings of the George H. Cady Symposium, Milwaukee, Wis.*, June 1970; M. Stacy and J. C. Tatlow, *Adv. Fluorine Chem.* **1**, 166 (1960).
4. E. A. Belmore, W. M. Ewalt, and B. H. Wojcik, *Ind. Eng. Chem.* **39**, 341 (1947).

5. R. D. Fowler and co-workers, *Ind. Eng. Chem.* **39**, 292 (1947).
6. Eur. Pat. Appl EP 281,784 (Sept. 14, 1988), W. Bailey and J. T. Lilack (to Air Products and Chemicals, Inc.).
7. U.S. Pat. 4,849,553 (July 18, 1989), W. T. Bailey, F. K. Schweighardt, and V. Ayala (to Air Products and Chemicals, Inc.).
8. Jpn. Kokai Tokkyo Koho JP 03 167,141 (July 19, 1991), H. Okajima and co-workers (to Kanto Denka Kaggo Co. Ltd.).
9. U.S. Pat. 2,670,387 (Feb. 23, 1954), H. B. Gottlich and J. D. Park (to E.I. du Pont de Nemours & Co., Inc.).
10. D. A. Rausch, R. A. Davis, and D. W. Osborn, *J. Org. Chem.* **28**, 494 (1963).
11. Ger. Pat. DD 287,478 (Feb. 28, 1991), W. Radeck and co-workers (to Akademie der Wissenschaften der DDR).
12. R. Hellman, Westinghouse Corp., Cincinnati, Ohio, private communication, Jan. 1989.
13. Jpn. Kokai Tokkyo Koho, JP 03,170,306 (July 23, 1991), S. Lizuka and co-workers (to Kanto Denka Kogyo Co. Ltd.).

DAYAL T. MESHRI
Advance Research Chemicals, Inc.

COPPER

Copper(II) Fluorides

Copper(II) forms several stable fluorides, eg, cupric fluoride [7789-19-7], CuF_2, copper(II) fluoride dihydrate [13454-88-1], $CuF_2 \cdot 2H_2O$, and copper hydroxyfluoride [13867-72-6], CuOHF, all of which are interconvertible. When CuF_2 is exposed to moisture, it readily forms the dihydrate, and when the latter is heated in the absence of HF, $CuOHF \cdot H_2O$ results. The colorless crystals of anhydrous CuF_2 are triclinic in structure and are moisture sensitive, turning blue when exposed to moist air. Physical properties of CuF_2 are listed in Table 1. CuF_2 reacts with ammonia to form $CuF_2 \cdot 5NH_3$.

Copper(I) fluoride is believed to be unstable (1) and no evidence for its existence has been found using mass spectrometry (2).

Manufacture. Several methods of synthesis for anhydrous CuF_2 have been reported, the most convenient and economical of which is the reaction of copper carbonate and anhydrous hydrogen fluoride to form the monohydrate, $CuF_2 \cdot H_2O$. Part of the water content from the monohydrate is removed by addition of excess HF. The excess HF is decanted and the remaining mass transferred to a Teflon-lined tray and dried under an atmosphere of hydrogen fluoride. The decanted material may also be dehydrated in a nickel or copper tray under an atmosphere of fluorine at 150–300°C. Both routes have successfully resulted in ultrapure (99.95%) white CuF_2 in good yields. The other method for the preparation of high purity anhydrous copper(II) fluoride is by the direct fluorination of commercially available CuOHF (3), or the action of a mixture of HF and BF_3 on $CuF_2 \cdot 2H_2O$ (4).

Uses. Copper(II) fluoride is used as a fluorinating reagent (5–7) in the fluorination of partially hydrogenated silanes; in superconductors (8–10); as a cathode material for high energy density primary and secondary batteries (qv)

Table 1. Physical Properties of CuF$_2$

Property	Value
molecular weight	101.54
melting point, °C	785 ± 10
boiling point, °C	1676
solubility, g/100 g	
water	4.75
anhydrous HF	0.01
aqueous 21.2% HF	12.1
density, g/cm^3	4.85
ΔH_f, kJ/mol[a]	−539
ΔG_f, kJ/mol[a]	−492
S, J/(mol·K)[a]	77.45
C_p, J/(mol·K)[a]	65.55

[a]To convert from J to cal, divide by 4.184.

(11–14); for the skeletal rearrangements of olefins (15); low temperature isomerization of pentane and hexane (16); as a selective herbicide (17); as a termite repellant (18); as a fungicide (19); in the manufacturing of conductive bicomponent fibers for electromagnetic shields (20); as a catalyst for the removal of nitrogen oxides from flue gases (21), and for the synthesis of heterocyclic tetraaromatics (22). The dihydrate is used in the casting of gray iron.

The high purity anhydrous copper(II) fluoride must be stored in a tightly closed or sealed container under an atmosphere of argon. The dihydrate may be stored in polyethylene-lined fiber drums. The ACGIH (1992–1993) adopted toxicity value for copper as Cu is 1 mg/m^3, and for fluorides a F$^-$, 2.5 mg/m^3.

In spite of the many applications for copper(II) fluoride, demand is restricted to 1 to 10 kg lots. It is available in the United States from Advance Research Chemicals, Aldrich Chemicals, Atomergic, Aesar, Johnson/Matthey, Cerac Corp., and PCR Corp. The 1993 price for the anhydrous copper(II) fluoride varied from $400 to $600/kg depending on the amount required. The dihydrate is available at $22/kg.

BIBLIOGRAPHY

"Copper Compounds" under "Fluorine Compounds, Inorganic" in *ECT* 1st ed., Vol. 6, p. 693, by F. D. Loomis; "Copper" under "Fluorine Compounds, Inorganic" in *ECT* 2nd ed., Vol. 9, pp. 583–584, by W. E. White; in *ECT* 3rd ed., Vol. 10, pp. 719–720, by D. T. Meshri, Advance Research Chemicals Inc.

1. I. G. Ryss, *Zh. Fiz. Khim.* **29**, 936 (1955).
2. R. K. Kent, J. D. McDonald, and J. Margrave, *J. Phys. Chem.* **70**, 874 (1966).
3. J. R. Lundquist, *Final Report Pacific Northwest Laboratories*, Seattle, Wash. NASA CR-72571, June 12, 1969; U.S. Pat. 3,607,015 (Sept. 21, 1971), J. R. Lundquist, R. Wash, and R. B. King (to NASA).
4. U.S. Pat. 2,782,099 (Feb. 19, 1957), D. A. McCaulay (to Standard Oil of Indiana).

5. Jpn. Kokai Tokkyo Koho, 02, 302,311 (Dec. 14, 1990), I. Harada, M. Aritsuka, and A. Yoshikawa (to Mitsui Tiatsu Chemicals).

6. B. Leng and J. H. Moss, *J. Flourine Chem.* **8**, 165 (1976).

7. J. H. Moss, R. Ottie, and J. B. Wilford, *J. Fluorine Chem.* **3** 317 (1973).

8. Jpn. Kokai Tokkyo Koho 01, 133,921 (Nov. 18, 1987), S. Aoki and co-workers (to Fujikura Ltd.).

9. Jpn. Kokai Tokkyo Koho, 63,313,426 (Dec. 21, 1988), Y. Tanaka, T. Shibata, and N. Uno (to Furukawa Electric Co. Ltd.).

10. Jpn. Kokai Tokkyo Koho, 63,288,943 (Nov. 25, 1988), T. Kyodo, S. Hirai, and K. Takahashi (to Sumitomo Electric Industries Ltd.).

11. Eur. Pat. 286,990 (Apr. 17, 1987), F. W. Dampier and R. M. Mank (to GTE Laboratories, Inc.).

12. Ger. Offen. 2,215,210 (Oct. 19, 1972), O. S. Savinovw (to Honeywell Inc.).

13. U.S. Pat. 3,953,232 (Apr. 27, 1976), W. L. Roth and G. C. Farrington (to General Electric Co.).

14. J. H. Kennedy and J. C. Hunter, *J. Electrochem. Soc.* **123**(1), 10 (1976).

15. U.S. Pat. 3,751,513 (Aug. 7, 1973), J. J. Tazuma (to Goodyear Tire & Rubber Co.).

16. Fr. Pat. 2,157,083 (July 6, 1973), T. Bernard (to Institut Francois du Petrole, des Carburnats et Lubricants).

17. I. G. Ryss, *The Chemistry of Fluorine and its Inorganic Compounds*, State Publishing House for Scientific and Chemical Literature, Moscow, 1956, Eng. Trans. ACE-Tr-3927, Vol. II, Office of Technical Services, U.S. Department of Commerce, Washington, D.C., 1960, p. 643.

18. G. N. Wolcott, *P. R. Agri. Exp. Stu. Bull.*, **73** (1947).

19. H. Martin, R. L. Wain, and E. H. Wilkinson, *Ann. Appl. Biol.* **29**, 412 (1942).

20. Jpn. Kokai Tokyo Koho, 01, 61,570 (Aug. 31, 1987), M. Oshida (to Tijin Ltd.).

21. Jpn. Kokai Tokkyo Koho, 63, 49,255 (Mar. 2, 1988), Y. Kawasaki (to Matsushita Electric Industrial Co. Ltd.).

22. T. Kaufmann and Z. R. Otter, *Angew. Chem.* **88**, 513 (1976).

DAYAL T. MESHRI
Advance Research Chemicals, Inc.

GERMANIUM

Germanium forms both a difluoride and a tetrafluoride. It also forms a stable hexafluorogermanate complex ion, GeF_6^{2-}, that is present in the aqueous acid and a number of salts.

Germanium Difluoride

Germanium difluoride [*13940-63-1*] is a white solid, mp 110°C, and $d_{23} = 3.7$ g/cm^3. This compound can be vacuum distilled. In a mass spectrometer, ions corresponding to $(GeF_2)_n^+$, where $n = 1-4$ have been observed at 361–403 K (1). At higher temperatures, GeF_2 disproportionates to GeF_4 (g), Ge (s), and GeF (g). Presumably the GeF [*39717-71-0*] formed is unstable and subsequently condenses and disproportionates. The initially reported (2) orange-red solid becomes red-brown going finally to the black of metallic Ge. The difluoride deliquesces (2) in moist air producing germanium(II) hydroxide (see GERMANIUM AND GERMANIUM COMPOUNDS). The difluoride is soluble in aqueous hydrofluoric acid and gives a solution having the reducing properties expected of divalent germanium. GeF_2 reacts with aqueous solutions of alkali metal fluorides to produce trifluoroger-manites (3), eg, cesium trifluorogermanite [72121-41-6], $CsGeF_3$, and potassium trifluorogermanite [72121-42-7], $KGeF_3$. The GeF_3^- ion is oxidized to GeF either by oxygen in neutral solution or by its reduction of H^+ to hydrogen in hydrofluoric acid solution. Germanium difluoride is soluble in ethanol (2), forms a reversible complex with diethyl ether (2), and forms a dimethyl sulfoxide complex [72121-40-5], $GeF_2 \cdot OS(CH_3)_2$, which is decomposed at 240°C (3).

Germanium difluoride can be prepared by reduction (2,4) of GeF_4 by metallic germanium, by reaction (1) of stoichiometric amounts of Ge and HF in a sealed vessel at 225°C, by Ge powder and HgF_2 (5), and by GeS and PbF_2 (6). GeF_2 has been used in plasma chemical vapor deposition of amorphous film (see PLASMA TECHNOLOGY; THIN FILMS) (7).

Germanium Tetrafluoride

Germanium tetrafluoride [7783-58-6] is a gas having a garlic-like odor, a reported (8) triple point of $-15°C$ and 404.1 kPa (4.0 atm), and a vapor pressure near 100 kPa (ca 1 atm) at $-36.5°C$. Germanium tetrafluoride fumes strongly in air and is hydrolyzed in solution to form GeF_6^{2-} ions. Germanium tetrafluoride can be prepared (8,9) by thermal decomposition of barium hexafluorogermanate [60897-63-4], $BaGeF_6$. Direct fluorination of germanium has been reported to give GeF_4 of higher purity (2). High purity GeF_4 is also manufactured by reaction of a finely powdered GeO_2 suspension in H_2SO_4 with UF_6 (10) or by the reaction of Ge metal or its oxide with F_2 or NF_3 (11). COF_2 has been used as a mild fluorinating agent to produce GeF_4 from GeO_2 (12). GeF_4 is used in ion implantation (qv) in semi-conductor chips (see SEMICONDUCTORS) (13,14). Germanium tetrafluoride acts as a Lewis acid (15,16) to form complexes with many donor molecules. The tetra-fluoride is commercially available.

Fluorogermanates

Fluorogermanic acid [16950-43-9] solutions, H_2GeF_6, are prepared by reaction of germanium dioxide and hydrofluoric acid or by hydrolysis of germanium tetra-fluoride. Addition of potassium fluoride, barium chloride (9), or other salts results in hexafluorogermanates such as potassium hexafluorogermanate [7783-73-5], K_2GeF_6, or $BaGeF_6$, both of which are stable at temperatures up to 500°C where $BaGeF_6$ starts to decompose to GeF_4 and BaF_2.

Germanium tetrafluoride produces hydrogen fluoride in aqueous acidic so-lutions. Hydrogen fluoride is toxic and very corrosive. The OSHA permissible ex-posure limit (17) and the American Conference of Governmental Industrial Hy-gienists' (ACGIH) TLV for fluoride is 2.5 mg/m^3 of air (18).

BIBLIOGRAPHY

"Germanium" under "Fluorine Compounds, Inorganic," in *ECT* 2nd ed., Vol. 9, pp. 584–585, by W. E. White, Ozark-Mahoning Co.; in *ECT* 3rd ed., Vol. 10, pp. 720–721, by C. B. Lindahl, Elf Atochem North America, Inc.

1. K. F. Zmbov, J. W. Hastie, R. Hauge, and J. L. Margrave, *Inorg. Chem.* **7**, 608 (1968).
2. N. Barlett and K. C. Yu. *Can. J. Chem.* **39**, 80 (1961).
3. E. L. Muetterties, *Inorg. Chem.* **1**, 342 (1962).
4. Jpn. Kokai Tokyo Koho 61111520 A2, (May 29, 1986), S. Ishihara, M. Hirooka, and S. Oono (to Showa).
5. P. Rivere, A. Castel, J. Stage, and C. Abdenhadheric, *Organometallics* **10**(5), 1227–1228 (1991).
6. S. M. Van der Kerk, *Polyhedron* **2**(6), 509–512 (1983).
7. EP 229707 A1 (July 22, 1987), S. Ishihara, M. Hirooka, J. Hanna, and I. Shimizu.
8. L. M. Dennis and A. W. Laubengayer, *Z. Phys. Chem.* **130**, 420 (1927).
9. C. J. Hoffman and J. S. Gutowsky, *Inorg. Syn.* **4**, 147 (1953).
10. Ger. Offen. DE 3841212 A1 (June 13, 1990), R. Doetzer.
11. EP 89-104364 (Mar. 11, 1989), I. Harada, Y. Yoda, N. Iwanaga, T. Nishitsuti, and A. Kikkawa.
12. S. P. Mallela, O. D. Gupta, and J. M. Shreeve, *Inorg. Chem.* **27**(1), 208–209 (1988).
13. A. Gottdang and co-workers, *Nucl. Instrum. Methods Phys. Res., Sect. B* **B55**(1–9), 310–313 (1991).
14. A. Ferreiro, J. DePontcharra, C. Jaussaud, and E. Lora-Tamayo, *Vacuum* **39**(7–8), 775–779 (1989).
15. E. L. Muetterties, *J. Am. Chem. Soc.* **82**, 1082 (1960).
16. R. C. Aggarwal and M. Onyszchuk, *J. Inorg. Nucl. Chem.* **30**, 3351 (1968).
17. *Code of Federal Regulation*, Title 29, Part 1910.1000, Washington, D.C., 1993.
18. *Threshold Limit Values for Chemical Substances and Physical Agents, 1992–1993*, The American Conference of Governmental Industrial Hygienists, Cincinnati, Ohio.

TARIQ MAHMOOD
CHARLES B. LINDAHL
Elf Atochem North America, Inc.

HALOGENS

The halogen fluorides are binary compounds of bromine, chlorine, and iodine with fluorine. Of the eight known compounds, only bromine trifluoride, chlorine trifluoride, and iodine pentafluoride have been of commercial importance. Properties and applications have been reviewed (1–7) as have the reactions with organic compounds (8). Reviews covering the methods of preparation, properties, and analytical chemistry of the halogen fluorides are also available (9).

The halogen fluorides are best prepared by the reaction of fluorine with the corresponding halogen. These compounds are powerful oxidizing agents; chlorine trifluoride approaches the reactivity of fluorine. In descending order of reactivity the halogen fluorides are chlorine pentafluoride [13637-63-3], ClF_5; chlorine trifluoride [7790-91-2], ClF_3; bromine pentafluoride [7789-30-2], BrF_5; iodine heptafluoride [16921-96-3], IF_7; chlorine monofluoride [7790-91-2], ClF; bromine trifluoride [7787-71-5], BrF_3; iodine pentafluoride [7783-66-6], IF_5; and bromine monofluoride [13863-59-7], BrF.

The halogen fluorides offer an advantage over fluorine in that the former can be stored as liquids in steel containers and, unlike fluorine, high pressure is not required. Bromine trifluoride is used as an oxidizing agent in cutting tools used in deep oil-well drilling, whereas chlorine trifluoride is used to convert uranium to UF_6 in nuclear fuel processing (see NUCLEAR REACTORS; PETROLEUM).

Except for iodine pentafluoride, the halogen fluorides have no commercial importance as fluorinating agents. Their extreme reactivity and the accompanying energy release of the reaction can be sufficient to disrupt C—C bonds and can result in explosive reactions or fires. In addition, both halogens are generally then introduced into organic compounds, giving rise to a complex mixture of products.

Physical Properties

The physical properties of the halogen fluorides are given in Table 1. Calculated thermodynamic properties can be found in Reference 24.

Bromine Monofluoride. Bromine monofluoride is red to red-brown (4) and is unstable, disproportionating rapidly into bromine and higher fluorides. Therefore, the measurement of its physical properties is difficult and the values reported in Table 1 are only approximate. The uv-absorption spectrum is available (25).

Bromine Trifluoride. Bromine trifluoride is a colorless liquid. The commercial grade is usually amber to red because of slight bromine contamination. The molecule has a distorted T structure (26). Infrared spectral data (26–30), the uv-absorption spectrum (31), and vapor pressure data (32) may be found in the literature.

Bromine Pentafluoride. Bromine pentafluoride is a colorless liquid having the molecular structure of a tetragonal pyramid (5). The index of refraction n_D is 1.3529 (33). Infrared spectra (13,34), the uv-absorption spectrum (35), and vapor pressure data (11) are all available.

Chlorine Monofluoride. Chlorine monofluoride is a colorless gas that condenses to a liquid with a slight yellow cast and freezes to a white solid. The infrared spectrum of gaseous chlorine monofluoride and the Raman spectrum of the liquid have been studied (36). The uv-absorption spectrum (37) and vapor pressure data are also available (11).

An equilibrium exists between chlorine trifluoride, chlorine monofluoride, and fluorine gas (38). The equilibrium constant may be expressed as

$$K_p = \frac{(P_{ClF})(P_{F_2})}{(P_{ClF_3})}$$

where P_X is the partial pressure of substance X. Values for K_p are

Temperature, °C	Pa (mm Hg)
250	30 (0.22)
300	240 (1.8)
350	1450 (10.9)

Table 1. Physical Properties of the Halogen Fluorides[a]

Property	BrF	BrF$_3$	BrF$_5$	ClF	ClF$_3$	ClF$_5$	IF$_5$	IF$_7$
boiling point, °C	20	125.7	40.9	−100.1	11.75	−13.1	102	5.5
melting point, °C	−33	8.8	−60.6	−155.6	−76.3	−103	8.5	4.5
liquid density at 25°C, g/mL		2.803	2.463[b]	1.620[b]	1.825[b]	1.790[c]	3.252	2.669
critical temperature, °C					154.5	142.6		
−ΔH_f (g) at 25°C, kJ/mol[d]	58.5	255.4	443.9	56.4	164.5	254.6	839.3	961.0
−ΔG_f (g) at 25°C, kJ/mol[d]	73.6	229.1	351.5	57.7	124.4	163.0	771.6	841.4
heat of vaporization, kJ/mol[d]		42.8	30.6	20.1	27.50	22.21	35.92	24.7
E_{diss}, kJ/mol[d]	254 260			253	105 160			122
heat of fusion, kJ/mol[d]		12.01	5.66		7.60		11.21	
specific heat, gas, J/(mol·K)[d]	32.9	66.5[e]		32.0	65.2		99.1	136.3
specific conductivity, liquid, at 25°C, W·cm	8.0 × 10^{-3}	9.1 × 10^{-8}		1.9 × 10^{-7f}	4.9 × 10^{-9}	1.25 × 10^{-9g}	5.4 × 10^{-6}	10^{-9}

[a]Compiled from References 8, 10–23.
[b]At boiling point.
[c]At 20°C.
[d]To convert J to cal, divide by 4.184.
[e]The specific heat of the liquid is 124.5 J/(mol·K) (29.8 cal/(mol·K)).
[f]At 145 K.
[g]At 256 K.

Chlorine Trifluoride.   Chlorine trifluoride is a pale yellow liquid or a colorless gas. It freezes to a white solid and undergoes a transition in the solid state at $-82.66°C$ (11). The infrared and Raman spectra have been studied (28,29,39,40) as has the uv absorption spectrum (41). Vapor pressure data are given in Reference 42. The viscosity of the liquid is 0.448 mPa·s($=cP$) at 290 K, and the surface tension is 26.6 mN/m ($=dyn/cm$) at 273 K (43). The density of the solid is 2.530 g/cm^3 at 153 K (44). The vapors of chlorine trifluoride are nonideal and this has been attributed to the following equilibrium:

$$2 \, ClF_3 \rightleftharpoons Cl_2F_6$$

The equilibrium constant at 24.2°C is 2.84 kPa (21.3 mm Hg) (11):

$$K_p = \frac{P_{Cl_2F_6}}{(P_{ClF_3})^2}$$

Chlorine Pentafluoride.   Chlorine pentafluoride is a colorless gas at room temperature. The ir and Raman spectra of the liquid and gas phase have been studied (34,39). The uv absorption spectrum (45) and vapor pressure data may be found in the literature (18).

Iodine Pentafluoride.   Iodine pentafluoride is a straw-colored liquid; the ir and Raman spectra of the gas phase have been studied (19,46,47); vapor pressure data are given in References 14 and 48.

Iodine Heptafluoride.   Iodine heptafluoride is a colorless liquid; the ir and Raman spectra of the gas have been studied (47,49); vapor pressure data are available (19).

Chemical Properties

Reactions With Metals.   All metals react to some extent with the halogen fluorides, although several react only superficially to form an adherent fluoride film of low permeability that serves as protection against further reaction. This protective capacity is lost at elevated temperatures, however. Hence, each metal has a temperature above which it continues to react. Mild steel reacts rapidly above 250°C. Copper and nickel lose the ability to resist reaction above 400 and 750°C, respectively.

Metals that form no protective fluoride film react readily with the halogen fluorides. Chlorine trifluoride reacts with Hg, As, Ca, Ti, Co, Pt, and Pb at elevated temperatures to give HgF_2, AsF_5, CaF_2, TiF_3, CoF_3, PtF_5, and PbF_{3-4}, respectively (50). Titanium alloys and molybdenum alloys and niobium metal react vigorously with ClF_3 (51). Molybdenum and tungsten react with BrF_3 to form the volatile MoF_6 and WF_6 (52,53). Chlorine trifluoride (54) converts Nb and Ta to pentafluorides, and Mo and Re to hexafluorides.

Uranium is converted by ClF_3, BrF_3, and BrF_5 to UF_6. The recovery of uranium from irradiated fuels has been the subject of numerous and extensive investigations sponsored by atomic energy agencies in a number of countries

(55–63). The fluorides of the nuclear fission products are nonvolatile; hence the volatile UF_6 can be removed by distillation (see NUCLEAR REACTORS; URANIUM AND URANIUM COMPOUNDS).

The rapid reaction of ClF_3 and BrF_3 with metals is the basis of the commercial use in cutting pipe in deep oil wells (64–68). In this application, the pipe is cut by the high temperature reaction of the halogen fluoride and the metal.

Reactions With Nonmetals. Few elements withstand the action of interhalogen compounds at elevated temperatures and many react violently at or below ambient temperatures. The oxidation of the element proceeds to its highest valence state, whereas the halogen other than fluorine is reduced either to the element or a lower valent interhalogen derivative. The oxidizing capacity of the interhalogens varies substantially from compound to compound. For example, chlorine trifluoride reacts vigorously with virtually every element at room temperature; on the other hand, iodine pentafluoride has a much milder oxidizing power. Thus chlorine trifluoride oxidizes xenon to xenon fluorides whereas iodine pentafluoride does not react (11). Furthermore, all stable halogen fluorides, except iodine pentafluoride, oxidize radon between -195 and $25°C$ (69) (see HELIUM-GROUP GASES, COMPOUNDS).

In general, reactions of halogens and halogen fluorides yield mixtures (4). Bromine pentafluoride reacts with iodine at ambient temperatures and with chlorine at 250–300°C, giving mixtures of interhalogen compounds. Bromine pentafluoride has been stored with bromine in steel cylinders at room temperature for extended periods of time without appreciable reaction. At elevated temperatures BrF_3 is formed. Bromine and chlorine trifluoride give bromine trifluoride (60–80% yields) and chlorine (70). Iodine and chlorine trifluoride produce IF_5 and ICl (70). Chlorine reacts with chlorine trifluoride to produce chlorine monofluoride. Bromine and iodine react with chlorine monofluoride to produce BrF_3 and IF_5, respectively. Bromine reacts with IF_5 on warming to give IBr and BrF_3 (4). Chlorine reacts with IF_7 to give ClF and interhalogens of iodine and chlorine (4).

Halogen fluorides react with sulfur, selenium, tellurium, phosphorus, silicon, and boron at room temperature to form the corresponding fluorides. Slight warming may be needed to initiate the reactions (4) which, once started, proceed rapidly to completion accompanied by heat and light. The lack of protective film formation allows complete reaction.

Reactions With Inorganic Compounds. In an investigation of the reactions of BrF_3 with oxides (71–73), little or no reaction was found with the oxides of Be, Mg, Ce, Ca, Fe, Zn, Zr, Cd, Sn, Hg, Th, and the rare earths, whereas the oxides of Mo and Re formed stable oxyfluorides. Manganese dioxide reacted incompletely but $KMnO_4$ released oxygen quantitatively. Complete replacement of oxygen took place with oxides of B, Ti, V, Cr, Cu, Ge, As, Se, Nb, Sb, Te, I, Ta, W, Tl, Pb, Bi, and U at 75°C.

Oxygen was partially replaced when P_2O_5, V_2O_5, and CrO_3 were dissolved in IF_5 to form POF_3, VOF_3, and CrO_2F_2. With WO_3 and MoO_3, $WO_3{\cdot}IF_5$ and $2\ MoO_3{\cdot}3IF_5$ complexes were formed (74). Reaction of excess IF_5 with $KMnO_4$ gives MnO_3F, IOF_3 [19058-78-7], and IO_2F [28633-62-7] (74).

Water reacts violently with all halogen fluorides. The hydrolysis process can be moderate by cooling or dilution. In addition to HF, the products may include oxygen, free halogens (except for fluorine), and oxyhalogen acids.

Fused silica and Pyrex glass (qv) are not significantly attacked by halogen fluorides up to 100°C if HF is absent.

Salts of halides other than fluorides react with halogen fluorides to produce the corresponding metal fluoride and release the free higher halogen. Filter paper moistened with KI solution darkens readily in the presence of ClF_3 and the bromine fluorides. This serves as a sensitive detector for leaks in equipment containing these halogen fluorides. If a metal exhibits more than one valence, reactions of halogen fluorides with halides (including fluorides) yields the fluoride in which the metal is usually at its highest valence. Chlorine trifluoride converts silver salts to AgF_2, cobalt compounds to CoF_3, and so on. Such reactions are useful in the preparation of fluorinated organic materials and also regenerate fixed-bed fluorinating agents without using fluorine itself (75). A comparison of the efficacy with which various halogen fluorides convert cobalt(II) chloride to cobalt(III) fluoride is indicated below (76).

fluorinating agent	ClF_3	BrF_5	BrF_3	IF_5
solid product	100% CoF_3	55% CoF_3	45% CoF_3	
		45% CoF_2	55% CoF_2	72% CoF_2

Reactions With Organic Compounds. Most organic compounds react vigorously exhibiting incandescence or even explosively with ClF_3 and BrF_3 (8,77,78). For this reason, only the less reactive iodine pentafluoride is used as a fluorinating agent to any extent. The reaction of iodine pentafluoride and various organic compounds is described in the literature (79–84).

Inert diluents in which the halogen fluorides are soluble, such as carbon tetrachloride and methylene dichloride (85), have been used for control in liquid-phase reactions. Anhydrous hydrogen fluoride is a good diluent, because it does not react with halogen fluorides but rather is miscible in all proportions with them (86). Control of vapor-phase reactions may be improved by diluting the halogen fluoride with an inert gas such as nitrogen or argon. However, any reaction of the halogen fluorides with an organic compound in either the gas or liquid phase should be approached with extreme caution.

Bromine trifluoride in bromine solution reacts smoothly with bromofluoro-ethanes to give a clean, progressive substitution of the bromine by fluorine with no replacement of the hydrogen (87). The relative ease of replacement of bromine in various groups is $CBr_3 > CBr_2F > CHBr_2 > CF_2Br > CHBrF > CH_2Br$.

The reactions with IF_5 are more amenable to control giving good yields of identifiable products and lower losses from oxidative fragmentation. The reaction of IF_5 and iodine with tetrafluoroethylene produces the telomer perfluoroethyl iodide [354-64-3] in yields that exceed 98% based on CF_2=CF_2 using SbF_5 as a catalyst (88).

$$2\, I_2 + IF_5 + 5\, CF_2{=}CF_2 \rightarrow 5\, CF_3CF_2I$$

Aryl and alkyl isothiocyanates are converted in good yields by IF_5 in pyridine to thiobis(N-trifluoromethylamines) (89,90):

$$RN{=}C{=}S + IF_5 \rightarrow RN\underset{\underset{\displaystyle CF_3}{|}}{\overset{\overset{\displaystyle CF_3}{|}}{S}}NR$$

Fluorination of aromatic isothiocyanates occurs much more readily than that of alkyl isothiocyanates. Alcohols treated with IF_5 in DMF give 30–70% yields of their respective formates.

Iodine pentafluoride fluorinates CCl_4 at room temperature to give $CClF_3$ and traces of CCl_2F_2 (91). It reacts with CHI_3 to yield CHF_3 and $CHIF_2$ (92) and with CI_4 to form C_2F_2 (92) and CIF_3 (93). With CBr_4 at 90°C, IF_5 forms 83% CBr_2F_2 and minor amounts of CBr_3F and $CBrF_3$.

Liquid Halogen Fluorides as Reaction Media. Bromine trifluoride and iodine pentafluoride are highly dimerized and behave as ionizing solvents:

$$(BrF_3)_2 \rightleftharpoons BrF_2{}^+ + BrF_4{}^-$$

$$(IF_5)_2 \rightleftharpoons IF_4{}^+ + IF_6{}^-$$

Antimony pentafluoride dissolves in each to form $BrF_2{}^+SbF_6{}^-$ and $IF_4{}^+SbF_6{}^-$ which act as acids. Potassium fluoride likewise forms $KBrF_4$ [15705-87-0] and KIF_6 [20916-97-6] which are both stable, white, crystalline solids (3,94,95). These compounds dissociate at 200°C to KF and the corresponding halogen fluoride. Other salts are formed similarly (71,95–99). Some of the acids and bases of these systems are listed in Table 2.

Table 2. Acids and Bases Derived from Halogen Fluorides

Name	CAS Registry Number	Formula	Reference
Acid			
difluorobromine hexafluoroantimonate	[19379-47-6]	BrF_2SbF_6	73
bis(difluorobromine) hexafluorostannate	[72229-86-8]	$(BrF_2)_2SnF_6$	73
difluorobromine hexafluoroniobate	[72229-87-9]	BrF_2NbF_6	100
difluorobromine hexafluorotantalate	[35967-87-4]	BrF_2TaF_6	100
difluorobromine hexafluorobismuthate	[36608-81-8]	BrF_2BiF_6	100
tetrafluoroiodine hexafluoroantimonate	[41646-48-4]	IF_4SbF_6	101
Base			
potassium hexafluorobromate	[32312-22-4]	$KBrF_6$	95
silver tetrafluorobromate	[35967-89-6]	$AgBrF_4$	95
barium tetrafluorobromate	[35967-90-9]	$Ba(BrF_4)_2$	95
potassium hexafluoroiodate	[20916-97-6]	KIF_6	94

The use of ClF_3 and BrF_3 as ionizing solvents has been studied (102,103). At 100°C and elevated pressures, significant yields of $KClF_4$ [19195-69-8], $CsClF_4$ [15321-04-7], $RbClF_4$ [15321-10-5], $KBrF_6$ [32312-22-4], $RbBrF_6$ [32312-22-4], and $CsBrF_6$ [26222-92-4] were obtained. Chlorine trifluoride showed no reaction with lithium fluoride or sodium fluoride.

Manufacture

Bromine Trifluoride. Bromine trifluoride is produced commercially by the reaction of fluorine with bromine in a continuous gas-phase process where the ratio of fluorine to bromine is maintained close to 3:1. It is also produced in a liquid-phase batch reaction where fluorine is added to liquid bromine at a temperature below the boiling point of bromine trifluoride.

Chlorine Trifluoride. Chlorine trifluoride is produced commercially by the continuous gas-phase reaction of fluorine and chlorine in a nickel reactor at ca 290°C. The ratio of fluorine to chlorine is maintained slightly in excess of 3:1 to promote conversion of the chlorine monofluoride to chlorine trifluoride. Sufficient time in the reactor must be provided to maintain high conversions to chlorine trifluoride. Temperature control is also critical because the equilibrium shift of chlorine trifluoride to chlorine monofluoride and fluorine is significant at elevated temperatures.

Iodine Pentafluoride. Iodine pentafluoride is produced by the reaction of iodine and fluorine. Because iodine has a high melting point, the reaction is either performed in a solvent or the reaction is maintained at a temperature where the iodine is liquid. In a continuous process using a solvent (104), ca 1% I_2 is dissolved in IF_5 and passed to a reactor where it is contacted with F_2 gas. The IF_5 is continuously discharged from the reactor where a small portion is taken off as product and the larger portion of the stream is recycled.

In another process (105), fluorine gas reacts under pressure with liquid I_2 held above its melting point (113°C) but below a temperature (150°C) that would result in the formation of significant amounts of IF_7. Fluorine is added continuously until all the iodine has been converted and yields of IF_5 in excess of 95% are reported. The reaction pressure is ca 300 kPa (3 atm) so that the IF_5 produced in the reaction is maintained as a liquid.

Economic Aspects

U.S. production of bromine trifluoride is several metric tons per year mostly used in oil-well cutting tools. Air Products and Chemicals, Inc. is the only U.S. producer. The 1992 price was ca $80/kg.

U.S. chlorine trifluoride production is several metric tons per year. Most of the product is used in nuclear fuel processing. A large production plant for chlorine trifluoride was operated in Germany during World War II with a reported capacity of 5 t/d (106,107). As of 1993, Air Products and Chemicals, Inc. was the only U.S. producer. The 1992 price was ca $100/kg.

United States production of iodine pentafluoride is several hundred metric tons per year. The two U.S. producers are Air Products and Chemicals, Inc. and AlliedSignal, Inc. The 1992 price was ca $50/kg.

Shipping, Specifications, and Analytical Methods

Bromine trifluoride is commercially available at a minimum purity of 98% (108). Free Br_2 is maintained at less than 2%. Other minor impurities are HF and BrF_5. Free Br_2 content estimates are based on color, with material containing less than 0.5% Br_2 having a straw color, and ca 2% Br_2 an amber-red color. Fluoride content can be obtained by controlled hydrolysis of a sample and standard analysis for fluorine content. Bromine trifluoride is too high boiling and reactive for gas chromatographic analysis. It is shipped as a liquid in steel cylinders in quantities of 91 kg or less. The cylinders are fitted with either a valve or plug to facilitate insertion of a dip tube. Bromine trifluoride is classified as an oxidizer and poison by DOT.

Chlorine trifluoride is commercially available at 99% minimum purity (108) and is shipped as a liquid under its own vapor pressure in steel cylinders in quantities of 82 kg per cylinder or less. Chlorine trifluoride is classified as an oxidizer and poison by DOT.

Iodine pentafluoride is commercially available at a minimum purity of 98% (108). Iodine heptafluoride is the principal impurity and maintained at less than 2%. Free I_2 and HF are minor impurities. Iodine pentafluoride is shipped as a liquid in steel cylinders in various quantities up to 1350 kg cylinders. It is classified as an oxidizer and poison by DOT.

Volatile impurities, eg, F_2, HF, ClF, and Cl_2, in halogen fluoride compounds are most easily determined by gas chromatography (109–111). The use of Ftoroplast adsorbents to determine certain volatile impurities to a detection limit of 0.01% has been described (112–114). Free halogen and halide concentrations can be determined by wet chemical analysis of hydrolyzed halogen fluoride compounds.

Handling

The halogen fluorides are highly reactive compounds and must be handled with extreme caution (115–120). The more reactive compounds, such as bromine trifluoride and chlorine trifluoride, are hypergolic oxidizers and react violently and sometimes explosively with many organic and inorganic materials at room temperature. At elevated temperatures, these cause immediate ignition of most organic substances and many metals.

Materials of Construction. Nickel, Monel, copper, mild steel, 304 stainless steel, and aluminum have been found to be suitable metals of construction for handling halogen fluorides (51). Silver solder is acceptable; lead solder is not recommended. Nickel and Monel are more suitable for elevated temperatures. Steel is not dependable above 150°C. Gaskets may be made of soft copper or calcium fluoride-impregnated polytetrafluoroethylene. Packing and gasketing should have smooth surfaces and the surfaces should be free from organic greases and embedded impurities, which may ignite in the presence of halogen fluorides.

Equipment should be carefully and completely degreased and passivated with low concentrations of fluorine or the gaseous halogen fluoride before use. Special care should be taken that valves are completely disassembled and each part carefully cleaned.

Disposal. Moderate amounts of chlorine trifluoride or other halogen fluorides may be destroyed by burning with a fuel such as natural gas, hydrogen, or propane. The resulting fumes may be vented to water or caustic scrubbers. Alternatively, they can be diluted with an inert gas and scrubbed in a caustic solution. Further information on disposal of halogen fluorides is available (115–118).

Toxicity

The time-weighted average (TWA) concentrations for 8-h exposure to bromine trifluoride, bromine pentafluoride, chlorine trifluoride, chlorine pentafluoride, and iodine pentafluoride have been established by ACGIH on a fluoride basis to be 2.5 mg/m^3. NIOSH reports (121) the following inhalation toxicity levels for chlorine trifluoride: LC_{50} monkey, 230 ppm/h; LC_{50} mouse, 178 ppm/h; for chlorine pentafluoride: LC_{50} monkey, 173 ppm/h; mouse, 57 ppm/h.

No toxicity data have been reported on the other halogen fluorides, but all should be regarded as highly toxic and extremely irritating to all living tissue.

Uses

Chlorine trifluoride is utilized in the processing of nuclear fuels to convert uranium to gaseous uranium hexafluoride. Chlorine trifluoride has also been used as a low temperature etchant for single-crystalline silicon (122,123).

Bromine trifluoride and chlorine trifluoride are used in oil-well tubing cutters (65–68). Chemical cutter tools are commercially available for use in wells at any depth. The cutter consists of three tubular chambers with the top chamber carrying an explosive charge, the middle chamber containing the halogen fluoride, and the lower chamber containing a catalyst. At the extreme end is the cutter head which guides the halogen fluoride against the pipe to be cut. Arranged around the head is a row of evenly spaced orifices or nozzles. The cable used to lower the cutter into the hole also serves to carry an electric charge to set off the explosive. The force drives the chemical into the head where it jets out of the orifices under enormous pressure to impinge against the inner walls of the tube to be cut. The catalyst raises the temperature of the halogen fluoride to trigger a high speed reaction so that the tube is cut in a fraction of a second. The cut is clean and unflared. Tension on the pipe at the top of the well aids in completing the separation (see PETROLEUM).

Iodine pentafluoride is an easily storable liquid source of fluorine having little of the hazards associated with other fluorine sources. It is used as a selective fluorinating agent for organic compounds. For example, it adds iodine and fluorine to tetrafluoroethylene in a commercial process to produce a useful telomer (124).

BIBLIOGRAPHY

"Halogen Fluorides" under "Fluorine Compounds, Inorganic" in *ECT* 1st ed., Vol. 6, pp. 694–695, by H. S. Booth, Case Western Reserve University, and J. T. Pinkston, Harshaw Chemical Co.; in *ECT* 2nd ed., Vol. 9, pp. 585–598, by H. S. Halbedel, Harshaw Chemical Co.; "Halogens" in *ECT* 3rd ed., Vol. 10, pp. 722–733, by A. J. Woytek, Air Products and Chemicals, Inc.

1. H. S. Booth and J. T. Pinkston, *Chem. Rev.* **41**, 421 (1947).
2. H. J. Emeleus, in J. H. Simons, ed., *Fluorine Chemistry*, Vol. 2, Academic Press, Inc., New York, 1954, pp. 39–49.
3. A. G. Sharpe, *Q. Rev. Chem. Soc.* **4**, 115 (1950).
4. H. S. Booth and J. T. Pinkston, in J. H. Simons, ed., *Fluorine Chemistry*, Vol. 1, Academic Press, Inc., New York, 1950, pp. 189–224.
5. H. C. Clark, *Chem. Rev.* **58**, 869 (1958).
6. J. C. Bailer and co-workers, *Comprehensive Inorganic Chemistry*, Vol. 2, Pergamon Press, Compendium Publishers, Elmsford, N.Y., 1973, pp. 1054–1062.
7. F. A. Cotton and G. Wilkinson, *Advanced Inorganic Chemistry*, 5th ed., John Wiley & Sons, Inc., New York, 1988, pp. 572–574.
8. W. K. R. Musgrave, in M. Stacey, J. C. Tatlow, and A. G. Sharpe, eds., *Advances in Fluorine Chemistry*, Vol. 1, Academic Press, Inc., New York, 1960, pp. 1–28; L. S. Boguslavskaya and N. N. Chuvatkin, in L. German and S. Zemskov, eds., *New Fluorinating Agents in Organic Synthesis*, Springer-Verlag, Berlin, 1989, pp. 140–196.
9. K. R. Brower, *J. Org. Chem.* **52**, 798 (1987); N. S. Nikolaev and co-workers, *Khimiyn Galoidnykh*, Soldinerii Ftorn, Moscow, 1968 (U.S. translation NTIS no. *AD-702-974*).
10. J. A. Dean, ed., *Lange's Handbook of Chemistry*, 11th ed., McGraw-Hill Book Co., New York, 1973, p. 4:139.
11. L. Stein, in V. Gutmann, ed., *Halogen Chemistry*, Vol. 1, Academic Press, Inc., New York, 1967, p. 133.
12. D. Pilipovich and co-workers, *Inorg. Chem.* **6**, 1918 (1967).
13. *Mellor's Comprehensive Treatise on Inorganic and Theoretical Chemistry*, Suppl. 2, Part I, Longmans, Green and Co., London, 1956.
14. H. Selig, C. W. Williams, and G. J. Moody, *J. Phys. Chem.* **71**, 2739 (1967).
15. *National Bureau of Standards Technical Note 270-3*, U.S. Government Printing Office, Washington, D.C., 1968.
16. L. Stein, *J. Phys. Chem.* **66**, 288 (1962).
17. R. C. King and G. T. Armstrong, *J. Res. Natl. Bur. Stand.* **74A**, 769 (1970).
18. H. H. Rogers and co-workers, *J. Chem. Eng. Data* **13**, 307 (1968).
19. D. W. Osborne, F. Schreiner, and H. Selig, *J. Chem. Phys.* **54**, 3790 (1971).
20. O. Ruff and H. Krug, *Z. Anorg. Alleg. Chem.* **190**, 270 (1930).
21. R. C. King and G. T. Armstrong, *J. Res. Nat. Bur. Stand.* **74A**, 769 (1970).
22. A. A. Banks, H. J. Emeleus, and A. A. Woolf, *J. Chem. Soc.*, 2861 (1949).
23. M. T. Rogers, J. L. Speirs, and M. B. Panish, *J. Am. Chem. Soc.* **78**, 3288 (1956).
24. *JANAF Thermochemical Tables, NSRDA-NBS37*, 2nd ed., NBS, Washington D.C., 1971.
25. S. N. Buben and A. M. Chaikin, *Kinet. Katal.* **21**, 1591 (1980).
26. D. W. Magnuson, *J. Chem. Phys.* **27**(1), 233 (1957).
27. H. M. Haendler and co-workers, *J. Chem. Phys.* **22**, 1939 (1954).
28. H. Selig, H. H. Claassen, and J. H. Holloway, *J. Chem. Phys.* **52**, 3517 (1970).
29. R. A. Frez, R. L. Redington, and A. L. K. Aljiburty, *J. Chem. Phys.* **54**, 344 (1971).
30. Y. A. Rymarchuk and V. S. Ivanov, *Zh. Prikl. Spektrosk.* **22**, 950 (1975).
31. S. N. Buben and A. M. Chaikin, *Kinet. Katal.* **21**, 1591 (1980).

32. J. W. Grisard and G. D. Oliver, *The Vapor Pressure and Heat Vaporization of Bromine Trifluoride, K-25, Plant Report K-766*, U.C.C. Nuclear Co., Oak Ridge, Tenn., June 8, 1951.

33. R. D. Long, J. J. Martin, and R. C. Vogel, *Ind. Eng. Chem. Data Ser.* **3**, 28 (1958).

34. G. M. Begun, W. H. Fletcher, and D. F. Smith, *J. Chem. Phys.* **42**, 2236 (1965).

35. S. N. Buben and A. M. Chaikin, *Kinet. Katal.* **21**, 1591 (1980).

36. E. A. Jones, T. F. Parkinson, and T. G. Burke, *J. Chem. Phys.* **18**, 235 (1950).

37. Y. A. Rymarchuk and V. S. Ivanov, *Zh. Prikl. Spektrosk.* **22**, 950 (1975).

38. H. Schmitz and H. J. Schumacker, *Z. Naturforsch Teil 2*, 363 (1947).

39. E. A. Jones, T. F. Parkinson, and R. B. Murray, *J. Chem. Phys.* **17**, 501 (1949).

40. K. Schaefer and E. Wicke, *Z. Elektrochem.* **52**, 205 (1948).

41. Y. A. Rymarchuk and V. S. Ivanov, *Zh. Prikl. Spektrosk.* **22**, 950 (1975).

42. J. W. Grisard, H. A. Burnhardt, and G. D. Oliver, *J. Am. Chem. Soc.* **73**, 5725 (1951).

43. A. A. Banks, A. Davies, and A. J. Rudge, *J. Chem. Soc.*, 732 (1953).

44. R. D. Burbank and F. N. Bensey, *J. Chem. Phys.* **21**, 602 (1953).

45. Y. A. Rymarchuk and V. S. Ivanov, *Zh. Prikl. Spektrosk.* **22** (1975).

46. L. E. Alexander and I. R. Beattie, *J. Chem. Soc. A*, 3091 (1971).

47. H. H. Claassen, E. L. Gasner, and H. Selig, *J. Chem. Phys.* **49**, 1803 (1968).

48. C. J. Schack and co-workers, *J. Phys. Chem.* **72**, 4697 (1968).

49. R. K. Khanna, *J. Mol. Spectroscopy* **8**, 134 (1962).

50. W. Huckel, *Nachr. Akad. Wiss. Gottingen Math. Phys. Kl.*, 36 (1946).

51. J. C. Grigger and H. C. Miller, *Met. Protect.*, 33 (Sept. 1964).

52. B. Cox, D. W. A. Sharp, and A. G. Sharpe, *Proc. Chem. Soc.*, 1242 (1956).

53. N. S. Nikolaev and A. A. Opalovskii, *Zh. Neorgan. Khim.* **4**, 1174 (1959).

54. N. S. Nikolaev and E. G. Ippolitov, *Dokl. Akad. Nauk* **134**, 358 (1960).

55. H. A. Bernhardt, E. J. Barber, and R. A. Gustison, *Ind. Eng. Chem.* **51**, 179 (1959).

56. J. F. Ellis and L. H. Brooks, *U. K. Atomic Energy Authority* **8111-D** (1959).

57. R. A. Gustison and co-workers, in F. R. Bruce and co-eds., *Progress in Nuclear Chemistry, Series 3: Process Chemistry*, McGraw-Hill Book Co., Inc., New York, 1956, pp. 281–285.

58. U.S. Pat. 3,012,849 (Dec. 12, 1961), F. L. Horn (to U.S. Atomic Energy Commission).

59. W. J. Mechan and co-workers, *Chem. Eng. Prog.* **53**(2), 72-F-77F (1957).

60. L. Stein and R. Vogel, *Ind. Eng. Chem.* **48**, 418 (1956).

61. G. Strickland, F. L. Horn, and R. Johnson, *U.S. At. Energy Comm. BNL-471 (T-107)* (1957), *BNL-457 (1–21)* (1957).

62. U.S. Pat. 3,825,650 (July 23, 1974), R. A. Gustison and co-workers (to U.S. Atomic Energy Commission).

63. Ger. Pat. 2,209,628 (1971), P. Cousin and co-workers.

64. E. McGhee, *Oil Gas J.* **54**(14), 67 (1955).

65. Ger. Pat. 1,029,770 (May 14, 1958), W. G. Sweetman.

66. *Oil Forum*, 332 (Sept. 1955).

67. U.S. Pat. 4,315,797 (Feb. 16, 1982) J. M. Peppers (to Gearhart Industries, Inc.).

68. *Chemical Cutter Bulletin*, Pipe Recovery Systems, Houston, Tex., 1993.

69. L. Stein, *J. Am. Chem. Soc.* **91**, 5396 (1969).

70. R. Burnett and R. E. Banks, "Fluorine Chemistry," paper presented at *British Symposium*, Nov. 30, 1949.

71. H. J. Emeleus and A. A. Woolf, *J. Chem. Soc.*, 164 (1950).

72. H. R. Hoekstra and J. J. Katz, *Anal. Chem.* **25**, 1608 (1953).

73. A. A. Woolf and H. J. Emeleus, *J. Chem. Soc.*, 2865 (1949).

74. E. E. Aynsley, R. Nichols, and P. L. Robinson, *J. Chem. Soc.*, 623 (1953).

75. E. G. Rochow and I. Kukin, *J. Am. Chem. Soc.* **74**, 1615 (1952).

76. J. F. Gall and co-workers, "Interhalogen Compounds of Fluorine," papers presented at *Annual ACS Meeting*, New York, Sept. 1947.

77. S. Rosen and co-workers, *Accounts Chem. Res.* **21**, 307 (1988).

78. K. R. Brower, *J. Org. Chem.* **52**, 798 (1987).

79. H. J. Frohn and W. Pahlmann, *J. Fluorine Chem.* **24**, 219 (1984).

80. H. J. Frohn, *Chem. -Ztg.* **108**, 146 (1984).

81. H. J. Frohn and W. Puhlmann, *J. Fluorine Chem.* **26**, 243 (1984).

82. H. J. Frohn and W. Puhlmann, *J. Fluorine Chem.* **28**, 191 (1985).

83. H. J. Frohn and H. Maurer, *J. Fluorine Chem.* **34**, 73 (1986).

84. H. J. Frohn and H. Maurer, *J. Fluorine Chem.* **34**, 129 (1986).

85. U.S. Pat. 1,961,622 (June 5, 1934), H. S. Nutting and P. S. Petrie (to The Dow Chemical Co.).

86. U.S. Pat. 2,918,434 (Dec. 22, 1959), J. G. Gall and C. E. Inman (to Pennsalt Chemicals Corp.).

87. R. A. Davis and E. C. Larsen, *J. Org. Chem.* **32**, 3478 (1967).

88. U.S. Pat. 3,123,185 (May 5, 1964), R. E. Parsons (to E. I. du Pont de Nemours & Co., Inc.).

89. T. E. Stevens, *Tetrahedron Lett.*, (17), 16 (1959).

90. T. E. Stevens, *J. Org. Chem.* **26**, 3451 (1961).

91. O. Ruff and R. Keim, *Z. Anorg. Allg. Chem.* **201**, 245 (1931).

92. J. H. Simons, R. L. Bond, and R. E. McArthur, *J. Am. Chem. Soc.* **62**, 3477 (1940).

93. A. A. Banks and co-workers, *J. Chem. Soc.*, 2188 (1948).

94. H. J. Emeleus and A. G. Sharpe, *J. Chem. Soc.*, 2206 (1949).

95. A. G. Sharpe and H. J. Emeleus, *J. Chem. Soc.*, 2135 (1948).

96. A. A. Woolf and H. J. Emeleus, *J. Chem. Soc.*, 1050 (1950).

97. A. A. Woolf, *J. Chem. Soc.*, 1053 (1950).

98. A. G. Sharpe, *J. Chem. Soc.*, 2901 (1949).

99. A. G. Sharpe and A. A. Woolf, *J. Chem. Soc.*, 798 (1951).

100. V. Gutman and H. J. Emeleus, *J. Chem. Soc.*, 1046 (1950).

101. A. A. Woolf, *J. Chem. Soc.*, 3678 (1950).

102. U.S. Pat. 3,143,391 (Aug. 4, 1964), T. L. Hurley, R. O. MacLaren, and E. D. Whitney (to Olin Matheson Chemical Corp.).

103. E. D. Whitney and co-workers, *J. Am. Chem. Soc.* **86**, 2583 (1964).

104. U.S. Pat. 3,367,745 (Feb. 6, 1968), H. G. Tepp (to Allied Chemical Corp.).

105. U.S. Pat. 4,108,966 (Aug. 22, 1978), J. T. Lileck (to Air Products and Chemicals, Inc.).

106. *Chem. Ind.* **57**, 1084 (1945).

107. H. R. Neumark, *Trans. Electrochem. Soc.* **91**, 367 (1947).

108. *Inorganic Fluorine Compounds*, bulletin, Air Products and Chemicals, Inc., Allentown, Pa., 1974.

109. J. C. Million, C. W. Weber, and P. R. Kuehn, *Gas Chromatography of Some Corrosive Halogen-Containing Gases, Report No. K-1639*, Union Carbide Corp., Nuclear Division, New York, 1966.

110. J. F. Ellis and G. Iveson, *The Application of Gas-Liquid Chromatography to the Analysis of Volatile Halogen and Interhalogen Compounds, Gas Chromatography*, Butterworths, London, 1956 pp. 300–309.

111. U.S. Pat. 3,877,894 (Apr. 15 1975), L. G. Swope and E. A. Emory (to U.S. Atomic Energy Commission).

112. V. F. Sukhoverkhov and L. G. Podzolko, *Zh. Anal. Khim.* **38**, 715 (1983).

113. V. F. Sukhoverkhov and L. G. Podzolko, *Zh. Anal. Khim.* **45**, 1101 (1990).

114. V. F. Sukhoverkhov, L. G. Podzolko, and V. F. Garanin, *Zh. Anal. Khim.* **33**, 1360 (1978).

115. *Chlorine Trifluoride Handling Manual, AD266,121*, U.S. Department of Commerce, OTS, Washington, D.C., 1961.

116. R. L. Farrar, *Safe Handling of Chlorine Trifluoride*, report K-1416, U.C.C. Nuclear Company, Oak Ridge, Tenn., 1960.

117. J. M. Siegmund and co-workers, *NASA Document N62-14523, AD 281-818*, NASA, Washington, D.C., 1962.

118. "Chemical Rocket/Propellant Hazards," in *Liquid Propellant Handling, Storage and Transportation, CPLA Publication No. 194*, Vol. 3, CPLA, Silver Spring, Md., May 1970, Chapt. 8.

119. *Handling Hazardous Materials, NASA SP-5032*, NASA, Washington, D.C., 1965, Chapt. 4.

120. R. L. Farrar, Jr., and E. J. Barber, *Some Considerations in the Handling of Fluorine and the Chlorine Fluorides*, report K/ET-252, Oak Ridge Gaseous Diffusion Plant, Oak Ridge, Tenn., 1979.

121. *Registry of Toxic Effects of Chemical Substances*, 1985–1986 ed., U.S. Dept. of Health, Education, and Welfare, Washington, D.C., 1987.

122. Y. Saito, O. Yamaoka, and A. Yoshida, *J. Vac. Sci. Technol.* **B9**, 2503 (1991).

123. Y. Saito, M. Hirabaru, and A. Yoshida, *J. Vac. Sci. Technol.* **B10**, 175 (1992).

124. H. C. Fielding, in R. E. Banks ed., *Organofluorine Chemicals and their Industrial Applications*, Ellis Horwood Publishers, Chichester, UK, 1979.

General Reference

R. A. Rhein and M. H. Miles, *Bromine and Chlorine Fluorides: A Review*, Naval Weapons Center technical publication 6811, NWC, China Lake, Calif., 1988.

WEBB I. BAILEY
ANDREW J. WOYTEK
Air Products and Chemicals, Inc.

HYDROGEN

Hydrogen fluoride [7664-39-3], HF, is the most important manufactured fluorine compound. It is the largest in terms of volume, and serves as the raw material for most other fluorine-containing chemicals. It is available either in anhydrous form or as an aqueous solution (usually 70%). Anhydrous hydrogen fluoride is a colorless liquid or gas having a boiling point of 19.5°C. It is a corrosive, hazardous material, fuming strongly, which causes severe burns upon contact. Rigorous safety precautions are the standard throughout the industry, and in practice hydrogen fluoride can be handled quite safely.

Although it was known in the early nineteenth century, commercial use of hydrogen fluoride was limited. All early production was as aqueous solutions for uses such as glass etching, foundry scale removal, and production of chemicals such as sodium fluoride and sodium bifluoride. Some hydrogen fluoride was also produced for captive use in aluminum manufacture. Production of anhydrous hydrogen fluoride began in the early 1930s, but the demand at that time was limited to the small market for chlorofluorocarbons (see FLUORINE COMPOUNDS, ORGANIC).

World War II brought a revolution in the HF field. The need for high octane aviation fuels (see AVIATION AND OTHER GAS TURBINE FUELS), the birth of the nuclear industry requiring uranium hexafluoride (see NUCLEAR REACTORS; URANIUM AND URANIUM COMPOUNDS), and the rapid growth of the chlorofluorocarbon market all contributed to a steadily rising demand for hydrogen fluoride, especially in the anhydrous form. Whereas earlier anhydrous production had been solely via distillation of aqueous HF, technology emerged allowing direct production of anhydrous hydrogen fluoride.

Properties

Physical Properties.   Physical properties of anhydrous hydrogen fluoride are summarized in Table 1. Figure 1 shows the vapor pressure and latent heat of vaporization. The specific gravity of the liquid decreases almost linearly from 1.1 at $-40°C$ to 0.84 at 80°C (4). The specific heat of anhydrous HF is shown in Figure 2 and the heat of solution in Figure 3.

Table 2 summarizes the properties of the hydrogen fluoride–water system. The freezing and boiling point curves of this system are shown in Figures 4 and 5, respectively. Figure 6 gives the partial pressures of HF and H_2O in aqueous HF solutions. The specific gravity of the solutions at various temperatures is shown in Figure 7. Specific conductivity of this system is given (27,28).

HF, wt %	Conductivity at 0°C, $(\Omega \cdot cm)^{-1}$
70	7.9×10^{-1}
80	7.1×10^{-1}
85	6.3×10^{-1}
90	4.9×10^{-1}
92.5	3.8×10^{-1}
95	2.5×10^{-1}
96	1.95×10^{-1}
97	1.04×10^{-1}
98	9.4×10^{-2}
99	5.6×10^{-2}
99.5	3.4×10^{-2}
99.75	1.8×10^{-2}
99.9	5.7×10^{-3}
99.95	2.8×10^{-3}
100	$<1.6 \times 10^{-6}$

Hydrogen fluoride is unique among the hydrogen halides in that it strongly associates to form polymers in both the liquid and gaseous states. At high temperatures or low partial pressures, HF gas exists as a monomer. At lower temperatures and higher partial pressures hydrogen bonding leads to the formation of chains of increasing length, and molecular weights of 80 and higher are observed. Electron diffraction study of the gas (29) has shown the hydrogen to fluorine distances to be about 0.10 nm and 0.155 nm, the F—H . . . F distance to be

Table 1. Properties of Anhydrous Hydrogen Fluoride

Property	Value	Reference
formula weight	20.006	
composition, wt %		
H	5.038	
F	94.96	
boiling point at 101.3 kPa,[a] °C	19.54	1
critical pressure, MPa[a]	6.48	2
critical temperature, °C	188.0	2
critical density, g/mL	0.29	2
critical compressibility factor	0.117	2
melting point, °C	-83.55	3
density, liquid, 25°C, g/mL	0.958	4
heat of vaporization, 101.3 kPa,[a] kJ/mol[b]	7.493	1
heat of fusion, -83.6°C, kJ/mol[b]	3.931	5
heat capacity, constant pressure, liquid at 16°C, J/(mol·K)[b]	50.6	5
heat of formation, ideal gas, 25°C, kJ/mol[b]	-272.5	6
free energy of formation, ideal gas, 25°C, kJ/mol[b]	-274.6	6
entropy, ideal gas, 25°C, J/(mol·K)[b]	173.7	6
vapor pressure, 25°C, MPa[a]	122.9	7
viscosity, liquid, 0°C, mPa·s($=$cP)	0.256	8
surface tension, mN/m($=$dyne/cm), 0°C	10.2	9
refractive index, liquid, 25°C, 589.3 nm	1.1574	10
molar refractivity, cm^3	2.13	10
dielectric constant, at 0°C	83.6	11
dipole moment, C·m[c]	6.104×10^{-30}	12
thermal conductivity, at 25°C, J/(s·cm·°C)[b]		
liquid	4.1×10^{-3}	13
vapor	2.1×10^{-4}	13
cryoscopic constant, K_f, mol/(kg·°C)	1.52	3
ebullioscopic constant, K_b, mol/(kg·°C)	1.9	14

[a]To convert kPa to psi, multiply by 0.145.
[b]To convert J to cal, divide by 4.184.
[c]To convert C·m to debye, divide by 3.336×10^{-30}.

0.255 nm, and the polymer to have a linear zig-zag configuration. The angle H—F—H is reported to be about 120°. Monomeric HF has an H—F distance of 0.0917 nm. Cyclical polymers (possibly H_6F_6) also probably occur. In general, polymers of differing molecular weights are present in equilibrium at a given temperature and pressure, and an average molecular weight encompasses many different actual molecules. The apparent molecular weight of anhydrous HF vapor is shown in Figure 8.

This high degree of association results in highly nonideal physical properties. For example, heat effects resulting from vapor association may be significantly larger than the latent heat of vaporization (Fig. 9). Vapor heats of associ-

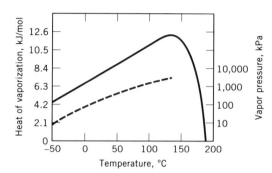

Fig. 1. (——) Latent heat of vaporization (1,7) and (— — —) vapor pressure (1,4,7,15) of anhydrous hydrogen fluoride. To convert kPa to psi, multiply by 0.145. To convert kJ to kcal, divide by 4.184.

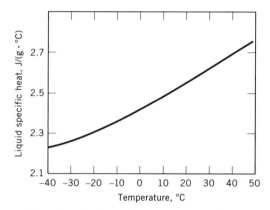

Fig. 2. Specific heat of liquid anhydrous hydrogen fluoride (5,16). To convert J to cal, divide by 4.184.

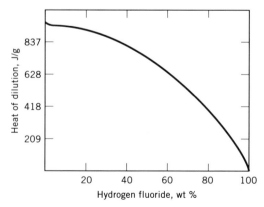

Fig. 3. Heat of solution per gram of anhydrous hydrogen fluoride in water when mixed to the final concentration shown in wt % of HF (16–18).

Table 2. Properties of 70% Aqueous Hydrogen Fluoride

Property	Value	Reference
boiling point at 101.3 kPa,[a] °C	66.4	19
freezing point, °C	−69	20
density, 0°C, g/mL	1.258	21
vapor pressure, 25°C, kPa[a]	20	22
viscosity, 25°C, mPa·s(=cP)	0.61	23
specific heat, 25°C, J/(g·°C)[b]	0.675	23

[a]To convert kPa to psi, multiply by 0.145.
[b]To convert J to cal, divide by 4.184.

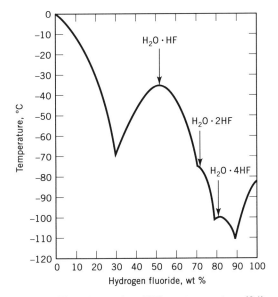

Fig. 4. Freezing point, HF–water system (24).

ation (ΔH_{assoc}) for HF to $(HF)_n$ per mole of $(HF)_n$ are as follows. To convert kJ to kcal, divide by 4.184.

n	$-\Delta H_{assoc}$, kJ
2	33.0
3	67.4
4	101.7
5	133.9
6	167.4
7	198.7
8	224.3

Chemical Properties. Hydrogen fluoride, characterized by its stability, has a dissociation energy of 560 kJ (134 kcal), which places HF among the most stable

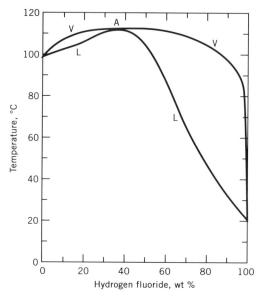

Fig. 5. Boiling point curve for the HF−water system, where A represents an azeotrope at 37.73 wt % HF, L is the liquid, and V the vapor (25,26).

diatomic molecules. Hydrogen fluoride is, however, highly reactive, and it has a special affinity for oxygen compounds, reacting with boric acid to form boron trifluoride and with sulfur trioxide and sulfuric acid to form fluorosulfonic acid. This last reaction demonstrates the dehydrating power of anhydrous hydrogen fluoride. HF belongs to the only class of compounds that readily react with silica and silicates, including glass (qv). With organic compounds, HF acts as a dehydrating agent, a fluorinating agent, a polymerizing agent, a catalyst for condensation reactions, and a hydrolysis catalyst. Hydrogen fluoride reacts with alcohols and unsaturated compounds to form fluorides and with alkylene oxides to give alkylene fluorohydrins.

The strong catalytic activity of anhydrous hydrogen fluoride results from the ability to donate a proton, as in the dimerization of isobutylene (see BUTYLENES):

$$CH_2{=}C(CH_3)_2 + HF \rightarrow (CH_3)_3C^+ + F^-$$

$$(CH_3)_3C^+ + CH_2{=}C(CH_3)_2 \rightarrow (CH_3)_3C{-}CH_2C^+(CH_3)_2 \rightarrow (CH_3)_3CCH{=}C(CH_3)_2 + H^+$$

Anhydrous hydrogen fluoride is an excellent solvent for ionic fluorides (Table 3). The soluble fluorides act as simple bases, becoming fully ionized and increasing the concentration of HF_2^-. For example,

$$HF + KF \rightarrow K^+ + HF_2^-$$

Because of the small size of the fluoride ion, F^- participates in coordination structures of high rank. Tantalum and niobium form stable hexafluorotantalate and hexafluoroniobate ions and hydrogen fluoride attacks these usually acid-

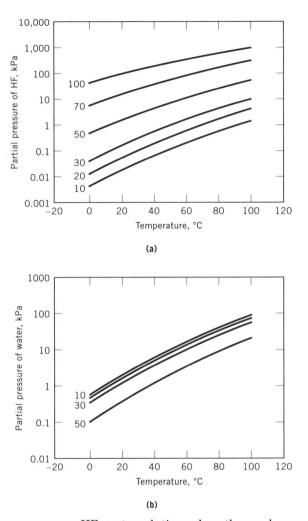

Fig. 6. Partial pressures over HF–water solutions where the numbers represent the quantity of HF in solution expressed as wt % (**a**) of HF and (**b**) of H_2O (22).

resistant metals. Hydrogen fluoride in water is a weak acid. Two dissociation constants are

$$K_1 = \frac{[H^+][F^-]}{[HF]} = 6.46 \times 10^{-4}\ M$$

$$K_2 = \frac{HF_2^-}{[HF][F^-]} = 5\ \text{to}\ 25\ M$$

Whereas hydrogen fluoride is a fairly weak acid as a solute, it is strongly acidic as a solvent. As the concentration of hydrogen fluoride increases in aqueous mixtures, the system becomes more acidic, with water acting as a very strong

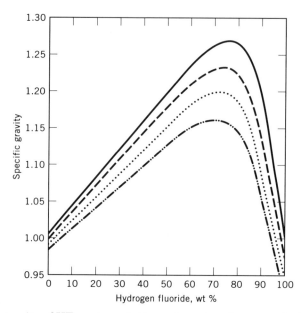

Fig. 7. Specific gravity of HF–water solutions where (——) represents 0°C, (— — —) 20°C, (····) 40°C, and (— ·· — ··) 60°C (21–23).

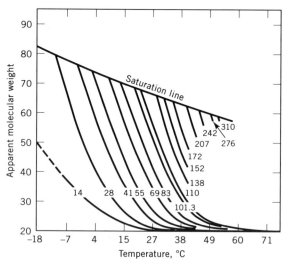

Fig. 8. Apparent vapor molecular weight of anhydrous HF where the numbers represent the partial pressure of HF in kPa. To convert kPa to psi, multiply by 0.145 (7,17,30–32).

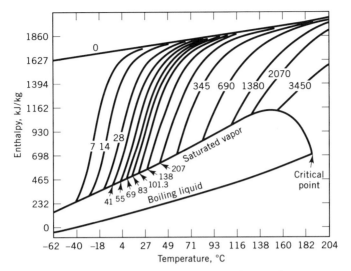

Fig. 9. Vapor-phase enthalpy of anhydrous HF where the numbers represent the partial pressure of HF in kPa (1,17,20,31,33). The critical point occurs at 188°C. To convert kPa to psi, multiply by 0.145. To convert kJ/kg to Btu/lb, multiply by 4.302×10^{-4}.

Table 3. Solubility of Metal Fluorides in Anhydrous Hydrogen Fluoride[a]

Fluoride	Temperature, °C	Solubility, g/100 g
LiF	12	10.3
NaF	11	30.1
KF	8	36.5
NH_4F	17	32.6
CaF_2	12	0.817
SrF_2	12	14.83
BaF_2	12	5.6
MgF_2	12	0.025
FeF_2	12	0.006
FeF_3	12	0.008
AlF_3	11	<0.002
SbF_5	25	miscible

[a]Ref. 8.

base. In dilute aqueous solution, an isolated hydrogen fluoride molecule donates a proton to an aggregate of water molecules and forms an aquated fluoride ion. When small amounts of water are present in the system, a proton is transferred to an isolated water molecule from polymeric hydrogen fluoride. The fluoride ion thus formed is part of a stable polymeric anionic complex. This difference in the solvation of the fluoride ion at the extremes of composition in the H_2O—HF system is probably the principal factor affecting the ease of proton transfer (14).

For anhydrous hydrogen fluoride, the Hammett acidity function H_0 approaches -11. The high negative value of H_0 shows anhydrous hydrogen fluoride to be in the class of superacids. Addition of antimony pentafluoride to make a 3 M solution in anhydrous hydrogen fluoride raises the Hammett function to -15.2, nearly the strongest of all acids (34).

Manufacture

Raw Materials. Essentially all hydrogen fluoride manufactured worldwide is made from fluorspar and sulfuric acid, according to the reaction:

$$CaF_2(s) + H_2SO_4 \rightarrow CaSO_4(s) + 2\ HF(g)$$

Generally, yields on both fluorspar and sulfuric acid are greater than 90% in commercial plants.

Fluorspar. A typical acid-grade fluorspar analysis gives:

Component	Composition, wt %
calcium fluoride	97.0
silica	0.7
calcium carbonate	1.0
organic (as carbon)	0.1
sulfur	
total	0.02
sulfide	0.01
phosphorus pentoxide	0.02
chloride, total	0.02

Also present are 0.3 wt % mixed metal oxides (R_2O_3) and 5 ppm of arsenic. Impurities in fluorspar may affect yield, plant operability, or product quality.

Silica. Silica, which has the greatest impact on yield losses, reacts with HF and is discharged from the manufacturing process as H_2SiF_6. Yield losses can be calculated based on the chemical stoichiometry:

$$SiO_2 + 3\ H_2SO_4 + 3\ CaF_2 \rightarrow H_2SiF_6 + 2\ H_2O + 3\ CaSO_4$$

or

$$6\ HF + SiO_2 \rightarrow H_2SiF_6 + 2\ H_2O$$

Calcium fluoride loss is equal to 3.9% for each 1% silica; sulfuric acid loss is equal to 4.9% for each 1% silica.

Mixed-Metal Oxides. Generally, iron oxide is the principal component of mixed-metal oxides. These affect the sulfuric and oleum consumption in HF production.

$$R_2O_3 + 3 H_2SO_4 \rightarrow R_2(SO_4)_3 + 3 H_2O$$

Sulfuric acid loss is approximately 1.84% H_2SO_4 for each percentage of R_2O_3. Oleum consumption is increased to consume the water that is formed. The metal sulfates are more stable than metal fluorides under furnace conditions and are discharged from the process with the residue.

Calcium Carbonate. Calcium carbonate, like R_2O_3, affects sulfuric and oleum consumption in the HF process. Sulfuric acid loss is approximately 0.98% H_2SO_4 for each percentage of $CaCO_3$. The carbon dioxide evolved by the reaction increases the noncondensable gas flow, and because it carries HF, contributes to yield losses in the vent stream.

Magnesium Oxide. Magnesium oxide behaves in a similar manner to other metal oxides. However, most spars contain practically no magnesium oxide, so it does not affect yield loss or plant operation.

Organic Carbon. Organic materials interfere with plant operation because these compounds react with sulfuric acid under furnace conditions to form sulfur dioxide. There is a reducing atmosphere in the furnace which may reduce sulfur dioxide to elemental sulfur, which results in sulfur deposits in the gas handling system.

Total Sulfur and Sulfide Sulfur. Total sulfur is predominately in the form of metal sulfate, and because sulfates act as inerts, these materials have little impact on the process. Sulfide sulfur compounds, on the other hand, react and leave the furnace as a sulfur vapor, which may deposit in the gas handling system. A possible mechanism for this is the partial reaction of SO_2 to H_2S, followed by

$$2 H_2S + SO_2 \rightarrow 3 S + 2 H_2O$$

Phosphorus Pentoxide. Phosphorus compounds form PF or POF compounds in the furnace. Some may be hydrolyzed to higher boiling forms in downstream process operation. Some of the phosphorus compounds do appear in the final product. This is objectionable to some users.

Chloride. Chloride is known to significantly increase the rate of corrosion in acidic fluoride media. The level of chloride that can be tolerated in the HF process before corrosion hinders plant operation is quite low.

Arsenic and Boron. Arsenic and boron form volatile fluorides which are difficult to separate from high purity HF. Special equipment and techniques must be used to remove the arsenic.

Sulfuric Acid. Generally, sulfuric acid of 93–99% is used. The sulfuric values may be fed to the plant as H_2SO_4, oleum (20% SO_3), or even SO_3 (see SULFURIC ACID AND SULFUR TRIOXIDE). Commonly, both H_2SO_4 and oleum are used. The split between the two is determined by water balance. All water entering the process or produced by side reactions reacts with the SO_3 component of the oleum:

$$H_2O + SO_3 \rightarrow H_2SO_4$$

The ratio of fluorspar to sulfuric acid fed depends on the relative cost of each raw material. As of this writing, fluorspar is more expensive than sulfuric acid; thus, most often a slight excess of sulfuric acid is desirable. Too much sulfuric acid, however, yields a reaction mixture which becomes wet, sticky, corrosive, and hard to handle.

Technology. The key piece of equipment in a hydrogen fluoride manufacturing plant is the reaction furnace. The reaction between calcium fluoride and sulfuric acid is endothermic (1400 kJ/kg of HF) (334.6 kcal/kg), and for good yields, must be carried out at a temperature in the range of 200°C. Most industrial furnaces are horizontal rotating kilns, externally heated by, for example, circulating combustion gas in a jacket. Other heat sources are possible, eg, supplying the sulfuric acid value as SO_3 and steam (qv), which then react and condense, forming sulfuric acid and releasing heat.

Even at the small production rates involved with the earliest HF production, the fundamental technical problem of HF production was apparent. When finely ground spar and acid are mixed in the proper proportions, a thin, almost watery slurry is obtained. Little reaction occurs until the suspension is heated. Upon heating, the slurry thickens rapidly, passing into a sticky paste that can build up on the furnace walls, thus reducing heat transfer. Additionally, intimate mixing of spar and acid is required, and paste formation interferes with this. The fundamental problem in designing an HF furnace is thus to find a method to keep the heat-transfer surfaces clean enough to allow the reaction to proceed at a reasonable rate, and to keep the reaction mass from forming a sticky material.

Historically, internal scrapers or paddles were used in some designs, and loose rails were used in others to break up any caking material which formed. The nature of these designs mechanically limited the furnaces to relatively small sizes producing about 3000 t/yr. High maintenance costs were also involved.

In the 1960s the need for improved technology extendable to higher capacities became apparent, and several new approaches were commercialized. One example is the use of a heavy-duty mechanical mixer to partially react the spar and acid. This first reaction phase carries the reaction past the point where sticky material forms, enabling the reaction to be completed on a flowable solid material in a standard externally jacketed rotating kiln. A second technology employs an Archimedean screw fixed to the rotating shell, to bring sufficient quantities of hot, dry solid from the discharge end of the furnace to the feed end, such that any sticky material formed is absorbed by the dry solid. This method also serves to bring the reactants up to reaction temperature quickly, by contact with the hot recycled material (35). In both of these technologies, large (> 10,000 t/yr) furnace capacities are attainable, and the furnaces can be made largely from inexpensive carbon steel.

In all HF processes, the HF leaves the furnace as a gas, contaminated with small amounts of impurities such as water, sulfuric acid, SO_2, or SiF_4. Various manufacturers utilize different gas handling operations, which generally include scrubbing and cooling. Crude HF is condensed with refrigerant, and is further purified by distillation (qv). Plant vent gases are scrubbed with the incoming sulfuric acid stream to remove the bulk of the HF. The sulfuric acid is then fed to the furnace. Water or alkali scrubbers remove the remainder of the HF from the plant vent stream.

Some manufacturers recover by-products from the process. Fluosilicic acid [16961-83-4], which is used in water fluoridation, can easily be recovered from the plant vent gases, which contain SiF_4:

$$SiF_4 + 2 HF \rightarrow H_2SiF_6 \text{ (aq)}$$

The calcium sulfate [7778-18-9] discharged from the furnace can also be recovered. This is less the practice in the United States where natural gypsum is plentiful and inexpensive than in Europe, where $CaSO_4$ recovery for use in cement (qv) and self-leveling floors is common. Some $CaSO_4$ is recovered in the United States, primarily for lower end uses such as road aggregate.

Figure 10 is a schematic of a typical HF process.

Alternative Processes.   Because of the large quantity of phosphate rock reserves available worldwide, recovery of the fluoride values from this raw material source has frequently been studied. Strategies involve recovering the fluoride from wet-process phosphoric acid plants as fluosilicic acid [16961-83-4], H_2SiF_6, and then processing this acid to form hydrogen fluoride.

Numerous processes have been proposed, but none has been commercialized on a large scale (36). The overall reaction in such processes is

$$H_2SiF_6 + 2 H_2O \rightarrow 6 HF + SiO_2$$

However, this reaction does not take place in a single step, and multiple reactions must be used. One such route involves using sulfuric acid to decompose the H_2SiF_6:

$$H_2SiF_6 \xrightarrow{H_2SO_4} 2 HF + SiF_4$$

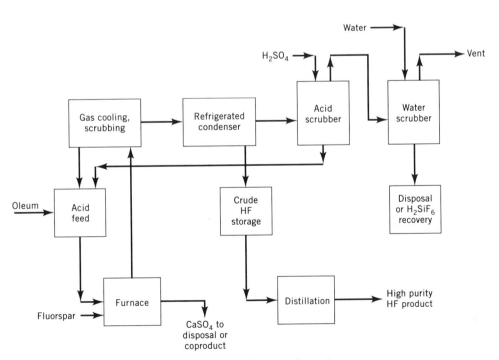

Fig. 10.   Hydrogen fluoride manufacturing process.

followed by hydrolysis of the SiF_4 in either the vapor or liquid phases:

$$3\ SiF_4(l)\ +\ 2\ H_2O\ \longrightarrow\ 2\ H_2SiF_6\ +\ SiO_2$$

$$SiF_4(g)\ +\ 2\ H_2O\ \xrightarrow[>600°C]{}\ 4\ HF\ +\ SiO_2$$

Other technologies proceeding via intermediates such as NH_4F or KHF_2 are also possible. A more recently developed process (36) involves the reaction of the H_2SiF_6 and phosphate rock, producing a calcium silicon hexafluoride ($CaSiF_6$) intermediate that can be converted to CaF_2 and then to HF by reaction of H_2SO_4. All of the processes produce silica. The quality of the silica varies greatly, and its value as a coproduct has a significant impact on the processes' economics.

The future for these technologies is uncertain. Economic comparisons with fluorspar-based processes indicate that as long as fluorspar supplies remain abundant, there is little justification to proceed with such processes.

Specifications, Shipping, and Analysis. Hydrogen fluoride is shipped in bulk in tank cars (specification 112S400W) and tank trucks (specification MC312). A small volume of overseas business is shipped in ISO tanks. Bulk shipments are made of anhydrous HF as well as 70% aqueous solutions. A small amount of aqueous solution may be shipped as 50%. Cars and trucks used for anhydrous HF transport are of carbon steel construction. It is possible to ship 70% aqueous in steel from a corrosion standpoint; however, rubber lining is commonly used to eliminate iron pickup, which is detrimental to product quality in a number of applications. Hydrogen fluoride of less than 60% strength must always be shipped in lined containers.

Anhydrous hydrogen fluoride is also available in cylinders, and aqueous hydrogen fluoride, either 50% or 70%, is also shipped in polyethylene bottles and carboys. Typical product specifications and analysis methods are given in Table 4.

Table 4. Hydrogen Fluoride Product Specifications[a]

Component	Specification[b]	Analytical method
Anhydrous HF		
HF, wt %	99.95[c]	difference
nonvolatile acid, ppm	100	evaporation/titration
sulfur dioxide, ppm	50	iodimetry
water, ppm	200	conductivity
arsenic, ppm	25	colorimetry
fluosilicic acid, ppm	100	colorimetry
Aqueous HF		
HF, wt %	70–72	titration
nonvolatile acid, ppm	200	evaporation/titration
sulfur dioxide, ppm	100	iodimetry
arsenic, ppm	18	colorimetry
fluosilicic acid, ppm	100	colorimetry

[a]Ref. 37.
[b]Values are the maximum allowable unless otherwise stated.
[c]Value is the minimum allowable.

Materials of Construction. Acceptable materials of construction for hydrogen fluoride handling are a function of such variables as temperature, hydrogen fluoride strength, and method of use. As examples of the latter, corrosion is greater for higher velocities as well as for metals used in reboiler heat-transfer surfaces. Mild steel is generally used for most anhydrous hydrogen fluoride applications, up to 66°C. Steel, in contact with HF, forms a passive film of iron fluoride, which then protects the metal against further corrosion. Any physical or chemical action which disrupts this passive film can lead to substantial increases in corrosion rate. In steel service, hydrogen blistering, caused by accumulation of hydrogen released by corrosion at laminations and inclusions in the steel, may occur and must be evaluated during periodic inspections.

At higher temperatures, Monel, a nickel–copper alloy, is suitable, as is Hastelloy-C, a nickel molybdenum–chromium alloy.

Aqueous hydrogen fluoride of greater than 60% may be handled in steel up to 38°C, provided velocities are kept low (< 0.3 m/s) and iron pickup in the process stream is acceptable. Otherwise, rubber or polytetrafluoroethylene (PTFE) linings are used. For all applications, PTFE or PTFE-lined materials are suitable up to the maximum use temperature of 200°C. PTFE is also the material of choice for gasketing. Alloy 20 or Monel is typically used for valve and pump applications. Materials unacceptable for use in HF include cast iron, type 400 stainless steel, hardened steels, titanium, glass, and silicate ceramics.

Economic Factors

Production. Global hydrogen fluoride production capacity in 1992 was estimated to be 875,000 metric tons. An additional 204,000 metric tons was used captively for production of aluminum fluoride. Worldwide capacity is tabulated in Table 5 (38). Pricing for hydrogen fluoride in 1990 was about $1.52/kg (39).

North America accounts for about 38% of the worldwide hydrogen fluoride production and 52% of the captive aluminum fluoride production. Table 6 (38) summarizes North American capacity for hydrogen fluoride as well as this captive capacity for aluminum fluoride production. In North America, HF is produced in the United States, Canada, and Mexico, but represents a single market, as well over 90% of the consumption is in the United States.

North American HF production capacity has declined since the early 1980s and several smaller producers, such as Harshaw and Essex, have closed plants. Production is expected to continue to decline in the short term because of chlorofluorocarbon (CFC) cutbacks, but is expected to rebound later in the 1990s as replacement hydrochlorofluorocarbons are introduced to the marketplace.

At least in the short term, European production is expected to be impacted by two trends: the move away from planned economics in the East should lead to more rapid demand growth; and the phaseout of CFCs, including, in Europe, aerosols (qv), should lead to overcapacity in the West. This excess capacity in western Europe could be used to supply the East.

Asian production of hydrogen fluoride is concentrated in Japan. The Japanese are leaders in the production of high quality HF. Hashimoto has the capacity

Table 5. Worldwide Hydrofluoric Acid Capacity[a]

Country	Hydrogen fluoride, t/yr	
	Market	Captive[b]
North America[c]	330,000	106,000
Brazil	18,000	
Venezuela	5,000	
Germany	90,000	
the Netherlands	7,000	
United Kingdom	73,000	
Spain	28,000	
Italy	32,000	20,000
France	45,000	53,000
Greece	5,000	
CIS	100,000	
Czechoslovakia	8,000	
Norway		25,000
South Africa	3,000	
India	9,000	
Australia	9,000	
Japan	113,000	
Total	*875,000*	*204,000*

[a]Ref. 38.
[b]Anydrous HF for the production of aluminum fluoride.
[c]See also Table 6.

Table 6. Hydrofluoric Acid Capacity in North America[a]

Producer	Market, t/yr	AlF₃ production
Alcan		55,000
Alcoa		45,000
Allied-Signal	142,000	
Du Pont	68,000	
Atochem	22,000	
Quimica Fluor	68,000	
Fluorex	18,000	
Industrias Quimica de Mexico	6,000	16,000
Quimobasicos	6,000	
Total	*330,000*	*116,000*

[a]Ref. 38.

for 3000 t/yr of ultrahigh purity product. For the future, increased production in many of the developing Asian nations is likely.

Fluorspar Supply. Production costs of hydrogen fluoride are heavily dependent on raw materials, particularly fluorspar, and significant changes have occurred in this area. Identified world fluorspar resources amount to approximately 400×10^6 metric tons of fluorspar (40). Of these 400×10^6 t, however, only 243×10^6 t are considered reserves and an additional 93×10^6 t is considered reserve base, ie, recoverable at higher market prices.

Fluorspar is marketed in three grades: acid, ceramic, and metallurgical. Metallurgical grade is commonly sold as lump or gravel, and ceramic-grade as a dried flotation filter cake or as briquettes or pellets. Acid-grade is used for HF manufacture and is the purest form, having a minimum CaF_2 content of 97%.

Based on previous splits in milling operations, about a 60% yield or 146×10^6 t of acid-grade spar could be expected. At the production rates of the early 1990s, this would be a 24-yr supply. Additional supplies are expected to be brought into production, however, and no decline in available reserves is expected through the year 2000.

Most of the acid-grade spar used for HF production in the United States is imported. More than two-thirds of the fluorspar consumed in the United States goes into production of HF; nearly 30% is consumed as a flux in steelmaking; and the remainder is consumed in glass manufacture, enamels, welding rod coatings, and other end uses or products (see FLUORINE COMPOUNDS, INORGANIC–CALCIUM).

A most significant development has been the bringing on line of mining capacity in the People's Republic of China. China now produces more than 1×10^6 t/yr of fluorspar, making it the world's largest supplier. Kenyan production has also increased substantially.

Purity is expected to become a significant concern as reserves are depleted. Higher levels of impurities in the fluorspar may require modifications to HF production technology to produce high quality hydrogen fluoride. This has already happened regarding high arsenic levels in some Mexican fluorspar. Both Fluorex and Allied-Signal have installed facilities to remove arsenic from the HF process. In addition, the new Kenyan fluorspar production contains high levels of phosphate impurity which must be dealt with.

Uses

In the North American HF market, approximately 70% goes into the production of fluorocarbons, 4% to the nuclear industry, 5% to alkylation processes, 5% to steel pickling, and 16% to other markets (41). This does not include the HF going to aluminum fluoride, the majority of which is produced captively for this purpose.

Fluorocarbons.   Fluorocarbons are a family of products that have properties which render them valuable as refrigerants, blowing agents, solvents, and sources of raw materials for production of fluoropolymer materials. Other specialty fluorine-containing, organic chemicals are also produced, some of which are used as anesthetics and fire extinguishants. Certain chlorofluorocarbon products are thought to be damaging to the ozone layer of the upper atmosphere and this market is expected to change to nonozone-damaging fluorocarbon products.

HF is used as a source of fluorine for production of all the various fluorocarbon products. HF reacts in the presence of a suitable catalyst and under the appropriate temperature and pressure conditions with various organic chemicals to yield a family of products. A by-product stream of hydrochloric acid may be coproduced.

Projection of HF requirements for this market segment is uncertain. The ultimate volume of this market segment is negatively impacted by replacement of fluorocarbons with nonfluorine-containing products for foam blowing and solvents. Sales of fluorocarbons would also be reduced by conservation, recovery, and recycle encouraged by high fluid cost, taxation, or regulation. The production of high growth fluorine-containing plastics would be unaffected.

It appears that the ultimate replacements for the high volume chlorofluorocarbon products are to be more highly fluorinated organic chemicals, thus requiring significantly higher volumes of HF in their manufacture.

Nuclear Industry. Technology in the manufacture of uranium reactor fuel for commercial electric power generation requires the uranium to be converted to gaseous uranium hexafluoride [7783-81-5], UF_6, so that enrichment may occur. UF_6, the only gaseous form of uranium, is the form used for the enrichment processes (see DIFFUSION SEPARATION METHODS). Following enrichment, reactor fuel elements are manufactured by converting to UO_2. Domestic U.S. nuclear power generation is not expected to grow. Most of the future demands for UF_6 are expected to be dictated by development of overseas reactor installations. Some other nuclear markets employing UF_6 involve weapons systems but usage and future growth is expected to be low.

To convert naturally occurring uranium oxide, yellow cake or U_3O_8, to the gaseous UF_6, hydrofluoric acid is first used to convert the U_3O_8 to UF_4. Further fluorination using fluorine (generated from more HF) is employed to convert the UF_4 to UF_6. The UF_6 is then processed at gaseous diffusion enrichment plants.

Alkylation. Petroleum and, to a lesser extent, detergent aklylation are processes which make use of the particular catalytic properties of anhydrous HF. Petroleum alkylation produces a very high octane gasoline blending component (C-7 or C-8 compounds) by condensation of C-3 or C-4 olefins obtained in the catalytic cracking process along with isobutane. Detergent alkylation generates a desirable biodegradable detergent intermediate. Although HF is used as a catalyst in these processes, the HF is slowly consumed because of side reactions, with impurities contained in the various feedstocks (qv). As of this writing there are 69 petroleum alkylation and two detergent alkylation units operating in North America.

In the petroleum alkylation process, liquid anhydrous HF is intimately contacted with isobutane [75-28-5] and mixed light olefins under pressure at an elevated (about 40°C) temperature to produce a branched-chain fuel having very high octane value. The mixture of HF and hydrocarbon is settled, the acid is recycled, and the alkylate is water-washed and dried. The HF catalyzes a broad range of desirable reactions in this process and although some HF is lost through reaction with impurities in the various feedstocks, regeneration of most of the HF is easily accomplished within the alkylation unit. The drawoff from this acid regeneration system is neutralized before disposal. This equates to the HF consumption of this process.

Increasing demand for higher octane, lead-free motor fuel having low volatility makes the alkylation process a proven way to maximize profitability. Alkylation using sulfuric acid as the catalyst competes with the HF process even though acid consumption and regeneration costs are much greater.

The choice between sulfuric and hydrofluoric acid-catalyzed processes for new alkylation capacity is influenced by proximity to sulfuric acid regeneration plants, energy costs, and the nature of the unit feed. HF produces a higher quality alkylate when the unit feed is rich in propenes or isobutene. Sulfuric acid is preferred when the feed is rich in pentenes or n-butene. Additionally, in some quarters the perception exists that the sulfuric acid alkylation process may be less hazardous and may present a lesser potential environmental threat. As a result, future installation of additional grass roots HF alkylation units in North America may be affected. Industry is working toward development of chemical additives which, when combined with HF, reduce the risks associated with an accidental release. The success of such work could have a significant impact on the alkylation market.

Chemicals.   Both organic and inorganic fluorine-containing compounds, most of which have highly specialized and valuable properties, are produced from HF. Typically these fluorinated chemicals are relatively complex, sometimes difficult to manufacture, and of high value. These materials include products used as fabric and fiber treatments, herbicide and pharmaceutical intermediates, fluoroelastomers, and fluorinated inert liquids. Other products include BF_3, SF_6, and fluoborates.

Many different processes using HF as a reactant or source of fluorine are employed in the manufacture of fluorinated chemical derivatives. In many cases the chemistry employed is complex and in some cases proprietary. Electrochemical fluorination techniques and gaseous fluorine derived from HF are used in some of these applications.

Some of the chemical derivatives, especially those tied to agricultural uses, tend to experience some cyclical demand. However, because of the specialized nature of many of the fluorinated chemicals, these products are positioned in strong, high performance market areas having above average growth rates.

Aqueous HF.   Aqueous solutions of hydrofluoric acid are used in stainless steel pickling (see METAL TREATMENTS), chemical milling, glass (qv) etching, exotic metals extraction, quartz purification, and a variety of other uses including metal coatings (qv) and other, small-volume, upgraded inorganic fluorine compounds. A substantial portion of aqueous HF is marketed through distributors as drummed or packaged product.

A small but significant use for aqueous HF is in the electronics industry (see ELECTRONIC MATERIALS). Aqueous HF (typically 49%) of extremely high purity is used as an etchant for silicon wafers (see ULTRAPURE MATERIALS).

Aluminum Industry.   Large amounts of HF are consumed in the production of aluminum fluoride [7784-18-1], AlF_3, and cryolite [15096-52-3] (sodium aluminum fluoride), used by the aluminum industry. Both of these compounds are used in the fused alumina bath from which aluminum is produced by the electrolytic method.

Most AlF_3 and cryolite producers have their own HF production facilities. HF vapor is reacted with alumina trihydrate to form AlF_3 in a fluid-bed reactor. HF is reacted with sodium hydroxide to form sodium fluoride, which is then used to produce cryolite. Producers who manufacture these products solely for use in the aluminum industry do not generally install liquid HF storage and handling facilities, and do not participate in the merchant HF market.

Health, Safety, and Environmental Aspects

Although it is widely recognized as a hazardous substance, large volumes of HF are safely manufactured, shipped, and used, and have been for many years. Excellent manuals describing equipment and procedures for the safe handling of hydrogen fluoride are available from manufacturers (16,17,42).

Mild exposure to HF via inhalation can irritate the nose, throat, and respiratory system. The onset of symptoms may be delayed for several hours. Severe exposure via inhalation can cause nose and throat burns, lung inflammation, and pulmonary edema, and can also result in other systemic effects including hypocalcemia (depletion of body calcium levels), which if not promptly treated can be fatal. Permissible air concentrations are (42) OSHA PEL, 3 ppm (2.0 mg/m^3) as F; OSHA STEL, 6 ppm (5.2 mg/m^3) as F; and ACGIH TLV, 3 ppm (2.6 mg/m^3) as F. Ingestion can cause severe mouth, throat, and stomach burns, and may be fatal. Hypocalcemia is possible even if exposure consists of small amounts or dilute solutions of HF.

Both liquid HF and the vapor can cause severe skin burns which may not be immediately painful or visible. HF can penetrate skin and attack underlying tissues, and large (over 160 cm^2) burns may cause hypocalcemia and other systemic effects which may be fatal. Even very dilute solutions may cause burns. Both liquid and vapor can cause irritation to the eyes, corneal burns, and conjunctivitis.

Unlike other acid burns, HF burns always require specialized medical care. The fluoride ion is extremely mobile and easily penetrates deeply into the skin. Immediate first aid consists of flushing the affected area with copious quantities of water for at least 20 minutes. Subsequently the area is immersed in iced 0.13% benzalkonium chloride solutions or massaged with 2.5% calcium gluconate gel. For larger burns, subcutaneous injection of 5% calcium gluconate solution beneath the affected area may be required. Eye exposure requires flushing with water for at least 15 minutes, and subsequent treatment by an eye specialist. Exposure to HF vapor should be treated by moving the victim to fresh air, followed by artificial respiration if required, and administration of oxygen if the victim is having difficulty breathing. As in other cases of exposure, a qualified physician must be called, and the victim should be held under observation for at least 24 hours. First aid for swallowed HF consists of drinking large quantities of water; milk or several ounces of milk of magnesia may be given. Vomiting should not be induced.

Hydrogen fluoride is not a carcinogen. However, HF is highly reactive, and heat or toxic fumes may be evolved. Reaction with certain metals may generate flammable and potentially explosive hydrogen (qv) gas.

The hydrogen fluoride industry has undertaken a significant effort to investigate the behavior of HF releases so as better to define the risks associated with an accidental spill, and to design effective mitigation systems. A series of tests conducted in the Nevada desert in 1986 showed that spills of pressurized, superheated HF under certain conditions could form a heavier-than-air vapor cloud consisting of flashed, cold HF vapor and an entrained aerosol of HF droplets. The HF did not form liquid pools as expected, reducing the effectiveness of diking in mitigating the effect of a release (43). The effect of water sprays in mitigating an

HF release was studied in detail as one of several components of the Industry Cooperative Hydrogen Fluoride Mitigation and Ambient Impact Assessment Program (ICHMAP). Water spray curtains or water monitors were found to remove between 25 and 95% of HF released in field tests. The removal efficiency depended primarily on the ratio of water to HF volume. The higher removal efficiency was obtained at a 50:1 ratio with a single spray curtain (44).

BIBLIOGRAPHY

"Hydrogen Fluoride" under "Fluorine Compounds, Inorganic," in *ECT* 1st ed., Vol. 6, pp. 695–708, by A. S. Woodard, Pennsylvania Salt Manufacturing Co.; in *ECT* 2nd ed., Vol. 9, pp. 610–625, by J. F. Gall, Pennsalt Chemicals Corp.; "Hydrogen" under "Fluorine Compounds, Inorganic," in *ECT* 3rd ed., Vol. 10, pp. 733–753, by J. F. Gall, Philadelphia College of Textiles and Science.

1. C. E. Vanderzee and W. W. Rosenberg, *J. Chem. Thermodyn.* **2**, 461 (1970).
2. J. F. Mathews, *Chem. Rev.* **72**(1), 85, 97 (1972).
3. R. J. Gillespie and D. A. Humphreys, *J. Chem. Soc.* **92**, 2311 (1970).
4. E. U. Franck, and W. Spalthoff, *Z. Electrochem.* **61**, 348 (1957).
5. J. H. Hu, D. White, and H. L. Johnston, *J. Am. Chem. Soc.* **75**, 1232 (1953).
6. D. R. Stull and H. R. Prophet, *JANAF Thermochemical Tables*, 2nd ed., National Bureau of Standards, NSRDS-NBS 37, U.S. Government Printing Office, Washington, D.C., 1971.
7. R. L. Jarry and W. Davis, Jr., *J. Phys. Chem.* **57**, 600 (1953).
8. H. H. Hyman and J. J. Katz, in T. C. Woddington, ed., *Non-Aqueous Solvent Systems*, Academic Press, Inc., London, 1965, pp. 47–81.
9. J. H. Simons and J. W. Bouknight, *J. Am. Chem. Soc.* **54**, 129 (1932).
10. A. J. Perkins, *J. Phys. Chem.* **68**, 654 (1964).
11. K. Fredenhagen and J. Dahmlos, *Z. Anorg. Allg. Chem.* **178**, 272 (1929).
12. S. I. Chan, D. Ikenberry, and T. P. Das, *J. Chem. Phys.* **41**, 2107 (1964).
13. C. L. Yaws and L. S. Adler, *Chem. Eng.*, 119 (Oct. 28, 1974).
14. T. A. O'Donnell, in J. C. Bailar and co-workers, eds., *Comprehensive Inorganic Chemistry*, Vol. 2, Pergamon Press, Oxford, UK, 1973, pp. 1038–1054.
15. W. H. Claussen and J. H. Hildebrand, *J. Am. Chem. Soc.* **56**, 1820 (1934).
16. *Hydrofluoric Acid, Anhydrous—Technical, Properties, Uses, Storage, and Handling*, E. I. du Pont de Nemours & Co., Inc., Wilmington, Del., 1984.
17. *Hydrofluoric Acid*, Allied-Signal Corp., Morristown, N.J., 1978.
18. G. K. Johnson, P. N. Smith, and W. N. Hubbard, *J. Chem. Thermodyn.* **5**, 793 (1973).
19. K. Fredenhagen, *Z. Anorg. Allg. Chem.* **210**, 210 (1933).
20. E. U. Franck and F. Meyer, *Z. Electrochem.* **63**, 571 (1959).
21. Hodgman, C. D., *Handbook of Chemistry and Physics*, 33rd ed., Chemical Rubber Publishing Co., Boca Raton, Fla., 1951, p. 1677.
22. P. A. Munter, O. T. Aepli, and R. A. Kossatz, *Ind. Eng. Chem.* **41**, 1504 (1949).
23. Allied-Signal Corp. data.
24. G. H. Cady and J. H. Hildebrand, *J. Am. Chem. Soc.* **52**, 3843 (1930).
25. P. A. Munter, O. T. Aepli, and R. A. Kossatz, *Ind. Eng. Chem.* **39**, 427 (1947).
26. N. Miki, M. Maeno, K. Maruhashi, and T. Ohmi, *J. Electrochem Soc.* **137**(3), 787 (1990).
27. K. Fredenhagen, *Z. Phys. Chem.* **128**, 1 (1927).
28. K. Fredenhagen and M. Wellman, *Z. Phys. Chem.* **162**, 454 (1932).
29. S. H. Bauer, J. Y. Beach, and J. H. Simons, *J. Am. Chem. Soc.* **61**, 19 (1939).
30. W. Strohmeier and G. Brieglab, *Z. Electrochem.* **57**(8), 662 (1953).

31. W. Spalthoff and E. U. Franck, *Z. Electrochem* **61**(8), 993 (1957).
32. R. W. Long, J. H. Hildebrand, and W. E. Morrell *J. Am. Chem. Soc.* **65**, 182 (1943).
33. R. M. Yabroff, J. C. Smith, and E. H. Lightcap *J. Chem. Eng. Data* **9**(2), 178 (1964).
34. M. Kilpatrick and J. G. Jones, in J. J. Lagowski, ed., *The Chemistry of Nonaqueous Solvents*, Vol. 2, Academic Press, Inc., New York, 1967, pp. 43–49.
35. U.S. Pat. 3,718,736 (Feb. 27, 1973), W. E. Watson and R. P. Troeger, (to Allied Chemical Corp.).
36. *Chem. Eng.*, 27 (Mar. 1993).
37. *Aqueous and Anhydrous Hydrogen Fluoride product specifications*, Allied-Signal Inc., Morristown, N.J., 1991.
38. *Chem-Intell Database*, Reed Telepublishing, London, update Feb. 1990.
39. *Chem. Mark. Rep.*, 23 (Aug. 10, 1992).
40. *Annual Mineral Industry Surveys.* U.S. Bureau of Mines, Washington, D.C., 1990.
41. *Chem. Mark. Rep.*, 30 (July 29, 1991).
42. *Hydrofluoric Acid, Anhydrous*, Product Safety Data Sheet, Allied-Signal Inc., Morristown, N.J., 1991.
43. *Industry Cooperative Hydrogen Fluoride Mitigation and Ambient Impact Assessment Program, Summary Report*, National Technical Information Service, Aug. 1989.
44. *Effectiveness of Water Spray Mitigation Systems for Accidental Releases of Hydrogen Fluoride, Summary Report*, National Technical Information Service, June 1989.

ROBERT A. SMITH
AlliedSignal Inc.

IRON

Iron(II) Fluoride

Anhydrous iron(II) fluoride [7789-28-8], FeF_2, is a white solid. The off-white to buff-colored appearance of the material is attributed to the partial oxidation of Fe^{2+} to Fe^{3+}. FeF_2 is highly stable and does not decompose when heated in the presence of nitrogen. It is sparingly soluble in water but the solubility can be increased by the addition of aqueous HF or any strong acid. Physical properties are listed in Table 1. FeF_2 holds great promise in the field of advanced magnets known as the iron–boron–rare-earth-alloy sintered magnets (1).

Table 1. Physical Properties of Iron Fluorides

Property	FeF_2	FeF_3
mol wt	93.84	112.84
density, g/cm^3	4.09	3.87
mp, °C	1100	1000^a
bp, °C	1837	
C_p, J/(mol·K)^b	68.12	+91.0

^{a}Sublimes.
^{b}To convert J to cal, divide by 4.184.

FeF_2 was first prepared by the action of gaseous hydrogen fluoride over $FeCl_2$ in an iron boat (2). The reaction of anhydrous $FeCl_2$, $FeCl_2\cdot4H_2O$, or $FeSO_4\cdot7H_2O$ and anhydrous HF in plastic reaction vessels such as vessels of polyethylene, polypropylene, or Teflon results in quantitative yields of very high purity FeF_2. The anhydrous salt has also been prepared from a solid-state reaction of a mixture of FeC_2O_4 and NH_4F (weight ratio 1:3) at 300°C and 13.3 Pa (0.1 torr) (3). Other methods of preparation are also available (4,5).

Colorless crystals of iron(II) fluoride tetrahydrate [13940-89-1], $FeF_2\cdot4H_2O$, can be obtained by dissolving metallic iron or the anhydrous salt in hydrofluoric acid. The crystals of $FeF_2\cdot4H_2O$ are sparingly soluble in water and decompose to Fe_2O_3 when heated in air.

The only reported industrial application for FeF_2 is its use in rust removal solutions based on oxalic acid (6). The anhydrous salt is commercially available in 100 g to 5 kg lots from Advance Research Chemicals, Aldrich Chemicals, Cerac, Johnson/Matthey, PCR, and other suppliers in the United States. As of 1993, the prices varied between $500 to $700/kg.

Toxicity of iron(II) fluoride has not been determined. FeF_2 is shipped as a nonhazardous material in plastic containers. The ACGIH has adopted (1991–1992) a TWA value of 1 mg/m^3 for iron as Fe, and 2.5 mg/m^3 for fluorides as F$^-$.

Iron(III) Fluoride

Iron(III) fluoride [7783-50-8], FeF_3, is the most widely known fluoride of iron. It is light greenish (lime green) in color and the crystals have a rhombic structure. Physical properties are listed in Table 1.

Anhydrous FeF_3 is prepared by the action of liquid or gaseous hydrogen fluoride on anhydrous $FeCl_3$ (see IRON COMPOUNDS). FeF_3 is insoluble in alcohol, ether, and benzene, and sparingly soluble in anhydrous HF and water. The pH of a saturated solution in water varies between 3.5 and 4.0. Low pH indicates the presence of residual amounts of HF. The light gray color of the material is attributed to iron oxide or free iron impurities in the product.

The most important industrial application of the iron(III) fluoride is in the manufacture of Fe–Co–Nd magnets. Other significant uses are as a hydrocracking catalyst (7), as a catalyst for the preparation of perfluoroacyl fluorides (8), as a catalyst for hydrorefining of lubricating oils (9), as a fluorinating agent (10), for pin-hole prevention in cast iron (11), as a catalyst for preparation of xenon–fluorine compounds (12), burning rate control catalyst, as a catalyst for aromatization, dealkylation, and polymerization, and conversion of vinylidene chloride to the fluoride (13), and in the manufacturing of flame-retardant polymers (14). The industrial market for iron(III) fluoride varies from 2000 kg/yr to 30,000 kg/yr and 1993 prices ranged from $25 to $100/kg. FeF_3 is available from Advance Research Chemicals, Aldrich Chemicals, Morrita Chemicals of Japan, and also Russian and European producers.

Hydrated Salts and Other Compounds

Hydrated iron(III) fluoride [15469-38-2], $FeF_3\cdot3H_2O$, is easily prepared from yellow Fe_2O_3 and hydrofluoric acid. Dehydration of $FeF_3\cdot3H_2O$ produces oxyfluorides of iron.

In the presence of excess HF, complex ions such as FeF_4^- and FeF_6^- are formed in solution. Neutralization using a base such as NaOH produces $NaFeF_4$ [*15274-99-4*] and Na_3FeF_6 [*20955-11-7*], respectively. The latter is used as a fluorinating agent (15).

A mixed valency pale yellow crystalline iron pentafluoride heptahydrate, $FeF_5 \cdot 7H_2O$, is prepared by dissolving iron powder in 40% HF in the presence of air (16). No applications have been reported for this material.

BIBLIOGRAPHY

"Iron Compounds" under "Fluorine Compounds, Inorganic," in *ECT* 1st ed., Vol. 6, p. 709, by F. D. Loomis; "Iron" under "Fluorine Compounds, Inorganic," in *ECT* 2nd ed., Vol. 9, pp. 625–626, by W. E. White; in *ECT* 3rd ed., Vol. 10, pp. 754–755, by D. T. Meshri, Advance Research Chemicals, Inc.

1. Jpn. Kokai Koho, 63,249,304 (Oct. 17, 1988), A. Kobayashi and T. Sato (to Hitachi Metals Ltd.).
2. C. Poulenc, *Compt. Rend. Hebd. Acad. Sci.* **115**, 942 (1980).
3. USSR Pat. 1,502,473 (Aug. 23, 1989), S. V. Petrov, N. I. Kuznetsova, D. D. Ikrami, S. Ganiev, and V. S. Sidorov.
4. G. Pourroy and P. Poix, *J. Fluorine Chem.* **42**(2), 257–263 (1989).
5. I. G. Ryss, *The Chemistry of Fluorine and its Inorganic Compounds*, State Publishing House for Scientific and Chemical Literature, Moscow, Russia, 1956; Eng. transl. ACE-Tr-3927, Vol. II, Office of Technical Services, U.S. Department of Commerce, Washington, D.C., 1960, p. 665.
6. U.S. Pat. 4,828,743 (May 7, 1989), S. Rahfield and B. Newman (to Boyle Midway Household Products Inc.).
7. U.S. Pat. 4,895,822 (Jan. 23, 1990), H. Okazaki, M. Adachi, and M. Ushio (to Nippon Oil Co. Ltd.).
8. Eur. Pat. Appl. 260,713 (Sept. 19, 1986), P. Cuzzato, A. Castellan, and A. Paquale (to Ausimont SPA).
9. Pol. Pat. 138,387 (Jan. 30, 1988), E. Zienkiewicz, J. Kudmierczyk, A. Kubacki, and K. Kowalczyk (to Gdanskse Zaklady Refineryine, Politechnika, Wroclawska).
10. S. Okazaki, *Nippon Kagaku Zasshi* **89**, 1054 (1968).
11. Jpn. Kokai, 75,17,173 (June 19, 1975), T. Kuska (to Hinoshita Rare Metal Institute).
12. B. Z. Slivnik, *Inorg. Nucl. Chem.*, 173 (1976).
13. U.S. Pat. 4,827,055 (Mar. 7, 1988), M. Elsheikh (to Pennwalt Corp.).
14. Ger. Offen. 2,531,816 (Feb. 12, 1976), E. Dorfman, R. R. Hindersim, and W. T. Schwatz (to Hooker Chemical Plastics Corp.).
15. B. Cornils, M. Rassch, and G. Shcieman, *Chem. Ztg. Chem. Appl.* **92**(5), 137 (1968).
16. K. J. Galagher and M. J. Ottaway, *J. Chem. Soc. Dalton*, 978 (1975).

DAYAL T. MESHRI
Advance Research Chemicals, Inc.

LEAD

Lead Difluoride

Lead difluoride [7783-46-2], PbF_2, has the highest melting and boiling points among all the dihalides of lead. Two colorless crystalline forms are known. The α-PbF_2 is orthorhombic in structure and is stable at ordinary temperatures. Upon heating to 200°C it transforms to the cubic β-form. Table 1 lists some of the physical properties of PbF_2.

PbF_2 is readily prepared by the action of hydrogen fluoride on lead hydroxide, lead carbonate, or α-lead oxide. It can also be obtained by precipitation from lead nitrate or lead acetate solutions using potassium fluoride, ammonium fluoride, or ammonium bifluoride.

PbF_2 exhibits very good electrical insulating properties and optical transparency. It is thus used in a variety of glass (qv) such as sealing glass (1), low melting glass (2), near infrared absorbing glass for fiber optics (qv) (3), weather-resistant glass (4,5), and glass for active optical fibers (6). It is also used in printing, photography (qv), brazing, scintillation counters (7), dielectric interference filters (see OPTICAL FILTERS) (8), as a mild fluorinating reagent, as a source material for PbF_4, and as an ingredient in lead–acid batteries (qv) (9).

High purity lead difluoride is available from Advance Research Chemicals, Aldrich Chemicals, Johnson/Matthey, Atomergic, Cerac, and other suppliers in the United States. The U.S. annual consumption varies between 500 to 2500 kg/yr. The 1993 price varied between $10–20/kg.

Table 1. Physical Properties of Lead Fluorides

Property	PbF_2	PbF_4
mol wt	245.19	283.2
density, g/cm^3	8.24	6.7
melting point, °C	855	600
boiling point, °C	1290	decomposes
solubility, g/100 g		
water	0.0641	a
anhydrous HF	2.628	
ΔH_f, kJ/mol^b	−677	
ΔG_f, kJ/mol^b	−631	
C_p, J/(mol·K)^b	72.3	
S, J/(mol·K)^b	−113	

^{a}Material hydrolyzes to PbO_2 and HF.
^{b}To convert J to cal, divide by 4.184.

Lead Tetrafluoride

Like all the lead tetrahalides, lead tetrafluoride [7783-59-7], PbF_4, is very reactive. It is relatively the most stable halide, however. PbF_4 is a white crystalline powder which is highly moisture sensitive, turning yellowish brown in moist air owing to hydrolysis. It should be handled in a dry box or under an atmosphere of dry nitrogen. Properties for PbF_4 are in Table 1.

PbF_4, produced by various routes including the *in situ* species, is a very effective fluorinating agent and also an oxidizing agent. It is classified as a hard fluorinating agent (10), replacing hydrogen with fluorine or adding fluorine to double bonds of both halogenated and hydrocarbon olefins to produce difluorocarbons (11,12).

$$CCl_2{=}CCl_2 + PbF_4 \rightarrow CCl_2FCCl_2F + PbF_2 \tag{1}$$

$$CF_3CCl{=}CCl_2 + PbF_4 \rightarrow CF_3CClFCCl_2F + PbF_2 \tag{2}$$

$$CHCl{=}CHCl + PbF_4 \rightarrow CHClFCHClF + PbF_2 \tag{3}$$

It is also used in the preparation of biologically active steroids where the fluorine is added in a cis configuration to the double bond (13,14).

Lead fluorides are highly toxic and should be handled with great care. The ACGIH adopted toxicity value for lead compounds as Pb is TWA 0.15 mg/m^3 and for fluorides as F^- 2.5 mg/m^3. PbF_4 is prepared by the action of elemental fluorine on very dry PbF_2 at 280–300°C (15).

BIBLIOGRAPHY

"Lead Compounds" under "Fluorine Compounds, Inorganic" in *ECT* 1st ed., Vol. 6, pp. 709–710, by F. D. Loomis, Pennsylvania Salt Manufacturing Co.; "Lead" under "Fluorine Compounds, Inorganic" in *ECT* 2nd ed., Vol. 9, pp. 626–627 by W. E. White, Ozark-Mahoning Co.; in *ECT* 3rd ed., Vol. 10, pp. 756–757 by D. T. Meshri, Ozark Mahoning Co.

1. USSR Pat. 1,701,656 (Dec. 30, 1991), N. B. Knyazyan and co-workers.
2. V. A. Kulgin and K. Pan, *Otkrytiya, Izobret* **33**, 84(1991); *Chem. Abstr.* **116** 157385r; USSR Pat. 1,675,238 (Sept. 7, 1991).
3. Jpn. Kokai Tokkyo Koho, JP 03, 32,735 (Oct. 16, 1991), N. Matsui and W. Takahashi (to Toshiba Glass Co. Ltd.).
4. Jpn. Kokai Tokkyo Koho, JP 04 21,541 (Jan. 24, 1992), T. Osuga and T. Kawaguchi (to Asahi Glass Co. Ltd.).
5. Jpn. Kokai Tokkyo Koho, JP 03 40,935 (Feb. 21, 1991), T. Osuga and T. Kawaguchi (to Asahi Glass Co. Ltd.).
6. J. L. Adam and co-workers, *Proc. SPIE, Int. Soc. Opt. Eng.*, (*Glass Optoelectronics*), 1513 1991, pp. 150–157.
7. J. L. Pauley and M. K. Testerman, *J. Am. Chem. Soc.* **76**, 4220 (1954).
8. G. Honcia and K. Krebs, *Optik* **19**(3), 156 (1962).
9. Jpn. Kokai Tokkyo Koho JP 02,119,055 (May 7, 1990), M. Terada, S. Saito, and A. Miura (to Shin-Kobe Electric Machinery Co. Ltd.).
10. D. T. Meshri and W. E. White, *George H. Cady ACS Symposium*, Milwaukee, Wis., June 1970.
11. A. L. Henne and T. H. Newby, *J. Am. Chem. Soc.* **70**, 130 (1948).
12. A. L. Henne and T. P. Waalkes, *J. Am. Chem. Soc.* **67**, 1639 (1945).
13. A. Bowers and co-workers, *J. Am. Chem. Soc.* **84**, 1050 (1962).
14. Ger. Pat. 1,167,828 (Apr. 16, 1964), K. Bruckner and H. J. Mannhardt (to E. Merck AG).
15. J. Bornstein and L. Skarlas, *J. Am. Chem. Soc.* **90**, 5046 (1968).

DAYAL T. MESHRI
Advance Research Chemicals, Inc.

LITHIUM

Lithium Fluoride

Properties.   Lithium fluoride [7789-24-4], LiF, is a white nonhygroscopic crystalline material that does not form a hydrate. The properties of lithium fluoride are similar to the alkaline-earth fluorides. The solubility in water is quite low and chemical reactivity is low, similar to that of calcium fluoride and magnesium fluoride. Several chemical and physical properties of lithium fluoride are listed in Table 1. At high temperatures, lithium fluoride hydrolyzes to hydrogen fluoride when heated in the presence of moisture. A bifluoride [12159-92-1], LiF·HF, which forms on reaction of LiF with hydrofluoric acid, is unstable to loss of HF in the solid form.

Table 1. Properties of Lithium Fluoride

Property	Value	Reference[a]
melting point, °C	848	1
boiling point, °C	1681	1
solubility, g/100 g solvent		
water, 25.4°C	0.133	2
water, 81.8°C	0.150	2
acetic acid, 25°C	0.084	3
acetic acid, 50°C	0.152	3
liquid HF, 12°C	10.3	4
tetrahydrofuran, 25°C	0.6	5
crystalline form	cubic (NaCl)	6
a_0, nm	0.401736	
density at 20°C, g/cm^3	2.635	6
index of refraction	1.3915	6
lattice energy, kJ/mol[b]	1020 ± 10	7
standard heat of formation, kJ/mol[b]	−613.0	1
standard entropy, J/(mol·K)[b]	35.9	1
heat capacity, C_p, J/(mol·K)[b]	42.01	1
heat of fusion, kJ/mol[b]	27.09	1
heat of vaporization, kJ/mol[b]	213	8
heat of hydration, kJ/mol[b]	1.023	8
debye temperature, °C	449	9

[a]Properties listed in this table can be supplemented by the comprehensive collections in References 10 and 11.
[b]To convert kJ to kcal, divide by 4.184.

Manufacture.   Lithium fluoride is manufactured by the reaction of lithium carbonate or lithium hydroxide with dilute hydrofluoric acid. If the lithium carbonate is converted to the soluble bicarbonate, insolubles can be removed by filtration and a purer lithium fluoride can be made on addition of hydrofluoric acid (12). High purity material can also be made from other soluble lithium salts such as the chloride or nitrate with hydrofluoric acid or ammonium bifluoride (13).

Optical crystals of high purity lithium fluoride are grown by use of the Stockbarger process (10) in sizes to 25 cm dia $\times$ 25 cm high (14). Typical commercial material contains 99.2% LiF; typical impurities include Li_2CO_3 and Fe_2O_3 at <0.1% levels, and SO_4^{2-}, PO_4^{3-}, and heavy metals as Pb at <0.01% levels. The price during 1991 was \$10.91/kg in truckload quantities. Annual production is probably less than 100 metric tons. Lithium fluoride toxicity relative to use in thermoluminescent dosimetry is discussed in Reference 15; 10 mg/d is proposed as the maximum permissible daily intake for the average human body. Ingestion of 200 mg/kg of body weight is lethal to guinea pigs (16).

Uses. Lithium fluoride is used primarily in the ceramic industry to reduce firing temperatures and improve resistance to thermal shock, abrasion, and acid attack (see CERAMICS). Another use of LiF is in flux compositions with other fluorides, chlorides, and borates for metal joining (17) (see SOLDERS).

Lithium fluoride is an essential component of the fluorine cell electrolyte; 1% LiF in the KF·2HF electrolyte improves the wettability of the carbon anodes and lowers the tendency of the cells to depolarize (18). Thermoluminescent radiation dosimeters used in personnel and environmental monitoring and in radiation therapy contain lithium fluoride powder, extruded ribbons, or rods (19).

Molten lithium fluoride is used in salt mixtures for an electrolyte in high temperature batteries (qv) (FLINAK) (20), and as a carrier in breeder reactors (FLIBE) (21) (see NUCLEAR REACTORS).

Large high purity crystals are cut into windows and refracting components for use in x-ray monochromators (14), and in the vacuum uv, uv, visible, and ir ranges.

BIBLIOGRAPHY

"Lithium Fluoride" in *ECT* 1st ed., under "Fluorine Compounds, Inorganic," Vol. 6, pp. 709–710, by F. D. Loomis, Pennsylvania Salt Manufacturing Co.; in *ECT* 2nd ed., Vol. 9, p. 627, by G. C. Whitaker, The Harshaw Chemical Co.; "Lithium" under "Fluorine Compounds, Inorganic," in *ECT* 3rd ed., Vol. 10, pp. 757–759, by H. S. Halbedel and T. E. Nappier, The Harshaw Chemical Co.

1. *JANAF Thermochemical Tables*, Clearinghouse for Federal Scientific and Technical Information, U.S. Dept. of Commerce, Springfield, Va., Dec. 1963.
2. C. B. Stubblefield and R. O. Bach, *J. Chem. Eng. Data* **17**, 491 (1972).
3. J. Emsley, *J. Chem. Soc. A*, 2511 (1971).
4. A. W. Jache and G. W. Cady, *J. Phys. Chem.* **56**, 1106 (1952).
5. Brit. Pat. 787,771 (Dec. 18, 1957), (to Metropolitan Vickers Electric Co.).
6. C. A. Hutchison and H. L. Johnson, *J. Am. Chem. Soc.* **62**, 3165 (1940).
7. D. F. C. Morris, *Acta Crystallogr.* **9**, 197 (1956).
8. M. C. Ball and A. A. Norbury, *Physical Data for Inorganic Chemists*, Longman, Inc., New York, 1974.
9. W. W. Scales, *Phys. Rev.* **112**, 49 (1958).
10. *Gmelins Handbuch der Anorganischen Chemie*, 8th ed., Deutsche Vol. 6, Verlag-Chemie, Weinheim/Bergstrasse, 1960, pp. 305–327.
11. J. W. Mellor, *Comprehensive Treatise on Inorganic and Theoretical Chemistry*, Vol. 2, Suppl. 2, Longman, Green and Co., New York, 1961, pp. 174–178.
12. D. C. Stockbarger, *Rev. Sci. Instrum.* **7**, 133 (1936).

13. U.S. Pat. 3,132,922 (May 12, 1954), R. D. Goodenough and T. G. Cook (to Dow Chemical Co.).
14. *Harshaw Optical Crystals*, Harshaw Chemical Co., Solon, Ohio, 1967, pp. 32–33.
15. N. C. Spoor, *Ann. Occup. Hyg.* **11**(1), 23 (1968).
16. H. C. Hodge and F. N. Smith, in J. H. Simon, ed., *Fluorine Chemistry*, Vol. 4, Academic Press, New York, 1965, p. 199.
17. U.S. Pat. 3,958,979 (May 25, 1976), A. R. Valdo (to Ethyl Corp.).
18. J. T. Pinkston, *Ind. Eng. Chem.* **39**, 255 (1947).
19. F. M. Cox, *Proc. 2nd Int. Conf. on Luminescence Dosimetry*, Oak Ridge National Laboratory, CONF 680920, Oak Ridge, Tenn., Sept. 1968; F. M. Cox, A. C. Lucas, and B. M. Kaspar, *Health Phys.* **30**, 135 (1976); J. F. Valley, C. Pache, and P. Lerch, *Helv. Phys. Acta* **49**(2), 171 (1976).
20. G. L. Green, J. B. Hunt, and R. A. Sutula, *U.S. Nat. Tech. Inform. Serv.*, A.D. Rep. 1973, No. 758001.
21. C. D. Scott and W. L. Carter, *AEC Accession No. 43422, Rept. No. ORNL-3791*, Oak Ridge National Laboratory, Oak Ridge, Tenn., 1966.

JOHN R. PAPCUN
Atotech

MAGNESIUM

Magnesium Fluoride

Properties. Magnesium fluoride [7783-40-6], MgF_2, is a fine white crystalline powder with low chemical reactivity. This relative inertness makes possible some of its uses, eg, stable permanent films to alter light transmission properties of optical and electronic materials. The reaction with sulfuric acid is so sluggish and incomplete that magnesium fluoride is not a suitable substitute for calcium fluoride in manufacturing hydrogen fluoride. Magnesium fluoride resists hydrolysis to hydrogen fluoride up to 750°C (1). Bimetallic fluorides, such as $KMgF_3$ [28042-61-7], are formed on fusion of MgF_2 alkali metal and ammonium fluorides (2). Chemical and physical properties are listed in Table 1. MgF_2 is birefringent and only mildly affected by high energy radiation, making possible optics for the uv region.

Manufacture. Magnesium fluoride is manufactured by the reaction of hydrofluoric acid and magnesium oxide or carbonate:

$$MgO + 2\,HF \rightarrow MgF_2 + H_2O$$
$$MgCO_3 + 2\,HF \rightarrow MgF_2 + CO_2 + H_2O$$

Formation of a gelatinous precipitate that is difficult to filter can be avoided by addition of magnesium oxide to the acid solution. In order to increase particle size it is often necessary to keep the solution hot for several hours; however, this problem is avoided by heating an intimate mixture of ammonium bifluoride with magnesium carbonate to 150–400°C (11). Particles of MgF_2 similar in size to those of the magnesium carbonate are obtained.

Table 1. Chemical and Physical Properties

Property	Value	Reference
melting point, °C	1263	3
boiling point, °C	2227	3
standard heat of formation, kJ/mol[a]	−112.4	4
standard entropy of formation, J/(mol·K)[a]	178	5
heat of fusion, kJ/mol[a]	58.2	3
heat of vaporization, kJ/mol[a]	264	5
lattice energy, kJ/mol[a]	2920	5
heat capacity, 25°C, J/(mol·K)[a]	61.59	3
free energy of solution, kJ/mol[a]	40.2	5
density, g/cm^3	3.127	6
index of refraction[b]		
n_o[c]	1.37770	7
n_e	1.38950	7
crystalline form (sellaite)	tetragonal	6
a, nm	0.4623	
c, nm	0.3052	
solubility, g/100 g of solvent		
water, 25°C	0.013	8
hydrogen fluoride, 12°C	0.025	9
acetic acid, 25°C	0.681	10

[a]To convert kJ to kcal, divide by 4.184.
[b]o, ordinary; e, extraordinary.
[c]At 589 nm.

$$MgCO_3 + NH_4HF_2 \xrightarrow{150-400°C} MgF_2 + NH_3 + CO_2 + H_2O$$

The same results are obtained by adding magnesium carbonate to an aqueous solution of ammonium bifluoride and ammonium hydroxide and warming to 60°C (12). The resulting precipitate is ammonium magnesium fluoride [35278-29-6] which settles rapidly.

$$MgCO_3 + 3 NH_4F \text{ (aq)} \longrightarrow NH_4MgF_3 + (NH_4)_2CO_3$$

$$NH_4MgF_3 \xrightarrow[4\text{ h}]{620°C} MgF_2 + NH_3 + HF$$

Magnesium fluoride is a by-product of the manufacture of metallic beryllium and uranium. The beryllium or uranium fluorides are intimately mixed with magnesium metal in magnesium fluoride-lined crucibles. On heating, a Thermite-type reaction takes place to yield the desired metal and MgF_2 (13). Part of the magnesium fluoride produced in this reaction is then used as a lining for the crucibles used in the process.

$$BeF_2 + Mg \rightarrow Be + MgF_2$$

A commercial grade of magnesium fluoride containing approximately 96–98% MgF_2 is manufactured by Advance Research Chemicals and the Bicron Co. in the United States. Imported technical grades suitable for fluxes containing 94–96% magnesium fluoride are available from Atomergic Chemetals Co., Fine Chemical Co., and Magnesium Elektron.

Magnesium fluoride optical crystals are made by hot-pressing (14) high quality MgF_2 powder. The optical quality powder is made by the NH_4HF_2 method described (11) or by reaction of magnesium bicarbonate and hydrofluoric acid (15). Lead fluoride can also be used in purification of MgF_2 for optical crystals (16). Such optical crystals are manufactured by Bicron Co.

Toxicity. The lethal dose of MgF_2 to guinea pigs by ingestion is 1000 mg/kg (17).

Uses. Established uses of magnesium fluoride are as fluxes in magnesium metallurgy and in the ceramics industry. A proposed use is the extraction of aluminum from arc-furnace alloys with Fe, Si, Ti, and C (18). The molten alloy in reacting with magnesium fluoride volatilizes the aluminum and magnesium which are later separated above the melting point of MgF_2. A welding (qv) flux for aluminum (19) as well as fluxes for steel (20) contain MgF_2.

Optical windows of highly purified magnesium fluoride which transmit light from the vacuum ultraviolet (140 nm) into the infrared (7) are recommended for use as ultraviolet optical components for use in space exploration.

BIBLIOGRAPHY

"Magnesium Fluoride" under "Fluorine Compounds, Inorganic," in *ECT* 1st ed., Vol. 6, p. 709, by F. D. Loomis, Pennsylvania Salt and Manufacturing Co.; in *ECT* 2nd ed., Vol. 9, pp. 627–628 by G. C. Whitaker, The Harshaw Chemical Co.; "Magnesium" under "Fluorine Compounds, Inorganic," Vol. 10, pp. 760–762, by T. E. Nappier and H. S. Halbedel, The Harshaw Chemical Co.

1. D. R. Messier, *J. Am. Ceram. Soc.* **48**, 452, 459 (1965).
2. I. G. Ryss, *The Chemistry of Fluorine and Its Inorganic Compounds*, State Publishing House for Scientific and Technical Literature, Moscow, 1956; Engl. transl. by F. Haimson for the U.S. Atomic Energy Commission, *AEC-tr-3927*, Washington, D.C., 1960, p. 812.
3. *JANAF Thermochemical Tables*, Clearinghouse for Federal Scientific and Technical Information, U.S. Dept. of Commerce, Springfield, Va., 1966.
4. E. Ruelzitis, H. M. Fedar, and W. N. Hubbard, *J. Chem. Phys.* **68**, 2978 (1964).
5. M. C. Ball and A. A. Norbury, *Physical Data for Inorganic Chemicals*, Longman, Inc., New York, 1974.
6. C. Palache, H. Berman, and C. Frandel, *Danas System of Mineralogy*, 7th ed., Vol. 2, John Wiley & Sons, Inc., New York, 1951, p. 38.
7. Technical data, The Harshaw Chemical Co., Crystal and Electronics Dept., Solon, Ohio.
8. D. D. Ikrami, A. S. Paramzin, and A. Kubr, *Russ. J. Inorg. Chem.* **16**, 425 (1971).
9. A. W. Jache and G. W. Cady, *J. Phys. Chem.* **56**, 1106 (1952).
10. C. J. Emsley, *J. Chem. Soc. (A)*, 2511 (1971).
11. U.S. Pat. 3,357,788 (Dec. 12, 1967), J. F. Ross (to General Electric Co.).
12. U.S. Pat. 3,848,066 (Nov. 12, 1974), C. D. Vanderpool and M. B. MacInnis (to G.T.E. Sylvania).

13. H. E. Thayer, *Proc. of the 2nd U. M. Internat. Conference, Peaceful Uses of Atomic Energy, (Geneva),* **4**, 22 (1958); W. E. Dennis and E. Proudfoot, *U.K. At. Energy R&D* **B**(C) TN-88 (1954).
14. U.S. Pat. 3,294,878 (Dec. 29, 1960), E. Carroll and co-workers (to Eastman Kodak Co.).
15. U.S. Pat. 3,920,802 (Nov. 18, 1975), R. H. Moss, C. F. Swinehart, and W. F. Spicuzza (to Kewanee Oil Co.).
16. U.S. Pat. 2,498,186 (Feb. 21, 1950), D. C. Stockbarger and A. A. Blanchard (to Research Corp.).
17. H. C. Hodges and F. A. Smith, in J. H. Simons, ed., *Fluorine Chemistry*, Vol. 4, Academic Press, Inc., New York, 1965, p. 199.
18. G. S. Layne and co-workers, *Light Met. Age* **30**(3,4), 8 (1972).
19. U.S. Pat. 2,552,104 (May 8, 1951), M. A. Miller and W. E. Haupin (to Alcoa).
20. Jpn. Pat. 7,556,339 (Sept. 19, 1973), T. Tanigaki, T. Koshio, and T. Enomoto (to Nippon Steel Corp.).

JOHN R. PAPCUN
Atotech

MERCURY

Mercury(I) Fluoride

Mercury(I) fluoride [*13967-25-4*], Hg_2F_2, also known as mercurous fluoride, is a light-sensitive golden yellow material decomposing in water at 15°C. Some of the physical properties are listed in Table 1. Hg_2F_2 resembles AgF in activity with the exception that the former does not form complex compounds or mixed halogen fluoride salts. Consequently, almost an equivalent amount of Hg_2F_2 is sufficient for halogen exchange reactions (1). A mixture of Hg_2F_2 and I_2 (1:1 molar ratio) is much more effective than Hg_2F_2 alone (2). Mercury(I) fluoride is classified as a soft fluorinating reagent (3). Reactions of Hg_2F_2 with monobromides and monoiodides produce fairly good yields, but in polybromides and polyiodides only one halogen is replaced at 120–140°C. Loss of hydrogen halide gives the corresponding olefins (4).

Table 1. Properties of Mercury Fluorides

Property	Hg_2F_2	HgF_2
mol wt	439.22	238.61
density, g/cm^3	8.73	8.95
melting point, °C	>570 (dec)	645
ΔH_f, kJ/mol^a	−485	−405
ΔG_f, kJ/mol^a	−469	−362
C_p, J/(mol·K)^a	+100.4	+74.86
S, J/(mol·K)^a	161	134.3

^{a}To convert J to cal, divide by 4.184.

Several preparatory methods for the manufacture of Hg_2F_2 have been reported (5). Whereas no commercial applications for Hg_2F_2 have been reported, it is available from Advance Research Chemicals and Aldrich Chemicals in the United States. As of 1993, the U.S. market was a few kilograms per year at a price of $1500/kg.

Mercury(II) Fluoride

Mercury(II) fluoride [7783-39-3], HgF_2, also known as mercuric fluoride, is a white, hygroscopic solid which turns yellow instantly on exposure to moist air. It must be handled in a dry box or under an atmosphere of dry nitrogen. Some of its physical properties are listed in Table 1. Whereas HgF_2, classified as a moderate fluorinating reagent (3), is superior to both AgF and Hg_2F_2, it has been replaced by anhydrous potassium fluoride [7789-23-3], KF, owing to the toxicity of mercury and the disposal regulations issued by the EPA (see MERCURY COMPOUNDS). HgF_2 is an excellent reagent for the addition of fluorine to olefins (1).

Mercury(II) fluoride is easily prepared by passing pure elemental fluorine over predried $HgCl_2$ at 100–150°C until all the chloride ions have been replaced. It is also produced *in situ* by condensing anhydrous HF over HgO (6) or over $HgCl_2$ (10).

Mercury(II) fluoride has been used in the process for manufacture of fluoride glass (qv) for fiber optics (qv) applications (11) and in photochemical selective fluorination of organic substrates (12). It is available from Advance Research Chemicals, Aldrich Chemicals, Johnson/Matthey, Aesar, Cerac, Strem, and PCR in the United States. The 1993 annual consumption was less than 50 kg; the price was $800–1000/kg.

Mercury salts are highly toxic and must be handled carefully. It is necessary to consult the material safety data sheet prior to handling. Strict adherence to OSHA/EPA regulations is essential. The ACGIH adopted (1991–1992) TLV for mercury as inorganic compounds is TWA 0.1 mg/m^3 and for fluorides as F^- 2.5 mg/m^3.

BIBLIOGRAPHY

"Mercury Fluorides" under "Fluorine Compounds, Inorganic" in *ECT* 1st ed., Vol. 6, pp. 747–748, by E. T. McBee and O. R. Pierce, Purdue University; "Mercury" under "Fluorine Compou ds, Inorganic" in *ECT* 2nd ed., Vol. 9, p. 628, by W. E. White, Ozark-Mahoning Co.; in *ECT* 3rd ed., Vol. 10, pp. 763–764 by D. T. Meshri, Ozark-Mahoning Co.

1. F. Swarts, *Bull. Acad. Roy. Belg.* (5)7, 438 (1921).
2. A. L. Henne and M. W. Renoll, *J. Am. Chem. Soc.* **60**, 1060 (1938).
3. D. T. Meshri and W. E. White, *George H. Cady ACS Symposium*, Milwaukee, Wis., June 1970.
4. R. N. Hazeldine and B. R. Steele, *J. Chem. Soc.*, 1199 (1953).
5. I. G. Ryss, *The Chemistry of Fluorine Compounds*, State Publishing House for Scientific and Chemical Literature, Moscow, Russia, 1956, English tranl. ACE-Tr-3927, Vol. II, Office of Technical Services, U.S. Dept. of Commerce, Washington D.C., 1960, pp. 634–635.

6. A. L. Henne and M. W. Renoll, *J. Am. Chem. Soc.* **60**, 1960 (1938).
7. A. L. Henne and T. Midgley, Jr., *J. Am. Chem. Soc.* **58**, 882 (1936).
8. A. L. Henne and M. W. Renoll, *J. Am. Chem. Soc.* **58**, 887 (1936).
9. J. B. Dicky and co-workers, *Ind. Eng. Chem.* **46**, 2213 (1954).
10. O. Ruff and co-workers, *Chem. Ber.* **51**, 1752 (1918)
11. Jpn. Kokai Tokkyo Koho, JP 63239,137 (Oct. 5, 1988), N. Mitachi, Y. Ooishi, and S. Sakaguchi (to Nippon Telegraph and Telephone Co.).
12. M. H. Habibi and T. E. Mallouk, *J. Fluorine Chem.* **51**(2), 291–294 (1991).

DAYAL T. MESHRI
Advance Research Chemicals, Inc.

MOLYBDENUM

Molybdenum Hexafluoride

Molybdenum hexafluoride [7783-77-9], MoF_6, is a volatile liquid at room temperature. It is very moisture sensitive, hydrolyzing immediately upon contact with water to produce HF and molybdenum oxyfluorides. MoF_6 should therefore be handled in a closed system or in a vacuum line located in a chemical hood. The crystals possess a body-centered cubic structure that changes to orthorhombic below $-96°C$ (1,2). The known physical properties are listed in Table 1.

Molybdenum hexafluoride can be prepared by the action of elemental fluorine on hydrogen-reduced molybdenum powder (100–300 mesh (ca 149–46 μm)) at 200°C. The reaction starts at 150°C. Owing to the heat of reaction, the tem-

Table 1. Physical Properties of MoF_6

Property	Value
mol wt	209.93
melting point, °C	17.4
boiling point, °C	35.0
solubility, g/100 g	[a]
density, g/cm^3	
liquid	2.544
solid	2.888
ΔH_f, kJ/mol[b]	
liquid	-1626
gas	-1558
ΔG_f, kJ/mol[b]	
liquid	-1511
gas	-1468
S, J/(mol·K)[b]	259.69

[a]Hydrolyzes in water.
[b]To convert J to cal, divide by 4.184.

perature of the reactor rises quickly but it can be controlled by increasing the flow rate of the carrier gas, argon, or reducing the flow of fluorine.

Molybdenum hexafluoride is used in the manufacture of thin films (qv) for large-scale integrated circuits (qv) commonly known as LSIC systems (3,4), in the manufacture of metallized ceramics (see METAL-MATRIX COMPOSITES) (5), and chemical vapor deposition of molybdenum and molybdenum–tungsten alloys (see MOLYBDENUM AND MOLYBDENUM ALLOYS) (6,7). The latter process involves the reduction of gaseous metal fluorides by hydrogen at elevated temperatures to produce metals or their alloys such as molybdenum–tungsten, molydenum–tungsten–rhenium, or molybdenum–rhenium alloys.

Molybdenum hexafluoride is classified as a corrosive and poison gas. It is stored and shipped in steel, stainless steel, or Monel cylinders approved by DOT. Electronic and semiconductor industries prefer the cylinders equipped with valves which have Compressed Gas Association (CGA) 330 outlets. This material is produced on pilot-plant scale and the U.S. annual consumption is less than 50 kg/yr. As of 1993, the price was $1500/kg. It is available from Advance Research Chemicals Inc., Aldrich Chemicals, Atomergic, Cerac, Johnson/Matthey, Pfaltz & Bauer, and Strem Chemicals.

Other Molybdenum Fluorides

Three other binary compounds of molybdenum and fluorine are known to exist: molybdenum trifluoride [20193-58-2], MoF_3, molybdenum tetrafluoride [23412-45-5], MoF_4, and molybdenum pentafluoride [13819-84-6], MoF_5. Also known are the two oxyfluorides, molybdenum dioxydifluoride [13824-57-2], MoO_2F_2, and molybdenum oxytetrafluoride [14459-59-7], $MoOF_4$. The use of these other compounds is limited to research applications.

BIBLIOGRAPHY

"Molybdenum" under "Fluorine Compounds, Inorganic" in *ECT* 3rd ed., Vol. 10, pp. 764–765, by A. J. Woytek, Air Products & Chemicals, Inc.

1. M. Carles, *L. Hexafluorure de Molybdene, MoF₆*, Commissariat a L'Energie Atomique, Pierelatte, France; *Report CEA-BIB-124, NASA Technical Translation TT-F-12, 702,* Sept. 1968, pp. 1–25.
2. S. Siegel and D. A. Northrup, *Inorg. Chem.* **5**, 2187 (1966).
3. Jpn. Kokai Tokkyo Koho JP 61 224,313 (Oct. 6, 1986), S. Tsujiku and co-workers.
4. Ger. Offen. 3,639,080 (May 21, 1987), Y. S. Liu and C. P. Yakmyshyn.
5. Eur. Pat. Appl. EP 443,277 (Aug. 28, 1991), P. Jalby and co-workers.
6. J. G. Donaldson and H. Kenworthy, *Electrodepostion Surf. Treat.* **2**, 435 (1973–1974).
7. A. M. Shroff and G. Delval, *High Temp. High Pressure* **3**, 695 (1971).

DAYAL T. MESHRI
Advance Research Chemicals Inc.

NICKEL

Nickel Fluoride Tetrahydrate

Nickel fluoride tetrahydrate [*13940-83-5*], $NiF_2 \cdot 4H_2O$, and its anhydrous counterpart, nickel fluoride [*10028-18-9*], NiF_2, are the only known stable binary compounds of nickel and fluorine. The former is a greenish light yellow crystal or powder prepared by the addition of nickel carbonate to 30–50% aqueous HF solution. The nickel fluoride formed first goes into solution and then precipitates out as the tetrahydrate as the concentration of nickel fluoride increases and that of HF decreases. When the addition of nickel is complete, the solution and the precipitates are dried at 75–100°C until all the water is expelled. The tetrahydrate has high solubility in aqueous HF, eg, 13.3 wt % in 30% HF. It is slightly soluble in water and insoluble in alcohol and ether.

Historically, the annual consumption of nickel fluoride was on the order of a few metric tons. Usage is dropping because nickel fluoride is listed in the EPA and TSCA's toxic substance inventory. Nickel fluoride tetrahydrate is packaged in 200–500-lb (90.7–227-kg) drums and the 1993 price was $22/kg. Small quantities for research and pilot-plant work are available from Advance Research Chemicals, Aldrich Chemicals, Johnson/Matthey, Pfaltz and Bauer, PCR, and Strem Chemicals of the United States, Fluorochem of the United Kingdom, and Morita of Japan.

Nickel Fluoride, Anhydrous

Anhydrous nickel fluoride, a light yellow colored powder, is prepared by the action of anhydrous HF on anhydrous $NiCl_2$, or nickel fluoride tetrahydrate at 300°C. It is also prepared by heating a mixture of NH_4HF_2 and $NiF_2 \cdot 4H_2O$. The other methods include the fluorination of metal salts using excess SF_4 (1) or using ClF_3 (2) at elevated temperatures, or the reaction of $NiCO_3$ and anhydrous HF at 250°C (3).

Nickel fluoride is used in marking ink compositions (see INKS), for fluorescent lamps (4) as a catalyst in transhalogenation of fluoroolefins (5), in the manufacture of varistors (6), as a catalyst for hydrofluorination (7), in the synthesis of XeF_6 (8), and in the preparation of high purity elemental fluorine for research (9) and for chemical lasers (qv) (10).

The 1993 price of high purity anhydrous nickel difluoride was $0.55/g in 100- or 250-g quantities. Small quantities are stored and shipped in polyethylene bottles, whereas large amounts are shipped in fiber board drums with polyethylene liners.

All nickel compounds are considered as suspected carcinogens and are listed in the EPA and TSCA's toxic substances inventory. LD_{50} (mice iv) for NiF_2 is 130 mg/kg (11–13).

Physical Properties.   Anhydrous nickel fluoride has a mol wt of 96.71; mp, 1450°C; bp, 1740°C; solubility in water of 4.0 g/100 g; density, g/mL, of 4.72; ΔH_f

of -651.5 kJ/mol (-135.3 kcal/mol); ΔG_f of -604.2 kJ/mol (-144.4 kcal/mol); S of 73.6 J/(mol·K) (17.6 cal/(mol·K)); and C_p, 75.3 J/(mol·K) (18.0 cal/(°C·mol)).

Other Nickel Fluorides. Nickel trifluoride has been observed during the electrolysis of the $NiF_2 \cdot HF$ system as a brownish solid capable of liberating iodine from KI solution and turning into yellow powder (14). It has also been observed during the fluorination of NiF_2 at 200°C. A black substance obtained by the addition of AsF_5 to $K_2NiF_6 \cdot HF$ solution and decomposing to NiF_2 during the purification process is also believed to be impure NiF_3 (15).

Nickel Fluoride Complexes

Nickel tetrafluoroborate [*14708-14-6*], $Ni(BF_4)_2 \cdot xH_2O$, can be prepared by dissolving nickel carbonate in tetrafluoroboric acid [*16872-11-0*], HBF_4. Nickel tetrafluoroborate, commercially available as a hydrated solid, and also as a 50% solution, plays an important role in the electroplating (qv) and electronics industries. Its consumption is several hundred metric tons a year and its 1993 price was $4.25/kg. It is available from Advance Research Chemicals, Aldrich Chemicals, Aesar Chemicals, Johnson/Matthey, Harshaw M & T Chemicals, and from various other sources.

The complex hexafluoronickelates, M_2NiF_6 (M = Na [*21958-95-2*], K [*17218-47-2*], Rb [*17218-48-3*], Cs [*17218-49-4*]) and M_3NiF_6 (M = Na [*22707-99-9*], K [*14881-07-3*], Rb [*72151-96-3*], and Cs [*72138-72-8*]), are prepared by reaction of elemental fluorine, chlorine trifluoride, or xenon difluoride and a mixture of nickel fluoride and alkali metal fluorides or other metal halides (16,17). If the fluorination is carried out using mixed fluorides, a lower temperature can be used, yields are quantitative, and the final products are of high purity. Bis(tetrafluoroammonium) hexafluoronickelate [*63105-40-8*], $(NF_4)_2NiF_6$, prepared from Cs_2NiF_6 and NF_4SbF_6 by a metathesis in anhydrous HF, is also known (18).

These hexafluoronickelates can be used as fluorinating reagents (15), as a source of high purity elemental fluorine (9,10), and as high energy solid propellant oxidizers (see EXPLOSIVES AND PROPELLANTS) (18).

BIBLIOGRAPHY

"Nickel Fluoride" under "Fluorine Compounds, Inorganic," in *ECT* 2nd ed., Vol. 6, p. 710, by F. D. Loomis; "Nickel" under "Fluorine Compounds, Inorganic," in *ECT* 2nd ed., Vol. 9, pp. 628–629, by W. E. White; in *ECT* 3rd ed., Vol. 10, pp. 766–767, by D. T. Meshri, Advance Research Chemical, Inc.

1. USSR Pat. 495,279 (Dec. 15, 1975), A. P. Kostyuk and L. M. Yagupolskii (to Odessa Polytechnic Inst.).
2. Y. I. Nikonorou and co-workers, *Izv. Sib. Otd. Akad. Nauk. SSSR Ser. Khim. Nauk*, (3), 88 (1976).
3. U.S. Pat. 3,836,634 (Sept. 17, 1974), J. Soldick (to FMC Corp.).
4. Jpn. Kokai Tokkyo Koho 61,278,576 (Dec. 9, 1986), M. Nakono (to Hitachi Ltd.).

5. Eur. Pat. 203,807 (Dec. 3, 1986), F. J. Weigert (to E.I. du Pont de Nemours & Co. Inc.).
6. Jpn. Kokai, 75,139,395 (Nov. 7, 1975), M. Matsuura and co-workers (to Matsushita Electrical Industrial Co.).
7. USSR Pat. 466,202 (Apr. 5, 1975), Kh. U. Usmanov and co-workers (to Tashkent State University).
8. B. Zemva and J. Slivnik, *Vestn. Slov. Kem. Drus.* **19**(1–4), 43 (1972).
9. L. B. Asprey, *J. Fluorine Chem.* **7**(1–3), 359 (1976).
10. U.S. Pat. 4,711,680 (Dec. 8, 1987), K. O. Christe (to Rockwell International Corp.).
11. *Metal Toxicity in Mammals*, Vol. 2, Plenum Publishing Corp., New York, 1978, p. 293.
12. *The Merck Index*, 11th ed., Merck & Co., Inc., Rahway, N.J., 1989, p. 1028.
13. N. I. Sax, *Dangerous Properties of Industrial Materials*, 6th ed., Van Nostrand Reinhold Co., New York, 1984, p. 1993.
14. L. Stein and co-workers, *Inorg. Chem.* **8**, 247 (1969).
15. T. L. Court and M. F. A. Dove, *J. Chem. Soc. Chem. Commun.*, 726 (1971).
16. S. V. Zemskov and co-workers, *Izv. Sib. Otd. Akad. Nauk. S.S.S.R. Ser. Khim. Nauk*, (3), 83 (1976).
17. W. Klemm and E. Huss, *Z. Anorg. Chem.* **258**, 221 (1949).
18. U.S. Pat. 4,108,965 (Aug. 22, 1978), K. O. Christe (to Rockwell International Corp.).

DAYAL T. MESHRI
Advance Research Chemicals, Inc.

NITROGEN

Nitrogen has four binary fluorides: nitrogen trifluoride [7783-54-2], NF_3; tetrafluorohydrazine [10036-47-2], N_2F_4; difluorodiazine [10578-16-2], N_2F_2; and fluorine azide [14986-60-8], FN_3. There are numerous other nitrogen fluorine compounds, the most significant of which are the perfluoroammonium salts based on the NF_4^+ cation. Of all the nitrogen fluorine compounds, only NF_3 has been of commercial importance. Nitrogen trifluoride is used as an etchant gas in the electronics industry and as a fluorine source in high power chemical lasers.

Nitrogen Trifluoride

Physical Properties. Nitrogen trifluoride, NF_3, is a colorless gas, liquefying at 101.3 kPa (1 atm) and −129.0°C, and solidifying at −206.8°C (1). High purity NF_3 has little odor, but material contaminated with traces of active fluorides may have a pungent, musty odor. NF_3, a pyramidal molecule with C_{3v} point group symmetry, has a structure similar to ammonia. The N–F bonds are 0.137 nm and the F–N–F bond makes a 102.1° angle. Selected physical properties of NF_3 are given in Table 1. An extensive tabulation and correlation of all the physical properties of NF_3 is available (2). The infrared (3), Raman (4), and ultraviolet spectra (5) of NF_3 have been investigated.

Chemical Properties. NF_3 can be a potent oxidizer, especially at elevated temperature. At temperatures up to ca 200°C, its reactivity is comparable to oxygen. At higher temperatures, the homolysis of the N–F bond into NF_2 and F free radicals becomes significant. The F radical reacts with organic compounds and

Table 1. Physical Properties of Nitrogen Trifluoride

Property	Value	Reference
boiling point, °C	−129.0	1
liquid density at −129°C, g/mL	1.533	6
heat of vaporization, kJ/mol[a]	11.59	1
triple point, °C, 0.263 Pa[b]	−206.8	7
heat of fusion, J/mol[a]	398	1
solid transition point, °C	−216.5	2
heat of transition, kJ/mol[a]	1.513	2
critical temperature, °C	−39.25	6
critical pressure, kPa[b]	4530 (44.7 atm)	6
critical volume, cm^3/mol	123.8	8
heat of formation, kJ/mol[a]	−131.5	9
heat capacity at 25°C, J/(mol·K)[a]	53.39	9
water solubility, 101.3 kPa,[b] 25°C	1.4×10^{-5} mol NF_3/mol H_2O	10
vapor pressure equation, P in kPa[c]	$\log P = 5.90445 - \dfrac{501.913}{T - 15.37}$	6

[a]To convert kJ to kcal, divide by 4.184.
[b]To convert kPa to mm Hg, multiply by 7.5.
[c]For P in mm Hg, $\log P = 6.77966 - \dfrac{501.913}{T - 15.37}$

certain metals, liberating heat and causing further dissociation of the NF_3. At temperatures above 400°C, the reactivity of NF_3 becomes more like that of fluorine. The thermal dissociation of NF_3 has been studied by a number of investigators (11–14) and was found to peak in the temperature range of 800 to 1200°C.

Nitrogen trifluoride acts primarily upon the elements as a fluorinating agent, but is not a very active one at lower temperatures. At elevated temperatures, NF_3 pyrolyzes with many of the elements to produce N_2F_4 and the corresponding fluoride. The element used in this reaction scavenges the fluorine radical, allowing the NF_2 radicals to combine. The pyrolysis of NF_3 over copper turnings produces N_2F_4 in a 62–71% yield at 375°C (15). Pyrolysis over carbon at 400–500°C is more favorable (16,17). This process was the basis for the commercial production of N_2F_4 in the early 1960s for use in rocketry.

Hydrogen and hydrides react with NF_3 with the rapid liberation of large amounts of heat. This is the basis for the use of NF_3 in high energy chemical lasers (qv). The flammability range of NF_3–H_2 mixtures is 9.4–95 mol % NF_3 (17), whereas the flammability range of NF_3–SiH_4 mixtures is even broader at 4.7–99.34 mol % NF_3 (18). Nitrogen trifluoride reacts with organic compounds but generally an elevated temperature is required to initiate the reaction. Under these conditions, the reaction often proceeds explosively and great care must be exercised when exposing NF_3 to organic compounds. Therefore, NF_3 has found little use as a fluorinating agent for organic compounds. The reactions of NF_3 with the elements, and various inorganic and organic substances, are summarized in Reference 2.

Although NF_3 is an amine, it exhibits virtually no basic properties and is not protonated by the HSO_3F–SbF_5–SO_3 superacid medium at 20°C (19). Commercial scrubbing systems for unwanted NF_3 are available (20) and work on the principle of pyrolysis of the NF_3 over reactive substrates at high temperatures.

NF_3 reacts with F_2 and certain Lewis acids under heat or uv light to form the corresponding NF_4^+ salts. For example, when NF_3, SbF_5, and F_2 are mixed in a 1:1:1.5 ratio in a Monel vessel and heated to 200°C for 50 hours, a 41% yield of NF_4SbF_6 [*16871-76-4*] is obtained (21). NF_4^+ salts had been thought to be too unstable to be synthesized (22). Preparation methods have been described which use other sources of energy to initiate the reaction. NF_4AsF_6 [*16871-75-3*] (23) and NF_4BF_4 [*15640-93-4*] (24) were prepared using a low temperature glow discharge. NF_4PF_6 [*58702-88-8*] and NF_4GeF_5 [*58702-86-6*] were prepared by uv photolysis (25). This series of compounds has been further extended by metathesis between NF_4SbF_6 and other salts in HF solution. For example, $(NF_4)_2TiF_6$ [*61128-92-5*] was prepared by the metathesis of Cs_2TiF_6 and NF_4SbF_6 in HF (26).

The salts rapidly hydrolyze to form NF_3 and O_2, and react with glass at temperatures above 85°C to form NF_3 and SiF_4 (21). The NF_4^+ salts are stable in dry atmospheres to 200°C, but rapidly decompose above 300°C to yield NF_3, F_2, and the corresponding Lewis acid. Therefore, these salts are solid sources of NF_3 and F_2, free of atmospheric contaminants and HF.

Manufacture and Economics.   Nitrogen trifluoride can be formed from a wide variety of chemical reactions. Only two processes have been technically and economically feasible for large-scale production: the electrolysis of molten ammonium acid fluoride; and the direct fluorination of the ammonia in the presence of molten ammonium fluoride. In the electrolytic process, NF_3 is produced at the anode and H_2 is produced at the cathode. In a divided cell of 4 kA having nickel anodes, extensive dilution of the gas streams with N_2 was used to prevent explosive reactions between NF_3 and H_2 (17).

In the direct process, NF_3 is produced by the reaction of NH_3 and F_2 in the presence of molten ammonium acid fluoride (27). The process uses a specially designed reactor (28). Because H_2 is not generated in this process, the hazards associated with the reactions between NF_3 and H_2 are eliminated.

As a result of the development of electronic applications for NF_3, higher purities of NF_3 have been required, and considerable work has been done to improve the existing manufacturing and purification processes (29). N_2F_2 is removed by pyrolysis over heated metal (30) or metal fluoride (31). This purification step is carried out at temperatures between 200–300°C which is below the temperature at which NF_3 is converted to N_2F_4. Moisture, N_2O, and CO_2 are removed by adsorption on zeolites (29,32). The removal of CF_4 from NF_3, a particularly difficult separation owing to the similar physical and chemical properties of these two compounds, has been described (33,34).

Production of NF_3 is less than 100 t/yr in the United States. Air Products and Chemicals, Inc. is the only commercial producer in the United States. The 1992 price ranged from $400–$800/kg depending on the grade.

Specifications and Analysis.   Nitrogen trifluoride is shipped as a high pressure gas at 10 MPa (1450 psig) and is available in tube trailers and cylinders. Table 2 shows NF_3 specifications for a commercial grade typically used in chemical laser applications, and two higher grades of NF_3 used by the electronics industry. Analysis of NF_3 for impurities can be performed on the gas chromatograph (35).

Table 2. Specifications of NF_3^a

Impurity	Commercial, ppmv	VLSI, ppmv	Megaclass, ppmv
total fluorides as HF	3,900	1	1
CO_2	130	16	4
CO	330	25	0.5
CF_4	1,200	560	25
N_2	19,000	130	10
O_2 + Ar	22,000	100	6
SF_6	50	25	1
N_2O	500	16	2
H_2O	1	1	1

[a]Commercial grades offered by Air Products and Chemicals, Inc.

Active fluorides are determined by scrubbing with a basic solution and wet-chemical analysis for fluoride content (35).

Handling and Toxicity. Nitrogen trifluoride gas is noncorrosive to the common metals at temperatures below 70°C and can be used with steel, stainless steel, and nickel. The corrosion rate of NF_3 on these materials significantly increases with moisture or HF. Nitrogen trifluoride is compatible with the fluorinated polymers such as Teflon, Kel-F, and Viton at ambient conditions. Extensive data on the corrosion rates of gaseous and liquid NF_3 to a variety of metals and nonmetals are compiled in Reference 2. In systems handling high pressure NF_3, precautions should be taken to avoid any sudden heating of the NF_3 which can occur during adiabatic compression of the gas during the introduction of NF_3 rapidly from a high pressure point to a low pressure dead-end space (20).

Nitrogen trifluoride is a toxic substance and is most hazardous by inhalation. NF_3 induces the production of methemoglobin which reduces the level of oxygen transfer to the body tissues. At the cessation of NF_3 exposure methemoglobin reverts back to hemoglobin. The OSHA permissible exposure limits is set as a TLV–TWA of 29 mg/kg or 10 ppm (36). Because NF_3 has very little odor, it cannot be detected by its odor at concentrations within the TLV. Therefore, adequate personnel protection or monitoring must be provided when handling NF_3. Commercially available monitors detect NF_3 by either pyrolysis to HF or NO_2 followed by electrochemical quantification, or by infrared absorption (20). All of the monitors can detect at least 1 ppm or 10 times below the TLV. The pyrolysis–electrolytic monitors are generally less expensive, but other halogen-containing compounds can interfere with the NF_3 detection.

The inhalation toxicity of NF_3 on animals has been studied extensively (37–40). These studies provide the basis of emergency exposure limits (EEL) that have been proposed for NF_3. The NAS–NRC Committee on Toxicology recommends that the EEL for NF_3 be 10 min at 2250 ppm, 30 min at 750 ppm, and 60 min at 375 ppm. Gaseous NF_3 is considered to be innocuous to the skin and a minor irritant to the eyes and mucous membranes. NF_3 does give a weakly positive metabolically activated Ames test but only at concentrations greater than 2% or 10 times the 10 minute EEL.

Environmental impact studies on NF_3 have been performed. Although undiluted NF_3 inhibits seed growth, no effect on plant growth was observed when exposed to 6,000 ppm·min of NF_3 and only minor effects were observed at the 60,000 ppm·min exposure level (41). Exposure of microbial populations to 25% NF_3 in air for seven hours showed normal growth. NF_3 is not an ozone-depleting gas (20).

Uses.   The principal use of NF_3 is as a fluorine source in the electronics industry. The use of NF_3 as a dry chemical etchant has been reviewed (20,42–44). The advantages of using NF_3 as an etchant over traditional carbon-based etchants include high etch rates, high selectivities for nitride-over-oxide etching and single-crystal silicon over thermally grown oxide, and the production of only volatile reaction products resulting in an etch with no polymer or fluoride residues. *In situ* plasma or thermal cleaning of chemical vapor deposition (CVD) reactors is also a use of NF_3. Residual coatings are deposited on the internal surfaces of CVD reactors during deposition processes. A plasma of NF_3 can remove these deposits as volatile fluorides in minutes at the process temperature eliminating the need to remove the internal CVD reactor components to be cleaned by acid tank immersion.

Another use of NF_3 is as a fluorine source for the hydrogen and deuterium fluoride (HF/DF) high energy chemical lasers (qv). The HF/DF lasers are the most promising of the chemical lasers under development because a substantial fraction (ca 25%) of the energy of the reaction between H_2/D_2 and F_2 can be released as laser radiation (45,46). The use of NF_3 is preferred to F_2 because of its comparative ease of handling at ambient temperatures. Storage and handling of NF_3 do not require the precautions necessary for the large-scale use of fluorine.

BIBLIOGRAPHY

"Nitrogen Trifluoride" under "Fluorine Compounds, Inorganic" in *ECT* 1st ed., Vol. 6, p. 710; "Nitrogen" under "Fluorine Compounds, Inorganic" in *ECT* 2nd ed., Vol. 9, pp. 629–630 by W. E. White, Ozark-Mahoning Co.; in *ECT* 3rd ed., Vol. 10, pp. 768–772, by A. J. Woytek, Air Products and Chemicals, Inc.

1. L. Pierce and E. L. Pace, *J. Chem. Phys.* **23**, 551 (1955).
2. R. E. Anderson, E. M. Vander Wall, and R. K. Schaplowsky, "Nitrogen Trifluoride," *USAF Propellant Handbook*, AFRPL-TR-77-71, Contract F04611-76-C-0058, Aerojet Liquid Rocket Co., Sacramento, Calif., 1977.
3. P. N. Schatz and I. W. Levin, *J. Chem. Phys.* **29**, 475 (1958).
4. M. Gilbert, P. Nectoux, and M. Drifford, *J. Chem. Phys.* **68**, 679 (1978).
5. V. Legasov and co-workers, *J. Fluorine Chem.* **11**(2), 109 (1978).
6. R. L. Jarry and H. C. Miller, *J. Phys. Chem.* **60**, 1412 (1956).
7. *DIPPR on-line database on STN*, American Institute of Chemical Engineers, New York, 1992.
8. D. N. Seshadri, D. S. Viswanath, and N. R. Kuloor, *Indian J. Technol.* **8**(5), 153 (1970).
9. D. Wagman and co-workers, *J. Phys. Chem. Ref. Data*, Suppl. No. 2 (1982).
10. C. R. S. Dean, A. Finch, and P. J. Gardner, *J. Chem. Soc., Dalton Trans.* **23**, 2722 (1973).
11. K. O. McFadden and E. Tschuikow, *J. Phys. Chem.* **77**, 1475 (1973).
12. G. L. Schott, L. S. Blair, and J. D. Morgan, Jr., *J. Phys. Chem.* **77**, 2823 (1973).

13. E. A. Karko and co-workers, *J. Chem. Phys.* **63**, 3596 (1975).

14. P. J. Evans and E. Tschuikow-Roux, *J. Chem. Phys.* **65**, 4202 (1976).

15. C. B. Colburn and A. Kennedy, *J. Am. Chem. Soc.* **80**, 5004 (1958).

16. *Chem. Eng. News* **38**, 85 (Sept. 19, 1960).

17. U.S. Pat. 3,235,474 (Feb. 15, 1966), J. F. Tompkins (to Air Products and Chemicals, Inc.).

18. Y. Urano, K. Tokuhashi, S. Kondo, S. Horiguchi, and M. Iwsaka, *20th Safety Engineering Symposium*, Tokyo, Japan, June 1990.

19. R. J. Gillespie and G. P. Pez, *Inorg. Chem.* **8**, 1233 (1969).

20. *Nitrogen Trifluoride: Safety, Applications, and Technical Data Manual*, Air Products and Chemicals, Inc., Allentown, Pa., 1992.

21. W. E. Tolberg and co-workers, *Inorg. Chem.* **6**, 1156 (1967).

22. W. C. Price, T. R. Passmore, and D. M. Roessler, *Discussions Faraday Soc.* **35**, 201 (1963).

23. J. P. Guertin, K. O. Christie, and A. E. Pavlath, *Inorg. Chem.* **5**, 1921 (1966).

24. S. M. Sinel'nikov and V. Ya. Rosolovskii, *Dokl. Akad. Nauk SSSR* **194**, 1341 (1970).

25. K. O. Christie, C. J. Schack, and R. D. Wilson, *Inorg. Chem.* **15**, 1275 (1976).

26. K. O. Christie and C. J. Schack, *Inorg. Chem.* **16**, 353 (1977).

27. U.S. Pat. 4,091,081 (May 23, 1978), A. J. Woytek and J. T. Lileck (to Air Products and Chemicals, Inc.).

28. *Chem. Eng.* **84**(26), 116 (1977).

29. A. J. Woytek and P. B. Henderson, in *Proceedings of the Institute of Environmental Sciences 37th Annual Meeting*, IES, Mount Prospect, Ill., 1990, p. 570.

30. U.S. Pat. 4,193,976 (Mar. 18, 1980), J. T. Lileck, J. Papinsick, and E. J. Steigerwalt (to Air Products and Chemicals, Inc.).

31. U.S. Pat. 4,948,571 (Dec. 3, 1988), I. Harada, H. Hokonohara, and T. Yamaguchi (to Mitsui Toatsu Chemicals, Inc.).

32. Eur. Pat. Appl. 366 078 A2, (May 2, 1990), M. Aritsuka and N. Iwanaga (to Mitsui Toatsu Chemicals, Inc.).

33. U.S. Pat. 5,069,690 (Dec. 3, 1991), P. B. Henderson, C. G. Coe, D. E. Fowler, and M. S. Benson (to Air Products and Chemicals, Inc.).

34. U.S. Pat. 5,069,887 (Dec. 3, 1991), T. Suenaga, T. Fujii, and Y. Kobayashi (to Central Glass).

35. L. A. Dee, *Analysis of Nitrogen Trifluoride*, AFRPL-TR-76-20 (AD-A022887), Air Force Rocket Propulsion Laboratory, Edward Air Force Base, Calif., Apr. 1976.

36. J. O. Accrocco and M. Cinquanti, eds., *Right-To-Know Pocket Guide for Laboratory Employees*, Genium, Schenectady, N.Y., 1990.

37. E. H. Vernot and C. C. Haun, *Acute Toxicology and Proposed Emergency Exposure Limits of Nitrogen Trifluoride*, AMRL-TR-69-130-Paper No. 13 (AD 710062), Contract F33615-70-V-1046, Syste-Med Corp., Dayton, Ohio, Dec. 1969. Reprinted from the *Proceedings of the 5th Annual Conference on Atmospheric Contamination in Conhned Spaces, Sept. 16–18, 1969*, AMRL-TR-69-130 (AD 709994), Dayton, Ohio, Dec. 1969, pp. 165–171.

38. F. N. Dost and co-workers, "Metabolism and Pharmacology of Inorganic and Fluorine Containing Compounds," *Final Report (July 1, 1964–June 30, 1967)*, AMRL-TR-67-224 (AD 681-161), Contract AF 33(615)-1799, Oregon State University, Corvallis, Oreg., Aug. 1968.

39. F. N. Dost, D. J. Reed, and C. H. Wang, *Toxicol. Appl. Pharmacol.* **17**, 585 (1970).

40. G. L. Coppoc and S. J. Leger, "Effect of Nitrogen Trifluoride on Plasma Concentrations of Lactate, Methemoglobin, and Selected Enzymes," *ApriWune 1968*, SAM-TR-70-42 (AD 711044), School of Aerospace Medicine, Brooks Air Force Base, Texas, July 1970.

41. D. J. Reed, F. N. Dost, and C. H. Want, "Inorganic Fluoride Propellant Oxidizers, Vol. 1: Their Effects Upon Seed Germination and Plant Growth," *Interim Report (May 15, 1964–May 15, 1966)*, AMRLTR-66-187, Vol. I (AD 667-556), Contract AF 33(615)-1767, Oregon State University, Corvallis, Oreg., Nov. 1967.

42. A. J. Woytek, J. T. Lileck, and J. A. Barkanic, *Solid State Technol.* **27**, 109 (1979).

43. J. A. Barkanic and co-workers, *Solid State Technol.* **32**, 172 (1984).

44. J. M. Parks, R. J. Jaccodine, J. G. Langan, and M. A. George, in *Proceedings of the 8th Symposium on Plasma Processing*, Electrochemical Society, Montreal, Canada, May 6, 1990, p. 701.

45. E. R. Schulman, W. G. Burwell, and R. A. Meinzer, "Design and Operation of Medium Power CW HF/DF Chemical Lasers," *AIAA 7th Fluid and Plasma Dynamics Conference*, Palo Alto, Calif., AIAA Paper No. 74-546, June 1974.

46. P. J. Klass, *Aviat. Week Space Technol.* **107**, 34 (Aug. 1976).

General Reference

C. J. Hoffman and R. G. Neville, *Chem. Reo.* **62**, 1 (1962); A. W. Jache and W. E. Shite, in C. A. Hunpel, ed., *Encyclopedia of Electrochemistry*, Reinhold Publishing Corp., New York, 1964, pp. 856–859; R. E. Anderson, E. M. Vander Wall, and R. K. Schaplowsky, "Nitrogen Trifluoride," *USAF Propellant Handbook*, AFRPL-TR-77-71, Contract F0461 1-76-C-0058, Aerojet Liquid Rocket Co., Sacramento, Calif., 1977.

PHILIP B. HENDERSON
ANDREW J. WOYTEK
Air Products and Chemicals, Inc.

OXYGEN

Oxygen Difluoride

Oxygen difluoride [7783-41-7], OF_2, is the most stable binary compound of oxygen and fluorine. Under ambient conditions, it is a colorless gas that condenses to a pale yellow liquid at $-145°C$ (1) and freezes at $-224°C$ (2). Oxygen difluoride is a powerful oxidizer that has attracted considerable attention as an ingredient of high energy rocket propellant systems (see EXPLOSIVES AND PROPELLANTS). Several comprehensive reviews of the physical and chemical properties of OF_2 (3–5) and its handling (6) are available.

Physical Properties. An extensive tabulation of the physical properties of OF_2 is available (4). Selected data are mp $-224°C$ (2); bp, $-145°C$ (1); critical temperature $-58°C$ (7); density of liquid, in g/mL from -145 to $-153°C$, t in K, $d = 2.190-0.00523\,t$ (8); heat of formation 31.8 kJ/mol (7.6 kcal/mol) (9); and heat of vaporization 11.1 kJ/mol (2.65 kcal/mol) (10).

Spectroscopic investigations have shown that OF_2 is bent and has equivalent O—F bonds. The O—F distance is 0.139–0.141 nm and the FOF angle is 103–104° (11–13). Measurements of the dipole moment have yielded values of $0.6-1.3 \times 10^{-30}$ C·m (0.18–0.40 D) (12,14,15). The ir (16–18), uv (19), mass (20), and nmr (21) spectra of OF_2 have been reported.

Chemical Properties. The kinetics of decomposition of OF_2 by pyrolysis in a shock tube are different, as a result of surface effects, from those obtained by conventional decomposition studies. Dry OF_2 is stable up to 250°C (22).

Reactions with Metals. Many common metals react with OF_2, but the reaction stops after a passive metal fluoride coating is formed (3,4).

Reactions with Nonmetallic Elements and Inorganic Compounds. Mixtures of OF_2 with carbon, CO, CH_4, H_2, or H_2O vapor explode when ignited with an electrical shock. Elemental B, Si, P, As, Sb, S, Se, and Te react vigorously on slight warming to produce fluorides and oxyfluorides. Oxides such as CrO_3, WO_3, As_2O_3, and CaO react with OF_2 to form fluorides. The corresponding chlorides react with OF_2 to form the respective fluorides and liberate free chlorine in the process (3,4).

In aqueous solution, OF_2 oxidizes HCl, HBr, and HI (and their salts), liberating the free halogens. Oxygen difluoride reacts slowly with water and a dilute aqueous base to form oxygen and fluorine. The rate of this hydrolysis reaction has been determined (23).

Nitric oxide and OF_2 inflame on contact; emission and absorption spectra of the flame have been studied (24). Oxygen difluoride oxidizes SO_2 to SO_3, but under the influence of uv irradiation it forms sulfuryl fluoride [2699-79-8], SO_2F_2, and pyrosulfuryl fluoride [37240-33-8], $S_2O_5F_2$ (25). Photolysis of SO_3–OF_2 mixtures yields the peroxy compound FSO_2OOF [13997-94-9] (25,26).

Oxygen Difluoride as a Source of the OF Radical. The existence of the ·OF radical [12061-70-0] was first reported in 1934 (27). This work was later refuted (28). The ·OF radical was produced by photolysis of OF_2 in a nitrogen or argon matrix at 4 K. The existence of the ·OF species was deduced from a study of the kinetics of decomposition of OF_2 and the kinetics of the photochemical reaction (25,26):

$$OF_2 + SO_3 \xrightarrow[\text{350 nm}]{h\nu} FSO_2OOF$$

The existence of the ·OF radical was further established by use of ^{17}O-labeled compounds and ^{17}O nmr studies to verify the mechanism (29):

$$OF_2 + h\nu \rightarrow F\cdot + \cdot OF$$
$$F\cdot + SO_3 \rightarrow FSO_3\cdot$$
$$FSO_3\cdot + \cdot OF \rightarrow FSO_2OOF$$

The ·OF radical has also been detected by CO_2 laser magnetic resonance (30). The O—F bond length is 0.135789 nm.

Carbonyl fluoride, COF_2, and oxygen difluoride react in the presence of cesium fluoride catalyst to give bis(trifluorylmethyl)trioxide [1718-18-9], CF_3OOOCF_3 (31). CF_3OOF has been isolated from the reaction in the presence of excess OF_2 (32).

Reactions with Organic Compounds. Tetrafluoroethylene and OF_2 react spontaneously to form C_2F_6 and COF_2. Ethylene and OF_2 may react explosively, but under controlled conditions monofluoroethane and 1,2-difluoroethane can be recovered (33). Benzene is oxidized to quinone and hydroquinone by OF_2. Methanol and ethanol are oxidized at room temperature (4). Organic amines are extensively degraded by OF_2 at room temperature, but primary aliphatic amines in a fluorocarbon solvent at $-42°C$ are smoothly oxidized to the corresponding nitroso compounds (34).

The reaction of OF_2 and various unsaturated fluorocarbons has been examined (35,36) and it is claimed that OF_2 can be used to chain-extend fluoropolyenes, convert functional perfluorovinyl groups to acyl fluorides and/or epoxide groups, and act as a monomer for an addition-type copolymerization with diolefins.

Preparation. The synthesis of OF_2 was first achieved by the electrolysis of molten KHF_2 in the presence of water (37). The electrolysis of aqueous HF in the presence of O_2 and O_3 was also found to produce OF_2 (38–40).

The most satisfactory method of OF_2 generation is probably the fluorination of aqueous NaOH (3,22,41–45):

$$2 F_2 + 2 NaOH \rightarrow OF_2 + 2 NaF + H_2O$$

Yields of greater than 60% are obtained (46). This method has been used for the commercial production of OF_2 (8). The NaOH concentration, however, must be kept low to avoid the loss of product by a secondary reaction:

$$OF_2 + 2 OH^- \rightarrow O_2 + F^- + H_2O$$

An economic study of the preparation of OF_2 is available (47).

Analytical Procedures. Oxygen difluoride may be determined conveniently by quantitative application of ir, nmr, and mass spectroscopy. Purity may also be assessed by vapor pressure measurements. Wet-chemical analyses can be conducted either by digestion with excess NaOH, followed by measurement of the excess base (2) and the fluoride ion (48,49), or by reaction with acidified KI solution, followed by measurement of the liberated I_2 (4).

Handling and Safety Factors. Oxygen difluoride can be handled easily and safely in glass and in common metals such as stainless steel, copper, aluminum, Monel, and nickel, from cryogenic temperatures to 200°C (4). At higher temperatures only nickel and Monel are recommended. The compatibility of OF_2 with process equipment depends largely on the cleanliness of the equipment; contaminants such as dirt, moisture, oil, grease, scale slag, and pipe dope must be avoided. Equipment should be passivated with elemental fluorine before contact with OF_2.

Oxygen difluoride must be regarded as a highly poisonous gas, somewhat more toxic than fluorine. It has a foul odor with a limit of detectability of 0.1–0.5 ppm. Repeated exposure of rats to 0.5 ppm OF_2 produced death; repeated exposure to 0.1 ppm, however, caused no discernible effects.

Dioxygen Difluoride

Dioxygen difluoride [7783-44-0], O_2F_2, prepared by passing a 1:1 mixture of O_2 and F_2 through a high voltage electric discharge tube cooled by liquid nitrogen, has also been prepared by uv irradiation of O_2 and F_2 (50,51) and by radiolysis of liquid mixtures of O_2 and F_2 at 77 K using 3 MeV bremsstrahlung (52). Heating an O_2/F_2 mixture to 700°C in stainless steel tubes followed by rapid cooling produces O_2F_2 (53). This compound is also obtained in high yield by subjecting a flowing gas mixture of F_2 to microwave, then downstream and outside of the region of discharge, introducing molecular oxygen (54).

Physical Properties. Because O_2F_2 is unstable, it is difficult to purify. Consequently, some of the reported physical properties are open to question. Selected data are density, in g/mL, from -87 to -156°C, $d = 2.074-0.00291\,t$ (50); heat of formation 19.8 kJ/mol (4.73 kcal/mol) (55); and heat of vaporization 19.2 kJ/mol (4.58 kcal/mol) at -57°C (55).

The structure of O_2F_2 is that of a nonlinear FOOF chain, having the following molecular constants (56,57): O—O distance, 0.122 nm; OOF angle, 109°30′; dihedral angle, 87°30′; dipole moment, 4.8×10^{-30} C·m (1.44 D). Additional physical and spectral data are summarized in References 4 and 58.

Chemical Properties. The bond distance of O—O is relatively short (121.7 $\pm$ 0.3 pm) and that of O—F is relatively long (157.5 $\pm$ 0.3 pm) (56). The weakest bond in O_2F_2 is thus the O—F bond and the mechanisms of reaction of O_2F_2 can probably be explained by the formation of F· and ·OOF and not two ·OF radicals. The ·OOF radical [15499-23-7] is a feasible intermediate as it has been shown to exist at low temperatures (56,59–61). If O_2F_2 is allowed to react quickly with other compounds, simple fluorination usually results. The controlled reactions of O_2F_2, however, yield products that appear to be formed via an ·OOF intermediate.

Simple Fluorination Reactions. Some examples (62) of O_2F_2 acting mainly as a fluorinating agent are

$$Xe \xrightarrow{\;O_2F_2\;} XeF_4$$

$$ClF_3 \xrightarrow{\;O_2F_2\;} ClF_5$$

$$Ag + ClF_5 \xrightarrow{\;O_2F_2\;} AgF_3$$

$$PuF_4 \xrightarrow{\;O_2F_2\;} PuF_6$$

Reactions Involving an ·OOF Intermediate. In controlled reactions of O_2F_2 and various compounds, ^{17}O tracer studies and other techniques have shown that the first step in the reaction appears to be

$$FOOF \rightarrow \cdot OOF + F\cdot$$

For example:

$$SO_2 + O_2F_2 \rightarrow FSO_2OOF$$

where the proposed mechanism (63) is

$$SO_2 + F\cdot \rightarrow FSO_2\cdot$$
$$FSO_2\cdot + \cdot OOF \rightarrow FSO_2OOF$$

Also:

$$2\ CF_3CF{=}CF_2 + 2\ O_2F_2 \rightarrow CF_3CF(OOF)CF_3 + CF_3CF_2CF_2OOF$$

in which the proposed mechanism (64) involves the transfer of an OOF group.

The formation of a new class of compounds, dioxygenyls, containing O_2^+, is also thought to take place via an ·OOF intermediate (65).

$$O_2F_2 \rightarrow \cdot OOF + F\cdot$$
$$\cdot O_2F + BF_3 \rightarrow O_2^+BF_4^-$$

A number of fluorides have been shown to form O_2^+ compounds upon reaction with O_2F_2.

Uses

Oxygen difluoride is mainly a laboratory chemical. It has been suggested as an oxidizer for rocket applications and has been used for small tests in this area.

Dioxygen difluoride has found some application in the conversion of uranium oxides to UF_6 (66), in fluorination of actinide fluorides and oxyfluorides to AcF_6 (67), and in the recovery of actinides from nuclear wastes (68) (see ACTINIDES AND TRANSACTINIDES; NUCLEAR REACTION, WASTE MANAGEMENT).

Higher Oxygen Fluorides

Several higher oxygen fluorides, O_3F_2 [16829-28-0] (50,69), O_4F_2 [12020-93-8] (70), O_5F_2 [12191-79-6] (71), and O_6F_2 [12191-80-9] (71), and radicals such as ·O_3F (72,73) have been reported. Only ·OF, OF_2, O_2F_2, ·OOF, and O_4F_2, however, have been satisfactorily characterized. From cryogenic mass spectroscopy, it appears that O_3F_2 consists of loosely bonded ·O_2F and ·OF radicals (74). The ^{19}F nmr spectrum of O_3F_2 suggests an O_3F_2 model consisting of O_2F_2 and interstitial oxygen (75). However, ^{19}F and ^{17}O nmr (7,76), and other studies have shown that O_3F_2, as reported in the literature, is actually a mixture of O_4F_2 and O_2F_2.

Little is known about O_4F_2. It has been reported to behave similarly to O_2F_2 in that it can act as a fluorinating agent or a source of the ·OOF radical. In fact, it appears to be a better source of the ·OOF radical than O_2F_2 in its reactions with SO_2 and BF_3.

BIBLIOGRAPHY

"Oxygen Compounds" under "Fluorine Compounds, Inorganic," in *ECT* 1st ed., Vol. 6, pp. 710–711; "Oxygen" under "Fluorine Compounds, Inorganic," in *ECT* 2nd ed., Vol. 9, pp. 631–635, W. B. Fox and R. B. Jackson, Allied Chemical Corp.; in *ECT* 3rd ed., Vol. 10, pp. 773–778, by I. J. Solomon, IIT Research Institute.

1. J. Schnitzlstein and co-workers, *J. Phys. Chem.* **56**, 233 (1952).
2. O. Ruff and K. Clusius, *Z. Anorg. Allgem. Chem.* **190**, 267 (1930).
3. H. R. Leech, in Mellor, ed., *Comprehensive Treatise on Inorganic and Theoretical Chemistry*, Suppl. II, Part I, Longmans, Green & Co., Inc., New York, 1956, pp. 186–193.
4. A. G. Streng, *Chem. Rev.* **63**, 607 (1963).
5. R. B. Jackson, *Oxygen Difluoride Handling Manual*, Report No. NASA-CR-72401, Allied Chemical Corp., Morristown, N.J., Dec. 1970.
6. R. F. Muraca, J. Neff, and J. S. Whittick, *Physical Properties of Liquid Oxygen Difluoride and Liquid Diborane—A Critical Review*, Report No. NASA-CR-88519, SRI-951581-4, Jet Propulsion Lab., Calif. Inst. of Tech., Pasadena, Stanford Research Inst., Menlo Park, Calif., July 1967.
7. R. Anderson and co-workers, *J. Phys. Chem.* **56**, 473 (1952).
8. *Oxygen Difluoride*, Product Data Sheet, General Chemical Division, Allied Chemical Corp., Morristown, N.J.
9. W. Evans, T. Munson, and D. Wagman, *J. Res. Natl. Bur. Std.* **55**, 147 (1955).
10. O. Ruff and W. Menzel, *Z. Anorg. Chem.* **198**, 39 (1931).
11. A. Hilton and co-workers, *J. Chem. Phys.* **56**, 473 (1952).
12. L. Pierce, R. Jackson, and N. Dicianni, *J. Chem. Phys.* **35**, 2240 (1961).
13. J. Ibers and V. Schomaker, *J. Phys. Chem.* **57**, 699 (1953).
14. J. Bransford, A. Kunkel, and A. Jache, *J. Inorg. Nucl. Chem.* **14**, 159 (1960).
15. R. Dodd and R. Little, *Nature* **188**, 737 (1960).
16. H. Bernstein and J. Powling, *J. Chem. Phys.* **18**, 685 (1960).
17. E. Jones and co-workers, *J. Chem. Phys.* **19**, 337 (1951).
18. A. Nielsen, *J. Chem. Phys.* **19**, 379 (1951).
19. A. Glissman and H. Schumacher, *Z. Physik. Chem.* **324**, 328 (1934).
20. V. Dibeler, R. Reese, and J. Franklin, *J. Chem. Phys.* **27**, 1296 (1957).
21. H. Agahigian, A. Gray, and G. Vickers, *Can. J. Chem.* **40**, 157 (1962).
22. G. Brauer, *Handbuch der Preparativen Anorganischen Chemie*, Ferdinand Enke, Stuttgart, 1954.
23. S. N. Misra and G. H. Cady, *Kinetics of Hydrolysis of Oxygen Difluoride*, Report No. TR-70, University of Washington Department of Chemistry, Seattle, Jan. 1972.
24. P. Goodfriend and H. Woods, *J. Chem. Phys.* **39**, 2379 (1963).
25. G. Franz and F. Neumayr, *Inorg. Chem.* **3**, 921 (1964).
26. R. Gath and co-workers, *Angew. Chem.* **75**, 137 (1963).
27. O. Ruff and W. Z. Menzel, *Z. Anorg. Allg. Chem.* **217**, 85 (1934).
28. P. Frisch and H. J. Schumacher, *Z. Anorg. Allg. Chem.* **229**, 423 (1936); (Leipzig) **B34**, 322 (1936); **B37**, 18 (1937).
29. I. J. Solomon, A. J. Kacmarek, and J. Raney, *J. Phys. Chem.* **72**, 2262 (1968).
30. A. R. W. McKellar, *Can. J. Phys.* **57**, 2106 (1979).
31. L. R. Anderson and W. B. Fox, *J. Am. Chem. Soc.* **89**, 431B (1967).
32. I. J. Solomon and co-workers, *Inorg. Chem.* **11**, 195 (1972).
33. R. Rhein and G. Cady, *Inorg. Chem.* **3**, 1644 (1964).
34. R. Merritt and J. Ruff, *J. Am. Chem. Soc.* **86**, 1342 (1964).
35. M. S. Toy, *Utilization of Oxygen Difluoride for Syntheses of Fluoropolymers*, Report No. Patent-3,931,132, Pat. Appl.-45,549, NASA, Pasadena Office, Calif., Jan. 1976.
36. M. Dos Santos Afonso, E. Castellano, and H. J. Schumacher, *An. Asoc. Quim. Argent.* **74**, 465 (1986).
37. P. Lebeau and A. Damiens, *Compt. Rend.* **185**, 652 (1927).
38. A. Englebrecht and E. Nachbaur, *Monatsh. Chem.* **90**, 367 (1959).
39. J. A. Donohue, T. D. Nevitt, and A. Zletz, *Adv. Chem. Ser.* **54**, 192 (1966).

40. D. Hass and P. Wolter, *Z. Anorg. Allg. Chem.* **463**, 91 (1980).
41. G. Rohrbach and G. H. Cady, *J. Am. Chem. Soc.* **69**, 677 (1947).
42. D. Yost, *Inorg. Synth.* **1**, 109 (1939).
43. W. Koblitz and H. Schumacher, *Z. Physik. Chem.* **B25**, 283 (1934).
44. P. Lebeau and A. Damiens, *Compt. Rend.* **188**, 1253 (1929).
45. A. Borning and K. E. Pullen, *Inorg. Chem.* **8**, 1791 (1969).
46. G. H. Cady, *J. Am. Chem. Soc.* **57**, 246 (1935).
47. F. L. Hyman and J. F. Tompkins, *An Economic Study of Oxygen Difluoride*, Final Report No. NASA-CR-117317, Air Products and Chemicals, Inc., Allentown, Pa., June 1970.
48. H. Willard and C. Horton, *Anal. Chem.* **22**, 1190 (1950).
49. H. Willard and C. Horton, *Anal. Chem.* **24**, 862 (1952).
50. S. Aoyama and S. Sakuraba, *J. Chem. Soc. Japan* **59**, 1321 (1938).
51. A. Kirshenbaum, A. Grosse, and J. Astor, *J. Am. Chem. Soc.* **81**, 6398 (1959).
52. C. D. Wagner and co-workers, *J. Am. Chem. Soc.* **91**, 4702 (1969).
53. T. R. Mills, *J. Fluorine Chem.* **52**, 267 (1991).
54. U.S. Pat. Appl. 6,696,548 (Jan. 1986), W. H. Beattie (to U.S. Dept. of Energy).
55. A. Streng, *J. Am. Chem. Soc.* **85**, 1380 (1963).
56. R. Jackson, *J. Chem. Soc.*, 4585 (1962).
57. L. Hedberg and co-workers, *Inorg. Chem.* **27**, 232 (1988).
58. K. C. Kim and G. M. Campbell, *J. Mol. Struct.* **129**, 263 (1985).
59. R. W. Fessenden and R. H. Schuler, *J. Chem. Phys.* **44**, 434 (1966).
60. A. Arkell, *J. Am. Chem. Soc.* **87**, 4057 (1965).
61. R. D. Sprately, J. J. Turner, and G. C. Pimentel, *J. Chem. Phys.* **44**, 2063 (1966).
62. J. B. Nielsen and co-workers, *Inorg. Chem.* **29**, 1779 (1990); S. A. Kinkead, L. B. Asprey, and P. G. Eller, *J. Fluorine Chem.* **29**, 459 (1985); Yu. M. Kiselev and co-workers, *Zh. Neorg. Khim.* **33**, 1252 (1988); J. G. Malm, P. G. Eller, and L. B. Asprey, *J. Am. Chem. Soc.* **106**, 2726 (1984).
63. I. J. Solomon, A. J. Kacmarek, and J. M. McDonough, *Chem. Eng. Data* **13**, 529 (1968).
64. I. J. Solomon, A. J. Kacmarek, and J. Raney, *Inorg. Chem.* **7**, 1221 (1968).
65. I. J. Solomon and co-workers, *J. Am. Chem. Soc.* **90**, 6557 (1968).
66. L. B. Asprey, S. A. Kinkead, and P. G. Eller, *Nucl. Technol.* **73**, 69 (1986).
67. U.S. Pat. Appl. 6,636,656 (Oct. 1985), P. G. Eller, J. G. Malm, and R. A. Penneman (to U.S. Dept. of Energy).
68. U.S. Pat. Appl. 6,649,626 (Oct. 1985), L. B. Asprey and P. G. Eller (to U.S. Dept. of Energy).
69. J. N. Keith and co-workers, *Inorg. Chem.* **7**, 320 (1968).
70. A. D. Kirshenbaum and A. V. Grosse, *J. Am. Chem. Soc.* **81**, 1277 (1959).
71. A. V. Grosse, A. G. Streng, and A. D. Kirshenbaum, *J. Am. Chem. Soc.* **83**, 1004 (1961).
72. A. G. Streng and A. V. Grosse, *J. Am. Chem. Soc.* **88**, 169 (1966).
73. A. D. Kirshenbaum and A. V. Grosse, *Production, Isolation, and Identification of the ·OF, ·O₂F, and ·O₃F Radicals*, Research Institute, Temple University, Philadelphia, Pa., June 1964.
74. T. J. Malone and H. A. McGee, *J. Phys. Chem.* **71**, 3060 (1967).
75. J. W. Nebgen, F. I. Metz, and W. B. Rose, *J. Am. Chem. Soc.* **89**, 3118 (1967).
76. I. J. Solomon and co-workers, *J. Am. Chem. Soc.* **89**, 2015 (1967).

I. J. SOLOMON
IIT Research Institute

JEAN'NE M. SHREEVE
University of Idaho

PHOSPHORUS

The majority of the fluorine in the earth's crust is present in the form of the phosphorus fluoride fluoroapatite [1306-05-4], $Ca_5(PO_4)_3F$. Phosphate rock deposits contain an average concentration of 3.5 wt % fluorine. During phosphate processing these fluorine values are partially recovered as by-product fluorosilicic acid. The amount of fluorosilicic acid recovered has grown steadily, in part because of environmental requirements (see PHOSPHORIC ACID AND THE PHOSPHATES).

The compounds phosphorus trifluoride [7783-55-3], PF_3; phosphorus pentafluoride [7647-19-0], PF_5; phosphorus oxyfluoride [13478-20-1], POF_3; and phosphorus thiofluoride [2404-52-6], PSF_3, were prepared prior to 1900. The most widely studied of these are PF_5 and PF_3. Physical properties are given in Table 1. The mixed chlorofluorides PCl_xF_y where $x + y = 3$ and 5 have also been studied. Diphosphorus tetrafluoride [13537-32-1], P_2F_4, was first reported in 1966 (1–3).

Table 1. Properties of Phosphorus Fluorides[a]

Property	PF_5	POF_3	PF_3	PSF_3
melting point, °C	-91.6	-39.1	-151.5	0.15
boiling point, °C	-84.8	-39.7	-101.8	-0.5
density, liquid, at bp, g/mL			1.6	
critical temperature, °C	>25	73.3	-2.05	-0.73
critical pressure, MPa[b]		4.23	4.33	
heat of fusion, kJ/mol[c]	12.1	14.9		
heat of vaporization, kJ/mol[c]	16.7	23.2[d]	16.5	
heat of formation, $-\Delta H_f$, kJ/mol[c]	1210[e]		946[f]	

[a]Refs. 4–6 unless otherwise noted.
[b]To convert MPa to atm, divide by 0.101.
[c]To convert kJ to kcal, divide by 4.184.
[d]Ref. 7.
[e]Ref. 8.
[f]Ref. 9.

Phosphorus Pentafluoride

Phosphorus pentafluoride was first prepared in 1876 through fluorination of phosphorus pentachloride using arsenic trifluoride (1). Other routes to PF_5 have included fluorination of PCl_5 by HF, AgF, benzoyl fluoride, SbF_3, PbF_2, or CaF_2 (10). It can also be made by the reaction of PF_3 and fluorine, chlorine (11), or chlorine in contact with calcium fluoride (12); by the reaction of FSO_3H on fluoride and phosphate-containing rocks (13); by the reaction of SF_6 and PF_3 at high (~400°C) temperature (14); VF_5 and POF_3 (15), and Ca_3P_2 and NF_3 (16); by the reaction of alkali or alkaline-earth metal fluorides or fluorosulfonate fluorides with P_2O_5 or H_3PO_4 at 180–200°C (17); by the reaction of POF_3 and HF at 60–80°C (18); $HPF_6 \cdot XH_2O$ and SO_3 or HSO_3F (19); by reaction of $POF_3 \cdot SO_3$ adduct with HF (20); and by the reaction of carbonyl fluoride with POF_3 (21). From 1968–1973 a series of patents (22–25) reported the production of PF_5, PF_3, POF_3, and the hexa-, di-, and monofluorophosphoric acids by fluorination of phosphoric and phosphorus acids using calcium fluorosulfate fluoride, $CaFSO_3F$, produced by reaction of CaF_2 and SO_3 (22–25). Based on these patents a pilot plant was established to produce fluorine–phosphorus chemicals, but the activity was not commercially successful and was terminated in 1972.

Phosphorus pentafluoride is a colorless gas which fumes in contact with moist air and reacts immediately with water to hydrolyze, first to POF_3 and then to the fluorophosphoric acids. Although PF_5 is probably stable in completely dry glass, glass (qv) is not recommended for storage as even a trace of moisture generates HF which regenerates moisture converting the glass to SiF_4 and PF_5 to POF_3.

Phosphorus pentafluoride behaves as a Lewis acid showing electron-accepting properties. It forms complexes, generally in a ratio of 1:1 with Lewis bases, with amines, ethers, nitriles, sulfoxides, and other bases. These complexes are frequently less stable than the similar BF_3 complexes, probably owing to stearic factors. Because it is a strong acceptor, PF_5 is an excellent catalyst especially in ionic polymerizations. Phosphorus pentafluoride is also used as a source of phosphorus for ion implantation (qv) in semiconductors (qv) (26).

Phosphorus Trifluoride

Phosphorus trifluoride was prepared by Moissan in 1884 by reaction of copper phosphide and lead fluoride [7783-55-3]. It is usually prepared by fluorination of PCl_3 with CaF_2 (11), AsF_3, SbF_3, AgF, PbF_2, ZnF_2, or NaF; reaction of fluorosulfonate, $CaF(FSO_3)$, using molten H_3PO_3 (27); by the reaction of phosphorus oxide and F_2 or NF_3 gas (28) or reaction of PH_3 and NF_3 (29); by the reaction of KHF_2 and PCl_3 or PBr_3 (30). PF_3 can be purified for semiconductor devices by contacting it with nickel silicide (31) or with Cu arsenide, phosphide, or silicide (32). Laboratory-scale syntheses have been published (33,34) and PF_3 is commercially available.

Phosphorus trifluoride is an almost odorless gas that does not fume in air and reacts slowly with water but rapidly with base. It may be very toxic, and great care should be taken in handling it. It reacts with ferrohemoglobin to form an unstable complex (35). Because the action may be similar to carbon monoxide poisoning, oxygen may be helpful in treatment. Phosphorus trifluoride acts as a Lewis base and forms many complexes by donating the lone electron pair on the phosphorus atom. Complexes similar to those of CO (see CARBONYLS) are $Ni(PF_3)_4$ [13859-65-9] (36), $Fe(PF_3)_5$ [13815-34-4] (37), $(PF_3)_2PtCl_2$ [15977-33-0] (38), $Mo(PF_3)_6$ [15339-46-5] (39), and $Cr(PF_3)_6$ [26117-61-3] (39). Although AsF_3 and SbF_3 have acceptor properties and function as Lewis acids, no evidence has been found for molecular complex formation by PF_3 as a Lewis acid nor for stable fluorophosphites (PF_4^-).

Phosphorus Oxyfluoride

Phosphorus oxyfluoride is a colorless gas which is susceptible to hydrolysis. It can be formed by the reaction of PF_5 with water, and it can undergo further hydrolysis to form a mixture of fluorophosphoric acids. It reacts with HF to form PF_5. It can be prepared by fluorination of phosphorus oxytrichloride using HF, AsF_3, or SbF_3. It can also be prepared by the reaction of calcium phosphate and ammonium fluoride (40), by the oxidization of PF_3 with NO_2Cl (41) and NOCl (42); in the presence of ozone (43); by the thermal decomposition of strontium fluorophosphate hydrate (44); by thermal decomposition of $CaPO_3F\cdot2H_2O$ (45); and reaction of SiF_4 and P_2O_5 (46).

Phosphorus Thiofluoride

Phosphorus thiofluoride can be prepared at a low temperature by uv radiation of OCS and PF_3 (47); by the reaction of PF_5 and $(C_2H_5)_4NSH$ in acetonitrile (48); by the reaction of PF_3 and SF_6 at elevated temperature (49), or with H_2S (50); by the reaction of $PSCl_3$ and NaF (51); and by the high temperature reaction of PF_3 and S (52).

Fluorophosphoric Acids and the Fluorophosphates

The three primary fluorophosphoric acids, monofluorophosphoric acid [13537-32-1], H_2O_3PF (1), difluorophosphoric acid [13779-41-4], HO_2PF_2 (2), and hexafluorophosphoric acid [16940-811], HPF_6 (3), were discovered in 1927 (53). In 1963 sym-difluorodiphosphoric acid [44801-72-1] (4) was reported (54). The three primary acids can be prepared by reaction of phosphoric acid or phosphoric anhydride (55) using varying amounts of HF or phosphorus oxyfluoride (23–26) and HF or water, or both. The reaction of anhydrous hydrogen fluoride and phosphoric anhydride is extremely violent. These acids are in equilibrium with each other, HF, and phosphoric acid. The interrelationships of the acids are

All three fluorophosphoric acids are commercially available. The mono- and difluoro acids can be made as anhydrous or hydrated liquids. Commercial hexafluorophosphoric acid is an aqueous solution. Anhydrous hexafluorophosphoric acid may be prepared at reduced temperatures and pressures but it dissociates rapidly into PF_5 and HF at 25°C (56). When diluted with water all the fluorophosphoric acids hydrolyze producing orthophosphoric acid. The hexafluoro acid is the most stable of the three fluorophosphoric acids.

A number of salts of the monofluoro- and hexafluorophosphoric acids are known and some are commercially important. The salts of difluorophosphoric acid are typically less stable toward hydrolysis and are less well characterized. Sodium monofluorophosphate [7631-97-2], the most widely used dentifrice additive for the reduction of tooth decay, is best known (see DENTIFRICES). Several hexafluorophosphates can be prepared by neutralization of the appropriate base using hexafluorophosphoric acid. The monofluorophosphates are usually prepared by other methods (57) because neutralization of the acid usually results in extensive hydrolysis.

Because HF is present in the aqueous acids and can be generated by hydrolysis of the anhydrous acids, glass should be avoided in handling or processing. Teflon or other plastics can generally be used for laboratory work although there is evidence for migration of the acids through both thin plastic and plastic bottle closures. The acids are generally shipped and stored in United Nations (UN) 6HA1 heavy plastic drums with steel overpacks. Aluminum is also satisfactory for storage and use of concentrated solutions of the difluoro acid and hexafluoro acid.

Experimentation with test animals and laboratory and plant experience indicate that the fluorophosphoric acids are less toxic and dangerous than hydrogen fluoride (58). However, they contain, or can hydrolyze to, hydrofluoric acid and must be treated with the same care as hydrofluoric acid. Rubber gloves and face shields are essential for all work with these acids, and full rubber dress is necessary for handling larger quantities. The fumes from these acids contain HF.

Monofluorophosphoric Acid. Monofluorophosphoric acid (**1**) is a colorless, nonvolatile, viscous liquid having practically no odor. On cooling it does not crystallize but sets to a rigid glass at $-78°C$. It has a density of $d_{25} = 1.818$ g/mL. Little decomposition occurs up to 185°C under vacuum but it cannot be distilled. An aqueous solution shows the normal behavior of a dibasic acid; the first neutralization point in 0.05 N solution is at pH 3.5 and the second at pH 8.5. Conductance measurements, however, indicate H_2PO_3F behaves as a monobasic acid in aqueous solution (59). The permanent end point shows the stability of PO_3F^{2-} ions at this pH. Slow hydrolysis occurs at low pH to give orthophosphoric and hydrofluoric acids. These kinetics have been studied (60,61). Equilibrium concentrations of H_2PO_3F at varying acid strengths have been determined by nmr studies (62). The anhydrous acid causes rapid swelling of cellulose, and acts as a polymerization catalyst. It has no oxidizing character.

Monofluorophosphoric acid is one of the hydrolysis products of POF_3. It is the primary product of reaction between theoretical amounts of phosphoric anhydride and aqueous hydrogen fluoride. However, the product of this reaction usually contains up to 20% each of orthophosphoric and difluorophosphoric acids. Reaction of P_2O_5 and 40% aqueous hydrogen fluoride gives more complete conversion to monofluorophosphoric acid. Although difluorophosphoric acid is formed initially, it hydrolyzes to give an aqueous solution of primarily monofluorophosphoric acid, some orthophosphoric acid, and HF.

Difluorophosphoric Acid. Difluorophosphoric acid (**2**) is a mobile, colorless liquid. It fumes on contact with air, probably owing to HF aerosol formation. The mp of anhydrous difluorophosphoric acid has been reported to be $-96.5 \pm 1°C$ (63) and $-91.3 \pm 1°C$ (64). The density at 25°C is 1.583 g/mL. It partially decomposes on heating above 80–100°C. An extrapolated normal boiling point is 116°C (63) although it boils at 107–111°C (64,65) with decomposition. Vapor pressure data are available (65,66), however, the data above ca 80°C may be unreliable.

A freshly made solution behaves as a strong monobasic acid. Neutralized solutions slowly become acidic because of hydrolysis to monofluorophosphoric acid and hydrofluoric acid. The anhydrous acid undergoes slow decomposition on distillation at atmospheric pressure, reacts with alcohols to give monofluorophosphoric acid esters, and is an alkylation (qv) and a polymerization catalyst.

The commercially available difluorophosphoric acid has a formula approximating $2\,HPO_2F_2 \cdot H_2O$ and contains some monofluorophosphoric acid (**1**) and some hexafluorophosphoric acid (**3**) as well as HF. It is primarily used in catalytic applications.

Hexafluorophosphoric Acid. Hexafluorophosphoric acid (**3**) is present under ambient conditions only as an aqueous solution because the anhydrous acid dissociates rapidly to HF and PF at 25°C (56). The commercially available HPF_6 is approximately 60% HPF_6 based on PF_5^- analysis with HF, HPO_2F_2, HPO_3F, and H_3PO_4 in equilibrium equivalent to about 11% additional HPF_6. The acid is a colorless liquid which fumes considerably owing to formation of an HF aerosol. Frequently, the commercially available acid has a dark honey color which is thought to be reduced phosphate species. This color can be removed by oxidation with a small amount of nitric acid. When the hexafluorophosphoric acid is diluted, it slowly hydrolyzes to the other fluorophosphoric acids and finally phosphoric acid. In concentrated solutions, the hexafluorophosphoric acid establishes equilibrium with its hydrolysis products in relatively low concentration. Hexafluorophosphoric acid hexahydrate [40209-76-5], $HPF_6 \cdot 6H_2O$, mp ca 31.5°C, also forms (66). This compound has been isolated in good yield when a concentrated acid solution is cooled rapidly to prevent renewed equilibration. The hexahydrate forms hard, coarse crystals having a cubic structure (67) which are very hygroscopic. However, upon melting, an equilibrium is again established. Liquid $HPF_6 \cdot (C_2H_5)_2O$ [4590-57-2], boiling at 114°C and melting at $-35°C$, has been reported (68). Hexafluorophosphoric acid is used catalytically, as a fluorinating reagent, as a fungicide, and extensively in preparation of numerous hexafluorophosphates (see FUNGICIDES, AGRICULTURAL).

The PF_6 ion can be determined by precipitation with nitron or tetraphenylarsonium chloride (69).

Monofluorophosphates. Monofluorophosphates are probably the best characterized series of fluoroxy salts. The PO_3F^{2-} ion is stable in neutral or slightly alkaline solution. The alkali metal and ammonium monofluorophosphates are soluble in water but the alkaline-earth salts are only slightly soluble, eg, $CaPO_3F$ is not water-soluble and precipitates as the dihydrate.

Monofluorophosphates of ammonium, lithium, sodium, potassium, silver, calcium, strontium, barium, mercury, lead, and benzidine have been described (70) as have the nickel, cobalt, and zinc salts (71), and the cadmium, manganese, chromium, and iron monofluorophosphates (72). Many of the monofluorophosphates are similar to the corresponding sulfates (73).

The monofluorophosphates can be prepared by neutralization of monofluorophosphoric acid (**1**). Sodium monofluorophosphate [*7631-97-2*] is prepared commercially (57) by fusion of sodium fluoride and sodium metaphosphate, and the potassium monofluorophosphate [*14104-28-0*] can be prepared similarly. Insoluble monofluorophosphates can be readily prepared from reaction of nitrate or chloride solutions with sodium monofluorophosphate. Some salts are prepared by metathetical reactions between silver monofluorophosphate [*66904-72-1*] and metal chlorides.

Molten alkali metal monofluorophosphates are reactive and corrosive, hydrolyzing to generate HF and reacting with many metals and ceramics. They readily dissolve metal oxides and are effective metal surface cleaners and fluxes (see METAL SURFACE TREATMENTS). They also have bactericidal and fungicidal properties (74). However, the main commercial application among monofluorophosphates is of sodium monofluorophosphate in dentifrices.

Sodium monofluorophosphate, mp 625°C, is soluble in water to the extent of 42 g/100 g solution. The pH of a 2% solution is between 6.5 and 8.0. Dilute solutions are stable indefinitely in the absence of acid or cations that form insoluble fluorides.

Sodium monofluorophosphate is used in most dentifrices at a concentration of 0.76 wt % which produces the desired fluoride level of 1000 ppm although one extra strength dentifrice has 1.14 wt % and 1500 ppm F. Although the mechanism of its efficacy in reducing dental decay is not completely understood (75), it almost certainly reacts with the apatite of the tooth converting it to fluoroapatite which is less soluble in mouth acids (see DENTIFRICES).

The *United States Pharmacopeia* (76) specifications for sodium monofluorophosphate require a minimum of 12.1% fluoride as PO_3F^{2-} (theoretical 13.2%) and a maximum of 1.2% fluoride ion reflecting unreacted sodium fluoride. Analysis for PO_3F^{2-} is by difference between total fluoride in the product less fluoride ion as determined by a specific ion electrode. The oral LD_{50} of sodium monofluorophosphate in rats is 888 mg/kg.

Sodium monofluorophosphate decahydrate [*7727-73-3*], $Na_2PO_3F \cdot 10H_2O$, melts at 9°C. Commercial sodium monofluorophosphate can be purified using this salt. The commercial salt is dissolved in about 10% more than the theoretical water and filtered. The filtrate is cooled to 0°C and allowed to stand. If crystals do not form, the solution may be seeded with $Na_2PO_3F \cdot 10H_2O$ or $Na_2SO_4 \cdot 10H_2O$. The resulting crystals are filtered cold and washed with small quantities of ice water. Attempts to remove the water of hydration by heating lead to hydrolysis of the sodium monofluorophosphate. The water can be extracted by multiple extractions with ethyl alcohol (77).

Calcium, strontium, and barium monofluorophosphates can be precipitated from aqueous solutions using sodium monofluorophosphate. The salts obtained are $CaPO_3F \cdot 2H_2O$ [*37809-19-1*], $SrPO_3F \cdot H_2O$ [*72152-36-4*], and $BaPO_3F \cdot xH_2O$ [*58882-62-5*] where $x < 1$. The solubility of $CaPO_3F \cdot 2H_2O$ is 0.417 g/100 mL solution at 27°C (78). It can be partially dehydrated to the hemihydrate [72152-38-6] but further attempts cause hydrolysis to CaF_2 and phosphate. Heating $SrPO_3F \cdot H_2O$ to 450°C gives anhydrous $SrPO_3F$ [66546-46-1] but $BaPO_3F \cdot xH_2O$ retains water even at 500°C (79).

Difluorophosphates. Difluorophosphates have limited applications largely because of hydrolytic instability of the $PO_3F_2^-$ ion. The ammonium salt can be prepared from ammonium fluoride and phosphoric anhydride.

$$3\ NH_4F\ +\ P_2O_5 \rightarrow NH_4PO_2F_2\ +\ (NH_4)_2PO_3F$$

The $NH_4PO_2F_2$ can be extracted from the solid reaction product with boiling methanol (80). Alkali metal difluorophosphates are prepared from the hexafluorophosphates by one of the following fusion reactions (81):

$$KPF_6\ +\ 2\ KPO_3 \rightarrow 3\ KPO_2F_2$$
$$3\ KPF_6\ +\ 2\ B_2O_3 \rightarrow 3\ KPO_2F_2\ +\ 4\ BF_3$$

Even though the $PO_2F_2^-$ is considered to be hydrolytically unstable, hydrolysis is slow in a neutral solution. However, in a solution initially 0.1 N in NaOH, at 70°C, $NaPO_2F_2$ is quantitatively hydrolyzed to give the PO_3F^{2-} and F^- ions within 10 min (82).

A number of organic nitrogen-containing basic compounds give insoluble difluorophosphates. Among these is nitron which can be used for the gravimetric determination of $PO_2F_2^-$ (83). Potassium and other metal difluorophosphates have been reported as stabilizing agents in chloroethylene polymers (84).

Hexafluorophosphates. There is a great deal of interest in the hexafluorophosphate anion [1691-18-8], mostly as organic hexafluorophosphates for catalysis in photopolymerization. A number of the compounds are diazonium compounds (see PHOTOREACTIVE POLYMERS).

The hexafluorophosphates are among the most stable halogen complexes known. The highly symmetrical PF_6^- ion is stable to boiling aqueous alkali and is decomposed only slowly in acidic solutions at ambient temperatures. It does hydrolyze rapidly at elevated temperatures in acid (85). The stability of the PF_6^- ion can be compared to the isoelectronic SF_6. The hexafluorophosphates can be decomposed yielding PF_5 although in many cases only at temperatures where the PF_5 reacts with the metal containers. Benzenediazonium hexafluorophosphate can be decomposed to PF_5, N_2, and fluorobenzene at 120°C (86) and is a convenient source for laboratory amounts of PF_5 as well as a frequently used catalyst.

Many of the organic and inorganic hexafluorophosphates can be prepared by reaction of hexafluorophosphoric acid and the appropriate base. Another method involves reaction of the appropriate chloride and PCl_5 with anhydrous HF (87).

Potassium hexafluorophosphate [17084-13-8], KPF_6 and ammonium hexafluorophosphate [16941-11-0], NH_4PF_6, are the most readily available PF_6^- salts. The KPF_6 salt melts with slow decomposition to PF_5 and KF at ca 565°C. The density of KPF_6 is 2.55 g/mL, and its solubility in water is 3.56 g/100 g solution at 0°C, 8.35 g/100 g solution at 25°C and 38.3 g/100 g solution at 100°C (88). A solution of KPF_6 is neutral and stable against hydrolysis unless the pH is reduced to about 3. The salt has a minimum intraperitoneal lethal dose of 1120 mg/kg for female albino mice (89). Potassium hexafluorophosphate is a soluble neutral salt which can be used in syntheses of other PF_6^- salts as a substitute for HPF_6 (**3**). Ammonium hexafluorophosphate has a water solubility of 60.4 g/100 g solution at 25°C (90).

Anhydrous silver hexafluorophosphate [*26042-63-7*], $AgPF_6$, as well as other silver fluorosalts, is unusual in that it is soluble in benzene, toluene, and *m*-xylene and forms 1:2 molecular crystalline complexes with these solvents (91). Olefins form complexes with $AgPF_6$ and this characteristic has been used in the separation of olefins from paraffins (92). $AgPF_6$ also is used as a catalyst. Lithium hexafluorophosphate [*21324-40-3*], $LiPF_6$, as well as KPF_6 and other PF_6^- salts, is used as electrolytes in lithium anode batteries (qv).

Substituted ammonium hexafluorophosphates are of decreasing water solubility corresponding to the greater number of hydrogens on the ammonium group replaced by the organic radicals. Unusual thermal stability is found in the quaternary compounds when the R groups are the low alkyl radicals. For example, tetramethylammonium hexafluorophosphate [*558-32-7*], $(CH_3)_4NPF_6$, is stable up to about 400°C.

Fluorophosphate Esters

The esters of monofluorophosphoric acid are of great interest because of their cholinesterase inhibiting activity which causes them to be highly toxic nerve gases and also gives them medical activity (see ENZYME INHIBITORS). The most studied is the bis(1-methylethyl)ester of phosphorofluoridic acid also known as diisopropyl phosphorofluoridate [*55-91-4*], DFP (**5**), and as the ophthalmic ointment or solution Isoflurophate USP. It is used as a parasympathomimetic agent, and as a miotic in glaucoma and convergent strabismus. Developed during World War II as a nerve gas, (93) it is prepared by reaction of PCl_3 and isopropanol, followed by chlorination and conversion to the desired product using NaF (94).

In the nerve gas known as Sarin [*107-44-8*] or GB (**6**), one of the isopropoxy groups of DFP is replaced by a methyl group. Sarin is more toxic than DFP; oral LD_{50} in rats are 550 μg/kg and 6 mg/kg (95), respectively (see CHEMICALS IN WAR).

The esters of monochloro- and dichlorophosphoric acids having polyfluoroalkyl groups Rf, eg, Rf = $(CF_3)_2CH_3C—$, $CF_3(CH_3)_2C—$, $CF_3CH_2—$, $CH_2-(CF_2CH_2)_n—$, n = 2, 4, have been synthesized by the reaction of PCl_3 and the lithium salt of the respective polyfluoroalkyl alcohols followed by oxidation with

N_2O_4 (96,97). Toxicity data of these compounds are not available. The esters have been hydrolyzed to acid phosphates. These acids have potential as fuel cell electrolytes (see FUEL CELLS).

$$RfOLi + PCl_3 \longrightarrow RfOPCl_2 + LiCl$$

$$2 RfOPCl_2 + N_2O_4 \longrightarrow 2 RfOP(O)Cl_2 + 2 NO$$

$$RfP(O)Cl_2 \xrightarrow[2 \text{ HCl}]{H_2O} RfOP(O)OH_2$$

The perfluoroalkylphosphorus(V) acids and bis(perfluoroalkyl)phosphorus-(V) acids, $RfPO(OH)_2$ and $(Rf)_2P(O)(OH)$, where $Rf = CF_3$, C_2F_5, C_3F_7, and C_4F_9 have also been reported (98).

BIBLIOGRAPHY

"Phosphorus Compounds" under "Fluorine Compounds, Inorganic" in *ECT* 1st ed., Vol. 6, pp. 711–721, by W. E. White, Ozark-Mahoning Co.; "Phosphorus" under "Fluorine Compounds, Inorganic" in *ECT* 2nd ed., Vol. 9, pp. 635–649, by W. E. White and C. Pupp, Ozark-Mahoning Co., in *ECT* 3rd ed., Vol. 10, pp. 779–788, by C. B. Lindahl, Elf Atochen North America, Inc.

1. R. W. Rudolph, R. C. Taylor, and R. W. Parry, *J. Am. Chem. Soc.* **88**, 3729 (1966).
2. M. Lustig, J. K. Ruff, and C. B. Colburn, *J. Am. Chem. Soc.* **88**, 3875 (1966).
3. L. B. Centofanti and R. W. Rudolph, *Inorg. Syn.* **12**, 281 (1970).
4. S. Johnson, Ph.D. dissertation, Purdue University, Lafayette, Ind., 1953.
5. T. D. Farr, *Phosphorus—Properties of the Element and Some of its Compounds*, Chemical Engineering Report No. 8, Tennessee Valley Authority, Part XI, U.S. Government Printing Office, Washington, D.C., 1950.
6. *Handbook Volume*: P: Mvol. C, 1965, p. 589.
7. G. Tarbutton, E. P. Egan, Jr., and S. G. Frary, *J. Am. Chem. Soc.* **63**, 1783 (1941).
8. C. J. Hoffman, *Phosphorus—Fluorine Oxidizers*, PF-150613-1, Part 7, Propulsion Chemistry Part II, Lockheed Aircraft Corp., Missiles and Space Div., Burbank, Calif., 1959.
9. H. C. Duus and D. P. Mykytiuk, *J. Chem. Eng. Data* **9**, 585 (1964).
10. E. L. Muetterties and co-workers, *J. Inorg. Nucl. Chem.* **16**, 52 (1960).
11. Brit. Pat. 822,539 (Oct. 28, 1959), K. C. Brinker (to E. I. du Pont de Nemours & Co., Inc.).
12. U.S. Pat. 2,810,629 (Oct. 22, 1957), E. L. Muetterties (to E. I. du Pont de Nemours & Co., Inc.).
13. Fr. Pat. 2,476,054, A1, (Aug. 21, 1981), W. C. Cannon and N. R. Hall.
14. A. P. Hagen and D. L. Terrell, *Inorg. Chem.* **20**(4), 1325–1326 (1981).
15. B. R. Fowler and K. C. Moss, *J. Fluorine Chem.* **15**(1), 67–73 (1980).
16. A. Tasaka and O. Gleuser, *Dashisha Daigaku, Rikogaku Kenkyu Hokoku*, **14**(3), 175–196 (1973).
17. U.S. Pat. 3,769,387 (Oct. 30, 1973), R. A. Wiesboeck and J. D. Nickerson.
18. Fr. Pat. 2,082,502 (Jan. 14, 1972).
19. Ger. Offen. DE 2013858. R. A. Wiesboeck.
20. U.S. Pat. 3,584,999 (Jan. 15, 1971), R. A. Wiesboeck.
21. S. P. Mallela, O. D. Gupta, and J. M. Shreeve, *Inorg. Chem.* **27**(1), 208–209 (1988).

22. U.S. Pat. 342,019 (Sept. 17, 1968), H. L. Bowkley and R. B. Thurman (to Armour Agricultural Chemical Co.).
23. U.S. Pat. 3,428,422 (Feb. 18, 1969), R. A. Wiesboeck (to USS Agri-Chemicals, Inc.).
24. U.S. Pat. 3,634,034 (Jan. 11, 1972), J. D. Nickerson and R. A. Wiesboeck (to United States Steel Corp.).
25. U.S. Pat. 3,728,435 (Apr. 17, 1973), R. A. Wiesboeck (to United States Steel Corp.).
26. R. G. Wilson and D. M. Jamba, *Appl. Phys. Lett.* **22**, 176 (1973).
27. V. S. Zuev, L. D. Mikheev, and I. V. Pogorelskii, *Kvantovanya Elektron (Moscow)* (2), 394–400 (1974).
28. Eur. Pat. 333,084 Aw (Sept. 20, 1989), I. Harada and co-workers.
29. J. K. McDonald and R. W. Jones, *Proceedings of the SPIE-International Society of Opt. Engineering, 1986*, pp. 99–104, and 669.
30. R. G. Kalbandkeru and co-workers, *Indian J. Chem.* **23**(A)(12), 990–991 (1984).
31. Jpn. Pat. 03,164,429 A2 (July 16, 1991), K. Kitahara, T. Shimada, and K. Iwata (to Heisei).
32. Jpn. Pat. 03,178,313 A2 (Aug. 2, 1991), K. Kitahara, T. Shimada, and K. Iwata (to Heisei).
33. C. J. Hoffman, *Inorg. Syn.* **4**, 149 (1953).
34. A. A. Williams, *Inorg. Syn.* **5**, 95 (1957).
35. G. Wilkinson, *Nature (London)* **168**, 514 (1951).
36. G. Wilkinson, *J. Am. Chem. Soc.* **73**, 5501 (1951).
37. R. J. Clark, *Inorg. Chem.* **3**, 1395 (1964).
38. J. Chatt and A. A. Williams, *J. Chem. Soc.*, 3061 (1951).
39. Th. Kruck, *Z. Naturforsch* **196**, 164 (1964).
40. Zh. K. Dzhanmuldaeva and co-workers, *Aktual. Vap. Poluch., Fosfora. Soedin. Ego OSU*, 99–101 (1990).
41. S. H. P. Kuma and K. D. Padma, *J. Fluorine Chem.* **49**(3), 301–311 (1990).
42. G. R. Kalbandkeri and co-workers, *Indian J. Chem. Sect. A* **20**(A)(12), 83–84 (1981).
43. B. W. Moores and L. Andrews, *J. Phys. Chem.* **93**(5), 1902–1907 (1989).
44. H. D. Menz and co-workers, *Z. Anorg. Allg. Chem.*, 540–541, 191–197 (1986).
45. K. Heide, H. D. Menz, and C. Schmidt, *K. Kolditz*, **32**(8), 520 (1985).
46. S. B. Suresh and K. D. Padma; *J. Fluorine Chem.* **24**(4), 399–407 (1984).
47. M. Hawkins, M. J. Almond, and A. J. Downs, *J. Phy. Chem.* **89**(15), 3325–3334 (1985).
48. L. Kolditz and co-workers, *Dokl. Akad. Nauk. SSSR* **267**(6), 1392–1395 (1982).
49. A. P. Hagen and D. L. Terrl, *Inorg. Chem.* **20**(4), 1325–1326 (1981).
50. A. P. Hagen and B. W. Callaway, *Inorg. Chem.* **17**(3), 554–555 (1978).
51. K. D. Padma, S. K. Vijayalashmi, and A. R. Vasudevamurthy, *J. Fluorine Chem.* **8**(6), 461–465 (1976).
52. A. P. Hagen and E. A. Elphingston, *Inorg. Chem.* **12**(2), 478–480 (1973).
53. W. Lange, *Chem. Ber.* **60**, 962 (1927).
54. K. B. Boerner, C. Stoelzer, and A. Simon, *Ber.* **96**, 1328 (1963).
55. L. C. Mosier and W. E. White, *Ind. Eng. Chem.* **43**, 246 (1951).
56. U.S. Pat. 2,718,456 (Sept. 20, 1955), A. J. Mulder and W. C. B. Smithuysen (to Shell Development Co.).
57. U.S. Pat. 2,481,807 (Sept. 13, 1949), C. O. Anderson (to Ozark-Mahoning Co.).
58. J. M. Godwin and W. E. White, *Chem. Eng. News* **28**, 2721 (1950).
59. Y.-T. Chen, T.-C Li, and K.-C. Yin, *Sci Sinica (Peking)* **13**, 1719 (1952).
60. I. G. Ryss and V. B. Tul'chimskii, *Dokl. Akad, Nauk SSSR* **142**, 141 (1962).
61. L. N. Devonshire and H. H. Rowley, *Inorg. Chem.* **1**, 680 (1962).
62. D. P. Ames and co-workers, *J. Am. Chem. Soc.* **81**, 6350 (1959).
63. W. Lange and R. Livingston, *J. Am. Chem. Soc.* **72**, 1280 (1950).

64. A. S. Lenskii, A. D. Shaposhnikova, and A. S. Allilueva, *Zh. Prikl. Khim. (Leningrad)* **35**, 760 (1962).
65. G. Tarbutton, E. P. Egan, Jr., and S. G. Frary, *J. Am. Chem. Soc.* **63**, 1782 (1941).
66. U.S. Pat. 2,488,298 (Nov. 15, 1949), W. Lange and R. Livingston (to Ozark-Mahoning Co.).
67. H. Bode and G. Teufer, *Acta Cryst.* **8**, 611 (1955).
68. Ger. Pat. 812,247 (Aug. 27, 1951), H. Jonas (to Farbenfabriken Bayer A.G.).
69. H. E. Affsprung and U. S. Archer, *Anal. Chem.* **35**, 1912 (1963).
70. W. Lange, *Chem. Ber.* **62**, 793 (1929).
71. H. C. Goswami, *J. Indian Chem. Soc.* **14**, 660 (1937).
72. E. B. Singh and P. C. Sinha, *J. Indian Chem. Soc.* **41**, 407 (1964).
73. I. V. Mardirosova and co-workers, *Izv. Akad, Nauk SSSR Neorg. Mater.* **9**, 970 (1973).
74. J. M. Godwin, Masters Thesis, University of Tulsa, Tulsa, Oklahoma, 1950.
75. E. J. Duff, *Caries Res.* **7**, 79 (1973).
76. *The United States Pharmacopeia XIX-National Formulary XIV*, 4th Suppl., United States Pharmacopeial Convention, Inc., Rockville, Md, Jan. 31, 1978, USP XX-NF XV, 1980.
77. G. F. Hill and L. F. Audrieth, *Inorg. Syn.* **3**, 108 (1950).
78. H. H. Rowley and J. E. Stuckey, *J. Am. Chem. Soc.* **78**, 4262 (1956).
79. Vu Quang Kinh, *Bull Soc. Chim. Fr.*, 1466 (1962).
80. W. Lange, in W. C. Fernelius, ed., *Inorganic Syntheses*, Vol. II, McGraw-Hill Book Co., Inc., New York, 1946, pp. 155–158.
81. Ger. Pat. 813–848 (Sept. 17, 1951), H. Jonas (to Farbenfabriken Bayer A.G.).
82. I. G. Ryss and V. B. Tul'chimskii, *Zh. Neorgna. Khim.* **7**, 1313 (1962); *Ibid.*, **9**, 831 (1964).
83. W. Lange, *Chem. Ber.* **62**, 786 (1929).
84. U.S. Pat. 2,846,412 (Aug. 5, 1958), C. B. Havens (to The Dow Chemical Co.).
85. W. Lange and E. Mueller, *Chem. Ber.* **63**, 1058 (1930).
86. R. Schmutzler, in M. Stacey, J. D. Tatlow, and A. G. Sharpe, eds., *Advances in Fluorine Chemistry*, Vol. 5, Butterworth & Co., Inc., Washington, D.C., 1965, pp. 31–287.
87. M. M. Woyski, in L. F. Audrieth, ed., *Inorganic Syntheses*, Vol. III, McGraw-Hill Book Co., Inc., New York, 1950, pp. 111–117.
88. J. N. Sarmousakis and M. J. D. Law, *J. Am. Chem. Soc.* **77**, 6518 (1955).
89. F. A. Smith and co-workers, *Toxicol. Appl. Pharmacol.* **2**, 54 (1960).
90. A. V. Nikolaev and co-workers, *Izv. Sib. otd. Akad. Nauk SSSR Ser. Khim. Nauk*, 48 (1976).
91. D. W. A. Sharp and A. G. Sharpe, *J. Chem. Soc.*, 1855 (1956).
92. U.S. Pat. 3,189,658 (June 15, 1965), H. W. Quinn (to The Dow Chemical Co.).
93. *Chemical Warfare Service TDMR 832*, Edgewood Arsenal, Apr. 1944.
94. U.S. Pat. 2,409,039 (Oct. 8, 1946), E. E. Hardy and G. M. Kosolapoff (to Monsanto Chemical Co.).
95. *Registry of Toxic Effects of Chemical Substances*, 1977 ed., Vol. II, NIOSH, U.S. Department of Health, Education, and Welfare, Washington, D.C., pp. 667, 686.
96. T. Mahmood and J. M. Shreeve, *Inorg. Chem.* **25**, 3830–3837 (1986).
97. Ref. 96, pp. 4081–4084.
98. Ref. 96, pp. 3128–3131.

General References

R. Schmutzler, in M. Stacey, J. D. Tatlow, and A. G. Sharpe, eds., *Advances in Fluorine Chemistry*, Vol. 5, Butterworth & Co., Inc., Washington, D.C., 1965, pp. 31–287.
G. I. Drozd, *Usp. Khim.* **39**, 3 (1970).

W. Lange, in J. H. Simons, ed., *Fluorine Chemistry*, Vol. I, Academic Press, Inc., New York, 1950, pp. 125–188.

E. L. Muetterties and co-workers, *J. Inorg., Nucl. Chem.* **16**, 52 (1960).

S. Johnson, "Some Chemical and Physical Properties of Phosphorus Pentafluoride," Ph.D. dissertation, Purdue University, Lafayette, Ind., 1953.

CHARLES B. LINDAHL
TARIQ MAHMOOD
Elf Atochen North America, Inc.

POTASSIUM

The two stable salts of potassium and fluorine of commercial significance are the normal fluoride [7789-23-3], KF, and potassium bifluoride [7789-29-9], KHF_2.

Potassium Fluoride

Properties. Anhydrous potassium fluoride [7789-23-3] is a white hygroscopic salt that forms two hydrates, $KF \cdot 2H_2O$ [13455-21-5] and $KF \cdot 4H_2O$ [34341-58-7]. The tetrahydrate exists at temperatures below 17.7°C. The dihydrate is stable at room temperature and starts to lose water above 40°C. Temperatures on the order of 250–300°C are required to remove the last few percent of water in a reasonable period of time. Potassium fluoride does not pyrohydrolyze at temperatures as high as 1000°C (1). Chemical and physical properties of KF are summarized in Table 1.

Halogen exchange with KF is not successful in acetic acid (10). Hydrogen bonding of the acid hydrogen with the fluoride ion was postulated to cause acetate substitution for the halide; however, the products of dissolved KF in acetic acid are potassium acetate and potassium bifluoride (11). Thus KF acts as a base rather than as a fluorinating agent in acetic acid.

Manufacture. Commercial KF is manufactured from potassium hydroxide and hydrofluoric acid followed by drying in a spray dryer or flaking from a heated drum. The KF assay is typically 97–99%; impurities are $KF \cdot 2H_2O$ and either potassium carbonate or potassium bifluoride. The 1992 price of the anhydrous salt was $4.68/kg and that of the reagent-grade dihydrate ranged from $8–$14/kg. Potassium fluoride can be purified by passing anhydrous HF through the melt (12).

Toxicology. By ingestion, the lethal dose of potassium fluoride in guinea pigs is 250 mg/kg body weight. The LD_{50} orally for rats is 245 mg/kg body weight (13). Ingestion of potassium fluoride may cause vomiting, abdominal pains, and diarrhea.

Uses. Potassium fluoride is used in the manufacture of silver solder fluxes and in fluxes for various metallurgical operations (see SOLDERS AND BRAZING ALLOYS). In tin deposition from halogen plating baths, KF is used to complex tetravalent tin to form K_2SnF_6 [16893-93-9] which may be filtered from the so-

Table 1. Physical and Chemical Properties of Potassium Fluoride

Property	Value	Reference
melting point, °C	857	2
boiling point, °C	1505	2
specific gravity at 25°C	2.48	2
solubility, g/100 g		
in H_2O, 25°C	49.6	3
in liquid HF, 8°C	36.5	4
in acetic acid, 25°C	28	5
in methanol, 25°C	2.3	6
standard heat of formation, kJ/mol[a]		
KF	-567.4	2
$KF \cdot 2H_2O$	-1159	7
heat of fusion, kJ/mol[a]	28.2	2
heat of vaporization, kJ/mol[a]	173	7
heat of dehydration, $KF \cdot 2H_2O$, kJ/mol[a]	58.2	8
entropy at 25°C, J/(mol·K)[a]	66.6	2
lattice energy, kJ/mol[a]	813.8	9
specific heat, J/(kg·K)[a]		
at 0°C	833	2
at 50°C	854	2

[a]To convert kJ to kcal, divide by 4.184.

lution (see ELECTROPLATING) (14). Highly purified KF is formed into single crystals by the Stockbarger process (15). The principal use of the crystals is in the studies of fundamental properties and defects in alkali halide crystals used in introducing fluorine in organic synthesis.

For many types of replacement of halogens by fluorine in organic compounds, KF is the most frequently used fluoride. As a fluorinating agent, KF must be as anhydrous as possible and very finely divided. The potassium chloride or bromide by-products from reactions with organic chlorides or bromides deposit on the potassium fluoride crystal surfaces, significantly retarding the reaction. Polar solvents such as acetonitrile, dimethyl sulfoxide, or formamide, and rapid stirring are useful in overcoming this drawback. Ball-milling the reaction mixture is also helpful in speeding up the reaction. This fluorination process is used commercially in the manufacture of sodium fluoroacetate, a useful rat poison sold as "1080" (see FLUORINE COMPOUNDS, ORGANIC–FLUORINATED ACETIC ACID). An alkyl haloacetate is converted to an alkyl fluoroacetate which is then hydrolyzed using sodium hydroxide to the final product. Similarly, fluoroacetamide, a systemic insecticide, is made from chloroacetamide. Organic fluorides (16) that may be prepared from corresponding bromides or chlorides using potassium fluoride include monofluoroalkanes; α-fluoroesters, -ethers, and -alcohols; acyl fluorides; sulfonyl fluorides; and 1-fluoro-2,4-dinitrobenzene. Potassium fluoride behaves as a base when used as a catalyst for reactions such as dehydrohalogenation (17), Michael addition (18), and the Knoevenagel reaction (19). Polyurethane foams can be formed when KF is used as a catalyst for the reaction of adiponitrile carbonate and a polyester polyol (20) (see URETHANE POLYMERS).

Potassium Bifluoride

Properties. Other names for potassium bifluoride are potassium hydrogen difluoride and potassium acid fluoride. This white crystalline salt is a soft, waxy solid. The crystal forms of potassium bifluoride are tetragonal and cubic (21). The bifluoride ion in KHF_2 averages 0.2292 nm between fluoride ions in the F–H–F group (22). At elevated temperatures, potassium bifluoride exhibits an appreciable vapor pressure of HF. At 440°C, KHF_2 is decomposed completely to KF and HF; this decomposition is a convenient means of obtaining very pure HF. Other chemical and physical properties are summarized in Table 2.

Manufacture. Potassium bifluoride is produced from potassium hydroxide or potassium carbonate and hydrofluoric acid. The concentrated solution is cooled and allowed to crystallize. The crystals are separated centrifugally and dried. The commercial product consists typically of 99.7% KHF_2 and 0.2% KF. Potassium bifluoride is available in the United States in 180-kg drums at $4.04/kg (1992).

Toxicology and Handling.   The lethal dose by ingestion in guinea pigs is 150 mg/kg body weight (13). The TLV for KHF_2 is 2.5 mg/m^3 (25). Potassium bifluoride crystals may break down to a fine white powder that is readily airborne. In this form, the salt is quite irritating to the nasal passages, eyes, and skin. Therefore, the hands and eyes should be protected and acid dust masks should be worn while handling, as an acid fluoride KHF_2 can cause superficial hydrofluoric acid-type burns. Areas of skin that have been in contact with potassium bifluoride should be washed as soon as possible with mildly alkaline soaps or borax-containing hand cleaners. If there has been contact with the eyes, they should be washed well with water and a physician should be consulted.

Uses. A primary use for potassium bifluoride is in the electrolyte for cells in fluorine manufacture (26). Sufficient hydrogen fluoride is dissolved with KHF_2 to bring the total HF content up to 40–42 wt %. This mixture approximates the formula KF·2HF; it is molten at 90°C, the operating temperature of the cell. Fluxes for a wide variety of metal joining applications utilize KHF_2, usually in combination with potassium pentaborate, boric acid, and other fluorides and chlo-

Table 2. Properties of Potassium Bifluoride, KHF$_2$

Property	Value	Reference
melting point, °C	238.8	23
specific gravity	2.37	24
solubility in H_2O, 20°C, g/100 g	39.2	
crystal transition temperature, °C	196.7	23
standard heat of formation, kJ/mol^a	−920.4	7
heat of fusion, kJ/mol^a	6.6	23
heat of dissociation to KF + HF, 226.8°C, kJ/mol^a	77.5	24
entropy, 25°C, J/(mol·K)^a	104.3	7
heat capacity, C_p, 25°C, J/(mol·K)^a	76.86	7
lattice energy, kJ/mol^a	641.8	24

^{a}To convert kJ to kcal, divide by 4.184.

rides such as $ZnCl_2$ (27). Solutions of KHF_2 and citric acid are used to etch aluminum prior to coating with an acrylic copolymer emulsion; this opaque white coating resists chipping and is useful for automotive, architectural, and decorative applications (28). Tetrahydrofuran is polymerized to poly(tetramethylene glycol) with fuming sulfuric acid and potassium bifluoride (29).

BIBLIOGRAPHY

"Potassium Compounds" under "Fluorine Compounds, Inorganic," in *ECT* 1st ed., Vol. 6, pp. 721–722, by D. C. Whitaker, The Harshaw Chemical Co.; "Potassium" under "Fluorine Compounds, Inorganic," in *ECT* 2nd ed., Vol. 9, pp. 649–650, by D. C. Whitaker, The Harshaw Chemical Co.; in *ECT* 3rd ed., Vol. 10, pp. 789–792, by H. S. Halbedel and T. E. Nappier, The Harshaw Chemical Co.

1. D. L. Deadman, J. S. Machin, and A. W. Allen, *J. Am. Ceram. Soc.* **44**(3), 105 (1961).
2. *JANAF Thermochemical Tables*, Clearinghouse for Federal, Scientific, and Technical Information, U.S. Dept. of Commerce, Springfield, Va., 1964.
3. J. H. Simons, *Fluorine Chemistry*, Vol. 1, Academic Press, Inc., New York, 1950, p. 28.
4. A. W. Jache and G. W. Cady, *J. Phys. Chem.* **56**, 1106 (1952).
5. J. Emsley, *J. Chem. Soc. A*, 2511 (1971).
6. R. E. Harner, J. B. Sydnor, and E. S. Gilbreath, *J. Chem. Eng. Data* **8**, 411 (1963).
7. F. D. Rossini and co-workers, *NBS, Circ. 500*, U.S. Government Printing House, Washington, D.C., 1952.
8. J. Bell, *J. Chem. Soc.*, 72 (1940).
9. H. Vaino and M. Kanko, *Ann. Univ. Turku. Ser. AI*, (40), 3 (1960).
10. J. H. Clark and J. Emsley, *J. Chem. Soc. Dalton Trans.*, 2129 (1975).
11. V. Kazakov and V. G. Kharchuk, *Zh. Obshch. Khim.* **45**, 2744 (1975).
12. H. Kojima, S. G. Whiteway, and C. R. Masson, *Can. J. Chem.* **46**, 2968 (1968).
13. H. C. Hodge and F. A. Smith, in J. H. Simons, ed., *Fluorine Chemistry*, Vol. 4, Academic Press, Inc., New York, 1965, p. 200.
14. I. Rajagonal and K. S. Rajams, *Met. Finish.* **76**(4), 43 (1978).
15. D. C. Stockbarger, *J. Opt. Soc. Am.* **14**, 448 (1927).
16. A. R. Basbour, L. F. Belf, and M. W. Bruxton, in M. Stacy and co-eds., *Advances in Fluorine Chemistry*, Vol. 3, Butterworth, Washington, D.C., 1963, pp. 181–250.
17. F. Naso and L. Ronzini, *J. Chem. Soc. Perkin Trans. 1*, 340 (1974); J. H. Clark and J. M. Miller, *J. Am. Chem. Soc.* **99**, 498 (1977); J. H. Clark, J. Emsley, and O. P. A. Hoyta, *J. Chem. Soc. Perkin Trans. 1*, 1091 (1977).
18. I. Belski, *Chem. Commun.*, 237 (1977).
19. L. Rand, J. V. Swisher, and C. J. Cromin, *J. Org. Chem.* **27**, 3505 (1962); L. Rand, D. Haidukewych, and R. J. Dohinski, *J. Org. Chem.* **31**, 1272 (1966).
20. U.S. Pat. 3,766,147 (July 31, 1972), L. G. Walgemuth (to Atlantic Richfield Co.).
21. R. Kruh, K. Fuwa, and T. E. McEver, *J. Am. Ceram. Soc.* **78**, 4526 (1956).
22. H. L. Carrell and J. Donohue, *Isr. J. Chem.* **10**(2), 195 (1972).
23. M. L. Davis and E. F. Westrum, Jr., *J. Phys. Chem.* **65**, 338 (1961).
24. T. C. Waddington, *Trans. Faraday Soc.* **54**, 25 (1958).
25. *Proceedings of American Conference Governing Individual Hygiene*, Cincinnati, Ohio, 1977.
26. S. P. Vavalides and co-workers, *Ind. Eng. Chem.* **50**(2), 178 (1958).

27. U.S. Pat. 2,829,078 (Apr. 1, 1958), H. B. Aull and A. S. Cross; U.S. Pat. 3,958,979 (May 29, 1976), A. R. Valdo (to Ethyl Corp.); Jpn. Kokai 75 113,449 (Sept. 5, 1975), K. Motoyoshi, M. Kume, and Y. Amano (to Sumitomo Electric Industries).
28. U.S. Pat. 3,849,208 (Aug. 13, 1973), M. N. Marosi (to Convertex, Ltd.).
29. Jpn. Kokai 73 01,100 (Jan. 9, 1973), K. Matsuzawa, Y. Suzuki, and K. Ohya (to Mitsubishi Chemical Industries Co.).

JOHN R. PAPCUN
Atotech

RHENIUM

Rhenium Hexafluoride

Rhenium hexafluoride [10049-17-9], ReF_6, is a pale yellow solid at 0°C, but a liquid at ambient temperature. In the presence of moisture it hydrolyzes rapidly forming HF, ReO_2, and $HReO_4$ (see RHENIUM AND RHENIUM COMPOUNDS). It is not safe to store ReF_6 in a glass trap or glass-lined container. Leaks in the system can initiate hydrolysis and produce HF. The pressure buildup causes the system to burst and an explosion may result.

Properties.   Some physical properties of ReF_6 are mol wt, 300.19; mp, 18.5°C; bp, 33.7°C; solubility in HF, 52.5 g/100 g; specific gravity, 3.58; and vapor pressure at 20.3°C, 61 kPa (458 mm Hg). The transition point has been reported as −3.45 (1) and −1.9°C (2). The compound can be handled in dry metal vacuum lines made of copper, nickel, stainless steel, or Monel. It forms a passive fluoride film on the surface which protects these metals from further corrosion. Reaction with nitric oxide yields nitrosonium hexafluororhenate [60447-76-9], $NOReF_6$ (3), and with potassium fluoride yields potassium octafluororhenate [57300-90-0], K_2ReF_8 (4). Reaction with alkali metal iodides dissolved in SO_2 results in the reduction to rhenium(IV) complex salts, M_2ReF_6 (M = Na [12021-61-3], K [16962-12-2], Rb [16962-13-3], and Cs [16962-14-4]) (5).

Rhenium hexafluoride is readily prepared by the direct interaction of purified elemental fluorine over hydrogen-reduced, 300 mesh (ca 48 μm) rhenium powder at 120°C. The reaction is exothermic and temperature rises rapidly. Failure to control the temperature may result in the formation of rhenium heptafluoride. The latter could be reduced to rhenium hexafluoride by heating with rhenium metal at 400°C.

Rhenium hexafluoride is used for the deposition of rhenium metal films for electronic, semiconductor, laser parts (6–8), and in chemical vapor deposition (CVD) processes which involve the reduction of ReF_6 by hydrogen at elevated (550–750°C) temperatures and reduced (< 101.3 kPa (1 atm)) pressures (9,10).

Rhenium hexafluoride is a costly (ca $3000/kg) material and is often used as a small percentage composite with tungsten or molybdenum. The addition of rhenium to tungsten metal improves the ductility and high temperature properties of metal films or parts (11). Tungsten–rhenium alloys produced by CVD processes exhibit higher superconducting transition temperatures than those alloys produced by arc-melt processes (12).

Rhenium hexafluoride (99.5% pure) is commercially available from Advance Research Chemicals, Atomergic, Atochem, Spectra Gases, and Matheson Gas of the United States, Fluorochem of the United Kingdom, and other sources. The 1993 price for small quantities varied from $3000 to $3500/kg. Larger quantities were available at $2000 to $2500/kg depending on the price of rhenium metal. U.S. production is less than 100 kg/yr. Because of its high irritating and corrosive nature it is classified as corrosive, poisonous liquid and shipped in steel, stainless steel, or Monel cylinders. Upon exposure to air it hydrolyzes producing HF fumes that are corrosive to the lower respiratory tract, skin, and eyes. Prolonged exposure to fumes may cause pulmonary edema. ACGIH (1992–1993) adopted TLV for fluorides as F^- is 2.5 mg/m^3; therefore great care should be taken while handling ReF_6. Personnel working with this material should use vacuum lines or closed systems located in a chemical hood. All precautions must be taken to avoid breathing of vapors or contact with skin.

Other Rhenium Fluoride Compounds

Rhenium heptafluoride [17029-21-9], ReF_7, is obtained by the direct interaction of elemental fluorine with hydrogen-reduced rhenium powder at 400°C and slightly over atmospheric pressure of fluorine. It is a pale yellow solid, mol wt 319.19; mp, 48.3°C; and bp, 73.7°C.

Rhenium pentafluoride [30937-52-1], ReF_5, is obtained along with rhenium tetrafluoride [15192-42-4], ReF_4, when reduction of ReF_6 is carried out with metal carbonyls (qv). ReF_5 is a greenish yellow solid with mp 48°C. Its ready thermal decomposition and magnetic properties suggest that it may be $ReF_4^+ReF_6^-$ (13). ReF_4, best prepared by the reduction of rhenium hexafluoride with hydrogen at 200°C, is a pale blue solid melting at 124.5°C and boiling at 795°C.

Rhenium also forms several important oxyfluorides: rhenium oxytetrafluoride [17026-29-8], $ReOF_4$; rhenium oxypentafluoride [23377-53-9], $ReOF_5$; rhenium dioxytrifluoride [57246-89-6], ReO_2F_3; and perrhenyl fluoride [25813-73-4], ReO_3F. All are solids at room temperature. Properties are summarized in Table 1.

Table 1. Rhenium Oxyfluorides[a]

Compound	$ReOF_5$	ReO_2F_3	ReO_3F	$ReOF_4$	$ReOF_3$
preparative route	$ReO_2 + F_2$	$ReO_2 + F_2$	$KReO_4 + IF_5$	$ReF_6 + M(CO)_x$	$ReF_6 + M(CO)_x$
color	cream	pale yellow	yellow	blue	black
mp, °C	40.8[b]	90	71	107.8[b]	> 200

[a]Refs. 14–16.
[b]Transition point.

BIBLIOGRAPHY

"Rhenium" under "Fluorine Compounds, Inorganic" in *ECT*, 3rd ed., Vol. 10, pp. 792–794, by A. J. Woytek, Air Products and Chemicals, Inc.

1. G. H. Cady and C. B. Hargreaves, *J. Chem. Soc.*, 1563–1568 (1961).
2. J. G. Malm and H. Selig, *J. Inorg. Nucl. Chem.* **20**, 189 (1961).
3. N. Bartlett, S. P. Beaton, and N. K. Jha, *Chem. Commun.*, 168 (1966).
4. E. G. Ippolitono, *J. Inorg. Chem., USSR* **7**, 485 (1962).
5. R. D. Peacock, *J. Chem. Soc.*, 467 (1957).
6. Jpn. Kokai Tokkyo Koho 03,293,725 (Dec. 25, 1991), T. Tsutsumi (to Mitsubishi Electric Corp.).
7. PCT Int. Appl. 9,201,310 (Jan. 23, 1992), T. Ohmi.
8. Jpn. Kokai Tokkyo Koho 03,36,734 (Feb. 18, 1991), T. Ooba (to Fujitsu Ltd.).
9. U.S. Pat. 3,565,676 (Feb. 23, 1971), R. A. Holzl (to Fanstell Metallurgical Corp.).
10. F. W. Hoertel and J. G. Donaldson, *J. Electrodep. Surf. Treat.* **2**, 343 (1974).
11. W. L. Roberts, *High Temp. Mater. Pap. Plansee Semin. 6th*, 880–884 (1969).
12. D. S. Easton and co-workers, *Philos Mag.* **30**, 1117 (1974).
13. F. A. Cotton and G. Wilkinson, *Advanced Inorganic Chemistry, A Comprehensive Text*, 3rd ed., Interscience Publishers, Division of John Wiley & Sons, New York, 1972, p. 977.
14. O. Ruff and W. Kwansik, *Z. Anorg. Chem.* **219**, 65 (1934).
15. E. E. Aynsley and M. L. Hair, *J. Chem. Soc.*, 3747 (1958).
16. A. Englebrecht and A. V. Grosse, *J. Am. Chem. Soc.* **76**, 2042 (1954).

DAYAL T. MESHRI
Advance Research Chemicals, Inc.

SILVER

Silver Subfluoride

Pure silver subfluoride [*1302-01-8*], Ag_2F, is a greenish shiny crystalline material, or yellowish green solid if contaminated with AgF. It decomposes in water but is stable in alcohol and saturated solutions of AgF. Ag_2F disproportionates to Ag and AgF when heated above 100°C (1,2).

Silver subfluoride is prepared by heating a concentrated solution of silver fluoride with metallic silver powder. It may also be obtained by electrolysis of a solution containing AgF, aqueous HF, and NH_4F at 50°C in a platinum dish using a silver electrode. Another process for the manufacture of Ag_2F involves heating a mixture of a saturated solution of AgF and fine Ag powder. This process yields a yellowish green solid mass. Silver subfluoride is a reagent of laboratory curiosity. No commercial applications are known.

Silver Fluoride

Anhydrous silver fluoride [*7775-41-9*], AgF, is a golden yellow solid in its pure form and is classified as a soft fluorinating agent (3). Several solid phases of the solvated species are reported for AgF in the system AgF–HF–H$_2$O (4), eg, AgF·2H$_2$O [72214-21-2], AgF·4H$_2$O [2242-42-6], 3AgF·2HF [72318-57-1], AgF·2HF [12444-84-7], AgF·3HF [12444-85-8], AgF·5HF [12444-86-9], and AgF·7HF·2H$_2$O [72318-56-0]. In addition, AgF·3HF is formed in the absence of water at 0°C. When this last is warmed to 25°C under an atmosphere of dry nitrogen it dissociates into AgF·HF or AgHF$_2$ [12249-52-4] and HF.

Preparation. Silver fluoride can be prepared by dissolving Ag$_2$O or Ag$_2$CO$_3$ in anhydrous hydrogen fluoride or aqueous hydrofluoric acid, evaporating to dryness, and then treating with methanol or ether.

Properties. Silver fluoride is light sensitive and has a specific gravity of 5.852. It melts at 435°C into a black liquid which boils at 1150°C. Unlike the other halides, it is extremely soluble (182 g/100 g) in water and in anhydrous hydrogen fluoride (83.2 g/100 g at 11.9°C). It is only slightly soluble in absolute methanol (1.5 g/100 mL). Its heat of formation, ΔH_f, is -204.6 kJ/mol (-48.9 kcal/mol); heat capacity, C_p, is 51.92 J/(mol·K) (12.4 cal/(mol·K)); and entropy, S, 83.7 J/(mol·K) (20.0 cal/(mol·K)).

Silver fluoride forms explosive adducts with ammonia (qv) (5,6), and therefore all of the reactions involving liquid or gaseous ammonia should be carried out with extreme precautions.

Uses. Silver fluoride has found many laboratory and special industrial applications. It is used as a soft (mild) fluorinating agent for selective fluorination (7–17), as a cathode material in batteries (qv) (18), and as an antimicrobial agent (19). Silver fluoride is commercially available from Advance Research Chemicals, Inc., Aldrich Chemicals, Cerac Corp., Johnson/Matthey, PCR, Atochem, and other sources in the United States. The U.S. price of silver fluoride in 1993 was $1000–$1400/kg and the total U.S. consumption was less than 200 kg/yr.

Silver Difluoride

Silver difluoride [7783-95-1], AgF$_2$, is a black crystalline powder. It has been classified as a hard fluorinating agent (3) which liberates iodine from KI solutions and ozone from dilute aqueous acid solutions on heating. It spontaneously oxidizes xenon gas to Xe(II) in anhydrous hydrogen fluoride solutions (20).

AgF$_2$ is prepared by the action of elemental fluorine on AgF or AgCl at 200°C. Both processes result in quantitative yields. Silver difluoride should be stored in Teflon, passivated metal containers, or in sealed quartz tubes.

Properties. Silver difluoride melts at 690°C, boils at 700°C, and has a specific gravity of 4.57. It decomposes in contact with water. Silver difluoride may react violently with organic compounds, quite often after an initial induction period. Provisions must be made to dissipate the heat of the reaction. Small-scale experiments must be run prior to attempting large-scale reactions.

Uses. AgF$_2$ is a powerful fluorinating agent and is used for substitution of hydrogen by fluorine in hydrohalocarbons (21), preparation of perfluorocompounds (21–23), purification of perfluoromorpholines containing partially fluorinated hydrocarbons (24), fluorination of compounds containing triple bonds (25), addition of fluorine to unsaturated halocarbons (25), and conversion of carbon monoxide into carbonyl fluoride.

Silver difluoride, commercially available from the same sources as those of AgF, had a 1993 price between $1000–$1400/kg. In spite of the technical success in laboratory experiments, silver fluorides have found limited use on a large scale mainly because of the high cost of the reagents. Demand for silver difluoride is less than 100 kg/yr.

Silver Trifluoride

The existence of diamagnetic salts of AgF_4^- was first reported in 1957 (26), but little was known about their properties. In 1988 (27) it was claimed that AgF_3 was prepared by a reaction of Ag metal and O_2F_2 in ClF_5. Silver trifluoride [91899-63-7], AgF_3, has since been prepared (28) from anhydrous HF solutions of AgF_4^- salts by addition of BF_3, PF_5, or AsF_5.

If excess AsF_5 is added, silver(III) is reduced and the $AgFAsF_6$ salt is produced.

$$AgF_3 + AsF_5 \rightarrow AgFAsF_6 + 1/2\ F_2$$

The red precipitates of AgF_3 are diamagnetic and isostructural with AuF_3. Silver trifluoride is a powerful oxidizing agent and thermodynamically unstable. Its powerful oxidizing properties result from the tight binding of its valence shell d-orbital electrons. No commercial source is available.

Silver Fluorocomplexes

The silver fluorocomplexes, ie, silver hexafluoroantimonate [26042-64-8], $AgSbF_6$; silver hexafluorophosphate [26042-63-7], $AgPF_6$; silver tetrafluoroborate [14104-20-2], $AgBF_4$; and other salts such as silver trifluoromethane sulfonate [2923-28-6], CF_3SO_3Ag, and silver trifluoroacetate [2966-50-9], CF_3COOAg, play an important role in the synthesis of organic compounds and have gained potential industrial importance.

These compounds perform a dual function in synthesis procedures. The introduction of a complex anion assists in the stabilization of the desired product and the generation of unique intermediates by chloride displacement, eg, silver hexafluorophosphate, $AgPF_6$, forms adducts with neutral diamagnetic organometallics which can act as controlled sources of highly reactive cations (29). Silver hexafluoroantimonate, $AgSbF_6$, is an electrophilic bromination catalyst (30) and is also used in promoting chlorination of reactive alkanes (31). Silver trifluoromethane sulfonate, CF_3SO_3Ag, is an excellent precursor to a number of derivatives useful as alkylating agents for aromatic compounds (32).

Silver fluorocomplexes are also used in the separation of olefin–paraffin mixtures (33), nitration (qv) of aromatic compounds (34), in the synthesis of o-bridged bicyclics (35), pyrroles (36), cyclo-addition of vinylbromides to olefins (37), and in the generation of thiobenzoyl cations (38).

These complex salts are very hygroscopic and light sensitive. They are slightly soluble in anhydrous hydrogen fluoride, very soluble in water, and soluble in organic solvents such as acetonitrile, benzene, toluene, and *m*-xylene. Except for the melting point of CF_3COOAg (257–260°C), not many other physical properties are known. Most of the salts decompose at higher temperatures.

These salts are corrosive and are to be considered toxic because of the presence of Ag^+ ions. The American Conference of Government Industrial Hygienists (ACGIH) (1992–1993) has adopted TWA values of 0.01 mg/m^3 for silver metal and 0.01 mg/m^3 for soluble silver salts. TWA for fluorides as F^- ions is 2.5 mg/m^3. The MSDS should be consulted prior to use. Skin contact and inhalation should be avoided.

These salts are commercially available. Worldwide consumption of fluoro-complex salts varies between 100 to 300 kg/yr. The most popular salt is $AgBF_4$. Prices vary between $1000 and $1400/kg.

BIBLIOGRAPHY

"Silver Compounds" under "Fluorine Compounds, Inorganic," in *ECT* 1st ed., Vol. 6, pp. 730–731, by F. D. Loomis, Pennsylvania Salt Manufacturing Co.; "Silver" under "Fluorine Compounds, Inorganic," in *ECT* 2nd ed., Vol. 9, pp. 661–662, by W. E. White, Ozark-Mahoning Co.; in *ECT* 3rd ed., Vol. 10, pp. 795–797, by D. T. Meshri, Ozark-Mahoning Co.

1. R. Scholder and K. Traulsen, *Z. Anorg. Allg. Chem.* **197**, 57 (1931).
2. X. L. Wang, *J. Phys. Soc. Jpn.* **60**(4), 1398–1405 (1991).
3. D. T. Meshri and W. E. White, *George H. Cady ACS Symposium*, Milwaukee, Wis., June 1970.
4. H. J. Thomas and A. W. Jache, *J. Inorg. Nucl. Chem.* **13**, 54 (1960).
5. L. J. Olner and M. Dervin, *Compt. Rend.* **175**, 1085 (1922).
6. W. Blitz and E. Rahlfs, *Z. Anorg. Allg. Chem.* **166**, 351 (1927).
7. D. T. Meshri and W. T. Miller, *158th ACS National Meeting*, New York, Sept. 1969, abstract 14.
8. W. T. Miller, R. A. Snider, and D. T. Meshri, *4th Winter Fluorine Conference*, Daytona Beach, Fla., Feb. 1979.
9. E. D. Bergmann and I. Shalhok, *J. Chem. Soc.*, 1418 (1959).
10. H. J. Emeleus and D. E. McDuffe, *J. Chem. Soc.*, 2597 (1961).
11. J. H. Simons, D. F. Herman, and W. H. Pearson, *J. Am. Chem. Soc.* **68**, 1672 (1946).
12. A. D. Britt and W. B. Moniz, *J. Am. Chem. Soc.* **91**, 6204 (1969).
13. A. Takashi, D. G. Cork, M. Fujta, K. Takahide, and T. Tatsuno, *Chem. Lett.* **11**, 187–188 (1988).
14. B. C. Jago and J. Gittins, *Am. Mineral* **74**(7–8), 936–937 (1989).
15. L. F. Chen, J. Mohthasham, and G. L. Gard, *J. Fluorine Chem.* **49**(3), 331–347 (1990).
16. C. M. Wang and T. E. Mallouk, *J. Am. Chem. Soc.* **112**(5), 2016–2018 (1990).
17. Jpn. Kokai Tokkyo Koho 02,169,523 (June 29, 1990), I. Yuji, A. Nakahara, and J. Nakajima.
18. Jpn. Kokai 75,131,034 (Oct. 16, 1961), T. M. Saaki, (to Japan Storage Battery Co. Ltd.).
19. *Chem. Week* **52** (Jan. 14, 1961).
20. B. Zemva, and co-workers, *J. Am. Chem. Soc.* **112**(12), 4849–4849 (1990).
21. W. B. Burford, III and co-workers, *Ind. Eng. Chem.* **39**, 379 (1947).
22. E. T. McBee and co-workers, in Ref. 21, p. 310.
23. Jpn. Kokai Tokkyo Koho 03,167,141 (July 19, 1991), H. Okajima, T. Hiroshi, I. Fuyuhiko, S. Masamichi, and S. Shiro (to Kanto Denka Kogyo Co. Ltd.).
24. Ger. Pat. 287,477 (Feb. 28, 1991), W. Radeck, S. Ruediger, A. V. Dimitrov, and H. Stewig (to Akademie der Wissenschaften der DDR).

25. D. A. Rausch, R. A. Davis, and D. W. Osborne, *J. Org. Chem.* **28**, 494 (1963).
26. R. Hoppe, *Z. Anorg. Allg. Chem.* **292**, 28 (1957).
27. Yu. M. Kiselev, A. I. Popov, A. A. Timakov, and K. V. Bukharin, *Zh. Neorg. Khim.* **33**(5), 1252–1256 (1988).
28. B. Zemva, and co-workers, *J. Am. Chem. Soc.* **113**, 4192–4198 (1991).
29. N. G. Connelly, A. R. Lucy, and A. M. R. Galas, *Chem. Commun.*, 43 (1981).
30. G. A. Olah and P. Schilling, *J. Am. Chem. Soc.* **95**, 7680 (1973).
31. G. A. Olah, R. Renner, P. Schilling, and Y. K. Mo, *J. Am. Chem. Soc.* **95**, 7686 (1973).
32. B. L. Booth, R. N. Hazeldine, and K. Laak, *J. Chem. Soc., Perkin Trans.*, 2887 (1980).
33. U.S. Pat. 3,189,658 (June 15, 1965), H. W. Quinn (to The Dow Chemical Co.).
34. G. A. Olah and S. Kahn, *J. Am. Chem. Soc.* **83**, 4564 (1961).
35. J. Mann and A. A. Usman, *Chem. Commun.*, 119 (1980).
36. J. E. Baeckvall and J. E. Nystroem, *Chem. Commun.*, 59 (1981).
37. M. Hanock, I. Harder, and K. R. Balinger, *Tetrahedron Lett.* **22**, 553 (1981).
38. G. A. Olah, G. K. S. Prakash, and T. Nakajima, *Angew. Chem., Int. Ed. Engl.* **19**, 812 (1980).

DAYAL T. MESHRI
Advance Research Chemicals, Inc.

SODIUM

Sodium has two fluorides, sodium fluoride [*7722-88-5*] and sodium bifluoride [*1333-83-1*].

Sodium Fluoride

Sodium fluoride, NaF, is a white, free-flowing crystalline powder, mp 992°C, bp 1704°C, with a solubility of 4.2 g/100 g water at 10°C, and 4.95 g/100 g water at 93.3°C. The purity of the commercial material is about 98%.

Sodium fluoride is normally manufactured by the reaction of hydrofluoric acid and soda ash (sodium carbonate), or caustic soda (sodium hydroxide). Control of pH is essential and proper agitation necessary to obtain the desired crystal size. The crystals are centrifuged, dried, sized, and packaged. Reactors are usually constructed of carbon brick and lead-lined steel, with process lines of stainless, plastic or plastic-lined steel; diaphragm, plug cock, or butterfly valves are preferred.

The salt is packaged in 45-kg multiwall bags or fiber drums of 45, 170, or 181 kg. It is available in both powdered and granular forms with densities of 1.04 and 1.44 g/cm³ (65 and 90 lb/ft³), respectively. Only the powdered grade is authorized by and registered with the EPA for use in pesticide formulations, with the further proviso that it must be tinted blue or green, or otherwise discolored. The word poison appears on all labels together with first-aid information.

Both sodium fluoride and sodium bifluoride are poisonous if taken internally. Dust inhalation and skin or eye contact may cause irritation of the skin, eyes, or respiratory tract, and should be avoided by the use of proper protective equipment (1).

Fluoridation of potable water supplies for the prevention of dental caries is one of the principal uses for sodium fluoride (see WATER, MUNICIPAL WATER TREATMENT). Use rate for this application is on the order of 0.7 to 1.0 mg/L of water as fluoride or 1.5 to 2.2 mg/L as NaF (2). NaF is also applied topically to teeth as a 2% solution (see DENTIFRICES). Other uses are as a flux for deoxidizing (degassing) rimmed steel (qv), and in the resmelting of aluminum. NaF is also used in the manufacture of vitreous enamels, in pickling stainless steel, in wood preservation compounds, casein glues, in the manufacture of coated papers, in heat-treating salts, and as a component of laundry sours.

Sodium Bifluoride

Sodium bifluoride (sodium acid fluoride, sodium hydrogen fluoride), $NaHF_2$ or $NaF \cdot HF$, is a white, free-flowing fine granular material. Its solubility in water is 3.7 g/100 g solution at 20°C, and 16.4 g/100 g at 80°C. It decomposes at temperatures above 160°C to give sodium fluoride and hydrogen fluoride. Commercial material is ca 99% pure. To prevent the formation of irritating dust, wetted products, containing 85–90% $NaHF_2$ and 10–15% water, are also in use.

The same reactants are used for manufacture as for sodium fluoride. An excess of acid is required to crystallize the bifluoride. The crystals are dewatered, dried, sized, and packaged. Cooling of the reaction is necessary to avoid overheating and decomposition. Reactors and auxiliary equipment are the same as for sodium fluoride.

The dried salt is shipped in 45-kg multiwall bags and in 57-, 170-, and 180-kg fiber drums. Densities range from ca 0.70 g/cm^3 (44 lb/ft^3) to 1.2 g/cm^3 (75 lb/ft^3) for crystalline material.

Sodium bifluoride, by itself or in conjunction with other materials, is a good laundry sour because, in the concentrations used, it does not create a pH below 4.0 and thus causes no damage to textile fibers, although it removes iron stains. Leather (qv) bleaching and cleaning of stone and brick building faces are other uses for this material (3).

BIBLIOGRAPHY

"Sodium Fluoride" under "Fluorine Compounds, Inorganic," in *ECT* 1st ed., Vol. 6, p. 731, by F. D. Loomis, Pennsylvania Salt Manufacturing Co.; "Sodium Bifluoride" under "Fluorine Compounds, Inorganic," in *ECT*, 1st ed., Vol. 6, pp. 731–732, by J. E. Dodgen, Pennsylvania Salt Manufacturing Co.; "Sodium Fluoride" and "Sodium Bifluoride," under "Fluorine Compounds, Inorganic," in *ECT* 2nd ed., Vol. 9, pp. 662–663, by J. Griswold, Allied Chemical Corp.; "Sodium" under "Fluorine Compounds, Inorganic" in *ECT* 3rd ed., Vol. 10, pp. 797–798, by K. Wachter, Olin Corp.

1. N. I. Sax, *Dangerous Properties of Industrial Materials*, 4th ed., Van Nostrand Reinhold, New York, 1975.

2. *Fluoridation Engineering Manual*, EPA, Office of Water Programs, Washington, D.C., 1972.
3. M. Windholz, ed., *Merck Index*, 9th ed., Merck & Co., Inc., Rahway, N.J. 1976.

WERNER H. MUELLER
Hoechst-Celanese Corporation

SULFUR

SULFUR FLUORIDES

The known binary compounds of sulfur and fluorine range in character from ephemeral to rock-like and provide excellent examples of the influence of electronic and structural factors on chemical reactivity. These marked differences are also reflected in the diversified technological utility.

Sulfur Hexafluoride

Sulfur hexafluoride [2551-62-4], SF_6, molecular weight 146.07, is a colorless, odorless, tasteless gas. It is not flammable and not particularly reactive. Its high chemical stability and excellent electrical characteristics have led to widespread use in various kinds of electrical and electronic equipment such as circuit breakers, capacitors, transformers, microwave components, etc (see ELECTRONIC MATERIALS). Other properties of the gas have led to limited usage in a variety of unique applications ranging from medical applications to space research.

Sulfur hexafluoride was first prepared in 1902 (1). The discovery in 1937 that its dielectric strength is much higher than that of air (2) led to its use as an insulating material for cables, capacitors (3), and transformers (4) (see INSULATION, ELECTRIC). Sulfur hexafluoride has been commercially available as AccuDri, SF_6 (AlliedSignal Inc.) since 1948. It is also produced by Air Products and Chemicals in the United States and by others in Germany, Italy, Japan, and Russia.

Properties.   Sulfur hexafluoride is a good dielectric because a high gas density can be maintained at low temperatures. Properties are given in Table 1.

The vapor pressure of the liquid for the range -50 to $45.6°C$ and a standard % deviation of ± 0.18 is calculated as (10):

Table 1. Physical Properties[a] of Sulfur Hexafluoride

Property	Value	References
sublimation point, °C	−63.9	5
heat of sublimation, kJ/mol[b]	23.59	6
triple point, °C	−50.52	7
pressure at triple point, kPa[c]	225.31	7
critical temperature, °C	45.55	8, 9
critical pressure, MPa[d]	3.759	9, 10
critical density, g/cm^3	0.737	10–12
density, g/cm^3		
solid at −195.2°C	2.863	13
liquid	1.336	10, 14
gas	6.0886×10^{-3}	15
vapor pressure of saturated liquid, MPa[d]	2.3676	15
heat of formation, kJ/mol[b]	−1221.66	
free energy of formation, kJ/mol[b]	−1117.73	16
heat of vaporization, kJ/mol[b]	9.6419	15
entropy, kJ/(mol·K)[b]	291.874	16
heat capacity, J/(mol·K)[b]		
liquid at −43°C	119.5	17
gas	97.234	15
surface tension at −20°C, mN/m(=dyn/cm)	8.02	18
viscosity, mPa·s(=cP)		
liquid	0.277	14
gas	0.01576	17
thermal conductivity, W/(m·K)		
liquid	0.0583	19
gas	0.01415	20
sound velocity, gas, m/s	136	14
refractive index, n_D	1.000783	21
dielectric constant		
liquid	1.81	22
gas	1.00204	23
loss tangent (liquid)	0.001	22

[a]All data refer to 25°C and 101.3 kPa (1 atm), unless otherwise stated.
[b]To convert J to cal, divide by 4.184.
[c]To convert kPa to mm Hg, multiply by 7.5.
[d]To convert MPa to atm, divide by 0.101.

$$\log P_{\mathrm{kPa}} = 0.87652594 - 816.48995/T + 0.029287342T$$
$$- 0.40107549 \times 10^{-4}\, T^2$$
$$+ 0.7142667 \frac{(319.802 - T)}{T} \log (319.802 - T)$$

where T is in Kelvin. Equations for the calculation of sublimation pressure are available (5,24). Heats of vaporization, calculated from the Clausius-Clapeyron equation (15), are

Temperature, °C	ΔH_{vap}, kJ/mol (kcal/mol)	
45.6	0	
40	5.608	(1.340)
20	9.80w	(2.344)
0	12.23	(2.923)
−20	13.58	(3.246)
−40	14.94	(3.571)

Chemical Properties. With few exceptions, SF_6 is chemically inert at ambient temperature and atmospheric pressure. Thermodynamically SF_6 is unstable and should react with many materials, including water, but these reactions are kinetically impeded by the fluorine shielding the sulfur. Sulfur hexafluoride does not react with alkali hydroxides, ammonia, or strong acids.

At elevated temperatures SF_6 forms the respective fluorides and sulfides with many metals (25). In quartz, it starts to decompose at 500°C (1); in copper or stainless steel, it is less stable (26). The stability of SF_6 at 200 and 250°C in the presence of aluminum, copper, silicon steel, and mild steel is shown in Table 2 (14). Careful exclusion of moisture from the system improves the stability of sulfur hexafluoride in the presence of most materials.

Sulfur hexafluoride is more stable in arcs (27) than fluorocarbons such as C_2F_6, or refrigerants such as CCl_2F_2, but less stable than CF_4, BCl_3, or SiF_4. Exposed to 1000°C temperatures, SF_6 decomposes to SOF_2 and SF_4 to the extent of 10 mol %. In spite of its decomposition, the dielectric strength of SF_6 remains the same.

The main products of SF_6 arc decomposition in the presence of air are SOF_2, SF_4, and SOF_4 plus metal fluorides and sulfides (28).

Electrical Properties. The electrical properties of SF_6 stem primarily from its effectiveness as an electron scavenger. To accomplish electrical breakdown in a dielectric gas, primary electrons must gain sufficient energy to generate appreciable numbers of secondary electrons on molecular impact. Sulfur hexafluoride interferes with this process by capturing the primary electrons, resulting in the formation of SF_6^- or SF_5^- ions and F atoms (29):

Table 2. Stability of Sulfur Hexafluoride in Various Materials of Construction[a]

Material	Decomposition, %/yr	
	200°C	250°C
aluminum		0.006
copper	0.18	1.4
silicon steel	0.005	0.01[b]
mild steel	0.2	ca 2

[a]Ref. 14.
[b]Estimated value.

$$SF_6 \rightarrow (SF_6^-)^* \rightarrow SF_6^-$$
$$(SF_6^-)^* \rightarrow SF_5^- + F$$

where $(SF_6^-)^*$ represents an activated complex. This complex is stable against autodissociation during tens of microseconds and thus can be stabilized by collisions.

Although production of SF_6^- is the primary process, formation of SF_5^- ions increases with temperature, and at 200°C the $SF_5^-:SF_6^-$ ratio is 1:25 (30). In addition to high dielectric strength, SF_6 can rapidly interrupt heavy currents at high voltages. At 550 or 756 kV, circuit breaker ratings can be 38,000 and 5,000 MVA, respectively (31). Both SF_6 and C_2F_6 have a more rapid arc recovery than nitrogen (32); six S—F bonds, mean energy 3.4–3.8 eV, are available for fast energy absorption. Both SF_6 and SF_4 have high stability in arcs (27).

Paschen's Rule and Breakdown Voltage. As pressure decreases to vacuum conditions, the breakdown voltage (BDV) first decreases, then increases, resulting in a minimum as shown in Figure 1. Table 3 gives BDV data for SF_6 and other

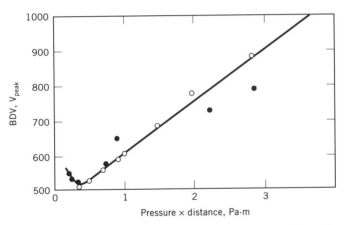

Fig. 1. Paschen's rule plot for SF_6 in uniform electric fields, 60 Hz or dc at 25°C, pressures ≤ 200 kPa, and a gap ≤ 0.3 mm. BDV = breakdown voltage. Compiled from References 33 (○) and 34 (●). To convert Pa to μm Hg, multiply by 7.50.

Table 3. Sulfur Hexafluoride Compared with other Dielectrics[a,b]

Compound	Molecular weight	Bp, °C	Relative BDV[c]
SF_6	146.05	− 64 (sub)	2.7
CF_2Cl_2	120.9	− 29	2.8
CF_3Cl	104.5	− 81	1.44
CF_4	88.0	− 128	1.14
N_2	28	− 194	1.0

[a]Ref. 35.
[b]Conditions: 60 Hz, 0.5 cm gap, 5 cm spheres, at 101.3 kPa (1 atm) and ca 25°C.
[c]BDV = breakdown voltage.

dielectrics. For optimum utility of a dielectric, a compromise is needed between low boiling point and high BDV. At 300–400 kPa (3–4 atm), the BDV of SF$_6$ gas has been shown to be equivalent to that of transformer oil under uniform field conditions (35). The BDV of SF$_6$ deviates from Paschen's rule as pressure increases. This rule is obeyed well only in the case of uniform fields at pressures up to ca 400 kPa (4 atm).

The description of SF$_6$ electrical properties needed for practical applications is more complex than knowledge of BDV. Corona-onset voltages (COV) must be considered, particularly for the more usual nonuniform fields. Figure 2 illustrates this for a point-to-plane electrode configuration, where extensive prebreakdown corona occurs before sparkover (36).

The theory and application of SF$_6$ BDV and COV have been studied in both uniform and nonuniform electric fields (37). The ionization potentials of SF$_6$ and electron attachment coefficients are the basis for one set of correlation equations. A critical field exists at 89 kV/(cm·kPa) above which coronas can appear. Relative field uniformity is characterized in terms of electrode radii of curvature. Peak voltages up to 100 kV can be sustained. A second BDV analysis (38) also uses electrode radii of curvature in rod-plane data at 60 Hz, and can be used to correlate results up to 150 kV. With d-c voltages (39), a similarity rule can be used to treat BDV in fields up to 500 kV/cm at pressures of 101–709 kPa (1–7 atm). It relates field strength, SF$_6$ pressure, and electrode radii to coaxial electrodes having 2.5-cm gaps. At elevated pressures and large electrode areas, a fall-off from this rule appears. The BDV properties of liquid SF$_6$ are described in the literature (40–41).

High Frequency Dielectric Strength. Dielectric strength at high frequency is important in microwave power uses such as radar (see MICROWAVE TECHNOLOGY). Because SF$_6$ has zero dipole moment, its dielectric strength is substantially constant as frequency increases. At 1.2 MHz, SF$_6$ has been shown to have a dielectric strength of 2.3–2.5 relative to N$_2$ (42). At 3 GHz, SF$_6$ has

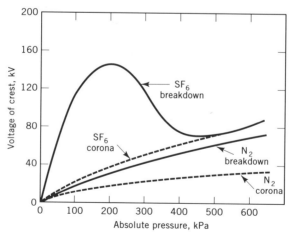

Fig. 2. (——) Sparkover and (– – –) corona onset voltages for SF$_6$ and N$_2$ (36). To convert kPa to atm, divide by 101.

about 10 times the power-carrying capacity of air (43), whereas at 9.375 GHz in a waveguide assembly the power-carrying capacity was 7.5 times that of air (42).

Particle Contamination. In assembling large, high voltage equipment such as coaxial lines, contamination by metal particles may occur which may decrease the dielectric strength under various conditions by 5 to 10-fold (44–45). Metal needles are the worst contaminants and electrostatic traps or adhesive areas have been designed to cope with them (46).

In some cases particles have been added to electrical systems to improve heat removal, for example with an SF_6-fluidized particulate bed to be used in transformers (47). This process appears feasible, using polytetrafluoroethylene (PTFE) particles of low dielectric constant. For a successful application, practical problems such as fluidizing narrow gaps must be solved.

Manufacture and Quality Control. Sulfur hexafluoride is manufactured by combining sulfur vapor and pure elemental fluorine (48,49). It is then given a preliminary scrubbing with caustic. Any disulfur decafluoride, S_2F_{10}, formed is decomposed by heating the product to 400°C, to give SF_4 and SF_6. The SF_4 and any remaining lower fluorides of sulfur are removed by a second caustic scrubber. The gas is then dried in a sulfuric acid tower, distilled, and packaged. A routine manufacturing quality-control test establishes the absence of toxic impurities. In this test, mice are exposed to an atmosphere of 80% sulfur hexafluoride and 20% oxygen for 16–20 h and must show no visible effects from the gas.

Economic Aspects and Shipping. Consumption of SF_6 has increased gradually as dielectric uses have broadened. The estimated worldwide annual consumption for 1992 was about 6000 metric tons. The 1992 U.S. price was ca $8–15/kg.

Sulfur hexafluoride is packaged as a liquefied gas in DOT 3AA 2015 steel cylinders containing 52 kg. Larger quantities are available in tube trailers containing ca 11,000 kg.

Specifications and Analytical Methods. Sulfur hexafluoride is made to rigid specifications. Per ASTM D2472-81 (reapproved 1985) (50), the only permissible impurities are traces of air, carbon tetrafluoride (0.05 wt % max), and water (9 ppm by wt max; dew point −45°C max).

Sulfur hexafluoride may be analyzed chromatographically using a molecular sieve or a Porapak QS column. Using an electron-capture detector, a sensitivity of 10^{-3} to 10^{-4} ppb is possible (51–53).

Health and Safety Factors. Sulfur hexafluoride is a nonflammable, relatively unreactive gas that has been described as physiologically inert (54). The current OSHA standard maximum allowable concentration for human exposure in air is 6000 mg/m³ (1000 ppm) TWA (55). The Underwriters Laboratories classification is Toxicity Group VI. It should be noted, however, that breakdown products of SF_6, produced by electrical decomposition of the gas, are toxic. If SF_6 is exposed to electrical arcing, provision should be made to absorb the toxic components by passing the gas over activated alumina, soda-lime, or molecular sieves (qv) (56).

Because of concerns about the production of the highly toxic S_2F_{10} in arced SF_6 gas, an electrical industry-supported research program was begun in 1992. An interim report indicated that S_2F_{10} may be present in SF_6 gas exposed to electrical discharge, but as of this writing, this could not be precisely quantified

(57). The effects of SF_6 on the environment and the Greenhouse effect have been discussed (58).

Applications. For use as a gaseous dielectric, other specific properties are needed in addition to high breakdown strength, and a compromise must be made between electrical and mechanical requirements. Desirable properties include low toxicity, thermal stability toward materials of construction, good heat transfer, and absence of electrically conducting carbon. Sulfur hexafluoride has a good balance of properties (see Table 1), good dew points, and chemical attributes.

Reviews of gas-phase kinetics (59) and ionization energies (60) have also listed some of the advantages SF_6 enjoys in service as a gaseous dielectric.

Circuit Breakers. Current interruption is essential in high voltage equipment when overloads or other emergencies occur. Circuit breakers consist of relays having contacts equipped with SF_6 jets and high voltage transformers holding fault-sensing coils that can activate a relay contact for each phase of current (61). At 60 Hz, a 765 kV root-mean-square (rms) breaker can have a rating of 50,000 MVA (31). High (1.7 MPa (17 atm)) pressure SF_6 jets extinguish the arcs generated on opening the contacts. Designs of circuit breakers called puffer breakers have encouraged the design of smaller devices, which has extended the use of these types of units (62,63). Sulfur hexafluoride can also act as the insulation for the sensing transformer and supply pressure to activate the relay contacts. At higher pressures and winter temperatures, heaters serve to prevent SF_6 condensation.

High Voltage Coaxial Lines. Sulfur hexafluoride is the main insulation for power transmision in high voltage coaxial lines which can move power above ground, underground, or underwater (64–66). Voltage ratings range between 65 and 500 kV and higher. Above ground, powers of 300 MVA have been transported. Generally each phase needs one coaxial line, but three-phase lines in one envelope have been developed (64). Lines now handle a-c power, and d-c lines have been tested. To reduce costs, SF_6–N_2 mixtures can be used (67). A long coaxial line having a dielectric constant very near unity is possible before reactive compensation is needed. Usually lines of only a few hundred meters are used, especially where land values are high or where passage under multilane highways or other power lines is necessary.

Mini-Substations. Development of SF_6-insulated lines and circuit breakers made possible development of a compact electric substation that requires one-tenth the land area of conventional designs (64,68). Other advantages of the mini-station are freedom from effects of weather and pollution, and reduced risk of vandalism (69). Substation units in which SF_6 insulates 400 and 500 kV d-c equipment have been developed (70).

Transformers. Units insulated with SF_6 are used mainly in circuit breakers. Some SF_6 power transformers have been designed, but the temperature limitation of ca 200°C restricts this use. The problem in high power units is heat transfer through the gas; this can be alleviated with a fluidized particulate bed (47). Although this approach appears feasible, it has yet to be commercialized.

Other Electrical Uses. Using SF_6 insulation, waveguides can transport 7 to 10-fold more microwave power, which results in doubling radar ranges (21,42,43). Voltage ratings in Van de Graaf generators and linear accelerators are also increased by replacing N_2 or air with SF_6 (21). Incorporation of SF_6 in polyethylene-insulated cables increases the starting voltage for tree-formation breakdown processes in the polymer (71,72).

Nonelectrical Uses.   Because of its inertness under normal conditions, SF_6 has been used as a tracer for a variety of studies such as air flow patterns (73), underground pipe leak detection (74), and dispersion of air pollutants (75). It has also been proposed as a refrigerant, either alone (76) or with $CHClF_2$ (77) or CHF_3 as an azeotrope (78). Owing to its low sound velocity, it can improve the performance of loudspeakers at lower pitch registers (79).

At elevated temperatures or under laser radiation, SF_6 becomes a source of fluorine atoms. In the operation of a chemical laser, it abstracts hydrogen from suitable molecules (80–81). SF_6 is also being used in etching of semiconductor surfaces (82,83). Mixed with air or CO_2 in the amount of a few tenths of a percent (0.22 vol %), SF_6 protects molten magnesium during its casting process, resulting in reduced slag and an improved metal surface (84,85). It has also been found useful in removing hydrogen and other gases from aluminum melts (86).

Additional uses include (*1*) filling the enclosed space in double-pane window units to reduce noise and heat transmission (87,88); (*2*) acting as a source of power from the reaction with lithium to produce heat (89,90); and (*3*) pressurizing recreation ball packages such as tennis balls to give improved shelf life (91).

Sulfur Tetrafluoride

Sulfur tetrafluoride [*7783-60-0*], SF_4, molecular weight 108.06, is a highly reactive colorless gas that fumes in moist air and has an irritating odor that resembles sulfur dioxide. Interest in this compound as a fluorinating agent was spurred by its unique ability to replace oxygen in compounds containing carbonyl groups. It was first reported in 1929 (92).

Physical Properties.   Sulfur tetrafluoride has the structure of a distorted trigonal bipyramid, the sulfur having hybrid sp^3d orbitals and an unshared electron pair (93). The FSF bond angles have been found to be 101° and 187°, and the bond distances 0.1646 and 0.1545 nm (94).

Selected physical properties are given in Table 4. The nmr data (97) and ir and Raman spectra (98) have also been determined. Thermodynamic functions have been calculated from spectral data (99).

Chemical Properties.   Sulfur tetrafluoride reacts rapidly with water to give hydrofluoric acid and thionyl fluoride [*7783-42-8*]:

$$SF_4 + H_2O \rightarrow SOF_2 + 2\ HF$$

With alcohols, mixtures of alkyl fluorides and alkyl ethers are obtained (100). Alcohols bearing electron-withdrawing groups can be converted to the corresponding fluorides in high yield (101). Sulfur tetrafluoride replaces the carbonyl oxygen with fluorine (100,102).

$$\begin{array}{c} R \\ \diagdown \\ \diagup \\ R' \end{array} C{=}O + SF_4 \rightarrow \begin{array}{c} R \\ \diagdown \\ \diagup \\ R' \end{array} CF_2 + SOF_2$$

Table 4. Physical Properties of Sulfur Tetrafluoride

Property	Value	Reference
molecular weight	108.055	
melting point, °C	−121.0	95
boiling point, °C	−38	96
critical temperature, °C	90.9	96
surface tension at −73°C, mN/m($=$dyn/cm)	257	95
density, liquid, at −73°C, g/mL	1.9190	95
vapor pressure at 25°C, MPa[a]	2.0219	95
heat of vaporization, kJ/mol[b]	26.4	95
heat of formation at 25°C, kJ/mol[b]	−781.1	16
free energy of formation at 25°C, kJ/mol[b]	−740.4	16
entropy at 25°C, J/(mol·K)[b]	300.7	16
dipole moment, C·m[c]	2.11×10^{-30}	94

[a]To convert MPa to atm, divide by 0.101.
[b]To convert J to cal, divide by 4.184.
[c]To convert C·m to debye, divide by 3.3366×10^{-30}.

Sulfur tetrafluoride reacts with most inorganic oxides and sulfides to give the corresponding fluorides (103):

$$SF_4 + SnS_2 \rightarrow SnF_4 + 3\,S$$
$$5\,SF_4 + I_2O_5 \rightarrow 2\,IF_5 + 5\,SOF_2$$
$$3\,SF_4 + UO_3 \rightarrow UF_6 + 3\,SOF_2$$

Extensive reviews of SF_4 in organic fluorination are available (104,105).

Preparation. In the laboratory, sulfur tetrafluoride is made by combining SCl_2 and NaF suspended in acetonitrile at ca 77°C (106). For commercial production, SF_4 is made by direct combination of sulfur with elemental fluorine (107). Commercial applications of SF_4 are limited. It is available from Air Products and Chemicals.

Toxicity. Sulfur tetrafluoride has an inhalation toxicity comparable to phosgene. The current OSHA standard maximum allowable concentration for human exposure in air is 0.4 mg/m^3 (TWA) (54). On exposure to moisture, eg, on the surface of skin, sulfur tetrafluoride liberates hydrofluoric acid and care must be taken to avoid burns. One case of accidental exposure of electrical workers to decomposed SF_6 gas containing SF_4 has been cited (108).

Other Sulfur Fluorides

Although eight other binary sulfur fluorides have been synthesized and characterized, proof of the existence of several members of this group was dependent on modern instrumental methods of analysis because of extreme instability. SF_5 and S_2F_{10} are stable, however, the latter is noted for its extreme toxicity. All sulfur fluorides other than SF_6 must be considered extremely toxic.

As a group, these materials have no technological utility because of instability, toxicity, and difficulty of preparation. An excellent review of many of these compounds is available (109).

Sulfur Pentafluoride. Sulfur pentafluoride [10546-01-7] is thought to be formed during the electrical breakdown of SF_6 and also to be present in plasma reactions involving SF_6. A number of theoretical studies have been reported (110–113).

Disulfur Decafluoride. Disulfur decafluoride [5714-22-7], S_2F_{10}, is an extremely toxic, colorless, volatile liquid (114). Electron diffraction studies show the molecule to be composed of two octahedral SF_5 groups joined by a sulfur—sulfur bond. The S—F bond distance is 0.156 nm, almost identical to that of SF_6, and the S—S distance is 0.221 nm (115). Table 5 summarizes the known physical properties.

Disulfur decafluoride does not react rapidly with water, mercury, copper, or platinum at ambient temperatures. There is evidence that it slowly decomposes on various surfaces in the presence of water when stored in the vapor state (118). It is decomposed by molten KOH to give a mixture of potassium compounds of sulfur and fluorine. The gas reacts vigorously with many other metals and silica at red heat (114). At ca 156°C it combines with Cl_2 or Br_2 to form SF_5Cl or SF_5Br (119,120). At ca 200°C, S_2F_{10} is almost completely thermally decomposed into the hexa- and tetrafluoride (121).

In the laboratory, S_2F_{10} is prepared by the photochemical reduction of SF_5Cl in the presence of hydrogen (122).

$$2\ SF_5Cl + H_2 \xrightarrow[h\nu]{} S_2F_{10} + 2\ HCl$$

The OSHA standard maximum allowable concentration for human exposure in air is 0.10 mg/m^3 (TWA) (55). No commercial uses for this compound have developed.

Table 5. Physical Properties of S_2F_{10}

Property	Value	Reference
molecular weight	254.13	
melting point, °C	−55	116
boiling point,[a] °C	28.7	114
critical temperature,[b] °C	165	14
density, liquid, at 25°C, g/cm^3	2.08	114
surface tension at 25°C, mN/m($=$dyn/cm)	13.9	114
heat of vaporization, kJ/mol[c]	29.18	114
heat of formation[b,d] at 25°C, kJ/mol[c]	−2.08	117
entropy[b,d] at 25°C, J/(mol·K)[c]	444.93	117
Trouton constant	3.0	114
dielectric constant at 35°C	2.042	114

[a]Vapor pressure calculated.
[b]Estimated.
[c]To convert J to cal, divide by 4.184.
[d]Ideal gas.

Because of the extreme toxicity of this material and the possibility it could be present in failed circuit breakers containing decomposed SF_6, several studies have been carried out to develop analytical methods and investigate possible ways to minimize environmental exposure. One method capable of determining S_2F_{10} in the ppb range has been reviewed (123).

Studies on the kinetics of formation of S_2F_{10} and reviews of applicable literature have been reported (124–126). Other work has concentrated on the use of cell culture evaluation methods for assessing cytotoxic activity of SF_6 decomposition products (127,128). Several laboratories seek to provide methods for accurately determining S_2F_{10} in operating electrical units (57).

Thiothionyl Fluoride and Difluorodisulfane. Thiothionyl fluoride [1686-09-9], $S{=}SF_2$, and difluorodisulfane [13709-35-8], FSSF, are isomeric compounds which may be prepared as a mixture by the action of various metal fluorides on sulfur vapor or S_2Cl_2 vapor. Chemically, the two isomers are very similar and extremely reactive. However, in the absence of catalytic agents and other reactive species, FSSF is stable for days at ordinary temperatures and $S{=}SF_2$ may be heated to 250°C without significant decomposition (127). Physical properties of the two isomers are given in Table 6. The microwave spectrum of $S{=}SF_2$ has been reported (130).

Difluoromonosulfane and Difluorodisulfane Difluoride. Difluoromonosulfane [13814-25-0] (sulfur difluoride), SF_2, and its dimer, disulfane tetrafluoride [27245-05-2], SF_3SF, are both extremely unstable compounds which have only a fleeting existence except under rigorously controlled laboratory conditions. These compounds may be prepared by passing SCl_2 vapor over HgF_2 at 150°C (131). Electronic and nmr examinations of SF_2 have been reported (132,133).

Other Fluorosulfanes. Difluorotrisulfane [31517-17-6], FSSSF, and difluorotetrasulfane [31517-18-7], FSSSSF, have been identified as the constituents of the yellow oil obtained when sulfur vapor reacts with AgF. Their existence was demonstrated by nmr and mass spectroscopy (134,135).

Table 6. Physical Properties of S_2F_2 Isomers[a]

Property	Value	
	FSSF	SSF_2
molecular weight	102.13	102.13
melting point, °C	−133	−164.6
boiling point, °C	15	−10.6
heat of vaporization, kJ/mol[b]	24.80	22.72
entropy of vaporization, J/(mol·K)[b]	86.67	78.08
heat capacity at 25°C, J/(mol·K)[b]	64.0	41.50
entropy[c] at 25°C, J/(mol·K)[b]	299	290.77
heat content at 25°C, kJ/mol[b]	14.017	13.342
free energy, kJ/(mol·K)[b]	72.48	73.26

[a]Ref. 129.
[b]To convert J to cal, divide by 4.184.
[c]Ideal gas.

BIBLIOGRAPHY

"Sulfur Compounds" under "Fluorine Compounds, Inorganic" in *ECT* 1st ed., Vol. 6, pp. 732–734, by H. C. Miller, Pennsylvania Salt Manufacturing Co.; "Sulfur Hexafluoride" in *ECT* 1st ed., Suppl. 2, pp. 793–802, by W. Mears, Allied Chemical & Dye Corp.; "Sulfur" under "Fluorine Compounds, Inorganic" in *ECT* 2nd ed., Vol. 9, pp. 664–676, by J. A. Brown, Allied Chemical Corp.; "Sulfur Fluorides" under "Fluorine Compounds, Inorganic–Sulfur" in *ECT* 3rd ed., Vol. 10, pp. 799–811, by R. E. Eibeck and W. Mears, Allied Chemical Corp.

1. H. Moissan and P. Lebeau, *Ann. Chim. Phys.* **26**, 145 (1902).
2. E. E. Charlton and F. S. Cooper, *Gen. Electr. Rev.* **40**, 438 (1937).
3. B. M. Hokhberg and co-workers, *J. Tech. Phys. (USSR)* **12**, 3 (1942).
4. Brit. Pat. 532,670 (Jan. 29, 1941), (to The British Thompson-Houston Co., Ltd.).
5. W. E. Schumb and E. L. Gamble, *J. Am. Chem. Soc.* **52**, 4302 (1930).
6. D. M. Yost and H. Russell, Jr., *Systematic Organic Chemistry*, Prentice-Hall, Inc., Englewood Cliffs, N.J., 1944, pp. 297–309.
7. V. P. Borisoylekskii, L. I. Strokovskii, and I. S. Zhizuleva *Zh. Fiz. Khim.* **48**, 119 (1974).
8. L. A. Makarevich, E. S. Sokolova, and G. A. Sorena, *Zh. Fiz. Khim.* **42**, 22 (1968).
9. K. E. MacCormack and W. G. Schneider, *Can. J. Chem.* **29**, 699 (1951).
10. W. H. Mears, E. Rosenthal, and J. V. Sinka, *J. Phys. Chem.* **73**, 2254 (1969).
11. L. A. Makarevich and O. N. Sokolova, *Zh. Fiz. Khim.* **47**, 763 (1973).
12. D. Balzarini and P. Palffy, *Can. J. Phys.* **52**, 2007 (1974).
13. T. G. Pearson and R. L. Robinson, *J. Chem. Soc.*, 1427 (1933).
14. Unpublished data, Allied Chemical Corp., Morristown, N.J., 1993.
15. E. Rosenthal, *Sulfur Hexafluoride—Thermodynamic Properties*, unpublished study, Specialty Chemicals Div., Allied Chemical Corp., Morristown, N.J., 1969.
16. *JANAF Thermochemical Tables*, NSRDS-NBS-37, 2nd ed., National Bureau of Standards, Washington, D.C., June 1971.
17. A. Eucken and E. Schroder, *Z. Phys. Chem.* **B41**, 307 (1938).
18. J. Neudorffer, *Ann. Chim. (Paris)* **8**, 501 (1953).
19. P. Grassman and W. Tauscher, *Allied Chemical Research Contract*, Institute of Heat and Engineering, EIDG Technical University, Zurich, Switzerland, 1967.
20. W. A. Tauscher, *Kaltetech.-Klimatisierin* **24**, 67 (1972).
21. *Technical Bulletin SFBR-1 (Sulfur Hexafluoride)*, Specialty Chemicals Division, Allied Chemical Corp., Morristown, N.J., 1973.
22. D. Berg, *J. Chem. Phys.* **31**, 572 (1959).
23. H. E. Watson, G. C. Rao, and K. L. Ramaswamy, *Proc. R. Soc. Ser. A* **132**, 569 (1931).
24. B. Genot, *J. Chim. Phys. Physiochem. Biol.* **68**, 111 (1971).
25. F. A. Cotton and G. Wilkinson, *Advanced Inorganic Chemistry*, 3rd ed., Wiley-Interscience, New York, 1962, p. 419.
26. D. K. Padma and A. R. Vasuderamurthy, *J. Fluor. Chem.* **5**, 181 (1975).
27. J. P. Manion, J. A. Philosophos, and M. B. Robinson, *IEEE Trans., Trans. Electr. Insul.* **E1-2**(1), 1 (Apr. 1967).
28. C. Boudene and co-workers, *Rev. Gen. Electr.* (Special No.), 45 (June 1974).
29. R. L. Champion, *Gaseous Dielectric 6*, Proceedings of the 6th International Symposium, 1990, (Pub. 1991), pp. 1–8.
30. F. C. Fehsenfeld, *J. Chem. Phys.* **53**, 2000 (1970).
31. R. N. Yerkley and C. F. Cromer, *IEEE Trans., Power Appar. Syst.* **89**(8), 2065 (1970).
32. M. Hudis, *CP74-090-7, IEEE Power Engineering Society, Winter Meeting*, New York, Jan. 27–Feb. 1, 1974.

33. G. Luxa and co-workers, *Item D. Paschen Curve for SF₆, 1975 CIGRE International Conference on Large High Tension Electrical Systems.*
34. S. Schreier, *IEEE Trans., Power Appar. Syst.* **83**, 468 (1964).
35. P. R. Howard, *Proc. Inst. Elect. Eng.* **104**(A), 123 (1957).
36. C. N. Works and T. W. Dakin, *Trans. AIEE* **72**(1), 682 (1953).
37. T. Nitta and Y. Shibuya, *IEEE Trans., Power Appar. Syst.* **90**, 1965 (1971).
38. A. A. Azer and P. P. Comsa, *IEEE Trans. Electr. Insul.* **8**(4), 136 (1973).
39. I. M. Bortnik and C. M. Cooke, *IEEE Trans., Power Appar. Syst.* **91**, 2196 (1972).
40. C. N. Works, T. W. Dakin, and R. W. Rogers, *N.A.S.N.R.C. Publ.* (1080), 69 (1963).
41. Y. V. Torshin, Conference Paper 118, *Third International Conference on Gas Discharges,* IEE, London, Sept. 9–12, 1974.
42. T. Anderson and co-workers, *AIEE Conf. Paper 52-82,* 1957.
43. J. W. Sutherland, *Electron. Eng.,* 538 (1955).
44. C. M. Cook, R. E. Wootton, and A. H. Cookson, *IEEE Trans., Power Appar. Syst.* **96**(3), 768 (1977).
45. A. H. Cookson, O. Farish, and G. M. Sommerman, *IEEE Trans., Power Appar. Syst.* **9**(4), 1329 (1972).
46. J. G. Trump, *IEEE Trans. Nucl. Sci.* **14**, 113 (1962).
47. *Gas Insulated Fluidized Bed Transformer, Final Report EL-302,* Project 479-1, Electric Power Research Inst. (EPRI), Buffalo, N.Y., May 1977.
48. U.S. Pat. 3,336,111 (Aug. 15, 1967), W. E. Watson, H. G. Tepp, and M. H. Cohen (to Allied Chemical Corp.)
49. E. P. 87338 (Aug. 31, 1983) M. Jaccaud and A. J. F. Ducouret (to PCUK-Ugine Kuhlmann-Atochem).
50. *ASTM D2472-81* (reapproved 1985), American Society of Testing and Materials, Philadelphia, Pa., 1981.
51. J. E. Lovelock and S. B. Lipsky, *J. Am. Chem. Soc.* **82**, 860 (1960).
52. *Ibid.,* 431 (1960).
53. P. G. Simonds and co-workers, *Anal. Chem.* **44**, 860 (1972).
54. D. Lester and L. A. Greenberg, *Arch. Ind. Hyg. Occup. Med.* **2**, 348 (1950).
55. *Threshold Limit Values for Chemical Substances and Physical Agents,* American Conference of Governmental Industrial Hygienists, Cincinnati, Ohio, 1990–1991.
56. W. C. Schumb, J. G. Trump, and G. L. Priest, *Ind. Eng. Chem.* **41,** 1348 (1949).
57. D. R. James, Technical Note No. 1, *Cooperative Research and Development Agreement (CRADA), Investigation of S₂F₁₀ Production and Mitigation in Compressed SF₆-Insulated Power System,* Oak Ridge National Laboratory, Oak Ridge, Tenn., Dec. 28, 1992.
58. L. Niemayer, F. Y. Chu, *IEEE Trans. Elec. Insul.* **27**(1), 184–187 (Feb. 1992).
59. J. T. Herron, *Int. J. Chem. Kinet.* **19**(2) 129-42 (1987).
60. M. L. Lanferd and co-workers, *Int. J. Mass. Spectrum Ion Processes,* **98**(2) 147-53 (1990).
61. T. Ushio, I. Shimura, and G. Tominago, *IEEE Trans., Power Appar. Syst.* **89**, 2615 (1970).
62. U.S. Pat. 5,059,753 (Oct. 22, 1991), S. R. Hamm (to Cooper Industries).
63. U.S. Pat. 4,752,860 (June 21, 1988), A. Giboulet and P. Romanet (to Merlin-Gerin SA).
64. S. D. Barrett, *The A to Z of SF₆, Special Report Electric Light and Power,* TID ed., Dec. 1972.
65. *Electr. World Trans. Dist.* **54**, (Feb. 15, 1972).
66. B. O. Pedersen, H. C. Doepken, and D. C. Bolin, *IEEE Trans., Power Appar. Syst.* **90**, 2631 (1971).

67. R. Nakata and co-workers, *An Underground High Voltage Direct Current Transmission Line, IEEE Underground Distribution and Transmission Conference, Dallas, Tex., Apr. 1–5, 1974.*

68. *Electr. World* **61**, (Feb. 23, 1976).

69. *IEEE Trans. Power Appar. Syst.* **94**, (1975).

70. E. E. Fischer, A. Glassanos, and N. G. Hingorani, *Electr. World Trans. Dist.* **40**, (Feb. 1, 1977).

71. T. Kojima and co-workers, *Showa Wire Cable Rev.* **221**, 11 (1972).

72. U.S. Pat. 4,783,576 (Nov. 8, 1988), W. G. Lawson and D. A. Silver (to Pirelli Cable Co.)

73. P. J. Drivas and F. H. Shair, *Atmos. Environ.* **8**, 1155 (1974).

74. U.S. Pat. 5,046,353 (Sept. 10, 1991), G. M. Thompson (to Tracer Research Corp.)

75. B. K. Lamb, D. E. Stock, HTD (*Am. Soc. Mech. Eng.*) **152** (*Mixed Convert Enveron Flows*) 55–59, (1990).

76. R. Plank, *Kaltetechnik* **8**, 302 (1956).

77. U.S. Pat. 3,642,639 (Feb. 15, 1972), K. P. Murphy and R. F. Stahl (to Allied Chemical Corp.).

78. U.S. Pat. 3,719,603 (Mar. 6, 1973), R. F. Stahl (to Allied Chemical Corp.).

79. U.S. Pat. 2,797,766 (July 2, 1957), H. W. Sullivan (to D. Bogen and Co.).

80. D. N. Kaye, *New Sci.* **14**, 65 (1971).

81. D. J. Spenser and co-workers, *Int. J. Chem. Kinet.* **1**, 493 (1969).

82. U.S. Pat. 4,680,087 (July 14, 1987), S. M. Bobbio (to AlliedSignal).

83. U.S. Pat. 4,980,018 (Dec. 25, 1990), Mu Xiao-Chun and Multani Jagin (to Intel Corp.).

84. J. W. Fruehling and J. D. Hanawalt, *Am. Foundry Soc. Trans.* **16**, 159 (1969).

85. O. Schlem, *Giesserei* **19**, 558 (1971).

86. U.S. Pat. 4,959,010 (Sept. 25, 1990), R. R. Corns and co-workers (to AGA AB).

87. U.S. Pat. 4,800,693 (Jan. 31, 1989), I. Fasth and J. Karlsen (to Barrier HB).

88. Fr. Pat. 2,529,609 (Jan. 6, 1984), M. Rehfeld (to Saint-Gobain Vitrage).

89. U.S. Pat. 4,959,566 (Sept. 25, 1990) Dobran Flavio

90. S. H. Chan and co-workers, *23rd Symposium on Internal Combustion. Proceedings, 1990* (Pub. 1991), pp. 1139–1146.

91. U.S. Pat. 4,358,111 (Nov. 11, 1982), J. J. Oransky and co-workers (to Air Products and Chemicals).

92. J. Fischer and W. Jaenckner, *Z. Angew. Chem.* **42**, 810 (1929).

93. F. H. Cotton, J. W. George, and J. S. Waugh, *J. Chem. Phys.* **28**, 994 (1958).

94. W. M. Tolles and W. D. Gwinn, *J. Chem. Phys.* **36**, 1119 (1962).

95. E. Brown and P. L. Robinson, *J. Chem. Soc.*, 3147 (1955).

96. *Sulfur Tetrafluoride, Tech. Bulletin, 2B*, E. I. du Pont de Nemours & Co., Inc., Wilmington, Del., 1946.

97. J. Bacon and R. J. Gillespie, *Can. J. Chem.* **41**, 1016 (1963).

98. K. O. Christe and co-workers, *Spectrochim. Acta* **32A**, 1141 (1976).

99. M. Radharkrishnan, Z. Naturforsch. **18a**, 103 (1963).

100. W. R. Hasek, W. C. Smith, and V. A. Engelhardt, *J. Am. Chem. Soc.* **82**, 543 (1960).

101. U.S. Pat. 2,980,740 (Apr. 18, 1961), W. R. Hasek and A. C. Haven, Jr. (to E. I. du Pont de Nemours & Co., Inc.).

102. U.S. Pat. 2,859,245 (Nov. 4, 1958), W. C. Smith (to E. I. du Pont de Nemours & Co., Inc.).

103. A. L. Oppegard and co-workers, *J. Am. Chem. Soc.* **82**, 3835 (1960).

104. C. L. J. Wang, *Org. React. (N.Y.)* **34**, 319–400 (1985).

105. W. Dmowski, *J. Fluorine Chem.* **32**(3) 255–282 (1986).

106. F. S. Fawcett and C. W. Tullock in J. Kleinberg, ed., *Inorganic Syntheses*, Vol. 7, McGraw-Hill Book Co., Inc., New York, 1963, pp. 119–124.

107. U.S. Pat. 3,399,036 (Aug. 27, 1968), S. Kleinberg and J. F. Tompkins, Jr. (to Air Products and Chemicals, Inc.).
108. A. Kraut, R. Lilis, *Br. J. Ind. Med.* **47**(12) 829–832 (1990).
109. F. Seel in H. J. Emeleus and G. G. Sharpe, eds., *Advances in Inorganic and Radio Chemistry*, Vol. 16, Academic Press, Inc., New York, 1974, pp. 297–333.
110. J. T. Herron, *Proceedings of the 5th International Symposium on Gaseous Dielectrics*, 1987, pp. 199–204.
111. M. Ticky and co-workers, *Int. J. Mass. Spectrum Ion Processes*, **76**(3) 231–235 (1987).
112. W. L. Sieck and P. J. Ausloos, *J. Chem. Phys.* **93**(11) 8374-8 (1990).
113. I. C. Plant and K. R. Ryan, *Plasma Chem. Plasma Process.*, **6**(3) 247–258 (1986).
114. K. B. Denbigh and R. W. Gray, *J. Chem. Soc.*, 1346 (1934).
115. R. B. Harvey and S. H. Bauer, *J. Am. Chem. Soc.* **75**, 2840 (1953).
116. N. R. S. Hollies and R. L. McIntosh, *Can. J. Chem.* **29**, 494 (1951).
117. G. Pass, *J. Appl. Chem.* **19**, 77 (1969).
118. J. K. Olthoff and co-workers, *Conference Record of the IEEE International Symposium on Electrical Insulation*, 1990, pp. 248–252.
119. B. Cohen and A. G. MacDiarmid, *Inorg. Chem.* **4**, 1782 (1965).
120. T. A. Kovacina, A. D. Berry, and W. B. Fox., *J. Fluor. Chem.* **7**, 430 (1976).
121. W. R. Trost and R. L. McIntosh, *Can. J. Chem.* **29**, 508 (1951).
122. H. L. Roberts, *J. Chem. Soc.*, 3183 (1962).
123. J. K. Olthoff and co-workers, *Anal. Chem.* **63**(7) 726 (1991).
124. F. Y. Chu and co-workers, *Conference Record of the IEEE International Symposium on Electrical Insulation*, 1988, pp. 131–134.
125. J. T. Herron, *Int. J. Chem. Kinet.* **19**(2) 129–142 (1987).
126. J. T. Herron, *J. Phys. Chem. Ref. Data* **16**(1) 1–6 (1987).
127. G. D. Griffin and co-workers, *Toxicol. Environ. Chem.* **9**(2) 139–166 (1984).
128. G. D. Griffin, *IEE Proc. Part A*, **137**(4) 221–227 (1990).
129. F. Seel, *Chimia* **22**, 79 (1968).
130. R. W. Davis, *J. Mol. Spectrosc* **116**(2) 371–383 (1986).
131. F. Seel, H. Heinrich, and W. Gombler, *Chimia* **23**, 73 (1969).
132. W. Gombler and co-workers, *Inorg. Chem.* **29**(14) 2697-8 (1990).
133. R. J. Glinski and co-workers, *J. Chem. Phys. Chem.* **94**(16) 6196–6201 (1990).
134. F. Seel and co-workers, *Z. Anorg. Allgem. Chem.* **380**, 262 (1971).
135. F. Seel, R. Budenz, and D. Werner, *Ber.* **97**, 1369 (1964).

FRANCIS E. EVANS
GANPAT MANI
AlliedSignal Inc.

FLUOROSULFURIC ACID

Fluorosulfuric acid [7789-21-1], HSO_3F, is a colorless-to-light yellow liquid that fumes strongly in moist air and has a sharp odor. It may be regarded as a mixed anhydride of sulfuric and hydrofluoric acids. Fluorosulfuric acid was first identified and characterized in 1892 (1). It is a strong acid and is employed as a catalyst and chemical reagent in a number of chemical processes, such as alkylation (qv), acylation, polymerization, sulfonation, isomerization, and production of organic fluorosulfates (see FRIEDEL-CRAFTS REACTIONS).

Properties. Selected physical properties of fluorosulfuric acid are shown in Table 1. Fluorosulfuric acid is soluble in acetic acid, ethyl acetate, nitrobenzene,

Table 1. Physical and Chemical Constants of Fluorosulfuric Acid

Property	Value[a]	References
molecular weight	100.07	
boiling point, °C	162.7	1, 2
freezing point, °C	−88.98	3
density, g/mL	1.726	2
viscosity, mPa·s(=cP)	1.56	2
dielectric constant	ca 120	2
specific conductance, $(\Omega \cdot m)^{-1}$	1.085×10^{-6}	2
heat of formation,[b] ΔH_f, kJ/mol[c]	792.45	4

[a]All values at 25°C.
[b]From SO_3 and HF.
[c]To convert kJ to kcal, divide by 4.184.

and diethyl ether, and insoluble in carbon disulfide, carbon tetrachloride, chloroform, and tetrachloroethane. Many inorganic and organic materials dissolve in fluorosulfuric acid; the physical and chemical properties of such solutions have been extensively investigated (5–8). The structure of fluorosulfuric acid has been determined (9), and the ir, Raman, and nmr spectra have been reported (10). The solution of antimony pentafluoride [7783-70-2] in fluorosulfuric acid results in a superacid possessing protonating power orders of magnitude greater than 100% sulfuric acid. Extensive studies on the properties of superacid compositions containing fluorosulfuric acid have been published (11,12).

Fluorosulfuric acid is stable to heat up to decomposition at about 900°C (13), where vapor-phase dissociation into hydrogen fluoride and sulfur trioxide probably occurs. Reviews of the chemistry and properties of fluorosulfuric acid have been published (14–16).

Reactions. The reaction of fluorosulfuric acid and water is violent and exothermic; it proceeds as follows:

Fast hydrolysis	$HSO_3F + H_2O \rightleftharpoons H_2SO_4 + HF$
Ionization	$HSO_3F + H_2O \rightleftharpoons H_3O^+ + SO_3F^-$
Slow hydrolysis	$SO_3F^- + H_2O \rightleftharpoons HSO_4^- + HF$

The extent of the initial hydrolysis depends on temperature and how the water is added. Hydrolysis is reduced at slower addition rates and lower temperatures. The hydrolysis subsequent to the initial fast reaction is slow, presumably because part of the acid is converted to fluorosulfate ions which hydrolyze slowly even at elevated temperatures. The hydrolysis in basic solution has also been studied (17). Under controlled conditions, hydrates of HSO_3F containing one, two, and four molecules of water have been observed (18,19).

The pure acid does not react in the cold with sulfur, selenium, tellurium, carbon, silver, copper, zinc, iron, chromium, or manganese, but slowly dissolves mercury and tin (20). At higher temperatures, lead, mercury, tin, and sulfur react rapidly, eg:

$$S + 2\ HSO_3F \rightarrow 3\ SO_2 + 2\ HF$$

Precipitated (hydrated) silica reacts vigorously with fluorosulfuric acid to give silicon tetrafluoride [7783-61-1] (21), but glass (qv) is not attacked in the absence of moisture (20). Alkali and alkaline-earth metal chlorides are readily converted to fluorosulfates by treatment with fluorosulfuric acid (7,13,22,23).

Electrolysis of fluorosulfuric acid produces either $S_2O_6F_2$ [13709-32-5] (24) or SO_2F_2 [13036-75-4] plus OF_2 (25), depending on specific conditions. Various reactions of fluorosulfuric acid with inorganic compounds are shown in Table 2, and with organic compounds in Table 3.

Other studies which have been reported describe unusual chemistry such as HSO_3F–$Nb(SO_3F)_5$ systems (42). Also the unique properties of fluorosulfuric acid have been found to provide unusual solvent systems, which can vary properties such as acidity, heats of solution, enthalpy, and heats of neutralization (43).

Table 2. Reactions of Fluorosulfuric Acid and Inorganic Compounds

Reactant	Product		References
	Name	Formula	
nitrogen oxides	nitrosyl fluorosulfate,	FSO_3NO,	26, 27
	nitryl fluorosulfate	FSO_3NO_2	
H_3BO_3	boron trifluoride	BF_3	28
$KClO_4$	perchloryl fluoride	ClO_3F	29
$KMnO_4$	manganese(VII) fluoride trioxide	MnO_3F	30
As_2O_3	arsenic trifluoride	AsF_3	31
As_2O_5	arsenic pentafluoride	AsF_5	31
CrO_3	difluorodioxochromium(VI)	CrO_2F_2	32
P_4O_{10}	phosphoryl fluoride	POF_3	31
H_2S in ethanol	monothiosulfuric acid	$H_2S_2O_3$	33

Table 3. Reactions of Fluorosulfuric Acid and Organic Compounds

Reactant	Product		References
	Name	Formula	
benzene[a]	benzenesulfonic acid	$C_6H_5SO_3H$	20, 34
	diphenyl sulfone	$(C_6H_5)_2SO_2$	34
	benzenesulfonyl fluoride	$C_6H_5SO_2F$	35
aliphatic amines	amidosulfuric acids	R_2NSO_3H or	36
		$RNHSO_3H$	
aromatic compound	arylsulfonyl fluorides[b]	$ArSO_2F$	
carboxylic acids	acid fluorides	$RCOF$	37
alcohol or alkene	alkyl fluorosulfates	$ROSO_2F$	20, 38, 39
perhaloolefins or perhaloalkyl iodides	perhaloalkyl fluorosulfates	R_fOSO_2F	40, 41

[a] Product is dependent on reaction conditions and proportions of reagents.
[b] Ar represents an aryl group.

Fluorosulfuric acid may be used to prepare diazonium fluorosulfates, $ArN_2^+SO_3F^-$ (44), which decompose on heating to give aryl (Ar) fluorosulfates (36,45). Aryl fluorosulfates are also obtained from arylsulfonyl chlorides and fluorosulfuric acid (35). Alkyl and other organofluorosulfates form during electrolysis of fluorosulfuric acid in the presence of organic species (46,47).

Preparation and Manufacture. Fluorosulfuric acid, first prepared by combining anhydrous HF and cooled, anhydrous SO_3 in a platinum container (1), has also been prepared from ionic fluorides or fluorosulfates and sulfuric acid (20,48). The reaction of chlorosulfuric acid (qv) with ionic fluorides also gives fluorosulfuric acid (49).

Commercially, fluorosulfuric acid is made by processes utilizing the product as a solvent. Solutions of HF and SO_3 in fluorosulfuric acid are mixed in stoichiometric quantities, or SO_3 and HF are separately introduced into a stream of fluorosulfuric acid to produce essentially pure HSO_3F. Some of the product is then recycled (50,51).

Fluorosulfuric acid can be very corrosive. A study of the corrosive properties of fluorosulfuric acid during preparation and use showed carbon steel to be acceptable up to 40°C, stainless steel up to 80°C, and aluminum alloys up to 130°C (52).

Economic Aspects. U.S. manufacturers of fluorosulfuric acid are AlliedSignal and Du Pont. These companies have a combined annual capacity estimated at 20,000 metric tons, most of which is used internally although some merchant sales exist. Fluorosulfuric acid is shipped in tank cars.

Specifications and Analysis. Commercial fluorosulfuric acid contains approximately 98% HSO_3F and approximately 1% H_2SO_4 and lesser amounts of sulfur trioxide and dioxide. No free HF is present.

The free sulfur trioxide can be titrated with water; the end point is determined conductimetrically. The sulfuric acid content is determined from the specific conductivity of the liquid at the point in the titration where no free SO_3 or excess water is present. If the presence of HF is suspected, a known amount of SO_3 is added to the acid and the excess SO_3 is determined as above. The content of another common impurity, SO_2, may be determined iodometrically in a dilute, aqueous solution.

Health and Safety Factors. Fluorosulfuric acid is a strong acid capable of causing severe burns similar to those experienced with sulfuric and hydrofluoric acids. In addition, the fumes of fluorosulfuric acid are extremely irritating, and breathing of the fumes is to be avoided. Precautions and first aid measures generally observed in handling strong sulfuric acid and hydrofluoric acid are applicable to fluorosulfuric acid. Small containers of fluorosulfuric acid should be well cooled before opening and precautions taken to relieve any gas pressure that may have developed. In the laboratory, fluorosulfuric acid may be handled in glass if water is not present; otherwise, containers of inert polymers or platinum should be used. For larger-scale equipment, iron or carbon steel (not stainless steel) may be used. Material safety data sheets and other literature from manufacturers describe additional precautions in handling large quantities of fluorosulfuric acid.

Uses. Fluorosulfuric acid serves as catalyst in the alkylation (qv) of branched-chain paraffins (53–58) and aromatic compounds (59), and in the polymerization of monoolefins (60) and rosin (61). Addition of strong Lewis acids, such as SbF_5, TaF_5, and NbF_5, to fluorosulfuric acid markedly increases the system acidity and catalytic activity (62–69). Other examples which show the marked catalytic effect of fluorosulfuric acid alone or in systems including SbF_5, etc, are the synthesis of methyl *tert*-butyl ether (70), the stereospecific formation of 2-naphthol (71), formation of aromatic aldehydes using CO (72), and polymerization of tetrahydrofuran (73,74) (see CATALYSIS).

As a reagent, fluorosulfuric acid has been employed in the preparation of boron trifluoride (28), silicon tetrafluoride (75,76), alkyl fluorosulfates (20,38), arenesulfonyl fluorides (35), acyl fluorides (77), sulfamic acid (78), and diazonium fluorosulfates (44). Among its other uses are the removal of small amounts of organic fluorides from petroleum alkylate made by the hydrogen fluoride process (79), the removal of HF from process exhaust gases (80), the removal of HF from F_2 (81), and as a constituent of baths for electropolishing metals (82,83) and glass polishing (84).

Derivatives. The nonmetallic inorganic derivatives of fluorosulfuric acid are generally made indirectly, although complex fluorosulfates of the Group 15 (V) elements and of xenon can be made directly (85,86), as can the NO^+ and NO_2^+ salts (26,27).

Peroxydisulfuryl difluoride [*13709-32-5*], FSO_2OOSO_2F, prepared from fluorine and SO_3 (87), is a ready source of fluorosulfate radicals, $FSO_2O\cdot$, (88) which react with many substances to form stable fluorosulfates (89,90). By using the route

$$X_2 + n\,S_2O_6F_2 \rightarrow 2\,X(OSO_2F)_n$$

where $n = 1$ or 3, compounds of the type $FOSO_2F$ [*13536-85-1*] (91), $BrOSO_2F$ [*13997-93-8*], and $I(OSO_2F)_3$ [*13709-37-0*] (92) have been prepared. Fluorosulfates of most metallic elements have been prepared but none have any commercial significance. The physical properties of some fluorosulfates are summarized in Table 4.

Table 4. Physical Properties of Some Fluorosulfates

Salt	CAS Registry Number	Appearance	Mp, °C	Solubility Water[a]	Other solvents[a]
NH_4SO_3F	[*13446-08-7*]	long colorless needles	245	s	sl s ethanol; v s methanol
$LiSO_3F$	[*13453-75-3*]	white powder	360	v s	v s ethanol, ether, acetone, amyl alcohol, ethyl acetate; i ligroin
$LiSO_3F\cdot3H_2O$		long shiny needles	60–61		
$NaSO_3F$	[*14483-63-7*]	shiny leaflets, hygroscopic		s	s ethanol, acetone; i ether
KSO_3F	[*13455-22-6*]	short white prisms	311	6.9^b	sl s methanol
$RbSO_3F$	[*15587-05-0*]	colorless needles	304	s	sl s methanol
$CsSO_3F$	[*13530-70-6*]	colorless rhombic	292	2.23^b	

^{a}i = insoluble; sl s = slightly soluble; s = soluble; v s = very soluble.
^{b}In g/100 mL water.

Ammonium fluorosulfate is produced from ammonium fluoride by reaction with sulfur trioxide, oleum, or potassium pyrosulfate, $K_2S_2O_7$ (48). Solutions of ammonium fluorosulfate show little evidence of hydrolysis and the salt may be recrystallized from hot water. Ammonium fluorosulfate absorbs anhydrous ammonia to form a series of liquid amines that contain 2.5–6 moles of ammonia per mole of salt (77).

Sodium fluorosulfate may be prepared by the action of fluorosulfuric acid on powdered, ignited sodium chloride (13) or of sulfur trioxide on sodium fluoride (48). In general, the alkali metal fluorosulfates may be prepared from the ammonium salt by evaporating a solution containing that salt and an alkali metal hydroxide (77). The solubilities of some Group 1 and 2 fluorosulfates in fluorosulfuric acid have been determined (93).

BIBLIOGRAPHY

"Fluosulfonic Acid" under "Fluorine Compounds, Inorganic" in *ECT* 1st ed., Vol. 6, pp. 734–738, by W. S. W. McCarter, Pennsylvania Salt Manufacturing Co.; "Fluorosulfuric Acid" under "Fluorine Compounds, Inorganic" in *ECT* 2nd ed., Vol. 9, pp. 676–681, by R. E. Eibeck, Allied Chemical Corp.; "Fluorosulfuric Acid" under "Fluorine Compounds, Inorganic–Sulfur" in *ECT* 3rd ed., Vol. 10, pp. 812–817, by R. E. Eibeck, Allied Chemical Corp.

1. T. E. Thorpe and W. Kirman, *J. Chem. Soc.*, 921 (1892).
2. J. Barr, R. J. Gillespie, and R. C. Thompson, *Inorg. Chem.* **3**, 1149 (1964).
3. R. J. Gillespie, J. B. Milne, and R. C. Thompson, *Inorg. Chem.* **5**, 468 (1966).
4. G. W. Richards and A. A. Woolf, *J. Chem. Soc. A*, 1118 (1967).
5. A. A. Woolf, *J. Chem. Soc.*, 2840 (1954); *ibid.*, 433 (1955).
6. R. J. Gillespie and co-workers, *Can. J. Chem.* **40**, 675 (1962); **41**, 148, 2642 (1963); **42**, 502, 1433 (1964).
7. R. J. Gillespie and co-workers, *Inorg. Chem.* **3**, 1149 (1964).
8. R. J. Gillespie and co-workers, *Inorg. Chem.* **4**, 1641 (1965); **8**, 63 (1969).
9. K. Bartmann and D. Mootz, *Acta Crystallogr., Sect. C* C46(2), 319–320 (1990).
10. R. J. Gillespie and E. A. Robinson, *Can. J. Chem.* **40**, 644, 675 (1962); R. Savoie and P. A. Giguere, *Can. J. Chem.* **42**, 277 (1964).
11. B. Carre and J. Devynck, *Anal. Chim. Acta* **159**, 149–158 (1984).
12. V. Gold and co-workers, *J. Chem. Soc., Perkin Trans.* 2(6) 859–864 (1985).
13. O. Ruff, *Chem. Ber.* **47**, 646 (1914).
14. R. J. Gillespie, *Accounts. Chem. Res.* **1**(7), 202 (1968).
15. R. C. Thompson in G. Nickless, ed., *Inorganic Sulphur Chemistry*, Elsevier, Amsterdam, the Netherlands, 1968, pp. 587–606.
16. A. W. Jache in H. J. Emeleus and A. G. Sharpe, eds., *Advances in Organic Chemistry and Radiochemistry*, Vol. 16, Academic Press, Inc., New York, 1974, pp. 177–200.
17. I. G. Ryss and A. Drabkina, *Kinet. Katal.* **7**, 319 (1966).
18. R. C. Paul, K. K. Paul, and K. C. Malhotra, *Inorg. Nucl. Chem. Lett.* **5**, 689 (1969).
19. D. Mootz, K. Bartmann, *Z. Anorg. Allg. Chem.* **592**, 171–178 (1991).
20. J. Meyer and G. Schramm, *Z. Anorg. Allg. Chem.* **206**, 24 (1932).
21. L. J. Belf, *Chem. Ind. (London)*, 1296 (1955).
22. P. Bernard, Y. Parent, and P. Vast, C. R. *Acad. Sci. Ser. C* **269**, 767 (1969).
23. E. Kemnitz and D. Hass, *Z. Chem.* **30**(7), 264–265 (1990).
24. J. M. Shreeve and G. H. Cady, *J. Am. Chem. Soc.* **83**, 4521 (1961).
25. H. Schmidt and H. D. Schmidt, *Z. Anorg. Allg. Chem.* **279**, 289 (1955).

26. D. R. Goddard, E. D. Hughes, and C. K. Ingold, *J. Chem. Soc.*, 2559 (1950).
27. W. Lange, *Chem. Ber.* **60B**, 967 (1927).
28. U.S. Pat. 2,416,133 (Feb. 18, 1947), D. Young and J. Pearson (to Allied Chemical Corp.).
29. G. Barth-Wehrenalp, *J. Inorg. Nucl. Chem.* **2**, 266 (1956).
30. A. Engelbrecht and A. V. Grosse, *J. Am. Chem. Soc.* **76**, 2042 (1954).
31. E. Hayek, A. Aignesberger, and A. Engelbrecht, *Monatch. Chem.* **86**, 470735 (1955).
32. A. Engelbrecht, *Angew. Chem. Int. Ed. Engl.* **4**, 641 (1965).
33. M. Schmidt and G. Talsky, *Z. Anorg. Allg. Chem.* **303**, 210 (1960).
34. J. H. Simons, H. J. Passino, and S. Archer, *J. Am. Chem. Soc.* **63**, 608 (1941).
35. W. Steinkopf and co-workers, *J. Prakt. Chem.* **117**, 1 (1927).
36. Ger. Pat. 532,394 (Aug. 8, 1930), W. Lange.
37. W. Traube and A. Krahmer, *Chem. Ber.* **B52**, 1293 (1919).
38. Ger. Pats. 342,898 (Oct. 25, 1921), 346,245 (Dec. 27, 1921), W. Traube.
39. G. Olah, J. Nishimura, and Y. Mo., *Synthesis* **4**(11), 661 (1973).
40. U.S. Pats 3,254,107 (May 31, 1966), 3,255,228 and 3,255,229 (June 7, 1966), M. Hauptschein and M. Braid (to Pennsalt Chemicals Corp.); 3,083,220 (Mar. 26, 1963), E. L. Edens (to E. I. du Pont de Nemours & Co., Inc.); 2,878,156 (Mar. 17, 1959), R. A. Davis (to The Dow Chemical Co.); 2,628,927 (Feb. 17, 1953), J. D. Calfee and P. A. Florio (to Allied Chemical Corp).
41. M. Hauptschein and M. Braid, *J. Am. Chem. Soc.* **83**, 2502 (1961).
42. W. V. Cicha and F. Aubke, *J. Am. Chem. Soc.* **111**(12), 4328–4331 (1989).
43. R. C. Paul, K. S. Dhindsa, *Proc. Indian Natl. Sci. Acad. Part A*, **47**(3), 357–372 (1981).
44. U.S. Pat. 1,847,513 (Mar. 1, 1932), W. Hentrich, M. Hardtmann, and H. Ossenbeck (to General Aniline Works).
45. W. Lange and E. Müller, *Chem. Ber.* **B63**, 2653 (1930).
46. J. P. Coleman and D. Pletcher, *Tetrahedron Lett.* (2), 147 (1974).
47. D. Pletcher and C. Smith, *Chem. Ind. (London)* **8**, 371 (1976).
48. W. Traube, *Chem. Ber.* **46**, 2525 (1913).
49. U.S. Pat. 2,312,413 (Mar. 2, 1943), R. K. Iler (to E. I. du Pont de Nemours & Co., Inc.).
50. U.S. Pats. 2,430,963 (Nov. 18, 1947), R. Stephenson and W. Watson (to Allied Chem. Corp.); U.S. Pat. 3,957,959 (May 18, 1976), R. Wheatley, D. Treadway, and R. Toennies (to E. I. du Pont de Nemours & CO., Inc.).
51. Jpn. Pat. 55126509 (Sept. 30, 1980), (to Akita Chem Co.).
52. V. I. D. Daritskii and co-workers, *Khim. Prom. (Moscow)*, (3) 183–184 (1991).
53. Br. Pat. 537,589 (June 27, 1941), (to Standard Oil Development Co.).
54. U.S. Pat. 2,313,103 (Mar. 9, 1943), C. L. Thomas (to Universal Oil Products Co.).
55. U.S. Pat. 3,778,489 (Dec. 11, 1973), P. T. Parker and I. Mayer (to Esso Research and Engineering Co.).
56. U.S. Pat. 3,922,319 (Nov. 25, 1975), J. W. Brockington (to Texaco, Inc.).
57. U.S. Pat. 3,928,487 (Dec. 23, 1975) D. A. McCauley (to Standard Oil Co.).
58. U.S. Pat. 4,008,178 (Feb. 15, 1977), J. W. Brockington (to Texaco, Inc.).
59. U.S. Pat. 2,428,279 (Sept. 3, 1947), V. N. Ipatieff and C. B. Linn (to Universal Oil Products Co.).
60. U.S. Pat. 2,421,946 (June 10, 1947), V. N. Ipatieff and C. B. Linn (to Universal Oil Products Co.).
61. U.S. Pat. 2,419,185 (Apr. 15, 1947), C. A. Braidwood and A. G. Hovey (to Reichold Chemicals).
62. U.S. Pat. 3,594,445 (July 20, 1971) P. T. Parker (to Esso Research and Engineering Co.).
63. U.S. Pat. 3,636,129 (Jan. 18, 1972), P. T. Parker and C. N. Kimerlin, Jr. (to Esso Research and Engineering Co.).
64. U.S. Pat. 3,678,120 (July 18, 1972), H. S. Bloch (to Universal Oil Products Co.).

65. U.S. Pat. 3,708,553 (Jan. 2, 1973), G. A. Olah (to Esso Research and Engineering Co.).
66. D. T. Roberts, Jr. and L. E. Calihan, *J. Macromol. Sci. Chem.* **7**, 1629 (1973).
67. U.S. Pat. 3,819,743 (June 25, 1974), D. A. McCauley (to Standard Oil Co.).
68. U.S. Pat. 3,925,495 (Dec. 9, 1975), P. G. Rodewald (to Mobil Oil Corp.).
69. U.S. Pat. 3,984,352 (Oct. 5, 1976), P. G. Rodewald (to Mobil Oil Corp.).
70. U.S. Pat. 5,081,318 (Mar. 4, 1991), J. F. Knifton (to Texaco).
71. G. A. Olah and co-workers, *J. Org. Chem.* **56**(21) 6148–6151 (1991).
72. Jpn. Pat. 01075442 A2 (Mar. 22, 1989), Y. Sama (to Agency of Industrial Science and Technology).
73. U.S. Pat. 4,544,774 (Oct. 1, 1985), R. Pick (to Du Pont).
74. U.S. Pat. 4,569,990 (Feb. 11, 1986), W. W. Kasper and co-workers (to Du Pont).
75. Can. Pat. 448,662 (May 25, 1948), A. C. Hopkins, Jr., R. M. Stephenson, and W. E. Watson (to Allied Chemical Corp.).
76. Brit. Pat. 755,692 (Aug. 22, 1956), A. J. Edwards (to National Smelting Co., Ltd.).
77. W. Traube, J. Horenz, and F. Wunderlich, *Chem. Ber.* **B52**, 1272 (1919).
78. W. Traube and E. Brehmer, *Chem. Ber.* **B52**, 1284 (1919).
79. U.S. Pat. 2,428,753 (Oct. 7, 1947), C. B. Linn (to Universal Oil Product Co.).
80. U.S. Pat. 2,434,040 (Jan. 6, 1948), B. F. Hartman (to Socony-Vacuum Oil Co.).
81. Brit. Pat. 824,427 (Dec. 2, 1959), H. R. Leech and W. H. Wilson (to Imperial Chemical Industries, Ltd.).
82. C. B. F. Young and K. R. Hesse, *Met. Finish.* **45**(2), 63, 84 (1947); **45**(3), 64 (1947).
83. Ger. Pat. DE3438433 A1 (May 15, 1985), K. Tajiri, H. Nomura (to Mitsubishi Heavy Ind. Ltd.).
84. A. Kaiser, *Glastech Ber.* **62**(4) 127–134 (1989).
85. R. C. Paul and co-workers, *J. Inorg. Nucl. Chem.* **34**, 2535 (1972).
86. D. D. Des Marteau and M. Eisenberg, *Inorg. Chem.* **11**, 2641 (1972).
87. F. B. Dudley and G. H. Cady, *J. Am. Chem. Soc.* **79**, 513 (1957).
88. *Ibid.* **85**, 3375 (1963).
89. F. Aubke and D. D. Des Marteau, *Fluorine Chem. Rev.* **8**, 74 (1977).
90. R. A. DeMarco and J. M. Shreeve, *Adv. Inorg. Chem. Radiochem.* **16**, 115 (1974).
91. J. E. Roberts and G. H. Cady, *J. Am. Chem. Soc.* **81**, 4166 (1959).
92. *Ibid.* **82**, 352 (1960).
93. R. Seeley and A. W. Jache, *J. Fluorine Chem.* **2**(3), 225 (1973).

FRANCIS E. EVANS
GANPAT MANI
AlliedSignal Inc.

TANTALUM

Tantalum Pentafluoride

Tantalum pentafluoride [*7783-71-3*], TaF_5, a white solid with a reported mp of 97°C and a bp of 229°C (1), is the only known binary fluoride. The vapor pressure of TaF_5 in kPa is given by the equation $\log P_{kPa} = 7.649 - 2834/T$ over the temperature range of 80–230°C and the heat of vaporization is 54.4 kJ/mol (13 kcal/mol) (1).

There are a number of methods of preparation for TaF_5. For example, tantalum pentafluoride has been produced by the reaction of F_2 or ClF_3 and Ta metal (2,3), by contacting Ta_2O_5 with excess HF in the presence of a dehydrating agent (4), by the reaction of Ta-containing ores and $HF-H_2SO_4$ followed by extraction with an organic solvent (5,6), by reaction of Ta_2O_5 and COF_2 (7), by heating ammonium hexafluorotantalate (8), by contacting fluorotantalic acid with a dehydrating agent containing C–Cl or C–Br bonds (9) and by halogen exchange of $TaCl_5$ with HF (10).

TaF_5 has been characterized by ir, Raman, x-ray diffraction, and mass spectrometry (3,11,12). TaF_5 has been used as a superacid catalyst for the conversion of CH_4 to gasoline-range hydrocarbons (qv) (12); in the manufacture of fluoride glass and fluoride glass optical fiber preforms (13), and incorporated in semiconductor devices (14). TaF_5 is also a catalyst for the liquid-phase addition of HF to polychlorinated ethenes (15). The chemistry of TaF_5 has been reviewed (1,16–19). Total commercial production for TaF_5 is thought to be no more than a few hundred kilograms annually.

BIBLIOGRAPHY

"Tantalum" under "Fluorine Compounds, Inorganic," in *ECT* 2nd ed., Vol. 9, pp. 681, W. E. White, Ozark-Mahoning Co.; in *ECT* 3rd ed., Vol. 10, p. 818, by A. J. Woytek, Air Products & Chemicals, Inc.

1. J. H. Canterford and R. Cotton, *Halides of the Second and Third Row Transition Metals*, John Wiley & Sons, Inc., New York, 1968.
2. J. K. Gibson, *J. Fluorine Chem.* **55**(3), 299–311 (1991).
3. B. Frlec, *Vestu. Slov. Kem. Drus* **16**(1–4), 47–50 (1969).
4. U.S. Pat. 5,091,168 A (Feb. 25, 1992), M. J. Nappa and J. Mario (to E. I. du Pont de Nemours and Co., Inc.).
5. Jpn. Kokai Tokkyo Koho 63236716 A2 (Oct. 3, 1988), M. Watanabe, M. Nanjo, and Y. Nishimura (to Solex Research Corp of Japan).
6. Jpn. Pat. 63147827 A2 (June 20, 1988), M. Watanabe, M. Nanjo, and Y. Nishimura (to Solex Research Corp. of Japan).
7. S. P. Mallela, O. D. Gupta, and J. M. Shreeve, *Inorg. Chem.* **27**(1), 208–209 (1988).
8. Eur. Pat. 85-301897 (Mar. 1985), M. Watanabe and S. Nishimura.
9. U.S. Pat. 77,864,687 (Dec. 27, 1977), C. J. Kim and D. Farcasiu.
10. S. Ruff, *Z. Anorg. Allgem. Chem.* **72**, 329 (1911).
11. A. I. Popov, V. F. Sukhoverkhov, and N. A. Chumae-Vskii, *Zh. Neorg. Khim.* **35**(5), 1111–1122 (1990).
12. I. R. Beattie, K. M. S. Livingston, G. A. Ozin, and D. J. Reynolds, *J. Chem. Soc. A.*, (6), 958–965 (1969).

13. U.S. Pat. 4,973,776 A (Nov. 27, 1990), V. M. Allenger and R. N. Pandey.
14. Eur. Pat. 331,483 A2, (Sept. 6, 1989), K. Fujiura, Y. Ohishi, M. Fujiki, T. Kanamori, and S. Takahashi.
15. A. E. Feiring, *J. Fluorine Chem.* **14**, 7 (1979).
16. Eur. Pat. 89-104364 (Mar. 11, 1989), I. Haroda and co-workers.
17. F. Fairbrother, in V. Gutmann, ed., *Halogen Chemistry*, Vol. 3, Academic Press, Inc., New York, 1966, p. 123.
18. F. Fairbrother, *The Chemistry of Niobium and Tantalum*, Elsevier Scientific Publishing Co., London, 1967.
19. D. Brown, in J. C. Barter, ed., *Comprehensive Inorganic Chemistry*, Vol. 3, Pergamon Press, Elmsford, N.Y., 1973, p. 565.

TARIQ MAHMOOD
CHARLES B. LINDAHL
Elf Atochem North America, Inc.

TIN

The main binary tin fluorides are stannous fluoride and stannic fluoride. Because the stannous ion, Sn^{2+}, is readily oxidized to the stannic ion, Sn^{4+}, most reported tin and fluorine complexes are of tin(IV) and fluorostannates. Stannous fluoroborates have also been reported.

Stannous Fluoride

Stannous fluoride [7783-47-3], SnF_2, is a white crystalline salt that has mp 215°C (1), bp 850°C, and is readily soluble in water and hydrogen fluoride. At 20°C stannous fluoride dissolves in water to a concentration of 30–39%; in anhydrous hydrogen fluoride to 72–82% (2–4).

The pH of a freshly prepared 0.4% solution of stannous fluoride is between 2.8 and 3.5. Initially clear aqueous solutions become cloudy on standing owing to hydrolysis and oxidation. The insoluble residue is a mixture containing stannous and stannic species, fluoride, oxide, oxyfluorides, and hydrates.

Stannous fluoride probably was first prepared by Scheele in 1771 and was described by Gay-Lussac and Thenard in 1809. Commercial production of stannous fluoride is by the reaction of stannous oxide and aqueous hydrofluoric acid, or metallic tin and anhydrous hydrogen fluoride (5,6). SnF_2 is also produced by the reaction of tin metal, HF, and a halogen in the presence of a nitrile (7).

Stannous fluoride is used widely in dentifrices (qv) and other dental preparations because of its anticaries effect (8). The chemistry (9) involved in cavity prevention is thought to be reaction of stannous fluoride and the hydroxyapatite, $Ca_5(PO_4)_3OH$, of the tooth to form the more insoluble fluoroapatite, $Ca_5(PO_4)_3F$. More concentrated solutions of stannous fluoride react with hydroxyapatite to produce $Sn_3F_3PO_4$ [12592-27-7] (10). The role of SnF_2 in reducing acidogenicity of dental plaque *in vivo* has also been studied (11). On heating stannous fluoride under nitrogen with stannic fluoride, Sn_7F_{16}, Sn_3F_8, Sn_2F_6, and $Sn_{10}F_{34}$ are formed (12).

Other uses of SnF_2 are in the synthesis of fluorophosphate glasses having low melting temperatures (13–15), in formation of transparent film (16), and in the preparation of optically active alcohols (17).

Fluorostannites and Fluorostannates

Complexes of the type $MSnF_3$, where M is NH_4 [15660-29-4], Na [13782-22-4], K [13782-23-5], and Cs [13782-25-7], have been crystallized from aqueous solutions (18–20). Solutions of these salts deposit tin(II) oxide crystals indicating hydrolysis but not oxidation. From molten mixtures of SnF_2 and NaF, RbF, and CsF, both the $MSnF_3$ (M = Na [13782-22-4], K [13782-23-5], Rb [13782-24-6], and Cs [13782-25-7]) and the fluorostannate salts, $MSnF_5$ (M = Na [58179-42-3], K [58179-40-1], Rb [72264-75-6], and Cs [72264-76-7]) have been obtained (21). Complexes of the type $Cd(H_2O)_6 \cdot (SnF_3)_2$ [125445-76-3], $Zn(H_2O)_6 \cdot SnF_3$ [125445-75-2] (22), $SnCl_3 \cdot SnF_3$ [108632-61-7], $N_2H_6 \cdot (SnF_3)_2$ [99625-93-1] (23), $N_2H_5 \cdot SnF_3$ [73953-53-4] (24), $Ca(SnF_3)_2$ [69244-56-0] (25), $Ni(SnF_3)_2$ [26442-44-4] (26), $Co(SnF_3)_2$ [26442-43-3] (26) have also been reported.

Stannic Fluoride

Stannic fluoride [7783-62-2], SnF_4, is a white solid that sublimes at 705°C and hydrolyzes in water to form insoluble stannic acid. It can be prepared by reaction of fluorine and probably ClF_3 or BrF_3 with virtually any tin(II) or tin(IV) compound, eg, Sn, SnO, SnO_2, SnS, SnS_2 (27), and $SnCl_2$ (28). Reaction of $SnCl_4$ and HF (29) forms $SnCl_4 \cdot SnF_4$ which can be decomposed by heating to 750°C where pure SnF_4 sublimes. Stannic fluoride forms numerous complexes as a Lewis acid. The other methods of preparation for stannic fluoride include the oxidation of SnF_2 by a halogen in acetonitrile (30,31); the reaction of NF_3O and Sn (32); and the reaction of COF_2 and SnO_2 (33). Stannic fluoride is used in the manufacture of glass (qv) (34).

Stannous Fluoroborate

Stannous fluoroborate [13814-97-6], $Sn(BF_4)_2$, is prepared in electrochemical cells using tin and fluoroboric acid (35,36), by reaction of 80% HF and H_3BO_3 followed by reaction with $Sn(OH)_2$ (37); and from the reaction of mossy tin and 30–70% HBF_4 (38). The main use of stannous fluoroborate is in electroplating (qv) (39).

Hexafluorostannates

The hexafluorostannate anion [21340-04-5], SnF_6^{2-}, forms readily and is stable over a wide pH range. Numerous hexafluorostannates have been prepared by dissolving stannates in excess hydrofluoric acid, dissolving stannic acid in excess HF and neutralizing, or by reaction of salts and SnF_4. Many of these stable and generally water-soluble hexafluorostannates were prepared as early as 1857. Spectral studies of the SnF_6^{2-} anion have been reported (40). Some of the newer hexafluorostannates are $K \cdot NaSnF_6$ [*112813-21-5*] (41), $CsNa \cdot SnF_6$ [*112813-23-7*], $Rb \cdot NaSnF_6$ [*112813-22-6*] (42), and $N_2H_6 \cdot SnF_6$ [*128493-43-6*] (43).

Safety, Handling, and Toxicity

Stannous fluoride is used in dentifrices and dental preparations. The OSHA permissible exposure limit (44) and ACGIH (45) established TLV for fluoride is 2.5 mg/m^3 of air.

BIBLIOGRAPHY

"Tin Fluoride" under "Tin Compounds," in *ECT* 1st ed., Vol. 4, p. 160, by H. Richter, Metal & Thermit Corp.; "Tin" under "Fluorine Compounds, Inorganic" in *ECT* 2nd ed., Vol. 9, pp. 682–683, by W. E. White, Ozark-Mahoning Co.; in *ECT* 3rd ed., Vol. 10, pp. 819–820, by C. B. Lindahl and D. T. Meshri, Ozark-Mahoning Co.

1. J. J. Dudash and A. W. Searcy, *High Temp. Sci.* **1**, 287 (1969).
2. W. H. Nebergall, J. C. Muhler, and H. G. Day, *J. Am. Chem. Soc.* **74**, 1604 (1952).
3. J. B. Bearl, Jr., Ph.D. dissertation, Texas A&M University, College Station, 1963.
4. J. E. Gilliland, M.S. thesis, Oklahoma State University, Stillwater, 1960.
5. U.S. Pats. 2,924,508 (Feb. 9, 1960); 2,955,914 (Oct. 11, 1960); 3,097,063 (July 9, 1963), J. E. Gilliland, R. Ray, and W. E. White (to Ozark-Mahoning Co.).
6. I. V. Murin, S. V. Chernov, and M. Yu. Vlasov, *Zh. Prikl. Khim. (Leningrad)* **58**(10), 2340–2342 (1985).
7. U.S. Pat. 4,034,070 (July 21, 1975) J. A. Wojtowicz and D. F. Gavin.
8. J. C. Muhler and co-workers, *J. Am. Dent. Assoc.* **50**, 163 (1955).
9. E. J. Duff, *Caries Res.* **7**, 79 (1973).
10. S. H. Y. Wei, *J. Dent. Res.* **53**, 57 (1974).
11. J. E. Ellingsen, B. Svatun, and G. Roella; *Acta Odontol. Scand.* **38**(4), 219–222 (1980).
12. R. Sabtier, A. M. Hebrard, J. D. Cousseinsv, and C. R. Hebd, *Seances Acad. Sci. Ser. C* **279**(26), 1121–1123 (1974).
13. J. Leissner, K. Sebastian, H. Roggendorf, and H. Schmidt, *Mater. Sci. Forum* **67–68**, 137–142 (1991).
14. C. M. Shaw and J. E. Shelby, *Phys. Chem. Glasses* **29**(2), 49–53 (1988).
15. N. Sakamoto and K. Morinaga; *Sogo Rikogaku Hokoku (Kyushu Daigaku Daigakuin)* **12**(3), 283–289 (1990).
16. Jpn. Pat. 633147B A2 (Dec. 22, 1988) N. Sonoda and N. Sato (to Showa).
17. Jpn. Pat. 62158222 A2 (July 14, 1987) M. Mukoyama, N. Minowa, T. Oriyama, and K. Narasaka (to Showa).
18. E. L. Muetterties, *Inorg. Chem.* **I**, 342 (1962).
19. W. B. Schaap, J. A. Davis, and W. N. Nebergall, *J. Am. Chem. Soc.* **76**, 5226 (1954).
20. J. D. Donaldson and J. D. O'Donoghue, *J. Chem. Soc.*, 271 (1964).
21. J. D. Donaldson, J. D. O'Donoghue, and R. Oteng, *J. Chem. Soc.*, 3876 (1965).
22. Yu. V. Kokunov, and co-workers, *Dokl. Akad. Nauk. SSSR* **307**(5), 1126–1130 (1989).
23. V. Kancic and co-workers, *Acta. Crystallogr., Sect. C: Crystal Struct. Commun.* **C44**(8), 1329–1331 (1988).

24. W. Granier and M. Lopez, *Calorim. Anal. Therm.* **16**, 358–371 (1985).
25. F. Babcock, K. David, C. Jeana, and T. H. Jordan, *J. Dent. Res.* **57**(9–10), 933–938 (1978).
26. J. D. Donaldson and R. Oteng, *J. Chem. Soc. A*, (18), 2696–2699 (1969).
27. H. M. Haendler and co-workers, *J. Am. Chem. Soc.* **76**, 2179 (1954).
28. A. A. Woolf and H. J. Emeleus, *J. Chem. Soc.*, 2864 (1949).
29. H. J. Emeleus, in J. H. Simons, ed., *Fluorine Chemistry*, Vol. 1, Academic Press, New York, 1950, pp. 1–76.
30. D. Tudela and F. Rey, *Z. Anorg. Allg. Chem.* **575**, 202–208 (1989).
31. O. D. Gupta, R. L. Kirchmeier, and J. M. Shreeve, *Inorg. Chem.* **29**(3), 573–574 (1990).
32. S. P. Mallela, O. D. Gupta, and J. M. Shreeve, *Inorg. Chem.* **27**(1), 208–209 (1988).
33. EP 156617 A (Oct. 22, 1985) M. Watanabe and S. Nishimura.
34. G. D. Lukiyanchuk, V. K. Gonsharuk, E. V. Merkulov, and T. I. Usol'tseva, *Fiz. Khim. Stekla*, **18**(2), 141–145 (1992).
35. DD 293,609 A5 (Sept. 5, 1991), D. Ohms and co-workers.
36. C. J. Chen and C. C. Wan, *Electrochim. Acta* **30**(10), 1307–1312 (1985).
37. RO 67814 (Jan. 15, 1980), V. Grigore and N. Cretu.
38. U.S. Pat. 3,432,256 (Mar. 11, 1969), H. P. Wilson.
39. EP 45,471 A1 (Feb. 10, 1982), H. Willenberg, W. Becher, and K. H. Hellberg.
40. H. Kreigsmann and G. Kessler, *Z. Anorg. Allegm. Chem.* **318**, 266–276 (1962).
41. A. V. Gerasimenko, S. B. Antokhina, and V. I. Sergienko, *Koor. Khim.* **18**(2), 129–132 (1992).
42. V. I. Sergienko, V. Ya. Kavun, and L. N. Ignat'eva, *Zh. Neorg. Khim.* **36**(5), 1265–1268 (1991).
43. A. Rahten, D. Gantar, and I. Leban, *J. Fluorine Chem.* **46**(3), 521–528 (1990).
44. *Code of Federal Regulations*, Title 29, Part 1910.1000, Washington, D.C.
45. *Threshold Limit Values for Chemical Substances and Physical Agents, 1992–1993*, The American Conference of Governmental Industrial Hygienists, Cincinnati, Ohio.

CHARLES B. LINDAHL
TARIQ MAHMOOD
Elf Atochem North America, Inc.

TITANIUM

Titanium(III) Fluoride

Titanium trifluoride [*13470-08-1*], TiF_3, is a blue crystalline solid that undergoes oxidation to TiO_2 upon heating in air at 100°C (see TITANIUM COMPOUNDS). In the absence of air, disproportionation occurs above 950°C to give TiF_4 and titanium metal. TiF_3 decomposes at 1200°C, has a density of 2.98 g/cm^3, and is insoluble in water but soluble in acids and alkalies. The magnetic moment is 16.2 × 10^{-24} J/T (1.75 μB).

Titanium trifluoride is prepared by dissolving titanium metal in hydrofluoric acid (1,2) or by passing anhydrous hydrogen fluoride over titanium trihydrate at 700°C or over heated titanium powder (3). Reaction of titanium trichloride and anhydrous hydrogen fluoride at room temperature yields a crude product that can be purified by sublimation under high vacuum at 930–950°C.

Titanium trifluoride can be stored in tightly closed polyethylene containers for several years. Shipping regulations classify the material as a corrosive solid and it should be handled in a fully ventilated area or in a chemical hood. The ACGIH adopted toxicity values (1992–1993) for TiF_3 is as TWA for fluorides as F^- 2.5 mg/m³.

This material is available from Advance Research Chemicals, Inc., Aldrich Chemical Company, Inc., Aesar, Johnson/Matthey, Cerac, PCR, and Pfaltz & Bauer in the United States, Fluorochem of the United Kingdom, and Schuchardt of Germany. Its 1993 price was approximately $500/kg. No commercial applications have been reported.

Titanium(IV) Fluoride

Titanium tetrafluoride [7783-63-3], TiF_4, has potential for use in dental hygiene products. It is used in infrared transmitting halide glass.

TiF_4 is a colorless, very hygroscopic solid and is classified as a soft fluorinating reagent (4), fluorinating chlorosilanes to fluorosilanes at 100°C. It also forms adducts, some of them quite stable, with ammonia, pyridine, and ethanol. TiF_4 sublimes at 285.5°C, and melts at temperatures >400°C. It is soluble in water, alcohol, and pyridine, hydrolyzing in the former, and has a density of 2.79 g/mL.

Titanium tetrafluoride may be prepared by the action of elemental fluorine on titanium metal at 250°C (5) or on TiO_2 at 350°C. The most economical and convenient method is the action of liquid anhydrous HF on commercially available titanium tetrachloride in Teflon or Kynar containers. Polyethylene reacts with $TiCl_4$ and turns dark upon prolonged exposure. The excess of HF used is boiled off to remove residual chloride present in the intermediates.

Titanium(IV) fluoride dihydrate [60927-06-2], $TiF_4 \cdot 2H_2O$, crystals can be prepared by the action of aqueous HF on titanium metal. The solution is carefully evaporated to obtain the crystals. Neutral solutions when heated slowly hydrolyze and form titanium(IV) oxyfluoride [13537-16-1], $TiOF_2$ (6). Upon dissolution in hydrogen fluoride, TiF_4 forms hexafluorotitanic acid [17439-11-1], H_2TiF_6.

The most promising application of titanium tetrafluoride is for use in topical applications for prevention of dental caries (7–13). It is being evaluated and compared to NaF, MFP, and SnF_2 used in these applications. The other use is in mixed optical halide glass (14–16), and in the preparation of fluorotitanates (17–19).

Total consumption of TiF_4 in both the United States and Europe is less than 500 kg/yr. TiF_4 is available from Advance Research Chemicals, Inc., Aldrich, Aesar, Johnson/Matthey, Cerac, PCR, and Pfaltz & Bauer of the United States, Fluorochem of the United Kingdom, and Schuchardt of Germany. Its 1993 price varied between $300 to $400/kg.

Fluorotitanates

Hexafluoroanions of Group 4 (IVB) are octahedral crystals that are quite stable in acidic media. Solutions having pH > 4 tend to hydrolyze forming the metal dioxides. All three hexafluoroacids are known, ie, hexafluorotitanic acid, hexa-fluorozirconic acid [12021-95-3], H_2ZrF_6, and hexafluorohafnic acid [12021-47-5], H_2HfF_6. These acids exist only in aqueous media in the presence of excess hydro-fluoric acid. Alkali, alkaline-earth, and other metal salts of these acids, M_2XF_6, where X = Ti and M = Li [19193-50-1], Na [17116-13-1], K [16919-27-0], Rb [16962-41-7], Cs [16919-28-1], NH_4 [16962-40-6], and Tl [26460-00-4], have been isolated as stable solids at ambient temperatures (20). Maximum concentration of hexafluorotitanic acid is found to be 63% in the presence of 0.5% excess HF. Its salts, NH_4^+, Li^+, and Na^+ are quite soluble in water, whereas those of K^+, Rb^+, and Cs^+ are only slightly soluble.

Fluorotitanic acid is used as a metal surface cleaning agent, as a catalyst, and as an aluminum finishing solvent (see METAL SURFACE TREATMENTS). Fluoro-titanates are used in abrasive grinding wheels and for incorporating titanium into aluminum alloys (see ABRASIVES; ALUMINUM AND ALUMINUM ALLOYS).

Although titanium compounds are considered to be physiologically inert (21), fluorides in general are considered as toxic above 3 ppm level and extreme care should be taken in handling large amounts of titanium salts as well as hexafluoro-titanic acid. The ACGIH adopted (1992–1993) toxicity limits are as TWA for fluorides as F^- 2.5 mg/m^3.

The total U.S. consumption of H_2TiF_6 is 20 t/yr. The 1993 price varied be-tween \$2.80 to \$7.50/kg depending on quantity and specifications. It is packaged in DOT approved polyethylene-lined drums and the salts in polyethylene-lined fiber board drums.

BIBLIOGRAPHY

"Titanium Fluorides" under "Titanium Compounds," in *ECT* 1st ed., Vol. 14, p. 217, by L. R. Blair, H. H. Beecham, and W. K. Nelson; "Titanium" under "Fluorine Compounds, In-organic," in *ECT* 2nd ed., Vol. 9, pp. 683–684, by W. E. White; in *ECT* 3rd ed., Vol. 10, pp. 821–822, by D. T. Meshri, Ozark-Mahoning Co.

1. M. E. Straumenis and J. I. Ballas, *Z. Anorg. Chem.* **278**, 33 (1955).
2. P. H. Woods and L. D. Cockrell, *J. Am. Chem. Soc.* **80**, 1534 (1958).
3. P. Ehrlich and G. Pietzka, *Z. Anorg. Chem.* **275**, 121 (1954).
4. D. T. Meshri and W. E. White, "Fluorinating Reagents in Inorganic and Organic Chem-istry" in the *Proceedings of George H. Cady Symposium*, Milwaukee, Wis., June 1970.
5. H. M. Haendler, *J. Am. Chem. Soc.* **76**, 2177 (1954).
6. K. S. Vorres and F. B. Dutton, *J. Am. Chem. Soc.* **77**, 2019 (1955).
7. B. Regolati and co-workers, *Helv. Odontol. Acta* **18**(2), 92 (1974).
8. A. S. Mundorff, M. F. Little, and B. G. Bibby, *J. Dent. Res.* **51**, 1567 (1972).
9. A. J. Reed and B. G. Bibby, *J. Dent. Res.* **55**, 357 (1976).
10. L. Skartveit, K. A. Selvig, S. Myklebust, and A. B. Tveit, *Acta Odontol. Scand.* **48**(3), 169–174 (1990).
11. L. Skartveit, A. B. Tveit, B. Klinge, B. Toetdal, and K. A. Selvig, *Acta Odontol. Scand.* **47**(2), 65–68 (1989).
12. L. Skartveit, A. B. Tveit, B. Toetdal, and K. A. Selvig, *Acta Odontol. Scand.* **47**(1), 25–30 (1989).
13. A. B. Tveit, K. Bjorn, B. Toetdal, and K. A. Selvig, *Scand. J. Dent. Res.* **96**(6), 536–540 (1988).

14. A. Jha and J. M. Parker, *Phys. Chem. Glasses.* **32**(1), 1–2 (1991).
15. B. Boulard and C. Jacoboni, *Mater. Res. Bull.* **25**(5), 671–677 (1990).
16. Eur. Pat. EP 331,483 (Sept. 6, 1989), K. Fujiura and co-workers (to Nippon Telegraph & Telephone Corp.).
17. Pol. Pat. PL 153,066 (May 11, 1988), L. Stoch, S. Mocydlarz, M. Laczka, and I. Waclawska (to Akademia Gorniczo Hutnicza).
18. Pol. Pat. PL 153,702 (May 31, 1991), I. Kustra, A. Chajduga, J. Konczal, and M. Jarzynowski (to Instytut Chemï Nieorganicznej).
19. B. N. Chernyshov and co-workers, *Zh. Neorg. Khim.* **34**(9), 2179–2186 (1989).
20. B. Cox and A. G. Sharpe, *J. Chem. Soc.*, 1783 (1953).
21. N. I. Sax, *Dangerous Properties of Industrial Materials*, 6th ed., Van Nostrand Reinhold Co., New York, 1984, p. 2585.

DAYAL T. MESHRI
Advance Research Chemicals, Inc.

TUNGSTEN

Tungsten has three readily prepared binary fluorides, tungsten hexafluoride [7783-82-6], tungsten pentafluoride [19357-83-6], and tungsten tetrafluoride [13766-47-7] (1,2). The three lower oxidation state tungsten binary fluorides have been observed only in high energy systems (3). Several complex oxyfluorides are known including WOF_4 [13520-79-1] and WO_2F_2 [14118-73-1] (4). Only tungsten hexafluoride is made commercially. Tungsten hexafluoride is used as a tungsten source in chemical vapor deposition (CVD) for very large-scale integration (VSLI) devices.

Tungsten Hexafluoride

Physical Properties. Tungsten(VI) fluoride [7783-82-6], WF_6, is a colorless gas that condenses at ca 100 kPa (1 atm) and 17.1°C to a water-white liquid that may be colored owing to metallic impurities. Below 2°C it forms a white solid. Tungsten hexafluoride has a symmetrical octahedral structure at near room temperature and a phase of lower symmetry below −8.5°C (5). The Raman and uv spectra (6), as well as the ir spectrum (7) have been studied. The physical properties of tungsten hexafluoride are given in Table 1.

Chemical Properties. Tungsten hexafluoride is readily hydrolyzed by water to give tungsten trioxide and hydrogen fluoride. It is a strong fluorinating agent and reacts with many metals at room temperature. Tungsten hexafluoride reacts with the alkali fluorides KF, RbF, and CsF to form the complex salts K_2WF_8 [57300-87-5], Rb_2WF_8 [57300-88-6], and Cs_2WF_8 [57300-89-7], respectively (10). The alkali iodides and WF_6 react in sulfur dioxide to form the tungsten(V) compounds $NaWF_6$ [55822-76-9], KWF_6 [34629-85-1], $RbWF_6$ [53639-97-7], and $CsWF_6$ [19175-38-3] (11). Tungsten hexafluoride reacts with hydrogen and hy-

Table 1. Physical Properties of Tungsten Hexafluoride

Property	Value	References
boiling point, °C	17.2	8
triple point, °C, 55.1 kPa[a]	2.0	9
liquid density at 15°C, g/mL	3.441	8
transition point, °C, 32.0 kPa[a]	−8.2	9
heat of vaporization, kJ/mol[b]	26.5	9
heat of fusion, kJ/mol[b]	1.76	9
heat of transition, kJ/mol[b]	5.86	9
heat of sublimation, kJ/mol[b]		
above transition	32.4	9
below transition	38.3	9
entropy of vaporization, J/(mol·K)[b]	91.2	9
entropy of fusion, J/(mol·K)[b]	6.07	9
entropy of transition, J/(mol·K)[b]	22.1	9
specific heat at 25°C, J/(mol·K)[b]	118.92	9
vapor pressure equation	$\log P_{kPa} = a - b/T$	
liquid		9
T, °C	2.0 to 17.1	
a	6.760[c]	
b	1380.5	
solid		9
T, °C	−8.2 to 2.0	
a	7.883[d]	
b	1689.9	
T, °C	−60 to −8.2	
a	9.076[e]	
b	2006.0	

[a]To convert kPa to mm Hg, multiply by 7.5.
[b]To convert kJ to kcal, divide by 4.184.
[c]For P in mm Hg, $a = 7.635$.
[d]For P in mm Hg, $a = 8.758$.
[e]For P in mm Hg, $a = 9.951$.

drogen-containing reducing agents at elevated temperature to form tungsten metal and hydrogen fluoride. This reaction is the basis of the primary use of tungsten hexafluoride in CVD (see THIN FILMS). The CVD chemistry of WF_6 has been reviewed (12,13). Reduction with hydrogen generally requires temperatures of 450–750°C and pressures of < 100 kPa (14.5 psi) (14). Other gaseous reductants include GeH_4 (15), SiH_2F_2 (16), SiH_4 (17), and diethyl silane (18).

Manufacture and Economics. Tungsten hexafluoride is produced commercially by the reaction of tungsten powder and gaseous fluorine at a temperature in excess of 350°C (19). Tungsten hexafluoride is the principal product of the reaction, and there are no by-products when high purity tungsten powder and fluorine are used. U.S. production is several metric tons per year. Essentially all of the product is used in CVD. Air Products and Chemicals, Inc. (Allentown, Pennsylvania) and Bandgap Technology Corp. (Broomfield, Colorado) are the only U.S. producers. The 1992 price ranges from \$300–\$850/kg, depending on the purity.

Because of the development of electronic applications for WF_6, higher purities of WF_6 have been required, and considerable work has been done to improve the existing manufacturing and purification processes (20). Most metal contaminants and gaseous impurities are removed from WF_6 by distillation. HF, which has a similar vapor pressure to WF_6, must be removed by adsorption (see ELECTRONIC MATERIALS; ULTRAPURE MATERIALS).

Specifications. The use of tungsten hexafluoride in CVD applications in the manufacture of high density silicon chips requires a high purity product, essentially free of all metallic contaminants. Several grades of WF_6 are available. Table 2 shows the specifications for three grades of WF_6.

Tungsten hexafluoride is shipped as a liquid under its own vapor pressure in nickel or steel cylinders in quantities of 45 kilograms per cylinder or less; however, it has been shown that the purity of WF_6 packaged in steel cylinders can degrade over time (21). It is classified as a corrosive liquid by the DOT.

Handling and Toxicity. Tungsten hexafluoride is irritating and corrosive to the upper and lower airways, eyes, and skin. It is extremely corrosive to the skin, producing burns typical of hydrofluoric acid. The OSHA permissible exposure limits is set as a time-weighted average of 2.5 mg/kg or 0.2 ppm (22).

Monel and nickel are the preferred materials of construction for cylinders and delivery systems; however, copper, brass, steel, and stainless steel can be used at room temperature, providing that these metals are cleaned, dried, and passivated with a fluoride film prior to use. Studies have shown that fluorine passivation of stainless steel and subsequent formation of an iron fluoride layer prior to WF_6 exposure prevents reaction between the WF_6 and the stainless steel surface (23).

Uses. The primary use of WF_6 is for blanket and selective deposition of tungsten and tungsten silicide films in the manufacture of VLSI electronic devices. The important aspects of this application have been reviewed (24). Addi-

Table 2. Specifications on Three Grades[a] of WF_6

Impurity	Electronic	VLSI	Megaclass
HF, ppmv	150	10	1
CO_2, CF_4, SF_6, SiF_4, ppmv each	10	0.5	0.5
N_2, ppmv	15	1	0.5
O_2 + Ar, ppmv	10	0.5	0.5
CO, ppmv	[b]	1	1
total metals, ppb[c]	1000	1000	1000
Cr, Fe, K, Na, ppb[c] each	10	10	10
U, ppb[c]	0.1	0.05	0.05
Th, ppb[c]	0.1	0.1	0.1

[a]Commercial grades offered by Air Products and Chemicals, Inc.
[b]No specifications given.
[c]By weight.

tionally, several conferences have been devoted to CVD using WF_6 (13,20). Non-electronic applications of tungsten hexafluoride include the CVD of tungsten to form hard tungsten carbide coatings on steel (25) and to fabricate solid tungsten pieces such as tubing or crucibles. Composite coatings of tungsten and rhenium are produced by the simultaneous chemical vapor deposition from these hexafluorides (26) and the addition of rhenium improves the ductility and high temperature properties of the deposit.

BIBLIOGRAPHY

"Tungsten" under "Fluorine Compounds, Inorganic" in *ECT* 3rd ed., Vol. 10, pp. 823–825, by A. J. Woytek, Air Products and Chemicals, Inc.

1. N. N. Greenwood and A. Earnshaw, *Chemistry of the Elements*, Pergamon Press, Elmsford, N.Y., 1984, pp. 1187–1191.
2. J. Schröeder and F. J. Grewe, *Angew. Chem., Int. Ed. Engl.* **7**, 132 (1968).
3. A. Bensaoula, E. Grossman, and A. Ignatiev, *J. Appl. Phys.* **62**, 4587 (1987).
4. J. C. Bailar and co-eds., *Comprehensive Inorganic Chemistry*, Vol. 3, Pergamon Press, Compendium Publishers, Elmsford, N.Y., 1973, pp. 749–763.
5. S. Siegal and D. A. Northrup, *Inorg. Chem.* **5**, 2187 (1966).
6. K. N. Tanner and A. B. F. Duncan, *J. Am. Chem. Soc.* **73**, 1164 (1951).
7. J. Gaunt, *Trans. Faraday Soc.* **49**, 1122 (1953).
8. J. A. Dean, ed., *Lange's Handbook of Chemistry*, 13th ed., McGraw-Hill Book Co., New York, 1985, pp. 4–125.
9. G. H. Cady and G. B. Hargreaves, *J. Chem. Soc.*, 1563 (1961).
10. B. Cox, D. W. Sharp, and A. G. Sharpe, *J. Chem. Soc.*, 1242 (1956).
11. G. B. Hargreaves and R. D. Peacock, *J. Chem. Soc.*, 4212 (1957).
12. M. L. Yu, K. Y. Ahn, and R. V. Joshi, *IBM J. Res. Dev.* **34**(6), 875 (1990).
13. V. V. S. Rana, R. V. Joshi, and I. Ohdomari, eds., *Advanced Metallization for ULSI Applications* and references therein, Materials Research Society, Pittsburgh, Pa., 1992.
14. U.S. Pat. 3,565,676 (Feb. 23, 1971), R. A. Holzl (to Fansteel Metallurgical Corp.).
15. C. A. Van der Jeugd, G. J. Leusink, G. C. A. M. Janssen, and S. Radelaar, *Appl. Phys. Lett.* **57**, 354 (1990).
16. Jpn. Pat. J63250463 (Oct. 18, 1988) T. Kusumoto and co-workers (to ULVAC Corp.).
17. H. L. Park and co-workers, *J. Electrochem Soc.* **137**, 3213 (1990).
18. D. A. Roberts and co-workers, in Ref. 13, p. 127.
19. H. F. Priest, *Inorg. Synth.* **3**, 181 (1950).
20. *Forum on Process Gases, TechWeek/East*, Semiconductor Equipment and Materials International, Cambridge, Mass., Sept. 29, 1992.
21. M. A. George and D. Garg, in *Proceedings of 1990 Microcontamination Conference*, Canon Communications, Santa Monica, Calif., 1990.
22. *Documentation of the Threshold Limit Values and Biological Exposure Indices*, 5th ed., American Conference of Governmental Industrial Hygienists, Inc., Cincinnati, Ohio, 1986.
23. D. A. Bohling and M. George, *Semicond. Int.* **14**, 104 (1991).
24. J. E. J. Schmitz, *Chemical Vapor Deposition of Tungsten and Tungsten Silicides for VLSI/ULSI Applications*, Noyes Publications, Park Ridge, N.J., 1992.
25. N. J. Archer, *Proceedings Conference Chemical Vapor Deposition, 5th International Conference*, Electrochemical Society, Princeton, N.J., 1975.

26. J. L. Federer and A. C. Schaffhauser, *9th Thermionic Conversion Specialist Conference,* Miami Beach, Fla., 1970, pp. 74–81.

General References

J. H. Simons, ed., *Fluorine Chemistry,* Vol. 5, Academic Press, Inc., New York, 1964.
J. W. Mellor, *A Comprehensive Treatise on Inorganic and Theoretical Chemistry,* Vol. 11, John Wiley & Sons, Inc., New York, 1962.

PHILIP B. HENDERSON
ANDREW J. WOYTEK
Air Products and Chemicals, Inc.

FLUORINE COMPOUNDS, INORGANIC (URANIUM).   See
FLUORINE; URANIUM AND URANIUM COMPOUNDS.

ZINC

Zinc Fluoride

Anhydrous zinc fluoride [7783-49-5], ZnF_2, melts at 872–910°C, has a solubility of only 0.024 g/100 g anhydrous HF at 14.2°C (1), and can be prepared by slowly drying zinc fluoride tetrahydrate [13986-18-0], $ZnF_2 \cdot 4H_2O$, in a current of anhydrous hydrogen fluoride to minimize hydrolysis and formation of the oxide. There is x-ray evidence for dihydrate formation during dehydration of the tetrahydrate (2). Anhydrous zinc fluoride can also be prepared from the reaction of Zn metal powder and pyridinium poly(hydrogen fluoride) at ambient temperature (3); by treating zinc hydroxycarbonate with NH_4F followed by thermal decomposition (4); by the reaction of NF_3O (5) or NH_4F and ZnO (6,7); by the thermal decomposition of $(NH_4)_2ZnF_4$ (8); by the reaction of SOF_2 and Zn (9); by the reaction of Zn and HF in the presence of acetonitrile (10); by the reaction of SF_6 and Zn (11); by the reaction of PF_3 and ZnO (12); and by the reaction of ZnO and hydrogen fluoride (13). Zinc fluoride of ca 96% purity is commercially produced for use as a flux in metallurgy (qv). Production is only on a small scale.

Zinc fluoride has been used as a mild fluorinating reagent in replacement of chlorine in halogenated hydrocarbons (14,15). It is also used as a catalyst in several applications including cyclization processes (15). High purity ZnF_2 is used in the synthesis of fluorophosphate glass (16,17), fluoride glass (18,19), high conducting oxyfluoride glass (20), as fluoride glass films (21), in the manufacture of fluoride glass optical fibers (22), and in the preparation of optical transmitting glass (23) (see GLASS; FIBER OPTICS).

The only reported toxicity data on zinc fluoride in the NIOSH RTECS file is a LD_{LO} of 280 mg/kg for subcutaneous administration in frogs. OSHA has a standard time-weighted average (TWA) of 2.5 mg/m^3 based on fluoride. NIOSH has issued a criteria document (24) on occupational exposure to inorganic fluorides.

Zinc Fluoride Tetrahydrate. Zinc fluoride tetrahydrate [13986-18-0] is prepared by reaction of ZnO and aqueous HF. $ZnF_2 \cdot 4H_2O$ has a water solubility of about 1.6 g/100 mL solution at 25°C. Addition of HF increases the solubility to 11.8 g/100 mL in a 29% HF solution. The tetrahydrate loses water at temperatures above 75°C.

Fluorozincates. Fluorozincates of the formula $MZnF_3$, where M = Na [18251-84-8], K [13827-02-6], Rb [29987-38-0], Cs [29507-53-7], NH$_4$ [14972-88-4], Ag [28667-89-2], N$_2$H$_5$ [63439-12-3], and Li [106207-44-7] (25–28); as well as M_2ZnF_4 where M = K [37732-22-2], Rb [35944-46-8], Cs [72161-48-9], and Li [155007-51-9]; Ba [13825-40-6], Sr [15154-47-9], and Ca [15246-41-0] (25–31), have been reported. Potassium fluorozincate [13827-02-6], $KZnF_3$, and sodium fluorozincate [18251-84-8], $NaZnF_3$, are used as catalysts in alginate dental impression materials (see DENTAL MATERIALS) (32).

BIBLIOGRAPHY

"Zinc Fluoride" under "Fluorine Compounds, Inorganic," in *ECT* 1st ed., Vol. 6, p. 738, by F. D. Loomis, Pennsylvania Salt Manufacturing Co.; "Zinc" under "Fluorine Compounds, Inorganic," in *ECT* 2nd ed., Vol. 9, pp. 684–685, by W. E. White, Ozark-Mahoning Co.; in *ECT* 3rd ed., Vol. 10, p. 826, by C. B. Lindahl, Elf Atochem North America, Inc.

1. A. W. Jache and G. H. Cady, *J. Phys. Chem.* **56**, 1106 (1952).
2. E. A. Secco and R. R. Martin, *Can. J. Chem.* **43**, 175 (1965).
3. K. R. Muddukrishna, R. N. Singh, and D. K. Padma, *J. Fluorine Chem.* **57**(1–3), 155–158 (1992).
4. USSR Pat. 1,590,433 (Sept. 7, 1990), R. Okhunov., N. N. Levina, and D. D. Ikrami.
5. O. D. Gupta, R. L. Kirchmeier, and J. M. Shreeve, *Inorg. Chem.* **29**(3), 573–574 (1990).
6. G. Pourroy, and P. Poix, *J. Fluorine Chem.* **42**(2), 257–263 (1989).
7. G. A. Lopatkina and co-workers, *Khim. Prom-st. (Moscow)*, (11), 846–847 (1978).
8. Eur. Pat. 156,617 A2 (Oct. 2, 1985), M. Watanabe and S. Nishimura.
9. D. K. Padma and co-workers, *J. Inorg. Nucl. Chem.* **43**(12), 3099–3101 (1981).
10. U.S. Pat. 597,546 (July 21, 1975), J. A. Wojtowicz and D. F. Gavin.
11. A. A. Opalovskii and co-workers, *Izv. Sib. Otd. Akad. Nauk SSSR, Ser. Khim. Nauk.*, (6), 83–86 (1974).
12. M. Chaigneau and M. Santarromana, *C. R. Acad. Sci., Ser. C* **278**(25), 1453–1455 (1974).
13. USSR Pat. 265,091 (Mar. 9, 1970), G. A. Lopatkina, T. N. Kolosova, and O. S. Suslova.
14. A. Sekiya and N. Ishikawa, *Bull. Chem. Soc. Jpn.* **51**, 1267 (1978).
15. U.S. Pat. 3,728,405 (Sept. 14, 1970), J. Allan (to E. I. du Pont de Nemours & Co., Inc.).
16. J. Leissner and co-workers, *Mater. Sci. Forum*, 67–68, 137–142 (1991).
17. M. Matecki and M. Poulain, *J. Non-Cryst. Solids* **56**(1–3) (1983).
18. Y. Wang, *J. Non-Cryst. Solids* **142**(1–2), 185–188 (1992).
19. K. Zhang, *J. Chem.*, (2), 136–140 (1990).
20. K. Hirao, A. Tsujimura, and N. Soga, *Zairyo* **39**(438), 283–286 (1990).
21. B. Boulard and C. Jacoboni, *Mater. Res. Bull.* **25**(5), 671–677 (1990).
22. Eur. Pat. 331,483 (Sept. 6, 1989), K. Fujiura and co-workers.
23. M. Poulain and Y. Messaddeq, *Mater. Sci. Forum*, 32–33, 131–136 (1988).

24. *Criteria for a Recommended Standard-Occupational Exposure to Inorganic Fluorides, PB 246 692, NIOSH 76-103*, U.S. Dept. of Health, Education, and Welfare, Washington, D.C., 1975.
25. O. Schmitz-Dumont and A. Bornefeld, *Z. Anorg. Allg. Chem.* **287**, 120 (1956).
26. J. Portier, A. Tressaud, and J. L. Dupin, *C. R. Acad. Sci., Ser. C* **270**(2), 216–218 (1970).
27. J. Slivnik and co-workers, *Vestn. Slov. Kem. Drus.* **26**(1), 19–26 (1979).
28. S. H. Pulcinelli and co-workers, *Rev. Chem. Miner* **23**(2), 238–249 (1986).
29. P. A. Rodnyi, M. A. Terekhin, and E. N. Melchakov, *J. Lumin.* **47**(6), 281–284 (1991).
30. H. G. Von Schnering, D. Vu, and K. Peters, *Z. Kristallogr.* **165**(1–4), 305–308 (1983).
31. Yu. Wan-Lun, and Z. Min-Guang, *J. Phys. Chem.* **17**(20), L525–L527 (1984).
32. U.S. Pat. 2,769,717 (Nov. 6, 1956), J. Cresson (to L. D. Caulk Co.).

CHARLES B. LINDAHL
TARIQ MAHMOOD
Elf Atochem North America, Inc.

ZIRCONIUM

Three binary zirconium fluorides ZrF_2, ZrF_3, and ZrF_4, are known to exist. The most important compounds industrially are zirconium tetrafluoride, ZrF_4, and fluorozirconic acid [*12021-95-3*], H_2ZrF_6, and its salts (see ZIRCONIUM AND ZIRCONIUM COMPOUNDS).

Zirconium Difluoride

Zirconium difluoride [*7783-49-5*], ZrF_2, has been isolated in rare gas–solid matrices (1). ZrF_2, prepared by Knudsen cell techniques, is not commercially available.

Zirconium Trifluoride

Zirconium trifluoride [*13814-22-7*], ZrF_3, was first prepared by the fluorination of ZrH_2 using a mixture of H_2 and anhydrous HF at 750°C (2). It can also be prepared by the electrolysis of Zr metal in KF–NaF melts (3). Zirconium trifluoride is stable at ambient temperatures but decomposes at 300°C. It is slightly soluble in hot water and readily soluble in inorganic acids. This compound is of academic interest rather than of any industrial importance.

Zirconium Tetrafluoride

Zirconium tetrafluoride [*7783-64-4*], ZrF_4, is one of the many important inorganic fluorides that have played a role in the development of heavy-metal fluoride glass (HMFG) technology (see GLASS). Table 1 summarizes some of the physical properties of zirconium tetrafluoride. Zirconium tetrafluoride monohydrate [*14956-*

Table 1. Properties of ZrF$_4$

Property	Value
mol wt	167.21
sublimation point, °C	912
specific gravity	4.54
solubility in water at 25°C	1.388
ΔH_f, kJ/mol[a]	−1911
ΔG_f, kJ/mol[a]	−1807
S, J/(mol·K)[a]	104.6
C_p, J/(mol·K)[a]	103.7

[a]To convert J to cal, divide by 4.184.

11-3], ZrF$_4$·H$_2$O, and the trihydrate [14517-16-9], ZrF$_4$·3H$_2$O, also exist. The hydrated forms can be prepared by dissolving zirconium hydroxy carbonate, commonly known as basic zirconium carbonate, in aqueous hydrogen fluoride and evaporating the solution to dryness. This produces the trihydrate ZrF$_4$·3H$_2$O. The monohydrate, ZrF$_4$·H$_2$O, is prepared by removing two moles of water from the trihydrate under dynamic vacuum at 70°C. The monohydrate can also be prepared by dissolving the anhydrous salt in aqueous hydrofluoric acid and evaporating the solution to dryness.

The anhydrous salt is prepared by several methods, eg, by reacting ZrCl$_4$ with liquid anhydrous HF. It is necessary to use an excess of HF which also acts as a wetting agent. The reaction is instantaneous and is carried out in a polyethylene jar or carboy. When the evolution of HCl ceases, the material is transferred to a tray and dried under an atmosphere of nitrogen. By proper selection of equipment, purification of raw material, and drying conditions, materials of spectrographic purity can be produced (4).

Other methods of preparation of anhydrous ZrF$_4$ include the decomposition of (NH$_4$)$_3$ZrF$_7$ [17250-81-6] at 297°C (5). NH$_4$F sublimes and leaves the flow reactor.

The principal application of ZrF$_4$ has been in the manufacture of HMFGs of which the most widely investigated is the system composed of Zr, Ba, La, Al, and Na, also popularly known as the ZBLAN glasses. This system has revolutionized the optics industry because of the significantly superior qualities of these glasses over conventional silica glasses. The theoretical transmission losses for fluoride glass fibers are calculated to be 0.001 dB/km at 3.2 μm and 0.005 dB/km at 3.5 μm (6) (see FIBER OPTICS). Transmission loss rates have been observed as low as 0.65 dB/km (4). Fluoride glasses, unlike the silica glasses, are expected to transmit light in the range of 2–4 micrometers or in the infrared range. ZBLAN fibers transmit light at a 5 to 10 times lower optical loss than that of silica fibers, and can be used for long distance data transmission, for use in mid-ir and multispectral optical components, ir domes, laser windows, laser hosts, for medical applications such as surgery and cauterization, and for nuclear radiation resistant transmitting devices (6–16).

Other applications of zirconium tetrafluoride are in molten salt reactor experiments; as a catalyst for the fluorination of chloroacetone to chlorofluoroacetone (17,18); as a catalyst for olefin polymerization (19); as a catalyst for the conversion of a mixture of formaldehyde, acetaldehyde, and ammonia (in the ratio of 1:1:3:3) to pyridine (20); as an inhibitor for the combustion of NH_4ClO_4 (21); in rechargeable electrochemical cells (22); and in dental applications (23) (see DENTAL MATERIALS).

High purity ZrF_4 is available in the United States from Advance Research Chemicals, Inc., Air Products and Chemicals, Inc., Johnson–Matthey/AESAR group, Aldrich Chemical, and EM Industries, Inc. Ultrahigh purity (99.999%) material is available only from Air Products and Chemicals, Inc. of Allentown, Pennsylvania. The price varies depending on the purity of the material from $25/kg (99%) through $250/kg (99.99%), to $1500–2500/kg (99.999%). Consumption of ZrF_4 in the United States is less than 5000 kg/yr.

Fluorozirconic Acid and Fluorozirconates

Hexafluorozirconic acid [12021-95-3], H_2ZrF_6, is formed by dissolving freshly prepared oxide, fluoride, or carbonate of zirconium in aqueous HF. This acid is produced commercially in a concentration range of 10 to 47%. The acid can be stored at ambient temperatures in polyethylene or Teflon containers without decomposition for at least two years. By neutralization of the acid, several stable fluorozirconates can be produced, such as $(NH_4)_2ZrF_6$ [16919-31-6], Na_2ZrF_6 [16925-26-1], K_2ZrF_6 [16923-06-8], Rb_2ZrF_6 [16923-95-8], and Cs_2ZrF_6, $CaZrF_6$ [30868-51-0]. The acid is regulated as a corrosive liquid NOS for shipping purposes (UN ID 1760) and packaging requires a corrosive label. The toxicity of zirconium compounds in general is represented by the TWA 5 mg/m^3 as Zr. The TWA for fluorides is 2.5 mg/m^3.

Hexafluorozirconic acid is used in metal finishing and cleaning of metal surfaces, whereas the fluorozirconates are used in the manufacture of abrasive grinding wheels, in aluminum metallurgy, ceramics industry, glass manufacturing, in electrolytic cells, in the preparation of fluxes, and as a fire retardant (see ABRASIVES; METAL SURFACE TREATMENTS).

High purity hexafluorozirconic acid and its salts are produced by Advance Research Chemicals of the United States, and Akita and Moritta of Japan. The technical-grade green-colored material is supplied by Cabot Corp. of the United States. In 1993, the U.S. market for fluorozirconic acid was about 250,000 kg/yr; the world market was less than 500,000 kg/yr. A principal part of this production is consumed by the wool, garment, and upholstery industries. The 1993 price varied between $2.4 to $6.6/kg depending on the quality and quantity required. Potassium fluorozirconate [16923-95-8], K_2ZrF_6, is commercially important; the world market is about 750,000 kg/yr. The most important application is as a fire-retardant material in the wool (qv) industry, for the manufacture of garments, upholstery for aeroplane industry, and children's clothes (see FLAME RETARDANTS). The 1993 unit price was between $5.0 and $6.6/kg.

BIBLIOGRAPHY

"Zirconium Fluorides" under "Zirconium Compounds" in *ECT* 1st ed., Vol. 15, pp. 297–298, by W. B. Blumenthal, National Lead Co.; "Zirconium" under "Fluorine Compounds, Inorganic", *ECT* 2nd ed., Vol. 9, pp. 685–686, by W. E. White, Ozark-Mahoning Co.; in *ECT* 3rd ed., Vol. 10, pp. 827–828, by D. T. Meshri, Ozark-Mahoning Co.

1. R. H. Hange, J. L. Margrave, and J. W. Hastie, *High Temp. Sci.* **5**(2), 89 (1973).
2. P. Ehrlich, F. Ploeger, and E. Kotch, *Z. Anorg. Allg. Chem.* **333**, 209 (1964).
3. Y. V. Baimakov, *Freiberg. Forschbungsh. B.* **118**, 43 (1967).
4. H. P. Withers Jr., V. A. Monk, and G. A. Cooper, *Proceedings of the SPIE International Society of Optical Engineers*, Vol. 1048, 1989, pp. 72–77; U.S. Pat. 4,983,372 (Jan. 8, 1991), H. P. Withers, Jr. and V. A. Monk (to Air Products & Chemicals, Inc.).
5. H. M. Haendler, C. M. Wheeler, Jr., and D. W. Robinson, *J. Am. Chem. Soc.* **74**, 2352 (1952).
6. K. Ohsawa and T. Shibata, *J. Lightwave Technol.* **LT-2**(5), 602–606 (1984).
7. J. M. Jewell and I. D. Aggarwal, *Proceedings of the SPIE International Society of Optical Engineers*, Vol. 1327, 1990, pp. 190–197.
8. G. C. Devyatykh and M. F. Churbanov, *Z. Anorg. Allg. Chem.* **576**, 25–32 (1989).
9. R. Mossadegh, P. M. Kutty, N. J. Garrito, and D. C. Tran, Proceedings of the SPIE International Society of Optical Engineers, Vol. 1112, 1989, pp. 40–46.
10. L. J. Moore, D. R. MacFarlane, and P. J. Newman, *J. Non-Cryst. Solids* **140**(1–3), 159–165 (1992).
11. L. A. Bursill, J. Peng, and J. R. Sellar, *Mater. Sci. Forum* **14**(1), 41–59 (1990).
12. R. N. Schwartz, M. Robinson, and G. L. Tangonan, *Mater. Sci. Forum*, **19–20** 275–285 (1987).
13. H. W. Schneider, A. Schoberth, and A. Standt, *Glasstech. Ber.* **60**(6), 205–210 (1987).
14. Y. Dai, T. Kawaguchi, K. Suzuki, S. Suzuki, and K. Yamamoto, *J. Non-Cryst. Solids* **142**(1–2), 159–164 (1992).
15. W. A. Sibley, *Mater. Sci. Forum* **6**, 611–616 (1985).
16. T. Iqbal, M. Shahriari, G. Merberg, and G. H. Sigel, *J. Mater. Res.* **6**(2), 401–406 (1991).
17. U.S. Pat. 2,807,646 (Sept. 24, 1957), C. B. Millen and C. Woolf (to Allied Chemicals and Dye Corp.).
18. U.S. Pat. 2,805,121 (Sept. 3, 1957), C. Woolf (to Allied Chemicals and Dye Corp.).
19. U.S. Pat. 3,165,504 (Jan. 12, 1965), J. P. Hogan (to Philips Petroleum Co.).
20. Jpn. Kokai 7,663,176 (June 1, 1976), W. Yasuo and co-workers (to Koci Chemical Co., Ltd.).
21. A. P. GalzKova, *Dokl. Akad. Nauk. USSR* **213**, 622 (1973).
22. U.S. Pat. 3,725,128 (Nov. 26, 1974), S. Senderoff (to Union Carbide).
23. R. M. Shrestha, S. A. Mundorff, and B. G. Bibby, *J. Dent. Res.* **51**, 1561 (1972).

DAYAL T. MESHRI
Advance Research Chemicals, Inc.

FLUORINE COMPOUNDS, ORGANIC

INTRODUCTION

Organic fluorine compounds were first prepared in the latter part of the nineteenth century. Pioneer work by the Belgian chemist, F. Swarts, led to observations that antimony(III) fluoride reacts with organic compounds having activated carbon–chlorine bonds to form the corresponding carbon–fluorine bonds. Preparation of fluorinated compounds was facilitated by fluorinations with antimony(III) fluoride containing antimony(V) halides as a reaction catalyst.

It was the 1930s before the direction of organic fluorine chemistry turned commercial (1). Facing the problem of replacing methyl chloride and ammonia in household refrigerators, researchers found that dichlorodifluoromethane (CFC-12) was the best alternative as a safe, stable gas whose liquefied state had low compressibility. In addition, it was not flammable. General Motors and E.I. du Pont de Nemours and Co. jointly led the early application of chlorofluorocarbons (CFCs) as refrigerants. Later, other diverse CFC applications were commercialized as cleaning agents, as blowing agents to make foam products, and as sources of monomers for many fluoropolymers.

Another impetus to expansion of this field was the advent of World War II and the development of the atomic bomb. The desired isotope of uranium, ^{235}U, in the form of UF_6 was prepared by a gaseous diffusion separation process of the mixed isotopes (see FLUORINE). UF_6 is extremely reactive and required contact with inert organic materials as process seals and greases. The wartime Manhattan Project successfully developed a family of stable materials for UF_6 service.

These early materials later evolved into the current fluorochemical and fluoro-polymer materials industry. A detailed description of the fluorine research performed on the Manhattan Project has been published (2).

Concern arose during the 1970s about cumulative CFC emissions into the atmosphere with progressive depletion of the stratospheric ozone layer by Cl atoms and led to the formation and global support of a multinational forum, called the Montreal Protocol on Substances That Deplete the Ozone Layer. As a result, CFC production has been dramatically decreased and will likely be totally phased out before the year 2000 (3) (see FLUORINATED ALIPHATICS). If hydrogen atoms are introduced into the CFC structure to lower the chlorine content, the resulting hydrochlorofluorocarbon (HCFC) is more susceptible to degradation in the lower atmosphere before it can reach the stratosphere. However when a hydrogen atom is introduced into a one-carbon compound, the boiling point is lowered and may be too low for the same CFC application. Therefore two-carbon compounds bearing some hydrogen are more attractive substitutes than the one-carbon modified CFCs. As hydrogen content increases, there is a counter effect of increasing flammability, which in turn limits some HCFC applications.

Physical Properties

Substitution of fluorine for hydrogen in an organic compound has a profound influence on the compound's chemical and physical properties. Several factors that are characteristic of fluorine and that underlie the observed effects are the large electronegativity of fluorine, its small size, the low degree of polarizability of the carbon–fluorine bond and the weak intermolecular forces. These effects are illustrated by the comparisons of properties of fluorocarbons to chlorocarbons and hydrocarbons in Tables 1 and 2.

The replacement of chlorine by fluorine results in a nearly constant boiling point (bp) drop of approximately 50°C for every chlorine atom that is replaced (see Table 1). In Table 2, a similar boiling point effect with hydrocarbons is apparent, even though the molecular weight of the fluorocarbon is much higher than the corresponding hydrocarbon analogue. An analogous drop in the corresponding fluorocarbon freezing point results in a widened liquid range for applications like lubricating fluids and greases. One other significant property difference, attributed to weak intermolecular forces, can be found in the very low surface tensions of fluorocarbons as compared to hydrocarbons and water (Table 3).

Table 1. Boiling Points of Halomethanes

Chloro-hydrocarbon	Bp, °C	Fluoro-hydrocarbon	CAS Registry Number	Bp, °C	Difference per F atom, °C
CH_3Cl	-24	CH_3F	[593-53-3]	-78	54
CH_2Cl_2	40	CH_2F_2	[75-10-5]	-52	46
$CHCl_3$	61	CHF_3	[75-46-7]	-83	48
CCl_4	77	CF_4	[75-73-0]	-128	51

Table 2. Boiling Points of Hydrocarbons and Fluorocarbons

Hydrocarbon	Bp, °C	Fluorocarbon	CAS Registry Number	Bp, °C
CH_4	-161	CF_4	[75-73-0]	-128
C_2H_6	-88	C_2F_6	[76-16-4]	-78
C_3H_8	-45	C_3F_8	[76-19-7]	-38
C_4H_{10}	0.6	C_4F_{10}	[355-25-9]	-5
C_7H_{16}	98	C_7F_{16}	[335-57-9]	82
C_6H_6	80	C_6F_6	[392-56-3]	80
C_6H_{12} (cyclic)	81	C_6F_{12} (cyclic)	[355-68-0]	52

Table 3. Surface Tensions of Selected Fluids

Compound	CAS Registry Number	Surface tension, 20 °C, mN/m (= dyn/cm)
perfluoroheptane	[335-57-9]	13.6
perfluoromethylcyclohexane	[355-02-2]	15.4
perfluoro-1,4-dimethylcyclohexane	[374-77-6]	16.3
octane	[111-65-9]	21.8
benzene	[71-43-2]	28.9
methyl bromide	[74-83-9]	41.5
water	[7732-18-5]	72.8

The low surface tension of highly fluorinated organic compounds is commercially important for their application in surfactants, antisoiling textile treatments, lubricants, and specialty wetting agents.

In contrast, the viscosities of fluorocarbons are higher than those of the corresponding hydrocarbons. This can be explained by the greater stiffness of the fluorocarbon chain arising from the large replusive forces between molecules, and from the greater density imparted by the more massive fluorine atoms (vs hydrogen). The fluorocarbon viscosity drops rapidly with increasing temperature and is accompanied by a simultaneous large decrease in density.

The refractive indexes and dielectric constants for the fluorocarbons are both lower than that for the corresponding hydrocarbon analogue.

Preparation

There are many known ways to introduce fluorine into organic compounds, but hydrogen fluoride [7664-39-3], HF, is considered to be the most economical source of fluorine for many commercial applications.

Halogen Exchange.   The exchange of another halogen atom in an organic compound for a fluorine atom is the most widely used method of fluorination. The relative ease of replacement follows the general order I > Br > Cl. Commonly used fluorinating agents are the fluorides of the alkali metals (especially KF), antimony, and mercury.

Antimony trifluoride [*7783-56-4*], SbF_3, can be used in the following preparations (4):

$$RCOCl \xrightarrow{SbF_3} RCOF \tag{1}$$

$$RSO_2Cl \xrightarrow{SbF_3} RSO_2F \tag{2}$$

$$\diagup\!\!\!C{=}C(R){-}CCl_3 \xrightarrow{SbF_3} \diagup\!\!\!C{=}C(R){-}CF_3 \tag{3}$$

$$RCCl_2R \xrightarrow{SbF_3} RCF_2R \tag{4}$$

$$CHCl_2OCH_2Cl \xrightarrow{SbF_3} CHF_2OCH_2F \tag{5}$$

The limitations of this reagent are several. It cannot be used to replace a single unactivated halogen atom with the exception of the chloromethyl ether (eq. 5) to form difluoromethyl fluoromethyl ether [*461-63-2*]. It also cannot be used to replace a halogen attached to a carbon–carbon double bond. Fluorination of functional group compounds, eg, esters, sulfides, ketones, acids, and aldehydes, produces decomposition products caused by scission of the carbon chains.

The effectiveness of antimony fluoride is increased if it is used in conjunction with chlorine or with antimony pentachloride. The formation of either $SbCl_2F_3$ or a complex of SbF_3 and $SbCl_5$ probably accounts for the increased activity (4).

Antimony pentafluoride [*7783-70-2*], SbF_5, is a highly active fluorinating agent and is generally used to fluorinate only completely halogenated compounds, since those containing hydrogen as well as halogen undergo decomposition. In the case of halogenated olefins or aromatic compounds (5), addition as well as substitution occurs. Thus hexachlorobenzene forms a fluorochlorocyclohexene [*27458-17-9*]:

$$C_6Cl_6 \xrightarrow{SbF_5} C_6Cl_2F_8 \tag{6}$$

Potassium fluoride [*7789-23-3*], KF, is the most frequently used of the alkali metal fluorides, although reactivity of the alkali fluorides is in the order CsF > RbF > KF > NaF > LiF (6). The preference for KF is based on cost and availability traded off against relative reactivity. In its anhydrous form it can be used to convert alkyl halides and sulfonyl halides to the fluorides. The versatility makes it suitable for halogen exchange in various functional organic compounds like alcohols, acids and esters (7). For example, 2,2-difluoroethanol [*359-13-7*] can be made as shown in equation 9 and methyl difluoroacetate [*433-53-4*] as in equation 10.

$$RX \xrightarrow{KF} RF \tag{7}$$

$$RSO_2Cl \xrightarrow{KF} RSO_2F \tag{8}$$

$$CHCl_2CH_2OH \xrightarrow{KF} CHF_2CH_2OH \tag{9}$$

$$CHCl_2COOCH_3 \xrightarrow{KF} CHF_2COOCH_3 \tag{10}$$

The preparation of fluoroaromatics by the reaction of KF with perhaloaromatics, primarily hexachlorobenzene, has received considerable attention. Two methods were developed and include either the use of an aprotic, polar solvent, such as *N*-methylpyrrolidinone (8), or no solvent (9). These methods plus findings that various fluoroaryl derivatives are effective fungicides (10) prompted development of a commercial process for the production of polyfluorobenzenes (11). The process uses a mixture of sodium and potassium fluorides or potassium fluoride alone in aprotic, polar solvents such as dimethyl sulfoxide or sulfolane.

Mercuric fluoride [*7783-39-3*], HgF_2, is used in the following conversions where $X = Br, I$ (4,12):

$$RX \xrightarrow{\text{HgF}_2} RF \tag{11}$$

Because alkyl chlorides react very slowly, chloroform or methylene chloride can be used as solvents in the above reaction. Polyhalides are known to react to form polyfluoride products. Oxygen-containing compounds, eg, esters, ethers, and alcohols, appear to inhibit this fluorination.

Mercurous fluoride [*13967-25-4*], Hg_2F_2, is less effective than HgF_2. The addition of chlorine or iodine to the reagent increases its reactivity owing to the formation of a complex between HgF_2 and HgX_2 (4,12).

Hydrogen fluoride, HF, when used alone is a comparatively ineffective exchange agent and replaces only active halogens (13), eg, acyl fluorides from acyl chlorides and benzotrifluoride [*98-08-8*] from benzotrichloride (eq. 12).

$$C_6H_5CCl_3 \xrightarrow{\text{HF}} C_6H_5CF_3 \tag{12}$$

When used with antimony pentachloride, the reactivity of HF is comparable to $SbCl_2F_3$ alone. Therefore a continuous fluorination exchange process is possible where antimony is the fluoride carrier from HF to the organic fluoride.

$$SbCl_5 + 3\ HF \rightarrow SbCl_2F_3 + 3\ HCl \tag{13}$$

$$SbCl_2F_3 + CCl_4 \rightarrow SbCl_4F + CCl_2F_2 \tag{14}$$

$$SbCl_4F + 2\ HF \rightarrow SbCl_2F_3 + 2\ HCl \tag{15}$$

Since antimony halides serve as fluorine carriers, the actual fluorination agent is HF. This process is the principal one used in the production of CFC-11, CFC-12, and HCFC-22 as well as many others. The application is well suited to the fluorination of one-carbon through three-carbon containing organic halides. In practice, the HF and organic halide enter by separate feeds into the process vessel. Under pressure, they react in the liquid phase with heating to form HCl and the organic fluoride. By suitable control of the feed ratios, temperature, pressure, and residence time, the degree of fluorination can be controlled. As the reaction progresses, vapors of HCl and the organic fluoride are continuously vented off for separation and recovery. Antimony salts stay in the process vessel and are periodically reactivated by treatment with chlorine to form the desired antimony(V) halide salts prior to reuse.

Heterogeneous vapor-phase fluorination of a chlorocarbon or chlorohydrocarbon with HF over a supported metal catalyst is an alternative to the liquid phase process. Salts of chromium, nickel, cobalt or iron on an AlF_3 support are considered viable catalysts in pellet or fluidized powder form. This process can be used to manufacture CFC-11 and CFC-12, but is hampered by the formation of over-fluorinated by-products with little to no commercial value. The most effective application for vapor-phase fluorination is where all the halogens are to be replaced by fluorine, as in manufacture of 3,3,3-trifluoropropene [677-21-4] (14) for use in polyfluorosilicones.

$$CCl_3CH_2CH_2Cl \xrightarrow[\text{excess}]{\text{HF}} CF_3CH{=}CH_2 \tag{16}$$

Another use of hydrogen fluoride, although not in halogen exchange, is the reaction with ethylenes or acetylenes to form the addition products, 1,1-difluoroethane [75-37-6] and vinyl fluoride [75-02-5]:

$$HC{\equiv}CH + HF \rightarrow CH_3CHF_2 + CH_2{=}CHF \tag{17}$$

Reaction conditions must be controlled since HF is also an excellent polymerization catalyst. Controlled reaction conditions can alternatively lead to vinyl fluoride or to HFC-152a (CH_3CHF_2). The latter can be thermally cracked to form vinyl fluoride.

Sulfur tetrafluoride [7783-60-0], SF_4, replaces halogen in haloalkanes, haloalkenes, and aryl chlorides, but is only effective (even at elevated temperatures) in the presence of a Lewis acid catalyst. The reagent is most often used in the replacement of carbonyl oxygen with fluorine (15,16). Aldehydes and ketones react readily, particularly if no alpha-hydrogen atoms are present (eg, benzal fluoride [455-31-2] from benzaldehyde), but acids, esters, acid chlorides, and anhydrides are very sluggish. However, these reactions can be catalyzed by Lewis acids (HF, BF_3, etc).

$$C_6H_5CHO + SF_4 \rightarrow C_6H_5CHF_2 \tag{18}$$

$$C_6H_5COOH + SF_4 \rightarrow C_6H_5CF_3 \tag{19}$$

Halogen Fluorides. These include compounds such as IF_3, IF_5, ClF, etc, of which only a few, ClF, ClF_3, BrF_3, and IF_5, are used to some extent. They act both as halogen exchange agents and, in the case of the monofluorides, as addition agents to unsaturated bonds (17).

Replacement of Hydrogen. Three methods of substitution of a hydrogen atom by fluorine are (1) reaction of a C–H bond with elemental fluorine (direct fluorination, (2) reaction of a C–H bond with a high valence state metal fluoride like AgF_2 or CoF_3, and (3) electrochemical fluorination in which the reaction occurs at the anode of a cell containing a source of fluoride, usually HF.

Direct Fluorination. The principal disadvantage of the use of elemental fluorine as a fluorinating agent is the high heat of reaction. A considerable degree of carbon–carbon bond scission can occur as well as polymer formation. In order to prevent these complications, fluorine is diluted with nitrogen and the reaction zone is constructed such that good heat conductivity is possible. Low temperatures are favored to achieve maximum selectivity and yield (18). Fluorine is also effective for the replacement of residual hydrogen in a highly fluorinated organic molecule to produce the corresponding fluorocarbon.

The fluorination reaction is best described as a radical-chain process involving fluorine atoms (19) and hydrogen abstraction as the initiation step. If the molecule contains unsaturation, addition of fluorine also takes place (17). Complete fluorination of complex molecules can be conducted using this method (see FLUORINE COMPOUNDS, ORGANIC–DIRECT FLUORINATION).

Reaction with a Metal Fluoride. A second technique for hydrogen substitution is the reaction of a higher valence metal fluoride with a hydrocarbon to form a fluorocarbon:

$$2 \ CoF_2 + F_2 \rightarrow 2 \ CoF_3 \tag{20}$$

$$2 \ CoF_3 + RH \rightarrow RF + 2 \ CoF_2 + HF \tag{21}$$

The principal advantage to this method is that the heat evolved for each carbon–fluorine bond formed, 192.5 kJ/mol (46 kcal/mol), is much less than that obtained in direct fluorination, 435.3 kJ/mol (104 kcal/mol). The reaction yields are therefore much higher and less carbon–carbon bond scisson occurs. Only two metal fluorides are of practical use, AgF_2 and CoF_3.

The reactivity of the metal fluoride appears to be associated with the oxidation potential of the metal. For example, AgF replaces halogen in organic compounds, whereas AgF_2 replaces hydrogen.

The reaction is conducted by passing fluorine through a bed of AgCl or $CoCl_2$ at an elevated temperature to form the higher valence state fluorides. The organic reactant is then passed through the bed to realize a semicontinuous fluorination process. In general, the method is used for the preparation of fluorocarbons since any unsaturation or functionality in the reactant is removed. The process can also be used to fluorinate polychlorohydrocarbons, whereby replacement of both chlorine and hydrogen occurs (2).

Electrochemical Fluorination. The electrochemical fluorination (ECF) of highly fluorinated organic compounds (20) involves the electrolysis of an organic reactant in liquid anhydrous HF at a voltage below that for liberation of fluorine. The reaction is limited by temperature (usually done at 0°C) and by the solubility of the reactant in HF. Electrical conductivity is required for current to flow and the reaction to proceed. Current density is 10–20 mA/cm^2. Fluorination takes place at the nickel anode by a free-radical process not involving the intermediate formation of elemental fluorine. Hydrogen is liberated at the cathode. The method is used to fluorinate acyl halides, sulfonyl halides, ethers, carboxylic acids, and amines. The product is a fluorocarbon having no residual hydrogen. Olefins and carbocyclics, as well as heterocyclic compounds, become saturated. Side reactions, resulting in reduced yields, include cleavage of carbon–carbon bonds and polymer formation. The electrochemical yields decrease with increasing number of carbon atoms in the structure.

ECF is successfully used on a commercial scale to produce certain perfluoroacyl fluorides, perfluoroalkylsulfonyl fluorides, perfluoroalkyl ethers, and perfluoroalkylamines. The perfluoroacyl fluorides and perfluoroalkylsulfonyl fluorides can be hydrolyzed to form the corresponding acid and acid derivatives. Examples include perfluorooctanoyl fluoride [*335-66-0*], perfluorooctanoic acid [*335-67-1*], perfluorooctanesulfonyl fluoride [*307-35-7*], perfluorooctanesulfonic acid [*763-23-1*], and tris(perfluoro-*n*-butyl)amine [*311-89-7*].

$$C_7H_{15}COCl \xrightarrow[\text{HF}]{\text{ECF}} C_7F_{15}COF \rightarrow C_7F_{15}COOH \tag{22}$$

$$C_8H_{17}SO_2Cl \xrightarrow[\text{HF}]{\text{ECF}} C_8F_{17}SO_2F \rightarrow C_8F_{17}SO_3H \tag{23}$$

$$(C_4H_9)_3N \xrightarrow[\text{HF}]{\text{ECF}} (C_4F_9)_3N \tag{24}$$

Telomer Formation. Fluorinated compounds with active C–Br or C–I bonds can add to fluoroolefins to form addition products in high yield. The olefin most often used is tetrafluoroethylene [*116-14-3*]. Telomerization involves reaction of a telogen, or addition agent like $CBrF_3$, CF_3I [*2314-97-8*], or C_2F_5I [*354-64-3*], with the olefin to form longer chain addition products called telomers. The reaction is initiated by thermolysis, photolysis, peroxides, and other free-radical initiators, certain metal complexes, and various redox chemicals. By control of the stoichiometry and reaction conditions, a simple addition product or telomers with high fluorine content can be formed. The yield is higher than that seen with ECF production methods. The route does suffer from a distribution of adducts formed vs formation of one specific reaction product. The usual adducts are perfluoroalkyl iodides having up to 14 carbon atoms in the alkyl chain.

$$C_2F_5I + x\ CF_2{=}CF_2 \rightarrow C_2F_5(CF_2CF_2)_xI$$

Often used as mixtures, the telomers are subsequently converted to commercial surfactants and stain-resistant fiber finishes through functionalizing steps using standard chemical reactions of the C–I bond.

Aromatic Ring Fluorination. The formation of an aryl diazonium fluoride salt, followed by decomposition, is a classical reaction (the Schiemann reaction) for aryl fluoride preparation (21). This method has been adapted to the production-scale manufacture of fluorobenzene [*462-06-6*] (22) where HF is the source of the fluoride.

$$C_6H_5NH_2 \xrightarrow[\text{HF}]{\text{NaNO}_2} C_6H_5N_2{}^+F^- \xrightarrow{-N_2} C_6H_5F \tag{25}$$

Chemical Properties and Applications

Substitution of fluorine into an organic molecule results in enhanced chemical stability. The resulting chemical reactivity of adjacent functional groups is drastically altered due to the large inductive effect of fluorine. These effects become more pronounced as the degree of fluorine substitution is increased, especially on the same carbon atom. This effect demonstrates a maximum in fluorocarbons and their derivatives.

Fluorinated Alkanes. As the fluorine content increases, the chemical reactivity decreases until complete fluorination is achieved, after which they are inert to most chemical attack, including the highly reactive element fluorine. Their lack of reactivity leads to their use in certain commercial applications where stability is valued when in contact with highly reactive chemicals.

Fluorinated Olefins. In electrophilic addition reactions, the reactivity of the unsaturated linkage is reduced by the inductive effect of fluorine. Nucleophilic additions are enhanced by this same effect. Amines, phenols, alcohols, and many other nucleophiles, including fluoride ion, add to the carbon–carbon double bond of highly fluorinated olefins (23). Free-radical addition of halogen halides proceeds easily using either peroxide or thermal initiation. Some halides, especially those derived from iodine, eg, ICl, react by an ionic mechanism. Fluorinated olefins also undergo free-radical polymerization producing a wide range of valuable fluoropolymers.

Certain CFCs are used as raw materials to manufacture key fluorinated olefins to support polymer applications. Thermolysis of HCFC-22 affords tetrafluoroethylene and hexafluoropropylene [116-15-4] under separate processing conditions. Dechlorination of CFC-113 forms chlorotrifluoroethylene [79-38-9]. Vinylidene fluoride [75-38-7] is produced by the thermal cracking of HCFC-142b.

Fluorinated Aromatic Hydrocarbons. Many aromatic fluorocarbon derivatives, eg, hexafluorobenzene, pentafluorotoluene [771-56-2], and perfluoronaphthalene [313-72-4] are examples of compounds that readily undergo nucleophilic ring substitution reactions with loss of one or more fluorine substituents. This is in sharp contrast to the fluorine substituents in perfluoroalkanes. Perfluoroalkyl substitution on the aromatic ring has a strong inductive effect, making the ring more susceptible to nucleophilic attack. Fluorine hyperconjugative effects are considered to be unimportant (24) in these reactivity patterns, leaving inductive effects as the primary factor to describe the substituent effect.

Fluorinated Heterocyclic Compounds. Heterocyclic compounds containing the CF_3 group are prepared by methods similar to those used in the fluorination of aliphatic compounds. The direct action of fluorine on uracil yields the cancer chemotherapy agent, 5-fluorouracil [51-21-8], as one special example of a selective fluorination on a commercial scale (25).

Fluorinated Acids. This class of compounds is characterized by the strength of the fluorocarbon acids, eg, CF_3COOH, approaching that of mineral acids. This property results from the strong inductive effect of fluorine and is markedly less when the fluorocarbon group is moved away from the carbonyl group. Generally, their reactions are similar to organic acids and they find applications, particularly trifluoroacetic acid [76-05-1] and its anhydride [407-25-0], as promotors in the preparation of esters and ketones and in nitration reactions.

Fluorinated Biologically Active Compounds. Many biologically active compounds are prepared from fluorobenzene, difluorobenzene, benzotrifluoride and fluorinated steroids. The preparation of fluorinated compounds for use in medicine has increased rapidly (26,27) since the 1950s. The strong interest in such substances is based on the following considerations: (1) fluorine most closely resembles bioactive hydrogen analogues with respect to steric requirements at the receptor sites; (2) fluorine alters electronic effects, owing to its high electronegativity; (3) fluorine imparts improved oxidative and thermal stability to the parent molecule; and (4) fluorine imparts lipid solubility, thereby increasing the *in vivo* absorption and transport rates in membranes.

Many fluorinated, biologically active agents have been developed and successfully used in the treatment of diseases. The biological property of fluorinated organics has been further extended to applications in the agrochemical and pest management fields.

Analgesics. Four examples of antiinflammatory agents are Sulindac [38194-50-2], based on a monofluoro indene derivative; diflunisal [22494-42-4], based on a substituted difluorobenzene; and dexamethasone [50-02-2] and fluocinonide [356-12-7], based on monofluorinated and difluorinated steroids, respectively.

Antiviral Agents. Trifluridine [70-00-8] is a trifluoromethyl substituted heterocyclic antiviral agent.

Appetite Depressants. Fenfluramine hydrochloride [404-82-0] is an anorexiant based on a meta-substituted benzotrifluoride.

Tranquilizers. Fluphenazine hydrochloride [146-56-5], trifluoperazine hydrochloride [440-17-5], and triflupromazine [146-54-3] are all trifluoromethyl substituted phenothiazine chemicals useful in the management of psychotic disorders.

Diuretics. The diuretic and antihypertensive agent bendroflumethiazide [73-48-3] is a benzotrifluoride-based pharmaceutical.

Inhalation Anesthetics. Examples of highly fluorinated halocarbons and ethers are halothane [151-67-7], fluroxene [406-90-6], enflurane [13838-16-9], methoxyflurane [76-38-0], sevoflurane [28523-86-6], desflurane [57041-67-5], isoflurane [26675-46-7], and pure enantiomers (28) of isoflurane as potentially more effective anesthetic compounds than the racemic mixture. Isoflurane is the leading inhalation anesthetic marketed in North America since its introduction in 1981. Desflurane is being newly introduced into the United States market. Enflurane is largely used in Europe while sevoflurane is used in Japan and Korea. Methoxyflurane is used in veterinary applications only.

Herbicides. Fluometuron [2164-17-2] is a fluorophenyl-substituted urea effective against grassy and broadleaf weeds in bean, grain, fruit, and cotton crops. Trifluralin [1582-09-8] is a trifluoromethyl-substituted dinitroaniline used as pre-emergence control against weeds in cotton and soybean crops. Profluralin [26399-36-0] and benfluralin [1861-40-1] are structural analogues of trifluralin. Fluorodifen [15457-05-3] is a trifluoromethyl-substituted diphenyl ether used for weed control in bean and rice crops. Fluroxypyr [81406-37-3] is a fluoropyridine compound used on cereals for post-emergent control of broadleaf weeds.

Insecticides. Diflubenzuron [35367-38-5] is a difluorobenzoyl urea. It inhibits insect chitin formation during larval molting. Application is with management of fruit, bean, and cotton crops. Perfluoroalkylsulfonamides of carbon chain length equal to 6 or 8 show good control against fire ants.

Fungicides. Flusilazole [85509-19-9] is a newer broad-spectrum foliar fungicide containing two fluorophenyl substituents with application on cereal, fruit, and vegetable crops. Flutriafol [76674-21-0] is another fluorobenzene derivative useful on small grain cereal diseases.

Economic Aspects

The CFC commodity application is undergoing significant change due to environmental pressures. Development of acceptable, alternative fluorinated compounds is extremely expensive. As the largest global supplier, Du Pont plans to spend $1 billion by the mid-1990s to develop CFC alternatives. The five-year toxicity testing program alone has been estimated at up to $5 million for each candidate compound tested.

The HCFC and HFC refrigeration alternatives are estimated to be two to five times higher in price, and some of the viable alternatives demonstrate a lower heat-transfer efficiency than the current CFCs. Total production will continue to drop due to conservation in use and elimination of emissive uses along with substitutions in refrigeration applications. The global CFC market is estimated at $4 billion with one-half of that being in the United States. The largest supplier is Du Pont with 25% of the global market share.

The 1990 United States production for CFCs, also generically called fluorocarbons in some sources, was 417,009 metric tons (29). Global production is approximately three times this value, with roughly one-third of the total produced in the North American area, one-third in Europe, and the remaining one-third being spread over all remaining global areas. The peak production year was in 1988 with a total of 624,550 t of CFCs. Production of CFC-11 and CFC-12 has been dropping due to environmental pressures while HCFC-22 has been growing modestly to pick up some of the refrigeration growth not addressed by CFC-11 or CFC-12 use. The continued refrigeration use of HCFC-22 is based on its very low relative ozone depletion potential. Some of the HCFC-22 production growth is attributable to growing captive conversion to tetrafluoroethylene for polymer manufacture at the same HCFC-22 production sites. Global poly(tetrafluoroethylene) demand has been currently growing at 4–5% per year, which in turn grows HCFC-22 demand.

United States CFC production is spread over different sites with a variety of companies. Allied-Signal Inc. has 143,180 t of capacity for CFC-11, CFC-12, HCFC-22, CFC-113, CFC-114, and HCFC-141b at its three sites in Baton Rouge, La., Danville, Ill., and El Segundo, Calif.; Ausimont USA, Inc. has 11,360 t of capacity for HCFC-141b and HCFC-142b at their Thorofare, N.J. site, where vinylidene fluoride is produced. Both HCFC-141b and HCFC-142b can be used as alternative feedstocks for vinylidene fluoride manufacture. Elf Atochem North America (former Pennwalt Corp. merged with Elf Atochem) can produce 72,730 t of combined CFC-11, CFC-12, HCFC-22, HFC 143a, HCFC-141b, and HCFC-142b at Calvert City, Ky., and another 36,360 t of combined CFC-11, CFC-12, and HCFC-22 capacity at their Wichita, Kans. site. Du Pont has an estimated 318,180 t of combined capacity for CFC-11, CFC-12, CFC-13, CFC-14, HCFC-22, HFC-23, CFC-113, CFC-114, CFC-115, CFC-116, HFC 125, HFC-134a, HFC 143a, and HFC-152a at the Antioch, Calif., Corpus Christi, Tex., Deepwater, N.J., Louisville, Ky., and Montague, Mich. sites. La Roche Chemicals Inc. has a plant in Gramercy, La. with 36,360 t per year capacity for the combined CFC-11, CFC-12, and HCFC-22. Great Lakes Chemical Co. has a Halon production facility at their El Dorado, Ark. site with unspecified capacity.

In Western Europe, the CFC producers are equally varied. The following is a partial list of the larger companies with total CFC production capacity (10^3 t) at all sites shown in parentheses: Atochem SA (148.5, France and Spain), Hoescht AG (102.0, Germany), Kali-Chemie AG (66.0, Germany and Spain), Montefluos SpA (100.0, Italy), and ICI Chemicals and Polymers Ltd. (> 113.6, United Kingdom). These producers account for over 80% of the Western European CFC production.

In Japan, the primary suppliers are Asahi Glass Co., Ltd. (60,000 t), Daikin Industries, Ltd. (115,000 t), and Du Pont-Mitsui Fluorochemicals Co. (64,000 t). Together these three producers account for over 90% of the Japanese CFC production.

Investments in plants to produce refrigeration alternatives have been announced by a variety of companies. ICI has an HFC-134a plant built in St. Gabriel, La. and is scheduled for 1993 startup. Their HFC-134a plant in Runcorn, UK is operational; another one is planned for the Mihara, Japan site. ICI's new HFC-32 pilot plant at Widnes, UK will produce this low temperature refrigerant beginning in 1992. Mixtures of HFC-32 and HFC-134a are also being pursued as a refrigerant. Initial HFC-32 supplies will support toxicology testing required prior to commercialization. ICI has also announced it will close its last Halon-1211 production facility in Manchester, UK by the end of 1993. In 1990, Du Pont built a commercial HFC-134a plant in Corpus Christi, Tex. The Du Pont HFC-134a facilities at Chiba, Japan, and Dordrecht, the Netherlands are targeted for 1995 completion. Another Du Pont plant in Maitland, Canada commercially manufactures HCFC-123, and a conversion of their Montague, Michigan plant is targeted at manufacture of HCFC-141b. Du Pont reports their global HFC 134a capacity at 55,000 t per year. Exact capacity of these facilities has not been confirmed but each is assumed to be well over 5000 t per year.

The Minnesota Mining and Manufacturing Co., or 3M, manufactures specialty perfluorochemicals using mainly electrochemical fluorination methods at their St. Paul, Minn., Decatur, Ala., and Cordova, Ill. sites. Their capacity is not reported, but is estimated at over 5000 t as fluorinated inert fluids, surfactants, and fire extinguishment chemicals. Asahi, Du Pont, and Hoescht all use fluoroolefin telomerization technology at a variety of their sites to manufacture a line of perfluorinated specialty chemicals for stain-resistant treatment and surfactant applications. Globally, these telomer-based fluorochemicals are estimated to be over 5000 t per year with Du Pont having one-third of the total. Daikin and 3M recently formed a U.S. joint venture called MDA Manufacturing. This venture will build a plant at the Decatur, Ala. site for production of about 4550 t of HCFC-22 and other unspecified fluorochemical intermediates. The HCFC-22 production will be dedicated to the manufacture of tetrafluoroethylene for Daikin use at their own proposed Decatur plant and of hexafluoropropylene to be used by 3M at its existing Decatur plant.

Aromatic fluorine compounds are varied in kind and in volume. Mallinckrodt Specialty Chemicals division of the Imcera Group claim their continuous process capacity for fluorobenzene is 1200 t per year from their St. Louis, Mo. site. EniChem reports a 910 t per year capacity for fluorinated aromatics at their Trissino, Italy site using a novel continuous diazotization process coupled with electrofluorination technology. Hoescht AG has announced plans to double its unspecified capacity at Griesheim, Germany for fluoroaromatics. ICI in the United Kingdom and Du Pont at their Deepwater, N.J. facility also have fluoroaromatic capabilities. Other smaller suppliers also manufacture fluorinated aromatic compounds for specialty applications, but their capacities are again unreported. Fluoroaromatics are basic intermediate building blocks leading toward the more advanced aromatic fluorine intermediates (AFI) including fluorinated aniline, quinoline, biphenyl, and phenol compounds. AFI uses are in surfactants, pharmaceuticals, agrochemicals, electronics, and biomedical applications. The worldwide AFI demand is estimated at 4000 t growing to 10,000 t by 1994. The 1994 market value is projected at about $400 million (30).

Safe Handling Aspects

The safety of fluorine compounds is possibly as varied as the numbers of compounds known which bear fluorine substituents. Most compounds bearing the C–F bond are synthetic and therefore not normally encountered in nature. Aerosol or vapor inhalation is the most likely route of exposure where adverse health effects may occur. As such all new fluorine compounds should be handled with caution as one would with any potentially hazardous substance until full toxicological properties are known. Existing fluorine compounds cover the range from biologically inert materials like fluorocarbon fluids suitable for potential blood substitutes (31) (see BLOOD, ARTIFICIAL) through biologically active materials like the very highly toxic octafluoroisobutylene [382-21-8]. The toxicity of one chemical vs another chemical is not predictable based on the number or the site of fluorine substituents. The main commercial fluorinated compounds, like the CFCs, exhibit a very low order of toxicity (32). The potential cardiotoxicity from inhalation of bronchiodilator aerosols using CFCs as propellants is well documented in the medical literature.

Many new fluorinated drug and agrochemical agents were discovered based initially on the properties that a C–F bond imparts to a molecule. Its size similarity to the C–H bond analogue allows entry into a binding site for a subsequent biological effect. The more stable linkage allows for longer term effects with slower metabolism and excretion. Some compounds possess toxic effects attributable to fluoride ion toxicity and irritation of the respiratory system may be a common response, but many others demonstrate a more complex biological behavior. Recent long-term inhalation studies with HCFC-123 as a CFC alternative have now shown nonmalignant tumors in male rats (33). This behavior is characteristic of this specific compound, and while unexpected, it is not representative of the family of fluorinated two-carbon compounds as a whole. Many of the other members of this family show excellent safety from inhalation toxicity testing. Fluorinated inhalation anesthetics require inhalation efficacy and safety acquired only through trial and error testing. Much of the work on anesthetics heralds back to the early refrigerant development work and the safety studies that were conducted to, in turn, identify this important property (34).

BIBLIOGRAPHY

"Nomenclature, Physical Properties, Reactions and Methods of Fluorination" under "Fluorine Compounds, Organic," in *ECT* 1st ed, Vol. 6, pp. 735–751, by E. T. McBee, O. R. Pierce, and W. F. Edgell, Purdue University; "Introduction" in *ECT* 2nd ed., under "Fluorine Compounds, Organic," Vol. 9, pp. 686–704, by E. T. McBee, C. J. Norton, and T. Hodgins, Purdue University; in *ECT* 3rd ed., Vol. 10, pp. 829–839, by O. R. Pierce, Dow Corning Corp.

1. U.S. Pat. 1,833,847 (Nov. 24, 1931), T. Midgley, Jr., A. L. Henne, and R. McNary (to Frigidaire Corp.) and related patents.
2. C. Slesser and S. R. Schram, *Preparation, Properties and Technology of Fluorine and Organic Fluorine Compounds*, McGraw-Hill Book Co., Inc., New York, 1951.
3. P. S. Zurer, *Chem. Eng. News*, 7–13 (July 24, 1989); 7–13 (June 22, 1992).
4. A. L. Henne, in R. Adams and co-workers, eds., *Organic Reactions*, Vol. II, John Wiley & Sons, Inc, New York, 1944, pp. 49–93.
5. E. T. McBee, P. A. Wiseman, and G. B. Bachman, *Ind. Eng. Chem.* **39**, 415–417 (1947).
6. N. N. Vorozhtsov and G. G. Yacobson, *Khim. Nauka i Prom.* **3**, 403 (1958).
7. E. Gryszkiewicz-Trochimowski, A. Sporzynski, and J. Wnuk, *Rec. Trav. Chim.* **66**, 413–418 (1947).
8. J. T. Maynard, *J. Org. Chem.* **28**, 112–115 (1963).
9. N. N. Vorozhtsov, V. E. Platonov, and G. G. Yakobson, *Izvest. Akad. Nauk SSSR, Ser. Khim.*, (8), 1524 (1963).
10. G. C. Finger, F. H. Reed, and L. R. Tehon, *Ill. State Geol. Surv. Circ.* **199**, 1–15 (1955).
11. W. Prescott, *Chem. Ind. (London)*, (2), 56–63 (1978).
12. W. Bockemuller, in *Newer Methods of Preparative Organic Chemistry*, rev. ed., Interscience Publishers, Inc., New York, 1948, pp. 229–248.
13. K. Wiechert, in Ref. 12, pp. 315–368.
14. U.S. Pat. 4,465,786 (Aug. 14, 1984), M. F. Zimmer, W. E. Smith, and D. F. Malpass (to General Electric Co.); U.S. Pat. 4,798,818 (Jan. 17, 1989), W. X. Bajzer and co-workers (to Dow Corning Corp.).
15. G. W. Parshall, *J. Org. Chem.* **27**, 4649–4651 (1962).
16. W. C. Smith, *Angew Chem. Int. Ed. Engl.* **1**, 467–475 (1962).
17. W. K. R. Musgrave, in M. Stacey and co-workers, eds., *Advances in Fluorine Chemistry*, Vol. 1, Butterworth & Co. Ltd., London, 1960, pp. 1–28.
18. W. E. Jones and E. G. Skolnik, *Chem. Reviews* **76**, 563–592 (1976); R. J. Lagow and J. L. Margrave, in S. J. Lippard, ed., *Progress in Inorganic Chemistry*, Vol. 26, John Wiley & Sons, Inc., New York, 1979, pp. 161–210.
19. S. T. Purrington, B. S. Kagen, and T. B. Patrick, *Chem. Reviews* **86**, 997–1018 (1986).
20. J. H. Simons and co-workers, *J. Electrochem. Soc.* **95**, 47–67 (1949); J. Burdon and J. C. Tatlow, in Ref. 17, pp. 129–165; S. Nagase, in P. Tarrant, ed., *Fluorine Chemistry Reviews*, Vol. 1, Marcel Dekker, Inc., New York, 1967, pp. 77–106.
21. A. Roe, in R. Adams and co-workers, eds., *Organic Reactions*, Vol. V, John Wiley & Sons, Inc., New York, 1949, pp. 193–228.
22. U.S. Pat. 4,822,927 (Apr. 18, 1989), N. J. Stepaniuk and B. J. Lamb (to Mallinckrodt, Inc.).
23. R. D. Chambers and R. H. Mobbs, in M. Stacey and co-workers, eds., *Advances in Fluorine Chemistry*, Vol. 4, Butterworth & Co., Ltd., London, 1965, pp. 50–112.
24. D. Holtz, *Chem. Reviews* **71**, 139–145 (1971).
25. U.S. Pat. 4,082,752 (Apr. 4, 1978), S. Misaki and T. Takahara (to Daikin Kogyo Co. Ltd.); U.S. Pat. 4,029,661 (June 14, 1977), R. Anderson, P. Schuman, and G. Westmoreland (to PCR Inc.); U.S. Pat. 3,954,758 (May 4, 1976), P. Schuman and co-workers, (to PCR Inc.).

26. R. Filler, in R. E. Banks, ed., *Organofluorine Chemicals and Their Industrial Applications*, Ellis Horwood Ltd., Chichester, UK, 1979.
27. *AMA Drug Evaluation*, 3rd ed., Publishing Sciences Group, Inc., Littleton, Mass., 1977.
28. C. G. Huang, in *Abstracts of the 203rd National Meeting of the American Chemical Society*, abstract #FLUO-018, San Francisco, Apr. 5–10, 1992.
29. United States International Trade Commission (USITC), *Synthetic Organic Chemicals–United States Production and Sales, 1990*, publication #2470, Washington, D.C., Dec. 1991; R. F. Bradley, A. Leder, and Y. Sakuma, *Fluorocarbons*, in *Chemical Economics Handbook*, SRI International, Menlo Park, Calif., 1990, sections 543.7000–543.7003, plus 1992 supplemental data.
30. *Chem. Eng. News*, 13 (Jan. 4, 1988); 27 (Nov. 27, 1989).
31. J. G. Riess and M. LeBlanc, in K. C. Lowe, ed., *Blood Substitutes: Preparation, Physiology and Medical Applications*, Ellis Horwood, Ltd., Chichester, UK, 1988, pp. 94–129.
32. G. D. Clayton and F. E. Clayton, eds., *Patty's Industrial Hygiene and Toxicology*, Vol. 2B, 3rd rev. ed., John Wiley & Sons, Inc., New York, 1981.
33. *Chem. Eng. News*, 26 (July 8, 1991); P. Zurer, *Chem. Eng. News*, 21 (July 22, 1991).
34. E. R. Larsen, in P. Tarrant, ed., *Fluorine Chemistry Reviews*, Vol. 3, Marcel Dekker, Inc., 1969, pp. 1–44.

General References

R. E. Banks, *Preparation, Properties and Industrial Applications of Organofluorine Compounds*, Ellis Horwood Ltd., Chichester, UK 1982.

R. E. Banks and M. G. Barbour, *Fluorocarbon and Related Chemistry*, Vols. 1–3, The Chemical Society, Burlington House, London, 1971–1976.

R. E. Banks, ed., *Organofluorine Chemicals and Their Industrial Applications*, Ellis Horwood, Ltd., Chichester, UK, 1979.

R. D. Chambers, *Fluorine in Organic Chemistry*, John Wiley & Sons, Inc., New York, 1973.

R. Filler and Y. Kobayashi, *Biochemical Aspects of Fluorine Chemistry*, Kodansha Scientific Books, Tokyo, and Elsevier Biomedical Press, Amsterdam, 1982.

M. Hudlicky, *Chemistry of Organic Fluorine Compounds*, 2nd ed., Ellis Horwood, Ltd., Chichester, UK, 1976.

I. L. Knunyants and G. G. Yakobson, *Synthesis of Fluoroorganic Compounds*, Springer-Verlag, New York, 1985.

J. F. Liebman, A. Greenberg, and W. R. Dolbier, Jr., eds., *Fluorine-Containing Molecules: Structure, Reactivity, Synthesis and Applications*, VCH Publishers, Inc., New York, 1988.

A. M. Lovelace, D. A. Rausch, and W. Postelnek, *Aliphatic Fluorine Compounds*, Reinhold Publishing Corp., New York, 1954.

G. A. Olah, R. D. Chambers, and G. K. S. Prakash, eds., *Synthetic Fluorine Chemistry*, John Wiley & Sons, Inc., New York, 1992.

A. E. Pavlath and J. E. Leffler, *Aromatic Fluorine Compounds*, Reinhold Publishing Corp., New York, 1962.

G. Siegemund and co-workers, *Fluorine Compounds, Organic*, in W. Gerhartz and co-workers, eds., *Ullmann's Encyclopedia of Industrial Chemistry*, 5th rev. ed., Vol. A11, VCH Publishers, New York, 1988.

J. H. Simons, ed., *Fluorine Chemistry*, Vols. 1–5, Academic Press, Inc., New York, 1950–1964.

W. A. Sheppard and C. M. Sharts, *Organic Fluorine Chemistry*, Benjamin, Inc., New York, 1969.

P. Tarrant, ed., *Fluorine Chemistry Reviews*, Vols. 1–8, Marcel Dekker, Inc., New York, 1967–1977.

J. C. Tatlow and co-workers, eds., *Advances in Fluorine Chemistry*, Vols. 1–7, W. A. Benjamin, Inc., New York, 1960–1973.

J. T. Welch and S. Eswarakrishnan, *Fluorine in Bioorganic Chemistry*, John Wiley & Sons, Inc., New York, 1991.

J. T. Welch, *Selective Fluorination in Organic and Bioorganic Chemistry*, ACS Symposium Series #456, American Chemical Society, Washington, D.C., 1991.

WILLIAM X. BAJZER
YUNG K. KIM
Dow Corning Corporation

DIRECT FLUORINATION

Organic compounds containing fluorine are well known for their special properties, especially their inertness, very low boiling points, comparatively high melting points, and high thermal stability. Binary compounds of carbon and fluorine, ie, fluorocarbons, can be traced back to Moissan who, in 1886, discovered and isolated fluorine (1). The simplest solid fluorocarbon (2) is poly(fluoromethylidyne) $(CF)_n$, CFX (MarChem), which has a layered structure (Fig. 1). It can be synthesized by the reaction of fluorine with graphite at 300–600°C and fluorine pressures up to 101 kPa (1 atm) (3,4). Other solids (C_4F, C_2F, etc) have also been reported from the reaction of F_2 with graphite (5,6). This material is a grayish to white powder of variable stoichiometry and stable up to 600°C in air. Fluorinated graphite has a very low coefficient of friction, similar to Teflon, and is an excellent lubricant that can be used up to 600°C, surpassing Teflon which softens and decomposes above 350°C (see FLUORINE COMPOUNDS, ORGANIC–POLYTETRAFLUOROETHYLENE). When $(CF_x)_n$ decomposes above 650°C, it yields mainly car-

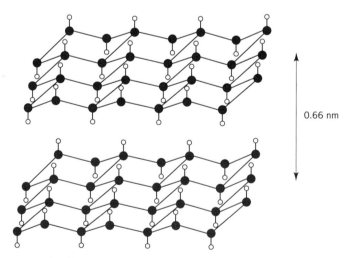

Fig. 1. Structure of poly(carbon monofluoride). ●, carbon; ○, fluorine. The interstitial space is 0.66 nm.

bon black (finely divided soot) and the inert gas carbon tetrafluoride (7). Thus it can be used safely at high temperatures without risk of poisonous gases being produced on decomposition (2,8).

Initial attempts at reactions between fluorine and hydrocarbons were described as similar to combustion and the reaction products contained mostly carbon tetrafluoride and hydrogen fluoride:

$$C_xH_y + \left(\frac{4x + y}{2}\right) F_2 \rightarrow x\ CF_4 + y\ HF + \text{energy}$$

This reaction has often reached explosive proportions in the laboratory. Several methods were devised for controlling it between 1940 and 1965. For fluorination of hydrocarbons of low (1–6 carbon atoms) molecular weight at room temperature or below by these methods, yields as high as 80% of perfluorinated products were reported together with partially fluorinated species (9–11). However, fluorination reactions in that era involving elemental fluorine with complex hydrocarbons at elevated temperatures led to appreciable cleavage of the carbon–carbon bonds and the yields invariably were only a few percent.

Before the LaMar process was developed in 1969, the use of direct fluorination was usually considered the classical method of fluorination (12,13) and other approaches were regarded as modern methods. Now only telomerization reactions using tetrafluoroethylene and reactions in hydrogen fluoride-based electrochemical cells are more widely used than direct fluorination on a commercial scale; however, this may change in the future. Direct fluorination not only gives higher yields in most cases but preparation in this manner is applicable to a wider range of organofluorine compounds and classes of compounds inaccessible by these more established technologies. Many compounds are uniquely prepared in the laboratory by direct fluorination, and ton quantities of various fluorocarbon materials are available from 3M Co. manufactured by new direct fluorination technology.

Metal Fluoride Method

Before the LaMar process was developed, it was generally believed that because the reaction of elemental fluorine with hydrocarbons is highly exothermic, the carbon–carbon bonds are first ruptured, creating free radicals and eventually leading to polymerization. Therefore in most instances the expected product was a complex mixture of polymers and degradation products, including pure carbon and various fragments of the original hydrocarbon material.

In the 1940s researchers developed diverse fluorination methods that avoid these problems by using certain metal fluorides, inorganic fluorides, or halogen fluorides and electrochemical or indirect methods using fluorides as reagents. Hydrogen fluoride has been successfully used as a fluorinating agent and is, in fact, still widely used industrially.

Fluorination of organic compounds using high valency metallic fluorides (14) may be represented as follows:

Exchange of halogen with fluorine of the metal fluorides, MF_n:

$$\diagdown \!\!\!\! \overset{\diagup}{\underset{\diagup}{C}}\!\!-\!X + M^+F^- \rightarrow \diagdown \!\!\!\! \overset{}{\underset{\diagup}{C}}\!\!-\!F + M^+X^- \tag{1}$$

where X = Cl, Br, or I and M = K, Sb, $AgHg_2$, or Hg

Replacement of hydrogen with the fluorine of metal fluorides:

$$\diagdown \!\!\!\! \overset{}{\underset{\diagup}{C}}\!\!-\!H + 2\,MF_n \rightarrow \diagdown \!\!\!\! \overset{}{\underset{\diagup}{C}}\!\!-\!F + HF + 2\,MF_{n-1} \tag{2}$$

Addition to double bonds:

$$\overset{\diagdown}{\underset{\diagup}{}}C\!\!=\!\!C\overset{\diagup}{\underset{\diagdown}{}} + 2\,MF_n \rightarrow F\!\!-\!\!\overset{\diagdown}{\underset{\diagup}{C}}\!\!-\!\!\overset{\diagup}{\underset{\diagdown}{C}}\!\!-\!\!F + 2\,MF_{n-1} \tag{3}$$

High valency metallic fluorides are very reactive compounds and most decompose in water. They include CoF_3, AgF_2, MnF_3, CeF_4, PbF_4, and possibly BiF_5 and UF_6. As shown in equations 1–3, at elevated temperatures, usually 100–400°C, they react with organic compounds producing the fluoro derivatives and the corresponding lower fluorides such as CoF_2 and AgF. Using cobalt trifluoride [10026-18-3] at 200–400°C, all hydrogen atoms of a hydrocarbon can be replaced by fluorine if the product is thermally stable. Similar addition of fluorine to unsaturated linkages and to aromatic nuclei takes place.

The requirement that organic compounds be vaporized at temperatures averaging 280°C across the bed of cobalt trifluoride or silver difluoride causes serious limitations to the broad applicability of the synthesis of organofluorine compounds using metal fluoride technology. There are at least two companies, Imperial Smelting, Ltd. of Britain and Air Products and Chemicals of Allentown, Pennsylvania, still active in this field; the number of organic compounds that can be prepared effectively with this technique numbers approximately 100. Fused-ring aromatic compounds are the most able to survive these harsh fluorination conditions. Fluorination of polymers is almost impossible using this method because few polymers are easily volatilized. The most effective of these processes utilizing metal fluorides, the cobalt trifluoride technology, is much less flexible than the 3M electrochemical cell (Simons' cell) or the Du Pont tetrafluoroethylene telomerization technology. In general, cobalt trifluoride fluorination has been more successful in the vapor phase than the liquid phase; yields tend to be low with extensive thermal degradation occurring at high temperatures.

Hydrogen Fluoride Electrochemical Cell Methods

Direct fluorination using hydrogen fluoride electrochemical cell methods is mechanistically similar in some regards to direct fluorination with F_2. This method uses an electrolytically activated fluoride ion produced by a Simons' designed hydrogen fluoride electrochemical cell as its primary means of fluorination. The Simons' electrochemical cell fluorination technology is practiced widely by Minnesota Mining & Manufacturing Co. (3M) of St. Paul, Minnesota. In this method, organic precursors are dissolved in liquid hydrogen fluoride and a voltage slightly under the voltage required for generation of elemental fluorine is applied across carbon electrodes. This technique, invented by J. H. Simons (15–17), has been a successful source of organofluorine compounds, functional fluids, and low molecular weight perfluorocarbon acids and diacids. There are also a number of companies in Japan and Europe that now use electrochemical fluorination for production of fluorocarbons.

The principal disadvantage to electrochemical fluorination is the requirement that the organic material be at least somewhat soluble in the polar liquid hydrogen fluoride. Therefore 3M product lines are generally based on perfluoro amines and functionalized materials such as carboxylic acids or sulfonic acids which are soluble in hydrogen fluoride (see FLUORINE COMPOUNDS, ORGANIC– FLUORINATED HIGHER CARBOXYLIC ACIDS; FLUOROETHERS AND FLUORAMINES; PERFLUOROALKANESULFONIC ACIDS). Even so this technology is not capable of producing high molecular weight functional products beyond C-8 or C-10 in yields which make the process an economically viable technique. 3M uses the electrochemical cell technology to produce a well-known line of fluids known as Fluorinerts, largely based on perfluoro amines.

Other limitations of electrochemical fluorination are that compounds such as ethers and esters are decomposed by hydrogen fluoride and cannot be effectively processed. Branching and cross-linking often take place as a side reaction in the electrochemical fluorination process. The reaction is also somewhat slow because the organic reactant materials have to diffuse within 0.3 nm of the surface of the electrode and remain there long enough to have all hydrogen replaced with fluorine. The activated fluoride is only active within 0.3 nm of the surface of the electrode.

Fluorocarbons produced by electrochemical fluorination often have small quantities (1–5%) of up to 20 by-products produced by rearrangement. Rearrangement is not characteristic of modern direct fluorination technology using elemental fluorine. By-product formation is a particular disadvantage for applications such as production of fluorocarbon oxygen carriers, fluorocarbon blood, and other biomedical fluorocarbon products where high purity materials are required. Single compound materials are essential for advantageous consideration by the U.S. Food and Drug Administration (FDA) because each often requires individual FDA approval before the mixture receives approval.

Direct Fluorination Using Elemental Fluorine

Kinetic as well as thermodynamic problems are encountered in fluorination. The rate of reaction must be decelerated so that the energy liberated may be absorbed or carried away without degrading the molecular structure. The most recent advances in direct fluorination are the LaMar process (18–20) and the Exfluor process (21–24), which is practiced commercially by 3M.

Thermochemistry. Thermodynamic considerations are of utmost importance in fluorinations. Table 1 is based on JANAF data (25) for CH_4, which indicate an average carbon–hydrogen bond strength of 410.0 kJ/mol (98 kcal/mol) based on the atomization energy of CH_4.

Table 1. Thermodynamic Data[a] for Fluorination of CH_4, kJ/mol[b]

Step	Reaction	ΔH_{25}	ΔH_{325}	ΔG_{25}	ΔG_{325}
initiation					
1a	$F_2 \rightarrow 2\,F^{\cdot}$	157.7	161.0	123.6	87.4
1b	$F_2 + RH \rightarrow R^{\cdot} + HF + F^{\cdot}$	16.3	21.3	−24.4	−79.1
propagation					
2a	$RH + F^{\cdot} \rightarrow R^{\cdot} + HF$	−141.4	−139.7	−151.2	−156.9
2b	$R^{\cdot} + F_2 \rightarrow RF + F^{\cdot}$	−289.1	−290.8	−284.9	−268.4
termination					
3a	$R^{\cdot} + F^{\cdot} \rightarrow RF$	−446.8	−451.8	−407.9	−356.0
3b	$R^{\cdot} + R^{\cdot} \rightarrow R-R$	−350.6	−347.5	−294.1	−240.6
overall reaction	$RH + F_2 \rightarrow RF + HF$	−430.5	−430.5	−432.6	−430.9

[a]Based on JANAF table data (25).
[b]To convert J to cal, divide by 4.184.

The limiting parameter to be considered in attempting to develop a satisfactory method for controlling reactions of elemental fluorine is the weakest bond in the reactant compound. For hydrocarbons the average carbon–carbon single-bond strength is 351.5–368.2 kJ/mol (84–88 kcal/mol). The overall reaction for the replacement of hydrogen by fluorine is exothermic enough [$\Delta G_{25} = -432.6$ kJ/mol (−103.4 kcal/mol)] for a fracture of carbon–carbon bonds if it were to occur via a concerted mechanism or on several adjacent carbon atoms simultaneously. This energy must be dissipated so as to avoid the fragmentation of the molecular skeleton. The comparison of 359.8 kJ/mol (86 kcal/mol) vs 430.9 kJ/mol (103 kcal/mol) has been cited in many previous discussions as an obvious basis to predict the failure of direct fluorination methods. For rapid reaction rates, which were employed in most previous experiments, this is a valid argument.

It can be seen from Table 1 that there are no individual steps that are exothermic enough to break carbon–carbon bonds except the termination of step 3a of −407.9 kJ/mol (−97.5 kcal/mol). Consequently, procedures or conditions that reduce the atomic fluorine concentration or decrease the mobility of hydrocarbon radical intermediates, and/or keep them in the solid state during reaction, are desirable. It is necessary to reduce the reaction rate to the extent that these hydrocarbon radical intermediates have longer lifetimes permitting the advantages of fluorination in individual steps to be achieved experimentally. It has been demonstrated by electron paramagnetic resonance (epr) methods (26) that, with high fluorine dilution, various radicals do indeed have appreciable lifetimes.

The two possible initiations for the free-radical reaction are step 1b or the combination of steps 1a and 2a from Table 1. The role of the initiation step 1b in the reaction scheme is an important consideration in minimizing the concentration of atomic fluorine (27). As indicated in Table 1, this process is spontaneous at room temperature [$\Delta G_{25} = -24.4$ kJ/mol (−5.84 kcal/mol)] although the enthalpy is slightly positive. The validity of this step has not yet been conclusively established by spectroscopic methods which makes it an unsolved problem of prime importance. Furthermore, the fact that fluorine reacts at a significant rate with some hydrocarbons in the dark at temperatures below −78°C indicates that step 1b is important and may have little or no activation energy at RT. At extremely low temperatures (ca 10 K) there is no reaction between gaseous fluorine and CH_4 or C_2H_6 (28).

A simple equilibrium calculation reveals that, at 25°C and atmospheric pressure, fluorine is less than 1% dissociated, whereas at 325°C an estimated 4.6% dissociation of molecular fluorine is calculated. Obviously, less than 1% of the collisions occurring at RT would result in reaction if step 1a were the only important initiation step. At 325°C the fluorine atom initiation step should become more important. From the viewpoint of energy control, as shown in Table 1, it would be advantageous to have step 1b predominate over step 2a and promote attack by molecular rather than atomic fluorine. Ambient or lower temperatures keep the atomic fluorine concentration low.

In the addition of fluorine to double bonds, the energetic situation is less severe, ie, the addition of fluorine to double bonds is only 251.4–292.9 kJ/mol (60–70 kcal/mol) exothermic per carbon–carbon bond. This energy is not sufficient to fracture the carbon skeleton if care is taken to keep addition from occurring on several adjacent carbon atoms simultaneously. Here, as in the case of hydrogen removal, the individual steps are less exothermic than the overall reaction. It has been established experimentally that less fragmentation occurs and, correspondingly, a higher yield is obtained with most conventional fluorination processes when an unsaturated rather than a saturated hydrocarbon is the starting material. This is owing to the greater exothermicity of the reaction with hydrogen, ie, 434.7 kJ/mol (103.9 kcal/mol) per saturated carbon atom as compared with 207–289.9 kJ/mol (50–70 kcal/mol) per unsaturated carbon atom. In the case of addition of fluorine to double bonds, the initiation step (eq. 4) is probably exothermic by 20.7–190.4 kJ/mol (5–46 kcal/mol) and thus plays an important role.

$$R_2C{=}CR_2 + F_2 \rightarrow R_2\dot{C}{-}CFR_2 + F^{\cdot} \qquad (4)$$

A second possibility is that a concerted mechanism (eq. 5) which is exothermic by 207.0–283.2 kJ/mol (50–58.4 kcal/mol) per carbon atom, is important.

$$R_2C{=}CR_2 + F_2 \rightarrow R_2CF{-}CFR_2 \qquad (5)$$

Steric Factors.   Initially, most of the collisions of fluorine molecules with saturated or aromatic hydrocarbons occur at a hydrogen site or at a π-bond (unsaturated) site. When collision occurs at the π-bond, the double bond disappears but the single bond remains because the energy released in initiation (eq. 4) is insufficient to fracture the carbon–carbon single bond. Once carbon–fluorine bonds have begun to form on the carbon skeleton of either an unsaturated or alkane system, the carbon skeleton is somewhat sterically protected by the sheath of fluorine atoms. Figure 2, which shows the crowded helical arrangement of fluorine around the carbon backbone of polytetrafluoroethylene (PTFE), is an example of an extreme case of steric protection of carbon–carbon bonds (29).

The nonbonding electron clouds of the attached fluorine atoms tend to repel the oncoming fluorine molecules as they approach the carbon skeleton. This reduces the number of effective collisions, making it possible to increase the total

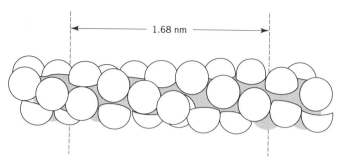

Fig. 2. The steric protection of the carbon backbone by fluorine of a polytetrafluoroethylene chain. The helical configuration with a repeat distance of 1.68 nm results from the steric crowding of adjacent fluorine.

number of collisions and still not accelerate the reaction rate as the reaction proceeds toward completion. This protective sheath of fluorine atoms provides the inertness of Teflon and other fluorocarbons. It also explains the fact that greater success in direct fluorination processes has been reported when the hydrocarbon to be fluorinated had already been partially fluorinated by some other process or was prechlorinated, ie, the protective sheath of halogens reduced the number of reactive collisions and allowed reactions to occur without excessive cleavage of carbon–carbon bonds or runaway exothermic processes.

Kinetic Control. In direct fluorination processes, concentration, time, and temperature can be controlled. In most previous work, the fluorine was diluted with an inert gas such as nitrogen, helium, or even carbon dioxide. However, the concentration of fluorine in the reactor was kept at a constant level, usually 10% or greater, by premixing the inert gas with fluorine in the desired proportion and then introducing this mixture into the reactor. The rate of reaction between a hydrocarbon compound and a 10% fluorine mixture is relatively high, and this very exothermic process can lead to fragmentation and, in some cases, to combustion. The initial stages of reaction are most critical; nearly all the fragmentation occurs at this time. An initial concentration of 10% fluorine or more is, for most compounds, much too high for nondestructive fluorination.

Molecular relaxation processes such as vibrational or rotational relaxations or thermal conduction make it possible to dissipate the energy released during fluorination. Such relaxation processes can minimize the chances that the energy required to break the weakest bond is appropriately localized if the reaction sites are widely distributed over the system. Therefore, in the initial stages of fluorination, it is necessary to reduce the probability of adjacent reaction sites simultaneously occurring in the same molecule, or in adjacent molecules in a crystal, by diluting the reactants and relying on relaxation processes to distribute the energy over the entire system and thus avoid fragmentation.

Reactant molecules are able to withstand more fluorine collisions, as they become more highly fluorinated, without decomposition because some sites are sterically protected, ie, collisions at carbon–fluorine sites are obviously nonreactive. The fluorine concentration may therefore be increased as the reaction proceeds to obtain a practical reaction rate. Actual dilution schemes to achieve successful fluorination must be individually tailored for specific reaction systems and may, in some cases, include a stepwise procedure. In Figure 3, the horizontal line at about 10 kPa (10%) represents the 1940s approach to direct fluorination. High initial concentrations result in extensive fragmentation. The curved lines that asymptotically approach 101.3 kPa (1 atm) of fluorine pressure or 50.6 kPa (0.5 atm) of fluorine pressure, etc, represent the controlled approach of the LaMar process (18–20).

To achieve the very low initial fluorine concentration in the LaMar fluorination process initially a helium or nitrogen atmosphere is used in the reactor and fluorine is bled slowly into the system. If pure fluorine is used as the incoming gas, a concentration of fluorine may be approached asymptotically over any time period (Fig. 3). It is possible to approach asymptotically any fluorine partial pressure in this manner. The very low initial concentrations of fluorine in the system greatly decreases the probability of simultaneous fluorine collisions on the same molecules or on adjacent reaction sites.

Thus, for a successful fluorination process involving elemental fluorine, the number of collisions must be drastically reduced in the initial stages; the rate of fluorination must be slow enough to allow relaxation processes to occur and a heat sink must be provided to remove the reaction heat. Most direct fluorination reactions with organic compounds are performed at or near room temperature unless reaction rates are so fast that excessive fragmentation, charring, or decomposition occurs and a much lower temperature is desirable.

Low temperature fluorination techniques ($-78°C$) are promising for the preparation of complex fluorinated molecules, especially where functional groups are present (30), eg, fluorination of hexamethylethane to perfluorohexamethylethane [39902-62-0], of norbornane to perfluoro- (C_7F_{12}) and 1-hydro undecafluoronorbornane [4934-61-6], C_7HF_{11}, and of adamantane to 1-hydropentadecafluoroadamantane [54767-15-6].

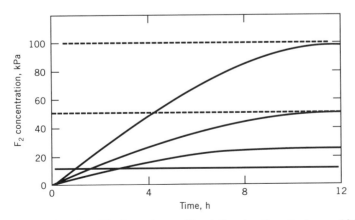

Fig. 3. Fluorine dilution scheme direct fluorination. 1 atm = 101.3 kPa.

Experimental Techniques

In early reaction systems (9,10,31,32) the vaporized hydrocarbon was combined with nitrogen in a reactor and mixed with a nitrogen–fluorine mixture from a preheated source. The jet reactor (11) for low molecular weight fluorocarbons was an important improvement. The process takes place at around 200–300°C, and fluorination is carried out in the vapor state.

At 200–300°C many compounds, both organic and inorganic, are marginally stable, and certainly not in their lowest vibrational states. They may even undergo some pyrolytic decomposition. Thus the addition of extra energy produced by the interaction of fluorine with these compounds is likely to produce substantial fragmentation. Vaporization of high molecular weight hydrocarbons is rather difficult, and their fluorination was not successful in the early jet fluorination studies.

The typical fluorination apparatus used in the LaMar process for these reactions is simple in design (Fig. 4) (33). It is essential that the materials of construction are resistant to fluorine (34). The presence of even traces of oxygen or moisture can have a deleterious effect and, therefore, extreme precautions must be taken to eliminate these contaminants.

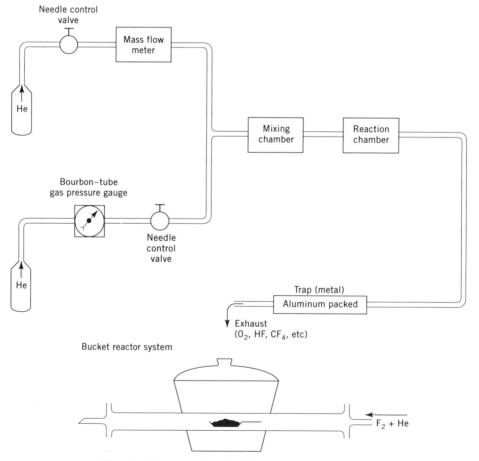

Fig. 4. Diagram of typical fluorination apparatus.

The connections are conveniently made of 0.635 cm OD copper tubing. When the fluorocarbon produced in the reaction is volatile, a cold finger-type trap can be placed between the reaction chamber and the trap to catch volatile products. The temperature of the trap must be high enough to pass unreacted or excess fluorine and nitrogen but cold enough to condense the reaction products. Before the reaction is started, the whole system is purged with helium or nitrogen for ca 30 minutes and then 0.5–2 mL of fluorine and 50–100 mL of nitrogen per minute are passed through the system. Solid material has to be ground to a very fine powder (37 mm (~100 mesh)) to achieve complete fluorination.

A special cryogenic reactor (35) in which the reactions of fluorine with liquid and gaseous samples can be controlled at very low temperatures is shown in Figure 5. Reactants are volatilized into the reaction zone of the cryogenic reactor from the heated oil evaporator prior to initiation of the reaction. The main reaction chamber is a nickel tube, 2.54 cm in diameter, packed with copper turnings. The compartments (10.1 × 10.1 × 20.2 cm) are constructed of stainless steel and insulated with urethane foam and act as heat sinks. All connections are made of 0.635 cm copper or aluminum tubing. A sodium fluoride trap is used to remove the hydrogen fluoride from the reaction products. By cooling or warming the compartments, they can be used to create a temperature gradient along the reaction tube. Because the products are highly fluorinated, they are usually volatile and tend to move through the reactor tube rapidly, depending on the temperature gradient. This provides a continually renewed surface of reactant at the optimum temperature for fluorination. Fluorinated copper turnings effectively increase the surface area of the compound exposed to fluorine. The individual zones of the reactor may be cooled with various solvent–solid carbon dioxide or with solvent–liquid nitrogen slushes. Preferably, the temperature is precisely regulated with

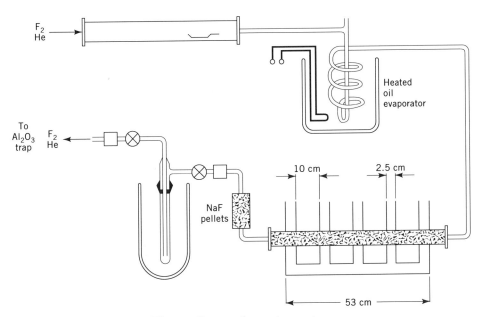

Fig. 5. Cryogenic reactor system.

an automatic liquid nitrogen temperature controller. In addition to the four-zone reactor shown in Figure 5, a multizone reactor can also be used; an eight-zone reactor has been found to be particularly efficient (36). Internal Freon cooling is effective for controlling the temperatures of the various compartments (37).

Oxygen or moisture has to be excluded because the presence of oxygen leads to cross-linking, presumably with epoxy bridges, to carbonyl groups, which give acid fluorides, and to peroxides (38). Cross-linking obviously decreases the yield of pure perfluorocarbon. It can be detected by infrared absorption in the 1600–2000 cm^{-1} region and by noting the polymeric nature of the products, ie, high melting points, low vapor pressures, etc. On the other hand, oxyfluorination is a technique offering unique possibilities for the functionalization of fluorocarbons and for the preparation of functional fluorocarbon membranes (39).

Aerosol-Based Direct Fluorination.   A technology that works on liter and half-liter quantities has been introduced (40–42). This new aerosol technique, which functions on principles similar to LaMar direct fluorination (Fig. 5), uses fine aerosol particle surfaces rather than copper filings to maintain a high surface area for direct fluorination. The aerosol direct fluorination technique has been shown to be effective for the synthesis of bicyclic perfluorocarbon such as perfluoroadamantane, perfluoroketones, perfluoroethers, and highly branched perfluorocarbons.

Modern Direct Fluorination.   Direct fluorination technology has been scaled up at Exfluor Research Corp. of Austin, Texas (21–24). The synthesis of perfluoroethers by this method is licensed to 3M Co. where it is practiced on a commercial (multiton) scale. Using direct fluorination it is possible to produce almost any desired fluorocarbon structure for which there is a hydrocarbon or organic structural precursor (22–24,43). There are two basic approaches to controlling direct fluorination: the LaMar method where the rate of fluorine addition is the limiting factor, and the Lagow-Exfluor method in which the rate of addition of the hydrocarbon is the limiting factor. A highly effective solvent fluorination technology with rapid heat transfer has been developed on this principle by Exfluor Research Corp. Multikilogram quantities of new fluorocarbons are produced in yields ranging from 95 to 99% with this technology (21). This technique is capable of producing very high molecular weight perfluoro acids and diacids that are precursors for new fluorocarbon copolymers. This is accomplished rapidly and on a commercial scale with essentially no branching or rearrangement.

Applications

In 1954 the surface fluorination of polyethylene sheets by using a solid CO_2 cooled heat sink was patented (44). Later patents covered the fluorination of PVC (45) and polyethylene bottles (46). Studies of surface fluorination of polymer films have been reported (47). The fluorination of polyethylene powder was described (48) as a fiery intense reaction, which was finally controlled by dilution with an inert gas at reduced pressures. Direct fluorination of polymers was achieved in 1970 (8,49). More recently, surface fluorinations of poly(vinyl fluoride), polycarbonates, polystyrene, and poly(methyl methacrylate), and the surface fluorination of containers have been described (50,51). Partially fluorinated poly(ethylene terephthalate) and polyamides such as nylon have excellent soil release properties as well as high wettability (52,53). The most advanced direct fluorination technology in the area of single-compound synthesis and synthesis of high performance fluids is currently practiced by 3M Co. of St. Paul, Minnesota, and by Exfluor Research Corp. of Austin, Texas.

The following companies manufacture organic fluorine compounds by direct fluorination techniques: 3M; Exfluor; Air Products and Chemicals, Inc., Allentown, Pennsylvania; MarChem, Inc., Houston, Texas; Ozark-Mahoning, Inc., Tulsa, Oklahoma; and PCR, Inc., Gainesville, Florida.

Simple and Complex Organic Molecules. Using modern direct fluorination technology, the synthesis of even the most complex perfluorocarbon structures from hydrocarbon precursors is now possible. For example, syntheses of the first perfluoro crown ethers, perfluoro 18-crown-6, perfluoro 15-crown-5, and perfluoro 12-crown-4 (54) have been reported. Perfluoro crown ethers (54,55) are becoming important as the molecules of choice for many ^{19}F-nmr imaging applications (56) in humans and are particularly effective in brain and spinal diagnostics when

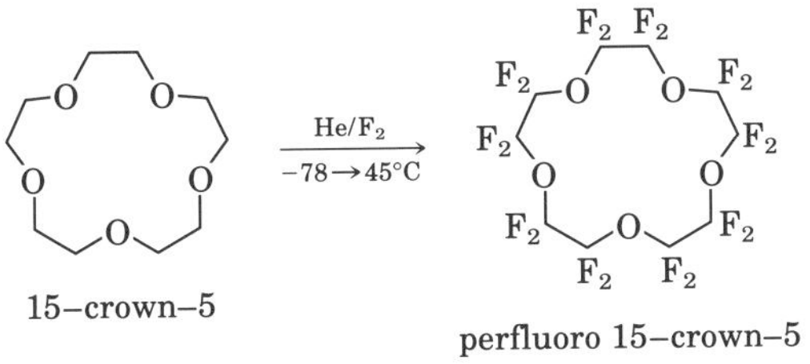

15–crown–5

perfluoro 15–crown–5

administered to the cerebrospinal fluid compartment. Synthesis scale-up of perfluoro 15-crown-5 (54,55) and plans for commercialization are underway while research is being conducted on other biological applications of these new compounds (57). In collaboration with Air Products, excellent brain imaging scans have been obtained by infusing the perfluoro 15-crown-5 in spinal fluids. Toxicology reports on these are very favorable with essentially no toxic effects physiologically in several different animals.

Perfluoro crown ethers from the hydrocarbon dibenzo crown ethers have also been synthesized (58) and the first perfluorocryptand molecule [2.2.2] has been reported (59). The perfluorocryptand is a stable, inert, high boiling clear oil.

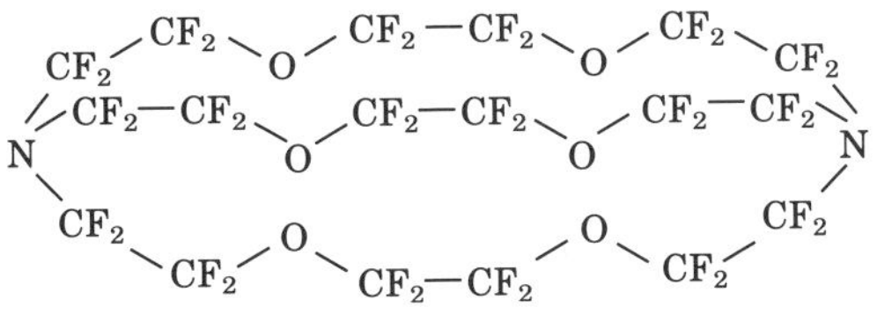

Hydrocarbon crown ethers coordinate cations; however, both the perfluoro crown ethers and the perfluorocryptands coordinate anions. For example, perfluoro crown ethers and perfluorocryptands tenaciously encapsulate O_2^- and F^- (60,61).

The first several perfluoro spiro compounds have also been synthesized (62). An example of this technology is the synthesis and crystal structure of perfluoro-1,4,9,12-tetraoxadispiro[4.2.4.2]tetradecane.

Many novel small molecule perfluoropolyethers have been made using direct fluorination technology. For example, even branched ethers such as perfluoro(pentaerythritol tetramethyl ether) can be prepared:

$$C{-}(CH_2{-}O{-}CH_3)_4 \xrightarrow{F_2} C{-}(CF_2{-}O{-}CF_3)_4$$

Very good low temperature fluids are obtained by direct fluorination of trialkylorthoformates (63):

$$HC(OCH_2CH_3)_3 \xrightarrow{F_2/He} FC(OCF_2CF_3)_3$$

Perfluoropolyethers emerged on the market in the early 1970s; however, for the next 15 years there were only two basic structures known. The first perfluoropolyether was the homopolymer of hexafluoropropylene oxide produced by Du Pont having the structure

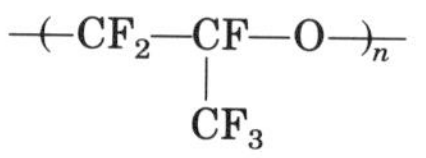

$$-(CF_2{-}\underset{\underset{CF_3}{|}}{CF}{-}O)_n{-}$$

Du Pont called this new lubricant material Krytox (64,65) and initially it had such extraordinary properties that it sold for \$200/kg (\$187/kg ca 1993). Krytox was and is used in most of the vacuum pumps and diffusion oil pumps for the microelectronics industry in this country and in Japan because it produces no hydrocarbon (or fluorocarbon) vapor contamination. It has also found important applications in the lubrication of computer tapes and in other data processing applications as well as military and space applications.

Materials similar in high temperature properties to the Du Pont material with better low temperature properties have been synthesized using direct fluorination. The first was produced by reaction of fluorine with inexpensive hydrocarbon polyethers such as poly(ethylene oxide). In the simplest case, poly(ethylene oxide) is converted to the perfluoroethylene oxide polymer:

$$HO{-}(CH_2CH_2O)_n{-}H \xrightarrow{F_2/He} R_f(OCF_2CF_2)_nOR_f$$

This simple reaction chemistry was first reported in 1978 (66).

Other interesting perfluoro ether structures can be obtained by copolymerization of hexafluoroacetone with ethylene oxide, propylene oxide, and trimethylene oxide with subsequent fluorination to yield the following structures (67):

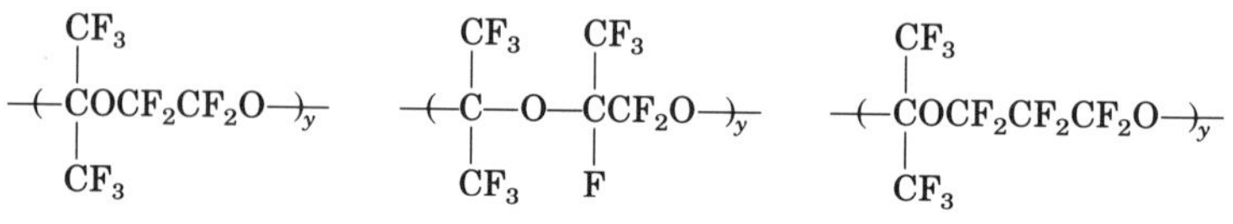

$$-(\underset{\underset{CF_3}{|}}{\overset{\overset{CF_3}{|}}{C}}OCF_2CF_2O)_y- \qquad -(\underset{\underset{CF_3}{|}}{\overset{\overset{CF_3}{|}}{C}}{-}O{-}\underset{\underset{F}{|}}{\overset{\overset{CF_3}{|}}{C}}CF_2O)_y- \qquad -(\underset{\underset{CF_3}{|}}{\overset{\overset{CF_3}{|}}{C}}OCF_2CF_2CF_2O)_y-$$

Two of the perfluoropolyether fluid structures yet to be commercialized are interesting. The first structure is a strictly alternating copolymer of ethylene oxide and methylene oxide, which has the longest liquid range of any molecule containing carbon (40). The second structure is the perfluoromethylene oxide polyether which has low temperature liquid properties down to $-120°C$:

$$-\!\!\left(\!CF_2O\!-\!CF_2CF_2O\!\right)_{\!\overline{n}} \qquad -\!\!\left(\!CF_2O\!-\!CF_2\!-\!O\!\right)_{\!\overline{n}}$$

Other perfluoropolyether structures that have been synthesized are (24,43)

$$-\!\!\left(\!\underset{\underset{\displaystyle CF_2CF_3}{|}}{CF_2CFO}\!\right)_{\!\overline{n}} \qquad -\!\!\left(\!\underset{\underset{\displaystyle CF_2Cl}{|}}{CF_2CFO}\!\right)_{\!\overline{n}}$$

Hydrocarbon Polymers.   It is difficult to produce perfluorocarbon polymers by the usual methods. Many monomers, such as hexafluoropropylene, polymerize only slowly because of the steric hindrance of fluorine. Furthermore, some monomers are not very stable and are difficult to synthesize. Direct fluorination can be used for the direct synthesis of fluorocarbon polymers (68–70) and for producing fluorocarbon coatings on the surfaces of hydrocarbon polymers (8,29,44–47,49,68–71).

Thus fluorocarbon polymers can be produced with chemical compositions similar to polytetrafluoroethylene by the direct reaction of fluorine with polyethylene and the perfluoro analogues of polypropylene and polystyrene can be prepared. These fluorocarbon polymers differ from the more familiar linear structures because carbon–carbon cross-linking occurs to a significant extent during fluorination. Most of these fluoropolymers are white solids with high thermal stability; some are stable in air as much as 200°C above the ignition temperatures of their corresponding hydrocarbon precursors. Hydrocarbon polymers such as polyethylene and the new surfaces formed by direct fluorination have been studied by esca which shows that the surface is truly converted to a fluorocarbon polymer (72). Most of these fluorocarbon surfaces are inert and many of them have good lubricant properties. The fluorination of PVC has been followed with esca and $(CF_x)_n$ polymer was also identified (73).

Surface Fluorination of Polymers.   Fluorocarbon-coated objects have many practical applications because the chemically adherent surface provides increased thermal stability, resistance to oxidation and corrosive chemicals and solvents, decreased coefficient of friction and thus decreased wear, and decreased permeability to gas flow. Unusual surface effects can be obtained by fluorinating the polymer surfaces only partially (74).

Natural and Synthetic Rubber.   Fluorination of natural or synthetic rubber creates a fluorocarbon coating (29,75,76) which is very smooth and water repellent (see WATERPROOFING). Rubber articles such as surgical gloves, O-rings, gaskets, and windshield wiper blades can be fluorinated on the surface while the interior retains the elastic, flexible properties of the natural rubber. Fluorinated O-rings can be used without extra lubricant in corrosive atmospheres since the fluorocarbon is unreactive. In food-processing equipment, grease or lubricants are eliminated and do not contaminate the food products. Fluorinated O-rings have smooth surfaces, very low frictional coefficients, and enhanced thermal stabilities. Fluorinated windshield wiper blades have a very low coefficient of friction, run smoother with less squeak, their surface is more resistant to the sun's uv radiation and attack by ozone, and they require less electrical energy for operation.

Many applications of this technique are apparent in medicine, such as surgical rubber gloves, rubber sheets, drain tubes, catheters, etc. Since talcum or other lubricating powder often used with surgical gloves can cause allergic reactions, thin powderless gloves are desirable. Teflon or silicone films have proved unsuccessful for surgical gloves. However, an excellent direct fluorination process has been developed (76) whereby the inside surface is fluorinated under expanded conditions at elevated temperatures. A very smooth surface is obtained and powder is not required. At the same time, the outside surface remains rough so that surgical instruments can be held firmly without slippage. The tactile sensitivity of the tips of the fingers seems to be increased, and the problem of powder forming lumps or a mud-like slush inside the glove is eliminated.

Blow-Molded Containers. A surface-fluorination process (Airopak) has been developed by Air Products & Chemicals for the blow-molding industry to produce solvent-resistant polyolefin containers. In this application, the air that is normally used to blow-mold containers is replaced by a low concentration of fluorine in nitrogen. Airopak containers produced by this process show outstanding resistance to nonpolar solvents (50,51,77,78) and such blow-molding fluorination procedures have been widely used for the last 10 years by Ford Motor Co. and many European auto manufacturers to produce low cost–high performance gas tanks for cars and trucks. A similar technology is now practiced by Fluoroseal, Inc. (Houston, Texas). This group has generated a successful product line based on post-treating containers and other objects with elemental fluorine.

BIBLIOGRAPHY

"Direct Fluorination" under "Fluorine Compounds, Organic" in *ECT* 3rd ed., Vol. 10, pp. 840–855, by J. L. Margrave. R. H. Hauge, and R. B. Badachhape, Rice University, and R. J. Lagow, University of Texas.

1. H. Moissan, *Le Fluor et ses Composes*, Steinbeil, Paris, 1900.
2. P. Kamarchik and J. L. Margrave, *Acc. Chem. Res.* **11**, 296 (1978); U.S. Pat. 3,519,657 (July 7, 1970), G. A. Olah (to Dow Chemical Co.,); L. B. Ebert, J. I. Brauman, and R. A. Huggins, *J. Am. Chem. Soc.* **96**, 7841 (1974).
3. O. Ruff and O. Bretschneider, *Z. Anorg. Allgem. Chem.* **217**, 1 (1937); W. Rudorff and G. Rudorff, *Z. Anorg. Allgem. Chem.* **253**, 281 (1947); W. Rudorff and G. Rudorff, *Chem. Ber.* **80**, 413 (1947); W. Rudorff and K. Brodersen, *Z. Naturforsch* **12b**, 575 (1957); W. Rudorff, *Adv. Inorg. Chem. Radiochem.* **1**, 230 (1959).
4. A. K. Kuriakose and J. L. Margrave, *J. Phys. Chem.* **69**, 2772 (1965); U.S. Pat. 3,674,432 (July 4, 1972), J. L. Margrave, R. J. Lagow, and co-workers (to R. I. Patents, Inc.); J. L. Margrave, R. J. Lagow, and co-workers, *J. Am. Chem. Soc.* **96**, 1268, 2628 (1974); R. B. Badachhape, V. K. Mahajan, and J. L. Margrave, *Inorg. Nucl. Chem. Lett.* **10**, 1103 (1974).
5. N. Watanabe and M. Ishii, *J. Electrochem. Soc. Jpn.* **29**, 364 (1961); N. Watanabe, Y. Koyama, and S. Yoshizawa, *J. Electrochem. Soc. Jpn.* **31**, 756 (1963); N. Watanabe and K. Kumon, *J. Electrochem. Soc. Jpn.* **35**, 19 (1967); H. Imoto and N. Watanabe,

Bull. Chem. Soc. Jpn. **49**, 1736 (1976); M. Takashima and N. Watanabe, *Nippon Kagaku Kaishi*, 1222 (1976); N. Watanabe, Y. Kita, and T. Kawaguchi, *Nippon Kagaku Kaishi*, 191 (1977).

6. Y. Kita, N. Watanabe, and Y. Fujii, *J. Am. Chem. Soc.* **101**, 3823 (1979); R. J. Lagow and co-workers, *Inorg. Metalorg. Chem.* **2**, 145 (1972).

7. P. Kamarchik and J. L. Margrave, *J. Therm. Anal.* **11**, 259 (1977).

8. J. L. Margrave, R. J. Lagow, and co-workers, *IR-100 Award for CFX, a Lubricant Powder*, Rice University, Houston, Tex., 1970; *Ind. Res.* **12**, 47 (1970).

9. G. H. Cady and co-workers, *Ind. Eng. Chem.* **39**, 290 (1947).

10. R. N. Haszeldine and F. Smith, *J. Chem. Soc.*, 2689, 2787 (1950).

11. L. A. Bigelow, in J. H. Simons, ed., *Fluorine Chemistry*, Vol. 1, Academic Press, Inc., New York, 1970, p. 373; E. A. Tyczkowski and L. A. Bigelow, *J. Am. Chem. Soc.* **77**, 3007 (1955).

12. C. M. Sharts, *J. Chem. Ed.* **45**, 3 (1968).

13. R. E. Banks, *Fluorocarbons and Their Derivatives*, Oldbourne Press, London, 1964, p. 87.

14. W. A. Shepherd and C. M. Sharts, *Organic Fluorine Chemistry*, W. A. Benjamin, New York, 1969; M. Hudlicky, *Chemistry of Organic Fluorine Compounds*, 2nd ed., Halsted Press, a division of John Wiley & Sons, Inc., New York, 1976; M. Stacey and J. C. Tatlow, in *Advances in Fluorine Chemistry*, Vol. 1, Butterworths Scientific Publications, London, 1960, pp. 166–198.

15. J. H. Simons, *Trans. Electrochem. Soc.* **95**, 47 (1949).

16. J. H. Simons, *Fluorine Chemistry*, Vol. 1, Academic Press, Inc., New York, 1950, p. 401.

17. U.S. Pat. 2,490,099 (Dec. 6, 1949), J. H. Simons (to Minnesota Mining & Manufacturing Co.).

18. R. J. Lagow and J. L. Margrave, *Proc. Natl. Acad. Sci.* **67**(4)8A (1970).

19. R. J. Lagow, Ph.D. dissertation, Rice University, Houston, Tex., 1970.

20. B. Fegley, *MIT Technol. Eng. News*, 13 (Apr. 1973).

21. U.S. Pat. 5,093,432 (Mar. 3, 1992), T. R. Bierschenk, R. J. Lagow, T. J. Juhlke, and H. Kawa (to Exfluor Research Corporation).

22. U.S. Pat. 4,760,198 (July 26, 1988), T. R. Bierschenk, T. J. Juhlke, and R. J. Lagow (to Exfluor Research Corp.).

23. U. S. Pat. 4,827,042 (May 2, 1989), T. R. Bierschenk, T. J. Juhlke, and R. J. Lagow (to Exfluor Research Corp.).

24. U.S. Pat. Appl. Ser. No. 07/982,030 (Nov. 24, 1992), T. R. Bierschenk, R. J. Lagow, T. J. Juhlke, and H. Kawa (to Exfluor Research Corp.).

25. D. R. Stull and co-eds, *JANAF Thermochemical Tables*, 2nd ed., *NSRDS-NBS37*, U. S. Government Printing Office, Washington, D.C., June 1971, and subsequent revisions.

26. R. E. Florin and L. A. Wall, "Electron Spin Resonance Studies on Fluorination of Polymers," abstract no. 8, Fluorine Chemistry Div., *165th American Chemical Society Meeting*, Dallas, Tex., Apr. 10, 1973.

27. W. T. Miller, S. D. Koch, and F. W. McLafferty, *J. Am. Chem. Soc.* **78**, 4992 (1956).

28. R. H. Hauge, J. Wang, and J. L. Margrave, paper presented at the *First Winter Fluorine Conference*, St. Petersburg, Fla., Jan. 1972.

29. C. W. Bunn and E. R. Howell, *Nature* **174**, 549 (1954).

30. N. J. Maraschin and R. J. Lagow, *J. Am. Chem. Soc.* **94**, 8601 (1972).

31. O. Ruff, *Die Chemie des Fluors*, Springer, Berlin, 1920.

32. W. Bockemuller, *Organische Fluorverbindungen*, F. Enke, Stuttgart, 1936.

33. U.S. Pat. 3,758,450 (Sept. 11, 1973), R. J. Lagow and J. L. Margrave (to R. I. Patents, Inc.); U.S. Pat. 3,775,489 (Nov. 27, 1973), R. J. Lagow and J. L. Margrave (to R. I. Patents, Inc.)

34. H. F. Priest and A. V. Grosse, *Ind. Eng. Chem.* **39**, 279 (1947); R. Landau and R. Rosen, *Ind. Eng. Chem.* **39**, 281 (1947).

35. N. J. Maraschin and co-workers, *J. Am. Chem. Soc.* **97**, 513 (1975).

36. U.S. Pat. 3,904,501 (Sept. 9, 1975), R. J. Lagow and co-workers (to Massachusetts Institute of Technology).

37. U.S. Pat. 4,281,119 (July 28, 1981), R. J. Lagow and co-workers (to Massachusetts Institute of Technology).

38. U.S. Pat. 3,480,667 (Nov. 25, 1969), W. R. Siegart and W. D. Blackley.

39. S. Inoue, J. L. Adcock, and R. J. Lagow, *J. Am. Chem. Soc.* **100**, 1948 (1978).

40. J. L. Adcock, K. Horita, and E. B. Renh, *J. Am. Chem. Soc.* **103**, 6932 (1981).

41. J. L. Adcock and M. L. Robin, *J. Org. Chem.* **49**, 1442 (1984).

42. J. L. Adcock, *J. Fluorine Chem.* **33**, 327 (1986).

43. U.S. Pat. 4,931,199 (June 5, 1990), T. R. Bierschenk, R. J. Lagow, T. J. Juhlke, and H. Kawa (to Exfluor Research Corp.).

44. Brit. Pat. 710,523 (June 16, 1954), A. J. Rudge (to Exfluor Research Corp.).

45. U.S. Pat. 2,497,046 (Feb. 7, 1950), E. L. Kopra (to American Cyanamid Co.).

46. U.S. Pat. 2,811,468 (Oct. 29, 1957), S. P. Joffre (to Shulton, Inc.).

47. H. Schonhorn and R. Hansen, *J. Appl. Polym. Sci.* **12**, 1231 (1968).

48. M. Okade and K. Makuuchi, *Ind. Eng. Chem. Prod. Res. Dev.* **8**, 334 (1969).

49. *Chem. Eng. News* **48**, 40 (Jan. 12, 1970); *Chemistry* **43**(4), 30 (1970).

50. U.S. Pat. 3,862,284 (Jan. 21, 1975), D. D. Dixon, D. G. Manly, and G. W. Recktenwald (to Air Products and Chemicals, Inc.).

51. *Ind. Res. Dev.* **12**, 102 (1978).

52. U.S. Pats. 3,988,491 (Oct. 26, 1976), and 4,020,223 (Apr. 26, 1977), D. D. Dixson and L. J. Hayes (to Air Products and Chemicals, Inc.).

53. L. J. Hayes, *J. Fluorine Chem.* **8**, 69 (1976).

54. W. H. Lin, W. I. Bailey, Jr., and R. J. Lagow, *J. Chem. Soc., Chem. Commun.*, 1550 (1985).

55. U.S. Pat. 4,570,005 (Feb. 11, 1986), W. H. Lin and R. J. Lagow (to University of Texas System).

56. U.S. Pat. 4,838,274 (June 13, 1989), F. K. Schweighardt and J. A. Rubertone (to Air Products and Chemicals, Inc.).

57. T. Y. Lin, L. C. Clark, Jr., and R. J. Lagow, to be published, 1993.

58. T. Y. Lin and R. J. Lagow, *J. Chem. Soc., Chem. Commun.*, 12 (1991).

59. W. D. Clark and R. J. Lagow, *J. Org. Chem.* **55**, 5933 (1990).

60. J. Brodbelt, R. J. Lagow, and co-workers, *J. Am. Chem. Soc.* **113**, 5913 (1991).

61. J. Brodbelt, R. J. Lagow, and co-workers, *J. Chem. Soc., Chem. Commun.*, 1705 (1991).

62. T. Y. Lin and R. J. Lagow, in print, 1994.

63. R. J. Lagow, T. E. Mlsna and co-workers, *Eur. J. Solid State and Inorg. Chem.* **29**, 907 (1992).

64. J. T. Hill, *J. Macromol. Sci., Chem.* **8**, 499 (1974).

65. H. S. Eleuterio, *J. Macromol. Sci., Chem.* **6**, 1027 (1972).

66. G. E. Gerhardt and R. J. Lagow, *J. Org. Chem.* **43**, 4505 (1978).

67. D. F. Persico and R. J. Lagow, *Macromolecules* **18**, 1383 (1985).

68. J. L. Margrave and R. J. Lagow, *J. Polym. Sci. Polym. Lett. Ed.* **12**, 177 (1974).

69. A. J. Otsuka and R. J. Lagow, *J. Fluorine Chem.* **4**, 371 (1974).

70. R. J. Lagow, H. Kawa and co-workers, *J. Polym. Sci., Polym. Lett. Ed.* **28**, 297 (1990).

71. U.S. Pat. 3,647,613 (Mar. 7, 1972), J. L. Scotland (to British Resin Products Ltd.).

72. D. T. Clark and co-workers, *J. Polym. Sci. Polym. Chem. Ed.* **13**, 857 (1975); D. T. Clark and co-workers, in L. H. Lee, ed., *Advances in Friction Wear*, Vol. 5A, Plenum Press, New York, 1975, p. 373.

73. G. Parks, Ph.D. dissertation, Rice University, Houston, Tex., 1976.

74. J. Pederson, M.A. thesis, Rice University, Houston, Tex., 1979.
75. Can. Pat. 1,002,689 (Dec. 28, 1976), R. J. Lagow and J. L. Margrave (to DAMW Associates); Brit. Pat. 1,440,605 (Oct. 20, 1976), R. J. Lagow and J. L. Margrave (to DAMW Associates).
76. U.S. Pat. 3,992,221 (Nov. 16, 1976), R. B. Badachhape, C. Homsy, and J. L. Margrave (to Vitek Inc. and MarChem, Inc.).
77. A. J. Woytek and J. F. Gentilecore, "A New Blow Molding Process to Resist Solvent Permeation of Polyolefin Containers," paper no. 13, presented at *Advances in Blow Molding Conference*, Rubber and Plastics Institute, London, Dec. 6, 1977.
78. J. F. Gentilecore, M. A. Triolo, and A. J. Woytek, *Plast. Eng.* **34**, 23 (1978).

RICHARD J. LAGOW
University of Texas at Austin

FLUORINATED ALIPHATIC COMPOUNDS

The hydrogen atoms in alkanes can be partially or completely replaced by fluorine. Partially fluorinated alkanes are commonly called hydrofluorocarbons (HFCs) and the fully fluorinated derivatives are perfluorocarbons (PFCs). Alkanes whose hydrogens are replaced by both fluorine and chlorine are designated chlorofluorocarbons (CFCs), or hydrochlorofluorocarbons (HCFCs) if the replacement is incomplete. Similar designations are used for other halogenated fluorocarbons. Fluorinated aliphatics are further identified by a series of numbers related to the formula of the compound. In this numbering system for methane and ethane derivatives, the first digit on the right is the number of fluorine atoms in the compound, and the second digit from the right is one more than the number of hydrogen atoms. The third digit from the right is one less than the number of carbon atoms, but when this digit is zero, it is omitted. The remaining available positions in the compound are taken by chlorine atoms unless specified otherwise. For example, CCl_3F, $CHClF_2$, CF_3CHF_2, and CF_3CF_3 are designated CFC-11, HCFC-22, HFC-125, and PFC-116, respectively. When bromine is present, the same rules apply except that the letter B is used, followed by a number that indicates the number of chlorine atoms replaced by bromine. For example, CF_3Br and $CHBrF_2$ are coded BFC-13B1 and HBFC-22B1, respectively. In the fire extinguishing trade, the brominated derivatives are usually called Halons and have a different numbering system in which the digits from right to left are respectively the number of bromine, chlorine, fluorine, and carbon atoms. Any remaining available positions are hydrogen atoms. The above BFC and HBFC become H-1301 and H-1201 in this system. This numbering system has been extended systematically to both acyclic and cyclic compounds with more than two carbon atoms, but the code for distinguishing various isomers becomes rather complex (1).

Perfluorocarbons and Hydrofluorocarbons

Properties. Aliphatic PFCs have an unusual combination of physical properties relative to their hydrocarbon counterparts (2–5). The volatilities of PFCs are much higher than expected based on their molecular weights. For example, tetrafluoromethane, mol wt 88, boils at $-128°C$, whereas n-hexane, mol wt 86, boils at $+69°C$. Perfluorocarbons containing up to four carbon atoms boil somewhat higher than the corresponding hydrocarbons; the reverse is true of PFCs with more carbon atoms. Liquid PFCs are two to three times as dense as hydrocarbons with the same carbon skeleton, and aliphatic PFCs have among the lowest dielectric constants, refractive indexes, and surface tensions of any liquids at room temperature. The compressibilities and absolute viscosities of PFCs are considerably higher than those of hydrocarbons. Aliphatic PFCs are poor solvents for all materials except for those with low cohesive energies, such as gases and other PFCs. They are practically insoluble in water and only slightly soluble in hydrocarbons.

The extremely nonpolar character of PFCs and very low forces of attraction between PFC molecules account for their special properties. Perfluorocarbons boil only slightly higher than noble gases of similar molecular weight, and their solvent properties are much more like those of argon and krypton than hydrocarbons (2). The physical properties of some PFCs are listed in Table 1.

The physical properties of hydrofluorocarbons reflect their polar character, and possibly the importance of intermolecular hydrogen bonding (3). Hydrofluorocarbons often boil higher than either their PFC or hydrocarbon counterparts. For example, $1\text{-}C_6H_{13}F$ boils at $91.5°C$ compared with $58°C$ for $n\text{-}C_6F_{14}$ and $69°C$ for $n\text{-}C_6H_{14}$. Within the series of fluorinated methanes, the boiling point reaches a maximum for CH_2F_2, which contains an equal number of hydrogen and fluorine atoms for maximum hydrogen bonding. The methane boiling points, however, also parallel their dipole moments, which reflect relative polar character: CH_3F ($\mu = 1.85D$), CH_2F_2 (1.97D), CHF_3 (1.65D), CF_4 (0.0D) ($1D = 3.336 \times 10^{-30}$ C·m).

Hydrofluorocarbons invariably have higher refractive indexes, dielectric constants, and surface tensions, but lower densities than their PFC counterparts. The physical properties of some HFCs are listed in Table 2. Because of their very strong carbon–fluorine and carbon–carbon bonds (10), the chemical and thermal stability of PFCs is considerably higher in general than that of the corresponding hydrocarbons (3,11). Perfluorocarbons normally are significantly less reactive than hydrocarbons toward all chemical reagents except alkali metals. Molten alkali metals or alkali–metal hydrocarbon complexes degrade most fluorocarbons, and this reaction is used for their chemical analysis (3). Perfluorocarbons are not affected by acids or oxidizing agents and are not hydrolyzed below 500°C. Carbon tetrafluoride decomposes only slowly at carbon arc temperatures, and it does not react with Cu, Ni, W, or Mo at 900°C. Perfluorocarbons of higher molecular weight are less thermally stable, but temperatures approaching 1000°C are still required to decompose C_2F_6 or $n\text{-}C_3F_8$, and most PFCs are stable below 300°C. Partially fluorinated hydrocarbons are less stable and more reactive, especially when only one fluorine atom is present. Hydrogen fluoride can be eliminated by chemical or thermal action from hydrofluorocarbons.

Manufacture. The direct fluorination of hydrocarbons with elemental fluorine is extremely exothermic and difficult to control. Special methods including metal packing techniques, jet reactors, and high dilution have been developed to control the reaction, but currently they have limited industrial importance (12). Poly(carbon monofluoride), $(CF)_x$, is one product that is made commercially by direct fluorination (of graphite) (13). The disadvantages of direct fluorination have

Table 1. Physical Properties of Aliphatic Perfluorocarbons (PFCs)[a]

PFC number	Formula	CAS Registry Number	Molecular weight	Boiling point, °C	Melting point, °C	Liquid density, g/mL at °C	Liquid refractive index, n_D at °C	Critical temp, °C	Critical pressure, MPa[b]
14	CF_4	[75-73-0]	88.01	-128.1	-183.6	1.613_{-130}	1.151_{-73}	-45.6	3.74
116	CF_3CF_3	[76-16-4]	138.02	-78.2	-100.6	1.600_{-80}	1.206_{-73}	19.7	2.99
218	$CF_3CF_2CF_3$	[76-19-7]	188.03	-36.7	-183	1.350_{20}		71.9	2.68
31-10	$CF_3(CF_2)_2CF_3$	[355-25-9]	238.04	-2.2	-128	1.543_{20}	1.217_{25}	113.2	2.32
C-318	cyclo-C_4F_8	[115-25-3]	200.04	-5.9	-41.4	1.500_{25}	1.242_{15}	115.2	2.78
41-12	$CF_3(CF_2)_3CF_3$	[678-26-2]	288.05	29.2	-126	1.620_{20}	1.251_{22}	149	2.04
51-14	$CF_3(CF_2)_4CF_3$	[355-42-0]	338.07	58	-86	1.680_{25}	1.262_{20}	174.5	1.90
61-16	$CF_3(CF_2)_5CF_3$	[335-57-9]	388.08	82.5	-51	1.733_{20}	1.290_{25}	201.6	1.62
PP3[c]	cyclo-$C_6F_{10}(CF_3)_2$[d]	[335-27-3]	400.09	102	f	1.828_{25}	1.313_{25}	241.5	1.88
PP6[c]	cyclo-$C_{10}F_{18}$[e]	[306-94-5]	462.11	142	-70	1.917_{25}	1.320_{25}	292.0	1.75
PP9[c]	cyclo-$C_{10}F_{17}(CF_3)$[g]	[306-92-3]	512.12	160	-70	1.972_{25}	1.335_{25}	313.4	1.66
PP11[c]	cyclo-$C_{14}F_{24}$[h]	[306-91-2]	624.15	215	-20	2.03_{25}		377[i]	1.46[i]

[a] Refs. 6–8.
[b] To convert MPa to psi, multiply by 145.
[c] Flutec number (trademark of Rhône-Poulenc, Inc., RTZ Chemicals, ISC Division).
[d] Perfluoro-1,3-dimethylcyclohexane.
[e] Perfluorodecalin, cis/trans mixture.
[f] −11.2 to 18.0°C, depending on cis/trans ratio.
[g] Perfluoro-2-methyldecalin.
[h] Perfluorotetradecahydrophenanthrene.
[i] Estimated values.

Table 2. Physical Properties of Aliphatic Hydrofluorocarbons (HFCs)[a]

HFC number	Formula	CAS Registry Number	Molecular weight	Boiling point, °C	Melting point, °C	Liquid density, g/mL at °C	Liquid refractive index, n_D at °C	Critical temp, °C	Critical pressure, MPa[b]
23	CHF$_3$	[75-46-7]	70.01	−82.2	−155.2	1.442$_{-80}$	1.215$_{-73}$	25.7	4.83
32	CH$_2$F$_2$	[75-10-5]	52.02	−51.6	−136	1.200$_{-50}$	1.190$_{20}$		
41	CH$_3$F	[593-53-3]	34.03	−78.3	−141.8	0.884$_{-80}$	1.1727$_{20}$	44.6	5.86
125	CHF$_2$CF$_3$	[354-33-6]	120.02	−48.5	−103	1.53$_{-48.5}$	1.5012$_{19}$	72.4	3.52
134	CHF$_2$CHF$_2$	[359-35-3]	102.03	−19.7	−89		1.250$_{20}$		
134a	CH$_2$FCF$_3$	[811-97-2]	102.03	−26.5	−101	1.21$_{25}$		101.1	4.14
143	CHF$_2$CH$_2$F	[430-66-0]	84.04	5.0	−84			71.2	
143a	CH$_3$CF$_3$	[420-46-2]	84.04	−47.4	−111.3	1.176$_{-50}$	1.22$_{25}$	73.1	3.76
152	CH$_2$FCH$_2$F	[624-72-6]	66.05	30.7		0.913$_{19}$	1.28$_{25}$	107.5	
152a	CH$_3$CHF$_2$	[75-37-6]	66.05	−25.8	−117	1.023$_{-30}$	1.3011$_{-72}$	113.5	4.50
161	CH$_3$CH$_2$F	[353-36-6]	48.06	−37.4	−143.2	0.818$_{-37}$	1.3033$_{-37}$	102.2	4.72
227ea	CF$_3$CFHCF$_3$[c]	[431-89-0]	170.03	−18	−129.5	1.407$_{25}$		101.7	2.91
245ca	CHF$_2$CF$_2$CH$_2$F	[679-86-7]	134.05	26	−82				
245cb	CF$_3$CF$_2$CH$_3$	[1814-88-6]	134.05	−18			1.30$_{15}$	106.9	
254fb	CH$_2$FCH$_2$CF$_3$	[460-36-6]	116.06	29.4		1.2584$_{25}$	1.2765$_{25}$		
272ca	CH$_3$CF$_2$CH$_3$	[420-45-1]	80.08	−0.4	−104.8	0.9205$_{20}$	1.2904$_{20}$		
272fa	CH$_2$FCH$_2$CH$_2$F	[462-39-5]	80.08	41.6		1.0057$_{25}$	1.3190$_{26}$		
281ea	CH$_3$CHFCH$_3$	[420-26-8]	62.09	−10	−133.4	0.7238$_{-20}$	1.3075$_{-10}$		
281fa	CH$_2$FCH$_2$CH$_3$	[460-13-9]	62.09	−2.5	−159	0.7956$_{20}$	1.3115$_{20}$		

[a]Refs. 5, 7–9.
[b]To convert MPa to psi, multiply by 145.
[c]Unpublished data, Great Lakes Chemical Co.

been overcome by the use of fluorine carriers, in particular, high valence metal fluorides such as cobalt trifluoride, CoF_3, or potassium tetrafluorocobaltate, $KCoF_4$. These reagents replace hydrogen and halogen atoms by fluorine and add fluorine to double bonds and aromatic systems (12).

$$2\ CoF_3 + RH \longrightarrow RF + HF + 2\ CoF_2$$

$$\underset{F_2}{\underline{\qquad\qquad\qquad\qquad\qquad}}$$

Cobalt trifluoride is generated *in situ* by passing fluorine over cobalt difluoride contained in a horizontal, mechanically agitated steel reactor. The compound to be fluorinated is passed through the reactor at 150–300°C as a vapor in a stream of nitrogen. After the reaction is completed, the CoF_3 is regenerated by adding fluorine. Advances in process control technology have allowed the process to be run continuously by simultaneously introducing the fluorine and hydrocarbon into the cobalt fluoride bed (6). Principally, cyclic and higher molecular weight acyclic fluorocarbons are prepared by this method.

Fluorocarbons are made commercially also by the electrolysis of hydrocarbons in anhydrous hydrogen fluoride (Simons process) (14). Nickel anodes and nickel or steel cathodes are used. Special porous anodes improve the yields. This method is limited to starting materials that are appreciably soluble in hydrogen fluoride, and is most useful for manufacturing perfluoroalkyl carboxylic and sulfonic acids, and tertiary amines. For volatile materials with little solubility in hydrofluoric acid, a complementary method that uses porous carbon anodes and $HF \cdot 2KF$ electrolyte (Phillips process) is useful (14).

Hydrofluorocarbons are also prepared from acetylene or olefins and hydrogen fluoride (3), or from chlorocarbons and anhydrous hydrogen fluoride in the presence of various catalysts (3,15). A commercial synthesis of 1,1-difluoroethane, a CFC alternative and an intermediate to vinyl fluoride, is conducted in the vapor phase over an aluminum fluoride catalyst.

$$HC{\equiv}CH + 2\ HF \xrightarrow{\text{catalyst}} CH_3CHF_2$$

$$CH_3CHF_2 + heat \longrightarrow CH_2{=}CHF + HF$$

Perfluorocyclobutane is prepared by the thermal cyclodimerization of tetrafluoroethylene [*116-14-3*].

Health and Safety Factors.   Completely fluorinated alkanes are essentially nontoxic (16). Rats exposed for four hours to 80% perfluorocyclobutane and 20% oxygen showed only slight effects on respiration, but no pathological changes in organs. However, some fluorochemicals, especially functionalized derivatives and fluoroolefins, can be lethal. Monofluoroacetic acid and perfluoroisobutylene [*382-21-8*] are notoriously toxic (16).

Uses.   The chemical inertness, thermal stability, low toxicity, and nonflammability of PFCs coupled with their unusual physical properties suggest many useful applications. However, the high cost of raw materials and manufacture has limited commercial production to a few, small-volume products. Carbon tetrafluoride and hexafluoroethane are used for plasma, ion-beam, or sputter etching of semiconductor devices (17) (see ION IMPLANTATION). Hexafluoroethane and octafluoropropane have some applications as dielectric gases, and perfluorocyclobutane is used in minor amounts as a dielectric fluid. Perfluoro-1,3-dimethylcyclohexane is used as an inert, immersion coolant for electronic equipment, and perfluoro-2-methyldecalin is used for pin-hole leak testing of encapsulated electronic devices (6,18). Perfluoroperhydrophenanthrene has several diverse applications, ranging from a vapor-phase soldering agent for fabrication of printed circuits to a substitute for internal eye fluid in remedial eye surgery (6,19).

Medical applications of PFC emulsions for organ perfusion and intravenous uses have received much attention in recent years. The first commercial blood substitute (Fluosol DA 20%, trademark of the Green Cross Corp.) employed perfluorodecalin, and improved, second generation products based on this PFC, or perfluorooctylbromide, are now under development (20,21). The relatively high oxygen dissolving capability of PFCs underlies these applications (see BLOOD, ARTIFICIAL).

Poly(carbon monofluoride) is used as a high temperature lubricant and as a cathode material in high energy lithium batteries (13), but the fluorocarbons of greatest commercial interest and volume are the high molecular weight fluoroplastics and elastomers derived from tetrafluoroethylene, hexafluoropropylene, vinylidene fluoride, and vinyl fluoride (22). Poly(tetrafluoroethylene) [9002-84-0], Teflon (trademark of E. I. du Pont de Nemours & Co.), and its copolymer [25067-11-2] with hexafluoropropylene, Teflon FEP, are fluorinated plastics, particularly notable for their outstanding chemical and thermal stability, electrical inertness, and nonflammability (23). Teflon has service temperatures in the range of -196 to 260°C. The copolymers [9011-17-0] of hexafluoropropylene and vinylidene fluoride, and the terpolymers [25190-89-0] of hexafluoropropylene, tetrafluoroethylene, and vinylidene fluoride (Viton fluoroelastomers, trademark of E. I. du Pont de Nemours & Co.) are rubbers with excellent thermal, chemical, and oxidative stability (24). They remain useful elastomers for indefinite periods of continuous exposure in air up to about 230°C (see ELASTOMERS, SYNTHETIC–FLUORO-CARBON ELASTOMERS).

Chlorofluorocarbons and Hydrochlorofluorocarbons

Properties.   The physical properties of aliphatic fluorine compounds containing chlorine are similar to those of the PFCs or HFCs (3,5). They usually have high densities and low boiling points, viscosities, and surface tensions. The irregularity in the boiling points of the fluorinated methanes, however, does not appear in the chlorofluorocarbons. Their boiling points consistently increase with the number of chlorines present. The properties of some CFCs and HCFCs are shown in Tables 3 and 4.

Although the CFCs and HCFCs are not as stable as the PFCs, they still can be rather stable compounds (3,11). Dichlorodifluoromethane, CCl_2F_2, is stable at 500°C in quartz; CCl_3F and $CHClF_2$ begin to decompose at 450 and 290°C, re-

Table 3. Physical Properties of Aliphatic Chlorofluorocarbons (CFCs)[a]

CFC number	Formula	CAS Registry Number	Molecular weight	Boiling point, °C	Melting point, °C	Liquid density, g/mL at °C	Liquid refractive index, n_D at °C	Critical temp, °C	Critical pressure, MPa[b]
11	CCl_3F	[75-69-4]	137.36	23.8	-111	1.476_{25}	1.374_{25}	198.0	4.41
12	CCl_2F_2	[75-71-8]	120.91	-29.8	-158	1.311_{25}	1.287_{25}	112.0	4.11
13	$CClF_3$	[75-72-9]	104.46	-81.4	-181	1.298_{-30}	1.199_{-73}	28.9	3.87
111	CCl_3CCl_2F	[354-56-3]	220.29	137	100	1.740_{25}			
112	CCl_2FCCl_2F	[76-12-0]	203.82	92.8	26	1.634_{30}	1.413_{25}	278	3.44
112a	CCl_3CClF_2	[76-11-9]	203.82	91.5	40.6	1.649_{20}			
113	CCl_2FCClF_2	[76-13-1]	187.38	47.6	-35	1.565_{25}	1.354_{25}	214.1	3.41
113a	CCl_3CF_3	[354-58-5]	187.38	45.8	14.2	1.579_{20}	1.361_{20}		
114	$CClF_2CClF_2$	[76-14-2]	170.92	3.8	-94	1.456_{25}	1.288_{25}	145.7	3.26
114a	CCl_2FCF_3	[374-07-2]	170.92	3.6	-94	1.455_{25}	1.309_{0}	145.6	3.29
115	$CClF_2CF_3$	[76-15-3]	154.47	-39.1	-106	1.291_{25}	1.214_{25}	80.0	3.12

[a]Refs. 5, 7–9.
[b]To convert MPa to psi, multiply by 145.

505

Table 4. Physical Properties of Aliphatic Hydrochlorofluorocarbons (HCFCs)[a]

HCFC number	Formula	CAS Registry Number	Molecular weight	Boiling point, °C	Melting point, °C	Liquid density, g/mL at °C	Liquid refractive index, n_D at °C	Critical temp, °C	Critical pressure, MPa[b]
21	$CHCl_2F$	[75-43-4]	102.92	8.92	−135	1.366_{25}	1.354_{25}	178.5	5.17
22	$CHClF_2$	[75-45-6]	86.47	−40.75	−160	1.194_{25}	1.256_{25}	96.0	4.97
31	CH_2ClF	[593-70-4]	68.48	−9.1	−133	1.271_{20}			
121	$CHCl_2CCl_2F$	[354-14-3]	185.84	116.6	−82.6	1.622_{20}	1.4463_{20}		
121a	$CHClFCCl_3$	[354-11-0]	185.84	116.5	−95.4	1.625_{20}	1.4525_{20}		
122	$CClF_2CHCl_2$	[354-21-2]	169.39	71.9	−140	1.5447_{25}	1.3889_{20}		
122a	$CHClFCCl_2F$	[354-15-4]	169.39	72.5		1.5587_{20}	1.3942_{20}		
122b	CHF_2CCl_3	[354-12-1]	169.39	73		1.566_{20}	1.3979_{20}		
123	$CHCl_2CF_3$	[306-83-2]	152.93	28.7	−107	1.475_{15}	1.3332_{15}	185	3.79
123a	$CHClFCClF_2$	[354-23-4]	152.93	28.2	−78	1.498_{10}	1.327_{20}		
124	$CHFClCF_3$	[2837-89-0]	136.48	−12	−199	1.364_{25}		122.2	3.57
124a	CHF_2CClF_2	[354-25-6]	136.48	−10.2	−117	1.379_{20}			
131	$CHCl_2CHClF$	[359-28-4]	151.40	102.5	−140.7	1.5497_{17}	1.4390_{20}		
131a	CH_2ClCCl_2F	[811-95-0]	151.40	88	−155	1.4227_{20}	1.4248_{20}		
132	$CHFClCHFCl$	[431-06-1]	134.94	59		1.46_{20}	1.391_{20}		
132a	CHF_2CHCl_2	[471-43-2]	134.94	60		1.4945_{17}	$1.3830_{16.4}$		
132b	$CH_2ClCClF_2$	[1649-08-7]	134.94	46.8	−101	1.416_{20}	1.362_{20}	222	
133	$CHClFCHF_2$	[431-07-2]	118.49	17.2		1.365_{10}			
133a	CH_2ClCF_3	[75-88-7]	118.49	6.1	−105.5	1.389_0	1.309_0	153.0	
141	$CH_2ClCHClF$	[430-57-9]	116.95	75.7	−60	1.3814_{20}	1.4113_{20}		
141b	CCl_2FCH_3	[1717-00-6]	116.95	32	−103.5	1.2500_{10}	1.3600_{10}	210.3	4.640
142	CHF_2CH_2Cl	[338-65-8]	100.50	35.1		1.312_{15}	1.3528_{15}		
142a	$CHClFCH_2F$	[338-64-7]	100.50	35			1.3416_{20}		
142b	CH_3CClF_2	[75-68-3]	100.50	−9.2	−130.8	1.113_{25}		137.1	4.12
151	CH_2FCH_2Cl	[762-50-5]	82.50	53.2	<−50	1.1675_{25}	1.3752_{20}	237.6	
225ca	$CF_3CF_2CHCl_2$[c]	[422-56-0]	202.94	51.1	−94	1.55_{25}	1.326_{20}		
225cb	$CClF_2CF_2CHClF$[c]	[507-55-1]	202.94	56.1	−97	1.56_{25}	1.3262_{25}		

[a] Refs. 5, 7–9.
[b] To convert MPa to psi, multiply by 145.
[c] Physical data from Ref. 25.

spectively (7). The pyrolysis of $CHClF_2$ at 650–700°C in metal tubes is the basis of a commercial synthesis of tetrafluoroethylene:

$$2 \text{ CHClF}_2 + \text{heat} \rightarrow \text{CF}_2{=}\text{CF}_2 + 2 \text{ HCl}$$

The chlorofluorocarbons react with molten alkali metals and CCl_2F_2 reacts vigorously with molten aluminum, but with most metals they do not react below 200°C. An exception is the dechlorination of chlorofluorocarbons with two or more carbon atoms in the presence of Zn, Mg, or Al in polar solvents. A commercial synthesis of chlorotrifluoroethylene [79-38-9] employs this reaction:

$$\text{CClF}_2\text{CCl}_2\text{F} + \text{Zn} \xrightarrow{\text{alcohol}} \text{CF}_2{=}\text{CClF} + \text{ZnCl}_2$$

Most chlorofluorocarbons are hydrolytically stable, CCl_2F_2 being considerably more stable than either CCl_3F or $CHCl_2F$. Chlorofluoromethanes and ethanes disproportionate in the presence of aluminum chloride. For example, CCl_3F and CCl_2F_2 give $CClF_3$ and CCl_4; $CHClF_2$ disproportionates to CHF_3 and $CHCl_3$. The carbon–chlorine bond in most chlorofluorocarbons can be homolytically cleaved under photolytic conditions (185–225 nm) to give chlorine radicals. This photochemical decomposition is the basis of the prediction that chlorofluorocarbons that reach the upper atmosphere deplete the earth's ozone shield.

Manufacture.   The most important commercial method for manufacturing CFCs and HCFCs is the successive replacement of chlorine by fluorine using hydrogen fluoride (3,15). The traditional, liquid-phase process uses antimony pentafluoride or a mixture of antimony trifluoride and chlorine as catalysts. Continuous vapor-phase processes that employ gaseous hydrogen fluoride in the presence of heterogenous chromium, iron, or fluorinated alumina catalysts also are widely used. Carbon tetrachloride, chloroform, and hexachloroethane (or tetrachloroethylene plus chlorine) are commonly used starting materials for one- and two-carbon chlorofluorocarbons. The extent of chlorine exchange can be controlled by varying the hydrogen fluoride concentration, the contact time, or the reaction temperature.

$$\text{CHCl}_3 \xrightarrow{\text{HF}} \text{CHCl}_2\text{F} + \text{CHClF}_2 + \text{CHF}_3$$

$$\text{CCl}_4 \xrightarrow{\text{HF}} \text{CCl}_3\text{F} + \text{CCl}_2\text{F}_2 + \text{CClF}_3$$

$$\text{CCl}_3\text{CCl}_3 \xrightarrow{\text{HF}} \text{CCl}_2\text{FCCl}_3 + \text{CCl}_2\text{FCCl}_2\text{F} + \text{CClF}_2\text{CCl}_2\text{F} + \text{CClF}_2\text{CClF}_2$$

$$\text{CH}_3\text{CCl}_3 \xrightarrow{\text{HF}} \text{CH}_3\text{CCl}_2\text{F} + \text{CH}_3\text{CClF}_2 + \text{CH}_3\text{CF}_3$$

The direct chlorination of hydrofluorocarbons and fluoroolefins has also been used commercially, eg, in the preparations of CH_3CClF_2 from CH_3CHF_2 and $CClF_2CClF_2$ from tetrafluoroethylene.

Economic Aspects.   The estimated worldwide production of important industrial CFCs is shown in Table 5. Trichlorofluoromethane, dichlorodifluoromethane, and trichlorotrifluoroethane account for over 95% of the total produc-

Table 5. Worldwide Production of Important Chlorofluorocarbons,[a] 10^3 t

Product	1986 Market	1991 Market	Percentage of 1986 market
CFC-11	415	263	63%
CFC-12	441	259	59%
CFC-113	241	143	59%
CFC-114	18	5	30%
CFC-115	13	11	85%

[a]Ref. 26.

tion. Between 1986 and 1991 the production of CFCs has decreased dramatically due to global adherence to the provisions of the Montreal Protocol and eventually will be phased out entirely. Estimates of the distribution by use in 1986 and subsequent reductions in use are shown in Table 6.

In 1990, approximately 115,000 t of CFCs were used as propellants, which represents a 58% decrease from the 1986 level. Most of this market segment likely will move to using hydrocarbon propellants, but the optimal choice of alternative will depend on the particular application. No suitable replacement for CFC-12 in pharmaceutical metered dose inhalers has been found. This single application represents a 6000 t/yr market. It also has been difficult to replace CFC-12 as the inert propellant for the potent sterilant ethylene oxide, which is a market for approximately 20,000 t per year.

Worldwide use of CFCs for refrigeration, air conditioning, and heat pumps totaled 260,000 t in 1991 (see REFRIGERATION) which is a dropoff of only 7% from 1986 levels.

Cleaning agent and solvent use has decreased 41% from 1986 levels, but in 1990, over 178,000 t of CFC-113 were still used in electronics, metal, precision, and dry cleaning (see SOLVENTS, INDUSTRIAL).

The use of CFCs as foam blowing agents has decreased 35% from 1986 levels. Polyurethanes, phenolics, extruded polystyrenes, and polyolefins are blown with CFCs, and in 1990 the building and appliance insulation markets represented about 88% of the 174,000 t of CFCs used in foams (see FOAMED PLASTICS).

Health and Safety Factors. The toxicity of aliphatic CFCs and HCFCs generally decreases as the number of fluorine atoms increases (16), as shown in

Table 6. Worldwide Estimates of CFC Use by Industry, 1991 vs 1986[a]

Application	1986 Total uses, %	Reduction since 1986, %
propellants	28	58
refrigerants	23	7
cleaning	21	41
foam blowing agents	26	35
other uses	2	

[a]Ref. 26.

Table 7, but there are exceptions as in the case of 141b vs 142b. Also, some derivatives like HCFC-132b can have low acute but high chronic toxicities (29).

Chlorofluorocarbons and Stratospheric Ozone Destruction. In 1971, it was shown (30) that CFCs were accumulating in the atmosphere, and three years later a relationship between CFCs and stratospheric ozone destruction, wherein the longlived CFCs that migrated to the upper stratosphere were being photolyzed by the intense uv radiation from the sun to form chlorine atoms, was proposed

Table 7. Toxicity of Selected Halocarbons[a]

Compound	Formula	TLV,[b] ppm	ALC,[c] ppm
BFC-12B2	CBr_2F_2	100	7,102[d]
BCFC-12B1	$CBrClF_2$		131,000
BFC-13B1	$CBrF_3$	1000	800,000
HBCC-30B1	CH_2BrCl	200	28,800[e]
HBFC-22B1	$CHBrF_2$		108,000
CC-10	CCl_4	10	8,000
CFC-11	CCl_3F	1000	26,200
CFC-12	CCl_2F_2	1000	800,000
CFC-13	$CClF_3$	1000	>600,000[f]
HCC-20	$CHCl_3$	10[g]	8,861
HCFC-21	$CHCl_2F$	10	49,900
HCFC-22	$CHClF_2$	1000	220,000
HFC-23	CHF_3	1000[h]	>663,000
HCC-30	CH_2Cl_2	100	22,669[i]
CFC-113	CCl_2FCF_2Cl	1000	52,500
CFC-114	CF_2ClCF_2Cl	1000	720,000[i]
CFC-115	CF_3CF_2Cl	1000	>800,000
HCFC-123	CF_3CHCl_2	10[h]	32,000
HCFC-124	CF_3CHFCl	500[h]	>230,000
HFC-125	CF_3CF_2H	1000[h]	>709,000
HFC-134a	CF_3CH_2F	1000[h]	567,000
HCFC-141b	CH_3CFCl_2	500[h]	61,647
HCFC-142b	CH_3CF_2Cl	1000[h]	128,000
HFC-152a	CH_3CHF_2	1000[h]	383,000
HCFC-225ca	$CF_3CF_2CHCl_2$		31,000[j]
HCFC-225cb	CF_2ClCF_2CHFCl		31,000[j]
HFC-227ea	CF_3CHFCF_3		>800,000[k]
PFC-31-10	$CF_3CF_2CF_2CF_3$		>800,000[l]

[a]Refs. 7 and 27.
[b]Except for CO_2, no compound has a higher TLV than 1000 ppm.
[c]Approximate lethal concentration, inhalation by rats, 4 h exposure unless noted otherwise.
[d]15 min exposure, mouse.
[e]15 min exposure.
[f]2 h exposure.
[g]Suspected carcinogen in humans.
[h]Du Pont Allowable Exposure Limit.
[i]30 min exposure, mouse.
[j]Ref. 28.
[k]Unpublished results, Great Lakes Chemical Co.
[l]Unpublished results, 3M, Inc.

(31). These chlorine atoms then participated in an ozone destruction cycle, and it has been estimated that one chlorine atom destroys 10,000 ozone molecules before getting trapped as inactive HCl (32). Bromine also participates in a destruction cycle, but it is about 10 times more efficient than chlorine in destroying ozone (33).

The possibility that CFCs and Halons can deplete the earth's ozone layer has had a significant impact on the fluorochemicals industry. Also, CFCs have been cited as contributors to global warming owing to their absorption of infrared irradiation and long atmospheric lifetimes (34). Because aerosol products containing CFC-11 and CFC-12 had accounted for the biggest release of CFCs, their manufacture for this use was banned in 1978. The discovery of the Antarctic ozone hole in 1985 (35) has prompted the world community to take much more extensive action toward halting the release of CFCs.

The Montreal Protocol. In response to the growing scientific consensus that CFCs and Halons would eventually deplete the ozone layer, the United Nations Environmental Programme (UNEP) began negotiations in 1981 aimed at protecting the ozone layer. In March 1985, the Vienna Convention for the Protection of the Ozone Layer was convened and provided a framework for international cooperation in research, environmental monitoring, and information exchange. In September of 1987, the Montreal Protocol on Substances that Deplete the Ozone Layer was signed by 24 nations and took force on January 1, 1989. This treaty called for (1) limiting production of specified CFCs, including 11, 12, 113, 114, and 115, to 50% of 1986 levels by 1998, (2) freezing production of specified Halons 1211, 1301, and 2402 at 1986 levels starting in 1992, and (3) convening the signatories yearly to reevaluate and update the Protocol articles in light of recent developments. By 1988 the Ozone Trends Panel issued a report based on new scientific evidence that conclusively linked CFCs to ozone depletion in the stratosphere. In June of 1990, the parties to the Montreal Protocol met in London and amended the Protocol to strengthen the controls on ozone depleting chemicals, expand the list of chemicals to include carbon tetrachloride and 1,1,1-trichloroethane, and specify stepped-up timetables for total phaseout of ozone depleting chemicals by the year 2000.

In April of 1991, the U.S. National Aeronautics and Space Administration concluded that ozone depletion was occurring even faster than had been estimated, and at the third meeting of the parties to the Montreal Protocol in June of 1991, an earlier phaseout of controlled substances was proposed. An assessment of the technical and economic consequences of a 1997 phaseout is currently underway, and further acceleration of the phaseout schedule to as soon as 1995 seems likely. Many countries already have unilaterally banned or curbed the use of controlled substances well ahead of the Montreal Protocol timetable. As of early July 1992, there were 81 parties to the Protocol.

Chlorofluorocarbon Alternatives

Properties. The ideal substitute should have identical or better performance properties than the CFC it replaces. The ideal CFC substitute must not harm the ozone layer, and must have a short atmospheric lifetime to ensure a low greenhouse warming potential (GWP). It also must be nontoxic, nonflammable, thermally and chemically stable under normal use conditions, and manufacturable at a reasonable price. The chemical industry has found substitutes that match many but not all of these criteria.

The general strategy has been to incorporate at least one hydrogen atom in the proposed CFC substitute's structure which provides a means for its destruction via hydrogen atom abstraction by tropospheric hydroxyl radicals. The haloalkylradicals thus formed are then rapidly degraded to acids and CO_2, which are both removed from the atmosphere by natural processes. Since fluorine does not participate in the ozone destruction cycle, a substitute composed of only hydrogen, fluorine, and carbon would be ideal, but HFCs for every application have not yet been identified. In some applications, HCFCs and even PFCs have been suggested as transitional replacements to accelerate phaseout of the much more harmful CFCs. Trade-offs will likely be required in most applications, and the alternatives that have been identified for the various markets are listed in Table 8.

The physical and environmental properties of the leading commercial CFCs and their proposed substitutes are compared in Tables 9 and 10. The HCFCs have relatively small but non-zero ozone depletion potentials (ODP) and low global warming potentials (GWP). Recent results indicate even these values may be 15% too high (37). The HFCs have zero ODPs and low-to-moderate GWPs. Perfluorocarbons also have zero ODPs, but very large GWPs.

Manufacture.   The manufacture of CFC alternatives is a far more complex challenge than production of the CFCs themselves (38). The very design feature

Table 8. Alternatives to CFCs

CFC	Application	Near-term substitute	Long-term substitute
CFC-11	blowing agents and refrigerants	HCFC-123 HCFC-22 HCFC-141b HCFC-142b	HFCs HFC-152a blends
CFC-12	refrigerants	HFC-134a HCFC-22	HFC-134a HFC-152a blends
CFC-113	cleaning agents	blends/azeotropes HCFC-225ca/cb	HFCs
CFC-114	blowing agents and refrigerants	HCFC-124 HCFC-142b blends/azeotropes	HFCs
CFC-115	refrigerants	HFC-125 blends/azeotropes	HFC-125
H-1301	fire extinguishant	HFC-23 HFC-125 PFC-31-10	HFC-23 HFC-227ea
H-1211	fire extinguishant	HCFC-123 HBFC-22B1 HBFC-124B1 PFC-51-14	HFCs

Table 9. Physical Property Comparisons of CFCs and Their HCFC or HFC Substitutes

Property	CFC	Substitutes				CFC	Substitutes	
	CFC-11	HCFC-123	HCFC-22	HCFC-141b	HCFC-142b	CFC-113	HCFC-225ca[a]	HCFC-225cb[a]
molecular weight	137.37	152.9	86.47	116.95	100.47	187.38	202.94	202.94
boiling point, °C	23.8	27.9	−40.7	32	−9.8	47.6	51.1	56.1
freezing point, °C	−111.1	−107	−157.4	−103.5	−130.8	−35.0	−94	−97
critical temperature,°C	198.0	185	96.0	210	137.1	214.2		
critical pressure, kPa[b]	4408	3789	4977	4641	4123	3415		
viscosity,[c] gas, mPa·s(= cP)	0.0105_{24}	0.0136_{60}	0.0122_{0}	0.0129_{60}	0.01099_{25}	0.0108_{49}		
viscosity,[c] liquid, mPa·s(= cP)	0.43_{20}	0.449_{25}		0.409_{25}	$0.453_{-20.9}$	$0.497_{48.9}$	0.58_{25}	0.60_{25}
surface tension,[c] mN/m(= dyn/cm)	18_{25}		8_{25}			17.3_{25}	15.8_{25}	16.7_{25}
heat capacity,[c] kJ/(kg·K)[d]								
liquid	0.870_{25}	1.0174_{25}	1.0962_{-40}	1.1556_{25}	1.2979_{25}	0.912_{25}		
vapor at 101.3 kPa[b]	0.565_{25}	0.7201_{25}		0.7913_{25}	0.8792_{25}	0.674_{60}		
latent heat of vaporization, kJ/kg[d]	180.3	174.17	233.84	223.15	223.15	146.73		
solubility in water at 101.3 kPa,[b] wt %	0.11_{25}	0.39	0.30	0.021	0.14	0.017		
atmospheric lifetime, years[e]	55	1.71	15.8	10.8	22	110	2.8	8.0
ozone depletion potential (ODP)[e]	1.00	0.02	0.055	0.11	0.065	1.07	0.025	0.033
global warming potential (GWP)[e]	1.00	0.02	0.39	0.14	0.44	1.80		

[a]Ref. 25.
[b]To convert kPa to atm, divide by 101.3.
[c]At the temperature (°C) indicated by subscripts.
[d]To convert J to cal, divide by 4.184.
[e]Ref. 36.

512

Table 10. Environmental Property Comparisons of CFCs and Substitutes

Property	CFC	Substitute	CFC	Substitutes		CFC	Substitute
	CFC-12	HFC-134a	CFC-114	HCFC-124	HCFC-142b	CFC-115	HFC-125
viscosity,[a] gas, mPa·s (=cP)	$0.0117_{4.4}$	0.0152_{60}	0.0118_{25}	0.0138_{60}	0.01099_{25}	0.0125_{25}	0.015_{25}
viscosity,[a] liquid, mPa·s (=cP)	0.398	0.205_{25}	0.485_{0}	0.0314_{25}	$0.453_{-20.9}$	0.193_{25}	0.104_{25}
heat capacity,[a] kJ/(kg·K)[b]							
liquid	0.971_{25}	1.428_{25}	1.016_{25}	1.130_{25}	1.298_{25}	1.192_{25}	1.260_{25}
vapor at 101.3 kPa[c]	0.607_{25}	0.854_{25}	0.711_{25}	0.741_{25}	0.879_{25}	0.686_{25}	0.707_{25}
latent heat of vaporization, kJ/kg[b]	165.1	219.8	136.0	167.9	223.2	126.0	159.0
solubility in water[a] at 101.3 kPa,[c] wt %	0.028_{25}	0.15_{25}	0.013_{25}	1.71_{24}	0.14	0.006_{25}	0.09_{25}
atmospheric lifetime, years[d]	116	15.6	220	7.0	22.4	550	40.5
ozone depletion potential (ODP)[d]	1.00	0.00	0.8	0.022	0.065	0.52	0.00
global warming potential (GWP)[d]	2.90	0.29	4.1	0.11	0.44	6.1	0.86

[a] At the temperature (°C) indicated by subscripts.
[b] To convert J to cal, divide by 4.184.
[c] To convert kPa to atm, divide by 101.3
[d] Ref. 36.

513

which makes the alternatives tropospherically labile, the hydrogen atom substituent, also significantly complicates their manufacture because of potential by-product formation or catalyst inactivation. At least a dozen different routes to HFC-134a have been identified, but a simple, single-step process is very unlikely (39). A two-step process that has been commercialized first involves reaction of trichloroethylene with HF in the vapor or liquid phase to form HCFC-133a (40,41), which is then separated and reacts again with HF to form HFC-134a.

$$CCl_2{=}CHCl + 3\ HF \rightarrow CF_3CH_2Cl + 2\ HCl$$

$$CF_3CH_2Cl + \text{excess HF} \rightarrow CF_3CH_2F + HCl$$

The HCFC-123 alternative to CFC-11 is made by the fluorination of tetrachloroethylene with either liquid or gaseous HF (41). Further reaction of HCFC-123 with HF provides the HCFC-124 and HFC-125 alternatives.

$$CCl_2{=}CCl_2 + 3\ HF \longrightarrow CF_3CHCl_2 + 2\ HCl$$

$$CF_3CHCl_2 \xrightarrow{\ HF\ } CF_3CHClF + CF_3CHF_2$$

The HCFC-225 isomers designed to replace CFC-113 are manufactured by Lewis acid promoted addition of HCFC-21 to tetrafluoroethylene (25,42).

$$CHCl_2F + CF_2{=}CF_2 \xrightarrow{\ AlX_3\ } CF_3CF_2CHCl_2 + CClF_2CF_2CHClF$$

Since HFC-134a likely will be the single largest volume CFC alternative produced, many manufacturers around the world are in the process of or have plans to commercialize it, each under their own trade name. United States and foreign trademarks and manufacturers of CFC alternatives are listed in Table 11.

Economic Aspects.   Manufacturing facilities for CFC alternatives are just now coming on line. The size of the markets for the alternatives is estimated to be quite large (several thousand t/yr), but it will not be as large as the prior markets for CFCs themselves. This is largely because of the higher cost of the alternatives, typically 3–5 times that of the incumbents. Low value-in-use applications which cannot support the cost of the alternatives will disappear or will switch to not-in-kind alternatives such as hydrocarbons for foam blowing.

Health and Safety Factors.   The toxicity of CFC alternatives is the subject of intense study. Fifteen fluorocarbon producers have formed the Program for Alternative Fluorocarbon Toxicity testing (PAFT) to share the costs associated with determining safe operating and handling procedures for the proposed CFC alternatives. Long-term chronic toxicity studies are still underway and results to date generally look encouraging, although prolonged exposure to HCFC-123 produced benign tumors in rats (43). The approximate lethal concentrations (ALC) are shown in Table 7.

Hydrofluorocarbons generally are less toxic than HCFCs, with the notable exception of HFC-152, CH_2FCH_2F, which apparently can be metabolically converted to monofluoroacetic acid and is therefore quite toxic (44).

Table 11. Trademarks and Manufacturers of CFC Alternatives

Country and trademark	Manufacturer
France	
Forane	Elf Atochem
Belgium	
Solkane	Solvay
Germany	
Frigen	Farbwerke Hoechst
Fridohna	Hüls
Italy	
Algogrene	Montedison
Japan	
Asahiflon	Asahi Glass
Daiflon	Daikin Kogyo
Suva	Du Pont Mitsui Fluorochemicals
United Kingdom	
Klea	Imperial Chemical Industries
Isceon	Rhône-Poulenc
United States	
Genetron	AlliedSignal Corp.
Suva	Du Pont
FM	Great Lakes Chemical Co.
3M Brand	Minnesota Mining and Manufacturing Co.

Fluorocarbons Containing Other Halogens

Properties. The physical and chemical properties of bromo- and iodofluorocarbons are similar to those of the chlorofluorocarbons except for higher densities and generally decreased stability. The stability of these compounds decreases as the ratio of bromine or iodine to fluorine increases. The reactivity of carbon–halogen bonds toward exchange by fluorine or homolytic cleavage increases in the order C–Cl, C–Br, C–I. Iodofluorocarbons and most bromofluorocarbons readily lose iodine or bromine radicals under photolytic, thermal, or radical initiation to give the corresponding carbon-centered radical. The physical properties of several examples are shown in Tables 12 and 13.

Manufacture. Brominated fluoromethanes are prepared industrially by the halogen exchange of tetrabromomethane or by the bromination of CH_2F_2 or CHF_3 at elevated temperatures (3). Other bromo- or iodofluorocarbons can be prepared by halogenating suitable fluorocarbons, including fluoroolefins, or by halogen exchange of perfluoroiodocarbons (47).

$$CF_3CH_2Cl + Br_2 \xrightarrow{500°C} CF_3CHBrCl$$

$$CI_4 + IF_5 \longrightarrow CF_3I$$

$$CF_2{=}CF_2 + I_2 \longrightarrow ICF_2CF_2I$$

$$5\ CF_2{=}CF_2 + IF_5 + 2\ I_2 \longrightarrow 5\ CF_3CF_2I$$

$$CF_3(CF_2)_6CF_2I + Br_2 \xrightarrow{100–290°C} CF_3(CF_2)_6CF_2Br$$

Table 12. Physical Properties of Aliphatic Hydrobromofluorocarbons (HBFCs) and Hydroiodofluorocarbons (HIFCs)[a]

B(I)FC number	Formula	CAS Registry Number	Molecular weight	Boiling point, °C	Melting point, °C	Liquid density, g/mL at °C	Refractive index, n_D at °C
11B1	CBrCl$_2$F	[353-58-2]	181.82	52	−106	1.9317$_{20}$	1.4304$_{20}$
11B2	CBr$_2$ClF	[353-55-9]	226.28	80		2.3172$_{20}$	1.4750$_{20}$
11B3	CBr$_3$F	[353-54-8]	270.74	106	−75	2.765$_{20}$	1.5256$_{20}$
12B1	CBrClF$_2$	[353-59-3]	165.37	−3.9	−161	1.850$_{15}$	
12B2	CBr$_2$F$_2$	[75-61-6]	209.83	24.5	−110	2.306$_{15}$	
13B1	CBrF$_3$	[75-63-8]	148.92	−57.8	−168	1.538$_{25}$	1.238$_{25}$
21B1	CHBrClF	[593-98-6]	147.38	−45	36.1	1.977$_0$	1.4144$_{25}$
21B2	CHBr$_2$F	[1868-53-7]	191.84	64.9	26.5	2.421$_{20}$	1.4685$_{20}$
22B1	CHBrF$_2$	[1511-62-2]	130.92	−15.5	−145	1.825$_{20}$	
113aB1	CBrCl$_2$CF$_3$	[354-50-7]	231.83	69.2		1.950$_{20}$	1.3977$_{20}$
114B2	CBrF$_2$CBrF$_2$	[124-73-2]	259.85	47.3	−110	2.163$_{25}$	1.367$_{25}$
115B1	CBrF$_2$CF$_3$	[354-55-2]	198.92	−21		1.810$_0$	1.2966$_{29.8}$
123B1	CHBrClCF$_3$	[151-67-7]	197.39	50.2		1.860$_{20}$	1.3700$_{20}$
123aB1a	CHClFCBrF$_2$	[354-06-3]	197.39	52.5		1.864$_{25}$	1.3685$_{25}$
124B1	CF$_3$CHFBr[b]	[124-72-1]	180.94	8.6	−80	1.85$_{20}$	
124aB1	CF$_2$BrCF$_2$H	[354-07-4]	180.94	10.8		1.900$_{15}$	1.321$_{15}$
132bB1a	CH$_2$BrCClF$_2$	[421-01-2]	179.40	68.4	−76	1.830$_{20}$	1.4018$_{20}$
133aB1	CF$_3$CH$_2$Br	[421-06-7]	162.94	26	−94	1.7881$_{20}$	1.3331$_{20}$
142B1	CH$_2$BrCHF$_2$	[359-07-9]	144.95	57.3	−75	1.824$_{18.5}$	1.3940$_{10.5}$
1I1	CCl$_2$FI	[420-48-4]	228.82	90.0	−107	2.313$_{20}$	1.510$_{20}$
13I1	CF$_3$I	[2314-97-8]	195.91	−22.5		2.361$_{-32}$	1.379$_{-42}$
21I2	CHFI$_2$	[1493-01-2]	285.83	100.3	−34.5	3.197$_{22}$	
22I1	CHF$_2$I	[1493-03-4]	177.92	21.6	−122	3.238$_{-19}$	
31I1	CH$_2$FI	[373-53-5]	159.93	53.4		2.366$_{20}$	1.491$_{20}$
113I1	CClFICClF$_2$	[354-61-0]	278.83	99		2.196$_{25}$	1.447$_{25}$
133aI1	CH$_2$ICF$_3$	[353-83-3]	209.94	55		2.142$_{25}$	1.3981$_{25}$

[a]Refs. 8 and 9. [b]Ref. 45.

Table 13. Physical Properties of Telomer Iodides[a]

Formula	CAS Registry Number	Molecular weight	Boiling point, °C	Melting point, °C	Liquid density, g/mL at °C	Refractive index, n_D at °C
CF_3CF_2I	[354-64-3]	245.92	13		2.072_{28}	$1.3378_{0.5}$
$CF_3(CF_2)_3I$	[423-39-2]	345.91	67		2.0424_{25}	1.3258_{25}
$CF_3(CF_2)_5I$	[355-43-1]	445.92	118	−46	$2.028_{26.4}$	1.3220_{20}
$CF_3(CF_2)_7I$	[507-63-1]	545.96	163	20.8	2.008_{25}	
$CF_3(CF_2)_9I$	[423-62-1]	645.98	195	65.5	1.9400_{70}	$1.3350_{25.5}$
$CF_3(CF_2)_2I$	[754-34-7]	295.93	41.2	−95.0	2.0026_{20}	1.3281_{20}
$CF_3(CF_2)_4I$	[638-79-9]	395.94	94.4	−50.0	$2.0349_{27.8}$	$1.3389_{0.5}$
$CF_3(CF_2)_6I$	[335-58-0]	495.96	137.5			1.3230_{30}
$CF_3(CF_2)_8I$	[558-97-4]	595.97	181			
$I(CF_2)_2I$	[354-65-4]	353.82	112		2.629_{25}	1.4895_{25}
$I(CF_2)_4I$	[375-50-8]	453.82	150	−9.0	2.4739_{27}	$1.4273_{26.2}$

[a]Refs. 9 and 46.

The higher molecular weight perfluoroalkyl iodides are prepared by telomerization of tetrafluoroethylene with lower molecular weight perfluoroalkyl iodides (46,48).

$$CF_3CF_2I + m\ CF_2{=}CF_2 \xrightarrow{\text{cat. SbF}_5/\text{IF}_5} CF_3CF_2(CF_2CF_2)_mI$$

$$ICF_2CF_2I + n\ CF_2{=}CF_2 + \text{heat} \xrightarrow{\hspace{3cm}} ICF_2CF_2(CF_2CF_2)_nI$$

Health and Safety Factors. Fluorocarbons containing bromine or iodine are more toxic than the corresponding chloro compounds. When the ratio of the fluorine to other halogens is high, the toxicity can be quite low, especially for bromofluorocarbons. Perfluoro-1-bromooctane [423-55-2] has an LD_{50} of greater than 64 mL/kg when administered into the gastrointestinal tract, and has little effect when instilled into the lungs (49). Other examples are included in Table 7.

Uses. The most important industrial products of this class have been the fire-extinguishing agents $CBrClF_2$ and $CBrF_3$. The latter is considerably more effective than CO_2, and is nontoxic as well as its decomposition products (50). It is used in commercial aircraft for in-flight engine fires and in portable fire extinguishers for both military and civilian markets. Both of these Halons have very high ozone depletion potentials, and their production is scheduled to be completely phased out by the year 1994 (51). Suitable replacements are being sought, and the properties of some leading candidates are compared in Table 14.

Halothane, $CF_3CHClBr$, was a widely used anesthetic which has the advantages of nonflammability, high anesthetizing power, and general lack of post-narcotic effects, but it has lost its leading market share to the fluorinated ethers enflurane, $CHClFCF_2OCHF_2$, and isofluorane, $CF_3CHClOCHF_2$ (52,53) (see ANESTHETICS).

Perfluorooctyl bromide [423-55-2], which has one of the highest oxygen-dissolving capabilities among fluorinated liquids and is readily emulsified, shows great promise in various medical applications such as tissue oxygenation, chemotherapy, and radiographic imaging (20,21,49).

Table 14. Physical Property Comparisons of BFCs and Their Substitutes

Property	BFC	Substitutes			BFC	Substitutes		
	H-1301	HFC-23	HFC-227ea	PFC-31-10	H-1211	HCFC-123	HBFC-22B1	PFC-51-14
molecular weight	148.91	70.01	170.03	238.03	165.4	152.9	130.92	338.0
boiling point, °C	−57.8	−82.0	−16.4	−2.0	−4	27.9	−15.5	56
freezing point, °C	−168.0	−155.2	−131	−128.2	−160.5	−107	−145	−90
critical temperature, °C	67.0	25.7	101.7	113.2	153.8	185	138.8	178.0
critical pressure, kPa[a]	3964	4810	2909	2323	4104	3789	5129	1835
critical density, kg/cm³	745	525	621	630	713			
viscosity,[b] gas, mPa·s	0.00016_{25}	0.0144_{25}	0.0132_{25}		0.0138_{25}	0.0136_{60}	0.0153_{25}	
viscosity,[b] liquid, mPa·s	0.32_{-40}	0.368_{-80}	0.184_{25}	0.607_{25}	0.292_{25}	0.449_{25}	0.269_{25}	0.700_{25}
heat capacity, kJ/(kg·K)[c]								
liquid	0.828_{0}	1.211_{-100}	1.102_{25}	1.045_{25}	0.742_{25}	1.017	0.814_{25}	
vapor at 101.3 kPa[d]	0.474_{25}	0.736_{25}	0.777_{25}	0.805_{25}	0.474_{25}	0.720	0.478_{25}	
latent heat of vaporization, kJ/kg[c]	118.7	251.2_{-100}	132.7	96.3	149.35	174.17	172.08	88.4
relative dielectric strength, $N_2 = 1$	1.83	1.04		5.25				
solubility in water[b] at 101.3 kPa, wt %	0.03_{25}	0.10_{25}		0.001_{25}		0.39_{25}		0.001_{25}
atmospheric lifetime, years[e]	67	310^{f}		>500	19	1.71	5.6	>500
ozone depletion potential (ODP)[e]	16.0	0.00	0.00	0.00	4.0	0.02	1.40	0.00
global warming potential (GWP)[e]	1.6	8		>8		0.02		>8

[a]To convert kPa to atm, divide by 101.3. [b]At the temperature (°C) indicated by subscripts. [c]To convert J to cal, divide by 4.184.
[d]101.3 kPa = 1 atm. [e]Ref. 36. [f]Estimated value.

The use of α,ω-diiodoperfluoroalkanes as chain-transfer agents in the manufacture of fluoroelastomers (54) is the only direct commercial application of iodofluorocarbons, although several telomer iodides, such as $CF_3(CF_2)_7I$ [*507-63-1*] and $CF_3(CF_2)_7CH_2CH_2I$ [*2043-53-0*], are intermediates in the manufacture of oil and water repellants, surfactants, and fire-extinguishing foams (55). The alcohols derived from perfluoroalkane carboxylic acids, eg, $CF_3(CF_2)_nCH_2OH$, or from the telomer iodides, eg, $CF_3(CF_2)_nCH_2CH_2OH$, are used to prepare fluorinated acrylate esters. Aqueous dispersions of the fluoroacrylate polymers are used as soil, water, and oil repellents for fabrics (56) (see WATERPROOFING). They are sold under the Scotchgard (3M) and Zepel or Zonyl (Du Pont) trade names.

BIBLIOGRAPHY

"Aliphatic Fluorinated Hydrocarbons" under "Fluorine Compounds, Organic" in *ECT* 1st ed., Vol. 6, pp. 752–757, by A. F. Benning, E. I. du Pont de Nemours & Co., Inc., and L. J. Hals and W. H. Pearlson, Minnesota Mining & Manufacturing Co.; "Fluorinated Hydrocarbons" under "Fluorine Compounds, Organic" in *ECT* 2nd ed., Vol. 9, pp. 739–751, by R. C. Downing, E. I. du Pont de Nemours & Co., Inc.; "Fluorinated Aliphatic Compounds" under "Fluorine Compounds, Organic" in *ECT* 3rd ed., Vol. 10, pp. 856–870, by B. E. Smart, E. I. du Pont de Nemours & Co., Inc.

1. *Refrig. Eng.* **65**, 49 (1957), ASRE Standard 34.
2. A. Maciejewski, *J. Photochem. Photobio., A: Chemistry* **51**, 87 (1990).
3. M. Hudlicky, *Chemistry of Organic Fluorine Compounds*, 2nd ed., Ellis Horwood Ltd., Chichester, UK, 1976.
4. H. G. Bryce, in J. H. Simons, ed., *Fluorine Chemistry*, Vol. 5, Academic Press, Inc., New York, 1965, pp. 297–492.
5. T. M. Reed, in Ref. 4, pp. 133–236.
6. B. D. Joyner, *J. Fluorine Chem.* **33**, 337 (1986).
7. *Freon Fluorocarbons, B-2*, E. I. du Pont de Nemours & Co., Inc., Wilmington, Del., 1969.
8. *Selected Values of Properties of Chemical Compounds*, Thermodynamics Research Center Data Project, Texas Engineering Experimental Sta., Texas A&M University, College Station, 1977.
9. A. M. Lovelace, D. A. Rausch, and W. Postelnek, *Aliphatic Fluorine Compounds*, Reinhold, Inc., New York, 1958.
10. B. E. Smart, in J. F. Liebman and A. Greenberg, eds., *Molecular Structure and Energetics*, VCH Publishers, Inc., Deerfield Beach, Fl., 1986, pp. 141–191.
11. R. E. Banks, *Fluorocarbons and Their Derivatives*, Macdonald, Ltd., London, 1970.
12. G. G. Furin, *Sov. Sci. Rev. B. Chem.* **16**, 1 (1991).
13. N. Watanabe, T. Nakajima, and H. Touhara, *Graphite Fluorides*, Elsevier, Ltd., Oxford, 1988.
14. W. V. Childs and co-workers, in C. H. Lund and M. M. Baizer, eds., *Organic Electrochemistry*, 3rd ed., Marcel Dekker, Inc., New York, 1991, Chap. 6, p. 1103.
15. A. K. Barbour, L. J. Belf, and M. W. Buxton, *Adv. Fluorine Chem.* **3**, 181 (1963).
16. J. W. Clayton, Jr., *Fluorine Chem. Rev.* **1**, 197 (1967).
17. C. M. Melliar-Smith and C. J. Mogab, in J. L. Vossen and W. Kern, eds., *Thin Film Processes*, Academic Press, Inc., New York, 1978.
18. D. S. L. Slinn and S. W. Green, in R. E. Banks, ed., *Preparation, Properties, and Industrial Applications of Organofluorine Compounds*, Ellis Horwood, Chichester, UK, 1982, Chap. 2, pp. 45–82.

19. *Flutec*, Rhône-Poulenc, Inc., RTZ Chemicals, ISC Division, Princeton, N.J., 1989.
20. K. C. Lowe, *Adv. Mater.* , 87 (1991).
21. T. M. S. Chang and R. P. Geyer, eds., *Blood Substitutes*, Marcel Dekker, Inc., New York, 1989.
22. L. A. Wall, ed., *Fluoropolymers*, John Wiley & Sons, Inc., New York, 1972.
23. S. V. Gangal, in J. I. Kroschwitz, ed., *Encyclopedia of Polymer Science and Engineering*, Vol. 16, John Wiley & Sons, Inc., New York, 1989, pp. 577–642.
24. A. L. Logothetis, *Prog. Polym. Sci.* **14**, 251 (1989).
25. M. Yamabe, "HCFC-225s as CFC-113 Substitutes" in *Symposium on Progress on the Development and Use of Chlorofluorocarbon (CFC) Alternatives, 200th ACS National Meeting*, Abstract No. 22, Washington, D.C., Aug. 28, 1990.
26. United Nations Environment Programme (UNEP), *Report of the Technology and Economic Assessment Panel*, Dec. 1991.
27. *Threshold Limit Values for Chemical Substances and Physical Agents and Biological Exposure Indices*, American Conference of Governmental Industrial Hygienists, Cincinnati, Ohio, 1991.
28. S. R. Frame, M. C. Carakostas, and D. B. Warheit, *Fundam. Appl. Toxicol.* **18**, 590 (1992).
29. *Du Pont Toxicology Information System*, Vol. 1.02, Du Pont Co., 1988.
30. J. E. Lovelock, *Nature* **230**, 379 (1971).
31. F. S. Rowland and M. J. Molina, *Nature* **249**, 810 (1974).
32. L. T. Molina and M. J. Molina, *J. Phys. Chem.* **91**, 433, (1987).
33. M. B. McElroy and co-workers, *Nature* **321**, 759 (1986).
34. A. Lacis and co-workers, *Geophys. Res. Lett.* **8**, 1035 (1981).
35. J. C. Farman, B. G. Gardiner, and J. D. Shanklin, *Nature* **315**, 207 (1985).
36. *Scientific Assessment of Ozone Depletion: 1991*, Report No. 25, World Meteorological Organization, Global Ozone Research and Monitoring Project, Geneva, 1991.
37. R. K. Talukar and co-workers, *Science* **257**, 227 (1992).
38. L. E. Manzer, *Science* **249**, 31 (1990).
39. L. E. Manzer, *Catalysis Today*, **13**, 13 (1992).
40. E. Chynowyth, *European Chem. News*, 8 (Apr. 17, 1991); 21 (July 15, 1991).
41. G. W. Parshall and S. D. Ittel, *Homogeneous Catalysis*, 2nd ed., John Wiley & Sons, Inc., New York, 1992, pp. 305–308.
42. Eur. Pat. Appl. EP 456,841 (Nov. 21, 1991), K. Ohnishi and co-workers (to Asahi Glass).
43. H. J. Trochimowicz, *Toxicol. Lett.* **68**, 25 (1993).
44. P. H. Lieder and D. A. Keller, *Chem. Eng. News* **70**, 2 (1992).
45. C. N. Fletcher, P. Jones, and M. Winterton, "Clean Agent Fire Extinguishant: Break-Down Products," *1990 International Conference on CFC and Halon Alternatives*, Baltimore, Md., 1990.
46. U. S. Pat. 3,234,294 (Feb. 8, 1966) and 3,132,185 (May 5, 1964), R. E. Parsons (to Du Pont).
47. Ger. Offen. DE 4,116,361 (Jan. 2, 1992), B. Felix and H. Katezenberger (to Hoechst AG).
48. C. D. Bedford and K. Baum, *J. Org. Chem.* **45**, 347 (1980).
49. D. M. Long and co-workers, in R. Filler, ed., *Biochemistry Involving Carbon–Fluoride Bonds*, American Chemical Society, Washington, D.C., 1976, pp. 171–189.
50. *Du Pont Freon FE 1301 Fire Extinguishing Agent*, E. I. du Pont de Nemours & Co., Inc., Wilmington, Del., 1969.
51. *Handbook for the Montreal Protocol on Substances that Deplete the Ozone Layer*, 3rd ed., Ozone Secretariat, United Nations Environmental Program, Nairobi, 1993.
52. D. Noble and L. Martin, *Anaesthesia* **45**, 339 (1990).
53. J. Tarpley and P. Lawler, *Anaesthesia* **44**, 596 (1989).

54. M. Oka and M. Tatemoto, *Contemporary Topics in Polymer Science*, Vol. 4, Plenum Press, New York, 1984, p. 763.
55. H. C. Fielding, in R. E. Banks, ed., *Organofluorine Chemicals and their Industrial Applications*, Ellis Horwood Ltd., Chichester, U.K., 1979, pp. 214–234.
56. M. J. Owen, in J. I. Kroschwitz, ed., *Encyclopedia of Polymer Science and Engineering*, Vol. 14, John Wiley & Sons, Inc., New York, 1988, pp. 411–421.

BRUCE E. SMART
RICHARD E. FERNANDEZ
E. I. du Pont de Nemours & Co., Inc.

FLUOROETHANOLS

Ethanol may be readily fluorinated at C-2. Replacement of H by F at C-1 would lead to unstable compounds which readily form carbonyls by loss of HF.

Monofluoro Derivative

2-Fluoroethanol [*371-62-0*] (ethylene fluorohydrin, β-fluoroethyl alcohol), FCH_2CH_2OH, is a colorless liquid with an alcohol-like odor; mp, $-26.45°C$; bp, $103.55°C$; d_4, 1.1297; n_D^{18}, 1.13647; heat of combustion, -1214.0 kJ/mol (-290.16 kcal/mol) (1,2). It is miscible with water, stable to distillation, and low in flammability. It is the least acidic of the fluoroethanols, although more acidic than ordinary alcohols with a pK_a value of 14.42 ± 0.04 in aqueous solution (3). Its most notable difference from the other fluoroethanols is its extreme toxicity (4–12). In mice an LD_{50} of 10 mg/kg has been measured (10–12). The toxicity is due to its facile oxidation in animals to derivatives of fluoroacetic acid [*144-49-0*], a known inhibitor of the tricarboxylic acid cycle of respiration. No effective antidote to this poisoning is known, although ethanol appears to diminish the lethal effects of 2-fluoroethanol in rats and monkeys (9).

In its chemical reactions, 2-fluoroethanol behaves like a typical alcohol. Oxidation (12) yields fluoroacetaldehyde [*1544-46-3*] or fluoroacetic acid; reaction with phosphorus tribromide (12) gives 1-bromo-2-fluoroethane [*762-49-2*]; addition to olefins results in ethers (13); and additions to isocyanates give carbamates (14). The alcohol can be prepared in 50% yield by the reaction of potassium fluoride with 2-chloroethanol at 175°C in high boiling glycol solvents (15). Alternatively, the addition of hydrogen fluoride to ethylene oxide (16) or the fluorination of 2-bromoethyl or 2-chloroethyl acetate with silver, potassium, or mercuric fluoride followed by hydrolysis gives the alcohol (1,2). 2-Fluoroethanol is not currently produced in commercial quantities, although (in 1992) it was available in research quantities for ca $3/g.

Because of its high toxicity, special procedures should be followed by users of 2-fluoroethanol. Suggested precautions include working with it in sealed reactors at subatmospheric pressure and careful monitoring to ensure that contamination of the surroundings is minimized (17). Another potential hazard is the formation of the alcohol as a minor by-product in reactions such as those involving boron trifluoride and ethylene oxide (18). Despite these problems, several potential uses for the alcohol and its derivatives have been reported. The alcohol has been used to control rodent populations (19) and, when labeled with ^{18}F, as a radiodiagnostic agent (20). Various derivatives have shown promise as herbicides or as agents to control mites and other plant pests (14,21–24).

Difluoro Derivative

2,2-Difluoroethanol [359-13-7], F_2CHCH_2OH, is a colorless liquid with an alcohol-like odor; mp, 28.2°C, bp, 96°C; d_1^{17}, 1.3084; n_D^{17}, 1.3320; heat of combustion, −1026 kJ/mol (−245.3 kcal/mol). It is stable to distillation and miscible with water and many organic solvents. As expected, its acidity lies between that of 2-fluoroethanol and 2,2,2-trifluoroethanol both in the gas phase (25) and in 50% aqueous ethanol solution (26), where its K_a of 1.0 × 10⁻¹² is about 4.8 times smaller than that of trifluoroethanol.

2,2-Difluoroethanol is prepared by the mercuric oxide catalyzed hydrolysis of 2-bromo-1,1-difluoroethane with carboxylic acid esters and alkali metal hydroxides in water (27). Its chemical reactions are similar to those of most alcohols. It can be oxidized to difluoroacetic acid [381-73-7] (28); it forms alkoxides with alkali and alkaline-earth metals (29); with alkoxides of other alcohols it forms mixed ethers such as 2,2-difluoroethyl methyl ether [461-57-4], bp 47°C, or 2,2-difluoroethyl ethyl ether [82907-09-3], bp 66°C (29). 2,2-Difluoroethyl difluoromethyl ether [32778-16-8], made from the alcohol and chlorodifluoromethane in aqueous base, has been investigated as an inhalation anesthetic (30,31) as have several ethers made by addition of the alcohol to various fluoroalkenes (32,33). Methacrylate esters of the alcohol are useful as a sheathing material for polymers in optical applications (34). The alcohol has also been reported to be useful as a working fluid in heat pumps (35). The alcohol is available in research quantities for ca $6/g (1992).

Trifluoroethanol

2,2,2-Trifluoroethanol [75-89-8], CF_3CH_2OH, is a colorless liquid with an ethanol-like odor; mp, −45°C; bp, 73.6°C; d_4^{25}, 1.3823; n_D^{20}, 1.2907; flash point (open cup), 41°C; flash point (closed cup), 33°C; no fire point (36); heat of combustion, −886.6 kJ/mol (−211.9 kcal/mol) (36); and dielectric constant (25°C), 26.14 (37). Many other physical and thermodynamic properties of the alcohol and its solutions have been published (36,38–41). It is the most acidic fluoroethanol with an ionization constant of 4.3 × 10⁻¹³ (42). It is stable to distillation and miscible with water and many organic solvents. It has the unusual property of dissolving most polyamides, both nylons (43) and polypeptides (44), at room temperature. Because of its excellent combination of physical and thermodynamic properties, 2,2,2-trifluoroethanol–water mixtures (also known as fluorinols) have application as working fluids in Rankine-cycle engines for recovering energy from waste heat sources (36,45,46). Its high ionizing power and low specific conductance make the alcohol useful as a solvent for ionic reactions and conductometric titrations (47), and basic research into solvolysis mechanisms (48–50).

Chemically, 2,2,2-trifluoroethanol behaves as a typical alcohol. It can be converted to trifluoroacetaldehyde [75-90-1] or trifluoroacetic acid [76-05-1] by various oxidizing agents such as aqueous chlorine solutions (51) or oxygen in the presence of a vanadium pentoxide catalyst (52). Under basic conditions, it adds to tetrafluoroethylene and acetylene to give, respectively, 1,1,2,2-tetrafluoroethyl 2′,2′,2′-trifluoroethyl ether [406-78-0] (53) and 2,2,2-trifluoroethyl vinyl ether [406-90-6] which was used as the inhalation anesthetic Fluroxene. Its alkoxides react with bromoethane to give trifluoroethyl ethyl ether [461-24-5], bp 50.3°C. Similarly prepared is bis(trifluoroethyl) ether used as the convulsant drug Flurothyl as a substitute for electric shock therapy. As the trichlorosulfonate ester, trifluoroethanol is used to introduce the trifluoroethyl group into the anxiolytic drug Halazepam [23092-17-3] (54). 2,2,2-Trifluoroethanol is also the starting material for the anesthetic Isoflurane (1-chloro-2,2,2-trifluoroethyl difluoromethyl ether [26675-46-7]) (55,56) and Desflurane (2-difluoromethoxy-1,1,1,2-tetrafluoroethane [57041-67-5]) (57).

Trifluoroethanol was first prepared by the catalytic reduction of trifluoroacetic anhydride [407-25-0] (58). Other methods include the catalytic hydrogenation of trifluoroacetamide [354-38-1] (59), the lithium aluminum hydride reduction of trifluoroacetyl chloride [354-32-5] (60) or of trifluoroacetic acid or its esters (61,62), and the acetolysis of 2-chloro-1,1,1-trifluoroethane [75-88-7] followed by hydrolysis (60). More recently, the hydrogenation of 2,2,2-trifluoroethyl trifluoroacetate [407-38-5] over a copper(II) oxide catalyst has been reported to give the alcohol in 95% yield (63).

The largest producer of trifluoroethanol is Halocarbon Products Corp. Other producers include Japan Halon and Rhône-Poulenc. Commercial quantities sell for approximately $20/kg (1992).

Toxicity studies on trifluoroethanol show acute oral LD_{50}, 240 mg/kg; acute dermal LD_{50}, 1680 mg/kg; and acute inhalation $L(ct)_{50}$, 4600 ppm·h. Long-term subchronic inhalation exposure to 50–150 ppm of the alcohol has caused testicular depression in male rats, but no effects were noted at the 10 ppm level (32). Although the significance of the latter observations for human safety is unknown, it is recommended that continuous exposure to greater than 5 ppm or skin contact with it be avoided.

BIBLIOGRAPHY

"Fluoroethanols" under "Fluorine Compounds, Organic," in *ECT* 1st ed., Vol. 6, pp. 760–762, by J. F. Nobis, Xavier University; in *ECT* 2nd ed., Vol. 9, pp. 751–752, by L. L. Ferstandig, Halocarbon Products Corp.; in *ECT* 3rd ed., Vol. 10, pp. 871–874, by G. Astrologes, Halocarbon Products Corp.

1. F. Swarts, *Rec. Trav. Chim.* **33**, 252 (1914).
2. F. Swarts, *J. Chem. Soc.* **106**, 475 (1914).
3. W. L. Mock and J. Z. Zhang, *Tetrahedron Lett.*, 5687 (1990).
4. E. Gryszkiewicz-Trochimowski, *Rec. Trav. Chim.* **66**, 427 (1947).
5. H. McCombie and B. C. Saunders, *Nature* **158**, 382 (1946).

6. E. V. Avdeeva and N. M. Dukel'skaya, *Vestn. Mosk. Univ. Ser. VI* **21**(4), 49 (1966).
7. F. R. Johannsen and C. O. Knowles, *Comp. Gen. Pharmacol.* **5**(1), 101 (1974).
8. E. O. Dillingham and co-workers, *J. Pharm. Sci.* **62**, 22 (1973).
9. D. I. Peterson, J. E. Peterson, and M. G. Hardinge, *J. Pharm. Pharmacol.* **20**, 465 (1968).
10. F. L. M. Pattison, *Toxic Aliphatic Fluorine Compounds*, Elsevier Publishing Co., New York, 1959, p. 65.
11. F. L. M. Pattison and co-workers, *J. Org. Chem.* **21**, 739 (1956).
12. B. C. Saunders, G. J. Stacey, and I. G. E. Wilding, *J. Chem. Soc.*, 773 (1949).
13. R. J. Koshar, T. C. Simmons, and F. W. Hoffmann, *J. Am. Chem. Soc.* **79**, 1741 (1957).
14. East Ger. Pat. 111,149 (Feb. 5, 1975), H. G. Werchan and co-workers.
15. F. W. Hoffmann, *J. Am. Chem. Soc.* **70**, 2596 (1948).
16. I. L. Knunyants, O. V. Kil'disheva, and I. P. Petrov, *J. Gen. Chem. USSR, Eng. Transl.* **19**, 95 (1949).
17. G. DiDrusco and F. Smai, *Quad. Ing. Chim. Ital.* **9**(11), 156 (1973).
18. C. T. Bedford, D. Blair, and D. E. Stevenson, *Nature* **267**, 335 (1977).
19. A. I. Kryl'tsov and co-workers, *Tr. Kaz. Nauch. Issled Inst. Zashch. Rast.* **11**, 171 (1972).
20. G. D. Robinson, Jr., *Radiopharm. Label Compounds, Proc. Symp. 1*, 423 (1973).
21. U.S. Pat. 4,022,609 (May 10, 1977), D. E. Hardies and J. K. Rinehart (to PPG Industries, Inc.).
22. U.S. Pat. 4,960,884 (Oct. 2, 1990), D. M. Roush and co-workers (to FMC Corp.).
23. U.S. Pat. 3,852,464 (Dec. 3, 1974), D. E. Hardies and J. K. Rinehart (to PPG Industries, Inc.).
24. Fr. Pat. 1,604,978 (July 30, 1971), G. Rossi, G. Michieli, and P. Paolucci (Montecatini Edison S.p.A.).
25. J. H. J. Dawson and K. R. Jenning, *Int. J. Mass Spectrom. Ion Phys.* **25**(1), 47 (1977).
26. R. N. Haszeldine, *J. Chem. Soc.*, 1757 (1953).
27. Jpn. Kokai 62 273,925 (Nov. 28, 1987), T. Komatsu and Y. Asai (to Asahi Chemical Industry Co., Ltd.).
28. F. Swarts, *Chem. Zentr.* **II**, 709 (1903).
29. F. Swarts, *Bull. Soc. Chim. Belg.* **11**, 731 (1902).
30. U.S. Pat. 3,769,433 (Oct. 30, 1973), R. C. Terrell (to Airco, Inc.).
31. U.S. Pat. 3,896,178 (July 22, 1975), R. C. Terrell (to Airco, Inc.).
32. U.S. Pat. 3,746,769 (July 17, 1973), R. C. Terrell (to Airco, Inc.).
33. U.S. Pat. 3,862,240 (Jan. 21, 1975), R. C. Terrell (to Airco, Inc.).
34. Jpn. Kokai 63 066154 (Mar. 24, 1988), T. Ide and T. Komatsu (to Asahi Chemical Industry Co., Ltd.).
35. Jpn. Kokai 62 013481 (Jan. 22, 1987), M. Sagami and H. Matsuo (to Asahi Glass Co.).
36. *Trifluoroethanol Brochure*, Halocarbon Products Corp., Hackensack, N.J., 1979.
37. J. M. Mukherjee and E. Grunwald, *J. Phys. Chem.* **62**, 1311 (1958).
38. J. Murto and E. Heino, *Suom. Kemistil.* **39**, 263 (1966).
39. C. H. Rochester and J. R. Symonds, *J. Fluorine Chem.* **4**, 141 (1974).
40. A. Kivenen, J. Murto, and M. Lehtonen, *Suom. Kemistil.* **B 41**, 359 (1968).
41. C. H. Rochester and J. R. Symonds, *J. Chem. Soc. Faraday Trans. I* **69**, 1274 (1973).
42. P. Ballinger and F. A. Long, *J. Am. Chem. Soc.* **81**, 1050 (1959).
43. Ger. Pat. 1,017,782 (Oct. 17, 1957), P. Schlack (to Farbwerke Hoechst AG).
44. M. Goodman, I. G. Rosen, and M. Safdy, *Biopolymers* **2**, 503,519,537 (1964).
45. U.S. Pat. 3,722,211 (Mar. 27, 1973), R. C. Conner and L. L. Ferstandig (to Halocarbon Products Corp.).
46. D. G. Shepherd, *Hydrocarbon Proc.*, 141 (Dec. 1977).
47. N. Paetzold, *J. Polym. Sci. Part B* **1**, 269 (1963).
48. V. J. Shiner, Jr. and co-workers, *J. Am. Chem. Soc.* **91**, 4838 (1969).

49. D. A. daRoza, L. J. Andrews, and R. M. Keefer, *J. Am. Chem. Soc.* **95**, 7003 (1973).
50. D. S. Noyce, R. L. Castenson, and D. A. Meyers, *J. Org. Chem.* **37**, 4222 (1972).
51. U.S. Pat. 3,088,896 (May 7, 1963), M. Braid (to Pennsalt Chemicals Corp.).
52. U.S. Pat. 3,038,936 (June 12, 1962), M. Braid (to Pennsalt Chemicals Corp.).
53. A. L. Henne and M. A. Smook, *J. Am. Chem. Soc.* **72**, 4378 (1950).
54. M. Steinman and co-workers, *J. Med. Chem.* **16**, 1354 (1973).
55. U.S. Pat. 3,535,425 (Oct. 20, 1970), R. C. Terrell (to Air Reduction Co., Inc.).
56. U.S. Pat. 3,637,477 (Jan. 25, 1972), L. S. Croix (to Air Reduction Co., Inc.).
57. U.S. Pat. 4,762,856 (Aug. 9, 1988), R. C. Terrell (to BOC, Inc.).
58. F. Swarts, *Compt. Rend.* **197**, 1201 (1933).
59. H. Gilman and R. G. Jones, *J. Am. Chem. Soc.* **70**, 1281 (1948).
60. A. L. Henne, R. M. Alm, and M. Smook, *J. Am. Chem. Soc.* **70**, 1968 (1948).
61. K. N. Campbell, J. O. Knobloch, and B. K. Campbell, *J. Am. Chem. Soc.* **72**, 4380 (1950).
62. D. R. Husted and A. H. Ahlbrecht, *J. Am. Chem. Soc.* **74**, 5422 (1952).
63. U.S. Pat. 4,072,726 (Feb. 7, 1978), H. R. Nychka and co-workers (to Allied Chemical Corp.).

ARTHUR J. ELLIOTT
Halocarbon Products Corporation

FLUOROETHERS AND FLUOROAMINES

Perfluoroaliphatic ethers and perfluorotertiary amines together with the perfluoroalkanes and cycloalkanes comprise a class of extremely unreactive materials known in the industry as inert fluids. These fluids are colorless, odorless, essentially nontoxic, nonflammable, dense, and extremely nonpolar. In the electronics industry, the lower molecular weight compounds find application in the areas of heat transfer, testing, and vapor-phase soldering. Higher molecular weight polymers and oligomers are used in a variety of applications, including hazardous-duty vacuum pump fluids, specialty greases, and various specialty cosmetics and lubricants.

Many perfluoroaliphatic ethers and tertiary amines have been prepared by electrochemical fluorination (1–6), direct fluorination using elemental fluorine (7–9), or, in a few cases, by fluorination using cobalt trifluoride (10). Examples of lower molecular weight materials are shown in Table 1. In addition to these, there are three commercial classes of perfluoropolyethers prepared by anionic polymerization of hexafluoropropene oxide [*428-59-1*] (11,12), photooxidation of hexafluoropropene [*116-15-4*] or tetrafluoroethene [*116-14-3*] (13,14), or by anionic ring-opening polymerization of tetrafluorooxetane [*765-63-9*] followed by direct fluorination (15).

Physical Properties

Perfluorinated compounds boil at much lower temperatures and have lower heats of vaporization than the corresponding hydrocarbon analogues even though they have considerably higher molecular weights. This holds true not only for the per-

Table 1. Physical Properties of Some Perfluorinated Liquids

Name	CAS Registry Number	Molecular formula	Bp, °C	d_4^{25}	n_D^{24}	Pour point,[a] °C
perfluoro-4-methylmorpholine	[382-28-5]	$CF_3N(CF_2)_2O(CF_2)_2$	51	1.70	1.267	−80
perfluoro-2-ethyltetrahydrofuran	[356-48-9]	$C_6F_{12}O$	56	1.69	1.263	−136
perfluorohexane	[355-42-0]	C_6F_{14}	58	1.68	1.252	−74
Galden HT 70[b]			70	1.73		−110
perfluorotriethylamine	[359-70-6]	$(C_2F_5)_3N$	71	1.73	1.262	−110
perfluoro-4-ethylmorpholine	[55716-11-5]	$C_2F_5N(CF_2)_2O(CF_2)_2$	72	1.74	1.273	−73
perfluoro-4-isopropylmorpholine	[1600-71-1]	$(CF_3)_2CFN(CF_2)_2O(CF_2)_2$	95	1.79	1.283	−112
perfluorobutyl ether	[308-48-5]	$(C_4F_9)_2O$	102	1.71	1.261	−100
Fluorinert FC-75	[11072-16-5]	$C_8F_{16}O^c$ (cyclic)	103	1.76	1.276	−93
perfluorooctane	[307-34-6]	C_8F_{18}	103	1.77	1.272	−42
perfluorononane	[375-96-2]	C_9F_{20}	123	1.80	1.276	−16
perfluorotripropylamine	[338-83-0]	$(C_3F_7)_3N$	130	1.82	1.279	−52
perfluorobis(2-butoxyethoxy)methane	[130085-23-3]	$(C_4F_9OC_2F_4O)_2CF_2$	178	1.76		<−110
perfluorotributylamine	[311-89-7]	$(C_4F_9)_3N$	178	1.86	1.291	−50
perfluoro(diethylamino)ethyl ether	[108709-75-7]	$[(C_2F_5)_2NC_2F_4]_2O$	178			−80
perfluorohexyl ether	[424-20-4]	$(C_6F_{13})_2O$	181	1.81	1.278	−90
K7 fluid	[59884-34-3]	$C_3F_7O[CF(CF_3)CF_2O]_5C_2F_5$	250	1.82		−80

[a] ASTM D97.
[b] A mixture of perfluorinated polyethers marketed by Montefluos.
[c] $C_8F_{16}O$ represents a mixture of isomers of cyclic perfluoroaliphatic ethers, primarily perfluoro-2-butyltetrahydrofuran [335-36-4].

fluoroalkanes and cycloalkanes but for the perfluorinated ethers and tertiary amines as well. The latter compounds have boiling points very close to the perfluoroalkanes having the same number of carbon atoms; the heteroatoms contribute little polarity to the molecules. Catenary oxygen and nitrogen atoms have marked effects on the freezing points and on the viscosity at low temperatures. This is illustrated by the C_{12}-perfluorohexyl ether and the C_{12}-perfluorotributylamine in Table 1 that freeze at -90 and $-50°C$, respectively, in contrast to the 32°C melting point of perfluorodecane. Note also the extremely low pour point of the Du Pont K7 fluid despite its high molecular weight. This effect is believed to be due to the increased flexibility that catenary oxygen or nitrogen atoms contribute to the perfluorinated chain (16) and is also observed for the higher molecular weight perfluoropolyethers that have relatively low pour points at relatively high molecular weights.

Many of the unusual properties of the perfluorinated inert fluids are the result of the extremely low intermolecular interactions. This is manifested in, for example, the very low surface tensions of the perfluorinated materials (on the order of 9–19 mN/m = dyn/cm) at 25°C which enables these liquids to wet any surface including polytetrafluoroethene. Their refractive indexes are lower than those of any other organic liquids, as are their acoustic velocities. They have isothermal compressibilities almost twice as high as water. Densities range from 1.7 to 1.9 g/cm^3 (17).

The absolute viscosities of the perfluorinated inert liquids are higher than the analogous hydrocarbons but the kinematic viscosities are lower due to the higher density of the perfluorinated compounds. The viscosity index, ie, the change in viscosity with temperature, is generally higher for the perfluorinated liquids than for hydrocarbons.

Thermal Stabilities. The perfluoroethers have thermal stabilities comparable to those of the perfluoroalkanes. Typically, although this depends somewhat on structure, they do not undergo significant decomposition until about 400°C. Perfluorotertiary amines are less stable thermally and begin to decompose at temperatures of about 250°C (18). Generally this slight instability is not a significant problem during use.

Electrical Properties. The low polarizability of perfluorinated liquids makes them excellent insulators. Their dielectric strengths are about 40 kV (ASTM D877); dissipation factors are about 0.0001 at 1 MHz; dielectric constants are about 1.8; volume resistivities are about 1×10^{15} ohm·cm (ASTM D257) (17).

Chemical Properties

The inert character of the perfluoroethers and tertiary amines is demonstrated by their lack of basicity or reactivity as compared with their hydrocarbon analogues. Both classes of compounds are nonflammable. The perfluorotertiary amines do not form salts with any protic acid nor do they form complexes with boron trifluoride. Neither class reacts with most oxidizing or reducing agents nor with strong acids or bases. As with the perfluoroalkanes, perfluoroethers and tertiary amines may, under some conditions, react violently with fused alkali metals. In contrast to the perfluoroalkanes, both classes of compounds react with aluminum chloride or bromide at elevated temperatures (ca 90–200°C) (19). The $-OCF_2O-$ linkage is especially vulnerable to this attack (20,21).

Solvent Properties. In comparison to the more familiar hydrocarbon systems, the solvent properties of the perfluorinated inert liquids are also unusual due to their nonpolar nature and low intermolecular forces. They are generally very poor solvents for most organic compounds. Water and hydrocarbon alcohols are nearly completely insoluble in them. Lower aliphatic hydrocarbons, lower molecular weight ethers, and some highly chlorinated solvents such as carbon tetrachloride are relatively soluble. As the molecular weight of the perfluorinated compound increases, the hydrocarbon solubility decreases. Partially fluorinated compounds such as benzotrifluoride or bis(trifluoromethyl)benzene are soluble in perfluorinated liquids. Aliphatic highly fluorinated compounds containing hydrogen such as $C_7F_{15}H$ are also miscible with perfluorinated liquids. The chlorofluorocarbons (CFCs) such as CFC-113, $CF_2ClCFCl_2$, or the hydrochlorofluorocarbons (HCFCs) such as HCFC-123, CF_3CHCl_2, HCFC-141b, or CH_3CFCl_2, are miscible in all proportions with the perfluoro compounds at 25°C. Typical solubility data are shown in Table 2.

The solubilities of gases such as oxygen, nitrogen, and carbon dioxide are generally high as shown in Table 3 (22). The oxygen and carbon dioxide solubilities of related compounds has led to their use as specialized synthetic blood substitutes.

These solubility relationships are consistent with the predictions based on the Hildebrand solubility parameter (23). For perfluorinated liquids, the solubility

Table 2. Solubility Relationships for $C_8F_{16}O$, $C_5F_{11}NO$, and $(C_4F_9)_3N^a$

| | Solubility at 20°C, mL/100 mL | | | | | |
Solvent	$C_8F_{16}O$ in solvent	Solvent in $C_8F_{16}O$	$C_5F_{11}NO$ in solvent	Solvent in $C_5F_{11}NO$	$(C_4F_9)_3N$ in solvent	Solvent in $(C_4F_9)_3N$
acetone	4.8	1.2	2.9	1.6	0.9	0.6
benzene	2.6	3.8	6.0	2.7	0.3	0.2
benzotrifluoride	miscible	miscible	miscible	miscible	miscible	miscible
benzyl alcohol	0.2	0.4	0.1	0.1	insoluble	insoluble
2-butanone	2.2	1.9	5.1	2.6	0.3	0.7
carbon tetrachloride	20.2	36.5	48.0	58.9	2.4	15.0
chlorobenzene	1.8	3.2	1.6	2.3	0.3	0.4
chloroform	7.7	4.5	13.2	8.1	1.2	5.4
cyclohexane	7.7	8.4	8.9	6.4	1.8	2.2
ethyl ether	miscible	miscible	miscible	miscible	miscible	5.0
ethyl acetate	7.5	6.5	8.8	5.5	2.2	2.4
heptane	25.5	11.6	25.1	10.1	6.4	3.4
isopropyl alcohol	4.1	1.3	4.7	0.7	insoluble	insoluble
methanol	1.0	0.1	1.6	0.6	insoluble	insoluble
petroleum ether	miscible	miscible	miscible	miscible	33.2	7.0
toluene	2.9	4.1	3.1	3.6	0.4	2.0
turpentine	5.3	1.0	6.4	0.4	0.9	insoluble
xylene	3.0	3.0	3.1	2.6	0.2	1.0
water	insoluble	insoluble	insoluble	insoluble	insoluble	insoluble

$^a C_8F_{16}O$ denotes a mixture of cyclic perfluoroaliphatic ethers, primarily perfluoro-2-butyltetrahydrofuran; $C_5F_{11}NO$ denotes perfluoro-4-methylmorpholine.

parameters are on the order of $10–12$ $J^{1/2}/cm^{3/2}$ ($5–6$ $cal^{1/2}/cm^{3/2}$) which are the lowest known values for liquids.

Table 3. Solubility of Gases in Perfluorinated Liquids[a]

	Solubility	
Gas	mL gas/100 mL $C_8F_{16}O$ (cyclic)	mL gas/100 mL $(C_4F_9)_3N$
oxygen	48.8	38.9
nitrogen	33.4	28.4
carbon dioxide	192.0	152.0
air	40.5	33.1

[a]At 25°C and 101 kPa (1 atm).

Methods of Preparation

Electrochemical Fluorination. In the Simons electrochemical fluorination (ECF) process the organic reactant is dissolved in anhydrous hydrogen fluoride and fluorinated at the anode, usually nickel, of an electrochemical cell. This process has been reviewed (6). Essentially all hydrogen atoms are substituted by fluorine atoms; carbon–carbon multiple bonds are saturated. The product phase is heavier than the HF phase and insoluble in it and is recovered by phase separation.

$$(C_4H_9)_3N \ + \ 27\ HF \ \xrightarrow{\ ECF\ } \ (C_4F_9)_3N \ + \ 27\ H_2$$

With an amine reactant, it has been shown that roughly one-third of the current passed makes liquid product, one-third gas, and one-third goes to HF-soluble polyfluorinated products (24). The ether perfluoro(2-butyltetrahydrofuran) [335-36-4] is made from a cyclization process during the ECF of perfluorooctanoyl chloride; other cyclic ethers have been prepared from certain ester reactants by a similar cyclization (25). Perfluoroaminoethers have been prepared by ECF (26).

Electrochemical fluorination leads to fragmentation, coupling, and rearrangement reactions as well as giving the perfluorinated product. In addition, small amounts of hydrogen can be retained in the crude product. The products are purified by treatment with base to remove the hydrogen-containing species and subsequently distilled.

Direct Fluorination. This is a more recently developed method for the synthesis of perfluorinated compounds. In this process, fluorine gas is passed through a solution or suspension of the reactant in a nonreactive solvent such as trichlorotrifluoroethane (CFC-113). Sodium fluoride may also be present in the reaction medium to remove the coproduct hydrogen fluoride. There has been enormous interest in this area since the early 1980s resulting in numerous journal publications and patents (7–9) (see FLUORINE COMPOUNDS, ORGANIC–DIRECT FLUORINATION). Direct fluorination is especially useful for the preparation of perfluoroethers.

Multiple ether oxygen atoms can be present in the molecule. Cleavage and coupling reactions occur with direct fluorination although to a lesser extent than with ECF. This allows the direct fluorination of acid-sensitive materials, such as the formal shown below, which would not survive ECF (8).

$$(C_4H_9OC_2H_4O)_2CH_2 + F_2 \rightarrow (C_4F_9OC_2F_4O)_2CF_2$$

As opposed to ECF, direct fluorination affords a much lower degree of isomerization so that the carbon skeleton of the reactant remains intact in the perfluorinated product. Direct fluorination is also complementary to ECF in the significantly higher yields observed for the direct fluorination of ethers. As with ECF the products are purified by treatment with base and subsequent distillation.

Polymerization. The higher molecular weight perfluoropolyethers are prepared by distinctly different technology. The anionic polymerization of hexafluoropropene oxide is carried out using cesium or potassium fluoride as catalyst in a polar aprotic solvent such as diglyme (11). This leads to repeating units (n can vary widely) of the perfluoroisopropoxy group in the oligomeric chain which is terminated with an acyl fluoride.

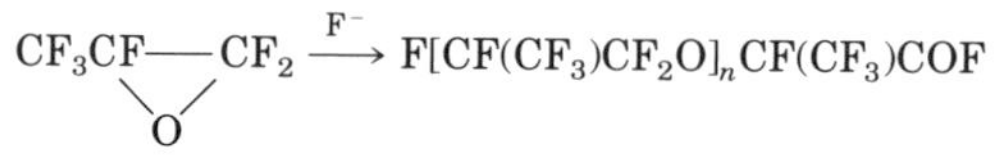

$$CF_3CF\overset{\diagdown \diagup}{\underset{O}{\quad}}CF_2 \xrightarrow{F^-} F[CF(CF_3)CF_2O]_n CF(CF_3)COF$$

This reactive end group must be removed to render the final compound inert. There are several methods to do this. Deacylation using antimony pentafluoride or aluminum fluoride as catalyst leads to the stable OC_2F_5 end group. The acyl fluoride may be converted to the corresponding acid salt by reaction with a base such as potassium hydroxide followed by thermal decarboxylation. This yields the hydrogen-containing $OCFHCF_3$ end group. The acid (27), or at higher temperatures the acyl fluoride, may be directly replaced by fluorine with either elemental fluorine or cobalt trifluoride. The lower molecular weight members of this family have been marketed by Du Pont as their K-Series fluids.

The photooxidation of hexafluoropropene or tetrafluoroethene or mixtures thereof leads to perfluoro polyether peroxides of varying molecular weights (14). The peroxidic links are subsequently decomposed by thermal treatment, the presence of base is optional, to give lower molecular weight fragments containing a variety of end groups. These include the CF_3O and the fluoroformyl FCO_2 groups. The polymers are then stabilized by removal of the functional group via direct fluorination. The final products of the photooxidation of hexafluoropropene have the following structure where the ratio q/p can vary between 0 and 0.1:

$$CF_3O[(CF(CF_3)CF_2O)_p(CF_2O)_q] CF_3$$

The lower molecular weight fractions from this process have been marketed by Montefluos under their trade name Galden.

The final products for the photooxidation of tetrafluoroethene have the following structure where the ratio m/n is between 0.6–1.5 and is typically about

0.8 (14). The average molecular weight of these polyethers is between ca 1,000 and 40,000.

$$CF_3O(C_2F_4O)_m(CF_2O)_nCF_3$$

Perfluoropolyethers with the linear perfluoropropoxy repeat unit have been commercialized (28). They are prepared by the anionic oligomerization of tetrafluorooxetane followed by direct fluorination to remove the acyl fluoride end group as well as to fluorinate the remaining CH_2 groups; n can vary widely.

$$\underset{\underset{CH_2-O}{|\qquad|}}{CF_2-CF_2} + F^- \rightarrow F(CH_2CF_2CF_2O)_nCH_2CF_2COF \xrightarrow{F_2} C_3F_7O(C_3F_6O)_nC_2F_5$$

All three processes give perfluoropolyethers with a broad distribution of molecular weights. They are typically separated into fractions by vacuum distillation.

Economic Aspects

Information on the production levels of the perfluoroethers and perfluorotertiary amines is not disclosed, but the products are available commercially and are marketed, for instance, as part of the Fluorinert Electronic Liquids family by 3M Co. (17). These liquids have boiling points of 30–215°C with molecular weights of about 300–800. They range in price from $26–88/kg. Perfluoropropene oxide polyethers are marketed by Du Pont with the trade name Krytox (29). The linear perfluoropropene oxide polyethers are marketed by Daikin under the trade name Demnum (28). The perfluoropolyethers derived from photooxidation are marketed by Montefluos under the trade name of Fomblin (30). These three classes of polyethers are priced from about $100–150/kg.

Environmental, Health, and Safety Factors

Over the years animal studies have repeatedly shown that perfluorinated inert fluids are nonirritating to the eyes and skin and practically nontoxic by ingestion, inhalation, or intraperitoneal injection (17,22). Thermal degradation can produce toxic decomposition products including perfluoroisobutene which has a reported LC_{50} of 0.5 ppm (6 hr exposure in rats) (31). This decomposition generally requires temperatures above 200°C.

Perfluorinated ethers and perfluorinated tertiary amines do not contribute to the formation of ground level ozone and are exempt from VOC regulations (32). The commercial compounds discussed above have an ozone depletion potential of zero because they do not contain either chlorine or bromine which take part in catalytic cycles that destroy stratospheric ozone (33).

Uses

The unique combination of properties of the perfluorinated fluids makes them useful in a variety of applications in the electronics industry (34). The lower molecular weight materials are used in three principal areas in this industry: direct contact cooling of electronic components, testing, and reflow-soldering. Recently perfluorinated liquids have been used as the total immersion coolant for a new generation supercomputer. This fluid is used to cool the power supplies, memory boards, logic circuits, and main processors. Vapor-phase reflow-soldering has been described (35). Testing applications include liquid burn-in testing, gross and fine leak testing, and electrical environmental testing. In the early 1990s, these perfluorinated fluids have found use in some specialized applications as replacements for the chlorofluorocarbons (CFCs) which have come under tight regulation. Perfluorinated liquids have replaced CFCs in several applications including coolant for ion implanters, secondary blanket in vapor-phase soldering, solvent for magnetic media lubricants, and coolant for large rectifiers. They have also found use in blowing agent systems for polyurethane foams (36) and in the removal of water from precision parts that have been cleaned using an aqueous rinse (37). They are under investigation as potential replacements for the Halon fire-extinguishing agents (qv) (38).

The higher molecular weight perfluoropolyethers are useful as specialty lubricants. They provide good lubrication under boundary conditions in systems in which the mechanical parts are exposed to high temperatures or aggressive chemical environments. They are typically used as the working fluid in hazardous duty vacuum pumps used in plasma etching. Specialty greases, used in high temperature environments in which a hydrocarbon-based grease fails, have also been formulated by blending perfluoropolyethers with fluorinated polymers. Additionally, perfluoropolyethers are used as lubricants for magnetic media, lubricant and sealing agent for oxygen service, inert hydraulic fluids, etc (14). They have also found application in cosmetics (39) and as a protective coating for outdoor stone art and masonry (40).

Perfluorinated compounds are also potentially useful as inert reaction media, particularly when one of the reactants is gaseous. The high solubility of oxygen and carbon dioxide in perfluorinated liquids has allowed their use as blood substitutes (41) and as oxygenation media for biotechnology (42). One product, Fluosol DA (43) (Green Cross Corp.), has been commercialized, and there is an abundant patent art in this area (see BLOOD, ARTIFICIAL).

BIBLIOGRAPHY

"Fluoro Ethers and Amines" under "Fluorine Compounds, Organic" in *ECT* 1st ed., Vol. 6, pp. 762–763, by L. J. Hals and W. H. Pearlson, Minnesota Mining & Manufacturing Co.; in *ECT* 2nd ed., Vol. 9, pp. 753–754, by L. J. Hals and W. H. Pearlson, Minnesota Mining & Manufacturing Co.; in *ECT* 3rd ed., Vol. 10, pp. 874–881, by R. D. Danielson, Minnesota Mining & Manufacturing Co.

1. J. H. Simons and co-workers, *J. Electrochem. Soc.* **95**, 47 (1949).
2. J. H. Simons, ed., *Fluorine Chemistry*, Vol. 1, Academic Press, Inc., New York, 1950, pp. 414–420.

3. S. Nagase, in P. Tarrant, ed., *Fluorine Chemistry Reviews*, Vol. 1, Marcel Dekker, Inc., New York, 1967, pp. 77–106.

4. T. Abe and S. Nagase, in R. E. Banks, ed., *Preparation, Properties and Industrial Applications of Organofluorine Compounds*, Ellis Horwood, Chichester, U.K., 1982, p. 19.

5. I. N. Rozhkov, in M. M. Baizer and H. Lund, eds., *Organic Electrochemistry*, 2nd ed., Marcel Dekker, Inc., New York, 1983, p. 805.

6. W. V. Childs and co-workers, in H. Lund and M. M. Baizer, eds., *Organic Electrochemistry*, 3rd ed., Marcel Dekker, Inc., New York, 1991, p. 1103.

7. U.S. Pat. 5,093,432 (Mar. 3, 1992), T. R. Bierschenk and co-workers (to Exfluor Research Corp.).

8. World Pat. 90,06296 (June 14, 1990), M. G. Costello and G. G. I. Moore (to Minnesota Mining & Mfg. Co.).

9. R. J. Lagow, *Prog. Inorg. Chem.* **26**, 161–210 (1979).

10. R. D. Chambers and co-workers, *J. Fluorine Chem.* **29**, 323 (1985).

11. J. T. Hill, *J. Macromol. Sci.-Chem.* **A8**(3), 499 (1974); P. Tarrant and co-workers, in P. Tarrant, ed., *Fluorine Chemistry Reviews*, Vol. 5, Marcel Dekker, Inc., New York, 1971, pp. 77–113.

12. H. S. Eleuterio, *J. Macromol. Sci.-Chem.* **A6**(6), 1027 (1972).

13. D. Sianesi and co-workers, *Chim. Ind. (Milan)* **55**(2), 208 (1973).

14. G. Caporiccio, in R. E. Banks, D. W. A. Sharp, and J. C. Tatlow, eds., *Fluorine: The First Hundred Years (1886–1986)*, Elsevier Sequoia, New York, 1986, pp. 314–320.

15. Y. Ohsaka, *J. Jpn. Pet. Inst.* **8**(9), 2 (1985).

16. U.S. Pat. 3,810,874 (May 14, 1974), R. A. Mitsch and J. L. Zollinger (to Minnesota Mining & Mfg. Co.).

17. *Fluorinert Liquids*, technical notebook, 3M Co., St. Paul, Minn., 1987.

18. H. G. Bryce, in J. H. Simons, ed., *Fluorine Chemistry*, Vol. 5, Academic Press, Inc., New York, 1964.

19. G. V. D. Tiers, *J. Am. Chem. Soc.* **77**, 4837, 6703, 6704 (1955).

20. P. Kasai and P. Wheeler, *Appl. Surface Sci.* **52**, 91 (1991).

21. P. Kasai, W. Tang, and P. Wheeler, *Appl. Surface Sci.* **51**, 201 (1991).

22. J. W. Sargent and R. J. Seffl, *Fed. Proc.* **29**, 1699 (1970).

23. J. H. Hildebrand, J. M. Prausnitz, and R. L. Scott, *Regular and Related Solutions*, Van Nostrand Reinhold, New York, 1970, p. 207.

24. A. Dimitrov, St. Rudiger, and M. Bartoszek, *J. Fluorine Chem.* **47**, 23 (1990).

25. T. Abe and co-workers, *J. Fluorine Chem.* **12**, 359 (1978).

26. G. G. I. Moore and co-workers, *J. Fluorine Chem.* **32**, 41 (1986).

27. U.S. Pat. 4,847,427 (July 11, 1989), M. J. Nappa (to E. I. du Pont de Nemours & Co., Inc.).

28. *Demnum, Demnum Grease Technical Bulletin*, Daikin Industries, Ltd., Osaka, Japan, Jan. 1987.

29. *Krytox, Technical Bulletins No. E74823*, E. I. du Pont de Nemours & Co., Inc., Wilmington, Del., Aug. 1985.

30. *Fomblin y Fluorinated Fluids*, Montedison SpA, Milan, Italy, 1971.

31. J. W. Clayton, Jr., in Ref. 3, pp. 225–232.

32. *Fed. Reg.* **57**(22), 3945 (Feb. 3, 1992).

33. M. J. Molina and F. S. Rowland, *Nature* **249**, 810 (1974).

34. D. Slinn and S. Green, in Ref. 4, pp. 45–82.

35. W. H. Pearlson, *J. Fluorine Chem.* **32**, 29 (1986).

36. U.S. Pat. 4,972,002 (Nov. 20, 1990), O. Volkert (to BASF AG).

37. U.S. Pat. 5,089,152 (Feb. 18, 1992), R. M. Flynn, D. A. Johnson, and J. G. Owens (to Minnesota Mining & Mfg. Co.).

38. R. E. Tapscott, *1992 Halon Alternatives Technical Working Conference*, Albuquerque, N.M., May 12–14, 1992.
39. Eur. Pat. Appl. 390,206 (Oct. 3, 1990), F. Brunetta and G. Pantini (to Ausimont SRL).
40. *Fomblin MET Technical Bulletin*, Montefluos, Milan, Italy, Oct. 1987.
41. *Science* **206**, 205 (1979).
42. B. Mattiasson and P. Adlercreut, *Trends Biotechnol.* **5**, 250 (1987).
43. *Chem. Eng. News*, 12 (Nov. 26, 1979).

RICHARD M. FLYNN
3M Company

PERFLUOROEPOXIDES

Perfluoroepoxides were first prepared in the late 1950s by Du Pont Co. Subsequent work on these compounds has taken place throughout the world and is the subject of a number of reviews (1–5). The main use of these epoxides is as intermediates in the preparation of other fluorinated monomers. Although the polymerization of the epoxides has been described (6–12), the resulting homopolymers and their derivatives are not significant commercial products. Almost all the work on perfluoroepoxides has been with three compounds: tetrafluoroethylene oxide (TFEO), hexafluoropropylene oxide (HFPO), and perfluoroisobutylene oxide (PIBO). Most of this work has dealt with HFPO, the most versatile and by far the most valuable of this class of materials (4).

Physical Properties

In general, the perfluoroepoxides have boiling points that are quite similar to those of the corresponding fluoroalkenes. They can be distinguished easily from the olefins by ir spectroscopy, specifically by the lack of olefinic absorption and the presence of a characteristic band between 1440 and 1550 cm^{-1}. The nmr spectra of most of the epoxides have been recorded. Little physical property data concerning these compounds have been published (Table 1). The structure of HFPO by electron diffraction (13) as well as its solubility and heats of solution in some organic solvents have been measured (14,15).

Chemical Properties

There are three general reactions of perfluoroepoxides: pyrolyses (thermal reactions), electrophilic reactions, and by far the most important, reactions with nucleophiles and bases.

Thermal Reactions. Those perfluoroepoxides that contain a CF_2 group in the epoxide ring undergo a smooth decomposition at relatively mild, neutral conditions (140–220°C) to give a perfluorocarbonyl compound and difluorocarbene (16,17) (eq. 1).

Table 1. Physical Properties of Perfluorocarbon Epoxides

Material	CAS Registry Number	Molecular structure	Bp,[a] °C	Infrared absorption, μm	Reference
tetrafluoroethylene oxide	[694-117-7]	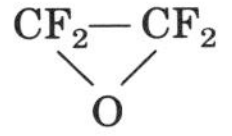 CF_2—CF_2 with O	-63.5^b	6.21	67,68
hexafluoropropylene oxide	[428-59-1]	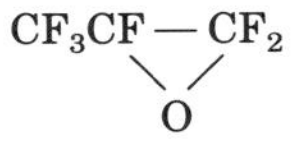 CF_3CF—CF_2 with O	-27.4	6.43	69
trifluoroglycidyl fluoride	[24419-82-7]	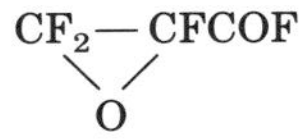 CF_2—$CFCOF$ with O	16^c (extrap.)	6.61 / 5.35 (COF)	70
perfluoro-1,2-epoxybutane	[3709-90-8]	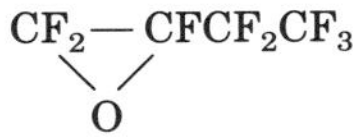 CF_2—$CFCF_2CF_3$ with O		6.48	21
perfluoro-2,3-epoxybutane	[773-29-5]	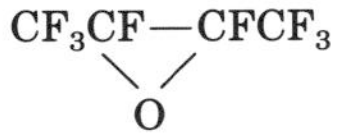 CF_3CF—$CFCF_3$ with O	0–1	6.63	32
perfluoroisobutylene oxide	[707-13-1]	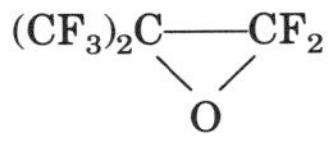 $(CF_3)_2C$—CF_2 with O	3^d	6.66	69
perfluoro-1,2-epoxycyclobutane	[13324-28-2]	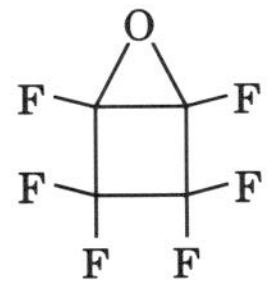		6.3	16
perfluoro-4,5-epoxy-1-pentene	[15453-08-4]	 CF_2—$CFCF_2CF$=CF_2 with O	37	6.45 / 5.65 (C=C)	72
perfluoro-1-oxaspiro[2,3]hexane	[53389-66-5]	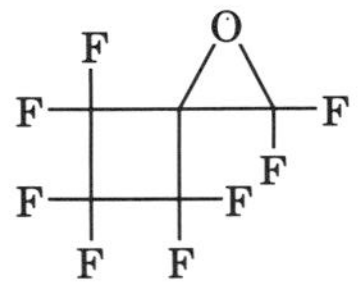	18–21	6.55	73

Table 1. (Continued)

Material	CAS Registry Number	Molecular structure	Bp,[a] °C	Infrared absorption, μm	Reference
perfluoro-1,2-epoxycyclopentane	[710-70-3]	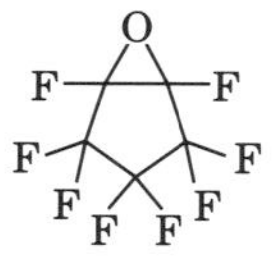	26.5	6.55	61
perfluoro-4,5-epoxyvaleryl fluoride	[140173-04-2]	CF₂—CFCF₂CF₂COF	37–38	6.49	71
perfluoro-1,2-epoxyhexane	[72804-48-9]	CF₂—CFCF₂CF₂CF₂CF₃	55–56	6.45	32
perfluoro-1,2-epoxy-6-oxaheptane[e]	[71877-16-2]	CF₂—CFCF₂CF₂CF₂OCF₃	61–62	6.49	74
perfluoro-2-methyl-2,3-epoxypentane	[788-67-0]	(CF₃)₂C—CFCF₂CF₃	57	6.85	69
perfluoro-2,3-epoxy-4-methylpentane	[788-50-1]	CF₃CF—CFCF(CF₃)₂	53	6.61	69
2,3-bis(trifluoromethyl)perfluoro-2,3-epoxybutane	[1708-78-7]	(CF₃)₂C—C(CF₃)₂	53–54		75
perfluoro-5,6-epoxy-1-hexene	[15453-10-8]	CF₂—CFCF₂CF₂CF=CF₂	57–58	6.45 5.59 (C=C)	71
perfluoro-1,2;5,6-dienoxyhexane	[140173-03-1]	CF₂—CFCF₂CF₂CF—CF₂	57–58	6.49	71

536

Compound	CAS Number	Structure	Bp, °C	n_D^{20}	Refs.
perfluoro-1,2-epoxy-5-methyl-4-oxahexane	[84424-45-3]	 CF₂—CFCF₂OCF(CF₃)₂ with epoxide O	58.5–59	6.47	78
perfluoro-1,2-epoxycyclohexane	[5927-67-3]	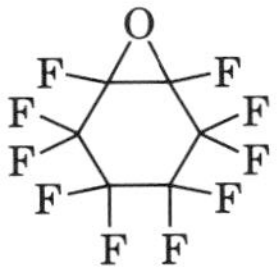 (perfluorocyclohexane epoxide structure)	54	6.71	79
perfluoro-1-methyl-1,2:4,5-diepoxycyclohexane	[130482-35-8]	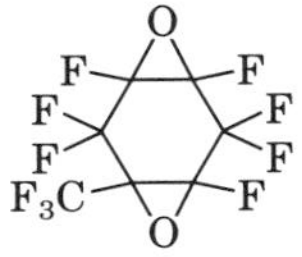 (perfluoromethyl diepoxycyclohexane structure)	77	6.69 / 6.92	80
perfluoro-7,8-epoxy-1-octene	[72264-78-9]	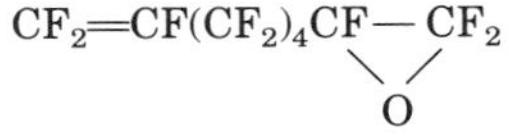 CF₂=CF(CF₂)₄CF—CF₂ with epoxide O	105		76
perfluoro-1,2:7,8-diepoxyoctane[f]	[13714-88-0]	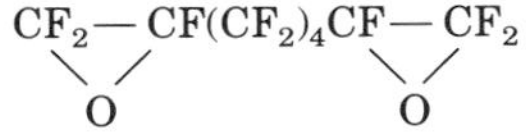 CF₂—CF(CF₂)₄CF—CF₂ with two epoxide O	104	6.45	76,77
perfluorophenylglycidyl ether	[84329-68-0]	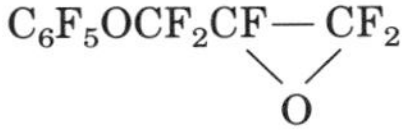 C₆F₅OCF₂CF—CF₂ with epoxide O	61–64 (4 kPa)[g]	6.47	78
perfluoro-3,4-epoxy-2,3,5-trimethylhexane	[2355-27-3]	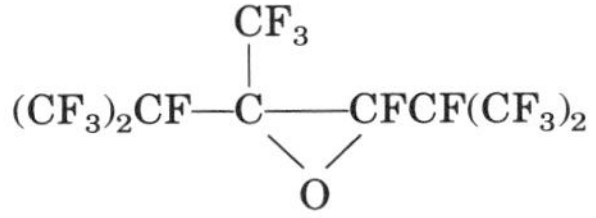 (CF₃)₂CF—C(CF₃)—CFCF(CF₃)₂ with epoxide O	36–39 (0.13 kPa)[g]	7.21	69
perfluoro-1,2:9,10-diepoxydecane[h]	[13714-90-4]	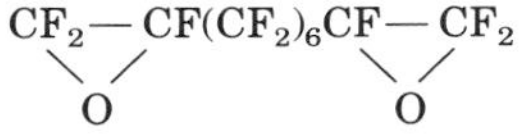 CF₂—CF(CF₂)₆CF—CF₂ with two epoxide O	88 (5.3 kPa)[g]		76

[a]At 101.3 kPa = 1 atm unless otherwise noted in parentheses. [b]Mp, −118°C. [c]Heat of vaporization = 28.9 kJ/mol (6.9 kcal/mol). [d]Mp, 122°C. [e]n_D^{20} 1.2560; d_4^{20} 1.6441. [f]n_D^{20} 1.2900; d_4^{20} 1.7220. [g]To convert kPa to mm Hg, multiply by 7.5. [h]n_D^{20} 1.3030; d_4^{20} 1.8260.

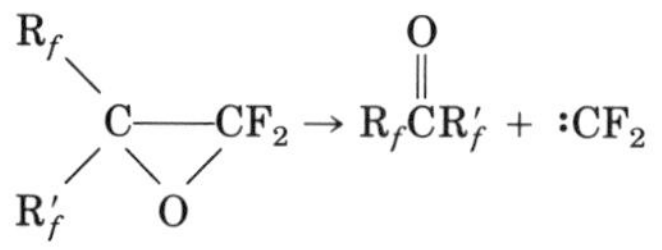

$$\underset{R_f'}{\overset{R_f}{\diagdown}} C \underset{O}{\overset{}{\diagup}} CF_2 \rightarrow R_f\overset{O}{\overset{\|}{C}}R_f' + :CF_2 \tag{1}$$

where R_f = perfluoroalkyl; R_f' = perfluoroalkyl or fluorine

The difluorocarbene produced in this way may react with a variety of compounds (18). Epoxides of internal olefins which do not contain a CF_2 group have much greater stability (19).

Electrophilic Reactions. Perfluoroepoxides are quite resistant to electrophilic attack. However, they react readily with Lewis acids, for example SbF_5, to give ring-opened carbonyl compounds (20–22) (eq. 2).

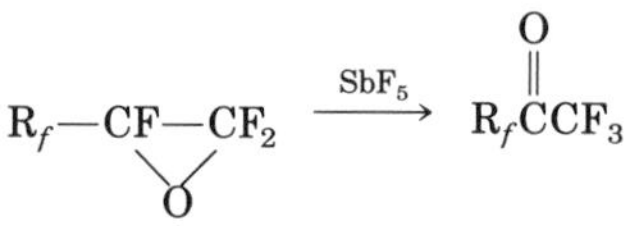

$$R_f\overset{}{-}CF\overset{}{-}CF_2 \xrightarrow{SbF_5} R_f\overset{O}{\overset{\|}{C}}CF_3 \tag{2}$$

The structure of the ketones produced from unsymmetrical internal perfluoroepoxides has been reported (5). The epoxide ring may also be opened by strong protic acids such as fluorosulfonic acid or hydrogen fluoride at elevated temperatures (23–25). The ring opening of HFPO by sulfur trioxide at 150°C has been interpreted as an example of an electrophilic reaction (26) (eq. 3).

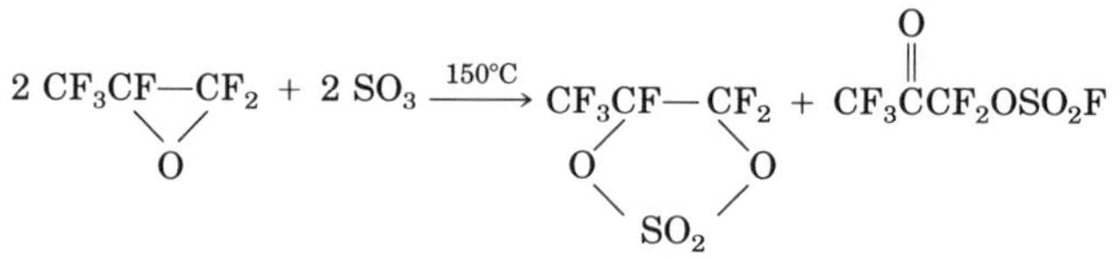

$$2\ CF_3CF\overset{}{-}CF_2 + 2\ SO_3 \xrightarrow{150°C} CF_3CF\overset{}{-}CF_2 + CF_3\overset{O}{\overset{\|}{C}}CF_2OSO_2F \tag{3}$$

Nucleophilic Reactions. The strong electronegativity of fluorine results in the facile reaction of perfluoroepoxides with nucleophiles. These reactions comprise the majority of the reported reactions of this class of compounds. Nucleophilic attack on the epoxide ring takes place at the more highly substituted carbon atom to give ring-opened products. Fluorinated alkoxides are intermediates in these reactions and are in equilibrium with fluoride ion and a perfluorocarbonyl compound. The process is illustrated by the reaction of methanol and HFPO to form methyl 2,3,3,3-tetrafluoro-2-methoxypropanoate (eq. 4).

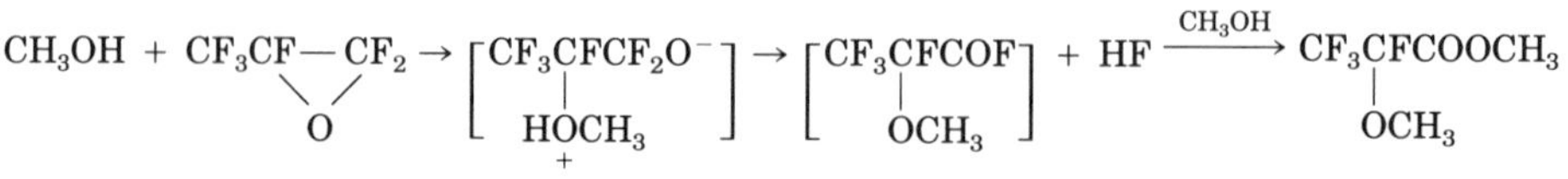

$$CH_3OH + CF_3CF\overset{}{-}CF_2 \rightarrow \left[\underset{HOCH_3}{\overset{}{\underset{+}{CF_3CFCF_2O^-}}} \right] \rightarrow \left[\underset{OCH_3}{CF_3CFCOF} \right] + HF \xrightarrow{CH_3OH} \underset{OCH_3}{CF_3CFCOOCH_3} \tag{4}$$

TFEO is by far the most reactive epoxide of the series. However, all the reported perfluoroepoxides undergo similar ring-opening reactions. The most important reactions of these epoxides are those with the fluoride ion or perfluoroalkoxides. The reaction of PIBO and the fluoride ion is an example (27). It also illustrates the general scheme of oligomerization of perfluoroepoxides (eq 5).

$$(CF_3)_2C\!\!-\!\!CF_2 \xrightarrow{F^-} (CF_3)_2CFCF_2O^- \leftrightarrows (CF_3)_2CFCOF + F^- \xrightarrow{PIBO} (CF_3)_2CFCF_2O\overset{\overset{\displaystyle CF_3}{|}}{\underset{\underset{\displaystyle CF_3}{|}}{C}}CF_2O^- \leftrightarrows etc$$

$$\underset{\diagdown\diagup}{}$$
$$O$$

(5)

The direction of nucleophilic ring opening of unsymmetrical perfluoroepoxides has been shown to be a function of the nature of the nucleophile and the solvent (23,28). Although many oligomeric products have been prepared by this procedure and variations of it, no truly high polymers have been obtained (9).

Preparation

A large number of methods have been used to prepare perfluoroepoxides (5). All of these methods must contend with the great chemical reactivity of the epoxide product, especially with subsequent ionic and thermal reactions which result in the loss of the desired epoxide.

The reaction of perfluoroalkenes with alkaline hydrogen peroxide is a good general method for the preparation of the corresponding epoxides with the exception of the most reactive of the series, TFEO (eq. 6).

$$R_fCF\!\!=\!\!CF_2 + H_2O_2 \xrightarrow[solvent]{OH^-} R_fCF\!\!-\!\!CF_2 + H_2O$$
$$\underset{O}{\diagdown\diagup}$$

(6)

The alkene is allowed to react at low temperatures with a mixture of aqueous hydrogen peroxide, base, and a co-solvent to give a low conversion of the alkene (29). These conditions permit reaction of the water-insoluble alkene and minimize the subsequent ionic reactions of the epoxide product. Phase-transfer techniques have been employed (30). A variation of this scheme using a peroxycarbimic acid has been reported (31).

Reaction of perfluoroalkenes and hypochlorites has been shown to be a general synthesis of perfluoroepoxides (32) (eq. 7). This appears to be the method of choice for the preparation of epoxides from internal fluoroalkenes (38). Excellent yields of HFPO from hexafluoropropylene and sodium hypochlorite using phase-transfer conditions are claimed (34).

$$R_fCF\!\!=\!\!CFR_f + {}^-OCl \rightarrow R_fCF\!\!-\!\!CFR_f + Cl^-$$
$$\underset{O}{\diagdown\diagup}$$

(7)

The direct oxidation of fluoroalkenes is also an excellent general synthesis procedure for the preparation of perfluoroepoxides (eq. 8). This method exploits the low reactivity of the epoxide products to both organic and inorganic free radicals.

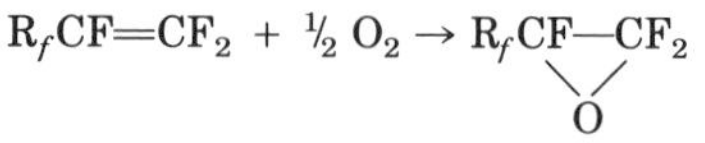

$$R_fCF\!=\!CF_2 + \tfrac{1}{2}\,O_2 \rightarrow R_fCF\!\!-\!\!CF_2 \quad\underset{O}{\diagdown\!\!\diagup} \tag{8}$$

The oxidation may be carried out with an inert solvent thermally (35), with a sensitizer such as bromine (36), with uv radiation (37), or over a suitable catalyst (38). Principal by-products of all these oxidation processes are the acyl fluoride products derived from oxidative cleavage of the perfluoroalkene (eq. 9).

$$R_fCF\!=\!CF_2 + O_2 \rightarrow R_fCOF + COF_2 \tag{9}$$

Perfluoroepoxides have also been prepared by anodic oxidation of fluoroalkenes (39), the low temperature oxidation of fluoroalkenes with potassium permanganate (40), by addition of difluorocarbene to perfluoroacetyl fluoride (41) or hexafluoroacetone (42), epoxidation of fluoroalkenes with oxygen difluoride (43) or peracids (44), the photolysis of substituted 1,3-dioxolan-4-ones (45), and the thermal rearrangement of perfluorodioxoles (46).

Tetrafluoroethylene Oxide

TFEO has only been prepared by a process employing oxygen or ozone because of its extreme reactivity with ionic reagents. This reactivity may best be illustrated by its low temperature reaction with the weak nucleophile, dimethyl ether, to give either of two products (47) (eq. 10).

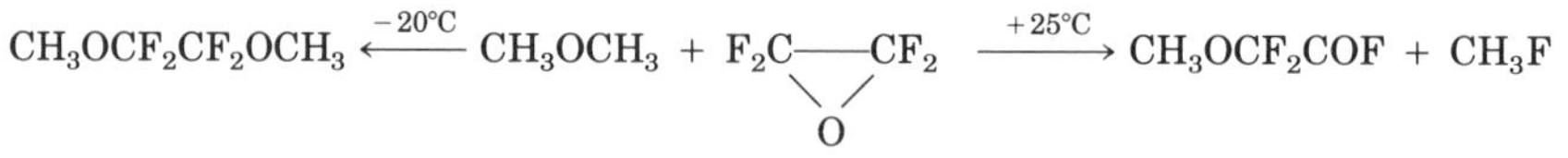

$$CH_3OCF_2CF_2OCH_3 \xleftarrow{-20^\circ C} CH_3OCH_3 + F_2C\!\!-\!\!CF_2 \;\underset{O}{\diagdown\!\!\diagup}\; \xrightarrow{+25^\circ C} CH_3OCF_2COF + CH_3F \tag{10}$$

Reaction of TFEO with acid fluorides and the fluoride ion yields oligomers with the structure $R_fCF_2O(CF_2CF_2O)_nCF_2COF$ (47,48). The epoxide yields a waxy solid polymer when exposed to high energy radiation (47,49) or when treated with amines at low temperature (47,50). The extreme chemical reactivity and facile rearrangement to perfluoroacetyl fluoride have been deterrents to the large-scale development of TFEO. The structure of TFEO has been measured using microwave spectroscopy (51).

Hexafluoropropylene Oxide

HFPO is the most important of the perfluoroepoxides and has been synthesized by almost all of the methods noted. Many attempts have been made to polymerize HFPO (6,8). The most successful has been the reaction of HFPO with fluoride ion at low temperature to give a series of oligomeric acid fluorides which have been end capped to yield stable fluids (eq. 11, where X = H,F).

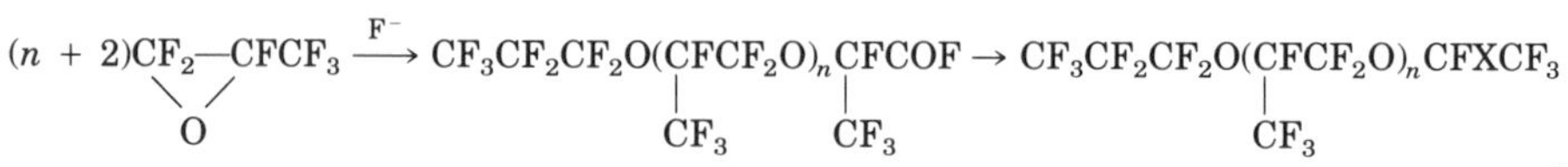

$$(n+2)CF_2\!\!-\!\!CFCF_3 \;\underset{O}{\diagdown\!\!\diagup}\; \xrightarrow{F^-} CF_3CF_2CF_2O(\underset{CF_3}{CFCF_2}O)_n\underset{CF_3}{CFCOF} \rightarrow CF_3CF_2CF_2O(CFCF_2O)_n\underset{CF_3}{CFXCF_3} \tag{11}$$

Materials of this type have been sold by Du Pont Co. under the Freon E and Krytox trademarks. Perfluorinated materials structurally similar to those in equation 11 have been prepared by Ausimont by the low temperature irradiation of either hexafluoropropylene or tetrafluoroethylene with oxygen followed by heating and/ or irradiation and have been sold as Fomblin liquids (52). An isomeric polyether, Demnum, prepared by the oligomerization of 2,2,3,3-tetrafluorooxetane followed by fluorination has been commercialized by Daikin (eq. 12).

$$CH_2\!-\!CF_2 \rightarrow FCH_2CF_2CF_2O(CH_2CF_2CF_2O)_nOCH_2CF_2COF \rightarrow CF_3CF_2CF_2O(CF_2CF_2CF_2O)_nCF_2CF_2$$
$$O\!-\!\!-\!\!-CF_2$$

$$(12)$$

Higher molecular weight HFPO-based materials have been prepared by reaction with both ends of a perfluorodiacyl fluoride followed by coupling through triazine rings (9). Lower molecular weight HFPO oligomers have been coupled to give inert perfluorinated ethers which are sold as Hostinert liquids by Hoechst-Celanese (eq. 13).

$$2\ CF_3CF_2CF_2O(CFCF_2O)_nCFCOF \rightarrow C_3F_7O(CFCF_2O)_nCF\!-\!CF(OCF_2CF)_nOC_3F_7 \quad (13)$$
$$CF_3 \qquad CF_3 \qquad\qquad CF_3 \qquad CF_3\ CF_3 \qquad CF_3$$

HFPO reacts with a large number of acyl fluorides in a general reaction to give 2-alkoxytetrafluoropropionyl fluorides which in turn may be converted to trifluorovinyl ethers (eq. 14).

$$R_fCOF + F_2C\!-\!\!-\!\!-CFCF_3 \xrightarrow{\ F^-\ } R_fCF_2OCCFCOF \rightarrow R_fCF_2OCF\!=\!CF_2 \quad (14)$$
$$O \qquad\qquad\qquad CF_3$$

These ethers readily copolymerize with tetrafluoroethylene and other fluoroalkenes to commercially significant plastics, elastomers, and ion-exchange resins such as Teflon PFA, Kalrez, and Nafion (see FLUORINE COMPOUNDS, ORGANIC–TETRAFLUOROETHYLENE–PERFLUOROVINYL ETHER COPOLYMERS; ELASTOMERS, SYNTHETIC–FLUOROCARBON ELASTOMERS; IONIC POLYMERS).

Publications have described the use of HFPO to prepare acyl fluorides (53), fluoroketones (54), fluorinated heterocycles (55), as well as serving as a source of difluorocarbene for the synthesis of numerous cyclic and acyclic compounds (56). The isomerization of HFPO to hexafluoroacetone by hydrogen fluoride has been used as part of a one-pot synthesis of bisphenol AF (57). HFPO has been used as the starting material for the preparation of optically active perfluorinated acids (58). The nmr spectrum of HFPO is given in Reference 59. The molecular structure of HFPO has been determined by gas-phase electron diffraction (13).

Perfluoroisobutylene Oxide

PIBO has been prepared primarily by the addition of difluorocarbene to hexafluoroacetone or by the reaction of alkaline hydrogen peroxide with perfluoroisobutylene. The small amount of published work on PIBO deals with its oligomerization (60), isomerization to perfluoroisobutyryl fluoride (61), conversion to perfluoro-*t*-butyl alcohol (62), and reaction with nucleophiles (63). PIBO has been reported to be as toxic as perfluoroisobutylene (64). The nmr spectrum of PIBO is reported in Reference 65.

Other Epoxides

Large numbers of epoxides have been reported that contain only fluorine and carbon bound to the oxirane ring but which contain other halogens, oxygen, hydrogen, and other functional groups in one of the carbon side chains. Although these are not true perfluoroepoxides their syntheses and reactions of their epoxide rings are virtually identical to those of the perfluoro analogues. One example is the reaction of 3-chloropentafluoropropylene oxide with nucleophiles such as fluoride ion (66) (eq. 15).

$$CF_2\text{—}CFCF_2Cl \xrightarrow{F^-} CClF_2CF_2CF_2O(CFCF_2O)_n CFCOF \qquad (15)$$
$$\underset{O}{\diagdown\diagup} \qquad\qquad\qquad \underset{CF_2Cl}{|} \quad \underset{CF_2Cl}{|}$$

BIBLIOGRAPHY

"Perfluoroepoxides" under "Fluorine Compounds, Organic" in *ECT* 3rd ed., Vol. 10, pp. 956–962, by P. R. Resnick, E. I. du Pont de Nemours & Co., Inc.

1. D. Sianesi and co-workers, *Gazz. Chem. Ital.* **98**, 265, 277, 290 (1968).
2. P. Tarrant and co-workers, *Fluorine Chem. Rev.* **5**, 77 (1971).
3. N. Ishikawa, *Yuki Gosei Kagaku Kyokai Shi* **35**, 131 (1977).
4. H. Millauer, W. Schwertfeger, and G. Siegemund, *Angew. Chem. Int. Ed. Engl.* **24**, 161 (1985).
5. L. F. Sokolov, P. I. Valov, and S. V. Sokolov, *Usp. Khim.* **53**, 1222 (1984).
6. H. S. Eleuterio, *J. Macromol. Sci. Chem.* **A6**, 1027 (1972).
7. J. T. Hill and J. P. Erdman, *Am. Chem. Soc. Polym. Prepr.* **18**, 100 (1977).
8. J. T. Hill, *J. Macromol. Sci.* **A8**, 499 (1974).
9. Jpn. Kokai JP 02-202919 (Aug. 13, 1990), M. Ikeda and A. Aoshima (to Asahi Chemical).
10. Y. Duan, D. Ni, and Y. He, *Kao Fen Tzu T'ung Hsun*, 139 (1981).
11. P. P. Shpakov and co-workers, *Zh. Prikl. Khim.* **54**, 2726 (1981).
12. Eur. Pat. 154,297A (Sept. 11, 1985), T. R. Darling (to E. I. du Pont de Nemours & Co., Inc.).
13. B. Beagley, R. G. Pritchard, and R. E. Banks, *J. Fl. Chem.* **18**, 159 (1981).
14. V. A. Gubanov and co-workers, *Zh. Fiz. Khim.* **48**, 2084 (1974)
15. R. G. Makitra, T. I. Politanskaya, and F. B. Moin, *Zh. Prik. Khim.* **52**, 2623 (1979); **56**, 2724 (1983).
16. D. P. Carlson and A. S. Milian, *4th International Fluorine Symposium*, Estes Park, Colo., July 1967.
17. R. C. Kennedy and J. B. Levy, *J. Fluorine Chem.* **7**, 101 (1976).
18. P. B. Sargeant, *J. Am. Chem. Soc.* **91**, 3061 (1969); V. V. Karpov, V. E. Platonov, and G. G. Yakobson, *Izv. Akad. Nauk SSSR, Ser. Khim.*, 981 (1975); R. N. Beauchamp, C. W. Gillies, and N. C. Craig, *J. Am. Chem. Soc.* **109**, 1696 (1987).

19. P. L. Coe, A. W. Mott, and J. C. Tatlow, *J. Fluorine Chem.* **30**, 297 (1985).
20. Jpn. Kokai 78 25,512 (Mar. 9, 1978), Y. Osaka and H. Takashi (to Daiken Kogyo).
21. U.S. Pat. 3,213,134 (Oct. 19, 1965), D. E. Morin (to 3M).
22. U.S. Pat. 4,302,608 (Nov. 24, 1981), E. N. Squire (to E. I. du Pont de Nemours & Co., Inc.).
23. A. Ya. Zapevalov and co-workers, *Zh. Org. Khim.* **22**, 93 (1986).
24. T. I. Filyakova and co-workers, *Izv. Akad. Nauk SSSR, Ser. Khim.*, 1878 (1979).
25. U.S. Pat. 4,400,546 (Aug. 23, 1983), P. -P. Rammelt and G. Siegemund (to Hoechst).
26. I. L. Knunyants, V. V. Shokina, and E. I. Mysov, *Isv. Akad. Nauk SSSR Ser. Khim.*, 2725 (1973).
27. J. T. Hill, *J. Fl. Chem.* **9**, 97 (1972).
28. A. Ya. Zapevalov and co-workers, *Zh. Org. Khim.* **25**, 492 (1989).
29. U.S. Pat. 3,358,003 (Dec. 12, 1967), H. S. Eleuterio and R. W. Meschke (to E. I. du Pont de Nemours & Co., Inc.).
30. U.S. Pat. 4,954,643 (Sept. 4, 1990), G. Bornengo and co-workers.
31. Ger. Offen. 2,557,655 (June 30, 1977), R. A. Sulzbach and F. Heller (to Hoescht).
32. I. P. Kolenko and co-workers, *Izv. Akad. Nauk SSSR, Ser. Khim.*, 2509 (1979).
33. T. I. Filyakova and co-workers, *Zh. Org. Khim.* **24**, 371 (1988).
34. Eur. Pat. 64,293 (Dec. 10, 1986), M. Ikeda, M. Miura, and A. Aoshima (to Asahi Kasei Kogyo K. K.).
35. S. V. Kartsov and co-workers, *Izv. Akad. Nauk SSSR, Ser. Khim.*, 2230 (1975); U.S. Pat. 3,536,733 (Oct. 27, 1970), D. P. Carlson (to E. I. du Pont de Nemours & Co., Inc.).
36. Brit. Pat. 931,857 (July 17, 1963), (to E. I. du Pont de Nemours & Co., Inc.).
37. V. Caglioti, M. Lenzi, and A. Mele, *Nature* **201**, 610 (1964); D. Sianesi, A. Pasetti, and C. Corti, *Makromol. Chem.* **86**, 308 (1965).
38. U.S. Pat. 3,775,438 (Nov. 27, 1973, R. J. Cavanaugh (to E. I. du Pont de Nemours & Co., Inc.); U.S. Pat. 3,775,439 (Nov. 27, 1973), G. M. Atkins, Jr. (to E. I. du Pont de Nemours & Co., Inc.); U.S. Pat. 3,775,440 (Nov. 27, 1973), R. J. Cavanaugh and G. M. Atkins, Jr. (to E. I. du Pont de Nemours & Co., Inc.).
39. U.S. Pat. 4,014,762 (Mar. 29, 1977), H. Millauer (to Hoescht).
40. I. L. Knunyants and co-workers, *Izv. Akad. Nauk SSSR Ser. Khim.*, 2780 (1967).
41. W. Mahler and P. R. Resnick, *J. Fluorine Chem.* **3**, 451 (1973/74).
42. U.S. Pat. 3,338,978 (Aug. 29, 1967), E. P. Moore (to E. I. du Pont de Nemours & Co., Inc.).
43. U.S. Pat. 3,622,601 (Nov. 23, 1971), J. W. Dale (to Monsanto); U.S. Pat. 3,639,429 (Feb. 1, 1972), V. Weinmayr (to E. I. du Pont de Nemours & Co., Inc.).
44. Jpn. Kokai 77 108,914 (Sept. 12, 1977), Y. Oda, K. Uchida, and S. Morikawa (to Asahi Glass).
45. T. S. Croft, *J. Fluorine Chem.* **7**, 438 (1976).
46. M. -H. Hung and P. R. Resnick, *J. Am. Chem. Soc.* **112**, 9671 (1990).
47. J. L. Warnell in Ref. 16.
48. U.S. Pat. 3,250,806 (May 10, 1966), J. L. Warnell (to E. I. du Pont de Nemours & Co., Inc.).
49. P. Barnaba and co-workers, *Chim. Inc. (Milan)* **47**, 1060 (1965).
50. Can. Pat. 778,490 (Feb. 13, 1968), J. L. Warnell (to E. I. du Pont de Nemours & Co., Inc.).
51. J. W. Agopovich and co-workers, *J. Am. Chem. Soc.* **106**, 2251 (1984).
52. D. Sianesi, *Am. Chem. Soc. Polym. Prepr.* **12**, 411 (1971); D. Sianesi and co-workers, *Chim. Ind. (Milan)* **55**, 208 (1973).
53. N. Ishikawa and S. Sasaki, *Chem. Lett.* **483**, 1407 (1976).
54. T. Martini, *Tetrahedron Lett.*, 1857, 1861 (1976).
55. N. Ishikawa and S. Sasaki, *Bull. Chem. Soc. Jpn.* **50**, 2164 (1977).

56. V. M. Karpov, V. E. Platonov, and G. G. Yacobsen, *Izv. Akad. Nauk SSSR Ser. Khim.*, 2295 (1976).

57. Eur. Pat. Appl. EP54227A (June 23, 1982), P. P. Rammelt and G. Siegemund (to Hoechst).

58. N. Ishikawa, *J. Fl. Chem.* **25**, 17 (1984); H. Kawa and N. Ishikawa, *Chem. Lett.*, 843 (1980).

59. J. K. Ruff and R. F. Merritt, *J. Org. Chem.* **30**, 3968 (1965); K. W. Jolley, L. H. Sutcliffe, and K. L. Williamson, *Spectrochemica Acta* **30A**, 1455 (1974).

60. J. T. Hill, *8th International Fluorine Symposium*, Kyoto, Japan, Aug. 1977.

61. U.S. Pat. 3,321,515 (May 23, 1967), E. P. Moore and A. S. Milian (to E. I. du Pont de Nemours & Co., Inc.).

62. U.S. Pat. 3,385,904 (May 28, 1968), F. J. Pavlik (to Minnesota Mining and Manufacturing Co.).

63. I. L. Knunyants and co-workers, *Izv. Akad. Nauk SSSR Ser. Khim.*, 1133 (1972).

64. I. L. Knunyants, V. V. Shokina, and I. V. Salakhov, *Khim. Geterotsikl. Soedin.* **2**, 873 (1966).

65. F. J. Pavlik and P. E. Toren, *J. Org. Chem.* **35**, 2054 (1970).

66. Eur. Pat. Appl. 72579 (Feb. 23, 1983), B. R. Ezzell, W. P. Carl, and W. A. Mod (to Dow Chemical Co.).

67. Brit. Pat. 931, 857 (July 17, 1963), (to E. I. du Pont de Nemours & Co., Inc.).

68. V. Caglioti and co-workers, *J. Chem. Soc.*, 5430 (1964).

69. Brit. Pat. 904,877 (Sept. 5, 1962), (E. I. du Pont de Nemours & Co., Inc.).

70. W. Stuckey, J. Heicklen, and V. Knight, *Can. J. Chem.* **47**, 2329 (1969).

71. T. I. Filyakova, R. E. Ilatovskii, and A. Ya. Zapevalov, *Zh. Org. Khim.* **27**, 2055 (1991).

72. U.S. Pat. 3,366,610 (Jan. 30, 1968), B. C. Anderson (to E. I. du Pont de Nemours & Co., Inc.).

73. R. J. DePasquale, K. B. Baucom, and J. R. Patton, *Tetrahedron Lett.*, 1111 (1974).

74. V. V. Berenblit and co-workers, *Zh. Org. Khim.* **15**, 1417 (1979).

75. E. M. Rokhlin and co-workers, *Dokl. Akad. Nauk SSSR* **161**, 1356 (1965).

76. I. L. Knunyants and co-workers, *Zh. Obsch. Khim.* **36**, 1981 (1966).

77. USSR Pat. 545,645 (Mar. 25, 1977) D. S. Rondarev and co-workers.

78. U.S. Pat. 4,360,645 (Nov. 23, 1982), C. G. Krespan and A. P. King (to E. I. du Pont de Nemours & Co., Inc.).

79. P. L. Coe, A. W. Mott, and J. C. Tatlow, *J. Fl. Chem.* **20**, 243 (1982).

80. P. L. Coe, A. W. Mott, and J. C. Tatlow, *J. Fl. Chem.* **49**, 21 (1990).

PAUL R. RESNICK
E. I. du Pont de Nemours & Co., Inc.

FLUORINATED ACETIC ACIDS

Fluoroacetic acid [*144-49-0*], FCH_2COOH, is noted for its high toxicity to animals, including humans. It is sold in the form of its sodium salt as a rodenticide and general mammalian pest control agent. The acid has mp, 33°C; bp, 165°C; heat of combustion, -715.8 kJ/mol (-171.08 kcal/mol) (1); enthalpy of vaporization, 83.89 kJ/mol (20.05 kcal/mol) (2). Some thermodynamic and transport properties of its aqueous solutions have been published (3), as has the molecular structure of the acid as determined by microwave spectroscopy (4). Although first prepared in 1896 (5), its unusual toxicity was not published until 50 years later (6). The acid is the toxic constituent of a South African plant *Dichapetalum cymosum*, better known as gifblaar (7). At least 24 other poisonous plant species are known to contain it (8).

Chemically, fluoroacetic acid behaves like a typical carboxylic acid, although its acidity is higher (K_a = 2.2 × 10^{-3}) than the average (9). It can be prepared from the commercially available sodium salt by distillation from sulfuric acid (10).

Sodium Fluoroacetate. Sodium fluoroacetate [*62-74-8*], FCH_2COONa, known as Compound 1080, is a hygroscopic white solid, mp, 200–202°C, which decomposes when heated above the melting point. Its solubility at 25°C in g/100 g solvent is water, 111; methanol, 5; ethanol, 1.4; acetone, 0.04; and carbon tetrachloride, 0.004. Because its carbon–fluorine bond is unreactive under most conditions, this salt can be converted by standard procedures to typical carboxylic acid derivatives such as fluoroacetyl esters (11,12), fluoroacetyl chloride [*359-06-8*] (13), fluoroacetamide (14), or fluoroacetonitrile [*503-20-8*] (14).

Sodium fluoroacetate is usually made by displacing the halogen from an ester of bromo- or chloroacetic acid with potassium fluoride or, in one instance, antimony fluoride, followed by hydrolysis with aqueous sodium hydroxide (15–17). A commercial process for its manufacture from ethyl chloroacetate and potassium fluoride has been described (18). The ester, purified by distillation to remove traces of acid and water, is treated with oven-dried, finely powdered potassium fluoride in a well-stirred autoclave at 200°C for 11 hours. The resulting ethyl fluoroacetate [*459-72-3*] is then distilled into an agitated tank containing sodium hydroxide dissolved in methanol. The solid product is isolated by centrifugation, followed by vacuum drying. Through this process, all liquids are handled in a closed system of pipes and vessels, carefully inspected for leaks before each run. This is important since the intermediate fluoroacetate is highly toxic, and the starting chloroacetate is a lacrimator.

Toxicity. Sodium fluoroacetate is one of the most effective all-purpose rodenticides known (18). It is highly toxic to all species of rats tested and can be used either in water solution or in bait preparations. Its absence of objectionable taste and odor and its delayed effects lead to its excellent acceptance by rodents. It is nonvolatile, chemically stable, and not toxic or irritating to the unbroken skin of workers. Rats do not appear to develop any significant tolerance to this compound from nonlethal doses. However, it is extremely dangerous to humans, to common household pets, and to farm animals, and should only be used by experienced personnel. The rodent carcasses should be collected and destroyed since they remain poisonous for a long period of time to any animal that eats them.

The unusually high toxicity of fluoroacetic acid and of other monofluorinated organic compounds that can be metabolized to fluoroacetate has stimulated much research into the mechanism of this toxicity (8,19–23). Fluoroacetate mimics acetate by being incorporated into the tricarboxylic acid cycle of cellular respiration where it becomes converted into fluorocitric acid. This acid inhibits the enzyme, aconitate hydratase, which normally catalyzes the dehydration of citric acid. As a result, citric acid accumulates in the organism and the energy-producing cycle is interrupted. Because of the time it takes for the fluorocitrate to form and accumulate, there is usually a latent time of at least an hour before the appearance of symptoms of fluoroacetate poisoning, eg, ventricular fibrillation or convulsions. This is advantageous in its use as a pesticide. One characteristic of fluoroacetate toxicity is the wide range in lethal doses for different species ranging from (LD_{50}, mg/kg) 0.06 in dogs, 0.2 in cats, 0.4 in sheep or rabbits, 2–10 in humans, 5 in rats, 7 in mice, to about 400 in toads (20,24). The only suggested antidotes for the poisoning are 1,2,3-propanetriol monoacetate (20,23), acetamide (20), and other acetate donors, but these only have an effect if administered before significant amounts of fluoroacetate have been converted to fluorocitrate. To determine if fluoroacetate poisoning has occurred, it is often desirable to detect the presence of small amounts of the poison in animal tissue. Although difficult, this can be done by spectrochemical methods (25), processes involving ion-selective fluoride electrodes (21,26), or gas chromatography often combined with mass spectrometry (27). A microbial detection of fluoroacetate utilizing DNA technology and bioluminescence has been reported (28).

Fluoroacetamide.   Fluoroacetamide [*640-19-7*], FCH_2CONH_2, is a white water-soluble solid having mp 108°C (14). It has been used as a rodenticide and has been reported to have a better acceptability to rats than sodium fluoroacetate (29). However, like the latter compound, its misuse has caused deaths to farm animals and pets (20).

Tull Chemical Co. (Oxford, Alabama) is the only producer of sodium fluoroacetate. It is sometimes colored with the black dye nigrosine. It is usually packed in 8 oz (227 g) or 5 kg cans and is almost exclusively exported. There is very limited use in the United States.

Difluoroacetic Acid

Difluoroacetic acid [*381-73-7*], $F_2CHCOOH$, is a colorless liquid with a sharp odor; mp, 35°C; bp, 134°C; d_4^{10}, 1.539 g/mL; n_D^{20}, 1.3428 (30,31); flash point, 78°C (95% aqueous solution); enthalpy of vaporization, 67.82 kJ/mol (16.21 kcal/mol); and enthalpy of solution in water of the undissociated acid, -7.03 kJ/mol (-1.68 kcal/mol) (2). It is a moderately strong acid; determinations of its acid dissociation constant are 4.6×10^{-2} (32) and 3.5×10^{-2} (33). Its molecular structure in the gas phase has been determined by electron diffraction studies (34). Details of the acid's ir (35) and nmr (36) spectra also have been reported.

Difluoroacetic acid undergoes reactions typical of a carboxylic acid such as forming an ester when heated with an alcohol and sulfuric acid. Typical esters are methyl difluoroacetate [*433-53-4*], bp, 85.2°C, and ethyl difluoroacetate [*454-31-9*], bp, 99.2°C. It can also be photochemically chlorinated to chlorodifluoroacetic acid [*76-04-0*] or brominated in the presence of iron to bromodifluoroacetic acid [*667-27-6*] (37,38).

The acid can be synthesized in several different ways. The reaction of tetrafluoroethylene with ammonia to give 2,4,6-tris(difluoromethyl)-*s*-triazine, followed by its alkaline hydrolysis, has been reported to give the acid in 80% overall yield (31). The addition of diethylamine to tetrafluoroethylene gives, after partial hydrolysis, a 49% yield of amide *N,N*-diethyldifluoroacetamide [*56425-08-2*], $F_2CHCON(C_2H_5)_2$, which can be hydrolyzed in excellent yield to the acid (39). The same amide can be prepared in 60% yield by the addition of diethylamine to chlorotrifluoroethylene followed by hydrolysis and fluorination with KF in diethylene glycol. Another method that gives the acid in 86% yield is the permanganate oxidation of $CHF_2CH=CCl_2$ (40).

Difluoroacetic acid is much less toxic than fluoroacetic acid ($LD_{50} = 180$ mg/kg mouse iv) (41). It is available in research quantities for about $5/g (1992).

Trifluoroacetic Acid

Physical Properties. Trifluoroacetic acid [76-05-1], CF_3COOH, is a colorless liquid with a sharp odor resembling that of acetic acid. Its physical properties are shown in Table 1. It is a strong carboxylic acid with an acid dissociation constant at 25°C of 0.588 (9) or 0.32 (32). It is miscible with water, fluorocarbons, and most common organic solvents including methanol, benzene, carbon tetrachloride, acetone, ether, and hexane. Compounds with limited solubility in the acid include alkanes with more than six carbon atoms and carbon disulfide. It is a good solvent for proteins (50) and polyesters. The viscosities, densities, and conductivities of solutions of the acid in acetic acid, water, and several other liquids have been studied (46).

Chemical Properties. Trifluoroacetic acid undergoes reactions typical of a carboxylic acid. The trifluoromethyl group is inert to most common reducing agents, including lithium aluminum hydride, which give trifluoroacetaldehyde [75-90-1] and 2,2,2-trifluoroethanol [75-89-8] (51,52). Common oxidizing agents do not attack the acid at room temperature except for potassium permanganate, which slowly oxidizes the anhydrous acid to carbon dioxide and other products (53). The acid is also slowly attacked by boiling 25% aqueous sodium hydroxide to yield oxalate and fluoride ions (44). Although the acid is stable to temperatures above 250°C, its sodium salt decomposes above 205°C to give sodium fluoride, trifluoroacetyl fluoride [354-34-7], carbon monoxide, carbon dioxide, and other products (44). In ethylene glycol solution at 180°C the sodium salt can be made to decompose quantitatively to trifluoromethane [75-46-7] and carbon dioxide if a boric acid buffer is present (54). Except for a few instances like these, the reactions of trifluoroacetic acid closely parallel those of other carboxylic acids, but there are

Table 1. Physical Properties of Trifluoroacetic Acid

Property	Value	References
freezing point, °C	-15.36	42
boiling point, °C	71.8	43
water azeotrope (20.6% H_2O) bp, °C	105.5	44
density at 25°C, g/mL	1.4844	45
heat of vaporization, kJ/mol^a	33.26	43
viscosity at 25°C, mPa·s($=$cP)	0.813	46
dielectric constant at 25°, ϵ	42.1	47
conductivity at 25°C, $1/\Omega \cdot cm$	2600	46
surface tension at 25°C mN/m($=$dyn/cm)	13.44	48
heat of formation, liquid, kJ/mol^a	-1060 ± 2	49

^{a}To convert kJ to kcal, divide by 4.184.

important differences: eg, its amides and esters are more easily hydrolyzed than is typical for carboxylic acids. This has led to the use of the acid and its anhydride [407-25-0] (55) in making derivatives of carbohydrates (56), amino acids (57), and peptides (57) from which the trifluoroacetyl protective group can be removed with relative ease. Peroxytrifluoroacetic acid [359-48-8], formed from the reaction of trifluoroacetic anhydride and hydrogen peroxide, is a stronger oxidizing agent than other peroxycarboxylic acids and gives better yields of epoxides from alkenes (58), esters from ketones (59), and nitrobenzenes from anilines (60). Trifluoro-acetic acid and its anhydride are also useful as catalysts for reactions involving other carboxylic acids such as esterifications of alcohols or acylations of aromatic or other unsaturated compounds (45). The acid has been reported to be superior to sulfuric acid as a catalyst for the Beckmann rearrangement of oximes to amides (61). Owing to its low nucleophilicity, the acid has been used as a solvent for basic research into solvolysis mechanisms (62).

Preparation.   Because of its stability to further oxidation, trifluoroacetic acid can be prepared by the oxidation of compounds containing a trifluoromethyl group bonded to carbon. Although first prepared in 1922 by the oxidation of tri-fluoromethylcyclohexane or 3-aminobenzotrifluoride, later better results were obtained from the alkaline permanganate oxidation of olefins such as 1,1,2-tri-chloro-3,3,3-trifluoropropene (40), or more economically, 2,3-dichlorohexafluoro-2-butene which gives an 87% yield of the acid (63). The acid has been prepared by photochemical oxidation of ethanes such as 2-chloro-1,1,1-trifluoroethane or 2,2-dichloro-1,1,1-trifluoroethane with oxygen to give high yields of trifluoroacetyl chloride [354-32-5] which easily hydrolyzes to the acid (64,65). Another process involves the trimerization of trichloroacetonitrile to a triazine which can be fluor-inated with a mixture of SbF_3 and $SbCl_5$ and then hydrolyzed to the acid (66). The reaction of trichloroacetyl chloride with hydrogen fluoride at 320°C over a chromium and nickel oxide catalyst has been reported to give trifluoroacetic acid in 92% yield (67).

Trifluoroacetic acid was produced commercially by 3M Co. by the electrolysis of mixtures containing acetyl fluoride, hydrogen fluoride, and sodium fluoride to give trifluoroacetyl fluoride, which upon hydrolysis gave the acid (68). Although a 71% yield is claimed, isolation of the low boiling acid fluoride product from by-product hydrogen is costly. Improvements in this process have been patented (69,70) as well as processes involving the electrochemical fluorination of 2-chloroethanol (71) or chloroacetyl fluoride (72).

Health and Safety Factors.   Unlike fluoroacetic acid, trifluoroacetic acid presents no unusual toxicity problems. However, owing to its strong acidity, its vapors can be irritating to tissue, and the liquid acid can cause deep burns if allowed to contact the skin. The acid can be safely stored in containers made of glass or common corrosion-resistant alloys and metals such as stainless steel or aluminum.

Economic Aspects.   Halocarbon Products Corp. is the largest producer of trifluoroacetic acid. The commercial grade is of very high purity with the main impurity being ca 0.2% water. A grade, which has a low residue specification, intended for use in protein synthesis (Biograde) is available. Other producers include Rhône-Poulenc and Solvay. The 1992 price was ca $15/kg.

BIBLIOGRAPHY

"Monofluoroacetic Acid" under "Fluorine Compounds, Organic," in *ECT* 1st ed., Vol. 6, pp. 764–766, by E. E. Hardy and J. H. Saunders, Monsanto Chemical Co.; "Difluoroacetic Acid" under "Fluorine Compounds, Organic," in *ECT* 1st ed., Vol. 6, pp. 766–767, by M. G. Gergel and M. Revelise, Columbia Organic Chemicals Co.; "Trifluoroacetic Acid" under "Fluorine Compounds, Organic," in *ECT* 1st ed., Vol. 6, pp. 767–768, by M. G. Gergel and M. Revelise, Columbia Organic Chemicals Co.; "Monofluoroacetic Acid" under "Fluorine Compounds, Organic," in *ECT* 2nd ed., Vol. 9, pp. 767–770, by E. Hardy, Monsanto Research Corp., J. H. Saunders, Mobay Chemical Co., and J. B. Hynes, Hynes Chemical Research Corp.; "Difluoroacetic Acid" under "Fluorine Compounds, Organic," in *ECT* 2nd ed., Vol. 9, pp. 770–771, by J. B. Hynes, Hynes Chemical Research Corp.; "Trifluoroacetic Acid" under "Fluorine Compounds, Organic," in *ECT* 2nd ed., Vol. 9, pp. 771–772, by C. Woolf, Allied Chemical Corp.; "Fluorinated Acetic Acids" under "Fluorine Compounds, Organic," in *ECT* 3rd ed., Vol. 10, pp. 891–896, by G. Astrologes, Halocarbon Products Corp.

1. F. Swarts, *Bull. Acad. R. Belg.* **35**, 849 (1898).
2. P. Haberfield and A. K. Rakshit, *J. Am. Chem. Soc.* **98**, 4393 (1976).
3. M. V. Kaulgud and G. H. Pandya, *Indian J. Chem. Sect. A.* **14A**(2), 91 (1976).
4. B. P. Van Eijck, P. Brandts, and J. P. M. Maas, *J. Mol. Struct.* **44**, 1 (1978)
5. F. Swarts, *Bull. Acad. R. Belg.* **31**, 675 (1896).
6. F. L. M. Pattison, *Toxic Aliphatic Fluorine Compounds*, Elsevier Publishing Co., New York, 1959, p. 16.
7. M. J. J. Meyer and N. Grobbelaar, *J. Plant Physiol.* **138**, 122 (1991).
8. G. W. Miller, M. H. Yu, and M. Psenak, *Fluoride* **6**(3), 203 (1973).
9. A. L. Henne and C. J. Fox., *J. Am. Chem. Soc.* **73**, 2323 (1953).
10. F. L. M. Pattison, J. B. Stothers, and R. G. Woolford, *J. Am. Chem. Soc.* **78**, 2255 (1956).
11. F. L. M. Pattison, S. B. D. Hunt, and J. B. Stothers, *J. Org. Chem.* **21**, 883 (1956)
12. C. C. Price and W. G. Jackson, *J. Am. Chem. Soc.* **69**, 1065 (1947).
13. F. L. M. Pattison and co-workers, *Can. J. Technol.* **34**, 21 (1956).
14. F. J. Buckle, R. Heap, and B. C. Saunders, *J. Chem. Soc.*, 912 (1949).
15. B. C. Saunders and G. J. Stacey, *J. Chem. Soc.*, 1773 (1948).
16. E. D. Bergmann and I. Blank, *J. Chem. Soc.*, 3786 (1953).
17. Ref. 6, pp. 21–22.
18. Ref. 6, p. 167.
19. D. D. Clarke, *Neurochem. Res.* **16**, 1055 (1991).
20. R. A. Peters, *Fluoride* **6**(3), 189 (1973).
21. M. N. Egyed, *Fluoride* **6**(3), 215 (1973).
22. P. Buffa, V. Guarriero-Bobyleva, and R. Costa-Tiozzo, *Fluoride* **6**, 224 (1973).
23. Ref. 6, pp. 27–56, 208–210.
24. Ref. 6, pp. 3–4.
25. I. Schoenfeld and M. Lidji, *J. Forsensic Sci.* **13**, 267 (1968).
26. J. A. Peters and K. J. Baxter, *Bull. Environ. Contam. Toxicol.* **11**(2), 177 (1974).
27. H. M. Stahr, W. B. Buck, and P. F. Ross, *J. Assn. Off. Anal. Chem.* **57**, 405 (1974).
28. S. Lee and co-workers, *Anal. Chim. Acta.* **244**, 201 (1991).
29. T. Kusano, *J. Fac. Agric. Tottori Univ.* **10**, 15 (1975).
30. F. Swarts, *Bull. Soc. Chim. Fr.*, 597 (1903).
31. A. L. Henne and R. L. Pelley, *J. Am. Chem. Soc.* **74**, 1426 (1952).
32. J. L. Kurz and J. M. Farrar, *J. Am. Chem. Soc.* **91**, 6057 (1969).
33. M. M. Kreevoy and co-workers, *J. Am. Chem. Soc.* **89**, 1201 (1967).
34. J. M. Bijen and J. L. Derissen, *J. Mol. Struct.* **27**, 233 (1975).
35. J. R. Barcelo and C. Otero, *Spectrochim. Acta* **18**, 1231 (1962).

36. V. Barboiu, *Rev. Roum. Chim.* **19**, 363 (1974).
37. F. Swarts, *Chem. Zentr.* **II**, 709 (1903).
38. F. Swarts, *Chem. Zentr.* **I**, 1237 (1906).
39. N. N. Yarovenko and co-workers, *Obschei Khim.* **27**, 2246 (1957).
40. A. L. Henne, T. Alderson, and M. S. Newman, *J. Am. Chem. Soc.* **67**, 918 (1945).
41. Ref. 6, pp. 62–63.
42. H. H. Cady and G. E. Cady, *J. Am. Chem. Soc.* **76**, 915 (1954).
43. M. D. Taylor and M.,B. Templeman, *J. Am. Chem. Soc.* **78**, 2950 (1956).
44. F. Swarts, *Bull. Acad. R. Belg. Classe Sci.* **8**, 343 (1922).
45. *Trifluoracetic Acid Brochure*, Halocarbon Products Corp., Hackensack, N.J., 1967.
46. Y. Y. Fialkov and V. S. Zhikarev, *Zh. Obshch. Khim.* **33**, 3466, 3471, 3790 (1963).
47. J. H. Simons and K. E. Lorentzen, *J. Am. Chem. Soc.* **72**, 1426 (1950).
48. J. J. Jasper and H. L. Wedlick, *J. Chem. Eng. Data* **9**, 446 (1964).
49. V. P. Kolesov, G. M. Slavutskaya, and T. S. Papino, *Zh. Fiz. Khim.* **46**, 815 (1972).
50. J. J. Katz, *Nature* **174**, 509 (1954).
51. M. Braid, H. Iserson, and F. E. Lawlor, *J. Am. Chem. Soc.* **76**, 4027 (1954).
52. O. R. Pierce and T. G. Kane, *J. Am. Chem. Soc.* **76**, 300 (1954).
53. G. S. Fujioka and G. H. Cady, *J. Am. Chem. Soc.* **79**, 2451 (1957).
54. I. Auerbach, F. H. Verhoek, and A. L. Henne, *J. Am. Chem. Soc.* **72**, 299 (1950).
55. J. M. Tedder, *Chem. Rev.* **55**, 787 (1955).
56. E. J. Bourne and co-workers, *J. Chem. Soc.*, 2976 (1949).
57. F. Weygand, *Bull. Soc. Chim. Biol.* **43**, 1269 (1961).
58. W. D. Emmons and A. S. Pagano, *J. Am. Chem. Soc.* **77**, 89 (1955).
59. W. D. Emmons and G. B. Lucas, *J. Am. Chem. Soc.* **77**, 2287 (1955).
60. W. D. Emmons, *J. Am. Chem. Soc.* **76**, 3470 (1954).
61. U.S. Pat. 2,721,199 (Oct. 18, 1955), M. L. Huber (to E. I. du Pont de Nemours & Co., Inc.).
62. P. E. Peterson and co-workers, *J. Am. Chem. Soc.* **87**, 5169 1965).
63. A. L. Henne and P. Trott, *J. Am. Chem. Soc.* **69**, 1820 (1947).
64. R. N. Haszeldine and F. Nyman, *J. Am. Chem. Soc.*, 387 (1959).
65. U.S. Pat. 3,883,407 (May 13, 1975), A. L. Dittman (to Halocarbon Products Corp.).
66. T. R. Norton, *J. Am. Chem. Soc.* **72**, 3527 (1950).
67. Ger. Offen. 2,221,849 (Nov. 16, 1972), Ramanadin (to Rhone-Progil).
68. U.S. Pat. 2,717,871 (Sept. 13, 1965), H. M. Scholberg and H. G. Bryce (to Minnesota Mining and Manufacturing Co.).
69. U.S. Pat. 4,022,824 (May 10, 1977), W. V. Childs (to Phillips Petroleum Co.).
70. Jpn. Kokai 75 30,827 (Mar. 27, 1975), T. Suzuki and S. Yahara (to Mitsubishi Gas Chemical Co., Inc.).
71. USSR Pat. 329,165 (Feb. 9, 1972), N. M. Arakelyan and S. E. Isabekyan.
72. Czech. Pat. 119,682 (Sept. 15, 1966), D. Frantisek.

ARTHUR J. ELLIOTT
Halocarbon Products Corporation

FLUORINATED HIGHER CARBOXYLIC ACIDS

Perfluorinated carboxylic acids are corrosive liquids or solids. The acids are completely ionized in water. The acids are of commercial significance because of their unusual acid strength, chemical stability, high surface activity, and salt solubility characteristics. The perfluoroalkyl acids with six carbons or less are liquids; the higher analogues are solids (Table 1).

The higher members of the series decrease the surface tension of aqueous solutions well below the point possible with any type of hydrocarbon surfactant, although in practice because of their strong acid character and solubility characteristics, more commonly salts and other derivatives are employed. A 0.1% solution of $C_9F_{19}COOH$ has a surface tension of only 19 mN/m (dyn/cm) at 30°C (6).

Table 1. Properties of Perfluoroalkylcarboxylic Acids, $C_nF_{2n+1}COOH^a$

Acid	CAS Registry Number	Bp, °C	Mp, °C	Density at 20°C, g/mL	Reference
perfluoropropanoic	[422-64-0]	96		1.561	1
perfluorobutanoic	[375-22-4]	120	−17.5	1.641	1,2
perfluoropentanoic	[2706-90-3]	139		1.713	1,3
perfluorohexanoic	[307-24-4]	157		1.762	1,3
perfluorocyclohexane carboxylic^b	[374-88-9]	168		1.789	1,2
perfluoroheptanoic	[375-85-9]	175		1.792	1,2
perfluorooctanoic	[335-67-1]	189	52–54	1.792	1,2,4
perfluorodecanoic	[335-76-2]	218			1,2
perfluorotetradecanoic		270			1,5

^{a}Except where noted.
^{b}Cyclo-$C_6F_{11}COOH$.

Preparation

There are five methods for the preparation of long-chain perfluorinated carboxylic acids and derivatives: electrochemical fluorination, direct fluorination, telomerization of tetrafluoroethylene, oligomerization of hexafluoropropylene oxide, and photooxidation of tetrafluoroethylene and hexafluoropropylene.

Many of the perfluoroalkyl carboxylic acids were first prepared by the electrochemical fluorination (ECF) of the corresponding carboxylic acids (7). In ECF acid chlorides are converted to the corresponding perfluoroacid fluorides as shown in equation 1 for octanoyl chloride.

$$\underset{\substack{\|\\C_7H_{15}C-Cl}}{O} + 16\,HF \xrightarrow{ECF} \underset{\substack{\|\\C_7F_{15}C-F}}{O} + C_7F_{16} + \underset{C_4F_9}{\overset{O}{\diamondsuit}}_F + \overset{O}{\diamondsuit}_{F}^{C_3F_7} + HCl + H_2 \tag{1}$$

The principal by-products are cyclic perfluoroethers; the fluorine in the center of the ring denotes a perfluorinated ring structure. Octanoyl chloride gives only 10–15% of the perfluorooctanoyl fluoride (8), although the yield of fluorination can be improved by running under different cell conditions (9). As the chain length increases, formation of cyclic ethers and cleavage products becomes more prominent. In addition to cleavage and cyclization products, ECF results in isomerization of the carbon backbone in the product. The relative weight % of C_7F_{15} isomers in $C_7F_{15}COOH$ commercially produced by ECF is 78% linear, 9% internal branched, and 13% terminal (isopropyl) branched (4). Hydrolysis of the acid fluoride followed by removal of the fluoride ion and distillation yields the fluorinated acid.

Perfluorinated acid fluorides containing heteratoms are also accessible by ECF. Long-chain perfluorinated acid fluorides produced by ECF containing nitrogen (10–12), oxygen (13), and sulfur (14,15) have been reported. The fluorinated mixed sulfonic acid–carboxylic acid precursors are also known. ECF of hydrocarbon sultones has led to formation of $FSO_2(CF_2)_nCOF$, where $n = 2,3$ (16).

Direct fluorination involves the treatment of an appropriate hydrocarbon precursor dissolved in an inert liquid with fluorine gas to yield a perfluorinated precursor to a long-chain carboxylic acid. Equations 2 and 3 illustrate the process for perfluorooctadecanoic acid (17).

$$C_{18}H_{37}O\overset{\overset{\displaystyle O}{\|}}{-}C-CF_3 + 37\ F_2 \rightarrow C_{18}F_{37}O\overset{\overset{\displaystyle O}{\|}}{-}C-CF_3 + 37\ HF \qquad (2)$$

$$C_{18}F_{37}O\overset{\overset{\displaystyle O}{\|}}{-}C-CF_3 + 2\ H_2O \rightarrow C_{17}F_{35}COOH + 2\ HF + HO\overset{\overset{\displaystyle O}{\|}}{-}C-CF_3 \qquad (3)$$

Unlike ECF, direct fluorination does not alter the carbon backbone; preparation of isomerically pure acids is possible (18). Both direct fluorination and ECF permit a great variety of structures to be made, but each method is better at certain types of structures than the other. Ether acids are produced in good yields, by direct fluorination (17), while ECF of ether-containing acids is fair to poor depending on the substrate. Despite much industrial interest, the costs and hazards of handling fluorine gas have prevented commercial application of this process.

Fluorinated carboxylic acids are also prepared by telomerization of tetrafluoroethylene, followed by oxidation (19–21).

$$C_2F_5I + n\ CF_2{=}CF_2 \rightarrow C_2F_5(CF_2CF_2)_nI \xrightarrow{SO_3} C_2F_5(CF_2CF_2)_{n-1}CF_2COOH \qquad (4)$$

This process yields a purely straight-chain acid of even carbon number. Typically, the value of n varies from two to six, and distillation yields the pure components. Du Pont pioneered the development of this technology. Allied has used hexafluoroacetone to produce telomer iodides containing the perfluoroisopropoxy end group, eg, $(CF_3)_2CFO(CF_2CF_2)_nI$ (22,23). Dichromate oxidation (24) or ozonolysis (25) of $C_2F_5(CF_2CF_2)_nCH{=}CH_2$, derived from reaction of the telomer iodide shown in equation 4, gives the acid $C_2F_5(CF_2CF_2)_nCOOH$.

Fluoride ion-catalyzed oligomerization of fluorinated epoxides leads to long-chain ether-containing acids. Equation 5 exemplifies this reaction for hexafluoropropylene oxide (HFPO). Tetrafluoroethylene oxide can undergo similar ring-opening reactions, but is seldom used because of its chemical instability. HFPO, however, is stable at room temperature in an anhydrous atmosphere, and in the absence of acid and base (26). Thermal decomposition occurs only at temperatures of 150°C or higher (27).

$$F^- + (n+2) \quad \begin{array}{c} F \\ F \end{array} \diagup\!\!\!\diagup\!\!\!\!\overset{O}{\diagdown}\diagdown\!\!\!\begin{array}{c} F \\ CF_3 \end{array} \quad \longrightarrow \quad C_3F_7O(\underset{CF_3}{C}FCF_2O)_{\overline{n}}\overset{O}{\overset{\|}{C}}F\underset{CF_3}{C}-F + F^- \tag{5}$$

Hydrolysis of the acid fluoride, removal of fluoride ion, and distillation yield the perfluorinated acid. The value of n typically varies from one to six, depending on reaction conditions. Higher values of n are possible by employing more rigorous conditions. These acids are marketed by Du Pont under the trade name of Krytox acids. This process yields perfluoroether acids containing regular repeat units of perfluoroisopropoxy group and terminated by an alpha-branched carboxylate. The C_3F_7O end of the molecule can be varied by fluoride ion condensation of a perfluorinated acid fluoride with HFPO (26).

Photooxidation of tetrafluoroethylene (TFE) and hexafluoropropylene (HFP) yield peroxides that can be decomposed to esters and ultimately long-chain ether-containing carboxylic acids. Equation 6 shows a simplified version of what occurs during photooxidation and workup (TFE R = F, HFP R = CF_3) (28,29).

$$CFR{=}CF_2 + O_2 \xrightarrow{\text{uv light}} CF_3O(CF_2O)_n(\underset{R}{C}FCF_2O)_m[O(CF_2O)_n(\underset{R}{C}FCF_2O)_m]_zO(\underset{R}{C}FCF_2O)_m(CF_2O)_nCOF$$

$$\rightarrow CF_3O(CF_2O)_n(\underset{R}{C}FCF_2O)_{m-1}CF_2CO_2CH_3 + CH_3OCOCF_2(CF_2O)_n(\underset{R}{C}FCF_2O)_{m-2}CF_2CO_2CH_3 \tag{6}$$

The acid is obtained by saponification of the ester. Photooxidation of TFE in practice yields mostly difunctional ether acids. Photooxidation of HFP yields mostly the monofunctional ether acid, but yields significant quantities of inert materials that are difficult to separate out. The segments $(CFRCF_2O)$ and (CF_2O) are randomly arranged in the chain. The random arrangement of these units in the chain is thought to be responsible for the exceptional low temperature properties of polymers and other materials derived from this route (30). Molecular weight of the resulting materials is determined by process control parameters and subsequent distillation.

Derivatives

In general, the reactions of the perfluoro acids are similar to those of the hydrocarbon acids. Salts are formed with the ease expected of strong acids. The metal salts are all water soluble and much more soluble in organic solvents than the salts of the corresponding hydrocarbon acids. Esterification takes place readily with primary and secondary alcohols. Acid anhydrides can be prepared by distillation of the acids from phosphorus pentoxide. The amides are readily prepared by the ammonolysis of the acid halides, anhydrides, or esters and can be dehydrated to the corresponding nitriles (31).

The ammonium salts, $C_nF_{2n+1}COONH_4$, where n equals 7 and larger, are particularly useful as emulsifiers in the polymerization of fluorinated olefin monomers such as tetrafluoroethylene or vinylidene fluoride. Their surface activity, low rate of chain transfer, and the highly fluorochemical nature of micelles formed in aqueous media are unique features that give them broad utility in the emulsification and polymerization of fluorochemical monomers (32).

Amines of the formula $C_nF_{2n+1}CH_2NH_2$ can be prepared by the lithium aluminum hydride reduction of the corresponding amide, hydrogenolysis of the nitrile, or diborane reduction of the amide (33). The analogous alcohols, $C_nF_{2n+1}CH_2OH$, can be prepared by high pressure hydrogenation of an appropriate ester (ie, $C_nF_{2n+1}COOCH_3$) over a barium oxide stabilized chromite catalyst, lithium aluminum hydride reduction of the acid or esters, or sodium borohydride reduction of the esters (34,35) or acid fluorides (36,37). Acrylate esters of the dihydroalcohols copolymerize with hydrocarbon acrylates and other monomers and have been found to impart resistance to a variety of organic liquids at elevated temperatures (38).

$$C_nF_{2n+1}CH_2O\overset{\overset{\displaystyle O}{\|}}{C}C(R){=}CH_2$$

The surface energy of the homopolymers of these acrylates, 10.6 mN/m(dyn/cm), is among the lowest ever recorded (39), lower even than Teflon, 18.5 mN/m(dyn/cm) (40).

Perfluorodicarboxylic Acids

The lowest members of the series of perfluoroalkanedicarboxylic acids have been prepared and are stable compounds. They have been synthesized by oxidation of the appropriate chlorofluoroolefin as well as by electrochemical fluorination and direct fluorination. Perfluoromalonic acid is an oxidation product of $CH_2{=}CHCF_2CH{=}CH_2$ (21). Perfluorosuccinic acid has been produced by oxidation of the appropriate olefin (see eq. 7) (5) or by electrochemical fluorination of succinyl chloride or butyrolactone (41) and subsequent hydrolysis.

$$\boxed{F}\begin{matrix} {-}Cl \\ {-}Cl \end{matrix} + \begin{matrix} \text{basic} \\ KMnO_4 \end{matrix} \longrightarrow \begin{matrix} CF_2COOH \\ | \\ CF_2COOH \end{matrix} \qquad (7)$$

Table 2 lists some typical properties of perfluoroalkanedicarboxylic acids and their esters along with references to their synthesis.

Table 2. Properties of Perfluroalkanedicarboxylic Acids, $HOOC(CF_2)_nCOOH$

n	Acid	CAS Registry Number	Mp, °C	Bp, °C$_{kPa^a}$	Bp of ester,[b] °C$_{kPa^a}$	Reference
1	perfluoromalonic	[1514-85-5]	117		$58-58_{1.2}$	35,42
2	perfluorosuccinic	[377-35-8]	115–116	150_2	173	35
3	perfluoroglutaric	[376-73-8]	78–88	$134_{0.4}$	$100_{4.5}$	5
4	perfluoroadipic	[336-08-3]	134		$108-110_4$	35,43,44
6	perfluorosuberic	[678-45-5]	154–158		$156-159^c_{3.6}$	45
8	perfluorosebacic	[307-78-8]			$102-113_{0.005}$	44
12	perfluorotetradecanedioic		191			35

^{a}To convert kPa to mm Hg, multiply by 7.5.
^{b}Methyl ester unless otherwise noted.
^{c}Ethyl ester.

Fluorinated ether-containing dicarboxylic acids have been prepared by direct fluorination of the corresponding hydrocarbon (17), photooxidation of tetra-fluoroethylene, or by fluoride ion-catalyzed reaction of a diacid fluoride such as oxalyl or tetrafluorosuccinyl fluorides with hexafluoropropylene oxide (46,47). Equation 8 shows the reaction of oxalyl fluoride with HFPO. A difunctional ether-containing acid fluoride derived from HFPO contains regular repeat units of per-fluoroisopropoxy group and is terminated by two alpha-branched carboxylates.

$$F{-}\overset{O}{\overset{\|}{C}}{-}\overset{O}{\overset{\|}{C}}{-}F + F^- + (n+2)\ \ \underset{F}{\overset{F}{\diagdown}}\!\!\overset{\overset{O}{\diagup\diagdown}}{\diagup}\!\!\underset{CF_3}{\overset{F}{\diagdown}} \longrightarrow$$

$$F\overset{O}{\overset{\|}{C}}{-}\underset{CF_3}{\overset{}{C}}FOCF_2CF_2O(\underset{CF_3}{\overset{}{C}}FCF_2O)_{\overline{n}}\underset{CF_3}{\overset{}{C}}F\overset{O}{\overset{\|}{C}}{-}F + F^- \quad (8)$$

Fluorinated diacids offer a convenient method for introducing a perfluoro moiety into organic molecules. They are of potential interest in the preparation of polyamides and other fluorinated polymers. A detailed description of the per-fluorocarboxylic acids and their derivatives has been published (1), and a review article on polyfluorinated linear bifunctional compounds has appeared (35).

Derivatives similar to those mentioned for the monofunctional fluorinated carboxylic acids have been prepared: tetrahydrodiols, tetrahydrodiamines, di-amides, and diesters.

Toxicology and Safety. Because of their strong acidity, the perfluorinated carboxylic acids themselves are corrosive to the skin and eyes. Protective clothing, ie, gloves and face shield/safety glasses, should be employed when handling them (48,49). Although perfluorooctanoic acid has been found to be corrosive to the eyes and severely irritating to the skin on contact, it is only slightly toxic on oral con-tact. The acute LD_{50} (rat) of perfluorooctanoic acid was found to be slightly less than 1000 mg per kg of body weight. However, in feeding studies the test animals suffered erosion of the gastric mucous membrane because of corrosivity of the acid

(50). Perfluorodecanoic acid has been assessed for its genotoxic activity. It was found to test negative in the Ames test, Chinese hamster ovary gene mutation assay, sister chromatid exchange assay, chromosomal aberration assay, and *in vivo/in vitro* unscheduled DNA synthesis. Chromosomal aberrations were observed only when the S-9 fraction was incubated with perfluorodecanoic acid in the S-phase DNA synthesis assay (51).

The salts of the perfluorinated acids are not corrosive, so one is in a better position to discuss toxicity not related to corrosivity. The toxicity of the salts varies depending on the exact structure. The ammonium salt of perfluorooctanoic acid is nonirritating to the skin and moderately irritating to the eyes. Its oral toxicity is rated at moderate; the LD_{50} is 540 mg per kg of body weight (52). There has been some concern in the past that ammonium perfluorooctanoate was teratogenic. More recent results indicate that it is neither embryotoxic nor teratogenic (52,53). It was not found to be mutagenic in either the Ames assay or one employing *Saccharomyces cerevisiae* D4 yeast (52). It also did not cause cell transformation in a mammalian cell transformation assay (53). Although ammonium perfluorooctanoate was fed to albino rats for two years, no compound-induced carcinogenicity was found in the study. There were statistically significant compound-related benign testicular tumors (52,53). Prolonged or repeated exposure can cause liver damage which results in jaundice or tenderness of the upper abdomen (53). The dust from the ammonium salts of the perfluorinated acids is irritating to breathe and should only be handled in a well-ventilated area or preferably a hood.

BIBLIOGRAPHY

"Heptafluorobutyric Acid" under "Fluorine Compounds, Organic" in *ECT* 1st ed., Vol. 6, p. 769, by M. G. Gergel and M. Revelise, Columbia Organic Chemicals Co., Inc.; "Other Perfluorocarboxylic Acids" under "Fluorine Compounds, Organic" in *ECT* 1st ed., Vol. 6, pp. 769–771, by L. J. Hals and W. H. Pearlson, Minnesota Mining & Manufacturing Co.; "Heptafluorobutyric Acid" under "Fluorine Compounds, Organic" in *ECT* 2nd ed., Vol. 9, p. 773, by M. G. Gergel, Columbia Organic Chemicals Co., Inc.; "Other Perfluorocarboxylic Acids" under "Fluorine Compounds, Organic" in *ECT* 2nd ed., Vol. 9, pp. 773–775, by L. J. Hals and W. H. Pearlson, Minnesota Mining & Manufacturing Co.; "Fluorinated Higher Carboxylic Acids" under "Fluorine Compounds, Organic" in *ECT* 3rd ed., Vol. 10, pp. 897–900, by R. A. Guenthner, 3M Co.

1. A. M. Lovelace, W. Postelnek, and D. A. Rausch, *Aliphatic Fluorine Compounds*. ACS monograph 138 Reinhold Publishing Co., New York, 1958.
2. U.S. Pat. 2,567,011 (Sept. 4, 1951), A. R. Diesslin, E. A. Kauck, and J. H. Simons (to 3M Co.).
3. E. A. Kauck and A. R. Diesslin, *Ind. Eng. Chem.* **43**, 2332 (1952).
4. *Fluorad Fluorochemical Acid FC-26*, Minnesota Mining & Manufacturing Co., St. Paul, Minn., 1986.
5. A. L. Henne and W. J. Zimmerscheid, *J. Am. Chem.* **69**, 281 (1947).
6. H. M. Scholberg, "Surface Chemistry of Fluorocarbons and Their Derivatives," in *Abstracts of Papers, 116th Meeting, Am. Chem. Soc.* Atlantic City, N.J., Sept. 1949, p. 36K.
7. J. H. Simons and co-workers, *J. Electrochem. Soc.* **95**, 47–67 (1949).

8. H. C. Fielding, in R. E. Banks, ed., *Organofluorine Chemicals and their Application*, Ellis Howard, 1979, p. 216.
9. U.S. Pat. 3,919,057 (Nov. 11, 1975), E. Plattner, C. Comninellis, and P. Javet (to CIBA-GEIGY AG).
10. U.S. Pat. 3,471,484 (Oct. 7, 1969), R. A. Guenthner (to 3M Co.).
11. T. Abe and co-workers, *J. Fluorine Chem.* **48**, 257–279 (1990).
12. A. Dimitrov and S. T. Rudiger, *J. Fluorine Chem.* **50**, 197–205 (1990).
13. U.S. Pat. 2,826,564 (Mar. 11, 1958), F. A. Bovey, and J. F. Abere (to 3M Co.).
14. T. J. Brice, R. I. Coon, and W. A. Severson, "Properties of Some Fluorocarbon Derivatives of Sulfur Hexafluoride," paper presented at *American Chemical Society*, Minneapolis, Minn., 1955.
15. Eur. Pat. Appl. 444,822 (Feb. 28, 1991), J. C. Hansen and P. M. Savu (to 3M Co.).
16. Eur. Pat. Appl. 058,466 (Dec. 18, 1985), F. E. Behr and R. J. Koshar (to 3M Co.).
17. WO Pat. 90/06,296 (June 14, 1990), M. G. Costello and G. G. I. Moore (to 3M Co.).
18. R. M. Flynn, T. A. Kestner, and G. G. I. Moore, "Stereochemistry of the Direct Fluorination Process," *Tenth Winter Fluorine Conference*, St. Petersburg, Fla., 1991.
19. U.S. Pat. 3,132,185 (May 5, 1964), R. E. Parsons (to E. I. du Pont de Nemours and Co., Inc.); U.S. Pat. 3,226,449 (Dec. 28, 1965), W. A. Blanchard and J. C. Rhode (to E. I. du Pont de Nemours and Co., Inc.); U.S. Pat. 3,234,294 (Feb. 8, 1966), R. E. Parsons (to E. I. du Pont de Nemours and Co., Inc.); U.S. Pat. 4,425,199 (Jan. 10, 1984), M. Hamada, J. Ohmura, and F. Muranaka (to Asahi Kasei Kogyo Kabushi Kaisha).
20. M. Hauptschein, *J. Am. Chem. Soc.* **83**, 2500 (1961).
21. Brit. Pat. 1,004,575 (Sept. 15, 1962), M. Hauptschein and Parris (to Pennsalt Chem Corp.).
22. Brit. Pat. 1,165,912 (Oct. 1, 1969), Evans and Litt (to Allied Chem Corp.); U.S. Pat. 3,558,721 (Jan. 26, 1971), C. C. Y. Yao (to Allied Chem Corp.).
23. F. W. Evans and co-workers, *J. Org. Chem.* **33**, 1839 (1968).
24. U.S. Pat. 3,525,758 (Aug. 25, 1970), A. Katsushima (to Daikin Kogyo Kabushi Kaisha).
25. U.S. Pat. 4,138,417 (Feb. 6, 1979), H. Ukiihashi and co-workers, (to Asahi Glass Co.).
26. H. Millauer, W. Schwertfeger, and G. Siegemund, *Ang. Chem. Int. Ed. Eng.* **24**, 161–179 (1985).
27. H. S. Eleuterio, *J. Macmol. Sci.-Chem.* **A6**, 1027 (1972).
28. U.S. Pat. 3,442,942 (May 6, 1966), D. Sianesi and co-workers (to Montecatini Edison sPa).
29. U.S. Pat. 3,847,978 (Nov. 12, 1974), D. Sianesi and G. Caporiccio (to Montecatini Edison sPa).
30. U.S. Pat. 3,810,874 (May 14, 1974), R. A. Mitsch and J. L. Zollinger (to 3M Co.).
31. D. R. Husted and A. H. Diesslin, *J. Am. Chem. Soc.* **75**, 1605 (1953).
32. *Fluorad Fluorochemical Surfactant FC-143*, Minnesota Mining & Manufacturing Co., St. Paul, Minn., 1987.
33. T. Takakura and N. Sugiyama, *Asahi Garasu Kenkyu Hokosu* **37**, 257–262 (1987).
34. U.S. Pat. 4,156,791 (May 29, 1979), W. V. Childs (to Phillips Petroleum).
35. I. L. Knunyants, L. Chih-yuan, and V. V. Shokina, *Advances in Chem. (Uspekhi Khimi)* **32**, original 1502, Eng. trans. 461–476 (1963); translation RSIC-165, Redstone Information Center.
36. U.S. Pat. 3,293,306 (1966), R. E. Bleu and J. H. Fasenacht (to E. I. du Pont de Nemours and Co., Inc.).
37. U.S. Pat. 3,574,770 (Apr. 13, 1971), E. C. Stump, Jr., and S. E. Rochow (to NASA).
38. F. A. Bovey and co-workers, *J. Polym. Sci.* **XV**, 520–536 (1955).
39. M. K. Bernett and W. A. Zisman, *J. Chem. Physics* **66**, 1207 (1962).
40. H. W. Fox and W. A. Zisman, *J. Colloid Sci.* **7**, 109 (1952).

41. Jpn. Kokai Tokkyo Koho, JP 59,177,384 [87,177,384] (Oct. 8, 1984), M Hamada and F. Muranaka (to Asahi Chem. Industry Co.).
42. A. L. Henne and H. G. DeWitt, *J. Am. Chem. Soc.* **70**, 1548 (1948).
43. E. T. McBee, P. A. Wiseman, and G. B. Bachman, *Ind. Eng. Chem.* **39**, 415 (1947).
44. U.S. Pat. 5,093,432 (Mar. 3, 1992), T. R. Bierschenk and co-workers (to Exfluor Research Co.).
45. U.S. Pat. 2,606,206 (Aug. 5, 1952), R. Guenthner (to 3M Co.).
46. J. T. Hill, *J. Macromol. Sci-Chem* **A8**, 499–520 (1974).
47. U.S. Pat. 4,647,413 (Mar. 3, 1987), P. M. Savu (to 3M Co.).
48. *Material Safety Data Sheet Fluorad Brand Fluorochemical Acid FC-23*, Minnesota Mining & Manufacturing Co., St. Paul, Minn., July 22, 1992.
49. *Material Safety Data Sheet Fluorad Brand Fluorochemical Acid FC-26*, Minnesota Mining & Manufacturing Co., St. Paul, Minn., July 6, 1992.
50. *Product Toxicity Summary Sheet Fluorad Brand Fluorochemical Acid FC-26*, Minnesota Mining & Manufacturing Co., St. Paul, Minn., Dec. 1989.
51. C. S. Godin and co-workers, *Fundam. Appl. Toxicol.* **18**, 45431–0009 (1992).
52. *Product Toxicity Summary Sheet Fluorad Brand Fluorochemical Acid FC-143*, Minnesota Mining & Manufacturing Co., St. Paul, Minn., Apr. 26, 1990.
53. *Material Safety Data Sheet Fluorad Brand Fluorochemical Acid FC-143*, Minnesota Mining & Manufacturing Co., St. Paul, Minn., June 8, 1992.

Patricia M. Savu
3M Company

PERFLUOROALKANESULFONIC ACIDS

Perfluoroalkanesulfonic acids and their derivatives are of commercial significance because of their unusual acid strength, chemical stability, and the surface activity of the higher members of the series (eight carbons and larger).

Preparation

The perfluoroalkane sulfonic acids were first reported in 1954. Trifluoromethanesulfonic acid was obtained by the oxidation of bis(trifluoromethyl thio) mercury with aqueous hydrogen peroxide (1). The preparation of a series of perfluoroalkanesulfonic acids derived from electrochemical fluorination (ECF) of alkane sulfonyl halides was also disclosed in the same year (2). The synthetic operations employed when the perfluoroalkanesulfonic acid is derived from electrochemical fluorination, which is the best method of preparation, are shown in equations 1–3.

$$R_hSO_2F + HF \rightarrow R_fSO_2F + H_2 \tag{1}$$

where R_h is an alkyl group and R_f is a perfluoroalkyl group

$$R_fSO_2F + KOH \rightarrow R_fSO_3K + HF \tag{2}$$

$$R_fSO_3K + H_2SO_4 \rightarrow R_fSO_3H + KHSO_4 \tag{3}$$

Perfluorosulfonyl fluorides can also be prepared by the electrochemical fluorination of saturated or unsaturated cyclic sulfones (3–5). Perfluorobutanesulfonyl fluoride can be prepared in 40–48% yield from sulfolane (eq. 4) (6).

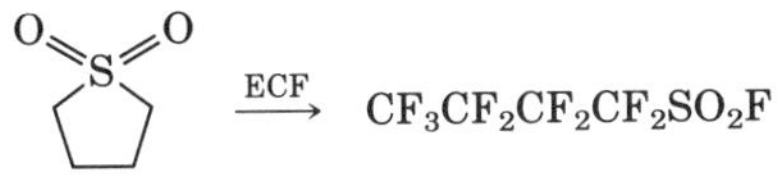

$$\text{(cyclic sulfone)} \xrightarrow{\text{ECF}} CF_3CF_2CF_2CF_2SO_2F \qquad (4)$$

Yields of sulfonyl fluorides prepared by ECF vary depending on the particular structure. Chain degradation becomes more important as the chain length increases (6). Yields can vary from 96% for perfluoromethanesulfonyl fluoride (7) to 43–50% for perfluorooctanesulfonyl fluoride (8).

Trifluoromethanesulfonic acid can be prepared via trifluoromethanesulfenyl chloride as shown in equations 5–7 (9).

$$CF_3S\text{–}SCF_3 + Cl_2 \rightarrow 2\ CF_3SCl \qquad (5)$$

$$CF_3SCl + 2\ HOCl \rightarrow CF_3SO_2Cl + 2\ HCl \qquad (6)$$

$$CF_3SO_2Cl + NaOH \rightarrow CF_3SO_3Na + HCl \qquad (7)$$

Other preparations of trifluoromethanesulfonic acid include oxidation of methyltrifluoromethyl sulfide under a variety of conditions (10,11). Perfluorosulfonyl fluorides have also been prepared by reaction of fluoroolefins with sulfuryl fluoride (12,13). Chinese chemists have published numerous papers on the conversion of telomer-based alkyl iodides to sulfonyl fluorides (14,15) (eqs. 8 and 9):

$$2\ R_fI + Na_2S_2O_4 + NaHCO_3 \rightarrow 2\ R_fSO_2Na \qquad (8)$$

$$R_fSO_2Na + Cl_2 \rightarrow R_fSO_2Cl + NaCl \qquad (9)$$

Perfluorosulfonyl fluorides have also been prepared by direct fluorination, although in general yields are lower than preparation by ECF. Perfluoromethanesulfonyl fluoride has been produced in 15% yield from direct fluorination of dimethyl sulfone (16). Perfluoro-2-propanesulfonyl fluoride was prepared in 29% yield from propanesulfonyl fluoride (17). Direct fluorination of tetramethylene sulfone leads to the intact perfluorinated sulfone in 28% yield and the ring-opened product (perfluorobutanesulfonyl fluoride) in 10% yield (eq. 10) (17).

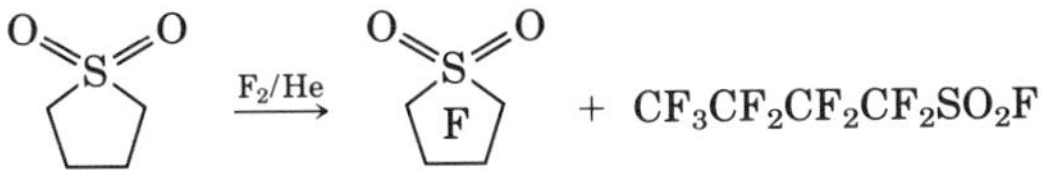

$$\text{(cyclic sulfone)} \xrightarrow{F_2/He} \text{(perfluoro cyclic sulfone, F)} + CF_3CF_2CF_2CF_2SO_2F \qquad (10)$$

Currently, the commercially important methods of preparations of perfluorinated sulfonic acid derivatives are electrochemical fluorination and sulfur trioxide addition to tetrafluoroethylene with subsequent ring opening.

The boiling points of a series of perfluoroalkanesulfonic acids are listed in Table 1 (2).

Table 1. Boiling Points of Perfluoroalkanesulfonic Acids

Compound	CAS Registry Number	Bp, °C/kPa[a]	Bp, °C[b]
CF_3SO_3H	[1493-13-6]	60/0.4	166
$C_2F_5SO_3H$	[354-88-1]	81/2.9	175[c]
$C_4F_9SO_3H$	[59933-66-3]	76–84/0.017	200[c]
$C_5F_{11}SO_3H$	[3872-25-1]	110/0.67[d]	212[c,e]
$C_6F_{13}SO_3H$	[355-46-4]	95/0.47	225[c]
$C_8F_{17}SO_3H$	[1763-23-1]	133/0.8	249
4-CF_3(cyclo-C_6F_{10})SO_3H	[374-62-9][f]	120/0.4	241
4-C_2F_5(cyclo-C_6F_{10})SO_3H	[335-24-0][f]		254

[a]To convert kPa to mm Hg, multiply by 7.5.
[b]At 101.3 kPa = 1 atm.
[c]Estimated.
[d]The hydrate, $C_5F_{11}SO_3H \cdot H_2O$.
[e]$C_5F_{11}SO_3H$ anhydrous.
[f]Potassium salt.

Trifluoromethanesulfonic Acid

The first member of the series, CF_3SO_3H, has been extensively studied. Trifluoromethanesulfonic acid [1493-13-6] is a stable, hydroscopic liquid which fumes in air. Addition of an equimolar amount of water to the acid results in a stable, distillable monohydrate, mp 34°C, bp 96°C at 0.13 kPa (1 mm Hg) (18). Measurement of conductivity of strong acids in acetic acid has shown the acid to be one of the strongest protic acids known, similar to fluorosulfonic and perchloric acid (19).

Trifluoromethanesulfonic acid is miscible in all proportions with water and is soluble in many polar organic solvents such as dimethylformamide, dimethylsulfoxide, and acetonitrile. In addition, it is soluble in alcohols, ketones, ethers, and esters, but these generally are not suitably inert solvents. The acid reacts with ethyl ether to give a colorless, liquid oxonium complex, which on further heating gives the ethyl ester and ethylene. Reaction with ethanol gives the ester, but in addition dehydration and ether formation occurs.

Alkyl esters of trifluoromethanesulfonic acid, commonly called triflates, have been prepared from the silver salt and an alkyl iodide, or by reaction of the anhydride with an alcohol (18,20,21). Triflates of the 1,1-dihydroperfluoroalkanols, $CF_3SO_2OCH_2R_f$, can be prepared by the reaction of perfluoromethanesulfonyl fluoride with the dihydroalcohol in the presence of triethylamine (22,23). Triflates are important intermediates in synthetic chemistry. They are among the best leaving groups known, so they are commonly employed in anionic displacement reactions.

The metallic salts of trifluoromethanesulfonic acid can be prepared by reaction of the acid with the corresponding hydroxide or carbonate or by reaction of sulfonyl fluoride with the corresponding hydroxide. The salts are hydroscopic but can be dehydrated at 100°C under vacuum. The sodium salt has a melting point of 248°C and decomposes at 425°C. The lithium salt of trifluoromethanesulfonic acid [33454-82-9], CF_3SO_3Li, commonly called lithium triflate, is used as a battery electrolyte in primary lithium batteries because solutions of it exhibit high electrical conductivity, and because of the compound's low toxicity and excellent chemical stability. It melts at 423°C and decomposes at 430°C. It is quite soluble in polar organic solvents and water. Table 2 shows the electrical conductivities of lithium triflate in comparison with other lithium electrolytes which are much more toxic (24).

Due to the strong ionic nature of lithium trifluoromethanesulfonate, it can increase the conductivity of coating formulations, and thereby enhance the dissipation of static electricity in nonconducting substrates (see ANTISTATIC AGENTS) (25).

Trifluoromethanesulfonic acid anhydride, bp 84°C, is prepared by refluxing the acid over an excess of phosphorous pentoxide (18,26). The anhydride reacts instantaneously with ammonia or amines to form trifluoromethanesulfonamides. The anhydride reacts with most polar organic solvents. It polymermizes THF to give a living polyether having cationic activity at each chain end (27).

Several excellent review articles (28–31) cover the chemistry of the acid and its derivatives in great detail. Trifluoromethanesulfonic acid is available from the 3M Co. as Fluorochemical Acid FC-24; the lithium salt is available as Fluorochemical Specialties FC-122, FC-123, and FC-124 (32).

Table 2. Comparative Electrical Conductivity^a of Lithium Salts

Concentration, M	CF_3SO_3Li	$LiClO_4$	$LiAsF_6$
0.5	24.4	29.9	26.3
0.1	5.81	7.4	6.94
0.05	3.12	3.83	3.57
0.01	0.70	0.83	0.79

^{a} In water at 25°C, $ohm^{-1}cm^{-1} \times 10^{-3}$.

Higher Perfluoroalkanesulfonic Acids

The longer perfluoroalkanesulfonic acids are hydroscopic oily liquids. Distillation of the acid from a mixture of its salt and sulfuric acid gives a hydrated mixture with melting points above 100°C. These acids show the same general solubilities as trifluoromethanesulfonic acid, but are insoluble in benzene, heptane, carbon tetrachloride, and perfluorinated liquids. All of the higher perfluoroalkanesulfonic acids have been prepared by electrochemical fluorination (20).

The longer-chain acids and their salts, particularly $C_8F_{17}SO_3H$ and higher, are surface-active agents in aqueous media. They reduce the surface tension of water to levels not possible with hydrocarbon surfactants. The surfactant $C_8F_{17}SO_2N(C_2H_5)CH_2COOK$ [2991-51-7] lowers the surface tension of water to 17 mN/m(= dyn/cm) at 0.2 weight percent (33) and exhibits outstanding thermal and chemical stabilities. The potassium salt of perfluorooctanesulfonic acid [2795-39-3], $C_8F_{17}SO_3K$, or perfluoroethylcyclohexanesulfonic acid [335-24-0], C_2F_5-cyclo-(C_6F_{10})-SO_3K, can form a stable foam in hostile media such as chromium trioxide and sulfuric acid where conventional hydrocarbon and silicone surfactants would be destroyed (34). The ability of these materials to foam concentrated sulfuric acid is utilized to prevent sulfuric acid from aerosoling into the air in industrial situations where chrome plating is done. Instead of forming an aerosol, the sulfuric acid forms a foam blanket on top of the plating bath. The foam derived from $C_8F_{17}SO_3K$ is generally more stable and dense than that derived from C_2F_5-cyclo-(C_6F_{10})-SO_3K. As a result these fluorochemical surfactants are often used in combination to produce desired wetting and foaming activity.

Generally, derivatives of the longer-chain perfluoroalkanesulfonic acids have a number of unique surface-active properties and have formed a basis for a number of commercial products. Derivatives of N-alkyl perfluorooctanesulfonamidoethanol, $C_8F_{17}SO_2N(R)CH_2CH_2OH$, and polymers of N-alkyl perfluorooctanesulfonamidoethyl methacrylate, $C_8F_{17}SO_2N(R)CH_2CH_2OCOC(CH_3)$=$CH_2$, impart soil, oil, and water repellency to treated fabrics and paper; this forms the basis for 3M's Scotchguard and Scotchban products (35). Polymers of N-alkyl perfluorooctanesulfonamidoethyl acrylates, $C_8F_{17}SO_2N(R)CH_2CH_2OCOC(R')$=$CH_2$, with certain hydrocarbon acrylates and methacrylates have also been found to be surface-active agents in organic solvents and water. These polymers have applications in the areas of secondary crude oil recovery and wetting, leveling, and flow control agents (36,37).

Higher perfluoroalkanesulfonates are slightly more reactive than triflates toward nucleophilic displacements. The rate constants for acetolysis of methyl nonafluorobutanesulfonate [6401-03-2], methyl trifluoromethanesulfonate [333-27-7], and methyl toluenesulfonate [80-48-8] are 1.49×10^{-4}, 7.13×10^{-5}, and 3.1×10^{-9} s^{-1}, respectively. This means that the relative reactivities for nonafluorobutanesulfonate, trifluoromethanesulfonate, and toluenesulfonate are 48,000/22,900/1 (38).

Difunctional Perfluoroalkanesulfonic Acids

Alpha, omega-perfluoroalkanedisulfonic acids were first prepared by aqueous alkali permanganate oxidation of the bis-sulfone, $RSO_2(CF_2CF_2)_nSO_2R$ (39). Disulfonyl fluorides of the formula $FSO_2(CF_2)_nSO_2F$ have also been prepared by electrochemical fluorination where $n = 1$ to 5. These disulfonyl fluorides have been converted to the cyclic anhydrides by basic hydrolysis of the disulfonyl fluoride, acidification, and dehydration with phosphorous pentoxide (40). The alpha, omega-perfluoroalkanedisulfonic acids can also be prepared by the action of sodium dithionate on the diiodides to form the disulfinate salt, followed by chlorination and hydrolysis to give the disulfonic acids (14,15).

Carbonyl sulfonyl fluorides of the formula $FCO(CF_2)_nSO_2F$ have been prepared by electrochemical fluorination of hydrocarbon sultones (41,42). More commonly in a technology pioneered by Du Pont, perfluoroalkanecarbonyl sulfonyl fluorides are prepared by addition of SO_3 to tetrafluoroethylene followed by isomerization with a tertiary amine such as triethylamine (43).

$$CF_3{=}CF_2 + SO_3 \rightarrow \underset{O}{\overset{O\ F_2C{-\!-\!-}CF_2}{\underset{\|}{\overset{|\qquad\quad|}{S{-\!-}O}}}} \xrightarrow{(CH_3CH_2)_3N} \underset{\overset{\|}{O}}{\overset{O}{F{-}\underset{}{\overset{\|}{S}}{-}CF_2}}\overset{O}{\overset{\|}{C}}{-}F \qquad (11)$$

Fluorosulfonyldifluoroacetyl fluoride [677-67-8] is an important industrial intermediate used in the production of Du Pont's Nafion ion-exchange membrane. Nafion is an ion-exchange membrane used under the extreme conditions in electrolytic cells, especially for the electrolysis of sodium chloride to produce chlorine and caustic soda. Other commercial fluorinated ion-exchange resins utilize perfluorinated carboxylate groups instead of sulfonate groups in the monomer in order to make the polymer conductive. Shown in equation 12 are the synthetic operations used to convert fluorosulfonyldifluoroacetyl fluoride to its polymerizable form.

$$F-\overset{\overset{\displaystyle O}{\|}}{\underset{\underset{\displaystyle O}{\|}}{S}}-\overset{\overset{\displaystyle O}{\|}}{CF_2C}-F + 2\; \underset{F}{\overset{F}{\diagup}}\!\!\triangle\!\!\underset{CF_3}{\overset{F}{\diagdown}} \xrightarrow{F^-} F-\overset{\overset{\displaystyle O}{\|}}{\underset{\underset{\displaystyle O}{\|}}{S}}-CF_2CF_2OCFCF_2O-\overset{\overset{\displaystyle O}{\|}}{\underset{\underset{\displaystyle CF_3}{|}}{CFC}}-F \xrightarrow{Na_2CO_3}$$

$$F-\overset{\overset{\displaystyle O}{\|}}{\underset{\underset{\displaystyle O}{\|}}{S}}-CF_2CF_2O\underset{\underset{\displaystyle CF_3}{|}}{CFCF_2}O-CF=CF_2 + 2\,NaF + 2\,CO_2 \quad (12)$$

The vinyl ether in the latter part of the equation is copolymerized with tetrafluoroethylene, and then the sulfonyl fluoride group is hydrolyzed under basic conditions in order to produce the ion-exchange membrane (44–46).

BIBLIOGRAPHY

"Perfluoroalkane Sulfonic Acids" under "Fluorine Compounds, Organic" in *ECT* 3rd ed., Vol. 10, pp. 952–955, by R. A. Guenthner, 3M Co.

1. R. N. Hazeline and J. M. Kidd, *J. Chem. Soc.*, 4228 (1954).
2. P. W. Trott and co-workers, *126th National Meeting of the American Chemical Society*, New York, 1954, abstract p. 42-M.
3. U.S. Pat. 3,623,963 (1971), P. Voss (to Bayer AG); U.S. Pat. 3,951,762 (1972), P. Voss (to Bayer AG).
4. Brit. Pat. 1,099,240 (1968) (to Dow Corning).
5. I. N. Rozhkow, A. V. Bukhtiarov, and I. L. Knunyants, *Ikv. Akad. Nauk SSR, Ser. Khim.* **4**, 945 (1969).
6. T. Abe and S. Nagase, in R. E. Banks, ed., *Preparation, Properties, and Industrial Applications of Organofluorine Compounds*, Ellis Howard, 1982, p. 37.
7. T. Gramstad and R. N. Hazeldine, *J. Chem. Soc.*, 173 (1956); *Ibid.*, 2640 (1957).
8. Ger. Offen. 2,201,649 (1973), P. Voss, H. Niederprum, and M. Wechsberg; Ger. Offen. 2,234,837 (1974), P. Heinze and M. Schwarzmann; Ch. Comninellis, Ph. Javet, and E. Platterner, *J. Appl. Electrochem.* **4**, 287 (1974).
9. R. N. Hazeltine and J. M. Kidd, *J. Chem. Soc.*, 2901 (1955).
10. R. N. Hazeltine and co-workers, *Chem. Commun.*, 249 (1972).
11. R. B. Ward, *J. Org. Chem.* **30**, 3009 (1965).
12. U.S. Pat. 3,542,864 (Nov. 24, 1970), R. J. Koshar (to 3M Co.).
13. Brit. Pat. 1,189,561 (1970) (to Du Pont).
14. H. Weiyuan and C. Qingyun, *Chemistry* **2**, 31–76 (1987).
15. W. Y. Huang, *J. Fluorine Chem.* **32**, 179–195 (1986).

16. R. J. Lagow and co-workers, *J. C. S. Perkin I* **11**, 2675–2678 (1979).
17. H. Huang, H. Roesky, and R. J. Lagow, *Inorg. Chem.* **30**, 789–794 (1991).
18. T. Gramstad and R. N. Hazeldine, *J. Chem. Soc.*, 4069 (1957).
19. T. Gramstad, *Tidsskr, Kremi, Bergres, Metall.* **19**, 62 (1959).
20. U.S. Pat. 2,732,398 (Jan. 12, 1956), T. J. Brice and P. W. Trott (to 3M Co.).
21. P. G. Gassman and C. K. Harrington, *J. Org. Chem.* **49**, 2258–2273 (1984).
22. K. A. Epstein and co-workers, "Fluorinated Ferroelectric Liquid Crystals: Overview and Synthesis," *Eleventh Winter Fluorine Conference*, St. Petersburg, Fla., 1993.
23. U.S. Pat. 3,419,595 (Dec. 31, 1968), R. L. Hansen (to 3M Co.).
24. *Fluorad Fluorochemical Specialties FC-122*, and *Fluorad Fluorochemical Specialties FC-124*, Minnesota Mining & Manufacturing Co., St. Paul, Minn., 1986.
25. *Fluorad Fluorochemical Specialties FC-123*, Minnesota Mining & Manufacturing Co., St. Paul, Minn., 1992.
26. J. Burden and co-workers, *J. Chem. Soc.*, 2574 (1957).
27. S. Smith and A. J. Hubin, *J. Macromol. Sci.-Chem.* **A7**, 1399–1413 (1973).
28. A. Senning, *Chem. Rev.* **65**, 385 (1965).
29. R. D. Howells and J. D. McCown, *Chem. Rev.* **77**, 69 (1977).
30. P. J. Stang and M. R. White, *Aldrichimica Acta* **16**, 15 (1983).
31. P. J. Stang, M. Hanack, and L. R. Subramanian, *Synthesis*, 85 (1982).
32. *Fluorad Fluorochemical Acid FC-24*, Minnesota Mining & Manufacturing Co., St. Paul, Minn., 1978.
33. *Fluorad Fluorochemical Specialties FC-129*, Minnesota Mining & Manufacturing Co., St. Paul, Minn., 1991.
34. *Fluorad Fluorochemical Specialties FC-95*, Minnesota Mining & Manufacturing Co., St. Paul, Minn., 1987.
35. H. C. Fielding, in Ref. 6, pp. 226–232.
36. *Fluorad Fluorochemical Specialties FC-430/431*, Minnesota Mining & Manufacturing Co., St. Paul, Minn., 1990.
37. *Fluorad Fluorochemical Specialties FC-740*, Minnesota Mining & Manufacturing Co., St. Paul, Minn., 1987.
38. R. L. Hansen, *J. Org. Chem.* **30**, 4322–4324 (1965).
39. U.S. Pat. 3,346,606 (Oct. 10, 1967), R. B. Ward (to Du Pont).
40. U.S. Pat. 4,329,478 (May 11, 1982), F. E. Behr (to 3M Co.).
41. U.S. Pat. 4,332,954 (June 1, 1982), R. J. Koshar (to 3M Co.).
42. Eur. Pat. Appl. 062,430 (1982), M. Hamada, J. Ohmura, and F. Muranaka.
43. U.S. Pat. 2,852,554 (Sept. 16, 1958), D. C. England and H. Oak (to Du Pont).
44. Y. Yen, C. Nieh, and L. C. Hsu, *Fluorinated Polymers, Process Economics Report #166*, SRI International, Menlo Park, Calif., 1983, p. 165.
45. A. Eisenberg and H. L. Yeager, eds., *Perfluorinated Ionomer Membranes*, ACS Series *180*, American Chemical Society, Washington, D.C., 1982.
46. L. A. Ultracki and R. A. Weiss, eds., *Multiphase Polymers: Blends and Ionomers*, ACS *Symposium Series 395*, American Chemical Society, Washington, D.C., 1989.

Patricia Savu
3M Center

FLUORINATED AROMATIC COMPOUNDS

Aromatic fluorine compounds have been known for nearly a century, but numerous applications have surfaced only in recent years. The special properties conferred by fluorine justify the higher costs required to produce fluoroaromatics. The unusual physiochemical and biological properties that fluorine imparts to aromatics result from the small size of fluorine (it is bioisosteric with both the hydrogen atom and the hydroxyl group) and from its striking electronic properties, including high electronegativity and the ability to alter polarity of adjacent groups, as well as to donate electrons by resonance. Other significant properties are the enhanced stability of the C—F bond in the absence of activating groups, high lipid solubility, hydrogen-bonding potential (acceptor role), and enzyme inhibition. The carbon–fluorine link in fluoroaromatics has been of considerable value as a label for metabolic, mechanistic, and structural studies.

Depending on which substituents are present, fluoroaromatic intermediates can be converted into fluorinated or fluorine-free products. Fluorine substitution can affect the biological spectrum of the parent aromatic or heterocyclic compound by enhancement of desired properties or by suppression of undesired properties. Fluorine-containing aromatics have been incorporated into drugs (hypnotics, tranquilizers, antiinflammatory agents, analgesics, antibacterials, etc) and into crop protection chemicals (herbicides, insecticides, fungicides). Liquid crystals, positron emission tomography, and imaging systems are newer use areas for fluoroaromatics and fluoroheterocyclics.

For fluorine-free products, the lability of fluorine in fluoronitrobenzenes and other activated molecules permits it to serve as a handle in hair-dye manufacturing operations, high performance polymers such as polyetheretherketone (PEEK), production of drugs such as diuretics, and fiber-reactive dyes. Labile fluorine has also been used in analytical applications and biological diagnostic reagents.

Preparative Methods

Ring-Fluorinated Aromatics and Heterocyclics. In contrast with other molecular halogens (Cl_2, Br_2), early attempts at direct aromatic substitution with fluorine (F_2) gave violent reactions involving ring scission, addition, coupling, and polymerization. Consequently, indirect fluorination techniques based on diazotization of anilines or exchange fluorination of activated haloaromatics were developed. Recent advances in synthetic methods include discoveries of new fluorinating agents and modifications of known methods. Some of these efforts were stimulated by objectives to effect selective fluorination of natural or biologically active compounds. The need to prepare ^{18}F-labeled pharmaceuticals for use in positron emission tomography also accelerated the need for improved aromatic fluorination techniques (1).

Substitutive Aromatic Fluorination.   The search for improved substitutive aromatic fluorination tools based on tamed fluorine continues (2). These reagents include elemental fluorine (F_2) (3–7), chlorine trifluoride (8) and pentafluoride (9), xenon fluorides (eg, XeF_2) (10), silver difluoride (11), cesium fluoroxysulfate (12), trifluoromethyl hypofluorite (CF_3OF) (13), bis(fluoroxy)difluoromethane ($CF_2(OF)_2$) (13), and acetyl hypofluorite (CH_3CO_2F) (14,15). Substitutive aromatic fluorination with elemental fluorine is commercially practiced for the manufacture of the antineoplastic, 5-fluorouracil (5-FU) from uracil. Nitrogen–fluorine reagents can also effect substitutive aromatic fluorination, for example, *N*-fluorobis[(trifluoromethyl)sulfonyl]imide (16); *N*-fluorobenzenesulfonimides (NFSi) (17); *N*-fluoropyridinium trifluoromethanesulfonates (18); and 1-alkyl-4-fluoro-1,4-diazoniabicyclo[2·2·2]octane salts, marketed as SELECTOFLUOR reagents (19,20).

$(C_6H_5SO_2)_2NF$

$BF_4^-;\ CF_3SO_3^-$

Diazotization Routes.   Conventional Sandmeyer reaction conditions are not suitable to make fluoroaromatics. Phenols primarily result from high solvation of fluoride ion in aqueous media.

Fluoroaromatics are produced on an industrial scale by diazotization of substituted anilines with sodium nitrite or other nitrosating agents in anhydrous hydrogen fluoride, followed by *in situ* decomposition (fluorodediazoniation) of the aryldiazonium fluoride (21). The decomposition temperature depends on the stability of the diazonium fluoride (22,23). A significant development was the addition of pyridine (24), tertiary amines (25), and ammonium fluoride (or bifluoride) (26,27) to permit higher decomposition temperatures (>50°C) under atmospheric pressure with minimum hydrogen fluoride loss.

The Balz-Schiemann reaction is a useful laboratory and industrial method for the preparation of fluoroaromatics. The water-insoluble diazonium fluoroborate is filtered, dried, and thermally decomposed to give the aryl fluoride, nitrogen, and boron trifluoride (28–30).

$+ N_2 + BF_3$

Extreme caution must be exercised in the handling of nitroaryldiazonium fluoroborates because of unruly decomposition (29,30). Water-insoluble aryl diazonium hexafluorophosphates, ArN_2PF_6, frequently give higher yields of the fluoroaromatic (31). Substitution of aqueous sodium nitrite by nitrite esters–boron trifluoride in organic solvents gives high yields of aryl diazonium fluoroborate (32,33). A variant of the Balz-Schiemann reaction features diazotization by nitrosonium tetrafluoroborate, $NO^+BF_4^-$, in organic solvents followed by *in situ* decomposition to give high yields of aryl fluoride (34). A single fluorine atom can be introduced sequentially by the Balz-Schiemann reaction (via successive nitration, reduction, and diazotization) for a total of up to four fluorine atoms, eg, 1,2,4,5-tetrafluorobenzene. The Balz-Schiemann process is used to manufacture fluoroaromatics, eg, o- and p-difluorobenzene, not readily accessible by standard aniline–hydrogen fluoride diazotization or exchange-fluorination (Halex) routes. Estimates of producer capacities utilizing Balz-Schiemann technology range from 50–100 t/yr (35) to hundreds of t/yr (36). A continuous feed aryl diazonium fluoroborate decomposition step has been patented (37) and commercial details described (38).

The discovery of the Balz-Schiemann reaction in 1927 replaced the earlier Wallach procedure (1886) based on fluorodediazoniation of arenediazonium piperidides (aryltriazenes) in aqueous hydrogen fluoride (39,40). The Wallach aryltriazene fluorodediazoniation technique has found new utility in agrochemicals (41), pharmaceuticals (42), and positron emission tomography (43). This is illustrated in the synthesis of 2,4-dichloro-5-fluorotoluene [86522-86-3], an intermediate to the fluoroquinolone antibacterial ciprofloxacin, by heating N-(2,4-dichloro-5-methylphenyl)-N',N'-dimethyltriazene in anhydrous fluoride (42).

Exchange Fluorination. Fluorobenzene cannot be made from chlorobenzene and potassium fluoride because of absence of substrate activation. The halogen exchange (Halex) reaction of activated haloaromatics and haloheterocyclics with potassium fluoride is a primary industrial fluoroaromatics synthesis tool (44,45). Early work featured preparation of o- and p-fluoroaromatics activated by nitro or cyano groups by exchange fluorination in dipolar aprotic solvents (46). Features of this technique include good fluorine utilization (1:1 stoichiometry), facile product separation, and potential recycling of potassium chloride (as KF) by treatment with hydrogen fluoride or fluorine. Aprotic solvents permit less solvation of fluoride ion (as compared with protic solvents), a kinetically significant amount of fluoride ion in solution, and greater insolubility of potassium chloride which, in turn, provides a further reaction driving force.

$$p\text{-ClC}_6\text{H}_4\text{NO}_2 \ + \ \text{KF} \xrightarrow{\text{solvent}} p\text{-FC}_6\text{H}_4\text{NO}_2$$

The degree of fluorination can be limited by the thermal stability of the solvent or by its reaction with basic potassium fluoride through proton abstraction. Such solvent-derived by-products can subsequently react with the starting material and/or main product.

Of the alkali metal fluorides, potassium fluoride offers the best compromise between cost and effectiveness. Although cesium fluoride generally gives higher yields, its higher cost may be a potential drawback as an industrial fluorination tool except for those substrates resistant to potassium fluoride. In contrast, inexpensive sodium fluoride consistently gives lower yields than potassium fluoride. Tetra-n-butylammonium fluoride (TBAF) (47), tetra-n-butylphosphonium hydrogen difluoride, and dihydrogen trifluoride (48) have been successfully employed in Halex reactions. Mode of potassium fluoride preparation (spray-, calcine- or freeze-dried) can affect exchange fluorination activity (49,50). Enhanced fluorination rates are associated with decreasing particle size and increasing surface area.

Halex rates can also be increased by phase-transfer catalysts (PTC) with widely varying structures: quaternary ammonium salts (51–53); 18-crown-6-ether (54); pyridinium salts (55); quaternary phosphonium salts (56); and poly(ethylene glycol)s (57). Catalytic quantities of cesium fluoride also enhance Halex reactions (58).

The inertness of chlorine in the meta position in Halex reactions is of commercial value. For example, 3,4-dichloronitrobenzene [99-54-7] forms 3-chloro-4-fluoronitrobenzene [350-30-1], which is then reduced to 3-chloro-4-fluoroaniline [367-21-5] for incorporation in the herbicide flamprop–isopropyl or the fluoroquinolone antibacterials, norfloxacin and pefloxacin.

Activating groups other than nitro or cyano have extended the versatility of exchange-fluorination reactions: —CHO (59,60); —COCl (60); —CO$_2$R (61); —(CONRCO)— (62); —SO$_2$Cl (63); and —CF$_3$ (56,64).

Explosions have been reported during preparation of fluoronitroaromatics by the Halex reaction on a laboratory or industrial scale: o-fluoronitrobenzene (65); 2,4-dinitrofluorobenzene (66); 2,4-difluoronitrobenzene (67); and 1,5-difluoro-2,4-dinitrobenzene (68).

Fluorodenitration of nitroaromatics represents an exchange fluorination technique with commercial potential. For example, m-fluoronitrobenzene [402-67-5] from m-dinitrobenzene [99-65-0] and KF in the presence of various promoters can be realized (69–72). This is not feasible under Halex conditions with m-chloronitrobenzene [121-73-3]

Saturation–Rearomatization. The first commercial route to perfluorinated aromatics such as hexafluorobenzene, octafluorotoluene [434-64-0] and fused-ring polycyclics was based on a multistage saturation–rearomatization process (73). In the first stage, benzene is fluorinated by a high valency oxidative metal fluoride (cobalt trifluoride) to give a mixture of polyfluorocyclohexanes. The latter is subjected to a combination of dehydrofluorination (with alkali) and/or defluorination (with heated iron, iron oxide, or nickel packing) to give hexa-, penta-, and tetrafluorobenzenes. Modifications of the first stage include the use of complex metal fluorides, eg, potassium tetrafluorocobaltate(III), that have been found to be milder and more selective fluorinating agents than cobalt trifluoride (74). A related process features successive treatment of hexachlorobenzene with chlorine trifluoride (75) or fluorine (76), followed by dehalogenation with iron powder at 300°C. More emphasis is now given to Halex processes for perfluoroaromatics manufacture rather than saturation–rearomatization routes.

Fluoroaliphatic Thermolytic Routes. The reaction of difluorocarbene (generated from CHClF$_2$ at 600°C) with cyclopentadiene to give fluorobenzene (70% yield) has been scaled up in a pilot-plant/semiworks facility (capacity = several dozen t/yr) (77,78). The same process can now be effected under liquid-phase conditions in the presence of phase-transfer catalysts (79,80).

Miscellaneous Methods. Exhaustive evaluation of the decarbonylation of benzoyl fluorides, ArCOF, by Wilkinson's catalyst [14694-95-2], Rh[(C$_6$H$_5$)$_3$P]$_3$Cl, to give aryl fluorides has established (81) that previous claims (82) cannot be reproduced.

One approach to aryl fluorides (83) based on phenolic derivatives features more moderate thermal decarboxylation of phenyl fluoroformates employing alumina-impregnated platinum group catalysts (84). Treatment of phenyl chloroformates with hydrogen fluoride using Lewis acid catalysts to give aryl fluorides may have potential industrial importance (85,86).

Fluorodesulfonylation represents a complementary extrusion technique to aryl fluorides (87) which has attracted interest (88,89). For example, 2-fluorobenzonitrile [394-47-8] was obtained in 84% yield from 2-cyanobenzenesulfonyl fluoride and potassium fluoride in sulfolane (88).

The electrochemical route to fluoroaromatics (90) based on controlled potential electrolysis in the absence of hydrogen fluoride (platinum anode, $+2.4$ V; acetonitrile solvent; tetraalkylammonium fluoride electrolyte) has not been commercialized. However, considerable industrial interest in the electrochemical approach still exists (91–93).

The single-step p-fluoroaniline [31-40-4] process based on fluorodeoxygenation of nitrobenzene (via *in situ* generation of N-phenylhydroxylamine) in anhydrous hydrogen fluoride (94–96) has not been commercialized primarily due to concurrent formation of aniline, as well as limited catalyst life. The potential attractiveness of this approach is evidenced by numerous patents (97–101). Concurrent interest has been shown in the two-step process based on N-phenylhydroxylamine (HF-Bamberger reaction) (102–104).

Side-Chain Fluorinated Aromatics and Heterocyclics. Benzotrifluorides generally are prepared from trichloromethylaromatics with metal fluorides or hydrogen fluoride. Industrial processes feature reaction with hydrogen fluoride under high pressure, atmospheric pressure, or vapor-phase conditions. A potential simplification is the single-step conversion of toluene to benzotrifluoride employing chlorine–hydrogen fluoride (CCl_4 diluent, 460°C) (105).

Sulfur Tetrafluoride and Aromatic Carboxylic Acids. Benzotrifluorides also are prepared from aromatic carboxylic acids and their derivatives with sulfur tetrafluoride (SF_4) (106,107). Hydrogen fluoride is frequently used as a catalyst. Two equivalents of sulfur tetrafluoride are required:

$$ArCOOH + SF_4 \xrightarrow{HF} ArCOF + HF + SOF_2$$
$$ArCOF + SF_4 \longrightarrow ArCF_3 + SOF_2$$

The high cost of SF_4 and the incomplete use of fluorine justify its use only for inaccessible benzotrifluorides. The related liquid S–F reagent, (diethylamino)sulfur trifluoride (DAST), $(C_2H_5)_2NSF_3$, also effects similar transformations with aromatic carboxylic acids (108).

Perfluoroalkylation. A significant technical advance features perfluoroalkylation of aromatics (devoid of electron-withdrawing groups) with carbon tetrachloride–hydrogen fluoride to give high selectivity of benzotrifluorides (109,110). Hydrogen fluoride performs a threefold role: solvent, Friedel-Crafts alkylation catalyst, and fluorinating agent.

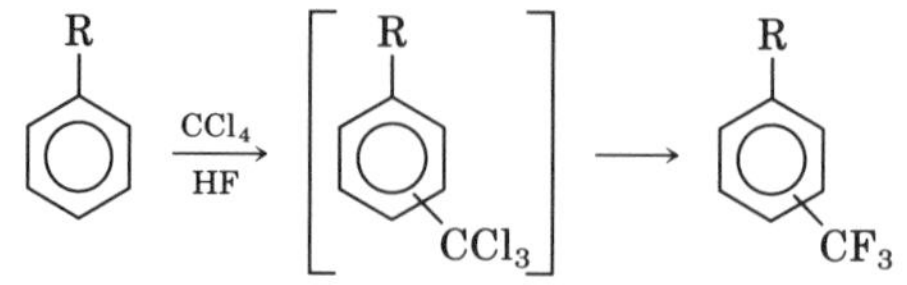

$$R\text{—}C_6H_5 \xrightarrow[\text{HF}]{\text{CCl}_4} \left[R\text{—}C_6H_4\text{—}CCl_3 \right] \longrightarrow R\text{—}C_6H_4\text{—}CF_3$$

Aromatic perfluoroalkylation can be effected by fluorinated aliphatics via different techniques. One category features copper-assisted coupling of aryl halides with perfluoroalkyl iodides (eg, CF_3I) (111,112) or difluoromethane derivatives such as CF_2Br_2 (Burton's reagent) (113,114), as well as electrochemical trifluoromethylation using CF_3Br with a sacrificial copper anode (115). Extrusion of spacer groups attached to the fluoroalkyl moiety, eg, CF_3COONa and higher perfluorocarboxylated salts (116,117), CF_3SO_2Na (118), and esters such as $CF_2ClCOOCH_3$ (119) or $FSO_2CF_2COOCH_3$ (120), represents a novel trifluoromethylation concept.

$$\text{ArI} + CF_3COONa \xrightarrow{\text{CuI/NMP}} \text{Ar—CF}_3$$

Aromatic perfluoroalkylation can also be performed in the absence of copper employing $(CF_3COO)_2$ (121) or $R_fI(C_6H_5)OSO_2CF_3$ (FITS reagents) (122). Aluminum chloride-catalyzed alkylation of fluorobenzene with hexafluoroacetone, CF_3COCF_3, gave 66% yield of p-fluoro-α,α-bis(trifluoromethyl)benzyl alcohol [2402-74-6] (123).

Oxidative Fluorination of Aromatic Hydrocarbons. The economically attractive oxidative fluorination of side chains in aromatic hydrocarbons with lead dioxide or nickel dioxide in liquid HF stops at the benzal fluoride stage (67% yield) (124).

$$p\text{-}CH_3C_6H_4NO_2 \xrightarrow{\text{PbO}_2/\text{HF}} p\text{-}CHF_2C_6H_4NO_2$$

Cyclization. Construction of benzotrifluorides from aliphatic feedstocks represents a new technique with economic potential. For example, 1,1,1-trichloro-2,2,2-trifluoroethane [354-58-5] and dimethyl itaconate [617-52-7] form 4-methoxy-6-trifluoromethyl-2H-pyran-2-one [101640-70-4], which is converted to methyl 3-(trifluoromethyl)benzoate [2557-13-3] with acetylene or norbornadiene (125).

Ring-Fluorinated Benzenes

FLUOROBENZENE

Properties. Fluorobenzene [462-06-6] (monofluorobenzene), C_6H_5F, has a molecular weight of 96.1, and is a colorless mobile liquid with a pleasant aromatic odor (Table 1). Its thermal stability is of a high order; fluorobenzene undergoes no detectable decomposition when kept at 350°C for 24 h at pressures of up to

Table 1. Physical Properties of Fluorobenzene

Property	Value
melting point, °C	−42.22
boiling point, °C	84.73
density, 25°C, g/mL	1.0183
coefficient of expansion	0.00116
refractive index, n_D^{25}	1.4629
viscosity, mPa·s(= cP)	
9.3°C	0.653
19.9°C	0.585
80.9°C	0.325
surface tension, mN/m(= dyn/cm)	
9.3°C	28.49
20.0°C	27.71
34.5°C	25.15
latent heat of fusion, J/mol[a]	11,305.2
latent heat of vaporization, 25°C, J/mol[a]	34,576.6
specific heat, 25°C, J/mol[a]	146.3
critical temperature, °C	286.94
critical pressure, kPa[b]	4550.9
critical density, g/mL	0.269
dielectric constant, 30°C	5.42
dipole moment, C·m[c]	4.90×10^{-30}
heat of combustion, J/g[a]	−32,273.3
heat of formation, kJ/mol[a]	
vapor	−110.5
liquid	−145.2
solubility in water, 30°C, g/100 g	0.154
solubility of water in fluorobenzene, 25°C, g/100 g	0.031
boiling point of binary azeotrope, °C	
with 31 wt % *tert*-butyl alcohol	76.0
with 32 wt % methanol	59.7
with 30 wt % isopropyl alcohol	74.5
flash point (Tag open cup), °C[d]	−13
vapor pressure, in °C and kPa[b]	Antoine equation[e]

[a]To convert J to cal, divide by 4.184.
[b]To convert kPa to mm Hg, multiply by 7.5; log kPa = log mm Hg − 0.895.
[c]To convert C·m to debye (D), divide by 3.336×10^{-30}.
[d]Ref. 126.
[e]$\text{Log}_{10}P = 6.07687 - \dfrac{1248.083}{(t + 221.827)}$.

40.5 MPa (400 atm). Toxicity: oral (rat), $LD_{50} > 4$ g/kg; inhalation (mouse), LD_{50} 45 g/m^3 (2 h) (127).

    Reactions. *Electrophilic Substitution.* Fluorobenzene electrophilic substitution reactions are more para directing than are the same chlorobenzene reactions (128). Nitration of fluorobenzene with concentrated nitric and sulfuric acid gives a 92:8 mixture of *p*- and *o*-fluoronitrobenzene [*1493-27-2*] which can be separated by distillation. The other commercial route to *o*- and *p*-fluoronitrobenzene

[350-46-9] is based on exchange fluorination (KF) in a polar solvent; phase-transfer catalysts are frequently employed.

The Friedel-Crafts ketone synthesis is of commercial importance in upgrading fluorobenzene for drug, polymer, and electronic applications (Table 2).

Nucleophilic Displacement Reactions. The presence of activating groups, eg, *o*, *p* nitro groups, makes aromatic fluorine reactive in nucleophilic displacement reactions. This has been demonstrated by determination of the relative fluorine–chlorine displacement ratios from the reaction of halonitrobenzenes with sodium methoxide in methanol (137); F is displaced 200–300 times more readily than Cl.

Numerous applications have been developed based on the lability of fluoronitroaromatics; 4-fluoro-3-nitroaniline [364-76-1] and 4-fluoro-3-nitro-*N*,*N*-bis(hydroxyethyl)aniline [29705-38-2], commercial hair dye intermediates (138–140); 2,4-dinitrofluorobenzene [70-34-8] (Sanger's reagent), for amino acid characterization (141); 4-fluoro-3-nitrophenyltrimethylammonium iodide [39508-27-5], a protein solubilizing reagent (142); and 4-fluoro-3-nitrophenylazide [28166-06-5], an antibody tagging reagent (143) also used for industrial immobilization of enzymes (144). Other examples of biochemical applications (amino acid or peptide characterization, protein cross-linking reagent) include 2,4-dinitro-5-fluoroaniline [361-81-7] (Bergmann's reagent) (145); 4-fluoro-3-nitrobenzoates (146,147); 4-fluoro-3-nitrobenzenesulfonic acid [349-05-3] (148); 4-fluoro-3-nitrophenyl sulfone [51451-34-4] (149); 1,5-difluoro-2,4-dinitrobenzene [327-92-4] (150); 3,5-dinitro-2-fluoroaniline [18646-02-1] (151); 4-fluoro-7-nitrobenzofurazan [29270-56-2] (NBD-F) (152); and 1-fluoro-2,4-dinitrophenyl-5-L-alanine amide

Table 2. Friedel-Crafts Ketone Synthesis with Fluorobenzene[a]

Acylating agent	Reference	Product	CAS Registry Number	End use
4-chlorobutyryl chloride	129	4-chloro-4'-fluorobutyr-ophenone	[3874-54-2]	haloperidol[b] (tranquilizer)
acetyl chloride or acetic anhydride	130	4-fluoroacetophenone	[403-42-9]	flazalone[b] (anti-inflammatory)
4-fluorobenzoyl chloride	131	4,4'-difluorobenzo-phenone	[345-92-6]	polyetheretherketone (PEEK), a high performance thermoplastic
isophthaloyl chloride	132, 133	1,3-bis(4-fluorobenz-oyl)benzene	[108464-88-6]	poly(arylene ethers) (PAE); polyimides
oxalyl chloride	134	4-fluorobenzoyl chloride	[403-43-0]	liquid crystal intermediate
chloroacetyl chloride	135	2-chloro-4'-fluoroaceto-phenone	[456-04-2]	flutriafol[c] (fungicide)
2,3-naphthalenedicar-boxylic anhydride	136	3-(4-fluorobenzoyl)-2-naphthalenecarbox-ylic acid	[91786-16-2]	organo-selenium metallic conductors

[a]AlCl₃ catalyst.
[b]See Table 4.
[c]See Table 3.

[*95713-52-3*] (Marfey's reagent) (153). Labile fluorine in 4-fluoronitrobenzene can be used to form piperidinylimino-linked polar chromophores for nonlinear optical (NLO) materials (154).

Examples of commercial reactive fluoroaromatics are not restricted to fluoronitrobenzenes. The fluorine-free diuretic, furosemide [*54-31-9*], is prepared in 85% yield from 2-fluoro-4-chloro-5-sulfamoylbenzoic acid and furfurylamine at 95°C for 2 h (155).

Cyclothiazide [*2259-96-3*] is another example of a fluorine-free pharmaceutical (diuretic, antihypertensive) based on *m*-chlorofluorobenzene [*625-98-9*] where fluorine activation is subsequently provided by two sulfonamide groups (156).

Another commercial application of nucleophilic reactions of nitro-free fluoroaromatics is the manufacture of polyetheretherketone (PEEK) high performance polymers from 4,4′-difluorobenzophenone [*345-92-6*], and hydroquinone [*121-31-9*] (131) (see POLYETHERS, AROMATIC).

Polyether sulfones (PES) prepared from 4,4′-difluorodiphenyl sulfone and bisphenol A (potassium salt, DMSO) react faster than the corresponding reaction with 4,4′-dichlorodiphenyl sulfone (157) (see POLYMERS CONTAINING SULFUR, POLYSULFONES). Poly(ether sulfone)s prepared from sodium 4-fluorobenzene-thiolate, α,ω-diiodoperfluoroalkanes, and bisphenol A exhibit good permeability and selectivity for O_2–N_2 gas separations (158,159). Fluorine-free membranes based on 2,6-difluorobenzonitrile and bisphenol A can also be used to separate gas mixtures (160,161).

Less activated substrates such as fluorohalobenzenes also undergo nucleophilic displacement and thereby permit entry to other useful compounds. Bromine is preferentially displaced in *p*-bromofluorobenzene [*460-00-4*] by hydroxyl ion under the following conditions: calcium hydroxide, water, cuprous oxide catalyst, 250°C, 3.46 MPa (500 psi), to give *p*-fluorophenol [*371-41-5*] in 79% yield (162,163). This product is a key precursor to sorbinil, an enzyme inhibitor (aldose reductase).

Fluoroaryl Organometallics. Fluorobenzene does not form a Grignard reagent with magnesium (164). 4-Bromofluorobenzene [*460-00-4*] can be selectively converted to 4-fluorophenylmagnesium bromide [*352-13-6*] for subsequent incorporation into the silicon-containing fungicide, flusilazole [*85509-19-9*] (165).

This represents the first large-scale application of a fluoroaryl organometallic. Other silicon-containing aryl fluorides such as pentafluorophenyldimethyl silanes, $C_6F_5Si(CH_3)_2X$ (X = Cl; NH_2; $N(C_2H_5)_2$), are offered commercially as Flophemsyl reagents for derivatization of sterols in chromatographic analysis (166).

Phenyllithium cannot be formed from fluorobenzene. Instead, the electronegativity of fluorine makes the ortho hydrogen sufficiently acidic to permit reaction with n-butyllithium in tetrahydrofuran at $-50°C$ to give 2-fluorophenyllithium [348-53-8]. An isomer, 4-fluorophenyllithium [1493-23-8], was reported to be explosive in the solid state (167).

The chelate, cobalt bis(3-fluorosalicylaldehyde)ethyleneimine [6220-65-5] (fluomine) had been under active evaluation for an oxygen-regenerative system in aircraft (168). Boron-containing fluoroaromatics are commercially offered as laboratory reagents: tetrakis(4-fluorophenyl)boron sodium·2H₂O, a titration agent for nonionic surfactants (169); and 4-fluorobenzeneboronic acid [1765-93-1], a glc reagent for derivatization of diols.

Biotransformation Reactions. Enzymatic oxygenation of aryl fluorides without ring opening provides a new production tool to fluoroaromatic fine chemicals. Microbial oxidation of fluorobenzene forms 3-fluoro-*cis*-1,2-dihydrocatechol, followed by chemical rearomatization to give 3-fluorocatechol [363-52-0] (170–172). This technique represents a significant improvement over the standard four-step chemical route based on 3-fluoroanisole [456-49-5]. Dehydration (acid pH) of the fluorodihydrocatechol also provides a new route to 2-fluorophenol [367-12-4]. Biological oxidation of fluoroaromatics has been demonstrated at the tonnage scale in up to 20-m³ reactors (171).

Manufacture. Fluorobenzene is produced by diazotization of aniline in anhydrous hydrogen fluoride at 0°C, followed by *in situ* decomposition of benzenediazonium fluoride at 20°C (21). According to German experience during World War II, the yield for 750-kg batches was 75–77%. Aryldiazonium fluoride–hydrogen fluoride solutions can also be decomposed by continuous feed through a heated reaction zone (173,174) or under super atmospheric pressure conditions (175).

The spent hydrogen fluoride layer, which contains water and sodium bifluo-ride, from this process is treated with sulfur trioxide or 65% oleum, and hydrogen fluoride is distilled for recycle to the next batch (176,177).

Nitrosyl chloride (178), nitrosyl chloride–hydrogen fluoride (NOF·3HF, NOF·6HF) (179), nitrous acid–hydrogen fluoride solutions (180,181), or nitrogen trioxide (prepared *in situ* from nitric oxide and oxygen) (27) can be used in place of sodium nitrite in the diazotization step.

Firms producing fluorobenzene and other ring-fluorinated aromatics by the diazotization of anilines in hydrogen fluoride include Rhône-Poulenc, ICI, Du Pont, Mallinckrodt, MitEni, and Riedel de Haën/Hoechst (182). With announce-ments of plant expansions and entry of new manufacturers, surplus capacity in basic fluoroaromatics exists. Prices (1991) are quoted at \$15–18/kg delivered in the United States for basic intermediates such as fluorobenzene and the fluoro-toluenes (182). Emphasis has now been placed on higher value downstream de-rivatives development programs.

Applications. *Crop Protection Chemicals.* The fluorinated analogue of DDT (GIX, DFDT), 1,1-bis(4-fluorophenyl)-2,2,2-trichloroethane [475-26-3], was produced from chloral and fluorobenzene as an insecticide in Germany during World War II. Other agricultural applications did not subsequently materialize since lower manufacturing costs of chlorinated aromatic crop-protection chemicals represented an advantage over ring-fluorinated analogues. However, chloroaro-matics pose ecological problems such as pesticide persistency, toxicity, etc. Be-cause fluoroaromatics offer agronomic advantages, eg, dosage, selectivity, and crop safety, significant commercialization of these compounds as crop protection chemicals (herbicides, fungicides, and insecticides) has occurred. Table 3 lists rep-resentative examples.

Drugs. Ring-fluorinated aromatics have found broad pharmaceutical appli-cations, eg, in tranquilizers, hypnotics, sedatives, antibacterial agents (qv), etc. Representative monofluorinated drugs are listed in Table 4. Arprinocid [5579-18-15] is a fluoroaromatic-based veterinary drug that has found wide acceptance as a coccidiostat for chicken feed.

Other Medical Applications. Positron emission tomography, a noninvasive technique for monitoring biochemical functions in humans, represents a signifi-cant advance in medical diagnosis. Synthetic methods have been developed for incorporation of the ^{18}F isotope ($t_{1/2}$, 109.27 ± 0.06 min) into numerous biologi-cally active radiopharmaceuticals (1). One example is the stereospecific and re-giospecific synthesis of 6-[^{18}F]fluoroDOPA (6-fluoro-3,4-dihydroxyphenylalanine) by fluorodemetallation of a trimethylsilyl precursor using $^{18}F_2$ for Parkinson's disease research.

$$\text{ArCH=N}\diagdown \underset{\underset{\underset{\text{CH}_3\text{O}}{\text{CH}_3\text{O}}}{\overset{\text{CH}_2}{|}}}{\text{CH}}\diagup \text{COOC}_2\text{H}_5 \quad \xrightarrow[\text{2. HBr}]{\text{1. }^{18}\text{F}_2} \quad \text{H}_2\text{N}\diagdown \underset{\underset{\underset{\text{HO}}{\text{HO}}}{\overset{\text{CH}_2}{|}}}{\text{CH}}\diagup \text{COOH}$$

Labeling aromatics with fluorine using ^{19}F-nmr as a probe for product iden-tification has been a useful analytical tool (183) which has been extended to med-

Table 3. Monofluoroaromatic Crop Protection Chemicals

Common name	CAS Registry Number	Structure	Application
flamprop-isopropyl	[52756-22-6]		post-emergent herbicide
fluoronitrofen	[13738-63-1]		post-emergent herbicide
fluoroimide	[41205-21-4]		fungicide
cyfluthrin	[68359-37-5]		insecticide
flutriafol	[76674-21-0]		fungicide

ical diagnosis. Magnetic resonance imaging (mri) is a noninvasive technique complementary to x-ray contrast agents, ultrasound devices, and computerized tomography, without the need of radioisotopes (184). For example, the mri technique has been applied to the interaction of fluoroquine, a fluorine analogue of the antimalarial drug, chloroquine, with DNA and t-RNA (185).

Liquid Crystals.   Based on worldwide patent activity, numerous compounds containing fluoroaromatic moieties have been synthesized for incorporation into liquid crystals. For example, fluoroaromatics are incorporated in ZLI-4792 and

Table 4. Monofluorinated Aromatic Drugs

Common name	CAS Registry Number	Structure	Application
haloperidol	[52-86-8]	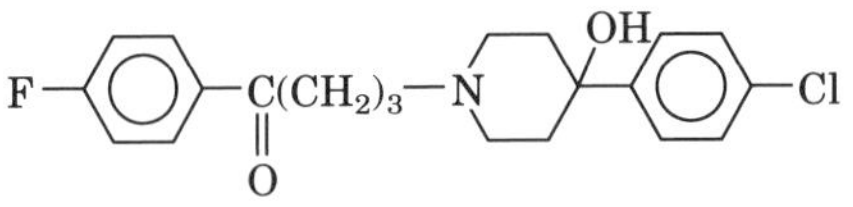	tranquilizer
flurazepam hydrochloride	[1172-39-5]	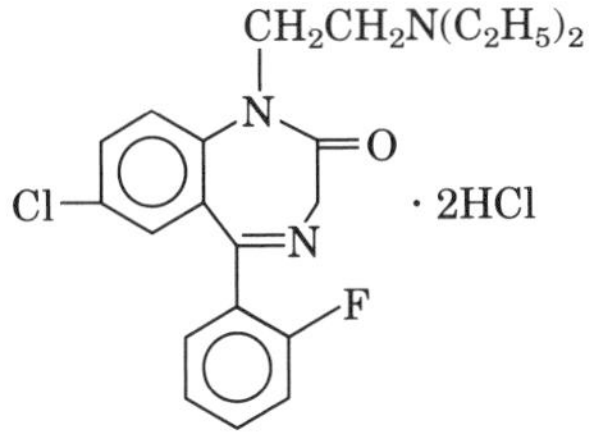	hypnotic, sedative
floxacillin	[5250-39-5]	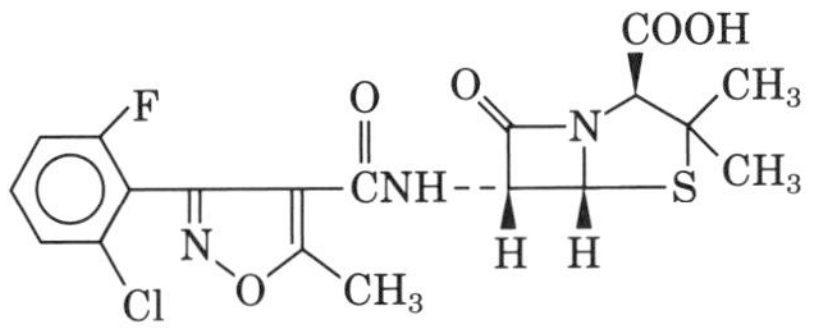	antibacterial
flazalone	[21221-18-1]	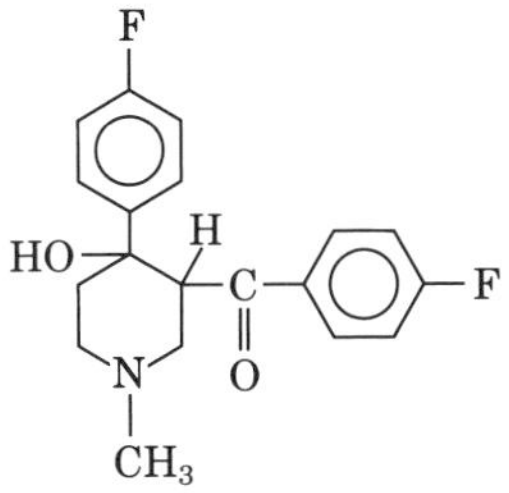	anti-inflammatory
fluspirilene	[1841-19-6]	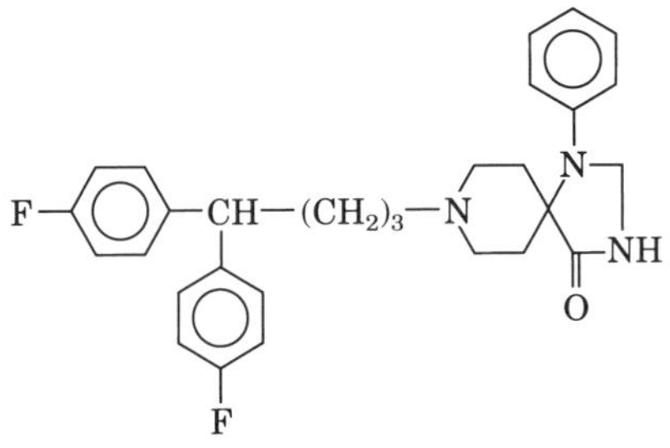	tranquilizer
lidoflazine	[3416-26-0]	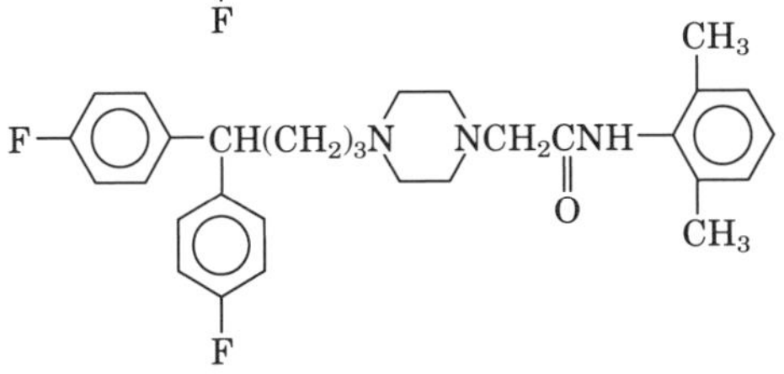	vasodilator (coronary)
sorbinil	[68367-52-2]	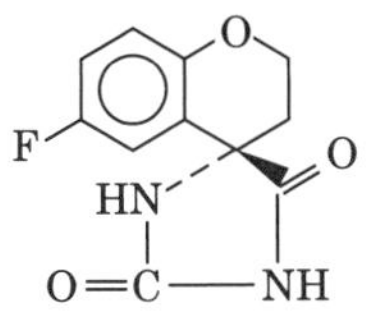	enzyme inhibitor (aldose reductase)

ZLI-4801-000/-100 for active matrix displays (AMD) containing super fluorinated materials (SFM) (186,187). Representative structures are as follows.

Photoconductive Imaging. Considerable attention has been placed on the xerographic properties of fluorosquaraines based on N,N-dimethyl-3-fluoroaniline and other 3-fluoroaniline derivatives for imaging applications (188–191). A typical structure of a fluorosquaraine is as follows:

Dyes. In contrast to benzotrifluorides and fluoropyrimidines, limited commercialization has developed for dyes containing a fluoroaromatic group. Fluorophenylhydrazines have been converted to (fluorophenyl)pyrazolones, which are disperse dyes for cellulose acetate and nylon (192).

DIFLUOROBENZENES

Interest in the commercialization of difluoroaromatics in crop protection chemicals and drugs (Table 5) continues to be strong. Numerous liquid crystals containing the 1,2-difluorobenzene moiety have been synthesized. Table 6 lists physical properties of commercially significant intermediates such as o-, m-, and p-difluorobenzene, 2,4-difluoroaniline and 2,6-difluorobenzonitrile. The LD_{50} values for the three isomeric difluorobenzenes are identical: 55 g/m^3 for 2 h (inhalation, mouse) (127).

1,2-Difluorobenzene. Tetrazotization-fluorination of o-phenylenediamine [95-54-5] in hydrogen fluoride or by the Balz-Schiemann reaction is not a practical route to 1,2-difluorobenzene but this product can be prepared from 2-fluoroaniline [348-54-9] by the Balz-Schiemann reaction (193); heating the diazonium fluoroborate in organic solvents increases the yield to 78% (194). Electrophilic substitution reactions are site-specific: nitration gives 3,4-difluoronitrobenzene [369-34-6], and bromination forms 3,4-difluorobromobenzene [348-61-8], a precursor to dicyclohexylethylene liquid crystals (195). Vicinal metallation (n-butyllithium, $-78°C$) of 1,2-difluorobenzene is also employed to prepare *trans*-4-alkyl cyclohexyl-substituted 2,3-difluorobiphenyls for liquid crystal applications (196).

1,3-Difluorobenzene. This isomer has been prepared in 78% yield by tetrazotization-fluorination of m-phenylenediamine [108-45-2] in pyridine–hydrogen fluoride at 100°C (23,197). Balz-Schiemann yields for the corresponding reaction vary from 31 to 49% (198,199). Diazotization of m-fluoroaniline [372-19-0] in the presence of ammonium bifluoride, tertiary amines, or dimethyl sulfoxide gave 46–

Table 5. Difluoroaromatic Applications

Common name	CAS Registry Number	Structure
		Crop protection chemicals[a]
diflubenzuron	[35367-38-5]	
diflufenican[b]	[83164-33-4]	
flufenoxuron	[101463-69-8]	
teflubenzuron	[83121-18-0]	
		Drugs
diflunisal[c]	[22494-42-4]	
fluconazole[d]	[86386-73-4]	

[a]Insecticide unless otherwise noted.
[b]Herbicide.
[c]Analgesic; antiinflammatory.
[d]Antifungal.

Table 6. Properties of Fluorinated Aromatic Compounds[a]

Component	CAS Registry Number	Mol wt	Mp, °C	Bp, °C[b]	Refractive index, n_D^t	Specific gravity, d_4^t	Surface tension, 20°C mN/m (=dyn/cm)	Flash point,[c] °C
$C_6H_4F_2$								
Difluorobenzenes								
1,2-difluorobenzene	[367-11-3]	114.09	−34	91–92	1.4452[20]	1.1496[25]		7.2[d]
1,3-difluorobenzene	[372-18-9]	114.09	−59.3	82–83	1.4410[20]	1.1572[20]	25.93	−11.1[d]
1,4-difluorobenzene	[540-36-3]	114.09	−13	88–89	1.4421[20]	1.1716[20]	27.05	−11.7[d]
$C_6H_3F_3$								
Trifluorobenzenes								
1,2,3-trifluorobenzene	[1489-53-8]	132.08		94–95	1.4230[20]	1.280[20]		−3
1,2,4-trifluorobenzene	[367-23-7]	132.08		88	1.4230[20]	1.264[20]	26.2	−5[d]
1,3,5-trifluorobenzene	[372-38-3]	132.08	−5.5	75.5	1.4140[20]	1.277[20]	27.16	−7
$C_6H_2F_4$								
Tetrafluorobenzenes								
1,2,3,4-tetrafluorobenzene	[551-62-2]	150.08	−42	95	1.4069[20]	1.422[25]		20
1,2,3,5-tetrafluorobenzene	[2367-82-0]	150.08	−48	83	1.4011[25]	1.393[20]	23.99	4
1,2,4,5-tetrafluorobenzene	[327-54-8]	150.08	4	90	1.4045[20]	1.424[25]	24.9	16
C_6HF_5								
Pentafluorobenzene								
	[363-72-4]	168.07	−48	85	1.3881[25]	1.531[20]		13

	CAS	MW	mp	bp	n_D	density	fp
Fluorotoluenes							
2-fluorotoluene[e]	[95-52-3]	110.13		113–114	1.4704^{25}	1.001^{20}	12
3-fluorotoluene	[352-70-5]	110.13		115	1.4691^{20}	0.991^{20}	9
4-fluorotoluene	[352-32-9]	110.13		116	1.4690^{20}	1.000^{20}	17
2-chloro-6-fluorotoluene	[443-83-4]	144.58		155	1.5026^{20}	1.129^{20}	48
Fluoroanilines							
$C_6H_4F(NH_2)$							
2-fluoroaniline[f]	[348-54-9]	111.12	−29	175	1.5406^{25}	1.152^{25}	60
3-fluoroaniline[g,h]	[372-19-0]	111.12		186	1.5445^{25}	1.152^{25}	77
4-fluoroaniline[e,f,i]	[371-40-4]	111.12	−1.9	187	1.5375^{25}	1.158^{25}	73
$C_6H_3F_2(NH_2)$							
2,4-difluoroaniline	[367-25-9]	129.11	−7.5	169.5	1.5043^{20}	1.268^{20}	62
$C_6H_3(Cl)(F)(NH_2)$							
3-chloro-4-fluoroaniline	[367-21-5]	145.57	44–47	227–228		1.42^{20} (solid) 1.3^{60} (liquid)	110
Fluorobenzonitriles							
2,6-difluorobenzonitrile	[1897-52-5]	139.11	30–32	99^j	1.4875^{25}	1.236^{40}	80

[a] Colorless unless otherwise noted.
[b] At 101.1 kPa = 1 atm unless otherwise noted.
[c] Closed cup (ASTM Procedure D3278) unless otherwise noted.
[d] Open cup.
[e] LD_{50} = 100 mg/kg (oral, wild bird) (127).
[f] Pale yellow.
[g] Amber.
[h] LD_{50} = 56 mg/kg (oral, wild bird) (127).
[i] LD_{50} = 50 mg/kg (oral, rat) (127).
[j] At 2.67 kPa = 20 mm Hg.

73% yields of 1,3-difluorobenzene (25,26). The latter can also be made by reductive-dediazoniation of 2,4-difluoroaniline [367-25-9] in 77% yield from sodium nitrite, hydrochloric acid, and hypophosphorus acid (200). A 95% yield was realized by treatment of 2,4-difluorobenzenediazonium fluoroborate with copper powder in the presence of 18-crown-6 ether in dichloromethane (201).

Nitration of 1,3-difluorobenzene at 0°C forms 2,4-difluoronitrobenzene [446-35-5] in 92% yield. The latter can also be prepared from 2,4-dichloronitrobenzene and potassium fluoride in polar solvents (46,202); phase-transfer catalysts, eg, quaternary ammonium salts, serve to both lower reaction temperature and enhance fluorination rates (203). Reduction gives 2,4-difluoroaniline, a precursor to the analgesic/antiinflammatory diflunisal, and the herbicide diflufenican.

1,4-Difluorobenzene.   This compound has been prepared in 65% yield by tetrazotization-fluorination of *p*-phenylenediamine [106-50-3] in pyridine–hydrogen fluoride at 120°C (23,197); 27–40% yields are obtained by the Balz-Schiemann reaction with *p*-phenylenediamine or *p*-fluoroaniline [371-40-4] (198).

TRIFLUOROBENZENES

Table 6 lists physical properties of representative trifluorobenzenes.

1,2,3-Trifluorobenzene.   This compound is formed in low yield (13–24%) from 1,2,3-trichlorobenzene or 2,3-difluorochlorobenzene and KF/CsF in dimethyl sulfone (204). Likewise, low yields are realized when the Balz-Schiemann reaction is applied to 2,3-difluoroaniline or 2,6-difluoroaniline (205). Pyrolysis (520°C, iron gauze) of 1*H*, 2*H*, 3*H*-pentafluorocyclohexa-1,3-diene forms 1,2,3-trifluorobenzene (206). Derivatives such as 2,3,4-trifluoronitrobenzene [393-79-3] and 2,3,4-trifluoroaniline [3862-73-5] have been used to prepare fluoroquinolone antibacterials such as ofloxacin (207) and lomefloxacin (208), respectively.

1,2,4-Trifluorobenzene.   This isomer can be prepared in good yield from 2,4-difluoroaniline by the standard Balz-Schiemann route (209) or modifications using nitrite esters–boron trifluoride (210). Its ionization potential is 9.37 V. Electrophilic substitution reactions of 1,2,4-trifluorobenzene provide useful routes to 2,4,5-trifluorobenzoic acid [446-17-3], a key precursor to fluoroquinolone antibacterials: bromination forms 1-bromo-2,4,5-trifluorobenzene [327-52-6] (211), followed by exchange cyanation–hydrolysis (212); acetylation gives 2,4,5-trifluoroacetophenone [129322-83-4], followed by oxidation with commercial bleach (213). New routes to 2,4,5-trifluorobenzoic acid are also based on exchange-fluorination of chloroaromatic feedstocks such as 3,4,6-trichlorophthalic acid (62) and tetrachloroisophthalonitrile (214). Other 1,2,4-trifluorobenzene derivatives such as 3-chloro-2,4,5-trifluorobenzoic acid have also been converted to fluoroquinolone antibacterials (215).

1,3,5-Trifluorobenzene.   This isomer, *s*-trifluorobenzene, has been prepared in 63% yield by the Balz-Schiemann reaction with 3,5-difluoroaniline [372-39-4] (216). By modification of exchange fluorination conditions, tetrachloroisophthalonitrile [1897-45-6] was converted to 1,3,5-trifluorobenzene by a four-step process (217).

TETRAFLUOROBENZENES

Interest in tetrafluoroaromatics includes crop protection and as intermediates to fluoroquinolone antibacterials. Physical properties of tetrafluorobenzenes are listed in Table 6. A useful compilation of recipes for 35 tetrafluorinated aromatics has been published (199).

1,2,3,4-Tetrafluorobenzene. This compound has been prepared by fluorination of benzene with cobalt trifluoride and subsequent combination of the dehydrofluorination and defluorination steps. Its ionization potential is 9.01 V. Nitration gives 2,3,4,5-tetrafluoronitrobenzene [5580-79-0] in 75% yield, an intermediate to fluoroquinolone antibacterials (218).

Halex technology has also been employed to prepare 1,2,3,4-tetrafluorobenzene derivatives, eg, tetrachlorophthalic anhydride [117-08-8] was converted to 2,3,4,5-tetrafluorobenzoic acid [1201-31-6] for use in fluoroquinolone antibacterials (219,220).

1,2,3,5-Tetrafluorobenzene. This isomer has been prepared from 2,3,5-trifluoroaniline [363-80-4] in 43% yield by the Balz-Schiemann reaction.

1,2,4,5-Tetrafluorobenzene. This compound has been prepared from 2,4,5-trifluoroaniline [57491-45-9] by the Balz-Schiemann reaction in 38–46% yield or from pentafluorophenylhydrazine [828-39-9] with aqueous sodium hydroxide in 90–95% yield (221). Its ionization potential is 9.39 V.

Derivatives of 1,2,4,5-tetrafluorobenzene such as 2,3,5,6-tetrafluorobenzoic acid have been converted into fluoroquinolone antibacterials (222–224). The synthetic pyrethroid, tefluthrin [795-38-2], is prepared from 2,3,5,6-tetrafluoro-4-methylbenzyl alcohol.

Isomeric dichlorotetrafluorobenzenes have been studied for Rankine-cycle external combustion engines (225).

PENTAFLUOROBENZENE AND PENTAFLUOROPHENYL COMPOUNDS

Pentafluorobenzene. Pentafluorobenzene has been prepared by several routes: multistage saturation–rearomatization process based on fluorination of benzene with cobalt trifluoride; reductive dechlorination of chloropentafluorobenzene with 10% palladium-on-carbon in 82% yield (226,227); and oxidation of pentafluorophenylhydrazine in aqueous copper sulfate at 80°C in 77% yield (228). Its ionization potential is 9.37 V. One measure of toxicity is $LD_{50} = 710$ mg/kg (oral, mouse) (127).

Nucleophiles react with pentafluorobenzene to give para-substituted (relative to the hydrogen atom) tetrafluorophenyl products, p-XC_6F_4H (X = H, NH_2, $NHNH_2$, SH, OCH_3, SC_6H_5, OH). Nitration of pentafluorobenzene with concentrated nitric acid and boron trifluoride in sulfolane gave pentafluoronitrobenzene [880-78-4] in 82% yield (229).

Pentafluoroaniline. Pentafluoroaniline [771-60-8] has been prepared from amination of hexafluorobenzene with sodium amide in liquid ammonia or with ammonium hydroxide in ethanol (or water) at 167–180°C for 12–18 h. It is weakly basic ($pK_a = 0.28$) and dissolves only in concentrated acids. Liquid crystals have been prepared from Schiff bases derived from pentafluoroaniline (230).

Pentafluorophenol. This compound has been prepared from the reaction of hexafluorobenzene with potassium hydroxide in *t*-butyl alcohol. Pentafluorophenyl esters prepared from pentafluorophenol [*771-61-9*] illustrate the key features of a rapid stepwise peptide synthesis technique (231). Commercial high performance elastomers based on copolymerization of tetrafluoroethylene, perfluoro(methyl vinyl ether), and a third monomer incorporating a pentafluorophenoxy group as a cure site, give vulcanizates with good chemical and fluid resistance and high temperature oxidative resistance (232,233).

Pentafluorotoluene. Pentafluorotoluene [*771-56-2*] has been prepared from the reaction of methyllithium with hexafluorobenzene or from pentafluorophenylmagnesium bromide with dimethyl sulfate. Derivatives such as 2,3,4,5,6-pentafluorobenzyl bromide [*1765-40-8*] are used to derivatize organic acids as esters for determination by electron-capture gas chromatography (234). The synthetic pyrethroid, fenfluthrin [*75867-00-4*], is an insecticide containing a pentafluorobenzyl group.

Bromopentafluorobenzene. Aluminum bromide-catalyzed bromination of pentafluorobenzene in 20% oleum gives bromopentafluorobenzene [*344-04-7*]. It is readily converted to pentafluorophenylmagnesium bromide [*879-05-0*]; the latter undergoes conventional Grignard reactions (qv). Pentafluorophenyllithium [*1076-44-4*] can be synthesized from bromopentafluorobenzene and *n*-butyllithium or lithium amalgam in ether at 0°C. The preferred route is metallation of pentafluorobenzene with *n*-butyllithium at −65°C (235). A serious explosion has been reported during hydrolysis (D_2O) of pentafluorophenyllithium (236).

Pentafluorophenylmagnesium bromide or lithium can be converted to other pentafluorophenyl organometallics by reaction with the corresponding metal chloride (237). Bis(pentafluorophenyl)phenylphosphine [*5074-71-5*] (Ultramark 443), $(C_6F_5)_2C_6H_5P$, is offered commercially as a marker for mass spectral standardization (238).

Pentafluorobenzoic Acid. Standard routes to pentafluorobenzoic acid [*602-94-8*] include chloropentafluorobenzene (*n*-butyllithium or magnesium, carbonation, hydrolysis); pentafluorobenzene (phosgene, hydrolysis); octafluorotoluene (hydrolysis). Of potential economic significance is a new route based on benzonitrile: chlorination to pentachlorobenzonitrile, exchange fluorination, and hydrolysis (239). Pentafluorobenzoyl chloride [*2251-50-8*] has been used to derivatize anticonvulsants such as ethosuximide, carbamazepine, and primidone in electron-capture gas chromatography (240,241).

Pentafluorobenzaldehyde. Pentafluorobenzaldehyde [*653-37-2*] can be prepared by reaction of *N*-methylformanilide with pentafluorophenylmagnesium bromide or pentafluorophenyllithium. One process is based on the catalytic hydrogenation of pentafluorobenzonitrile (239). Pentafluorobenzaldehyde is used as a reagent for gas chromatographic assay of biological amines such as catecholamines by conversion to the pentafluorobenzylimine–trimethylsilyl derivatives (242). Catalytic hydrogenation gives pentafluorobenzyl alcohol [*440-60-8*]. Derivatives of the latter are employed in gas chromatography (electron capture): *O*-(2,3,4,5,6-pentafluorobenzyl)hydroxylamine hydrochloride [*57981-02-9*] (Florox reagent) $C_6F_5CH_2ONH_2 \cdot HCl$, for assay of ketosteroids (243,244); 2,3,4,5,6-pentafluorobenzyl chloroformate [*53526-74-2*], for assay of physiologically active tertiary amines (245).

Numerous examples for the incorporation of the pentafluorophenyl group in chromatographic derivatization of biologically active compounds have been compiled in a monograph (246). A review on the effects of the pentafluorophenyl group on the reactivity of organic compounds has been published (247).

HEXAFLUOROBENZENE

The development of commercial routes to hexafluorobenzene [392-56-3] included an intensive study of its derivatives. Particularly noteworthy was the development of high temperature lubricants, heat-transfer fluids, and radiation-resistant polymers (248).

Hexafluorobenzene. Hexafluorobenzene [392-56-3] C_6F_6, is a colorless liquid with a sweet odor. Hexafluorobenzene (perfluorobenzene) has a good thermal stability; slight decomposition occurs at 500°C in Nimonic 75 (alloy containing 85% nickel and 20% chromium) after three weeks. Toxicity: inhalation (mouse), LD_{50}-95 g/m^3 (2 h) (127). Physical properties of hexafluorobenzene are given in Table 7.

Manufacture. One commercial process features a three-stage saturation–rearomatization technique using benzene and fluorine gas as raw materials (73). Principal problems with this method are the complex nature of the process, its dependence on fluorine gas which is costly to produce, and the poor overall utili-

Table 7. Physical Properties of Hexafluorobenzene

Property	Value
mol wt	186.06
melting point, °C	5.10
boiling point, °C	80.261
density, 25°C, g/mL	1.60682
refractive index, n_D^{25}	1.3761
latent heat of fusion, kJ/mol[a]	11.59
latent heat of vaporization	
at 25°C, kJ/mol[a]	35.69 ± 0.084
at bp, kJ/mol[a]	32.69
specific heat, 23°C, kJ/mol·°C)[a]	0.221
critical temperature, °C	243.57 ± 0.03
critical pressure, MPa[b]	3.304 ± 0.005
coefficient of cubical expansion at 25°C	0.001412
heat of combustion, kJ/mol[a]	−2444.0 ± 1.2
heat of formation at 25°C, kJ/mol[a]	
liquid	−958.30
gas	−922.15
vapor pressure in °C and kPa[c]	Antoine equation[d]

[a]To convert J to cal, divide by 4.184.
[b]To convert MPa to atm, divide by 0.101.
[c]To convert kPa to mm Hg, multiply by 7.5; log kPa = log mm Hg −0.875.
[d]$\text{Log}_{10}P = 6.1422 - \dfrac{1219.410}{(t + 214.525)}$.

zation of fluorine, because nearly one-half of the input fluorine is removed during the process.

An alternative hexafluorobenzene process features exchange fluorination (KF) of hexachlorobenzene in the presence of polar solvents (226,249) or under solvent-free conditions (450–540°C, autoclave) (250). Intermediates such as chloropentafluorobenzene can be further fluorinated to hexafluorobenzene (42–51% yield) by cesium fluoride in sulfolane (226,249).

Pyrolytic routes to hexafluorobenzene have also attracted attention but have not been commercialized. Pyrolysis of tribromofluoromethane [353-54-8], CBr_3F, at 630–640°C in a platinum tube gives hexafluorobenzene in 55% yield (251–253). The principal disadvantage of this process is the low weight yield of product; 90% of the costly CBr_3F that is charged is lost as bromine. Of economic potential is the related copyrolysis of dichlorofluoromethane [75-43-4] and chlorofluoromethane [593-70-4] (254,255).

Reactions.   Hexafluorobenzene is susceptible to attack by nucleophilic agents to give pentafluorophenyl compounds of the general formula C_6F_5X, where X is OCH_3, NH_2, OH, SH, $NHNH_2$, $NHCH_3$, $N(CH_3)_2$, H, C_6H_5, CH_3, $CH_3CH=CH$, n-C_4H_9, C_6H_5S, etc (256).

FLUOROBIPHENYLS

Fluorobiphenyls are incorporated into the analgesic and antiinflammatory drugs diflunisal [22494-42-4] and flurbiprofen [5104-49-4]. The first is a difluoro compound and the other monofluoro.

Fluorinated biphenyls have been incorporated into numerous liquid crystal structures as attested by patents and publications from the following organizations: E. Merck GmbH (257,258); Sharp (258); Hoffmann-LaRoche (259); Kanto Chemical (260); U.K. Defence Secretariat (261); Dainippon (262); Chisso (263); Sanyo Chemical (264); and the University of Hull (265). Seiko Epson has also patented fluorinated terphenyls for liquid crystal applications (266).

Fluorinated biphenyls can be synthesized by diazotization–fluorination, Gomberg-Bachmann arylation, or Ullmann coupling reactions. Mono- and difluorophenyls can be prepared by the Balz-Schiemann reaction (or modification in HF), eg, 4,4′-difluorobiphenyl was formed in 80% yield from 4,4′-diaminobiphenyl by the Balz-Schiemann reaction (267). 2,4-Difluorobiphenyl [2285-28-1], a key precursor to diflunisal, is formed by successive diazotization of 2,4-difluoroaniline and coupling with benzene (268). Similar diazotization-coupling of 4-bromo-2-fluoroaniline [367-24-8] with benzene gives 4-bromo-2-fluorobiphenyl [41604-19-7], a key intermediate to flurbiprofen (269).

Decafluorobiphenyl [434-90-2], $C_6F_5C_6F_5$ (mol wt, 334.1; mp, 68°C; bp, 206°C), can be prepared by Ullmann coupling of bromo- [344-04-7], chloro- [344-

07-0], or iodopentafluorobenzene [*827-15-6*] with copper. This product shows good thermal stability; decafluorobiphenyl was recovered unchanged after 1 h below 575°C (270). Decafluorobiphenyl-based derivatives exhibit greater oxidative stability than similar hydrocarbon compounds (271). Thermally stable poly-(fluorinated aryl ether) oligomers prepared from decafluorobiphenyl and bisphenols show low dielectric constant and moisture absorption which are attractive for electronic applications (272).

Fluoronaphthalenes and Other Fused-Ring Fluoroaromatics

Few applications for fluoronaphthalenes and related polycyclic structures have materialized. The fused-ring bicyclic, sulindac [*38194-50-2*], a monofluorinated indene-3-acetic acid, is used as an antiinflammatory agent.

1-Fluoronaphthalene [*321-38-0*] is prepared from 1-naphthylamine by the Balz-Schiemann reaction in 52% yield or by diazotization in anhydrous hydrogen fluoride in 82% yield. Electrophilic substitution occurs at the 4-position, eg, nitration with fuming nitric acid in acetic acid gave 88% yield of 1-fluoro-4-nitro-naphthalene [*341-92-4*].

2-Fluoronaphthalene [*323-09-1*] is prepared in 54–67% yield from 2-naphthylamine by the Balz-Schiemann reaction or in 51% yield by pyrolysis of indene and chlorofluoromethane at 600°C (77).

1,4-Difluoronaphthalene [*315-52-6*] is prepared from 4-fluoro-1-naphthyl-amine by the Balz-Schiemann reaction. 1,4-Difluoronaphthalene is used in chemical carcinogenesis studies as a synthon for highly condensed difluoro–polycyclic aromatic hydrocarbons (273).

Octafluoronaphthalene [*313-72-4*] is prepared in 53% yield by defluorination of perfluorodecahydronaphthalene [*306-94-5*] over iron or nickel at 500°C. Exchange fluorination of octachloronaphthalene with KF in sulfolane (235°C) gave 60% yield of octafluoronaphthalene. This product exhibits good stability to ionizing radiation (274).

Fused-ring polycyclic fluoroaromatics can be made from the corresponding amino fused-ring polycyclic or from preformed fluoroaromatics, eg, 4-fluorophenyl-acetonitrile [*459-22-3*] (275). Direct fluorination techniques have been successfully applied to polycyclic ring systems such as naphthalene, anthracene, benzanthracenes, phenanthrene, pyrene, fluorene, and quinolines with a variety of fluorinating agents: xenon fluorides (10), acetyl hypofluorite (276), cesium fluoroxysulfate (277), and electrochemical fluorination (278,279).

Side-Chain Fluorinated Aromatics

Trifluoromethyl aromatics are used widely in the production of drugs, crop-protection chemicals, germicides, dyes, etc.

General Properties.   The trifluoromethyl group is stable under different reaction conditions, eg, the multistep classical transformation of benzotrifluoride to trifluoroacetic acid features successive nitration, reduction, and oxidation.

Thermal Stability. Benzotrifluoride is stable at 350°C in the presence of iron or copper. Working fluids for external combustion engines (Rankine cycle) must exhibit thermal stability with engine materials at high temperatures. Some of the promising working fluids include 1,3-bis(trifluoromethyl)benzene [402-31-3] (280) or a mixture of perfluorotoluene [434-64-0], $CF_3C_6F_5$, and hexafluorobenzene [392-56-3] (281). The stability of the isomeric CF_3-substituted anilines has been established by differential thermal analysis (dta) (282): m-$CF_3C_6H_4NH_2$ [98-15-7], 223°C > o-$CF_3C_6H_4NH_2$ [88-17-5], 187°C > p-$CF_3C_6H_4NH_2$ [455-14-1], 155°C.

Hydrolytic Stability. The trifluoromethyl group is sensitive to hydrolysis in acidic media. Benzotrifluoride is hydrolyzed to benzoic acid by heating with hydrobromic, hydrofluoric, or > 80% sulfuric acid. Reaction conditions and structural features for this reaction have been summarized (283). Benzotrifluorides are generally stable to base. Benzotrifluoride was recovered unchanged after heating (120–130°C) with sodium hydroxide. Although m-hydroxybenzotrifluoride [98-17-9] is stable to refluxing 50% sodium hydroxide, cold dilute alkali polymerizes the para isomer [402-45-9] (284). Similar polymers are formed from 2,3,5,6-tetra-fluoro-4-trifluoromethylphenol [2787-79-3] (285). Photohydrolysis of hydroxy- and aminobenzotrifluorides in dilute acid and alkali, respectively, gives the corresponding hydroxy- and aminobenzoic acid in high yield (286). Benzotrifluoride is inert under these photohydrolytic conditions.

Oxidative Stability. Benzotrifluoride resists ring oxidation. In contrast, chromic acid readily oxidizes 3-aminobenzotrifluoride to trifluoroacetic acid in 95% yield (287).

Stability to Reducing Agents. The trifluoromethyl group is inert to numerous reducing agents. Catalytic hydrogenation (platinum black) of benzotrifluoride gives trifluoromethylcyclohexane [401-75-2] (288). Benzotrifluoride was not reduced by lithium aluminum hydride (289). However, o-trifluoromethylbenzoic acid [433-97-6] and m-trifluoromethylbenzoic acid [454-92-2] are catalytically reduced (Raney nickel or cobalt alloys) to o- and m-toluic acid, respectively (290).

Instability of Trifluoromethylphenyl Organometallics. Care must be exercised in handling trifluoromethylphenyl organometallics. o-Trifluoromethylphenyl-lithium [49571-35-5] has exploded during reflux in diethyl ether under nitrogen (291). Both m- [368-49-0] and p-trifluoromethylphenyllithium [2786-01-8] are explosive in the solid state (292). Explosions have been reported in the preparation of o- [395-47-1], m- [402-26-6], and p-trifluoromethylphenylmagnesium bromide [402-51-7] (292,293). A violent explosion accompanied by loss of life and destruction of a chemical plant during preparation of p-trifluoromethylphenylmagnesium chloride [2923-41-3] has been reported (294). A compilation of reactive chemical hazards of trifluoromethylphenyl organometallics was published in 1990 (68).

Reactions. Benzotrifluoride undergoes electrophilic substitution reactions, eg, halogenation, nitration, typical of an aromatic containing a strong electron-withdrawing group. The trifluoromethyl group (sometimes referred to as a pseudohalogen) is meta directing.

Halogenation. Liquid-phase monochlorination of benzotrifluoride gives pronounced meta orientation (295); in contrast, vapor-phase halogenation favors para substitution (296). Sealed tube, photochemical, or dark chlorination (radical initiator) forms hexachloro(trifluoromethyl)cyclohexane; thermal dehydrochlorination (550°C) gives 2,4,6-trichlorobenzotrifluoride [567-59-9] (297). Liquid-phase bromination (Br_2) provides 3-bromobenzotrifluoride [401-78-5] in 60% yield. Catalyst performance decreases in the order $FeCl_3 > Fe > SbCl_5 > I_2$ (298). Silica gel (299) and calcium chloride (300) serve as hydrogen fluoride scavengers to suppress corrosion of glass reactors during halogenation. Bromination of benzotrifluoride can also be accomplished with bromine chloride in the presence of a halogen carrier, eg, antimony pentachloride; this technique permits complete utilization of bromine (301).

Hydrogen peroxide–hydrochloric acid reagent converts 2-aminobenzotrifluoride to 2-amino-5-chlorobenzotrifluoride [445-03-4], a dye intermediate (CI Azoic Diazo Component 17), without contamination by the 3-chloro isomer such as is observed with molecular chlorine (Cl_2) (302).

Nitration. Nitration of benzotrifluorides is an important industrial reaction. Mononitration of benzotrifluoride gives pronounced meta-orientation: 91% meta [98-46-4]; 6% ortho [384-22-5]; and 3% para [402-54-0] (296). Further nitration to 3,5-dinitrobenzotrifluoride [401-99-0] can be effected under forcing conditions at 100°C.

Alkylation. Benzotrifluoride can also be alkylated, eg, chloromethyl methyl ether–chlorosulfonic acid forms 3-(trifluoromethyl)benzyl chloride [705-29-3] (303,304), which can also be made from *m*-xylene by a chlorination–fluorination sequence (305). Exchange cyanation of this product in the presence of phase-transfer catalysts gives 3-(trifluoromethylphenyl)acetonitrile [2338-76-3] (304,305), a key intermediate to the herbicides flurtamone [96525-23-4] (306) and fluridone [59756-60-4].

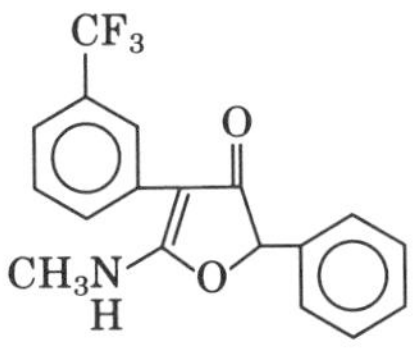

flurtamone 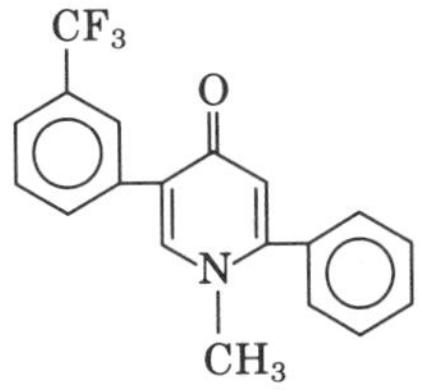   fluridone

Nucleophilic Displacement Reactions. The strong electron-withdrawing effect of a trifluoromethyl group activates ortho and para halogen toward nucleophilic attack. Such chlorine lability is utilized in the manufacture of crop control chemicals containing trifluoromethyl and nitro groups.

Reactions Involving the Trifluoromethyl Group. Aluminum chloride effects chlorinolysis of benzotrifluoride to give benzotrichloride (307). High yields of volatile acid fluorides are formed from benzotrifluoride and perfluorocarboxylic acids (308).

$$HCF_2CF_2CO_2H + C_6H_5CF_3 \xrightarrow{\text{Lewis acid}} HCF_2CF_2\overset{\|}{\underset{O}{C}}F + C_6H_5\overset{\|}{\underset{O}{C}}F + HF$$

4,4′-Difluorobenzophenone, the key precursor to PEEK high performance resins, can be prepared by sequential (*1*) Friedel-Crafts coupling of 3,4-dichlorobenzotrifluoride, (2) exchange fluorination using KF/CH$_3$SO$_2$CH$_3$, and (3) reductive dechlorination with HCOONa/Pd-C (309).

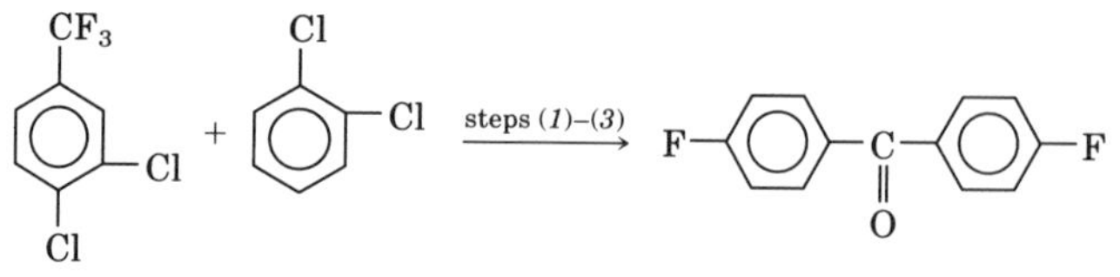

Electroreductive coupling of benzotrifluorides with sacrificial aluminum or magnesium anodes in the presence of acetone, carbon dioxide, or N,N-dimethylformamide provides a novel route to ArCF$_2$-derivatives (310).

$$C_6H_5CF_3 \ + \ CH_3CCH_3 \xrightarrow{\text{Al anode}} C_6H_5CF_2C(CH_3)_2OH$$
$$\underset{O}{\|}$$
$$(80\%)$$

Biotransformation.   Enzymatic oxidation of benzotrifluoride forms 3-trifluoromethyl-*cis*-1,2-dihydrocatechol; dehydration (acid pH) provides a novel route to 3-hydroxybenzotrifluoride [98-17-9] (171).

Benzotrifluoride.   Benzotrifluoride   [98-08-8]   (α,α,α-trifluorotoluene), C$_6$H$_5$CF$_3$ (mol wt, 146.11), is a colorless liquid (Table 8). Toxicity: oral (rat), LD$_{50}$, 1500 mg/kg; oral (mouse), LD$_{LO}$, 10,000 mg/kg; subcutaneous (frog), LD$_{LO}$, 870 mg/kg; intraperitoneal (mouse), LD$_{LO}$, 100 mg/kg (127).

Benzotrifluoride was first synthesized in 1898 via the reaction of benzotrichloride and antimony trifluoride (313). Benzotrifluoride can be produced by the high pressure reaction of benzotrichloride with anhydrous hydrogen fluoride (AHF). Typical conditions include a 4:1 AHF–benzotrichloride mole ratio at 80–110°C and 1.52–1.55 MPa (220–225 psi) for 2–3 h to give 70–75% yields of benzotrifluoride (314,315). The pressure fluorination can be performed continuously in a series of autoclaves (316) or through a nickel reaction tube at 90–130°C at 3–5 MPa (435–725 psi) (317). Batch liquid-phase catalyzed processes at atmospheric pressure (318,319) and continuous processes have been developed (320). High temperature vapor-phase fluorination processes have also been described (321).

Benzotrifluoride Derivatives.   Laboratory recipes for 45 benzotrifluorides have been published (322). Physical properties and toxicity of commercially significant benzotrifluoride derivatives are listed in Table 9; the amino compounds are colorless to yellow and other derivatives are colorless.

2-Chlorobenzotrifluoride.   This compound is produced from 2-chlorobenzotrichloride and anhydrous hydrogen fluoride under atmospheric or high pressure conditions. Nitration forms 2-chloro-5-nitrobenzotrifluoride [777-37-7], a dye and germicide precursor.

4-Chlorobenzotrifluoride.   This isomer is produced from 4-chlorobenzotrichloride and anhydrous hydrogen fluoride. Nitration provides either 4-chloro-3-

Table 8. Physical Properties of Benzotrifluoride[a]

Property	Value
mol wt	146.11
color	colorless
melting point, °C	−29.02
boiling point, °C	102.05
density, 25°C, g/mL	1.1814
coefficient of expansion (30–40°C)	0.00121
refractive index, n_D^{25}	1.4114
viscosity, mPa·s($=$cP)	
38°C	0.488
99°C	0.282
surface tension, mN/m($=$dyn/cm)	
20°C	23.39
latent heat of fusion, J/mol[b]	13,782.1
latent heat of vaporization, 102.05°C	
J/mol[b]	32.635.2
specific conductivity at 25°C, S/cm	1×10^{-7}
dielectric constant, 30°C	9.18
dipole moment, C·m[c]	8.54×10^{-30}
heat of combustion, J/g[b]	23,064.3
heat of formation, kJ/mol[b]	
vapor	−580.7
liquid	−618.4
boiling point of binary azeotrope, °C with 96.7 mol %	58.1
bromine	
solubility in water, g/100 g at room temperature	0.045
flash point (Cleveland open cup), °C[d]	15.6
fire point (Cleveland open cup), °C[d]	15.6
vapor pressure, in °C and kPa[e]	Antoine equation[f]

[a]Other properties of this compound and its derivatives have been reviewed (311).
[b]To convert J to cal, divide by 4.184.
[c]To convert C·m to debye (D), divide by 3.336×10^{-30}.
[d]Ref. 312.
[e]To convert kPa to mm Hg, multiply by 7.5; log kPa = log mm Hg − 0.875.
[f]$\mathrm{Log}_{10} P = 6.0939 - \dfrac{1305.509}{(t + 217.280)}$.

nitrobenzotrifluoride [*121-17-5*] (one-step) or 4-chloro-3,5-dinitrobenzotrifluoride [*393-75-9*] (two-step) for use in crop protection applications. Dinitration can also be accomplished in one step (85% yield) with 90% nitric acid/20% oleum (325). Single-step dehalogenation-reduction of 4-chloro-3,5-dinitrobenzotrifluoride provides a 96% yield of 3,5-diaminobenzotrifluoride [*368-53-6*] (326), an intermediate to specialty polymers.

2,4-Dichlorobenzotrifluoride. This dichloro compound is produced from 2,4-dichlorobenzotrichloride and hydrogen fluoride. One commercial application is the manufacture of the pre-emergent herbicide, dinitramine [*29091-05-2*].

3,4-Dichlorobenzotrifluoride. This compound is produced by chlorination of 4-chlorobenzotrifluoride and exhibits sufficient activation to undergo nucleophilic

Table 9. Properties of Benzotrifluoride Derivatives

Component	CAS Registry Number	Mol wt	Mp, °C	Bp, °C$_{kPa}$[a]	Refractive index n_D^t	Specific gravity d_4^t	Flash point,[b] °C	Toxicity, LD$_{50}$
H$_2$NC$_6$H$_4$CF$_3$				*Aminobenzotrifluoride*				
2-aminobenzotrifluoride	[88-17-5]	116.13	34	174–175$_{100.4}$	1.4800^{25}	1.290^{25}	55	440^c,
3-aminobenzotrifluoride	[98-16-3]	116.13	5–6	187–188	1.4788^{20}	1.305^{25}	85	690^{d,e}
				86$_{2.67}$				220^f
CF$_3$C$_6$H$_4$Cl				*Monochlorobenzotrifluorides*				
2-chlorobenzotrifluoride	[88-16-4]	180.56		152.5	1.4550^{20}	1.367^{20}	98	>6.8^g
4-chlorobenzotrifluoride	[98-56-6]	180.56	−36	139	1.4444^{25}	1.338^{25}	110	>2.7^h
				29.5$_{1.33}$				
CF$_3$C$_6$H$_3$Cl$_2$				*Dichlorobenzotrifluorides*				
2,4-dichlorobenzotrifluoride	[320-60-5]	215.00	−26	177.5	1.4793^{25}	1.501$^{15.5}$	72	2900^i
3,4-dichlorobenzotrifluoride	[328-84-7]	215.00	−12.4	173.5	1.4736^{25}	1.478^{25}	65	>2^j

[a]To convert kPa to mm Hg, multiply by 7.5.
[b]Closed cup (ASTM procedure D3278).
[c]Inhalation (rat), LC$_{50}$ mg/m^3 for 4 h (127).
[d]Inhalation (mouse), g/m^3 for 2 h (127).
[e]LC$_{50}$.
[f]Oral (mouse), mg/kg (127).
[g]Acute oral (rat), g/kg (323).
[h]Acute dermal (rabbit), g/kg (323).
[i]Acute oral (rat), mg/kg (324).
[j]Acute dermal (rabbit), g/kg (324).

displacement with phenols to form diaryl ether herbicides, eg, acifluorofen sodium [62476-59-9].

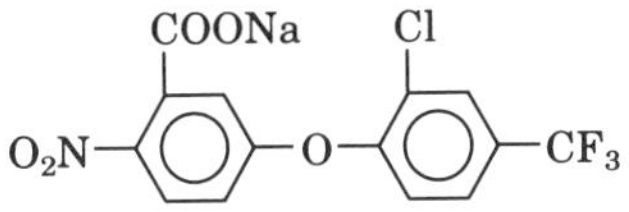

3-Aminobenzotrifluoride. The standard manufacturing route to 3-amino-benzotrifluoride involves nitration of benzotrifluoride to 3-nitrobenzotrifluoride [98-46-4], followed by hydrogenation. A comprehensive study on materials of construction to minimize corrosion during catalytic hydrogenation led to the recommendation of Cr–Ni–Mo steel (with ≥ 3% Mo) (327). Gas chromatographic details for monitoring this two-step process have been described (328), as well as analytical methods for assay of the 2- and 4-isomer impurities in 3-aminobenzo-trifluoride (329). Environmental aspects of the manufacture of this product in Germany have been published (330). A novel process based on the one-step *in situ* fluorination-reduction of 3-nitrobenzotrichloride with ammonium bifluoride–hydrogen fluoride has been described (331,332).

The amine group of 3-aminobenzotrifluoride can be replaced by Cl, Br, I, F, CN, or OH groups by standard diazotization reactions. Phosgenation gives 3-tri-fluoromethylphenylisocyanate [329-01-1], which can then be converted to the se-lective herbicide fluometuron [2164-17-2], a substituted urea.

Application. *Crop Protection Chemicals.* Benzotrifluoride derivatives have gained wide acceptance as herbicides, insecticides, and fungicides (Table 10).

Drugs. Trifluoromethyl-based pharmaceuticals had been limited to pheno-thiazine tranquilizers and benzothiadiazine 1,1-dioxide diuretics (qv). However, new drugs have been developed (Table 11). One of the key properties of the CF_3 group is its high lipophilicity; it increases the lipid solubility of the pharmaceutical and thus accelerates absorption and transport within the host organism.

Germicides. Benzotrifluoride derivatives have also found wide use as anti-microbial agents in soaps, eg, the brominated and chlorinated materials, fluoro-salan [4776-06-1] and cloflucarban [369-77-7].

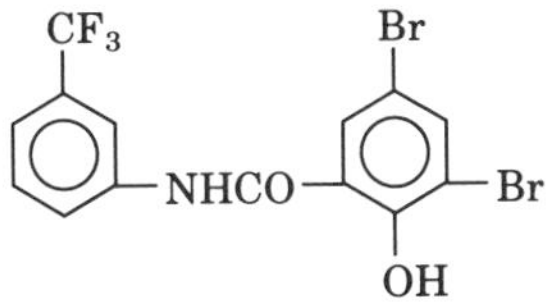

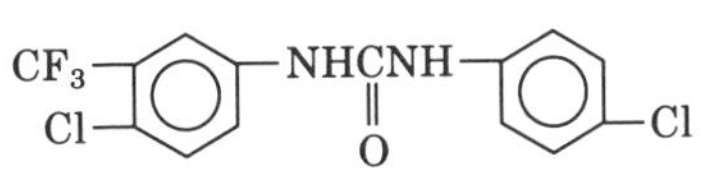

Other Biological Applications. 4-Nitro-3-(trifluoromethyl)phenol [88-30-2] (TFM) is still employed by the Canadian Bureau of Fisheries and the U.S. Fish and Wildlife Service as a lampricide for the control of parasitic sea lamprey in the Great Lakes (see AQUACULTURE).

Dyes. Several reviews on fluorine-containing dyes have been published (333–335). The relative accessibility of benzotrifluorides has reflected the wide incorporation of the trifluoromethylphenyl group into azo, anthraquinone, and triphenylmethane dyes (qv). The trifluoromethyl group is claimed to improve the

Table 10. Benzotrifluoride Crop-Protection Chemicals

Common name	CAS Registry Number	Structure	Application
trifluralin	[1582-09-8]	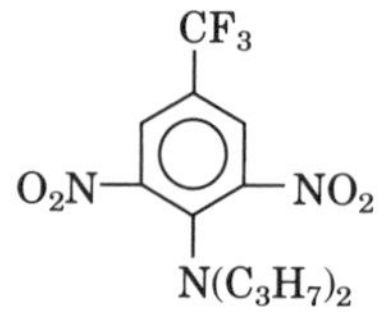	pre-emergent herbicide
lactofen	[77501-63-4]	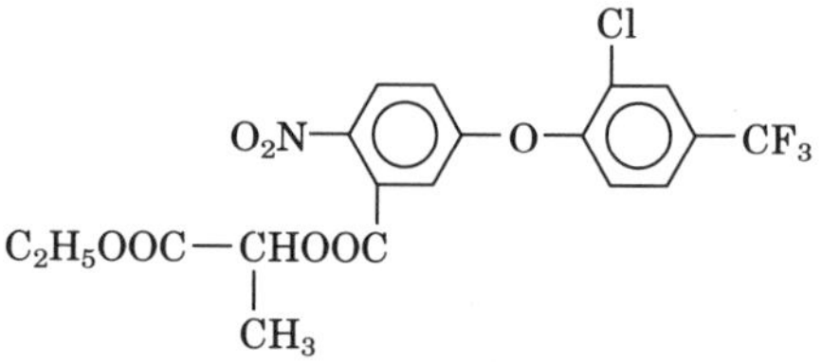	herbicide
norflurazon	[27314-13-2]	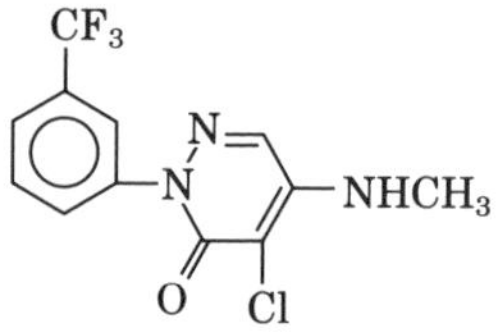	selective herbicide
fluvalinate	[69409-94-5]	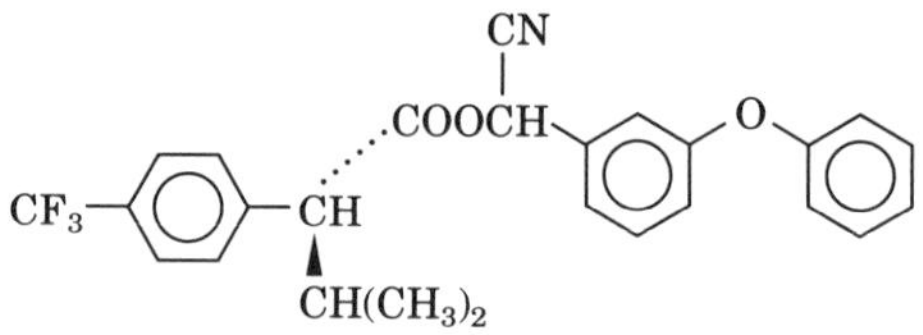	insecticide
hydramethylnon	[67485-29-4]	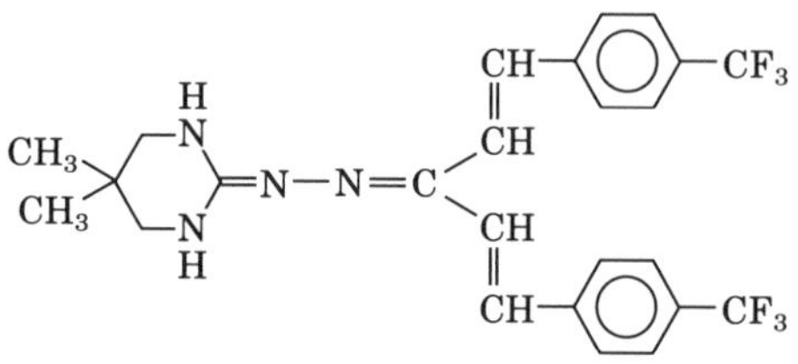	insecticide
fentrifanil	[62441-54-7]	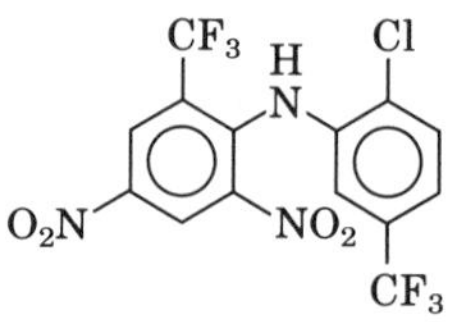	acaricide
triflumizole	[68694-11-1]	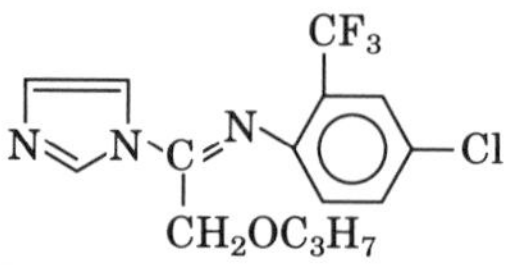	fungicide

Table 11. Benzotrifluoride Drugs

Common name	CAS Registry Number	Structure	Application
triflupromazine	[146-54-3]	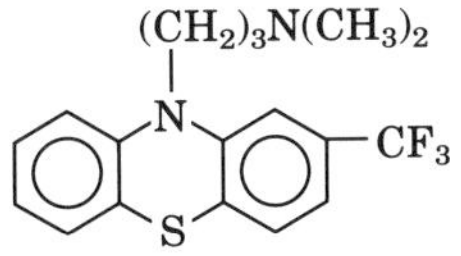	tranquilizer
flumethiazide	[148-56-1]	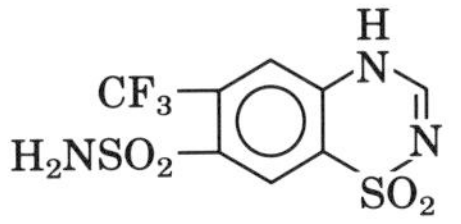	diuretic
fenfluramine	[404-82-0]	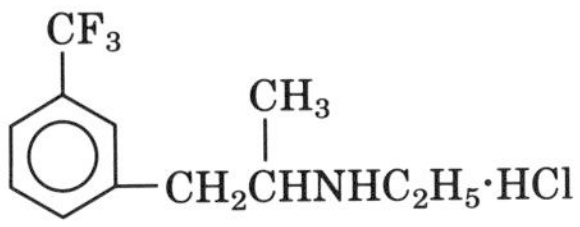	anorexigen
flufenamic acid	[53-78-9]	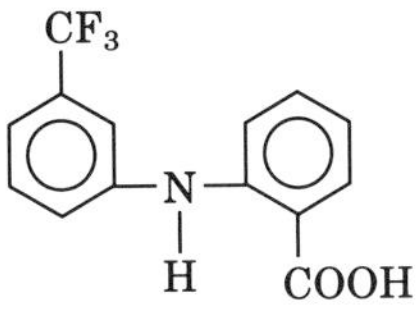	analgesic
trifluperidol	[749-13-3]	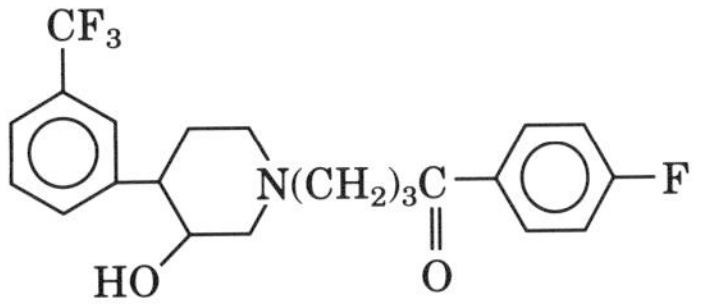	tranquilizer
flumetramide	[7125-73-7]	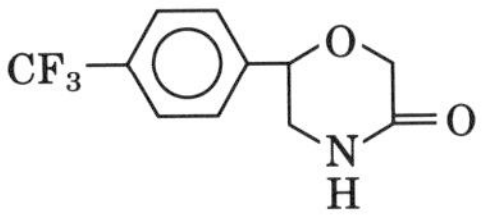	muscle relaxant
fluoxetine	[54910-89-3]	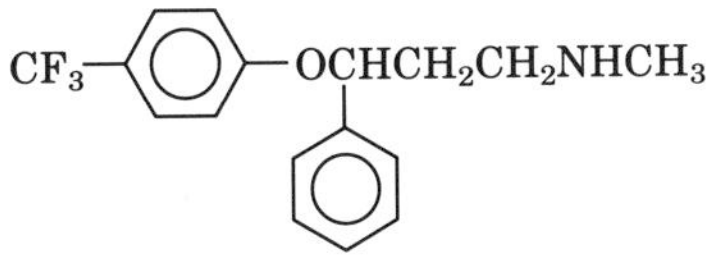	antidepressant

brightness, tinctorial values, and lightfastness of dyes. The electron-withdrawing effects of this group also tend to modify the absorption of light by a dye in the visible and uv region. In azo dyes (qv), aminobenzotrifluorides (Fast Base) are diazotized for subsequent coupling: 2-aminobenzotrifluoride CI Pigment Yellow 154 [88-17-5]; 3,5-bis(trifluoromethyl)aniline [328-74-5], CI Azoic Diazo Component 16; 2-amino-5-chlorobenzotrifluoride [445-03-4], CI Azoic Diazo Component 17; 3-amino-4-ethylsulfonylbenzotrifluoride [382-85-4], CI Azoic Diazo Component 19; and 3-amino-4-chlorobenzotrifluoride, CI Azoic Diazo Component 49

[*121-50-6*]. 3-Trifluoromethylbenzoyl halides are used to make anthraquinone vat dyes, eg, Indanthrene blue CLB [*6942-78-0*] (see DYES, ANTHRAQUINONE).

Miscellaneous Applications.   Benzotrifluoride derivatives have been incorporated into polymers for different applications. 2,4-Dichlorobenzotrifluoride or 2,3,5,6-tetrafluorobenzotrifluoride [*651-80-9*] have been condensed with bisphenol A [*80-05-7*] to give benzotrifluoride aryl ether semipermeable gas membranes (336,337). 3,5-Diaminobenzotrifluoride [*368-53-6*] and aromatic dianhydrides form polyimide resins for high temperature composites (qv) and adhesives (qv), as well as in the electronics industry (338,339).

Photoresist applications in the microelectronics industry have also been disclosed (340). Thermally stable benzyl sulfonate esters based on 2-methyl-3-nitrobenzotrifluoride [*6656-49-1*] can serve as nonionic photoacid generators to promote a cascade of reactions during irradiation of the resist.

Liquid crystal applications include esters based on *m*- or *p*-hydroxybenzotrifluoride (341,342), hydroxytrifluoromethylbiphenyls (343) or hydroxytrifluoromethylphenyl Schiff bases (344).

Inorganic analytical applications for benzotrifluoride derivatives include sodium tetrakis[3,5-bis(trifluoromethyl)phenyl]borate (Kobayashi's reagent) (345), and 4-(2,6-dinitro-4-trifluoromethylphenyl)aminobenzo-15-crown-5 (modified Takagi reagent (346).

Benzotrifluoride has been recommended as a fuel additive for internal combustion engines (347).

Arylfluoroalkyl Ethers

α,α,α-Trifluoromethoxybenzene [*456-55-3*], $C_6H_5OCF_3$, and other arylfluoroalkyl ethers and thioethers (HCF_2O—, HCF_2CF_2O—, CF_3CH_2O—, and CF_3S—), are assuming greater importance as crop-protection chemicals and pharmaceuticals.

Properties.   The trifluoromethoxy (CF_3O—) group in $ArOCF_3$ exhibits unusual stability to strong acids and bases (including organometallic reagents), as well as to strong oxidizing and reducing conditions (348). The thermal stability is exceptional; extensive degradation in the gas phase was not observed (mass spectroscopy) in a sealed nickel tube until 600°C (348). Nuclear chlorinated trifluoromethoxy and bis(trifluoromethoxy)benzenes have exhibited moderate thermal stability for use in transformers and Rankine cycle engines (349).

Reactions.   The CF_3O— group exerts predominant para orientation in electrophilic substitution reactions such as nitration, halogenation, acylation, and alkylation (350).

Trifluoromethoxybenzenes ($ArOCF_3$).   Trifluoromethoxybenzene (α,α,α-trifluoroanisole, phenyl trifluoromethyl ether [*456-55-3*]), $C_6H_5OCF_3$ (mol wt 162.11), is a colorless liquid, bp 102°C, mp −50°C, n_D^{20} 1.4060, d_4^{25} 1.226, flash point (closed cup), 12°C.

Depending on the ring substituent, trifluoromethoxybenzenes can be made by the sequential chlorination–fluorination of anisole(s) (351–354). A one-step process with commercial potential is the BF_3 (or SbF_3)-catalyzed reaction of phenol with carbon tetrachloride/hydrogen fluoride (355). Aryl trifluoromethyl ethers, which may not be accessible by the above routes, may be made by fluorination of aryl fluoroformates or aryl chlorothioformates with sulfur tetrafluoride (348) or molybdenum hexafluoride (356).

$$\underset{O}{(ArOCF)} \qquad \underset{S}{(ArOCCl)}$$

Trifluoromethylthioaromatics (ArSCF$_3$).   Trifluoromethylthioaromatics (aryl trifluoromethyl sulfides) can be made by sequential chlorination–fluorination (SbF$_3$ or HF) of the corresponding thioanisole (351,357). In the case of 4-trifluoromethylmercaptophenol [825-83-2], 4-CF$_3$C$_6$H$_4$SH, used for the production of the coccidiostat toltrazuril [69004-03-1] (Table 12), protection of the phenolic group as the carbonate ester is required prior to chlorination (358). Coupling of aryl halides with trifluoromethylthiocopper, CF$_3$SCu, provides an alternative entry to trifluoromethylthioaromatics (359,360).

Difluoromethoxyaromatics (ArOCHF$_2$) and Sulfur Analogues (ArSCHF$_2$). Difluoromethyltion of phenol (or thiophenols) with chlorodifluoromethane, CHClF$_2$, and aqueous caustic in dioxane gives good yields of aryldifluoromethyl ethers (361). A modification features the use of phase-transfer catalysts such as tris(3,6-dioxaheptyl)amine (TDA-1) (362).

Tetrafluoroethoxyaromatics (ArOCF$_2$CF$_2$H).   Tetrafluoroethoxyaromatics are produced by base-catalyzed addition of tetrafluoroethylene to phenols (348,363).

Aryltrifluoroethyl Ethers (ArOCH$_2$CF$_3$).   2,2,2-Trifluoroethoxybenzenes are obtained from the reaction of activated haloaromatics with sodium 2,2,2-trifluoroethoxide in polar solvents (364); phase-transfer catalysts are also employed (365). Nitro groups can also be displaced by the fluoroalkylation technique, eg, 4-nitrobenzonitrile was converted to 4-(2,2,2-trifluoroethoxy)benzonitrile (366). Trifluoroethoxybenzene pharmaceutical intermediates can be prepared by the base-catalyzed reaction of 2,2,2-trifluoroethyl trifluoromethanesulfonate (a trifluorethyl-transfer agent), CF$_3$SO$_3$CH$_2$CF$_3$, with phenols (367).

Applications.   Table 12 lists crop-protection chemicals and pharmaceuticals containing the aryl fluoroalkyl ether group. Trifluoromethoxybenzene (ArOCF$_3$) derivatives (186,187) and related arylfluoroalkyl ethers, ArO(CH$_2$)$_n$R$_f$ (n = 1–6) (368,369) are of use in liquid crystal applications.

Fluorinated Nitrogen Heterocyclics

Ring- or side-chain fluorinated nitrogen heterocyclics have been incorporated into crop-protection chemicals, drugs, and reactive dyestuffs. Key intermediates include fluorinated pyridines, quinolines, pyrimidines, and triazines. Physical properties of some fluorinated nitrogen heterocyclics are listed in Table 13.

RING-FLUORINATED PYRIDINES

Exchange fluorination of chloropyridines is the principal tool for production of ring-fluorinated pyridines. Diazotization of aminopyridines in pyridine–hydrogen fluoride (Olah's reagent) (370) or ammonium fluoride–hydrogen fluoride (371) has also been used. An emerging synthesis tool is the use of fluorinated aliphatic

Table 12. Aryl Fluoroalkyl Ether Applications

Common name	CAS Registry Number	Structure
		Insecticides
triflumuron	[64628-44-0]	
toltrazuril[a]	[69004-03-1]	
flucythrinate	[70124-77-5]	
hexaflumuron	[86479-06-3]	
		Herbicides
primsulfuronmethyl	[86209-51-0]	
tetrafluron	[27954-37-6]	
		Other
flurprimidol[b]	[56425-91-3]	
flecainide acetate[c]	[54143-56-5]	

[a]Coccidiostat.
[b]Plant growth regulator.
[c]Cardiac depressant.

Table 13. Properties of Miscellaneous Fluorinated Heterocyclic Compounds

Component	CAS Registry Number	Mol wt	Mp, °C	Bp, °C/kPa[a]	Refractive index, n_D^t	Specific gravity, d_4^t	Flash point, °C
Fluoropyridines							
C_5H_4FN							
2-fluoropyridine	[372-48-5]	97.09		$126_{100.4}$	1.4678^{20}	1.1281^{20}	24^b
3-fluoropyridine	[372-47-4]	97.09		$105–107_{100.3}$	1.4700^{20}	1.125^{25}	13^b
4-fluoropyridine	[694-52-0]	97.09	100^c	$108_{100.0}$	1.4730^{20}		
$C_5H_3F_2N$							
2,4-difluoropyridine	[3491-90-7]	115.08	$134–135^c$	104–105			
2,6-difluoropyridine	[1513-65-1]	115.08		$124.5_{99.1}$	1.4349^{25}	1.265^{25}	32^b
$C_5H_2F_3N$							
2,4,6-trifluoropyridine	[3512-17-2]	133.07		94–95			
C_5HF_4N							
2,3,5,6-tetrafluoropyridine	[2875-18-5]	151.06		102			
$C_5H_5N^d$							
pentafluoropyridine	[70-16-3]	169.05	–41.5	83.3	1.3856^{20}		
Perfluoroalkylpyridines							
2-trifluoromethylpyridine	[368-48-9]	147.10		$143_{99.4}$	1.4144^{25}		
3-trifluoromethylpyridine	[3796-23-4]	147.10		113–115	1.4150^{25}		
4-trifluoromethylpyridine	[3796-24-5]	147.10		108–110	1.4144^{25}		
2-chloro-5-trifluoromethylpyridine	[52334-81-3]	181.55	32–34	152_{100}		1.417^{20}	110^e
Fluoropyrimidines							
2,4,6-trifluoropyrimidine	[696-82-2]	134.06		$98;60_{24}$	1.4015^{25}		
2,4,5,6-tetrafluoropyrimidine	[767-79-3]	152.06		89	1.3875^{25}		
5-chloro-2,4,6-trifluoropyrimidine	[697-83-6]	168.51		114.5	1.4390^{20}		
Fluorotriazines							
2,4,6-trifluoro-1,3,5-triazine	[675-14-9]	135.05	–38	$72.4_{101.67}$	1.3844^{24}	1.60^{25}	
2,4,6-*tris*-(trifluoromethyl)-1,3,5-triazine	[368-66-1]	285.07	–24.8	95–96	1.3161^{25}	1.593^{25}	

[a] To convert kPa to mm Hg, multiply by 7.5. [b] Tag closed cup. [c] Mp of HCl salt. [e] Closed cup (ASTM procedure D3278).
[d] Trouton's constant, 24.9; latent heat of vaporization, 36, 338 J/mol.

building blocks to make fluoropyridines. Early studies on the substitutive fluorination (F_2) of pyridine gave 2-fluoropyridine in low yields and posed severe reaction hazards (372,373). Modifications featuring low temperature fluorination of substituted pyridines (alkyl, halogen, ester, or ketone functions) in 1,1,2-trichloro-1,2,2-trifluoroethane, $CF_2ClCFCl_2$, solvent give good yields of the corresponding 2-fluoropyridine (374,375). Tamed fluorine reagents such as xenon difluoride (376) and cesium fluoroxysulfate (377) can also fluorinate pyridine. A promising substitutive fluorination technique is the base-catalyzed decomposition of N-fluoropyridinium salts, BF_4^-, PF_6^-, or SbF_6^-, to give high yields of the substituted 2-fluoropyridine (378). The salt is made in 80% yield; the decomposition yields 72–91% of product.

Monofluoropyridines.   *2-Fluoropyridine.*   Diazotization of 2-aminopyridine in anhydrous hydrogen fluoride forms 2-fluoropyridine in high yield (178,370). Modifications include fluorodediazonization of substituted 2-aminopyridines in ammonium fluoride–hydrogen fluoride (371) or pyridine–hydrogen fluoride (370) media. Exchange-fluorination of 2-chloropyridine with potassium fluoride in polar solvents is sluggish (210°C for 21 d; 50–58% yield) (379). The solvent-free exchange-fluorination employing potassium bifluoride (KHF_2) and 2-chloropyridine (315°C/4 h) gave 2-fluoropyridine in 74% yield (380). A new development features exchange fluorination of 2-chloropyridine with hydrogen fluoride–γ-collidine at 150–200°C to give 94% yield of product (381).

Fluorine at the 2 position ($pK_a = -0.44$) significantly reduces pyridine ($pK_a = 5.17$) basicity more than at the 3 position ($pK_a = 2.97$) (382). 2-Fluoropyridine is readily hydrolyzed to 2-pyridone in 60% yield by reflux in 6 N hydrochloric acid (383). It is quite reactive with nucleophiles. For example, the halogen mobility ratio from the comparative methoxydehalogenation of 2-fluoropyridine and 2-chloropyridine was 85.5/1 at 99.5°C (384). This lability of fluorine has been utilized to prepare fluorine-free O-2-pyridyl oximes of 3-oxo steroids from 2-fluoropyridine for possible use as antifertility agents (385).

2-Fluoropyridine is a useful reagent for synthetic applications. It reacts with methyl p-toluenesulfonate to give 1-methyl-2-fluoropyridinium p-toluenesulfonate [*58086-67-2*] (Mukaiyama's reagent). Such onium salts are used in the oxidation of alcohols, cross-coupling of Grignard reagents, Beckmann rearrangements of ketoximes, nonenzymatic biogenetic-like synthesis of terpenes (386), and preparation of new synthetic penicillins (387).

Metallation of 2-fluoropyridine with lithium diisopropylamide (LDA) gives 2-fluoro-3-lithiopyridine, thereby providing entry to 3-substituted pyridines (388). This technique has been used to make fluorine analogues of the antitumor ellipticines (389).

$$\underset{N}{\bigcirc}-F \xrightarrow[-70°C]{LDA} \left[\underset{N}{\bigcirc}-\underset{F}{Li}\right] \xrightarrow[2.\ H_2O]{1.\ R_1COR_2} \underset{N}{\bigcirc}-\underset{F}{CH(OH)R_1R_2}$$

3-Fluoropyridine. Diazotization of 3-aminopyridine(s) in hydrogen fluoride (390), pyridine–hydrogen fluoride (370), or ammonium fluoride–hydrogen fluoride (371) can be effected in good yield. 3-Fluoropyridine can also be made by the Balz-Schiemann technique in 50% yield. Earlier warnings concerning the instability of 3-pyridyldiazonium fluoroborate (391) were confirmed by later reports on detonations involving this salt (392). Related compounds such as 2-chloro-3-pyridyldiazonium fluoroborate also decomposed with explosive violence (393).

3-Fluoropyridine derivatives can be constructed from fluoroaliphatic feedstocks. 5-Fluoro-2,6-dihydroxynicotinamide [*655-13-0*], a precursor to the antibacterial, enoxacin [*74011-58-8*], was prepared in 63% yield from ethyl fluoroacetate [*459-72-3*], ethyl formate [*109-94-4*], and malonamide [*108-13-4*] (394).

$$FCH_2COOC_2H_5 \xrightarrow[NaOC_2H_5]{HCOOC_2H_5} \left[\begin{array}{c} F\diagdown_{\displaystyle C}\diagup COOC_2H_5 \\ \parallel \\ H\diagup^{\displaystyle C}\diagdown ONa \end{array}\right] \xrightarrow{CH_2(CONH_2)_2} \underset{HO-\underset{N}{\bigcirc}-OH}{F-\underset{}{\bigcirc}-CONH_2}$$

A complementary cyclization technique based on dichlorofluoroacetonitrile [*83620-05-7*], $FCCl_2CN$, was employed to form 2-chloro-3-fluoro-5-methylpyridine [*34552-15-3*] (395).

Derivatives such as 3-fluoro-4-nitropyridine [*13505-01-6*] (396) or the 1-oxide [*769-54-0*] (397) have been used to characterize amino acids and peptides. 5-Fluoro-3-pyridinemethanol [*22620-32-2*] has been patented as an antilipolytic agent (398). A promising antidepressant, 1-(3-fluoro-2-pyridyl)piperazine hydrochloride [*85386-84-1*] is based on 2-chloro-3-fluoropyridine [*17282-04-1*] (399).

4-Fluoropyridine. This isomer can be prepared in 54–81% yield by diazotization of 4-aminopyridine in anhydrous hydrogen fluoride (370,371,400). Free 4-fluoropyridine readily undergoes self-quaternization to give pyridyl pyridinium salts (401); stabilization can be effected as the hydrochloride salt (371,400). Numerous 4-fluoropyridinium salts, eg, 4-fluoro-1-methylpyridinium iodide, have been converted to novel penicillins (387,402).

Difluoropyridines. 2,4-Difluoropyridine can be prepared (26% yield) from 2,4-dichloropyridine and potassium fluoride in sulfolane and ethylene glycol initiator (403). The 4-fluorine is preferentially replaced by oxygen nucleophiles to give 2-fluoro-4-hydroxypyridine derivatives for herbicidal applications (404).

Fluorination of 2,6-diaminopyridine in anhydrous hydrogen fluoride gave a 62% yield of 2,6-difluoropyridine (26,371,405). 2,6-Difluoropyridine is also prepared in 52% yield (200°C, 100 h) from 2,6-dichloropyridine and potassium fluoride in dimethyl sulfone or sulfolane (406). The reaction can be performed in dimethyl sulfoxide with shorter reaction times (9 h, 186°C) (407); addition of tetramethylammonium chloride (TMAC) catalyst lowers reaction temperature to 150°C, thereby minimizing solvent degradation (408). Solvent-free exchange-fluorination (KF) at 400°C (16 h) gave 80% yield of 2,6-difluoropyridine which attests to its high thermal stability (409).

Displacement reactions with oxygen nucleophiles are of potential commercial interest. Alkaline hydrolysis provides 2-fluoro-6-hydroxypyridine [*55758-32-*

2], a precursor to 6-fluoropyridyl phosphorus ester insecticides (410–412). Other oxygen nucleophiles such as bisphenol A and hydroquinone have been used to form aryl–pyridine copolymers (413).

Nitration with mixed nitric and sulfuric acids provides 79% yield of 3-nitro-2,6-difluoropyridine [5860-02-1], bp 218–220°C (414).

3-Bromo or chloro-2,6-difluoropyridines can be prepared in 50% yield by diazotization of the corresponding 3-halo-2,6-diaminopyridine in ammonium fluoride–hydrogen fluoride solvent (371). 5-Chloro-2,3-difluoropyridine [89402-43-7], a precursor to the herbicide pyroxofop [105512-06-9], was synthesized by a multistep sequence based on allyl chlorodifluoroacetate [118337-48-7], $ClCF_2CO_2$-$CH_2CH=CH_2$ (415).

Tri-, Tetra-, and Pentafluoropyridines. 2,4,6-Trifluoropyridine can be prepared in 75% yield by catalytic hydrogenolysis (palladium-on-carbon, 280°C) of 3,5-dichloro-2,4,6-trifluoropyridine [1737-93-5] (416). The latter is synthesized by exchange fluorination of pentachloropyridine with potassium fluoride in polar solvents such as N-methylpyrrolidinone (417,418). 3,5-Dichloro-2,4,6-trifluoropyridine is used to prepare the herbicides haloxydine [2693-61-0] and fluroxypyr-(1-methylheptyl) [81406-37-3].

2,3,5,6-Tetrafluoropyridine can be prepared in 75% yield from the hydrogenation of pentafluoropyridine under free-radical (catalytic) or nucleophilic (lithium aluminum hydride) conditions (416,419). No practical uses for 2,3,5,6-tetrafluoropyridine are known.

Pentafluoropyridine was first synthesized in 1960 in 27% yield by the defluorination of undecafluoropiperidine (prepared in low yield by the electrochemical fluorination of pyridine) over a nickel or iron surface at 560–610°C (420,421). The preferred route is the solvent-free exchange-fluorination (KF) of pentachloropyridine at 480–500°C to give 69–83% yields of pentafluoropyridine (418,422). Pentafluoropyridine is a weak base, does not form a hydrochloride salt, and is more volatile (bp 83.3°C) than pyridine (bp 115°C). Pentafluoropyridine readily undergoes reaction with nucleophilic agents to give 4-substituted-2,3,5,6-tetrafluoropyridines. More than 30 examples of these 4-substitution reactions have been compiled (423). Derivatives of 4-hydroxytetrafluoropyridine [2693-66-5] and related compounds exhibit herbicidal properties (424,425).

The nucleophilic equivalent of the Friedel-Crafts reaction of pentafluoropyridine with hexafluoropropene–potassium fluoride in sulfolane gave perfluoro-(4-isopropyl)pyridine in 94% yield (426).

Pentafluoropyridine–hexafluorobenzene working fluids show the requisite stability at 382°C for automotive Rankine-cycle power units (427). Hydroxyl and related functions in steroids can be selectively protected as tetrafluoro-4-pyridyl ethers by pentafluoropyridine (428).

Applications.   Until recently, haloxydine, a herbicide, was one of the few early examples of crop-protection chemicals containing ring-fluorinated pyridines. Fluroxypyr-(1-methylheptyl) and pyroxofop are new herbicides that are being commercialized (Table 14).

Several 3-fluoropyridine derivatives are employed to produce enoxacin, tosufloxacin, and other naphthyridine antibacterials (Table 14). Examples of such intermediates include 2,6-dichloro-5-fluoronicotinonitrile (429), ethyl 2,6-dichloro-5-fluoronicotinate (430), 2-chloro-3-fluoropyridine (393), 6-acetyl-2-(4-acetyl-1-piperazinyl)-3-fluoropyridine (431), and 5-fluoro-2,6-dihydroxynicotinamide (394).

Table 14. Applications of Ring-Fluorinated Pyridines

Common name	CAS Registry Number	Structure
Herbicides		
haloxydine	[2693-61-0]	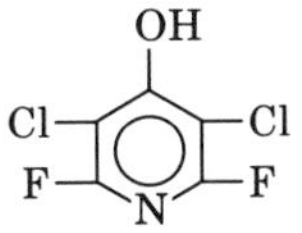
fluroxypyr-(1-methylheptyl)	[81406-37-3]	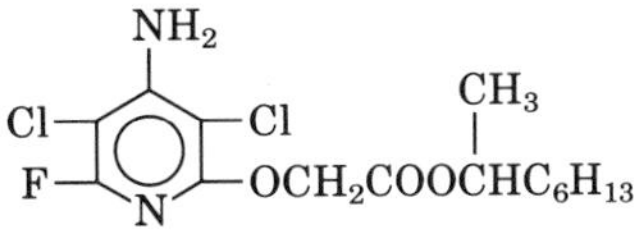
pyroxofop	[105512-06-9]	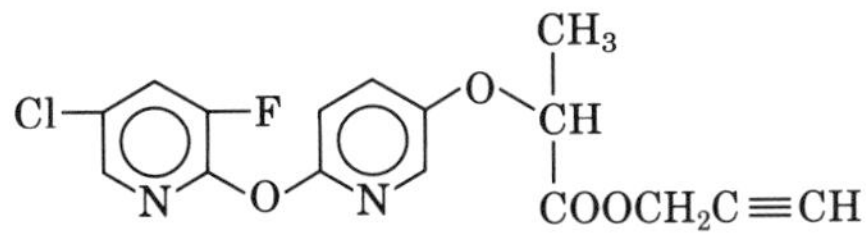
Antibacterials		
enoxacin	[74011-58-8]	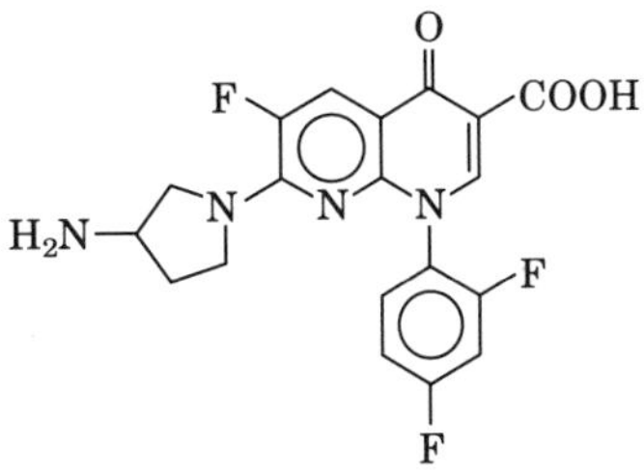
tosufloxacin	[108138-46-1]	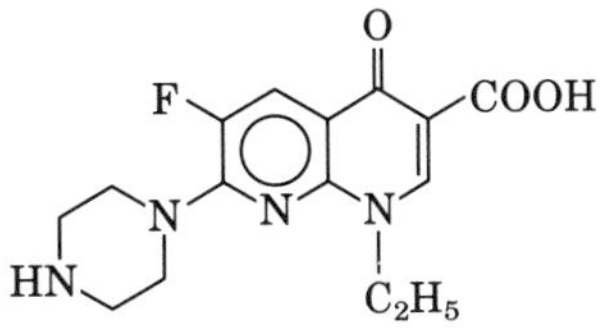

PERFLUOROALKYLPYRIDINES

New developments in trifluoromethylpyridine technology are associated with the commercialization of numerous crop-protection chemicals as herbicides, fungicides, and insecticides (Table 15). Physical properties for representative trifluoromethylpyridines are listed in Table 13.

The standard synthesis method features side-chain chlorination of a methylpyridine (picoline), followed by exchange-fluorination with hydrogen fluoride or antimony fluorides (432,433). The fluorination of pyridinecarboxylic acids by sul-

Table 15. Trifluoromethylpyridine-Based Crop-Protection Chemicals[a]

Common name	CAS Registry Number	Structure
dithiopyr	[97886-45-8]	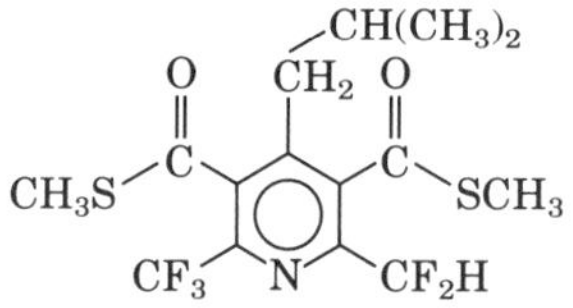
fluazilop–butyl	[69806-50-4]	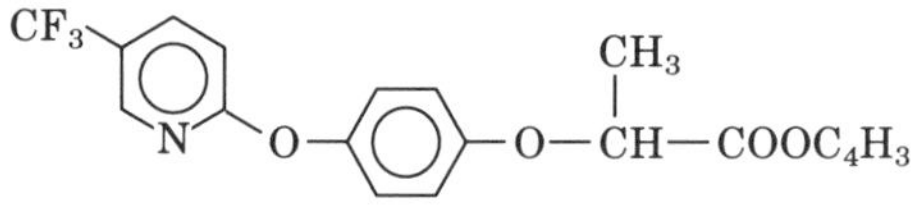
haloxyfop–methyl	[69806-40-2]	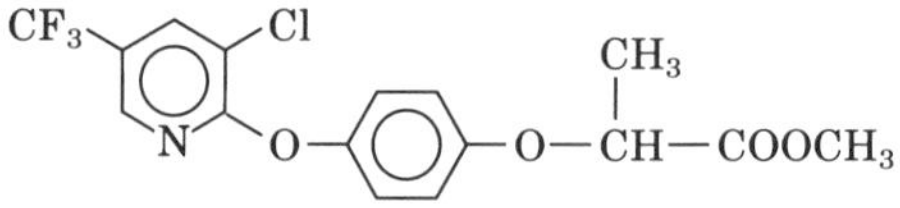
flazasulfuron	[104040-78-0]	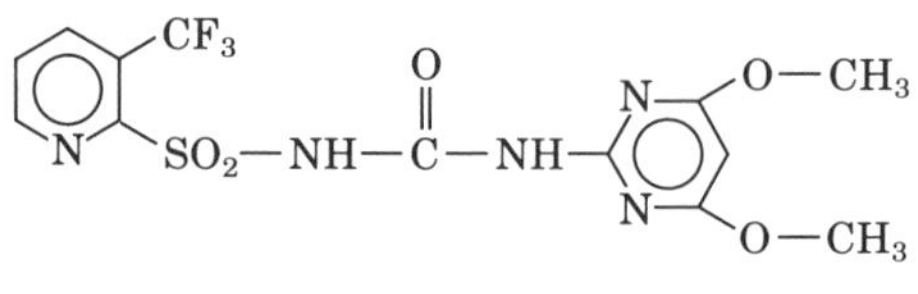
chlorfluazuron[b]	[71422-67-8]	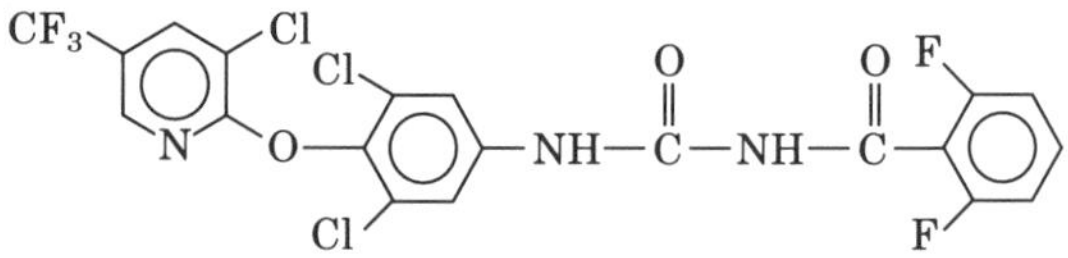
fluazinam[c]	[79622-59-6]	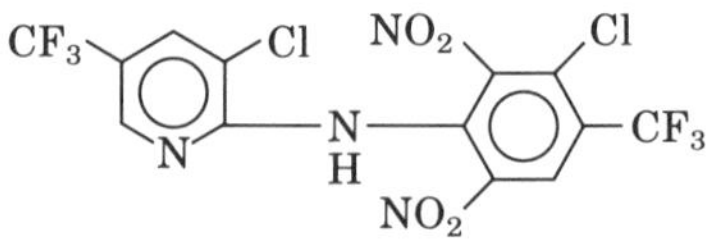

[a]Herbicide unless otherwise noted.
[b]Insecticide.
[c]Fungicide.

fur tetrafluoride (434) or molybdenum hexafluoride (435) is of limited value for high volume production operations due to high cost of fluorinating agent.

A significant development in trifluoromethylpyridine synthesis strategy is the use of fluorinated aliphatic feedstocks for the ring-construction sequence. Examples include the manufacture of the herbicide dithiopyr, utilizing ethyl 4,4,4-trifluoroacetoacetate [372-31-6], $CF_3COCH_2COOC_2H_5$ (436,437). 2,3-Dichloro-5-trifluoromethylpyridine [69045-84-7], a precursor to several crop-protection chemicals (see Table 15), can be prepared by conversion of 1,1,1-trichloro-2,2,2-trifluoroethane [354-58-5], CF_3CCl_3, to 2,2-dichloro-3,3,3-trifluoropropionaldehyde [82107-24-2], CF_3CCl_2CHO, followed by cyclization with acrylonitrile [107-13-1] (415).

2-Trifluoromethylpyridine can be prepared in 54% yield from picolinic acid and sulfur tetrafluoride–hydrogen fluoride (434). 2-Trifluoromethylpyridine is a weak base; no hydrochloride salt is formed. However, 2-trifluoromethylpyridine 1-oxide [22253-71-0] (bp 132–133°C/2.7 kPa (20 mm Hg)) is prepared in 81% yield using 30% hydrogen peroxide–acetic acid (438).

3-Trifluoromethylpyridine can be prepared in 25–65% yield from nicotinic acid and sulfur tetrafluoride (434,439). An alternative method is the passage of chlorine into a mixture of β-picoline and hydrogen fluoride in an autoclave (190°C, 3 MPa) (440). 4-Trifluoromethylpyridine is prepared in 57% yield from isonicotinic acid and sulfur tetrafluoride.

2-Chloro-5-trifluoromethylpyridine, an intermediate to the herbicide fluazilop–butyl, can be made from β-picoline by two processes. β-Picoline is chlorinated to 2-chloro-5-trichloromethylpyridine [69405-78-9], followed by fluorination with hydrogen fluoride under pressure (200°C, 10 h) (441) or vapor-phase (350°C, CCl_4 diluent) conditions (442). An alternative process features the single-step vapor-phase reaction of β-picoline with chlorine–hydrogen fluoride (400°C, N_2 or CCl_4 diluent) (443).

FLUOROQUINOLINES

The standard routes to monofluoroquinolines have been the Balz-Schiemann reaction from the corresponding aminoquinoline or the Skraup reaction from glycerol and a fluoroaniline. Exchange-fluorination also has been used. 2-Chloroquinoline and potassium fluoride in dimethyl sulfone gave 60% yield of 2-fluoroquinoline [580-21-2], C_9H_5FN; bp 133°C at 4 kPa (30 mm Hg); 75°C at 0.3 kPa (2 mm Hg); n_D^{25} 1.5827 (406). Likewise, heptachloroquinoline and potassium fluoride at 470°C for 17 h gave a 71% yield of heptafluoroquinoline [13180-38-6], C_9F_7N; mp 95–95.5°C, bp 205°C (444).

The main preparative techniques used to make all seven trifluoromethylquinoline isomers include copper-assisted coupling of the haloquinoline with trifluoromethyl iodide (112); quinolinecarboxylic acid with sulfur tetrafluoride–hydrogen fluoride (434,445); and aminobenzotrifluoride and glycerol (Skraup reaction) (446,447).

Commercial trifluoromethylquinoline-based products are mefloquine [53230-10-7], an antimalarial, and floctafenine [23779-99-9], an analgesic. The cyclization step to construct the 2,8-bis(trifluoromethyl)quinoline nucleus in mefloquine employs 2-aminobenzotrifluoride [*88-17-5*] and ethyl 4,4,4-trifluoro-acetoacetate [*372-31-6*] (448).

mefloquine   floctafenine

FLUOROQUINOLONES

A primary development has been the rapid commercialization of fluoroquinolone antibacterials (427). The single-fluorinated quinolones (second generation) constitute 10% of the worldwide prescriptions for antibiotics: norfloxacin [*70458-96-7*], enoxacin [*74011-58-8*], perfloxacin [*70458-92-3*], ciprofloxacin [*85721-33-1*], and ofloxacin [*82419-36-1*] (449) (see ANTIBACTERIAL AGENTS, QUINOLONES). Annual sales (1992) were estimated at $800 million. As new agents are introduced, fluoroquinolones are expected to maintain an average growth rate of 30% during the 1991–1997 time period (450). New synthesis strategy includes multiple-fluorinated quinolones (third generation) such as lomefloxacin [*98079-51-7*], fleroxacin [*79660-72-3*], temafloxacin hydrochloride [*105784-61-0*], and tosufloxacin [*108138-46-1*].

The discovery of new broad spectrum antibiotics has been accompanied by the development of processes for fluorinated feedstocks: ring-fluorinated aromatics for those quinolones containing a fluorobenzopyridone group, and fluorinated pyridine precursors for those antibiotics containing a naphthyridine nucleus (enoxacin, tosufloxacin) (see Table 14).

FLUOROPYRIMIDINES

Fluoropyrimidines find diverse use in cancer chemotherapy and other drug applications, as well as in fiber-reactive dyes. Table 13 lists physical properties of representative fluoropyrimidines.

5-Fluoropyrimidine derivatives are of tremendous importance in cancer chemotherapy, eg, 5-fluorouracil [*51-21-8*] (5-FU). The original 5-fluorouracil process featured a multistep low yield route based on ethyl fluoroacetate (451). Direct fluorination (fluorine) of uracil [*66-22-8*] gives high yields of 5-FU (452–455). This process has now been commercialized.

Other monofluoropyrimidines of biological interest are 5-fluorocytosine [2022-85-7], an antifungal agent; 2'-deoxy-5-fluorouridine [50-91-9] (5-FUDR), an antiviral and antineoplastic agent; 5-fluoroorotic acid [703-95-7] (5-FOA), used in yeast molecular genetics (456); and tegafur [17902-23-7] (Ftorafur), an antineoplastic agent which releases 5-FU *in vivo* (see CHEMOTHERAPEUTICS, ANTICANCER).

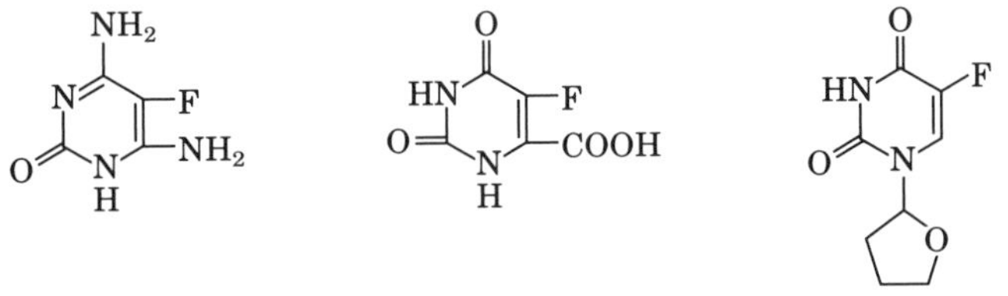

5-fluorocytosine   5-FOA   tegafur

Exchange fluorination is the main synthetic tool to prepare polyfluoro-pyrimidines. It was established that choice of fluorinating agent permits selectivity during exchange fluorination of nuclear- and side-chain chlorinated pyrimidines: NaF and KF fluorinate only in the heterocyclic nucleus; HF in the nucleus and in the chlorinated methyl group; and SbF$_3$ only in the chlorinated methyl group (457).

2,4,6-Trifluoropyrimidine can be prepared in 85% yield from 2,4,6-trichloropyrimidine [3764-01-0] and potassium fluoride in sulfolane or solvent-free conditions (458,459). Derivatives such as 1,1,1-trichloro-3-[5-(2,4,6-trifluoropyrimidyl)]-3,4-epoxybutane [121058-68-2] have been prepared as potential herbicides (460).

2,4,5,6-Tetrafluoropyrimidine has been prepared by direct fluorination of 2,4,6-trifluoropyrimidine with silver difluoride in perfluorobutylamine solvent (461,462). A more direct route (85% yield) is the reaction of tetrachloropyrimidine and potassium fluoride in an autoclave at 480°C for 42 h (463).

Tetrafluoropyrimidine was converted to the antineoplastic 5-fluorouracil (5-FU) by a novel process based on the sequence: partial exchange chlorination (61% yield), selective hydrogenolysis in triethylamine (71% yield) and hydrolysis (85–93% yield) (464).

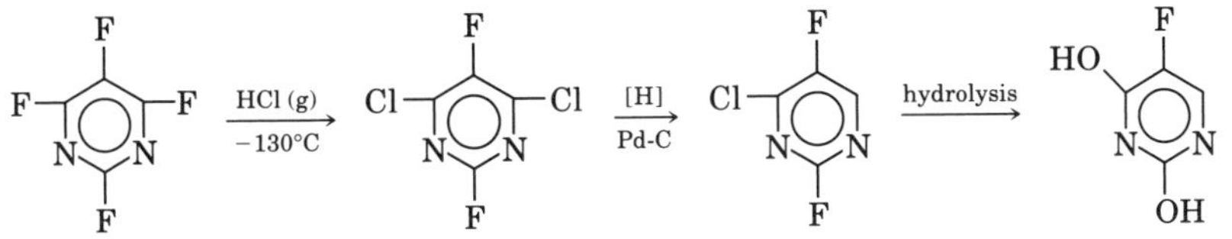

5-Chloro-2,4,6-trifluoropyrimidine [697-83-6] has gained commercial importance for the production of fiber-reactive dyes (465,466). It can be manufactured by partial fluorination of 2,3,5,6-tetrachloropyrimidine [1780-40-1] with anhydrous hydrogen fluoride (autoclave or vapor phase) (467) or sodium fluoride (autoclave, 300°C) (468). 5-Chloro-2,4,6-trifluoropyrimidine is condensed with amine chromophores to provide the 5-chloro-2,4-difluoropyrimidyl group; the fluorine atom of the latter then reacts with a nucleophilic site in the fabric. Commercial reactive dyes for cottons and cellulosics include Levafix EA and PA Dyestuffs and Drimarene K and R Dyestuffs. For wool, the following 5-chloro-2,4-difluoropyrimidyl reactive dyes are offered: Verofix Dyestuffs and Drimalene Dyestuffs (see DYES, REACTIVE).

FLUOROTRIAZINES

Ring-fluorinated triazines are used in fiber-reactive dyes. Perfluoroalkyl triazines are offered commercially as mass spectral markers and have been intensively evaluated for elastomer and hydraulic fluid applications. Physical properties of representative fluorotriazines are listed in Table 13. Toxicity data are available. For cyanuric fluoride, LD_{50} = 3.1 ppm for 4 h (inhalation, rat) and 160 mg/kg (skin, rabbit) (127).

2,4,6-Trifluoro-1,3,5-Triazine. Cyanuric fluoride [675-14-9] can be produced from 2,4,6-trichloro-s-1,3,5-triazine [108-77-0] (cyanuric chloride) with hydrogen fluoride under autoclave (469,470) or vapor-phase (471) conditions. Sodium fluoride (in sulfolane solvent) can also be used to manufacture cyanuric fluoride (472,473).

Cyanuric fluoride is readily hydrolyzed to 2,4,6-trihydroxy-1,3,5-triazine [108-80-5] (cyanuric acid). Cyanuric fluoride reacts faster with nucleophilic agents such as ammonia and amines than cyanuric chloride.

Fiber-reactive dyes containing the fluorotriazinyl group are based on the condensation of chromophores containing amino groups with 6-substituted-2,4-difluorotriazines. The latter can be prepared from cyanuric fluoride or from the reaction of alkali metal fluorides with 6-substituted-2,4-dichlorotriazines. Comparative advantages of monofluorotriazinyl dyes over commercial monochlorotriazinyl analogues have been reviewed (466).

Cyanuric fluoride has been employed as a specific reagent for tyrosine residues in enzymes (474). Cyanuric fluoride can also serve as a fluorinating agent in fluorodehydroxylation reactions, eg, the conversion of 2-hydroxypyridine [142-08-5] to 2-fluoropyridine (475). This technique was subsequently extended to the preparation of acid fluorides from the corresponding carboxylic acid (476). It has found application in peptide synthesis from amino acids through the corresponding acid fluoride (477).

2,4,6-tris-(Trifluoromethyl)-1,3,5-Triazine. This compound can be prepared by trimerization of trifluoroacetonitrile (478) or fluorination of 2,4,6-*tris*-(trichloromethyl)-1,3,5-triazine with hydrogen fluoride−antimony pentachloride or antimony trifluoride−antimony pentafluoride (479). LC_{50} = 1400 ppm for 4 h (inhalation, rat) (127).

tris-(Trifluoromethyl)-*s*-triazine [368-66-1], as well as the *tris*-perfluoroethyl [858-46-8], propyl [915-22-9], heptyl [21674-38-4], and nonyl [57104-59-4] *s*-triazines are commercially offered as mass spectrometry internal reference standards for a wide mass range, 285−1485. The perfluoroalkylene (perfluoroalkyl)-*s*-triazines and perfluoroalkylene(perfluoroalkyloxy)-*s*-triazines were found to be suitable nonflammable hydraulic fluids in the −25 to +300°C temperature range. Numerous laboratories have investigated the synthesis and properties of perfluoroalkylene elastomers containing the *s*-triazine functionality (480,481).

MISCELLANEOUS FLUORINATED NITROGEN HETEROCYCLICS

Two reviews (1981, 1990) include nitrogen heterocyclics not covered in the present survey (482,483). The 1990 review dealing with four-, five-, and six-membered ring heterocyclic compounds emphasizes biological properties (482).

BIBLIOGRAPHY

"Fluorinated Aromatic Compounds" under "Fluorine Compounds, Organic," in *ECT* 2nd ed., Vol. 9, pp. 775–802, by A. K. Barbour, M. W. Buxton, and G. Fuller, Imperial Smelting Corp., Ltd.; in *ECT* 3rd ed., Vol. 10, pp. 901–936, by M. M. Boudakian, Olin Chemicals; "Perfluoroalkylene Triazines" under "Fluorine Compounds, Organic," *ECT* 3rd ed., Vol. 10, pp. 948–951, by W. R. Griffen, Air Force Materials Laboratory.

1. M. R. Kilbourn, *Fluorine-18 Labeling of Radiopharmaceuticals*, National Academy Press, Washington, D.C., 1990.
2. G. Furin, in L. German and S. Zemskov, eds., *New Fluorinating Agents in Organic Synthesis*, Springer-Verlag, Berlin, 1989, Chapt. 9, p. 35.
3. V. Grakauskas, *J. Org. Chem.* **35**, 723 (1970); *Ibid.*, **34**, 2835 (1969).
4. N. B. Kaz'mina and co-workers, *Dokl. Akad. Nauk SSSR* **194**, 1329 (1970).
5. F. Cacace and co-workers, *J. Am. Chem. Soc.* **102**, 3511 (1980).
6. S. T. Purrington and D. L. Woolard, *J. Org. Chem.* **56**, 142 (1991).
7. S. Misaki, *J. Fluorine Chem.* **21**, 191 (1982); *Ibid.*, **17**, 159 (1981).
8. J. Ellis and W. K. R. Musgrave, *J. Chem. Soc.*, 3608 (1950).
9. M. M. Boudakian and G. A. Hyde, *J. Fluorine Chem.* **25**, 435 (1984).
10. R. Filler, *Isr. J. Chem.* **17**, 71 (1978).
11. A. Zweig and co-workers, *J. Org. Chem.* **45**, 3597 (1980).
12. S. Stavber and M. Zupan, *J. Org. Chem.* **50**, 3609 (1985).
13. M. J. Fifolt and co-workers, *J. Org. Chem.* **50**, 4576 (1985).
14. S. Rozen and co-workers, *J. Org. Chem.* **49**, 806 (1984); *Ibid.*, **46**, 4629 (1981).
15. D. Hebel and co-workers, *Tetrahedron Lett.* **31**, 619 (1990).
16. S. Singh and co-workers, *J. Am. Chem. Soc.* **109**, 7194 (1987).
17. E. Differding and H. Ofner, *SYNLETT*, 187 (Mar. 1991); *Chem. Eng. News*, 33 (Sept.7, 1992).
18. T. Umemoto and co-workers, *J. Am. Chem. Soc.* **112**, 8563 (1990).
19. A. G. Gilicinski and co-workers, *J. Fluorine Chem.* **59**, 157 (1992).
20. R. E. Banks and co-workers, *J. Chem. Soc., Chem. Commun.*, (8), 595 (1992).
21. Ger. Pat. 600,706 (July 30, 1934), P. Osswald and O. Scherer (to I. G. Farbenind.); F. O. Robitschek and B. H. Wilcoxon, *FIAT Final Report*, No. 998, Mar. 31, 1947.
22. R. L. Ferm and C. A. VanderWerf, *J. Am. Chem. Soc.* **72**, 4809 (1950).
23. T. Fukuhara and co-workers, *Synth. Commun.* **17**, 685 (1987).
24. G. A. Olah and J. Welch, *J. Am. Chem. Soc.* **97**, 208 (1975).
25. U.S. Pat. 4,096,196 (June 20, 1978), M. M. Boudakian (to Olin).
26. U.S. Pat. 4,075,252 (Feb. 21, 1978), M. M. Boudakian (to Olin).
27. U.S. Pat. 4,912,268 (Mar. 27, 1990), M. H. Krackov and C. H. Ralston (to Du Pont).
28. D. T. Flood, *Org. Synth., Coll. Vol.* **II**, 295 (1943).
29. A. Roe, in R. Adams, ed., *Organic Reactions*, Vol. 5, John Wiley & Sons, Inc., New York, 1949, Chapt. 4, p. 193.
30. H. Suschitzky, in M. Stacey, J. C. Taylor, and A. G. Sharpe, eds., *Advances in Fluorine Chemistry*, Vol. 4, Butterworths, Washington, D.C., 1964, p. 1.
31. K. G. Rutherford and co-workers, *J. Org. Chem.* **26**, 5149 (1961).
32. M. P. Doyle and W. J. Bryker, *J. Org. Chem.* **44**, 1572 (1979).

33. U.S. Pat. 4,476,320 (Oct. 9, 1984), H. Diehl, H. Pelster, and H. Habetz (to Bayer).
34. D. J. Milner, *Synth. Commun.* **22**, 73 (1992).
35. L. Treschanke, Riedel-de Haën, personal communication, Feb. 5, 1992.
36. J. P. Regan, *Specialty Chemicals* **21**, 56 (Jan. 1986).
37. U.S. Pat. 2,705,730 (Apr. 5, 1955), J. D. Head (to Dow).
38. *Chem. Eng. News*, 65 (Apr. 24, 1961).
39. O. Wallach, *Ann. Chem.* **235**, 255 (1886).
40. O. Wallach and F. Heusler, *Ann. Chem.* **243**, 219 (1888).
41. U.S. Pat. 4,194,054 (Mar. 18, 1980), H. Förster and co-workers (to Bayer).
42. Ger. Pat. 3,142,856 (May 11, 1983), E. Klauke and K. Grohe (to Bayer).
43. T. J. Tewson and M. Welch, *J. Chem. Soc., Chem. Commun.*, 1149 (1979).
44. W. Prescott, *Chem. Ind. London*, 56 (1978); L. Dolby-Glover, *Chem. Ind. London*, 518 (1986).
45. Y. Kimura, *J. Synth. Org. Chem., Japan* **47**, 258 (1989).
46. G. C. Finger and co-workers, *J. Am. Chem. Soc.* **78**, 6034 (1956); *Chem. Ind. London*, 1328 (1962).
47. A. J. Beaumont and J. H. Clark, *J. Fluorine Chem.* **32**, 295 (1991).
48. Y. Uchibori and co-workers, *SYNLETT*, 345 (Apr. 1992).
49. N. Ishikawa and co-workers, *Chem. Lett.*, 761 (1981).
50. Y. Kimura and H. Suzuki, *Tetrahedron Lett.* **30**, 1271 (1989).
51. U.S. Pat. 4,069,262 (Jan. 17, 1978), R. A. Kunz (to Du Pont).
52. U.S. Pat. 4,642,399 (Feb. 10, 1987), C. R. White (to Mallinckrodt).
53. U.S. Pat. 4,287,374 (Sept. 1, 1981), R. A. North (to Boots).
54. C. Liotta and H. P. Harris, *J. Am. Chem. Soc.* **96**, 2250 (1974).
55. U.S. Pat. 4,642,398 (Feb. 10, 1987), G. L. Cantrell (to Mallinckrodt).
56. S. Kumai and co-workers, *Reports Res. Lab. Asahi Glass Co.* **39**, 317 (1989).
57. J. Deutsch and H.-J. Niclas, *Synth. Commun.* **21**, 205 (1991).
58. U.S. Pat. 4,429,365 (Oct. 21, 1980), H. -G. Oeser and co-workers (to BASF).
59. Y. Yoshida and Y. Kimura, *J. Fluorine Chem.* **44**, 291 (1989).
60. R. E. Banks and co-workers, *J. Fluorine Chem.* **46**, 529 (1990).
61. Y. Yoshida and co-workers, *J. Fluorine Chem.* **53**, 301 (1991).
62. N. C. O'Reilly and co-workers, *SYNLETT*, 609 (Oct. 1990).
63. Y. Yoshida and co-workers, *J. Fluorine Chem.* **53**, 335 (1991).
64. U.S. Pat. 4,937,396 (June 26, 1990), R. G. Pews and co-workers (to Dow).
65. U.S. Pat. 3,240,824 (Mar. 15, 1966), M. M. Boudakian and E. R. Shipkowski (to Olin).
66. *Chem. Eng. News* **29**, 2666 (1951).
67. D. G. Mooney, *Inst. Chem. Eng. Symp. Series* **124**, 381 (1991); A. T. Cates, *J. Hazard. Mater.* **32**, 1 (1992).
68. L. Bretherick, *Handbook of Reactive Chemical Hazards*, 4th ed., Butterworths, London, 1990.
69. G. Bartoli and co-workers, *J. Chem. Soc. Perkin Trans.* **1**, 2671 (1972).
70. F. Effenberger and W. Streicher, *Chem. Ber.* **124**, 157 (1991).
71. S. Kumai and co-workers, *Reports Res. Lab. Asahi Glass Co.* **35**, 153 (1985).
72. U.S. Pat. 4,568,781 (Feb. 4, 1986), F. Effenberger and W. Streicher (to BASF); H. Suzuki and co-workers, *Bull. Chem. Soc. Jpn.* **63**, 2010 (1990).
73. M. Stacey and J. C. Tatlow, in M. Stacey, J. C. Tatlow, and A. G. Sharpe, eds., *Advances in Fluorine Chemistry*, Vol. 1, Butterworths, London, 1960, p. 166.
74. P. A. Coe and co-workers, *J. Chem. Soc. C.*, 1060 (1969).
75. R. D. Chambers and co-workers, *Tetrahedron* **19**, 891 (1963).
76. G. M. Brooks and co-workers, *J. Chem. Soc.*, 729 (1964).
77. U.S. Pat. 3,499,942 (Mar. 10, 1970), O. M. Nevedov and A. Ivashenko.

78. N. Ishikawa, *Senryo to Yakuhin* **26**(6), 106 (1981); *J. Synth. Org. Chem., Jpn.* **40**(2), 158 (1982).

79. N. V. Volchkov and co-workers, *Bull. Acad. Sci. USSR* **38**, 1782 (1989). (Engl. trans.).

80. U.S. Pat. 4,390,740 (June 28, 1983), I. Tabushi and co-workers (to Daikin Kogyo).

81. R. E. Ehrenkaufer and co-workers, *J. Org. Chem.* **47**, 2489 (1982).

82. G. A. Olah and P. Kreienbühl, *J. Org. Chem.* **32**, 1614 (1967).

83. K. O. Christe and A. E. Pavlath, *J. Org. Chem.* **31**, 559 (1966).

84. D. P. Ashton and co-workers, *J. Fluorine Chem.* **27**, 263 (1985); U.S. Pat. 4,745,235 (May 17, 1988), D. P. Ashton and co-workers (to ICI).

85. Fr. Demande Pat. 2,647,106 (Nov. 23, 1990), L. Gilbert and co-workers (to Rhône-Poulenc).

86. Eur. Pat. Appl. 427,603 (May 15, 1991), H. Garcia and co-workers (to Rhône-Poulenc).

87. G. G. Yakobson and co-workers, *Zh. Vses. Khim. Obshchestva im. D. I. Mendeleeva* **10**, 466 (1965).

88. M. Van Der Puy, *J. Org. Chem.* **53**, 4398 (1988); U.S. Pat. 5,081,275 (Jan. 14, 1992) (to AlliedSignal).

89. Y. Yazawa and co-workers, *Chem. Lett.*, 2213 (1989).

90. I. N. Rozhkov and co-workers, *Isv. Akad. Nauk SSSR Ser. Khim.*, 1130 (1972); *Russian Chem. Rev.* **45**, 615 (1976) (Engl. trans.).

91. Ger. Offen. 2,516,355 (Oct. 28, 1976), F. Beck (to BASF).

92. Jpn. Kokai Tokkyo Koho JP 63,111,192 (May 16, 1988), A. Shimizu and K. Yamatka (to Asahi Chemical Ind.); Jpn. Kokai Tokkyo Koho 62,127,488 (June 9, 1987), A. Shimizu and K. Yamatka (to Asahi Chemical Ind.).

93. J. H. H. Meurs and co-workers, *Angew. Chem. Int. Ed. Engl.* **28**, 927 (1989).

94. U.S. Pat. 2,884,458 (Apr. 28, 1959), D. A. Fidler (to Olin).

95. D. A. Fidler, J. S. Logan, and M. M. Boudakian, *J. Org. Chem.* **26**, 4014 (1961).

96. P. H. Scott and co-workers, *Tetrahedron Lett.*, 1153 (1970); U.S. Pat. 3,558,707 (Jan. 26, 1971), J. W. Churchill and E. H. Kober (to Olin).

97. Brit. Pat. Appl. 2,241,952 (Sept. 18, 1991), D. Levin and J. S. Moillet (to ICI).

98. Jpn. Kokai Tokkyo Koho 01,238,560 (Sept. 22, 1989), J. Negishi and T. Kawai (to Central Glass).

99. Eur. Pat. 248,746 (Apr. 25, 1990), M. Desbois (to Rhône-Poulenc).

100. U.S. Pat. 3,910,985 (Oct. 7, 1975), P. P. Montijn (to Shell Oil).

101. Span. ES 493,977 (Aug. 1, 1981), A. L. Porta (to Dr. Andreu S. A.).

102. A. I. Titov and A. N. Baryshnikova, *J. Gen. Chem. (USSR)* **23**, 346 (1953) (Engl. trans.).

103. U.S. Pat. 4,391,991 (July 5, 1983), P. F. Mundhenke and M. J. Fifolt (to Occidental Chemical).

104. PCT Int. Appl. WO 91 17,138 (Nov. 14, 1991), T. F. Braish (to Pfizer).

105. U.S. Pat. 4,367,350 (Jan. 4, 1983), U. Hiramatsu and co-workers (to Daikin Kogyo); U.S. Pat. 4,400,563 (Apr. 23, 1983), Y. Ohsaka and co-workers (to Daikin Kogyo).

106. G. A. Boswell and co-workers, *Organic Reactions* **21**, 1 (1974).

107. C. -L. Wang, *Organic Reactions* **34**, 319 (1985).

108. U.S. Pat. 3,976,691 (Aug. 24, 1976), W. J. Middleton (to Du Pont).

109. U.S. Pat. 4,533,777 (Aug. 6, 1985), A. Marhold and E. Klauke (to Bayer).

110. A. Marhold and E. Klauke, *J. Fluorine Chem.* **18**, 281 (1981).

111. V. C. R. McLaughlin and J. Thrower, *Tetrahedron* **25**, 5921 (1969).

112. Y. Kobayashi and I. Kumadaki, *Tetrahedron Lett.*, 4095 (1969); *Chem. Pharm. Bull.* **18**, 2334 (1970).

113. D. M. Wiemers and D. J. Burton, *J. Am. Chem. Soc.* **108**, 832 (1986).

114. J. H. Clark and co-workers, *J. Fluorine Chem.* **50**, 411 (1950).

115. J. M. Paratian and co-workers, *J. Chem. Soc., Chem. Commun.*, 53 (1992).

116. K. Matsui and co-workers, *Chem. Lett.*, 1719 (1981).
117. J. N. Freskos, *Synth. Commun.* **18**, 965 (1988).
118. B. R. Langlois and co-workers, *Tetrahedron Lett.* **32**, 7525 (1991).
119. J. G. MacNeil and D. J. Burton, *J. Fluorine Chem.* **55**, 225 (1991).
120. Q.-Y. Chen and co-workers, *J. Fluorine Chem.* **55**, 291 (1991).
121. H. Sawada and co-workers, *J. Fluorine Chem.* **46**, 423 (1990).
122. T. Umemoto and co-workers, *Chem. Lett.*, 1663 (1981).
123. W. A. Sheppard, *J. Am. Chem. Soc.* **87**, 2410 (1965).
124. A. F. Feiring, *J. Fluorine Chem.* **10**, 375 (1977); *J. Org. Chem.* **44**, 1252 (1979).
125. P. Martin, *Tetrahedron* **41**, 4057 (1985).
126. *Handbook of Aromatic Fluorine Compounds*, Olin Corp., Stamford, Conn., 1976.
127. D. V. Sweet, ed., *Registry of Toxic Effects of Chemical Substances-1985/1986 Edition*, Vols. 1–5, DHHS (NIOSH) publication no. 87-114, Washington, D.C., 1987.
128. G. A. Olah and co-workers, *J. Chem. Soc.*, 1823 (1957).
129. P. A. Janssen and co-workers, *J. Med. Pharm. Chem.* **1**, 281 (1959).
130. H. C. Brown and G. Marino, *J. Am. Chem. Soc.* **84**, 1658 (1962).
131. P. A. Staniland, in G. Allen and J. C. Bevington, eds., *Comprehensive Polymer Science*, Vol. 5, Pergamon Press, New York, 1989, Chapt. 29, p.490.
132. P. M. Hergenrother and co-workers, *Polymer* **29**, 358 (1988).
133. P. M. Hergenrother and co-workers, *J. Polym. Sci. Part A: Polym. Chem.* **29**, 1483 (1991).
134. M. E. Neubert and D. L. Fishel, *Mol. Cryst. Liq. Cryst.* **53**, 101 (1979).
135. P. A. Worthington, in D. R. Baker and co-workers, eds., *Synthesis and Chemistry of Agrochemicals*, ACS Symposium Series no. 355. American Chemical Society, Washington, D.C. 1987, Chap. 27, p.302.
136. U.S. Pat. 4,522,754 (June 11, 1985), B. Hilti and co-workers (to Ciba-Geigy).
137. R. E. Parker, in M. Stacey, J. C. Tatlow, and A. G. Sharpe, eds., *Advances in Fluorine Chemistry*, Vol. 3, Butterworths, Washington, D.C., 1963, p.63.
138. M. Bil, *J. Appl. Chem. Biotechnol.* **22**, 853 (1972); *Chem. Ind. London*, 656 (1971).
139. U.S. Pat. 3,959,377 (May 25, 1976) and U.S. Pat. 3,632,582 (Jan. 4, 1972), M. Bil (to Clairol).
140. Ger. Offen. 3,534,369 (Apr. 3, 1986), J. F. Grollier and co-workers (to Oreal).
141. F. Sanger, *Biochem. J.* **39**, 507 (1945).
142. S. E. Drewes and co-workers, *J. Chem. Soc., Perkin Trans.* **1**, 1283 (1975).
143. G. W. J. Fleet and R. Porter, *Nature*, 511 (1970).
144. K. J. Skinner, *Chem. Eng. News*, 23 (Aug. 18, 1975).
145. E. D. Bergmann and M. Bentov, *J. Org. Chem.* **26**, 1480 (1961).
146. R. W. Holley and A. D. Holley, *J. Am. Chem. Soc.* **74**, 1110 (1952).
147. F. F. Micheel and co-workers, *Ann. Chem.* **581**, 238 (1953).
148. H. Zahn and K. H. Lebkücher, *Biochem. Z.* **334**, 133 (1961).
149. H. Zahn and H. Zuber, *Ber.* **86**, 172 (1953).
150. H. Zahn and J. Meinhoffer, *Makromol. Chem.* **26**, 126 (1958).
151. K. L. Kirk and L. A. Cohen, *J. Org. Chem.* **34**, 395 (1969).
152. Y. Watanabe and Y. Imai, *J. Chromatogr.* **239**, 723 (1982).
153. P. Marfey, *Carlsberg Res. Commun.* **49**, 591 (1984).
154. A. E. Katz and co-workers, *Polym. Prepr. (Am. Chem. Soc., Div. Polym. Chem.)* **33**(3), 144 (1991).
155. K. Sturm and co-workers, *Chem. Ber.* **99**, 328 (1966).
156. U.S. Pats. 3,668,248 (June 6, 1972) and 3,419,552 (Dec. 31, 1968), C. W. Whitehead and J. J. Traverso (to Eli Lilly).
157. R. N. Johnson and co-workers, *J. Polym. Sci. Part A-1* **5**, 2375 (1967).
158. A. E. Feiring and co-workers, *J. Polym. Sci. Part A: Polym. Chem.* **28**, 2809 (1990).

159. U.S. Pat. 5,084,548 (Jan. 28, 1992), A. E. Feiring and S. D. Archer (to Du Pont).
160. U.S. Pat. 5,034,034 (July 23, 1991), E. S. Sanders and T. L. Parker (to Dow).
161. H. R. Kricheldorf and M. Bergahn, *Makromol. Chem., Rapid Commun.* **12**, 529 (1991).
162. M. M. Boudakian and co-workers, *J. Org. Chem.* **26**, 4641 (1961).
163. G. G. Yakobson and co-workers, *Dokl. Akad. Nauk SSSR* **141**, 1395 (1961).
164. S. H. Yu and E. C. Ashby, *J. Org. Chem.* **36**, 2123 (1971).
165. W. K. Moberg and co-workers, in Ref. 135, Chapt. 26, p. 288.
166. E. D. Morgan and C. F. Poole, *J. Chromatogr.* **104**, 351 (1975).
167. L. Bretherick, *Chem. Ind. London*, 1017 (1971).
168. A. J. Adduci, *Chemtech*, 575 (Sept. 1976).
169. M. Tsubouchi and co-workers, *Anal. Chem.* **57**, 783 (1985).
170. D. W. Ribbons and co-workers, *J. Fluorine Chem.* **37**, 299 (1987).
171. S. C. Taylor and co-workers, *Performance Chemicals*, 18 (Nov. 1986); *Specialty Chemicals* **8**(3), 236 (June 1988).
172. Eur. Pat. Appl. 253,438 (Jan. 20, 1988), J. A. Schofield and co-workers (to Shell Int. Research).
173. Ger. Offen. 3,520,316 (Dec. 11, 1986), E. Begemann and H. Schmand (to Riedel-de Haën).
174. U.S. Pat. 4,822,927 (Apr. 18, 1989), N. J. Stepaniuk and B. J. Lamb (to Mallinckrodt).
175. U.S. Pat. 4,812,572 (Mar. 14, 1989), M. S. Howarth and D. M. Tomkinson (to ICI).
176. U.S. Pat. 2,939,766 (June 7, 1960), J. W. Churchill (to Olin).
177. U.S. Pat. 5,032,371 (July 16, 1991), H. J. Buehler (to Mallinckrodt).
178. U.S. Pat. 2,563,796 (Aug. 7, 1951), W. H. Shenk, Jr. and G. R. Mellon (to Harshaw Chemical).
179. U.S. Pat. 3,160,623 (Dec. 8, 1964), L. G. Anello and C. Woolf (to Allied Chemical).
180. F. Seel, *Angew. Chem. Int. Ed. Engl.* **4**, 635 (1965).
181. Jpn. Kokai 74 81,330 (Aug. 6, 1974), S. Misaki and M. Okamoto (to Daikin Kogyo).
182. D. Hunter, *Chem. Week* **148**(27), 36 (July 24, 1991); *Ibid*, **146**(9), 41 (Mar. 1, 1989).
183. F. J. Weigert and W. A. Sheppard, *J. Org. Chem.* **25**, 4006 (1976).
184. R. Filler and S. M. Naqvi, in R. Filler and Y. Kobayashi, eds., *Biomedicinal Aspects of Fluorine Chemistry*, Kodansha Biomedical Press, Tokyo, 1982, Chapt. 1, p. 28.
185. P. H. Bolton and co-workers, *Biopolymers* **20**, 435 (1981).
186. H. J. Plach, *New LC Mixtures for STN and Active Matrix Displays Using New Terminally Fluorinated Compounds*, technical literature, E. Merck & Co., Darmstadt, Oct. 1990.
187. R. Neef, *The Merck Group Liquid Crystal Newsletter*, (7) (Feb. 1991), (8) (Sept. 1991), Darmstadt.
188. U.S. Pat. 5,077,160 (Dec. 31, 1991), K. -Y. Law and F. C. Bailey (to Xerox).
189. K.-Y. Law and F. C. Bailey, *J. Imaging Sci.* **31**, 172 (1987).
190. K.-Y. Law and co-workers, *Dyes and Pigments* **9**, 187 (1988).
191. P. M. Kazmeier and co-workers, *J. Imaging Sci.* **32**, 1 (1988).
192. H. Hopff, *Chimia* **15**, 193 (1961).
193. J. T. Minor and C. A. VanderWerf, *J. Org. Chem.* **17**, 1425 (1952).
194. Jpn. Kokai Tokkyo Koho JP 03,232,828 (Oct. 16, 1991), K. Momota and T. Yonezawa (to Morita Kagaku).
195. U.S. Pat. 5,055,220 (Oct. 8, 1991), M. Uchida and co-workers (to Chisso).
196. M. Hird and co-workers, *Liquid Crystals* **11**, 531 (1992).
197. Jpn. Kokai Tokkyo Koho JP 61 63,627 (Apr. 1, 1986), N. Yoneda and co-workers (to Mitsubishi Chemical).
198. G. Schiemann and R. Pillarsky, *Chem. Ber.* **62**, 3035 (1929).
199. G. Furin and co-workers, in I. L. Knunyants and G. G. Yakobson, eds., *Syntheses of Fluoroorganic Compounds*, Springer-Verlag, Berlin, 1985, Chapt. 2, p. 109.

200. Jpn. Kokai Tokkyo Koho JP 03 34,944 (Feb. 14, 1991), T. Mogi (to Tokemu Product K.K.).
201. G. D. Hartman and S. E. Biffar, *J. Org. Chem.* **42**, 1468 (1977).
202. Brit. Pat. 1,514,082 (June 14, 1978), G. Fuller (to ISC).
203. Brit. Pat. Appl. 2,058,067 (Apr. 8, 1988), D. Wotton (to ISC).
204. R. H. Shiley, D. R. Dickerson and G. C. Finger, *J. Fluorine Chem.* **2**, 19 (1972/73).
205. A. M. Roe, R. A. Burton, and D. R. Reavill, *J. Chem. Soc., Chem. Commun.*, 582 (1965).
206. W. J. Feast and R. Stephens, *J. Chem. Soc.*, 3502 (1965).
207. I. Hayakawa and co-workers, *Chem. Pharm. Bull.* **32**, 4907 (1984).
208. U.S. Pat. 4,528,287 (July 9, 1985), Y. Itoh and co-workers (to Hokuriku Pharmaceutical).
209. G. C. Finger and R. E. Oesterling, *J. Am. Chem. Soc.* **78**, 2593 (1956).
210. T. F. Braish and D. E. Fox, *Org. Prep. Proced. Int.* **23**, 655 (1991).
211. G. C. Finger and co-workers, *J. Am. Chem. Soc.* **73**, 145 (1951).
212. J. P. Sanchez and co-workers, *J. Med. Chem.* **31**, 983 (1988); *Ibid.* **35**, 361 (1992).
213. D. T. W. Chu and co-workers, *Can. J. Chem.* **70**, 1323 (1992).
214. U.S. Pat. 5,021,605 (June 4, 1991), H. Kobayashi and M. Shimizu (to SDS Biotech).
215. U.S. Pat. 4,885,386 (Dec. 5, 1989), J. N. Wemple and co-workers (to Warner-Lambert).
216. G. C. Finger and co-workers, *J. Am. Chem. Soc.* **73**, 153 (1951).
217. Eur. Pat. Appl. 460,639 (Dec. 11, 1991), Y. Kobayashi and co-workers (to SDS Biotech).
218. T. Ishizaki and co-workers, *Nippon Kagaku Kaishi*, 2054 (1985).
219. U.S. Pat. 4,782,180 (Nov. 1, 1988), J. N. Wemple and co-workers (to Warner-Lambert).
220. U.S. Pat. 5,047,553 (Sept. 10, 1991), D. M. Novak and H. C. Lin (to Occidental Chemical).
221. D. G. Holland and co-workers, *J. Org. Chem.* **29**, 3042 (1964).
222. U.S. Pat. 4,952,695 (Aug. 28, 1990), K. Grohe and co-workers (to Bayer).
223. D. T. W. Chu and co-workers, *J. Med. Chem.* **30**, 504 (1987).
224. H. Egawa and co-workers, *Chem. Pharm. Bull.* **34**, 4098 (1986).
225. S. K. Ray, *Adv. Energy Convers.* **6**, 89 (1966).
226. G. Fuller, *J. Chem. Soc.*, 6264 (1965).
227. Brit. Pat. 996,498 (June 30, 1965), G. Fuller (to ISC).
228. J. M. Birchall and co-workers, *J. Chem. Soc.*, 4966 (1962).
229. P. L. Coe and co-workers, *J. Chem. Soc. C*, 2323 (1966).
230. M. M. Murza and co-workers, *Zh. Org. Khim.* **13**, 1046 (1977).
231. I. Kisfaludy and co-workers, *Tetrahedron Lett.*, 1785 (1974).
232. G. H. Kalb and co-workers, *Appl. Polym. Symposium*, (22), 127 (1973).
233. G. H. Kalb and co-workers, in N. A. J. Platzer ed., *Polymerization Reactions and New Polymers*, Advances in Chemistry Series no. 129, American Chemical Society, Washington, D.C., 1973, Chapt. 2, p. 13.
234. F. K. Kawahara, *Anal. Chem.* **40**, 2073 (1968).
235. R. J. Harper and co-workers, *J. Org. Chem.* **29**, 3042 (1964).
236. E. Kinsella and A. G. Massey, *Chem. Ind. London*, 1017 (1971).
237. S. C. Cohen and A. G. Massey, in J. C. Tatlow, R. D. Peacock, and H. H. Hyman, eds., *Advances in Fluorine Chemistry*, Vol. 6, CRC Press, Cleveland, Ohio, 1970, p. 83.
238. J. W. Eichelberger and co-workers, *Anal. Chem.* **47**, 995 (1975).
239. T. Nakamura and O. Kaieda, *J. Synth. Org. Chem., Jpn.* **47**, 20 (1989).
240. J. Wallace and co-workers, *Clin. Chem.* **25**, 252 (1979); *Ibid* **24**, 895 (1978).
241. J. Wallace and co-workers, *Anal. Chem.* **49**, 903 (1977).
242. J. C. Lhuguenot and B. F. Baume, *J. Chromatogr. Sci.* **12**, 411 (1974).
243. K. T. Koshy and co-workers, *J. Chromatogr. Sci.* **13**, 97 (1975).

244. T. Nambara and co-workers, *J. Chromatogr.* **114**, 81 (1975).

245. U.S. Pat. 3,946,063 (May 23, 1976), J. Vessman and co-workers (to Pierce Chemical).

246. K. Blau and G. S. King, eds., *Handbook of Derivatives for Chromatography*, Heyden, London, 1977.

247. R. Filler, *Fluorine Chem. Rev.* **8**, 1 (1977).

248. L. A. Wall, ed., *Fluoropolymers*, Vol. 25, Wiley-Interscience, New York, 1972.

249. G. W. Holbrook, L. A. Loree, and O. R. Pierce, *J. Org. Chem.* **31**, 1259 (1966).

250. N. N. Vorozhtsov, Jr. and co-workers, *Isv. Akad. Nauk SSSR, Ser. Khim.* (8), 1524 (1963).

251. Y. Désirant, *Bull. Acad. Sci. Roy. Belg.* **41**, 759 (1955).

252. J. M. Birchall and R. N. Haszeldine, *J. Chem. Soc.*, 13 (1959).

253. L. A. Wall and co-workers, *J. Research NBS* **65A**, 239 (1961).

254. R. A. Falk, *Sperry Eng. Rev.* **16**(3), 24 (Fall 1963).

255. U.S. Pat. 3,158,657 (Nov. 24, 1964), F. R. Callihan and C. L. Quateta (to Sperry Rand).

256. L. S. Kobrina, *Fluorine Chem. Rev.* **7**, 1 (1974).

257. U.S. Pat. 5,087,764 (Feb. 11, 1992), V. Reiffenrath and J. Krause (to E. Merck GmbH).

258. U.S. Pat. 5,064,567 (Nov. 12, 1991), F. Funada and co-workers (to Sharp and E. Merck GmbH).

259. A. Villiger and F. Leenhouts, *Mol. Cryst. Liq. Cryst.* **209**, 297 (1991).

260. Jpn. Kokai Tokkyo Koho JP 03,197,438 (Aug. 28, 1991), N. Satake and co-workers (to Kanto Chemical).

261. Brit. Pat. Appl. 2,232,416 (Dec. 12, 1990), D. Coates and co-workers (to U.K. Secretary of State for Defence).

262. Jpn. Kokai Tokkyo Koho JP 03,176,445 (July 31, 1991), T. Kuriyama and co-workers (to Dainippon).

263. U.S. Pat. 5,032,314 (July 16, 1991), M. Ushioda and co-workers (to Chisso).

264. Jpn. Kokai Tokkyo Koho JP 04 29,951 (Jan. 31, 1992), T. Watanabe and M. Sato (to Sanyo Chemical).

265. M. Hird and co-workers, *Liquid Crystals* **11**, 531 (1992).

266. U.S. Pat. 5,061,400 (Oct. 29, 1991), T. Obikawa (to Seiko Epson).

267. G. Schiemann and E. Bolstead, *Chem. Ber.* **61B**, 1403 (1928).

268. J. Hannah and co-workers, *J. Med. Chem.* **21**, 1093 (1978).

269. U.S. Pat. 4,443,631 (Apr. 13, 1984), A. G. Padilla (to Upjohn).

270. W. J. Pummer and L. A. Wall, *J. Chem. Eng. Data* **6**, 76 (1961).

271. M. W. Buxton and co-workers, *J. Fluorine Chem.* **2**, 387 (1972/73).

272. F. Mercer and co-workers, *J. Polym. Sci., Part A: Polym. Chem.* **30**, 1767 (1992); U.S. Pat. 5,114,780 (May 19, 1992), F. Mercer and co-workers (to Raychem).

273. M. S. Newman, in P. O. P. Ts'o and J. D. DiPaolo, eds., *Chemical Carcinogenesis*, Marcel Dekker, Inc., New York, 1974, p. 177.

274. D. R. MacKenzie and co-workers, *J. Phys. Chem.* **69**, 2526 (1965).

275. E. Boger and co-workers, *J. Fluorine Chem.* **8**, 513 (1976).

276. O. Lerman and co-workers, *J. Org. Chem.* **49**, 806 (1984).

277. S. Stavber and M. Zupan, *J. Fluorine Chem.* **17**, 597 (1981).

278. I. N. Rozhkov and co-workers, *Dokl. Akad. Nauk SSSR, Chem. Sect.* **193**, 618 (1970) (Engl. trans.).

279. R. F. O'Malley and co-workers, *J. Org. Chem.* **46**, 2816 (1981); *J. Electrochem. Soc.* **130**, 2170 (1983).

280. U.S. Pat. 3,707,843 (Jan. 2, 1973), R. C. Connor and L. L. Ferstandig (to Halocarbon Products).

281. U.S. Pat. 3,753,345 (Aug. 21, 1973), F. H. Cassidy and R. Garcia (to Aerojet-General).

282. E. Kühle and E. Klauke, *Angew. Chem. Int. Ed. Engl.* **16**, 735 (1977).

283. R. Filler and H. Novar, *Chem. Ind. London*, 1273 (1960).

284. R. G. Jones, *J. Am. Chem. Soc.* **69**, 2346 (1947).

285. V. C. R. McLoughlin and J. Thrower, *Chem. Ind. London*, 1557 (1964).

286. R. Grinter and co-workers, *Tetrahedron Lett.*, 3845 (1968).

287. R. Wächter, *Angew. Chem.* **67**, 305 (1955).

288. F. Swarts, *Bull. Sci. Acad. Roy. Belg.*, 399 (1920).

289. Y. Kobayashi and I. Kumadaki, *Acc. Chem. Res.* **11**, 197 (1978).

290. N. P. Buu-Hoi and co-workers, *Compt. Rend.* **257** (21), 3182 (1963).

291. *Product List No. 4*, Marshallton Research Labs, West Chester, Pa., Jan. 1973, p. 27.

292. L. Bretherick, *Chem. Ind. London*, 1017 (1971).

293. I. C. Appleby, *Chem. Ind. London*, 120 (1971).

294. E. C. Ashby and D. M. Al-Fekri, *J. Organometallic Chem.* **390**, 275 (1990).

295. U.S. Pat. 3,234,292 (Feb. 8, 1966), S. Robota and E. A. Belmore (to Hooker Chemical).

296. R. J. Albers and E. C. Kooyman, *Rec. Trav. Chim.* **83**, 930 (1964).

297. A. A. Ushakov and co-workers, *Zh. Org. Khim.* **12**, 158 (1963).

298. K. Inukai and T. Ueda, *Kogyo Kagaku Zasshi* **64**, 2156 (1961).

299. Jpn. Kokai 75 76,029 (June 21, 1975), S. Misaki and co-workers (to Daikin Kogyo).

300. U.S. Pat. 4,401,623 (Aug. 30, 1983), T. Giacobbe and G. Tsien (to Rhône-Poulenc).

301. E. T. McBee and co-workers, *J. Am. Chem. Soc.* **72**, 1651 (1950).

302. U.S. Pat. 4,008,278 (Feb. 15, 1977), M. M. Boudakian (to Olin).

303. Ger. Pat. 1,568,938 (Dec. 12, 1974), H. Treiber (to Knoll).

304. U.S. Pat. 4,144,265 (Mar. 13, 1979), W. Dowd and T. H. Fisher (to Dow).

305. U.S. Pat. 4,966,988 (Oct. 30, 1990), W. I. Schinski and P. Denisevich (to Chevron).

306. C. E. Ward and co-workers, in Ref. 135, Chapt. 6, p.65.

307. A. L. Henne and M. S. Newman, *J. Am. Chem. Soc.* **60**, 1697 (1938).

308. W. Schwertfeger and G. Siegemund, *J. Fluorine Chem.* **36**, 237 (1987).

309. U.S. Pat. 4,978,798 (Dec. 18, 1990), J. S. Stults (to Occidental Chemical).

310. C. Saboureau and co-workers, *J. Chem. Soc., Chem. Commun.*, 1138 (1989).

311. R. Filler, in Ref. 237, p. 1.

312. *Benzotrifluoride*, data sheet no. 778-D, Hooker Chemical Corp., Niagara Falls, N.Y., 1969.

313. F. Swarts, *Bull. Acad. Roy. Belg. Cl. Sci.* **35**, 375 (1898).

314. A. K. Barbour, L. J. Belf, and M. W. Buxton, in M. Stacey, J. C. Tatlow, and A. G. Sharpe, eds., *Advances in Fluorine Chemistry*, Vol. 3, Butterworths, London, 1963, p. 181.

315. J. H. Brown and co-workers, *J. Chem. Soc. Suppl. Issue* **1**, S95 (1949).

316. U.S. Pat. 3,966,832 (June 26, 1978), R. Lademann and co-workers (to Hoechst).

317. Ger. Pat. 1,618,390 (Sept. 28, 1972), O. Scherer and co-workers (to Hoechst).

318. U.S. Pat. 4,183,873 (Jan. 15, 1980), Y. Baxamusa and S. Robota (to Hooker Chemical).

319. U.S. Pat. 4,130,594 (Dec. 19, 1978) and U.S. Pat. 4,129,602 (Dec. 12, 1978), L. Sendlak (to Hooker Chemical).

320. U.S. Pat. 4,462,937 (July 31, 1984), A. Ramanadin and L. Seigneurin (to Rhône-Poulenc).

321. U.S. Pat 3,859,372 (Jan. 7, 1975), S. Robota (to Hooker Chemical).

322. Y. A. Fialkov and L. M. Yagupolski, in Ref. 199, Chapt. 3, p.233.

323. *para-Chlorobenzotrifluoride (OXSOL 100)*, data sheet no. 784 588, MSDS No. M29358, Occidental Chemical Co., Niagara Falls, N.Y., Dec. 26, 1991.

324. *3,4-Dichlorobenzotrifluoride (OXSOL 1000)*, data sheet no. 343, 588, MSDS No. M29478, Occidental Chemical Co., Niagara Falls, N.Y., Jan. 16, 1992.

325. U.S. Pat. 4,110,405 (Aug. 29, 1978), M. Bornengo (to Montedison).

326. Eur. Pat. Appl. 490,115 (June 17, 1992) and Eur. Pat. Appl. 484,767 (May 13, 1992), R. Krishnamurti and co-workers (to Occidental Chemical).

327. A. Pozdeeva, *Zh. Prikl. Khim. (Leningrad)* **52**, 1119 (1979).

328. G. A. Mandrov and co-workers, *Khim. Prom-st., Ser.: Metody Anal. Kontrolya Kach. Prod. Khim. Prom-sti.* (8), 10 (1981).

329. B. W. Lawrence and co-workers, *Manuf. Chem. Aerosol News* **41**, 37 (Jan. 1970).

330. *3-(Trifluoromethyl)aniline, BUA Substance Report 44 (Sept. 1989)*, Advisory Board for Environmental Relevant Waste of the German Chemical Society, VCH Publishers, Weinheim, Germany, 1990, 58 pp.

331. U.S. Pat. 4,582,935 (Apr. 15, 1986), M. M. Boudakian (to Olin).

332. M. M. Boudakian, *J. Fluorine Chem.* **36**, 283 (1987).

333. L. M. Yagupolski and co-workers, *Russian Chem. Rev.* **52**, 993 (1983) (Engl. trans.).

334. N. Ishikawa, *Senryo to Yakuhin* **28**, 52 (1983).

335. G. Wulfram, in R. E. Banks, ed., *Organofluorine Chemicals and Their Industrial Applications*, E. Horwood, Chichester, U.K., 1979, Chapt. 10, p.208.

336. U.S. Pat. 5,030,252 (July 9, 1991), E. Sanders and T. L. Parker (to Dow).

337. U.S. Pat. 5,082,921 (Jan. 21, 1992), T. L. Parker (to Dow).

338. U.S. Pat. 4,876,329 (Oct. 24, 1989), W. L. Chiang and co-workers (to Amoco).

339. U.S. Pat. 5,021,540 (June 4, 1991), A. Leone-Bay and co-workers (to American Cyanamid).

340. F. M. Houlihan and co-workers, *Chem. Mater.* **3**, 462 (1991).

341. S. Misaki and co-workers, *Mol. Cryst. Liq. Cryst.* **66**, 443 (1981).

342. Jpn. Kokai Tokkyo Koho JP 59,16750 (Sept. 21, 1984) (to Chisso).

343. Jpn. Kokai Tokkyo Koho JP 03,200,737 (Sept. 2, 1991), S. Takehara and co-workers (to Dainippon).

344. A. C. Griffin and N. W. Buckley, *Mol. Cryst. Liq. Cryst.* **41**, 141 (1978).

345. H. Kobayashi and co-workers, *Chem. Lett.*, 1185 (1982).

346. G. E. Pacey and co-workers, *Analyst* **106**, 636 (1981).

347. U.S. Pat. 3,947,257 (Mar. 30, 1976), R. S. Johannsen and W. M. Moyer (to Raychem).

348. W. A. Sheppard, *J. Org. Chem.* **29**, 1 (1964).

349. F. E. Herkes, *J. Fluorine Chem.* **9**, 113 (1977).

350. G. A. Olah and co-workers, *J. Am. Chem. Soc.* **109**, 3708 (1987).

351. B. Langlois and M. Desbois, *Ann. Chim. Fr.*, 729 (1984).

352. Brit. Pat. 765,527 (Jan. 9, 1957) (to Hoechst).

353. U.S. Pat. 4,620,040 (Oct. 28, 1986), D. J. Alsop (to Occidental Chemical).

354. L. M. Yagupolski, *Dokl. Akad. Nauk (SSSR)* **105**, 100 (1955).

355. A. E. Feiring, *J. Org. Chem.* **44**, 2907 (1979).

356. F. Mathey and J. Bensoam, *Tetrahedron Lett.*, 2253 (1973).

357. L. M. Yagupolski and V. V. Orda, *Zh. Obsch. Khim.* **34**, 1979 (1964).

358. R. P. J. Braden and E. Klauke, *Pesticide Sci.* **17**, 418 (1986).

359. L. M. Yagupolski and co-workers, *Synthesis*, 721 (1975).

360. D. C. Remy and co-workers, *J. Org. Chem.* **41**, 1644 (1976).

361. T. G. Miller and J. W. Thanasi, *J. Org. Chem.* **25**, 2009 (1960).

362. B. R. Langlois, *J. Fluorine Chem.* **41**, 247 (1988).

363. D. C. England and co-workers, *J. Am. Chem. Soc.* **82**, 5116 (1960).

364. J. P. Idoux and co-workers, *J. Org. Chem.* **48**, 3771 (1983).

365. J. T. Gupton and co-workers, *Synth. Commun.* **15**, 431 (1985).

366. J. P. Idoux and co-workers, *J. Org. Chem.* **50**, 1876 (1985).

367. E. H. Banitt and co-workers, *J. Med. Chem.* **18**, 1130 (1975); *Ibid.* **20**, 821 (1977).

368. U.S. Pat. 5,082,587 (Jan. 21, 1992), E. P. Janulis (to 3M Co.).

369. World Pat. 91/00897 (Jan. 24, 1991). A. Wächtler and co-workers (to E. Merck GmbH).

370. N. Yoneda and co-workers, *J. Fluorine Chem.* **38**, 435 (1988).

371. M. M. Boudakian, *J. Fluorine Chem.* **18**, 497 (1981).

372. U.S. Pat. 2,447,717 (Aug. 24, 1948), J. H. Simons (to Minnesota Mining and Manufacturing).

373. H. Meinert, *Z. Chem.* **5**, 64 (1965).

374. M. Van der Puy, *Tetrahedron Lett.* **28**, 255 (1987).

375. U.S. Pat. 4,786,733 (Nov. 22, 1988), M. Van der Puy and R. E. Eibeck (to Allied-Signal).

376. S. P. Anand and R. Filler, *J. Fluorine Chem.* **7**, 179 (1976).

377. S. Stavber and M. Zupan, *Tetrahedron Lett.* **31**, 775 (1990).

378. T. Umemoto and G. Tomizawa, *J. Org. Chem.* **54**, 1726 (1989).

379. G. C. Finger and co-workers, *J. Org. Chem.* **28**, 1666 (1963).

380. M. M. Boudakian, *J. Heterocyclic Chem.* **4**, 381 (1967); U.S. Pat. 3,296,269 (Jan. 3, 1967), M. M. Boudakian (to Olin).

381. Jpn. Kokai Tokkyo Koho JP 04,124,176 (Apr. 24, 1992), T. Fukuhara and co-workers (to Tohkem Products).

382. H. C. Brown and D. McDaniel, *J. Am. Chem. Soc.* **77**, 3752 (1955).

383. L. Bradlow and C. VanderWerf, *J. Org. Chem.* **14**, 509 (1949).

384. G. Bressan and co-workers, *J. Chem. Soc. B.*, 225 (1971).

385. U.S. Pats. 3,873,701 (Mar. 25, 1975), and 3,816,406 (June 11, 1974), A. F. Hirsch (to Ortho Pharmaceuticals).

386. T. Mukaiyama, *Angew. Chem. Int. Ed. Engl.* **18**, 707 (1979).

387. J. Hannah and co-workers, *J. Med. Chem.* **25**, 457 (1982).

388. T. Güngör, *J. Organometallic Chem.* **215**, 139 (1981).

389. F. Marsais and co-workers, *J. Org. Chem.* **57**, 565 (1992).

390. R. D. Beaty and W. K. R. Musgrave, *J. Chem. Soc.*, 875 (1952).

391. A. Roe and C. Hawkins, *J. Am. Chem. Soc.* **69**, 2443 (1947).

392. *Chem. Eng. News*, 44 (Oct. 16, 1967); *Ibid.*, 8 (Dec. 8, 1967).

393. Span. Pat. ES 548,693 (May 16, 1986), J. C. Verde.

394. T. Miyamoto and co-workers, *Chem. Pharm. Bull.* **35**, 2280 (1987).

395. R. G. Pews and Z. Lysenko, *J. Org. Chem.* **50**, 5115 (1985).

396. T. Talik and Z. Talik, *Rocz. Chem.* **40**, 1187 (1966).

397. *Ibid.*, **38**, 785 (1964).

398. U.S. Pat. 3,637,714 (Jan. 25, 1972), L. A. F. Carlsson and co-workers (to Aktiebolaget Astra).

399. W. S. Saari and co-workers, *J. Med. Chem.* **26**, 1696 (1983).

400. U.S. Pat. 3,703,521 (Nov. 21, 1972), M. M. Boudakian (to Olin).

401. P. B. Desai, *J. Chem. Soc. Perkin Trans.* **1**, 1866 (1973).

402. U.S. Pats. 4,283,397 (Aug. 11, 1981), 4,255,424 (Mar. 10, 1981), 4,241,062 (Dec. 23, 1980), 4,263,306 (Apr. 21, 1981), and 4,282,219 (Aug. 4, 1981), J. Hannah (to Merck).

403. Ger. Offen. 2,128,540 (Dec. 16, 1971), A. Nicolson (to ICI).

404. Ger. Offen. 2,425,239 (May 24, 1973), D. W. R. Headford and co-workers (to ICI).

405. U.S. Pat. 3,798,228 (Mar. 19, 1974), M. M. Boudakian (to Olin).

406. J. Hamer and co-workers, *Rec. Trav. Chim.* **81**, 1059 (1962).

407. U.S. Pat. 4,071,521 (Jan. 31, 1978), T. G. Muench (to Dow).

408. U.S. Pat. 4,031,100 (June 21, 1977), T. L. Giacobbe (to Dow).

409. M. M. Boudakian, *J. Heterocyclic Chem.* **5**, 683 (1968).

410. U.S. Pat. 4,320,122 (Mar. 16, 1982), H. Theobold and co-workers (to BASF).

411. U.S. Pat. 3,810,902 (May 14, 1974), R. H. Rigterink (to Dow).

412. U.S. Pat. 4,115,557 (Sept. 19, 1978), C. E. Pawloski (to Dow).

413. H. R. Kricheldorf and co-workers, *Makromol. Chem.* **189**, 2255 (1988); *Ibid.* **191**, 2027 (1990).

414. F. Mutterer and C. D. Weis, *Helv. Chim. Acta* **59**, 229 (1976).

415. E. Differding and co-workers, *Bull. Soc. Chim. Belg.* **99**, 647 (1990).

416. R. D. Chambers and co-workers, *J. Chem. Soc.*, 5045 (1965).

417. U.S. Pat. 4,746,744 (May 24, 1988), C. A. Wilson and A. Fung (to Dow).

418. R. E. Banks and co-workers, *J. Chem. Soc.*, 594 (1965).
419. *Ibid.*, 575 (1965).
420. *Ibid.*, 1740 (1961).
421. J. A. Burdon and co-workers, *Nature* **196**, 231 (1960).
422. R. D. Chambers and co-workers, *J. Chem. Soc.*, 3573 (1964).
423. G. G. Yakobson and co-workers, *Fluorine Chem. Rev.* **7**, 115 (1974).
424. U.S. Pat. 3,850,943 (Nov. 26, 1974), R. D. Bowden and co-workers (to ICI).
425. U.S. Pat. 4,063,926 (Dec. 20, 1977), C. D. S. Tomlin and co-workers (to ICI).
426. R. D. Chambers and co-workers, *J. Chem. Soc. C*, 2221 (1968); *J. Chem. Soc., Chem. Commun.*, 384 (1966).
427. L. A. Mitscher and co-workers, *Chemtech*, 50 (Jan. 1991); *Ibid.*, 249 (Apr. 1991).
428. M. Jarman and R. McCague, *J. Chem. Soc., Chem. Commun.*, 125 (1984).
429. T. Miyamoto and co-workers, *J. Heterocyclic Chem.* **24**, 1333 (1987).
430. D. T. W. Chu and co-workers, *J. Med. Chem.* **29**, 2363 (1986).
431. J. Matsumoto and co-workers, *J. Heterocyclic Chem.* **21**, 673 (1984).
432. E. T. McBee and co-workers, *Ind. Eng. Chem.* **39**, 389 (1947).
433. U.S. Pat. 2,516,402 (July 25, 1950), E. T. McBee and E. M. Hodnett (to Purdue Research Foundation).
434. M. S. Raasch, *J. Org. Chem.* **27**, 1406 (1962).
435. I. D. Shustov and co-workers, *Zh. Obshch. Khim.* **53**, 103 (1983).
436. L. F. Lee and co-workers, *J. Org. Chem.* **55**, 2872 (1990).
437. L. F. Lee and co-workers, in D. R. Baker, J. G. Fenyes, and W. K. Moberg, eds., *Synthesis and Chemistry of Agrochemicals, II.* ACS Symposium Series No. 443, American Chemical Society, Washington, D.C., 1991, Chapt. 16, p.195.
438. Y. Kobayashi and I. Kumadaki, *Chem. Pharm. Bull.* **17**, 510 (1969).
439. Y. Kobayashi and E. Chinen, *Chem. Pharm. Bull.* **17**, 510 (1967).
440. U.S. Pat. 4,259,496 (Mar. 31, 1981), G. Whittaker (to ICI).
441. U.S. Pat. 4,324,627 (Apr. 13, 1982), D. Cartwright (to ICI).
442. U.S. Pat. 4,266,064 (May 5, 1981), R. Nishiyama and co-workers, (to Ishihara).
443. U.S. Pats. 4,417,055 (Nov. 22, 1983), and 4,288,599 (Sept. 8, 1981), R. Nishiyama and co-workers (to Ishihara).
444. R. D. Chambers and co-workers, *J. Chem. Soc. C*, 2328 (1966).
445. Y. Kobayashi and co-workers, *Chem. Pharm. Bull.* **17**, 2335 (1969).
446. H. Gilman and D. Blume, *J. Am. Chem. Soc.* **65**, 2467 (1943).
447. E. Pouterman and A. Girardet, *Helv. Chim. Acta* **30**, 107 (1947).
448. C. J. Ohnmacht and co-workers, *J. Med. Chem.* **14**, 926 (1971).
449. H. Vergin and R. Metz, *Drugs of Today* **27**(3), 177 (1991).
450. *Chem. Mktg. Rep.*, 5 (Dec. 15, 1992).
451. R. Duschinsky and co-workers, *J. Am. Chem. Soc.* **79**, 4559 (1957).
452. U.S. Pat. 3,682,917 (Aug. 8, 1972), I. L. Knunyants and co-workers.
453. U.S. Pat. 3,846,429 (Nov. 5, 1974), S. A. Giller and co-workers.
454. U.S. Pats. 3,954,758 (May 4, 1976), and 4,113,949 (Sept. 12, 1978), P. D. Schuman and co-workers (to PCR).
455. U.S. Pat. 4,082,752 (Apr. 4, 1978), T. Takahara and S. Misaki (to Daikin Kogyo).
456. F. Winston, *PCR Research Chemicals Catalog, 1990–91.* Gainesville, Fla., p.174.
457. E. Klauke and co-workers, *J. Fluorine Chem.* **21**, 495 (1982).
458. U.S. Pat. 3,314,955 (Apr. 18, 1967), M. M. Boudakian and C. W. Kaufman (to Olin).
459. U.S. Pat. 3,280,124 (Oct. 18, 1966), M. M. Boudakian and co-workers (to Olin).
460. R. G. Pews and W. E. Puckett, *J. Fluorine Chem.* **42**, 179 (1989).
461. H. Schroeder, *J. Am. Chem. Soc.* **82**, 4115 (1960).
462. H. Schroeder and co-workers, *J. Org. Chem.* **27**, 2580 (1962).
463. R. E. Banks and co-workers, *J. Chem. Soc. C*, 1822 (1967).

464. B. Baasner and E. Klauke, *J. Fluorine Chem.* **45**, 417 (1989).

465. D. R. Hildebrand, *Chemtech*, 224 (Apr. 1978).

466. W. Harms, in Ref. 335, Chapt. 9, p. 188.

467. U.S. Pat. 3,694,444 (Sept. 26, 1972), E. Klauke and H.-S. Bien (to Bayer); Brit. Pat. 1,273,914 (June 12, 1970), H.-U. Alles and co-workers (to Bayer).

468. Can. Pat. 844,625 (June 16, 1970), E. Klauke and H.-S. Bien (to Bayer).

469. Brit. Pat. 873,251 (July 19, 1961), A. Dorlars (to Bayer).

470. Ger. Offen. 2,729,762 (Jan. 18, 1979), F. Kysela and co-workers (to Bayer).

471. U.S. Pat. 4,332,939 (June 1, 1982), G. Seifert and S. Stäubli (to Ciba-Geigy).

472. C. W. Tullock and D. D. Coffman, *J. Org. Chem.* **25**, 2016 (1960).

473. U.S. Pat. 4,329,458 (May 11, 1982), E. Klauke and co-workers (to Bayer).

474. K. Kuriha and co-workers, *Biochem. Biophys. Acta* **384**, 127 (1975).

475. U.S. Pat. 2,975,179 (Mar. 14, 1961), A. Dorlars (to Bayer).

476. G. A. Olah and co-workers, *Synthesis*, 487 (1973).

477. L. A. Carpino and co-workers, *J. Am. Chem. Soc.* **112**, 9651 (1990); *J. Org. Chem.* **56**, 2611 (1991).

478. W. L. Reilly and H. C. Brown, *J. Org. Chem.* **22**, 698 (1967).

479. E. T. McBee and co-workers, *Ind. Eng. Chem.* **39**, 391 (1947).

480. S. Smith, in R. E. Banks, ed., *Preparation, Properties and Industrial Applications of Organofluorine Compounds*, E. Horwood, Chichester, U.K., 1982, Chapt. 8, p. 234.

481. J. A. Young, in Ref. 248, Chapt. 9.

482. K. Tanaka, *J. Synth. Org. Chem., Japan* **48**, 16 (1990).

483. R. D. Chambers and C. R. Sargent, in A. R. Katritzky and A. J. Boulton, eds., *Advances in Heterocyclic Chemistry*, Academic Press, Inc., New York, 1981, Vol. 28, p. 1.

General References

A. E. Pavlath and A. L. Leffler, *Aromatic Fluorine Compounds*, ACS Monograph No. 155, Reinhold, N.Y., 1962.

W. A. Sheppard and C. M. Sharts, *Organic Fluorine Chemistry*, W. A. Benjamin, N.Y., 1969.

M. Hudlicky, *Chemistry of Organic Fluorine Compounds*, 2nd ed., Ellis Horwood, Chichester, U.K., 1976.

G. Schiemann and B. Cornils, *Chemie und Technologie Cyclischer Fluorverbindungen*, F. Enke Verlag, Stuttgart, 1969.

Handbook of Aromatic Fluorine Compounds, Olin Corporation, Stamford, Conn., 1976.

R. D. Chambers, *Fluorine in Organic Chemistry*, John Wiley & Sons, Inc., New York, 1973.

E. Forche, "Houben-Weyl, Methoden der Organischen Chemie," in E. Mueller, ed., *Halogen-Verbindungen*, Georg Thieme Verlag, Stuttgart, 1962, pp. 1–502.

R. Filler, in J. C. Tatlow, R. D. Peacock, and H. H. Hyman, eds., *Advances in Fluorine Chemistry*, Vol. 6, CRC Press, Cleveland, Ohio, 1970, p.1.

G. Fuller, *Manuf. Chem. Aerosol News* **35**, 43 (May 1964); *Ibid.* **35**, 45 (June 1964).

A. K. Barbour, L. J. Belf, and M. W. Buxton, in M. Stacey, J. C. Tatlow, and A. G. Sharpe, eds., *Advances in Fluorine Chemistry*, Vol. 3, Butterworths, Washington, D.C., 1963, p.181.

R. H. Shiley, D. R. Dickerson, and G. C. Finger, *Aromatic Fluorine Chemistry at the Illinois State Geological Survey*, circular 501, Urbana, Ill., 1978.

M. M. Boudakian, "Halopyridines," in R. A. Abramovitch, ed., *The Chemistry of Heterocyclic Compounds*, Vol. 14, supplement part 2, *Pyridine and Its Derivatives*, Wiley-Interscience, New York, 1974, Chapt. 6, p. 407.

A. Roe, in R. Adams ed., *Organic Reactions*, Vol. 5, John Wiley & Sons, Inc., New York, 1949, Chapt. 4, p. 193.

H. Suschitzky, in M. Stacey, J. C. Tatlow, and A. G. Sharpe, eds., *Advances in Fluorine Chemistry*, Vol. 4, Butterworths, Washington, D.C., 1964, p. 1.

R. Filler and Y. Kobayashi, eds., *Biomedical Aspects of Fluorine Chemistry*, Kodansha, Tokyo, 1982.

R. E. Banks, D. W. A. Sharp, and J. C. Tatlow, eds., *Fluorine: The First Hundred Years*, Elsevier, Lausanne, 1986.

L. German and S. Zemzkov, eds., *New Fluorinating Agents in Organic Synthesis*, Springer-Verlag, Berlin, 1989.

I. L. Knunyants and G. G. Yakobson, eds., *Syntheses of Organic Fluorine Compounds*, Springer-Verlag, Berlin, 1985.

S. B. Walker, ed., *Fluorine in Agrochemicals*, Fluorochem, Ltd., Old Glossop, U.K., 1990.

R. E. Banks, ed., *Organofluorine Chemicals and Their Industrial Applications*, E. Horwood, Chichester, U.K., 1979.

R. E. Banks, ed., *Preparation, Properties and Industrial Applications of Organofluorine Compounds*, Ellis Horwood, Chichester, U.K., 1982.

MAX M. BOUDAKIAN
Chemical Consultant

POLYTETRAFLUOROETHYLENE

Polytetrafluoroethylene [*9002-84-0*] (PTFE), more commonly known as Teflon (Du Pont), a perfluorinated straight-chain high polymer, has a most unique position in the plastics industry due to its chemical inertness, heat resistance, excellent electrical insulation properties, and low coefficient of friction over a wide temperature range. Polymerization of tetrafluoroethylene (TFE) monomer gives this perfluorinated straight-chain high polymer with the formula $-(CF_2-CF_2)_n-$. The white to translucent solid polymer has an extremely high molecular weight, in the 10^6-10^7 range, and consequently has a viscosity in the range of 1 to 10 GPa·s ($10^{10}-10^{11}$ P) at 380°C. It is a highly crystalline polymer and has a crystalline melting point. Its high thermal stability results from the strong carbon–fluorine bond and characterizes PTFE as a useful high temperature polymer.

The discovery of PTFE (1) in 1938 opened the commercial field of perfluoropolymers. Initial production of PTFE was directed toward the World War II effort, and commercial production was delayed by Du Pont until 1947. Commercial PTFE is manufactured by two different polymerization techniques that result in two different types of chemically identical polymer. Suspension polymerization produces a granular resin, and emulsion polymerization produces the coagulated dispersion that is often referred to as a fine powder or PTFE dispersion.

Because of its chemical inertness and high molecular weight, PTFE melt does not flow and cannot be fabricated by conventional techniques. The suspension-polymerized PTFE polymer (referred to as granular PTFE) is usually fabricated by modified powder metallurgy techniques. Emulsion-polymerized PTFE behaves entirely differently from granular PTFE. Coagulated dispersions are processed by a cold extrusion process (like processing lead). Stabilized PTFE dispersions, made by emulsion polymerization, are usually processed according to latex processing techniques.

Manufacturers of PTFE include Daikin Kogyo (Polyflon), Du Pont (Teflon), Hoechst (Hostaflon), ICI (Fluon), Ausimont (Algoflon and Halon), and the CIS (Fluoroplast). India and The People's Republic of China also manufacture some PTFE products.

Monomer

Preparation.   The manufacture of tetrafluoroethylene [116-14-3] (TFE) involves the following steps (2–9). The pyrolysis is often conducted at a PTFE manufacturing site because of the difficulty of handling TFE.

$$CaF_2 + H_2SO_4 \longrightarrow CaSO_4 + 2\ HF$$

$$CH_4 + 3\ Cl_2 \longrightarrow CHCl_3 + 3\ HCl$$

$$CHCl_3 + 2\ HF \xrightarrow{\ SbF_3\ } CHClF_2 + 2\ HCl$$

$$2\ CHClF_2 \xrightarrow{\ \Delta\ } CF_2{=}CF_2 + 2\ HCl$$

Pyrolysis of chlorodifluoromethane is a noncatalytic gas-phase reaction carried out in a flow reactor at atmospheric or subatmospheric pressure; yields can be as high as 95% at 590–900°C. The economics of monomer production is highly dependent on the yields of this process. A significant amount of hydrogen chloride waste product is generated during the formation of the carbon–fluorine bonds.

A large number of by-products are formed in this process, mostly in trace amounts; more significant quantities are obtained of hexafluoropropylene, perfluorocyclobutane, 1-chloro-1,1,2,2-tetrafluoroethane, and 2-chloro-1,1,1,2,3,3-hexafluoropropane. Small amounts of highly toxic perfluoroisobutylene, $CF_2{=}C(CF_3)_2$, are formed by the pyrolysis of chlorodifluoromethane.

In this pyrolysis, subatmospheric partial pressures are achieved by employing a diluent such as steam. Because of the corrosive nature of the acids (HF and HCl) formed, the reactor design should include a platinum-lined tubular reactor made of nickel to allow atmospheric pressure reactions to be run in the presence of a diluent. Because the pyrolysate contains numerous by-products that adversely affect polymerization, the TFE must be purified. Refinement of TFE is an extremely complex process, which contributes to the high cost of the monomer. Inhibitors are added to the purified monomer to avoid polymerization during storage; terpenes such as d-limonene and terpene B are effective (10).

Tetrafluoroethylene was first synthesized in 1933 from tetrafluoromethane, CF_4, in an electric arc furnace (11). Since then, a number of routes have been developed (12–18). Depolymerization of PTFE by heating at ca 600°C is probably the preferred method for obtaining small amounts of 97% pure monomer on a laboratory scale (19,20). Depolymerization products contain highly toxic perfluoroisobutylene and should be handled with care.

Properties.   Tetrafluoroethylene (mol wt 100.02) is a colorless, tasteless, odorless, nontoxic gas (Table 1). It is stored as a liquid; vapor pressure at

Table 1. Physical Properties of Tetrafluoroethylene[a]

Property	Value
boiling point at 101.3 kPa,[b] °C	−76.3
freezing point, °C	−142.5
liquid density at t °C, g/mL	
$-100 < t < -40$	$1.202 - 0.0041\,t$
$-40 < t < 8$	$1.1507 - 0.0069\,t\text{-}0.000037\,t^2$
$8 < t < 30$	$1.1325 - 0.0029\,t\text{-}0.00025\,t^2$
vapor pressure at T K, kPa[c]	
$196.85 < T < 273.15$	$\log_{10} P_{kPa} = 6.4593\text{--}875.14/T$
$273.15 < T < 306.45$	$\log_{10} P_{kPa} = 6.4289\text{--}866.84/T$
critical temperature, °C	33.3
critical pressure, MPa[d]	39.2
critical density, g/mL	0.58
dielectric constant at 28°C	
at 101.3 kPa[b]	1.0017
at 858 kPa[b]	1.015
thermal conductivity at 30°C, mW/(m·K)	15.5
heat of formation for ideal gas at 25°C, ΔH, kJ/mol[e,f]	−635.5
heat of polymerization at 25°C to solid polymer ΔH, kJ/mol[e,g]	−172.0
flammability limits in air at 101.3 kPa,[c] vol %	14–43

[a]From Ref. 21, unless otherwise stated.
[b]To convert kPa to atm, multiply by 0.01.
[c]To convert kPa to psi, multiply by 0.145.
[d]To convert MPa to atm, divide by 0.101.
[e]To convert J to cal, divide by 4.184.
[f]Ref. 22.
[g]Ref. 23.

−20°C = 1 MPa (9.9 atm). It is usually polymerized above its critical temperature and below its critical pressure. The polymerization reaction is highly exothermic.

Tetrafluoroethylene undergoes addition reactions typical of an olefin. It burns in air to form carbon tetrafluoride, carbonyl fluoride, and carbon dioxide (24). Under controlled conditions, oxygenation produces an epoxide (25) or an explosive polymeric peroxide (24). Trifluorovinyl ethers, RO—CF=CF$_2$, are obtained by reaction with sodium salts of alcohols (26). An ozone–TFE reaction is accompanied by chemiluminescence (27). Dimerization at 600°C gives perfluorocyclobutane, C$_4$F$_8$; further heating gives hexafluoropropylene, CF$_2$=CFCF$_3$, and eventually perfluoroisobutylene, CF$_2$=C(CF$_3$)$_2$ (28). Purity is determined by both gas–liquid and gas–solid chromatography; the ir spectrum is complex and therefore of no value.

Uses. Besides polymerizing TFE to various types of high PTFE homopolymer, TFE is copolymerized with hexafluoropropylene (29), ethylene (30), perfluorinated ether (31), isobutylene (32), propylene (33), and in some cases it is used as a termonomer (34). It is used to prepare low molecular weight polyfluorocarbons (35) and carbonyl fluoride (36), as well as to form PTFE *in situ* on metal surfaces (37). Hexafluoropropylene [*116-15-4*] (38,39), perfluorinated ethers, and other oligomers are prepared from TFE.

In the absence of air, TFE disproportionates violently to give carbon and carbon tetrafluoride; the same amount of energy is generated as in black powder explosions. This type of decomposition is initiated thermally and equipment hot spots must be avoided. The flammability limits of TFE are 14–43%; it burns when mixed with air and forms explosive mixtures with air and oxygen. It can be stored in steel cylinders under controlled conditions inhibited with a suitable stabilizer. The oxygen content of the vapor phase should not exceed 10 ppm. Although TFE is nontoxic, it may be contaminated by highly toxic fluorocarbon compounds.

Manufacture of PTFE

Engineering problems involved in the production of TFE seem simple compared with those associated with polymerization and processing of PTFE resins. The monomer must be polymerized to an extremely high molecular weight in order to achieve the desired properties. The low molecular weight polymer does not have the strength needed in end use applications.

Polytetrafluoroethylene is manufactured and sold in three forms: granular, fine powder, and aqueous dispersion; each requires a different fabrication technique. Granular resins are manufactured in a wide variety of grades to obtain a different balance between powder flows and end use properties (Fig. 1). Fine pow-

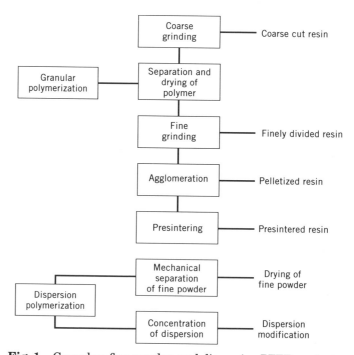

Fig. 1. Granular, fine powder, and dispersion PTFE products.

ders that are made by coagulating aqueous dispersions also are available in various grades. Differences in fine powder grades correspond to their usefulness in specific applications and to the ease of fabrication. Aqueous dispersions are sold in latex form and are available in different grades. A variety of formulation techniques are used to tailor these dispersions for specific applications.

Polymerization. In aqueous medium, TFE is polymerized by two different procedures. When little or no dispersing agent is used and vigorous agitation is maintained, a precipitated resin is produced, commonly referred to as granular resin. In another procedure, called aqueous dispersion polymerization, a sufficient dispersing agent is employed and mild agitation produces small colloidal particles dispersed in the aqueous reaction medium; precipitation of the resin particles is avoided. The two products are distinctly different, even though both are high molecular weight PTFE polymers. The granular product can be molded in various forms, whereas the resin produced by the aqueous dispersion cannot be molded, but is fabricated by dispersion coating or conversion to powder for paste extrusion with a lubricant medium. Granular resin cannot be paste extruded or dispersion coated.

Granular Resins. Granular PTFE is made by polymerizing TFE alone or in the presence of trace amounts of comonomers (40,41). An initiator, a small amount of dispersing agent, and other additives (42) may be present; an alkaline buffer is occasionally used (43). In the early stages of polymerization, an unstable dispersion is formed, but lack of dispersing agent and vigorous agitation cause the polymer to partially coagulate; the remainder of the process is fairly complex. The polymerized product is stringy, irregular, and variable in shape. The dried granular polymer is ground to different average particle sizes, depending on the product requirements, eg, the flow and other properties. Coarser fabrication of particles leaves a higher void in the sintered article. A better balance between handleability and moldability (ability to mold and sinter in the absence of voids) is achieved by agglomerating the finely divided resin to ca 400–800 μm (44). For ram extrusion of granular resin into long tubes and rods, a partially presintered resin is preferred. Granular PTFE resin is nonflammable.

Fine Powder Resins. Fine powder resins are made by polymerizing TFE in an aqueous medium with an initiator and emulsifying agents (45). The polymerization mechanism is not a typical emulsion type, but is subject to some of the principles of emulsion polymerization. The process and ingredients have a significant effect on the product. It is extremely important that the dispersion remains sufficiently stable throughout polymerization, avoiding premature coagulation (46), but unstable enough to allow subsequent coagulation into a fine powder. Gentle stirring ensures dispersion stability. The amount of emulsifying agent in the polymerization process is usually less than its critical micelle concentration. The rate of polymerization and the particle shape are influenced by the amount of the emulsifying agent (47–50). The particle structure can be influenced by the polymerization process. Most of the particles are formed in the early stages of the polymerization process and the particles grow as the batch progresses; hence, the radial variation in molecular weight and polymer composition within the dispersion particle can be achieved by controlling the polymerization variables, including ingredients and operating conditions (51–57).

The thin dispersion rapidly thickens into a gelled matrix and coagulates into a water-repellent agglomeration that floats on the aqueous medium as the mechanical agitation is continued. The agglomeration is dried gently; shearing must be avoided.

Aqueous Dispersions.   The dispersion is made by the polymerization process used to produce fine powders of different average particle sizes (58). The most common dispersion has an average particle size of about 0.2 μm, probably the optimum particle size for most applications. The raw dispersion is stabilized with a nonionic or anionic surfactant and concentrated to 60–65 wt % solids by electrodecantation, evaporation, or thermal concentration (59). The concentrated dispersion can be modified further with chemical additives. The fabrication characteristics of these dispersions depend on polymerization conditions and additives.

Filled Resins.   Fillers such as glass fibers, graphite, asbestos, or powered metals are compounded into all three types of PTFE. Compounding is achieved by intimate mixing. Coagulation of the polymer with a filler produces a filled fine powder.

Properties

The properties described herein are related to the basic structure of polytetrafluoroethylene and are exhibited by both granular and fine powder products. The carbon–carbon bonds, which form the backbone of the PTFE chain, and the carbon–fluorine bonds are extremely strong and are the key contributors in imparting an outstanding combination of properties. The fluorine atoms form a protective sheath over the chain of carbon atoms. If the atoms attached to the carbon-chain backbone were smaller or larger than fluorine, the sheath would not form a regular uniform cover. This sheath shields the carbon chain from attack and confers chemical inertness and stability. It also reduces the surface energy resulting in low coefficient of friction and nonstick properties.

Polytetrafluoroethylene does not dissolve in any common solvent; therefore, its molecular weight cannot be measured by the usual methods. A number-average molecular weight has been estimated by determining the concentration of end groups derived from the initiator. Earlier estimates, based on an iron bisulfite system containing radioactive sulfur, ^{35}S, ranged from 142×10^3 to 534×10^3 for low molecular weight polymer. The same technique applied to polymers of industrial interest gave molecular weights of 389×10^3 to 8900×10^3 (60,61). In the absence of a normal molecular weight determination method, an estimated relative molecular weight is used for all practical purposes. It is obtained by measuring the specific gravity following a standardized fabricating and sintering procedure (ASTM D1457-83). Because the rate of crystallization decreases with increasing molecular weight, samples prepared from the high molecular weight polymer and cooled from the melt at a constant slow rate have lower standard specific gravities than those prepared from low molecular weight polymer cooled at the same rate (62). The correlation between number-average molecular weight (M_n) based on end group estimations, and standard specific gravity (SSG) is given by

$$SSG = 2.612 - 0.058 \log_{10} M_n$$

The SSG procedure assumes absence of voids (or constant void content). Voids depress the values of the measured specific gravity. The inaccuracies that result from voids can be corrected by applying ir techniques (63).

Melting and recrystallization behavior of virgin PTFE has been studied by dsc (64). A quantitative relationship was found between M_n and the heat of crystallization (ΔH_c) in the molecular weight range of 5.2×10^5 to 4.5×10^7, where H_c is heat of crystallization in J/g, which is independent of cooling rates of 4–32°C/min.

$$M_n = 2.1 \times 10^{10} \cdot \Delta H_c^{-5.16}$$

At ca 342°C, virgin PTFE changes from white crystalline material to almost transparent amorphous gel. Differential thermal analysis indicates that the first melting of virgin polymer is irreversible and that subsequent remeltings occur at 327°C, which is generally reported as the melting point. Most of the studies reported in the literature are based on previously sintered (ie, melted and recrystallized) polymer; very little work is reported on the virgin polymer. Melting is accompanied by a volume increase of ca 30%. Because the viscosity of the polymer at 380°C is 10 GPa·s (10^{11} P), the shape of the melt is stable. The melting point increases with increasing applied pressure at the rate of 1.52°C/MPa (0.154°C/atm) (65).

Virgin PTFE has a crystallinity in the range of 92–98%, which indicates an unbranched chain structure. The fluorine atoms are too large to allow a planar zigzag structure, which would permit chain flexibility; therefore the chains are rigid (66). Electron micrographs and diffraction patterns (67) of PTFE dispersion particles indicate that the rod-like particles present in virgin PTFE dispersions are fully extended chain crystals containing few defects. The spherical particles appear to be composed of similar rod-like entities that are wrapped around themselves in a more or less random fashion.

Between 50 and 300°C, PTFE obeys the relationship between stress τ and the apparent shear rate γ: $\tau = K\gamma^{1/4}$. Melting of PTFE begins near 300°C. Above this temperature, the shear stress at constant shear rate increases and the rheological exponent rises from 0.25 toward 0.5 at the final melting point (68).

Transitions. Transitions observed by various investigators (69–74), their interpretation, and the modes of identification are shown in Table 2. Besides the transition at the melting point, the transition at 19°C is of great consequence because it occurs at ambient temperature and significantly affects the product behavior. Above 19°C, the triclinic pattern changes to a hexagonal unit cell. Around 19°C, a slight untwisting of the molecule from a 180° twist per 13 CF_2 groups to a 180° twist per 15 CF_2 groups occurs. At the first-order transition at 30°C, the hexagonal unit cell disappears and the rod-like hexagonal packing of the chains in the lateral direction is retained (69). Below 19°C there is almost perfect three-dimensional order; between 19 and 30°C the chain segments are disordered; and above 30°C, the preferred crystallographic direction is lost and the molecular segments oscillate above their long axes with a random angular orientation in the lattice (70,71).

Table 2. Transitions in Polytetrafluoroethylene

Temperature, °C	Region affected	Technique	Reference
	1st order		
19	crystalline, angular displacement causing disorder	thermal methods, x-ray, nmr	70
30	crystalline, crystal disordering	thermal methods, x-ray, nmr	70
90 (80 to 110)	crystalline	stress relaxation, Young's modulus, dynamic methods	73
	2nd order		
−90 (−110 to −73)	amorphous, onset of rotational motion around C—C bond	thermal methods, dynamic methods	74
−30 (−40 to −15)	amorphous	stress relaxation, thermal expansion, dynamic methods	73
130 (120 to 140)	amorphous	stress relaxation, Young's modulus, dynamic methods	73

The dynamic mechanical properties of PTFE have been measured at frequencies from 0.033 to 90 Hz. Abrupt changes in the distribution of relaxation times are associated with the crystalline transitions at 19 and 30°C (75). The activation energies are 102.5 kJ/mol (24.5 kcal/mol) below 19°C, 510.4 kJ/mol (122 kcal/mol) between the transitions, and 31.4 kJ/mol (7.5 kcal/mol) above 30°C.

Polytetrafluoroethylene transitions occur at specific combinations of temperature and mechanical or electrical vibrations. Transitions, sometimes called dielectric relaxations, can cause wide fluctuations in the dissipation factor.

Mechanical Properties. Mechanical properties of PTFE depend on processing variables, eg, preforming pressure, sintering temperature and time, cooling rate, void content, and crystallinity. Properties, such as the coefficient of friction, flexibility at low temperatures, and stability at high temperatures, are relatively independent of fabrication. Molding and sintering conditions affect flex life, permeability, stiffness, resiliency, and impact strength. The physical properties of PTFE have been reviewed and compiled (72,76,77) (Table 3).

A marked change in volume of 1.0–1.8% is observed for PTFE in the transition zone from 18 to 25°C. An article that has been machined on either side of this zone changes dimensions when passing through the transition zone; hence, the final operating temperature of a precision part must be accurately determined. Articles fabricated of PTFE resins exhibit high strength, toughness, and self-lubrication at low temperatures. They are useful from 5 K and are highly flexible from 194 K. They tend to return to their original dimensions after a deformation. At sintering temperature, they rapidly recover their original shapes. For most

Table 3. Typical Mechanical Properties of Molded and Sintered PTFE Resins[a]

Property	Granular resin	Fine powder	ASTM method
tensile strength at 23°C, MPa[b]	7–28	17.5–24.5	D638-61T
elongation at 23°C, %	100–200	300–600	D628-61T
flexural strength at 23°C, MPa[b]	does not break		D790-61
flexural modulus at 23°C, MPa[b]	350–630	280–630	D747-61T
impact strength, J/m[c]			
21°C	106.7		D256-56
24°C	160		
77°C	>320		
hardness durometer, D	50–65	50–65	D1706-59T
compression stress, MPa[b]			
at 1% deformation at 23°C	4.2		D695-52T
at 1% offset at 23°C	7.0		D695-52T
coefficient of linear thermal expansion per °C, 23–60°C	12×10^{-5}		D696-44
thermal conductivity, 4.6-mm thickness, W/(m·K)	0.24		Cenco-Fitch
deformation under load, at 26°C, 24 h, %			D621-59
6.86 MPa[b]		2.4	
13.72 MPa[b]	15		
water absorption, %	<0.01	<0.01	D570–54T
flammability	nonflammable		D635-56T
static coefficient of friction with polished steel	0.05–0.08		

[a]Ref. 77.
[b]To convert MPa to psi, multiply by 145.
[c]To convert J/m to ft·lbf/in., divide by 53.38.

applications no special precautions are necessary because decomposition rates below the recommended maximum service temperature of 260°C are very low. Impact strength is excellent over a wide range of temperatures. Static friction decreases with an increase in load. Static coefficient of friction is lower than the dynamic coefficient and therefore reduces stick-slip problems.

The surface of PTFE articles is slippery and smooth. Liquids with surface tensions below 18 mN/m(=dyn/cm) are spread completely on the PTFE surface; hence, solutions of various perfluorocarbon acids in water wet the polymer (78). Treatment with alkali metals promotes the adhesion between PTFE and other substances (79) but increases the coefficient of friction (80).

Filled Resins. Filled compositions meet the requirements of an increased variety of mechanical, electrical, and chemical applications. Physical properties of filled granular compounds are shown in Table 4 (81).

Chemical Properties. Vacuum thermal degradation of PTFE results in monomer formation. The degradation is a first-order reaction (82). Mass spectroscopic analysis shows that degradation begins at ca 440°C, peaks at 540°C, and continues until 590°C (83).

Table 4. Properties of Filled PTFE Compounds[a]

Property	Unfilled	Glass fiber, wt %		Graphite, 15 wt %	Bronze, 60 wt %
		15	25		
specific gravity	2.18	2.21	2.24	2.16	3.74
tensile strength, MPa[b]	28	25	17.5	21	14
elongation, %	350	300	250	250	150
stress at 10% elongation, MPa[b]	11	8.5	8.5	11	14
thermal conductivity, mW/ (m·K)	0.244	0.37	0.45	0.45	0.46
creep modulus, kN/m[c]	2	2.21	2.1	3.4	6.2
hardness, Shore durometer, D	51	54	57	61	70
Izod impact, J/m[d]	152	146	119		
PV,[e] (kPa·m)/s[f]	0.70	106	177	52	281
wear factor, 1/Pa[g]	5×10^{-14}	28×10^{-17}	26×10^{-17}	100×10^{-17}	12×10^{-17}
coefficient of friction					
static, 3.4 MPa[b] load	0.08	0.13	0.13	0.10	0.10
dynamic at					
PV = 172, (kPa·m)/s[f]		0.15–0.24	0.17	0.15	0.15
V = 900 m/s	0.01		−0.24	−0.18	−0.22

[a]Ref. 81.
[b]To convert MPa to psi, multiply by 145.
[c]To convert kN/m to lbf/in., divide by 0.175.
[d]To convert J/m to ft·lbf/in., divide by 53.38.
[e]PV = pressure × velocity. For 0.13-mm radial wear in 1000 h, unlubricated.
[f]To convert kPa to psi, multiply by 0.145.
[g]To convert 1/Pa to (in.^{3}·min)/(ft·lbf·h), divide by 2×10^{-7}.

Radiation Effects. Polytetrafluoroethylene is attacked by radiation. In the absence of oxygen, stable secondary radicals are produced. An increase in stiffness in material irradiated in vacuum indicates cross-linking (84). Degradation is due to random scission of the chain; the relative stability of the radicals in vacuum protects the materials from rapid deterioration. Reactions take place in air or oxygen and accelerated scission and rapid degradation occur.

Crystallinity has been studied by x-ray irradiation (85). An initial increase caused by chain scission in the amorphous phase was followed (above 3 kGy or 3×10^5 rad) by a gradual decrease associated with a disordering of the crystallites. The amorphous component showed a maximum of radiation-induced broadening in the nmr at 7 kGy (7×10^5 rad).

In air, PTFE has a damage threshold of 200–700 Gy ($2 \times 10^4 - 7 \times 10^4$ rad) and retains 50% of initial tensile strength after a dose of 10^4 Gy (1 Mrad), 40% of initial tensile strength after a dose of 10^5 Gy (10^7 rad), and ultimate elongation of 100% or more for doses up to 2–5 kGy (2×10^5–5×10^5 rad). During irradiation, resistivity decreases, whereas the dielectric constant and the dissipation factor increase. After irradiation, these properties tend to return to their preexposure values. Dielectric properties at high frequency are less sensitive to radiation than are properties at low frequency. Radiation has very little effect on dielectric strength (86).

Absorption, Permeation, and Interactions.   Polytetrafluoroethylene is chemically inert to industrial chemicals and solvents even at elevated temperatures and pressures (87). This compatibility is due to the strong interatomic bonds, the almost perfect shielding of the carbon backbone by fluorine atoms, and the high molecular weight of the polymer. Under some severe conditions PTFE is not compatible with certain materials. It reacts with molten alkali metals, fluorine, strong fluorinating agents, and sodium hydroxide above 300°C. Shapes of small cross section burn vertically upward after ignition in 100% oxygen. Because gases may be evolved, the weight loss during sintering of a blend of PTFE and white asbestos is many times greater than loss from pure PTFE. Finely divided aluminum and magnesium thoroughly mixed with finely divided PTFE react vigorously after ignition or at high temperatures (87).

Absorption of a liquid is usually a matter of the liquid dissolving in the polymer; however, in the case of PTFE, no interaction occurs between the polymer and other substances. Submicroscopic voids between the polymer molecules provide space for the material absorbed; which is indicated by a slight weight increase and sometimes by discoloration. Common acids or bases are not absorbed up to 200°C. Aqueous solutions are scarcely absorbed at atmospheric pressure. Even the absorption of organic solvents is slight, partially resulting from the low wettability of PTFE. Since absorption of chemicals or solvents has no substantial effect on the chemical bond within the fluorocarbon molecule, absorption should not be confused with degradation; it is a reversible physical process. The polymer does not suffer loss of mechanical or bulk electrical properties unless subjected to severely fluctuating conditions (87).

Dynamic mechanical measurements were made on PTFE samples saturated with various halocarbons (88). The peaks in loss modulus associated with the amorphous relaxation near -90°C and the crystalline relaxation near room temperature were not affected by these additives. An additional loss peak appeared near -30°C, and the modulus was reduced at all higher temperatures. The amorphous relaxation that appears as a peak in the loss compliance at 134°C is shifted to 45–70°C in the swollen samples.

The sorption behavior of perfluorocarbon polymers is typical for nonpolar partially crystalline polymers (89). The weight gain strongly depends on the solubility parameter. Little sorption of substances such as hydrocarbons and polar compounds occurs.

As an excellent barrier resin, PTFE is widely used in the chemical industry. However, it is a poor barrier for fluorocarbon oils because similarity in the chemical composition of a barrier and a permeant increases permeation. Most liquids and gases (other than fluorocarbons) do not permeate highly crystalline PTFE. Permeabilities at 30°C (in mol/(m·s·Pa) $\times$ 10^{15}) are as follows: CO_2, 0.93; N_2, 0.18; He, 2.47; anhydrous HCl, < 0.01 (89).

Gases and vapors diffuse through PTFE more slowly than through most other polymers (Table 5). The higher the crystallinity, and the less space between polymer molecules, the slower the permeation. Voids greater than molecular size cause an increase in permeability. However, the permeability of the finished article can be controlled by molding the resin to low porosity and high density. The

Table 5. Permeability of PTFE Resin to Vapors

Permeant	Permeability constant,[a,b] $\mathrm{mol/(m \cdot s \cdot Pa)} \times 10^{15}$	
	23°C	30°C
benzene	1.81	2.93
carbon tetrachloride	0.13	
ethanol	1.88	
HCl, 20%	<0.71	
piperidine	0.96	
H_2SO_4, 98%	54.20	
water		20.70

[a]Ref. 87. Test method ASTM E96-35T (at vapor pressure; for 25.4 μm film thickness). Values are averages only and not for specification purposes.
[b]Original data converted to SI units using vapor pressure data from Ref. 90.

optimum specific gravity for low permeability and good flexural properties is 2.16–2.195. Permeability increases with temperature as a result of the increase in activity of the solvent molecules and because of the increase in vapor pressure of the liquids. Swelling of PTFE resins and film is very low.

Electrical Properties.   Polytetrafluoroethylene is an excellent electrical insulator because of its mechanical strength and chemical and thermal stability as well as excellent electrical properties (Table 6). It does not absorb water and volume resistivity remains unchanged even after prolonged soaking. The dielectric constant remains constant at 2.1 for a temperature range of −40 to 250°C and a frequency range of 5 Hz to 10 GHz.

Articles fabricated according to standard practice should have dielectric constants in the range of 2.05 ± 0.5 when tested at RT. The dielectric constant varies with density and factors that affect density. Machined components can be fabricated to a predetermined dielectric constant by controlling the rod density during processing by adjusting the preforming pressure on the resin and cooling after

Table 6. Electrical Properties of Polytetrafluoroethylene[a]

Property	Granular	Fine powder	ASTM method
dielectric strength, short time, 2-mm thickness, V/mm	23,600	23,600	D149-55T
surface arc-resistance, s	>300	>300	D495-55T
volume resistivity, Ω·cm	>10^{18}	>10^{18}	D257-57T
surface resistivity at 100% rh, Ω/sq	>10^{16}		D257-57T
dielectric constant, at 60 to 2×10^9 Hz	2.1	2.1	D150-59T
dissipation factor, at 60 to 2×10^9 Hz	0.0003		D150-59T

[a]Ref. 77.

sintering. The dielectric constant and the density have a linear relationship. Predictable variations in the dielectric constant result from density changes that accompany thermal expansion occuring with increasing temperature. The dielectric constant did not change over two to three years of measurements.

The dissipation factor (the ratio of the energy dissipated to the energy stored per cycle) is affected by the frequency, temperature, crystallinity, and void content of the fabricated structure. At certain temperatures and frequencies, the crystalline and amorphous regions become resonant. Because of the molecular vibrations, applied electrical energy is lost by internal friction within the polymer which results in an increase in the dissipation factor. The dissipation factor peaks for these resins correspond to well-defined transitions, but the magnitude of the variation is minor as compared to other polymers. The low temperature transition at $-97°C$ causes the only meaningful dissipation factor peak. The dissipation factor has a maximum of 10^8-10^9 Hz at RT; at high crystallinity (93%) the peak at 10^8-10^9 Hz is absent.

As crystallinity increases, the internal molecular friction and the dissipation factor decrease. Voids reduce the dissipation factor in proportion to the percentage of microvoids present. Certain extruded shapes utilize air to reduce the effective dielectric constant and dissipation factor of a coaxial cable. The dielectric strength of these resins is high and is unaffected by thermal aging at 200°C. Frequency has a marked effect on the dielectric strength because corona discharge becomes more continuous as frequency increases. If the voltage stress is not high enough to cause corona ignition, a very long dielectric life is anticipated at any frequency. Corona discharges on the surface or in a void initiate dielectric breakdown (91). Surface arc resistance of these resins is high and not affected by heat aging. The resins do not track or form a carbonized conducting path when subjected to a surface arc in air. Polytetrafluoroethylene resins are capable of continuous service up to 260°C and can withstand much higher temperatures for limited periods of time. They do not melt or flow and retain some strength even in the gel state which begins at 327°C.

Fabrication

Granular Resins.   These resins are sold in different forms; an optimum balance between handleability and product properties is desired. A free-flowing resin is used in small and automatic moldings. A finely divided resin is more difficult to handle but it distributes evenly in large moldings and has superior properties in sintered articles; it is used for large billet- and sheet-molding operations. A presintered resin with low crystallinity and superior handleability is highly suitable for ram extrusion.

Virgin PTFE melts at about 342°C; viscosity, even at 380°C, is 10 GPa·s (10^{11} P). This eliminates processing by normal thermoplastic techniques, and other fabrication techniques had to be developed: the dry powder is compressed into handleable form by heating above the melting point. This coalesces the particles into a strong homogeneous structure; cooling at a controlled rate achieves the desired degree of crystallinity.

Molding. Many PTFE manufacturers give detailed descriptions of molding equipment and procedures (92–98). Round piston molds for the production of solid or hollow cylinders are the most widely used. Because preforming usually takes place below 100°C, carbon steel is a suitable material of construction. The compression ratio (ie, the bulk volume of the powder to the specific volume of the unsintered molding) for granular resins is 3:1 to 6:1. For large-area, thin-walled moldings of unfilled polymer, a short-stroke press with a working capacity of 19.6–34.3 MPa (194–339 atm) mold pressure is sufficient; for tall moldings of filled compounds with a small cross-sectional area, a long-stroke press with a low thrust is required. The powder should be evenly distributed and leveled in the mold (92). To ensure adequate compression uniformly throughout the preform, maximum pressure should be maintained for a sufficient length of time, and then be released slowly.

Automatic molding permits high speed mass production; it is preferable to machining finished material. Automatic presses can be operated mechanically, pneumatically, or hydraulically. The mold is filled by means of a special metering system from a storage hopper containing a free-flowing resin. Loading buckets that shuttle back and forth over the single-cavity mold are also used. Because automatic molding requires short cycles, the powder is usually compressed at high speed with a high preform pressure. Small articles such as rings, bushings, washers, gaskets, and ball-valve seats can be molded by this technique.

Isostatic molding allows uniform compression from all directions. A flexible mold is filled with a free-flowing granular powder and evacuated, tightly sealed, and placed in an autoclave containing a liquid that can be raised to the pressure required for preforming. The moldings require subsequent finishing because close tolerance cannot be achieved.

Sintering. Electrical ovens with air circulation and service temperatures up to 400°C are satisfactory for sintering. In free sintering, the cheapest and most widely used process, a preformed mold is placed in an oven with a temperature variation of ± 2°C. In pressure sintering, the preform is not removed from the mold; instead the mold containing the preform is heated in an oven until the sintering temperature is reached. During sintering and cooling, the mold is again placed under pressure but lower than the preform pressure. Pressure-sintered products have internal stresses that can be relieved by subsequent annealing. In the pressure-cooling process, pressure is applied on the molded article after it has reached sintering temperature and is maintained throughout the cooling period. The final product has a lower void content than the free-sintered mold.

To improve homogeneity, the preformed article is heated to 370–390°C. The time required for heating and sintering depends on the mold dimensions; cooling, which affects the crystallinity and product properties, should be slow.

Free-sintered articles do not have the same dimensions as the mold cavity because they shrink at right angles to the direction of the preform pressure and grow in the direction of the applied pressure.

For processing after sintering, in the least expensive method for sintered PTFE tape or sheet, a large billet is skived on a lathe after it has been sintered and cooled. High precision articles are machined from ram-extruded rods.

Articles that are too complicated to be made by machining are made by coining. A sintered molding is heated to its melting point, transferred to a mold, and quickly deformed at low pressure, where it is held until it has cooled sufficiently to retain the improved shape. However, the coined molding, if reheated to a high temperature, returns to its original shape, and hence there is a limit on the maximum temperature to which coined moldings can be heated.

Ram Extrusion. Compression molding is not suitable for the manufacture of continuous long moldings such as pipes or rods. In ram extrusion, a small charge of PTFE powder is preformed by a reciprocating ram and sintered. Subsequent charges are fused into the first charge, and this process continues to form homogeneous long rods (92,99–101). The die tube, which is made of a corrosion-resistant material, is heated by resistance heating. Good temperature control is essential, and the melted and compacted powder must not pass any constrictions in its path. Thermal expansion and friction produce great resistance to movement, and as a result, a considerable force is required to push the polymer through the tube. A high quality surface finish on the inside of the tube reduces the pressure. If adequate bond strength between successive charges is not developed, the extrudate may break at the interface (poker chipping). Free-flowing powders and presintered resins are preferred for ram extrusion. Ram-extruded rods are used for automatic screw machining. Tubing is used as pipe liners or stock from which seals, gaskets, and bellows are machined.

Fine Powder Resins. Fine powder PTFE resins are extremely sensitive to shear. They must be handled gently to avoid shear, which prevents processing. However, fine powder is suitable for the manufacture of tubing and wire insulation for which compression molding is not suitable. A paste-extrusion process may be applied to the fabrication of tubes with diameters from fractions of a millimeter to about a meter, walls from thicknesses of 100–400 μm, thin rods with up to 50-mm diameters, and cable sheathing. Calendering unsintered extruded solid rods produces thread-sealant tape and gaskets.

The paste-extrusion process includes the incorporation of ca 16–25 wt % of the lubricant (usually a petroleum fraction); the mixture is rolled to obtain uniform lubricant distribution. This wetted powder is shaped into a preform at low pressure (2.0–7.8 MPa or 19–77 atm) which is pushed through a die mounted in the extruder at ambient temperature. The shear stress exerted on the powder during extrusion confers longitudinal strength to the polymer by fibrillation. The lubricant is evaporated and the extrudate is sintered at ca 380°C.

The exact amount of lubricant required for extrusion depends on the design of the extruder, the reduction ratio (ie, ratio of the cross-sectional preform area to the cross-sectional area in the die), and the quality of the lubricant. A low lubricant content results in a high extrusion pressure, whereas a high lubricant content causes a poor coalescence and generates defects in the extrudate.

Fine powder resins can be colored with pigments that can withstand the sintering temperature. The pigment should be thoroughly mixed with the powder by rolling the mixture before adding the lubricant. Detailed design parameters of the paste extruder are given in Reference 102–108.

The extrudate is dried and sintered by passing it through a multistage oven located immediately after the extruder. Pipes and rods may be heated up to 380°C. The throughput rate depends on the length of the sintering oven. Residence time varies from a few seconds for thin-walled insulations on a wire to a few minutes for large diameter tubing. For short residence times temperatures may be as high as ca 480°C. The extrusion pressure depends on the reduction ratio, the extrusion rate, the lubricant content, and the characteristics of the extruder.

To produce unsintered tape by paste extrusion, the fine powder is lubricated and preformed according to the procedure described above. The preform is extruded in the form of rods, which are calendered on hot rolls to the desired width and thickness (109,110).

Different resins have been developed for use in different reduction–ratio application ranges (111,112). The powders suitable for high reduction–ratio applications, such as wire coatings, are not necessarily suitable for the medium reduction–ratio applications, such as tubings, or the low reduction–ratio applications, such as thread-sealant tapes or pipe liners. Applications and processing techniques are being used, which utilize the unique combination of properties offered by PTFE in fine powder form (113–115).

Dispersion Resins. Polytetrafluoroethylene dispersions in aqueous medium contain 30–60 wt % polymer particles and some surfactant. The type of surfactant and the particle characteristics depend on the application. These dispersions are applied to various substrates by spraying, flow coating, dipping, coagulating, or electrodepositing.

Aqueous dispersion is sprayed on metal substrates to provide chemical resistance, nonstick, and low friction properties. The coated surface is dried and sintered. Impregnation of fibrous or porous materials with these dispersions combines the properties of the materials with those of PTFE. Some materials require only a single dipping, eg, asbestos. The material is usually dried after dipping. For high pressure sealing applications, sintering at 380–400°C increases strength and dimensional stability. For film castings, the dispersion is poured on a smooth surface; the formed film is dried and sintered and peeled from the supporting surface.

Aqueous dispersions are used for spinning PTFE fibers. The dispersion is mixed with a matrix-forming medium (116,117) and forced through a spinneret into a coagulating bath. The matrix material is removed by heating and the fibers are sintered and drawn molten to develop their full strength.

Effects of Fabrication on Physical Properties of Molded Parts. The physical properties are affected by molecular weight, void content, and crystallinity. Molecular weight can be reduced by degradation but not increased during processing. These factors can be controlled during molding by the choice of resin and fabricating conditions. Void distribution (or size and orientation) also affects properties; however, it is not easily measured.

Preforming primarily affects void content, sintering controls molecular weight, and cooling determines crystallinity. Voids caused by insufficient consolidation of particles during preforming may appear in the finished articles. Densities below 2.10 g/cm^3 indicate a high void content. Electrical and chemical applications require a minimum density of 2.12–2.14 g/cm^3. Particle size, shape, and porosity are also important in determining void content. Although void content is determined largely by particle characteristics and preforming conditions, sintering conditions can also have an effect. Temperatures too high or too low increase void content. Excessively high sintering temperature can decrease the molecular weight. The final crystallinity of a molding depends on the initial molecular weight of the polymer, the rate of cooling of the molding, and to a lesser extent on sintering conditions. The degree of crystallinity of moldings is affected by the cooling or annealing conditions.

Flexural modulus increases by a factor of five as crystallinity increases from 50 to 90% with a void content of 0.2%; however, recovery decreases with increasing crystallinity. Therefore, the balance between stiffness and recovery depends on the application requirements. Crystallinity is reduced by rapid cooling but increased by slow cooling. The stress–crack resistance of various PTFE insulations is correlated with the crystallinity and change in density due to thermal mechanical stress (118).

Applications

Consumption of PTFE increases continuously as new applications are being developed. Electrical applications consume half of the PTFE produced; mechanical and chemical applications share equally the other half. Various grades of PTFE and their applications are shown in Table 7.

Electrical Applications. The largest application of PTFE is for hookup and hookup-type wire used in electronic equipment in the military and aerospace industries. Coaxial cables, the second largest application, use tapes made from fine powder resins and some from granular resin. Interconnecting wire applications include airframes. Other electrical applications include computer wire, electrical tape, electrical components, and spaghetti tubing.

Mechanical Applications. Seals and piston rings, basic shapes, and antistick uses constitute two-thirds of the resin consumed in mechanical applications. Bearings, mechanical tapes, and coated glass fabrics also consume a large amount of PTFE resins. Seals and piston rings, bearings, and basic shapes are manufactured from granular resins, whereas the dispersion is used for glass–fabric coating and antistick applications. Most pressure-sensitive mechanical tapes are made from granular resins.

Chemical Applications. The chemical processing industry uses large amounts of granular and fine powder PTFE. Soft packing applications are manufactured from dispersions, and hard packings are molded or machined from stocks and shapes made from granular resin.

Overbraided hose liners are made from fine powder resins by paste extrusion, and thread-sealant tapes are produced from fine powder by calendering. Fabricated gaskets are made from granular resins and pipe liners are produced from fine powder resins. Fibers and filament forms are also available.

Highly porous fabric structures, eg, Gore-Tex, that can be used as membranes have been developed by exploiting the unique fibrillation capability of dispersion-polymerized PTFE (113).

Micropowders. The PTFE micropowders, also called waxes, are tetrafluoroethylene homopolymers with molecular weights significantly lower than that of normal PTFE. The molecular weight for micropowders varies from 2.5×10^4 to 25×10^4, whereas that of normal PTFE is of the order of 10×10^6. Micropowders are generally white in color and are friable. The average agglomerate particle size is between 5 to 10 μm and is composed of smaller, "as polymerized" primary particles which are approximately 0.2 μm in diameter. The dsc curves of

Table 7. Applications of Polytetrafluoroethylene Resins

Resin grade	Processing	Description	Main uses
Granular			
agglomerates	molding, preforming, sintering, ram extrusion	free-flowing powder	gaskets, packing seals, electronic components, bearings, sheet, rod, heavy-wall tubing; tape and molded shapes for nonadhesive applications
coarse	molding, preforming, sintering	granulated powder	tape, molded shapes, nonadhesive applications
finely divided	molding, preforming, sintering	powder for highest quality, void-free moldings	molded sheets, tape, wire wrapping, tubing, gaskets
presintered	ram extrusion	granular, free-flowing powder	rods and tubes
Fine powder			
high reduction ratio	paste extrusion	agglomerated powder	wire coating, thin-walled tubing
medium reduction ratio	paste extrusion	agglomerated powder	tubing, pipe, overbraided hose, spaghetti tubing
low reduction ratio	paste extrusion	agglomerated powder	thread-sealant tape, pipe liners, tubing, porous structures
Dispersion			
general-purpose	dip coating	aqueous dispersion	impregnation, coating, packing
coating	dip coating	aqueous dispersion	film coating
stabilized	coagulation	aqueous dispersion	bearings

lower molecular weight micropowder show a higher heat of crystallization and melting (second heating) than normal PTFE. This is due to the higher crystallinity of the micropowder.

The production of micropowders involves the scission of the high molecular weight PTFE chain by gamma or electron beam irradiation at a variety of dosage levels. An increase in dosage reduces the molecular weight. The irradiated low molecular weight material is ground to a particle size ranging from 1 to 25 μm in the final product.

Economic Aspects

Polytetrafluoroethylene homopolymers are more expensive than most other thermoplastics because of high monomer refining costs. For extremely high molecular weights, ingredients and manufacturing process must be free of impurities, which increases costs. In the United States, the 1992 list prices from primary producers were between 16.3 and 23.5 $/kg, depending on the resin type. For example, granular PTFE resins cost 16.3–18.0 $/kg supplied in 22.5-kg containers. The coagulated fine powders cost 19.10–22.10 $/kg packaged in 22.5-kg containers. Formulated dispersions are 20.00–23.5 $/kg in 19-L or 113-L containers. Although fine powder sales have increased in recent years, the sales of granular PTFE are the highest on a worldwide basis. Most of the resin is consumed in the United States (ca 9000 t in 1991), followed by Europe and Japan.

Testing and Standards

A description of PTFE resins and their classification are given in ASTM D1457-83. A comprehensive listing of industrial and military specifications covering mechanical, electrical, and chemical applications of PTFE can be found in Reference 119.

Health and Safety

Exposure to PTFE can arise from ingestion, skin contact, or inhalation. The polymer has no irritating effect to the skin, and test animals fed with the sintered polymer have not shown adverse reactions. Dust generated by grinding the resin also has no effect on test animals. Formation of toxic products is unlikely. Only the heated polymer is a source of a possible health hazard (120).

Because PTFE resins decompose slowly, they may be heated to a high temperature. The toxicity of the pyrolysis products warrants care where exposure of personnel is likely to occur (120). Above 230°C decomposition rates become measurable (0.0001% per hour). Small amounts of toxic perfluoroisobutylene have been isolated at 400°C and above; free fluorine has never been found. Above 690°C the decomposition products burn but do not support combustion if the heat is removed. Combustion products consist primarily of carbon dioxide, carbon tetrafluoride, and small quantities of toxic and corrosive hydrogen fluoride. The PTFE resins are nonflammable and do not propagate flame.

Prolonged exposure to thermal decomposition products causes so-called polymer fume fever, a temporary influenza-like condition. It may be contracted by smoking tobacco that has been contaminated with the polymer. It occurs several hours after exposure and passes within 36–48 hours; the temporary effects are not cumulative.

Large quantities of PTFE resins have been manufactured and processed above 370°C. In various applications they are heated above the recommended use temperatures. No cases of serious injury, prolonged illness, or death have been reported resulting from the handling of these resins. However, when high molecular weight PTFE is converted to micropowder by thermal degradation, highly toxic products result.

Micropowders are added to a wide variety of material used in industry, where they provide nonstick and sliding properties. They are incorporated into the product by blending and grinding. To disperse well, the powder must have good flow properties. Conditions that make the powder sticky should be avoided.

The PTFE micropowders are commonly used in plastics, inks, lubricants, and finishes such as lacquer. Lubricants containing micropowders are used for bearings, valve components, and other moving parts where sliding friction must be minimized or eliminated. Nonstick finishes that require good release properties, for example, in the food and packaging industry, commonly use PTFE micropowders.

In some applications the high heat stability of the micropowder can be utilized over a reasonably wide temperature range. A maximum service temperature is normally 260°C, provided the crystalline melting point is between 320 and 335°C. Exposure above 300°C leads to degradation and possible evolution of toxic decomposition products.

The particulate morphology of PTFE micropowder in printing inks provides desirable gloss to the printed product. Its inherent lubricity results in good wear and slip properties and surface smoothness. The chemical resistance of the micropowder is as high as that of high molecular weight PTFE. It is therefore used in applications requiring service in strong or corrosive chemical environments such as concentrated mineral acids and alkalies.

BIBLIOGRAPHY

"Tetrafluorethylene Resins" in *ECT* 1st ed., Vol. 11, pp. 687–691, by B. E. Ely, E.I. du Pont de Nemours & Co., Inc.,; "Polytetrafluoroethylene" under "Fluorine Compounds, Organic" in *ECT* 2nd ed., Vol. 9, pp. 805–831, by S. Sherratt, Imperial Chemical Industries, Ltd., Plastics Division; in *ECT* 3rd ed., Vol. 11, pp. 1–24, by S. V. Gangal, E.I. du Pont de Nemours & Co., Inc.

1. U. S. Pat. 2,230,654 (Feb. 4, 1941), R. J. Plunkett (to Kinetic Chemicals, Inc.).
2. J. D. Park and co-workers, *Ind. Eng. Chem.* **39**, 354 (1947).
3. J. M. Hamilton, in M. Stacey, J. C. Tatlow, and A. G. Sharpe, eds., *Advances in Fluorine Chemistry*, Vol. 3, Butterworth & Co., Ltd., Kent, U.K., 1963, p. 117.
4. J. W. Edwards and P. A. Small, *Nature* **202**, 1329 (1964); J. W. Edwards and P. A. Small, *Ind. Eng. Chem. Fundam.* **4**, 396 (1965).
5. F. Gozzo and C. R. Patrick, *Nature* **202**, 80 (1964).
6. Jpn. Pat. 60 15,353 (Oct. 14, 1960), M. Hisazumi and H. Shingu.
7. U.S. Pat. 2,994,723 (Aug. 1, 1961), O. Scherer and co-workers (to Farbewerke Hoechst).
8. Brit. Pat. 960,309 (June 10, 1964), J. W. Edwards, S. Sherratt, and P. A. Small (To ICI).
9. U.S. Pat. 3,459,818 (Aug. 5, 1969), H. Ukahashi and M. Hisasne (to Asahi Glass Co.).
10. U.S. Pat. 2,407,405 (Sept. 10, 1946), M. A. Dietrich and R. M. Joyce (to E.I. du Pont de Nemours & Co., Inc.).
11. O. Ruff and O. Bretschneider, Z. *Anorg. Allg. Chem.* **210**, 173 (1933).
12. E. G. Locke, W. R. Brode, and A. L. Henne, *J. Am. Chem. Soc.* **56**, 1726 (1934).

13. O. Ruff and W. Willenberg, *Chem. Ber.* **73**, 724 (1940).
14. L. T. Hals, T. S. Reid, and G. H. Smith, *J. Am. Chem. Soc.* **73**, 4054 (1951); U.S. Pat. 2,668,864 (Feb. 9, 1954), (to Minnesota Mining and Manufacturing Co.).
15. U.S. Pat. 3,009,966 (Nov. 21, 1961), M. Hauptschein and A. H. Fainberg (to Pennsalt Chemical Corp.).
16. U.S. Pat. 3,471,546 (Oct. 7, 1969), G. Bjornson (to Phillips Petroleum Co.).
17. U.S. Pat. 3,662,009 (May 9, 1972), W. M. Hutchinson (to Phillips Petroleum Co.).
18. U.S. Pat. 3,799,996 (Mar. 26, 1974), H. S. Bloch (to Universal Oil Products).
19. E. E. Lewis and M. A. Naylor, *J. Am. Chem. Soc.* **69**, 1968 (1947).
20. U.S. Pat. 3,832,411 (Aug. 27, 1974), B. C. Arkles and R. N. Bonnett (to Liquid Nitrogen Processing Co.).
21. M. M. Renfrew and E. E. Lewis, *Ind. Eng. Chem.* **38**, 870 (1946).
22. H. C. Duus, *Ind. Eng. Chem.* **47**, 1445 (1955).
23. W. M. D. Bryant, *J. Polym. Sci.* **56**, 277 (1962).
24. A. Pajaczkowski and J. W. Spoors, *Chem. Ind.* **16**, 659 (1964).
25. Brit. Pat. 931,587 (July 17, 1963), H. H. Gibbs and J. L. Warnell (to E.I. du Pont de Nemours & Co., Inc.).
26. U.S. Pat. 3,159,609 (Dec. 1, 1964), J. F. Harris, Jr., and D. I. McCane (E.I. du Pont de Nemours & Co., Inc.).
27. F. S. Toby and S. Toby, *J. Phys. Chem.* **80**, 2313 (1976).
28. B. Atkinson and V. A. Atkinson, *J. Chem. Soc. Part II*, 2086 (1957).
29. U.S. Pat. 2,946,763 (July 26, 1960), M. I. Bro and B. W. Sandt (to E.I. du Pont de Nemours & Co., Inc.).
30. U.S. Pat. 3,847,881 (Nov. 12, 1974), M. Mueller and S. Chandrasekaran (to Allied Chemicals Co.).
31. U.S. Pat. 3,528,954 (Sept. 15, 1970), D. P. Carlson (to E.I. du Pont de Nemours & Co., Inc.).
32. U.S. Pat. 3,475,391 (Oct. 28, 1969), J. N. Coker (to E.I. du Pont de Nemours & Co., Inc.).
33. U.S. Pat. 3,846,267 (Nov. 5, 1974), Y. Tabata and G. Kojima (to Japan Atomic Energy Research Institute).
34. U.S. Pat. 3,467,636 (Sept. 16, 1969), A. Nersasian (to E.I. du Pont de Nemours & Co., Inc.).
35. U.S. Pat. 3,403,191 (Sept. 24, 1968), D. P. Graham (to E.I. du Pont de Nemours & Co., Inc.).
36. U.S. Pat. 3,404,180 (Oct. 1, 1969), K. L. Cordes (to E.I. du Pont de Nemours & Co., Inc.).
37. U.S. Pat. 3,567,521 (Mar. 2, 1971), M. S. Toy and N. A. Tiner (to McDonnell Douglas).
38. U.S. Pat. 3,446,858 (May 27, 1969), H. Shingu and co-workers (to Daikin Kogyo Co.).
39. U.S. Pat. 3,873,630 (Mar. 25, 1975), N. E. West (to E.I. du Pont de Nemours & Co., Inc.).
40. U.S. Pat. 3,855,191 (Dec. 17, 1974), T. R. Doughty, C. A. Sperati, and H. Un (to E.I. du Pont de Memours & Co., Inc.).
41. U.S. Pat. 3,655,611 (Apr. 11, 1972), M. B. Mueller, P. O. Salatiello, and H. S. Kaufman (to Allied Chemicals Co.).
42. U.S. Pat 4,189,551 (Feb. 19, 1980), S. V. Gangal (to E.I. du Pont de Nemours & Co., Inc.).
43. U.S. Pat. 3,419,522 (Dec. 31, 1968), P. N. Plimmer (to E.I. du Pont de Nemours & Co., Inc.).
44. U.S. Pat. 3,766,133 (Oct. 16, 1973) R. Roberts and R. F. Anderson (to E.I. du Pont de Nemours & Co., Inc.).

45. U.S. Pat. 2,612,484 (Sept. 30, 1952), S. G. Bankoff (to E.I. du Pont de Nemours & Co., Inc.).
46. U.S. Pat 4,186,121 (Jan. 29, 1980), S. V. Gangal (to E.I. du Pont de Nemours & Co., Inc.).
47. U.S. Pat. 4,725,644 (1988), S. Malhotra (to E.I. du Pont de Nemours & Co., Inc.).
48. T. Folda and co-workers, *Nature* **333**, 55 (1988).
49. B. Luhmann and A. E. Feiring, *Polymer* **30**, 1723 (1989).
50. B. Chu, C. Wu, and W. Buck, *Macromolecules* **22**, 831 (1989).
51. U.S. Pat. 4,576,869 (Mar. 18, 1986), S. C. Malhotra (to E.I. du Pont de Nemours & Co., Inc.).
52. U.S. Pat. 4,363,900 (Dec. 14, 1982), T. Shimizu and S. Koizumi (to Daikin Kogyo Co.).
53. U.S. Pat. 4,766,188 (Aug. 23, 1988), T. E. Attwood and R. F. Bridges (to ICI).
54. U.S. Pat. 4,036,802 (July 19, 1977), R. V. Poirier (to E.I. du Pont de Nemours & Co., Inc.).
55. U.S. Pat. 4,129,618 (Dec. 12, 1978), J. M. Downer, W. G. Rodway, and L. S. J. Shipp (to ICI).
56. U.S. Pat. 4,840,998 (June 6, 1989), T. Shimizu and K. Hosokawa (to Daikin Kogyo Co.).
57. U.S. Pat. 4,879,362 (Nov. 7, 1979), R. A. Morgan (to E.I. du Pont de Nemours & Co., Inc.).
58. U.S. Pat. 4,342,675 (Aug. 3, 1982), S. V. Gangal (to E.I. du Pont de Nemours & Co., Inc.).
59. U.S. Pat. 2,478,229 (Aug. 9, 1949), K. L. Berry (to E.I. du Pont de Nemours & Co., Inc.).
60. K. L. Berry and J. H. Peterson, *J. Am. Chem. Soc.* **73**, 5195 (1951).
61. R. C. Doban and co-workers, paper presented at *130th Meeting of the American Chemical Society*, Atlantic City, N. J., Sept. 1956.
62. C. A. Sperati and H. W. Starkweather, *Fortschr. Hochpolym. Forsch.* **2**, 465 (1961).
63. R. E. Moynihan, *J. Am. Chem. Soc.* **81**, 1045 (1959).
64. T. Suwa, M. Takehisa, and S. Machi, *J. Appl. Polym. Sci.* **17**, 3253 (1973).
65. P. L. McGeer and H. C. Duus, *J. Chem. Phys.* **20**, 1813 (1952).
66. C. W. Bunn, *J. Polym. Sci.* **16**, 332 (1955).
67. H. D. Chanzy, P. Smith, and J. Revol, *J. of Polym. Sci. Polym. Lett. Ed.* **24**, 557 (1986).
68. H. W. Starkweather, Jr., *J. Polym. Sci. Polym. Phys. Ed.* **17**, 73–79 (1979).
69. R. H. H. Pierce and co-workers, in Ref. 61.
70. E. S. Clark and L. T. Muus, paper presented at *133rd Meeting of the American Chemical Society*, New York, Sept. 1957.
71. E. S. Clark, paper presented at *Symposium on Helices in Macromolecular Systems*, Polytechnic Institute of Brooklyn, Brooklyn, N.Y., May 16, 1959.
72. C. A. Sperati, in J. Brandrup and E. H. Immergut, eds., *Polymer Handbook*, 2nd ed., John Wiley & Sons, Inc., New York, 1975 pp. V-29–36.
73. Y. Araki, *J. Appl. Polym. Sci.* **9**, 3585 (1965).
74. N. G. McCrum, *J. Polym. Sci.* **34**, 355 (1959).
75. H. W. Starkweather, Jr., *Macromolecules* **19**, 2541 (1986).
76. J. T. Milek, *A Survey Materials Report on PTFE Plastics*, AD 607798, U.S. Dept. of Commerce, Washington, D.C., Sept. 1964.
77. *Mechanical Design Data, Teflon Fluorocarbon Resins*, bulletin, E.I. du Pont de Nemours & Co., Inc. Wilmington, Del. Sept. 1964.
78. M. K. Bernett and W. A. Zisman, *J. Phys. Chem.* **63**, 1911 (1959).
79. U.S. Pat. 2,871,144 (Jan. 27, 1959), R. C. Doban (to E.I. du Pont de Nemours & Co., Inc.).
80. A. J. G. Allan and R. Roberts, *J. Polym. Sci.* **39**, 1 (1959).

81. *J. Teflon (Du Pont)* **13**(2), 3 (1972).

82. J. C. Siegle and co-workers, *J. Polym. Sci.* **Part A2**, 391 (1964).

83. G. P. Shulman, *Polym. Lett.* **3**, 911 (1965).

84. L. A. Wall and R. E. Florin, *J. Appl. Polym. Sci.* **2**, 251 (1959).

85. W. M. Peffley, V. R. Honnold, and D. Binder, *J. Polym. Sci.* **4**, 977 (1966).

86. *J. Teflon (Du Pont)* **10**(1), (Jan.–Feb. 1969).

87. *J. Teflon (Du Pont)* **11**(1), (Jan.–Feb. 1970).

88. H. W. Starkweather, Jr., *Macromolecules* **17**, 1178 (1984).

89. *Ibid.*, **10**, 1161 (1977).

90. D. W. Green, ed., *Perry's Chemical Engineers' Handbook*, 6th ed., McGraw-Hill Book Co., Inc., New York, 1984.

91. J. C. Reed, E. J. McMahon, and J. R. Perkins, *Insulation (Libertyville, Ill.)* **10**, 35 (1964).

92. *Hostaflon*, TF product information booklet, American Hoechst Corp., Somerville, N.J., 1970.

93. *Soreflon*, products information booklet, Ugine Kuhlmann, France.

94. *Teflon TFE-Fluorocarbon Resins Molding Techniques*, 2nd ed., bulletin, E.I. du Pont de Nemours & Co., Inc., Wilmington, Del., 1966.

95. *The Moulding of Granular Polymers*, technical service note, ICI, Wilmington, Del., 1966.

96. J. A. Ross, *High Speed Molding of Teflon Tetrafluoroethylene Resins*, technical release, E.I. du Pont de Nemours & Co., Inc., Wilmington, Del., 1963.

97. *Automatic Molding with "Halon" TFE*, Allied Chemical Corp., Morristown, N.J., 1967.

98. *Isostatic Molding of Teflon TFE-Fluorocarbon Resins*, preliminary information bulletin, E.I. du Pont de Nemours & Co., Wilmington, Del., 1969.

99. R. J. Dahlen, *The Ram Extrusion of Teflon*, technical release, E.I. du Pont de Nemours & Co., Inc., Wilmington, Del., 1969.

100. *The Granular Extrusion of "Halon" TFE*, Allied Chemical Corp., Morristown, N.J., 1968.

101. *The Extrusion of Granular Polymers*, 2nd ed., technical service note, ICI, Wilmington, Del., 1966.

102. V. Adamec, *Nature* **200**, 1196 (1963).

103. Technical brochure, *Teflon 62 Hose and Tubing*, H-11959, E. I. du Pont de Nemours & Co., Inc., Wilmington, Del., Feb. 1991.

104. J. F. Lontz and co-workers, *Ind. Eng. Chem.* **44**, 1805 (1952).

105. *TFE-Fluorocarbon Resins Paste Extrusion of Wire Insulations*, bulletin, E.I. du Pont de Nemours & Co., Inc., Wilmington, Del., 1961.

106. *The Extrusion Coating of Wire*, 2nd ed., technical service note, Wilmington, Del., 1969.

107. *Teflon TFE-Fluorocarbon Resins Extrusion of Thin-Walled Tubing*, bulletin, E.I. du Pont de Nemours & Co., Inc., Wilmington, Del., 1957.

108. *The Extrusion of Thin Sections*, 3rd ed., technical service note, ICI, Wilmington, Del., 1969.

109. R. C. Ribbans, *Unsintered Tape Manufactured Calendering Round Rods*, technical release, E.I. du Pont de Nemours & Co., Inc., Wilmington, Del., 1966.

110. *The Manufacture of Unsintered Tape*, 2nd ed., technical data, ICI, Wilmington, Del., 1966.

111. U.S. Pat. 3,142,665 (July 28, 1964), A. J. Cardinal, W. L. Edens, and J. W. Van Dyk (to E.I. du Pont de Nemours & Co., Inc.).

112. U.S. Pat. 4,038,231 (July 26, 1977), J. M. Douner, W. G. Rodway, and L. S. J. Shipp (to ICI).

113. U.S. Pat. 3,962,153 (June 8, 1976), R. W. Gore (to W. L. Gore and Assoc.).

114. U.S. Pat. 3,993,584 (Nov. 23, 1976), J. E. Owen and J. W. Vogt (to Kewanee Oil Co.).

115. U.S. Pat. 3,704,171 (Nov. 28, 1972), H. P. Landi (to American Cyanamid Co.).
116. U.S. Pat. 3,051,545 (Aug. 28, 1962), W. Steuber (to E.I. du Pont de Nemours & Co., Inc.).
117. P. E. Frankenburg, in *Ullmann's Encyclopedia of Industrial Chemistry*, Vol. A-10, 5th ed., VCH Publishing, Inc., New York, 1987, pp. 649–650.
118. R. L. Baillie, J. J. Bednarczyk, and P. M. Mehta, paper presented at *35th International Wire and Cable Symposium*, Nov. 18–20, 1986.
119. *J. Teflon (Du Pont)* **8**, 6 (Nov. 1967).
120. *Teflon Occupational Health Bull.* **17**(2), (1962) (published by Information Service Division, Dept. of National Health and Welfare, Ottawa, Canada).

Subhash V. Gangal
E.I. du Pont de Nemours & Co., Inc.

PERFLUORINATED ETHYLENE–PROPYLENE COPOLYMERS

Perfluorinated ethylene–propylene (FEP) resin [25067-11-2] is a copolymer of tetrafluoroethylene [116-14-3] (TFE) and hexafluoropropylene [116-15-4] (HFP); thus its branched structure contains units of $-CF_2-CF_2-$ and units of $-CF_2-CF(CF_3)-$. It retains most of the desirable characteristics of polytetrafluoroethylene (PTFE) but with a melt viscosity low enough for conventional melt processing. The introduction of hexafluoropropylene lowers the melting point of PTFE from 325°C to about 260°C.

The desire for a resin with polytetrafluoroethylene properties yet capable of being fabricated by conventional melt processing led to the discovery of this product (1). It allows melt extrusion of wire insulations of longer continuous lengths than the batchwise paste extrusion of PTFE as well as the injection molding of intricately shaped parts. The FEP polymer is melt-fabricable without severe sacrifice in mechanical properties because the perfluoromethyl side groups on the main polymer chain reduce crystallinity, which varies between 30 and 45%. This change in the crystallinity causes FEP and other copolymer particles to behave differently form PTFE particles; they do not fibrillate like PTFE particles and therefore do not agglomerate easily.

As a true thermoplastic, FEP copolymer can be melt-processed by extrusion and compression, injection, and blow molding. Films can be heat-bonded and sealed, vacuum-formed, and laminated to various substrates. Chemical inertness and corrosion resistance make FEP highly suitable for chemical services; its dielectric and insulating properties favor it for electrical and electronic service; and its low frictional properties, mechanical toughness, thermal stability, and nonstick quality make it highly suitable for bearings and seals, high temperature components, and nonstick surfaces.

Mechanical properties are retained up to 200°C, even in continuous service, which is better than with most plastics. At high temperatures, these copolymers react with fluorine, fluorinating agents, and molten alkali metals. They are commercially available under the Du Pont trademark Teflon FEP fluorocarbon resin. A similar product is manufactured by Daikin Kogyo of Japan and sold under the trademark Neoflon. The People's Republic of China also manufactures some FEP products.

Monomers

Preparation.   The preparation, properties, and uses of tetrafluoroethylene have been described (see FLUORINE COMPOUNDS, ORGANIC–POLYTETRAFLUORO-ETHYLENE).

Hexafluoropropylene (HFP) was initially prepared by pyrolysis of PTFE (2,3) and by fluorination of 1,2,3-trichloropropane followed by dehalogenation (4). A number of other routes are described in the patent literature (5–10). Hexafluoro-propylene can be prepared in high yield by thermally cracking TFE at reduced pressure at 700–800°C (11,12). Pyrolysis of PTFE at 860°C under vacuum gives a 58% yield of HFP (13). Fluorination of 3-chloropentafluoro-1-propene [79-47-0] at 200°C over activated carbon catalyst yields HFP (14). Decomposition of fluo-roform [75-46-7] at 800–1000°C in a platinum-lined nickel tube is another route (15). The thermal decomposition of sodium heptafluorobutyrate [2218-84-4], $CF_3CF_2CF_2CO_2Na$ (16), and copyrolyses of fluoroform and chlorotrifluoroethylene [79-38-9] (17), and chlorodifluoromethane [75-45-6] and 1-chloro-1,2,2,2-tetra-fluoroethane [2837-89-0] (18) give good yields of HFP.

Properties and Reactions.   The properties of HFP are shown in Table 1. It does not homopolymerize easily and hence can be stored as a liquid. It undergoes many addition reactions typical of an olefin. Reactions include preparation of lin-ear dimers and trimers and cyclic dimers (21,22); decomposition at 600°C with

Table 1. Properties of Hexafluoropropylene[a]

Property	Value
molecular weight	150.021
boiling point at 101 kPa,[b] °C	− 29.4
freezing point, °C	− 156.2
critical temperature, °C	85
critical pressure, kPa[b]	3254
critical density, g/cm^3	0.60
vapor pressure at K, kPa[b]	
243.75 < T < 358.15	$\log P(\text{kPa}) = 6.6938 - 1139.156/T$
liquid density, g/cm^3	
60°C	1.105
20°C	1.332
0°C	1.419
− 20°C	1.498
heat of formation for ideal gas at 25°C, ΔH, kJ/mol[c,d]	− 1078.6
flammability limits in air at 101 kPa[b]	nonflammable for all mixtures of air and hexafluoropropylene
heat of combustion, kJ/mol[c,d]	879
toxicity, LC_{50} (rat), 4 h, ppm[e]	3000

[a]Ref. 4.
[b]To convert kPa to mm Hg, multiply by 7.5.
[c]To convert kJ to kcal, divide by 4.184.
[d]Ref. 19.
[e]Ref. 20.

subsequent formation of octafluoro-2-butene and octafluoroisobutylene (23); oxidation with formation of an epoxide (24), an intermediate for a number of perfluoroalkyl perfluorovinyl ethers (25,26); and homopolymerization to low molecular weight liquids (27,28) and high molecular weight solids (29,30). Hexafluoropropylene reacts with hydrogen (31), alcohols (32), ammonia (33), and the halogens and their acids, except I_2 and HI (31,34−36). It is used as a comonomer to produce elastomers and other copolymers (37−41). The toxicological properties are discussed in Reference 42.

Copolymers

Hexafluoropropylene and tetrafluoroethylene are copolymerized, with trichloracetyl peroxide as the catalyst, at low temperature (43). Newer catalytic methods, including irradiation, achieve copolymerization at different temperatures (44,45). Aqueous and nonaqueous dispersion polymerizations appear to be the most convenient routes to commercial production (1,46−50). The polymerization conditions are similar to those of TFE homopolymer dispersion polymerization. The copolymer of HFP−TFE is a random copolymer; that is, HFP units add to the growing chains at random intervals. The optimal composition of the copolymer requires that the mechanical properties are retained in the usable range and that the melt viscosity is low enough for easy melt processing.

Hexafluoropropylene−tetrafluoroethylene copolymers are available in low melt viscosity, extrusion grade, intermediate viscosity, high melt viscosity, and as dispersions. The low melt viscosity (MV) resin can be injection molded by conventional thermoplastic molding techniques. It is more suitable for injection molding than other FEP resins (51).

The extrusion grade is suitable for tubing, wire coating, and cable jacketing. It is less suitable for injection molding than the low MV resin because of its relatively high melt viscosity. The intermediate MV (Teflon FEP-140) resin is used for insulation of wires larger than AWG 12 (American wire gauge) and applications involving smaller wire sizes, where high current loads or excessive thermal cycling may occur. It is also ideal for jacketing wire braid construction, such as coaxial cables, and for heater cable jackets.

The high MV resin is used as liners for process equipment. Its melt viscosity is significantly higher than that of other resins and therefore it is unsuitable for conventional injection molding. Stress-crack resistance and mechanical properties are superior to those of the other three products (52) (Table 2).

Both high and low color concentrates are available for pigmenting extruded coatings of FEP resins. The concentrates are prepared for melt dispersion in extrusion applications. The pigments are purified, thermally stable, and carefully selected to meet electrical, mechanical, and thermal end use specifications. Color concentrate pellets are easily dispersed among clear pellets by conventional tumbling. The ratio of concentrate to natural resin varies, depending on the wire size, insulation thickness, and color intensity desired.

An FEP copolymer dispersion is available as a 55-wt % aqueous dispersion containing 6% nonionic surfactant (on a solids basis) and a small amount of anionic dispersing agent. Its average particle size is ca 0.2 μm.

Table 2. Properties of Teflon FEP Fluorocarbon Resin[a]

Mechanical property	ASTM method	Teflon 110	Teflon 100	Teflon 140	Teflon 160
melt flow number, g/10 min	D2116		7.0	3.0	1.5
specific gravity	D792	2.13–2.17	2.13–2.17	2.13–2.17	2.13–2.17
tensile strength,[b] MPa[c]	D1708	20	23	30	31
elongation,[b] %	D1708	300	325	325	305
compressive strength, MPa[c]	D695		21	21	23
flexural strength,[b] MPa[c]	D790		18	18	18
impact strength,[b] J/m[d]	D256		no break	no break	no break
flexural modulus,[b] MPa[c]	D790	655	620	620	586
hardness durometer, Shore D	D2240	55	56	56	57
coefficient of friction, metal–film	D1894		0.27	0.27	0.235
deformation under load,[e] %	D621	1.8	0.5	0.5	0.5
water absorption, 24 h, %	D570	<0.01	0.004	0.004	0.004
linear coefficient of expansion per °C $\times 10^{-5}$	E381				
0–100°C			13.5	13.9	7.6
100–150°C			20.8	21.2	11.5
150–200°C			26.6	27.0	14.2

[a]Compression-molded specimens; property data on extruded wire specimens are similar.
[b]At 23°C
[c]To convert MPa to psi, multiply by 145.
[d]To convert J/m to ft·lbf/in., divide by 53.38.
[e]At 23°C, 6.9 MPa,[c] 23 h.

Properties.   The crystallinity of FEP polymer is significantly lower than that of PTFE (70 vs 98%). The structure resembles that of PTFE, except for a random replacement of a fluorine atom by a perfluoromethyl group (CF_3). The crystallinity after processing depends on the rate of cooling the molten polymer. The presence of HFP in the polymer chain tends to distort the highly crystallized structure of the PTFE chain and results in a higher amorphous fraction.

In the free-radical polymerization of FEP copolymers, chain termination occurs by binary coupling of chain ends, thus contributing to high molecular weights. Linear viscoelastic properties of these polymers in the amorphous melts were measured by dynamic rheometry. The FEP samples had high molecular weights and were found to verify the relation of zero shear viscosity vs (mol wt)^3 predicted by the reptation theory. At lower molecular weights, the empirical relation of viscosity vs (mol wt)$^{3.4}$ holds (53).

Transitions and Relaxations.   Only one first-order transition is observed, the melting point. Increasing the pressure raises mp. At low pressure, the rate of increase in the melting point is ca 1.74°C/MPa (0.012°C/psi); at high pressures this rate decreases to ca 0.725°C/MPa (0.005°C/psi). Melting increases the volume by 8%. In the presence of the HFP comonomer, crystal distortion occurs with an increase in intramolecular distance that, in turn, reduces the melting point (54).

The relaxation temperature appears to increase with increasing HFP content. Relaxation involves 5–13 of the chain carbon atoms. Besides α and γ relaxations, one other dielectric relaxation was observed below $-150°C$, which did not vary in temperature or in magnitude with comonomer content or copolymer density (55). The α relaxation (also called Glass I) is a high temperature transition (157°C) and γ relaxation (Glass II) (internal friction maxima) occurs between -5 and 29°C.

Thermal Stability. The polymer is thermally stable and can be processed at ca 270°C. Thermal degradation is a function of temperature and time, and the stability is therefore limited. The melt-flow rate (thermal degradation) increases significantly for short periods above 280°C, and degradation occurs at lower temperatures with longer hold times. The hourly weight loss is 0.0004% at 230°C, 0.001% at 260°C, 0.01% at 290°C, 0.02% at 320°C, 0.08% at 340°C, and 0.3% at 370°C. Degradation is not significant if the change in melt-flow rate during molding is $< 10\%$. Physical strength decreases after prolonged exposure above 205°C, which accounts for the lower temperature rating of FEP resins (56).

Radiation Effects. The primary effect of radiation is the degradation of large molecules to small molecules. Molecular weight reduction can be minimized by excluding oxygen. If FEP is lightly irradiated at elevated temperatures in the absence of oxygen, cross-linking offsets molecular breakdown (55,57).

The degree to which radiation exposure affects FEP resins is determined by the energy absorbed, regardless of the type of radiation. Changes in mechanical properties depend on total dosage, but are independent of dose rate. The radiation tolerance of FEP in the presence or absence of oxygen is higher than that of PTFE by a factor of 10:1.

Mechanical Properties. Extensive lists of the physical properties of FEP copolymers are given in References 58–63. Mechanical properties are shown in Table 3. Most of the important properties of FEP are similar to those of PTFE; the main difference is the lower continuous service temperature of 204°C of FEP compared to that of 260°C of PTFE. The flexibility at low temperatures and the low coefficients of friction and stability at high temperatures are relatively independent of fabrication conditions. Unlike PTFE, FEP resins do not exhibit a marked change in volume at room temperature, because they do not have a first-order transition at 19°C. They are useful above $-267°C$ and are highly flexible above $-79°C$ (64).

Static friction decreases with an increase in load, and the static coefficient of friction is lower than the dynamic coefficient. The tendency to creep must be considered carefully in FEP products designed for service under continuous stresses. Creep can be minimized by suitable fillers. Fillers are also used to improve wear resistance and stiffness. Compositions such as 30% bronze-filled FEP, 20% graphite-filled FEP, and 10% glass-fiber-filled FEP offer high PV values ($\sim400(kPa\cdot m)/s$) and are suitable for bearings.

Articles fabricated from FEP resins can be made bondable by surface treatment with a solution of sodium in liquid ammonia, or naphthalenyl sodium in tetrahydrofuran (64) to facilitate subsequent wetting. Exposing the surface to corona discharge (65) or amines at elevated temperatures in an oxidizing atmosphere (66) also makes the resins bondable. Some of the more recent work is described in References 67–69.

Table 3. Mechanical Properties of FEP[a]

Property	Value	ASTM method
specific gravity	2.14–2.17	D792-50
thermal conductivity, W/(m·K)		Cenco-Fitch
−129 to 182°C	2.4	
−253°C	1.4	
water absorption in 24 h, 3.175-mm thick sample		D570-547
% wt increase	<0.1	
dimensional change at 23°C	none	
coefficient of thermal expansion per °C		D696-44
>23°C	9.3×10^{-5}	
<23°C	5.7×10^{-5}	
specific heat, kJ/(kg·K)[b]		
20°C	1.09	
100°C	1.17	
260°C	1.30	
heat distortion, °C		D648-56
455 kPa[c]	70	
1820 kPa[c]	51	
tensile yield strength, av, MPa[d]		D638-527
−251°C	165	
−73°C	62	
23°C	12	
121°C	3.5	
tensile modulus, MPa[d]		
−251°C	57	
−73°C	24	
23°C	4	
100°C	1	
tensile elongation, %		D638-527
−251°C	4	
−73°C	200	
23°C	350	
flexural modulus, MPa[d]		D747-50
−251°C	5300	
−101°C	3200	
23°C	660	
55°C	340	
compressive strength, MPa[d]		D695
−251°C	251	
23°C	15	
100°C	3.4	
Izod impact strength, notched, J/m[e]		D256-56
23°C	no break	
hardness, Durometer		D2240-T
23°C	D59	
Taber abrasion, g/MHz, 100-g load		
CS-17 wheel	7.5	

[a]Measured on Teflon FEP T-100. [b]To convert kJ to kcal, divide by 4.184.
[c]To convert kPa to atm, multiply by 0.01. [d]To convert MPa to psi, multiply by 145.
[e]To convert J/m to ft·lbf/in., divide by 53.38.

Vibration-dampening properties at sonic and ultrasonic frequencies are excellent. However, the thickness of the resin must be sufficient to absorb the energy produced; this is usually determined experimentally.

Electrical Properties.   Because of excellent electrical properties, FEP is a valuable and versatile electrical insulator. Within the recommended service temperature range, PTFE and FEP have identical properties as electrical insulators. Volume resistivity, which is $> 10^{17}$ Ω/cm, remains unchanged even after prolonged soaking in water; surface resistivity is $> 10^{15}$ Ω/sq.

At low frequencies, the dielectric constant of FEP remains the same ($\sim$2). However, at > 100 MHz the constant drops slightly with increasing frequency. As a true thermoplastic, FEP has a void content of zero and most of the fabricated material has a density of 2.14–2.17 g/cm^3. The National Bureau of Standards has selected Teflon FEP resins for dielectric reference specimens because of the stability of their dielectric constant. The dissipation factor has several peaks as a function of temperature and frequency (3×10^{-4} at 100 kHz; 7×10^{-4} at 1 MHz). The magnitude of the dissipation factor peak is greater for FEP than for PTFE because the molecular structure of the former is less symmetrical. The dissipation factor is hardly affected by irradiation annealing (70) and unaffected by humidity. The dielectric strength is high (80 GV/mm for 0.25 mm film at 23°C) and unaffected by thermal aging at 200°C. At high frequencies, the dielectric properties deteriorate in the presence of corona. If the voltage stress is not high enough to cause corona ignition, an infinitely long dielectric life is expected at any frequency. Corona discharges on the surface or in a void initiate dielectric breakdown (71). The FEP resins are recommended for continuous service up to 205°C. Although they begin to melt flow at 270°C, they retain some structural integrity up to 250°C (70).

Chemical Properties.   The FEP resin is inert to most chemicals and solvents, even at elevated temperatures and pressures. However, it reacts with fluorine, molten alkali metal, and molten sodium hydroxide. Acids or bases are not absorbed at 200°C and exposures of one year. The absorption of organic solvents is less than 1% at elevated temperatures and long exposure times. Absorption of chemicals or solvents has no effect on the chemical integrity of the FEP molecule and is a reversible physical process.

Gases and vapors permeate FEP resin at a rate that is considerably lower than that of most plastics. Because FEP resins are melt processed, they are void-free and permeation occurs only by molecular diffusion. Variation in crystallinity and density is limited, except in unusual melt-processing conditions.

Because of its low permeability, FEP polymer is used extensively in the chemical industry. Its permeation characteristics are similar to those of PTFE (Table 4). An inverse relationship between permeability and film thickness applies to FEP.

Weathering.   Articles fabricated from FEP are unaffected by weather, and their resistance to extreme heat, cold, and uv irradiation suits them for applications in radar and other electronic components. For example, after 15 years of solar exposure in Florida, the tensile strength (73) and light transmission (96%) of a 25-μm thick film was unchanged and the film remained crystal clear. Elongation increased slightly for the first 5 to 7 years of outdoor exposure, probably as a result of stress relaxation. Beyond 10 years, a small decrease was observed.

Table 4. Permeability of FEP Fluorocarbon Resins to Liquid Vapors and Gases

Permeant	Permeability constant,[a,b] $mol/(m \cdot s \cdot Pa) \times 10^{15}$		
	23°C	35°C	50°C
Liquid vapors			
acetic acid		9.07	
acetone	0.37		3.23
benzene	0.75		
carbon tetrachloride	0.24	0.41	
decane	112.18		33.48
dipentene	23.50		10.67
ethyl acetate	0.27	2.06	4.09
ethanol	1.61	4.66	
H_2SO_4, 98%	21.70		
toluene	5.38		
water	8.14	20.32	18.26
Gases[c]			
oxygen	18.69		
helium	113.47		
nitrogen	6.10		
hydrogen	40.15		
methane	3.17		

[a]Ref. 60. Test method ASTM E96-35T (at vapor pressure; for 25.4-μm film thickness). Values are averages only and not for specification purposes.
[b]Original data converted to SI units using vapor pressure data from Ref. 72.
[c]At 20°C.

Optical Properties.   Teflon FEP fluorocarbon film transmits more ultraviolet, visible light, and infrared radiation than ordinary window glass. The refractive index of FEP film is 1.341–1.347 (74).

Fabrication

Standard thermoplastic processing techniques can be used to fabricate FEP. Thermal degradation must be avoided, and a homogeneous structure and good surface quality must be maintained.

Injection Molding.   Compared to most thermoplastic products, even the low MV resin has a significantly higher melt viscosity and therefore requires higher processing temperatures, slower injection rates, special mold design, and corrosion-resistant material of construction. When the flow velocity in melt processing exceeds a critical value, melt fracture occurs. The critical shear rate of FEP is much lower than that of other thermoplastics. Recommendations for materials of construction and the screw design, valves, smear heads, nozzle, operating conditions, and mold design are given in Reference 52.

Pigments (thermally stable at processing temperature) are dry blended with the resin before molding. At loadings of 0.1–1%, pigments have no appreciable effect on the dielectric strength, dielectric constant, or mechanical properties. The dissipation factor of pigmented resin varies with the pigment and its amount (75).

Extrusion.   Conventional melt-extrusion equipment is used in processing FEP resins. Commercial pigments are mixed with the resin before extrusion into wire coating, tubing, rods, molding, beading channels, etc. Coating thicknesses of 0.076–2.54 mm have been extruded over such materials as silicone rubber, poly(vinyl chloride), glass braid, metal-shielded cables, twisted conductors, and parallel multiconductor cables.

For primary insulation or cable jackets, high production rates are achieved by extruding a tube of resin with a larger internal diameter than the base wire and a thicker wall than the final insulation. The tube is then drawn down to the desired size. An operating temperature of 315–400°C is preferred, depending on holdup time. The surface roughness caused by melt fracture determines the upper limit of production rates under specific extrusion conditions (76). Corrosion-resistant metals should be used for all parts of the extrusion equipment that come in contact with the molten polymer (77).

Tubing is made in a wide range of sizes and is used as slip-on electrical insulation, instrument tubing, and for hoses. Small tubing, called spaghetti tubing, can be produced by a free-extrusion technique, whereas hose-size tubing is produced by conventional forming-box techniques; FEP also is extruded into films.

Dispersion Processing.   The commercial aqueous dispersion of FEP contains 55 wt % of hydrophobic, negatively charged FEP particles and ca 6 wt % (based on FEP) of a mixture of nonionic and anionic surface-active agents. The average particle size is ca 0.2 μm. The dispersion is processed by the same technique used for PTFE dispersion. For example, the fabric is coated with FEP dispersion, the water is evaporated from the coating, the wetting agent is removed, and the FEP layer is fused with the fabric.

Dispersion is used as a coating for glass fabric, chemical barriers, and wire-insulating tapes; as adhesive coatings for bonding seals and bearings of PTFE to metallic and nonmetallic components; and as antifriction or antistick coatings for metals. The fusion of FEP to provide a continuous film depends on a time–temperature relationship; 1 min at 400°C or 40 min at 290°C are sufficient to achieve good fusion (78).

Other Techniques.   The FEP resin is bonded to metal surfaces by the application of heat and pressure; it can be heat sealed or hot-gas welded. Heating FEP at 260°C and allowing it to cool slowly results in stress relieving, or annealing. The FEP film is used to weld PTFE-coated surfaces.

Effects of Fabrication on Product Properties.   Extrusion conditions have a significant effect on the quality of the product (77). Contamination can be the result of corrosion, traces of another resin, or improper handling. Corrosion-resistant Hastelloy C parts should be used in the extruder. Surface roughness is the result of melt fracture or mechanical deformation. Melt fracture can be eliminated by increasing the die opening, die temperature, and the melt temperature and reducing the extrusion rate. Bubbles and discoloration are caused by resin degradation, air entrapment, or condensed moisture. Excessive drawdown, resin degradation, or contamination can result in pinholes, tears, and cone breaks. The blisters are caused by degassing of primary coatings, and loose coatings are caused by rapid cooling and long cones.

Testing and Standards.   Requirements for extrusion and molding grades are cited in ASTM specifications (79) and in Federal specification LP-389A of May 1964 (80). For fabricated shapes, FEP film and sheet are covered by Aeronautical Material Specifications (AMS) 3647 and LP-523 (81). Besides the specifications covered by the Fluorocarbons Division of the Society of the Plastics Industry, Inc. (82), other specifications are listed in Reference 83.

Economic Aspects

Because of the high cost of hexafluoropropylene, FEP is more expensive than PTFE. In the United States in 1992, FEP sold at prices up to $28.3 kg, depending on the type and quantity. Most grades are marketed in a colorless, translucent, extruded pellet form. The dispersion containing about 55% solids is priced at ca $33 kg. During the 1980s FEP sales increased rapidly because of usage in plenum cable, but since there are other polymers that can be used in this application the growth rate for FEP is expected to slow down.

Health and Safety

The safety precautions required in handling TFE–HFP copolymers are the same as those applied to handling PTFE. Large quantities have been processed safely by many different fabricators in a variety of operations. With proper ventilation, the polymer can be processed and used at elevated temperatures without hazard. The fumes from heated FEP or its thermal decomposition products are toxic in high concentrations, like the fumes or decomposition products of other polymers. Ventilation should be provided in areas where the resin is at processing temperature (270–400°C). At ambient temperatures, FEP resin is essentially inert. Inhalation of fumes given off by heated FEP resin may result in influenza-like symptoms. They may occur several hours after exposure and disappear within 35–48 hours, even in the absence of treatment; the effects are not cumulative (52). Such attacks usually follow exposure to vapors evolved from the polymer without adequate ventilation or from smoking tobacco or cigarettes contaminated with the polymer. Toxicology study of the particulates and fumes is reported in Reference 84.

Applications

The principal electrical applications include hook-up wire, interconnecting wire, coaxial cable, computer wire, thermocouple wire, plenum cable, and molded electrical parts. Principal chemical applications are lined pipes and fittings, overbraided hose, heat exchangers, and laboratory ware. Mechanical uses include antistick applications, such as conveyor belts and roll covers. A recent development of FEP film for solar collector windows takes advantage of light weight, excellent weatherability, and high solar transmission. Solar collectors made of FEP film are efficient, and installation is easy and inexpensive.

BIBLIOGRAPHY

"Fluorinated Ethylene–Propylene Copolymers," under "Fluorine Compounds, Organic," in *ECT* 3rd ed., Vol. 11, pp. 24–35, by S. V. Gangal, E. I. du Pont de Nemours & Co., Inc.

1. U.S. Pat. 2,946,763 (July 26, 1960), M. I. Bro and B. W. Sandt (to E. I. du Pont de Nemours & Co., Inc.).
2. U.S. Pat. 2,394,581 (Feb. 12, 1946), A. F. Benning, F. B. Dowing, and J. D. Park (to Kinetic Chemicals, Inc.).
3. E. G. Young and W. S. Murray, *J. Am. Chem. Soc.* **70**, 2814 (1949).
4. A. L. Henne and T. P. Waalkes, *J. Am. Chem. Soc.* **68**, 496 (1946).
5. U.S. Pat. 3,446,858 (May 27, 1969), H. Shinzu and co-workers (to Daikin Kyogo Co.).
6. U.S. Pat. 3,459,818 (Aug. 15, 1969), H. Ukihashi and M. Hisasue (to Asahi Glass Co.).
7. U.S. Pat. 3,873,630 (Mar. 25, 1975), N. E. West (to E. I. du Pont de Nemours & Co., Inc.).
8. U.S. Pat. 5,043,491 (Aug. 27, 1991), J. Webster and co-workers, (to E. I. du Pont de Nemours & Co., Inc.).
9. U.S. Pat. 5,057,634 (Oct. 15, 1991), J. Webster and co-workers (to E. I. du Pont de Nemours & Co., Inc.).
10. U.S. Pat. 5,068,472 (Nov. 26, 1991), J. Webster and co-workers (to E. I. du Pont de Nemours & Co., Inc.).
11. U.S. Pat. 3,758,138 (Aug. 7, 1956), D. A. Nelson (to E. I. du Pont de Nemours & Co., Inc.).
12. B. Atkinson and A. B. Trenwith, *J. Am. Chem. Soc., Pt. II*, 2082 (1953).
13. U.S. Pat. 2,759,983 (Aug. 21, 1956), J. S. Waddell (to E. I. du Pont de Nemours & Co., Inc.).
14. U.S. Pat. 3,047,640 (July 31, 1962), R. F. Sweeny and C. Woolf (to Allied Chemical Corp.).
15. U.S. Pat. 3,009,966 (Nov. 21, 1961), M. Hauptschein and A. Fainberg (to Pennsalt Chemicals Corp.).
16. L. T. Hals, T. S. Reid, and G. H. Smith, *J. Am. Chem. Soc.* **73**, 4054 (1951); U.S. Pat. 2,668,864 (Feb. 9, 1954), J. T. Hals, T. S. Reid, and G. H. Smith (to Minnesota Mining and Manufacturing Co.).
17. G. Pass, *J. Am. Chem. Soc., Pt. I*, 824 (Jan. 1965); Fr. Pat. 1,399,414 (May 14, 1965), (to Imperial Chemical Industries, Ltd.).
18. Ger. Pat. 1,236,497 (Mar. 16, 1967), W. Oese, H. Dude, and F. Reinke (to VEB Fluor-werke Dohma).
19. H. C. Duus, *Ind. Eng. Chem.* **47**, 1445 (1955).
20. J. W. Clayton, *Occup. Med.* **4**, 262 (1962).
21. U.S. Pat. 2,918,501 (Dec. 22, 1959), W. J. Brehm and co-workers (to E. I. du Pont de Nemours & Co., Inc.).
22. U.S. Pat. 3,316,312 (Apr. 25, 1967), D. I. McCane and I. M. Robinson (to E. I. du Pont de Nemours & Co., Inc.).
23. R. A. Matula, *J. Phys. Chem.* **72**, 3054 (1968).
24. U.S. Pat. 3,358,003 (Dec. 12, 1967), H. S. Eleuterio and R. W. Meschke (to E. I. du Pont de Nemours & Co., Inc.).
25. U.S. Pat. 3,180,895 (Apr. 27, 1965), J. F. Harris, Jr. and D. I. McCane (to E. I. du Pont de Nemours & Co., Inc.).

26. U.S. Pat. 3,291,843 (Dec. 13, 1966), C. G. Fritz and S. Selman (to E. I. du Pont de Nemours & Co., Inc.).

27. E. V. Volkova and A. E. Skobina, *Vysokomol. Soedin.* **6**(5), 964 (1964).

28. Fr. Pat. 1,524,571 (May 10, 1968), S. W. Osborn and E. Broderich (to Thiokol Chemical Corp.).

29. U.S. Pat. 2,983,764 (May 9, 1961), D. F. Knaack (to E. I. du Pont de Nemours & Co., Inc.).

30. U.S. Pat. 2,958,685 (Nov. 1, 1960), H. S. Eleuterio (to E. I. du Pont de Nemours & Co., Inc.).

31. I. L. Knunyants, E. I. Mysov, and M. P. Krasuskaya, *Izv. Akad. Nauk. SSSR Otd. Khim. Nauk.*, 906 (1958).

32. I. L. Knunyants, A. E. Shchekotikhin, and A. V. Fakin, *Izv. Akad. Nauk. SSSR Otd. Khim. Nauk.*, 282 (1953).

33. I. L. Knunyants, L. S. German, and B. L. Dyatkin, *Izu. Akad. Nauk. SSSR Otd. Khim. Nauk.*, 1353 (1956).

34. R. N. Haszeldine and B. R. Steele, *J. Chem. Soc.*, 1592 (1953).

35. W. T. Miller, Jr., E. Bergman, and A. H. Fainberg, *J. Am. Chem Soc.* **79**, 4159 (1957).

36. I. L. Knunyants, V. V. Shokina, and N. D. Kuleshova, *Izv. Akad. Nauk. SSSR Otd. Khim. Nauk.*, 1936 (1960).

37. U.S. Pat. 3,467,636 (Sept. 16, 1969), A. Nersasian (to E. I. du Pont de Nemours & Co., Inc.).

38. U.S. Pat. 3,536,683 (Oct. 27, 1970), F. V. Bailor and J. R. Cooper (to E. I. du Pont de Nemours & Co., Inc.).

39. U.S. Pat. 3,790,540 (Feb. 5, 1974), J. E. Dohany and A. C. Whiton (to Pennwalt Corp.).

40. U.S. Pat. 3,817,951 (June 18, 1974), D. N. Robinson (to Pennwalt Corp.).

41. U.S. Pat. 3,868,337 (Feb. 25, 1975), P. Gros (to Society Superflexit).

42. G. L. Kennedy, Jr., *Crit. Revs. Toxicol.* **21**(2), 149–170 (1990).

43. U.S. Pat. 2,598,283 (May 27, 1952), W. T. Miller (to U.S. Atomic Energy Commission).

44. R. A. Naberezhnykh and co-workers, *Dokl. Akad. Nauk. SSSR* **214**, 149 (1974).

45. A. S. Kabankin, S. A. Balabanova, and A. M. Markevich, *Vysokomol. Soedin. Ser. A* **12**, 267 (1970).

46. Br. Pat. 781,532 (Aug. 21, 1957), C. G. Krespan (to E. I. du Pont de Nemours & Co., Inc.).

47. U.S. Pat. 3,132,124 (May 5, 1964), M. J. Couture, D. L. Schindler, and R. B. Weiser (to E. I. du Pont de Nemours & Co., Inc.).

48. U.S. Pat. 4,380,618 (1983), A. Khan and R. Morgan (to E. I. du Pont de Nemours & Co., Inc.).

49. U.S. Pat. 4,384,092 (1983), J. Herison (to Ugine Kuhlmann).

50. U.S. Pat. 4,861,845 (1989), E. Slocum, A. Sobrero, and R. Wheland (to E. I. du Pont de Nemours & Co., Inc.).

51. R. S. Atland, *Modern Plast.* **62**, 200 (1985).

52. *Teflon-FEP Fluorocarbon Resin, Techniques for Injection Molding*, information bulletin 95d, E. I. du Pont de Nemours & Co., Inc., Wilmington, Del., 1969.

53. S. Wu, *Macromolecules* **18**, 2023–2030 (1985).

54. R. K. Eby, *J. Appl. Phys.* **34**, 2442 (1963).

55. R. K. Eby and F. C. Wilson, *J. Appl. Phys.* **33**, 2951 (1962).

56. *Safe Handling Guide, Teflon Fluorocarbon Resins, du Pont Materials for Wire and Cable*, bulletin E-85433, E. I. du Pont de Nemours & Co., Inc., Wilmington, Del., 1986.

57. R. Y. M. Huang and P. J. F. Kanitz, *Polym. Prepr. Am. Chem. Soc. Div. Polym. Chem.* **10**(2), 1087 (1969).

58. R. J. Diamond, *Plastics* **27**, 109 (1962).

59. J. Frados, ed., *Modern Plastics Encyclopedia*, Vol. 46, No. 10A, McGraw-Hill Book Co., Inc., New York, 1969, p. 974.
60. *J. Teflon* **11**(1), 8 (1970).
61. J. A. Brydson, *Plastics Materials*, Iliffe Books, Ltd., London, 1966, Chapt. 10, pp. 203–218.
62. *Teflon 100 FEP–Fluorocarbon Resin—Melt Processible Resin*, information bulletin X-90a, E. I. du Pont de Nemours & Co., Inc., Wilmington, Del., 1960.
63. *Teflon Fluorocarbon Resins, Mechanical Design Data*, 2nd ed., E. I. du Pont de Nemours & Co., Inc., Wilmington, Del., 1965.
64. A. A. Benderly, *J. Appl. Polym. Sci.* **6**, 221 (1962).
65. Brit. Pat. 890,466 (Feb. 28, 1962), D. L. Ryan (to E. I. du Pont de Nemours & Co., Inc.).
66. U. S. Pat. 3,063,882 (Nov. 13, 1962), J. R. Chesire (to E. I. du Pont de Nemours & Co., Inc.).
67. R. R. Rye and G. W. Arnold, *Langmuir* **5**, 1331 (1989).
68. D. T. Clark and D. R. Hutton, *J. Polym. Sci., Polym. Chem. Ed.* **25**, 2643 (1987).
69. R. C. Bening and J. J. McCarthy, *Polym. Prep.* **29**, 336 (1988).
70. *Electrical/Electronic Design Data for Teflon*, E. I. du Pont de Nemours & Co., Inc., Wilmington, Del.
71. J. C. Reed, E. J. McMahon, and J. R. Perkins, *Insulation (Libertyville, Ill.)* **10**, 35 (1964).
72. D. W. Green, ed., *Perry's Chemical Engineers' Handbook*, 6th ed., McGraw-Hill Book Co., New York, 1984.
73. *Teflon Solar Film for Solar Collectors*, E. I. du Pont de Nemours & Co., Inc., Wilmington, Del.
74. *Teflon FEP–Fluorocarbon Film*, bulletin T-5A, Optical, E. I. du Pont de Nemours & Co., Inc., Wilmington, Del.
75. L. H. Gillespe, D. O. Saxton, and F. M. Chapman, *New Design Data for Teflon*, E. I. du Pont de Nemours & Co., Inc., Wilmington, Del., 1960.
76. *J. Teflon* **18**(1), 8(1977).
77. *Teflon FEP–Fluorocarbon Resin—Techniques for Processing by Melt Extrusion*, 2nd ed., information bulletin X-82, E. I. du Pont de Nemours & Co., Inc., Wilmington, Del., 1960.
78. *Properties and Processing Techniques for Teflon 120 FEP–Fluorocarbon Resin Dispersion*, preliminary information bulletin no. 20, E. I. du Pont de Nemours & Co., Inc., Wilmington, Del., 1961.
79. *ASTM Annual Book of ASTM Standards*, Vol. 08.01, American Society for Testing and Materials, Philadelphia, Pa., 1993.
80. *Federal Supply Service Bureau Specification L-P-389A*, Section SW, 470E L'Enfant Plaza, Washington, D.C.
81. Technical data, SAE International, Inc., Warrendale, Pa., 1993.
82. *ASTM Annual Book of ASTM Standards*, Vol. 08.03, American Society for Testing and Materials, Philadelphia, Pa., 1993.
83. Technical data, Society of the Plastics Industry, Inc., Fluorocarbon Division, Washington, D.C., 1993; *J. Teflon* **15**(1), 10 (1974).
84. K. P. Lees and W. C. Seidal, *Inhalation Toxicol.* **3**(3), 237 (1991).

General References

S. V. Gangal, "Tetrafluoroethylene Polymers, Tetrafluoroethylene–Hexafluoropropylene Copolymers," in J. I. Kroschwitz, ed., *Encyclopedia of Polymer Science and Engineering*, 2nd ed., Vol. 16, John Wiley & Sons, Inc., New York, 1989, pp. 601–613.

SUBHASH V. GANGAL
E. I. du Pont de Nemours & Co., Inc.

TETRAFLUOROETHYLENE–ETHYLENE COPOLYMERS

Copolymers of ethylene [74-85-1] and tetrafluoroethylene [116-14-3] (ETFE) have been a laboratory curiosity for more than 40 years. These polymers were studied in connection with a search for a melt-fabricable PTFE resin (1–5); interest in them fell with the discovery of TFE–HFP (FEP) copolymers (6). In the 1960s, however, it became evident that a melt-fabricable fluorocarbon resin was needed with higher strength and stiffness than those of PTFE resins. Earlier studies indicated that TFE–ethylene copolymers [11939-51-6] might have the right combination of properties. Subsequent research efforts (7) led to the introduction of modified ethylene–tetrafluoroethylene polymer [25038-71-5] (Tefzel) by E. I. du Pont de Nemours & Co., Inc. in 1970.

Modified ethylene–tetrafluoroethylene copolymers are the products of real commercial value because they have good tensile strength, moderate stiffness, high flex life, and outstanding impact strength, abrasion resistance, and cut-through resistance. Electrical properties include low dielectric constant, high dielectric strength, excellent resistivity, and low dissipation factor. Thermal and cryogenic performance and chemical resistance are good. These properties, combined with elasticity, make this material an ideal candidate for heat-shrinkable film and tubing. This family of copolymers can be processed by conventional methods such as melt extrusion, injection molding, transfer molding, and rotational molding. The properties of the copolymers vary with composition; polymers containing 40–90% tetrafluoroethylene (by weight) soften between 200 and 300°C, depending on composition (1). The tetrafluoroethylene segments of the molecules account for >75% of the weight of an approximately 1:1 mole ratio copolymer. The two monomers combine readily into a nearly 1:1 alternating structure. Such polymers exhibit a unique combination of mechanical, chemical, and electrical properties as well as excellent weatherability. However, thermal stress-crack resistance is poor. The copolymer can be modified with a termonomer that undergoes free-radical polymerization and does not cause undesirable chain transfer or termination during polymerization. The modified copolymer exhibits almost the identical physical, chemical, and electrical properties characteristic of the 1:1 alternating copolymer, but retains high ultimate elongation up to 200°C.

Ethylene and tetrafluoroethylene are copolymerized in aqueous, non-aqueous, or mixed medium with free-radical initiators. The polymer is isolated and converted into extruded cubes, powders, and beads, or a dispersion. This family of products is manufactured by Du Pont, Hoechst, Daikin, Asahi Glass, and Ausimont and sold under the trade names of Tefzel, Hostaflon ET, Neoflon EP, Aflon COP, and Halon ET, respectively.

Monomers

Tetrafluoroethylene of purity suitable for granular or dispersion polymerizations is acceptable for copolymerization with ethylene. Polymerization-grade ethylene is suitable for copolymerization with tetrafluoroethylene. Modifying termonomers, eg, perfluorobutylethylene and perfluoropropylene, are incorporated by free-radical polymerization.

Manufacture

Tetrafluoroethylene–ethylene copolymers have tensile strengths two to three times as high as the tensile strength of polytetrafluoroethylene or of the ethylene homopolymer (1). Because these copolymers are highly crystalline and fragile at high temperature, they are modified with a third monomer, usually a vinyl monomer free of telegenic activity. The termonomer provides the copolymer with side chains of at least two carbon atoms, such as perfluoroalkylvinyl or vinylidene compounds, perfluoroalkyl ethylenes, and perfluoroalkoxy vinyl compounds. For high tensile properties and cut-through resistance, a molar ratio of ethylene and tetrafluoroethylene between 60:40 and 40:60 is required (8,9).

Copolymerization is effected by suspension or emulsion techniques under such conditions that tetrafluoroethylene, but not ethylene, may homopolymerize. Bulk polymerization is not commercially feasible, because of heat-transfer limitations and explosion hazard of the comonomer mixture. Polymerizations typically take place below 100°C and 5 MPa (50 atm). Initiators include peroxides, redox systems (10), free-radical sources (11), and ionizing radiation (12).

Purely aqueous polymerization systems give copolymers that are not wetted by the reaction medium. The products agglomerate and plug valves, nozzles, and tubing, and adhere to stirrer blades, thermocouples, or reactor walls. These problems do not occur in organic media or mixtures of these with water.

Aqueous emulsion polymerization is carried out using a fluorinated emulsifier, a chain-transfer agent to control molecular weight, and dispersion stabilizers such as manganic acid salts and ammonium oxalate (13,14).

Reactivity ratios of ethylene and tetrafluoroethylene are

Temperature, °C	r_{TFE}	r_E
-35	0.014 ± 0.008	0.010 ± 0.02
65	0.045 ± 0.010	0.14 ± 0.03

These values indicate strong alternation tendencies that decrease with increasing temperature. Computations show that 1:1 ETFE copolymers obtained at -30 and 65°C should have about 97 and 93%, respectively, of alternating sequences (15).

Properties

The equimolar copolymer of ethylene and tetrafluoroethylene is isomeric with poly(vinylidene fluoride) but has a higher melting point (16,17) and a lower dielectric loss (18,19) (see FLUORINE COMPOUNDS, ORGANIC–POLY(VINYLIDENE FLUORIDE)). A copolymer with the degree of alternation of about 0.88 was used to study the structure (20). Its unit cell was determined by x-ray diffraction. Despite irregularities in the chain structure and low crystallinity, a unit cell and structure was derived that gave a calculated crystalline density of 1.9 g/cm³. The unit cell is believed to be orthorhombic or monoclinic (a = 0.96 nm, b = 0.925 nm, c = 0.50 nm; γ = 96°).

H H F F
| | | |
—C—C—C—C—
| | | |
H H F F

ethylene–tetrafluoroethylene unit

H F H F
| | | |
—C—C—C—C—
| | | |
H F H F

poly(vinylidene fluoride) segment

The molecular conformation is that of extended zigzag. Molecular packing appears to be orthorhombic, each molecule having four nearest neighbors with the CH_2 groups of one chain adjacent to the CF_2 groups of the next. The x-ray spectrum of a 1:1 copolymer has two main peaks at $Z_0 = 19.63°$ and $Z_0 = 21.00°$, corresponding to Bragg distances of 0.45 and 0.42 nm, respectively. Compression-molded samples are 50–60% crystalline; however, crystallinity is greatly affected by composition, quench rate, and temperature.

Alternation is usually above 90%. Nearly perfect alternation of isomeric units in a ca 1:1 monomer ratio has been confirmed by infrared spectroscopy. Bands at 733 and 721 cm^{-1} have an intensity proportional to the concentration of $(CH_2)_n$ groups ($n = 4$ and <6, respectively) present in a copolymer containing 46 mol % tetrafluoroethylene; intensity decreases with increasing concentration of fluorinated monomer.

The molecular weight and its distribution have been determined by laser light scattering, employing a new apparatus for ETFE dissolution and solution clarification at high temperature; diisobutyl adipate is the solvent at 240°C. The molecular weight of molten ETFE is determined by high temperature rheometry (21).

This polymer can be dissolved in certain high boiling esters at temperatures above 230°C (22), permitting a weight-average molecular weight determination by light scattering. Solution viscosity data suggest that the polymer exists as a slightly expanded coil under similar conditions (23).

Transitions.   Samples containing 50 mol % tetrafluoroethylene with ca 92% alternation were quenched in ice water or cooled slowly from the melt to minimize or maximize crystallinity, respectively (19). Internal motions were studied by dynamic mechanical and dielectric measurements, and by nuclear magnetic resonance. The dynamic mechanical behavior showed that the α relaxation occurs at 110°C in the quenched sample; in the slowly cooled sample it is shifted to 135°C. The β relaxation appears near −25°C. The γ relaxation at −120°C in the quenched sample is reduced in peak height in the slowly cooled sample and shifted to a slightly higher temperature. The α and γ relaxations reflect motions in the amorphous regions, whereas the β relaxation occurs in the crystalline regions. The γ relaxation at −120°C in dynamic mechanical measurements at 1 Hz appears at −35°C in dielectric measurements at 10⁵ Hz. The temperature of the α relaxation varies from 145°C at 100 Hz to 170°C at 10⁵ Hz. In the mechanical measurement, it is 110°C. There is no evidence for relaxation in the dielectric data.

The activation energy is 318.1 kJ/mol (76 kcal/mol) for the α relaxation and 44.3 kJ/mol (10.6 kcal/mol) for the γ relaxation. These relaxations are attributed to the motion of long and short segments in the amorphous regions, respectively. As ETFE copolymer is isomeric with poly(vinylidene fluoride) (18), the γ relaxations occur at about the same temperature. Activation energies are similar and are attributed to the motion of short amorphous segments. The β relaxation in PVF_2 is considered to be the main-chain amorphous relaxation and is analogous to the α relaxation in the ethylene–tetrafluoroethylene copolymer. However, the arrangement of dipoles in the all-trans conformation is more symmetrical.

Physical and Mechanical Properties. Modified ethylene–tetrafluoroethylene copolymer has a good combination of mechanical properties, including excellent cut-through and abrasion resistance, high flex life, and exceptional impact strength. As wire insulation, it withstands physical abuse during and after installation. Lightweight wire constructions are designed with a minimum diameter and are useful as single, general-purpose insulation and for multiple or composite constructions.

Modified ETFE is less dense, tougher, and stiffer and exhibits a higher tensile strength and creep resistance than PTFE, PFA, or FEP resins. It is ductile, and displays in various compositions the characteristic of a nonlinear stress–strain relationship. Typical physical properties of Tefzel products are shown in Table 1 (24,25). Properties such as elongation and flex life depend on crystallinity, which is affected by the rate of crystallization; values depend on fabrication conditions and melt cooling rates.

Light transmittance of 25-μm films in the visible-to-ir range varies from 91 to 95% for Tefzel 200 and from 89 to 93% for Tefzel 280. In the uv range transmittance increases from 50% at 200 nm to 90% at 400 nm.

Thermal Properties. Modified ETFE copolymer has a broad operating temperature range up to 150°C for continuous exposure (24). Cross-linking by radiation improves the high temperature capability further. However, prolonged exposure to higher temperatures gradually impairs the mechanical properties and results in discoloration.

The thermodynamic properties of Tefzel 200 and 280 are shown in Table 2; the annual rate of loss of weight with thermal aging for Tefzel 200 ranges from 0.0006 g/g at 135°C to 0.006 g/g at 180°C after an initial loss of absorbed gases of 0.0013 g/g at elevated temperature. The excellent thermal stability of ETFE is demonstrated by aging at 180°C; at this temperature, the annual weight loss of six parts per 1000, or a 1% weight loss, takes almost two years.

Friction and Bearing Wear of the Glass-Reinforced Copolymer. Glass reinforcement improves the frictional and wear properties of modified ETFE resins (HT-2004). For example, the dynamic coefficient of friction (689.5 kPa (100 psi) at > 3 m/min) for Tefzel 200 is 0.4, which drops to 0.3 for the 25% glass-reinforced product at these conditions (24). The wear factor also improves from 12×10^{-14} to 32×10^{-17} 1/Pa (6000×10^{-10} to 16×10^{-10} in.$^3 \cdot$ min/ftlbf·h). These frictional and wear characteristics, combined with outstanding creep resistance, indicate suitability for bearing applications. Glass-reinforced ETFE is less abrasive on mating surfaces than most glass-reinforced polymers. Its static coefficient of friction depends on bearing pressure; for Tefzel HT-2004 the coefficient of friction changes from 0.51 at 68 Pa to 0.34 at 3.43 kPa (0.5 psi).

Dynamic friction depends on pressure and rubbing velocity (PV). The generation of frictional heat depends on the coefficient of friction and the PV factor. For the glass-reinforced product, temperature buildup begins at about PV 10,000 and thermal runaway occurs just below PV 20,000. High wear rates begin above

Table 1. Typical Properties of Tefzel[a]

Property	ASTM method	Tefzel 200, 280	Tefzel[b] HT-2004
ultimate tensile strength, MPa[c]	D638	44.8	82.7
ultimate elongation	D887-64T	200[d]	8
compressive strength, MPa[c]	D695	48.9	68.9
shear strength, MPa[c]		41.3	44.8
heat deflection temp, °C	D648		
at 0.45 MPa		104	265
at 1.8 MPa		74	210
max continuous use temp, no load, °C		150[e]	200
low temp embrittlement	D746	below −100°C	
tensile modulus, MPa[c]	D638	827	8270
flexural modulus, MPa[c]	D790	965	6550
impact strength notched Izod	D256		
at −54°C, J/m[f]		>1067	373
at 23°C		no break	485
deformation under load, 13.7 MPa[c] at 50°C, %	D621	4.11	0.68
coefficient of linear expansion per °C × 10[−5]	D696-70		
20–30°C		9	3
50–90°C		9.3	1.7
104–180°C		14	3.2
specific gravity	D792	1.70	1.86
refractive index, n_D		1.4028	
flammability	UL 94	94 V-0	94 V-0
	D635	ATB[g] <5 s	
		ALB[g] 10 mm	
melting point, dta peak, °C		270	270
water absorption at saturation, %	D570	0.029	0.022
hardness			
Rockwell	D785	R50	R74
Durometer D		D75	
coefficient of friction[h]			
dynamic, 689 kPa (at >3 m/min)		0.4	0.3
static, 689 kPa			0.3

[a]At 23°C and 50% rh, unless otherwise specified.
[b]Reinforced with 25 wt % glass fiber.
[c]To convert MPa to psi, multiply by 145.
[d]Elongations between 100 and 300% are achieved with varying methods of sample fabrication.
[e]Long-term heat-aging tests on Tefzel 280 are in progress. It is expected that its continuous-use temperature will be above 150°C.
[f]To convert J/m to ftlbf/in., divide by 53.38.
[g]ATB, average time of burning to nearest 5 s; ALB: average length of burn to nearest 5 mm. Test bar thickness, 2.9 mm.
[h]Mating material AISI 1018 Steel, Rc20, 16AA; 689 kPa = 100 psi.

Table 2. Thermodynamic Properties of Modified ETFE

Property	Tefzel 200 and 280
melting point, °C	270
specific heat, J/(mol·K)[a]	0.46–0.47
heat of sublimation, kJ/mol[a]	50.2
heat of fusion,[b] J/g[a]	46.0
heat of combustion, kJ/g[a]	13.72
thermal conductivity, W/(m·K)	0.238
critical surface tension of molten resin, mN/m(= dyn/cm)	22

[a]To convert J to cal, divide by 4.184.
[b]Little dependence on temperature.

PV 15,000. The wear rate depends on the type of metal rubbing surface and finish, lubrication, and clearances. Lubrication, hard shaft surfaces, and high finishes improve wear rates. Table 3 gives wear factors for steel and aluminum. Because the wear rate of both ETFE and the metal is much higher for aluminum than for steel, an anodized surface is preferred with aluminum.

Electrical Properties. Modified ethylene–tetrafluoroethylene is an excellent dielectric (Table 4). Its low dielectric constant confers a high corona-ignition voltage. The dielectric constant does not vary with frequency or temperature. Both dielectric strength (ASTM D149) and resistivity are high. The loss characteristics

Table 3. Bearing Wear Rate[a] of Tefzel HT-2004

Pressure, kPa[b]	Velocity, cm/s	Wear factor, $K \times 10^{-17}$, 1/Pa[c]	
		Tefzel	Metal
On steel[d]			
6.8	2.5	32	8
6.8	5.1	28	12
6.8	7.6	38	26
6.8	8.9	60	32
6.8	10.2	fail	
On aluminum[e]			
2.0	5.1	2400	2400
0.68	25.4	960	780

[a]Thrust-bearing tester, no lubricants ambient air temperature, metal finish 406 nm.
[b]To convert kPa to psi, multiply by 0.145.
[c]To convert 1/Pa to (in.^{3}·min)/(ftlbf·h), divide by 2×10^{-7}.
[d]AISI 1018.
[e]LM24M (English).

Table 4. Electrical Properties of ETFE Resins

Property	ASTM test	ETFE	Reinforced
dissipation factor, Hz	D150		
10^2		0.0006	0.004
10^3		0.0008	0.002
10^4			0.002
10^5			0.003
10^6		0.005	0.005
10^9		0.005	
10^{10}		0.010	0.012
volume resistivity, $\Omega\cdot$cm	D257	$>10^{16}$	10^{16}
surface resistivity, Ω/sq	D257	5×10^4	10^{15}
arc resistance, s		75	110

are minimum; the dissipation factor, although low, increases at higher frequencies. Glass reinforcement increases losses and the dielectric constant rises from 2.6 to 3.4 (from 10^2 to 10^{10} Hz); the dissipation factor is increased by tenfold. Exposure to radiation also increases losses. Dielectric strength is not reduced by thermal aging, unless a physical break occurs in the material. The short-time test of ASTM D149 gives values of 16–20 kV/mm with 3-mm thick specimens to 160–200 kV/mm with films 25–75 μm thick. Tracking resistance is about 70 s by ASTM D495. This is comparable to materials considered to be nontracking; under unusual conditions tracking occurs. When these resins are foamed they provide insulation with even lower dielectric constant (26).

Chemical Resistance and Hydrolytic Stability. Modified ethylene–tetrafluoroethylene copolymers are resistant to chemicals and solvents (Table 5) that often cause rapid degradation in other plastic materials. Performance is similar to that of perfluorinated polymers (27), which are not attacked by strong mineral acids, inorganic bases, halogens, and metal salt solutions. Organic compounds and solvents have little effect. Strong oxidizing acids, organic bases, and sulfonic acids at high concentrations and near their boiling points affect ETFE to varying degrees.

Physical properties remain stable after long exposure to boiling water. Tensile strength and elongation of Tefzel 200 are unaffected after 3000 h in boiling water. The higher molecular weight ETFE behaves similarly, whereas the glass-reinforced product shows a reduction of 25–35% in tensile strength with loss of reinforcement.

Water absorption of Tefzel is low (0.029% by weight), which contributes to its outstanding dimensional stability as well as to the stability of mechanical and electrical properties regardless of humidity.

High temperature resistance of ETFE and other fluoropolymers in automotive fuels and their permeation resistance have been discussed (28,29).

The ETFE copolymer can be cross-linked by radiation (30), despite the high content of tetrafluoroethylene units. Cross-linking reduces plasticity but en-

Table 5. Tefzel Resistance to Chemicals after Seven Days Exposure[a]

Chemical	Bp, °C	Test temperature, °C	Retained properties, %		
			Tensile strength	Elongation	Weight gain
organic acids and anhydrides					
acetic acid (glacial)	118	118	82	80	3.4
acetic anhydride	139	139	100	100	0
trichloroacetic acid	196	100	90	70	0
hydrocarbons					
mineral oil		180	90	60	0
naphtha		100	100	100	0.5
benzene	80	80	100	100	0
toluene	110	110			
amines					
aniline	185	120	81	99	2.7
aniline	185	180	95	90	
N-methylaniline	195	120	85	95	
N,N-dimethylaniline	190	120	82	97	
n-butylamine	78	78	71	73	4.4
di-n-butylamine	159	120	81	96	
di-n-butylamine	159	159	55	75	
tri-n-butylamine	216	120	81	80	
pyridine	116	116	100	100	1.5
solvents					
carbon tetrachloride	78	78	90	80	4.5
chloroform	62	61	85	100	4.0
dichloroethylene	77	32	95	100	2.8
methylene chloride	40	40	85	85	0
Freon 113	46	46	100	100	0.8
dimethylformamide	154	90	100	100	1.5
dimethyl sulfoxide	189	90	95	95	1.5
Skydrol		149	100	95	3.0
Aerosafe		149	92	93	3.9
A-20 stripper solution		140	90	90	
ethers, ketones, esters					
tetrahydrofuran	66	66	86	93	3.5
acetone	56	56	80	83	4.1
acetophenone	201	180	80	80	1.5
cyclohexanone	156	156	90	85	0
methyl ethyl ketone	80	80	100	100	0
n-butyl acetate	127	127	80	60	0
ethyl acetate	77	77	85	60	0
other organic compounds					
benzyl alcohol	205	120	97	90	
benzoyl chloride	197	120	94	95	
o-cresol	191	180	100	100	
decalin	190	120	89	95	
phthaloyl chloride	276	120	100	100	

Table 5. (*Continued*)

Chemical	Bp, °C	Test temperature, °C	Retained properties, %		
			Tensile strength	Elongation	Weight gain
inorganic acids					
hydrochloric (conc)	106	23	100	90	0
hydrobromic (conc)	125	125	100	100	
hydrofluoric (conc)		23	97	95	0.1
sulfuric (conc)		100	100	100	0
nitric, 70%	120	120	0	0	
chromic	125	125	66	25	
phosphoric (conc)		100			
halogens					
bromine (anhy)	59	23	90	90	1.2
chlorine (anhy)		120	85	84	7
bases, peroxides					
ammonium hydroxide		66	97	97	0
potassium hydroxide, 20%		100	100	100	0
sodium hydroxide, 50%		120	94	80	0.2
hydrogen peroxide, 30%		23	99	98	0
other inorganic compounds					
ferric chloride, 25%	104	100	95	95	0
zinc chloride, 25%	104	100	100	100	0
sulfuryl chloride	68	68	86	100	8
phosphoric trichloride	75	75	100	98	
phosphoric oxychloride	104	104	100	100	
silicon tetrachloride	60	60	100	100	

[a]Changes in properties <15% are considered insignificant; test performed on 250–1250-μm microtensile bars; tensile strength, elongation, and weight gain determined within 24 h after termination of exposure.

hances high temperature properties and nondrip performance. The irradiated resin withstands a 400°C solder iron for 10 min without noticeable effect.

Modified ETFE copolymer has excellent weather resistance; tensile strength and elongation are not affected. On the other hand, tensile and elongation properties of the glass-reinforced compound show a significant reduction.

Modified ETFE films are used as windows in greenhouses and conservatories due to their high transparency to both uv and visible light and excellent resistance to weathering (31).

Vacuum Outgassing and Permeability. Under vacuum, modified ethylene–tetrafluoroethylene copolymers give off little gas at elevated temperatures. The loss rate is about one-tenth of the acceptable maximum rates for spacecraft uses. Exposing 750-μm specimens for 24 h at 149°C to a high vacuum results in a maximum weight loss of 0.12%; volatile condensible material is less than 0.02%.

The following permeability values were determined on Tefzel film (100-μm, ASTM D1434) at 25°C (1 nmol/m·s·GPa = 0.5 cc·mil/100 in.^{2} d·atm):

Material	nmol/m·s·GPa
carbon dioxide	500
nitrogen	60
oxygen	200
helium	1800
water vapor (ASTM E96)	3.3

Fabrication

Modified ethylene–tetrafluoroethylene copolymers are commercially available in a variety of physical forms (Table 6) and can be fabricated by conventional thermoplastic techniques. Commercial ETFE resins are marketed in melt-extruded cubes, that are sold in 20-kg bags or 150-kg drums. In the United States, the 1992 price was $27.9–44.2/kg, depending on volume and grade; color concentrates are also available.

Like other thermoplastics, they exhibit melt fracture (32) above certain critical shear rates. In extrusion, many variables control product quality and performance (33).

Melt Processing. Articles are made by injection molding, compression molding, blow molding, transfer molding, rotational molding, extrusion, and coating. Films can be thermoformed and heat sealed (24). Because of high melt viscosity, ETFE resins are usually processed at high (300–340°C) temperatures.

Injection-molded articles shrink about 1.5–2.0% in the direction of resin flow and about 3.5–4.5% in the transverse direction under normal molding conditions. A 25% glass-reinforced composition shrinks only about 0.2–0.3% in the flow direction and about 3.0% in the transverse direction. Although shrinkage depends on shape and processing conditions, uniformity is excellent.

Molten ETFE polymers corrode most metals, and special corrosion-resistant alloys are recommended for long-term processing equipment; short-term prototype runs are possible in standard equipment.

Table 6. Forms of Modified ETFE Resins

Tefzel grade	Form	Melt flow,[a] g/10 min	Application
210	extruded cubes	45	injection molding, thin coating
200	extruded cubes	8	general-purpose, insulation, tubing, fasteners
280	extruded cubes	3	chemical resistance, jacketing, heavy-wall logging cables
HT-2000	compacted powder	8	compounded products
HT-2010	compacted powder	3	compounded products, coating lining
HT-2010	compacted powder	45	coating

[a]At 297°C and 45 N (5 kg) load.

Forming and Machining.   Articles can be formed below the melting point with conventional metal-forming techniques. Tetrafluoroethylene–ethylene copolymers are readily machined with the same tools and feed rates as are used for nylon and acetal. For best dimensional stability, the article should be annealed at the expected use temperature before the final machine cut.

Coloring and Decorating.   Commercial pigments that are thermally stable at the resin processing temperature may be used. Pigments may be dry-blended with the resin, or ETFE pellets may be blended with color concentrates, which are available in pellet form.

Nontreated surfaces can be hot-printed with special foils in a manner similar to a typewriter ribbon. The type is heated to about 321°C, and a printing pressure of 172–206 kPa (25–30 psi) is applied for about 0.25 s; no further treatment is required.

Stripes may be applied to wire coated with ETFE fluoropolymer over Du-Lite 817-5002 fluoropolymer clear enamel or other bases. Thermally stable pigments are required. Stripes may be applied by gravure-wheel-type applicators and oven-cured in-line.

Assembly.   The success of many applications depends on the ability of ETFE fluoropolymer to be economically assembled.

Screw Assembly.   Self-tapping screws are used for joining ETFE parts. For maximum holding power, the boss diameter should be about double the screw diameter, and the engagement length about 2.5 times the screw diameter; lubricants should be avoided. Threaded inserts can be molded in place, pressed in, or driven in ultrasonically.

Snap-Fit and Press-Fit Joints.   Snap-fit joints offer the advantage that the strength of the joint does not diminish with time because of creep. Press-fit joints are simple and inexpensive, but lose holding power. Creep and stress relaxation reduce the effective interference, as do temperature variations, particularly with materials with different thermal expansions.

Cold or Hot Heading.   Rivets or studs can be used in forming permanent mechanical joints. The heading is made with special tools and preferably with the rivet at elevated temperatures. Formed heads tend to recover part of their original shape if exposed to elevated temperatures, resulting in loose joints. Forming at elevated temperature reduces recovery.

Spin Welding.   Spin welding is an efficient technique for joining circular surfaces of similar materials. The matching surfaces are rotated at high speed relative to each other and then brought into contact. Frictional heat melts the interface and, when motion is stopped, the weld is allowed to solidify under pressure.

Ultrasonic Welding.   Ultrasonic welding has been applied to Tefzel with weld strength up to 80% of the strength of the base resin. Typical conditions include a contact pressure of 172 kPa (25 psi) and 1–2 s cycle time. The two basic designs, the shear and butt joints, employ a small initial contact area to concentrate and direct the high frequency vibrational energy.

Potting.   Potting of wire insulated with Tefzel has been accomplished with the aid of a coating of a colloidal silica dispersion. The pots produced with a polysulfide potting compound meeting MIL-S-8516C Class 2 standards exhibit pullout strengths of 111–155 N (25–35 lbf).

Bonding. Surface treatment, such as chemical etch, corona, or flame treatments, is required for adhesive bonding of Tefzel. Polyester and epoxy compounds are suitable adhesives.

Ethylene–tetrafluoroethylene copolymers respond well to melt bonding to untreated aluminum, steel, and copper with peel strengths above 3.5 kN/m (20 lbf/in.). For melt bonding to itself, hot-plate welding is used. The material is heated to 271–276°C, and the parts are pressed together during cooling.

The plasma surface treatment of ETFE to improve adhesion has been studied (34).

Health and Safety

Large quantities of Tefzel have been processed and used in many demanding service applications. No cases of permanent injury have been attributed to these resins, and only limited instances of temporary irritation to the upper respiratory tract have been reported (35).

As with other melt-processable fluoropolymers, trace quantities of harmful gases, including hydrogen fluoride, diffuse from the resin even at room temperature. Therefore, the resins should be used in well-ventilated areas. Even though the resin is physiologically inert and nonirritating to the skin, it is recommended that spills on the skin be washed with soap and water. These resins are stable at 150°C and are recommended for continuous use at this temperature. Degradation, as measured by weight loss, is insignificant up to the melting point of 270°C. At processing temperatures sufficient quantities of irritating and toxic gases are generated to require removal of the gases by exhaust hoods over the die and at the hopper heater. For extrusion into water, a quench tank or partially filled container for purging is recommended. In extrusion operations proper procedures must be maintained to control temperature and pressure. The weight loss with increasing temperature is as follows:

Temperature, °C	Hourly weight loss, %
300	0.05
330	0.26
350	0.86
370	1.60

To remove all decomposition products, a "total-capture" exhaust hood is recommended.

Under normal processing conditions at 300–350°C, Tefzel resins are not subject to autocatalytic degradation. However, extended overheating can result in "blow-backs" through extruder feed hopper or barrel front.

Prolonged soldering in confined spaces with restricted air circulation requires ventilation. A small duct fan is recommended for hot-wire stripping. Tefzel articles should not be exposed to welding conditions.

The limiting oxygen index of Tefzel as measured by the candle test (ASTM D2863) is 30%. Tefzel is rated 94 V-0 by Underwriters' Laboratories, Inc., in their burning test classification for polymeric materials. As a fuel, it has a comparatively low rating. Its heat of combustion is 13.7 MJ/kg (32,500 kcal/kg) compared to 14.9 MJ/kg (35,000 kcal/kg) for poly(vinylidene fluoride) and 46.5 MJ/kg (110,000 kcal/kg) for polyethylene.

Bulk quantities of Tefzel fluoropolymer resins should be stored away from flammable materials. In the event of fire, personnel entering the area should have full protection, including acid-resistant clothing and self-contained breathing apparatus with a full facepiece operated in the pressure-demand or other positive-pressure mode. All types of chemical extinguishers may be used to fight fire involving Tefzel resins. Large quantities of water may be used to cool and extinguish the fire.

The Du Pont Haskell Laboratory for Toxicology and Industrial Medicine has conducted a study to determine the acute inhalation toxicity of fumes evolved from Tefzel fluoropolymers when heated at elevated temperatures. Rats were exposed to decomposition products of Tefzel for 4 h at various temperatures. The approximate lethal temperature (ALT) for Tefzel resins was determined to be 335–350°C. All rats survived exposure to pyrolysis products from Tefzel heated to 300°C for this time period. At the ALT level, death was from pulmonary edema; carbon monoxide poisoning was probably a contributing factor. Hydrolyzable fluoride was present in the pyrolysis products, with concentration dependent on temperature.

Testing and Standards

A description of modified ethylene–tetrafluoroethylene copolymers and their classification is given by the American Society for Testing and Materials under the designation D3159-83 (36). A comprehensive listing of industrial and military specifications is available (37).

Applications

Tefzel 200 is a general-purpose, high temperature resin for insulating and jacketing low voltage power wiring for mass transport systems, wiring for chemical plants, and control and instrumentation wiring for utilities. In injection-molded form, it is used for sockets, connectors, and switch components (38). Because of excellent mechanical properties it provides good service in seal glands, pipe plugs, corrugated tubing, fasteners, and pump vanes. In chemical service, it is used for valve components, laboratory ware, packing, pump impellers, and battery and instrument components.

Tefzel 210, the high melt-flow resin, provides a high speed processing product for use in coating of fine wire and injection molding of thin-walled or intricate shapes. It is also used for other fine-wire applications requiring high line speeds and mechanical strength, but where harsh environmental conditions are not anticipated.

For high temperature wiring with mechanical strength and stress-crack and chemical resistance, Tefzel 280 is preferred. Rated by UL at 150°C, it is widely used for insulating and jacketing heater cables and automotive wiring and for other heavy-wall application where temperatures up to 200°C are experienced for short periods of time or where repeated mechanical stress at 150°C is encountered. It is also suitable for oil-well logging cables and is used in transfer moldings and extrusions for lined chemical equipment. It is injection molded into articles with metal inserts, thick sections, and stock shapes.

BIBLIOGRAPHY

"Tetrafluoroethylene Copolymers With Ethylene" under "Fluorine Compounds, Organic" in *ECT* 3rd ed., Vol. 11, pp. 35–41, by R. L. Johnson, E. I. du Pont de Nemours & Co., Inc.

1. U.S. Pat. 2,468,664 (Apr. 26, 1949), W. E. Hanford and J. R. Roland (to E. I. du Pont de Nemours & Co., Inc.).
2. U.S. Pat. 2,479,367 (Aug. 16, 1949), R. M. Joyce, Jr. (to E. I. du Pont de Nemours & Co., Inc.).
3. Brit. Pat. 1,166,020 (Oct. 1, 1969), M. Modena and co-workers (to Montecatini Edison, SpA).
4. Jpn. Kokai 64 22,586 (Sept. 12, 1964), K. Hirose and co-workers (to Nitto Chemical Industry, Co., Ltd.).
5. Belg. Pat 725,356 (Feb. 14, 1969), Z. Kenkyusho (to Asahi Glass, Ltd.).
6. U.S. Pat. 2,946,763 (July 26, 1960), M. I. Bro and B. W. Sandt (to E. I. du Pont de Nemours & Co., Inc.).
7. D. P. Carlson, *Development of Tefzel Fluoropolymer Resins*, unpublished paper.
8. U.S. Pat. 3,624,250 (Nov. 30, 1971), D. P. Carlson (to E. I. du Pont de Nemours & Co., Inc.).
9. U.S. Pat. 4,123,602 (Oct. 31, 1978), H. Ukihashi and M. Yamake (to Asahi Glass Co.).
10. Brit. Pat. 1,353,535 (May 22, 1974), R. Hartwimmer (to Farbwerke Hoechst, AG).
11. U.S. Pat. 3,401,155 (Sept. 10, 1986), G. Borsini and co-workers (to Montecatini Edison SpA).
12. Y. Tabata, H. Shibano, and H. Sobue, *J. Polym. Sci. Part A* **2**(4), 1977 (1964).
13. U.S. Pat. 3,960,825 (June 1, 1976), D. N. Robinson and co-workers (to Pennwalt Corporation).
14. U.S. Pat. 4,338,237 (July 1982), R. A. Sulzbach and co-workers (to Hoechst Aktiengesellschaft).
15. M. Modena, C. Garbuglio, and M. Ragazzini, *Polym. Lett.* **10**, 153 (1972).
16. M. Modena, C. Garbuglio, and M. Ragazzini, *J. Polym. Sci. Part B* **10**, 153 (1972).
17. F. S. Ingraham and D. F. Wooley, Jr., *Ind. Eng. Chem.* **56**(9), 53 (1964).
18. S. Yano, *J. Polym. Sci. Part A-2* **8**, 1057 (1970).
19. H. W. Starkweather, *J. Polym. Sci. Part A-2* **11**, 587 (1973).
20. F. C. Wilson and H. W. Starkweather, *J. Polym. Sci. Part A-2* **11**, 919 (1973).
21. B. Chu and Chi Wu, *Macromolecules* **20**, 93–98 (1987).
22. B. Chu, C. Wu, and W. Buck, *Macromolecules* **22**, 371 (1989).
23. Z. Wang, A. Tontisakis, W. Tuminello, W. Buck, and B. Chu, *Macromolecules* **23**, 1444 (1990).
24. *Tefzel Fluoropolymer, Design Handbook*, E. I. du Pont de Nemours & Co., Inc., Wilmington, Del., 1973.
25. *Tefzel—Properties Handbook*, E-31301-3, E. I. du Pont de Nemours & Co., Inc., Wilmington, Del., Dec. 1991.
26. S. K. Randa, C. R. Frywald, and D. P. Reifschneider, *Proceedings of the 36th International Wire and Cable Symposium*, Arlington, Va., 1987, pp. 14–22.

27. Can. Pat. 900,075 (May 9, 1972), D. P. Carlson, J. A. Effenberger, and M. B. Polk (to E. I. du Pont de Nemours & Co., Inc.).
28. M. Carpenter, S. Chillons, and R. Will, Society of Automobile Engineers (SAE) International Technical Paper Series 910103, presented at the *International Congress and Exposition*, Detroit, Mich., Feb. 25–Mar. 1, 1991.
29. D. Goldsberry, S. Chillons and R. Will, SAE Technical Paper Series 910104, presented at the *International Congress and Exposition*, Detroit, Mich., Feb. 25–Mar. 1, 1991.
30. U.S. Pat. 3,738,923 (June 12, 1973), D. P. Carlson and N. E. West (E. I. du Pont de Nemours & Co., Inc.).
31. J. Emsley, *New Scientist*, 46 (April 22, 1989).
32. *Technical Information, Jacketing Rate Calculation*, Bulletin No. 10, Fluoropolymers Division Technical Service Laboratory, E. I. du Pont de Nemours & Co., Inc., Wilmington, Del., Aug. 1982.
33. *Extrusion Guide for Melt Processible Fluoropolymers*, Bulletin E-41337, E. I. du Pont de Nemours & Co., Inc., Wilmington, Del.
34. S. Kaplan, O. Kolluri, G. Hansen, R. Rushing, and R. Warren, technical paper presented at *Conference on Adhesives*, Society of Manufacturing Engineers, Atlanta, Ga., Sept. 12, 1989.
35. *Tefzel Fluoropolymers, Safe Handling Guide*, Bulletin E-85785, E. I. du Pont de Nemours & Co., Inc., Wilmington, Del., May 1986.
36. *Modified ETFE-Fluorocarbon Molding and Extrusion Materials*, ASTM D3159-83, American Society for Testing and Materials, Philadelphia, 1987.
37. *J. Teflon* **15**(1), 1974.
38. *Tefzel Fluoropolymers, Product Information* (Du Pont Materials for Wire and Cable), Bulletin E-81467, E. I. du Pont de Nemours & Co., Inc., Wilmington, Del., May 1986.

General Reference

S. V. Gangal, "Tetrafluoroethylene Polymers, Tetrafluoroethylene–Ethylene Copolymers," in J. I. Kroschwitz, ed., *Encyclopedia of Polymer Science and Engineering*, 2nd ed., Vol. 16, Wiley-Interscience, New York, 1989, pp. 626–642.

SUBHASH V. GANGAL
E. I. du Pont de Nemours & Co., Inc.

TETRAFLUOROETHYLENE–PERFLUOROVINYL ETHER COPOLYMERS

Perfluoroalkoxy (PFA) fluorocarbon resins are designed to meet industry's needs in chemical, electrical, and mechanical applications. These melt processible copolymers contain a fluorocarbon backbone in the main chain and randomly distributed perfluorinated ether side chains:

$$-CF_2-CF_2-CF-CF_2-$$
$$\mid$$
$$O$$
$$\mid$$
$$C_3F_7$$

A combination of excellent chemical and mechanical properties at elevated temperatures results in reliable, high performance service to the chemical processing and related industries. Chemical inertness, heat resistance, toughness and flexibility, stress-crack resistance, excellent flex life, antistick characteristics, little moisture absorption, nonflammability, and exceptional dielectric properties are among the characteristics of these resins.

The introduction of a perfluoromethyl side chain (Teflon FEP) greatly reduces the crystallinity of PTFE. Crystallinity is reduced even further by replacing the short side chain with a long side chain, such as perfluoropropyl ether. In contrast to Teflon FEP, only a small amount of vinyl ether is required to reduce crystallinity and develop adequate toughness.

Tetrafluoroethylene [116-14-3] and perfluorovinyl ether are copolymerized in aqueous (1,2) or nonaqueous (3) media. The polymer is separated and converted into various forms, such as extruded cubes, powders, beads, or dispersions. This family of products is manufactured by Du Pont, Daikin, and Hoechst and sold under the trade names of Teflon PFA, Neoflon AP, and Hostaflon TFA, respectively.

Monomers

Preparation.   The preparation of tetrafluoroethylene has been described previously. Perfluorovinyl ethers (4–7) are prepared by the following steps. Hexafluoropropylene [116-15-4] (HFP) is oxidized to an epoxide HFPO [428-59-1] (5) which, on reaction with perfluorinated acyl fluorides, gives an alkoxyacyl fluoride.

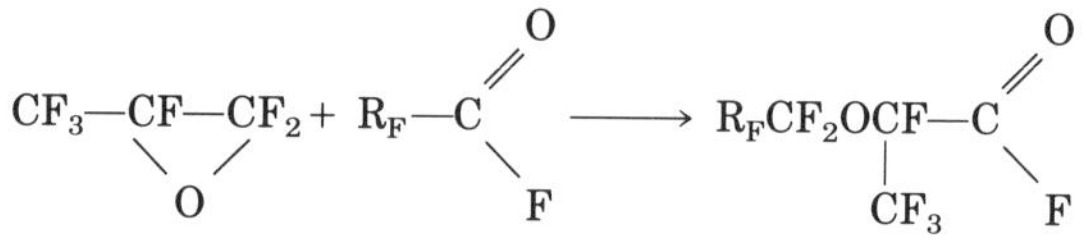

$$CF_3—CF—CF_2 + R_F—C\overset{O}{\underset{F}{\diagdown}} \longrightarrow R_FCF_2OCF—C\overset{O}{\underset{F}{\diagdown}}$$

HFPO

The alkoxyacyl fluoride is converted to vinyl ethers by treatment with base at ca 300°C (8).

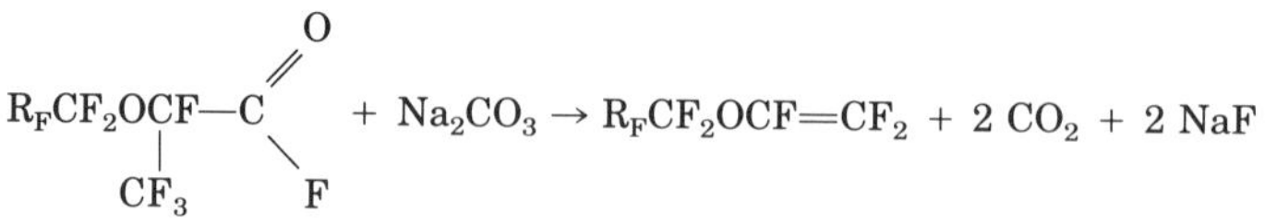

$$R_FCF_2OCF—C\overset{O}{\underset{F}{\diagdown}} + Na_2CO_3 \rightarrow R_FCF_2OCF{=}CF_2 + 2\ CO_2 + 2\ NaF$$

where $R_F = F(CF_2)_n$

Alkoxyacyl fluorides are also produced by an electrochemical process (9).

Properties.   Properties of perfluoropropyl vinyl ether [1623-05-8] (PPVE), a colorless, odorless liquid (mol wt 266) are shown in Table 1. Perfluoropropyl vinyl ether is an extremely flammable liquid and burns with a colorless flame. It is significantly less toxic than hexafluoropropylene; the average lethal concentration (ALC) is 50,000 ppm (10).

Table 1. Properties of Perfluoropropyl Vinyl Ether, $F_3C—CF_2—CF_2—O—CF=F_2$

Property	Value
critical temperature, K	423.58
critical pressure, MPa[a]	1.9
critical volume, cm³/mol	435
surface tension, mN/m(= dyn/cm)	9.9
boiling point, °C	36
specific gravity at 23°C	1.53
vapor density at 75°C, g/cm³	0.2
vapor pressure at 25°C, kPa[b]	70.3
solubility in water	0
odor	none
color	colorless
flash point, °C	−20
flammable limits in air,[c] % by vol	1

[a]To convert MPa to atm, divide by 0.1013.
[b]To convert kPa to psi (psia), multiply by 0.145.
[c]Extremely flammable.

Copolymerization

Tetrafluoroethylene–perfluoropropyl vinyl ether copolymers [26655-00-5] are made in aqueous (1,2) or nonaqueous media (3). In aqueous copolymerizations water-soluble initiators and a perfluorinated emulsifying agent are used. Molecular weight and molecular weight distribution are controlled by a chain-transfer agent. Sometimes a second phase is added to the reaction medium to improve the distribution of the vinyl ether in the polymer (11); a buffer is also added.

In nonaqueous copolymerization, fluorinated acyl peroxides are used as initiators that are soluble in the medium (12); a chain-transfer agent may be added for molecular weight control.

Temperatures range from 15 to 95°C, and the pressures from 0.45 to 3.55 MPa (65–515 psi). The temperatures used for the aqueous process are higher than those for the nonaqueous process.

Alkyl vinyl ethers tend to rearrange when exposed to free radicals (13). Temperatures must be kept low enough to prevent termination by free-radical coupling. In the aqueous process, temperatures below 80°C minimize the number of acid end groups derived from vinyl ether transfer. In the nonaqueous process, temperature must also be limited to avoid excessive vinyl ether transfer as well as reaction with the solvent. End groups are stabilized by treating the polymer with methanol, ammonia, or amines (14–16). Treatment of PFA with elemental fluorine generates CF_3 end groups and a very low level of contamination (17) which is important for the semiconductor industry (18).

The polymer is separated from the medium and converted to useful forms such as melt-extruded cubes for melt processible applications. Teflon PFA is also available as a dispersion, a fine powder, or in unmelted bead form.

Description and classification of PFA resins are given in Reference 19. Various specifications are given in Reference 20.

Properties

The melting point of commercial Teflon PFA is 305°C, ie, between those of PTFE and FEP. Second-order transitions are at -100, -30, and 90°C, as determined by a torsion pendulum (21). The crystallinity of the virgin resin is 65–75%. Specific gravity and crystallinity increase as the cooling rate is reduced. An ice-quenched sample with 48% crystallinity has a specific gravity of 2.123, whereas the press-cooled sample has a crystallinity of 58% and a specific gravity of 2.157.

Mechanical Properties. Table 2 shows the physical properties of Teflon PFA (22,23). At 20–25°C the mechanical properties of PFA, FEP, and PTFE are

Table 2. Properties of Teflon PFA

Property	ASTM method	Teflon 340	Teflon 350
nominal melting point, °C		302–306	302–306
specific gravity	D3307	10.6	1.8
continuous use temp, °C		260	260
tensile strength, MPa^a			
at 23°C	D1708	28	31
at 250°C		12	14
tensile yield, MPa^a			
at 23°C	D1708	14	15
at 250°C		3.5	4.1
ultimate elongation, %			
at 23°C		300	300
at 250°C	D1708	480	500
flexural modulus, MPa^a			
at 23°C	D790	655	690
at 250°C		55	69
creep resistance^b tensile modulus, MPa^a			
at 20°C	D695	270	270
at 250°C	D695	41	41
hardness Durometer	D2240	D60	D60
MIT folding endurance, 775–200 μm film thickness, cycles		50,000	500,000
water absorption, %	D570	0.03	0.03
coefficient of linear thermal expansion, per °C, $\times 10^{-5}$	D696		
20–100°C		12	12
100–150°C		17	17
150–210°C		20	20

^{a}To convert MPa to psi, multiply by 145.
^{b}Apparent modulus after 10 h: stress = 6.89 MPa at 20°C, 6.89 kPa at 250°C.

similar; differences between PFA and FEP become significant as the temperature is increased. The latter should not be used above 200°C, whereas PFA can be used up to 260°C. Tests at liquid nitrogen temperature indicate that PFA performs well in cryogenic applications (Table 3).

Unfilled Teflon PFA has been tested in mechanical applications using Teflon FEP-100 as a control (24). Tests were run on molded thrust bearings at 689.5 kPa (100 psi) against AISI 1080, Rc 20, 16AA steel, and at ambient conditions in air without lubrication. A limiting PV value of 5000 was found. Wear factors and dynamic coefficients of friction are shown in Table 4.

Table 3. Cryogenic Properties of Teflon PFA Resins

Property	ASTM method	At 23°C	At −196°C
yield strength, MPa[a]	D1708[b]	15	
ultimate tensile strength, MPa[a]	D1708[b]	18	129
elongation, %	D1708[b]	260	8
flexural modulus, MPa[a]	D790-71[c]	558	5790
impact strength, notched Izod, J/m[d]	D256-72a[e]	no break	64
compressive strength, MPa[a]	D695		414
compressive strain, %	D695		35
modulus of elasticity, MPa[a]	D695		4690

[a]To convert MPa to psi, multiply by 145.
[b]Crosshead speed B, 1.3 mm/min; used at both temperatures for more direct comparison.
[c]Method 1, procedure B.
[d]To convert J/m to ftlbf/in., divide by 53.38.
[e]Method A, head weight is 4.5 kg at 23°C and 0.9 kg at 160°C.

Table 4. Teflon PFA Fluorocarbon Resin Thrust-Bearing Wear-Test Results[a]

Velocity, m/min	Wear factor K × 10^{-17}, 1/Pa[b]	Dynamic coefficient of friction	Test duration, h
Teflon PFA TE-9704			
0.91	3.12	0.210	103
3.05	3.67	0.214	103
9.1	1.96	0.229	103
15.24	1.38	0.289	103
Teflon FEP-100			
0.91	3.71	0.341	104
3.05	2.19	0.330	104
9.1	3.16	0.364	104
15.24	1.60	0.296	103

[a]Mating surface: AISA 1018 steel, Rc 20, 16AA; contact pressure; 689 kPa; at 20°C in air; no lubricant.
[b]To convert 1/Pa to (in.³·min)/(ftlbf·h), divide by 2 × 10^{-7}.

Hardness (qv) is determined according to ASTM D2240 on $7.6 \times 12.7 \times 0.48$ cm injection-molded panels (25). Results on the D scale are 63–65 for Teflon PFA and 63–66 for Teflon FEP.

Chemical Properties. A combination of excellent chemical and mechanical properties at elevated temperatures result in high performance service in the chemical processing industry. Teflon PFA resins have been exposed to a variety of organic and inorganic compounds commonly encountered in chemical service (26). They are not attacked by inorganic acids, bases, halogens, metal salt solutions, organic acids, and anhydrides. Aromatic and aliphatic hydrocarbons, alcohols, aldehydes, ketones, ethers, amines, esters, chlorinated compounds, and other polymer solvents have little effect. However, like other perfluorinated polymers, they react with alkali metals and elemental fluorine.

Thermal Stability. Teflon PFA resins are very stable and can be processed up to 425°C. Thermal degradation is a function of temperature and time. A significant increase in melt flow rate indicates degradation after a short time above 425°C; at lower temperatures degradation takes longer. Degradation is not significant if the change in melt flow rate of the resin during molding is below 20%. Degradation is also indicated by the formation of small bubbles or discoloration; however, high stock temperatures may cause slight discoloration without adversely affecting properties.

Heat aging at 285°C, a temperature slightly below but near the melting point, increases the strength of Teflon PFA. Samples aged in a circulating air oven for 7500 h at 285°C show a decrease in melt flow number as defined by ASTM D2116. A decline in melt flow number indicates an increase in average molecular weight, which is also indicated by a 25% increase in tensile strength and enhanced ultimate elongation. Toughness is also measured by MIT flex life, which improves severalfold on heat aging at 285°C.

When exposed to fire, Teflon PFA contributes little in fuel value and is self-extinguishing when the flame is removed. The fuel value is approximately 5.4 MJ/kg (2324 Btu/lb). It passes the UL 83 vertical-flame test and is classified as 94VE-O according to UL 94. The limiting oxygen index (LOI) by ASTM D2863 is above 95%.

Electrical Properties. The electrical properties of Teflon PFA are given in Table 5. The dielectric constant of PFA resins is about 2.06 over a wide range of frequencies (10^2–2.4×10^{10} Hz), temperatures, and densities (ASTM D150). The values for PFA density vary only slightly, 2.13–2.17, and the dielectric constant varies only about 0.03 units over this range, among the lowest of all solid materials. Humidity has no measurable effect on the dielectric constant of PFA. The dielectric strength (short-term) of PFA resins is 80 kV/mm (0.25-mm films, ASTM D149); FEP films give similar results, whereas PTFE films are typically measured at 47 kV/mm. Like other fluoropolymer resins, PFA loses dielectric strength in the presence of corona discharge. The dissipation factor at low frequency (10^2–10^4 Hz) decreases with increasing frequency and decreasing temperature. Temperature and frequency have little influence on the dissipation factor over the frequency range 10^4–10^7 Hz. As frequencies increase to 10^{10} Hz, there is a steady increase in dissipation factor. Above 10^7 Hz, increases measured at room temperature are highest; a maximum at about 3×10^9 Hz is indicated. The higher dissipation factor with increasing frequency should be considered in electrical

Table 5. Electrical Properties of Teflon PFA

Property	Value
dielectric strength,[a] kV/m	79
volume resistivity,[b] ohm·cm	10^{18}
surface resistivity,[b] ohms/sq	10^{18}
dissipation factor[c]	
at 10^2 Hz	0.000027
at 10^6 Hz	0.000080
at 10^7 Hz	0.000145
at 10^9 Hz	0.00115
at 3×10^9 Hz	0.00144
at 1.4×10^{10} Hz	0.00131
at 2.4×10^{10} Hz	0.00124

[a]Short-term, 250-μm-thick sample.
[b]ASTM method D257.
[c]ASTM method D150.

insulation applications at high frequencies. The volume and surface resistivities of fluorocarbon resins are high and are not affected by time or temperature. When tested with stainless steel electrodes (ASTM D495), no tracking was observed for the duration of the test (180 s), indicating that PFA resin does not form a carbonized conducting path (27,28).

Optical Properties and Radiation Effects. Within the range of wavelengths measured (uv, visible, and near-ir radiation), Teflon PFA fluorocarbon film transmits slightly less energy than FEP film (29) (Table 6). In thin sections, the resin is colorless and transparent; in thicker sections, it becomes translucent. It is highly transparent to ir radiation; uv absorption is low in thin sections. Weather-O-Meter tests indicate unlimited outdoor life.

Like other perfluoropolymers, Teflon PFA is not highly resistant to radiation (30). Radiation resistance is improved in vacuum, and strength and elongation are increased more after low dosages (up to 30 kGy or 3 Mrad) than with FEP or PTFE. Teflon PFA approaches the performance of PTFE between 30 and 100 kGy

Table 6. Optical Properties of Teflon PFA Film

Property	ASTM method	Value
refractive index[a]	D542-50	1.350 ± 0.002
haze, %	D1003-52	4
light transmission, %		
uv,[b] 0.25–0.40 μm		55–80
visible, 0.40–0.70 μm		80–87
infrared, 0.70–2.1 μm		87–93

[a]Measured at 546 nm and 20°C.
[b]Cary Model Spectrophotometer.

Table 7. Effects of Radiation on Tensile Strength of PFA[a]

| Exposure, kGy[b] | ASTM D1708 | |
	Tensile strength, MPa[c]	Elongation, %
0	30.27	358
5	28.20	366
10	24.96	333
20	21.24	302
50	14.55	35
200		<5
500		<5

[a]Sample: 250-μm compression-molded films of Teflon PFA 340 from G.E. resonance transformer 2 MeV capacity, at a current of 1 mA.
[b]To convert kGy to Mrad, multiply by 0.1.
[c]To convert MPa to psi, multiply by 145.

(3–10 Mrad) and embrittles above 100 kGy (10 Mrads). At 500 kGy (50 Mrad) PTFE, FEP, and PFA are degraded. The effect of radiation on tensile strength and elongation is shown in Table 7.

Fabrication

Teflon PFA resins are fabricated by the conventional melt-processing techniques used for thermoplastics. Processing equipment is constructed of corrosion-resistant materials and can be operated at 315–425°C. A general-purpose grade, PFA 340, is designed for a variety of molding and extrusion applications, including tubing, shapes, and molded components, in addition to insulation for electrical wire and cables. Because of the excellent thermal stability of PFA 350, a wide range of melt temperatures can be used for fabrication. Extrusion temperatures are 20–26°C above the melting point.

Teflon PFA 440 HP is a chemically modified form of PFA 340 that provides additional benefits such as enhanced purity and improved thermal stability. This product is suitable for producing tubing, pipe linings for production of ultrapure chemicals, semiconductor components, and fluid handling systems for high performance filters (31).

Extrusion. Like other thermoplastics, Teflon PFA resin exhibits melt fracture above certain critical shear rates. For example, samples at 372°C and 5-kg load show the following behavior:

Teflon PFA	Melt flow, g/10 min	Critical shear rate, s^{-1}
340	14	50
310	6	16
350	2	6

Because Teflon PFA melt is corrosive to most metals, special corrosion-resistant alloys must be used for the extrusion equipment, such as Hastelloy C, Monel 400, and Xaloy 306. Barrels, liners, screws, adapters, breaker plates, and dies are made of corrosion-resistant metals (32). Corrosion is promoted by resin degradation and high processing temperatures, long residence times, or dead spots. Extruders used with Teflon FEP are also suitable for PFA resins. Heaters and controllers capable of accurate operation in the range of 330–425°C are required. Extruder barrels should have three or four independently controlled heating zones, each equipped with its own thermocouple and temperature-indicating control.

The screw consists of a feed section, a rapid transition section, and a metering section; a rounded forward end prevents stagnation. The breaker plate that converts the rotary motion of the melt into smooth, straight flow should have as many holes as possible; both ends of each hole should be countersunk for streamlined flow.

The temperature of the melt downstream from the breaker plate may exceed the front barrel temperature, because of the mechanical work transmitted to the resin by the screw; it varies with screw speed and flow rate. The melt temperature is measured by a thermocouple inserted into the melt downstream from the breaker plate. A hooded exhaust placed over the extruder die and feed hopper removes decomposition products when the extrudate is heated.

High melt strength of Teflon PFA 350 permits large reductions in the cross section of the extrudate by drawing the melt in air after it leaves the die orifice (33). At a given temperature, the allowable flow rates are limited at the low end by resin degradation and at the high end by the onset of melt fracture. A broader range of specific gravities (2.13–2.17) may be obtained in articles fabricated from PFA 350 than with FEP 160. Unlike with polytetrafluoroethylene, higher crystallinity in PFA seems to have little effect on flex life.

Injection Molding.   Any standard design plunger or reciprocating screw injection machine can be used for PFA 340, although a reciprocating screw machine is preferred (32). Slow injection into mold cavities avoids surface or internal melt fracture, and control of ram speed is important at low speed. Corrosion-resistant metals are used for parts in continuous contact with molten resin; Hastelloy C, and Xaloy 306 or 800 are recommended.

Because the mold is usually maintained at temperatures below the melting point of the resin, corrosion on the mold surface is less than in the molding machine. Nonreturn ball check valves and ring check valves are used; the latter is preferred for PFA. A streamlined flow must pass through the valve, preventing areas of stagnant flow or holdup and localized degradation.

A smear head causes less stagnation and overpacking than a nonreturn valve. A conventional-type reverse-tapered nozzle with the bore as large as possible without sudden changes in diameter is preferred. Independently controlled, zone-type heaters for heating the nozzle and at least two zones on the cylinder are used.

At a holdup time longer than 10–15 min at a high temperature, resin degradation is avoided by keeping the rear of the cylinder at a lower temperature than the front. At short holdup times (4–5 min), cylinder temperatures are the same in rear and front. If melt fracture occurs, the injection rate is reduced; pressures are in the range of 20.6–55.1 MPa (3000–8000 psi). Low back pressure and screw rotation rates should be used.

The cycle can usually be estimated on the basis of about 30 s/3 mm of thickness; most of it is devoted to ram-in-motion time (except for very thin sections). The mold temperature used with PFA 340 is often the highest temperature that allows the part to be ejected undamaged from the mold and retain its shape while cooling.

The resin must be of highest purity for optimum processing characteristics and properties. Degradation results in discoloration, bubbling, and change in melt flow rate.

Transfer Molding.   Valve and fitting liners are made by a transfer-molding process (33), with the valve or fitting serving as the mold. Melted resin is forced into the fitting at a temperature above the melting point of the resin. The melt may be produced by an extruder or an injection molding machine or melted cubes contained in a melt pot and transferred by applying pressure to a piston in the pot. After the resin transfer is completed, the fitting is cooled under pressure. Stock temperatures of 350–380°C and fitting temperatures of 350–370°C are used to process PFA 350. A slight adjustment in the cooling cycles may be required for transfer molding PFA 350 because it has higher melting and freezing points than FEP.

Rotocasting Teflon PFA Beads.   The resin has sufficient thermal stability for a commercial rotocasting operation; that is, TE-9738 has a melting point of about 303°C. In rotocasting trials, incoming flue gas temperatures of 355–365°C (34,35) and heat cycles of 90–180 min have been used. Conventional rotations for major and minor axes can be applied without modifications; Freecote 33 performs adequately as a mold release agent. Mold release instructions can be followed without modification. Heating cycles, including a preheat and a fusion stage, give consistent rotocasting. Preheating at 15–30°C below the fusion temperature takes 10–25 min. Heat-cycled Teflon PFA rotocastings are translucent white, often with bluish tinge. Rotocastings that have been heated too long may darken to a translucent brown. Uniform cooling is essential for undistorted, stress-free products; combinations of air and water are employed. The rotocasting is cooled below the resin melting temperature with air at ambient temperature and then with a water spray, and finally with a stream of air.

Dispersion Processing.   A commercial aqueous dispersion of Teflon PFA 335 contains more than 50 wt % PFA particles, about 5 wt % surfactants and fillers. This dispersion is processed by the same technique as for PTFE dispersion. It is used for coating various surfaces, including metal, glass, and glass fabrics. A thin layer of Teflon PFA coating can also serve as an adhesive layer for PTFE topcoat.

Powder Coating.   Teflon PFA is also available in a finely divided powder form. It can be used to produce thin layers on various surfaces by heating these surfaces above the melting point of PFA and then bringing the powder in contact with them. This allows a thin layer of the powder to melt on the surface of the substrate.

For some applications the powder is suspended in an aqueous medium or a solvent with the help of emulsifying agents and then sprayed onto the substrate.

The powder is also used as a filler to prepare sprayable compositions of PTFE dispersions, which then can be used to coat various substrates (36).

Pigmentation. Commercial color concentrates of Teflon PFA containing approximately 2% pigment can be easily dispersed in clear extruded cubes. The resin can also be dry-blended with stable inorganic pigments. At 0.1–1% concentration, the pigment has no appreciable effect on the dielectric strength and constant or mechanical properties. The dissipation factor of pigmented resin varies with the type and concentration of the pigment.

Pigment used for dry blending is dried overnight at 150°C in a vacuum oven to remove absorbed gases and moisture. It is screened through a 149-μm (100-mesh) screen directly onto the cubes, which are rolled or tumbled for at least 15 min. The pigmented resin is stored in an airtight container to prevent absorption of moisture.

Health and Safety

Safe practices employed for handling PTFE and FEP resins are adequate for Teflon PFA (37); adequate ventilation is required for processing above 330–355°C. In rotoprocessing, a vacuum (250–750 Pa or 1.8–5.6 mm Hg) in the oven ensures exhaust to the outside (36). Removal of end caps or opening of sealed parts in a well-ventilated area ensures ventilation of decomposition fumes. During rotoprocessing, molds should be vented.

Applications and Economic Aspects

The perfluorovinyl ether comonomer used for PFA is expensive, as is PFA. Most PFA grades are sold as extruded, translucent cubes in various colors at $47.9–60.3/kg. Some PFA types are also marketed in nonextruded forms.

Teflon PFA can be fabricated into high temperature electrical insulation and components and materials for mechanical parts requiring long flex life. Teflon PFA 350 is used as liner for chemical process equipment, specialty tubing, and molded articles for a variety of applications. Teflon PFA 340 is a general-purpose resin for tubing, shapes, primary insulation, wire and cable jacketing, injection- and blow-molded components, and compression-molded articles. Teflon PFA 440 HP is a chemically modified form of PFA-340 with enhanced purity and improved thermal stability while processing. This resin is suitable in semiconductor manufacturing, fluid handling systems for industry or life sciences, and instrumentation for precise measurements of fluid systems.

BIBLIOGRAPHY

"Tetrafluoroethylene Copolymers with Perfluorovinyl Ethers" under "Fluorine Compounds, Organic," in *ECT* 3rd ed., Vol. 11, pp. 42–49, by R. L. Johnson, E. I. du Pont de Nemours & Co., Inc.

1. U.S. Pat. 3,132,123 (May 5, 1964), J. F. Harris and D. I. McCane (to E. I. du Pont de Nemours & Co., Inc.).
2. U.S. Pat. 3,635,926 (Jan. 18, 1972) W. F. Fresham and A. F. Vogelpohl (to E. I. du Pont de Nemours & Co., Inc.).
3. U.S. Pat. 3,536,733 (Oct. 27, 1970), D. P. Carlson (to E. I. du Pont de Nemours & Co.).
4. U.S. Pat. 3,358,003 (Dec. 12, 1967), H. S. Eleuterio and R. W. Meschke (to E. I. du Pont de Nemours & Co., Inc.).
5. U.S. Pat. 3,180,895 (Apr. 27, 1965), J. F. Harris and D. I. McCane (to E. I. du Pont de Nemours & Co., Inc.).
6. U.S. Pat. 3,250,808 (Oct. 10, 1966), E. P. Moore, A. S. Milian, Jr., and H. S. Eleuterio (to E. I. du Pont de Nemours & Co., Inc.).
7. U.S. Pat. 4,118,421 (Oct. 3, 1978), T. Martini (to Hoechst Aktiengesellschaft).
8. U.S. Pat. 3,291,843 (Dec. 13, 1966), C. G. Fritz and S. Selman (to E. I. du Pont de Nemours & Co., Inc.).
9. U.S. Pat. 2,713,593 (July 1955), T. J. Brice and W. H. Pearlson (to Minnesota Mining and Manufacturing Co.).
10. A. H. Olson, E. I. du Pont de Nemours & Co., Inc., private communication, 1992.
11. U.S. Pat. 4,499,249 (Feb. 12, 1985), S. Nakagawa and co-workers (to Daikin Kogyo Co., Ltd.).
12. U.S. Pat. 2,792,423 (May 14, 1957), D. M. Young and W. N. Stoops (to Union Carbide and Carbon Corp.).
13. R. E. Putnam, in R. B. Seymour and G. S. Kirshenbaum, eds., *High Performance Polymers: Their Origin and Development*, Elsevier Scientific Publishing, Inc., New York, 1986, p. 279.
14. U.S. Pat. 3,674,758 (July 4, 1972), D. P. Carlson (to E. I. du Pont de Nemours & Co., Inc.).
15. U.S. Pat. 4,599,386 (July 8, 1986), D. P. Carlson and co-workers (to E. I. du Pont de Nemours & Co., Inc.).
16. PCT Int. Appl. WO 89,11,495 (1989), M. D. Buckmaster (to E. I. du Pont de Nemours & Co., Inc.).
17. U.S. Pat. 4,943,658 (1988), J. Imbalzano and D. Kerbow (to E. I. du Pont de Nemours & Co., Inc.).
18. C. J. Goodman and S. Andrews, *Solid State Technol.*, 65 (July 1990).
19. *PFA-Fluorocarbon Molding and Extrusion Materials*, ASTM 3307-86, American Society for Testing and Materials, Philadelphia, Pa., 1987.
20. *J. Teflon*, **15**(1) (1974).
21. R. A. Darby, E. I. du Pont de Nemours & Co., Inc., private communication, 1992.
22. *PFA Fluorocarbon Resins*, sales brochure, E08572, E. I. du Pont de Nemours & Co., Inc., Wilmington, Del.
23. M. I. Bro and co-workers, *29th International Wire and Cable Symposium*, Cherry Hill, N.J., Nov. 1980.
24. *Teflon PFA Fluorocarbon Resins: Wear and Frictional Data*, APD #2 bulletin, E. I. du Pont de Nemours & Co., Inc., Wilmington, Del., 1973.
25. *Teflon PFA Fluorocarbon Resins: Hardness*, APD #4 bulletin, E. I. du Pont de Nemours & Co., Inc., Wilmington, Del., 1973.
26. *Teflon PFA Fluorocarbon Resins: Chemical Resistance*, PIB #2 bulletin, E. I. du Pont de Nemours & Co., Inc., Wilmington, Del., 1972.
27. E. W. Fasig, D. I. McCane, and J. R. Perkins, paper presented at the *22nd International Wire and Cable Symposium*, Atlantic City, N.J., Dec. 1973.
28. *Handbook of Properties for Teflon PFA*, sales brochure, E46679, E. I. du Pont de Nemours & Co., Inc., Wilmington, Del., Oct. 1987.

29. *Teflon PFA Fluorocarbon Resins: Optical Properties*, APD #6 bulletin, E. I. du Pont de Nemours & Co., Inc., Wilmington, Del., 1973.

30. *Teflon PFA Fluorocarbon Resins: Response to Radiation*, APD #3 bulletin, E. I. du Pont de Nemours & Co., Inc., Wilmington, Del., 1973.

31. *Teflon PFA 440 HP*, product information, H-27760, E. I. du Pont de Nemours & Co., Inc., Wilmington, Del., 1990.

32. *Teflon PFA Fluorocarbon Resin: Injection Molding of Teflon PFA TE-9704*, PIB #4 bulletin, E. I. du Pont de Nemours & Co., Inc., Wilmington, Del., 1973.

33. *Teflon PFA Fluorocarbon Resins: Melt Processing of Teflon PFA TE-9705*, PIB #1 bulletin, E. I. du Pont de Nemours & Co., Inc., Wilmington, Del., 1973.

34. *Technical Information, No. 11, Processing Guidelines for Du Pont Fluoropolymer Rotocasting Powders of Tefzel and Teflon PFA*, E. I. du Pont de Nemours & Co., Inc., Wilmington, Del., 1982.

35. *Teflon PFA TE-9783 Rotation Molding Powder*, technical information, H-26600, E. I. du Pont de Nemours & Co., Inc., Wilmington, Del., June 1990.

36. Brit. Pat. 2,051,091B (Feb. 9, 1983), J. E. Bucino (to Fluorocoat Ltd.).

37. *Handling and Use of Teflon Fluorocarbon Resins at High Temperatures*, bulletin, E. I. du Pont de Nemours & Co., Inc., Wilmington, Del., 1961.

General Reference

S. V. Gangal, in J. I. Kroschwitz, ed., *Encyclopedia of Polymer Science and Engineering*, 2nd ed., Vol. 16, Wiley-Interscience, New York, 1989, pp. 614–626.

SUBHASH V. GANGAL
E. I. du Pont de Nemours & Co., Inc.

POLY(VINYL FLUORIDE)

Homopolymers and copolymers of vinyl fluoride are based on free-radical polymerization of vinyl fluoride and comonomers, usually under high pressure. Du Pont first commercialized a poly(vinyl fluoride)-based film in 1961 under the trade name Tedlar. Poly(vinyl fluoride) homopolymers and copolymers have excellent resistance to sunlight degradation, chemical attack, water absorption, and solvent, and have a high solar energy transmittance rate. These properties have resulted in the utilization of poly(vinyl fluoride) (PVF) film and coating in outdoor and indoor functional and decorative applications. These films are used where exceptional high temperature stability, outdoor longevity, stain resistance, adherence, and release properties are required.

Monomer

Vinyl fluoride [75-02-5] (VF) (fluoroethene) is a colorless gas at ambient conditions. It was first prepared by reaction of 1,1-difluoro-2-bromoethane [359-07-9] with zinc (1). Most approaches to vinyl fluoride synthesis have employed reactions of acetylene [74-86-2] with hydrogen fluoride (HF) either directly (2–5) or utilizing catalysts (3,6–10). Other routes have involved ethylene [74-85-1] and HF (11), pyrolysis of 1,1-difluoroethane [624-72-6] (12,13) and fluorochloroethanes (14–18), reaction of 1,1-difluoroethane with acetylene (19,20), and halogen exchange of vinyl chloride [75-01-4] with HF (21–23). Physical properties of vinyl fluoride are given in Table 1.

Table 1. Physical Properties of Vinyl Fluoride

Property	Value
molecular weight	46.04
boiling point, °C	−72.2
freezing point, °C	−160.5
critical temperature, °C	54.7
critical pressure, MPa[a]	5.1
critical density, g/cm^3	0.320
liquid density at 21°C, g/cm^3	0.636
vapor pressure at 21°C, MPa[a]	2.5
solubility in water at 80°C, g/100 g H$_2$O	
at 3.4 MPa[a]	0.94
at 6.9 MPa[a]	1.54

[a]To convert MPa to atm, divide by 0.101.

Polymerization

Vinyl fluoride undergoes free-radical polymerization. The first polymerization involved heating a saturated solution of VF in toluene at 67°C under 600 MPa (87,000 psi) for 16 h (24). A wide variety of initiators and polymerization conditions have been explored (25–27). Examples of bulk (28,29) and solution (25,28,30,31) polymerizations exist; however, aqueous suspension or emulsion methods are generally preferred (26,32–40). VF volatility dictates that moderately high pressures be used. Photopolymerizations, usually incorporating free-radical initiators, are also known (26,28,29,35).

The course of VF polymerizations is dominated by the high energy and hence high reactivity of the propagating VF radical. The fluorine substituent provides little resonance stabilization, leading to a propagating intermediate which is indiscriminate in its reactions. Monomer reversals, branching, and chain-transfer reactions are common. The reactivity of the vinyl fluoride radical limits the choice of polymerization medium, surfactants, initiators, or other additives and makes impurity control important. Species which can participate in chain transfer or incorporate in the polymer can depress molecular weight or degrade the thermal stability characteristics of the final polymer.

The combination of triisobutylborane [1116-39-8] and oxygen has been used to polymerize VF at reduced temperature and pressure (41). Polymerization temperature was varied from 0 to 85°C with a corresponding drop in melting point from about 230°C (0°C polymerization) to about 200°C (85°C polymerization). This dependance of melting temperature, and degree of crystallinity, have been interpreted in terms of variations in the extent of monomer reversals during polymerization (42). Copolymers of VF with vinylidene fluoride [75-38-7] and tetrafluoroethylene [116-14-3] also have been prepared with this initiation system. VF tends toward alternation with tetrafluoroethylene and incorporates preferentially in copolymerization with vinylidene fluoride (see FLUORINE COMPOUNDS, ORGANIC–POLYTETRAFLUOROETHYLENE; POLY(VINYLIDENE FLUORIDE)).

Copolymers of VF and a wide variety of other monomers have been prepared (6,41–48). The high energy of the propagating vinyl fluoride radical strongly influences the course of these polymerizations. VF incorporates well with other monomers that do not produce stable free radicals, such as ethylene and vinyl acetate, but is sparingly incorporated with more stable radicals such as acrylonitrile [107-13-1] and vinyl chloride. An Alfrey-Price Q value of 0.010 ± 0.005 and an e value of 0.8 ± 0.2 have been determined (49). The low value of Q is consistent with little resonance stability and the e value is suggestive of an electron-rich monomer.

Polymer Properties

Poly(vinyl fluoride) [24981-14-4] (PVF) is a semicrystalline polymer with a planar, zig-zag configuration (50). The degree of crystallinity can vary significantly from 20–60% (51) and is thought to be primarily a function of defect structures. Wideline nmr and x-ray diffraction studies show the unit cell to contain two monomer units and have the dimensions of a = 0.857 nm, b = 0.495 nm, and c = 0.252 nm (52). Similarity to the phase I crystal form of poly(vinylidene fluoride) suggests an orthorhombic crystal (53).

The relationship of polymer structure to melting point and degree of crystallinity has been the subject of controversy. Head-to-head regio irregularities in PVF are known (51,54,55) and the concentration of such units has been suggested as the source of variations in melting point (42,47,56). Commercial PVF contains approximately 12% head-to-head linkages by ^{19}F-nmr and displays a peak melting point of about 190°C (47,48,57,58). Both nmr and ir studies have shown PVF to be atactic (47,51,54,55,59–62) and, as such, variations in stereoregularity are not thought to be a contributor to variations in melting point.

PVF with controlled amounts of head-to-head units varying from 0 to 30% have been prepared (47,48) by using a chlorine substituent to direct the course of polymerization of chlorofluoroethylenes and then reductively dechlorinating the products with tributyltin hydride. This series of polymers shows melting point distributions ranging from about 220°C for purely head-to-tail polymer down to about 160°C for polymer containing 30% head-to-head linkages. This study, however, does not report the extent of branching in these polymers. Further work has shown that the extent of branching has a pronounced effect upon melting temperature (57,58). Change of polymerization temperature from 90 to 40°C produces a change in branch frequency from 1.35 to 0.3%, while the frequency of monomer reversals is nearly constant (12.5 ± 1%). The peak melting point for this series varies from 186°C (90°C polymerization) to 206°C (40°C polymerization).

PVF displays several transitions below the melting temperature. The measured transition temperatures vary with the technique used for measurement. T_g (L) (lower) occurs at −15 to −20°C and is ascribed to relaxation free from restraint by crystallites. T_g (U) (upper) is in the 40 to 50°C range and is associated with amorphous regions under restraint by crystallites (63). Another transition at −80°C has been ascribed to short-chain amorphous relaxation and one at 150°C associated with premelting intracrystalline relaxation.

PVF has low solubility in all solvents below about 100°C (61). Polymers with greater solubility have been prepared using 0.1% 2-propanol polymerization modifier and were characterized in *N,N*-dimethylformamide solution containing 0.1 *N* LiBr. M_n ranged from 76,000 to 234,000 (osmometry), and M_s from 143,000 to 654,000 (sedimentation velocity). Sedimentation velocity molecular weights can be related to intrinsic viscosity using the Mark-Houwink equation:

$$\eta_{\text{inh}} = KM^a$$

Using an *a* value of 0.80, which is typical of an extended polar polymer in good solvent, *K* is determined to be 6.52×10^{-5} (64).

The conformational characteristics of PVF are the subject of several studies (53,65). The rotational isomeric state (RIS) model has been used to calculate mean square end-to-end distance, dipole moments, and conformational entropies. ^{13}C-nmr chemical shifts are in agreement with these predictions (66). The stiffness parameter (δ) has been calculated (67) using the relationship between chain stiffness and cross-sectional area (68). In comparison to polyethylene, PVF has greater chain stiffness which decreases melting entropy, ie, $(\Delta S)_m = 8.58$ J/(mol·K) [2.05 cal/(mol·K)] versus 10.0 J/(mol·K)[2.38 cal/(mol·K)].

A solubility parameter of 24.5–24.7 MPa$^{1/2}$ [12.0–12.1 (cal/cm^3)$^{1/2}$] has been calculated for PVF using room temperature swelling data (69). The polymer lost solvent to evaporation more rapidly than free solvent alone when exposed to air. This was ascribed to reestablishment of favorable dipole–dipole interactions within the polymer. Infrared spectral shifts for poly(methyl methacrylate) in PVF have been interpreted as evidence of favorable acid–base interactions involving the H from CHF units (70). This is consistent with the greater absorption of pyridine than methyl acetate despite a closer solubility parameter match with methyl acetate.

PVF is more thermally stable than other vinyl halide polymers. High molecular weight PVF is reported to degrade in an inert atmosphere, with concurrent HF loss and backbone cleavage occurring at about 450°C (71,72). In air, HF loss occurs at about 350°C, followed by backbone cleavage around 450°C.

More recent work reports the onset of thermal degradation at lower temperatures and provides a clearer picture of the role of oxygen (73–75). In the presence of oxygen, backbone oxidation and subsequent cleavage reactions initiate decomposition. In the absence of oxygen, dehydrofluorination eventually occurs, but at significantly higher temperatures.

PVF is transparent to radiation in the uv, visible, and near ir regions, transmitting 90% of the radiation from 350 to 2,500 nm. Radiation between 7,000 and 12,000 nm is absorbed (76). Exposure to low dose γ irradiation produces crosslinks in PVF and actually increases tensile strength and etching resistance, whereas the degree of crystallinity and melting point are reduced (77). PVF becomes embrittled upon exposure to electron-beam radiation of 10 MGy (10^9 rad), but resists breakdown at lower doses. It retains its strength at 0.32 MGy (32×10^6 rad) while polytetrafluoroethylene is degraded at 0.02 MGy (2×10^6 rad) (78).

Fabrication and Processing

Commercial PVF is insoluble at room temperature because of the large number of hydrogen bonds and high degree of crystallinity. Some latent solvents solvate PVF at temperatures above 100°C. PVF is converted to thin films and coatings. Processing of PVF, eg, by melt extrusion, depends on latent solvation of PVF in highly polar solvents and its subsequent coalescence. An example is plasticized melt extrusion of PVF into thin films (79). Pigments, stabilizers, plasticizers, and other additives can be incorporated in the film by dispersing them with the polymer in the latent solvent. The solvent is recovered by evaporation after extrusion. The extruded film can be biaxially oriented to varying degrees.

Poly(vinyl fluoride) can be applied to substrates with solvent-based or waterborne dispersions, or by powder-coating techniques. Viscosity modifiers are often needed to obtain a coatable dispersion. Dispersions can be applied by spraying, reverse roll coating, dip coating, and centrifugal casting. Other methods include casting on a continuous belt, extrusion into a hot liquid (80), and dipping a hot article into the dispersion (81).

Table 2 lists properties of PVF films. Various multilayer cast PVF films have been reported (82). Physical and tensile properties of the film depend on the extent of its orientation (83).

Adherability of the film may be enhanced by its treatment with flame, electric discharge, boron trifluoride gas, activated gas plasma, dichromate sulfuric acid, and a solution of alkali metal in liquid ammonia (84–87). A coating of polyurethane, an alkyl polymethacrylate, or a chlorinated adhesive can be applied to PVF surfaces to enhance adhesion (80,88,89).

Economic Aspects

Poly(vinyl fluoride) is available from Du Pont both as a resin and as transparent and pigmented films under the trademark Tedlar PVF film. Films are available in nonoriented and oriented grade in several tensile modifications and thicknesses, with either adherable or nonadherable release-grade surfaces. The 1992 prices ranged from $30 to $70/kg, vs $24 to $62/kg in 1988. Prices for specially tailored films were significantly higher.

Health and Environment

Acute inhalation exposure of rats to 200,000 ppm VF for 30 minutes or more produced weak anaesthesia and no deaths (90). In rats VF is only slightly metabolized at a rate of one-fifth that of vinyl chloride (91–95). An extensive program of toxicity testing of vinyl fluoride is in progress (96,97).

Vinyl fluoride is flammable in air between the limits of 2.6 and 22% by volume. Minimum ignition temperature for VF and air mixtures is 400°C. A small amount, < 0.2%, of terpenes is added to VF to prevent spontaneous polymeriza-

Table 2. Properties of Poly(vinyl fluoride) Film

Property	Value	ASTM[a] test method
Physical and thermal properties		
bursting strength, kPa[b]	200–450[c]	D774
coefficient of friction with metal	0.18–0.21	D1894-78
density, g/cm³	1.38–1.72	weighed samples
impact strength, kJ/m[d]	43–90	D3420-80
refractive index, n_D	1.46	D542; Abbe refractometer, 30°C
tear strength, kJ/m[d]		
propagated	6–22	D1922-67
initial	129–196	D1004-66
tensile modulus, MPa[e]	44–110	D882
ultimate elongation, %	115–250	D882
ultimate yield strength, MPa[e]	33–41	D882
linear coefficient of expansion, cm/(cm·°C)	0.00005	air oven, 30 min
useful temperature range, °C		
continuous use	−70 to +107	
short cycle (1–2 h)	175	
zero strength, °C	260–300	hot bar
thermal conductivity (1°C/cm), W/(m·K)		
−30°C	0.14	
60°C	0.17	
self-ignition temperature, °C	390	D1929
solar energy transmittance, 359–2500 nm, %	90	E427-71
Permeability		
moisture absorption, %	0.5	D570-81
moisture vapor transmission,[f] nmol/(m²·s)[g]		
at 7.0 kPa,[h] 39.5°C	4.65–29.4	E96-58T
gas permeability,[f] nmol/(m·s·GPa)[i] at		
98 kPa,[h] 23°C		D1434
carbon dioxide	22.4	
helium	302	
hydrogen	117	
nitrogen	0.5	
oxygen	6.6	
vapor transmission rate,[j] nmol/(m²·s)[k] at 23.5°C		E96, modified
acetic acid	4.9	
acetone	1570	
benzene	13	
carbon tetrachloride	3.9	
ethyl acetate	13	
hexane	10	
water[l]	22	

Table 2. (*Continued*)

Property	Value	ASTM[a] test method
Electrical properties[m]		
corona endurance, h at 60 Hz, 40 V/μm	2.5–6.0	D2275
dielectric constant at 1 MHz, 23°C	6.2–7.7	D150-81
dielectric strength,		D150-81
short term ac, kV/μm	0.08–0.13	
short term dc, kV/μm	0.15–0.19	
dissipation factor, %		D150-81
1 MHz at 23°C	0.17–0.28	
1 MHz at 100°C	0.09–0.21	
10 kHz at 23°C	0.019–0.019	
10 kHz at 100°C	0.21–0.067	
volume resistivity for transparent film, GΩ·m		D257
23°C	2000–700	
100°C	0.7–2	
surface resistivity, GΩ·m		D257
23°C	60,000–20,000	
100°C	7–20	

[a]Unless otherwise noted.
[b]To convert kPa to psi, multiply by 0.145.
[c]Range dependent on composition and tensile modification.
[d]To convert kJ/m to ft·lbf/in., divide by 0.0534 (ASTM D256).
[e]To convert MPa to psi, multiply by 145.
[f]Measurements made on films of nominal 25 μm thickness.
[g]To convert nmol/(m²·s) to g/(m²·d), multiply by 1.94.
[h]To convert kPa to mm Hg, multiply by 7.5.
[i]To convert nmol/(m·s·GPa) to mL·mil/(m²·d·atm), multiply by 7.725.
[j]At partial pressure of vapor at given temperature.
[k]To convert nmol/(m²·s) to g/(m²·d), multiply by 1.94 and by the density.
[l]At 39.5°C.
[m]Range of electrical properties is given; the first value refers to 54.8-μm transparent film, and the second value to 54.8-μm white pigmented film.

tion. The U.S. Department of Transportation has classified the inhibited VF as a flammable gas.

The self-ignition temperature of PVF film is 390°C. The limiting oxygen index (LOI) for PVF is 22.6% (98), which can be raised to 30% in antimony oxide-modified film (99). Hydrogen fluoride and a mixture of aromatic and aliphatic hydrocarbons (100) are generated from the thermal degradation of PVF. Toxicity studies, ie, survival and time to incapacitation, of polymers, cellulosics (101,102), and airplane interior materials (103) expose mice to pyrolysis products and show PVF thermal degradation products to have relatively low toxicity.

Uses

The uses of PVF depend on its weatherability, strength over a wide range of temperatures, and inertness toward a wide variety of chemicals, corrosives, and staining agents. It finds wide use as a protective or decorative coating. It can be applied as a preformed film in a laminating step or from a dispersion in a coating step. It may be transparent or pigmented in a variety of colors. Poly(vinyl fluoride) film is laminated to cellulosics, flexible vinyls, plastics, rubbers, and resin-impregnated felt. These laminated products are applied to exterior wall panels for buildings (104,105), highway sound barriers (106), automobile trim, truck and trailer siding (107), vinyl awnings, backlit signs (108), pipe covering (109), stain-resistant wall coverings, and aircraft cabin interiors (110,111).

On metal or plastic, PVF surfaces serve as a primer coat for painting, eg, automobile parts, or where improved adhesion is desired (112). Because of its moisture impermeability and wide operating temperature range, PVF film is used to fabricate bags to contain glass fiber mats for insulating exterior airplane walls and cargo space, and air conditioning ducts. PVF has long been used to construct bags for sampling gases (113).

Fiber-reinforced panels covered with PVF have been used for greenhouses. Transparent PVF film is used as the cover for flat-plate solar collectors (114) and photovoltaic cells (qv) (115). White PVF pigmented film is used as the bottom surface of photovoltaic cells. Nonadhering film is used as a release sheet in plastics processing, particularly in high temperature pressing of epoxy resins for circuit boards (116–118) and aerospace parts. Dispersions of PVF are coated on the exterior of steel hydraulic brake tubes and fuel lines for corrosion protection.

BIBLIOGRAPHY

"Poly(vinyl fluoride)" under "Fluorine Compounds, Organic," in *ECT* 2nd ed., Vol. 9, pp. 835–840, by L. E. Wolinski, E. I. du Pont de Nemours & Co., Inc.; in *ECT* 3rd ed., Vol. 11, pp. 57–64, by D. E. Brasure, E. I. du Pont de Nemours & Co., Inc.

1. F. Swarts, *Bull. Clin. Sci. Acad. Roy. Belg.* **7**, 383 (1901); F. Swarts, *J. Chem. Soc. Abstr.* **82**, 129 (1902).
2. U.S. Pat. 1,425,130 (Aug. 8, 1922), H. Plauson (to Plauson's Ltd.).
3. U.S. Pat. 2,118,901 (May 31, 1938), J. Soll (to I. G. Farbenindustrie AG).
4. A. V. Grossee and C. B. Linn, *J. Am. Chem. Soc.* **64**, 2289 (1942).
5. A. L. Henne, *Organic Reactions*, Vol. 2, John Wiley & Sons, Inc., New York, 1944.
6. U.S. Pat. 2,419,010 (Apr. 15, 1947), D. D. Coffman and T. A. Ford (to E. I. du Pont de Nemours & Co., Inc.).
7. U.S. Pat. 2,437,307 (Mar. 9, 1948), L. F. Salisburg (to E. I. du Pont de Nemours & Co., Inc.).
8. U.S. Pat. 2,674,632 (Apr. 1, 1954), B. F. Skiles (to E. I. du Pont de Nemours & Co., Inc.).
9. U.S. Pat. 3,178,483 (Apr. 13, 1965), C. M. Christy and G. Teufer (to E. I. du Pont de Nemours & Co., Inc.).
10. U.S. Pat. 3,607,955 (Sept. 21, 1971), L. E. Gardner (to Phillips Petroleum Co.).
11. Jpn. Kokai 77 122,310 (Oct. 14, 1977), T. Kuroda and T. Yamamoto (to Onoda Cement Co., Ltd.).
12. U.S. Pat. 2,442,993 (June 8, 1948), O. W. Cass (to E. I. du Pont de Nemours & Co., Inc.).
13. U.S. Pat. 2,461,523 (Feb. 15, 1949), D. D. Coffman and R. D. Cramer (to E. I. du Pont de Nemours & Co., Inc.).
14. U.S. Pat. 3,621,067 (Nov. 16, 1971), J. Hamersma (to Atlantic Richfield Co.).

15. Jpn. Pat. 46 21,607 (71 21,607) (June 18, 1971), B. Tatsutani and co-workers (to Electro Chemical Industrial Co., Ltd.).
16. U.S. Pat. 3,642,917 (Feb. 15, 1972), J. Hamersma (to Atlantic Richfield Co.).
17. Jpn. Pat. 47 11,728 (72 11,728) (Apr. 12, 1972), B. Tatsutani and co-workers (to Electro Chemical Industrial Co., Ltd.).
18. Jpn. Pat. 51 13,123 (76 13,123) (Apr. 26, 1976), B. Ryutani and co-workers (to Electro Chemical Industrial Co., Ltd.).
19. U.S. Pat. 3,317,619 (May 2, 1967), T. E. Hedge (to Diamond Shamrock Corp.).
20. T. S. Sirlibaev and co-workers, *Zh. Prikl. Khim.* **58** (7), 1666–1668 (1985).
21. Jpn. Pat. 47 11,726 (72 11,726), (Apr. 12, 1972), B. Tatsutani, I. Kobayashi, and K. Yamamoto (to Electro Chemical Industrial Co., Ltd.).
22. T. S. Sirlibaev, A. Akramkhodzhaev, A. A. Yul'chibaev, and K. U. Usmanov, *Synthesis of Vinyl Fluoride and 1,1-difluoroethane from Vinyl Chloride*, Deposited Document, VINITI 48575, USSR, 1975.
23. T. S. Sirlibaev and co-workers, *Uzb. Khim. Zh.* (1), 29–31 (1980).
24. H. W. Starkweather, *J. Am. Chem. Soc.* **56**, 1870 (1934).
25. G. H. Kalb and co-workers, *J. Appl. Polym. Sci.* **4**, 55 (1960).
26. A. E. Newkirk, *J. Am. Chem. Soc.* **68**, 2467 (1946).
27. D. Sianesi and G. Caporiccio, *J. Polym. Sci. Part A-1* **6**, 335 (1968).
28. D. Raucher and M. Levy, *J. Polym. Sci.* **13**(6) 1339–1346 (1975).
29. K. U. Usmanov and co-workers, *Russ. Chem. Rev.* **46** (5), 462–478 1977; trans. from Usp. Khim. **46,** 878–906 (1977).
30. T. S. Sirlibaev and I. Tirkashev, *Uzb. Khim. Zh.* **2**, 40–42 (1983).
31. D. Raucher and co-workers, *J. Polym. Sci.* **17**, 2825–2832 (1979).
32. Jpn. Pat. 74 027108 (July 15, 1974), H. Iwamichi and Y. Adachi (to Electro Chemical Industries Co.).
33. U.S. Pat. 3,627,744 (Dec. 14, 1971), R. A. Bonsall and B. Hopkins (to Monsanto Co.).
34. Jpn. Pat. 74 028,907 (July 30, 1974), S. Yashida, K. Tamaiima, and H. Kurovama (to Electro Chemical Industries Co., Ltd.).
35. V. G. Klaydin and co-workers, *Zh. Prikl. Khim.* **56**(2), 462–465 (1983).
36. A. K. Gafurov and co-workers, *Sb. Nauchn. Tr. Tashk. Gos. Univ. im. V. I. Lenina* **667**, 16–23 (1981).
37. U.S. Pat. 2,510,783 (June 6, 1950), F. L. Johnston (to E. I. du Pont de Nemours & Co., Inc.).
38. U.S. Pat. 3,129,207 (Apr. 14, 1964), V. E. James (to E. I. du Pont de Nemours & Co., Inc.).
39. Brit. Pat. 1,161,958 (Aug. 20, 1969), J. G. Frielink (to Deutsche Solvay-Werke Gesellschaft mit Beschrankter Haftung).
40. U.S. Pat. 3,265,678 (Aug. 9, 1966), J. L. Hecht (to E. I. du Pont de Nemours & Co., Inc.).
41. G. Natta and co-workers, *J. Polym. Sci. Part A-1* **3**, 4263 (1965).
42. D. Sianesi and G. Caporiccio, *J. Polym. Sci. Part A-1* **6**, 335 (1968).
43. U.S. Pat. 2,406,717 (Aug. 27, 1946), C. A. Thomas (to Monsanto Chemical Co.).
44. U.S. Pat. 2,847,401 (Aug. 12, 1958), E. W. Gluesenkamp and J. D. Calfee (to Monsanto Chemical Co.).
45. U.S. Pat. 3,057,812 (Oct. 9, 1962), J. R. Straughan, R. Stickl, Jr., and W. F. Hill, Jr. (to Union Carbide Corp.).
46. F. Z. Yusupbekova and co-workers, *Uzb. Khim. Zh.* (3), 47–51 (1987).
47. R. E. Cais and J. M. Kometani, *Polymer* **29**, 168–172 (1988).
48. R. E. Cais and J. M. Kometani, in J. C. Randall, ed., *NMR and Macromolecules*, ACS Symposium. Ser. No. 247, American Chemical Society, Washington, D.C., 1984, p. 153.

49. T. Alfrey and C. C. Price, *J. Polym. Sci.* **2**, 101 (1947).
50. G. Natta, *Makromol. Chem.* **35**, 94 (1960).
51. M. Goerlitz and co-workers, *Angew. Makromol. Chem.* **29/30**(371), 137 (1973).
52. G. Natta, I. W. Bassi, and G. Allegra, *Atti Accad. Naz. Lincei Cl. Sci. Fis. Mat. Natur. Rend.* **31**, 350–356 (1961).
53. J. B. Lando, H. G. Olf, and A. Peterlin, *J. Polym. Sci. Part A-1* **4**, 941–951 (1966).
54. G. Caporiccio, E. Strepparola, and D. Sianesi, *Chim. Ind. (Milan)* **52**, 28–36 (1970).
55. C. W. Wilson III and E. R. Santee, Jr., *J. Polym. Sci. Part C*, **8**, 97–112 (1965).
56. M. D. Hanes, PhD. dissertation, Case-Western Reserve University, Cleveland, Ohio, 1991.
57. D. W. Ovenall and R. E. Uschold, *Macromolecules* **24**, 3235 (1991).
58. L. L. Burger and M. T. Aronson, in *Polymer* **34**(12), pp. 25, 46 (1993).
59. M. D. Bruch, F. A. Bovey, and R. E. Cais, *Macromolecules* **17**, 2547–2551 (1984).
60. F. J. Weigert, *Org. Magn. Resonance* **3**, 373–377 (1977).
61. J. L. Koenig and J. J. Mannion, *J. Polym. Sci. Part A-2*, **4**, 401–414 (1966).
62. G. Zerbi and G. Cortili, *Spectrochim. Acta* **26**, 733–739 (1970).
63. R. F. Boyer, *J. Polym. Sci. Part C* **50**, 189–242 (1975).
64. M. L. Wallach and M. A. Kabayama, *J. Polym. Sci. Part A-1* **4**, 2667–2674 (1966).
65. A. E. Tonelli, *Macromelecules* **13**, 734–741 (1980).
66. A. E. Tonelli, F. C. Schiling, and R. E. Cais, *Macromolecules* **14**, 560 (1982).
67. H. Tianbai, *Yingyong Huaxue (Chinese Journal of Applied Chemistry)* **2**, 15–18 (1985).
68. R. F. Boyer and R. L. Miller, *Macromolecules* **10**, 1167–1169 (1977).
69. A. Chapira, Z. Mankowski, and N. Schmitt, *J. Polym. Sci. Part A-1* **20**, 1791–1796 (1982).
70. F. M. Fowkes, D. O. Tischler, J. A. Wolfe, L. A. Lannigan, C. M. Ademu-John, and M. J. Halliwell, *J. Polym. Sci. Part A-1* **22**, 547–566 (1984).
71. D. Raucher and M. Levy, *J. Polym. Sci. Part A-1* **17**, 2675–2680 (1979).
72. G. Montaudo, C. Puglisi, E. Scamporrino, and D. Vitalini, *J. Polym. Sci. Part A-1* **24**, 301–316 (1986).
73. M. L. O'Shea, C. Morterra, and M. J. D. Low, *Mater. Chem. Phys.* **25**, 501 (1990).
74. B. F. Mukhiddinov and co-workers, *Doki. Akad. Nauk SSSR*, **316**, 165–168 (1991).
75. W. E. Farneth, M. T. Aronson, and R. E. Uschold *Macromolecules* **26**(18), 4765 (Feb. 1993).
76. Technical Information Bulletin TD-31, E. I. du Pont de Nemours & Co., Inc., Wilmington, Del., 1979.
77. Y. Rosenberg, A. Siegmann, M. Narkis, and S. Shkolnik, *J. Appl. Polym. Sci.* **45**, 783 (1992).
78. R. Timmerman and W. Greyson, *J. Appl. Polym. Sci.* **6**, 456 (1962).
79. U.S. Pat. 2,953,818 (Sept. 27, 1960), L. R. Bartron (to E. I. du Pont de Nemours & Co., Inc.).
80. U.S. Pat. 3,723,171 (Mar. 27, 1973), O. Fuchs (to Dynamite Nobel AG).
81. U.S. Pat. 4,645,692 (Feb. 24, 1987), E. Vassiliou (to E. I. du Pont de Nemours & Co., Inc.).
82. U.S. Pat. 4,988,540 (Jan. 29, 1991), R. F. Davis, C. G. Bragaw, and T. P. Cancannon (to E. I. du Pont de Nemours & Co., Inc.).
83. U.S. Pat. 3,139,470 (June 30, 1964), R. S. Prengle and R. L. Richards, Jr. (to E. I. du Pont de Nemours & Co., Inc.).
84. U.S. Pat. 3,145,242 (Aug. 18, 1964), W. L. Bryan (to E. I. du Pont de Nemours & Co., Inc.).

85. U.S. Pat. 3,274,088 (Sept. 20, 1966), L. E. Wolinski (to E. I. du Pont de Nemours & Co., Inc.).
86. U.S. Pat. 3,122,445 (Feb. 25, 1964), R. O. Osborn (to E. I. du Pont de Nemours & Co., Inc.).
87. C. A. L. Westerdahl and co-workers, *Activated Gas Plasma Surface Treatment of Polymers for Adhesive Bonding, Part III*, Technical Report 4279, Picatinny Arsenal, Dover, N.J., 1972.
88. U.S. Pat. 4,215,177 (July 29, 1980), A. Strassel (to Produits).
89. U.S. Pat. 3,880,690 (Apr. 29, 1975), O. Fuchs (to Dynamite Nobel AG).
90. D. Lester and L. A. Greenberg, *Arch. Ind. Hyg. Occup. Med.* **2**, 335 (1950).
91. M. E. Andersen, *Neurobehav. Toxicol. Teratol.* **3**, 383 (1981).
92. M. E. Andersen, *Drug Metab. Rev.* **13**(5), 799 (1982).
93. J. G. Filser and H. M. Bolt, *Arch. Toxicol.* **42**, 123 (1979).
94. H. M. Bolt, *Arbeitsmed. Sozialmed. Praventivmed.* **15**, 49 (1980).
95. H. M. Bolt, R. J. Laib, and K. P. Klein, *Arch. Toxicol.* **47**, 71 (1981).
96. *Fed. Reg.* **52**(109), 21516 (June 8, 1987).
97. J. R. Fiddle, *Occupat. Health Safety News Dig.* **4**, 3 (1988).
98. C. P. Fennimore and F. J. Martin, *Combust. Flame* **10**, 135 (1966).
99. U.S. Pat. 3,963,672 (June 15, 1976), D. E. Brasure (to E. I. du Pont de Nemours & Co., Inc.).
100. I. N. Einhorn and co-workers, *Final Report, FRC/UU-41*, ETEC 75-022, NASA Contract No. NAS2-8244, National Technical Information System (NTIS), Springfield, Va., Dec. 14, 1974.
101. C. H. Hilado and co-workers, *J. Combust. Toxicol.* **3**, 157 (1976).
102. *Ibid.*, p. 270.
103. J. C. Spurgeon, *Report No. FAA-RD-78-131*, Federal Aviation Administration Contract No. 181-521-100, NTIS, Springfield, Va., Nov. 1978.
104. R. D. Leaversuch, *Mod. Plast.* **64**, 52 (July 1987).
105. V. M. Cassidy, *Mod. Met.* **41**, 36 (May 1985).
106. *Public Works* **III**, 78 (1980).
107. *Du Pont Mag.* **82**, 8–11 (Mar.–Apr. 1988).
108. *Du Pont Mag.* **82**, 13 (May–June 1988).
109. *Energy Managmt. Tech.* **8**, 22–23 (July–Aug. 1984).
110. R. A. Anderson and G. A. Johnson, *J. Fire Flammabil.* **8**, 364–381 (1977).
111. U.S. Pat. 5,137,775 (Aug. 11, 1992), S. Ebnesajjad and R. F. Davis (to E. I. du Pont de Nemours & Co., Inc.).
112. *Tedlar*, Du Pont Technical Bulletin TD-40, E. I. du Pont de Nemours & Co., Inc., Wilmington, Del., Feb. 1988.
113. J. C. Pau, J. E. Knoll, and M. R. Midgett, *J. Air Waste Managmt. Assoc.* **41**(8), 1095–1097 (1991).
114. B. Baum and M. Binette, *Report DOE/CS/35359-T1* (DE84011488), U.S. Dept. of Energy Contract No. AC04-78CS35359, NTIS, Springfield, Va., June 1983.
115. R. S. Sugimura, D. H. Otth, R. G. Ross, Jr., J. C. Arnett, and G. T. Samuelson, *IEEE Conference Report*, IEEE, New York, Oct. 21–25, 1985.
116. G. L. Schmutz, *Circuits Mfg.* **23**, 51 (Apr. 1983).
117. J. L. Wilson, C. L. Long, D. L. Mathews, and M. L. Wilson, *BDX-613-1657*, U.S. Department of Energy Contract No. EY-76-C-04-0613, NTIS, Springfield, Va., Jan. 1978.
118. Jpn. Pat. 62 214,939 (Jan. 21, 1987), S. Suzuki and S. Onari (to Matsushita Electric Works).

General References

D. E. Brasure and S. Ebnesajjad, in J. I. Kroschwitz, ed., *Concise Encyclopedia of Polymer Science and Engineering*, John Wiley & Sons, Inc., New York, 1990, pp. 1273–1275.

D. E. Brasure and S. Ebnesajjad, in J. I. Kroschwitz, ed., *Encyclopedia of Polymer Science and Engineering*, 2nd ed., Vol. 17, John Wiley & Sons, Inc., New York, 1989, pp. 468–491.

S. EBNESAJJAD
L. G. SNOW
Du Pont Company

POLY(VINYLIDENE FLUORIDE)

Poly(vinylidene fluoride) [24937-79-9] is the addition polymer of 1,1-difluoroethene [75-38-7], commonly known as vinylidene fluoride and abbreviated VDF or VF_2. The formula of the repeat unit in the polymer is —CH_2—CF_2—. The preferred acronym for the polymer is PVDF, but the abbreviation PVF_2 is also frequently used. The history and development of poly(vinylidene fluoride) technology has been reviewed (1–3).

PVDF is a semicrystalline polymer that contains 59.4 wt % fluorine and 3 wt % hydrogen and is commercially polymerized in emulsion or suspension using free-radical initiators. The spatial arrangement of the CH_2 and CF_2 groups along the polymer chain accounts for the unique polarity, unusually high dielectric constant, polymorphism, and high piezoelectric and pyroelectric activity of the polymer. It has the characteristic resistance of fluoropolymers to harsh chemical, thermal, ultraviolet, weathering, and oxidizing or high energy radiation environments. Because of these characteristics it has many applications in wire and cable products, electronic devices, chemical and related processing fields, as a weather-resistant binder for exterior architectural finishes, and in many specialized uses. The polymer is readily melt-processed using conventional molding or extrusion equipment; porous membranes are cast from solutions, and finishes are deposited from dispersions using specific solvents. PVDF contains an extremely low level of ionic contamination and does not require additives for stabilization during melt-processing, thereby qualifying it for applications such as ultrapure water systems where high purity is demanded from materials of construction.

There is growing commercial importance and escalating scientific interest in PVDF. The World Patent database, including the United States, lists 678 patents that cite the term poly(vinylidene fluoride) for the period 1963–1980 and 2052 patents for the period 1981–1992; *Chemical Abstracts* files covering the years 1967–1992 contain 5282 references for the same term. Thirty years ago there was only one commercial producer of PVDF in the world; now there are two in the United States, two in Japan, and three in Europe.

Monomer

Properties. Vinylidene fluoride is a colorless, flammable, and nearly odorless gas that boils at $-82°C$. Physical properties of VDF are shown in Table 1. It is usually polymerized above its critical temperature of 30.1°C and at pressures above 3 MPa (30 atm); the polymerization reaction is highly exothermic.

Preparation. Thermal elimination of HCl from 1-chloro-1,1-difluoroethane (HCFC-142b) [75-68-3] is the principal industrial route to VDF covered by numerous patents (8–19). Dehydrohalogenation of 1-bromo-1,1-difluoroethane (20), or 1,1,1-trifluoroethane (HFC-143a) (21–25), or dehalogenation of 1,2-dichloro-1,1-difluoroethane (26–28) are investigated alternative routes (see FLUORINE COMPOUNDS, ORGANIC–FLUORINATED ALIPHATIC COMPOUNDS).

The commercially preferred monomer precursor HCFC-142b has been prepared by hydrofluorination of acetylene (29), vinylidene chloride (30–32), or 1,1,1-trichloroethane (33–39).

$$CH{\equiv}CH \xrightarrow{+2\,HF} CH_3{-}CHF_2 \xrightarrow{+Cl_2} CH_3{-}CClF_2 + HCl$$

$$CH_2{=}CCl_2 \xrightarrow{+2\,HF} CH_3{-}CClF_2 + HCl$$

$$CH_3{-}CCl_3 \xrightarrow{+2\,HF} CH_3{-}CClF_2 + 2\,HCl$$

The monomer can also be continuously prepared by the pyrolysis of trifluoromethane (CHF_3) in the presence of a catalyst and either methane or ethylene (40–43). Passing 1,1-difluoroethane ($CH_3{-}CHF_2$), oxygen, and CO_2 over a catalyst gives a mixture of VDF and vinyl fluoride (44). Using either methanol or dichloromethane as a source of the carbene moiety, VDF can be continuously prepared from chlorodifluoromethane (HCFC-22) (CHF_2Cl) (45,46). Pyrolysis of dichlorodifluoromethane (CFC-12) (CCl_2F_2) with either methane (47) or methyl chloride

Table 1. Properties of Vinylidene Fluoride

Property	Value	Reference
molecular weight	64.038	
boiling point, °C	-84	
freezing point, °C	-144	
vapor pressure at 21°C, kPa^a	3683	
critical pressure, kPa^a	4434	4
critical temperature, °C	30.1	
critical density, kg/m^3	417	
explosive limits, vol % in air	5.8–20.3	5
heat of formation at 25°C, kJ/mol^b	-345.2	6
heat of polymerization at 25°C, kJ/mol^b	-474.21	7
solubility in water, cm^3/100 g at 25°C, 10 kPa^a	6.3	

^{a}To convert kPa to atm, divide by 101.3.
^{b}To convert kJ to kcal, divide by 4.184.

yields the monomer (48,49). Copyrolysis of methane and either bromotrifluoro-($CBrF_3$) or chlorotrifluoromethane ($CClF_3$) yields VDF (50). Deuterated VDF has also been prepared (51).

Storage and Shipment. VDF or HFC-1132a is stored and shipped in gas cylinders or high pressure tube trailers without polymerization inhibitor and is placarded as flammable compressed gas. Terpenes or quinones can be added to inhibit polymerization. Elf Atochem North America, Inc. and Ausimont USA, Inc. supply VDF in the United States; other producers are in Japan and Europe.

Health and Safety Factors. VDF is a flammable gas; its combustion products are toxic. Liquid VDF on contact with the skin can cause frostbite. Acute inhalation toxicity of VDF is low; median lethal concentrations (LC_{50}) for rats were 128,000 ppm after a single 4-h exposure (52) and 800,000 ppm after a 30-min exposure (53). Cumulative toxicity is low; exposure of rats and mice at levels of up to 50,000 ppm for 90 days did not cause any systemic toxicity (54,55). No teratogenic or reproductive effects were found in rats. VDF was positive in bacterial gene mutation assay but negative in mammalian gene mutation, chromosomal aberration, and cell transformation assays. In 1979, a paper reported that rats developed lipomas after being given over 52 weeks' oral doses of VDF dissolved in olive oil (56). More relevant, lifetime (18 months) inhalation studies on rats and mice have not detected chronic or carcinogenic effects up to 10,000 ppm VDF (57,58). Additional information is available (59,60). Toxicology test data on VDF were submitted to the EPA pursuant to a final test rule and consent order under the Toxic Substances Control Act (TSCA) (61).

Uses. Vinylidene fluoride is used for the manufacture of PVDF and for copolymerization with many fluorinated monomers. One commercially significant use is the manufacture of high performance fluoroelastomers that include copolymers of VDF with hexafluoropropylene (HFP) (62) or chlorotrifluoroethylene (CTFE) (63) and terpolymers with HFP and tetrafluoroethylene (TFE) (64) (see ELASTOMERS, SYNTHETIC–FLUOROCARBON ELASTOMERS). There is intense commercial interest in thermoplastic copolymers of VDF with HFP (65,66), CTFE (67), or TFE (68). Less common are copolymers with trifluoroethene (69), 3,3,3-trifluoro-2-trifluoromethylpropene (70), or hexafluoroacetone (71). Thermoplastic terpolymers of VDF, HFP, and TFE are also of interest as coatings and film. A thermoplastic elastomer that has an elastomeric VDF copolymer chain as backbone and a grafted PVDF side chain has been developed (72).

Polymer

Polymerization. The first successful polymerizations of VDF in aqueous medium using peroxide initiators at 20–150°C and pressures above 30 MPa were described in a patent issued in 1948 (73). About a year later, the first copolymerizations of VDF with ethylene and halogenated ethylenes were also patented (74). After a hiatus of over 12 years a commercially feasible process was developed and PVDF was ready for market introduction (2).

PVDF is manufactured using radical initiated batch polymerization processes in aqueous emulsion or suspension; operating pressures may range from 1 to 20 MPa (10–200 atm) and temperatures from 10 to 130°C. Polymerization method, temperature, pressure, recipe ingredients, the manner in which they are added to the reactor, the reactor design, and post-reactor processing are variables that influence product characteristics and quality.

Emulsion polymerization of VDF is a heterogeneous reaction that requires, as is typical with most fluorine-containing monomers, addition of a polyfluoro-alkanoic acid salt as surfactant (75) to avoid radical scavenging reactions during polymerization. Sometimes chain-transfer agents or buffers, or both, are used in the emulsion process. Radical generators that initiate polymerization of VDF are either water-soluble, eg, persulfate salts (76–78), disuccinic acid peroxide (79), β-hydroxyalkyl peroxide (80,81), alkylperoxybutyric acid (82) or monomer soluble, eg, di-*tert*-butyl peroxide (83,84), dialkylperoxydicarbonate (85–88), or *tert*-butylperoxybutyrate (89). A radiotracer study found that the number of end groups formed in the polymer from primary radicals of the initiator decreased during emulsion polymerization of PVDF, whereas overall branching increased (90). Upon completion of the polymerization, the discharged reactor product is a milky white colloidal dispersion or latex that is subsequently filtered, coagulated, thoroughly washed and usually spray-dried to produce a very fine powder. It is typical of emulsion polymerization that the polymer solids in latex are spheres of about 250 nm in diameter and the dried powders contain agglomerates of about 2 to 5 μm in diameter. The powder is either packaged or processed as required for the intended use.

Suspension polymerization of VDF in water are batch processes in autoclaves designed to limit scale formation (91). Most systems operate from 30 to 100°C and are initiated with monomer-soluble organic free-radical initiators such as diisopropyl peroxydicarbonate (92–96), *tert*-butyl peroxypivalate (97), or *tert*-amyl peroxypivalate (98). Usually water-soluble polymers, eg, cellulose derivatives or poly(vinyl alcohol), are used as suspending agents to reduce coalescence of polymer particles. Organic solvents that may act as a reaction accelerator or chain-transfer agent are often employed. The reactor product is a slurry of suspended polymer particles, usually spheres of 30–100 μm in diameter; they are separated from the water phase thoroughly washed and dried. Size and internal structure of beads, ie, porosity, and dispersant residues affect how the resin performs in applications.

Solution polymerization of VDF in fluorinated and fluorochlorinated hydrocarbons such as CFC-113 and initiated with organic peroxides (99), especially bis(perfluoropropionyl) peroxide (100), has been claimed. Radiation-induced polymerization of VDF has also been investigated (101,102). Alkylboron compounds activated by oxygen initiate VDF polymerization in water or organic solvents (103,104). Microwave-stimulated, low pressure plasma polymerization of VDF gives polymer film that is <10 μm thick (105). Highly regular PVDF polymer with minimized defect structure was synthesized and claimed (106). Perdeuterated PVDF has also been prepared and described (107).

Polymer Properties.   PVDF is a tough, semicrystalline engineering polymer. Compared to the softer and mechanically less robust perfluorocarbon polymers, PVDF has high mechanical and impact strength, and excellent resistance to both creep under long-term stress and fatigue upon cyclic loading (108,109). PVDF also has excellent abrasion resistance and thermal stability, and resists damage from most chemicals and solvents, as well as from ultraviolet and nuclear radiation. Typical PVDF design properties are shown in Table 2.

Properties of PVDF depend on molecular weight, molecular weight distribution, chain configuration, ie, the sequence in which the monomer units are linked together, including side groups or branching, and crystalline form. The morphology of PVDF reflects differences in both the utilized polymerization procedure and the thermomechanical treatment that followed polymerization. During radical-initiated polymerization, the head-to-tail addition of VDF molecules predominates, in which —CF_2— is denoted as "head" and —CH_2— as "tail," but reversed monomeric addition leading to head-to-head and tail-to-tail defects does occur; the extent of defects is influenced by polymerization process conditions, particularly temperature (110). The incidence of these defects is best determined by high resolution ^{19}F nmr (111,112); infrared (113) and laser mass spectrometry (114) are alternative methods. Typical commercial polymers show 3–6 mol % defect content. Polymerization methods have a particularly strong effect on the sequence of these defects. In contrast to suspension polymerized PVDF, emulsion polymerized PVDF forms a higher fraction of head-to-head defects that are not

Table 2. Properties of Poly(vinylidene fluoride)

Property	Method	Value
specific gravity	ASTM D792	1.75–1.80
water absorption, 24 h at 23°C, %	ASTM D570	0.04
refractive index, n_D	ASTM D542	1.42
melting peak, T_m, °C	ASTM D3418	156–180
crystallization peak, T_c, °C	ASTM D3418	127–146
glass transition, T_g, °C	ASTM D2236	−40
brittleness temperature, °C	ASTM D746	−62 to −64
deflection temperature at 1.82 MPa,[a] °C	ASTM D648	84–115
specific heat, kJ/kg·K[b]	DSC	1.26–1.42
thermal conductivity, W/K·m	ASTM D433	0.17–0.19
tensile stress at yield, MPa[a]	ASTM D638	28–57
tensile stress at break, MPa[a]	ASTM D638	31–52
elongation at break, %	ASTM D638	50–250
compressive strength, MPa[a]	ASTM D695	55–110
flexural strength, MPa[a]	ASTM D790	59–94
modulus of elasticity, MPa[a]		
in tension	ASTM D882	1040–2600
in flexure	ASTM D790	1140–2500
impact strength at 25°C, J/m[c]	ASTM D256	
unnotched		800–4270
notched		107–214
limiting oxygen index, %	ASTM D2863	43
vertical burn	UL 94	V-0
sand abrasion, m³/mm	ASTM D968	4.0

[a]To convert MPa to psi, multiply by 145.
[b]To convert kJ to kcal, divide by 4.184.
[c]To convert J/m to ft·lbf/in., divide by 53.38.

followed by tail-to-tail addition (115,116). Crystallinity and other properties of PVDF or copolymers of VDF are influenced by these defect structures (117).

Crystallinity affects toughness and mechanical strength as well as impact resistance. PVDF crystals are seen in the optical microscope as spherulites that are lamellae of polymer chain segments, which are packed crystallographically; the interposed amorphous regions consist of disordered chains. The crystallinity can range between 35 and 70%. Various parameters, including molecular weight, molecular weight distribution, polymerization method, thermal history, and cooling rates influence crystallization kinetics (118).

Unlike other synthetic polymers, PVDF has a wealth of polymorphs; at least four chain conformations are known and a fifth has been suggested (119). The four known distinct forms or phases are alpha (II), beta (I), gamma (III), and delta (IV). The most common α-phase is the trans-gauche (tgtg') chain conformation placing hydrogen and fluorine atoms alternately on each side of the chain (120,121). It forms during polymerization and crystallizes from the melt at all temperatures (122,123). The other forms have also been well characterized (124–128). The density of the α polymorph crystals is 1.92 g/cm^3 and that of the β polymorph crystals 1.97 g/cm^3 (129); the density of amorphous PVDF is 1.68 g/cm^3 (130).

Relaxations of α-PVDF have been investigated by various methods including dielectric, dynamic mechanical, nmr, dilatometric, and piezoelectric and reviewed (3). Significant relaxation ranges are seen in the loss-modulus curve of the dynamic mechanical spectrum for α-PVDF at about 100°C (α'), 50°C (α''), −38°C (β), and −70°C (γ). PVDF relaxation temperatures are rather complex because the behavior of PVDF varies with thermal or mechanical history and with the testing methodology (131).

Suspension- and emulsion-polymerized PVDF exhibit dissimilar behavior in solutions. The suspension resin type is readily soluble in many solvents; even in good solvents, solutions of the emulsion resin type contain fractions of microgel, which contain more head-to-head chain defects than the soluble fraction of the resin (116). Concentrated solutions (15 wt %) and melt rheology of various PVDF types also display different behavior (132). The Mark-Houwink relation ($\eta = KM^a$) for PVDF in N-methylpyrrolidinone (NMP) containing 0.1 molar LiBr at 85°C, for the suspension (115) and emulsion (116) respectively is: $\eta = (4.5 \pm 0.3) \times 10^{-4} M^{0.70}$ and $\eta = 1.4 \times M^{0.96}$.

Unlike most crystalline polymers, PVDF exhibits thermodynamic compatibility with other polymers (133). Blends of PVDF and poly(methyl methacrylate) (PMMA) are compatible over a wide range of blend composition (134,135). Solid-state nmr studies showed that isotactic PMMA is more miscible with PVDF than atactic and syndiotactic PMMA (136). Miscibility of PVDF and poly(alkyl acrylates) depends on a specific interaction between PVDF and oxygen within the acrylate and the effect of this interaction is diminished as the hydrocarbon content of the ester is increased (137). Strong dipolar interactions are important to achieve miscibility with poly(vinylidene fluoride) (138). PVDF blends are the object of many papers and patents; specific blends of PVDF and acrylic copolymers have seen large commercial use.

PVDF cross-links readily when subjected to electron beam radiation (139) or gamma radiation (140). Cross-linking efficiency is proportional to molecular weight, molecular weight distribution, or extent of head-to-head chain defects (141). The cross-linked PVDF, when highly stressed or compressed above the melting point, exhibits thermodynamic and physical properties similar to polyethylene and polypropylene (142). Polyfunctional monomers having good solubility in PVDF increase the cross-linking rate (143,144). The effect of radiation on the structure and properties of PVDF has been reviewed (145).

Some electrical properties are shown in Table 3. Values of other parameters have been published (146). Polymorphism of the PVDF chains and the orientation of the two distinct dipole groups, $-CF_2-$ and $-CH_2-$, rather than trapped space charges (147) contribute to the exceptional dielectric properties and the extraordinarily large piezoelectric and pyroelectric activity of the polymer (146,148,149).

Prolonged exposure of PVDF to processing temperatures exceeding 300°C could lead to discoloration and chemical reactions that present hazards. The primary reaction at high temperature is loss of hydrogen fluoride (HF) that results in conjugation, $-CH=CF-CH=CF-$, along the chains; this explains the observed discoloration. The extent and rate of discoloration is not homogeneous among PVDF resins and may be commensurate with chain perfection, ie, percentage of head-to-tail repeat units in the chain (150); reversed repeat units may interrupt dehydrofluorination (151). If the temperature exceeds 375°C in air, rapid thermal decomposition takes place and HF gas evolves. After 70 wt % loss, at about 480°C, the residue is char that eventually burns completely at higher temperature. The charring phenomenon is considered basic to the superior performance of PVDF in severe fire tests such as the Underwriters Laboratories UL 910 Modified Steiner Tunnel Test (152).

Fabrication and Processing. PVDF is available in a wide range of melt viscosities as powder or pellets to fulfill typical fabrication requirements; latices are also commercially available.

PVDF is readily molded in conventional compression, transfer, and injection-molding equipment (153–155); typical molding temperatures for the cylinder and nozzle are 180–240°C and molds are at 50–90°C. PVDF resins do not require drying because the resin does not absorb moisture. As a crystalline polymer, it shows a relatively high mold shrinkage of ca 3%. To obtain a high dimensional

Table 3. Electrical Properties of Poly(vinylidene fluoride)

Property	Method	Value
volume resistivity, $\Omega \cdot cm$	ASTM D257	$1.5-5 \times 10^{14}$
surface arc resistance, s	ASTM D495	50–60
dielectric strength, kV/mm	ASTM D149	63–67
dielectric constant at 25°C	ASTM D150	
1 kHz		8.15–10.46
10 kHz		8.05–9.90
100 kHz		7.85–9.61
dissipation factor		
1 kHz		0.005–0.026
10 kHz		0.015–0.021
100 kHz		0.039–0.058

stability, carbon-filled, mica-filled, or carbon–fiber-reinforced (156) grades are used. To achieve best results and avoid warping or voids, it is essential to coordinate the cooling rate with the crystallization of the resin or anneal the part at 140–150°C. For compression or transfer molding the PVDF pellets are preheated in an oven to 210–240°C and transferred to the mold that is heated to 190–200°C. The resin in the filled mold is placed under sufficient pressure to complete flow and fusion. Sufficient time must be allowed to cool the molded part under pressure to 90°C to prevent vacuum voids and distortion.

Smooth PVDF profiles of all types—film, sheet, rod, profile, pipe, tubing, fiber, monofilament, wire insulation, and cable-jackets—can be extruded; no heat stabilizers are needed. In both molding and extrusion operations, care must be exercised to eliminate hang-up zones in the equipment where molten resin (at 230–260°C) can stagnate and thermally decompose with time. Equipment built with material of construction used for processing polyolefins or PVC is adequate; for long-term or high shear processing, a highly wear-resistant alloy such as Xaloy 306 for barrel liner and SAE 4140 steel for the screw is suggested. Gradual transition-type screws having L/D ratios at least 20:1, ample metering sections, and compression ratio of about 3 are recommended. Temperature profiles vary from 190 to 290°C depending on resin grade and shape being extruded. Water quenching is practiced for wire insulation, tubing, and pipe, whereas sheet and flat film are melt-cast on polished steel rolls operating at 65–150°C (157).

PVDF sheets can be backed during extrusion-calendering using fabrics of glass (158), polyamide, or polyester fibers; they can also be press-laminated with the fabrics at 185 to 200°C (159). Nonvulcanized rubber can also be press-laminated with PVDF sheet at 150°C (160). Melt-cast PVDF sheets can be oriented uniaxially or biaxially to produce films with vastly increased mechanical strength, specular transmission, or ferroelectric activity (161). Blown-film equipment typically used for HDPE can also be used for extrusion of blown PVDF film. Monofilaments are usually extruded or spun at 240–260°C into a 30–50°C water bath and then reheated to 130–160°C, oriented using draw ratios of 3:1 to 5:1, and heat-set at elevated temperatures to produce high strength filament having tenacities of 350–440 mN/tex (4–5 gf/den) (162,163). Coextrusion of PVDF with other polymers is the subject of several patents. Interlayer adhesion is critical, although matching the coefficients of thermal expansion and melt viscosities are other important considerations (164). To promote interleaf bond, an adhesive "tie-layer" consisting of a polymer that is partially compatible both with PVDF and the incompatible polymer layer, such as ABS (165) and polyolefins (166), has been used in coextrusion.

Semifinished PVDF products can be machined and processed by methods used for other thermoplastics (155). PVDF parts can be joined by standard welding methods. Pipe, fittings, or sheets can be welded using a hot-air gun with a welding rod or a heated tool for butt or socket welding. Films can be bonded by heat sealing, high frequency welding, or ultrasonic welding.

Manufactured PVDF parts can be cross-linked using high energy radiation to produce high temperature wire insulation, and heat-shrinkable tubing or film.

Organosol dispersions of PVDF used extensively for exterior architectural finishes can be produced from the very fine powder obtained only by the emulsion polymerization method. These dispersions include the very fine PVDF powder, pigments, acrylate or methacrylate copolymer, and selected solvents (167–170); comparable water-based coating compositions can also be prepared (171–174). These dispersions are factory applied by spray or roller to primed steel or aluminum surfaces and oven-fused at 230–260°C to form continuous films that adhere firmly to the substrata. Other organic dispersions of PVDF are formulated for spray applications of relatively thick coatings to protect metals from corrosive environments. Powders for electrostatic spraying, fluidized-bed deposition, or rotomolding are obtained by melt compounding PVDF with appropriate ingredients, cryogenic grinding, and classification to desirable particle-size range for the application.

Microporous filtration membranes from VDF polymers are made by casting a polymer solution on a rigidly supported backing belt, then passing the belt through a bath to form the membrane, followed by extraction of any residual solvent from and drying of the membrane (175). Formation of microporous PVDF membranes has been reviewed (176). To improve performance, PVDF membranes are often chemically modified (177–181). Hollow fibers useful for microfiltration are produced by extruding a spinning solution of PVDF from an annular spinning orifice into coagulating liquids (182). Porous structures can also be made by sintering very fine granules under controlled conditions (183), from extruded compounds that contain leachable additives that upon extraction leave voids in the product, or by extrusion of compounds containing chemical blowing agents (184,185).

Economic Aspects

Because of its excellent combination of properties, processibility, and relatively low price compared to other fluoropolymers, PVDF has become the largest volume fluoropolymer after PTFE; consumption in the United States has grown from zero in 1960 to about 6200 metric tons in 1991 (186). About 49% of the consumed volume is PVDF modified by copolymerization with 5–12-wt % HFP to enhance flexibility. In 1992, list price for homopolymer powders was $15.32/kg, and for pellets $15.42/kg; the reported market price was $14.09–14.22/kg (187). In the United States, almost all PVDF is supplied by Ausimont USA, Inc., Elf Atochem North America, Inc., and Solvay Polymers, Inc. Ausimont and Elf Atochem are producers; Solvay is an importer of the resin. Small amounts of resin are imported from Germany by Hüls America, Inc. and from Japan by Kureha Chemical Industry Co., Ltd. PVDF producers and their trademarks are listed in Table 4.

After 10 years of unabated rapid growth in the plenum wire and cable market, fluoropolymers including PVDF, primarily the flexible VDF/HFP copolymer, are beginning to lose market share to lower priced PVC-alloys. The loss of market share in the plenum market probably will be compensated by growth of PVDF in other fields; thus during the mid-1990s the total volume of PVDF may not grow (188).

Specifications and Standards.   Commercial PVDF resin types and standards are defined in ASTM D3222. A list of military and industrial specifications covering applications, material suppliers, and PVDF resin grades can be found in Reference 189.

Table 4. Producers and Trademarks of Poly(vinylidene fluoride)

Producer	Country	Trademark
Ausimont USA, Inc.	United States	Hylar
Elf Atochem North America, Inc.	United States	Kynar
Elf Atochem, SA	France	Foraflon
Solvay & Cie, SA	Belgium	Solef/Vidar
Hüls, AG	Germany	Dyflor
Daikin Kogyo Co., Ltd.	Japan	Neoflon
Kureha Chemical Industry Co., Ltd.	Japan	KF Polymer

Health and Safety Factors

PVDF is a nontoxic resin and may be safely used in articles intended for repeated contact with food (190). Based on studies under controlled conditions, including acute oral, systemic, subchronic, and subacute contact; implantation; and tissue culture tests, no adverse toxicological or biological response has been found in test animals (191,192). PVDF is acceptable for use in processing and storage areas in contact with meat or poultry products prepared under federal inspection and it complies with the 3-A sanitary standards for dairy equipment.

PVDF is not hazardous under typical processing conditions. If the polymer is accidentally exposed to temperatures exceeding 350°C, thermal decomposition occurs with evolution of toxic hydrogen fluoride (HF).

Some silica-containing additives such as glass and titanium dioxide lower the thermal stability of PVDF and should be used with caution. Processors should consult the resin producer about safe processing practice.

Uses

PVDF is used in many diverse industrial applications for products that require high mechanical strength and resistance to severe environmental stresses. The most important fields of application for PVDF resins include electric and electronic industry products, architectural and specialty finishes, products for the chemical and related industries, and rapidly growing specialized uses.

In the electric and electronics field the largest usage of PVDF is for plenum wire and cables, plenum being the space between the suspended and structural ceiling in high rise buildings. PVDF-insulated wire and cables jacketed with the flexible VDF–HFP copolymer pass the UL 910 specification for low smoke generation and flame spread and are approved for remote-control, signaling and power-limited circuits, fire protective signaling systems, and communication systems. Other important wire constructions include cross-linked PVDF jackets (193) for commercial aircraft, industrial power control, and cathodic protection wires and cables. Self-limiting strip heaters consisting of a cross-linked conductive PVDF core, which separates two parallel conductors, and a fluoropolymer jacket are useful for heating pipes or other process fluid-handling equipment (194,195).

Cross-linked heat-shrinkable PVDF tubings (196) are used as connector sleeves for wires and cables, or to coat ordnance (197). Some sleeves incorporate a ring of solder, forming a so-called solder sleeve for power control, electronic, aircraft, and communication wiring.

Uniaxially or biaxially oriented PVDF film upon metallization and poling under a high dielectric field is a flexible, tough, light, and active transducer for many piezo- and pyroelectric applications (198,199). Current applications include infrared detectors (200); audio devices, eg, stereo speakers, microphones, headphones, phonograph cartridges, hydrophones for long-range tracking in ocean depth; pressure or stress sensors (201); contactless keyboards (202); motion detectors (203); and medical devices, eg, detectors for heartbeat and breathing rate, or sensors for ultrasonic imaging.

The largest commercial application for PVDF homopolymer powder is as a base for long-lasting decorative finishes on aluminum and galvanized steel siding, curtain-wall panels, roofing systems, aluminum extrusions and other building components (204) that are used on power plants, schools, airport buildings, department stores, high rise office and hotel buildings, sports stadiums, and, to a lesser extent, residential buildings. These organosol finishes, available in many colors from paint companies throughout the world, are factory-applied by conventional state-of-the-art coil or spray-coating procedure to the primed base metal (205). Usually, the coating consists of a suitable primer layer up to about 5 μm thick and a 20–30 μm finishing layer of a PVDF topcoat. Accelerated weathering tests along with the experience with buildings erected since the 1960s prove that these finishes are unique in durability in terms of film integrity, color retention, corrosion resistance, flexibility, sand-abrasion resistance, and chemical resistance (206). Similar PVDF organosol dispersions are also being used for corrosion-protection coating of automotive break-line tubings. Pigmented thin film that is continuously cast from PVDF solutions or dispersions is used for decorative laminates and has been specified for body trim by principal automobile manufacturers. PVDF-based powders analogous in composition to the liquid finishes have been proposed as decorative protective coatings for metallic substrata (207). PVDF-based powder for rotomolding, eg, for tanks, valves, or fittings, and for fluidized-bed deposition and electrostatic spraying are available (208).

Fluid-handling systems in the chemical processing and related fields are also large users of PVDF products such as solid or lined pipes, valves, pumps, tower packing, and tank and trailer linings (209,210). Because PVDF is manufactured by methods that assure extremely low ionic contamination, it has qualified for use in ultrapure water systems (211–213), including WFI (water for injection) and *U.S. Pharmacopoeia* (USP) standards (214). Blow-molded PVDF bottles are used for shipping or storing high purity chemicals in the semiconductor industry (215). Extruded monofilament woven into coarse fabric is used widely for drum filtration during bleaching of wood pulp with chemicals, eg, sodium hypochlorite or chlorine dioxide and caustic soda. Like other fluoropolymers, PVDF is used as a binder for asbestos-fiber-based diaphragms used in cells for the electrolysis of brine to produce chlorine and caustic soda (216,217).

PVDF-based microporous filters are in use at wineries, dairies, and electro-coating plants, as well as in water purification, biochemistry, and medical devices. Recently developed nanoselective filtration using PVDF membranes is 10 times more effective than conventional ultrafiltration (UF) for removing viruses from protein products of human or animal cell fermentations (218). PVDF protein-sequencing membranes are suitable for electroblotting procedures in protein research, or for analyzing the phosphoamino content in proteins under acidic and basic conditions or in solvents (219).

Pigmented PVDF and ABS laminates manufactured by coextrusion with a tie-coat exhibit excellent weather resistance resulting from the protective PVDF cap layer; they are used in Europe for thermoformed automotive dash panels, trailer and tractor roofs, motorcycle gas tank housings, and lawn-mower blade guards (220). A PVDF alloy which is a blend of PVDF and alkyl methacrylate homo- or copolymer is coextruded with acrylate or methacrylate resin blend to form a sheet for hydrosanitary components (221). Similar blends of PVDF and compatible resins can be coextruded both with PVC, to form home siding panels with outstanding resistance to weather (222), and with an engineering resin, ie, polycarbonate, polyurethane, polyamide, polyester, or ABS, or their compounds (223).

In Japan, PVDF monofilament for fishing lines for both commercial and sport fishing is a specialty in demand (224–226) because it displays no water absorption, is not visible in water, and has high knot strength and high specific gravity. PVDF as a processing aid eliminates melt fracture and other flow-induced imperfections in blown LLDPE and HDPE films (227). Optical disk memory devices utilize the decrease in transmittance on crystallization of PVDF and thus provide an overwritable memory (228). The exceptional dielectric properties of PVDF are utilized in electrophotographic carrier (toner) compositions (229).

BIBLIOGRAPHY

"Poly(vinylidene fluoride) under "Fluorine Compounds, Organic" in *ECT* 2nd ed., Vol. 9, pp. 840–847, by W. S. Barnhart and N. T. Hall, Pennsalt Chemicals Corp.; in *ECT* 3rd ed., Vol. 11, pp. 64–74, by J. E. Dohany and L. E. Robb, Pennwalt Corp.

1. J. E. Dohany and J. S. Humphrey, in J. I. Kroschwitz, ed., *Encyclopedia of Polymer Science and Engineering*, 2nd ed., Vol. 17, p. 532.
2. J. E. Dohany, in R. B. Seymour and G. S. Kirshenbaum, eds. *High Performance Polymers: Their Origin and Development*, Elsevier Science Publishing Co., New York, 1986, p. 287.
3. A. J. Lovinger in G. C. Bassett, ed., *Developments in Crystalline Polymers*, Vol. 1, Applied Science Publishers, Ltd., Barking, UK, 1982, pp. 195–273.
4. W. H. Mears and co-workers, *Ind. Eng. Chem.* **47**(7), 1449–1454 (1955).
5. A. N. Baratov and V. M. Kucher, *Zh. Prikl. Khim.* **38**(5), 1068–1072 (1965).
6. D. R. Stull, E. F. Westrum, and G. C. Sinke, *The Chemical Thermodynamics of Organic Compounds*, John Wiley & Sons, Inc., New York, 1969, p. 502.
7. W. D. Wood, J. L. Lacina, B. L. DePrater, and J. P. McCullough, *J. Phys. Chem.* **68**(3), 579 (1964).
8. U.S. Pat. 2,551,573 (Aug. 5, 1951), F. B. Downing, A. F. Benning, and R. C. McHarness (to E. I. du Pont de Nemours & Co., Inc.).
9. U.S. Pat. 2,774,799 (Dec. 18, 1956), R. Mantell and W. S. Barnhart (to M. W. Kellogg Co.).
10. USSR Pat. 216,699 (Apr. 26, 1968), B. P. Zverev, A. L. Goldinov, Yu. A. Panshin, L. M. Borovnev, and N. S. Shirokova.

11. U.S. Pat. 3,246,041 (Apr. 12, 1966), M. E. Miville and J. J. Earley (to Pennwalt Corp.).
12. Ger. Pat. 1,288,085 (Jan. 30, 1969), F. Kaess, K. Lienhard, and H. Michaud (to Sueddeutsche Kalkstickstoff-Werke A.G.).
13. Ger. Pat. 1,288,593 (Feb. 6, 1969), F. Kaess, K. Lienhard, and H. Michaud (to Sueddeutsche Kalkstickstoff-Werke A.G.).
14. Jpn. Pat. 58 217,403 (Dec. 17, 1983), (to Pennwalt Corp.).
15. H. Mueller, G. Emig, and H. Hofmann, *Chem. Ing. Tech.* **56**(8), 626–628 (1984); *Chem. Abstr.* **101**(19), 170373v (1984).
16. J. Wolfrum, M. Schneider, *Proc. SPIE Int. Soc. Opt. Eng.* **458**, 46–52 (1984); *Chem. Abstr.* **101**(13), 110151 (1984).
17. J. Wolfrum, *Laser Chem.* **6**(2), 125–147 (1986).
18. Z. F. Dong, M. Schneider, J. Wolfrum, *Int. J. Chem. Kinet.* **21**(6), 387–397 (1989); *Chem. Abstr.* **111**(23), 213982 (1989).
19. Can. Pat. 2,016,691 (Dec. 28, 1990), M. Y. Elsheikh (to Elf Atochem North America, Inc.).
20. Fr. Pat. 1,337,360 (Sept. 13, 1963), Produits Chimique Pechiney Saint-Gobain.
21. Jpn. Pat. 68 29,126 (Dec. 13, 1968), H. Ukihashi and M. Ichimura (to Asahi Glass Co., Ltd.).
22. Jpn. Pat. 62 169,737 (July 25, 1987), (to Pennwalt Corp.).
23. U.S. Pat. 4,818,513 (Apr. 4, 1989), F. C. Trager, J. D. Mansell, and W. E. Wimer (to PPG Industries, Inc.).
24. Eur. Pat. Appl. 402,652 (Dec. 19, 1990), M. Y. Elsheikh and M. S. Bolmer (to Elf Atochem North America, Inc.).
25. Eur. Pat. Appl. 407,711 (Jan. 16, 1991), M. Y. Elsheikh (to Elf Atochem North America, Inc.).
26. U.S. Pat. 2,401,897 (June 11, 1946), A. F. Benning, F. B. Downing, and R. J. Plunkett (to E. I. du Pont de Nemours & Co., Inc.).
27. Jpn. Pat. 68 11,202 (May 11, 1968), Kureha Chem. Ind. Co., Ltd.
28. U.S. Pat. 2,734,090 (Feb. 7, 1956), J. C. Calfee and C. B. Miller (to Allied Chemical Corp.).
29. Ger. Pat. 2,659,712 (July 6, 1976), N. Schultz, P. Martens and H-J. Vahlensieck (to Dynamit Nobel AG).
30. U.S. Pat. 3,600,450 (Aug. 17, 1971), F. Kaess and H. Michaud (to Sueddeutsche Kalkstickstoff-Werke AG).
31. Eur. Pat. 3,723,549 (Mar. 27, 1973), F. Kaess, K. Lienhard, and H. Michaud (to Sueddeutsche Kalkstickstoff-Werke AG).
32. Eur. Pat. Appl. 361,578 (Apr. 4, 1990), J. Franklin and F. Janssens (to Solvay & Cie.).
33. E. T. McBee and co-workers, *Ind. Eng. Chem.* **39**(3), 409–412 (1947).
34. U.S. Pat. 3,833,676 (Sept. 3, 1974), R. Ukaji and I. Morioka (to Daikin Industries, Ltd.).
35. Jpn. Pat. 58 217,403 (Dec. 17, 1983), (to Pennwalt Corp.).
36. Eur. Pat. Appl. 297,947 (Jan. 4, 1989), B. Cheminal and A. Lantz (to Elf Atochem SA).
37. Eur. Pat. Appl. 407,689 (Jan. 16, 1991), D. W. Wright and B. L. Wagner (to Elf Atochem North America, Inc.).
38. Eur. Pat. Appl. 421,830 (Apr. 10, 1991), M. Bergougnan, J. M. Galland, and S. Perdieux (to Elf Atochem SA).
39. Jpn. Pat. 03 151,335 (June 27, 1991), M. Iwasaki and T. Yoshida (to Toa Gosei Industry Co., Ltd.).
40. U.S. Pat. 3,047,637 (July 31, 1962), F. Olstowski (to The Dow Chemical Co.).
41. Fr. Pat. 1,330,146 (June 2, 1963), A. E. Pavlath and F. H. Walker (to Stauffer Chemical Co.).

42. U.S. Pat. 3,188,356 (June 8, 1965), M. Hauptschein and A. H. Feinberg (to Pennwalt Corp.).
43. Jpn. Pat. 65 22,453 (Oct. 5, 1965), S. Okazaki and N. Sakauchi (to Kureha Chemical Industry Co., Ltd.).
44. Eur. Pat. Appl. 461,297 (Dec. 18, 1991), M. S. Bolmer and M. Y. Elsheikh (to Elf Atochem North America, Inc.).
45. U.S. Pat. 3,073,870 (Jan. 15, 1963), D. M. Marquis (to E. I. du Pont de Nemours & Co., Inc.).
46. Jpn. Pat. 68 10,602 (May 4, 1968), Y. Kometani and M. Takemoto (to Daikin Kogyo Co., Ltd.).
47. Eur. Pat. Appl. 313,254 (Apr. 26, 1989), D. W. Edwards (to Imperial Chemical Industries, PLC).
48. Ger. Pat. 42,730 (Jan. 5, 1966), H. Madai.
49. U.S. Pat. 3,428,695 (Feb. 18, 1969), J. R. Soulen and W. F. Schwartz (to Pennwalt Corp.).
50. U.S. Pat. 3,089,910 (May 14, 1963), F. Olstowski and J. D. Watson (to The Dow Chemical Company).
51. R. E. Cais and J. M. Kometami, *Macromolecules* **17**, 1887–1889 (1984).
52. C. P. Carpenter, U. C. Pozzani, and H. F. Smith, *J. Ind. Hyg. Toxicol.* **31**, 343 (1949).
53. L. A. Greenberg and O. Lester, *Arch. Ind. Hyg. Occ. Med.* **2**, 335 (1950).
54. Litton Bionetics, Inc., LBI Project No. 12199-02, National Toxicology Program, Contract No. NO1-ES-28, 1984.
55. Litton Bionetics, Inc., LBI Project No. 12199-03, National Toxicology Program, Contract No. NO1-ES-2, 1984.
56. C. Maltoni and D. Tovoli, *Med. Lavoro*, **5**, 353 (1979).
57. TNO Nutrition and Food Research Project No. B 84-1408, Report No. 91.039, Netherlands Institute for Applied Scientific Research, Delft, the Netherlands, 1991.
58. Bio/Dynamics, Inc., Project 87-8022, CMA Reference NO. FIG-3.3-ONCO-BIO, 1991.
59. G. L. Kennedy, Jr., *Crit. Rev. Toxicol.* **21**, 149 (1990).
60. *Vinylidene Fluoride*, Toxicology Data, Elf Atochem North America, Inc., Philadelphia, Pa., July 1992.
61. U.S. *Fed. Regist.* **57**(3), 409 (1992).
62. U.S. Pat. 3,051,677 (Aug. 28, 1964), D. R. Rexford (to E. I. du Pont de Nemours & Co., Inc.).
63. U.S. Pat. 2,738,343 (Mar. 13, 1956); U.S. Pat. 2,752,331 (June 26, 1956), A. Dittman, H. J. Passino, and W. O. Teeters (to M. W. Kellogg Co.).
64. U.S. Pat. 2,968,649 (Jan. 17, 1961), J. R. Pailthorp and H. E. Schroeder (to E. I. du Pont de Nemours & Co., Inc.).
65. U.S. Pat. 3,178,399 (Apr. 13, 1965), E. S. Lo (to 3M Co.).
66. Eur. Pat. Appl. 456,019 (Nov. 13, 1991), L. A. Barber (to Elf Atochem North America, Inc.).
67. U.S. Pat. 4,851,479 (Jul. 25, 1989), J. Blaise and P. Kappler (to Elf Atochem SA).
68. Brit. Pat. 827,308 (Feb. 3, 1960), (to 3M Co.).
69. Eur. Pat. Appl. 320,344 (June 14, 1989), P. Kappler (to Elf Atochem SA).
70. U.S. Pat. 3,706,723 (Dec. 19, 1972), S. Chandrasekaran and M. B. Mueller (to Allied Signal Corp.).
71. U.S. Pat. 4,591,616 (May 27, 1986), S. Miyata and S. Kobayashi (to Central Glass Co., Ltd.).
72. U.S. Pat. 4,472,557 (Sept. 18, 1984), C. Kawashima and T. Yasumura (to Central Glass Co., Ltd.).
73. U.S. Pat. 2,435,537 (Feb. 3, 1948), T. A. Ford and W. E. Hanford (to E. I. du Pont de Nemours & Co., Inc.).

74. U.S. Pat. 2,468,054 (Apr. 26, 1949), T. A. Ford (to E. I. du Pont de Nemours & Co., Inc.).

75. U.S. Pat. 2,559,752 (July 10, 1951), K. L. Berry (to E. I. du Pont de Nemours & Co., Inc.).

76. U.S. Pat. 3,714,137 (Jan. 30, 1973), K. Lienhard and D. Ulmschneider (to Suddeutsche Kalkstickstoff-Werke AG).

77. U.S. Pat. 4,025,709 (May 24, 1977), J. Blaise and E. Grimaud (to Elf Atochem SA).

78. Eur. Pat. Appl. 387,938 (Sept. 19, 1990), X. Bacque and P. Lasson (to Solvay & Cie.).

79. U.S. Pat. 3,245,971 (Apr. 12, 1966), H. Iserson (to Pennwalt Corp.).

80. U.S. Pat. 3,640,985 (Feb. 8, 1972), H. C. Stevens (to PPG Industries, Inc.).

81. U.S. Pat. 3,708,463 (Jan. 2, 1973), J. P. Stallings (to Diamond Shamrock Corp.).

82. U.S. Pat. 3,642,755 (Feb. 15, 1972), J. A. Baxter, C. O. Eddy, and H. C. Stevens (to PPG Industries, Inc.).

83. U.S. Pat. 3,193,539 (July 6, 1965), M. Hauptschein (to Pennwalt Corp.).

84. U.S. Pat. 4,076,929 (Feb. 28, 1978), J. E. Dohany (to Pennwalt Corp.).

85. U.S. Pat. 3,475,396 (Oct. 28, 1969), G. H. McCain, J. R. Semancik, and J. J. Dietrich (to Diamond Shamrock Corp.).

86. U.S. Pat. 3,857,827 (Dec. 31, 1974), J. E. Dohany (to Pennwalt Corp.).

87. U.S. Pat. 4,360,652 (Nov. 23, 1982), J. E. Dohany (to Pennwalt Corp.).

88. U.S. Pat. 4,569,978 (Feb. 11, 1986), L. A. Barber (to Pennwalt Corp.).

89. U.S. Pat. 3,598,797 (Aug. 10, 1971), Y. Kometani, M. Okuda, and C. Okuno (to Daikin Industries, Ltd.).

90. L. Y. Madorskaya and co-workers, *Vysokomol. Soedin., Ser. B* **31**(10), 737–742 (1989).

91. Eur. Pat. Appl. 215,710 (Mar. 25, 1987), J. Blaise (to Elf Atochem SA).

92. U.S. Pat. 3,553,785 (Jan. 12, 1971), Y. Amagi and N. Bannai (to Kureha Chemical Co.).

93. U.S. Pat. 3,781,265 (Dec. 25, 1973), J. E. Dohany (to Pennwalt Corp.).

94. U. S. Pat. 4,542,194 (June 18, 1985), J. Dumoulin (to Solvay & Cie.).

95. Jpn. Pat. 01 129,005 (May 29, 1989), K. Ihara, Y. Noda, and T. Amano (to Daikin Industries, Ltd.).

96. Jpn. Pat. 02 029,402 (Jan. 31, 1990), J. Watanabe (to Shin-Etsu Chemical Industry Co., Ltd.).

97. U.S. Pat. 3,780,007 (Dec. 18, 1973), J. F. Stallings (to Diamond Shamrock Corp.).

98. Eur. Pat. Appl. 417,585 (Mar. 20, 1991) and 423,097 (Apr. 17, 1991), P. Lasson (to Solvay & Cie.).

99. Brit. Pat. 1,057,088 (Feb. 1, 1967), Kali-Chemie AG.

100. Ger. Pat. 1,806,426 (May 16, 1969); Fr. Pat. 1,590,301 (Apr. 14, 1970), D. P. Carlson (to E. I. du Pont de Nemours & Co., Inc.).

101. W. W. Doll and J. B. Lando, *J. Appl. Polym. Sci.* **14**, 1767 (1970).

102. U.S. Pat. 3,616,371 (Oct. 26, 1971), H. Ukihashi and M. Ichimura (to Asahi Glass Co., Ltd.).

103. Brit. Pat. 1,004,172 (Sept. 8, 1965), Deutsche Solvay-Werke GmbH.

104. R. Liepins, J. R. Surles, N. Morosoff, V. T. Stannett, M. L. Timmons, and J. J. Wortman, *J. Polym. Sci. Part A-1* **16**, 3039 (1978).

105. Eur. Pat. Appl. 403,915 (Dec. 27, 1990), J. Kammermaier and G. Rittmayer (to Siemens AG).

106. U.S. Pat. 4,438,247 (Mar. 20, 1984), R. E. Cais (to AT&T Technologies).

107. R. E. Cais and J. M. Kometani, *Macromolecules*, **17**, 1887 (1984).

108. P. E. Bretz, Ph.D. dissertation, Lehigh University, Bethlehem, Pa., 1980.

109. P. E. Bretz, R. W. Hertzberg, and J. A. Manson, *Polymer* **22**, 1272–1278 (1981).

110. M. Gorlitz, R. Minke, W. Trautvetter, and G. Weisgerber, *Angew. Makromol. Chem.* **29/30** 137 (1973).

111. R. C. Ferguson and E. G. Baume, Jr., *J. Phys. Chem.* **83**, 1379 (1979).
112. R. C. Ferguson and D. W. Ovenall, *Polymer Preprints, Div. Polym. Chem. Am. Chem. Soc.* **25**(1), 340 (1984).
113. M. A. Bachmann, W. Gordon, J. L. Koenig, and J. B. Lando, *J. Appl. Phys.* **50**, 6106 (1979).
114. D. E. Mattern, L. Fu-Tyan, and D. M. Hercules, *Anal. Chem.* **56**, 2762–2769 (1984).
115. G. Lutringer and G. Weill, *Polymer* **32**(5), 877 (1991).
116. G. Lutringer, B. Meurer, and G. Weill, *Polymer* **32**(5), 884 (1991).
117. A. J. Lovinger, D. D. Davis, R. E. Cais, and J. M. Kometani, *Polymer* **28**, 617–626 (1987).
118. S. Russel, K. L. McElroy, and L. H. Judovits, *Polym. Eng. Sci.* **32**(17), 1300 (1992).
119. A. J. Lovinger, *Macromolecules* **15**, 40 (1982).
120. J. Herschinger, D. Schaefer, H. W. Spiess, and A. J. Lovinger, *Macromolecules* **24**, 2428 (1991).
121. M. A. Bachmann and J. B. Lando, *Macromolecules* **14**, 40 (1981).
122. A. J. Lovinger, *J. Polym. Sci. Part A-2* **18**, 793–809 (1980).
123. Y. S. Yadav and P. C. Jain, *J. Macromol. Sci. Phys.* **B25**(3), 335 (1986).
124. T. Mizuno, K. Nakamura, N. Murayama, and K. Okuda, *Polymer* **26**(6), 853 (1985).
125. A. J. Lovinger, *Polymer* **21**(11), 1317 (1980).
126. C. C. Hsu and P. H. Geil, *Polymer Comm.* **27**, 105 (1986).
127. W. M. Prest and D. J. Luca, *J. Appl. Phys.* **49**(10), 5042 (1978).
128. G. T. Davis, J. E. McKinney, M. G. Broadhurst, and S. C. Roth, *J. Appl. Phys.* **49**, 4998 (1978).
129. R. Hasegawa, Y. Takahashi, Y. Chatani, and H. Tadokoro, *Polymer J.* **3**, 600 (1972).
130. K. Nakagawa and Y. Ishida, *Kolloid Z. Z. Polym.* **251**, 103 (1973).
131. A. J. Lovinger and T. T. Wang, *Polymer* **20**, 725 (1979).
132. K. F. Auyeung, *Polym. Eng. Sci.* **30**(7), 394 (1990).
133. D. R. Paul and J. W. Barlow, *J. Macromol. Sci. Rev. Macromol. Chem.* **C18**, 109 (1980).
134. J. S. Noland, N. N.-C. Hsu, R. Saxon, and J. M. Schmitt, *Advan. Chem. Ser.* **99**, 15 (1971).
135. J. Mijovic, H.-L. Luo, and C. D. Han, *Polym. Eng. Sci.* **22**(4), 234 (1982).
136. A. P. A. M. Eijkelenboom and co-workers, *Macromolecules*, **25**(18), 4511 (1992).
137. D. C. Wahrmund, R. E. Bernstein, J. W. Barlow, and D. R. Paul, *Polym. Eng. Sci.* **18**, 677 (1978).
138. G. Guerra, F. E. Karasz, and W. J. MacKnight, *Macromolecules* **19**, 1935 (1986).
139. R. Timmerman and W. Greyson, *J. Appl. Polym. Sci.* **6**(22), 456 (1962).
140. T. Yoshida, R. E. Florin, and L. A. Wall, *J. Polymer Sci.* **A3**, 1685 (1965).
141. K. Makuuchi, M. Asano, and T. Abe, *Nippon Nogei Kagaku Kaishi* (4), 686 (1976).
142. S.-H. Hyon and R. Kitamaru, *Bull. Inst. Chem. Res. Kyoto Univ.* **57**(2), 193 (1979).
143. K. Makuuchi, F. Yoshii, and T. Abe, *Nippon Nogei Kagaku Kaishi*, (10), 1828 (1975).
144. V. S. Ivanov, I. I. Migunova, and A. I. Mikhailov, *Radiat. Phys. Chem.* **37**(1), 119 (1991).
145. A. J. Lovinger in R. L. Clough and S. W. Shalaby, eds., *Radiation Effects on Polymers*, American Chemical Society Symposium Series 475, ACS, Washington, D.C., 1991, p. 84.
146. R. G. Kepler, *Ann. Rev. Phys. Chem.* **29**, 497 (1978).
147. D. K. Das-Gupta, *Ferroelectrics* **118**, 165 (1991).
148. A. J. Lovinger, *Science* **220**, 1115 (1983).
149. T. T. Wang, J. M. Herbert, and J. M. Glass, eds., *The Applications of Ferroelectric Polymers*, Chapman and Hall, New York, 1988.
150. A. J. Lovinger and D. J. Freed, *Macromolecules* **13**, 889 (1980).

151. H. Ishii, *Kobunshi Kagaku* **27**(307), 858 (1970).
152. U.S. Pat. 4,401,845 (Aug. 30, 1983), J. W. Michaud and O. R. Odhner (to Pennwalt Corp.).
153. *Kynar PVDF*, Technical Brochure, Elf Atochem North America, Inc., Philadelphia, Pa., 1990.
154. *Hylar PVDF*, Technical Brochure, Ausimont USA, Inc., Morristown, N.J., 1991.
155. *Solef PVDF*, Technical Brochure, Solvay & Cie. SA, Brussels, Belgium, 1987.
156. U.S. Pat. 4,328,151 (May 4, 1982), D. N. Robinson (to Pennwalt Corp.).
157. *Extrusion of Kynar and Kynar Flex Poly(vinylidene Fluoride) (PVDF)*, Technical Data, Elf Atochem North America Inc., Philadelphia, Pa., Apr. 1990.
158. U.S. Pat. 3,922,186 (Nov. 25, 1975), M. Segawa, Y. Kawakami, and I. Itoh (to Kureha Chemical Co., Ltd.).
159. U.S. Pat. 4,208,462 (June 17, 1980), R. Dauphin and N. Maquet (to Solvay & Cie.).
160. *New Mater. Jpn.,* 2 (Feb. 1990).
161. U.S. Pat. 4,481,158 (Nov. 6, 1984), P. Georlette and N. Maquet (to Solvay & Cie.).
162. U.S. Pat. 4,264,555 (Apr. 28, 1981), E. Lang, W. Nachtigall, and J. Stark (to Dynamit Nobel AG).
163. U.S. Pat. 4,302,556 (Nov. 24, 1981), H. Endo, H. Ohhira, and T. Sasaki (to Kureha Chemical Co., Ltd.).
164. U.S. Pat. 4,051,293 (Sept. 27, 1977), D. F. Wiley (to Cosden Oil & Chemical Co.).
165. U.S. Pat. 4,317,860 (Mar. 2, 1982), A. Strassel (to Atochem SA).
166. Eur. Pat. Appl. 484,053 (May 6, 1992), T. Ozu, K. Hayama, K. Abe, and K. Hata (to Mitsubishi Petrochemical Co., Ltd.).
167. U.S. Pat. 3,340,222 (Sept. 5, 1967), J. C. Fang (to E. I. du Pont de Nemours & Co., Inc.).
168. U.S. Pat. 4,314,004 (Feb. 2, 1982), to R. L. Stoneberg (to PPG Industries, Inc.).
169. U.S. Pat. 4,400,487 (Aug. 23, 1983), R. L. Stoneberg and R. R. Stec (to PPG Industries, Inc.).
170. U.S. Pat. 4,656,768 (Apr. 14, 1987), A. J. Tortorello and C. A. Higginbotham (to DeSoto, Inc.).
171. U.S. Pat. 4,022,737 (May 10, 1977), K. Sekmakas and R. O. Yates (to DeSoto, Inc.).
172. U.S. Pat. 4,141,873 (Feb. 27, 1979), J. E. Dohany (to Pennwalt Corp.).
173. U.S. Pat. 4,309,328 (Jan. 5, 1982), D. W. Carson, R. C. Gray, and G. W. Luckock (to PPG Industries, Inc.).
174. U.S. Pat. 4,383,075 (May 10, 1983), P. T. Abel (to SCM Corp.).
175. U.S. Pat. 4,203,847 and 4,203,848 (May 20, 1980), J. D. Grandine (to Millipore Corporation).
176. A. Bottino, G. Camera-Roda, G. Capannelli, and S. Munari, *J. Membr. Sci.* **57**, 1 (1991).
177. U.S. Pat. 4,340,482 (July 20, 1982), S. Sternberg (to Millipore Corp.).
178. F. F. Stengaard, *J. Membr. Sci.* **36**, 257 (1988); *Desalination* **70**, 207 (1988).
179. U.S. Pat. 4,849,106 (July 18, 1989), L. Mir (to Koch Membrane Systems, Inc.).
180. U.S. Pat. 4,954,256 (Sept. 4, 1990), P. J. Degen, I. Rothman, and T. C. Gsell (to Pall Corporation).
181. U.S. Pat. 5,137,633 (Aug. 11, 1992), D. Wang (to Millipore Corp.).
182. U.S. Pat. 4,399,035 (Aug. 16, 1983), T. Nohmi and T. Yamada (to Asahi Kasei Kogyo Kabushiki Kaisha).
183. U.S. Pat. 3,896,196 (July 22, 1975), C. A. Dickey and J. E. McDaniel (to Glasrock Products, Inc.).
184. U.S. Pat. 4,425,443 (Jan. 10, 1984), P. Georlette and J. Leva (to Solvay & Cie.).

185. U.S. Pat. 4,615,850 (Oct. 7, 1986) and 4,675,345 (June 23, 1987), R. L. Pecsok (to Pennwalt Corp.).
186. M. J. Haley with A. Leder and Y. Sakuma, *Chemical Economics Handbook*, SRI International, Menlo Park, Calif., 1992.
187. L. Manolis Sherman, *Plastics Technol.* **38**(13), 77 (1992).
188. *Chem. Mark. Rep.* **240**(23), 7,26 (1991); *ibid.* **237**(6), 7,21 (1990).
189. R. J. Martino, ed., *Modern Plastics Encyclopedia for '93*, McGraw-Hill Book Co., Inc., New York, 1992, pp. 211–212.
190. U.S. Federal Regulations, Title 21, Chapt. I, Part 177.2510.
191. W. L. Guess, and J. Autian, *J. Oral Therapeut. Pharm.* **3**(2), 116 (1966).
192. D. J. Yturraspe, W. V. Lumb, S. Young, and H. G. Gorman, *J. Neurosurg.* **42**(1), 47 (1975).
193. U.S. Pat. 3,269,862 (Aug. 30, 1966), V. L. Lanza and E. C. Stivers (to Raychem Corp.).
194. *Plastics Design Forum*, (Nov-Dec. 1976).
195. U.S. Pat. 4,318,881 (Mar. 9, 1982); U.S. Pat. 4,591,700 (May 27, 1986), U. K. Sopory (to Raychem Corp.).
196. U.S. Pat. 3,582,457 (June 1, 1971), F. E. Bartell (to Electronized Chemicals Corp.).
197. M. D. Heaven, *Prog. Rubber Plast. Tech.* **2**, 16 (1986).
198. N. Murayama, *J. Polym. Sci. Part A-2* **13**, 929 (1975).
199. M. G. Broadhurst, S. Edelman, and G. T. Davis, *Am. Chem. Soc. Org. Coat. Plast. Chem.* **42**, 241 (1980).
200. H. Meixner and G. Mader, *Phys. Unserer Zeit* **21**(5), 210 (1990).
201. M. U. Anderson and D. E. Wackerbarth, Sandia National Laboratories Report SAND-88-2327; Order No. DE89010529, Albuquerque, N.M., 1988.
202. G. T. Pearman, J. L. Hokanson, and T. R. Meeker, *Ferroelectrics* **28**, 311 (1980).
203. B. Andre, J. Clot, E. Partouche, J. J. Simonne, and F. Bauer, *Sens. Actuators* **A33**, 111 (1992).
204. *Buildings*, 78 (Oct. 1978); *Building Design & Construction*, 134 (May 1983).
205. J. E. Dohany and N. P. Murray, in J. A. Wilkes, ed, *Encyclopedia of Architecture, Design, Engineering and Construction*, John Wiley & Sons, Inc, New York, 1988, p. 478.
206. American Architectural Manufacturers Association, Palatine, Ill., Specification No. AAMA 605.
207. U.S. Pat. 4,770,939 (Sept. 13, 1988) and U.S. Pat. 5,030,394 (July 9, 1991), W. Sietses, T. M. Plantenga, and J.-P. Dekerk (to Labofina SA).
208. *Pulp Pap.* **63**(13), 167 (1989).
209. N. L. Maquet, *Proceedings of the AESF Annual Technical Conference*, Vol. 73, American Electroplater's and Surface Finishers Society, Orlando, Fla., 1986, pp. 1–3.
210. D. K. Heffner, *Mater. Perform.* **31**(7), 33–36 (1992).
211. S. P. Daly, J. E. Dohany, and J. S. Humphrey, *Proceedings, 32nd Annual Conference Institute of Environmental Sciences*, Mt. Prospect, Ill., 1986, p. 397.
212. J. S. Humphrey, J. E. Dohany, and C. Ziu, *1st Annual High Purity Water Conference Proceedings*, Philadelphia, Pa., 1987, p. 135
213. J. M. De Berraly, *Ultrapure Water J.* **4**(4), 36 (1987); *CPI Equip. Reporter*, (July-Aug. 1988).
214. D. Spann, C. Mitchell, and D. A. Toy, *Chem. Process.* 26 (Mar. 1988).
215. *Chem. Mark. Rep.* **239**(4), 49 (1991).
216. U.S. Pat. 4,093,533 (June 6, 1978), R. N. Beaver and C. W. Becker (to The Dow Chemical Company).
217. U.S. Pat. 4,341,596 (July 27, 1982), P. R. Mucenieks (to FMC Corporation).
218. *Chem. Eng.* **98**(7), 17,19 (1991).

219. *Biotech. News*, **10**(12), 6 (1991); *ibid.*, **9**(12), 7 (1990).
220. A. Strassel, *Kunststoffe*, **78**(9), 801 (1988).
221. Eur. Pat. Appl. 419,166 (Mar. 27, 1991), C. Sempio, A. Anghileri, M. Binaghi, T. Ronchetti, and I. Vailati (to Vedril S.p.A.).
222. U.S. Pat. 4,585,701 (Apr. 29, 1986), E. J. Bartoszek and S. F. Mones (to Pennwalt Corp.).
223. U.S. Pat. 4,563,393 (Jan. 7, 1986), Y. Kitagawa, A. Nishioka, Y. Higuchi, T. Tsutsumi, T. Yamaguchi, and T. Kato (to Japan Synthetic Rubber Co., Ltd.).
224. H. Endo and S. Ohira, *Sen-i Gakkaishi* **47**(6), 333 (1991).
225. Jpn. Pat. 92 91,215 (Mar. 24, 1990), Y. Nishikawa, H. Nakada, and T. Sato (to Toray K. K.).
226. Jpn. Pat. 87 25,0217 (Oct. 31, 1987), K. Nakagawa, K. Toma, S. Murakami, and T. Eguchi (to Unitika Ltd.).
227. *Modern Plast.* **69**(6), 133 (1992); *Plast. World*, **50**(7), (1992).
228. Jpn. Pat. 03 13,383 (Jan. 22, 1991), A. Tanaka, Y. Kojima (to Fujitsu Ltd.).
229. Jpn. Pat. 03 01,164 (Jan. 7, 1991), H. Okuno, E. Tominaga, R. Kimura, M. Takeda, and T. Aokit

JULIUS E. DOHANY
Consultant

POLYCHLOROTRIFLUOROETHYLENE

Many challenging industrial and military applications utilize polychlorotrifluoroethylene [*9002-83-9*] (PCTFE) where, in addition to thermal and chemical resistance, other unique properties are required in a thermoplastic polymer. Such has been the destiny of the polymer since PCTFE was initially synthesized and disclosed in 1937 (1). The synthesis and characterization of this high molecular weight thermoplastic were researched and utilized during the Manhattan Project (2). The unique combination of chemical inertness, radiation resistance, low vapor permeability, electrical insulation properties, and thermal stability of this polymer filled an urgent need for a thermoplastic material for use in the gaseous UF_6 diffusion process for the separation of uranium isotopes (see DIFFUSION SEPARATION METHODS).

Properties

The physical properties of PCTFE are primarily determined by a combination of molecular weight and percent crystallinity. Because of the lack of suitable solvents, a correlation between the number average molecular weight and zero-strength time (ZST: typical values of 200 to 400 s) has been developed (3,4). The high molecular weight thermoplastic has a melt temperature (T_m) of 211–216°C, a glass-transition temperature (T_g) of 71–99°C (5), and is thermally stable up to 250°C. The useful operational temperature range is considered to be from -240 to 200°C although an increase in service temperature can be achieved through selected fiber filling of the polymer (fiber glass, from 1 to 20% weight of the fiber).

The theoretical specific gravity of PCTFE for the amorphous and crystalline polymers has been calculated to range from 2.075 to 2.185, respectively (6–12). In reality, PCTFE molded parts have exhibited ranges of crystallinity from approximately 45% (specific gravity of 2.10) for quick-quenched parts to 65% (specific gravity of 2.13) for slow-cooled parts. The use of the terms amorphous and crystalline are relative but can be significant in the application. Basically, two types of crystallinity, micro and macro, exist in the polymer as a result of the synthesis and processing. The higher crystalline forms are less transparent, have higher tensile modulus, lower elongation, and have more resistance to liquids and vapors. The less crystalline form is optically clear, tough, and ductile, exhibiting higher elongation and lower modulus.

The typical mechanical properties that qualify PCTFE as a unique engineering thermoplastic are provided in Table 1; the cryogenic mechanical properties are recorded in Table 2. Other unique aspects of PCTFE are resistance to cold flow due to high compressive strength, and low coefficient of thermal expansion over a wide temperature range.

Table 1. Mechanical Properties of Polychlorotrifluoroethylene

Property	Value
tensile strength, MPa[a]	32–39
compressive strength, MPa[a]	38
modulus of elasticity, MPa[a]	1400
hardness, Shore D	76
deformation under load, at 25°C, 24 h, 7 MPa[a], %	0.3
heat deflection temperature, at 0.46 MPa[a], °C	126

[a]To convert MPa to psi, multiply by 145.

The high fluorine content contributes to resistance to attack by essentially all chemicals and oxidizing agents; however, PCTFE does swell slightly in halogenated compounds, ethers, esters, and selected aromatic solvents. Specific solvents should be tested. PCTFE has the lowest water-vapor transmission rate of any plastic (14,15), is impermeable to gases (see also BARRIER POLYMERS), and does not carbonize or support combustion.

PCTFE plastic is compatible with liquid oxygen, remains ductile at cryogenic temperatures (16–22), and retains its properties when exposed to either uv or gamma radiation. PCTFE exhibits a refractive index of 1.43 (ASTM D542) and an amorphous sheet can provide over 90% transmittance.

PCTFE exhibits very good electrical properties in terms of high insulation resistance, minimal tracking, corona formation, and surface flashover due to the polymer's nonwettable surface and ultralow moisture absorption (Table 3).

Table 2. Cryogenic Mechanical Properties of Polychlorotrifluoroethylene[a]

Property	PCTFE, % crystallinity	Temperature, °C	Value[a]
tensile: ultimate strength, MPa[b]	40	25	38.6
		−129	150
		−252	200
elongation, %	40	25	140
		−129	9
		−252	5
modulus of elasticity, MPa[b]	40	25	1520
		−129	5500
		−252	8700
impact strength notched Izod, J/m[c]	60	25	13.7
		−196	12.8
		−252	13.7

[a]ASTM D1430–89 Type 1, Grade 2 (13).
[b]To convert MPa to psi, multiply by 145.
[c]To convert J/m to ft·lbf/in., divide by 53.38 (see ASTM D256).

Table 3. Electrical Properties of PCTFE

Property	ASTM method	Value
dielectric strength,[a] V/μm	D149	20
arc resistance, s	D495	360
volume resistivity,[b] ohm cm^2/cm	D257	10^{18}
surface resistivity,[b] ohm	D257	10^{15}

[a]Short time. To convert to V/mil, multiply by 25.
[b]Fifty percent rh at 25°C.

Manufacture and Processing

The synthesis of the high molecular weight polymer from chlorotrifluoroethylene [79-38-9] has been carried out in bulk (23–27), solution (28–30), suspension (31–36), and emulsion (37–41) polymerization systems using free-radical initiators, uv, and gamma radiation. Emulsion and suspension polymers are more thermally stable than bulk-produced polymers. Polymerizations can be carried out in glass or stainless steel agitated reactors under conditions (pressure 0.34–1.03 MPa (50–150 psi) and temperature 21–53°C) that require no unique equipment.

After polymerization, the polymer is isolated from the latex or suspension. The suspension polymer, already in powder form, is washed to remove initiator residues and then dried. The emulsion polymer is coagulated from the latex by freezing or by the addition of salts, acids, and solvents (see LATEX TECHNOLOGY) and separated from the aqueous phase. The isolated powder is then washed and dried. The dried powder from either process additionally can be chemically treated to remove trace impurities that can result in chain degradation during further processing. Treatment with carboxylic acids (42), ozone in air (43), or chlorine (44) improves thermal stability, color, and light transmission of the final polymer. The polymer product can then be processed by plastic fabrication techniques in powder or melt-extruded pellet forms.

The lower molecular weight oils, waxes, and greases of PCTFE can be prepared directly by telomerization of the monomer or by pyrolysis of the higher molecular weight polymer (45–54).

PCTFE plastics can be processed by the standard thermoplastic fabrication techniques, eg, extrusion, injection, compression, and transfer molding. Specific corrosion-resistant alloys or chrome or nickel plating are recommended for equipment parts in contact with the polymer melt, such as molds, barrels, screws, etc (see PLASTICS TECHNOLOGY). The control of processing temperatures is paramount since prolonged overheating (above 260°C) can result in degradation of the polymer causing discoloration, voids, blisters, and loss of properties. The plastic can be easily machined from billets or rod stock on standard machining equipment to fabricate more precise part geometries, but sharp tools should be employed.

Economic Aspects

Several worldwide commercial manufacturers of PCTFE and vinylidene fluoride-modified copolymers [9010-75-7] offer a variety of products as shown in Table 4. PCTFE plastics have selling prices in the range of $40–100/kg, depending on the molecular weight, grade, product form, and supplier. As a result, PCTFE thermoplastics are used in high technology, specialty engineering areas where the unique combination of properties and part reliability demands a high performance thermoplastic polymer.

Table 4. PCTFE Manufacturers and Products

Trademark	Manufacturer	Product forms
homopolymers		
Daiflon	Daikin Koygo, Osaka, Japan	molding powder, pellets, dispersion oils, and greases
Kel-F 81	3M Co., St. Paul, Minn.	molding powders and pellets
Voltalef	Ugine Kuhlmann, Pierre-Benite, France	molding powders and pellets
Halocarbon oil	Halocarbon Products Corp., River Edge, N.J.	oils, waxes, and greases
copolymers		
Aclon, Aclar	Allied-Signal Chemical, Morristown, N.J.	molding powders, pellets, and film
Kel-F 800	3M Co., St. Paul, Minn.	molding powders

Specifications and Test Methods

PCTFE plastic is available in products that conform to ASTM 1430-89 Type I (Grades 1 and 2) and is suitable for processing into parts that meet MIL-P 46036

(Federal Specification LP-385C was canceled 1988). Standards for fabricated forms are available for compression molded heavy sections (AMS-3645 Class C), thin-walled tubing, rod, sheet, and molded shapes (AMS-3650). PCTFE plastics have been approved for use in contact with food by the FDA (55).

The test methods employed are the determination of molecular weight as measured by ZST (ASTM D1430); specific gravity (ASTM D792); tensile strength, elongation, and modulus (ASTM D638); compressive strength and modulus (ASTM D621); heat deflection (ASTM D648); impact strength (ASTM D256); flammability (ASTM D2863); hardness (ASTM D2240 and D785); and coefficient of linear expansion (ASTM D696).

Health and Safety Factors

In general, the PCTFE resins have been found to be low in toxicity and irritation potential under normal handling conditions. Specific toxicological information and safe handling procedures are provided by the manufacturer of specified PCTFE products upon request.

Uses

The principal uses of PCTFE plastics remain in the areas of aeronautical and space, electrical/electronics, cryogenic, chemical, and medical instrumentation industries. Applications include chemically resistant electrical insulation and components; cryogenic seals, gaskets, valve seats (56,57) and liners; instrument parts for medical and chemical equipment (58), and medical packaging; fiber optic applications (see FIBER OPTICS); seals for the petrochemical/oil industry; and electrodes, sample containers, and column packing in analytical chemistry and equipment (59).

The lower molecular weight PCTFE oils, waxes, and greases are used as inert sealants and lubricants for equipment handling oxygen and other oxidative or corrosive media. Other uses include gyroscope flotation fluids and plasticizers for thermoplastics.

BIBLIOGRAPHY

"Polychlorotrifluoroethylene" under "Fluorine Compounds, Organic" in *ECT* 3rd ed., Vol. 11, pp. 49–54, by A. C. West, 3M Co.
1. Brit. Pat. 465,520 (May 3, 1937), (to I. G. Farbenindustrie).
2. U.S. Pat. 2,564,024 (Aug. 14, 1951), W. T. Miller (to USAEC).
3. H. S. Kaufman, C. O. Kroncke, and C. K. Giannotta, *Mod. Plast.* **32**, 146 (1954).
4. E. K. Walsh and H. S. Kaufman, *J. Polym. Sci.* **26**, 1 (1957).
5. Y. P. Khanna and R. Kumar, *J. Polymer Sci.* **32**(11) 2010–2013 (1991).
6. T. Hashimoto, H. Kawasaki, and H. Kawai, *J. Polymer Sci.* **16**(2) 271–288 (1978).
7. E. Sacher, *J. Polymer Sci.* **18**(5), 333–337 (1980).
8. H. Matsuo, *J. Polymer Sci.* **21**, 331 (1956).

9. H. Matsuo, *J. Polymer Sci.* **25**, 234 (1957).
10. H. Matsuo, *Bull. Chem. Soc. Jpn.* **30**, 593 (1957).
11. J. D. Hoffman, *J. Am. Chem. Soc.* **74**, 1696 (1952).
12. J. D. Hoffman and J. J. Weeks, *J. Res. Nat. Bur. Stand.* **60**, 465 (1958).
13. R. E. Mowers *Cryogenic Properties of Poly(Chlorotrifluoroethylene), Technical Document Report No. RTD-TDR-63-11*, Air Force Contract No. AF04(611)-6354, 1962.
14. J. D. Hoffman and J. J. Weeks, *J. Chem. Phys.* **37**, 1723 (1962).
15. A. W. Myers and co-workers, *Mod. Plast.* **37**, 139 (1960).
16. N. Brown, B. D. Metzger, and Y. Imai, *J. Polym. Sci.* **16**, 1085 (1978).
17. Y. Imai and N. Brown, *Polymer* **18**, 298 (1977).
18. N. Brown and S. Fischer, *J. Poly. Sci. Polym. Phys.* **13**, 1315 (1975).
19. J. L. Currie, R. S. Irani, and J. Sanders, *Factors Affecting the Impact Sensitivity of Solid Polymer Materials in Contact with Liquid Oxygen*, ASTM Spec. Tech. Publ. 986, ASTM, Philadelphia, Pa., 1988, pp. 233–247; S. Chandrasekaran, in J. I. Kroschwitz, ed., *Encyclopedia of Polymer Science and Engineering*, 2nd ed., Vol. 3, John Wiley & Sons, Inc., 1985, pp. 463–480.
20. N. Schmidt and co-workers, *Ignition of Nonmetallic Materials by Impact of High-pressure Oxygen*, ASTM Spec. Tech. Publ. 1040, ASTM, Philadelphia, Pa., 1989, pp. 23–37.
21. B. J. Lockhart, M. D. Hampton, and C. J. Bryan, *The Oxygen Sensitivity / Compatability Ranking of Several Materials by Different Test Methods*, ASTM Spec. Tech. Publ. 1040, ASTM, Philadelphia, Pa., 1989, pp. 93–105.
22. M. L. Reath and R. S. Britton, *Plastics Manufacture and Processing*, 37-5 (1991).
23. U.S. Pat. 2,586,550 (Feb. 19, 1952), W. T. Miller, A. L. Dittman, and S. K. Reed (to USAEC).
24. U.S. Pat 2,792,377 (May 14, 1957), W. T. Miller (to 3M).
25. U.S. Pat. 2,636,908 (Apr. 28, 1953), A. L. Dittman and J. M. Wrightson (to M. W. Kellogg Co.).
26. Brit. Pat. 729,010 (Apr. 27, 1955), (to Farbenfabriken Bayer AG).
27. Fr. Pat. 1,419,741 (Dec. 3, 1965), (to Kureha Chemical Co.).
28. U.S. Pat. 2,700,662 (Jan. 25, 1955), D. M. Young and B. Thompson (to Union Carbide Co.).
29. U.S. Pat. 2,820,027 (Jan. 14, 1958), W. F. Hanford (to 3M).
30. M. Lazar, *J. Polym. Sci.* **29**, 573 (1958).
31. U.S. Pat. 2,613,202 (Oct. 7, 1952), G. F. Roedel (to General Electric Co.).
32. U.S. Pat. 2,600,202 (June 10, 1952), D. W. Caird (to General Electric Co.).
33. Fr. Pat. 1,155,143 (Apr. 23, 1958), (to Society d'Ugine).
34. U.S. Pat. 2,842,528 (July 8, 1958), R. L. Herbst and B. F. Landrum (to 3M).
35. U.S. Pat. 2,689,241 (Sept. 14, 1954) A. L. Dittman, H. J. Passino, and J. M. Wrightson, (to M. W. Kellogg Co.).
36. J. M. Hamilton, *Ind. Eng. Chem.* **45**, 1347 (1953).
37. U.S. Pat. 2,569,524 (Oct. 21, 1951), J. M. Hamilton (to E. I. du Pont de Nemours & Co., Inc.).
38. U.S. Pat. 2,744,751 (Dec. 19, 1956), H. J. Passino and co-workers (to M. W. Kellogg Co.).
39. Brit. Pat. 840,735 (July 6, 1960), F. Fahnoe and B. F. Landrum (to 3M).
40. U.S. Pat. 2,559,749 (July 10, 1951), A. F. Benning (to E. I. du Pont de Nemours & Co., Inc.).
41. U.S. Pat. 2,559,749 (July 10, 1951), K. L. Berry (to E. I. du Pont de Nemours & Co., Inc.).
42. U.S. Pat. 2,751,376 (July 19, 1956), R. M. Mantell and W. S. Barnhart (to 3M).
43. U.S. Pat. 2,902,477 (Sept. 1, 1959), E. Fischer, K. Weissermel, and G. Bier (to 3M).

44. U.S. Pat. 3,045,000 (July 17, 1962), R. R. Divis.
45. U.S. Pat. 2,770,659 (Nov. 13, 1956), W. S. Barnhart (to M. W. Kellogg Co.).
46. U.S. Pat. 2,786,827 (Mar. 26, 1957), W. S. Barnhart (to M. W. Kellogg Co.).
47. U.S. Pat. 2,664,449 (Dec. 29, 1953), W. T. Miller (to USAEC).
48. U.S. Pat. 2,902,477 (Apr. 28, 1953), W. T. Miller (to M. W. Kellogg Co.).
49. U.S. Pat. 2,636,908 (Apr. 28, 1953), A. L. Dittman and J. M. Wrighton (to M. W. Kellogg Co.).
50. U.S. Pat. 2,706,715 (Apr. 19, 1955), R. C. Conner (to M. W. Kellogg Co.).
51. U.S. Pat. 2,716,141 (Aug. 23, 1955), W. T. Miller (to 3M).
52. U.S. Pat. 2,854,490 (Sept. 30, 1958), E. Fischer and H. Frey (to Farbwerke Hoechst).
53. U.S. Pat. 2,992,988 (July 18, 1961), C. D. Dipner (to 3M).
54. U.S. Pat. 3,076,765 (Feb. 5, 1963), F. W. West, R. J. Seffl, and L. J. Reilly (to 3M).
55. *Code of Federal Regulations*, Title 21, Paragraph 177.1380, U. S. Government Printing Office, Washington, D.C., revised Apr. 1, 1979, p. 604.
56. W. Broadway, "A Pressure Sensitivity and Temperature Response Butterfly Valve for Cryogenic Service," paper presented at *Energy Technology Conference and Exhibition*, Houston, Nov. 5–9, 1978.
57. W. Broadway, "Development of Cryogenic Butterfly Valve Seat," paper presented at *AIChE 71st Annual Meeting, Cryogenic Equipment Session*, Miami, Nov. 12–16, 1978.
58. J. E. Harrar and R. J. Sherry, *Anal. Chem.* **47**, 601 (1975).
59. S. L. Petersen and D. E. Tallman, *Anal. Chem.* **62**(5) 459–465 (1990).

General References

R. P. Bringer, "Influence of Unusual Environmental Conditions on Fluorocarbon Plastics," paper presented at *SAMPE (Society of Aerospace Material and Process Engineers) Symposium*, St. Louis, Mo., May 7–9, 1962.
R. P. Bringer and C. C. Solvia, *Chem. Eng. Prog.* **56**(10), 37 (1960).
R. E. Schawmm, A. F. Clark, and R. P. Reed, *A Compilation and Evaluation of Mechanical, Thermal and Electrical Properties of Selected Polymers*, NBS Report, AEC SAN-70-113, SANL 807 Task 7, SANL Task 6, National Technical Information Service, U.S. Dept. of Commerce, Springfield, Va., Sept. 1973, pp. 335–443.
C. A. Harper, ed., *Handbook of Plastics Elastomers, and Composit*, 2nd ed. McGraw-Hill Book Co., New York, 1992.
W. T. Miller, "General Discussion of Chlorotrifluoroethylene Polymers," in C. Slesser and S. R. Schram, eds., *Preparation, Properties and Technology of Fluorine and Organic Fluoro Compounds*, McGraw-Hill Book Co., New York, 1951.
L. A. Wall, ed., *Fluoropolymers*, Vol. 25, Wiley-Interscience, New York, 1992, Chapts. 15 and 16.
S. Chandrasekaran, "Chlorotrifluoroethylene Homopolymer" under "Chlorotrifluoroethylene Polymers," in J. I. Kroschwitz, ed., *Encyclopedia of Polymer Science and Engineering*, 2nd ed., Vol. 3, John Wiley & Sons, Inc., New York, 1985, pp. 463–480.

G. H. MILLET
J. L. KOSMALA
3M Company

BROMOTRIFLUOROETHYLENE

Bromotrifluoroethylene is a valuable reagent for the synthesis of trifluorovinylic compounds by means of its intermediate organometallic compounds.

Physical Properties

The monomer, bromotrifluoroethylene [598-73-2], CF_2=CFBr, is a colorless gas; bp $-3.0°C$ at 101 kPa (754 mm Hg); 58°C at 790 kPa (100 psig); and d_4^{25} 1.86 g/cm^3. Since it is spontaneously flammable in air, its odor is that of its oxidation products, mixed carbonyl halides. The olefin can be distilled, but it polymerizes on standing at ambient temperature unless an inhibitor such as 0.1% tributylamine is added. If desired, the inhibitor can be readily removed by passing the gas through silica gel. Higher temperatures or uv light increase the polymerization rate. The nmr (1,2), ir (3), uv (4), and photoelectron spectra (5) of the monomer have been reported, and some thermochemical data have been calculated for it (3). Its dipole moment has been determined to be 2.54×10^{-30} C·m (0.76 D) (6).

Chemical Properties

Many reactions of bromotrifluoroethylene have been studied. Under basic conditions it adds alcohols such as methanol (7,8) or ethanol (8,9), forming ethers with the general formula $ROCF_2CFBrH$. Similarly, diethylamine adds to it giving $(C_2H_5)_2NCF_2CFBrH$ (10). This addition is faster than diethylamine additions to tetrafluoroethylene or chlorotrifluoroethylene. Vapor-phase photochemical bromination of bromotrifluoroethylene gives the expected adduct, $CF_2BrCFBr_2$ (11). On the other hand, photochemical chlorination results in only 60% of the expected adduct and 40% scrambled bromo and chloro products. Hydrogenation of the double bond can be accomplished either catalytically (12) or with sodium borohydride (13). The former method also gives some trifluoroethylene and the latter method gives only a 37% yield of the pure product, $CF_2HCFBrH$. Other reagents that add across the double bond include S_2Cl_2, to give mainly a disulfide (14), and aqueous sodium nitrite (15).

Another class of reactions that bromotrifluoroethylene undergoes is cycloaddition with acetylenes (16) or olefins (17). When heated, the pure monomer, in addition to polymerizing, dimerizes to cis- and trans-1,2,-dibromohexafluorocyclobutane, which can be debrominated with zinc to perfluorocyclobutene [697-11-0] (18). Bromotrifluoroethylene is a valuable reagent for the synthesis of trifluorovinylic compounds by means of its intermediate organometallics. Trifluorovinylzinc (19) and trifluoromagnesium bromides (20,21) may be prepared directly using the metals, whereas trifluorovinyllithium [683-78-3] results from the metathesis metal–bromine exchange (22). The lithium and magnesium derivatives are thermally unstable but readily convert aldehydes, ketones, and carbon dioxide to alcohols (22,23) and trifluoroacrylic acid [433-68-1] (20). The zinc derivative transforms iodo- or bromobenzenes to α,β,β-trifluorostyrenes in the presence of zero valent palladium complexes (24) and participates in a variety of alkylation, coupling, and acylation reactions upon the addition of cuprous salts (25).

Manufacture

Although bromotrifluoroethylene was first prepared (9) by the dehydrobromination of 1,2-dibromo-1,1,2-trifluoroethane [354-04-1], it is more conveniently prepared from chlorotrifluoroethylene [79-38-9] by the following high yield steps (26):

$$CF_2{=}CFCl + HBr \rightarrow CF_2BrCFClH \xrightarrow{Zn} CF_2{=}CHF + ZnBrCl$$

$$CF_2{=}HF + Br_2 \rightarrow CF_2BrCHFBr \xrightarrow{KOH} CF_2{=}CFBr + KBr + H_2O$$

Bromotrifluoroethylene is manufactured and sold in commercial quantities with a purity of 99.9% by the Halocarbon Products Corp. for about $100/kg as of 1992.

Polymers

The olefin can be polymerized in trichlorofluoromethane solution at $-5°C$ for 7 days with a halogenated acetyl peroxide such as 0.037% trichloroacetyl peroxide as the initiator (27). Alternatively, it may be polymerized in an aqueous suspension with 2 parts by weight distilled water, 0.01 part ammonium persulfate, 0.004 part sodium bisulfite, and 0.001 part hydrated ferrous sulfate present for each part of the monomer. Mixing for 24.5 h at 20°C gives a 52% conversion of the monomer to the homopolymer (27). Prepared either way, the homopolymer [55157-25-0] is a white powder soluble in acetone and useful as a hard, chemically resistant coating for metal or fabric surfaces. The addition of small amounts of chain-transfer agents such as chloroform, carbon tetrachloride, bromotrichloromethane, and in particular, 0.01 to 0.3% 1-dodecanethiol to the polymerization mixture gives a lower molecular weight homopolymer that is softer and more soluble. Copolymers of bromotrifluoroethylene with many other monomers such as chlorotrifluoroethylene (28), tetrafluoroethylene (29), or trifluoronitrosomethane (30) have been reported. Neither the homopolymer nor the copolymers have any commercial utility.

Telomers. Bromotrifluoroethylene telomers have been prepared using chain-transfer agents such as CF_3SSCF_3 (31), C_2F_5I (32), CBr_4 (32), or CBr_3F (33). For example, when the olefin is slowly added to tribromofluoromethane under light from sunlamps, a liquid is obtained which, after saturation and distillation, has a viscosity of 510 mm²/s($=$cSt) and a density of 2.65 g/cm³ at 58.3°C. These and other bromotrifluoroethylene telomers are useful as flotation agents and damping fluids for gyroscopes and accelerometers in inertial guidance systems. These telomers are noncrystalline up to a higher degree of polymerization than those of chlorotrifluoroethylene, allowing the preparation of liquids of greater viscosity. The higher densities of the bromine-containing oils permit further miniaturization of the instruments floating in them. For these reasons, these oils complement their less expensive chloro analogues in this application. Commercially available bromotrifluoroethylene telomers have densities of 2.14–2.65 g/cm³ and viscosities of 2–4000 mm²/s(= cSt). These fluids are expensive but are made in small volume for the aerospace industry.

Toxicity

Rats exposed to 500 ppm of bromotrifluoroethylene died following a 4-h exposure. Since the monomer decomposes in air, the level of exposure to it was actually lower. The effects in rats of repeated exposure over a two-week period have been studied. At 50 ppm, the animals lost weight and renal damage was noted although the effect was reversible. Very mild testicular damage was seen at 50 but not 10 ppm. The amount of urinary fluoride excreted suggested that extensive metabolism was occurring (34).

BIBLIOGRAPHY

"Polybromotrifluoroethylene" under "Fluorine Compounds, Organic," in *ECT* 2nd ed., Vol. 9, pp. 833–835, by L. L. Ferstandig, Halocarbon Products Corp.; in *ECT* 3rd ed., Vol. 11, pp. 54–56, by G. Astrologes, Halocarbon Products Corp.

1. D. D. Elleman and S. L. Manatt, *J. Chem. Phys.* **36**, 1945 (1962).
2. J. Reuben, Y. Shvo, and A. Demiel, *J. Am. Chem. Soc.* **87**, 3995 (1965).
3. D. E. Mann, N. Acquista, and E. K. Plyler, *J. Chem. Phys.* **22**, 1199 (1954).
4. J. Schander and B. R. Russel, *J. Mol. Spectrosc.* **65**, 379 (1977).
5. K. Wittel and H. Bock, *Chem. Ber.* **107**, 317 (1974).
6. E. J. Gauss and T. S. Gilman, *J. Phys. Chem.* **73**, 3969 (1969).
7. U.S. Pat. 3,666,864 (May 30, 1972), R. C. Terrel (to Airco, Inc.).
8. A. Demiel, *J. Org. Chem.* **25**, 993 (1960).
9. F. Swarts, *Chem. Zentr.* **II**, 281 (1899).
10. R. N. Sterlin and co-workers, *Izv. Akad. Nauk SSSR Otd. Khim. Nauk.* **4**, 8,22 (1962).
11. J. R. Lacher, R. D. Burkhart, and J. D. Park, *Univ. Colo. Stud. Ser. Chem. Pharm.* **4**, 8,22 (1962).
12. J. R. Lacher and co-workers, *J. Phys. Chem.* **61**, 1125 (1957).
13. A. L. Anderson, R. T. Bogan, and D. J. Burton, *J. Fluorine Chem.* **1**, 121 (1971).
14. U.S. Pat. 2,451,411 (Oct. 12, 1948), M. S. Raasch (to E. I. du Pont de Nemours & Co., Inc.).
15. A. M. Krzhizhevskii, Y. A. Cheburkov, and I. L. Knunyants, *Izv. Akad. Nauk SSSR Ser. Khim.*, 2144 (1974).
16. J. C. Blazejewski, D. Cantacuzene, and C. Wakselman, *Tetrahedron Lett.*, 2055 (1974).
17. U.S. Pat. 3,954,893 (May 4, 1976), G. J. O'Neill, R. S. Holdsworth, and C. W. Simons (to W. R. Grace and Co.).
18. W. R. Cullen and P. Singh, *Can. J. Chem.* **41**, 2397 (1963).
19. S. W. Hansen, T. D. Spawn, and D. J. Burton, *J. Fluorine Chem.* **35**, 415 (1987).
20. J. L. Knunyants and co-workers, *Izv. Akad. Nauk SSSR Otdel Khim. Nauk.*, 1345 (1958).
21. H. D. Kaesz, S. L. Stafford, and F. G. A. Stone, *J. Am. Chem. Soc.* **81**, 6336 (1959).
22. P. Tarrant, P. Johncock, and J. Savory, *J. Org. Chem.* **28**, 839 (1963).
23. D. D. Denson, C. F. Smith, and C. Tamborski, *J. Fluorine Chem.* **3**, 247 (1974).

24. P. L. Heinze and D. J. Burton, *J. Org. Chem.* **53**, 2714 (1988).
25. D. J. Burton and S. W. Hansen, *J. Am. Chem. Soc.* **108**, 4229 (1986).
26. J. D. Park, W. R. Lycan, and J. R. Lacher, *J. Am. Chem. Soc.* **73**, 711 (1951).
27. U.S. Pat. 2,793,202 (May 21, 1957), J. M. Hoyt (to M. W. Kellogg Co.).
28. Brit. Pat. 593,605 (Oct. 21, 1947), (to E. I. du Pont de Nemours & Co., Inc.).
29. S. Afr. Pat. 69 05,518 (Mar. 3, 1970), J. Kuhls, H. Hahn, and A. Steininger (to Farbwerke Hoechst AG).
30. P. Tarrant, E. C. Stump, Jr., and C. D. Padgett, *Polym. Prepr. Am. Chem. Soc. Div. Polym. Chem.* **12**, 391 (1971).
31. R. E. A. Bear and E. E. Gilbert, *J. Fluorine Chem.* **4**, 107 (1974).
32. Ger. Pat. 2,235,885 (Feb. 7, 1974), J. Kuhls, H. Fitz, and P. Haasemann (to Farbwerke Hoechst AG).
33. U.S. Pat. 3,668,262 (June 6, 1972), A. L. Dittman (to Halocarbon Products Corp.).
34. G. L. Kennedy, Jr., *CRC Crit. Rev. Toxicol.* **21**, 149 (1990).

ARTHUR J. ELLIOTT
Halocarbon Products Corporation

POLY(FLUOROSILICONES)

The presence of carbon–fluorine bonds in organic polymers is known to characteristically impart polymer stability and solvent resistance. The poly(fluorosilicones) are siloxane polymers with fluorinated organic substituents bonded to silicon. Poly(fluorosilicones) have unique applications resulting from the combination provided by fluorine substitution into a siloxane polymer structure (see SILICON COMPOUNDS, SILICONES).

The incorporation of a single carbon–fluorine bond into a polymer cannot provide the stability and solvent resistance offered by multiple bonds or clusters of carbon–fluorine bonds available with substituents like the CF_3, C_2F_5, or C_3F_7 groups. Therefore, commercially interesting poly(fluorosilicones) have at least one CF_3 group per silicon in their structure. The proximity of silicon and fluorine in such compounds governs the stability of the structure. If fluorine is alpha to silicon the compounds are subject to thermal rearrangement (eq. 1). The thermodynamically more stable silicon–fluorine bond and difluoromethylene by-product form. The by-product undergoes further chemistry characteristic of divalent carbon compounds.

$$CF_3-\underset{\underset{R''}{|}}{\overset{\overset{R}{|}}{Si}}-R' \rightarrow F-\underset{\underset{R''}{|}}{\overset{\overset{R}{|}}{Si}}-R' + [:CF_2] \qquad (1)$$

When fluorine is beta to silicon, compounds undergo a facile elimination of an ethylenic compound and again form the stable silicon–fluorine bond (eq. 2).

$$CF_3CH_2-\underset{\underset{R''}{|}}{\overset{\overset{R}{|}}{Si}}-R' \rightarrow F-\underset{\underset{R''}{|}}{\overset{\overset{R}{|}}{Si}}-R' + CF_2=CH_2 \qquad (2)$$

Structures with the widest temperature range of demonstrated stability have fluorine in the gamma position relative to silicon (or further removed), as in $CF_3CH_2CH_2SiRR''R''''$. Longer hydrocarbon chains, with or without hetero atoms, are feasible, but oxidative stability is compromised and such materials are generally disfavored. Poly(3,3,3-trifluoropropyl)methylsiloxane [26702-40-9] demonstrates this structural principle. This polymer is one key member of the industrially important family of fluorosilicone materials.

Properties

Fluorosilicone elastomers can be formulated to provide specific durometer (hardness), tear strength, modulus, and solvent resistance properties (1). The specific gravity is 1.35 to 1.65. Durometer variation in the range of 20–80 Shore A-2 are attainable by formulation methods. Materials designed to resist tearing exhibit a tear strength of up to 52.5 kN/m (300 ppi). Modulus at 100% elongation can range from 0.4–6.2 MPa (60–900 psi). Elongation is 150–500%. Compression set as low as 6% (22 h/177°C) can be achieved, with the upper end being about 35%. Tensile strength in the range of 5.5–12.4 MPa (800–1800 psi) has been reported. Rubber resiliency (Bashore) is 14–30%.

Fluid and Chemical Resistance. Fluorosilicone elastomers and greases are especially suited for applications involving repeated exposure to fuels, oils, hydraulic fluids, and various chemicals (1). Fluid resistance is excellent to almost all solvents including alcohol–hydrocarbon mixtures currently being evaluated as alternative fuels. Even at elevated temperatures, prolonged immersion causes only slight elastomer swelling. Exceptions to this rule are highly polar solvents, such as esters and ketones.

Heat Resistance. Fluorosilicone elastomers have long-lasting dependability in static and dynamic applications over a wide range of temperature. Thermal cycling does not lead to embrittlement. Elastomer service temperatures range from −60 to 200°C. Some elastomers have service temperature excursions allowable (for hours) up to 250°C with little change in hardness; brief exposures up to 260°C result in retention of about 50% of the original tensile strength.

Low Temperature Properties. The property of solvent resistance makes fluorosilicone elastomers useful where alternative fluorocarbon elastomers cannot function. The ability to retract to 10% of their original extension after a 100% elongation at low temperature is an important test result. Fluorosilicones can typically pass this test down to −59°C. The brittle point is approximately −68°C.

Electrical Properties. Like unfluorinated silicone counterparts, fluorosilicone elastomers have inherently good electrical insulating properties. The dielectric properties remain relatively unchanged when the elastomer is exposed to severe environments.

Manufacture

Monomer Production. The key industrial monomer is 2,4,6-trimethyl-2,4,6-tris-(3,3,3-trifluoropropyl)cyclotrisiloxane [2374-14-3], which is produced by the hydrosilylation of 3,3,3-trifluoropropene [677-21-4] with methyldichlorosilane [75-54-7], catalyzed by various platinum and other noble metal compounds (eq. 3).

$$CF_3CH=CH_2 + CH_3SiHCl_2 \rightarrow CF_3CH_2CH_2Si(CH_3)Cl_2 \tag{3}$$

The preparation of 3,3,3-trifluoropropene in high yield has been described (2). The hydrosilylation reaction can also be conducted using peroxides, radiation, or photochemical means. The hydrosilylation product, 3,3,3-trifluoropropylmethyldichlorosilane [675-62-7], is hydrolyzed with water to form a hydrolyzate siloxane mixture of cyclic siloxanes and linear hydroxyl end blocked siloxanes (eq. 4). This hydrolyzate mixture is washed to remove residual acid and then made basic with sodium hydroxide or potassium hydroxide. Distillation of the resulting mixture under reduced pressure affords the cyclotrisiloxane monomer as the lowest boiling siloxane. Base catalysis continues to rearrange both linear and cyclic materials in the distillation pot to re-form additional cyclotrisiloxane, which can then be recovered until nearly all the siloxane material is converted to useable monomer.

$$CF_3CH_2CH_2Si(CH_3)Cl_2 \longrightarrow \text{hydrolyzate} \longrightarrow \tag{4}$$

Preparation of other fluorosilicone monomers follows methods similar to that described above. For example, 2,4,6-trimethyl-2,4,6-tris(3,3,4,4,5,5,6,6,6-nonafluorohexyl)cyclotrisiloxane [38521-58-3] is produced from 3,3,4,4,5,5,6,6,6-nonafluorohexene [19430-93-4] and methyldichlorosilane in three steps (3).

Polymerization. Cyclotrisiloxanes are strained ring compounds. Polymerization is driven by relief of this ring strain. Acid- or base-catalyzed equilibration reactions of cyclotrisiloxane with a measured amount of end-blocking agent lead to fluid polymers with predictable molecular weight distributions. The amount of added end blocking agent controls polymer chain length. Various polar, aprotic solvents are known promoters, including tetrahydrofuran, acetonitrile, and dimethylformamide. The reactions with metallic base catalysts (MB) adhere to the following order of decreasing rate: K > Na > Li. After neutralization or removal of the catalyst, fluid polymers are used as is or to formulate grease compounds. If a reactive end blocking group was incorporated during polymerization, the fluid polymer is then useful for making a coating, a sealant, or a liquid rubber product.

High molecular weight polymers or gums are made from cyclotrisiloxane monomer and base catalyst. In order to achieve a good peroxide-curable gum, vinyl groups are added at 0.1 to 0.6% by copolymerization with methylvinylcyclosiloxanes. Gum polymers have a degree of polymerization (DP) of about 5000 and are useful for manufacture of fluorosilicone rubber. In order to achieve the gum state, the polymerization must be conducted in a kinetically controlled manner because of the rapid depolymerization rate of fluorosilicone. The expected thermodynamic end point of such a process is the conversion of cyclotrisiloxane to polymer and then rapid reversion of the polymer to cyclotetrasiloxane [429-67-4]. Careful control of the monomer purity, reaction time, reaction temperature, and method for quenching the base catalyst are essential for reliable gum production.

Compounding. Fluorosilicone gums are compounded generally with fumed or precipitated silica fillers, hydroxy-containing low viscosity silicone oils, and readily available peroxides to produce various rubber products.

These rubber products are based on high molecular weight ($M_w > 700,000$) polymers, but because of the freedom of rotation of the silicon–oxygen bond they are soft and easily processed in conventional mixers, water-cooled mills and calenders. Gum plasticities are generally 2.3–3.6 mm and fully compounded bases are in the 2.5–5.1 mm range. Processing characteristics can be modified with hydroxy-containing low viscosity silicone oils, silicone gums, 1–10 parts of polydimethylsilicone softeners, and 1–50 parts of fumed or precipitated silica fillers. Alternatively, fillers based on diatomaceous earth or crushed quartz can be utilized. Most fluorosilicone rubbers can also be colored by the addition of pigments.

Vulcanization. Fluorosilicone elastomers can be peroxide-vulcanized by a free-radical mechanism using vinyl side groups that have been incorporated into the basic polymer structure during the initial polymerization process. Peroxide initiated cross-linking results in a carbon–carbon bond formed by the reaction of free radicals. The free radicals are generated in the polymer via the peroxide radicals, either by abstracting hydrogen from a methyl group or by adding to a vinyl group.

During the vulcanization, the volatile species formed are by-products of the peroxide. Typical cure cycles are 3–8 min at 115–170°C, depending on the choice of peroxide. With most fluorosilicones (as well as other fluoroelastomers), a post-cure of 4–24 h at 150–200°C is recommended to maximize long-term aging properties. This post-cure completes reactions of the side groups and results in an increased tensile strength, a higher cross-link density, and much lower compression set.

Another form of vulcanization is the "addition reaction" of one polymer containing a SiH functionality with a second polymer containing a vinyl functionality. These two polymers are made individually by copolymerization methods and then formulated prior to vulcanizing into two-part or inhibited one-part systems. The vulcanization reaction is usually catalyzed by platinum compounds or other noble metal compounds. Heat may or may not be employed, depending on the catalyst or inhibitor selected. The resulting —CH_2CH_2— linkage between siloxane chains vulcanizes the polymers into the desired elastomeric form. Many liquid silicone rubber (LSR) systems use this type of vulcanization for the rapid-cycle injection molding of fluorosilicone parts.

Fabrication. Fluorosilicones can be molded, extruded, or calendered by any of the conventional methods employed in the industry. Compression molding is the most widely used method and is ideal for a great many fabrications at 115–170°C and 5.5–10.3 MPa (800–1500 psi). Injection molding becomes increasingly important for high production operations and generally requires higher temperatures and pressures than compression molding. Transfer-press molding is particularly useful for molding complex parts in a multicavity press. Where dimensional accuracy of molded parts is important, shrinkage of the parts must be considered in the design of the mold. Linear shrinkage of most fluorosilicones is 2.5–3.5%.

Extrusion techniques are used to make tubes, rods, gaskets, preforms, etc. Standard rubber equipment may be used to extrude fluorosilicone elastomers. The green strength of fluorosilicones is less than that of typical fluorocarbon elastomers, and this should be considered when designing the feed system.

Calendering is used to produce long, thin sheets of fluorosilicone elastomers and to coat fluorosilicones on reinforcing substrates, eg, certain polymers and metals, to provide the protection of fluorosilicones at a minimal cost. When very thin films are desired or when the vulcanizing temperature of typical fluorosilicones is too high for the supporting substrate, room temperature vulcanizing (RTV) dispersions are commonly available. Dispersions of fluorosilicone gums, bases, and fully compounded stocks, with or without catalysts, can be used to coat many fibrous metal or polymeric substrates. It is necessary to maintain dispersion neutrality to maximize the physical properties of the resulting coating. Usually Silastic A-4040 Primer is recommended to prepare a dense polymeric or metal substrate surface for bonding.

Economic Aspects

Globally, there is a small number of basic fluorosilicone producers: General Electric Co. and Dow Corning Corp. in the United States, ShinEtsu in Japan, and Wacker Chemie in Germany. Prices tend to be about \$55–220/kg and higher depending on the physical form and the application.

Production capacity is not well understood because producers do not report their figures. Growth of fluorosilicones in automotive elastomer applications was estimated (3) at 15–20% per year through the 1980s. U.S. and western European fluorosilicone elastomer consumption (4) was estimated in 1982 to reach 1950 t by 1986, up from 1227 t in 1981. A 1989 market study (5) on elastomers reported its 1988 estimation of U.S. fluorosilicone production at 2770 t and U.S. capacity at 3270 t per year. Growth of U.S. fluorosilicone elastomers was estimated at an average of 7.5% per year for the period 1989 through 1993.

Health and Safety Factors

Information on fluorosilicone polymers is limited to safe handling information available in specific fluorosilicone product brochures. No known chronic health effects have been reported. Eye contact with fluorosilicone fluid materials may cause temporary eye discomfort with redness and dryness similar to wind burn. A single prolonged skin exposure (24–48 h) causes no known adverse effect. Small amounts transferred to the mouth by the fingers during incidental use should not cause injury. Swallowing large amounts of the fluid may cause digestive discomfort. Attempted inhalation of fluids showed no eye or respiratory passage irritation. Fluorosilicone sealant and rubber materials use a variety of curing agents. Curing to the final form usually releases small amounts of volatile by-products and unnecessary exposure during curing should be avoided. Ventilation to control vapor exposure is recommended. Some uses at elevated temperatures or in aerosol-spray applications may require added precautions.

Uses

Surface Protection. The surface properties of fluorosilicones have been studied over a number of years. The CF_3 group has the lowest known intermolecular force of polymer substituents. A study (6) of liquid and solid forms of fluorosilicones has included a comparison to fluorocarbon polymers. The low surface tensions for poly(3,3,3-trifluoropropyl)methylsiloxane and poly(3,3,4,4,5,5,6,6,6-nonafluorohexyl)methylsiloxane both resemble some of the lowest tensions for fluorocarbon polymers, eg, polytetrafluoroethylene.

Solutions of fluorosilicones impart oil and water repellent finishes to nylon–cotton fabrics. One series of C-1 through C-9 perfluoroalkyl substituents with varying structures were attached to silicon through amide or ether linkages. The fluorosilicones having perfluorinated straight-chain substituents with seven or more carbons gave the best repellencies (7) and exhibited durability toward repeated laundering, wear, and dry cleaning. The amide linkage to silicon was preferred over the ether linkage. Some nonfluorinated silicone can be tolerated in a fluorosilicone copolymer without affecting the repellency of the resulting treated fabric (see TEXTILES, FINISHING).

Foam Control. Whereas some silicones are known to be foam promoters, Dow Corning FS-1265 Fluid is a liquid fluorosilicone with effective antifoam properties. Petroleum industry application of fluids and dispersions in gas–oil separators on offshore drilling platforms has been successful. Their use peaked in the early 1980s, coinciding with constrained crude oil capacity and production. Diesel fuels are an excellent solvent for dimethylsilicones and render them ineffective as an antifoam. A new antifoam which does not require the use of added silica is formulated from a fluorosilicone copolymer. It has shown promise to antifoam (8) diesel fuel (see DEFOAMERS).

Fluids (Oils) and Greases. Fluorosilicone fluid polymers or oils are used as lubricants for pumps and compressors in harsh chemical service, such as those using acids, bases, and halogenated compounds and solvents. Unlike their hydrocarbon oil counterparts, fluorosilicones resist oxidation and degradation to form deposits, while providing a flatter viscosity–temperature profile than the hydrocarbon oils. Fluorosilicone fluids are chemically inert and do not corrode or react with most engineering materials of construction, including many metals, rubbers, and plastics. A serviceable temperature range is −40 to 204°C in open systems and up to 288°C in closed systems. The high autoignition temperature (above 480°C) makes air service possible over a wider temperature range with fire-resistant properties.

Gels. Fluorosilicone fluids with vinyl functionality can be cured using the platinum catalyst addition reactions. The cure can be controlled such that a gel or a soft, clear, jelly-like form is achieved. Gels with low (12% after 7 d) swell in gasoline fuel are useful (9) to protect electronics or circuitry from dust, dirt, fuels, and solvents in both hot (up to 150°C) and cold (down to −65°C) environments. Applications include automotive, aerospace, and electronic industries, where harsh fuel–solvent conditions exist while performance requirements remain high.

Sealants. Applications for sealants (qv) parallel the solvent-resistant applications cited below for rubbers. Sealants can be a one-part system utilizing a fluorosilicone with moisture-sensitive end groups like acetoxy groups. Hydrolysis of acetoxy end groups liberates acetic acid as a by-product. The compositions are called room temperature vulcanizing (RTV) sealants because they react with ambient moisture in air to hydrolyze the end group and cause cure by hydroxyl end group condensation of polymer ends. Cross-linking occurs through multifunctional moisture-sensitive cross-linker additives. Sealants can also be a two-part system with polymer cure depending on the reaction of a vinyl group with a silicon hydride. No volatile by-products are liberated. Application is primarily where thick section cure is desired.

Uses include both subsonic and supersonic aircraft fuel sealing applications, where temperatures of -57 to $232°C$ are experienced while in constant contact with aircraft fuel. Filleting sealants for integral fuel tanks built into the aircraft wings and noncurable channeling sealants comprise their primary uses. They are also used in bonding, sealing, caulking, encapsulating, and potting applications.

Rubber. Fluorosilicone rubber is used successfully as O-rings for fuel lines containing gasoline and aviation fuels. Its insulative properties allow its use as spark-plug boots and plug wire in transportation vehicles. The effect of alternative fuels on fluorosilicone rubbers has generated the largest interest in this class of materials. The use of alcohol–gasoline mixtures has been selected because it is a key transportation vehicle fuel of the future. Fluorosilicone rubbers exhibit stable physical properties (10) over the entire range of methanol–fuel blends for periods up to six months at $60°C$. They also return to their original physical properties after dryout following immersion. Immersion tests have utilized methanol–ASTM Reference Fuel C (RFC) blends and blends of methyl *tert*-butyl ether (MTBE) with RFC. The greatest fuel swell (33%) is seen at a methanol–RFC ratio of 25:75. This behavior is identical to that shown by a fluorocarbon reference material. Similar resistant effects were seen using oxidized gasoline or "sour" gasoline as the immersion medium. Its low temperature properties are demonstrated as follows: fluorosilicone rubbers can be formulated to show the smallest value for compression set after immersion in 25:75 methanol–RFC for 22 hours at $177°C$ and $-30°C$ (38% vs fluorocarbon reference at 100%). MTBE is shown to have less effect on fluorosilicones than methanol.

Heat and oil resistance coupled with its low swell have led automotive applications into laminated tubing and hoses (11) with this material. This resistance to the effects of ASTM No. 3 oil at service temperatures of $200°C$ makes it competitive with fluorocarbons and with the tetrafluoroethylene–propylene copolymer. Fluorosilicones are used to make exhaust gas recirculation (EGR) diaphragms for some passenger cars.

Hydraulic fluid resistance makes fluorosilicones the preferred military aircraft choice for the manufacture of the flexible bellows (12) between the hydraulic fluid reservoir and the suction pump on Northrop Corp.'s T-38 trainers and T-5 fighters. Its use allows for fluid continuity during normal and inverted flight attitudes.

Resiliency provides another opportunity for the rubber functioning as a cushion between stainless steel loop clamps and fuel–hydraulic fluid lines in aircraft. Pratt and Whitney F-100 military jet engine use (12) provides vibration damping without the clamp abrading the tube surfaces in normal service as well as at temperatures down to $-55°C$.

Electrically conductive rubber (13) can be achieved by incorporation of conductive fillers, eg, use of carbon or metal powders. These rubbers exhibit volume resistivities as low as 10^{-4} $\Omega\cdot$cm. Applications include use in dissipation of static charge and in conductive bridging between dissimilar electronic materials under harsh operating conditions.

BIBLIOGRAPHY

"Poly(fluorosilicones)" under "Fluorine Compounds, Organic" in *ECT* 3rd ed., Vol. 11, pp. 74–81 by Y. K. Kim, Dow Corning Corp.

1. D. J. Cornelius and C. M. Monroe, *Polym. Eng. Sci.* **25**(8), 467–473 (1985).
2. U.S. Pat. 4,465,786 (Aug. 14, 1984), M. F. Zimmer, W. E. Smith, and D. F. Malpass (to General Electric Co.); U.S. Pat. 4,798,818 (Jan. 17, 1989), W. X. Bajzer, R. L. Bixler, Jr., M. D. Meddaugh, and A. P. Wright (to Dow Corning Corp.).
3. T. Gabris, *Rubber World* **184**(1), 41, 59 (1981).
4. *Chem. Mark. Rep.*, 351 (Dec. 6, 1982).
5. J. L. Leone and J. P. Kelly, *Strategic Opportunities in Performance Elastomers U.S.A. 1989,* Kline & Company, Inc., Fairfield, N.J., 1989.
6. H. Kobayashi and M. J. Owen, *Macromolecules* **23**, 4929–4933 (1990).
7. J. W. Bovenkamp and B. V. LaCroix, *Ind. Eng. Chem. Prod. Res. Dev.* **20**, 130–133 (1981).
8. G. C. Sawicki and J. W. White, *Specialty Chemicals* **12**(2), 140–145 (1992).
9. M. T. Maxson and K. F. Benditt, SAE Technical Paper Series, Paper No. 88023, *SAEQ. Trans.* **97**(Part 2), 1–7 (1989).
10. M. S. Virant, L. D. Fiedler, T. L. Knapp, and A. W. Norris, SAE Technical Paper Series, Paper No. 910102, *SAEQ. Trans.* **100**(5), 37–48 (1991).
11. J. W. Horvath, *Rubber World* **197**(3), 21–29 (1987); R. E. Eggers, *Rubber World*, **204**(3), 24–35 (1991).
12. M. J. Dams, *Kautsch. Gummi Kunstst.* **38**(12), 1109–1113 (1985).
13. L. Kroupa, *Rubber World* **200**(3), 23–28 (1989).

WILLIAM X. BAJZER
YUNG K. KIM
Dow Corning Corporation

FLUORITE, FLUOROSPAR. See FLUORINE COMPOUNDS, INORGANIC–
CALCIUM FLUORIDE.

FLUOROALUMINATES, FLUOROBERYLLATES, FLUOROPHOSPHATES, FLUOROSILICATES, AND SIMILAR ENTRIES. See FLUORINE COMPOUNDS, INORGANIC.

FLUOROCHEMICALS. See FLUORINE COMPOUNDS, ORGANIC.

FOAMED PLASTICS

Foamed polymers, otherwise known as cellular polymers or polymeric foams, or expanded plastics have been important to human life since primitive people began to use wood, a cellular form of the polymer cellulose. Cellulose (qv) is the most abundant of all naturally occurring organic compounds, comprising approximately one-third of all vegetable matter in the world (1). Its name is derived from the Latin word *cellula*, meaning very small cell or room, and most of the polymer does indeed exist in cellular form as in wood, straws, seed husks, etc. The high strength-to-weight ratio of wood, good insulating properties of cork and balsa, and cushioning properties of cork and straw have contributed both to the incentive to develop and to the background knowledge necessary for development of the broad range of cellular synthetic polymers in use.

The first cellular synthetic plastic was an unwanted cellular phenol–formaldehyde resin produced by early workers in this field. The elimination of cell formation in these resins, as given by Baekeland in his 1909 heat and pressure patent (2), is generally considered the birth of the plastics industry. The first commercial cellular polymer was sponge rubber, introduced between 1910 and 1920 (3).

Many cellular plastics that have not reached significant commercial use have been introduced or their manufacture described in literature. Examples of such polymers are chlorinated or chlorosulfonated polyethylene, a copolymer of vinylidene fluoride and hexafluoropropylene, polyamides (4), polytetrafluoroethylene (5), styrene–acrylonitrile copolymers (6,7), polyimides (8), and ethylene–propylene copolymers (9).

Cellular polymers have been commercially accepted in a wide variety of applications since the 1940s (10–19). The total usage of foamed plastics in the United States has risen from 441×10^3 t in 1967 to 1.6×10^6 t in 1982, and has been projected to rise to about 2.8×10^6 t in 1995 (20).

Classification

A cellular plastic has been defined as a plastic the apparent density of which is decreased substantially by the presence of numerous cells disposed throughout its mass (21). In this article the terms cellular plastic, foamed plastic, expanded plastic, and plastic foam are used interchangeably to denote all two-phase gas–solid systems in which the solid is continuous and composed of a synthetic polymer or rubber.

The gas phase in a cellular polymer is distributed in voids, pores, or pockets called cells. If these cells are interconnected in such a manner that gas can pass from one to another, the material is termed open-celled. If the cells are discrete and the gas phase of each is independent of that of the other cells, the material is termed closed-celled.

The nomenclature of cellular polymers is not standardized; classifications have been made according to the properties of the base polymer (22), the methods of manufacture, the cellular structure, or some combination of these. The most comprehensive classification of cellular plastics, proposed in 1958 (23), has not been adopted and is not consistent with some of the common names for the more important commercial products.

One ASTM test procedure has suggested (24) that foamed plastics be classified as either rigid or flexible, a flexible foam being one that does not rupture when a 20 × 2.5 × 2.5 cm piece is wrapped around a 2.5 cm mandrel at a uniform rate of 1 lap/5 s at 15–25°C. Rigid foams are those that do rupture under this test. This classification is used in this article.

In the case of cellular rubber, the ASTM uses several classifications based on the method of manufacture (25,26). These terms are used here. Cellular rubber is a general term covering all cellular materials that have an elastomer as the polymer phase. Sponge rubber and expanded rubber are cellular rubbers produced by expanding bulk rubber stocks and are open-celled and closed-celled, respectively. Latex foam rubber, also a cellular rubber, is produced by frothing a rubber latex or liquid rubber, gelling the frothed latex, and then vulcanizing it in the expanded state.

The term structural/integral foam has been defined as flexible or rigid foams having a foamed core which gradually transforms to solid skins (27), but is used here to refer to those rigid foams produced at greater than about 320 kg/m³ density having holes in a foamed core with solid skins rather than a typical lower density structure of pentagonal dodecahedron type (28).

Theory of the Expansion Process

Foamed plastics can be prepared by a variety of methods. The most important process, by far, consists of expanding a fluid polymer phase to a low density cellular state and then preserving this state. This is the foaming or expanding process. Other methods of producing the cellular state include leaching out solid or liquid materials that have been dispersed in a polymer, sintering small particles, and dispersing small cellular particles in a polymer. The latter processes are relatively straightforward processing techniques but are of minor importance.

The expansion process consists of three steps: creating small discontinuities or cells in a fluid or plastic phase; causing these cells to grow to a desired volume; and stabilizing this cellular structure by physical or chemical means.

Initiation and Growth of Cells. The initiation or nucleation of cells is the formation of cells of such size that they are capable of growth under the given conditions of foam expansion. The growth of a hole or cell in a fluid medium at equilibrium is controlled by the pressure difference (ΔP) between the inside and the outside of the cell, the surface tension of the fluid phase γ, and the radius r of the cell:

$$\Delta P = 2\,\gamma/r \qquad (1)$$

The pressure outside the cell is the pressure imposed on the fluid surface by its surroundings. The pressure inside the cell is the pressure generated by the blowing agent dispersed or dissolved in the fluid. If blowing pressures are low, the radii of initiating cells must be large. The hole that acts as an initiating site can be filled with either a gas or a solid that breaks the fluid surface and thus enables blowing agent to surround it (29–32).

During the time of cell growth in a foam, a number of properties of the system change greatly. Cell growth can, therefore, be treated only qualitatively. The following considerations are of primary importance: (*1*) the fluid viscosity is changing considerably, influencing both the cell growth rate and the flow of polymer to intersections from cell walls leading to collapse; (*2*) the pressure of the blowing agent decreases, falling off less rapidly than an inverse volume relationship because new blowing agent diffuses into the cells as the pressure falls off according to equation 1; (*3*) the rate of growth of the cell depends on the viscoelastic nature of the polymer phase, the blowing agent pressure, the external pressure on the foam, and the permeation rate of blowing agent through the polymer phase; and (*4*) the pressure in a cell of small radius r_2 is greater than that in a cell of larger radius r_1. There is thus a tendency to equalize these pressures either by breaking the wall separating the cells or by diffusion of the blowing agent from the small to the large cells. The pressure difference ΔP between cells of radii r_1 and r_2 is shown in equation 2.

$$\Delta P = 2 \, \gamma \left(\frac{1}{r_2} - \frac{1}{r_1} \right) \tag{2}$$

Stabilization of the Cellular State.   The increase in surface area corresponding to the formation of many cells in the plastic phase is accompanied by an increase in the free energy of the system; hence the foamed state is inherently unstable. Methods of stabilizing this foamed state can be classified as chemical, eg, the polymerization of a fluid resin into a three-dimensional thermoset polymer, or physical, eg, the cooling of an expanded thermoplastic polymer to a temperature below its second-order transition temperature or its crystalline melting point to prevent polymer flow.

Chemical Stabilization.   The chemistry of the system determines both the rate at which the polymer phase is formed and the rate at which it changes from a viscous fluid to a dimensionally stable cross-linked polymer phase. It also governs the rate at which the blowing agent is activated, whether it is due to temperature rise or to insolubilization in the liquid phase.

The type and amount of blowing agent governs the amount of gas generated, the rate of generation, the pressure that can be developed to expand the polymer phase, and the amount of gas lost from the system relative to the amount retained in the cells.

Additives to the foaming system (cell growth-control agents) can greatly influence nucleation of foam cells, either through their effect on the surface tension of the system, or by acting as nucleating sites from which cells can grow. They can influence the mechanical stability of the final solid foam structure considerably by changing the physical properties of the plastic phase and by creating discontinuities in the plastic phase that allow blowing agent to diffuse from the cells to the surroundings. Environmental factors such as temperature and pressure also influence the behavior of thermoset foaming systems.

Physical Stabilization.   In physically stabilized foaming systems the factors are essentially the same as for chemically stabilized systems but for somewhat different reasons. Chemical composition of the polymer phase determines the temperature at which foam must be produced, the type of blowing agent required, and the cooling rate of the foam necessary for dimensional stabilization. Blowing agent composition and concentration controls the rate at which gas is released, the amount of gas released, the pressure generated by the gas, escape or retention of gas from the foam cells for a given polymer, and heat absorption or release owing to blowing agent activation.

Additives have the same effect on thermoplastic foaming processes as on thermoset foaming processes. Environmental conditions are important in this case because of the necessity of removing heat from the foamed structure in order to stabilize it. The dimensions and size of the foamed structure are important for the same reason.

Manufacturing Processes

Cellular plastics and polymers have been prepared by a wide variety of processes involving many methods of cell initiation, growth, and stabilization. The most convenient method of classifying these methods appears to be based on the cell growth and stabilization processes. According to equation 1, the growth of the cell depends on the pressure difference between the inside of the cell and the surrounding medium. Such pressure differences may be generated by lowering the external pressure (decompression) or by increasing the internal pressure in the cells (pressure generation). Other methods of generating the cellular structure are by dispersing gas (or solid) in the fluid state and stabilizing this cellular state, or by sintering polymer particles in a structure that contains a gas phase.

Foamable compositions in which the pressure within the cells is increased relative to that of the surroundings have generally been called expandable formulations. Both chemical and physical processes are used to stabilize plastic foams from expandable formulations. There is no single name for the group of cellular plastics produced by the decompression processes. The various operations used to make cellular plastics by this principle are extrusion, injection molding, and compression molding. Either physical or chemical methods may be used to stabilize products of the decompression process.

A summary of the methods for commercially producing cellular polymers is presented in Table 1. This table includes only those methods thought to be commercially significant and is not inclusive of all methods known to produce cellular products from polymers.

EXPANDABLE FORMULATIONS

Physical Stabilization Process. Cellular polystyrene [9003-53-6], the outstanding example; poly(vinyl chloride) [9002-86-2]; copolymers of styrene and acrylonitrile (SAN copolymers [9003-54-7]); and polyethylene [9002-88-4] can be manufactured by this process.

Polystyrene. There are two types of expandable polystyrene processes: expandable polystyrene for molded articles and expandable polystyrene for loose-fill packing materials.

Table 1. Methods for Production of Cellular Polymers

Type of polymer	Extrusion	Expandable formulation	Froth foam	Compression mold	Injection mold	Sintering
cellulose acetate[a]	X					
epoxy resin[b]		X	X			
phenolic resin		X				
polyethylene[a]	X	X		X	X	X
polystyrene	X	X			X	X
silicones		X				
urea–formaldehyde resin			X			
urethane polymers[b]		X	X		X	
latex foam rubber			X			
natural rubber	X	X		X		
synthetic elastomers	X	X		X		
poly(vinyl chloride)[a]	X	X	X	X	X	
ebonite				X		
polytetrafluoroethylene						X

[a]Also by leaching.
[b]Also by spray.

Expandable polystyrene for molded articles is available in a range of particle sizes from 0.2 to 3.0 mm, and in shapes varying from round beads to ground chunks of polymer. These particles are prepared either by heating polymer particles in the presence of a blowing agent and allowing the blowing agent to penetrate the particle (33) or by polymerizing the styrene monomer in the presence of blowing agent (34) so that the blowing agent is entrapped in the polymerized bead. Typical blowing agents used are the various isomeric pentanes and hexanes, halocarbons, and mixtures of these materials (35).

The fabrication of these expandable particles into a finished cellular-plastic article is generally carried out in two steps (36–39). First the particles are expanded by means of steam, hot water, or hot air into low density replicas of the original material, called prefoamed or pre-expanded beads. After proper aging enough of these prefoamed beads are placed in a mold to just fill it; the filled mold is then exposed to steam. This second expansion of the beads causes them to flow into the spaces between beads and fuse together, forming an integral molded piece. Stabilization of the cellular structure is accomplished by cooling the molded article while it is still in the mold. The density of the cellular article can be adjusted by varying the density of the prefoamed particles.

Expandable polystyrene for loose-fill packaging materials is available in various sizes and shapes varying from round disks to S-shaped strands. These particles can be prepared either by deforming the polystyrene under heat and impregnating the resin with a blowing agent in an aqueous suspension (40) or by the extrusion method with various die orifice shapes (41). The expansion of these particles into a product is usually carried out in two or three expansions by means of steam with at least one day of aging in air after each expansion (42). Stabilization is accomplished by cooling the polymer phase below its glass-transition temperature during the expansion process.

Poly(vinyl chloride).   Cellular poly(vinyl chloride) can be produced from several expandable formulations as well as by decompression techniques. Rigid or flexible products can be made depending on the amount and type of plasticizer used (43).

Polyethylene. Because polyethylene has a sharp melting point and its viscosity decreases rapidly over a narrow temperature range above the melting point, it is difficult to produce a low density polyethylene foam with nitrogen or chemical blowing agents because the foam collapses before it can be stabilized. This problem can be eliminated by cross-linking the resin before it is foamed, which slows the viscosity decrease above the melting point and allows the foam to be cooled without collapse of cell structure.

Cross-linking of polyethylene can be accomplished either chemically or by high energy radiation. Radiation cross-linking is usually accomplished by x-rays (44) or electrons (45,46). Chemical cross-linking of polyethylene is accomplished with dicumyl peroxide (47), di-*tert*-butyl peroxide (48), or other peroxides. Radiation cross-linking (49) is preferred for thin foams, and chemical cross-linking for the thicker foams.

Expandable polyethylene foam sheet can be made by a four-step process: (*1*) mixing of polyethylene, chemical blowing agent, and cross-linking agent (in the case of chemical cross-linking) at low or medium temperature (examples of decomposable blowing agents used for expandable polyethylene are azodicarbonamide, 4,4'-oxybis(benzenesulfonyl hydrazide), and dinitrosopentamethylenetetramine (35); (*2*) shaping at low or medium temperature; (*3*) chemical cross-linking at medium temperature or radiation cross-linking; and (*4*) heating and expanding at high temperature. Expansion of the cross-linked, expandable polyethylene sheet can be accomplished either by floating the sheet on the surface of a molten salt bath at 200–250°C and heating from above with ir heaters or by circulating hot air, or by expanding in the mold with a high pressure steam.

Chemical Stabilization Processes. This method is more versatile and thus has been used successfully for more materials than the physical stabilization process. Chemical stabilization is more adaptable for condensation polymers than for vinyl polymers because of the fast yet controllable curing reactions and the absence of atmospheric inhibition.

Polyurethane Foams. The most important commercial example of the chemical stabilization process is the production of polyurethane foams, which began in the mid-1950s. Depending on the choice of starting materials and processing techniques, it is possible to generate a wide variety of foams for such diverse uses as wood replacement in decorative cabinetwork or all-foam mattresses; to insulate portable coolers or for ultrasoft furniture cushions; as a sprayed-on insulating foam for pipes; or molded seat cushions for cars. Excellent summaries of the chemistry and technology of these polymers have been published (13,50,51) (see URETHANE POLYMERS).

The urethane forming ingredients in a polyurethane foam formulation are the isocyanate (**1**) and the polyol (**2**) as shown in equation 3.

$$\text{OCN—R—NCO} + \text{HO—R'—OH} \rightarrow \underset{(\textbf{3})}{-\!\!\left(\!\!\underset{\;}{\overset{\displaystyle O}{\overset{\|}{C}}}\text{—NH—R—NH—}\overset{\displaystyle O}{\overset{\|}{C}}\text{—O—R'—O}\!\!\right)\!\!-} \qquad (3)$$

$$\underset{(\textbf{1})}{} \quad \underset{(\textbf{2})}{}$$

Another useful reaction is the reaction of water with isocyanate to generate CO_2

and urea groups which modify the polymeric structure. This vigorous reaction is also a prime source of exothermic heat to drive equation 3 to completion.

$$OCN—R—NCO + 2 \ HOH \rightarrow NH_2—R—NH_2 + 2 \ CO_2 \tag{4}$$

Further reaction of the active hydrogens on nitrogen in the urethane groups (**3**) can occur with additional isocyanate (**1**) at higher temperatures to cause formation of allophanate structures. The active hydrogens in urea groups can also react with additional isocyanate to form disubstituted ureas which can still further react with isocyanate to form biurets (13).

The urethane-forming reaction (eq. 3) is known as the gelling reaction since it is the primary means of polymerizing the starting materials into long-chain polymer networks. The CO_2 forming reaction is known as the blowing reaction due to its contribution of CO_2 as an *in situ* blowing agent. The amount of blowing reaction is controlled by the water level of the formulation. The gelling and blowing reaction rates are determined by the catalyst choices. Typically, tertiary amines are used to foster the blowing reaction and organometallics are used to promote gellation although both contribute to both reactions. Urethane reactions often use a combination of catalysts to achieve the desired reactivity balance. Additional blowing may be obtained through the use of an auxiliary blowing agent such as methylene chloride, CFC-11, or HCFC-141b.

Silicone surfactants are used to assist in controlling cell size and uniformity through reduced surface tension and, in some cases, to assist in the solubilization of the various reactants (52,53).

The foam process may be described as follows: the materials are metered in appropriate quantities into a mixing chamber and thoroughly mixed. Tiny air (or gas) bubbles are generated in the liquid to effect nucleation. After a short induction period the blowing agents begin to diffuse into and enlarge the tiny nucleation bubbles causing a creamy appearance. The period from mixing to this point is known as the cream time which is normally about 6–15 s for flexible foams. As more blowing agents are generated the foaming mixture continues to expand and becomes more viscous as the polymerization occurs in the liquid phase. The total number of bubbles remains constant during the foam rise. The reduction of surface tension by the surfactant stabilizes the tender foaming mixture to prevent coalescing of the bubbles.

About 100–200 s after mixing, the blowing reaction ceases but the gelling reaction continues, strengthening the struts of the foam cells. The thin cell walls of a flexible foam then burst (blow-off) and the gases are released throughout the foam which has polymerized sufficiently to prevent collapse. The period from mixing to full rise (with blow-off in flexible foams) is known as rise time. The polymerization continues until the foam has gelled, usually 20–120 s after rise time. Loss of surface tackiness is known as tack free time. Rigid foams display a gel time prior to full rise. Additional cure time is necessary to achieve full polymer physical properties. This is a time/temperature characteristic which may vary from hours to days in duration.

The physical properties of the final foam can be varied broadly by controlling the degree of cross-linking in the final polymer as well as the structure of R and R′ in (**1**) and (**2**). The average molecular weight between cross-links is generally

400–700 for rigid polyurethane foams, 700–2500 for semirigid foams, and 2,500–20,000 for flexible foams (13). The structure of the diisocyanate is limited to some six or eight commercially available compounds (13,54). For this reason the variation between cross-links is controlled primarily by the polyol (**2**); it is common to use the equivalent weight (the ratio of molecular weight to hydroxyl units) as a criterion for the expected foam rigidity. The equivalent weights of polyhydroxy resins used for rigid foams are less than 300, for semirigids between 70 and 2000, and for flexibles from 500 to 3000.

Two general types of processes have been developed for producing polyurethanes on a commercial scale: the one-shot process and the prepolymer process. In the one-shot process, which is most widely used today, all primary streams (some of which may be premixed) are delivered to the foam mixing head at once for mixing and dispensing. In the prepolymer process the polyhydroxy component first reacts with isocyanate as shown in equation 5 to form an isocyanate terminated molecule, which can ultimately react with water to liberate CO_2 for foaming and obtain chain linkage via the urea groups. Use of excess isocyanate results in the formulation of an isocyanate/polyol adduct which contains a quantity of free isocyanate as well as a structured prepolymer. This adduct may be used as the source of isocyanate in a conventional system using additional polyol, catalysts, blowing agents, etc.

$$\text{OCN—R—NCO} + \text{HO—R'—OH} \rightarrow \text{OCN—R—NH—}\overset{\overset{\displaystyle O}{\|}}{C}\text{O—R'—O}\overset{\overset{\displaystyle O}{\|}}{C}\text{—NH—R—NCO} \qquad (5)$$

The foam forming ingredients are carefully metered to obtain the proper ratio of reactants, thoroughly mixed by either mechanical or impingement means, then applied as a liquid, a spray, or a froth with subsequent expansion and curing.

Polyisocyanurates. The isocyanurate ring formed by the trimerization of isocyanates is known to possess high thermal and flammability resistance as well as low smoke generation during burning (55–58). Cross-linking via the high functionality of the isocyanurates produces a foam with inherent friability. Modification of the isocyanurate system with a longer chain structure such as polyether polyols or terephthalate-based polyester polyols increases the abrasion resistance of the resultant foam. Aluminum foil-faced sheets of modified isocyanurate-based foams are now widely used as an insulation material. The manufacturing process for isocyanurate foams is similar to that for rigid polyurethane foams (see ISOCYANATES, ORGANIC).

Polyphenols. Another increasingly important example of the chemical stabilization process is the production of phenolic foams (59–62) by cross-linking polyphenols (resoles and novolacs) (see PHENOLIC RESINS). The principal features of phenolic foams are low flammability, solvent resistance, and excellent dimensional stability over a wide temperature range (59), so that they are good thermal insulating materials.

Most phenolic foams are produced from resoles and acid catalyst; suitable water-soluble acid catalysts are mineral acids (such as hydrochloric acid or sulfuric acid) and aromatic sulfonic acids (63). Phenolic foams can be produced from novolacs but with more difficulty than from resoles (59). Novolacs are thermoplastic and require a source of methylene group to permit cure. This is usually supplied by hexamethylenetetramine (64).

A typical phenolic foam system consists of liquid phenolic resin, blowing agent, catalyst, surface-active agent, and modifiers. Various formulations and composite systems (65–67) can be used to improve one or more properties of the foam in specific applications such as insulation properties (63,68–71), flammability (72–74), and open cell (76–78) (quality).

Several manufacturing processes can be used to produce phenolic foams (59,79): continuous production of free-rising foam for slabs and slab stock similar to that for polyurethane foam (61,80); foam-in-place batch process (61,81); sandwich paneling (63,82,83); and spraying (70,84).

Other Materials. Foams from epoxy resins (59,60,85,86) and silicone resins (32,60,87,88) can also be formed by a chemical stabilization process.

DECOMPRESSION EXPANSION PROCESSES

Physical Stabilization Process. Cellular polystyrene, cellulose acetate, polyolefins, and poly(vinyl chloride) can be manufactured by this process.

Polystyrene. The extrusion process for producing cellular polystyrene is probably the oldest method utilizing physical stabilization in a decompression expansion process (89). A solution of blowing agent in molten polymer is formed in an extruder under pressure. This solution is forced out through an orifice onto a moving belt at ambient temperature and pressure. The blowing agent then vaporizes and causes the polymer to expand. The polymer simultaneously expands and cools under such conditions that it develops enough strength to maintain dimensional stability at the time corresponding to optimum expansion. The stabilization is due to cooling of the polymer phase to a temperature below its glass-transition temperature by the vaporization of the blowing agent, gas expansion, and heat loss to the environment. Polystyrene foams produced by the decompression process are commercially offered in the density range of 23–53 kg/m^3 (1.4–3.3 lbs/ft^3) as well as at higher densities (90).

The extrusion of expandable polystyrene beads or pellets containing pentane blowing agent was originally used to produce low density foam sheet (91,92). The current method is to extrude polystyrene foam in a single-screw tandem line or twin-screw extruder and produce foam sheet by addition of pentane or fluorocarbon blowing agents into the extruder (93,94). For sheet thicknesses of less than 500 μm (20 mil), the blown-bubble method is normally used. This method involves blowing a tube from a round or annular die, collapsing the bubble, and then slitting the edges to obtain two flat sheets. For greater sheet thicknesses the sheet is pulled over a sizing mandrel and slit to obtain a flat sheet. Cooling of the expanded material by the external air is necessary to stabilize the foam sheet with a good skin quality.

Cellular polystyrene can also be produced by an injection-molding process. Polystyrene granules containing dissolved liquid or gaseous blowing agents are used as feed in a conventional injection-molding process (95). With close control of time and temperature in the mold and use of vented molds, high density cellular polystyrene moldings can be obtained.

Cellulose Acetate. The extrusion process has also been used to produce cellular cellulose acetate (96) in the density range of 96–112 kg/m^3 (6–7 lbs/ft^3). A hot mixture of polymer, blowing agent, and nucleating agent is forced through an orifice into the atmosphere. It expands, cools, and is carried away on a moving belt.

Polyolefins. Cellular polyethylene and polypropylene are prepared by both extrusion and molding processes. High density polyolefin foams in the density range of 320–800 kg/m^3 are prepared by mixing a decomposable blowing agent with the polymer and feeding the mixture under pressure through an extruder at a temperature such that the blowing agent is partially decomposed before it emerges from an orifice into a lower pressure zone. Simultaneous expansion and cooling take place, resulting in a stable cellular structure owing to rapid crystallization of the polymer, which increases the modulus of the polymer enough to prevent collapse of cell structure (29,39,97). This process is widely used in wire coating and structural foam products. These products can also be produced by direct injection of inert gases into the extruder (98,99).

Low density polyethylene foam products (thin sheets, planks, rounds, tubes) in the range of 32–160 kg/m^3 (2–10 lbs/ft^3) have been prepared by an extrusion technique using various gaseous fluorocarbon blowing agents (100,101). The techniques are similar to those described earlier for producing extruded polystyrene foam planks and foam sheets.

Thermoplastic Structural Foams. Structural foams having an integral skin, cellular core, and a high strength-to-weight ratio are formed by means of injection molding, extrusion, or casting, depending on product requirements (102–104). The two most widely used injection molding processes are the Union Carbide low pressure process (105) and the USM high pressure process (106).

In the low pressure process, a short shot of a resin containing a blowing agent is forced into the mold where the expandable material is allowed to expand to fill the mold under pressures of 690–4140 kPa (100–600 psi). This process produces structural foam products with a characteristic surface swirl pattern produced by the collapse of cells on the surface of molded articles.

In the high pressure process, a resin melt containing a chemical blowing agent is injected into an expandable mold under high pressure. Foaming begins as the mold cavity expands. This process produces structural foam products with very smooth surfaces since the skin is formed before expansion takes place.

Extruded structural foams are produced with conventional extruders and a specially designed die. The die has an inner, fixed torpedo located at the center of its opening, which provides a hollow extrudate. The outer layer of the extrudate cools and solidifies to form solid skin; the remaining extrudate expands toward the interior of the profile. One of the most widely used commercial extrusion processes is the Celuka process developed by Ugine-Kuhlmann (107).

Large structural foam products are produced by casting expandable plastic pellets containing a chemical blowing agent in aluminum molds on a chain conveyor. After closing and clamping the mold, it is conveyed through a heating zone where the pellets soften, expand, and fuse together to form the cellular products. The mold is then passed through a cooling zone. This process produces structural foam products with uniform, closed-celled structures but no solid skin.

Poly(vinyl chloride).   Cellular poly(vinyl chloride) is prepared by many methods (108), some of which utilize decompression processes. In all reported processes the stabilization process used for thermoplastics is to cool the cellular state to a temperature below its second-order transition temperature before the resin can flow and cause collapse of the foam.

A type of physical stabilization process, unique for poly(vinyl chloride) resins, is the fusion of a dispersion of plastisol resin in a plasticizer. The viscosity of a resin–plasticizer dispersion shows a sharp increase at the fusion temperature. In such a system expansion can take place at a temperature corresponding to the low viscosity; the temperature can then be raised to increase viscosity and stabilize the expanded state.

Extrusion processes have been used to produce high and low density flexible cellular poly(vinyl chloride). A decomposable blowing agent is usually blended with the compound prior to extrusion. The compounded resin is then fed to an extruder where it is melted under pressure and forced out of an orifice into the atmosphere. After extrusion into the desired shape, the cellular material is cooled to stabilize it and is removed by a belt.

Another type of extrusion process involves the pressurization of a fluid plastisol at low temperatures with an inert gas. This mixture is subsequently extruded onto a belt or into molds, where it expands (109,110). The expanded dispersion is then heated to fuse it into a dimensionally stable form.

Injection molding of high density cellular poly(vinyl chloride) can be accomplished in a manner similar to extrusion except that the extrudate is fed for cooling into a mold rather than being maintained at the uniform extrusion cross-section.

Chemical Stabilization Processes.   Cellular rubber and ebonite are produced by chemical stabilization processes.

Cellular Rubber.   This material is an expanded elastomer produced by expansion of a rubber stock, whereas latex foam rubber is produced from a latex. The following general procedure applies to production of cellular rubbers from a variety of types of rubber (111). A decomposable blowing agent, along with vulcanizing systems and other additives, is compounded with the uncured elastomer at a temperature below the decomposition temperature of the blowing agent. When the uncured elastomer is heated in a forming mold, it undergoes a viscosity change, as shown in Figure 1. The blowing agent and vulcanizing systems are chosen to yield open-celled or closed-celled cellular rubber. Although inert gases

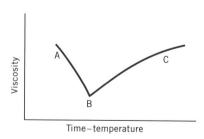

Fig. 1.   Viscosity of cellular rubber stock during a production cycle (111).

such as nitrogen have been pressurized into rubber and the rubber then expanded upon release of pressure, the current cellular rubbers are made almost entirely with decomposable blowing agents as exemplified by sodium bicarbonate [*144-55-8*], 2,2'-azobisisobutyronitrile [*78-67-1*], azodicarbonamide [*123-77-5*], 4,4'-oxy-bis(benzenesulfonyl hydrazide) [*80-51-3*], and dinitrosopentamethylenetetramine [*101-25-7*]. The compound named is the most important commercial compound in its particular class.

To produce open-celled cellular rubber the blowing agent is decomposed just prior to point A in Figure 1 so that the gas is released at the point of minimum viscosity. As the polymer expands the cell walls become thin and rupture; however, the connecting struts have developed enough strength to support the foam. This process is ordinarily carried out in one step inside a mold under pressure.

The timing for blowing-agent decomposition is more critical in making closed-celled cellular rubber; it must occur soon enough after point A to cause expansion of the elastomer but far enough past point A to allow the cell walls to become strong enough not to rupture under the blowing stress. The expansion of closed-celled rubber is often carried out in two main steps: a partial cure is carried out in a mold that is a reduced-scale replica of the final mold; removed from this mold, it expands partly toward its final form. It is then placed in an oven to complete the expansion and cure.

Most elastomers can be made into either open-celled or closed-celled materials. Natural rubber, SBR, nitrile rubber, polychloroprene, chlorosulfonated polyethylene, ethylene–propylene terpolymers, butyl rubbers, and polyacrylates have been successfully used (4,111,112).

A continuous extrusion process, as well as molding techniques, can be used as the thermoforming method. A more rapid rate of cure is then necessary to ensure the cure of the rubber before the cellular structure collapses. The stock is ordinarily extruded at a temperature high enough to produce some curing and expansion and then oven-heated to complete the expansion and cure.

A unique process for chemical stabilization of a cellular elastomer upon extrusion has been shown for ethylene–propylene rubber: the expanded rubber obtained by extrusion is exposed to high energy radiation to cross-link or vulcanize the rubber and give dimensional stability (9). EPDM is also made continuously through extrusion and a combination of hot air and microwaves or radio frequency waves which both activate the blow and accelerate the cure.

Polyurethane structural foam produced by reaction injection molding (RIM) is a rapidly growing product that provides industry with the design flexibility required for a wide range of applications. This process is more efficient than conventional methods in producing large area, thin wall, and load-bearing structural foam parts. In the RIM process, polyol and isocyanate liquid components are metered into a temperature controlled mold that is filled 20–60%, depending on the density of structural foam parts (113). When the reaction mixture then expands to fill the mold cavity, it forms a component part with an integral, solid skin and a microcellular core. The quality of the structural part depends on precise metering, mixing, and injection of the reaction chemicals into the mold.

Cellular Ebonite. Cellular ebonite is the oldest rigid cellular plastic. It was produced in the early 1920s by a process similar to the processes described for making cellular rubber. The formulation of rubber and vulcanizing agent is changed to produce an ebonite rather than rubber matrix (114).

DISPERSION PROCESSES

In several techniques for producing cellular polymers, the gas cells are produced by dispersion of a gas or liquid in the polymer phase followed, when necessary, by stabilization of the dispersion and subsequent treatment of the stabilized dispersion. In frothing techniques a quantity of gas is mechanically dispersed in the fluid polymer phase and stabilized. In another method, solid particles are dispersed in a fluid polymer phase, the dispersion stabilized, and then the solid phase dissolved or leached, leaving the cellular polymer. Still another method relies on dispersing an already cellular solid phase in a fluid polymer and stabilizing this dispersion. This results directly in cellular polymers called syntactic foams.

Frothing.   The frothing process for producing cellular polymers is the same process used for making meringue topping for pies. A gas is dispersed in a fluid that has surface properties suitable for producing a foam of transient stability. The foam is then permanently stabilized by chemical reaction. The fluid may be a homogeneous material, a solution, or a heterogeneous material.

Latex Foam Rubber.   Latex foam rubber was the first cellular polymer to be produced by frothing. (*1*) A gas is dispersed in a suitable latex; (*2*) the rubber latex particles are caused to coalesce and form a continuous rubber phase in the water phase; (*3*) the aqueous soap film breaks owing to deactivation of the surfactant in the water, breaking the latex film and causing retraction into the connecting struts of the bubbles; (*4*) the expanded matrix is cured and dried to stabilize it.

The earliest frothing process developed was the Dunlop process, which made use of chemical gelling agents, eg, sodium fluorosilicate, to coagulate the rubber particles and deactivate the soaps. The Talalay process, developed later, employs freeze-coagulation of the rubber followed by deactivation of the soaps with carbon dioxide. The basic processes and a multitude of improvements are discussed extensively in Reference 3. A discussion more oriented to current use of these processes is given in Reference 115.

Latex rubber foams are generally prepared in slab or molded forms in the density range 64–128 kg/m^3 (4–8 lbs/ft^3). Synthetic SBR latexes have replaced natural rubber latexes as the largest volume raw material for latex foam rubber. Other elastomers used in significant quantities are polychloroprene, nitrile rubbers, and synthetic *cis*-polyisoprene (115).

One method (116) of producing cellular polymers from a variety of latexes uses primarily latexes of carboxylated styrene–butadiene copolymers, although other elastomers such as acrylic elastomers, nitrile rubber, and vinyl polymers can be employed.

Urea–Formaldehyde Resins.   Cellular urea–formaldehyde resins can be prepared in the following manner: an aqueous solution containing surfactant and catalyst is made into a low density, fine-celled foam by dispersing air into it mechanically. A second aqueous solution consisting of partially cured urea–formaldehyde resin is then mixed into the foam by mechanical agitation. The catalyst in the initial foam causes the dispersed resin to cure in the cellular state. The resultant hardened foam is dried at elevated temperatures. Densities as low as 8 kg/m^3 can be obtained by this method (117).

Polyurethanes. Polyurethane foam systems have also been frothed using both low boiling dissolved materials and whipped-in air or other gas. Rigid polyurethane foam systems using a previously mixed polyol, surfactant, and catalyst system pressurized in a container with blowing agent are used for froth discharge into pour-in-place cavity filling (118). Flexible polyurethane foam is mechanically frothed by whipping dry gas such as air into the combined polyol and isocyanate. The thick, creamy froth is then doctored onto a carpet or textile back to form a variety of coatings ranging from a very thin unitary to a 1.8-cm thick resilient foam (119).

Syntactic Cellular Polymers. Syntactic cellular polymer is produced by dispersing rigid, foamed, microscopic particles in a fluid polymer and then stabilizing the system. The particles are generally spheres or microballoons of phenolic resin, urea–formaldehyde resin, glass, or silica, ranging 30–120 μm dia. Commercial microballoons have densities of approximately 144 kg/m^3 (9 lbs/ft^3). The fluid polymers used are the usual coating resins, eg, epoxy resin, polyesters, and urea–formaldehyde resin.

The resin, catalyst, and microballoons are mixed to form a mortar which is then cast into the desirable shape and cured. Very specialized electrical and mechanical properties may be obtained by this method but at higher cost. This method of producing cellular polymers is quite applicable to small quantity, specialized applications because it requires very little special equipment.

In a variation on the usual methods for producing syntactic foams (120,121), expandable polystyrene or styrene–acrylonitrile copolymer particles (in either the unexpanded or prefoamed state) are mixed with a resin (or a resin containing a blowing agent) which has a large exotherm during curing. The mixture is then placed in a mold and the exotherm from the resin cure causes the expandable particles to foam and squeeze the resin or foamed matrix to the surface of the molding. A typical example is Voraspan, expandable polystyrene in a flexible polyurethane foam matrix (122). These foams are finding acceptance in cushioning applications for bedding and furniture.

OTHER PROCESSES

Some plastics cannot be obtained in a low viscosity melt or solution that can be processed into a cellular state. For these cases two methods have been used to achieve the needed dispersion of gas in solid: sintering of solid plastic particles and leaching of soluble inclusions from the solid plastic phase.

Sintering has been used to produce a porous polytetrafluoroethylene (16). Cellulose sponges are the most familiar cellular polymers produced by the leaching process (123). Sodium sulfate crystals are dispersed in the viscose syrup and subsequently leached out. Polyethylene (124) or poly(vinyl chloride) can also be produced in cellular form by the leaching process. The artificial leather-like materials used for shoe uppers are rendered porous by extraction of salts (125) or by designing the polymers in such a way that they precipitate as a gel with many holes (126).

Phase Separation.   Microporous polymer systems consisting of essentially spherical, interconnected voids, with a narrow range of pore and cell-size distribution have been produced from a variety of thermoplastic resins by the phase-separation technique (127). If a polyolefin or polystyrene is insoluble in a solvent at low temperature but soluble at high temperatures, the solvent can be used to prepare a microporous polymer. When the solutions, containing 10–70% polymer, are cooled to ambient temperatures, the polymer separates as a second phase. The remaining nonsolvent can then be extracted from the solid material with common organic solvents. These microporous polymers may be useful in microfiltrations or as controlled-release carriers for a variety of chemicals.

Properties of Cellular Polymers

The mechanical properties of rigid foams vary considerably from those of flexible foams. The tests used to characterize these two classes of foams are, therefore, quite different, and the properties of interest from an application standpoint are also quite different. In this discussion the ASTM definition of rigid and flexible foams given earlier is used.

Several countries have developed their own standard test methods for cellular plastics, and the International Organization for Standards (ISO) Technical Committee on Plastics TC-61 has been developing international standards. Information concerning the test methods for any particular country or the ISO procedures can be obtained in the United States from the American National Standards Institute. The most complete set of test procedures for cellular plastics, and the most used of any in the world, is that developed by the ASTM; these procedures are published in new editions each year (128). There have been several reviews of ASTM methods and others pertinent to cellular plastics (32,59,129–131).

MECHANICAL PROPERTIES OF COMMERCIAL FOAMED PLASTICS

The properties of commercial rigid foamed plastics are presented in Table 2. The properties of commercial flexible foamed plastics are presented in Table 4. The definition of a flexible foamed plastic is that recommended by the ASTM Committee D 11. The data shown demonstrate the broad ranges of properties of commercial products rather than an accurate set of properties on a specific few materials. Specific producers of foamed plastics should be consulted for properties on a particular product (137,138,142).

The properties that are achieved in commercial structural foams (density >0.3 g/cm^3) are shown in Table 3. Because these values depend on several structural and process variables, they can be used only as general guidelines of mechanical properties from these products. Specific properties must be determined on the particular part to be produced. A good engineering guide has been published (103).

Structural Variables.   The properties of a foamed plastic can be related to several variables of composition and geometry often referred to as structural variables.

Polymer Composition.   The properties of foamed plastics are influenced both by the foam structure and, to a greater extent, by the properties of the parent

Table 2. Physical Properties of Commercial Rigid Foamed Plastics[a]

| | | | | Polystyrene | | | | | | | | | Polyurethane | | | |
| | | | | | | | | | | | | | | | Isocyanurate | |
Property	ASTM test	Cellulose acetate[b]	Phenolic[c]	Extruded plank[b,d]		Expanded plank[e,f]			Extruded sheet		PVC[g]		Polyether[h]		Bun[g]	Laminate[i]
density, kg/m³[j]		96–128	32–64	35	53	16	32	80	96	160	32	64	32–48	64–128	32	32
mechanical properties																
compressive strength, kPa[k] at 10%	D1621	862	138–620	310	862	90–124	207–276	586–896	290	469	345	1035	138–344	482–1896	210	117–206
tensile strength, kPa[k]	D1623	1172	138–379	517		145–193	310–379	1020–1186	2070–3450	4137–6900	551	1207	138–482	620–2000	250	248–290
flexural strength, kPa[k]	D790	1014	172–448	1138		193–241	379–517				586	1620	413–689	1380–2400		
shear strength, kPa[k]	C273	965	103–207	241			241				241	793	138–207	413–896	180	117
compression modulus, MPa[l]	D1621	38–90		10.3			3.4–14				13.1	35	2.0–4.1	10.3–31		
flexural modulus, MPa[l]	D790	38		41			9.0–26				10.3	36	5.5–6.2	5.5–10.3		
shear modulus, MPa[l]	C273		2.8–4.8	10.3			7.6–11.0				6.2	21	1.2–1.4	3.4–10.3		1.7
thermal properties																
thermal conductivity, W/(m·K)	C177	0.045–0.046	0.029–0.032	0.030		0.037	0.035	0.035	0.035	0.035	0.023		0.016–0.025	0.022–0.030	0.054	0.019
coefficient of linear expansion, 10^{-5}/°C	D696		0.9	6.3	6.3	5.4–7.2	5.4–7.2	5.4–7.2					5.4–7.2	7.2	7.2	
max service temperature, °C		177	132	74		74–80	74–80	74–80	77–80	80			93–121	121–149	149	149
specific heat, kJ/(kg·K)[m]	C351			1.1									ca 0.9	ca 0.9	ca 0.9	
electrical properties																
dielectric constant	D1673	1.12	1.19–1.20	<1.05	<1.05	1.02	1.02	1.02	1.27	1.28			1.05	1.1	1.4	
dissipation factor		20	0.028–0.031	<0.0004	<0.0004	0.0007	0.0007	0.0007	0.00011	0.00014			13	18		
moisture resistance																
water absorption, vol %	C272	4.5	13–51	0.02	0.05	1–4	1–4	1–4			15					
moisture vapor transmission, g/(m·s·GPa)[n]	E96			35		<120	35–120	23–35	86	56			35–230	50–120		230

[a]Data on epoxy resins can be found in Ref. 132; on urea–formaldehyde resins, Ref. 133. [b]Ref. 22. [c]Refs. 134 and 135. [d]Refs. 22 and 136. [e]Refs. 135 and 137. [f]Ref. 138.
[g]Ref. 139. [h]Ref. 140. [i]Ref. 141. [j]To convert kg/m³ to lb/ft³, multiply by 0.0624. [k]To convert kPa to psi, divide by 6.895. [l]To convert MPa to psi, multiply by 145,000.
[m]To convert kJ/(kg·K) to Btu/(lb·F), divide by 4.184. [n]To convert g/(m·s·GPa) to ..., multiply by 145.
To convert MPa to psi, multiply by 145.

polymer. The polymer phase description must include the additives present in that phase as well. The condition or state of the polymer phase (orientation, crystallinity, previous thermal history), as well as its chemical composition, determines the properties of that phase. The polymer state and cell geometry are intimately related because they are determined by common forces exerted during the expansion and stabilization of the foam.

Density. Density is the most important variable in determining mechanical properties of a foamed plastic of given composition. Its effect has been recognized since foamed plastics were first made and has been extensively studied.

Cell Structure. A complete knowledge of the cell structure of a cellular polymer requires a definition of its cell sizes, cell shapes, and location of each cell in the foam.

Cell size has been characterized by measurements of the cell diameter in one or more of the three mutually perpendicular directions (143) and as a measurement of average cell volume (144,145). Mechanical, optical, and thermal properties of a foam are all dependent upon the cell size.

Cell geometry is governed predominantly by the final foam density and the external forces exerted on the cellular structure prior to its stabilization in the expanded state. In a foam prepared without such external forces, the cells tend to be spherical or ellipsoidal at gas volumes less than 70–80% of the total volume, and they tend toward the shape of packed regular dodecahedra at greater gas volumes. These shapes have been shown to be consistent with surface chemistry arguments (144,146,147). Photographs of actual foam cells (Fig. 2) show a broad range of variations in shape.

In the presence of external forces, plastic foams in which the cells are elongated or flattened in a particular direction may be formed. This cell orientation can have a marked influence on many properties. The results of a number of studies have been reviewed (59,60).

The *fraction of open cells* expresses the extent to which the gas phase of one cell is in communication with other cells. When a large portion of cells are interconnected by gas phase, the foam has a large fraction of open cells, or is an open-celled foam. Conversely, a large proportion of noninterconnecting cells results in a large fraction of closed cells, or a closed-celled foam.

The nature of the opening between cells determines how readily different gases and liquids can pass from one cell to another. Because of variation in flow of different liquids or gases through the cell-wall openings, a single measurement of fraction open cells does not fully characterize this structural variable, especially in a dynamic situation.

Gas Composition. In closed-celled foams, the gas phase in the cells can contain some of the blowing agent (called captive blowing agent), gas components of air which have diffused in, or other gases generated during the foaming process. Such properties as thermal and electrical conductivity can be profoundly influenced by the cell gas composition. In open-celled foams the presence of air exerts only a minor influence on the static properties but does affect the dynamic properties such as cushioning.

Rigid Cellular Polymers. A separate class of high density, rigid cellular polymers has grown continually since the 1970s to become significant commercially. These are the structural foams with a density $>300 \text{ kg/m}^3$. They are treated here as a separate category of rigid foams.

Compressive strength and modulus are widely used as general criteria to characterize the mechanical properties of rigid plastic foams. Rigid cellular polymers generally do not exhibit a definite yield point when compressed but instead show an increased deviation from Hooke's law as the compressive load is increased (148,149). For precision the compressive strength is usually reported at some definite deflection (commonly 5 or 10%). The compressive modulus is reported as extrapolated to 0% deflection unless otherwise stated. Structural variables that affect the compressive strength and modulus of a rigid plastic foam are, in order of decreasing importance: plastic-phase composition, density, cell structure, and plastic state. The effect of gas composition is minor, with a slight effect of gas pressure in some cases.

Density and polymer composition have a large effect on compressive strength and modulus (Fig. 3). The dependence of compressive properties on cell size has been discussed (22). The cell shape or geometry has also been shown important in determining the compressive properties (22,59,60,153,154). In fact, the foam cell structure is controlled in some cases to optimize certain physical properties of rigid cellular polymers.

Strengths and moduli of most polymers increase as the temperature decreases (155). This behavior of the polymer phase carried over into the properties of polymer foams and similar dependence of the compressive modulus of polyurethane foams on temperature has been shown (151).

Tensile strength and modulus of rigid foams have been shown to vary with density in much the same manner as the compressive strength and modulus. General reviews of the tensile properties of rigid foams are available (22,59,60,131,156).

Those structural variables most important to the tensile properties are polymer composition, density, and cell shape. Variation with use temperature has also been characterized (157). Flexural strength and modulus of rigid foams both increase with increasing density in the same manner as the compressive and tensile properties. More specific data on particular foams are available from manufacturers literature and in References 22,59,60,131 and 156. Shear strength and modulus of rigid foams depend on the polymer composition and state, density, and cell shape. The shear properties increase with increasing density and with decreasing temperature (157).

Creep. The creep characteristic of plastic foams must be considered when they are used in structural applications. Creep is the change in dimensions of a material when it is maintained under a constant stress. Data on the deformation of polystyrene foam under various static loads have been compiled (158). There are two types of creep in this material: short-term and long-term. Short-term creep exists in foams at all stress levels; however, a threshold stress level exists below which there is no detectable long-term creep. The minimum load required to cause long-term creep in molded polystyrene foam varies with density ranging from 50 kPa (7.3 psi) for foam density 16 kg/m^3 (1 lb/ft^3) to 455 kPa (66 psi) at foam density 160 kg/m^3 (10 lb/ft^3).

The successful application of time–temperature superposition (159) for polystyrene foam is particularly significant in that it allows prediction of long-term

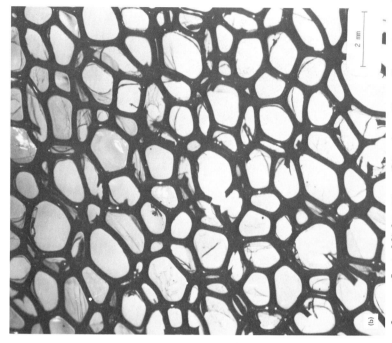

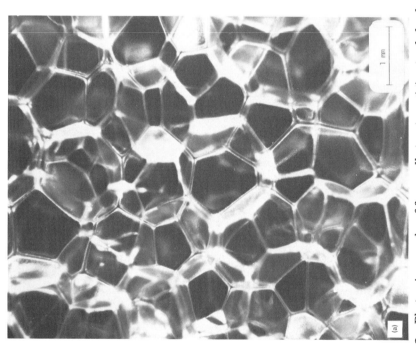

Fig. 2. Photomicrographs of foam cell structure: (**a**) extruded polystyrene foam, reflected light, 26 ×; (**b**) polyurethane foam, transmitted light, 26 ×; (**c**) polyurethane foam, reflected light, 12 ×; (**d**) high density plastic foam, transmitted light, 50 × (22). Courtesy of Van Nostrand Reinhold Publishing Corp.

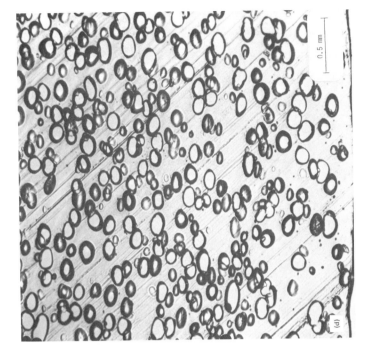

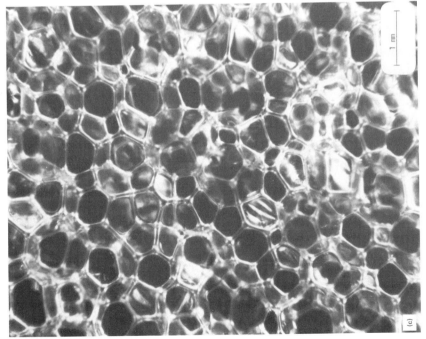

Fig. 2. (*Continued*)

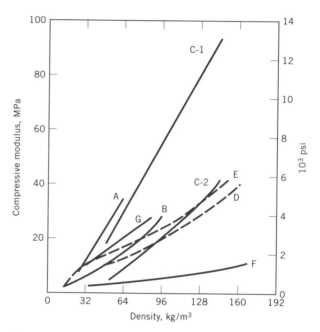

Fig. 3. Effect of density on compressive modulus of rigid cellular polymers. A, extruded polystyrene (131); B, expanded polystyrene (150); C-1, C-2, polyether polyurethane (151); D, phenol–formaldehyde (150); E, ebonite (150); F, urea–formaldehyde (150); G, poly(vinyl chloride) (152). To convert kg/m^3 to lb/ft^3, multiply by 0.0624.

behavior from short-term measurements. This is of interest in building and construction applications.

Structural Foams.   Structural foams are usually produced as fabricated articles in injection molding or extrusion processes. The optimum product and process match differs for each fabricated article, so there are no standard commercial products for one to characterize. Rather there are a number of foams with varying properties. The properties of typical structural foams of different compositions are reported in Table 3.

The most important structural variables are again polymer composition, density, and cell size and shape. Structural foams have relatively high densities (typically >300 kg/m^3) and cell structures similar to those in Figure 2d which are primarily comprised of holes in contrast to a pentagonal dodecahedron type of cell structure in low density plastic foams. Since structural foams are generally not uniform in cell structure, they exhibit considerable variation in properties with particle geometry (103).

The mechanical properties of structural foams and their variation with polymer composition and density has been reviewed (103). The variation of structural foam mechanical properties with density as a function of polymer properties is extracted from stress–strain curves and, owing to possible anisotropy of the foam, must be considered apparent data. These relations can provide valuable guidance toward arriving at an optimum structural foam, however.

Table 3. Typical Physical Properties of Commercial Structural Foams

Property	ASTM test	ABS	Noryl[a]	Nylon[b]	PC[c]	Polyester[d]	HDPE	Polypropylene	High impact polystyrene	Polyurethane[e]	PVC
glass-reinforced		yes	no	yes	no	30%	no	no 20%	no 20%	no no no	no
density, g/cm³		0.85	0.80	0.97	0.80	1.10	0.60	0.60 0.73	0.70 0.84	0.40 0.50 0.60	0.50
tensile strength, kPa[f]	D1623	18,600	22,700	101,000	37,900	76,000	8,900	13,800 20,700	12,400 34,500	11,000 17,200 23,400	6,900
compression strength, kPa[f] at 10% compression	D1621	6,900	34,500		51,700	76,000	8,900			5,500 12,400 19,300	
flexural strength, kPa[f]	D790	25,500	41,400	172,000	68,900	137,900	18,800	22,000 41,400	31,000 58,600	22,000 31,700 41,400	
flexural modulus, GPa[g]	D790	0.86	1.7	5.2	2.1	6.6	0.83	0.83 2.8	1.4 5.2	0.7 0.9 1.1	
max use temperature, °C		82	96	203	132	193	110	115			

[a]Noryl is an alloy of poly(2,6-dimethyl-1,4-phenylene ether) and polystyrene.
[b]Nylon-6,6 glass-reinforced.
[c]Polycarbonate.
[d]Thermoplastic polyester.
[e]Ref. 160.
[f]To convert kPa to psi, divide by 6.895.
[g]To convert GPa to psi, multiply by 145,000.

Table 4. Physical Properties of Commercial Flexible Foamed Plastics

Property	ASTM test	Expanded NR[a,b]	Expanded CR[a,b]	Expanded[b] SBR	Expanded[b] SBR	Latex foam rubber	PE extruded plank[c]	PE extruded plank[c]	PE extruded plank[c]	PE sheet Extruded[c]	PE sheet Cross-linked[d]
density, kg/m³[k]		56	320	192	72	80	35	96	144	43	26–28
cell structure		closed	closed	closed	closed	open	closed	closed	closed	closed	closed
compressive strength 25% deflection, kPa[l]	D3574, D3575			52			48	124	360		
tensile strength, kPa[l]	D3574		206	758	551	103	138	413	690	41	
tensile elongation, %	D3574			500		310	60	60	60	276	276–480
rebound resilience, %	D3574					73				50	
tear strength, (N/m)[n] × 10²	D3574						10.5	26	51	26	
max service temperature, °C		70	70	70	70		82	82	82	82	79–93
thermal conductivity, W/(m·K)	C177	0.036	0.043	0.065		0.030	0.053	0.058	0.058	0.040–0.049	0.036–0.040

Flexible Cellular Polymers. The application of flexible foams has been predominantly in comfort cushioning, packaging, and wearing apparel (161), resulting in emphasis on a different set of mechanical properties than for rigid foams. The compressive nature of flexible foams (both static and dynamic) is their most significant mechanical property for most uses (Table 4). Other important properties are tensile strength and elongation, tear strength, and compression set. These properties can be related to the same set of structural variables as those for rigid foams.

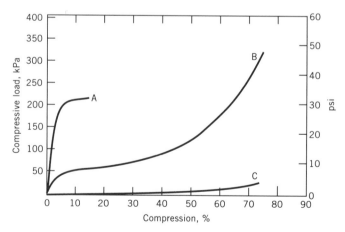

Fig. 4. Load vs compression for plastic foams (149). A, polystyrene, 32 kg/m³ (2 lbs/ft³); B, polyethylene, 32 kg/m³; C, latex rubber foam. To convert kg/m³ to lb/ft³, multiply by 0.0624.

Table 4. (continued)

Polypropylene			Polyurethane					PVC[b]		Silicone	
			Standard cushioning[f]		Carpet underlay[g]	High resilience type[h,i]				Liquid[j]	Sheet[d]
Unmodified[e]	Modified[e]	Sheet[d]									
64–96	64–96	10	16	24	34	26	40	112	96	272	160
closed	closed		open	open	open	open	open	closed	open	open	open
550	206	4.8	4.4	5.7	15.7	1.9	4.6				
830	344		88	118	258	79	103	24	3.4	36 at 20%	
1100	1380	138–275	160	205	135	200	160		220	227	310
25	75				40	65	62				
			3.3	4.4	3.7	2.6	2.4				
135	135	121								350	260
0.039	0.039	0.039						0.040		0.078	0.086

[a]NR = natural rubbber; CR = chloroprene rubber. [b]Ref. 131. [c]Ref. 162. [d]Ref. 135. [e]Ref. 163. [f]Ref. 164. [g]Ref. 165. [h]Ref. 166. [i]Ref. 167. [j]Ref. 168. [k]To convert kg/m^3 to lb/ft^3, multiply by 0.0624. [l]To convert kPa to psi, multiply by 0.145. [m]To convert N/m to lbf/in., divide by 1.75.

Compressive Behavior.　The most informative data in characterizing the compressive behavior of a flexible foam are derived from the entire load-deflection curve of 0–75% deflection and its return to 0% deflection at the speed experienced in the anticipated application. Various methods have been reported (3,161,169–172) for relating the properties of flexible foams to desired behavior in comfort cushioning. Other methods to characterize package cushioning have been reported. The most important variables affecting compressive behavior are polymer composition, density, and cell structure and size.

Polymer composition is the most important structural variable (Fig. 4). Although the polystyrene and polyethylene foams are approximately the same density and the open-celled latex foam significantly more dense, all three show markedly different compressive strengths. The compressive behavior of latex rubber foams of various densities (3,173) is illustrated in Figure 5. Similar relationships undoubtedly hold for vinyl and flexible polyurethane foams as well. In the case of polyurethane foams there are many variables in addition to density which heavily influence compressive behavior (32,50,60). For example the effects of reaction water content, polyol molecular weight, polymer polyol content, and isocyanate index on polymer tensile stiffness have been described (174). A further strong variation of flexible polyurethane foam compressive behavior can occur due to changes in the closed-cell content as measured by means of an airflow manometer described in ASTM method D3574.

Various geometric coring patterns in polyurethanes (171,175) and in latex foam rubber (176) exert significant influences on their compressive behavior. A good discussion of the effect of cell size and shape on the properties of flexible foams is contained in References 60 and 156. The effect of open-cell content is demonstrated in polyethylene foam (173).

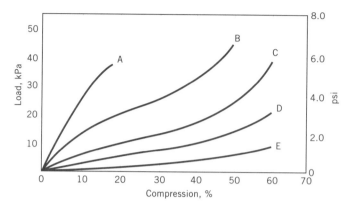

Fig. 5. Effect of load on compression for latex foams of different densities (3,173). A, 304 kg/m³; B, 208 kg/m³; C, 179 kg/m³; D, 139 kg/m³; E, 99 kg/m³. To convert kg/m³ to lb/ft³, multiply by 0.0624.

Tensile Strength and Elongation. The tensile strength of latex rubber foam has been shown to depend on the density of the foam (149,177) and on the tensile strength of the parent rubber (177,178). At low densities the tensile modulus approximates a linear relation with density but increases with a higher power of density at higher densities. Similar relations hold for polyurethane and other flexible foams (156,179,180).

The tensile elongation of solid latex rubber has been shown to correlate well with the elongation of foam from the latex (178). The elongation of flexible polyurethane has been related to cell structure (180,181).

Tear Strength. A relation for the tearing stress of flexible foams that predicts linear increase in the tearing energy with density and increased tearing energy with cell size has been developed (177). Both relationships are verified to a limited extent by experimental data.

Flex Fatigue. Considerable information on the measurement and cause of flex fatigue in flexible foams has been published (182–184). Changing compressive strength and volume upon repeated flexing over long periods of time is a significant deterrent to the use of polyurethane foam in many cushioning applications. For polyurethane foams these changes have been correlated mainly with changes in chemical structure.

Compression Set. The compression set is an important property in cushioning applications. It has been studied for polyurethane foams (185,186), and has been discussed in reviews (32,60,156). Compression set has been described as flex fatigue and creep as well.

OTHER PROPERTIES

The thermal, electrical, acoustical, and chemical properties of all cellular polymers are of such a similar nature that the discussions of these properties are not separated into rigid and flexible groups.

Thermal Properties. *Thermal Conductivity.* More information is available relating thermal conductivity to structural variables of cellular polymers than for any other property. Several papers have discussed the relation of the thermal conductivity of heterogeneous materials in general (187,188) and of plastic foams in particular (132,143,151,189–191) with the characteristic structural variables of the systems.

The following separation of the total heat transfer into its component parts, even if not completely rigorous, proves valuable to understanding the total thermal conductivity, k, of foams:

$$k = k_s + k_g + k_r + k_c \tag{6}$$

where k_s, k_g, k_r, and k_c are the components of thermal conductivity attributable to solid conduction, gaseous conduction, radiation, and convection, respectively.

As a good first approximation (187), the heat conduction of low density foams through the solid and gas phases can be expressed as the product of the thermal conductivity of each phase times its volume fraction. Most rigid polymers have thermal conductivities of 0.07–0.28 W/(m·K) and the corresponding conduction through the solid phase of a 32 kg/m^3 (2 lbs/ft^3) foam (3 vol %) ranges 0.003–0.009 W/(m·K). In most cellular polymers this value is determined primarily by the density of the foam and the polymer-phase composition. Smaller variations can result from changes in cell structure.

Although conductivity through gases is much lower than that through solids, the amount of heat transferred through the gas phase in a foam is generally the largest contribution to the total heat transfer because the gas phase is the principal part of the total value (ca 97 vol % in a 32 kg/m^3 foam). Table 5 lists values of the thermal conductivity for several gases that occur in the cells of cellular polymers. The thermal conductivities of the halocarbon gases are considerably less than those of oxygen and nitrogen. It has, therefore, proved advantageous to prepare cellular polymers using such gases that measurably lower the k of the polymer foam. Upon exposure to air the gas of low thermal conductivity in the cells can become mixed with air, and the k of the mixture of gases can be estimated by a mixing rule such as the Riblett equation 7.

$$k_m = \sum_i k_i M_i^{1/3} P_i / \sum_i M_i^{1/3} P_i \tag{7}$$

where k_m is the k of the gaseous mixture; k_i, M_i, and P_i are the component thermal conductivity, molecular weight, and partial pressure, respectively. Changes in total k calculated by equations 6 and 7 with change in gas composition agree well with experimental measurements (144,191,194,195).

There is ordinarily no measurable convection in cells of diameter less than about 4 mm (143). Theoretical arguments have been in general agreement with this work (151,191). Since most available cellular polymers have cell diameters smaller than 4 mm, convection heat transfer can be ignored with good justification. Studies of radiant heat transfer through cellular polymers have been made (143,151,191,196,197).

The variation in total thermal conductivity with density has the same general nature for all cellular polymers (143,189). The increase in k at low densities is owing to an increased radiant heat transfer; the rise at high densities to an increasing contribution of k_s.

Table 5. Thermal Conductivity at 20°C of Gases Used in Cellular Polymers[a]

Compound[b]	Thermal conductivity, W/(m·K)
trichlorofluoromethane (CFC-11)	0.0084
dichlorodifluoromethane (CFC-12)	0.0098
trichlorotrifluoroethane (CFC-113)	0.0072
dichlorotetrafluoroethane (CFC-114)	0.0104
dichlorofluoromethane (CFC-21)	0.0112
chlorodifluoromethane (HCFC-22)	0.0106
difluoromethane (HFC-32)	0.0163
2-chloro-1,1,1,2-tetrafluoroethane (HCFC-124)	0.0106
pentafluoroethane (HFC-125)	0.0131
1,1,1,2-tetrafluoroethane (HFC-134a)	0.0127
1,1-dichloro-1-fluoroethane (HCFC-141b)	0.0083
1-chloro-1,1-difluoroethane (HCFC-142b)	0.0108
trifluoroethane (HFC-143a)	0.0137
1,1-difluoroethane (HFC-152a)	0.0136
dichloromethane	0.0063
methyl chloride	0.0105
2-methylpropane	0.0161
carbon dioxide	0.0168
air	0.0259

[a]Refs. 192, 193.
[b]CFC = chlorofluorocarbon; HCFC = hydrochlorofluorocarbon; see FLUORINE COMPOUNDS, ORGANIC–ALIPHATIC.

The thermal conductivity of most materials decreases with temperature. When the foam structure and gas composition are not influenced by temperature, the k of the cellular material decreases with decreasing temperature. When the composition of the gas phase may change (ie, condensation of a vapor), then the relationship of k to temperature is much more complex (143,191,198).

The thermal conductivity of a cellular polymer can change upon aging under ambient conditions if the gas composition is influenced by such aging. Such a case is evidenced when oxygen or nitrogen diffuses into polyurethane foams that initially have only a fluorocarbon blowing agent in the cells (32,130,143,190,191, 198–201).

Thermal conductivity of foamed plastics has been shown to vary with thickness (197). This has been attributed to the boundary effects of the radiant contribution to heat-transfer.

Specific Heat. The specific heat of a cellular polymer is simply the sum of the specific heats of each of its components. The contribution of the gas is small and can be neglected in many cases.

Coefficient of Linear Thermal Expansion. The coefficients of linear thermal expansion of polymers are higher than those for most rigid materials at ambient temperatures because of the supercooled-liquid nature of the polymeric state, and this applies to the cellular state as well. Variation of this property with density and temperature has been reported for polystyrene foams (202) and for foams in general (22). When cellular polymers are used as components of large structures, the coefficient of thermal expansion must be considered carefully because of its magnitude compared with those of most nonpolymeric structural materials (203).

Maximum Service Temperature. Because the cellular materials, like their parent polymers (204), gradually decrease in modulus as the temperature rises rather than undergoing a sharp change in properties, it is difficult to precisely define the maximum service temperature of cellular polymers. The upper temperature limit of use for most cellular polymers is governed predominantly by the plastic phase. Fabrication of the polymer into a cellular state normally builds some stress into the polymer phase; this may tend to relax at a temperature below the heat-distortion temperature of the unfoamed polymer. Of course, additives in the polymer phase or a plasticizing effect of the blowing agent on the polymer affect the behavior of the cellular material in the same way as the unfoamed polymer. Typical maximum service temperatures are given in Tables 2, 3, and 4.

Flammability. The results of small-scale laboratory tests of plastic foams have been recognized as not predictive of their true behavior in other fire situations (205). Work aimed at developing tests to evaluate the performance of plastic foams in actual fire situations continues. All plastic foams are combustible, some burning more readily than others when exposed to fire. Some additives (131,135), when added in small quantities to the polymer, markedly improve the behavior of the foam in the presence of small fire sources. Plastic foams must be used properly following the manufacturers recommendations and any applicable regulations.

Moisture Resistance. Plastic foams are advantageous compared to other thermal insulations in several applications where they are exposed to moisture pickup, particularly when subjected to a combination of thermal and moisture gradients. In some cases the foams are exposed to freeze–thaw cycles as well. The behavior of plastic foams has been studied under laboratory conditions simulating these use conditions as well as under the actual use conditions.

In a study (206) of the moisture gain of foamed plastic roof insulations under controlled thermal gradients the apparent permeability values were greater than those predicted by regular wet-cup permeability measurements. The moisture gains found in polyurethane are greater than those of bead polystyrene and much greater than those of extruded polystyrene.

Moisture pickup and freeze–thaw resistance of various insulations and the effect of moisture on the thermal performance of these insulations has been reported (207). In protected membrane roofing applications the order of preference for minimizing moisture pickup is extruded polystyrene $\gg$ polyurethane $>$ molded polystyrene (207). Water pickup values for insulation in use for five years were extruded polystyrene 0.2 vol %, polyurethane without skins 5 vol %, and molded polystyrene 8–30 vol %. These correspond to increases in k of 5–265%. For below-grade applications extruded polystyrene was better than molded polystyrene or polyurethane without skins in terms of moisture absorption resistance and retention of thermal resistance. Increased water content has been related with increased thermal conductivity of the insulations (208–212).

Electrical Properties. Cellular polymers have two important electrical applications (22). One takes advantage of the combination of inherent toughness and moisture resistance of polymers along with the decreased dielectric constant and dissipation factor of the foamed state to use cellular polymers as electrical-wire insulation (97). The other combines the low dissipation factor and the rigidity of plastic foams in the construction of radar domes. Polyurethane foams have been used as high voltage electrical insulation (213).

Environmental Aging. All cellular polymers are subject to a deterioration of properties under the combined effects of light or heat and oxygen. The response of cellular materials to the action of light and oxygen is governed almost entirely by the composition and state of the polymer phase (22). Expansion of a polymer into a cellular state increases the surface area; reactions of the foam with vapors and liquids are correspondingly faster than those of solid polymer.

Foams prepared from phenol–formaldehyde and urea–formaldehyde resins are the only commercial foams that are significantly affected by water (22). Polyurethane foams exhibit a deterioration of properties when subjected to a combination of light, moisture, and heat aging; polyester-based foam shows much less hydrolytic stability than polyether-based foam (50,199).

A great deal of work has been done to develop additives that successfully eliminate environmental degradation (214). The best source of information on specific additives for specific foams is the individual manufacturer of the foam. The resistance to rot, mildew, and fungus of cellular polymers can be related to the amount of moisture that can be taken up by the foam (150). Therefore, open-celled foams are much more likely to support growth than are closed-celled foams. Very high humidity and high temperature are necessary for the growth of microbes on any plastic foam.

Miscellaneous Properties. The acoustical properties of polymers are altered considerably by their fabrication into a cellular structure. Sound transmission is altered only slightly because it depends predominantly on the density of the barrier (in this case, the polymer phase). Cellular polymers by themselves are, therefore, very poor materials for reducing sound transmission. They are, however, quite effective in absorbing sound waves of certain frequencies (150); materials with open cells on the surface are particularly effective. The combination of other advantageous physical properties with fair acoustical properties has led to the use of several different types of plastic foams in sound-absorbing constructions (215,216). The sound absorption of a number of cellular polymers has been reported (21,150,215,217).

The permeability of cellular polymers to gases and vapors depends on the fraction of open cells as well as the polymer-phase composition and state. The presence of open cells in a foam allows gases and vapors to permeate the cell structure by diffusion and convection flow, yielding very large permeation rates. In closed-celled foams the permeation of gases or vapors is governed by composition of the polymer phase, gas composition, density, and cellular structure of the foam (194,199,215,218,219).

The penetration of visible light through foamed polystyrene has been shown to follow approximately the Beer-Lambert law of light absorption (22). This behavior presumably is characteristic of other cellular polymers as well.

Comfort cushioning is the largest single application of cellular polymers; flexible foams are the principal contributors to this field. Historically, cushioning in particular and flexible foams in general have been the greatest volume of cellular polymers. However, the rapid growth rate of structural, packaging, and insulation applications has brought their volume over that of flexible foams during the past few years. Table 6 shows United States consumption of foamed plastics by resin and market (20).

Comfort Cushioning. The properties of greatest significance in the cushioning applications of cellular polymers are compression–deflection behavior, resilience, compression set, tensile strength and elongation, and mechanical and environmental aging; compression–deflection behavior is the most important. The broad range of compressive behavior of various types of flexible foam is one of the strong points of cellular polymers, since the needs of almost any cushioning application can be met by changing either the chemical nature or the physical structure of the foam. Flexible urethanes, vinyls, latex foam rubber, and olefins are used to make foamed plastic cushioning for automobile padding and seats, furniture, flooring, mattresses, and pillows. These materials compete with felt, fibers, innerspring, and other filling materials.

Latex foam rubber was initially accepted as a desirable comfort-cushioning material because of its softness to the touch and its resilience (equal to that of a steel spring alone but with better damping qualities than the spring).

Table 6. Market for Cellular Polymers,[a] 10^3 t

Item	1967	1982	1995[b]
By market			
insulation	58	261	472
flooring	20	98	154
other construction	9	136	288
cushioning	52	195	336
other furniture	40	103	175
packaging	43	177	311
transportation	76	140	238
consumer	44	136	225
bedding	18	57	113
appliances	14	40	61
other	68	225	408
Total	*441*	*1567*	*2781*
By resin			
flexible urethane	181	511	844
rigid urethane	68	248	449
styrene	125	410	699
vinyl	61	232	413
others	6	165	376
Total	*441*	*1567*	*2781*

[a]Ref. 20.
[b]Projected.

Cellular rubber has been used extensively as shoe soles, where its combination of cushioning ability and wear resistance, coupled with desirable economics, has led to very wide acceptance. In this case the cushioning properties are of minor importance compared with the abrasion resistance and cost. Other significant cushioning applications for cellular rubbers and latex foam rubbers are as carpet underlay and as cushion padding in athletic equipment.

Thermal Insulation. Thermal insulation is the second largest application of cellular polymers and the largest application for the rigid materials. The properties of greatest importance in determining the applicability of rigid foams as thermal insulants are thermal conductivity, ease of application, cost, moisture absorption and transmission permeance, and mechanical properties (see INSULATION, THERMAL). Plastic foams containing a captive blowing agent have considerably lower thermal conductivities than other insulating materials, whereas other rigid cellular plastics are roughly comparable with the latter.

Domestic Refrigeration. The very low thermal conductivity of polyurethanes plus the ease of application and structural properties of foamed-in-place materials gives refrigeration engineers considerable freedom of styling. This has resulted in an increasingly broad use of rigid polyurethane foams in home freezers and refrigerators which has displaced conventional rock wool and glass wool.

Commercial Refrigeration. Again, low thermal conductivity is important, as are styling and cost. Application methods and mechanical properties are of secondary importance because of design latitude in this area. For example, large institutional chests, commercial refrigerators, freezers, and cold storage areas, including cryogenic equipment and large tanks for industrial gases, are insulated with polystyrene or polyurethane foams. Polystyrene foam is still popular where cost and moisture resistance are important; polyurethane is used where spray application is required. Polystyrene foam is also widely used in load-bearing sandwich panels in low temperature space applications.

Refrigeration in Transportation. Styling is unimportant. The volume of insulation and a low thermal conductivity are of primary concern. Volume is not large, so application methods are not of prime importance. Low moisture sensitivity and permanence are necessary. The mechanical properties of the insulant are quite important owing to the continued abuse the vehicle undergoes. Cost is of less concern here than in other applications.

Residential Construction. Owing to rising energy costs, the cost and low thermal conductivity are of prime importance in wall and ceiling insulation of residential buildings. The combination of insulation efficiency, desirable structural properties, ease of application, ability to reduce air infiltration, and moisture resistance has led to use of extruded polymeric foam in residential construction as sheathing, as perimeter and floor insulation under concrete, and as a combined plaster base and insulation for walls.

Commercial Construction. The same attributes desirable on residential construction applications hold for commercial construction as well but insulation quality, permanence, moisture insensitivity, and resistance to freeze–thaw cycling in the presence of water are of greater significance. For this reason cellular plastics have greater application here. Both polystyrene and polyurethane foams are highly desirable roof insulations in commercial as in residential construction.

Cellular polymers are also used for pipe and vessel insulation. Spray and pour-in-place techniques of application are particularly suitable, and polyurethane and epoxy foams are widely used. Ease of application, fire properties, and low thermal conductivity have been responsible for the acceptance of cellular rubber and cellular poly(vinyl chloride) as insulation for smaller pipes.

The insulating value and mechanical properties of rigid plastic foams have led to the development of several novel methods of building construction. Polyurethane foam panels may be used as unit structural components (220) and expanded polystyrene is employed as a concrete base in thin-shell construction (221).

Packaging. Because of the extremely broad demands on the mechanical properties of packaging materials, the entire range of cellular polymers from rigid to flexible is used in this application. The most important considerations are mechanical properties, cost, ease of application or fabrication, moisture susceptibility, thermal conductivity, and aesthetic appeal.

The proper mechanical properties, particularly compressive properties, are the primary requirements for a cushioning foam (222,223). The reader is referred to the following sources for more specific information: package design (224); general vibration and shock isolation (225); protective package design (226); selection of cushioning material (222,227); and characterization of cellular polymers for cushioning applications (223,225,226,228).

Creep of a cushion packaging material when subjected to static stresses for long periods of storage or shipment is also an important consideration. Polystyrene foam shows considerable creep (158) at high static loadings but that creep is insignificant under loadings in the static stress region of optimum package design (22). The ability of polystyrene foam to withstand repeated impacts has also been studied (152,229).

The low density of most cellular plastics is important because of shipping costs for the cushioning in a package. Foams with densities ranging from 4 to 32 kg/m^3 are used in this application. The inherent moisture resistance of cellular plastics is of added benefit where packages may be subjected to high humidity or water. Many military applications require low moisture susceptibility. Foamed polystyrene is used as packaging inserts and as containers such as food trays, egg cartons, and drinking cups which require moisture resistance, rigidity, and shock resistance. Foamed polyurethane is also used as specialty packaging materials for expensive and delicate equipment.

The clean, durable, nondust-forming character of polyethylene foam has led to its acceptance in packaging missile parts (230). Polyethylene foam sheet has also displaced polystyrene foam sheet for packaging glass bottles and containers because of its greater resiliency and tear resistance.

Antistatic protection is an important consideration within the electronic industry and various antistatic agents are used commercially to alleviate this problem in cushion packaging materials.

Structural Components. In most applications structural foam parts are used as direct replacements for wood, metals, or solid plastics and find wide acceptance in appliances, automobiles, furniture, materials-handling equipment, and in construction. Use in the building and construction industry account for more than one-half of the total volume of structural foam applications. High impact polystyrene is the most widely used structural foam, followed by polypropylene, high density polyethylene, and poly(vinyl chloride). The construction industry offers the greatest growth potential for cellular plastics.

The sandwich-type structure of polyurethanes with a smooth integral skin produced by the reaction injection molding process provides a high degree of stiffness as well as excellent thermal and acoustical properties necessary for its use in housing and load-bearing structural components for the automotive, business machine, electrical, furniture, and materials-handling industry.

Buoyancy. The low density, closed-celled nature of many cellular polymers coupled with their moisture resistance and low cost resulted in their immediate acceptance for buoyancy in boats and floating structures such as docks and buoys. Since each cell in the foam is a separate flotation member, these materials cannot be destroyed by a single puncture.

The combination of structural strength and flotation has stimulated the design of pleasure boats using a foamed-in-place polyurethane between thin skins of high tensile strength (231). Other cellular polymers that have been used in considerable quantities for buoyancy applications are those produced from polyethylene, poly(vinyl chloride), and certain types of rubber. The susceptibility of polystyrene foams to attack by certain petroleum products that are likely to come in contact with boats led to the development of foams from copolymers of styrene and acrylonitrile which are resistant to these materials (6,7).

Electrical Insulation. The substitution of a gas for part of a solid polymer usually results in large changes in the electrical properties of the resulting material. The dielectric constant, dissipation factor, and dielectric strength are all generally lowered in amounts roughly proportional to the amount of gas in the foam.

For low frequency electrical insulation applications, the dielectric constant of the insulation is ideally as low as possible (see INSULATION, ELECTRICAL). The lower the density of the cellular polymer, the lower the dielectric constant and the better the electrical insulation. Dielectric strength is also reduced at lower density; the insulation is, therefore, susceptible to breakdown from voltage surges from such sources as lightning and short circuits. Because physical properties are also diminished proportionally to density, optimum density is determined by a compromise in properties. For many applications this compromise has been at an expansion of two or three volumes, mainly because the minimum physical properties required for fabrication and use are obtained at that point. Polyolefin foams have been most used as low frequency electrical insulation; poly(vinyl chloride) and polystyrene foams are used also. Producing a completely homogenous, closed-celled foam at lower densities in high speed wire-coating apparatus is difficult.

In high frequency applications, the dissipation factor is of greater importance. Coaxial cables using cellular polyolefins have been quite successfully used for frequencies in the megahertz range and above. Cellular plastics have also been used as structural materials in constructing very large radar-receiving domes (232). The very low dissipation factor of these materials makes them quite transparent to radar waves.

Space Filling and Seals. Cellular polymers have become common for gasketing, sealing, and space filling. Cellular rubber, poly(vinyl chloride), silicone (103), and polyethylene are used extensively for gasketing and sealing of closures in the automotive and construction trade (111). Most cellular materials must be predominantly closed-celled in order to provide the necessary barrier properties. The combination of chemical inertness, excellent conformation to irregular surfaces, and ability to be compressed to greater than 50% with relatively small pressures and still function satisfactorily contribute to the acceptance of cellular polymers in these applications.

In the construction industry, cellular polymers are used as spacers and sealant strips in windows, doors, and closures of other types, as well as for backup strips for other sealants.

Miscellaneous Applications. Cellular plastics have been used for display and novelty pieces from their early development. Polystyrene foam combines ease of fabrication with lightweight, attractive appearance, and low cost to make it a favorite in these uses. Phenolic foam has its principal use in floral displays. Its ability to hold large amounts of water for extended periods is used to preserve cut flowers. Cellular poly(vinyl chloride) is used in toys and athletic goods, where its toughness and ease of fabrication into intricate shapes have been valuable.

Cellular urea–formaldehyde and phenolic resin foams have been used to some extent in interior sound-absorbing panels and, in Europe, expanded polystyrene has been used in the design of sound-absorbing floors (233). In general, cost, flammability, and cleaning difficulties have prevented significant penetration of the acoustical tile market. The low percent of reflection of sound waves from plastic foam surfaces has led to their use in anechoic chambers (216).

Commercial Products and Processes

FLEXIBLE POLYURETHANE

These foams are produced from long-chain, lightly branched polyols reacting with a diisocyanate, usually toluene diisocyanate [1321-38-6] (TDI), to form an open-celled structure with free air flow during flexure. During manufacture these foams are closely controlled for proper density, ranging from 13 to 80 kg/m^3 (0.8–5 lbs/ft^3), to achieve the desired physical properties and cost.

In flexible polyurethane foams, the primary blowing agent is carbon dioxide, which is formed by the reaction of water and toluene diisocyanate. Softer foams with lower densities require an auxiliary blowing agent such as CFC-11, HCFC-141b, or methylene chloride. Since the load bearing characteristics of the foam are of great importance to the ultimate consumer this property is also closely controlled during manufacture.

Raw Materials. Polyether polyols are used in about 90% of polyurethane foams. The elastomeric polymer is provided additional toughness in the overall polymer matrix by the presence of hard segment urea-based polymers derived from the water/isocyanate reaction (see ISOCYANATES, ORGANIC; URETHANE POLYMERS). Intermolecular hydrogen bonding plays a further role in overall foam hardness. The polyols are typically trifunctional but di- and tetrafunctional polyols are also used. The polyol chain initiator determines the functionality of the final product; glycerol or trimethylolpropane are the most common triol initiators. Propylene oxide (PO) is then polymerized onto the initiator to form a long-chain triol with an equivalent weight of 1000 to 1500. PO chains are characterized by pendent methyl groups and terminal secondary hydroxyl groups which provide the lower level of reactivity used for slab foam manufacture. Ethylene oxide (EO) can be used in conjunction with PO to modify the polyether chain by reducing the pendent methyl groups. This is called a hetero polyol with the possibility of adding a mixed PO/EO feed to form a random hetero or a batch EO feed to form a block hetero polyol. Additionally, EO can be used at the end of the polyol polymerization to produce primary hydroxyl groups at chain termination. This is known as EO capping and results in polyols with considerably higher reactivities toward isocyanates which is the polyol type required for molded foam production.

Another type of polyol often used in the manufacture of flexible polyurethane foams contains a dispersed solid phase of organic chemical particles (234–236). The continuous phase is one of the polyols described above for either slab or molded foam as required. The dispersed phase reacts in the polyol using an addition reaction with styrene and acrylonitrile monomers in one type or a coupling reaction with an amine such as hydrazine and isocyanate in another. The solids content ranges from about 21% with either system to nearly 40% in the styrene–acrylonitrile system. The dispersed solids confer increased load bearing and in the case of flexible molded foams also act as a cell opener.

The isocyanates used in the manufacture of flexible foam are toluene diisocyanate (TDI) and polymeric 4,4'-methylenediphenyl diisocyanate [*101-68-8*] (MDI). Slab foam manufacturing is based almost entirely on TDI which is most often supplied as a blend of 80% 2,4 isomer and 20% 2,6 isomer by weight. There have been efforts to develop slab foaming technology using polymeric MDI in place of TDI (237–239). Polymeric MDI is often used in manufacturing molded foams usually blended with TDI, often at a 4 to 1 ratio of TDI to MDI by weight. The acidity and isomer distribution are key factors controlling the reactivity of these isocyanates. Foams are generally produced with a slight excess of isocyanate groups. The stoichiometric balance of a foam formulation is known as the foam index with 100 index as the balance point and 110 index indicating 110% isocyanate equivalents compared to active hydrogen equivalents.

Catalysis of the flexible polyurethane foaming operation is accomplished through the use of tertiary amine compounds, often using two different amines to balance the blowing and gelling reactions. Organometallic compounds, usually stannous salts, are also used to facilitate gelling and promote final cure.

Hydrolyzable or nonhydrolyzable siloxane compounds provide nucleating assistance for fine, uniform cells and surface tension depression for stabilization of the expanding cell walls prior to gelation of the polymer. The slab foam cells are mostly open after ultimate foam rise and blow off. Too much surfactant or too much tin gelation catalyst cause the foam to have a larger number of closed cells. This tight foam lets very little air pass through a cut block of foam. Tight foams are prone to shrink as the hot gas inside the closed cell cools thus producing less pressure and volume. Molded foams often need to be crushed after demolding to mechanically open closed cells and prevent shrinkage. Surfactants must be carefully chosen for use in flexible slab, high resilience (HR) slab, and HR or hot molded systems since most are not interchangeable.

Fillers (qv) are occasionally used in flexible slab foams; the two most commonly used are calcium carbonate (whiting) and barium sulfate (barytes). Their use level may range up to 150 parts per 100 parts of polyol. Various other ingredients may also be used to modify a flexible foam formulation. Cross-linkers, chain extenders, ignition modifiers, auxiliary blowing agents, etc, are all used to some extent depending on the final product characteristics desired.

Process and Equipment.   The critical requirements for urethane foam dispensing equipment are accurate metering of the ingredients to the mixing chamber, adequate short-cycle mixing, and proper dispensing ability. The polyol, isocyanate, and water must all be delivered at an accurate rate to maintain the desired stoichiometry which is essential for predicting final foam performance and properties. The other ingredients must also be precisely controlled to obtain optimum processing and performance. Thorough ingredient mixing is made more critical because the components are reactive and thus may not remain in the mixing chamber for more than a few seconds. There is also a wide range of component viscosities; low viscosity isocyanates are dispersed in fairly high viscosity polyols. Additionally the mixing head must deliver the foam ingredients in a smooth flowing manner to minimize air entrainment or splashing.

There are two basic metering/mixing systems (based on pressure) in wide use. Low pressure (less than 2,000 kPa) systems use positive displacement pumps to deliver material via a heat exchanger and recycle valve to a mixing chamber with a mechanically driven impeller. High pressure (2,000–20,000 kPa) systems use precision high pressure pumps to deliver material via flow adjusting valves and/or orifices to a cylindrical impingement mixing chamber. Following each use the impingement mixing chamber is cleared by advancing a piston that eliminates the need for solvent flushing as is required for low pressure machines.

The mixing head dispenses the foam in several ways depending on the particular foam production process. Flexible foam molding requires the head to be positioned over the open mold, moved in relation to the mold for the best pour pattern and to dispense material on a required quantity shot basis. After the ingredients are placed in the mold cavity the lid is closed and the mold heated. The materials foam and expand to fill the mold, then gel and cure. The mold is then opened, the foam part removed, and a fresh layer of mold release sprayed onto the mold. The foam object is crushed to enhance cell opening and then may be post-cured. The two basic processes for molding are the earlier developed hot process where the molds are subjected to a high temperature (204–371°C) and the cold process where the mold ranges from room temperature to about 120°C. The chemistry used for the cold process is called high resilience or HR foam system.

Flexible slab operations often use a traversing arrangement to dispense the foam ingredients back and forth on a layer of polyethylene film carried on a conveying belt. Side papers are brought up to the edge of the film and the assembly enters a tunnel fitted with an exhaust system. The liquid foam ingredients begin to react and the foam rises to full height within 3–4.5 m after entering the tunnel. As the slab bun exits the tunnel the side papers are pulled off; the bun is then cut into appropriate lengths and delivered to a cure area. Generally a minimum of 24 hours is required to cure the bun prior to cutting into blocks for shipping. This simplest form of slab foam manufacture leaves the top of the bun with a rounded cross-section much like the top of a loaf of bread. This rounding introduces a waste factor during subsequent cutting of the bun into rectangular blocks for final use as furniture cushions, mattresses, etc. Starting in the late 1970s a number of patented processes were introduced to provide a square block with less wasted foam. These include side paper lifting, top smoothing, and bottom dropping, in which case the foam ingredients are fed to an overflow trough and the expanding foam is allowed to grow down instead of up.

Alternative processes are block pouring into a large, open-topped box lined with plastic film from which the cured bun is subsequently removed, or a recently introduced vertical foaming operation. In the latter case the foam ingredients are fed to the bottom of an enclosed trough. As the foam expands vertically it is pulled up by side conveyors. At the top of the square conveyor the foam is cut into length (usually about 2 m) and laid on its side for further curing. Since one of the large foam markets, carpet underlayment, uses long thin sections of foam it is also desirable to generate cylinders of foam which can then be peeled using a long sharp blade. Round buns of foam are generated by proprietary techniques using conventional conveyors and also with the vertical foaming apparatus modified accordingly. Scrap foam is utilized by shredding into small pieces, adding a prepolymer glue, tumbling to mix, compressing into a mold, then curing with steam. This so-called rebond foam is prepared in a variety of density grades then cut, sliced, or peeled to proper form for a number of applications including carpet underlayment.

Applications.   Carpet underlayment as just described is a substantial market. Most furniture cushioning is made from blocks of slab-produced polyurethanefoam in the density range of 16 to 29 kg/m^3 (1.0–1.8 lbs/ft^3). A minor portion of the market, 9–14 thousands of metric tons (20–30 million pounds) uses 40 kg/m^3 (2.5 lbs/ft^3) high resilient (HR) foam for higher priced furniture cushions. The furniture market for polyurethane foams grew strongly until saturation occurred around 1979. Market use now tends to reflect the current economic trends.

For passenger car seating about 90% is made by the molded foam process. The transportation market has experienced a decline since 1979 due to decreased automotive production and also because U.S. cars have been downsized, resulting in the use of less polyurethane foam per car.

Consumption of polyurethane foam in bedding reached a maximum in 1978 and has since declined. The innerspring mattress has remained the standard in the United States whereas all-foam mattresses have gained a dominant market share in Europe.

Textile uses are a relatively stable area and consist of the lamination of polyester foams to textile products, usually by flame lamination or electronic heat sealing techniques. Flexible or semirigid foams are used in engineered packaging in the form of special slab material. Flexible foams are also used to make filters (reticulated foam), sponges, scrubbers, fabric softener carriers, squeegees, paint applicators, and directly applied foam carpet backing.

Economics.   Flexible polyurethane foam is generally sold by the board foot, 1 in. × 1 ft × 1 ft (0.083 ft^3 = 0.0024 m^3), in the United States. Typical densities are 18.5–32.0 kg/m^3 (1.15 to 2.0 lbs/ft^3) for conventional foams and 40.0 kg/m^3 (2.5 lbs/ft^3) for HR foam. Foam prices are usually double the cost of the chemicals for standard grades. Thus typical foam prices in 1992 were about $2.75/kg ($1.25 per pound). This provides a range of foam costs from $0.12 per board ft at 1.15 lbs/ft^3 to $0.21 per board ft at 2.0 lbs/ft^3 ($51–$88/m^3).

RIGID POLYURETHANE

These foams are characterized by closed-celled structure and very high compressive strength. They are produced by using a highly branched, short-chain polyol reacted with an aromatic isocyanate of two or more functionality which is often polymeric. Pour-in-place and free rise rigid polyurethane foams usually have a density in the region of 32.0 kg/m^3 (2.0 lbs/ft^3), although molded rigid foams have densities ranging up to 640 kg/m^3 (40 lbs/ft^3) in structural foams. Insulation effectiveness is one of the outstanding characteristics of rigid polyurethane foams which display thermal conductivities as low as 0.017 W/(m·K).

Raw Materials. The highly branched, short-chain polyols used for rigid foams can be initiated from amines such as diethylenetriamine to provide five functional sites or saccharides such as sorbitol or sucrose that have 6 or 8 functional sites, respectively. Subsequent polymerization of PO and/or EO at low levels further controls viscosity and reactivity of the resultant polyol. The level of oxide addition also contributes to the rigidity of the final foam product by controlling the molecular weight per branch point as well as influencing shrinkage resistance and moisture sensitivity. Amine-initiated polyols tend to be autocatalytic due to the tertiary amine groups residual in the molecule.

The isocyanates used with rigid foam systems are either polymeric MDI or specialty types of TDI. Both contain various levels of polymerized isocyanate groups which contribute to molecular weight per cross-link and also may affect reactivity due to steric hindrance of some isocyanate positions.

Surfactants for use with rigid foams are also silicone based but are quite different from those used for flexible foams. In this case it is more important for the surfactant to also act as a compatibilizer in assisting the intermixing of the isocyanate and polyol during the reaction period. Of course nucleation and cell stabilization during the early phase of foaming are also important functions of the surfactant. Water may also be used in rigid formulations but to a much lesser degree than in flexible foams.

Rigid polyurethane foams are normally foamed with CFC-11 or HCFC-141b as a blowing agent which has very low thermal conductivity. Because of the closed-celled structure of these foams and the low permeability of CFC-11 or HCFC-141b, the blowing agent is retained in the foam for a long period. Therefore, the superior insulating properties of these products is primarily due to the presence of CFC-11 or HCFC-141b in the foam cells.

Catalysis is usually accomplished through the use of tertiary amines such as triethylenediamine. Other catalysts such as 2,4,6-*tris*(*N,N*-dimethylaminomethyl)phenol are used in the presence of high levels of crude MDI to promote trimerization of the isocyanate and thus form isocyanurate ring structures. These groups are more thermally stable than the urethane structure and hence are desirable for improved flammability resistance (236). Some urethane content is desirable for improved physical properties such as abrasion resistance.

Miscellaneous chemicals are used to modify the final properties of rigid polyurethane foams. For example, halogenated materials are used for flammability reduction, diols may be added for toughness or flexibility, and terephthalate-based polyester polyols may be used for decreased flammability and smoke generation. Measurements of flammability and smoke characteristics are made with laboratory tests and do not necessarily reflect the effects of foams in actual fire situations.

Process and Equipment. Rigid polyurethane foam processes use the same high or low pressure pumping, metering, and mixing equipment as earlier described for flexible foams. Subsequent handling of the mixture is determined by the end product desired.

Lamination. Rigid foam boardstock with a variety of facer materials is commonly used for insulation in building construction. The boardstock is produced on a continuous basis by applying the polyurethane (or polyisocyanurate) forming mixture onto one facer sheet, allowing the mixture to begin foaming, applying the second facer on top, and passing the assembly into a fixed gap conveyor to provide heat for cure and control thickness. This is followed by edge trimming and cutting to board length. In this manner boardstock is produced with facer materials such as kraft paper, aluminum foil, tarpaper, etc, and a foam core thickness ranging up to 7.6 cm (3 in.).

Pour-in-Place. The polyurethane forming mixture can be poured into a cavity which will then be filled by the flowing, foaming reaction mixture. This method is used for such things as insulating refrigerator cabinets and filling hull cavities in boats and barges.

Molding. The reaction mixture can be discharged into a mold to flow out and fill the cavity. High density (about 320 kg/m^3 or 20 lbs/ft^3) moldings can be used for decorative furniture items such as drawer fronts or clock frames. The formulation can be adjusted to produce articles with a nonfoam skin layer and a cellular core which are known as structural foams.

Bun Stock. By pouring the reaction mixture on a continuous belt a long bun can be produced like the flexible slab foam previously described. After curing, the bun can be cut into slabs or blocks as required by the end use.

Box Foams. A measured quantity of the reaction mixture can be placed in an open-topped crate or box and allowed to foam in a free rise mode. The block is removed after gelling and is cut into end use pieces after curing.

Spray. In spray-on applications the reactive ingredients are impingement mixed at the spray head. Thickness of the foam is controlled by the amount applied per unit area and additional coats are used if greater than 2.5 cm (1.0 in.) thickness is required. This method is commonly used for coating industrial roofs or insulating tanks and pipes.

Applications. The principal use for rigid polyurethane foams is for insulation in various forms utilized by a variety of industries. Laminates for residential sheathing (1.2 to 2.5 cm thick with aluminum skins) and roofing board (2.5 to 10.0 cm thick with roofing paper skins) are the leading products with about 45 metric tons of liquid spray systems also in use. Metal doors insulated by a pour-in-place process constitute another substantial use.

Household refrigerator and freezer designs have been influenced by the increased cost of energy and the need to develop competitive units with comparable energy efficiency ratings. These factors have increased the use of rigid polyurethane foam as pour-in-place insulation in place of the fiber glass insulation now used in only about 30% of the market. Since CFC blown foam has much better insulating effectiveness, the cabinet wall thickness can be reduced from the former fiber glass-centered design. The pour-in-place cabinet insulating process is carried out in large-scale integrated operations. Commercial refrigeration applications are found in cold storage room insulation, reach-in coolers, and retail display cases. These markets are also using more insulation to offset the higher cost of energy.

The principal use of rigid foam in the transportation market is for insulation of refrigerated truck trailers and bodies as well as refrigerated rail cars. The liquid urethane ingredients are usually poured into large panels held in a fixture. These are then used as integral components: walls, roofs, or floors of the trailer or rail car. Additional uses are insulated truck bodies, recreational vehicles, and cargo containers.

Tanks, pipes, and ducts have been increasingly insulated due to the high cost of energy. For example oil storage tanks must be kept warm to maintain a moderate viscosity for pumping. The energy required to maintain this temperature can be sharply reduced by insulating the tank with rigid polyurethane foam. This type of insulation is often spray applied but may also be cut from boardstock.

Packaging constitutes another significant use and is often a foam-in-place operation to protect industrial equipment such as pumps or motors. Furniture articles molded from rigid foam are used in the form of decorative drawer fronts, clock cases, and simulated wooden beams. Flotation for barge repair and sport boats as well as insulation for portable coolers are a few other uses.

Economics.   Rigid foam systems are typically in the range of 32 kg/m^3 (2 lbs/ft^3) and, in 1992, had a foam price of about $3.63/kg ($1.65 per lb) with liquid foam systems at about $2.75/kg. Unit prices for pour-in-place polyurethane packaging systems fall between the competitive expandable polystyrene bead foam at $3.30/kg and low density polyethylene foams at $5.80/kg.

POLYSTYRENE

There are five basic types of polystyrene foams produced in a wide range of densities and employed in a wide variety of applications: (1) extruded polystyrene board; (2) extruded polystyrene sheet; (3) expanded bead molding; (4) injection molded structural foam; and (5) expanded polystyrene loose-fill packaging.

Expanded polystyrene (EPS) beadboard insulation is produced with expandable polystyrene beads. These beads are produced by impregnating with 5 to 8% pentane and sometimes with flame retardants such as hexabromocyclododecane, pentabromomonochlorocyclohexane, or a synergistic mixture of antimony trioxide and dicumyl peroxide during suspension polymerization. The beads are pre-expanded by fabricators with steam or vacuum and then allowed to age. The pre-expanded beads are fed to the steam heated block molds where further expansion and fusion of beads take place. The molded blocks are then sliced into various sizes needed for specific applications after curing. Block densities range from 13–48 kg/m^3 (0.8–3 lbs/ft^3) with 24 kg/m^3 (1.5 lbs/ft^3) most common for cushion packaging and 16 kg/m^3 (1.0 lb/ft^3) for insulation applications. The 1993 price of EPS beads was $1.58/kg ($0.72/lb) to the molder.

Expanded polystyrene bead molding products account for the largest portion of the drinking cup market and are used in fabricating a variety of other products including packaging materials, insulation board, and ice chests. The insulation value, the moisture resistance, and physical properties are inferior to extruded boardstock, but the material cost is much less.

Expanded polystyrene loose-fill packaging materials are produced normally by extrusion process followed by multiple steam expansions to give low density foam shapes that resemble "S", "8", and hollow shells. They are produced with either pentane or HCFC-141b or pentane/HCFC-141b mixed blowing agents. Expandable polystyrene loose-fill packaging material is also produced by suspension polymerization process with blowing agent incorporated into the polymer during the polymerization. These products are used as dunnage or space filling materials for cushion packaging. Under severe load conditions, vibrational settling may occur, resulting in a nonuniform cushioning protection throughout the package. They have good shock absorbency, excellent resiliency, and are odorless.

The light weight of these products reduces user's shipping costs and conserves energy in transportation. These products are reusable, a key property from economic, ecological, and energy conservation standpoints. Most products are available in bulk densities of 4.0 to 4.8 kg/m^3 (0.25 to 0.30 lb/ft^3). Average price is about $1.50 per pound from the manufacturer.

Extruded polystyrene board was first introduced in the early 1940s by Dow Chemical Co. with the tradename Styrofoam (89,240,241). The Styrofoam process consists of the extrusion of a mixture of polystyrene and volatile liquid blowing agent expanded through a die to form boards in various sizes. The continuous boards are then passed through the finishing equipment for further sizing.

In 1979, UC Industries, a joint venture between U.S. Gypsum and Condec Corp., began manufacture of a similar extruded polystyrene foam under the tradename FoamulaR. Its process is believed to consist of a single-screw tandem extrusion line (114.3-mm main extruder and 152.4-mm extruder as a cooler) and produce foam boardstock in a vacuum chamber connected to a barometric leg which acts as a vacuum seal (242,243). The continuous foam board coming out of a pool of water is then passed through the finishing equipment for sizing.

In 1982, Minnesota Diversified Products, Inc. started to produce extruded polystyrene foam insulation under the trade name Certifoam. This product is believed to be produced by the process (244) developed by LMP (Lavorazione Materie Plastiche) SpA, Turin, Italy. The LMP process for producing extruded polystyrene board consists of a corotating twin-screw extruder (132 mm diameter, 21:1 L/D) with a single-screw extension as a cooling section, a combination motionless mixer/homogenizer and heat exchanger, a flat die, and finishing equipment for sizing and curing.

In residential sheathing insulation, fiberboard is still the most widely used product, although the use of extruded and molded polystyrene foam and of foil-faced isocyanurate foam is increasing depending on the cost, the amount of insulation required, and compatibility of insulation with other construction systems. In cavity-wall insulation, mineral wool, polyurethane, urea–formaldehyde, and fiber glass are widely used, although fiber glass batt is the most economical insulation for stud-wall construction. In mobile and modular homes, cellular plastics are used widely because of their light weight and more efficient insulation value.

The list price of foam plastic sheathing is about $0.26 per board foot (> $100/m^3) for a typical 25 mm (1 in.) sheathing material. The foam density ranges between 23 kg/m^3 (1.4 lbs/ft^3) and 40 kg/m^3 (2.5 lbs/ft^3) depending on the process and blowing agent used to produce a typical 25 mm (1 in.) sheathing product.

Extruded polystyrene foam sheet is primarily produced in a single-screw tandem extrusion line consisting of a 114.3 mm (4.5 in.) primary extruder, screen changer, 152.4 mm (6.0 in.) secondary extruder as a cooler, and an annular die. Typical throughput rate for this size ranges from 340 to 450 kg/h. The sheet is normally extruded in thicknesses of about 0.4 to 6.5 mm, and at densities from about 50 to 160 kg/m^3. Polystyrene pellets and a nucleating agent such as talc or a combination of citric acid and sodium bicarbonate are fed to a primary extruder and melted. A blowing agent such as *n*-pentane, isopentane, HCFC-22, or HFC-152a is then injected into the primary extruder and mixed with the molten polymer. The mixture is passed through a secondary extruder to cool the mixture to appropriate foaming temperature. The cooled polymer gel is then passed through an annular die at which point foaming takes place. The foam bubble is pulled over a sizing mandrel and slit to obtain a flat sheet, which is then wound into a roll for storage and curing. The cured sheet is thermoformed into a finished product by either sheet manufacturers or fabricators. The raw material cost for the foam sheet is higher than that for the foam insulation boardstock because of its higher density. On the other hand, the capital cost for the foam sheet line is lower than that for the foam board because of its simpler finishing equipment. Primary application of foam sheet is as a packaging material in items such as disposable dishes and food containers, trays for meat, poultry and produce products, and egg cartons.

Injection molded structural foam is used widely for high density items such as picture frames, furniture, appliances, housewares, utensils, toys, pipes, and fittings. Most of these products are produced by injection molding or profile extrusion methods from impact modified polystyrene. Almost all high density foam products are produced with a chemical blowing agent that releases either nitrogen or carbon dioxide, typically sodium bicarbonate or azodicarbonamides. Medium density products can be produced with either a physical or chemical blowing agent, or a combination of both.

POLY(VINYL CHLORIDE)

Cellular poly(vinyl chloride) (PVC) foam is available in both flexible and rigid foams. Flexible PVC foams are primarily produced by spread coating and calendering of fluid plastisols by means of a chemical blowing agent or mechanical frothing with air. Flexible PVC foams also are made by the extrusion process. Rigid PVC foams are produced by the extrusion or injection molding processes. Blowing is achieved by a chemical blowing agent or gas injection into the extruder.

Raw Materials. PVC is inherently a hard and brittle material and very sensitive to heat; it thus must be modified with a variety of plasticizers, stabilizers, and other processing aids to form heat-stable flexible or semiflexible products or with lesser amounts of these processing aids for the manufacture of rigid products (see VINYL POLYMERS, VINYL CHLORIDE AND POLY(VINYL CHLORIDE)). Plasticizer levels used to produce the desired softness and flexibility in a finished product vary between 25 parts per hundred (pph) parts of PVC for flooring products to about 80–100 pph for apparel products (245). Numerous plasticizers (qv) are commercially available for PVC, although dioctyl phthalate (DOP) is by far the most widely used in industrial applications due to its excellent properties and low cost. For example, phosphates provide improved flame resistance, adipate esters enhance low temperature flexibility, polymeric plasticizers such as glycol adipates and azelates improve the migration resistance, and phthalate esters provide compatibility and flexibility (245).

In addition to modifying PVC with plasticizers, it is also necessary to incorporate heat stabilizers (qv) into the formulation in order to scavenge the evolved HCl at the processing temperatures, thereby reducing thermal degradation of the polymer. Typical heat stabilizers used for PVC are metallic compounds of barium, cadmium, zinc, lead, and tin; lead and zinc are the most common (245). Plasticizers containing epoxy linkages such as epoxidized soy bean oil or synergistic compounds such as dibasic lead phthalate and dibasic lead phosphite are also used to enhance heat stability. Other ingredients such as color pigments and fillers are added to the formulation for the desired coloration and cost reduction, respectively.

There are two principal PVC resins for producing vinyl foams: suspension resin and dispersion resin. The suspension resin is prepared by suspension polymerization with a relatively large particle size in the 30–250 μm range and the dispersion resin is prepared by emulsion polymerization with a fine particle size in the 0.2–2 μm range (245). The latter is used in the manufacture of vinyl plastisols which can be fused without the application of pressure. In addition, plastisol blending resins, which are fine particle size suspension resins, can be used as a partial replacement for the dispersion resin in a plastisol system to reduce the resin costs.

A most widely used decomposable chemical blowing agent is azodicarbonamide. Its decomposition temperature and rate of evolution of gaseous components are greatly influenced by the stabilizers containing zinc. Lead and cadmium are considered moderate activators for p,p'-oxybis benzenesulfonyl hydrazide (OBSH). OBSH can also be used as a blowing agent for PVC foams.

Process and Equipment. *Flexible Poly(vinyl chloride) Foam. Spread coating* is usually carried out by applying a thin coating of plastisol skin coat on a release paper which is then partially fused in a forced air convection oven in the range of 150°C to facilitate rolling and unrolling of the product. This product passes through the second coating head where a plastisol containing suitable chemical blowing agent is applied to the plastisol skin side of the laminate. The fabric is then adhered to the foam plastisol and passed through the final oven at 200–235°C for fusion and foaming. The paper is separated from the vinyl foam and both the paper and the product are taken away by separate winding rolls (245). The optimum oven temperatures depend on the residence time and the type of blowing agent used.

A *calender processing* is also used to produce substantial quantities of vinyl-fabric laminates. Raw materials are first blended in a Banbury mixer operated at either elevated or room temperatures to dissolve the plasticizer into the PVC resin. The blended materials are fluxed into a homogeneous mass of vinyl compound. The material is then discharged to a Banbury mill to cool the batch down.

The material can now be fed to an extruder and passed through the various nips between the calender rolls to obtain a sheet of well-controlled gauge. Vinyl foam-fabric laminates may be produced by combining a vinyl film to be used as the skin layer and a vinyl sheet containing blowing agent with fabric and activating the blowing agent by passing through a forced air convection oven.

The *chemical expansion method* is most widely used for the manufacture of flexible PVC foam. The three general methods used to produce flexible vinyl foam (246) are (*1*) the pressure molding technique, which consists of the decomposition of the blowing agent and fusion of the plastisol in a mold under pressure at elevated temperatures, cooling the mold, removing the molded part, and post expansion at some moderate temperature; (*2*) the one-stage atmospheric foaming method in which the blowing agent is decomposed in the hot viscosity range that lies between the gelation and complete fusion of the plastisol; and (*3*) the two-stage atmospheric foaming method in which the blowing agent is decomposed below the gelation of the plastisol, followed by heating at high temperature to fuse the foamed resin (247).

The *mechanical process* is used to produce low density, open-celled foam by expanding the plastisol before gelation and fusion. The three general methods (246) include the Dennis process (248), elastomer process (249), and the Vanderbilt process (250,251). The Dennis process utilizes a countercurrent adsorption technique by gravity feeding of the liquid plastisol through a packed absorption column under a low pressure (<690 kPa) of carbon dioxide in order to provide the largest surface area for absorption. The chilled plastisol mixture is pumped under pressure through a nozzle or tube and foams as it comes to atmospheric pressure. The wet foam is then gelled (170–182°C) in a conventional oven for thin sections or a high frequency oven for thick sections.

The elastomer process is very similar to the Dennis process. It involves a number of steps in which a gas, formerly carbon dioxide and now fluorocarbon, is mixed with a plastisol under pressure. When released to atmospheric pressure, the gas expands the vinyl compounds into a low density, open-celled foam which is then fused with heat.

The Vanderbilt process involves the mechanical frothing of air into a plastisol containing proprietary surfactants by means of an Oakes foamer or a Hobart-type batch whip. The resulting stable froth is spread or molded in its final form, then gelled and fused under controlled heat. The fused product is open-celled with fine cell size and density as low as 160 kg/m^3 (10 lbs/ft^3).

Rigid Poly(vinyl chloride) Foam. The techniques that have been used to produce rigid vinyl foams are similar to those for the manufacture of flexible PVC foams. The two processes that have reached commercial importance for the manufacture of rigid vinyl foams (246) are the Dynamit-Nobel extrusion process and the Kleber-Colombes Polyplastique process for producing cross-linked grafted PVC foams from isocyanate-modified PVC in a two-stage molding process.

The Dynamit-Nobel extrusion process (252) utilizes a volatile plasticizer such as acetone which is injected into the decompression section of a two-stage screw and is uniformly dispersed in the vinyl resin containing a stabilizer. The resulting PVC foam has low density and closed cells.

The Kleber-Colombes rigid PVC foam (253,254) is produced by compression molding vinyl plastisol to react and gel the compound, followed by steam expansion. The process involves mixing, molding, and expansion. The formulation consists of PVC, isocyanate, vinyl monomers such as styrene, anhydrides such as maleic anhydride, polymerization initiators, FC-11, and nucleators. The ingredients are mixed in a Werner-Pfleiderer or a Baker Perkins type of mixer and the resulting plastisol is molded under pressure. The initial temperature of the molds is 100–110°C which increases to 180–200°C due to exothermic polymerization of the vinyl monomers and anhydride. The mold is cooled and the partially expanded PVC is removed and then further expanded by steam. After the water treatment the foam is thermoset with a closed-celled structure and a relatively low thermal conductivity.

Applications.   Flexible cellular poly(vinyl chloride) was developed as a comfort cushioning material with compression–deflection behavior similar to latex rubber foam, and with the added feature of flame retardancy (43). It has a larger compression set than either latex rubber or polyurethane foams. The fact that the plasticizer in flexible vinyl foams can migrate to the surface restricts flexible vinyl foams in some applications. Furniture and motor vehicle upholstery is the largest market for flexible vinyl foams. Because of better aesthetics (leather-like plastics), comfort, and favorable pricing, they are expected to show good growth in upholstery, carpet backing, resilient floor coverings, outerwear, footwear, luggage, and handbags. The only application for flexible vinyl foams in protective packaging applications is for stretch pallet wraps. These wraps are produced by extrusion.

Rigid vinyl foams in construction markets have grown substantially due to improved techniques to manufacture articles with controlled densities and smooth outer surfaces. Wood molding substitute for door frames and other wood products is an area that has grown. Rigid vinyl foams are also used in the manufacture of pipes and wires as resin extenders and in sidings and windows as the replacement of wood or wood substitutes.

Economics.   The price of rubber modified flexible PVC foam ranges between about $2.00 to $3.00 per board foot ($800–1200/m^3) and that of unmodified, plasticized PVC foam is about $0.70 to $2.50 per board foot ($300–$1000/$m^3$) depending on the volume, thickness, and density of the product.

POLYETHYLENE

There are three basic types of polyethylene foams of importance: (1) extruded foams from low density polyethylene (LDPE); (2) foam products from high density polyethylene (HDPE); and (3) cross-linked polyethylene foams. Other polyolefin foams have an insignificant volume as compared to polyethylene foams and most of their uses are as resin extenders.

Extruded low density foam produced from LDPE is a tough, flexible, and resilient closed-celled foam used in a wide variety of applications such as cushion packaging and safety components. The resiliency of this product gives excellent energy absorption so important in cushion packaging, athletic pads, flotation devices, and occupant safety applications. Unlike other resilient products, uniform energy absorption can be achieved with low density polyethylene foam over an extremely wide temperature range from −54 to 71°C. The closed-celled nature of this product leads to negligible water pickup. This is important in military packaging where outdoor tropical storage or shipments in high humidity ship holds is common or where freeze–thaw arctic storage conditions are encountered. These products are produced with HCFC-142b, butane, or HFC-152a, or a mixture of these blowing agents depending on the foam thickness. Therefore, these blowing agents play a unique role in achieving the dimensional stability of the flexible low density polyethylene foam over the wide range of temperatures due to their close matching of permeability through LDPE with that of air.

HDPE foam is primarily used as a high density rigid product. Shipping pallets are a rapidly growing market at a projected growth rate of about 26% per year for the mid–1990s. Most of these products are produced by thermoforming sheet and injection molding.

Cross-linked polyethylene foams are produced by either radiation or chemical cross-linking of an extruded expandable sheet containing a chemical blowing agent. The cross-linked expandable sheet is subsequently passed over a molten salt bath or passed through hot air ovens. This process is somewhat complicated, expensive, and limited to the thin products in the continuous process but thicker foams can be produced in a more complicated batch process. A batch molding process utilizing expandable beads is also used to produce thicker foams. These products can be produced in a wide range of densities and thicknesses with fine cell size, having more flexibility, higher resiliency, and better thermoforming capability than the extruded foam products from LDPE. These products also have finer texture and a softer, more resilient feel than extruded low density polyethylene foams and are used in comfort cushioning and cushion packaging applications.

Kanegafuchi Chemical of Japan has introduced a chemical cross-linking process for producing PE foams by the bead technique similar to EPS. Their Eperan beads have been used to produce molded articles as cushioning materials, sound insulating panels, etc. Asahi-Dow and BASF have also been reported to have developed similar products.

The list price of 35 kg/m³ polyethylene foam boards is about $0.58/BF ($6.90/kg) and that of cross-linked polyethylene foam is about $0.33 to $0.88/kg more expensive than uncross-linked polyethylene foam depending on the density and thickness of the foam.

Health and Safety

Flammability. Plastic foams are organic in nature and, therefore, are combustible. They vary in their response to small sources of ignition because of composition and/or additives (255). All plastic foams should be handled, transported, and used according to manufacturers' recommendations as well as applicable local and national codes and regulations.

Virtually all plastic foams are blown with inert gases (CO_2, N_2, H_2O), chemical blowing agents that release inert gases, hydrocarbons containing 3–5 carbon atoms, chlorinated hydrocarbons, chlorofluorocarbons such as CFC-11, CFC-12,

CFC-113, and CFC-114, and hydrochlorofluorocarbons or hydrofluorocarbons such as HCFC-22, HCFC-141b, HCFC-142b, HFC-152a, and HFC-134a. Among these blowing agents, hydrocarbons and some of the HCFs and HFCs are flammable and pose a fire hazard in handling at the manufacturing plants (see FLUO-RINE COMPOUNDS, ORGANIC–FLUORINATED ALIPHATIC COMPOUNDS).

Atmospheric Emissions. Certain organic compounds are found to be smog generating substances because of their high photochemical reactivity at ambient conditions. Since fully or partially halogenated hydrocarbons are considered to have low reactivity in the lower atmosphere (troposphere), substitution of photochemically reactive compounds for the current blowing agents may reduce ozone depletion in the stratosphere, but has adverse impact on the indoor ambient air quality. Therefore, ozone/oxidant interaction with the total environment needs to be considered in developing environmentally acceptable alternative blowing agents (256).

Toxicity. The products of combustion have been studied for a number of plastic foams (257). As with other organics the primary products of combustion are most often carbon monoxide and carbon dioxide with smaller amounts of many other species depending on product composition and test conditions.

The presence of additives or unreacted monomers in certain plastic foams can limit their use where food or human contact is anticipated. Heavy metals can also be found in various additives. The manufacturers' recommendations or existing regulations again should be followed for such applications.

BIBLIOGRAPHY

"Foamed Plastics" in *ECT* 2nd ed., Vol. 9, pp. 847–884, by R. E. Skochdopole, The Dow Chemical Co.; in *ECT* 3rd ed., Vol. 11, pp. 82–126, by K. W. Suh and R. E. Skochcopole, The Dow Chemical Co.

1. E. Ott, H. M. Spulin, and M. W. Grafflin, eds., *Cellulose and Cellulose Derivatives*, 2nd ed., Part I, Interscience Publishers, Inc., New York, 1954, p. 9.
2. U.S. Pat. 942,699 (Dec. 7, 1909), L. H. Baekeland.
3. E. W. Madge, *Latex Foam Rubber*, John Wiley & Sons, Inc., New York, 1962.
4. A. Cooper, *Plast. Inst. Trans. J.* **29**, 39 (1961).
5. U.S. Pat. 3,058,166 (Oct. 16, 1962), R. T. Fields (to E. I. du Pont de Nemours & Co., Inc.).
6. A. R. Ingram, *J. Cell. Plast.* **1**(1), 69 (1965).
7. *TYRIL Foam 80*, technical data sheet no. 2-1, The Dow Chemical Co., Midland, Mich., Jan. 1964.
8. *Plast. Technol.* **28**(3) (Mar. 1982).
9. U.S. Pat. 3,062,729 (Nov. 6, 1962), R. E. Skochdopole, L. C. Rubens, and G. D. Jones (to The Dow Chemical Co.).
10. H. Junger, *Kunststoff-Rundschau* **9**, 437 (1962).
11. W. C. Goggin and O. R. McIntire, *Br. Plast.* **19**(223), 528 (1947).
12. A. F. Randolph, ed., *Plastics Engineering Handbook*, 3rd ed., Reinhold Publishing Corp., New York, 1960, pp. 137–138.
13. J. H. Saunders and K. C. Frisch, *Polyurethanes, Chemistry and Technology*, Vol. I, Wiley-Interscience, New York, 1963.

14. A. E. Lever, *Plastics (London)* **18**(193), 274 (1953).
15. *Plast. World*, 1 (May 1953).
16. *ETHAFOAM Brand Polyethylene Foam*, bulletin, The Dow Chemical Co., Midland, Mich., 1980.
17. U.S. Pat. 3,058,161 (Oct. 16, 1962), C. E. Beyer and R. B. Dahl (to The Dow Chemical Co.).
18. L. Nicholas and G. T. Gmitter, *J. Cell. Plast.* **1**(1), 85 (1965).
19. T. Wirtz, "Integral Skin Foam, A Progress in Urethane Molding," paper presented at the *2nd SPI International Cellular Plastic Conference*, Society of Plastics Industry, New York, Nov. 8, 1968.
20. G. P. Kratzschmer, *Plastic Trends, T44 Plastic Foams*, Predicasts Inc., Cleveland, Ohio, 1977; W. P. Weiser, *T71 U.S. Plastic Foam Markets, Part I–Industry Study*, Predicasts Inc., Cleveland, Ohio, Aug., 1983.
21. *ASTM D883-80C*, American Society for Testing and Materials, Philadelphia, Pa., 1982.
22. J. D. Griffin and R. E. Skochdopole, in E. Baer, ed., *Engineering Design for Plastics*, Reinhold Publishing Corp., New York, 1964.
23. A. Cooper, *Plast. Inst. (London) Trans.* **26** 299 (1958).
24. *ASTM D1566-82*, Vol. 37, ASTM, Philadelphia, Pa., 1982.
25. *ASTM D1056-78*, ASTM, Philadelphia, Pa., 1982.
26. *ASTM D1055-80*, ASTM, Philadelphia, Pa., 1982.
27. F. Shutov, *Integral/Structural Polymer Foams*, Springer-Verlag, Berlin, 1986.
28. K. W. Suh, in D. Klempner and K. C. Frisch, eds., *Polymeric Foams*, Hanser Publishers, New York, 1991.
29. R. H. Hansen, *SPE J.* **18**, 77 (1962).
30. A. R. Ingram and H. A. Wright, *Mod. Plast.* **41**(3), 152 (1963).
31. K. Hinselmann and J. Stabenow, *Ver. Dtsch. Ing.* **18**, 165 (1972).
32. K. C. Frisch and J. H. Saunders, *Plastic Foams*, Vol. 1, Part 1, Marcel Dekker, Inc., New York, 1972.
33. U.S. Pat. 2,681,321 (June 15, 1954), F. Stastny and R. Gaeth (to BASF).
34. U.S. Pat. 2,983,692 (May 9, 1961), G. F. D'Alelio (to Koppers Co.).
35. H. R. Lasman, *Mod. Plast.* **42**(1A), 314 (1964).
36. S. J. Skinner, S. Baxter, and P. J. Grey, *Trans. J. Plast. Inst.* **32**, 180 (1964).
37. *Ibid.*, p. 212.
38. S. J. Skinner and S. D. Eagleton, *Trans. J. Plast. Inst.* **32**, 321 (1964).
39. K. Goodier, *Br. Plast.* **35**, 349 (1962).
40. U.S. Pat. 3,697,454 (Oct. 10, 1972), R. L. Trimble (to Sinclair Koppers).
41. U.S. Pat. 3,066,382 (Dec. 4, 1962), M. L. Zweigle and W. E. Humbert (to The Dow Chemical Co.).
42. *PELASPAN-PAC Bulletin*, The Dow Chemical Co., Midland, Mich., 1979.
43. R. J. Meyer, *SPE J.* **18**, 678 (1962).
44. Jpn. Kokai 72,43,059 (June 22, 1972), J. Morita and co-workers (to Nitto Electric Industrial Co. Ltd.).
45. Jpn. Pat. 73,04,868 (Feb. 12, 1973), S. Nakada and co-workers (to Sekisui Chemical Co. Ltd.).
46. Brit. Pat. 1,333,392 (Oct. 10, 1973), S. Minami and co-workers (to Toray Industries, Inc.).
47. U.S. Pat. 3,812,225 (May 21, 1974), K. Hosoda and co-workers (to Furukawa Electric).
48. Fr. Pat. 1,446,187 (July 15, 1966), J. Zizlsperger and co-workers (to BASF).
49. A. Osakada and M. Koyama, *Jpn. Chem. Q.* **5**, 55 (1969).

50. J. H. Saunders and K. C. Frisch, *Polyurethanes, Chemistry and Technology*, Vol. II, Wiley-Interscience, New York, 1964.
51. G. Woods, *Flexible Polyurethane Foams*, Elsevier Applied Science Publishers Inc., New York, 1982.
52. F. O. Baskent and J. Pavlenyi, *Plastics Compounding* (Mar./Apr. 1983).
53. H. J. Kollmeier, H. Schator, and P. Zaeske, *Proceedings of the SPI, 26th Annual Conference, San Francisco*, Technomic Publishing Co., Stamford, Conn., Nov. 1981, p. 219.
54. *Bayer Pocket Book for the Plastics Industry*, 3rd ed., Farbenfabriken Bayer, AG, Leverkusen, Germany, Oct. 1963.
55. K. C. Frisch, K. J. Patel, and R. D. Marsh, *J. Cell. Plast.* **6**(5), 203 (1970).
56. R. Merten and co-workers, *J. Cell. Plast.* **4**(7), 262 (1968).
57. H. J. Papa, *Ind. Eng. Chem. Rod. Res. Dev.* **9**, 478 (1970).
58. R. H. Fish, *Proc. NASA Conf. Materials for Improved Fire Safety* **11**, 1 (1970).
59. K. C. Frisch and J. H. Saunders, Plastic Foams, Vol. 1, Part 2, Marcel Dekker, Inc., New York, 1973.
60. C. J. Benning, *Plastic Foams*, Vol. 1, Wiley-Interscience, New York, 1969.
61. H. Weissenfeld, *Kunststoffe* **51**, 698 (1961).
62. Ref. 12, p. 149.
63. U.S. Pat. 3,821,337 (June 28, 1974), E. J. Bunclark and co-workers (to Esso Research Engineering).
64. *Plastic Age*, 17th ed., Japan Plastics Industry Annual, Tokyo, 1974.
65. U.S. Pat. 3,726,708 (Apr. 10, 1973), F. Weissenfels and co-workers (to Dynamit Nobel).
66. Fr. Pat. 2,185,488 (Feb. 8, 1974), J. Dugelay.
67. U.S. Pat. 3,830,894 (Aug. 20, 1974), H. Juenger and co-workers (to Dynamit Nobel).
68. U.S. Pat. 3,807,661 (Mar. 11, 1975), P. J. Crook and S. P. Riley (to Pilkington Brothers, Ltd.).
69. U.S. Pat. 3,907,723 (Sept. 23, 1975), M. Pretot (to Certain-Teed Products Corp.).
70. Ger. Offen. 2,204,945 (Sept. 21, 1972), J. Tardy and co-workers.
71. Fr. Pat. 2,190,613 (Mar. 8, 1974), J. Tardy.
72. U.S. Pat. 3,766,100 (Oct. 16, 1973), H. A. Meyer-Stoll and co-workers (to Deutsche Texaco).
73. U.S. Pat. 3,741,920 (June 26, 1973), F. Weissenfels and co-workers (to Dynamit Nobel).
74. U.S. Pat. 3,694,387 (Sept. 26, 1972), H. Junger and co-workers (to Dynamit Nobel).
75. USSR Pat. 358,342 (Nov. 3, 1972), I. F. Ustinova and co-workers (to Kucherenko, V. A., Central Scientific Research Institute of Building Structures).
76. U.S. Pat. 3,101,242 (Aug. 20, 1963), J. M. Jackson, Jr. (to V. L. Smithers Manufacturing).
77. Jpn. Pat. 72,19,624 (June 5, 1972), M. Asaoka and co-workers (to Hitachi Chemical).
78. Brit. Pat. 1,268,440 (Mar. 29, 1972), D. J. Rush and co-workers (to Midwest Research Institute).
79. K. Murai, *Plast. Age* **18**(6), 93 (1972).
80. V. D. Valgin and co-workers, *Plast. Massy* **1**, 28 (1974).
81. V. C. Valgin, *Europlast. Mon.* **46**(7), 57 (July 1973).
82. U.S. Pat. 3,400,183 (Sept. 3, 1968), P. I. Vidal (to Rocma Anstalt).
83. U.S. Pat. 3,214,793 (Nov. 2, 1965), P. I. Vidal (to Rocma Anstalt).
84. U.S. Pat. 3,122,326 (Feb. 25, 1964), D. P. Cook (to Union Carbide).
85. Ref. 12, p. 140.
86. *Epon Foam Spray 175, A Low Temperature Insulation*, bulletin, Shell Chemical Co., Houston, Tex., Nov. 1963.

87. Ref. 12, p. 164.
88. H. Vincent and K. R. Hoffman, *Mod. Plast.* **40**(1A), 418 (1962).
89. U.S. Pat. 2,515,250 (July 18, 1950), O. R. McIntire (to The Dow Chemical Co.).
90. V. L. Gliniecki, *Seventh Ann. Tech. Conf., Proc.*, SPI, New York, Apr. 24–25, 1963.
91. *Mod. Plast.* **51**(1), 36, 40 (Jan. 1974).
92. T. P. Martens and co-workers, *Plast. Technol.* **12**(9), 46 (1966).
93. D. A. Knauss and F. H. Collins, *Plast. Eng.* **30**(2), 34 (1974).
94. F. H. Collins and co-workers, *Soc. Plast. Eng. Tech. Pap.* **19**, 643 (1973).
95. L. W. Meyer, *SPE J.* **18**, 1341 (1962).
96. Ref. 12, p. 139.
97. W. T. Higgins, *Mod. Plast.* **31**(7), 99 (1954).
98. U.S. Pat. 3,251,911 (May 17, 1966), R. H. Hansen (to Bell Telephone Lab.).
99. U.S. Pat. 3,268,636 (Aug. 23, 1966), R. G. Angell, Jr. (to Union Carbide Corp.).
100. U.S. Pat. 3,065,190 (Nov. 20, 1962), D. S. Chisholm and co-workers (to The Dow Chemical Co.).
101. U.S. Pat. 3,067,147 (Dec. 4, 1962), L. C. Rubens and co-workers (to The Dow Chemical Co.).
102. J. L. Throne, *J. Cell. Plast.* **12**(5), 264 (1976).
103. J. L. Throne and F. Shutov, in J. I. Kroschwitz, ed., *Encyclopedia of Polymer Science and Engineering*, 2nd ed., Vol. 15, John Wiley & Sons, Inc., New York, 1985, pp. 771–797.
104. R. W. Freund and co-workers, *Plast. Technol.*, 35 (Nov. 1973).
105. U.S. Pat. 3,436,446 (Apr. 1, 1969), R. G. Angell, Jr. (to Union Carbide Corp.).
106. *USM Foam Process, Technical Bulletin No. 653-A*, Farrel Co. Division, Ansonia, Conn.
107. Fr. Pat. 1,498,620 (Sept. 11, 1967), P. Botillier (to Ugine-Kuhlmann).
108. Ref. 12, p. 189.
109. U.S. Pat. 2,666,036 (Jan. 12, 1954), E. H. Schwencke (to Elastomer Chemical Co.).
110. U.S. Pat. 2,763,475 (Sept. 18, 1956), I. Dennis.
111. R. C. Bascom, *Rubber Age* **95**, 576 (1964).
112. L. Spenadel, *Rubber World* **150**(5), 69 (1964).
113. J. J. Kolb, in B. C. Wendle, ed., *Engineering Guide to Structural Foam*, Technomic Publishing Co., Inc., Westport, Conn., 1976, p. 161.
114. A. Cooper, *Plast. Inst. Trans.*, 51 (Apr. 1948).
115. T. H. Rogers, "Plastic Foams," paper presented at *Regional Tech. Conf.*, Palisades Section, Society of Plastics Engineers, New York, Nov. 1964.
116. *Dow Latex Foam Process*, bulletin, The Dow Chemical Co., Midland, Mich.
117. *Iporka*, bulletin, Badische Anilin- und Soda-Fabrik AG, Ludwigshafen am Rhein, Germany, July 1953.
118. T. H. Ferrigno, *Rigid Plastic Foams*, Reinhold Publishing Corp., New York, 1963, pp. 76–86.
119. T. E. Cravens, *Carpet Rug Ind.* (Oct. 1976).
120. *Dow Low Temperature Systems*, bulletin, 179-2086-77, The Dow Chemical Co., Midland, Mich., 1977.
121. U.S. Pat. 2,959,508 (Nov. 8, 1960), D. L. Graham and co-workers (to The Dow Chemical Co.).
122. J. B. Brooks and L. G. Rey, *J. Cell. Plast.* **9**(5), 232 (1973).
123. *Chem. Week* **19**(17), 43 (1962).
124. *Chem. Eng. News* **37**(36), 42 (1959).
125. U.S. Pat. 2,772,995 (Dec. 4, 1956), J. D. C. Wilson II (to E.I. du Pont de Nemours & Co., Inc.).
126. Can. Pat. 762,421 (July 4, 1967), M. E. Baguley (to Courtaulds Ltd.).

127. W. Worthy, *Chem. Eng. News*, 23 (Dec. 11, 1978).

128. *1993 Annual Book of ASTM Standards*, ASTM, Philadelphia, Pa., 1993.

129. W. H. Touhey, *J. Cell. Plast.* **4**(10), 395 (1968).

130. C. J. Hilado, *J. Cell. Plast.* **3**(11), 502 (1967).

131. K. W. Suh and D. D. Webb, in J. I. Kroschwitz, ed., *Encyclopedia of Polymer Science and Engineering*, 2nd ed., Vol. 3, John Wiley & Sons, Inc., New York, 1985, p. 1.

132. R. P. Toohy, *Chem. Eng. Prog.* **57**(10), 60 (1961).

133. C. A. Schutz, *J. Cell. Plast.* **4**(2), 37, (1968).

134. R. J. Bender, ed., *Handbook of Foamed Plastics*, Lake Publishing Co., Libertyville, Ill., 1965.

135. *Mod. Plast. Enc.* **54**(10A), 485 (1977–1978).

136. *STYROFOAM SI Brand Plastic Foam*, bulletin, The Dow Chemical Co., Midland, Mich., 1978.

137. Ref. 135, p. 776.

138. *Sweets Catalog File 7, Thermal and Moisture Protection*, Sweets Div., McGraw-Hill Information System Co., New York, 1978.

139. Y. Landler, *J. Cell. Plast.* **3**(9), 400 (1967).

140. R. K. Traeger, *J. Cell. Plast.* **3**(9), 405 (1967).

141. H. E. Reymove and co-workers, *J. Cell. Plast.* **11**(6), 328 (1975).

142. *U.S. Foamed Plastics Market and Directory*, Technomic Publishing Co., Stamford, Conn., 1973.

143. R. E. Skochdopole, *Chem. Eng. Prog.* **57**(10), 55 (1961).

144. R. H. Harding, *Mod. Plast.* **37**(10), 156 (1960).

145. D. M. Rice and L. J. Nunez, *SPE J.* **18**, 321 (1962).

146. *Br. Plast.* **35**, 18 (1962).

147. A. J. deVries, *Meded. Rubber Sticht. Delft* **326**, 11 (1957); *Rec. Trav. Chim.* **77**, 81, 209, 283, 383, 441 (1958).

148. T. L. Phillips and D. A. Lannon, *Br. Plast.* **34**, 236 (1961).

149. R. E. Skochdopole and L. C. Rubens, *J. Cell. Plast.* **1**(1), 91 (1965).

150. A. Cooper, *Plast. Inst. Trans.* **26**, 299 (1958).

151. D. J. Doherty, R. Hurd, and G. R. Lester, *Chem. Ind. (London)*, 1340 (1962).

152. J. F. Hawden, *Rubber Plast. Age* **44**, 921 (1963).

153. R. H. Harding, *J. Cell. Plast.* **1**(3), 385 (1965).

154. R. H. Harding, *Resinography of Cellular Materials*, ASTM Technical Publication 414, ASTM, Philadelphia, Pa., 1967.

155. R. J. Corruccini, *Chem. Eng. Prog.* **53**, 397 (1957).

156. E. A. Meinecke and R. E. Clark, *Mechanical Properties of Polymeric Foams*, Technomic Publishing Co., Stamford, Conn., 1972.

157. R. M. McClintock, *Adv. Cryog. Eng.* **4**, 132 (1960).

158. W. B. Brown, *Plast. Prog.* **1959**, 149 (1960).

159. G. M. Hart, C. F. Balazs, and R. B. Clipper, *J. Cell. Plast.* **9**(3), 139 (1973).

160. J. L. Eakin, *Plast. Eng.* **34**(7), 56 (1978).

161. G. H. Smith, *Rubber Plast. Age* **44**(2), 148 (1963).

162. *ETHAFOAM Brand Plastic Foam*, bulletin, The Dow Chemical Co., Functional Products and Systems Dept., Midland, Mich., 1976.

163. H. H. Lubitz, *J. Cell. Plast.* **5**(4), 221 (1969).

164. J. E. Knight, *Proceedings of the SPI 27th Annual Conference, Bal Harbour, Fla.*, Technomic Publishing Co., Stamford, Conn., Oct. 1982, p. 227.

165. W. R. Nicholson and J. E. Plevyak, *ibid.*, p. 306.

166. J. Pavlenyi and F. O. Baskent, *ibid.*, p. 295.

167. G. Woods, *Flexible Polyurethane Foams*, Elsevier Applied Science Publishers Inc., New York, 1982.

168. C. E. Lee and co-workers, *J. Cell. Plast.* **13**(1), 62 (1977).

169. G. J. Bibby, *Rubber Plast. Age* **45**(1), 52 (1964).

170. *Plast. Technol.* **8**(4), 26 (1962).

171. J. H. Saunders and co-workers, *J. Chem. Eng. Data* **3**, 153 (1958).

172. ·R. P. Marchant, *J. Cell. Plast.* **8**(2), 85 (1972).

173. T. H. Rogers and K. C. Hecker, *Rubber World* **139**, 387 (1958).

174. C. G. Seefried and co-workers, *J. Cell. Plast.* **10**(4), 171 (1974).

175. J. M. Buist and A. Lowe, *Trans. Plast. Inst.* **27**, 13 (1959).

176. J. Talaly, *Ind. Eng. Chem.* **46**, 1530 (1954).

177. A. N. Gent and A. G. Thomas, paper presented at *Prox. 7th Ann. Tech. Conf.*, Cellular Plastics Div., SPI, New York, Apr. 1963.

178. L. Talalay and A. Talalay, *Ind. Eng. Chem.* **44**, 791 (1952).

179. M. A. Mendelsohn and co-workers, *J. Appl. Polym. Sci.* **10**, 443 (1966).

180. E. A. Blair, *Resinography of Cellular Materials*, ASTM Tech. Publ. 414, ASTM, Philadelphia, Pa., 1967, p. 84.

181. J. H. Saunders, *Rubber Chem. Technol.* **33**, 1293 (1960).

182. B. Beals, F. J. Dwyer, and M. A. Kaplan, *J. Cell. Plast.* **1**(1), 32 (1965).

183. R. P. Kane, *J. Cell. Plast.* **1**(1), 217 (1965).

184. *Plast. Technol.* **8**(4), 26 (1964).

185. ·S. M. Terry, *J. Cell. Plast.* **7**(5), 229 (1971).

186. *Ibid.*, **12**(3), 156 (1976).

187. R. L. Gorring and S. W. Churchill, *Chem. Eng. Prog.* **57**(7), 53 (1961).

188. M. E. Stephenson, Jr. and M. Mark, *ASHRAE J.* **3**(2), 75 (1961).

189. F. O. Guenther, *SPE Trans.* **2**, 243 (1962); A. Cunningham and co-workers, *Cell. Plast.* **7**, 1 (1988).

190. R. E. Knox, *ASHRE J.* **4**(10), 3 (1962); D. W. Reitz and co-workers, *J. Cell. Plastics* **20**, 332 (1984).

191. R. H. Harding, *Ind. Eng. Chem. Proc. Des. Dev.* **3**(2), 117 (1964); J. R. Booth, in R. S. Graves and D. C. Wysocki, eds., *Insulating Materials: Testing and Applications*, ASTM 1116, ASTM, Philadelphia, Pa., 1991.

192. *Freon Technical Bulletin #B-2*, E. I. du Pont de Nemours & Co., Inc., Wilmington, Del., 1975; *Ibid.*, #AG-1, 1992.

193. J. H. Perry, *Chemical Engineers Handbook*, 4th ed., McGraw-Hill Book Co., New York, 1963.

194. F. J. Norton, *J. Cell. Plast.* **3**(1), 23 (1967).

195. R. M. Lander, *Refrig. Eng.* **65**(4), 57 (1957).

196. B. K. Larkin and S. W. Churchill, *AIChE J.* **5**, 467 (1959).

197. B. Y. Lao and R. E. Skochdopole, paper presented at *4th SPI International Cellular Plastics Conference*, Montreal, Canada, Nov. 1976.

198. G. A. Patten and R. E. Skochdopole, *Mod. Plast.* **39**(11), 149 (1962); I. R. Shankland, *Adv. Foam Aging* **1**, 60 (1986).

199. C. J. Hilado, *J. Cell. Plast.* **3**(4), 161 (1967); I. R. Shankland, *The Effect of Cell Structure on the Rate of Foam Aging*, International Workshop on Long Term Thermal Performance of Cellular Plastics, SPI, Canada, Oct. 1989.

200. R. R. Dixon, L. E. Edleman, and D. K. McLain, *J. Cell. Plast.* **6**(1), 44 (1970); C. F. Sheffield, *Description and Applications of a Diffusion Model for Rigid Closed-Cell Foams*, ORNL Symposium on Mathematical Modeling of Roofs, Oak Ridge National Laboratory, Oak Ridge, Tenn., Sept. 15, 1988.

201. G. W. Ball, R. Hurd, and M. G. Walker, *J. Cell. Plast.* **6**(2), 66 (1970); D. W. Reitz, and co-workers, *J. Cell. Plastics* **20**, 104 (1984).

202. L. Vahl, in Ref. 117, p. 267.

203. C. H. Wheeler, *Foamed Plastics*, U.S. Army Natick Labs and Committee on Foamed Plastics, U.S. Dept. Comm. Office Tech. Serv. PB Rept. 181576, Apr. 22–23, 1963, p. 164.
204. G. A. Patten, *Mater. Des. Eng.* **55**(5), 117 (1962).
205. *FTC Consent Agreement*, file #7323040, The Dow Chemical Co., Midland, Mich., 1974.
206. C. P. Hedlin, *J. Cell. Plast.* **13**(5), 313 (1977).
207. F. J. Dechow and K. A. Epstein, *Thermal Transmission Measurements of Insulation*, ASTM STP 660, ASTM, Philadelphia, Pa., 1978, p. 234.
208. M. M. Levy, *J. Cell. Plast.* **2**(1), 37 (1966).
209. I. Paljak, *Mater. Constr. (Paris)* **6**, 31 (1973).
210. H. Mittasch, *Plaste Kautsch,* **16**(4), 268 (1969).
211. J. Achtziger, *Kunststoffe* **23**, 3 (1971).
212. C. W. Kaplar, *CRREL Internal Report No. 207*, U.S. Army CRREL, Hanover, N.H., 1969.
213. P. J. Palmer, *J. Cell. Plast.* **9**(4), 182 (1973).
214. N. Z. Searle and R. C. Hirt, *SPE Trans.* **2**, 32 (1962).
215. A. Cooper, *Plastics* **29**(321), 62 (1964).
216. *Mod. Plast.* **39**(8), 93 (1962).
217. G. L. Ball, II, M. Schwartz, and J. S. Long, *Off. Dig. Fed. Soc. Paint Technol.* **32**, 817 (1960).
218. E. F. Cuddihy and J. Moacanin, *J. Cell. Plast.* **3**(2), 73 (1967).
219. C. E. Rogers, in E. Baer, ed., *Engineering Design for Plastics*, Reinhold Publishing Corp., New York, 1964.
220. S. C. A. Paraskevopoulos, *J. Cell. Plast.* **1**(1), 132 (1965).
221. *Forming Thin Shells*, bulletin, The Dow Chemical Co., Midland, Mich., 1962.
222. R. K. Stern, *Mod. Packag.* **33**(4), 138 (1959).
223. R. G. Hanlon and W. E. Humber, *Mod. Packag.* **35**(10), 158 (1962).
224. M. Bakker, ed., *The Wiley Encyclopedia of Packaging Technology*, John Wiley & Sons, Inc., New York, 1986, p. 228.
225. R. G. Hanlon and W. E. Humbert, *Package Eng.* **7**(4), 79 (1962).
226. K. Brown, *Package Design Engineering*, John Wiley & Sons, Inc., New York, 1959.
227. A. R. Gardner, *Prod. Eng.* **34**(25), 114 (1963).
228. N. C. Hilyard, and co-workers, *Mechanics of Cellular Plastics*, Macmillan Publishing Co., Inc., New York, 1982.
229. C. Kienzle, "Plastic Foams," paper presented at *Regional Technical Conference, Buffalo, N.Y.,* Society of Plastics Engineers, Inc., New York, Oct. 5, 1961, p. 93.
230. *Plast. World*, 15 (Mar. 1964).
231. *Mod. Plast.* **42**(1A), 294 (1964).
232. E. B. Murphy and W. A. O'Neil, *SPE J.* **18**, 191 (1962).
233. F. Stastny, *Baugewerbe* **19**, 648 (Apr. 1957).
234. U.S. Pat. 3,383,351 (May 14, 1968), P. Stamberger.
235. U.S. Pat. 3,652,639 (Mar. 28, 1972), L. C. Pizzini (to BASF).
236. U.S. Pat. 4,042,537 (Aug. 16, 1977), M. Dahm and co-workers.
237. S. C. Cohen and co-workers, in Ref. 51, p. 100.
238. R. J. Lockwood and co-workers, in Ref. 165, p. 196.
239. T. R. McClellan and co-workers, in Ref. 165, p. 204.
240. U.S. Pat. 2,023,204 (Dec. 3, 1935), C. G. Munters and co-workers (to C. G. Munters).
241. U.S. Pat. 3,960,792 (June 1, 1976), M. Nakamura (to The Dow Chemical Co.).
242. U.S. Pat. 3,704,083 (Nov. 28, 1972), A. L. Phipps.
243. U.S. Pat. 4,247,276 (Jan. 27, 1981), A. L. Phipps (to Condec Corp.).
244. *Proc. Eng. News*, 11 (July, 1978).
245. A. C. Werner, in Ref. 131.

246. Ref. 60, Chapt. 4.
247. C. S. Sheppard, H. N. Schnack, and O. L. Mageli, *J. Cell. Plast.* **2**, 97 (1966).
248. U.S. Pat. 2,763,475 (Sept. 18, 1956), I. Dennis.
249. U.S. Pat. 2,666,036 (Jan. 12, 1954), E. H. Schwencke (to Elastomer Chemical).
250. U.S. Pat. 3,301,798 (Jan. 31, 1967), R. R. Waterman and D. C. Morris (to R. T. Vanderbilt).
251. U.S. Pat. 3,288,729 (Nov. 29, 1966), R. R. Waterman and K. M. Deal (to R. T. Vanderbilt).
252. U.S. Pat. 3,020,248 (1962), (to Dynamit-Nobel Akt).
253. Brit. Pat. 901,118 (1960), (to Kleber-Colombes).
254. Brit. Pat. 993,763 (1963), (to Kleber-Colombes).
255. C. J. Hilado and R. W. Murphy, *Design of Buildings for Fire Safety*, ASTM Spec. Tech. Publ. STP 685, 16-105 ASTM, Philadelphia, Pa., 1979.
256. *Handbook for the Montreal Protocol on Substances that Deplete the Ozone Layer*, 2nd ed., Ozone Secretariat, UNEP, New York, Oct. 1991.
257. C. J. Hilado, H. J. Cumming, and C. J. Casey, *J. Cell. Plastic.* **15**(4), 205 (1979).

KYUNG W. SUH
The Dow Chemical Company

FOAMS

Foam is a nonequilibrium dispersion of gas bubbles in a relatively smaller volume of liquid. An essential ingredient in a liquid-based foam is surface-active molecules. These reside at the interfaces and are responsible for both the tendency of a liquid to foam and the stability of the resulting dispersion of bubbles. For instance, it is common experience that a relatively stable foam can be made by bubbling gas through soapy water, but not through pure water. Important uses for custom-designed foams vary widely from familiar examples of detergents, cosmetics, and foods, to fire extinguishing, oil recovery, and a host of physical and chemical separation techniques. Unwanted generation of foam, on the other hand, is a common problem affecting the efficiency and speed of a vast number of industrial processes involving the mixing or agitation of multicomponent liquids. In all cases, control of foam rheology and stability is desired. These physical properties, in turn, are determined by both the physical chemistry of their liquid–vapor interfaces and by the structure formed from the collection of gas bubbles.

Observed from a distance, foam made from a clear liquid appears homogeneous and white. When observed more closely, however, the intricate structure formed by the close packing of distinct gas bubbles becomes apparent. Figure 1 illustrates several features of this so-called microstructure which are common to many foams. The sample shown was photographed two hours after thoroughly shaking an aqueous solution of 5% by weight sodium dodecyl sulfate, a common

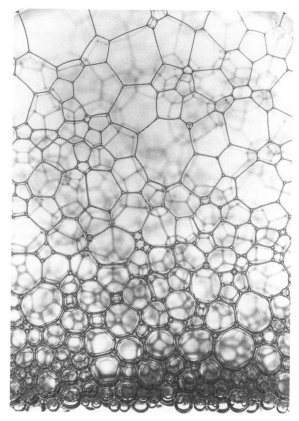

Fig. 1. Photograph illustrating the microstructure of the foam which still persists two hours after shaking an aqueous solution containing 5% sodium dodecylsulfate. The bubble shapes are more polyhedral near the top, where the foam is dry, and more spherical near the bottom, where the foam is wet. The average bubble size is approximately 2 mm.

surfactant. Near the top of the sample, most of the liquid has drained away leaving a dry foam consisting of nearly polyhedral gas bubbles separated by thin liquid films of uniform thickness. Near the bottom of the sample, by contrast, the foam is relatively wet and consists of bubbles that are more nearly spherical. Whether the bubbles are spherical, polyhedral, or in between, they typically have a distribution of sizes and pack together into a disordered, aperiodic structure. In Figure 1 the average bubble diameter is approximately 2 mm, but similar structures are also found in foams where the average bubble diameter is varied from 10 μm to 1 cm. In practice, the average bubble size and shape in a foam can be altered for a given liquid according to the production method, the surface-active ingredients, and other chemical additives such as viscosity modifiers or polymeric stabilizers.

The nonequilibrium nature of foams is revealed by the time evolution of their structures. The sample shown in Figure 1 was homogeneous immediately after shaking, but had evolved by the gravitational segregation of liquid downward and bubbles upward prior to being photographed. In addition to drainage, two other mechanisms by which foams evolve are by direct coalescence of neighboring bubbles via film rupture and by the diffusion of gas molecules through the liquid from small bubble to large bubbles. No matter which of these three processes dominates for a given foam, the liquid and vapor portions invariably consolidate and separate with time; in equilibrium there is no foam, only one region of liquid and one of vapor. The physical chemistry of the interfaces and the foam structure primarily determine the relative rates of the three aging mechanisms.

Foams that are relatively stable on experimentally accessible time scales can be considered a form of matter but defy classification as either solid, liquid, or vapor. They are solid-like in being able to support shear elastically; they are liquid-like in being able to flow and deform into arbitrary shapes; and they are vapor-like in being highly compressible. The rheology of foams is thus both complex and unique, and makes possible a variety of important applications. Many features of foam rheology can be understood in terms of its microscopic structure and its response to macroscopically imposed forces.

Physical Chemistry of Interfaces

The chemical composition, physical structure, and key physical properties of a foam, namely its stability and rheology, are all closely interrelated. Since there is a large interfacial area of contact between liquid and vapor inside a foam, the physical chemistry of liquid–vapor interfaces and their modification by surface-active molecules plays a primary role underlying these interrelationships. Thus the behavior of individual surface-active molecules in solution and near a vapor interface and their influence on interfacial forces is considered here first.

For aqueous solutions, the chemical constituents most commonly responsible for foaming are surfactants, ie, surface-active agents (1). Such molecules find wide use in other settings (see DETERGENCY; SURFACTANTS), and are distinguished by having both hydrophilic and hydrophobic regions. A typical example is the anionic surfactant sodium dodecylsulfate [151-21-3] (SDS). In spite of its hydrophobic hydrocarbon chain, SDS is readily soluble in water due to its polar head group. At concentrations higher than 8 mM (2), the so-called critical micelle concentration (CMC), SDS molecules form spherical micelles where the hydrophobic tails of approximately 64 molecules clump together so that only their hydrophilic heads are exposed to water (3). At still higher concentrations, even more exotic structures are formed in the bulk solution (1).

Reduced Surface Tension. Just as surfactants self-organize in the bulk solution as a result of their hydrophilic and hydrophobic segments, they also preferentially adsorb and organize at the solution–vapor interface. In the case of aqueous surfactant solutions, the hydrophobic tails protrude into the vapor and leave only the hydrophilic head groups in contact with the solution. The favorable energetics of the arrangement can be seen by the reduction in the interfacial free energy per unit area, or surface tension, σ. For example, the liquid–vapor surface tension reduces from about 80 to 33 mN/m(= dyn/cm) as the concentration of SDS increases from zero to the CMC (2). In most custom foams, the surfactant concentration in the base liquid is near or above the CMC. However, the reduced surface tension is not in itself responsible for the foaming; the primary benefit is that less mechanical energy need be supplied to create the large interfacial area in a foam.

The prevention of bubble coalescence needed for significant foaming is accomplished through other physical chemical mechanisms involving surfactants.

Gibbs Elasticity and Marangoni Flows. The reduction of surface tension with increasing surfactant adsorption gives rise to a nonequilibrium effect which can, in some cases, promote foaming. A sudden increase in the interfacial area by mechanical perturbation or thermal fluctuation results in a locally higher surface tension because the number of surfactant molecules per unit area simultaneously decreases. The Gibbs elasticity, E, is often used to quantify the instantaneous change in surface tension σ with area A, ie, $E = d\sigma/d\ln A$. If the film of liquid separating two neighboring bubbles in a foam develops a thin spot, the surface tension gradient in the vicinity of the thin spot will induce a Marangoni flow of liquid toward the direction of higher σ. This flow of liquid toward the thin spot helps heal the fluctuation and thus keeps the neighboring bubbles from coalescing. Note that if the time scale for surfactant diffusion and adsorption is shorter than the hydrodynamic time scale, then the Marangoni effect cannot improve stability. Thus the larger the elasticity E, and the longer it takes for surfactant molecules to diffuse to the new surface and reestablish the equilibrium surface excess concentration of surfactant, the more foaming is promoted. In practice, the Marangoni effect can cause severe foaming problems in industrial processes, but it alone never suffices to give a stable foam.

Interfacial Forces. Neighboring bubbles in a foam interact through a variety of forces which depend on the composition and thickness of liquid between them, and on the physical chemistry of their liquid–vapor interfaces. For a foam to be relatively stable, the net interaction must be sufficiently repulsive at short distances to maintain a significant layer of liquid in between neighboring bubbles. Otherwise two bubbles could approach so closely as to expel all the liquid and fuse into one larger bubble. Repulsive interactions typically become important only for bubble separations smaller than a few hundredths of a micrometer, a length small in comparison with typical bubble sizes. Thus attention can be restricted to the vapor–liquid–vapor film structure formed between neighboring bubbles, and this structure can be considered essentially flat.

van der Waals Interaction. The van der Waals force, also known as the London or dispersion force, always attracts adjacent bubbles together and is therefore destabilizing to foams. This attraction is ultimately of quantum mechanical origin, where like molecules are attracted through the electric fields associated with fluctuations in their instantaneous dipole moments. Summing up these molecular forces over all molecules in a flat film of thickness ℓ and infinite extent, the interaction energy per unit area is $V_{VDW}(\ell) = -A/12\pi\ell^2$ where A is the Hamaker constant. For a film of water in air, $A = 3.7 \times 10^{-21}$ J (4); for other liquids its value is still close to the thermal energy $k_B T$, which sets the scale for van der Waals interactions.

Electrostatic Double Layer Interaction. To prevent the film from thinning to zero thickness under influence of the van der Waals interaction, a balancing repulsive force is required. One possibility is the electrostatic double-layer interaction resulting from the mutual repulsion of charge clouds residing at each side of the film due to the dissociation of surface-adsorbed ionic surfactants. The thickness of the diffuse double layer of charges is roughly the debye screening length

$1/\kappa_D$, whose value is <u>determined</u> by electrolyte content. For 1:1 electrolytes such as NaCl, $1/\kappa_D \approx 1/\sqrt{10.8\rho}$ nm, where ρ is the molar bulk electrolyte concentration. The electrostatic double-layer interaction energy per unit area is given by $V_{DL}(\ell) \approx (64 k_B T \rho \gamma^2/\kappa_D)\exp(-\kappa_D \ell)$ where γ is a factor of order unity for highly dissociated ionic surfactants (4). In practice, the interaction strength may be estimated from knowledge of the liquid solution's pH and electrical conductivity.

The combined effect of van der Waals and electrostatic forces acting together was considered by Derjaguin and Landau (5) and independently by Vervey and Overbeek (6), and is therefore called DLVO theory. It predicts that the total interaction energy per unit area, also known as the effective interface potential, is given by $V(\ell) = V_{VDW}(\ell) + V_{DL}(\ell)$. In the absence of externally imposed forces, the equilibrium thickness of the liquid film separating two bubbles is found by minimizing $V(\ell)$ with respect to ℓ. This is demonstrated in Figure 2 for an aqueous film containing 1 mM of 1:1 electrolyte. For this case the minimum is located near 130 nm, but is too shallow to be seen in comparison with the energy barrier which keeps the film thickness away from the deep van der Waals minimum at zero thickness.

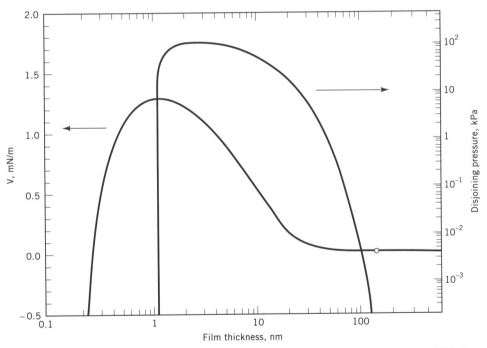

Fig. 2. Effective interface potential (left) and corresponding disjoining pressure (right) vs film thickness as predicted by DLVO theory for an aqueous soap film containing 1 mM of 1:1 electrolyte. The local minimum in $V(\ell)$, marked by $\circ$, gives the equilibrium film thickness in the absence of applied pressure as 130 nm; the disjoining pressure $\Pi = -dV/d\ell$ vanishes at this minimum. The minimum is extremely shallow compared with the stabilizing energy barrier. To convert kPa to atm, divide by 10^2. mN/m = dyn/cm = ergs/cm^2.

Other Interactions. In practical situations, quantitative application of DLVO theory may be uncertain due to uncontrolled Hamaker constants, screening lengths, or dissociation constants. Further complications and unanticipated behavior may also arise due to interactions neglected by DLVO, such as solvation forces. These are typically due to molecular structure in the liquid which plays a role when the film is thinner than several molecular diameters, and can be either attractive or repulsive. An example is the hydration force due to water hydrogen bonding with the polar head group of a surfactant molecule adsorbed to the liquid–vapor interface (7). Another important contribution is steric repulsion, which arises when the liquid contains polymers and the film thickness becomes thinner than the radius of gyration. Yet another example is a stabilizing depletion interaction which occurs if the solution is above the CMC and the film thickness is comparable to a few micelle diameters. In general, the more components the system contains, the more degrees of freedom exist for structure within the film which cannot be accounted for by DLVO.

Disjoining Pressure. A static pressure difference can be imposed between the interior and exterior of a soap film by several means including, for example, gravity. In such cases the equilibrium film thickness depends on the imposed pressure difference as well as on the effective interface potential. When the film thickness does not minimize $V(\ell)$, there arises a disjoining pressure $\Pi = -dV/d\ell$ which drives the system towards mechanical equilibrium. In response to a hydrostatic pressure, the film thickness thus adjusts itself so that the disjoining pressure balances the applied pressure and mechanical equilibrium is restored.

The disjoining pressure vs film thickness as predicted by DLVO theory for an aqueous film containing 1 mM of 1:1 electrolyte is shown along with the effective interface potential in Figure 2. The equilibrium thickness of a free film is where the effective interface potential is at a local minimum or, equivalently, where the disjoining pressure vanishes with a negative slope. If the same film is not free, but instead rises vertically from solution in the presence of the earth's gravitational field, its thickness will vary in response to the height dependence of the hydrostatic pressure. For example, at approximately 8 cm above the solution the hydrostatic pressure in the film drops by $\sim$ 10 kPa and, according to Figure 2, the film thickness at this height must decrease to 30 nm in order to be in equilibrium. Similar considerations are important for establishing the distribution of liquid around several bubbles packed together in a foam, and hence the bubble shapes.

Although the details of the interaction between neighboring films may not be accurately described by the simplest DLVO theory, it nevertheless captures the essential physics. There is a large energy barrier which prevents two films approaching too closely. This energy barrier may arise from electrostatic repulsion, as in the DLVO model, or it may arise from other interactions. However, its role is primarily to prevent two films from approaching sufficiently close that they fall into the deep attractive well. The degree to which the two films are forced together by external forces determines how high up the energy barrier they are forced; this is in turn parameterized by the disjoining pressure. Should the repulsive barrier be overcome, the films fall into the attractive minimum, whereupon they coalesce. Thus this repulsive barrier provides the essential stabilization of the foam.

Physical Properties of Foam

Based on the underlying physical chemistry of surfactants at interfaces, important features of foam structure, stability, rheology, and their interrelationships can be considered as ultimately originating in the molecular composition of the base liquid.

Structure.   *Very Wet Foam: Froth.*   Foam structure is characterized by the "wetness" of the system. Foams with arbitrarily large liquid to gas ratios can be generated by excessive agitation or by intentionally bubbling gas through a fluid. If the liquid content is sufficiently great, the foam consists of well-separated spherical bubbles that rapidly rise upwards displacing the heavier liquid. Such a system is usually called a froth, or bubbly liquid, rather than a foam. When the bubbles in a froth reach the surface, they may instantly burst, they may seethe and gradually burst, or they may collect together and form a more proper foam, all according to the quantity and nature of the surface active components in the liquid. This is familiar to anyone who has noticed the difference in opening agitated bottles of seltzer water and beer. There are no surface-active components in the former, and hence there are no interfacial forces or Marangoni effects to hinder the direct coalescence of bubbles.

Wet Foam: Spherical Bubbles.   If there are sufficiently strong repulsive interactions, such as from the electric double-layer force, then the gas bubbles at the top of a froth collect together without bursting. Furthermore, their interfaces approach as closely as these repulsive forces allow; typically on the order of 100 nm. Thus bubbles on top of a froth can pack together very closely and still allow most of the liquid to escape downward under the influence of gravity while maintaining their spherical shape. Given sufficient liquid, such a foam can resemble the random close-packed structure formed by hard spheres. With less liquid, depending on the distribution of bubble sizes, the bubbles must distort from their spherical shapes. For example, spheres of identical size can pack to fill at most $\pi/\sqrt{18} \approx 0.74$ of space; this occurs if they are packed into a crystalline lattice. A foam with a monodisperse size distribution but less than 26% liquid is thus composed of bubbles which are not spherical but are noticeably squashed together. Typical foams, as in Figure 1, have a fairly broad distribution of bubble sizes and can therefore maintain spherical bubbles with significantly less liquid. Empirically, foams with greater than about 5% liquid tend to have bubbles that are still approximately spherical, and are referred to as wet foams. Such is the case for the bubbles toward the bottom of the foam shown in Figure 1. Nevertheless, it is important to note that even in the case of these wet foams, some of the bubbles are deformed, if only by a small amount.

Dry Foam: Polyhedral Bubbles.   A dry foam, by contrast, is one with so little liquid that the bubbles are severely distorted into approximately polyhedral shapes. Typically this occurs for foams with less than 1% liquid by volume, as is the case for the bubbles toward the top of the foam shown in Figure 1. The structure of polyhedral foams is more appropriately described in terms of the liquid films separating neighboring bubbles rather than in terms of the packing of bubbles as individual units. Most of the interfacial area in a polyhedral foam is in the form of polygonal liquid films having uniform thickness and separating two adjacent gas bubbles. The structure formed by these films is seemingly random, but nevertheless possesses a certain regularity which follows from mechanical constraints. The first of these is that only three films can mutually intersect, and they must meet at an angle of 120°. The intersection of four films is unstable and breaks up into two sets of three because the surface tension of the films exerts a force which acts to minimize the total interfacial area. The region of intersection formed by three films is known as the Plateau border in honor of the French physicist J.A.F. Plateau, who first studied their properties. It is the Plateau borders, rather than the thin liquid films, which are apparent in the polyhedral foam shown toward the top of Figure 1. Lines formed by the Plateau borders of intersecting films themselves intersect at a vertex; here mechanical constraints imply that the only stable vertex is the one made from four borders. The angle between intersecting borders is the tetrahedral angle, $\cos^{-1}(-1/3) \approx 109.47°$. In terms of the arrangements of gas bubbles, the rules describing the structure of a polyhedral foam may be summarized as follows. First, only sets of four bubbles may be in mutual contact. All four bubbles share a common vertex, each of the four combinations of three bubbles share a common Plateau border, each of the six combinations of two bubbles share a common film, and the angles between pairs of films and between pairs of borders are respectively 120° and the tetrahedral angle.

These local structural rules make it impossible to construct a regular, periodic, polyhedral foam from a single polyhedron. No known polyhedral shape that can be packed to fill space simultaneously satisfies the intersection rules required of both the films and the borders. There is thus no ideal structure that can serve as a convenient mathematical idealization of polyhedral foam structure. Lord Kelvin considered this problem, and his minimal tetrakaidecahedron is considered the periodic structure of polyhedra that most nearly satisfies the mechanical constraints.

A real foam has further degrees of freedom available for establishing local mechanical equilibrium: the films and Plateau borders may curve. In fact, curvature can be readily seen in the borders of Figure 1. In order to maintain such curvature, there must be a pressure difference between adjacent bubbles given by Laplace's law according to the surface free energy of the film and the principle radii of curvature of the film: $\Delta P = \gamma_f \cdot (r_1^{-1} + r_2^{-1})$. Note that the pressure inside a bubble must be constant. The Laplace pressure is determined by the regions of greatest curvature. Thus, at the facets of the bubble where the surface is nearly flattened and the curvature is decreased, force balance is maintained by the effects of the disjoining pressure, which must balance the Laplace pressure in the regions of high curvature.

Even though pressure differences can exist between adjacent bubbles, and between the gas and the liquid, the pressure throughout the continuous liquid structure of films, borders, and vertices must be constant; otherwise, liquid flows until all pressure gradients vanish. Figure 3 shows cross sections of three films meeting in a Plateau border, and illustrates how pressure balance is achieved between liquid residing in a film and liquid residing in a border. Since the films are flat and opposite faces are parallel away from the border, the pressure inside the film equals the pressure in the gas minus the disjoining pressure $\Pi(\ell)$. In the border, by contrast, the pressure equals the gas pressure minus the Laplace pressure σ/r where r is the curvature of liquid–vapor interface at the Plateau border. The pressure balance is thus achieved by adjusting the distribution of liquid be-

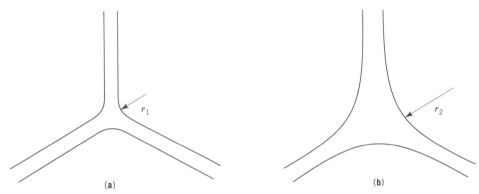

Fig. 3. Two-dimensional schematic illustrating the distribution of liquid between the Plateau borders and the films separating three adjacent gas bubbles. The radius of curvature r of the interface at the Plateau border depends on the liquid content and the competition between surface tension and interfacial forces. (**a**) Flat films and highly curved borders occur for dry foams with strong interfacial forces. (**b**) Nearly spherical bubbles occur for wet foams where the surface tension dominates the interfacial forces.

tween films and borders until the disjoining pressure equals the Laplace pressure $\Pi(\ell) = \sigma/r$. Figure 3 illustrates how flatter films with smaller, more highly curved Plateau borders are attained when interfacial forces dominate surface tension. The effect of more liquid being drawn into the Plateau border when surface tension dominates was first noticed by Plateau and is called border suction. For similar reasons, vertices are thicker than borders as can also be seen in the photograph in Figure 1. Thus the distribution of liquid between neighboring bubbles and the corresponding bubble shapes are determined not only by the ratio of gas to liquid, but by the competition between surface tension and interfacial forces as well.

Measurement. A complete characterization of the structure of a foam requires a characterization of the structure of the bubbles that comprise the foam. The total liquid content can be readily found from the mass densities of the foam and the liquid from which it was made. However, a more detailed determination of the bubble structure, including their average size, their shape, their structure and their size distribution is much more difficult, and is typically impeded by the problems in visualizing the interior of a foam. Even in the absence of any intrinsic optical absorption of the liquid, the strong mismatch in the indexes of refraction between the gas and the fluid results in a large scattering of light, usually precluding direct visualization of the interior structure of a foam. As a result, other, less direct, methods have been developed, and must be used, except in exceptional cases where the foam structure has been optimized for visualization.

To date, the most detailed characterization of the full three-dimensional structure of a foam is that by E. B. Matzke in 1946 (8). The method was to place by hand 1900 identical bubbles one at a time into a dish roughly 14 bubbles in diameter and then to image the structure with a dissecting microscope. After repeating this procedure sixteen times, the most abundant bubble shape was found to be a 13-sided polyhedron having 1 quadrilateral, 10 pentagonal, and 2 hexagonal faces. The number of faces per bubble ranges from 11 to 18, and is 13.7 on average. This can be compared with the theoretical result 13.39 for a statistical foam with isotropic cells of equal volume based on the structural rules given earlier and Euler's theorem (9,10). Such theory has recently been generalized to foams with curved faces (11).

Although informative, the technique of Matzke is restricted to use on the small class of very dry foams that have large cells and are essentially transparent to visible light. No such painstaking measurement and analysis have been carried out on foams with a more naturally broad bubble size distribution, like the one shown in Figure 1. However, one optical imaging technique that circumvents the problem of multiple light scattering and thus can be used more generally is to estimate the bubble size distribution from the area individual foam bubbles occupy at a glass surface. Such experiments, and the systematic differences between bulk and surface bubble distributions, have been reviewed (12). Another technique that also directly measures the bubble size distribution is the use of a Coulter counter, where individual bubbles are drawn through a small tube and counted (13). This yields a direct measure of the bubble size distribution, but it is invasive and cannot probe the structure of the foam.

One technique that does probe the foam structure directly is cryomicroscopy. The foam is rapidly frozen, and the solid structure is cut open and imaged with an optical or electron microscope (14). Such methods are widely applicable and provide a direct image of the foam structure; however, they destroy the sample and may also perturb the foam structure in an uncontrolled manner during the freezing.

Other methods attempt to probe the structure of the foam indirectly, without directly imaging it. For example, since the liquid portion of the foam typically contains electrolytes, it conducts electrical current, and much work has been done on relating the electrical conductivity of a foam to its liquid content, both experimentally (15) and theoretically (16). The value of the conductivity depends in a very complex fashion on not only the liquid content and its distribution between films and borders, but the geometrical structure of the bubble packing arrangement. Thus electrical measurements offer only a rather crude probe of the gas:liquid ratio, a quantity that can be accurately estimated from the foam's mass density.

Another nonimaging technique has been developed that exploits the strong multiple light scattering in foams, and provides a direct, noninvasive probe of bulk foam structure and dynamics (17). The time-averaged transmission of light through a foam gives a measure of the average bubble size, while temporal fluctuations in the scattered light intensity probe the motion of bubbles within the foam.

Stability.   Control of foam stability is important in all applications, whether degradation of a custom foam is to be minimized or whether excessive foaming is to be prevented. In all cases, the time evolution of the foam structure provides a natural means of quantifying foam stability. There are three basic mechanisms whereby the structure may change: by the gravitational segregation of liquid and bubbles, by the coalescence of neighboring bubbles via film rupture, and by the diffusion of gas across the liquid between neighboring bubbles.

Drainage. All foams and froths consist of liquid and vapor components that have very different mass densities, making them susceptible to gravitationally induced segregation. In very wet froths the vapor bubbles rapidly move upward while the liquid falls. In longer-lived foams, the gas fraction is higher and the bubbles are tightly packed. Nevertheless, the heavier fluid may still drain downward through the thin films and Plateau borders. In some cases, the addition of polymers or micelles in the liquid can increase its viscosity and slow the drainage. The influence of surface-active impurities on drainage through the thin film regions has been studied extensively through experiments on thin soap films. For example, thickness variations in a draining soap film can be observed by eye via the colors reflected under white light illumination. A large number of experiments has also been done on soap films pulled at constant speed from a soapy liquid (18–20). In addition to simple laminar flows set by film thickness, liquid viscosity, and the state of the adsorbed surfactant, whole regions of thick film can flow like a plug into the Plateau border and exchange for regions of thin film. This process is called marginal regeneration (18) and is believed to be important in foams as a means of bringing liquid from the films into the Plateau borders. Once in the Plateau borders the liquid can more rapidly drain downward. In wet long-lived foams, the bubbles are more nearly spherical so there is no distinction between flow within films vs Plateau borders.

Provided there is no rupture of the films, drainage proceeds until there develops a vertical, hydrostatic pressure gradient to offset gravity. The gradient is supported by the disjoining pressure due to the film thickness being too small to minimize the effective interface potential. Thus individual soap films in the foam decrease in thickness with increasing height. This results in a nonuniform gas:liquid volume fraction with the foam being more wet near the bottom of the container as in Figure 1. The formation of a macroscopic layer of liquid underneath a previously homogeneous foam is called gravitational syneresis, or creaming, and depends not only on the foam composition, but on the size and shape of the container as well (21).

Film Rupture. Another general mechanism by which foams evolve is the coalescence of neighboring bubbles via film rupture. This occurs if the nature of the surface-active components is such that the repulsive interactions and Marangoni flows are not sufficient to keep neighboring bubbles apart. Bubble coalescence can become more frequent as the foam drains and there is less liquid to separate neighbors. Long-lived foams can be easily formulated in which film rupture is essentially negligible, by ensuring that the surface-active agents provide a sufficiently large barrier that prevents the two films from approaching each other. Then film rupture is probably a thermally activated process in which a large, rare fluctuation away from equilibrium thickness and over an energy barrier is needed. Film rupture can also be enhanced by mechanical shock. Other external perturbations such as thermal cycling, mechanical shearing, composition change via evaporation or chemical or particulate additives, can also greatly affect the rate of film rupture (see DEFOAMERS).

Gas Diffusion. For very long-lived foams, film rupture is negligible and drainage slows to a stop as hydrostatic equilibrium is attained. Nevertheless, the foam is still not in thermodynamic equilibrium and continues to evolve with time. This occurs through an entirely different, though very general, means:

gas diffusion. Smaller bubbles have a greater interfacial curvature and hence, by Laplace's law, have a higher internal pressure than larger bubbles. This results in a diffusive flux of gas from smaller to larger bubbles. Thus with time small bubbles shrink while large bubbles grow. This process is known as coarsening, or ripening, and results in the net increase in the average bubble size over time. It is ultimately driven by surface tension and serves to decrease the total interfacial surface area with time. This process has many similarities to the phase separation of binary liquids and metal alloys (22–24).

As coarsening proceeds, the distribution of bubbles changes so that the average bubble size gets larger with time. The evolution of the bubble size distribution is typically expected to be self-similar; that is, the distribution is independent of time when scaled by the average bubble size which, in turn, grows as a power of time (22,23). Experimentally, the growth exponent of the average bubble radius is near one-half with slight dependence on the chemical composition and liquid content of the foam. The overall rate constant, as opposed to the growth exponent, depends directly on such material parameters as the surface tension and the solubility and diffusion constant of gas molecules in the interstitial liquid. Coarsening not only alters the size distribution, but alters the foam topology as well (9). Local stresses arise due to the change in packing conditions as small bubbles shrink and large bubbles grow. These stresses can be relieved by topology changes in which a bubble's nearest neighbors are changed (17).

Measurement.   Any means of characterizing foam structure can be used to study foam evolution provided that the measurement can be made noninvasively and sufficiently rapidly. Thus the measurement techniques that require the foam to be destroyed, such as cryomicroscopy, are not really suitable to study the stability of the foam. One technique that has been applied successfully is the measurement of the change in the pressure head over an evolving foam. This can be related to the total change in interfacial surface area (25,26), and provides a measure of evolution, and hence stability of the foam. Multiple light scattering can also be used to follow the time evolution of a foam (17). A common engineering technique for determining foam stability entails measuring the amount of foam produced. For defoaming applications, this is often a more important measure of stability than the foam structure.

Rheology.   The rheology of foam is striking; it simultaneously shares the hallmark rheological properties of solids, liquids, and gases. Like an ordinary solid, foams have a finite shear modulus and respond elastically to a small shear stress. However, if the applied stress is increased beyond the yield stress, the foam flows like a viscous liquid. In addition, because they contain a large volume fraction of gas, foams are quite compressible, like gases. Thus foams defy classification as solid, liquid, or vapor, and their mechanical response to external forces can be very complex.

One simple rheological model that is often used to describe the behavior of foams is that of a Bingham plastic. This applies for flows over length scales sufficiently large that the foam can be reasonably considered as a continuous medium. The Bingham plastic model combines the properties of a yield stress like that of a solid with the viscous flow of a liquid. In simple Newtonian fluids, the shear stress τ is proportional to the strain rate $\dot{\gamma}$, with the constant of proportionality being the fluid viscosity. In Bingham plastics, by contrast, the relation between stress and strain rate is $\tau = \tau_y + \mu_p\dot{\gamma}$ where τ_y is the yield stress below which there is no flow and μ_p is called the plastic viscosity. The effective viscosity is thus given by $\mu = \mu_p + \tau_y/\dot{\gamma}$ and is therefore shear thinning.

Consistent with this model, foams exhibit plug flow when forced through a channel or pipe. In the center of the channel the foam flows as a solid plug, with a constant velocity. All the shear flow occurs near the walls, where the yield stress has been exceeded and the foam behaves like a viscous liquid. At the wall, foams can exhibit wall slip such that bubbles adjacent to the wall have nonzero velocity. The amount of wall slip present has a significant influence on the overall flow rate obtained for a given pressure gradient.

While the Bingham plastic model is an adequate approximate description of foam rheology, it is by no means exact. More detailed models attempt to relate the rheological properties of foams to the structure and behavior of the bubbles. For very dry foams, the rheological properties are determined solely by the films separating the bubbles. The rheological properties of a set of randomly oriented films were determined first by Derjaguin (27) and later independently by Stamenovic and Wilson (28). These models set the scale for the elastic modulii of a foam. The bulk modulus of the foam is dominated by that of the gas in the bubbles, whereas the shear modulus is given by $G' = 4\gamma_f S/15V$ where S and V are the average surface area and volume of the bubbles. The bulk modulus of an ideal gas is equal to its pressure, so that for typical foams, the shear modulus is considerably weaker than the bulk modulus.

The model of randomly oriented thin films was made more precise through detailed micromechanical models that considered a two-dimensional array of hexagonal bubbles of equal sizes (29). Again only the surface tension of the thin films was considered. Because of the simpler geometry, the model could be solved exactly in two dimensions. Nevertheless, it provides considerable insight. It establishes the physical basis for the elastic modulus, which is the stretching of films with a shear deformation, as shown in Figure 4. The increased surface area results

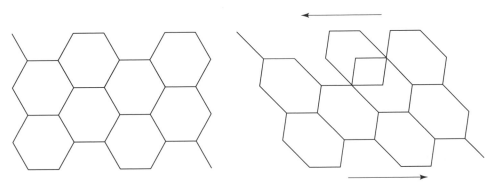

Fig. 4. Schematic representation of a two-dimensional model to account for the shear modulus of a foam. The foam structure is modeled as a collection of thin films; the Plateau borders and any other fluid between the bubbles is ignored. Furthermore, all the bubbles are taken to be uniform in size and shape. When shear is applied, the total area of the thin films increases, and the surface tension results in a restoring force, providing the shear modulus of the foam.

in a restoring force, providing the elastic model. It also suggests an origin for the yield stress; when the foam is sheared enough that the bubbles can slide over one another, the foam yields. This micromechanical model has been further refined through computer simulations, mainly restricted to two dimensions (30–32). This type of computer modeling has been extended to ordered three-dimensional structures (33).

Although all these models provide a description of the rheological behavior of very dry foams, they do not adequately describe the behavior of foams that have more fluid in them. The shear modulus of wet foams must ultimately go to zero as the volume fraction of the bubbles decreases. The foam only attains a solid-like behavior when the bubbles are packed at a sufficiently large volume fraction that they begin to deform. In fact, it is the additional energy of the bubbles caused by their deformation that must lead to the development of a shear modulus. However, exactly how this modulus develops, and its dependence on the volume fraction of gas, is not fully understood.

The viscous behavior of the foam once it begins to flow has also been investigated, both theoretically (34–36) and experimentally (37). The theoretical model assumes that all the viscous dissipation occurs in the thin films between the bubbles. However, the agreement between the predicted behavior and that observed is poor.

Measurement. To determine rheological parameters such as the yield stress and effective viscosity of a foam, commercial rheometers are available (38); rotational and continuous-flow-tube viscometry are most commonly employed (see RHEOLOGICAL MEASUREMENTS). However, obtaining reproducible results independent of the sample geometry is a difficult goal which arguably has not been achieved in most of the experiments reported in the scientific literature (38). One obvious difficulty is that the rheological properties depend sensitively on quantities such as the bubble size distribution and liquid content which are difficult to characterize and reproduce, and which tend to change with time. A more subtle and insidious difficulty is that nonuniform shear conditions due to wall slip and plug flow must be characterized and accounted for in the analysis of rheology experiments. For example, viscous dissipation in wall slip depends sensitively on the thickness of wetting layer of liquid which intrudes between the wall and the foam bubbles; this layer thickness varies greatly with surface chemistry and liquid composition and can also change with time as the foam drains. Furthermore, the extent of plug flow depends on the amount of wall slip and the sample geometry, and this can also greatly complicate the interpretation of the results.

Production

Several techniques are available for the generation of special-purpose foam with the desired properties. The simplest method is to disperse compressed gas directly into an aqueous surfactant solution by means of a glass frit. A variation of this method that allows for control of liquid content is to simultaneously pump gas and surfactant solution through a bead pack or steel wool, for example, at fixed rates. Less reproducible mechanical means of foam generation include brute force shaking and blending. For highly reproducible foams composed of small bubbles, such as shaving creams, the aerosol technique is especially suitable (39) (see AEROSOLS). Hydrocarbons or chlorofluorocarbons are liquefied at high pressure and then emulsified with the surfactant solution. When released to atmospheric pressure, the propellant droplets evaporate into tiny gas bubbles which aggregate into a foam.

Applications

Foams have a wide variety of applications that exploit their different physical properties. The low density, or high volume fraction of gas, enable foams to float on top of other fluids and to fill large volumes with relatively little fluid material. These features are of particular importance in their use for fire fighting. The very high internal surface area of foams makes them useful in many separation processes. The unique rheology of foams also results in a wide variety of uses, as a foam can behave as a solid, while still being able to flow once its yield stress is exceeded.

Foams are also widely encountered in circumstances where their presence is detrimental. Foams are common to many processes that entail agitation of fluids or bubbling of air through fluids. The presence of any type of surface-active ingredient, even in minute quantities, enhances the formation of foams in the processing. The presence of the foams increases the volume of the fluids and makes them more difficult to process and transport. Moreover, foams can have strong detrimental environmental effects. Thus, just as the production and stabilization of foams is important in some industrial processes, so the elimination of foams is crucial in many others. As a result, a wide variety of defoaming agents have been developed to eliminate or reduce the formation of foams (see DEFOAMERS).

Firefighting. Foams are widely used in firefighting applications (40). They are particularly useful in extinguishing flammable liquids, eg, gasoline. Whereas water simply agitates the gasoline, further spreading the fire, and then sinks to the bottom of the burning fluid, a foam is less dense than the burning liquid, and remains suspended on its surface. The collapsing bubbles cool the fluid near the surface, and reduce the amount of oxygen available to the flame, ultimately extinguishing it. A foam is also a more efficient use of the firefighting liquid, typically water, enabling it to be spread over a much larger area.

Foams for firefighting applications are typically made from a concentrated foaming agent diluted with water and then mixed with air. Rather than consider the volume fraction of air in the foam, firefighting foams are characterized by their expansion ratio, which is the increase in volume of the liquid after the foam is formed. Expansion ratios range from 5:1 to over 1000:1; ratios of 5:1 to 20:1 are called low expansion; ratios of 21:1 to 200:1, medium expansion; and ratios greater than 200:1, high expansion.

Low expansion foams are used most commonly. Because they are relatively more dense, they can more easily be sprayed larger distances, making them safer to use. In addition, because of the larger amount of liquid, they are more resistant to the heat of the fires, making them more effective as extinguishers. Their primary disadvantage is the relatively smaller area that they can cover due to their lower expansion ratio. Medium expansion foams are usually too light to be sprayed any distance, and instead must be formed very near to the flames. However, they can cover a much larger area of flame, and the low density reduces the probability of disrupting the surface of the burning fluid. They are less heat resistant and hence more easily destroyed than low expansion foams. However, they can cover a much greater area. High expansion foams cover the widest area, but suffer from even poorer heat resistance, and virtually no ability to be sprayed. Hence they are typically formed in place, and are often used to fill the places where a fire has already started, such as in the holds of ships, warehouses, or mines. They are also sometimes used in fighting forest fires in areas where water is scarce.

Most foam-forming concentrates used contain some form of protein, usually derived from animals. In addition, many contain fluorochemical surfactants to increase their foaming performance. Other foaming agents are comprised solely of synthetic surfactants. Most foams produced with either protein-based or synthetic foaming agents are susceptible to polar fluids, particularly alcohols, which are miscible in water and tend to destroy the firefighting foams. As a result, all-purpose foaming agents have been developed that produce foams which are not destroyed by alcohols, and are effective in fighting all types of fires. They typically contain natural polymers that are insoluble in polar solvents.

Food. Foams are common to a wide variety of food products. Whipped cream and meringue are essentially foams, and ice cream is comprised of a large amount of foam. These foams are stabilized by proteins; the two most important are egg white and milk proteins. For food products, it is desirable not only to achieve good foaming properties, but also to form stable foams (41–44). The ease with which foams are formed depends on the capacity of the proteins to rapidly adsorb onto the interface. The stability of the foams depends on the ability of the proteins to form an elastic membrane at the interface, which both prevents bubble coalescence and is sufficiently impermeable to reduce gas diffusion. One of the best food-foaming agents is egg white or egg albumen. It consists of a mixture of different proteins, each serving a particular function (45). Globulins are the most surface-active agents, leading to good foamability; drainage is retarded by the high viscosity caused by globulins and ovomucoids; the film strength is enhanced by surface complexes formed between lysozymes and ovomucins. Upon heating, thermal denaturation of ovalbumin and conalbumin results in a more permanent foam structure, leading to its widespread use in baked products. Ice cream is also a type of foam possessing varying amounts of air bubbles incorporated during an aeration step in the processing (46). These bubbles are initially stabilized by milk proteins, primarily β-casein, α-lactalbumin, and β-lactoglobulin (47). Further stabilization occurs due to the adsorption of fat globules on the interface.

Another important digestible foam is that on the top of a glass of freshly poured beer (qv). Although not as long lasting, it is nevertheless considered an important aesthetic quality of the beverage, and is thus the subject of considerable research (48,49). In addition, its aesthetic importance is somewhat dependent on location. For example, beer in the United Kingdom has traditionally possessed a higher and longer lasting head of foam than that in the United States. The foam in beer is usually formed by the dissolved CO_2, although dissolved nitrogen has also been used to improve the quality. The main stabilizer in the foam is proteins in the beer, although other components, such as trace metal ions, iso-α-acids, and propylene glycol alginate (PGA), also enhance the stability of beer foam. In fact, PGA is also sometimes added to beer to improve the foam (50).

Separations.   Foams have important uses in separations, both physical and chemical (51,52). These processes take advantage of several different properties of foams. The buoyancy and mechanical rigidity of foam is exploited to physically separate some materials. The large volume of vapor in a foam can be exploited to filter gases. The large surface area of a foam can also be exploited in the separation of chemicals with different surface activities.

Froth flotation (qv) is a significant use of foam for physical separations. It is used to separate the more precious minerals from the waste rock extracted from mines. This method relies on the different wetting properties typical for the different extracts. Usually, the waste rock is preferentially wet by water, whereas the more valuable minerals are typically hydrophobic. Thus the mixture of the two powders are immersed in water containing foam promoters. Also added are modifiers which help ensure that the surface of the waste rock is hydrophilic. Upon formation of a foam by bubbling air and by agitation, the waste rock remains in the water while the minerals go to the surface of the bubbles, and are entrapped in the foam. The foam rises, bringing the minerals to the surface with it. This can be collected, and the valuable minerals, now higher in purity, extracted.

Foam fractionation is a separation method that is chemical in origin (52). It relies on the preferential surface adsorption of some molecules and hence exploits the large surface area of foams. This is a commonly used method for separating surfactant molecules. It can be extended to other separations by coating the material to be separated by surfactant to enhance its adsorption at the surface. For example, foam fractionation has been applied to remove radioactive wastes. An advantage of this method is that the foam can be spread over a large contaminated volume, but the actual amount of material containing the radioactive waste is quite small, once the foam has been drained and collapsed. Foams can also be used to collect and separate small colloidal particles, if they are coated with a surfactant to bring them to the interface. Another use of foams are in the deinking processing to recycle waste paper. Air bubbles are used to remove the ink from the paper, and these are collected and separated as a foam.

Oil Recovery.   Foams find wide use in oil recovery, from the initial drilling of the bore holes, through the first recovery stage, and, increasingly, all the way to tertiary or enhanced oil recovery (see PETROLEUM). Again, this application exploits the unique features of foams, primarily the large interfacial area and the distinctive rheological properties of flowing foams.

In the drilling of oil wells, foam is sometimes used as the drilling fluid. Usually the drilling fluid is a clay or mud slurry, which is circulated down the bore hole to remove the waste generated by the drilling process, and to seal the well, preventing the expulsion of oil that is under pressure. However, in some wells, the pressure of the oil in the ground is less than the pressure head generated by a water column the height of the well. Thus a drilling mud would exert excessive pressure, and could contaminate the rock structure near the bore hole. In these cases, an aqueous foam is often used. It has all the features of a drilling mud, and can also aid in cleaning the drilling material from the well. However, since its density is lower, the pressure exerted by the drilling fluid on the oil is reduced.

Foams are also used in extracting the oil from the ground. One important use is in controlling the flow of fluids in the rock formation. The most common form of secondary oil recovery entails pushing the oil out of the ground by flooding the formation with water from one well and collecting the oil that is pushed out from an adjacent well. This process suffers from several problems that may be alleviated with foams. One problem is encountered if the rock formation contains channels of higher permeability, or lower resistance to the flow of fluids. These may arise, for example, from fractures in the formation. The water being forced through the formation flows more easily through these high permeability channels, bypassing the rest of the formation and greatly reducing the effectiveness of the recovery, or the sweep efficiency. One method that is sometimes used to alleviate this problem is the injection of foam-forming materials into the formations. The foam tends to go first into these larger channels, and then plugs them. By contrast, in the narrower channels, the shear stresses on the foam are larger, causing it to flow. This effectively blocks the high permeability regions and forces the pushing fluid to flow in the remainder of the formation, making it more effective in removing the oil. Another almost opposing use of foam is as a pressurizing agent to fracture the formation. This technique is used for very viscous oils, such as tar sands. These fractures then provide channels to allow the penetration of hot steam which is used to lower the viscosity of the heavy oil to enhance its flow. Sand is often added to fill the cracks and prop them open. When a foam is used as the pressurizing fluid, the settling of the sand during the injection is minimized (53).

Foam is also increasingly being considered in tertiary oil applications. Even with the water floods used in secondary oil recovery, typically 30–60% of the oil remains in the ground. Enhanced oil recovery, or tertiary recovery techniques, are used to try to extract this remaining oil. This is generally done by decreasing the interfacial tension of the oil, which is accomplished either through the use of a surfactant, or through the use of a miscible flood, where the pushing fluid is miscible in the oil. One type of miscible flooding uses the injection of CO_2, or other gases, which form a foam that displaces the oil (54). Foams also help control the mobility of the pusher. Often the oil that is being displaced has a higher viscosity, and thus a lower mobility, than the displacing fluid. Then the flow can become unstable, and instead of a uniform front of the displacing fluid advancing through the formation, narrow fingers are formed (55). This viscous fingering instability can greatly reduce the sweep efficiency. The occurrence of this instability can be reduced by decreasing the mobility of the pushing fluid. The use of foams are one technique for achieving this. This use exploits the rheological properties of the foams. More generally, the flow of foams in the very narrow pore spaces of an oil bearing formation is very complex, and is not completely understood (56). The behavior entails the motion of bubbles of air that can be comparable to the size of the pore spaces themselves. This can result in the bursting of the bubbles, which has the effect of introducing new interfaces into the flowing fluids. The flow of these interfaces, as well as the flow of the still intact bubbles themselves, in the restricted geometry of the pore spaces is a complex problem that is very sensitive to the nature of the rocks, the disorder of the pore spaces, and the local wetting properties of the formation. In addition, the flow of the interfaces, and the foam, at length scales larger than the pore space, but still smaller than the size of the whole oil field, remains a very poorly understood, and little investigated, problem. Thus, although foams have considerable promise for use in tertiary oil recovery, their applications have, to date, been limited.

Detergents. Foams are often associated with detergents, but they are generally not essential. Instead, foaminess is often a desirable trait more for its effect on the consumer than its function. In fact, excessive foaming can be detrimental to the cleaning if the volume of the foam is too high. One exception is the case where the cleansing action must be restricted to some particular region, and where excessive amounts of water must be avoided. For example, foams are often used to clean rugs, as they can spread the detergent on the surface of the rug, while avoiding excessively wetting the base of the rug.

Textiles. Foams are often encountered in the production of textiles (qv), which involves extensive interactions between the fibers, which have a large surface area, and a variety of aqueous treatments. In most cases, these foams are detrimental to the processing and fabrication of the textiles and measures are taken to reduce the foaming (57). These include modifications of the mechanical fabrication techniques, addition of foam inhibitors, and the use of low foaming surfactants in the processing. However, in other instances, foams can be used advantageously in textile processing, primarily for the application of screen printing, coatings, backings, and certain types of dyes (58). Foams can also be used to clean textiles as the foam helps wet between the fibers thereby more effectively spreading the detergent.

Cosmetics. Besides the esthetic appeal of foams, they have two properties that are exploited for cosmetic purposes. The first is their ability to retain different substances and distribute them as required, while using a relatively small quantity of fluid. An example of this application is the lather formed by some shampoos, which effectively spreads the detergent by wetting the surfaces of the hair, while avoiding excessive liquid that would otherwise fall off. The second application is to spread a moisturizer or lubricant, while still providing enough resiliency to hold the fluid in place. A prime example of this is shaving cream, which provides both moisturization and lubrication for shaving.

Other. Because a foam consists of many small, trapped gas bubbles, it can be very effective as a thermal insulator. Usually solid foams are used for insulation purposes, but there are some instances where liquid foams also find uses for insulation (see FOAMED PLASTICS; INSULATION, THERMAL). For example, it is possible to apply and remove the insulation simply by forming or collapsing the foam, providing additional control of the insulation process. Another novel use that is being explored is the potential of absorbing much of the pressure produced by an explosion. The energy in the shock wave is first partially absorbed by breaking the bubbles into very small droplets, and then further absorbed as the droplets are evaporated (53).

Safety, Health, and Environment

Foams play important roles in environmental issues, both beneficial and detrimental.

Natural Waters.   Many water systems have a natural tendency to produce foam upon agitation. The presence of pollutants exacerbates this problem. This was particularly severe when detergents contained surfactants that were resistant to biodegredation. Then, water near industrial sites or sewage disposal plants could be covered with a blanket of stable, standing foam (52,59). However, surfactant use has switched to biodegradable molecules, which has greatly reduced the incidence of these problems.

Wastewater Treatment.   The treatment of wastewater, either from sewage or from industrial processes, typically entails a preliminary filtration to remove the large volumes of solids, and then a slower settling to remove the sand and gravel (see WATER). The water is then treated by an activated sludge process to remove the remaining dissolved solids and organic colloidal particles. Activated sludge is a biomass that assists in the degradation of the organic waste in the water. The process entails a mixing and aeration of the wastewater with the activated sludge, which can lead to problems of foaming. The foams produced can be quite stable, resulting in additional problems for waste disposal. The foams produced in this process differ from those normally encountered in that the foam producing and stabilizing agents are microbial, primarily including *Nocardia* (60,61), *Microthrix parvicella* (62,63), and *Rhodococcus* (64,65). These foams are more difficult to treat with defoaming agents. Moreover, it is very difficult to predict the degree of foamability of the waste being treated (66). In other, more specialized wastewater treatments, these problems do not arise, and defoaming agents can be used effectively. For example, the wastewater remaining from the pulp used in the production of paper (qv) contains dissolved soaps from fatty acids and abietes, which can lead to foam problems. These can be controlled with mixtures of organic solvents and nonionic surfactants (67) or with gaseous sulfur dioxide (57).

Chlorofluorocarbon Alternatives.   There still is no completely satisfactory propellant for use in the aerosol method of foam production (39). Chlorofluorocarbons, still widely used, are harmful to atmospheric ozone and low molecular weight hydrocarbons, now popular, eg, in producing shaving cream, are explosive and promote the greenhouse effect (see FLUORINE COMPOUNDS, ORGANIC– ALIPHATIC COMPOUNDS). The difficulty is in creating a safe, stable liquid that can be readily emulsified and whose vapor pressure at room temperature is roughly 200–300 kPa (2–3 atm).

BIBLIOGRAPHY

"Foams" in *ECT* 1st ed., Vol. 6, pp. 772–778 by E. I. Valko, Polytechnic Institute of Brooklyn; in *ECT* 2nd ed., Vol. 9, pp. 884–901 by L. Shedlovsky, Colgate-Palmolive Co.; in *ECT* 3rd ed., Vol. 11, pp. 127–145 by S. Ross, Rensselaer Polytechnic Institute.

1. M. J. Rosen, *Surfactants and Interfacial Phenomena*, John Wiley & Sons, Inc., New York, 1989.
2. M. Dahanayake, A. W. Cohen, and M. J. Rosen, *J. Physical Chem.* **90**, 2413 (1986).
3. P. Lianos and R. Zana, *J. Colloid and Interface Sci.* **84**, 100 (1981).
4. J. N. Israelachvili, *Intermolecular and Surface Forces*, Academic Press Ltd., San Diego, 1991.

5. B. V. Derjaguin and L. Landau, *Acta Physicochim. USSR* **14**, 633 (1941).
6. E. J. W. Vervey and J. T. G. Overbeek, *Theory of the Stability of Lyophobic Colloids*, Elsevier, Amsterdam, 1948.
7. J. S. Clunie, J. F. Goodman, and P. C. Symons, *Nature* **216**, 1203 (1967).
8. E. B. Matzke, *Am. J. Bot.* **33**, 58 (1946).
9. D. Weaire and N. Rivier, *Contemp. Phys.* **25**, 55 (1984).
10. C. Isenberg, *The Science of Soap Films and Soap Bubbles*, Dover Publications, New York, 1992.
11. J. E. Avron and D. Levine, *Physical Review Letters* **69**, 208 (1992).
12. H. C. Cheng and R. Lemich, *Ind. Eng. Chem. Fundam.* **22**, 105 (1983).
13. A. Selecki and R. Wasiak, *J. Coll. Interface Sci.* **102**, 557 (1984).
14. A. J. Wilson, ed., *Foams: Physics, Chemistry, and Structure*, Springer-Verlag, New York, 1989, p. 69.
15. N. O. Clark, *Trans. Faraday Soc.* **44**, 13 (1948).
16. A. K. Agnihotri and R. Lemlich, *J. Colloid and Interface Sci.* **84**, 42 (1981).
17. D. J. Durian, D. A. Weitz, and D. J. Pine, *Science* **252**, 686 (1991).
18. K. J. Mysels, K. Shinoda, and S. Frankel, *Soap Films, Studies of Their Thinning and a Bibliography*, Pergamon Press, New York, 1959.
19. J. Lykema, P. C. Scholten, and K. J. Mysels, *J. Physical Chem.* **69**, 116 (1965).
20. I. B. Ivanov, *Thin Liquid Films: Fundamentals and Applications*, Marcel Dekker, Inc., New York, 1988.
21. H. M. Princen, *J. Coll. Interface Sci.* **134**, 188 (1989).
22. A. J. Markworth, *J. Coll. Interface Sci.* **107**, 569 (1984).
23. W. W. Mullins, *J. Appl. Phys.* **59**, 1341 (1986).
24. J. A. Glazier, S. P. Gross, and J. Stavans, *Phys. Rev. A* **36**, 306 (1987).
25. G. Nishioka and S. Ross, *J. Coll. Interface Sci.* **81**, 1 (1981).
26. A. Monsalve and R. S. Schechter, *J. Coll. Interface Sci.* **97**, 327 (1984).
27. B. V. Derjaguin, *Kolloid-Zeitschrift* **64**, 1 (1933).
28. D. Stamenovic and T. A. Wilson, *J. Appl. Mechan.* **51**, 229 (1984).
29. H. M. Princen, *J. Coll. Interface Sci.* **91**, 160 (1983).
30. F. Bolton and D. Weaire, *Phys. Rev. Lett.* **65**, 3449 (1990).
31. T. Herdtle and H. Aref, *J. Fluid Mech.* **241** 233 (1992).
32. T. Okuzono, K. Kawasaki, and T. Nagai, *J. Rheol.* **37**, 571 (1993).
33. D. A. Reinelt and A. M. Kraynick, *J. Coll. Interface Sci.* **159**, 460 (1993).
34. S. A. Khan and R. C. Armstrong, *J. Non-Newtonian Fluid Mech.* **25**, 61 (1987).
35. A. M. Kraynik and M. G. Hansen, *J. Rheol.* **31**, 175 (1987).
36. L. W. Schwartz and H. M. Princen, *J. Coll. Interface Sci.* **118**, 201 (1987).
37. H. M. Princen and A. D. Kiss, *J. Coll. Interface Sci.* **128**, 176 (1989).
38. J. P. Heller and M. S. Kuntamukkula, *Ind. Eng. Chem. Res.* **26**, 318 (1987).
39. P. A. Sanders, *Handbook of Aerosol Technology*, Van Nostrand Reinhold Co., New York, 1979.
40. F. Fitch, in Ref. 14, p. 207.
41. E. Dickinson, *Food Hydrocoll.* **1**, 3 (1986).
42. J. R. Mitchell, in B. J. F. Hudson, ed., *Developments in Food Proteins*, Elsevier Applied Science, London, 1986, p. 291.
43. S. Poole and J. C. Fry, in B. J. F. Hudson, ed., *Developments in Food Proteins*, Elsevier Applied Science, London, 1987, p. 257.
44. A. Prins, in E. D. a. G. Stainsby, ed., *Advances in Food Emulsions and Foams*, Elsevier Applied Science, London, 1988, p. 91.
45. W. D. Powrie, in W. J. S. a. O. J. Cotterill, ed., *Egg Science and Technology*, Avi Publishing, Westport, Conn., 1977, p. 61.
46. J. K. Madden, in Ref. 14, p. 185.

47. M. Anderson, B. E. Brooker, and E. C. Needs, in E. Dickinson, ed., *Food Emulsions and Foams*, Royal Society of Chemistry, London, 1987, p. 100.
48. C. W. Bamforth, *J. Inst. Brewing* **91**, 370 (1985).
49. P. K. Hegarty, in Ref. 14, p. 197.
50. P. T. Slack and C. W. Bamforth, *J. Inst. Brewing* **89**, 397 (1983).
51. R. Lemlich, *Adsorptive Bubble Separation Techniques*, Academic Press, Inc., New York, 1971.
52. J. J. Bikerman, *Foams*, Springer-Verlag, New York, 1973.
53. J. H. Aubert, A. M. Kraynik, and P. B. Rand, *Sci. Am.* **254**, 74 (1986).
54. F. I. Stalkup, Jr., *Miscible Displacement*, SPE, New York, 1983.
55. R. L. Chouke, C. van Meurs, and C. van der Pohl, *Pet. Tran. AIME* **216**, 188 (1959).
56. C. W. Nutt and R. W. Burley, in Ref. 14, p. 105.
57. R. Hofer, and co-workers, in W. Gerhartz, ed., *Ullmann's Encyclopedia of Industrial Chemistry*, VCH, Weinheim, 1988, p. 465.
58. G. M. Bryant and H. T. Walter, in *Handbook of Fiber Science and Technology*, Marcel Dekker, New York, 1983.
59. M. Raison, *Centre Belge Etude Doc. Eaux* **227**, 512 (1962).
60. H. Lemmer, *Korrespondenz Abwasser* **32**, 965 (1985).
61. O. J. Hao, P. E. Strom, and Y. C. Yu, *Water SA* **14**, 105 (1988).
62. J. R. Blackbeard, G. A. Ekama, and G. V. R. Marais, *Water Pollution Control* **85**, 90 (1986).
63. A. J. Goddard and C. F. Forster, *Enzyme Microbial Tech.* **9**, 164 (1987).
64. M. Segerer, *Korrespondenz Abwasser* **31**, 1073 (1984).
65. T. Mori and co-workers, *Environ. Tech. Lett.* **9**, 1041 (1988).
66. C. F. Forster, in Ref. 14, p. 167.
67. K. Roberts, C. Axberg, and R. Osterlund, *Emulsion foam Killers Containing Fatty and Rosin Acids*, Brunel University, Uxbridge, U.K. 1975.

General References

A. W. Adamson, *Physical Chemistry of Surfaces*, 4th ed., John Wiley & Sons, Inc., New York, 1982.
R. J. Akers, ed., *Foams: Proceedings of a Symposium Organized by the Society of Chemical Industry, Colloid and Surface Chemistry Group*, Brunel University, Uxbridge, U.K., Sept. 8–10, 1975, Academic Press, Inc., London, 1976.
G. R. Assar and R. W. Burley, in N. P. Cheremisinoff, ed., *Encyclopedia of Fluid Mechanics*, Vol. 3, p. 26, Gulf Publishing Co., Houston, 1986, p. 26.
J. H. Aubert, A. M. Kraynik, and P. B. Rand, *Sci. Am.* **254**, 74 (May 1986).
J. J. Bikerman, *Foams*, Springer-Verlag, New York, 1973.
C. V. Boys, *Soap Bubbles: Their Colors and the Forces Which Mold Them*, Dover Publications, New York, 1959; originally published by the Society for Promoting Christian Knowledge, London, 1890.
H. C. Cheng and T. E. Natan, in N. P. Cheremisinoff, ed., *Encyclopedia of Fluid Mechanics*, Vol. 3, Gulf Publishing Co., Houston, Tex., 1986, p. 3.
J. P. Heller and M. S. Kuntamukkula, *Ind. Eng. Chem. Res.* **26**, 318 (1987).
C. Isenberg, *The Science of Soap Films and Soap Bubbles*, Dover Publications, New York, 1992; originally published by Tieto, Clevedon, U.K., 1978.
J. N. Israelachvili, *Intermolecular and Surface Forces, With Applications to Colloidal and Biological Systems*, Academic Press, Inc., San Diego, 1985.
A. M. Kraynik, *Ann. Rev. Fluid Mech.* **20**, 325 (1988).
K. J. Mysels, K. Shinoda, and S. Frankel, *Soap Films, Studies of Their Thinning and a Bibliography*, Pergamon Press, New York, 1959.

M. J. Rosen, *Surfactants and Interfacial Phenomena*, 2nd ed., John Wiley & Sons, Inc., New York, 1989.

D. Weaire and N. Rivier, *Contemp. Phys.* **25**, 55 (1984).

A. J. Wilson, ed., *Foams: Physics, Chemistry and Structure*, Springer-Verlag, London, 1989.

Douglas J. Durian
UCLA

David A. Weitz
Exxon Research & Engineering Company

FOOD ADDITIVES

Food additives may improve nutritional composition, enhance flavor or eating quality, or prolong storage stability in food or beverage products. Very often, the role of the food additive is essential to the safety, effective distribution, and nutritional quality of the food supply. Nevertheless, the term food additive has a negative connotation to many consumers, and the use of food additives is often regarded with mistrust. Much of the critical scrutiny of food additives was engendered by the Delaney Clause, a portion of the 1958 *Food Additives Amendment to the U.S. Food, Drug and Cosmetic Act of 1938*, which bans the use in food of chemicals shown to produce cancer in humans or animals, at any level. Enforcement of the Delaney Clause led to a review of the safety of all additives, and the banning of some. Thus the addition of artificial, ie, unnatural, substances to foods and beverages is regarded by many consumers as unnecessary and unsafe, and the consumption of natural foods and beverages, those without artificial additives, has moved to the mainstream.

Whereas the concerns regarding food additives might be expected to lead to a dwindling market, in fact overall growth in additives has outpaced the growth of the food industry. This growth can be attributed to two factors. First, additive manufacturers have responded to consumer concerns by introducing many naturally derived alternatives to synthetic food additives, gradually shifting the technology of the industry away from synthetic chemistry, toward physical and genetic modification of agricultural materials. Second, consumer concerns about additives have been tempered by the desire for convenient foods and increased consumption of reduced calorie and reduced fat foods has occurred. Both food types have a higher requirement for certain additives. The market for food additives exceeded $10 billion in the developed world in 1992. This market can be expected to continue to provide growth opportunities well into the twenty-first century.

Definition and Regulatory Considerations

According to the *U.S. Code of Federal Regulations*, food additives may be defined as "substances . . . the intended use of which results or may reasonably be expected to result, directly or indirectly, either in their becoming a component of food or otherwise affecting the characteristics of food" (1). Canada and the European Community have adopted similar definitions. According to this broad definition, a food additive is synonymous to a food ingredient. In practice, however, the word additive is limited to substances that are used in small quantities.

Food additives may be categorized as

acidulants	flour bleaching agents and bread improvers
anticaking and free-flow agents	formulation aids
antifoaming agents	fumigants
antioxidants	gases
bulking agents	humectants
colors and coloring adjuncts	leavening agents
curing and pickling agents	lubricants and release agents
dietary fibers	nonnutritive sweeteners
emulsifiers	nutrient supplements
enzymes	preservatives
fat replacers	processing aids
firming agents	solvents and vehicles
flavors	stabilizers and thickeners
flavor enhancers	

For the most part, these categories are defined according to the functional or nutritional benefit provided to the food rather than the chemical identity of the additive. Some overlap exists among categories: for example, acidulants may be used as flavoring agents, preservatives, and leavening agents (see also BAKERY PROCESSES AND LEAVENING AGENTS; FLAVORS AND SPICES).

Most bulk food ingredients, eg, flour, fats and oils, and nutritive sweeteners (qv) such as sugar (qv), are excluded from the food additive category. In a few cases, substances that are used in relatively large quantities, eg, dietary fiber (qv) and bulking agents, are included herein because these have been the focus of market and technology developments.

In the United States, substances permitted in food and beverages are regulated by the U.S. Food and Drug Administration (FDA), an agency of the Department of Health and Human Services. Additives used in meat and poultry (see MEAT PRODUCTS) are regulated by the U.S. Department of Agriculture, and additives for alcoholic beverages are regulated by the Bureau of Alcohol, Tobacco and Firearms of the U.S. Department of Treasury (see BEER; BEVERAGE SPIRITS, DISTILLED; WINE). Premarketing approval is required.

The FDA recognizes two regulatory pathways for food additives and ingredients. For a new substance, a Food Additive Petition must be submitted. This document contains information regarding, among other things, the intended use of the substance, the usage levels at which it is efficacious, its safety, the manufacturing process, and environmental impact. The process of generating these data, developing the food additive petition, and responding to questions raised by the regulatory agency has become quite lengthy and financially burdensome. For example, demonstrating safety has come to involve issues not envisioned in the original framing of the petitioning process. The historical methods of demonstrating food additive safety involved animal testing in which the animals were fed quantities of the substance many times higher than expected maximum levels in humans. Some proposed new substances, such as synthetic fat substitutes, are designed to be macroingredients for which such methods are inapplicable (see FAT REPLACERS). Thus, as of the time of this writing, the FDA is in final stages of preparation of updated safety processes and protocols for food additive petitions.

The other pathway is a somewhat simpler route. This is the GRAS affirmation process by which a petitioner can affirm that a substance either has a long history of use in the food supply and/or is generally recognized as safe (GRAS) by experts in the field. The boundaries of this process are being tested by substances such as dietary fibers and bulking agents, which are physically modified forms of natural materials, and also have the potential to be used in foods in very high quantities. The GRAS affirmation process has also come under scrutiny, and is likely to undergo revision in the coming years.

In the United States, additional ramifications may be expected from FDA's announcement of final regulations for new food labeling requirements under the directive of the Nutrition Labeling and Education Act of 1990 (2). Among other things, these regulations limit health claims that can be made on food labels. They also require new information on nutrient content, and limit the use of descriptors such as low and free in association with calories, fat levels, and other food product characteristics.

In Europe, the formation of the European Economic Community has created a requirement to bring food additive approvals of the member nations into alignment, so as to eliminate differences in laws that hinder the movement of foodstuffs among these nations. Historically the member countries have differed widely in approaches to food additive approval and their tendency to approve new additives. At the time of this writing, a framework directive for food additives and several specific directives for various categories of additives are nearing completion (3).

Classes of Food Additives

Acidulants. Acidulants, the most versatile and widely used ingredients in the food industry, function well as flavoring agents. Many acids complement fruit and other flavors in carbonated beverages (qv), preserves, fruit drinks, and desserts (see FRUIT JUICES). Their ability to lower pH makes them useful as preservatives because an acid environment retards the growth of microorganisms responsible for spoilage and prevents enzymatic browning in fruit. They are also used to modify the acidity of wine (qv). In addition, acids are used in chemical leavening agents, as gelling agents, defoaming agents, emulsifiers, and in the production of cultured dairy products. In the choice of an appropriate acid, the effect of the acid on the overall flavor system, the rate and degree of solubility of the acid, its hygroscopicity, and its strength must all be considered.

Adipic Acid. Adipic acid (qv) [124-04-9], $C_6H_{10}O_4$, produced from cyclohexane, delivers a long-lasting flavor note, and is used in products that need lingering tartness. It is effective as a buffering agent and in preventing enzymatic browning in fruits. This acid improves the whipping properties of egg whites (see EGGS), enhances the melting characteristics of processed cheese and cheese spreads, and helps form gels in imitation jams and jellies. It can be used as a general-purpose acidulant in flavor emulsions, fruit-flavored hard candies, and for hot-filled canned foods (4,5).

Citric Acid. By far the most extensively used food acidulant is citric acid (qv) [77-92-9], $C_6H_8O_7$. This acid is favored because of its solubility, fresh flavor character, low cost, and low toxicity. It is commercially synthesized by fermentation (qv) of molasses by *Aspergillus niger* (6).

Citric acid is used in carbonated beverages to provide tartness, modify and enhance flavors, and chelate trace metals. It is often added to jams and jellies to control pH and provide tartness. It is used in cured and freeze-dried meat products to protect the amino acids (qv) and improve water retention. Bakers use it to improve the flavor of fruit fillings in baked goods. Because citric acid is a good chelator for trace metals, it is used as an antioxidant synergist in fats and oils, and as a preservative in frozen fish and shellfish (7) (see ANTIOXIDANTS).

Fumaric Acid. Fumaric acid [110-17-8], $C_4H_4O_4$, is unique in its low solubility in cold water and slow rate of solution, making it ideal for use in chilled biscuit leavening systems and for dry pudding mixes and beverage powders. It is also used for gelatin desserts, pie filling, fruit juices, and wine. Fumaric acid is produced by the acid-catalyzed isomerization of maleic acid (8,9) (see MALEIC ANHYDRIDE, MALEIC ACID, AND FUMARIC ACID).

Gluconolactone. Glucono delta lactone [90-80-2] (GDL), $C_6H_{10}O_6$, an inner ester of gluconic acid, is one of the less commonly used acidulants. In conjunction with reducing compounds, GDL accelerates the rate of development of cure color in smoked meats, which reduces the smoking time considerably. It may also be employed as a chemical leavening agent, and has been used for instant bread which needs no proofing (7).

Lactic Acid. The primary use of lactic acid [598-82-3], $C_3H_6O_3$, is in fermented foods and brine-packed products where it inhibits the growth of microorganisms. This acid has the most powerful preservative effect of all commonly used acidulants (10). There is a large market for lactic acid in cheese, where it is used as a pH-adjusting agent (see MILK AND MILK PRODUCTS). It is also used to adjust the pH of other dairy products, beer, and wine. Calcium lactate [814-80-2], $CaC_6H_{10}O_6$, is used as a firming agent for processed food and as a gelling salt for low methoxy pectin. Commercial lactic acid is produced by the esterification and hydrolysis of lactonitrile (8).

Malic Acid. Malic acid [6915-15-7], $C_4H_6O_5$, similar to citric acid in acidifying character and flavor, does not exhibit the initial burst of tartness that citric acid does. Malic acid is mostly used in fruit-flavored carbonated beverages, but its high solubility and low melting point make it ideal for hard candy applications. Malic acid is synthesized by hydrating maleic acid and fumaric acid in the presence of a catalyst, then separating malic from the mixture by equilibrium techniques (11).

Phosphoric Acid.   The only inorganic acid used for food applications is phosphoric acid [7664-38-2], H_3PO_4, which is second only to citric acid in popularity. The primary use of phosphoric acid is in carbonated beverages, especially root beer and cola. It is also used for its leavening, emulsification, nutritive enhancement, water binding, and antimicrobial properties. Food-grade phosphoric acid is produced by the furnace method. Elemental phosphorus is burned to yield phosphorus pentoxide which is then reacted with water to produce phosphoric acid (see PHOSPHORIC ACID AND THE PHOSPHATES) (12).

Tartaric Acid.   A by-product of the wine processing industry, tartaric acid [87-69-4], $C_4H_6O_6$, is a specialty acidulant used for neutralizing or adjusting pH. It is found in nature in grapes and is often used to augment natural and synthetic grape and other fruit flavors. It is used as a stabilizer in dry ground spices and exhibits synergistic activity with antioxidants as a chelating agent in fats and oils. In the form of potassium acid tartrate, or cream of tartar, it is used in baking powder leavening systems (13).

Anticaking Agents.   Anticaking agents, which must be insoluble in water and have the capacity to absorb water, are used to maintain free-flowing characteristics of granular and powdered forms of hygroscopic foods. These agents function by absorbing excess moisture, or by coating particles and making them water repellent. Calcium silicate [1344-95-2], $CaSiO_3$, used to prevent caking in baking powder, table salt, and other food products, absorbs oil in addition to water, and can be used in powdered mixes and spices that contain free oils. Calcium and magnesium salts of long-chain fatty acids, such as calcium stearate [1592-23-0], $C_{36}H_{70}CaO_4$, are used as anticaking agents in dehydrated vegetable products, salt, and onion and garlic powder. Other anticaking agents employed in the food industry include sodium silicoaluminate [1344-00-9], Na_2SiAlO_3, tricalcium phosphate [7758-87-4], $Ca_3(PO_4)_2$, magnesium silicate [1343-88-0], $MgSiO_3$, and magnesium carbonate [546-93-0], $MgCO_3$. Shredded or grated cheeses have a tendency to clump together, and microcrystalline cellulose is often used to prevent this (14).

Antifoaming Agents.   Foaming is a frequent problem in food manufacturing operations causing production inefficiencies. Polydimethylsiloxane [9016-00-6] [63148-62-9], or silicone, is used at a level of approximately 10 parts per million to control foam in food products. The silicone disperses itself throughout the liquid film that makes up the foam and causes it to collapse (15).

Antioxidants.   Oxygen is one of the most common causes of the deterioration and spoilage of food products. Oxidative rancidity, resulting in off-flavors and off-odors, may be accelerated by moisture, heat, light, air, metals, metal-containing compounds, and enzymes (16). Proper preparation, packaging, and refrigeration of food retards the onset of rancidity, but cannot prevent it. Antioxidants, which must be introduced into food prior to the onset of oxidation, are the most effective means of preventing or reducing this problem.

Oxidation begins with the breakdown of hydroperoxides and the formation of free radicals. These reactive peroxy radicals initiate a chain reaction that propagates the breakdown of hydroperoxides into aldehydes (qv), ketones (qv), alcohols, and hydrocarbons (qv). These breakdown products make an oxidized product organoleptically unacceptable. Antioxidants work by donating a hydrogen atom to the reactive peroxide radical, ending the chain reaction (17).

Antioxidants are added to products that are high in fat, such as shortenings, cooking oils, snack foods, nuts (qv), and salad dressings. They are also added to the packaging materials of some cereals (see FOOD PACKAGING). Even though cereal is low in fat, the fat it contains is highly unsaturated, and vulnerable to oxidation. Because it is difficult to get an antioxidant in contact with the fat phase of the cereal, the antioxidant is often added to the cereal box liner, where it slowly diffuses into the product (18).

Both synthetic and natural antioxidants exist. The most commonly used synthetic antioxidants include butylated hydroxyanisole [25013-16-5] (BHA), $C_{11}H_{16}O_2$, butylated hydroxytoluene [128-37-0] (BHT), $C_{15}H_{24}O$, propyl gallate [121-79-9] (PG), $C_{10}H_{12}O_5$, and *tert*-butylhydroquinone [1948-33-0] (TBHQ), $C_{14}H_{22}O_2$. In Europe, other gallate esters, such as octyl gallate [1034-01-1], $C_{15}H_{22}O_5$, and dodecyl gallate [1166-52-5], $C_{19}H_{30}O_5$, are also used. BHA and BHT, which are both fat soluble, are effective in protecting animal fat from oxidation, and are often added during the rendering process. Propyl gallate is also effective, but it has limited fat solubility, and turns bluish black in the presence of iron. It is typically used as a synergist in combination with BHA or BHT. TBHQ is most effective against oxidation in polyunsaturated vegetable oils (qv), and is often used in soybean oil (19).

FDA regulations permit the use of BHA, BHT, propyl gallate, and TBHQ singly or in combinations of two or more in food products at a maximum concentration of 0.02% based on the weight of the fat or oil in the food product (20). In the 1970s, the FDA proposed a restriction on the use of BHT as a food additive. Although BHT was never removed from the GRAS list, continuing concern over its safety has resulted in decreased usage.

The most popular natural antioxidants on the market are rosemary extracts and tocopherols. Natural antioxidants have several drawbacks which limit use. Tocopherols are not as effective in vegetable fats and oils as they are in animal fats. Herb extracts often impart undesirable colors or flavors in the products where used. In addition, natural antioxidants cost considerably more than synthetic ones. Despite this, the public's uncertainty of the safety of synthetic antioxidants continues to fuel the demand for natural ones (21).

Certain compounds, known as chelating agents (qv), react synergistically with many antioxidants. It is believed that these compounds improve the functional abilities of antioxidants by complexing the metal ions that often initiate free-radical formation. Citric acid and ethylenediaminetetraacetic acid [60-00-4] (EDTA), $C_{10}H_{16}N_2O_8$, are the most common chelating agents used (22).

Another group of compounds called oxygen scavengers retard oxidation by reducing the available molecular oxygen. Products in this group are water soluble and include erythorbic acid [89-65-6], $C_6H_8O_6$, and its salt sodium erythorbate [6381-77-7], $C_6H_8O_6Na$, ascorbyl palmitate [137-66-6], $C_{22}H_{38}O_7$, ascorbic acid [50-81-7], $C_6H_8O_6$, glucose oxidase [9001-37-0], and sulfites (23).

Bulking Agents and Bulking Sweeteners. Bulking agents are substances that add bulk to food products while contributing fewer calories than the ingredients they replace. In applications where sugar is replaced by a high intensity sweetener, bulking agents make up for the lost volume, and ideally provide some or all of the functional properties of the sugar. The most important properties of a bulking agent are reduced calorie content through limited digestibility, solubility, and minimal side effects (24).

Polydextrose [6824-04-4], a polymer of glucose that contains traces of sorbitol and citric acid, is the most widely used soluble bulking agent in the United States. It is approved for use in many applications, including bakery products, frozen desserts, candy and confectionery products, jams, chewing gums, salad dressings, gelatins, and puddings. The solubility of polydextrose makes it susceptible to action by microflora in the lower intestine, leading to flatulence and diarrhea. Accordingly, if a single serving of a food contains more than 15 grams of polydextrose, the label of the food must include a warning statement about this side effect (25).

Low calorie bulking agents represent an ingredient category having a great deal of potential, and several companies are developing products. The most common are naturally derived polymers of glucose and other sugars (polydextrose falls into this category), enantiomers of natural sugars, or synthetic polymers (26). Some of these newer developments in bulking agents are given (27).

Developer	Technology
Meiji Seika Kaisha (Japan)	fructooligosaccharides (neosugar)
Beghin Say (France)	
Coors Biotech (U.S.)	
NutraSweet (U.S.)	water-soluble hemicelluloses
	cellobiitol
	sugar amides
Raffinerie Tirlemontoise (France)	oligofructose
Novo Industri (Denmark)	beta-glucan oligomers
Dow Chemical (U.S.)	polyalkylene oxide polymers
Biospherics (U.S.)	D-tagatose

Bulking sweeteners provide a bulking effect, along with some of the sweetness and functional properties of sugar. They may be used alone to replace sugar in applications that can tolerate some reduction in sweetness. Products that fall into this category include mannitol [69-65-8], $C_6H_{14}O_6$, a sugar alcohol having 0.6 times the sweetness of sugar and half the calories; isomaltitol, which is up to 0.9 times as sweet as sugar and about half the calories; some L-sugars; and fructooligosaccharides. Palatinit, a hydrogenated isomaltulose which is about 0.5 times as sweet as sugar, is marketed throughout Europe by Palatinit Sussungsmittel (Germany), mainly for use in confectionery products (27) (see SUGAR ALCOHOLS).

Colorants. According to U.S. regulations, colorants are divided into two classes: certified and exempt (see COLORANTS FOR FOODS, DRUGS, COSMETICS, AND MEDICAL DEVICES). Batch samples of certified colors must be sent to the FDA for analysis and confirmation that the colorants comply with established specifications. Color manufacturers pay a small fee for each batch of color that is analyzed. The number of certified colors available to food technologists has declined. Several of the historical colorants were found to have carcinogenic effects. Table 1 shows the certified colors that are permissible for food use in the United States as of 1993.

Table 1. Certified Colors Permitted in the United States[a]

FDA name	Common name	CAS Registry Number	Molecular formula
FD&C Blue No. 1	Brilliant Blue	[2650-18-2]	$C_{37}H_{36}N_2O_9S_3 \cdot 2NH_3$
FD&C Blue No. 2	Indigotine	[860-22-0]	$C_{16}H_{10}N_2O_8S_2 \cdot 2Na$
FD&C Green No. 3	Fast Green	[12777-77-4]	$C_{34}H_{30}N_2O_{10}S_3 \cdot 2Na$
FD&C Yellow No. 5	Tartrazine	[1934-21-0]	$C_{16}H_{12}N_4O_9S_2 \cdot 3Na$
FD&C Yellow No. 6	Sunset Yellow	[2783-94-0]	$C_{16}H_{12}N_2O_7S_2 \cdot 2Na$
FD&C Red No. 3	Erythrosine	[16423-68-0]	$C_{20}H_8I_4O_5 \cdot 2Na$
FD&C Red No. 40	Allura Red	[25956-17-6]	$C_{18}H_{16}N_2O_8S_2 \cdot 2Na$
Orange B[b]		[15139-76-1]	$C_{22}H_{18}N_4O_9S_2 \cdot 2Na$
Citrus Red No. 2[c]		[6358-53-8]	$C_{18}H_{16}N_2O_3$

[a]Ref. 28.
[b]Allowed only on the surfaces of sausages and frankfurters at concentrations up to 150 ppm by weight.
[c]Allowed only on the skins of oranges, not intended for processing, at concentrations up to 2 ppm by weight.

The FD&C certified colors are all water-soluble dyes, but can be transformed into insoluble pigments known as lakes by precipitating the dyes with aluminum, calcium, or magnesium salts on a substrate of aluminum hydroxide. The lakes are useful in applications that require color whereas in dry form, such as cake mixes, or where water may be present and bleeding is a problem, such as food packaging. FD&C Red Lake No. 3 was delisted in February 1990 (29).

Exempt colors do not have to undergo formal FDA certification requirements, but are monitored for purity. Colors are not technically referred to as natural, although some of the exempt colors do come from natural plant and animal sources. In general, natural colors are costly, are only effective at high concentrations, and fade rapidly when exposed to light (30). The colorants exempt from FD&C certification are (30) annatto extract [8015-67-6], β-carotene [7235-40-7], beet powder, β-apo-8'-carotenal [1107-26-2], canthaxanthin [514-78-3], caramel [8028-89-5], carmine [1390-65-4], carrot oil, cochineal extract, cottonseed flour, ferrous gluconate [299-29-6], fruit juices, grape skin extract, paprika, paprika oleoresin, riboflavin [83-88-5], saffron, titanium dioxide [13463-67-7], turmeric, turmeric oleoresin, ultramarine blue, and vegetable juices.

Dietary Fiber. Dietary fiber (qv) is a broad term that encompasses the indigestible carbohydrate and carbohydrate-like components of foods that are found predominantly in plant cell walls (see CARBOHYDRATES). It includes cellulose [9004-34-6] (qv), lignin [9005-53-2] (qv), hemicelluloses [9034-32-6] (qv), pentosans, gums (qv), and pectins [9000-69-5]. Those fibers that have colligative properties, such as gums, are referred to as soluble fibers. They are often used to provide viscosity and texture in processed foods, and have been linked to lowered serum cholesterol [57-88-5] levels. Insoluble fibers, such as cereal brans and specialty flour ingredients, tend to cause a laxative effect when consumed in large quantities. Dietary fiber has become an important food additive owing to the link between high fiber intake and the lowering of serum cholesterol, the prevention of cancer, and the avoidance of digestive tract disease.

Dietary fibers are used in several food categories, including breakfast cereals, pasta, snack foods, and baked goods, as well as some pharmaceutical categories such as enteral nutritionals, bulk laxatives, and diet beverage mixes (31). The common dietary fiber additives and their sources are given (32).

Fiber	Source
wheat, corn, rice, oat bran	cereal grains
soy, pea, other leguminous fiber	edible legumes
tomato, apple, pear, citrus, sugar beet fiber	fruits and vegetables
gum arabic, tragacanth, ghatti	plant exudates
agar, alginates, carageenan, furcelleran	seaweed extracts
guar, locust bean gum	seed extracts
pectin	plant extracts, eg, citrus albeido
purified cellulose and derivatives	wood pulp, cereal husks and stems
microbial polysaccharides, eg, xanthan gum	microbial fermentation

Emulsifiers. The chemical structures of emulsifiers, or surfactants (qv), enable these materials to reduce the surface tension at the interface of two immiscible surfaces, thus allowing the surfaces to mix and form an emulsion (33). An emulsifier consists of a polar group, which is attracted to aqueous substances, and a hydrocarbon chain, which is attracted to lipids.

Emulsifiers are classified by the hydrophilic–lipophilic balance (HLB) system. This system indicates whether an emulsifier is more soluble in water or oil, and for which type of emulsion (water-in-oil or oil-in-water) it is best suited. Emulsifiers having a low HLB value are more oil soluble, and are better suited for water-in-oil applications such as margarine. Conversely, emulsifiers having a high HLB value are more water soluble, and function more effectively in oil-in-water emulsions such as ice cream (34). The use of this system is somewhat limited because the properties of emulsifiers are modified by the presence of other ingredients and different combinations of emulsifiers are needed to achieve a desired effect. The HLB values of some common emulsifiers are given (35).

Emulsifier	HLB value
glycerol monostearate	3.8
succinylated monoglyceride	5.3
propylene glycol monostearate	3.4
sodium stearoyl-2-lactylate	21.0
sorbitan monostearate	4.7
sorbitan monooleate	4.3
polyoxyethylene sorbitan monostearate	14.9
polyoxyethylene sorbitan monooleate	15.0

Mono- and Diglycerides. These glycerol esters are produced by heating triglycerides and glycerol with an alkaline catalyst. Commercial glycerol esters usually contain a mixture of mono-, di-, and triesters of fatty acids, but a concentrated form of monoglycerides can be produced by molecular distillation. Monoglycerides are the most commonly used food emulsifiers and find application in a wide range of products, including baked goods, margarine, confections, icings, toppings, and peanut butter. Many products that are derived from mono- and diglycerides have found uses in foods. These include acetylated monoglycerides, ethoxylated diglycerides, and lactylated, citrated, succinylated, and sodium sulfoacetate forms (36).

Lecithin. Lecithin [8002-43-5] (qv) is a mixture of fat-like compounds that includes phosphatidyl choline, phosphatidyl ethanolamines, inositol phosphatides, and other compounds (37). Commercial lecithin was originally obtained from egg yolks, but is now extracted from soybean oil. Lecithin is used in many products, including margarine, chocolate, ice cream, cake batter, and bread.

Propylene Glycol Esters. These emulsifiers are formed by an alcoholysis reaction of propylene glycol and fatty acids, and are predominantly used in cakes, prepared mixes, whipped toppings, and breads (36).

Lactylated Esters. Sodium (and calcium) stearoyl lactates are obtained when stearic acid and lactic acid are combined and converted to the calcium or sodium salts. These are highly hydrophilic emulsifiers and have strong starch-complexing abilities. They are used in starch puddings, whipped toppings, coffee whiteners, and cake icings. Lactylated esters are also used as dough conditioners in yeast-raised products (38).

Sorbitan and Sorbitol Esters. This group of emulsifiers is formed from the reaction of sorbitan and stearic acid. Sorbitan monostearate is often used in combination with polysorbate in ice cream, imitation dairy products, and baking applications (36).

Polysorbates. Polyoxyethylene sorbitan esters [9005-63-4] are formed from the reaction of sorbitol esters with ethylene oxide. These emulsifiers are almost always employed in combination with sorbitan esters and are used in the same applications (36).

Sucrose Esters. These newer emulsifiers, approved for direct addition in the United States in 1983 (35), are formed when sucrose is combined with various fatty acids and the resulting emulsion is dehydrated. These additives are odorless and tasteless, and can withstand the retort process. They are used in products when standards of identity do not preclude their use, such as baked goods, baking mixes, dairy product analogues, frozen dairy desserts and mixes, and whipped milk products (39). High price has limited use in the United States, but these compounds are used extensively in Japan as emulsifiers in baked goods (40).

In addition to surfactant properties, emulsifiers are sometimes used to enhance the fat-replacer properties of hydrocolloid systems. Off-flavors and legal restrictions limit use in this application to a level below 0.5% of the finished product (41).

Enzymes. One of the greatest advantages of using enzymes in food processing is their specificity. Select food components can be modified while others are not affected. In addition, enzymes are natural, ie, produced by living cells, and do not require extreme heat, pressure, or pH. In the food industry, the largest use of enzymes is in starch processing, cheese production, fruit and vegetable juice processing, baking, and brewing (42). Commercial enzyme preparations are obtained from animals and plants via extraction, or through cultivation of select microorganisms. Microbial enzyme production has become quite popular because the quality, quantity, and efficacy of enzymes obtained from the cultivation of selected organisms can be easily controlled (43) (see ENZYME APPLICATIONS).

Enzymes are divided into six main classes: hydrolases, isomerases, ligases, lyases, oxidoreductases, and transferases. Hydrolases catalyze the hydrolytic splitting of substrates; isomerases catalyze intramolecular rearrangements; ligases catalyze the joining together of two substrate molecules; lyases remove or add groups to their substrates; oxidoreductases catalyze oxidations or reductions; and transferases catalyze the shift of a chemical group from one substrate to another. Each class is further divided into a number of subclasses according to the reactions catalyzed (44). Table 2 lists the principal classes of enzymes used in the food industry and the commercial applications.

Fat Replacers. Fat has a ubiquitous presence in food and provides unique flavor, mouthfeel, and functional effects. At 9 kcal/g (38 kJ/g), fat can be a principal source of dietary calories, and excessive consumption has been correlated with the incidence of chronic disease and morbidity. Health officials have strongly urged consumers to reduce fat intake to no more than 30% of daily calories. Therefore, a demand for low fat versions of high fat foods has developed. Fat replacers (qv) are the ingredients that make these foods possible.

Two classes of fat replacers exist: mimetics, which are compounds that help replace the mouthfeel of fats but cannot substitute for fat on a weight for weight basis; and substitutes, compounds having physical and thermal properties similar to those of fat, that can theoretically replace fat in all applications (46). Because fats play a complex role in so many food applications, one fat replacer is often not a satisfactory substitute. Thus a systems approach to fat replacement, which relies on a combination of emulsifiers, gums, and thickeners, is often used.

Fat Mimetics. Existing fat mimetics are either carbohydrate-, cellulosic (fiber)-, protein-, or gum-based. These are used in a wide variety of applications including baked goods, salad dressings, frozen desserts, meats, confections, and dairy products. Table 3 lists some of the commercially available fat mimetics.

Fat Substitutes. As of this writing, only one fat substitute, caprenin, a triglyceride composed of capric acid [334-48-5], $C_{10}H_{20}O_2$, caprylic acid [124-07-2], $C_8H_{16}O_2$, and behenic acid [112-85-6], $C_{22}H_{44}O_2$, has had any commercial application. The ingredient, which is GRAS, has 5 kcal/g (21 kJ/g), and is manufactured in Denmark by Grindsted Products, Inc. via a joint venture agreement with its developer, Procter & Gamble. It has been used as a low calorie cocoa butter substitute in confectionery products. Most developmental fat substitutes are novel molecules and thus must undergo safety testing before they can receive regulatory approval.

Sucrose polyesters, which are made by esterifying sucrose with long-chain fatty acids, have the physical properties of fat, but are resistant to digestive enzymes (40). Olestra, a sucrose polyester developed by Procter & Gamble, was submitted for regulatory approval in May 1987. In order to facilitate the approval process, Procter & Gamble has since narrowed the scope of its food additive petition to include olestra's use only in savory and extruded snacks.

Many companies have received patents for development of fat substitutes (Table 4) but few have demonstrated strong intentions to move toward regulatory approval.

Table 2. Enzymes Used in the Food Industry[a]

Class	Enzymes	CAS Registry Number	Commercial applications
		Hydrolases	
carbohydrases	amylases		manufacture of dextrose from starch, maltose syrup, supplementation of flour to improve bread quality, saccharification of fermentation mashes in the brewing industry
	α-amylase	[9000-85-5]	
	β-amylase	[9000-91-3]	
	glucoamylase	[9032-08-0]	
	disaccharide-splitting enzymes		manufacture of artificial honey, invert sugar, liquid-center candies, prevent crystallization of lactose in ice cream, beet sugar refining
	invertase	[9001-57-4]	
	lactase	[9031-11-2]	
	maltase	[9001-42-7]	
	melibiase	[9025-35-8]	
	pectic enzymes		fruit juice clarification, citrus oil recovery, increase fruit juice yield, reduce viscosity of purees and concentrates
	pectin methylesterase	[9025-98-3]	
	polygalacturonase	[9032-75-1]	
	pectin lyase	[9033-35-6]	
	cellulases and hemicellulases		recovery of agar from seaweed, production of glucose from cellulosic plant waste
	β-glucosidase		
	β-glucanase		
	lysozyme		used in Europe to make cow milk more suitable for infant feeding
lipases			
	pregastric lipase		used to treat butterfat-containing products to produce flavors for dairy and confectionery products, enhance whipping qualities of egg whites
	pancreatic lipase		

proteases	rennin	[9001-98-3]	used in production of Japanese condiments, meat tenderizing, cheesemaking, breadmaking
	trypsin	[9002-07-7]	
	chymosin		
	papain	[9001-73-4]	
		Other enzymes	
isomerases	glucose isomerase	[9055-00-9]	convert glucose to fructose for large-scale production of high fructose corn syrup
oxidoreductases	lipoxidase	[9029-60-1]	improve dough properties, bleach natural pigments found in flour, remove glucose from egg albumin and whole egg prior to drying, remove oxygen from beverages, canned food products, beer, and mayonnaise
	glucose oxidase	[9001-37-0]	

[a]Ref. 45.

817

Table 3. Commercially Available Fat Mimetics[a]

Product	Raw material	Manufacturer
Carbohydrate-based mimetics		
Paselli-SA-2	potato	Avebe
Sta-Slim	corn	Staley
Stellar	corn	Staley
Maltrin	corn	Grain Processing Corp.
Remyline	rice	Remy
N-Oil	corn	National Starch
Cellulosic-based mimetics		
Avicel	microcrystalline cellulose carboxymethyl cellulose	FMC Corp.
Oatrim	oats	Rhône-Poulenc/Quaker Oats ConAgra
Protein-based mimetics		
Simplesse	egg and/or whey protein	NutraSweet
Lita	zein (corn protein)	Opta Food Ingredients, Pfizer
Trailblazer	whey/milk protein	Kraft General Foods
Gum-based mimetics		
Carrafat	carrageenan	FMC Corp.
Slendid	pectin	Hercules

[a]Courtesy of Arthur D. Little, Inc.

Table 4. Companies Having Patents on Fat Substitutes[a]

Company	Product	Date of original patent
Procter & Gamble	sucrose polyesters	1971
	branched-acid esters	1971
CPC International	trialkoxytricarballylates	1985
Frito-Lay (PepsiCo)	substituted malonate esters	1986
Dow Corning	polyorganosiloxanes	1986
Procter & Gamble	acylated glycerides	1986
Unilever	sucrose polyesters	1987
Atlantic Richfield	esterified propoxylated glycerols	1988
Curtice Burns	alkyl glycoside polyesters	1989
RJR Nabisco	carboxy/carboxylate esters	1989
Pfizer	encapsulated polydextrose	1990
Procter & Gamble	caprenin	1991
NutraSweet	polyhydroxyalkanoates	1992

[a]Ref. 47.

Firming Agents. During thermal processing and freezing, the bonds of pectic substances in plant walls that help to stabilize structure are modified, resulting in an unacceptably soft product. The cell wall structure of fruits and vegetables can be strengthened by adding polyvalent cations that promote the cross-linking of the free-carboxyl groups of pectic substances. Fruits such as tomatoes, berries, and apple slices are commonly firmed by added calcium salts prior to processing (see CALCIUM COMPOUNDS). The most common calcium salts used are calcium chloride [10043-52-4], $CaCl_2$, calcium citrate [813-94-5], $Ca_3(C_6H_5O_7)_2$, calcium sulfate [7778-18-9], $CaSO_4$, calcium lactate [814-80-2], $CaC_6H_{10}O_6$, and monocalcium phosphate [10031-30-8], $Ca(H_2PO_4)_2$. Acidic aluminum salts, such as sodium aluminum sulfate, $NaAl(SO_4)_2$, potassium aluminum sulfate [10043-01-3], $KAl(SO_4)_2$, ammonium aluminum sulfate, $(NH_4)Al(SO_4)_2$, and aluminum sulfate, $Al_2(SO_4)_3$, are added during the preparation of pickles and relishes to provide the same effect (48).

Flavors. Flavorings are used in the food industry to replace or enhance flavors that are lost during processing, to create flavor combinations that do not exist in nature, and to mask objectionable flavors (see FLAVORS AND SPICES). Over 6000 flavor ingredients exist.

Essential Oils. Essential oils (qv) are extracted from the flower, leaf, bark, fruit peel, or root of a plant to produce flavors such as mint, lemon, orange, clove, cinnamon, and ginger. These volatile oils are removed from plants either via steam distillation, or using the cold press method, which avoids heat degradation. Additional processing is sometimes employed to remove the unwanted elements from the oils, such as the terpenes in citrus oils which are vulnerable to oxidation (49,50).

Oleoresins. Oleoresins are prepared by passing a volatile solvent through a ground spice or herb. Along with the essential oils, resins and gums are extracted, which serve to protect the oil under high temperature processing conditions. Compared to essential oils, oleoresins are more heat stable and have better flavor characteristics. They are also more uniform and potent than the corresponding dry spice and require much less storage space. In addition, oleoresins are free from molds and fungi which are often present in whole spices. Some oleoresins, such as turmeric and paprika, are used for coloring rather than flavoring abilities (50).

Fruit Juices and Concentrates. Because fruit juices (qv) contain large amounts of water, they are often concentrated via evaporation (qv) followed by vacuum distillation. These compounds, especially ones of the citrus variety, are widely used by the beverage industry. Many fruit juices, because of weak flavor, are augmented with other natural flavors (WONF), and are labeled as such (51).

Botanical and Animal Extracts. Tinctures and fluid and solid extracts of items such as vanilla, coffee, cocoa, and licorice are produced by treating the raw materials with a solvent. Vanilla is by far the most widely used extract and is often found in chocolate products, baked goods, beverages, and frozen desserts (49,52).

Aroma Chemicals. Specialty chemicals produced either by extraction from natural sources or by synthesis, such as vanillin [121-33-5], $C_8H_8O_3$, diacetyl [431-03-8], $C_4H_6O_2$, and benzaldehyde [100-52-7], C_7H_6O, are aroma chemicals used by the food industry (53).

Compounded Flavors.   Liquid or dry blends of natural or synthetic flavor compounds are called compounded flavors. Most commercial preparations are available as water- and oil-soluble liquids, spray-dried and plated powders, emulsions, and carbohydrate-, protein-, and fat-based pastes. Compounded flavors are used throughout the food industry in confections, baked goods, snack foods, carbonated beverages, and processed foods (53).

The increasing demand for natural flavors has been the driving force behind the growing field of flavor biosynthesis. Much research and development is directed at enzyme and fermentation (qv) technologies, plant tissue culture, and genetic engineering (qv). In many cases, flavor houses are working with biotechnology firms to create more efficient ways to produce natural flavor compounds.

Flavor Enhancers.   Flavor enhancers have the ability to enhance flavors at a level below which they contribute any flavor of their own. Worldwide, the most popular flavor enhancers are monosodium L-glutamate [142-47-2] (MSG), $NaC_5H_8NO_4$, and the 5'-ribonucleotides: disodium 5'-inosinate [131-99-7] (IMP), $C_{10}H_{11}N_4O_8P \cdot 2Na$, and disodium 5'-guanylate [85-75-5] (GMP), $C_{10}H_{12}N_5O_8P \cdot 2Na$.

IMP and GMP are obtained either by degradation of RNA using 5'-phosphodiesterase to form 5'-nucleotides, or by fermentation which results in the production of nucleosides that are phosphorylated into 5'-nucleotides (54). MSG is the sodium salt of the amino acid, L-glutamic acid, which occurs naturally in plants and animals (see AMINO ACIDS). It is produced by fermentation of the glucose present in various vegetable substrates. Microorganisms such as *Corynebacterium* and *Brevibacterium* (55) can biosynthesize L-glutamic acid from a carbon source, such as glucose, fructose, sucrose, maltose, ribose, or xylose, and a nitrogen source, such as gaseous ammonia or a solution of urea. A synergistic effect exists between MSG and the ribonucleotides, and these are often used together in foods such as dried soups and broths, canned and frozen foods, nuts, sauces, spice blends, and other processed foods.

Ammonium glycyrrhizinate [53956-04-0] (AG), $C_{42}H_{65}NO_{16}$, is a flavor enhancer derived from licorice root. It is approximately 50 times sweeter than sucrose and is often used to enhance sweetness in a wide variety of food products (56). Maltol [118-71-8], $C_6H_6O_3$, and ethyl maltol [4940-11-8], $C_7H_8O_3$, are used as flavor enhancers in products such as cake mixes, confections, cookies, ice cream, fruit juices, puddings, and beverages (57).

Flour Bleaching Agents and Bread Improvers.   Freshly milled flour contains carotenoid pigments that cause the flour to have a yellow color. In addition, when the flour is made into dough the product is sticky and unmanageable. As the flour ages, a natural process takes place which turns the flour white and improves its baking qualities. Because the natural process takes quite a bit of time, additives are used to speed up the process.

Benzoyl peroxide [94-36-0], $C_{14}H_{10}O_4$, is a bleaching agent that is typically added at the flour mill at a level between 0.015 and 0.075%. This additive oxidizes the carotenoid pigments, resulting in a white flour. Benzoyl peroxide does not affect baking properties, however, and a number of other additives can be used for this effect. Gases that exert an effect on the flour upon immediate contact include chlorine gas, chlorine dioxide [10049-04-4], ClO_2, nitrosyl chloride [2696-92-6], NOCl, nitrogen oxides, N_xO_x, and nitrogen tetroxide [10544-72-6], N_2O_4.

Others that exert their effect when the flour is made into dough include potassium bromate [7758-01-2], KBrO$_3$, potassium iodate [7758-05-6], KIO$_3$, calcium iodate [7789-80-2], Ca(IO$_3$)$_2$, and calcium peroxide [1305-79-9], CaO$_2$ (58).

Formulation Aids. Formulation aids, which include carriers, binders, fillers (qv), plasticizers (qv), and film-formers, are ingredients used in processing to impart a particular physical state or textural characteristic. Table 5 gives an overview of the formulation aids used in the food industry.

Fumigants. Fumigants are volatile substances used for controlling insects or pests (60). Ethylene oxide [75-21-8], C$_2$H$_4$O, is used for the control of microorganisms and insects in ground spices and other natural seasonings (61). Propylene oxide [75-56-9], C$_3$H$_6$O, is used as a fumigant on cocoa, gums, processed spices, starch, and processed nutmeats (except peanuts) (62). Methyl bromide [74-83-9], CH$_3$Br, provides the same function in wheat and other cereal grains (qv).

Table 5. Formulation Aids Used in Food Processing[a]

Category	Function	Typical applications
Carriers		
starches dextrins cellulose compounds silicas	allow the addition of incompatible substances to a food product	cheese, dry mixes, flour, flavor compounds
Binders		
starches salts dextrins oils gums	used to hold food together, especially reformed products	prepared meat, fish, and poultry, chewing gum, confections
Fillers		
maltodextrin polydextrose starches	add bulk to food products	confections, dietary products, chewing gum, cereal mixes
Plasticizers		
oils waxes resins humectants	maintain the soft texture of food products	chewing gum, confections, margarine, cheese products
Film formers		
carnauba wax paraffin sodium caseinate mineral oil	increase palatability, preserve gloss, inhibit discoloration, and protect food surfaces	confections, snack foods, nuts, fresh and dried fruits and vegetables

[a]Ref. 59.

Gases. Gases provide three basic functions as food ingredients: preservation, carbonation, and aeration. Nitrogen or carbon dioxide gas is frequently used to fill the headspace of packaged foods, or is used to blanket foods, to prevent oxidative deterioration. Carbon dioxide is regularly added to provide carbonation to soft drinks or to supplement existing carbonation (in beer, for example). Nitrous oxide, nitrogen, and carbon dioxide are used to help dispense fluid food products from pressurized aerosol containers (see AEROSOLS) (63).

Humectants. In certain foods, it is necessary to control the amount of water that enters or exits the product. It is for this purpose that humectants are employed. Polyhydric alcohols (polyols), which include propylene glycol [57-55-6], $C_3H_8O_2$, glycerol [56-81-5], $C_3H_8O_3$, sorbitol [50-70-4], $C_6H_{14}O_6$, and mannitol [69-65-8], $C_6H_{14}O_6$, contain numerous hyroxyl groups (see ALCOHOLS, POLYHYDRIC). Their structure makes them hydrophilic and enables them to bind water in foods. Examples of products that use humectants include shredded coconut, cookies, glazed and dried fruit, gelatin products, and cakes. High dosages of polyhydric alcohols may cause a laxative effect and usage is somewhat limited as a result (64).

Leavening Agents. Many bakery products, such as self-rising flours, prepared baking mixes, and refrigerated doughs, rely on chemical leavening agents to produce the gas that gives them volume (see BAKERY PROCESSES AND LEAVENING AGENTS). Bicarbonates produce carbon dioxide in the presence of heat and moisture. Sodium bicarbonate [144-55-8], $NaHCO_3$, is the most commonly used product, but ammonium bicarbonate [1066-33-7], NH_4HCO_3, and potassium bicarbonate [298-14-6], $KHCO_3$, are used as well. When used alone, sodium bicarbonate reacts to give products a bitter, soapy flavor. Thus it is always combined with a leavening acid.

Leavening acids are classified according to the rate at which they release carbon dioxide from sodium bicarbonate. Some acids begin producing carbon dioxide as soon as they come into contact with water; others do not begin to react unless heat is present as well. The type of leavening acid needed depends on the product. For example, refrigerated dough products require limited carbon dioxide release initially so that they can be packed into containers, but need significant activity upon heating. A slow-acting leavening agent would be used for this product. Doughnuts, which must be leavened prior to being exposed to heat, require fast-acting leavening acids. Most products use both slow- and fast-acting leavening acids to obtain the appropriate volume (65).

Table 6 lists the leavening acids and the corresponding rates of reaction. The leavening acids most frequently used include potassium acid tartrate, sodium aluminum sulfate, δ-gluconolactone, and ortho- and pyrophosphates. The phosphates include calcium hydrogen phosphate [7757-93-9], $CaHPO_4$, sodium aluminum phosphate, and sodium acid pyrophosphate (66).

Lubricants and Release Agents. Lubricants and release agents are substances added to food processing equipment to prevent food ingredients and finished products from sticking to them. Ingredients that fall into this category include oils, lecithin [8002-43-5], starch, distilled acetylated monoglycerides, and magnesium silicate [1343-88-0], $MgSiO_3$ (68).

Nonnutritive Sweeteners. Consumer desire to reduce caloric intake and protect dental health has created an enormous market for nonnutritive sweet-

Table 6. Common Leavening Acids and their Reaction Rates[a]

Acid	CAS Registry Number	Molecular formula	Relative reaction rate[b]
sodium aluminum sulfate	[10102-71-3]	$AlNa(SO_4)_2$	slow
dicalcium phosphate dihydrate	[7789-77-7]	$CaHPO_4 \cdot 2H_2O$	none
monocalcium phosphate monohydrate	[10031-30-8]	$Ca(H_2PO_4)_2 \cdot H_2O$	fast
sodium aluminum phosphate	[7785-88-8]		slow
sodium acid pyrophosphate	[7758-16-9]	$Na_2H_2P_2O_7$	slow
potassium acid tartrate	[868-14-4]	$K_2C_4H_4O_6$	medium
δ-gluconolactone	[90-80-2]	$C_6H_{10}O_6$	slow

[a]Ref. 67.
[b]In the presence of sodium bicarbonate at room temperature.

eners (qv). As of this writing there are only three nonnutritive sweeteners approved for use in the United States.

Aspartame. Aspartame [53906-69-7], $C_{14}H_{18}N_2O_5$, was discovered by accident in 1965. It is a dipeptide of L-aspartic acid and L-phenylalanine, and is approximately 200 times as sweet as sucrose at a 4% concentration. It has a clean, sweet taste similar to sucrose. At high temperatures or during prolonged storage, aspartame undergoes hydrolytic degradation to its component amino acids and loses its sweetening ability. Therefore, aspartame cannot be used in products that undergo high temperatures during processing (69). Aspartame's largest use is in soft drinks, although it is also used in products such as yogurt, powdered drink mixes, confections, frozen desserts, and cereals.

Saccharin. Saccharin [81-07-2], $C_7H_5NO_3S$, which is approximately 300 times as sweet as sucrose in concentrations up to the equivalent of a 10% sucrose solution, has been used commercially as a nonnutritive sweetener since before 1900, predominantly in carbonated soft drinks, tabletop sweeteners, and dietetic foods marketed primarily to diabetics. In 1977, the FDA proposed a ban on saccharin because of its association with bladder cancer in laboratory animals. At the time, it was the only commercially available nonnutritive sweetener, and public outcry led to a delay of the ban, which was officially withdrawn in 1991. Instead, the FDA required that warning labels be placed on all foods that contained the ingredient. Although saccharin is heat stable, the public debate over its safety, as well as the fact that approximately one-third of the population perceives it to have a bitter aftertaste, has limited its use.

Acesulfame K. Acesulfame K [55589-62-3], $C_4H_5NO_4S \cdot K$, is an oxathiazine derivative approximately 200 times as sweet as sucrose at a 3% concentration in solution (70). It is approved for use as a nonnutritive sweetener in 25 countries (71), and in the United States has approval for use in chewing gum, confectionery products, dry mixes for beverages, puddings, gelatins, and dairy product analogues, and as a tabletop sweetener (72).

Other Sweeteners. Two other sweeteners, sucralose and cyclamates, are approved for use outside of the United States. Sucralose, a chlorinated derivative of sucrose which is 500–600 times as sweet as sugar, has received limited approval in Canada, and petitions for its approval are pending in the United States and Europe (71). Cyclamate sweeteners, once available in the United States, but now banned because they caused bladder cancer in animals, are still available in Canada and Europe. Table 7 gives several examples of nonnutritive sweeteners that have been developed.

Nutrients.   In the United States, foods are either restored, enriched, or fortified with nutrients. Restoration refers to the addition of nutrients to foods to replace those nutrients that are lost in processing. Enrichment is similar to restoration, but federal guidelines specify the exact amount and kinds of nutrients added to specific products. Fortification refers to the addition of nutrients that do not naturally occur in the food. This last is typically done to prevent diseases of nutritional deficiency.

The enrichment program followed in the United States is (1) the enrichment of flour, bread, and degerminated and white rice using thiamin [59-43-8], $C_{12}H_{17}N_5O_4S$, riboflavin [83-88-5], $C_{17}H_{20}N_4NaO_9P$, niacin [59-67-6], $C_6H_5NO_2$, and iron [7439-89-6]; (2) the retention or restoration of thiamin, riboflavin, niacin, and iron in processed food cereals; (3) the addition of vitamin D [67-97-0] to milk, fluid skimmed milk, and nonfat dry milk; (4) the addition of vitamin A [68-26-8], $C_{20}H_{30}O$, to margarine, fluid skimmed milk, and nonfat dry milk; (5) the addition of iodine [7553-56-2] to table salt; and (6) the addition of fluoride [16984-48-8] to areas in which the water supply has a low fluoride content (74).

Preservatives.   Without control of yeasts (qv), molds, and bacteria, the food industry would experience considerable economic losses each year owing to spoilage. Sugar, salt, and wood smoke have been used for centuries to preserve food.

Table 7. High Intensity Sweeteners[a]

Sweetener	Intensity[b]	Developer and/or manufacturer	Regulatory status
aspartame	180	NutraSweet	approved broadly
saccharin	300	many	approved in many markets, including U.S.
cyclamates	35	Abbott Laboratories	some approvals in Canada, Europe
acesulfame K	up to 200	Hoechst AG	broad approvals in Europe, limited in U.S.
sucralose	up to 600	Tate & Lyle, Johnson & Johnson	approved in Canada, pending elsewhere
alitame	up to 2,000	Pfizer	applications pending
sucrononic acid, other dipeptides	up to 200,000	Claude Bernard University, NutraSweet	not yet submitted
L-sugars	1[c]	Biospherics	not yet submitted

[a]Ref. 73.
[b]Values are times the sweetness of sucrose.
[c]L-Sugars are not absorbed.

These methods, however, are not compatible with all food products; thus preservatives, also known as antimicrobials, are used.

Most preservatives do not kill microorganisms present in food. Rather, they prevent further growth and proliferation of anything that is present by either lowering the water activity or increasing the pH of the foods in which they are used. Numerous factors, including the type of organism to be controlled, product pH, effect on product flavor, legal restrictions, and cost, all impact the selection of the proper preservative (75).

Benzoates. The sodium and potassium salts of benzoic acid [65-85-0], $C_7H_6O_2$, are most effective against yeast and mold. They are used in beverages, fruit products, chemically leavened baked goods, and condiments. Owing to their inhibitory effect on yeast, they cannot be used in yeast-leavened products. Potassium benzoate was developed for use in reduced-sodium products. Benzoates are permitted for use in foods up to a level of 0.1% (76).

Sorbates. The sodium and potassium salts of sorbic acid [110-44-1], $C_6H_8O_2$, are used as mold and yeast inhibitors in dairy products, chemically leavened baked goods, fresh and fermented vegetables, dried fruit, beverages, confections, and smoked meat and fish. They are widely used in Japan as preservatives in processed products made with fish paste. Sorbates have the ability to inhibit the growth of yeast at the surface of food during fermentation, but do not inhibit the organisms that are used in the fermentation process. Sorbates are typically added directly by dipping the food into a sorbate solution, or by spraying them on the surface of the food. Usage ranges from 0.025–0.2% (76).

Propionates. Propionic acid [79-09-4], $C_3H_6O_2$, and its calcium and sodium salts are effective mold inhibitors. They are particularly useful in yeast-leavened baked products because they do not affect the activity of yeast. In addition to being widely used in baked goods, they are used as mold inhibitors in cheese foods and spreads (77).

Organic Acids. One method of controlling the growth of microorganisms in food is to increase the acidity of the product. This can be accomplished by adding an organic acid such as acetic acid [64-19-7], $C_2H_4O_2$, citric acid [77-92-9], $C_6H_8O_7$, malic acid [6915-15-7], $C_4H_6O_5$, lactic acid [598-82-3], $C_3H_6O_3$, adipic acid [124-04-9], $C_6H_{10}O_4$, tartaric acid [87-69-4], $C_4H_6O_6$, or caprylic acid [124-07-2], $C_8H_{16}O_2$. Because acids can affect the functionality of other ingredients in food, care must be taken in selecting the appropriate one. These acids also function as acidulants (78).

Sulfur Dioxide and Sulfites. Sulfur dioxide [7446-09-5], SO_2, sodium bisulfite [15181-46-1], $NaHSO_3$, and sodium metabisulfite [23134-05-6] are effective against molds, bacteria, and certain strains of yeast. The wine industry represents the largest user of sulfites, because the compounds do not affect the yeast needed for fermentation. Other applications include dehydrated fruits and vegetables, fruit juices, syrups and concentrates, and fresh shrimp (79). Sulfites are destructive to thiamin, and cannot be used in foods, such as certain baked goods, that are important sources of this vitamin.

Sulfites and related compounds have been the subject of much controversy. It has been shown that these chemicals may cause serious allergic reactions in sensitive individuals. In the United States, sulfite-containing foods must be labeled as such, and sulfites are prohibited in areas such as salad bars where labeling is inappropriate.

Parabens. In the United States, the heptyl, methyl, and propyl esters of *para*-hydroxybenzoic acid [99-96-7], $C_7H_6O_3$, have been used as preservatives since the 1930s. Only the methyl and propyl esters are considered GRAS in the United States, and their use is limited to 0.1%. The heptyl esters are permitted in fermented beverages and some fruit-based beverages. The butyl and ethyl esters have been used in other countries since the 1920s. Unlike other antimicrobials, the parabens are active up to a pH of 8. They are used as mold and yeast inhibitors in baked goods, beverages, fruits, jams and jellies, syrups, olives, and pickles. Their high cost compared to some other preservatives has restricted their use in the food industry (78).

Sodium Nitrate and Sodium Nitrite. Nitrates and nitrites are used in meat-curing processes to prevent the growth of bacteria that cause botulism. Nitrates have been shown to form low, but possibly toxic, levels of nitrosamines in certain cured meats. For this reason, the safety of these products has been questioned, and use is limited (80).

Natamycin and Nisin. Concern over the safety of synthetic preservatives has led to the research and development of natural alternatives. Natamycin [7681-93-8], $C_{33}H_{47}NO_{13}$, an antibiotic produced by *Streptomyces natalensis*, has gained approval in the United States for use against molds on cured cheeses. Natamycin selectively inhibits molds while allowing the growth of bacteria needed for the ripening process (81). Nisin [1414-45-5], $C_{143}H_{230}N_{42}O_{37}S_7$, a polypeptide produced by the fermentation of a modified milk medium by *Lactococcus lactis*, is particularly effective against spore-forming gram-positive bacteria. It is used worldwide as a preservative in processed cheese, dairy products, canned foods, cured meat, and beer (82).

Processing Aids. Manufacturing aids, used to improve the appearance or performance of food products, include clarifying agents (flocculants), clouding agents, catalysts, and filter aids. Clarifying agents eliminate turbidity and particle suspension in products such as beer, wine, fruit juices, oils, and vinegar. Gelatin and lime are frequently used for this purpose. Clouding agents add a turbid appearance to products such as syrups, soft drinks, and powdered beverage mixes. Brominated vegetable oils [8016-94-2], gums, and citrus pulp are commonly used clouding agents. Catalysts, which are agents that facilitate a chemical reaction, are used for the hydrogenation of oil, transesterification of fats, modification of starches, and many enzyme reactions. Raney nickel [7440-02-0], Ni, sodium methoxide [124-41-4], CH_3ONa, and a variety of acids are typical catalysts. Filter aids break down or entrap undesired substances in fruit juices, wines, milk, oils, beer, and vinegar, thus making it easier to remove these substances by filtration (83).

Solvents. Solvents are generally used to either extract particular compounds, such as an essential oil from a plant, or to carry additives into a food system, such as a flavor into a powdered mix. Common solvents include ethanol [64-17-5], C_2H_6O, glycerine [56-81-5], $C_3H_8O_3$, propylene glycol [57-55-6], $C_3H_8O_2$, triethyl citrate [77-93-0], $C_{12}H_{20}O_7$, polyhydric alcohols, carbon dioxide [124-38-9], acetylated monoglycerides, hexane [110-54-3], C_6H_{14}, methylene chloride [75-

09-2], CH_2Cl_2, acetone [67-64-1], C_3H_6O, and trichloroethylene [79-01-6], C_2HCl_3 (68,84).

Stabilizers and Thickeners.   Many food products receive their textural properties from a group of compounds known as hydrocolloids. Hydrocolloids are high molecular weight polymers that are either extracted from plants, seaweed, or animal collagen, or are produced by microbial synthesis. They are widely used for their general thickening properties, as well as their ability to provide stabiliy for emulsions, suspensions, and foams. The structure of hydrocolloids gives them a slippery, creamy mouthfeel that mimics the organoleptic properties of fats and oils, thus making them useful as fat replacers. Many of the fat replacers that are on the market are based on one or a combination of hydrocolloids. Hydrocolloids fall into two classes: polysaccharides and proteins. Most stabilizers and thickeners are polysaccharides.

Locust Bean Gum.   Locust bean gum [9000-40-2], also known as carob seed gum, is a galactomannan extracted from the endosperm of the carob tree seed which is cultivated in the Mediterranean area. The primary use of locust bean gum is in dairy applications such as ice cream. It is often used in conjunction with carrageenan because the chemical structures of the two enable them to cross-link and form a gel (85).

Guar Gum.   Guar gum [9000-30-0], also a galactomannan, is extracted from the endosperm of the guar plant seed which is grown primarily in India and Pakistan. Guar hydrates rapidly in cold water and is used extensively in ice cream, cheese, and baked goods. Chemically, guar is closely related to locust bean gum and its use began as a result of a shortage of locust bean gum in the 1940s. The high price of guar gum has forced users to look to other materials for thickening agents (85).

Gum Arabic.   Gum arabic [9000-01-5] is an exudate of the Acacia tree, found in the Middle East. It dissolves readily in water to produce low viscosity solutions. It is used in confectionery products, bakery toppings, beverages, frozen dairy products, and dry drink mixes (86).

Carrageenan.   Carrageenan [9000-07-1] is extracted from red seaweeds found along the shores of the western United States, Nova Scotia, and the British Isles. The largest food use for carrageenan is in dairy products such as flavored milk and frozen desserts. Carrageenan is comprised of a mixture of polymers, but the most commonly used commercially are κ (gelling) and λ (nongelling). κ-Carrageenan is only soluble in hot conditions; λ-carrageenan is only soluble in cold conditions. Carrageenan interacts synergistically with other food gums and is often used in conjunction with locust bean gum (85).

Xanthan Gum.   Xanthan gum [11138-66-2] is produced by industrial fermentation of a carbohydrate under aerobic conditions by culturing the bacterium *Xanthomonas campestris*. It is unique in that it gives high viscosity solutions at low concentrations, exhibits little change in viscosity with variation in temperature, and is stable over a wide pH range. The greatest usage of xanthan gum is in salad dressing, but it is also used in baked goods, confectionery products, syrups, toppings, dry beverage mixes, frozen foods, and dairy products. Xanthan gum is synergistic with locust bean gum and guar gum, and they are often used together for enhanced gelation or viscosity (87).

Other Gums. Gellan gum, the first to receive FDA approval for food use since xanthan gum in 1968, is produced from *Pseudomonas elodea* by a pure culture fermentation process (88). In the United States, gellan gum received approval in November 1992 for use in all foods when a standard of identity does not preclude its use. In Japan, gellan gum is used in gelled desserts and jellies and approval is pending in Canada and Europe (87). Ghatti gum [*9000-28-6*], tragacanth gum [*9000-65-1*], and karaya gum [*9000-36-6*] are also used, but not as frequently as the others.

Cellulose. The principal structural component of plant cell walls is cellulose (qv). The most widely used cellulose derivative is the sodium salt of carboxymethylcellulose (CMC). It is made by treating cellulose with sodium hydroxide–chloroacetic acid. CMC is widely used in the food industry in products such as baked goods, icings, syrups, glazes, frozen dairy products, and dry drink mixes (89).

Methylcellulose, another useful cellulose derivative, is made by the reaction of cellulose and sodium hydroxide with methylchloride. It is used to prevent syneresis in frozen foods, and as a thickener and stabilizer in salad dressings (90).

Microcrystalline cellulose (MCC) is a partially depolymerized from of cellulose prepared by the hydrolysis of wood pulp with hydrochloric acid. MCC is used as a stabilizer in frozen desserts, meats, dairy products, baked goods, and aerosol toppings (91).

Agar. Agar [*9002-18-0*] is obtained from a variety of red marine algae found along the coast of Japan. Food applications include frozen desserts, confectionery products, and baked goods (92).

Starch. The most abundant natural, or unmodified, starch [*9005-25-8*] is that produced by the wet milling of corn. Other commercial starches include wheat, potato, sago, rice, and tapioca. Owing to low cost, unmodified starches are used extensively as thickening agents in the food industry. Starch is often pregelatinized so it can be used to impart thickening properties in foods that are not normally heated, such as instant pie fillings.

Starch is often modified by hyrolysis with hydrochloric or sulfuric acid (93). The resulting product is resistant to syneresis, keeps food in suspension after cooking, and exhibits much greater freeze–thaw stability than unmodified starch. Modified starch is commonly used in baby food, frozen prepared foods, pie fillings, meat products, and candy.

Pectin. Pectin [*9000-69-5*] is a mucilaginous substance extracted from the cell walls of citrus peel, sugar beet pulp, and apple pomace. Two classes of pectins exist. High methoxy pectins, which have a degree of esterification higher than 50%, are used in jams, jellies, and gummi candies. These pectins require the presence of sugar and acid in order to form a gel. Low methoxy pectins, which have a lower degree of esterification, form gels in the absence of sugar, but require the presence of calcium to aid with cross-linking. Low methoxy pectins are often used in low sugar jams and jellies (85).

Alginates. Alginates are extracted from brown seaweeds found along the rocky coasts of the north Atlantic coastline and the Pacific coast along southern California. They are unique because they are able to form irreversible gels by reaction with calcium salts without heat. They are used in ice cream, bakery products, puddings, dressing, and beer (for foam stabilization) (87).

Table 8. 1992 U.S. Sales and 1993 Price and Manufacturers of Food Additives[a]

Additive	Type	Sales, $ × 10⁶	Average price, $/kg	Manufacturer(s)
acidulants		225	1.60	
	citric			Haarmann & Reimer, Hoffmann-La Roche, Cargill
	fumaric			Pfizer, Monsanto
	lactic			Sterling, Takeda
	malic			Haarmann & Reimer, Takeda
	phosphoric			FMC, Monsanto, Rhône-Poulenc
antioxidants		75	6.6–33	
	BHA			Eastman Chemical, UOP
	BHT			PMC Specialties Group, Eastman Chemical
	erythorbates			Pfizer
	propyl gallate			Eastman Chemical
	tocopherols			Eastman Chemical, Henkel Corp., Hoffmann-La Roche
	TBHQ			Eastman Chemical, UOP
bulking agents	polydextrose	40	2.40	Pfizer
colorants		160	0.22–0.66	Universal Foods (Warner-Jenkinson, McCormick/Stange, Kohnstamn), Haarmann & Reimer, Hoffmann-La Roche, Colorcon, Quest/Biocon
dietary fibers		330	1.1–1.6	Protein Technologies International, Canadian Harvest, ConAgra, Opta Food Ingredients
emulsifiers		300	1.6–5.5	Van den Bergh Food Ingredients, Grindsted, Eastman Kodak, Lonza, Hunko, ADM Arkady, Karlshamn, ICI, Beatreme
enzymes		130	[b]	Novo, Solvay, Gist-Brocades, Cultor, Genecor, Christian Hansen's
fat replacers		80	2.2[c]–17.7[d]	NutraSweet Co., National Starch, Grain Processing Corp., Avebe, Staley, FMC, Hercules
flavors[e]		1000	[b]	International Flavors and Fragrances, Quest International, Givaudan, Takasago, Haarmann & Reimer

829

Table 8. (*Continued*)

Additive	Type	Sales, $ $\times 10^6$	Average price, $/kg	Manufacturer(s)
humectants		70	1.40	
	sorbitol			ICI, Lonza
	propylene glycol			Dow, Arco
	glycerin			Procter & Gamble, Quantum
leavening agents		50	1.20	Monsanto, Rhône-Poulenc, FMC
nonnutritive sweeteners		900		
	aspartame		66–132[f]	NutraSweet
	acesulfame K			Hoechst Celanese
	saccharin		4.4–8.8	PMC Specialties
nutrients	C, B	90	11–66	Hoffmann-La Roche, Takeda
	E			Eastman Kodak, Henkle
	K			Heterochemical
	niacin			Cambrex
preservatives		110		
	benzoates		2.00	Pfizer, Mallinckrodt, Kalamo Chemical, Southland Corp.
	sorbates		7.70	Eastman Kodak, Pfizer
	propionates			Mallinckrodt, Kalamo Chemical, Southland Corp.
thickeners				
	gums	280	1.50–27.00	Kelco, Rhône-Poulenc, FMC, Hercules, Sanofi, Meer, Polypro, Bulmer, Protan
	starches	210[f]	<0.50–3.50	National Starch, A. E. Staley, American Maize, CPC, Cargill, Archer Daniels Midland
	gelatin		2.75–5.30	General Foods, Unilever, Hormel

[a]Courtesy of Arthur D. Little, Inc.   [b]Price varies widely.
[c]Starch-based.   [d]Pectin-based.   [e]Includes flavor enhancers.
[f]Sales are for starches and gelatins.

Gelatin. Gelatin, a protein-based hydrocolloid, is a polymer made up of amino acids. It is obtained from beef or pork skin, hides, and bones by hydrolysis and extraction of collagen with hot water. The most popular usage of gelatin is in the preparation of gelatin desserts, but it is also used in marshmallows and other confectionery products, processed meats, frozen foods, and dairy products. Wine, beer, cider, and fruit juices use gelatin as a clarifying agent (94).

Market Overview

The U.S. market for food additives reached about $4 billion in 1992. The breakdown of this market and average prices in 1993 of common food additives are shown in Table 8.

The additives and ingredients industry remains quite fragmented. Competitors are drawn from a wide range of industries that include commodity grain and oilseed processors, other agricultural material processors, and bulk, specialty, and fine chemical suppliers. The large majority of participants have under $100 million in sales; only a handful have over $300 million. This relatively small size contrasts sharply with the multibillion dollar food companies that represent the customer base for ingredients and additives. Increasing demands are being placed on ingredient and additive suppliers by these customers, including applications support, favorable pricing, just-in-time deliveries, and conformance to standards.

Overall, the world market for food additives was approximately $10 billion in 1992. Besides the United States, the other principal markets were the European Community (about $3 billion) and Japan (about $2 billion). Growth rates are expected to average about 3% per year throughout most of the 1990s. Highest growth rates should continue to be realized in segments where the ingredients and additives are needed for low calorie, low fat, and other nutrition-oriented products. These include nonnutritive sweeteners, fat replacers, bulking agents, thickeners, and dietary fibers. Artificial flavors, colors, and preservatives are expected to experience slow or no growth.

BIBLIOGRAPHY

"Food Chemicals" in *ECT* 1st ed., Vol. 6, pp. 835–848, by M. B. Jacobs, Polytechnic Institute of Brooklyn; "Food Additives" in *ECT* 2nd ed., Vol. 10, pp. 1–22, by D. G. Chapman, Food and Drug Directorate, Department of National Health and Welfare, Canada, and Z. I. Kertesz, Food and Agriculture Organization of the United Nations; in *ECT* 3rd ed., Vol. 11, pp. 146–163, by T. Furia, Intechmark Corp.

1. *Code of Federal Regulations, Title 21 Part 170.3*, U.S. Government Printing Office, Washington, D.C., Apr. 1, 1990.
2. N. Mermelstein, *Food Technol.*, 81 (Feb. 1993).
3. C. Grjspaardtvink, *Food Technol.*, 106 (Mar. 1993).
4. R. D. McCormick, *Prepared Foods*, 106 (Apr. 1983).
5. *Monsanto Technical Literature*, MCFI-2002, St. Louis, Mo., 1992.
6. A. E. Humphrey and S. E. Lee, in J. A. Kent, ed., *Riegel's Handbook of Industrial Chemistry*, 8th ed., Van Nostrand Reinhold Co., New York, 1983, p. 656.
7. *Technical Literature*, Pfizer, Inc., New York, 1992.
8. Ref. 4, p. 107.
9. *Technical Literature*, Monsanto Co., FI-8504-LT, St. Louis, Mo., 1992.
10. *Technical Literature*, Purac Co., Lincolnshire, Ill.
11. Ref. 4, p. 110.
12. J. H. McLellan in Ref. 6, p. 242.
13. Ref. 4, p. 113.
14. R. C. Lindsay, in O. R. Fennema, ed., *Food Chemistry*, 2nd ed., Marcel Dekker, Inc., New York, 1985, pp. 665–666.

15. *Technical Literature*, Dow Corning, Midland, Mich.
16. C. Andres and D. Duxbury, *Food Processing* **51**, 100 (May 1990).
17. *Technical Literature*, Henkel, La Grange, Ill.
18. R. J. Sims and J. A. Fioriti, in T. E. Furia, ed., *Handbook of Food Additives*, Vol. 2, 2nd ed., CRC Press, Boca Raton, Fla., 1980, p. 20.
19. *Ibid.*, pp. 25, 30.
20. *The Food Chemical News Guide*, CRC Press, Inc., Washington, D.C., Feb. 17, 1992, p. 60.
21. B. F. Haumann, *Inform* **1**, 1002 (Dec. 1990).
22. J. D. Dzeziak, *Food Technol.* **40**, 101 (Sept. 1986).
23. *Ibid.*, pp. 97–101.
24. C. G. Greenwald, *Spectrum*, 2–52. (Sept. 1989).
25. *Ibid.*, p. 2–53.
26. *Ibid.*, p. 2–51.
27. C. G. Greenwald, "Sweeteners and Bulking Agents-Overlapping Technologies and Synergistic Markets," *Food Ingredients Europe Conference Proceedings*, Expoconsult Publishers, the Netherlands, 1990, p. 23.
28. F. J. Francis, in Ref. 14, p. 577.
29. J. B. Hallagan, *Cereal Foods World* **36**, 946 (Nov. 1991).
30. G. J. Lauro, *Cereal Foods World* **36**, 949 (Nov. 1991).
31. K. L. Wrick and A. B. Caragay, *Outlook for the U.S. Food Additives Industry to 1992*, Arthur D. Little Decision Resources, Burlington, Mass., Feb. 1989, pp. 59–79.
32. *Ibid.*, p. 61.
33. J. D. Dzeziak, *Food Technol.* **42**, 172 (Oct. 1988).
34. *Technical Literature*, Grindsted, Industrial Airport, Kans., May 1990.
35. W. W. Nawar, in Ref. 14, p. 171.
36. Ref. 33, pp. 178–179.
37. Ref. 33, p. 180.
38. N. H. Nash and L. M. Brickman, *J. Am. Oil Chem. Soc.* **49**, 458 (Aug. 1972).
39. Ref. 20, p. 440.1, Jan. 13, 1993.
40. *Food Engineering* **63**, 24 (Sept. 1991).
41. M. Glicksman, *Food Technol.* **45**, 94 (Oct. 1991).
42. J. D. Dzeziak, *Food Technol.* **45**, 78 (Jan. 1991).
43. L. A. Underkofler, in Ref. 18, p. 72.
44. *Ibid.*, p. 60.
45. *Ibid.*, pp. 80–111.
46. C. G. Greenwald, in C. Haberstroh and C. E. Morris, eds., *Fat and Cholesterol Reduced Foods*, Gulf Publishing Co., Texas, 1991, p. 27.
47. *Ibid.*, p. 28.
48. Ref. 15, pp. 660, 661.
49. A. J. Neilson and N. F. Smith, in Ref. 31, p. 27.
50. F. Fischetti, Jr., in Ref. 18, pp. 234, 242.
51. *Ibid.*, p. 249.
52. *Ibid.*, p. 251.
53. Ref. 49, p. 29.
54. Y. Sugita, in A. L. Branen, P. M. Davidson, and S. Salminen, eds., *Food Additives*, Marcel Dekker, Inc., New York, 1990, p. 266.
55. *Ibid.*, p. 265.
56. *Technical Literature*, MacAndrews and Forbes, Camden, N.J., 1992.
57. *Technical Literature*, Pfizer, Inc., New York, 1992.
58. Ref. 14, pp. 663–665.
59. E. S. Baranoski, in Ref. 54, pp. 512, 513.

60. *Code of Federal Regulations, Title 21*, section 170.3, U.S. Government Printing Office, Washington, D.C., Apr. 1, 1990, p. 8.
61. Ref. 20, p. 191, Sept. 28, 1992.
62. Ref. 20, p. 374.1, Jan. 18, 1993.
63. Ref. 14, pp. 666–667.
64. Ref. 14, pp. 658–659.
65. *Technical Literature*, Church & Dwight Co., Princeton, N.J., 1990.
66. Ref. 14, p. 634.
67. Ref. 14, p. 635.
68. Ref. 59, p. 514.
69. Ref. 14, p. 655.
70. Ref. 14, p. 656.
71. Ref. 27, p. 21.
72. Ref. 20, p. 2, Jan. 4, 1993.
73. Ref. 27, p. 22.
74. S. R. Tannenbaum, V. R. Young, and M. C. Archer, in Ref. 14, p. 486.
75. *Technical Literature*, Pfizer, Inc., New York, 1992.
76. C. Andres, *Food Processing* **46**, 27 (Mar. 1985).
77. J. D. Dzeziak, *Food Technol.* **40**, 105 (Sept. 1986).
78. Ref. 76, p. 28.
79. Ref. 77, p. 108.
80. Ref. 77, p. 109.
81. Ref. 14, p. 648.
82. J. Delves-Broughton, *Food Technol.* 100, 108 (Nov. 1990).
83. Ref. 59, pp. 513, 514.
84. Ref. 14, p. 672.
85. *Technical Literature*, Sanofi Bio-Industries, Waukesha, Wis., 1992.
86. J. D. Dzeziak, *Food Technol.* **45**, 120 (Mar. 1991).
87. *Technical Literature*, Kelco, San Diego, Calif., 1992.
88. Ref. 20, p. 210.2, Nov. 30, 1992.
89. R. L. Whistler and J. R. Daniel, in Ref. 14, p. 121.
90. Ref. 89, p. 122.
91. *Technical Literature G-34*, FMC Corp., Philadelphia, 1992, p. 3.
92. Ref. 86, p. 118.
93. Ref. 89, p. 118.
94. *Technical Literature*, Gelatin Manufacturers of America, New York, 1992 .

LESLIE J. FRIEDMAN
C. GAIL GREENWALD
Arthur D. Little, Inc.

FOOD PACKAGING

The principal functions of food packaging are to protect the food contents from physical damage, losses, or deterioration, and to facilitate distribution from processor to consumer. Food packaging also must attractively identify the product and must perform these functions at minimum system cost because the package itself has no intrinsic value to the consumer. In 1992, food packaging represented about 57% of the United States' more than $70 billion packaging industry.

Food packaging assists product preservation for distribution by reducing spoilage, infestation, contamination, and pilferage; makes economical use of warehouse space; conserves labor in both distribution and marketing; and permits distribution of identified products that can be effectively marketed through self-service retailing. Food packaging deals not only with the materials in contact with the product but also with secondary and tertiary (unitizing) packages, form, equipment, labor, consumer use, and systems costs.

The choice of packaging for a given food product is circumscribed by a variety of factors, ie, the food characteristics determine the protection needed to prevent deterioration, and the distribution system affects the product's shelf life and therefore imposes additional requirements for its protection. The food package must be attractive, convenient, and identifiable to the consumer. Its costs must be low, and the machinery to produce it must be available. As of the 1990s, packagers must actively consider the ultimate disposal of the used package in selecting food packaging.

Food deteriorates by biochemical, enzymatic, microbiological, and physical vectors. The biochemical vector is the result of interactions of food chemicals because of their proximity to each other. Enzymatic deterioration is biochemical deterioration catalyzed by enzymes naturally present in food. Microbiological deterioration from yeasts (qv), molds, and bacteria is a common food spoilage vector. Damage to food products not associated with biochemical, enzymatic, or microbial spoilage is usually physical, such as gain or loss of water. Elevated water activity in dry food products increases the rate of biochemical reactions. In food products with high water content, such as fresh produce or meat (see MEAT PRODUCTS), water loss alters physical characteristics and water gain can lead to favorable conditions for microbiological growth. Thus, much of the function of food packaging is to ensure against gain or loss of water. Almost all adverse reactions are accelerated by increasing temperatures.

Packaged Food Classification

Approximately half of all the food products in the United States are fresh or minimally processed. Fresh food products include meats, vegetables, and fruits that are unprocessed except for removal from the original environment and limited trimming and cleaning. Fresh foods are handled to retard deterioration, which is relatively rapid at ambient or higher temperature. Meats are chilled rapidly to below 10°C (50°F) and most vegetables and fruits are generally reduced to below 4.4°C (40°F) by low temperature air, water, or ice.

Minimally processed foods include those that have been altered to help retard deteriorative processes. For example, most dairy products must be refrigerated after pasteurization (see MILK AND MILK PRODUCTS); nitrite-cured meats must be kept refrigerated after processing to minimize microbial growth. Fully processed foods are intended for long-term shelf life at ambient temperature, and include almost all heat processed, dried, etc, foods (see FOOD PROCESSING).

Fresh Foods. *Meat, Poultry, Fish.* About a quarter of food products in the United States are meats, including beef, poultry, fish, lamb, veal, and pork. All of these are susceptible to microbiological, enzymatic, and physical changes.

The color of red meat depends on oxygen. The color of the meat pigment myoglobin is purple. The bright red color of the fresh-as-cut meat is from oxymyoglobin. To preserve red meat, the objectives are to retard spoilage and weight loss, and to deliver red color at the consumer level.

In distribution channels, most red meat is packaged under reduced oxygen in high oxygen–water vapor-barrier flexible packaging materials to retard deterioration. Fresh meats are transported to retail outlets at temperatures below 10°C (50°F) to retard deteriorative processes. At the retail level, exposure to air in gas-permeable packaging permits restoration of the bright red oxymyoglobin color. Oxygen-permeable flexible packaging, such as plasticized poly(vinyl chloride) (PVC) film, permits oxygen into the package while retarding the passage of water vapor.

Poultry, susceptible to microbiological deterioration, is an excellent substrate for *Salmonella*. Therefore, the temperature is reduced as rapidly as possible after slaughter. Packaging at factory level is in soft film, ie, low density polyethylene or plasticized PVC, which retards water-vapor loss and permits oxygen entry.

Fish is packaged to retard weight loss. Packaging for frozen fish generally has low water-vapor permeability to permit long-term frozen distribution without freezer burn or surface desiccation. Wrapped or polyethylene-coated paperboard cartons and water-vapor barrier flexible films such as polyethylene are employed to package frozen fish.

Produce. Fresh fruits and vegetables must be handled gently because of sensitive structures and the ubiquitous presence of microorganisms. Damage to the product surfaces provide channels through which microorganisms can enter to initiate spoilage.

Fresh fruit and vegetable packaging is often in bulk in a variety of traditional wooden boxes and crates, and corrugated fiberboard cases. At or near the retail level, bulk produce may be repackaged in oxygen-permeable flexible materials such as PVC, with or without a tray of foamed polystyrene.

Partially Processed Food Products. Partially processed food products have received more than minimal processing but still require refrigeration.

Nitrite-Cured Meats. Ham, bacon, sausage, bologna, etc, are cured to reduce water activity, are spiced for flavor, and usually have ingredients to maintain the desired red color. Curing agents include salt, sodium nitrite, and sodium nitrate. Cured meats maintained in an absence of oxygen have refrigerated shelf lives measured in weeks. Most processed meats are packaged under reduced oxygen on thermoform–vacuum–gas flush–seal systems, usually nylon-based, and are distributed under refrigeration. Small quantities of cured meats are packaged in oxygen barrier film pouches under inert atmosphere such as nitrogen (see also FOOD ADDITIVES).

Dairy Products. Dairy products are derived from milk, which must be treated to reduce microbial counts. Pasteurization, a low heat process that destroys disease microorganisms, does not destroy all microorganisms that cause spoilage. Pasteurized dairy products must be maintained under refrigeration. Nonreturnable packages such as blow-molded high density polyethylene bottles or polyethylene-coated paperboard gable-top packages are most often used for packaging and distributing milk under refrigeration in the United States. In Canada, flexible pouches for fluid milk packaging are made from medium density polyethylene.

In aseptic packaging, milk is sterilized, ie, rendered free of microorganisms. Simultaneously, high barrier paperboard–foil–plastic-lamination or all-plastic packaging material is sterilized. The two are assembled in a sterile environment and the package is sealed to produce sterile milk in a sterile package. The increased heat required for sterilization of the milk can lead to flavors different from those in pasteurized refrigerated milk. Aseptically packaged milk may be distributed at ambient temperature.

Cheese products generally must be maintained under refrigeration using closed flexible plastic, or plastic cups or tubs for packaging. Ice cream packaging is generally minimal, ie, lacquered or polyethylene extrusion-coated paperboard cartons, molded plastic tubs, or spiral wound composite paperboard tubs or cartons.

Fully Processed Foods. Fully processed foods are processed and packaged so that the ambient temperature shelf life can exceed three to six months.

Canned Foods. The canning process thermally destroys all microorganisms and enzymes and maintains sterility by hermetic sealing in oxygen- and water-vapor-impermeable packaging that excludes microorganisms. Whether a metal can or glass jar is used, the process begins with treating the food product prior to filling. Air that can cause oxidative damage is removed from the interior; however, air removal leads to an anaerobic condition which can foster the growth of pathogenic *Clostridia* organisms. The package is hermetically sealed and then subjected to heating. The package must be able to withstand heat up to 100°C (212°F) for high acid products and 127°C (260°F) for low acid products, which must receive added heat to destroy heat-resistant potentially pathogenic microbial spores. Packages containing low, above pH 4.5, acid foods must withstand pressure. The thermal process is calculated on the basis of time required for the most remote portion of the food within the package to achieve a temperature that destroys *Clostridia* spores. After reaching that temperature, the package must be cooled rapidly to retard further cooking.

The package must contain the product, exclude air, and withstand heat. It also must maintain a hermetic seal throughout distribution and ensure that no microorganisms can re-enter the package.

Attempts have been made to perform thermal retorting in a gas barrier flexible pouch or tray. The retort pouch, under development for many years, has a higher surface-to-volume ratio than a can and employs a heat seal rather than a mechanical closure. Similarly, plastic retort trays have higher surface-to-volume ratios and are usually heat seal closed. Plastic cans intended for microwave re-heating are composed of bodies fabricated from multilayer plastic including a high oxygen barrier material, plus double-seam aluminum closures.

Frozen Foods. Freezing reduces temperature to below the freezing point of water so that microbiological, enzymatic, and biochemical activities are virtually halted. In commercial freezing, the product temperature passes through the transition from liquid water to ice rapidly so that ice crystals are relatively small and do not physically disrupt food cells. The product may be frozen inside or outside of the package. Most freezing processes use high velocity cold air or liquid nitrogen to remove heat from bulk or individually quick frozen (IQF) unpackaged product. The frozen product is then packaged in polyethylene-coated paperboard cartons, or polyethylene or polyethylene-coated paper pouches.

Products frozen in the United States include precooked, processed entrees in meal-size portions packaged in microwaveable crystallized polyester trays with polyester film closures, and overpackaged in printed paperboard cartons.

Dry Foods. Dry products include those dried from liquid form and engineered mixes of dried components blended to become dry products. In the first category are instant coffee (qv), tea (qv), and milk. The liquid is spray-, drum-, or air-dried to remove water; the presence of water at above 1% can lead to browning. Engineered mixes include beverage mixes, eg, sugar, citric acid, color, flavor, etc, and soup mixes, ie, dehydrated soup stock, noodles, and some fat products which become a heterogeneous liquid on rehydration. Products having relatively high fat, such as bakery or soup mixes, must be packaged so that the fat does not interact with the packaging materials. Seasoning mixes that contain herbs and volatile flavoring components can interact with plastic packaging materials such as interior polyolefins. The package must be hermetically sealed, providing a total barrier against both access by water vapor and by oxygen for products susceptible to oxidation.

Fats and Oils. Cooking oils and hydrogenated vegetable shortenings contain no water and so are stable at ambient temperatures (see VEGETABLE OILS). Unsaturated fats and oils (see FATS AND FATTY OILS) are subject to oxidative rancidity; both usually are packaged under inert atmosphere such as nitrogen. Hydrogenated vegetable shortenings generally are packaged in composite paperboard cans with nitrogen to ensure against oxidative rancidity. Edible oils are packaged in blow-molded polyester bottles in the United States and both polyester and unplasticized PVC bottles in Europe.

Margarine and butter contain fat plus water and water-soluble ingredients, eg, salt and milk solids that impart flavor and color to the product. Generally these products are distributed at refrigerated temperatures to retain their quality. Greaseproof packaging, such as polyethylene-coated paperboard, aluminum foil/paper, parchment paper wraps, and polypropylene tubs, is used for butter and margarine (see DAIRY SUBSTITUTES).

Cereal Products. Breakfast cereals are susceptible to moisture absorption and require good water-vapor- and fat-barrier packaging that retains delicate flavors. Breakfast cereals are packaged in polyolefin coextrusion films in the form of pouches or bags within paperboard carton outer shells. Sugared cereals are often packaged in aluminum foil or barrier plastic, eg, ethylene vinyl alcohol,

laminations to retard water vapor and flavor transmission (see WHEAT AND
OTHER CEREAL GRAINS).

Soft baked goods such as breads, cakes, and pastries are highly aerated
structures and are subject to dehydration and staling. In moist environments,
baked goods are also subject to microbiological deterioration as a result of the
growth of mold and other microorganisms. To retard moisture loss, good water-
vapor barriers such as coextruded polyethylene film bags or polyethylene-coated
paperboard are used for packaging.

Hard baked goods such as cookies and crackers have a relatively low water
and high fat content. Water can be absorbed, and the product loses its desirable
texture and becomes subject to lipid rancidity. Packaging for cookies and crackers
includes polyolefin-coextrusion film pouches within paperboard carton shells, and
polystyrene trays overwrapped with polyethylene or oriented polypropylene film.
Soft cookies are packaged in high water-vapor-barrier laminations containing aluminum foil.

Salty Snacks. Salty snacks include dry grain or potato products such as
potato and corn chips, and roasted nuts (see NUTS). These snacks usually have
low water content and relatively high fat content. Snack packaging problems are
compounded by salt, a catalyst for lipid oxidation in the product formulations.
Snacks are often packaged in pouches derived from oriented polypropylene or
polyester that have low water-vapor transmission, relying on rapid and controlled
product distribution to obviate fat oxidation problems. Some salty snacks are
packaged under inert atmospheres in both pouches and rigid containers, such as
composite cans, to extend distribution. Generally, light which catalyzes lipid oxidation harms such products, and so opaque packaging is often, but not always,
employed.

Candy. Chocolate is subject to flavor or microbiological change. Inclusions
such as nuts and fillings such as caramel are susceptible to water gain or loss.
Chocolates, which are stable, are packaged in greaseproof papers and moisture/
fat barriers such as polypropylene film (see CHOCOLATE AND COCOA).

Hard sugar candies have very low moisture content. They are sealed in low
water vapor-transmission packaging such as aluminum foil or oriented polypropylene film.

Beverages. Beverages may be still or carbonated, alcoholic or nonalcoholic.
The largest quantity of packaging in the United States is for two carbonated beverages, ie, beer (qv) and soft drinks (see CARBONATED BEVERAGES). Both contain
dissolved carbon dioxide (qv) creating pressure within the package. The package
must be capable of withstanding the internal pressure of carbon dioxide. Coated
aluminum cans, and glass and polyester plastic bottles are the most used packaging for carbonated drinks.

Beer is more sensitive than other carbonated beverages to oxygen, loss of
carbon dioxide, off-flavor, and light. Most American beer undergoes thermal pasteurization performed after sealing in the package. Thus, the internal pressure
within the package can build to well over 690 kPa (100 psi) at 63°C (145°F), the
usual pasteurization temperature. Beer and other carbonated beverages are generally packaged at relatively high speeds; therefore, the packages must be extremely uniform, free of defects, and dimensionally stable.

Paper, Metal, and Glass Packaging

Paper. The largest volume packaging materials in the United States, accounting for 40% of all food packaging, are paper (qv) and paperboard. Paper consists primarily of cellulose (qv) fibers obtained from wood (qv) by pulping (see PULP). The packaging properties of paper, its strength, and mechanical properties depend on the treatment of the wood fibers and on the incorporation of fillers (qv) and binding materials at the paper mill. The physicochemical properties of paper and paperboard, such as permeability to vapor and gases, are derived from impregnation, coating, and/or laminating. Materials used to enhance barrier include plastics, such as polyethylene, and resins, such as urea–formaldehyde. Laminated and coated papers often warrant the designation of protective materials (see BARRIER POLYMERS). Many converted papers, however, offer little more than protection from light and mechanical damage.

Paper may be used as flexible packaging material components or as material for construction of rigid containers. Flexible packaging applications include bags, usually multiwall; liners; liner substrates; pouches; and overwraps. Some of the more important types of paper used for these purposes include kraft paper, greaseproof papers, glassines, and, infrequently, waxed papers. Glassine grades are made of tightly knit kraft paper fibers, highly supercalendered to deliver smooth surfaces capable of accepting coating and print. Greaseproof grades are similar to glassine except for the finishing, and so are not as readily printable. Both glassine and greaseproof papers have good oil resistance.

Rigid paperboard containers are made of paper greater than 0.254 mm (0.010 in) in caliper and include folding cartons, corrugated fiberboard cases, and spiral wound composite cans. Many paperboard cartons require the use of inner liners or overwraps, made of protective grades of paper, plastic, or aluminum foil laminations.

Folding Paperboard Cartons. Valued at more than $7 billion, paperboard folding cartons are produced in more than 450 plants, most of which are connected with paperboard mills. Nearly 40% of paperboard folding cartons are food and beverage packaging. Among the 10 largest producers, which account for half the industry sales, are Jefferson Smurfit, James River, Waldorf, Mead Packaging, International Paper, Rock-Tenn, Riverwood, and Packaging Corporation of America. Paperboard folding cartons protect food products from impact and crushing. Certain dense or easily flowing products can be held in shape better by the carton structure. Generally, folding paperboard cartons are used to contain less than 1.36 kg (3 lbs.) of product. Paperboard for noncontact food packaging has an interior flexible liner that prevents direct contact between the product and the paperboard of the carton. More than one-fourth of paperboard folding cartons are manufactured from recycled paper and paperboard, used newspapers, and some post-consumer waste papers (see RECYCLING, PAPER).

Carton liners, eg, coextruded polyolefin pouches, help to prevent loose product content sifting and moisture migration. The carton provides a surface for graphic decoration. Polyethylene extrusion coating or, in some cases, polyethylene blend hot melt coatings are used for the interior or exterior of folding cartons, especially if used to contain liquids (see COATINGS). Extrusion coatings can be used with cartons when higher levels of water/water-vapor protection are indicated and a film overwrap or inner liner is not desired. Coatings are usually less effective than an additional layer such as an inner liner.

Composite Paperboard Cans.   Composite cans usually consist of spiral or convolute wound paperboard or paperboard-laminated body, the ends of which have been formed to accept paper ends or flanged to seam a metal end. The package usually has a printed paper label covering the entire body. Composite paperboard cans using aluminum foil interior are used for packaging refrigerated cookie and biscuit doughs, snack foods, juice concentrates, and dry powders. Convolute wound cans are used for cocoa powder, roasted and ground coffee, and candy.

Shipping Containers.   A majority of food products are distributed in corrugated fiberboard case trays engineered to meet specifications under quasi-governmental regulations. Corrugated fiberboard is the material most widely used throughout the world for tertiary or distribution packaging. Printed, cut, and fabricated into a box or tray, corrugated fiberboard forms the shipping case. The largest single packaging material in the United States is corrugated fiberboard, valued at about $19 billion. Food packaging represents about 38–39% of the total (1). Most converters are linked to paperboard mills. Among the largest producers are Stone Container, Georgia Pacific, Weyerhauser, Inland, and Jefferson Smurfit.

Polyethylene shrink film wrapping of corrugated fiberboard trays is in common use outside of the United States. Equipment erects the trays, fills the trays with primary packages such as cans or jars, wraps the grouping in shrink film, and heat shrinks the combination. Shrink film wrapping keeps primary and secondary packaging materials clean and dry.

Metal Can Packaging.   Nearly 130 billion metal cans are produced annually in the United States, of which 100 billion are two-piece aluminum used for beer and carbonated beverage packaging, but with increasing numbers for still beverage packaging. About 30 billion of the cans are three- and two-piece steel for food containment. About one-third of all cans are self-manufactured by the packager. The leading merchant can manufacturers are Crown Cork and Seal, American National Can, and Ball.

Two-Piece Cans.   Almost all aluminum cans are two-piece drawn and ironed, but some are drawn and redrawn. Some steel cans are two-piece.

Draw and iron manufacturing starts with a coil of metal fed into a multiple cupping or blanking press which forms it into shallow cups. These cups are fed into an ironing press, where successive rings or a die stack form the can side wall. More metal is left near the top and bottom to give added strength. In addition to thinning the side wall of the shell, the ironing press imparts the bottom profile. Trimming of the top end follows to produce a can of uniform height. Aluminum and steel drawn and ironed cans are manufactured by essentially the same process.

Inside coatings to protect both the metal and the contents are applied to the can by an airless spray gun. After application, the cans are baked in an oven to remove the solvent and cure the coating.

Three-Piece Cans.   Most steel cans, whether or not tin-plated, are three-piece, ie, a body and two ends. In the past, solder was used to bond the longitudinal seam, but solder has been replaced by welded side seams.

For welded side-seam cans, sheets of steel are coated, baked, and slit into body blanks. The blanks are fed into the bodymaker, where the edges are cleaned to remove any interfering layer where the steel is welded. The blank is formed into a cylinder with the edges overlapped at the side seam and tack welded together. The cylinder is then passed between rotating wheel electrodes, which weld each side of the seam. Side-seam coating coverage is achieved by applying a powdered epoxy material to both sides of the seam immediately after welding. Residual heat from the weld fuses and cures the stripe.

The cylindrical body then travels to the flanger where the top and bottom edges are curled outward to form the flange. Roll or die-necking are incorporated to reduce the can body diameter on each end in combination with flanging.

When the flanged, or necked and flanged, bodies leave the flanger after having received the topcoat spray and bake, they proceed to the double seamer, where one end is applied.

Protective Coatings. The primary function of interior can coatings is to prevent interaction between the can and its contents. Exterior can coatings may be used to provide protection against the environment, or as decoration to give product identity as well as protection.

Enamel is applied to steel in the flat before fabrication. Cans manufactured by the draw and ironing operation must be coated internally after fabrication because of the metal deformation with surface disruption that takes place.

Types of internal enamel for food containers include oleoresins, vinyl, acrylic, phenolic, and epoxy–phenolic. Historically can lacquers were based on oleoresinous products. Phenolic resins have limited flexibility and high bake requirements, but are used on three-piece cans where flexibility is not required. Vinyl coatings are based on copolymers of vinyl chloride and vinyl acetate dissolved in ketonic solvents. These can be blended with alkyd, epoxy, and phenolic resins to enhance performance. Flexibility allows them to be used for caps and closures as well as drawn cans. Their principal disadvantage is high sensitivity to heat and retorting processes; this restricts their application to cans which are hot filled, and to beer and beverage products.

Epoxy phenolic coatings either are made by blending of a solid epoxy resin with a phenolic resin or are the products of the precondensation of a mixture of two resins. A three-dimensional structure is formed during curing which combines the good adhesion properties of the epoxy resin with the high chemical resistance properties of the phenolic resin. The balanced properties of epoxy phenolic coatings have made them almost universal in their application on food cans.

Vinyl organosol coatings, which incorporate a high molecular weight thermoplastic PVC organosol dispersion resin, are extremely flexible. Soluble thermosetting resins, including epoxy, phenolic, and polyesters, are added to enhance the film's product resistance and adhesion.

Two basic methods are used for the application of protective coatings to metal containers, ie, roller coating and spraying. Roller coating is used if physical contact is possible, eg, coating of metal in sheet and coil form. Spraying techniques are used if physical contact is not possible, eg, to coat the inside surface of two-piece drawn and ironed can bodies (see COATING PROCESSES).

Coatings that are applied wet must be dried after application by solvent removal, oxidation, and/or heat polymerization. Using powder coating, the resin is applied dry in the form of a fine powder, usually under the influence of an electrostatic field. Powder coating is used where heavy coatings are required, eg, in the protection of welded side seams where the bare metal that exists in the weld area must be covered. Curing is usually by infrared radiation or high frequency induction heating.

Polyester and polypropylene films have been laminated (ca 1991) to base steel sheet to impart protection to the metal. These laminated metals are used to make two-piece drawn cans.

Environmental Aspects.   More than two-thirds of aluminum cans are recaptured and returned for recycling into more cans. Because of the heat of melting, the use of post-consumer recycled cans is safe for beverage contents. Not only does recycling save on mass of materials, it also saves the energy of manufacture from aluminum ore (see RECYCLING, NONFERROUS METALS).

Because of the vast quantities of scrap steel available from automobiles and appliances, recycling of steel cans has been growing at a relatively modest rate. As of 1992 up to 30% of steel cans are returned and remelted (see RECYCLING, FERROUS METALS).

Glass Packaging.   Glass is used for carbonated beverages, beer, and still beverage packaging. This $4.5 billion industry is declining in importance because of weight, relative fragility, and energy requirements. Glass is recyclable, and so cullet, or crushed reused glass, is a part of every raw material batch (see RECYCLING, GLASS). Glass is virtually chemically inert and impermeable. The principal American glass bottle and jar makers include Owens-Illinois, Vitro, Ball, and American National Can.

Protective coatings capture the original strength of glass containers and delay deterioration. Coatings include hot end treatments in which newly formed hot bottles are subjected to an atmosphere of vaporized metallic compound. This atmosphere reacts with the glass surface, chemically bonding to form a primer that provides permanency to the cold end treatment. The second step of the protective coating, usually an emulsion of polyethylene, is applied after the cooling section of the annealing layer, usually at a bottle temperature of about 149°C (300°F). The second coating imparts lubricity to the container surface. This prevents abrasions or other surface damage from bottle-to-bottle or bottle-to-guide rail contact during normal package handling.

To produce glass bottles, the mixture is prepared in unit batches. Mixing is critical because complete homogeneity of the batch is necessary to produce quality glass. Cullet is added to the batch, usually at discharge from the mixer. The cullet must be of the same color and basic composition, and be free of contamination such as metal bottle caps and tramp metal scraps.

Molten glass moves through the melter to the refiner where the glass is conditioned for uniform temperature and for release of dissolved glasses to remove seeds and blisters. The molten glass is cooled and conditioned for equalization of temperature throughout the stream. The glass melting process, regardless of the type furnace or energy source, is designed to supply a definite tonnage of glass per unit time. Continuous regenerative type furnaces fired with either natural gas or fuel oil commonly are used for melting container glass. Electric melting furnaces also are utilized in glass container manufacturing.

At the gob feeder, the forming operation begins. The gob feeder delivers a glass gob shaped such that it enters the blank mold without excessive mold contact, distortion, or reshaping of the glass. Shears cut off a gob of glass. The gob is fed to either a blow and blow, press and blow, or press-only forming operation to produce a container.

In blow and blow operations, the gob of glass (parison) is delivered from the feeder to the blank mold. The gob drops through a guide funnel into the blank mold in the inverted position. Air is applied to settle the gob into the finish, and air is blown in to complete the parison shape.

Press and blow operations are used to produce wide mouth and some narrow neck containers, including beer bottles. The difference between the press and blow operation and the blow and blow operation is that the parison is pressed into shape by a plunger that fills the complete void in the parison.

The rapid transfer of heat and the mechanics of blowing the bottle creates both thermal and mechanically induced stresses in the newly formed bottle. To relieve the stresses, the newly formed bottles are put through an annealing process.

Plastic Packaging

Plastic materials represent less than 10% by weight of all packaging materials. They have a value of over $7 billion including composite flexible packaging; about half is for film and half for bottles, jars, cups, tubs, and trays. The principal materials used are high density polyethylene (HDPE) for bottles, low density polyethylene for film, polypropylene (PP) for film, and polyester for both bottles and films. Plastic resins are manufactured by petrochemical companies, eg, Union Carbide and Mobil Chemical for low density polyethylene (LDPE), Solvay for high density polyethylene, Himont for polypropylene, and Shell and Eastman for polyester.

Plastic packaging materials are thermoplastic, ie, reversibly fluid at high temperatures and solid at ambient temperatures. These materials may be modified by copolymerization, additives in the blend, alloying, and surface treatment and coating. Properties of principal plastic packaging materials are given in Table 1.

Materials. *Low Density Polyethylene.* LDPE film is slightly cloudy, having high tensile strength and good water-vapor barrier properties, but high gas transmission. It is employed for shrink and stretch bundling, as drum and case liners, and to package bread, fresh produce, and low sensitivity food products. The material may also be extrusion blow molded into soft sided bottles.

The density of LDPE for film ranges from <0.92 to approximately 0.93 g/cm^3. The resin is melted in a heated extruder and converted to film by extruding through a circular die. The resulting tube is collapsed and slit into film ranging in gauge from less than 0.25 mm to greater than 0.75 mm. Some monolayer and coextruded polyethylene films are produced by slot-die casting. Because of its extensibility, LDPE film is usually printed on central impression-drum flexographic presses; modern rotogravure presses permit finer printing.

Table 1. Properties of Plastic Packaging Materials

Material	Specific gravity	Tensile strength, MPa[a]	Elongation, %	Heat-seal range, °C	Water-vapor transmission rate, μmol/(m²·s)	Gas transmission, nmol/(m²·s)	Use temperature, °C Maximum	Use temperature, °C Minimum
ionomer	0.94–0.96	21–69	350–450	87–204	13–21	1800–3870	71	−73
nylon								
uncoated	1.13–1.14	48–124	250–500	176–260	240–260	21	176–232	−59
saran coated, one side	1.13–1.14	48–124	250–500	176–260	2	4	93	−40
polyester (PET)								
uncoated	1.35–1.39	172–228	120–140		13	40	204	−62
saran coated, one side	1.4	179–214	90–125	135–204	6	3.2	82[b]	−51
metallized	1.35–1.39	172–228	120–140		0.3–1.4	0.32	204	−62
polyethylene								
LDPE	0.910–0.925	6.9–2.4	225–600	121–176	12	2000–6700	65	−51
HDPE	0.941–0.965	21–52	10–500	135–154	3–6.5	260–2000	121	−51
LLDPE	0.915–0.935	24–55	400–800	121–176	12	2000–6700	76–82	−51
LDPE/12% EVA copolymer	0.94	21–35	300–500	93–148	39	4100–5200	60	−51
polypropylene								
nonoriented	0.88–0.90	21–62	400–800	162–204	5–6.5	670–3300	121	>0
oriented	0.905	172–207	60–100	[c]	3–4	880	121	−51
oriented and metallized	0.905	131[d]	50–400	[e]	1	24–80	121	−51
poly(vinyl chloride)	1.21–1.37	14–110	5–500	121–176	>40	40–12000	93	0
poly(vinylidene chloride)	1.64–1.71	55–138	40–100	121–148	0.5–3	0.64–14	ca 82[b]	>−18

[a]To convert MPa to psi, multiply by 1450.
[b]Coating softens.
[c]Requires coating or additive.
[d]Machine direction.
[e]Requires coating.

Linear Low Density Polyethylene.　Films from linear low density polyethylene (LLDPE) resins have 75% higher tensile strength, 50% higher elongation-to-break strength, and a slightly higher but broader heat-seal initiation temperature than do films from LDPE. Impact and puncture resistance are also improved over LDPE. Water-vapor and gas-permeation properties are similar to those of LDPE films.

Linear low density polyethylene films are used in many of the same packaging applications as LDPE. The greater film extensibility permits the printing of small bags by one or more of 800 American companies.

High Density Polyethylene.　High density polyethylene resins range in density from 0.93 to 0.96 g/cm^3. HDPE films are more translucent and stiffer than LDPE films, with lower elongation to break. Water-vapor and gas permeabilities are slightly higher than those of LDPE. Softening and melting points are high, and HDPE materials are not easily sealed on flexible packaging equipment. HDPE film is being used to replace paper sacks for retail grocery and department stores. The most important packaging use for HDPE is for extrusion blow molding of bottles for milk, water, and nonfoods such as liquid detergent, motor oil, etc, by companies such as Graham and Continental.

High density polyethylene coextrusions with ethylene–vinyl acetate copolymer or ionomer are widely used as liners in food cartons. High density polyethylene film is produced by blown-film extrusion methods, often blowing downward because the weight of the film could cause collapse of an upwardly blowing bubble, as used for low density polyethylene film.

Polypropylene.　Polypropylene film is produced as either unoriented (CPP) or oriented (OPP) films by companies such as Mobil Chemical and Hercules. Because of extremely poor cold-temperature resistance and a very narrow, short, heat-seal temperature range, CPP film is not widely used for packaging. In film form, the material is a good heat sealant for packages that have later high temperature requirements, eg, retorting. It is used as a transparent bag material for textile soft goods, twist wrap for candy, and other applications. Coextruded films of PP/PE and CPP are used to separate sliced cheese. Unoriented PP is made by extrusion through a slot die (see OLEFIN POLYMERS).

Oriented polypropylene film (OPP) may be classified as heat-set and non-heat-set, blown and tentered, coextruded and coated. Orientation improves the cold-temperature resistance and other physical properties. Heat-set biaxially oriented polypropylene film (BOPP) is the most widely used protective packaging film in the United States. It is used to wrap bakery products, as lamination plies for potato and corn chips, and for pastas and numerous other flexible pouch and wrapping applications. Nonheat-set OPP is used as a sparkling, transparent shrink-film overwrap for cartons of candy.

Oriented polypropylene film may be manufactured by blown or slot-die extrusion processes. In the slot-die tenter-frame process, polypropylene film is extruded through a flat-die and stretched or oriented in the machine direction. The film is then reheated, gripped along its edges, and stretched outwardly while in longitudinal motion to impart transverse directional orientation. To impart heat-seal properties, it may be coextruded or coated with acrylic or poly(vinylidene chloride) (PVDC). In the lesser-used double-bubble blown film process, the resin is extruded through a circular die, cooled, reheated, and blown again to produce a balanced, biaxially oriented, heat-set film. Film may be coextrusion blown to provide a heat-sealing coating. Film may also be coated with poly(vinylidene chloride) to impart water-vapor, gas-barrier, and heat-sealing properties.

OPP producers have expanded the core, creating a foam structure with lower density, greater opacity, and a stiffer, more paper-like feel. Vacuum metallization increases opacity and water-vapor barrier properties.

Oriented polypropylene film has excellent water-vapor barrier but poor gas barrier properties; excellent clarity, or opacity in newer forms; and good heat-seal properties in packaging applications.

Poly(vinyl chloride). To be converted into film, poly(vinyl chloride) [9002-86-2] (PVC) must be modified with heat stabilizers and plasticizers, which increase costs. Plasticized PVC film is highly transparent and soft, with a very high gas-permeation rate. Water-vapor transmission rate is relatively low. At present, PVC film is produced by blown-film extrusion, although casting and calendering are employed for heavier gauges (see VINYL POLYMERS).

The principal packaging use of PVC film is as a gas-permeable but water-vapor impermeable wrap for red meat, poultry, and produce. Sparkle and transparency, combined with the ability to transmit oxygen to maintain red-meat color, offer advantages in these applications.

Polyester. Poly(ethylene terephthalate) [25038-59-9] (PET) polyester film has intermediate gas- and water- vapor barrier properties, very high tensile and impact strengths, and high temperature resistance (see POLYESTERS, THERMOPLASTIC). Applications include use as an outer web in laminations to protect aluminum foil. It is coated with PVDC to function as the flat or sealing web for vacuum/gas flush packaged processed meat, cheese, or fresh pasta.

Polyester is manufactured by extrusion through a slot die and biaxially oriented by stretching first in a longitudinal and then in a transverse direction while still hot, ie, tentering. Polyester film cannot be heat-sealed by conventional methods and is either coextruded or coated for heat sealing. Being heat resistant and almost inextensible, polyester film is used as a substrate for vacuum metallizing and silica coating, processes that improve moisture and gas properties. Vacuum-metallized polyester film is used for packaging wine (qv) and bulk tomato and fruit products.

Nylon. Nylon is the designation for a family of thermoplastic polyamide materials which in film form are moderate-oxygen barriers. The gas-barrier properties are equal to odor and flavor barrier properties important in food applications. Nylon films are usually tough and thermoformable, but are only fair moisture barriers (see POLYAMIDES).

Nylon films are used in lamination or coated form to ensure heat sealability and enhance barrier properties. The largest uses are as thermoforming webs for twin-web processed meat and cheese packaging under vacuum or in an inert atmosphere. Other uses include bags for red meat, boil-in-bags, bag-in-box for wine, and as the outer protective layer for aluminum foil in cookie and vacuum coffee packages.

Poly(vinylidene chloride). Poly(vinylidene chloride) [9002-85-1] (PVDC), most of which is produced by Dow Chemical, is best known in its saran or PVC-copolymerized form (see VINYLIDENE CHLORIDE AND POLY(VINYLIDENE CHLORIDE)). As solvent or emulsion coating, PVDC imparts high oxygen, fat, aroma, and water-vapor resistance to substrates such as cellophane, oriented polypropylene, polyester, and nylon.

Of the common commercial resins and films, PVDC has the best water-vapor and oxygen-barrier properties. High crystallinity confers resistance to the permeation of odors and flavors, as well as to fat and oil. Because of its high chloride content, PVDC tends to corrode processing equipment, which increases manufacturing costs. Unlike other high oxygen-barrier materials, PVDC is almost insensitive to water and water vapor.

Copolymer film is produced by extrusion blowing followed by water quenching. In-line, the film is blown, crystallized, and oriented. PVDC copolymer film is difficult to produce.

Saran film is used to wrap cheese and occasionally for vertical form/fill/seal chub packaging of sausage and ground red meat. Mostly it is used as the high barrier component of laminations not containing aluminum foil. It is rarely used alone in commercial packaging because it is difficult to seal.

Polystyrene. Polystyrene [9003-53-6] packaging film has excellent clarity, stiffness, and dimensional stability. Because of high permeability to water vapor and gases, it is suited for packaging fresh fruit and vegetables requiring the presence of oxygen. Packaging applications are limited to folding-carton windows, overwraps for tomato trays, lettuce wrapping, etc.

Foaming polystyrene resin prepared by blending with gas delivers an opaque, low density sheet useful for beverage-bottle and plastic can labels as a water-resistant paper substitute (see STYRENE POLYMERS).

Other Films. Although commercially less important than polyethylenes and polypropylenes, a number of other plastic films are in commercial use or development for special applications, including ethylene–vinyl acetate, ionomer, and polyacrylonitrile [25014-41-9].

Ethylene vinyl acetate copolymer (EVA) forms a soft, tacky film with good water-vapor barrier but very poor gas-barrier properties. It is widely used as a low temperature initiation and broad-range, heat-sealing medium. The film also serves for lamination to other substrates for heat-sealing purposes.

Ionomers (qv), often known by their trade name Surlyn (Du Pont), are ionically cross-linked thermoplastics derived from ethylene–methacrylic acid copolymers. Ionomer films are tough, extensible, and impact resistant with excellent heat-seal and hot-tack characteristics. Extruded on other substrates, ionomer films are used as the heat-sealing component.

Polyacrylonitrile (PAN) films have outstanding oxygen and CO_2 barrier properties, but only modest water-vapor barrier properties. They are for processed-meat and fresh pasta packaging laminations where an oxygen barrier is required for vacuum or gas flush packaging.

Coextrusions. In coextrusion, two or more thermoplastic resin melts are extruded simultaneously from the same die. Coextrusion permits an intimate layering in precisely the quantities required to function. Incompatible plastic materials are bonded with thermoplastic adhesive layers. Coextruded films may be made by extrusion-blowing or slot-casting of two, three, or more layers, eg, AB or ABA. Slot-casting is capable of combining up to 11 layers, but most equipment is used for the simple AB, ABA, and ABC or ACBCA configurations where C is the adhesive layer. In the simplest combinations, oriented polypropylenes are coextruded with copolymer heat-seal layers. Low density polyethylenes are coextruded to impart toughness or slip characteristics.

In more complex combinations, HDPE, LDPE, and EVA resins are coextruded to produce stiff, heat-sealable films to be used as liners in cereal, cookie, and cracker cartons. Films of EVA and white-pigmented LLDPE are used for packaging of frozen vegetables and fruits. In these applications, one layer imparts toughness, opacity, or stiffness, and the other layer adds heat sealability. Coextrusions of nylon with polyethylene in five layers are used for thermoforming where high gas and water-vapor barrier are required, eg, medical packaging.

Plastic packaging materials may be classified into rigid and flexible, with an overlapping dividing line between rigid and flexible. Materials determined by caliper or gauge to have thicknesses less than 0.25 mm are generally regarded as being in the flexible category.

Flexible Packaging. Flexible packaging is composed of both single- and multilayer structures. The latter may be further subdivided into laminated, coated, and coextruded, or combinations of these.

More than half of flexible packaging is used for food. Within foods, candy, bakery products, and snack-type foods, such as potato and corn chips, use well over half of flexible packaging. Cheese, processed meat, shrink wraps, condiments, dry-drink mixes, fresh meats, and fresh produce represent smaller applications.

Over 1000 firms participate in the flexible packaging industry in the United States. Manufacturers process the plastic resin into film, a unit operation sometimes assumed by the converter, depending on the material. The processes of slitting, printing, coating, laminating, coextruding, and fabricating into preformed pouches and bags are performed by converters.

Fabrication. Flexible packaging materials may be mono- or multilayer. Monolayer materials are usually films that have been produced by polymer resin melting and extrusion.

Extrusion of polyethylene and some polypropylenes is usually through a circular die into a tubular form, which is cut and collapsed into flat film. Extrusion through a linear slot onto chilled rollers is called casting and is often used for polypropylene, polyester, and other resins. Cast, as well as some blown, films may be further heated and stretched in the machine or in transverse directions to orient the polymer within the film and improve physical properties such as tensile strength, stiffness, and low temperature resistance.

In coextrusion, two or more plastic melts from different extruders are combined into a single die in which the melts are joined. Coextrusion permits precise, small quantities of plastic materials to be intimately bonded to each other.

Free mono- and multilayer films may be adhesive- or extrusion-bonded in the laminating process. The bonding adhesive may be water- or solvent-based. Alternatively, a temperature-dependent polymer-based adhesive without solvent may be heated and set by cooling. In extrusion lamination, a film of a thermoplastic such as polyethylene is extruded as a bond between the two flat materials, which are brought together between a chilled and backup roll.

Flexible materials are printed in roll form by rotogravure, flexographic, and web offset printing. In rotogravure printing, a cylinder is engraved with minute depressions or wells that accept dilute solvent-based inks by capillarity. When contacted by the packaging material, the ink is drawn from the wells to the printing surface and the solvent is evaporated to set the ink. In flexographic printing, the design is elevated above the cylinder surface using rubber-like materials. In offset printing, oil-based ink is in a planographic attitude and printed or offset onto a rubber blanket from which the ink is later removed to the substrate to be printed. Rotogravure printing cylinders are usually considerably more expensive than flexographic printing plates. Offset printing plates are even less expensive. The detail produced by rotogravure and offset is finer than that produced by flexography. Rotogravure is usually used for long production runs and high resolution reproduction, whereas flexography is usually used for shorter runs and bolder design, and offset is applicable to short run fine design. Both flexography and rotogravure are used extensively to print flexible packaging in the United States. Coatings applied by printing or extrusion protect the printed surfaces.

Some flexible packaging is fabricated by converters into bags and pouches. Bag material is either small monolayer or large multiwall with paper as a principal substrate. Pouches are small and made from laminations. Bags usually contain a heat-sealed or adhesive-bonded seam running the length of the unit and a cross-seam bonded in the same fashion.

Applications.   Preformed bags are opened by the packager, filled with food product, and closed by adhesive, heat-sealing, clipping, stitching, etc.

A small quantity of flexible packaging material, usually oriented polypropylene, shrink polypropylene, or polyethylene, is used to overwrap paperboard cartons. The film is wrapped around the carton and sealed by heating. Products such as boxed chocolates, candies, and cookies are overwrapped, sometimes by a printed film.

Some heavier gauge flexible materials, usually containing nylon, are thermoformed, ie, heated and formed into three-dimensional shapes. Such structures are used to provide high gas-barrier, heat-sealable containment for processed meat or cheese.

Large quantities of flexible packaging materials are employed in horizontal form/fill/seal machines to enclose contents, sometimes with a hermetic seal. The web of material is unwound and folded so that its longitudinal edges are in contact. Meanwhile, the product is conveyed at the same speed as the flexible packaging material into the tube. The two edges are heat-sealed and transverse heat seals are formed between the product units. This system is used for overwrapping, as well as for unit packaging for candies, cookies, and crackers.

In a variant of the horizontal form/fill/seal operation, the material, moving in a horizontal direction, is folded on itself vertically. Vertical sections of the two faces are heat-sealed to each other to form a pouch, which may then be filled. The pouch, usually made from film or paper bonded to aluminum foil plus a plastic laminant and heat sealant, is closed by a heat seal. This type of pouch gives high moisture and oxygen protection and is used for moisture- and flavor-sensitive condiments and beverage mixes.

The largest volume of flexible packaging is used in vertical form/fill/seal applications for loose, flowable products such as potato and corn chips, nuts, and roasted and ground coffee. The roll of flexible materials is unwound over a forming collar forcing the web into a vertical, tubular shape. By heat-sealing the edges and a bottom transverse seal, an open-top tube is formed. The product is gravity filled, the web is drawn down, and the tube is closed by another heat seal. Vertical form/fill/seal operations use water-vapor barrier materials such as oriented poly-propylene film to package moisture-sensitive products.

In applications other than protective packaging, flexible materials are employed to unitize cans, bottles, cartons, or cases. Heat-shrinkable films, such as low density or linear low density polyethylene, are wrapped around groups of bottles or cans, sealed, and exposed to hot air to bind the contents tightly together within the film. In stretch wrapping, an extensible film cohesive to itself, eg, linear low density polyethylene, is wrapped very tightly around the contents. Surface cohesion causes the film wrap to hold to itself. As the stretched film attempts to revert to its original unstretched form, it binds the bundle more tightly.

Numerous variations and other applications are common for flexible packaging materials, eg, oxygen-permeable wraps for fresh red meat and produce; shrinkable, low oxygen permeability bags for meat; and rigid tray closures.

Rigid Containers. Most rigid plastic is used in bottles and jars. Jars have openings approximately the same as those of the body, whereas the neck diameter of bottles is significantly smaller than that of the body.

Fabrication Processes. Injection Molding. Matched metal molds are used in the fabrication of plastic closures, specialty packages, and bottle preforms. In conventional injection molding the plastic resin is melted in an extruder which forces a measured quantity or shot into a precision-machined chilled mold after which the nozzle of the extruder is withdrawn.

The pressure of the extruder forces uniform plastic distribution throughout the mold. Cooling the mold solidifies the plastic with slight shrinkage. The mold is maintained closed by mechanical or hydraulic pressure while the thermoplastic is injected and solidified.

Because cycle time to inject, flow, set, open, eject, and close is finite, and the face area or platen size is limited, the effective molding area is increased by increasing the number of mold cavities so that the number of finished pieces per cycle may be multiplied many times.

Injection molds are constructed of metal precision-machined with internal cooling, multiple cavities, and multifaces, and with devices for extraction or ejection of the molded piece.

Returnable distribution cases for carbonated-beverage and milk bottles are injection-molded from high density polyethylene or polypropylene–ethylene copolymer. HDPE and polypropylene–ethylene copolymer cups and tubs for dairy product and specialty frozen food applications are injection-molded. High and medium density polyethylenes are injection-molded as closures for metal and paperboard composite can closures. Because covers and closures have a high surface area-to-depth ratio, multilevel or stack molds are used to maximize unit output per cycle. Many bottle and jar closures are injection molded from polypropylene. High impact polystyrene is injection-molded for refrigerated dairy product and specialty food cup packaging.

Insert injection molding is used to manufacture snap closures for yogurt and ice cream cups and tubs and for breakfast cereal cans. In insert injection molding, a die-cut printed paperboard or other flat material is placed in the mold. The plastic is extruded around the insert to form a precision skeletal structure.

Injection-molded articles can be decorated by in-mold labeling or by post-mold decoration. In the former method, printed film is inserted into the mold cavity before injection. The plastic forms an intimate contact with the graphic material. Post-mold decoration includes hot stamping, dry offset printing, and decal printing.

Injection molding is used for the preparation of polyester preforms for blow molding into bottles, eg, carbonated beverages, and jars, eg, peanut butter. Resins such as polyester with narrow melt-temperature ranges require sequential fabrication steps. A blow-mold parison is injected at melt temperature and gently reheated to softening below melt temperature for stretching or blowing into a bottle or jar.

To enhance water-vapor- or gas-barrier properties, layers of different plastics may be injected together or sequentially. Multilayer injection-molded pieces may be prepared as packaging or for blowing into bottle or jar shapes.

Sheet Extrusion and Thermoforming. Sheet for thermoforming and analogous operations is usually formed by extruding the melt through a slot die onto a set of polished chill rolls. The sheet is usually approximately 150 cm wide. After rapid cooling, the web is coiled or cut into sheets. Polystyrene, PVC, polyethylene, polypropylene, and filled polypropylene are prepared in sheet form by extrusion.

Thermoforming includes the extrusion of sheets, thicker than 0.25 mm, followed by forming a reheated sheet in an open-face mold by pressure, vacuum, or both. Sheet of less than 0.25 mm thick is thermoformed in-line, and filled and sealed with contents such as processed meats, cheeses, and pastas.

Thermoformable sheet may be mono- or multilayer with the latter produced by lamination or coextrusion. Multilayers are employed to incorporate high oxygen-barrier materials between structural or high water-vapor barrier plastics. Both ethylene vinyl alcohol copolymers and poly(vinylidene chloride) (less often) are used as high oxygen-barrier interior layers with polystyrene or polypropylene as the structural layers, and polyolefin on the exterior for sealing.

Alternatively, gas may be introduced by blending thermally reactive chemicals which release gas into the resin at the extruder. Extrusion heat initiates the reaction to release gas and expand the melt.

Steam-Chest Expansion. In steam-chest expansion the resin beads in which gas is already present are poured into molds into which steam is injected. The steam increases the temperature close to the melting point and expands within the structure to create beads with food cushioning and insulating properties. Expanded polystyrene is widely used in this process for thermal insulation of frozen food packaging.

Three-Dimensional Packaging. *Thermoforming.* Thermoforming is the most common method of fabricating sheet into three-dimensional packaging. In conventional thermoforming, the sheet is heated to its softening point or just below the melting temperature. The softened plastic is forced by differential air pressure into an open-top mold to assume the shape of the female mold. The mold is chilled and the plastic sheet solidifies and is then removed from the mold.

The mold is usually prepared with orifices to permit air trapped between the sheet and the mold to escape and ensure uniform, close contact of the plastic with the mold surface. By clamping the sheet beyond the perimeter of the piece, plastic may be drawn from the peripheral areas into the mold, ensuring uniformity. Both pressure and vacuum are employed to force the softened plastic sheet into the mold.

Thermoforming is used for gauges above 6 mm in some nonpackaging applications; for packaging applications gauges are between 0.6 and 2.5 mm.

The material is stretched during formation, and so the greater the depth of draw, the higher must be the gauge of the original sheet. Depth of draw, defined as the ratio of piece height to cross-sectional surface area, is a measure of the ability to form deep articles. Although deep-drawn pieces are feasible, thermoforming is best suited to shallow profiles of up to 7.5 cm; deeper draws often result in thin side walls. Any fully or partially tapered shape with an opening equal to or larger than the body can be made.

In commercial practice, packaging is produced from continuous web on intermittent-motion thermoforming/die-cut machines. The web edge is clamped and conveyed into a heating box. If a plastic with a narrow softening temperature is used, heating is carefully controlled from top and bottom. The heated web is conveyed to the forming section where pressure or vacuum force the softened web into the mold. The mold opens and the web is conveyed to a die-cutting station.

The narrow softening temperature range and structural properties of polypropylene led to the development of special controlled-temperature forming processes called solid-phase pressure forming (SPPF). The polypropylene sheet temperature is brought to above the softening temperature but well below the melting point and subjected to forming pressures three to five times those used for fabrication of polystyrene.

Conventional thermoforming of polystyrene and PVC is the most widely used technique for making packages for dairy products and for disposable cups and trays.

Thermoforming may be integrated with filling and sealing on thermoform/fill/seal machines operated in-plant by food packagers. The base web is gripped and moved through heating and forming operations. The open-top pieces are filled and a second web of material is heat-sealed to the flange of the base by heated pressure bars. Cutting may take place during or after sealing.

Retortable plastic cans and trays, designed for low acid foods, may be hermetically sealed and thermally sterilized up to 125°C. High oxygen-barrier properties are usually required. These cans or trays are constructed of five-layer coextrusions of PVDC or ethylene–vinyl alcohol (EVOH) as the gas barrier layers and polypropylene, with extrudable adhesive layers. The trays must be thermoformed and aged before filling and sealing to minimize heat-seal distortion due to stress, unless melt-to-mold fabrication procedures are employed. Counterpressure and temperature during retorting must be carefully controlled to avoid heat seal distortion.

Melt-to-mold thermoforming overcomes the thermal stresses developed in forming polypropylene sheet. As the plastic is extruded from the slot die and while it is still hot, it is transferred into the mold. To maintain thermal and mass equilibrium, the molds move continuously past the slot from which the melt is discharged and the plastic is formed in the molds. Melt-phase thermoforming has been used to produce monolayer frozen food tubs and wide-mouth snap closures, for fabricating crystallized polyester trays for dual ovenable applications, and for coextruding multilayer structures incorporating EVOH for hot fill, aseptic, and retort packaging of foods.

In the cuspation–dilation thermoforming process developed in Australia, sheet formation is promoted by expanding blades extending into all areas and distributing the material uniformly throughout the mold. This process is claimed to deliver uniform distribution of high barrier components of sheet coextrusions and laminations. The process also permits almost vertical side walls to cups (2).

Blow Molding.   In conventional blow molding a single extruder pushes the plastic melt through an annular die to form a tubular parison which is delivered into the open mold. The mold closes, pinching off the bottom and gripping the top. An air-blow tube is inserted through the neck opening and pressurized air is blown in to expand the hot, soft plastic to the chilled walls of the mold. Upon setting, the parison drops or is extracted from the mold. Excess flash is mechanically trimmed.

Extrusion blow molding produces narrow-neck bottles from high and low density polyethylene. Bottles with adequate neck finishes are produced. Without good parison control, bottle wall thickness varies, ie, areas of greater diameter have thinner walls, whereas base, shoulder, and neck areas have thicker walls. These bottles may be fabricated with hollow handles. They may be decorated by in-molding labeling, with decals, by screen printing, or by hot stamp or post-mold labeling.

Conventional extrusion or coextrusion may be performed on vertical or horizontal rotary or shuttle mold configurations. In shuttle blow molding the extruder and die are in fixed horizontal and vertical position; two or more molds shuttle into and out of position beneath the die. By reciprocating in two planes, the mold may remove a parison and permit the extruder to function continuously.

Horizontal rotary machines employ multiple molds in a horizontal plane on a rotary turret. As each mold approaches the extruder die exit, it opens to accept the parison and then closes. The parison is then blown into the bottle shape. The extruder must extrude on an intermittent basis or be intermittently withdrawn to provide a parison for each passing mold.

Vertical rotary molds also employ multiple molds on a turret but rotate in a vertical plane. As each mold reaches the die exit, it grasps the parison and closes. Because of the vertical spacing between molds, intermittent extruder action is not required. Vertical wheels are used commercially for high volume applications.

Because high oxygen-barrier plastics are incompatible with other thermoplastics, extrudable adhesives must be extruded between the layers. Scrap can be included within the multilayer structure, provided an extrudable adhesive is incorporated.

Incorporating EVOH as high oxygen barrier with polypropylene is used for packaging tomato catsup, barbecue sauce, mayonnaise, pickle relish, and other foods. Bottles fabricated from internal and external layers of polypropylene contain EVOH as the principal high oxygen-barrier material.

Extrusion blow molding of polypropylene is not easy because of its narrow softening-to-melt temperature range. Bottle orientation enhances structural strength. For monolayer polypropylene bottles, a two-stage process produces a continuous extrusion of pipe which is cut into fixed parison lengths. The parisons are reheated and stretched longitudinally before circumferential blowing. Impact resistance, gloss clarity, and stiffness are improved, but barrier properties are not.

Materials with a very narrow melt–temperature range are formed into bottles by injection stretch blow molding. A test-tube shape is first formed by injection. This preform is transferred to the blow-molding machine and slowly heated to uniform temperature. The heated parison is placed in the blow mold in which it is stretched to induce vertical orientation and blown to shape. Blowing induces horizontal orientation to provide circumferential or hoop structural strength. Stretching and blowing reduce wall thickness, whereas orientation improves structural strength.

Injection stretch-blow molding may be performed on a single one-stage machine in sequence or on two independent sequential two-stage machines. PET carbonated beverage bottles are usually produced by injection stretch blow molding.

Multilayer injection stretch blow molding has been commercialized for both narrow neck and wide-mouth containers. The basic form is fabricated by injecting multiple layers, such as polypropylene and EVOH plus tie layers, and blowing the parison. Several high oxygen-barrier cans with plastic bodies and ends intended for metal end double seam closing have been introduced. Cans containing polypropylene and EVOH are retorted after filling to resist retort temperatures up to 125°C.

Heat-Stabilized Molding. Recognition of the merits of hot filling foods into plastic packaging, followed by sealing and cooling, has led to a need for high oxygen-barrier plastic containers capable of resisting temperatures up to 85–90°C for brief periods. Resistance to internal vacuum can be achieved by structural design. Plastics with distortion temperatures above 100°C, eg, polypropylene and high density polyethylene, may be filled with hot liquid without fear of thermal distortion, although vacuum collapse is an issue. Polyester requires physical modification to resist heat, drying to remove water, partial crystallization, and heat stabilization. In heat stabilization, the container is molded and briefly secured in the mold while at elevated temperature rather than chilled immediately. The crystalline structure of the material is thereby altered to resist moderately elevated temperatures. This technique is employed to produce PET bottles for filling with hot liquids.

Blow-Mold/Fill/Seal System. Some blow-molded bottles are produced in blow-mold/fill/seal operations designed for aseptic packaging. On blow-mold/fill/seal machines the parison is extruded through a multilayer annular die into a sterile space containing sterile molds. The parison is blown with sterile air and immediately filled with cooled product. After filling, the bottle is closed by heat-seal fusion within the mold and removed from mold and sterile chamber through a small opening protected against contamination by the pressure of sterile air.

This method is slow because of the multiple operations on a shuttle machine. The heat of extrusion sterilizes the bottle, which is not readily achieved after molding. Blow-mold/fill/seal systems are used commercially for beverages and for pharmaceutical packaging.

BIBLIOGRAPHY

"Packaging and Packages" in *ECT* 1st ed., Vol. 9, pp. 754–762, by R. D. Minteer, Monsanto Chemical Co.; in *ECT* 2nd ed., Vol. 14, pp. 432–443, by G. T. Stewart, The Dow Chemical

Co.; "Packaging Materials, Industrial" in *ECT* 3rd ed., Vol. 16, pp. 714–724, by S. J. Fraenkel, Container Corp. of America.

1. M. Bakker, ed., *The Wiley Encyclopedia of Packaging Technology*, John Wiley & Sons, Inc., New York, 1986.
2. G. L. Robertson, *Food Packaging-Principles and Practice*, Marcel Dekker, Inc., New York, 1993.

General References

A Processor's Guide to Establishment, Registration and Process Filing for Acidified and Low Acid Canned Foods, FDA, HHS publication 80-2126, U.S. Department of Health & Human Services, Washington, D.C., 1980.

A. C. Hersom and E. D. Hulland, *Canned Foods: An Introduction to Their Microbiology*, Chemical Publishing Co., New York, 1964.

A. L. Brody and E. P. Schertz, *Convenience Foods: Products, Packaging, Markets*, Iowa Development Commission, Des Moines, 1970.

A. L. Brody and L. M. Shepherd, *Modified/Controlled Atmosphere Packaging: An Emergent Food Marketing Revolution*, Schotland Business Research, Inc., Princeton, N.J., 1987.

A. L. Brody, ed., *Controlled/Modified Atmosphere/Vacuum Packaging of Foods*, Food & Nutrition Press, Inc., Trumbull, Conn., 1989.

A. L. Brody, *Flexible Packaging of Foods*, CRC Press, Inc., Boca Raton, Fla., 1972.

A. L. Brody, *International Conference on Controlled/Modified Atmosphere/Vacuum Packaging-CAP '87*, Schotland Business Research, Inc., Princeton, N.J., 1987.

A. L. Brody, *International Conference on Microwaveable Foods-Microready Foods, '88*, Schotland Business Research, Inc., Princeton, N.J., 1988.

A. L. Brody, *International Conference on Microwaveable Foods-Microready Foods, '89*, Schotland Business Research, Inc., Princeton, N.J., 1989.

A. L. Brody, *NOVA-PACK, International Conference on Packaging and Converting Advances*, Schotland Business Research, Inc., Princeton, N.J., 1988.

A. Lopez, ed., *A Complete Course in Canning: Book I Basic Information of Canning*, The Canning Trade, Inc., Baltimore, Md., 1987.

A. Lopez, ed., *A Complete Course in Canning: Book II Packaging; Aseptic Processing; Ingredients*, The Canning Trade, Inc., Baltimore, Md., 1987.

A. Lopez, ed., *A Complete Course in Canning: Book III Processing Procedures for Canned Food Products*, The Canning Trade, Inc., Baltimore, Md., 1987.

Anonymous, *Food Packaging Technology International 1991*, issue 4, Cornhill Publications Ltd., London, 1991.

Aseptic Processing and Packaging of Foods, IUFoST Symposium, Tylosand, Sweden, Sept. 1985.

C. J. Benning, *Plastic Films for Packaging*, Technomic Publishing Co., Inc., Lancaster, Pa., 1983.

C. M. Swalm, ed., *Chemistry of Food Packaging*, ACS Adv. Chem. Ser. **135**, American Chemical Society, Washington, D.C., 1974.

D. S. Hsu, *Ultra High Temperature Processing and Aseptic Packaging of Dairy Products*, Damana Tech., New York, 1979.

E. J. Stilwell and co-workers, *Packaging for the Environment-A Partnership for Progress*, American Management Association, Washington, D.C., 1991.

F. A. Paine and H. Y. Paine, *A Handbook of Food Packaging*, Leonard Hill Ltd., London, 1983.

F. A. Paine and H. Y. Paine, *Principles of Food Packaging*, Leonard Hill, Ltd., London, 1983.

F. A. Paine, ed., *Modern Processing, Packaging and Distribution Systems for Food*, Van Nostrand Reinhold Co., Inc., New York, 1987.

F. A. Paine, ed., *The Packaging Media*, John Wiley & Sons, Inc., New York, 1977.

Foodplas '89, The Plastics Institute of America, Inc., Hoboken, N.J., Mar. 1989.

Foodplas '90, The Plastics Institute of America, Inc., Hoboken, N.J., Mar. 1990.

Foodplas '91, The Plastics Institute of America, Inc., Hoboken, N.J., Mar. 1991.

Foodplas '92, The Plastics Institute of America, Inc., Hoboken, N.J., Mar. 1992.

G. A. Reineccius, "Flavor/Packaging Problems," presented at *Flavor Workshop III: Flavor Applications*, Department of Food Science and Nutrition, University of Minnesota, Sept. 1991.

Gorman's Food Conference, *8th Annual Conference on New Products: New Foods, New Reality, Lessons from the Past Decade*, Gorman Publishing, Chicago, 1990.

H. M. Broderick, *Beer Packaging*, Master Brewers' Association of America, Madison, Wis., 1982.

H. Reuter, ed., *Aseptic Packaging of Food*, Technomic Publishing Co., Inc., Lancaster, Pa., 1989.

J. A. Schlegel, *Barrier Plastics-The Impact of Emerging Technology*, American Management Association, Washington, D.C. 1985.

J. F. Hanlon, *Handbook of Package Engineering*, 2nd ed., McGraw-Hill Book Co., Inc., New York, 1984.

J. H. Briston and L. L. Katan, *Plastics Films*, 2nd ed., Longman Scientific & Technical in association with The Plastics and Rubber Institute, Essex, UK, 1983.

J. Hotchkiss, ed., *Food and Packaging Interactions*, ACS Symposium Series **365**, American Chemical Society, Washington, D.C., 1988.

J. I. Gray, B. R. Harte, and J. Miltz, eds., "Food Product–Package Compatibility," *Michigan State University School of Packaging Seminar Proceedings*, Technomic Publishing Co., Inc., Lancaster, Pa., 1987.

K. R. Osborn and W. A. Jenkins, *Plastic Films-Technology and Packaging Applications*, Technomic Publishing Co., Inc., Lancaster, Pa., 1992.

L. Erwin and L. Hall Healy, Jr., *Packaging and Solid Waste Management Strategies*, The American Management Association, Washington, D.C., 1990.

M. L. Troedel, ed., *Current Technologies in Flexible Packaging, ASTM Special Technical Publication* **912**, ASTM, Philadelphia, Pa., 1986.

P. E. Nelson, J. V. Chambers, and J. H. Rodriguez, eds., *Principles of Aseptic Processing and Packaging*, The Food Processors Institute, Washington, D.C., 1987.

Pack Alimentaire '89 Conference, Schotland Business Research, Inc., Princeton, N.J., 1989.

Pack Alimentaire '90 Conference, Schotland Business Research, Inc., Princeton, N.J., 1990.

Pack Alimentaire '91 Conference, Schotland Business Research, Inc., Princeton, N.J., 1991.

Packaging and Packaging Materials, United Nations, New York, 1969.

R. C. Griffin, S. Sacharow, and A. L. Brody, *Principles of Package Development*, 2nd ed., Avi Publishing Co., Inc., Westport, Conn., 1985.

R. Griffin and S. Sacharow, *Food Packaging*, 2nd ed., Avi Publishing Co., Inc., Westport, Conn., 1982.

S. E. M. Selke, *Packaging and the Environment-Alternatives, Trends and Solutions*, Technomic Publishing Co., Inc., Lancaster, Pa., 1990.

S. J. Risch and J. H. Hotchkiss, eds., *Food Packaging and Interaction II*, American Chemical Society Symposium Series 473, ACS, Washington, D.C., 1991.

S. Sacharow and A. L. Brody, *Packaging: An Introduction*, Harcourt Brace Jovanovich Publications, Inc., Duluth, Minn., 1987.

T. Kadoya, ed., *Food Packaging*, Academic Press, Inc., San Diego, Calif., 1990.

W. A. Jenkins and J. P. Harrington, *Packaging Foods with Plastics*, Technomic Publishing Co., Inc., Lancaster, Pa., 1991.

AARON L. BRODY
Rubbright-Brody, Inc.

FOOD PROCESSING

Food processing operations can be grouped into three categories: preparation, assembly, and preservation of foods. Preparation processes are used to convert raw plant or animal tissue into edible ingredients. This may include separation of inedible and hazardous components, extraction or concentration of nutrients, flavors, colors, and other useful components, and removal of water. Assembly processes are used to combine and form ingredients into consumer products. Preservation processes are used to prevent the spoilage of foods. Five sources of food spoilage must be addressed in order to deliver fresh, safe foods and ingredients: microbial contamination, including viruses; enzyme activity from enzymes in the food itself and from external enzymes such as from microbial activity; chemical deterioration such as oxidation and nonenzymatic browning; contamination from animals, insects, and parasites; and losses owing to mechanical damage such as bruising. Preservation processes can be used to extend the shelf life of fresh foods, such as produce, or to manufacture products for long-term storage where shelf lives are measured in years. The processing of foods is regulated by federal food laws that cover good manufacturing practices, nutritional content of foods, and food and ingredient standards (see also FOOD ADDITIVES; FOOD PACKAGING).

Plants and animals are the primary sources of food. The food processing industry devotes considerable research to the selection and improvement of plants and animals for raw materials. Genetic engineering (qv), as well as conventional breeding methods, are being used to improve the yield, color, flavor, texture, nutrient content, and resistance to diseases, insect loss, and climatic stress (see also FEEDS AND FEED ADDITIVES; FERTILIZERS; GROWTH REGULATORS). However, product quality can vary owing to weather, soil, growing practices, harvest methods, and post-harvest handling. Thus food processing unit operations must be designed to accept raw materials having a wide range of qualities. In addition, provision often must be made for profitable use of by-products and waste streams.

Regulations

Food processing operations are usually regulated and mandated by national and international laws, regulations, and standards which define nutritional requirements, certain ingredients, process conditions, and even the composition of some products. Food safety and toxicology regulations include standards for toxic and carcinogenic substances in foods; pathogenic microbes; and physical hazards. Chemical hazards include heavy metals, carcinogens such as aflatoxins and nitrosamines, pesticides and herbicides (qv), and other natural toxicants such as solanin and gossypol (see FOOD TOXICANTS, NATURALLY OCCURRING). Microbial hazards include *Clostridium botulinum, Listeria monocytogenes,* and *Salmonella* sp. Physical hazards include extraneous material such as metal, wood, pits, glass, insect fragments, and rodent hair. Food regulations are covered in the *Federal Code of Regulations* (1).

Process Optimization

A number of food processing unit operations, such as distillation (qv), filtration (qv), and crystallization (qv), are common to the chemical process industry. Mechanical operations such as size reduction (qv), materials handling (see CONVEYING), and mixing (see MIXING AND BLENDING) are also similar to those used in chemical processing.

Food processing operations can be optimized according to the principles used for other chemical processes if the composition, thermophysical properties, and structure of the food is known. However, the complex chemical composition and physical structures of most foods can make process optimization difficult. Moreover, the quality of a processed product may depend more on consumer sensory responses than on measurable chemical or physical attributes.

Food process optimization measurements may link a single chemical such as a vitamin, or a physical change such as viscosity, to process conditions and to consumer acceptance. Retention levels of ascorbic acid [50-81-7], $C_6H_8O_6$, or thiamine can often be used as an indicator of process conditions (see VITAMINS).

Particular food products have well-developed technologies associated with their preparation, processing, and packaging. Detailed discussions of processing technologies can be found in the general references.

Theoretical Basis. Food preservation theory has yielded mathematical models for predicting the heating times and temperatures needed to produce foods free of pathogenic or spoilage microbes (2). Mild heat treatments used to inactivate viruses, vegetative pathogenic bacteria such as *Salmonella* sp., and certain yeasts (qv) and molds, are referred to as pasteurization operations. Milk pasteurization treatments at 61.67°C for 30 min or 71.67°C for 15 s are examples (see MILK AND MILK PRODUCTS).

Spore-forming bacteria are among the most heat-resistant organisms known. For example, a population of 10,000 spores of *Bacillus stearothermophilus* must be held at 121°C for approximately 16 min to ensure complete inactivation. These spores are found as contaminants in many raw plant and animal ingredients and must be inactivated or prevented from germinating to prevent the spoilage of heat preserved foods. The spore-forming bacteria of greatest public health concern are the several types of *Clostridium botulinum*. Upon germination, *C. botulinum* releases a highly toxic cyclic polypeptide which is the most potent human toxin known, based on molecular weight. Research since the early 1920s has been directed toward the development of mathematical models to predict the rate of heat inactivation of *Clostridium botulinum* spores as a function of heating time and temperature, and the composition of the suspending media (2). Heat inactivation rates can be influenced by pH, water activity, salts such as nitrates, nitrites, and sodium chloride, and other chemicals. The polypeptide toxin itself can be inactivated by heating at 100°C for 10 min.

Spore germination can be inhibited by antibiotic substances produced by several types of lactic acid-producing bacteria (3). These substances, called bacteriosins, are finding increased use in preventing the growth of gram-positive bacteria. Nisin [1414-45-5], $C_{143}H_{230}N_{42}O_{37}S_7$, a particularly effective bacteriosin against *Clostridium botulinum*, is allowed in processed cheese food. Wider use is expected in the United States because nisin has already been approved for use in many foods in other countries.

Two other broad areas of food preservation have been studied with the objective of developing predictive models. Enzyme inactivation by heat has been subjected to mathematical modeling in a manner similar to microbial inactivation. Chemical deterioration mechanisms have been studied to allow the prediction of shelf life, particularly the shelf life of foods susceptible to nonenzymatic browning and lipid oxidation.

Water Activity. The rates of chemical reactions as well as microbial and enzyme activities related to food deterioration have been linked to the activity of water (qv) in food. Water activity, at any selected temperature, can be measured by determining the equilibrium relative humidity surrounding the food. This water activity is different from the moisture content of the food as measured by standard moisture tests (4).

At very low concentrations of water, or in foods held below the freezing point of water, physical conditions may be such that the available water may not be free to react. Under these conditions, the water may be physically immobilized as a glassy or plastic material or it may be bound to proteins (qv) and carbohydrates (qv). The water may diffuse with difficulty and thus may inhibit the diffusion of solutes. Changes in the structure of carbohydrates and proteins from amorphous to crystalline forms, or the reverse, that result from water migration or diffusion, may take place only very slowly.

When water activity is low, foods behave more like rubbery polymers than crystalline structures having defined domains of carbohydrates, lipids, or proteins. Water may be trapped in these rubbery structures and be more or less active than predicted from equilibrium measurements. As foods change temperature the mobility of the water may change. A plot of chemical activity vs temperature yields a curve having distinct discontinuities indicating phase changes in the structure of the food system and possible release or immobilization of water. An important phase change is that from a rubbery or glassy structure to a more crystalline one defined by the glassy point transition temperature. The glassy point transition temperature of a frozen food may be important for long-term storage stability. Similarly for baked or dried foods stored at room temperature, the glassy point transition temperature may indicate a temperature at which staling, loss of softness, or the development of an undesirable texture may take place. Glassy point transition temperatures and the nonequilibrium analysis of food structures have been studied (5).

Preservation of Foods

Preservation operations to reduce or eliminate food spoilage can be grouped into five categories: heat treatments; storage near or below the freezing point of water; dehydration and control of water activity; chemical preservation; and use of mechanical operations such as washing, peeling, filtration, centrifugation, grinding, ultrahigh hydrostatic pressure, and most importantly, the packaging. Most food preservation technologies use two or more preservation operations because virtually all processed foods are packaged.

The extension of the useful storage life of plant and animal products beyond a few days at room temperature presents a series of complex biochemical, physical, microbial, and economic challenges. Respiratory enzyme systems and other enzymes in these foods continue to function. Their reaction products can cause off-flavors, darkening, and softening. Microbes contaminating the surface of plants or animals can grow in cell exudates produced by bruises, peeling, or size reduction. Fresh plant and animal tissue can be contaminated by odors, dust, insects, rodents, and microbes. Packaging must be used to protect the food from these contaminants.

Short-Term Storage. *Controlled Atmosphere.* The composition of the gas atmosphere surrounding certain respiring fruits and vegetables can influence their rate of quality loss. For example, an atmosphere having only a few percent of oxygen, 1–2% carbon dioxide, and the remainder nitrogen, can prolong the useful storage life of selected varieties of apples in refrigerated storage. Using this atmosphere a shelf life extension from three to nine or ten months is possible (6). Studies using refrigerated meat and poultry products have shown that gas mixtures containing 50% oxygen, several percent carbon dioxide, and the remainder nitrogen can extend the useful shelf life of these products from days to weeks (see MEAT PRODUCTS). In this case the oxygen inhibits the growth of anaerobic spoilage bacteria under refrigerated conditions.

Controlled atmosphere storage, when used for blemish-free products, is an effective preservation method for clean, intact foods, such as fresh apples and pears. Packaging has been developed to take advantage of the respiration processes taking place in fresh foods. Packaged produce can metabolize oxygen and release carbon dioxide. Packaging materials can be designed to allow diffusion of oxygen from the atmosphere and carbon dioxide to the atmosphere to maintain a desirable ratio of these gases in the package. A ratio can be obtained that is sufficient to reduce the respiration rate of the food and inhibit the growth of aerobic microbes. Chemical scavengers can be included in the package to remove undesirable metabolic gases such as ethylene which can influence the rate and course of respiration.

Refrigeration. Foods contaminated with microbes from various unit operations may require additional preservation to ensure a useful shelf life. Refrigeration slows the growth of spoilage microbes (see REFRIGERATION AND REFRIGERANTS). The effectiveness of refrigeration can be improved by combining storage just above freezing with ionizing radiation or chemical preservatives, eg, food acids, sodium benzoate, and nisin; treatment with ultrahigh (ca 300–400 MPa (43,000–58,000 psi)) pressure (7); and for liquid foods, filtration or centrifugation to remove microbes. Holding pork at −15°C for up to 30 days can be used to inactivate *Trichina spirales*, the cause of trichinosis.

Vacuum packaging has been effective in extending the shelf life of refrigerated processed meats (see VACUUM TECHNOLOGY). The use of a combination of several preservation steps to inhibit spoilage and induce minimal changes in the quality of the fresh food has been likened to establishing a series of hurdles (8). In some cases a very mild heat treatment, such as holding at 60°C for several minutes, helps reduce microbial and possibly insect contamination without extensive damage to the quality of fresh plant or animal tissue.

Heat Treatment. Shellfish taken from sewage-contaminated waters and containing pathogenic virus can be made safe by thorough cooking. Foods contaminated with large numbers of pathogenic microbes, or containing heat-labile toxins, however, are generally unsuitable for human consumption even if heat could be used to render these foods safe. Foods containing heat-stable microbial toxins, but otherwise free of microbes, are particularly hazardous because these may have no detectable signs of spoilage. Canned mushrooms have been implicated in staphylococcal food poisoning (9). Staphylococcal enterotoxin may lose only 90% of its toxicity by heating 19 min at 121°C.

Preservatives. Bakery products represent an important category of minimally processed foods. Bread and other yeast and chemically leavened baked products can lose their fresh quality characteristics within 24 h at room temperature owing to chemical changes brought about by water migration and starch crystallization. This quality loss, evidenced by staling, represents a complex set of changes among the starch, protein, water, and lipid components of the product. Additives such as sodium stearoyl lactylate can inhibit staling in bread for up to several weeks. The potential for the growth of microbes, such as molds, which can tolerate low water activity conditions, can create the need for a hurdle preservation strategy for extending the shelf life of fresh bread. Propionates as mold inhibitors or packaging in an oxygen-free modified atmosphere can prevent mold growth.

Long-Term Storage. Inactivation of microbes and enzymes in foods and food ingredients is necessary to ensure a long useful packaged shelf life. This can be achieved by using one or more preservation operations such as applying heat; using storage temperatures below $-18°C$; drying to water activities below 0.65, that is, an equilibrium relative humidity surrounding the product below 65%; and by adding chemical preservatives such as organic acids (acetic or lactic) or table salt. Generally, heat is used to inactivate enzyme activity prior to other preservation treatments. A mild heat treatment to inactivate enzymes in foods prior to freezing, drying, or chemical preservation is known as blanching. A discussion of the methods and equipment for blanching is available (10).

Food processing firms producing heat-preserved, frozen, dehydrated, or chemically preserved foods may be classified by their finished products. Companies may be further grouped based on whether they process raw materials into ingredients, such as in poultry and meat processing plants, or whether they take these ingredients and convert them to ready-to-eat consumer products.

Thermal Preservation Technology. The heat preservation of foods can be accomplished by various combinations of heating times and temperatures depending on the number and type of heat-resistant spores present, the composition of the food, and the physical characteristics of the food and package. Physical characteristics of the food such as viscosity, size of particles, size of the package, and starting temperature influence the rate of heat penetration into the slowest heating point in the package.

The inactivation of heat-resistant spores appears to follow first-order kinetics. Thus if the rate of inactivation of a spore population is known at several temperatures, and the rate of heating of the slowest point in a package can be determined or calculated from heat-transfer principles, then the time needed to sterilize the package can be calculated for any external heating condition. Mathematical formulas for calculating times and temperatures needed to heat foods in cylindrical containers to achieve any desired level of microbial inactivation are discussed in Reference 2. Computer programs are available that can be used in an interactive mode to monitor steam retort operations in real time to adjust for process deviations. Process deviations are changes in the operating conditions of the retort, such as retort steam pressure, or initial conditions of the product to be heat sterilized, such as product temperature, from those established with the Food and Drug Administration as necessary to produce thermally processed foods free of *Clostridium botulinum.*

The establishment of safe thermal processes for preserving food in hermetically sealed containers depends on the slowest heating volume of the container. Heat-treated foods are called commercially sterile. Small numbers of viable, very heat-resistant thermophylic spores may be present even after heat treatment. Thermophylic spores do not germinate at normal storage temperatures.

Chemical changes in foods resulting from heating, such as the loss of pigments, flavors, and vitamins, can also be approximated by first-order kinetics. These reaction rates, however, are much less sensitive to temperature change than are the rates of microbial inactivation. The activation energies for chemical changes in foods are often lower by a factor of five than the activation energies for the inactivation of spores (11).

If food can be heated quickly to a temperature of 131°C a lethality equivalent to 6 min at 121°C can be accumulated in 36 s. This rapid heating and cooling of liquid foods, such as milk, can be performed in a heat exchanger and is known as high temperature–short time (HTST) processing. HTST processing can yield heat-preserved foods of superior quality because heat-induced flavor, color, and nutrient losses are minimized.

Equipment. Equipment and processes for thermal preservation depend on the physical form of the food and its pH. Foods having a pH < 4.5 often can be sterilized, for commercial purposes, at or near a temperature of 100°C. Commercial sterility for these products means that the product will not spoil owing to microbial growth as long as the pH remains at or below 4.5. Hydrogen-ion activity (qv) inhibits the germination of *Clostridium botulinum* spores as well as the spores of most other heat-resistant microbes. The spores of *Bacillus coagulans* are an important exception. This latter microbe is found in tomato products, and these products are often adjusted to a pH of 4.0 or lower, or given an additional heat treatment.

Acid foods generally require the simplest equipment for heat preservation. The food can be heated to 100°C and filled hot into suitable containers. The containers are sealed, inverted to sterilize the closure, held at the filling temperature for a short time to ensure that the package is thoroughly heated, and then cooled. Tomato sauces, jellies, fruits, fruit juices (qv), and pickles are routinely preserved in this fashion.

Low acid foods have a pH > 4.5, require sterilization at temperatures above 100°C, and thus require treatment in pressure vessels. Heat preservation processes above 100°C can be carried out in batch or continuous heat-exchange equipment. Batch retorts are simple pressure vessels in which the packaged food can be exposed to saturated steam (qv), water and air over pressure, water sprays and air over pressure, or mixtures of air and steam. Microwave energy can be used to heat suitable plastic or glass packages (see MICROWAVE TECHNOLOGY). Small cans serve as their own individual pressure vessels and thus can be heated by direct flames, fluidized beds, or electrical resistance heating. Packages containing fluid foods or slurries with particles can be agitated to increase internal heat transfer. Rigid cans may be filled with particulate foods under very high vacuum with very little liquid to promote the formation of a steam atmosphere inside the package during heat treatment. This atmosphere greatly increases internal heat transfer.

Batch process equipment has the advantages of low capital investment and flexibility. There is little restriction on the form or size of the package and length of heat treatment. Systems are available having fully automated process cycle controls and materials handling for ease of loading and unloading. Disadvantages of batch equipment are slow cycle times, because the system must be heated and cooled for each process cycle, and higher energy and labor costs. Materials handling costs are also higher. In general, batch heat process systems are useful in food processing operations that produce a mix of products, in a number of package sizes, with a limited numbers of cases required for any product style.

Continuous heat processing equipment can take the form of a heat exchanger for pumped foods or a materials handling system that introduces individual packages into a sealed steam pressure chamber. Equipment for handling individual packages can use either a mechanical valve system, such as rotating pockets, or hydrostatic legs of sufficient height to balance the internal steam pressure of the system. A hydrostatic cooker operating at 121°C with a gauge pressure of 0.1 MPa (14.5 psi) typically uses 20 m high hydrostatic inlet and outlet legs. This height is needed to provide a safety factor for possible changes in water level during operation. Packages are placed on carriers affixed to an endless chain which travels through the system. Package cooling takes place in the decompression leg. Package treatment rates of 1000 containers/min are possible. Treatment rates of several hundred packages/min are possible using mechanical valve systems.

Foods continuously heat sterilized and cooled in heat exchangers must be handled under aseptic conditions and filled aseptically in presterilized packages. Ohmic, microwave, and induction heating allow particulate foods to be heated more rapidly and uniformly than conduction heating by conventional or wiped-surface heat-exchanger technology. The filling of sterilized and cooled product into presterilized packages, under sterile conditions, is referred to as aseptic processing and has several advantages over in-package sterilization. One advantage is that heating and cooling for sterilization is independent of package size. Packages ranging from a few cubic centimeters to the size of bulk tank cars can be prepared with equal quality. Products are filled at room temperature. Thus packages can be constructed of plastic having a relatively low softening temperature. These packages can have lower costs and greatly reduced permeabilities because heat-resistant polymers are not needed. Additionally, a wider range of structures is possible.

Aseptic filling systems can accept a wide range of packages including metal cans and covers sterilized by superheated steam; paper, foil, and plastic laminates sterilized by hot hydrogen peroxide; ionizing radiation sterilized bags; and a variety of plastic and metal containers capable of being sterilized by high pressure steam. A novel aseptic filling system designed for acid foods, such as applesauce and fruit juices, uses hot, food-grade, organic or inorganic acid to produce commercially sterile packages as they enter the sterile filling area. Aseptic filling machines can use filtered air or nitrogen under a slight positive pressure to prevent the entrance of microbes into the sterile filling area where the sterile product is filled and sealed in presterilized packages. A nitrogen atmosphere is preferred to keep oxygen away from the product when the package is sealed (see STERILIZATION).

Aseptic processing systems have found wide use for packing juices and milk products for the retail market and for the bulk preservation of tomato paste and fruit slices for use as ingredients. Further information on aseptic processing can be found in the literature (2).

Freezing Preservation.   The rate of loss of color, flavor, texture, and nutrients, the growth of microbes, and the activity of enzymes and other life forms are all functions of temperature. Thus lower storage temperatures prolong the useful life of foods. Below 0°C, the free water in foods starts to form ice crystals. Ice crystal formation is a function of moisture content, solute composition, and storage temperature fluctuation. Supercooling can occur during the freezing process, but it is a transient phenomenon. Foods having high concentrations of solutes can behave as glassy materials in which water may exist in a noncrystalline state at freezing temperatures owing to inhibited diffusion and high soluble solids content. The conditions needed for establishing and maintaining the glassy state in frozen foods is discussed in detail elsewhere (5).

Ice formation is both beneficial and detrimental. Benefits, which include the strengthening of food structures and the removal of free moisture, are often outweighed by deleterious effects that ice crystal formation may have on plant cell walls in fruits and vegetable products preserved by freezing. Ice crystal formation can result in partial dehydration of the tissue surrounding the ice crystal and the freeze concentration of potential reactants. Ice crystals mechanically disrupt cell structures and increase the concentration of cell electrolytes which can result in the chemical denaturation of proteins. Other quality losses can also occur (12).

Equipment for food freezing is designed to maximize the rate at which foods are cooled to −18°C to ensure as brief a time as possible in the temperature zone of maximum ice crystal formation (12,13). This rapid cooling favors the formation of small ice crystals which minimize the disruption of cells and may reduce the effects of solute concentration damage. Rapid freezing requires equipment that can deliver large temperature differences and/or high heat-transfer rates.

Many formulated foods and certain animal products tolerate freezing and thawing well because their structures can accommodate ice crystallization, movement of water, and related changes in solute concentrations. Starches can be modified for freeze–thaw stability against gel breakdown through several cycles. By contrast, most fruits and vegetables lose significant structural quality on freezing and during storage because their rigid cell structures fail to accommodate to ice crystal formation. Frozen food storage equipment must be designed to minimize temperature fluctuations. It is not possible to store foods at temperatures low enough to ensure complete conversion of all water to ice. Commercial frozen food storage temperatures (−18 to −24°C) represent an economic balance between storage costs measured in time, energy, and capital investment, and desired shelf life and product quality.

Freezing can also disrupt tissue structures and allow cell contents to become mixed so that undesirable enzyme reactions can take place at significant rates even at storage temperatures of $-18°C$. These reactions can generate off-flavors, reduce nutrient concentrations, and cause changes in the structure and appearance of foods. The amount of free liquid or drip found after a freeze–thaw cycle is a good indication of structural damage. Heat treatment (blanching) prior to freezing is used to eliminate enzyme activity. Most enzymes responsible for quality loss in plant materials can be inactivated by exposure to a temperature of $100°C$ for one to five minutes. Enzymes in heat-sensitive fruits often can be inhibited using sulfur dioxide, sucrose, or combinations of citric acid, table salt, and ascorbic acid, preceded by vacuum removal of air if heat is not used.

Most frozen foods have a useful storage life of one year at $-18°C$. However, foods high in fat such as sausage products may become rancid after two weeks in frozen storage if not protected from oxygen by special packaging and antioxidants. The time–temperature tolerances of various frozen foods are discussed in Reference 14. Moisture migration and loss of moisture through packaging materials or defective seals can occur in frozen foods during storage. The process is similar to freeze-drying and is accelerated if the storage temperature is not constant. The heat-transfer surfaces used to maintain the storage temperature of a frozen storage room must be at a lower temperature than the storage area and frozen foods must be protected by moisture-proof packaging. Storage under varying temperature conditions favors the migration of water in the food from areas of high to areas of low concentration. In addition, foods susceptible to oxidative deterioration must be protected from air by hermetically sealed containers, by coating with an oxygen impermeable coating, or by incorporating an antioxidant in the product (see ANTIOXIDANTS). Water glazes have been used to protect fish during frozen storage. Edible barriers are being evaluated to limit the rate of moisture migration (15).

Equipment. Food freezing equipment can be classified by the method and medium of heat transfer used. High velocity air is the most common medium used for direct contact freezing of nonpackaged foods. Typically, foods are loaded on a continuous mesh or perforated belt and passed through air moving upward at five m/s at temperatures as low as $-40°C$. The air is recycled through cooling coils and fans located next to the conveyor and returned through the conveyor. Because cold air has a partial water vapor pressure lower than the warm food, evaporation to the extent of several percent of the weight of the food can occur from the surface. This water vapor condenses on the air cooling coils, making periodic defrosting necessary. Whereas air is a convenient and safe heat-transfer medium for food freezing, it has several drawbacks including its low heat capacity and poor heat-transfer characteristics. Rapid freezing requires high air velocities and low operating temperatures. For these and other reasons many foods are frozen in equipment using conduction or liquid heat-transfer methods. Capital and energy savings, reduced moisture loss, and elimination of defrosting are some advantages of these methods (see HEAT-EXCHANGE TECHNOLOGY).

Liquid heat-transfer media for immersion freezing include solutions of edible salts, sugars, alcohols, and esters. These heat-transfer agents offer high heat-transfer rates, reduced pumping costs, and allow operating at higher refrigerant temperatures. Not all foods are suitable for direct immersion in these freezing media. However, irregularly shaped foods such as whole turkeys can be shrink wrapped in plastic and thus can be adapted to immersion freezing.

Conduction freezing between chilled plates is a very cost-effective method of heat removal for products that can be packaged in a geometry to fit between refrigerated plates. Packages having a regular geometry, such as a semi-infinite slab, can be loaded automatically between platens through which a refrigerant is circulated. The platens are stacked to provide a large product-holding capacity and thus sufficient contact time to ensure complete freezing. As unfrozen packages are introduced, frozen packages are removed in a continuous fashion. Good conduction heat-transfer conditions are achieved by maintaining pressure on the stack of platens to ensure good contact between the package and heat-exchanger surfaces.

Cryogenic freezing equipment uses liquid nitrogen or carbon dioxide snow. These units have the advantage of portability and simplicity and can produce extremely fast freezing rates. The refrigerant can be sprayed directly on the product to ensure rapid heat transfer. Cryogenic freezing produces very high quality products and ensures that little product weight loss occurs. This is important for high unit value foods and ingredients such as meat, poultry, bakery, and seafood products.

The quality of a frozen food may be determined more by the temperature at which it is stored than by the method or rate of freezing. Storage temperatures may fluctuate as products move from manufacturing through distribution channels to the consumer's home freezer. The useful shelf life of a frozen food may be severely limited by exposure to storage temperatures above $-18°C$, even for a few hours.

Dehydration Processing.   Dehydration is one of the oldest means of preserving food. Microbes generally do not grow below a minimum water activity, A_w, of 0.65 defined as the equilibrium relative humidity surrounding food in a sealed container at a given temperature, ie, no microbes can grow at a water activity below 65% relative humidity at storage temperatures in the range of 0–40°C.

Each food or food ingredient shows a characteristic equilibrium relative humidity at a given moisture content and temperature. Thus as a food is dried and its moisture content is reduced from its fresh value where water activity is generally 1.0, to lower and lower values, the equilibrium water activity of the food decreases as a complex function of residual moisture. The shape of the equilibrium relative humidity–moisture content curve is set by the chemistry of the food. Foods high in fructose, for example, bind water and thus show lower water activities at high moisture contents. Dried prunes and raisins are examples. Drying can be terminated at any desired moisture content and hence any water activity.

Foods dried to water activities in the range of 0.65 to 0.85 are often referred to as intermediate moisture foods. These partially dried foods tend to be soft and to rehydrate easily. The remaining water acts as a plasticizer. Because molds and yeast may be able to grow in these partially dried products, they must be preserved by heat, vacuum, or modified atmosphere packaging, refrigeration, or chemical means. For example, bakery products can be filled or topped with intermediate moisture content fruit or cheese fillings which have a microbial stability matching the lower moisture baked pastry portion.

Foods high in sucrose, protein, or starch (qv) tend to bind water less firmly and must be dried to a low moisture content to obtain microbial stability. For example, grain and wheat flour can support mold growth at moisture contents above 15% (wet basis) and thus are stored at moisture contents below 14%. Stored grains and oil seeds must be kept at a water activity below 0.65 because certain molds can release aflatoxins as they grow. Aflatoxins are potent carcinogens (see FOOD TOXICANTS, NATURALLY OCCURRING).

Fresh plant and animal tissue when dried to a water activity much below 0.97 show irreversible disruption of metabolic processes. However, individual metabolic enzymes may retain activity almost to dryness. Foods are usually heat treated (blanched) prior to drying. Some foods, upon drying, are susceptible to rapid nonenzymatic browning owing to high concentrations of reducing sugars, ascorbic acid, and free amino acids (qv). Dry fruits can be treated with sulfur dioxide to prevent nonenzymatic browning. Products susceptible to oxidation and oxidative rancidity can be treated with antioxidants and vacuum or inert gas packed to minimize exposure to oxygen. Low temperature storage can further reduce the rate of chemical deterioration. Dehydrated ingredients must have the same water activity when mixed together in a blended product to prevent undesirable moisture migration.

Equipment. Continuous hot air driers are used to prepare most of the high quality, dried, piece-form fruits and vegetables produced in the United States (see DRYING). Optimum quality can be achieved by matching the rate of heat input to the food to the rate of moisture release from the food while carefully controlling the product temperature. The bed depth, dry-bulb temperature, relative humidity, air velocity, and direction of air movement through the bed are selected to maintain a wet-bulb temperature which minimizes product deterioration owing to heat-induced chemical reactions. Typically, continuous belt dryers are staged to provide three or more drying zones into which product is fed to give a very even bed depth. The drying product can be redistributed as a progressively deeper bed in each zone. Each zone is set to operate at a controlled wet- and dry-bulb temperature, up or down air-flow velocity, and bed depth which optimizes product quality, minimizes energy use, and maximizes product throughput. Excessive heating of the product during any stage of drying reduces product quality; thus food dryers must be designed to provide very uniform drying conditions. The successful operation of a continuous belt drying system depends on establishing a uniform feed rate to the belt to ensure a uniform bed depth. Liquids and pastes are commonly dried in spray, drum, or freeze dryers. Particulate foods can be dried in batch or continuous air-fluidized beds or freeze dryers. Many agricultural commodities are sun-dried when weather conditions at harvest provide low humidity, warm temperatures, and good air circulation.

Rehydration rate and extent of rehydration of a dried food are important quality factors. Instantized dried foods refer to products which rehydrate to approximate fresh appearance and eating quality in several minutes in the presence of either hot or cold water. Freeze-dried meats, seafood, vegetables, and specialty products are particularly useful for instant soups, sauces, meals, and garnishes. Instantized rice, potato, and carrot dice, and other vegetable and cereal products, can be made using puffing guns, centrifugal fluidized beds, or single- or double-screw extruders. These and other drying technologies are covered in the literature (16).

Chemical Preservation.   Food additives (qv) can enhance the effectiveness of food preservation by heat, refrigeration, and drying methods. The addition of a food-grade acid to a low acid food to shift the pH to a value below 4.5 allows heat preservation at a temperature of 100°C instead of in the range of 121°C. Antioxidants such as butylated hydroxyanisole [25013-16-5] (BHA) can be added to potato chips to reduce the need for expensive oxygen-impermeable flexible packaging. Sulfur dioxide is used in wine (qv) and in dry fruit and vegetable products to preserve colors and flavors and prevent nonenzymatic browning.

Food can be preserved by fermentation (qv) using selected strains of yeast, lactic acid-producing bacteria, or molds. The production of ethanol (qv), lactic and other organic acids, and antimicrobial agents in the food, along with the removal of fermentable sugars, can yield a product having an extended shelf life. Mild heating of foods, acidified by fermentation and packaged to prevent further contamination, results in a shelf stable product. Cucumber pickles and sauerkraut are examples. The course of the fermentation process can be controlled by the addition of sodium chloride to help provide optimum fermentation conditions for the lactic acid-producing bacteria present on the raw materials at harvest.

Lactic acid-producing bacteria associated with fermented dairy products have been found to produce antibiotic-like compounds called bacteriocins. Concentrations of these natural antibiotics can be added to refrigerated foods in the form of an extract of the fermentation process to help prevent microbial spoilage. Other natural antibiotics are produced by *Penicillium roqueforti*, the mold associated with Roquefort and blue cheese, and by *Propionibacterium* sp., which produce propionic acid and are associated with Swiss-type cheeses (3).

Ionizing radiation is considered to be a chemical preservation method and applications must be cleared by the Food and Drug Administration for use, not only on a product-by-product basis, but also on a dose basis. In the United States, up to 100 Gy (10,000 rad) may be used to inhibit potato sprouting, and up to 500 Gy may be used to kill insects and insect eggs in grain products. Fruit, poultry, and other fresh products can be pasteurized at doses up to 10,000 Gy to inactivate pathogenic bacteria, spoilage microbes, insects and larva, and to reduce total microbial counts to extend refrigerated shelf life. Packaging materials, spices, and medical devices can be sterilized at doses up to 60,000 Gy. The theory and applications of ionizing radiation in food preservation are discussed elsewhere (17).

Other Technologies.   Several technologies for the preservation of foods using a minimum of heat are being explored. The application of ultrahigh pressure to the preservation of foods has been investigated (7). Ultrahigh hydrostatic pressure is known to inactivate vegetative microbial cells and parasites in foods at a rate proportional to pressure at any temperature. Studies using a wide variety of fresh and processed foods have shown that these foods can be rendered free of vegetative bacteria, parasites, yeasts, and molds by subjecting the foods to hydrostatic pressures in the range of 300 to 600 MPa (43,000–87,000 psi) for times between one minute and one hour, at room temperature. Spores, viruses, and food spoilage enzymes, such as polyphenol oxidase, have been found to be quite resistant to inactivation by pressure at room temperature. However, combinations of pressures and temperatures up to 80–90°C have been used to inactivate spores and enzymes in low acid foods. Pressure resistant viruses are easily inactivated by heat.

Foods having a pH < 4.5 can be made commercially sterile using pressures in the range of 400 MPa (58,000 psi) because a pH ≤ 4.5 inhibits the germination of most bacterial spores. Yeasts and molds are much more susceptible to inactivation by pressure than are vegetative bacterial cells. Pressure appears to disrupt cell membranes and to denature proteins. Foods preserved by ultrahigh pressure processes must be stored at their intended storage temperature for a time sufficient to determine whether microbes can repair pressure damaged cell structures during storage. Inactivation and regeneration rates of pressure treated microbes are strongly influenced by the chemical composition of the media in which they are held.

Equipment for batch ultrahigh pressure preservation of foods has been adapted from units used in the metal and plastic industry for cold or warm isostatic pressing at operating pressures to 1000 MPa (145,000 psi). Operating volumes of several hundred liters are possible at these pressures using wire-wound vessels. Batch chambers operating from 300 to 600 MPa (43,000 to 87,000 psi) may be made from multiple shrink-fit cylindrical tubes which can be closed at either end using continuous thread, interrupted thread, pin, or yoke closures. It is possible to treat liquid foods in a continuous manner by pumping them, at treatment pressure, through a holding tube of sufficient length to provide an adequate treatment time, and returning the foods to atmospheric pressure. The treated foods must be stored and filled under aseptic conditions into presterilized packages.

Capacitance discharge has also been investigated as a means to pasteurize or commercially sterilize foods which can pass between plates sufficiently close together to allow an electric field of approximately 25,000 V/cm (18). The field is established preferably as a square-wave pulse of millisecond duration to avoid heating effects. The strong electrical field appears to disrupt or permeabilize cell membrane structures. The degree of disruption, and hence microbial inactivation, can be related to the number of pulses to which each element of food is subjected. Cell permeabilization by capacitance discharge can be used to improve the extraction of secondary metabolites from microbes and cells and to increase the yield of mechanically pressed fresh fruit and vegetable juices (18). Very high intensity flashes of visible light can be used to pasteurize fruit juices using a minimum of heating in a manner which appears to be similar to capacitance discharge.

Computer Integrated Manufacturing, Instrumentation, and Controls. Large food processing firms are exploring the use of computer integrated manufacturing (see PROCESS CONTROL). However, the diversity of products, the difficulty of measuring meaningful quality attributes on line, and the batch nature of many processes has tended to slow the application of this technology to food processing. However, thermal processing controls have been developed to the point where time and temperature process deviations can be corrected on line. Freezer, dryer, and vacuum evaporator operating conditions can be controlled and optimized using systems already available to the process industry.

An important aspect of food processing, common with other processing industries, is yield of finished product from starting raw materials for any shift and for specific unit operations. Computer integrated manufacturing can start with the measurement of material flows and build upon this information. Instrumentation for the on line measurement of specific food qualities of importance to the consumer such as food flavor, aroma, texture, and microbial content are under development. These quality factors are monitored using statistical quality control (qv) procedures using standard sampling plans and control strategies (see also ROBOTS, PROCESSING).

BIBLIOGRAPHY

"Food and Food Processing" in *ECT* 1st ed., Vol. 6, pp. 785–818, by Z. I. Kertesz, New York State Agricultural Experiment Station, Cornell University; in *ECT* 2nd ed., Vol. 10, pp. 23–61, by Z. I. Kertesz, Food and Agricultural Organization of the United Nations; "Food Processing" in *ECT* 3rd ed., Vol. 11, pp. 164–183, by D. F. Farkas, University of Delaware.

1. *Code of Federal Regulations (CFR)*, Title 7, subtitle B, Chapt. 1, parts 27–209; Title 9, Chapt. 3, parts 300–399; Title 21, Chapt. 1, subpart B, parts 100–199, U.S. Government Printing Office, Washington, D.C., 1992.
2. A. Teixeira, in D. R. Heldman and D. B. Lund, eds., *Handbook of Food Engineering*, Marcel Dekker, Inc., New York, 1992, Chapt. 11.
3. P. M. Davidson and A. L. Branen, eds, *Antimicrobials in Foods*, 2nd ed., Marcel Dekker, Inc., New York, 1993.
4. K. Helrich, ed., *Official Methods of Analysis*, 15th ed., Vols. 1 and 2, Association of Official Agricultural Chemists, A.O.A.C., Inc., Arlington, Va., 1990.
5. L. Slade and H. Lavine, *Beyond Water Activity: Recent Advances Based on an Alternative Approach to the Assessment of Food Quality and Safety*, CRC Critical Reviews in Food Science and Nutrition, CRC Press, Boca Raton, Fla., 1991, pp. 115–360.
6. D. K. Salunke, H. R. Bolin, and N. R. Reddy, *Storage, Processing, and Nutritional Quality of Fruits and Vegetables*, 2nd ed., Vol. 1, CRC Press, Boca Raton, Fla., 1991.
7. D. G. Hoover and co-workers, *Food Tech.* **43**(3), 99–107 (1989).
8. V. W. Leistner and W. Rodel, in R. Davies, G. G. Birch, and K. J. Parker, eds., *Intermediate Moisture Foods*, Applied Science Publishers, Ltd., London, 1976, pp. 112–147.
9. P. Hardt-English and co-workers, *Food Tech.* **44**(12) 74, 76–77 (1990).
10. N. N. Potter, *Food Science*, 4th ed., Van Nostrand Reinhold, New York, 1986.
11. R. Villota and J. G. Hawks, in Ref. 2, Chapt. 2.
12. O. R. Fennema, W. D. Powrie, and E. H. Marth, *Low-Temperature Preservation of Foods and Living Matter*, Marcel Dekker, Inc., New York, 1973.
13. D. K. Tressler, W. B. Van Arsdel, and M. J. Copley, eds., *The Freezing Preservation of Foods*, 4th ed., Vols. 1–4, Avi Publishing Co., Westport, Conn., 1979.
14. W. B. Van Arsdel, M. J. Copley, and R. L. Olson, eds., *Quality and Stability of Frozen Foods*, Wiley-Interscience, New York, 1969.
15. J. A. Torres, "Protein Functionality in Food Systems," in N. Hettiarachy and G. Ziegler, eds., *Proceedings of the 1993 Institute of Food Technologists Basic Symposium*, IFT/Marcel Dekker, Inc., New York, 1994.
16. M. R. Okos and co-workers, in Ref. 2, Chapt. 10.
17. E. S. Josephson and M. S. Peterson eds., *Preservation of Food by Ionizing Radiation*, Vol. 1–3, CRC Press, Inc., Boca Raton, Fla., 1983.
18. B. Mertens and D. Knorr, *Food Tech.* **46**(5), 124–133 (1992).

General References

M. D. Pierson and D. A. Corlett, Jr., eds., *HACCP-Principles and Applications*, Van Nostrand Reinhold Co., New York, 1992.

Canned Foods, Principles of Thermal Process Control, Acidification, and Container Closure Evaluation, 5th ed., The Food Processors Institute, Washington, D.C., 1988.

D. M. Considine and G. D. Considine, eds., *Foods and Food Production Encyclopedia*, Van Nostrand Reinhold Co., New York, 1982.

J. A. Troller, *Sanitation in Food Processing*, Academic Press, Inc., Orlando, Fla., 1983.

R. V. Decareau and R. E. Mudgett, *Microwaves in the Food Processing Industry*, Academic Press, Inc., Orlando, Fla., 1985.

D. A. Shapton and N. F. Shapton, eds., *Principles and Practices for the Safe Processing of Foods*, Butterworth-Heinemann Ltd., Oxford, U.K., 1991.

F. A. Paine, ed., *Packaging User's Handbook*, Van Nostrand Reinhold Co., New York, 1991.

DANIEL F. FARKAS
Oregon State University

FOODS, DIET. See DAIRY SUBSTITUTES; FAT REPLACERS; SWEETENERS.

FOODS, NONCONVENTIONAL

Nonconventional foods differ from the usual materials of plant and animal origin used for human food or animal feed (see FEEDS AND FEED ADDITIVES; FOOD PROCESSING). These materials can be produced from chemical feedstocks, eg, carbohydrates (qv), hydrocarbons (qv), or other industrial organics, by processes such as microbiological, enzymatic, or chemical synthesis, or from existing natural products, containing carbohydrates, proteins (qv), and fats, by physical, chemical, microbiological, or enzymatic modification.

Examples of nonconventional foods include single-cell proteins, ie, dried cells of microorganisms such as algae, bacteria, actinomycetes, yeasts (qv), molds, and higher fungi, or protein concentrates and isolates derived from them; derived plant and animal products, ie, leaf meals and leaf protein concentrates, seed meals and seed meal proteins (see NUTS; SOYBEANS AND OTHER OILSEEDS), concentrates and isolates of soy, cottonseed, peanut, etc, plant cells grown in tissue culture, fish, and meat protein concentrates and isolates (see AQUACULTURE CHEMICALS; MEAT PRODUCTS); synthetic products, ie, carbohydrates (qv), fats and fatty oils (qv), proteins (qv), peptides, amino acids (qv), and vitamins (qv) prepared by chemical, microbiological, or enzymatic synthesis; and manufactured or combination foods, ie, engineered, restructured, or textured foods, and formulated foods.

Each of these general classes of nonconventional foods has been developed to meet specific applications. For example, single-cell protein (SCP) products provide a source of protein for use in animal feeds in those regions of the world where conventional sources of protein feedstuffs, eg, soybean meal or fish meal, are periodically in short supply or available only at very high prices. SCP products have applications in human foods as protein sources or as functional food ingredients for their flavoring, water- or fat-binding, stabilizing, and thickening characteristics.

Derived plant and animal products make better use or upgrade the nutritional quality of already existing materials or products. Synthetic and manufactured products arose from knowledge of the functional properties of food ingredients and of human and animal nutrition that involved more precise definition of nutrient requirements for growth, reproduction, lactation, and body maintenance in both humans and domestic livestock. Food products have been developed to meet human needs under abnormal environments, eg, military rations for arctic, tropical, or desert environments, and special products for astronauts in space flights.

Numerous reviews have been published on various aspects of nonconventional foods (1–18).

Single-Cell Protein

Cells of microorganisms have constituted a portion of human food since ancient times. Yeast-leavened baked products contain the residual nutrients from the yeast cells destroyed during baking (see BAKERY PROCESSES AND LEAVENING AGENTS). Cultured dairy products, such as yogurt, buttermilk, and sour cream, contain up to 10^6 cells of lactic acid bacteria per gram (19) (see MILK AND MILK PRODUCTS). Other examples of fermented foods consumed since early times include fermented meats, fish, and soybean products.

Modern technology for producing microbial cells for human food or animal feed emerged in Germany during World War I. Baker's yeast, *Saccharomyces cerevisiae*, was grown in aerated tanks using incremental feeding of molasses as the carbon and energy source, and ammonium salts as the nitrogen source (20) (see FERMENTATION). Between World Wars I and II, processes were developed in Germany for producing fats from the sulfite waste liquor of paper (qv) manufacturing using *Endomyces vernalis*, and for producing *Saccharomyces cerevisiae* from wood hydrolysates for use as fodder yeast, ie, the Scholler-Tornesch process. Also during this period, the Heiskenskjold process for propagating *S. cerevisiae* from sulfite waste liquor was introduced in Finland (21,22).

During World War II, effort was undertaken in Germany to produce food and fodder yeast from waste products such as sulfite liquor. The yeast *Candida utilis*, ie, Torula yeast, grows on pentoses such as D-xylose and D-arabinose present in sulfite waste liquor, as well as on glucose. The Waldhof fermentor, introduced during this period, provided for both agitation and aeration. It was a significant advance in microbial cell production technology, and it enabled improved rates of oxygen transfer to growing cells to occur, resulting in faster growth rates than had been achieved previously.

Two broad classes of microorganisms are of interest (ca 1993) for single-cell protein (SCP) production, ie, photosynthetic organisms, including algae and certain bacteria; and nonphotosynthetic organisms, including bacteria, actinomycetes, yeasts, molds, and higher fungi. In addition, two different uses of SCP are distinguished, ie, food for humans and feed for animals.

Photosynthetic Organisms.   Mass cultivation of algae in ponds or tanks under photosynthetic conditions, using incident sunlight as the energy source and CO_2 as the carbon source, has been investigated in Japan, Taiwan, Mexico, Algeria, India (9,23–26), and in California in the United States (27,28). Artificial illumination systems have been used for experimental mass cultivation of algae (29) and in bioregenerative systems for converting CO_2 and human wastes into breathable oxygen and food as part of life-support systems for long-duration space exploration missions (30).

Research has been conducted on growth of blue-green algae heterotrophically in the dark, using organic carbon and energy sources such as glucose or acetate (31–35). The objective of these efforts has been to determine optimum conditions, including pH and substrate concentrations, for high specific growth rates and biomass yields.

Algal cultures must be agitated during growth to maintain cells in suspension and exposure to mutant sunlight, and to remove photosynthesis-inhibiting oxygen. Methods used for agitation include paddle wheels, raceways with arrays of foils to create vortices, and recirculation, either alone or in combination with injection of CO_2 (24,33) (see AERATION, BIOTECHNOLOGY). Table 1 shows that algal densities in culture ponds are in the range of 1–5 g/L, dry wt basis. Consequently, large volumes of water must be handled in harvesting, dewatering (qv), and drying algal cells.

Centrifugation; flocculation using $Al_2(SO_4)_3$, $Ca(OH)_2$, or cationic polymers; sedimentation; filtration; treatment with ion-exchange resins and drum; sand bed; and sun drying have been investigated for separating, concentrating, and drying algal cells (36,44–46). All of these methods add significantly to the cost of the product except for sun drying. However, sun drying is difficult to accomplish in humid climates such as in India and southeast Asia.

In California, *Spirulina* sp. grown in paddle-wheel-agitated open ponds with CO_2 is harvested through stainless steel screens, with recycling of the nutrient-rich water to the ponds. The wet *Spirulina* is spray-dried at 60°C for a few seconds to yield a food-grade product (47).

Yields of algae grown in outdoor pond cultures (Table 1) are on the order of 15–40 g/(m²·d) (24 short tons per acre per year). Higher yields can be obtained under artificial illumination, but growth of algae under these conditions is not economically feasible.

Product quality is an important consideration in producing algae for food or feed use. Algal cells must be dried using time–temperature combinations sufficient to destroy pathogenic bacteria and viruses that may be present in ponds, particularly in those culture systems based on sewage (see WATER, SEWAGE). The possible presence of heat-stable algal toxins must also be considered. The cyanobacteria (blue-green algae) *Anabaena flos-aquae* and *Microcystis* sp in algal waterblooms on ponds produce toxins which poison farm animals. These toxins are not destroyed by boiling or autoclaving the water (48).

The nitrogen requirements for algal growth can be met by either ammonium salts, urea, or nitrates. Bacterial action in sewage oxidation ponds may also liberate sufficient ammonia for algal growth. Most natural wastes supply sufficient

Table 1. Photosynthetic Microorganisms in SCP Production

Organism	Scale	Growth conditions[a]	pH	Yield,[b] g/(m²·d)	Reference
Algae					
Chlorella sp.	plastic ponds[c]	CO_2 or acetate	6.0–7.0	15–40	33
Chlorella ellipsoidea	200-m² ponds	outdoor sunlight, continuous CO_2, urea autotrophic or mixotrophic with acetate	6.0–7.0	18.7–27.5	34
Chlorella pyrenoidosa	200-m² ponds	outdoor sunlight, continuous CO_2, urea autotrophic or mixotrophic with acetate	6.0–7.0	19.0–30.5	34
	10-L tubular loop bioreactor	CO_2, 2 kPa (0.3 psi) pressure, sunlight, urea fed batch	6.6	26.0–30	35
Scenedesmus acutus	225-m² shallow tanks[d]	sunlight, CO_2, sugar cane, molasses (mixotrophic) urea	7.0–8.0	20–25	36
Spirulina maxima	700-m² pond	sunlight, 0.5% CO_2	9.0	15	37,38
Spirulina platensis	100-m² ponds[e]	$NaHCO_3$, KNO_3	9.5–10.0	22	39
Bacteria					
Rhodobacter (Rhodopseudomonas) capsulatus	waste ponds[f]	industrial waste substrates, sunlight			40,41
	2-L photobioreactor[g]	artificial light, calcium lactate, 30°C			
Rhodocyclus (Rhodopseudomonas)				10.41[h]	42
gelatinosus	14-L fermentor[i]	incandescent light, 3.0% wheat bran infusion, 30°C	6.7–7.5	4.33[h]	43

[a]Ambient temperature unless otherwise noted. [b]Dry wt. To convert g/(m²·d) to short tons per acre·year, multiply by 1.629. [c]Cell density of 2.5 g/L. [d]Strain 276-3A; 20-cm deep. [e]Fiber glass-stirred ponds, 13–15-cm deep. Cell density of 0.3 g/L. [f]Cell density of 1.2–2.0 g/L. [g]Continuous upflow photobioreactor. [h]g/L. [i]Continuous fermentor. Cell density of 3.15 g/L.

quantities of inorganic nutrients, but additional phosphorus may be needed in some regions for optimal SCP production.

Key factors influencing growth include temperature; pH; availability of CO_2, nitrogen, phosphorus, and other inorganic nutrients; and availability of sunlight as influenced by latitude, cloud cover, and depth of the culture pond or tank. Slow, erratic growth results from wide fluctuations between day and night temperatures in outdoor ponds, and from season to season. Significant amounts of algal biomass may be lost as a result of respiration during the night. In the case of *Spirulina* sp., this loss may be as great as 35% of the total biomass produced during the day (24).

The availability of CO_2 and the pH are intimately related because the preferred pH range for growth of many species, such as *Chlorella*, is pH 6.5–7.0 and most of the CO_2 is bound as bicarbonate (HCO_3^-) in solution. Additional CO_2 beyond that present in air (0.03%) must be provided to attain optimum growth.

At Lake Texcoco, Mexico, bicarbonate is available in the alkaline waters from soda ash [497-19-8] (sodium carbonate) deposits (see ALKALI AND CHLORINE PRODUCTS). This supply of carbon is adequate for growing *Spirulina maxima*, which tolerates alkaline pH values in the range 9–11 (37,38). Combustion gases have been used to grow this organism, but this carbon source is not available in many regions (49).

Sunlight availability is critical for algal growth. Outdoor algal cultivation is considered practical only in regions between latitudes 35° N and S, where cloud cover and variations in the length of day and night are minimized (50). Depth of the culture pond also affects availability of sunlight. Research in Israel has shown that as much as 80% of the algal cells in a pond may be in darkness because almost all of the solar irradiation is absorbed in the upper 2–3 cm of liquid depth (39).

Economic evaluations of algal production indicate that production costs vary from $0.15 to $4.00/kg of algal product, depending on type of bioreactor, culture technique, and operating conditions (51). For systems with controlled agitation and carbonation, including raceways and tubular reactors, production costs are estimated to range from $2.00 to $4.00/kg.

Tables 2, 3, and 4 list compositional and nutritional data of selected algae. More extensive compilations on algae are available (26,58). Algae tend to have lower contents of methionine than is desirable in human and animal nutrition

Table 2. Composition of Photosynthetically Grown Algae, %[a]

Organism	Nitrogen	Crude[b] protein	Fat	Ash	Reference
Chlorella sp.	9.3	58	9	3	26
Scenedesmus acutus	8.2–10.2	51.4–63.6	11.2–14.3	7.9–16.7	52
Spirulina sp.	8.8–11.2	55–70	4–7	5–10	47
Spirulina platensis	8.0	50.0	0.5	11.0	53

[a]Dry wt basis.
[b]Crude protein = % nitrogen × 6.25. Does not accurately reflect true protein content. Algal cells may contain nonprotein nitrogen substances, eg, 4–6% nucleic acids, dry wt basis.

Table 3. Amino Acid Composition of Photosynthetically Grown Algae, g/16 g Nitrogen

Amino acid	Chlorella sp.[a]	Scenedesmus sp.[b]	Spirulina sp.[c]	FAO reference pattern[d]
alanine	7.4	7.02	8.28	
arginine	5.74	6.94	7.43	
aspartic acid	8.18	8.34	9.95	
cysteine		1.17	0.93	
glutamic acid	9.74	10.02	13.81	
glycine	5.89	5.13	5.28	
isoleucine	4.00	3.32	6.14	4.2
leucine	8.18	7.11	9.26	4.8
lysine	5.39	5.73	4.93	4.2
methionine	2.26	11.95	2.65	2.2
phenylalanine	4.87	4.14	4.61	2.8
proline	4.35	3.78	4.46	
serine	3.48	3.74	6.30	
threonine	4.18	4.04	5.30	2.8
tryptophan	0.87		1.37	1.4
valine	5.56	4.89	7.00	4.2
nitrogen[e]	9.3	10.2	9.6	

[a]Has 1.91 g histidine/16 g nitrogen. Ref. 26.
[b]Ref. 52.
[c]Ref. 37.
[d]Ref. 54.
[e]Values given are percentages.

and supplementation with this amino acid is necessary with many species (Table 4).

There is considerable anecdotal information on the history of human consumption of *Spirulina maxima* as a source of protein in Mexico and in the region of Lake Chad in Africa. However, relatively few controlled human feeding studies have been conducted using algae as a significant source of protein in the diet. Consumption of 100 g/d of a mixture of *C. ellipsoidea* and *S. obliquus* resulted in gastrointestinal distress attributed to toxins in the algae (59). Partial substitution of proteins in eggs and fish with *C. pyrenoidosa* did not reduce human nitrogen retention but digestibility was low (60). In other studies, *Scenedesmus* was fed to human subjects at levels up to 20 g/d with no ill effects (61). *Spirulina* was incorporated into diets of hospitalized children in Mexico City and was well tolerated (38). However, consumption of algae in human diets is limited to the extent that nucleic acid intake should not exceed 2 g/d. Higher levels may lead to arthritis and gout.

In general, many species of algae have cell walls resistant to digestive enzymes, dark colors, and bitter flavor. All of these characteristics must be altered to make an acceptable food or feed product.

The principal interest in photosynthetic bacteria for their applicability to SCP production (Table 1) has been in Japan, where *Rhodobacter capsulatus* has been used to treat industrial wastes in sewage ponds (40,41). The product has

Table 4. Algae Protein Quality and Digestibility[a]

Organism	Treatment[b]	Algae protein,% in diet	digestibility	PER[c]	NPU[d]	Reference
Chlorella pyrenoidosa (Sorokiniana)		10	86	2.19		55
	0.20% L-methionine	10	86	2.90		55
Chlorella sp.[e,f]		7.5–15				56
Scenedesmus sp.[g,f]	0.10% DL-methionine	10				52
Spirulina sp.		10	8.4–8.5	2.2–2.6	53–61	37
Spirulina platensis			75.5		52.7 (68)	57
	0.2% DL-methionine		75.5		62.4 (82.4)	57

[a]Tests on rats unless otherwise noted.
[b]Dried plus addition of indicated compounds.
[c]Protein efficiency ratio (PER) = weight gain (g) for a 10% protein level in the diet of rats as compared to the standard of 2.5 for casein.
[d]Net protein utilization (NPU) = % digestibility (D) × biological value (BV); complete utilization = 100. Biological value is given in parentheses; BV = % of absorbed nitrogen retained in body tissue; complete retention = 100.
[e]Feed/gain ratio of 1.60–1.63.
[f]Tests on chicks.
[g]Feed/gain ratio of 2.0–2.3.

been evaluated as a protein supplement in laying hen rations for egg production with acceptable results (40).

Nonphotosynthetic Organisms. Nonphotosynthetic microorganisms of interest in SCP production include bacteria, actinomycetes, yeasts, molds, and higher fungi. Carbon and energy sources considered for growing these organisms include carbohydrates such as simple sugars, starches, and cellulose (qv); agricultural, forestry, pulp (qv), paper, and food processing wastes containing these carbohydrates; and hydrocarbons and chemicals derived from them, including alcohols and organic acids.

Commercial-scale operations are conducted in batch, fed-batch, or continuous culture systems. Fermentation vessels include the conventional baffled aerated tank, with or without impeller agitation, and the air-lift tower fermentors in which air is sparged into an annular space between the fermentor wall and internal cylinder (1–3). A corrosion-resistant grade of stainless steel (316 L) is usually used for fermentor construction; wood or concrete tanks have been used with agricultural or food wastes.

In batch systems, the concentration of the carbon and energy source for growth is 1–10%. In fed-batch and continuous culture systems, it is usually less than 1% and quantities of nutrients are limited to those required to meet nutritional requirements of the growing organisms. Suitable nitrogen sources include anhydrous ammonia or ammonium salts. Feed-grade phosphate is used as the source of phosphorus. Mineral-nutrient requirements vary among different organisms and are usually added to make up deficiencies in the water supply (see MINERAL NUTRIENTS). Sulfates are used rather than chlorides to minimize corrosion. Carbon–nitrogen ratios should be 7:1–10:1 for yeasts to favor high protein contents and minimize the fat synthesis in the cells that occurs at higher C:N ratios.

Temperature and pH conditions for optimum growth rates and productivities, ie, dry weight of cells per unit volume per unit time, vary widely but are generally 25–40°C and pH 3.0–7.0, respectively. It is desirable to use strains of microorganisms that tolerate higher temperatures in this range since considerable quantities of heat are liberated during aerobic growth of microorganisms on either carbohydrates or hydrocarbons. Typical values are 15–34 kJ/g (3.6–8.1 kcal/g) of dry wt cells, depending on yield from a given substrate. In many geographical regions, cooling water is not available at a temperature below 20°C and refrigeration must be provided to control the temperature in the fermentor.

Production of food-grade SCP products requires operation under aseptic conditions in which the air, growth medium, and equipment are sterilized. Feed-grade SCP can be produced under clean but nonsterile conditions provided that a pH of 3.0–4.5 and a large inoculum are used. Transfer of oxygen and substrates to and across the cell surface is an important factor affecting growth rate, yield, and productivity in SCP processes. For yeasts, oxygen requirements range from 1 g/g dry wt of cells with carbohydrates to 2 g/g dry wt of cells for hydrocarbons (62).

Several processes for bacterial SCP production have been developed but abandoned. Imperial Chemical Industries, Ltd. constructed a 50,000–75,000-t/yr plant for producing the bacterium *Methylophilus methylotrophus* from methanol (qv). This process employed an air-lift pressure cycle fermentor, and a proprietary system for separating the cells from the growth medium by agglomeration. This plant is no longer operating because the protein product, Pruteen, was not competitive as an animal feedstuff in west European markets (63).

Large-scale SCP production processes for growing yeasts of the genus *Candida* from hydrocarbon substrates were developed by British Petroleum Co., Ltd. and Kanegafuchi Chemical Industry, Ltd. of Japan (57). However, the 100,000-t/yr capacity plants based on these processes, and constructed in Sardinia and Italy, were abandoned because of regulatory agency questions regarding residual hydrocarbon contents of the products (2,3).

Table 5 presents typical operating conditions and cell production values for commercial-scale yeast-based SCP processes including (63) *Saccharomyces cerevisae*, ie, primary yeast from molasses; *Candida utilis*, ie, Torula yeast, from papermill wastes, glucose, or sucrose; and *Kluyveromyces marxianus* var. *fragilis*, ie, fragilis yeast, from cheese whey or cheese whey permeate. All of these products have been cleared for food use in the United States by the Food and Drug Administration (77).

S. cerevisiae is produced by fed-batch processes in which molasses supplemented with sources of nitrogen and phosphorus, such as ammonia, ammonium sulfate, ammonium phosphate, and phosphoric acid, are fed incrementally to meet nutritional requirements of the yeast during growth. Large (150 to 300 m^3) total

Table 5. SCP Production Processes Based on Nonphotosynthetic Microorganisms

Organism	Substrate	Scale	Fermentor	Temperature, °C	pH	Cell density,[a] g/L	Specific growth rate,[b] h⁻¹	Yield[c]	Reference
Bacterial processes									
Methylophilus (Pseudomonas) methylotrophus	methanol	75,000[d]	continuous[e]	35–40	6.0–7.0	30	0.38–0.50	0.50	64,65
Yeast processes									
Candida (Saccharomycopsis) lipolytica	n-alkanes	18,000[f]	continuous	32	5.5	23.6		0.88	66,67
Candida utilis	sulfite waste liquor	300,000[g]		30	4.5		0.5	0.50	5
	ethanol	4,450[d]	plant	30	4.6	6–7	0.3	0.80	68
Hansenula jadinii	sucrose	25,000[g]	continuous	32–35	3.5–4.5	12–150	0.13–0.15	0.52	69,70
Kluyveromyces marxianus var. *fragilis*	cheese whey (lactose)	56,781[g]	fed-batch	30	4.5	112.5	112.5	0.45–0.55	71,72
	cheese whey permeate	1,500[g]	continuous	37	4.6		0.1–0.3	0.45	73
Saccharomyces cerevisiae	molasses	150,000[g]	fed-batch	30	4.5–5.0	40–45	0.20	0.50–0.54	5
Mold and fungal processes									
Fusarium graminearum	glucose	1,300[g]	continuous	30	6.0	15–20	0.2	0.53	74
Morchella hortensis	glucose	7,570[g]	batch	25–30	6.5	24–30		0.48	62
Paecilomyces varioti (Pekilo)	spent sulfite liquor	360,000[g]		37	4.5	13	0.14–0.20	0.55	75,76

[a]Dry wt basis. [b]Dilution rate per h. [c]g/g of substrate utilized on dry wt basis. [d]Scale is in t/yr. [e]Air-lift pressure cycle fermentor. [f]Scale is in L. [g]Scale is in L, working volume.

volume aerated fermentors provided with internal coils for cooling water are employed in these processes (5). Substrates and nutrients are sterilized in a heat exchanger and then fed to a cleaned–sanitized fermentor to minimize contamination problems.

C. utilis yeast is produced by either fed-batch or continuous processes. Aerated–agitated fermentors range up to 300 m³ total capacity and are operated in the same manner as described for *S. cerevisiae* (2,5). *C. utilis* is capable of metabolizing both hexose and pentose sugars. Consequently, papermill wastes such as sulfite waste liquor that contain these sugars often are used as substrates.

The Provesteen process, developed by Phillips Petroleum Company, employs a proprietary 25,000-L continuous fermentor for producing *Hansenula jejunii*, the sporulating form of *C. utilis*, from glucose or sucrose at high cell concentrations up to 150 g/L. The fermentor is designed to provide optimum oxygen and heat transfer (69,70).

K. marxianus var. *fragilis*, which utilizes lactose, produces a food-grade yeast product from cheese whey or cheese whey permeates collected from ultrafiltration processes at cheese plants. Again, the process is similar to that used with *C. utilis* (2,63). The Provesteen process can produce fragilis yeast from cheese whey or cheese whey permeate at cell concentrations in the range of 110–120 g/L, dry wt basis (70,73).

Molds and higher fungi have been grown in aerated fermentors for food use utilizing a variety of carbohydrates as substrates. Mycelia of various species of mushrooms, such as *Agaricus, Lentinus, Morchella*, especially *Morchella crassipes*, and *M. hortensis* grow on simple sugars such as glucose or sucrose (62,63). A process was developed for growing the mycelium of these organisms on a commercial scale in the United States for use as a food-flavoring ingredient rather than as a source of protein. This process is no longer practiced because of the relatively high production costs as compared with the costs of imported dried mushrooms.

Other mold-based SCP processes that have been investigated include utilization of sulfite waste liquor by *Paecilomyces varioti*, conversion of carob bean waste by *Aspergillus niger*, corn- and pea-processing wastes by *Giotrichium* sp., and coffee-processing wastes by *Trichoderma harzianum* (62). However, none of these processes is practiced commercially.

A product called Myco-protein, based on the continuous aerobic culture of *Fusarium graminearum* with glucose as the substrate, has been developed (74). The nitrogen source fed to the fermentor is gaseous ammonia, which also is used to control pH. Mineral salts required as nutrients are sterilized with the glucose substrate before feeding to the fermentor. The mycelial product is used to form textured protein meat analogues which are sold (ca 1993) on a test-market basis in the United Kingdom.

Dry wt yields of bacteria and yeasts grown on hydrocarbons and methanol are ca 1.0 and 0.5 g/g substrate utilized, respectively. For yeasts, molds, and higher fungi grown on carbohydrate substrates, dry wt yields are 0.5–0.6 g/g substrate utilized. Yeast cells are harvested readily by centrifugation. Molds and higher fungi grow in either pellet or filamentary forms. These organisms can be separated from the growth medium and dewatered by screens, filter processes, or basket centrifuges. It is very costly to separate bacteria from the growth medium by centrifugation because of their small (1–2 μm) size and densities similar to that of water. Bacterial cells can be concentrated by agglomeration or electrocoagulation prior to centrifugation. The resulting wastewater and residual substrates are purified and recycled, particularly in processes based on hydrocarbons, methanol, or ethanol (78).

The product quality considerations for nonphotosynthetic microorganisms are similar to those for algae. Tables 6 and 7 present composition and amino acid analyses, respectively, for selected bacteria, yeasts, molds, and higher fungi produced on a large pilot-plant or commercial scale. Table 8 summarizes results of protein quality and digestibility studies.

Most of the bacteria, yeasts, molds, and higher fungi of interest for SCP production are deficient in methionine and must be supplemented with this amino acid to be suitable for animal feeding or human food applications. Also, lysine–arginine ratios should be adjusted in poultry rations in which yeast SCP is used (62). Human feeding studies have shown that only limited quantities of yeast such as *Candida utilis* can be added to food products without adverse effects on flavor (63).

Nucleic acid contents of SCP products, which range up to 16% in bacteria and 6–11% in yeasts, must be reduced by processing so that intakes are less than 2 g/d to prevent kidney stone formation or gout. Adverse skin and gastrointestinal reactions have also been encountered as a result of human consumption of some SCP products (87).

The FDA regulations provide for the use of dried cells of the yeasts *S. cerevisiae*, *K. marxianus* var. *fragilis*, and *C. utilis* in foods. Folic acid contents must not exceed 0.04 mg/g (88). Functional concentrates and isolates can be prepared from dried microbial cells by disrupting or removing the cell walls using mechanical means or acid, alkaline, and enzyme hydrolysis; or removing the cell walls and reducing nucleic acid contents by chemical or enzymatic methods. Also, microbial proteins can be spun into fibers (63). Baker's yeast protein concentrate has been approved by the FDA for use as a functional food additive (88).

Derived Plant and Animal Products

Leaf Protein Concentrates. Leaf protein concentrates (LPC) are prepared by crushing plant material, extracting the juices, and either using the juice per se or recovering the protein from the juice by heating or chemical precipitation. Dehydrated alfalfa (lucerne) has a long history of use as a source of plant protein for animal feeds. The leaves of alfalfa and other crops are a source of protein that can be extracted to give a concentrated product having increased protein and decreased fiber contents suitable for animal feeding. Plants, other than alfalfa, considered as sources of LPC include pea vines, clover, field beans, mustard, kale, fodder radish, banana leaves, and aquatic plants (89,90). LPC production requires crops having rapid growth and high yields or protein during the growing season, ie, 1600 kg/hm^2 (1430 lbs/acre); absence of mucilaginous sap which makes it difficult to separate juice from the fiber; absence of acidic or high tannin saps which prevent extraction of protein into the juice because of precipitation in the pulp; and absence of toxic materials such as cyanogenic glycosides, glucosinolates, and

Table 6. Composition of Nonphotosynthetic Microorganisms Grown on Various Substrates, g/100 g[a]

Organism	Substrate	Nitrogen	Protein[b]	Fat	Crude fiber	Ash	Reference
		Bacteria					
Methylophilus methylotrophus[c]	methanol	13	83	7	<0.05	8.6	64
		Yeasts					
Candida (Saccharomycopsis) lipolytica	*n*-alkanes	10	65	8.1		6	66,67
Candida utilis	sulfite liquor	8.3–8.8	52–55	4.6	2.6	7.3	79
	ethanol	8.3	52	7	5	8	80
Kluyveromyces marxianus var. *fragilis*	cheese whey	7.2–8.8	45–55	2		6–10	72
	cheese whey permeate	7.6	47	4.6		20	73
Saccharomyces cerevisae	molasses	8.3–8.8	52–55	4.1–5.3		7.1–8.4	5
		Molds and higher fungi					
Fusarium graminearum	glucose	9.6	60	73		6	74
Morchella hortensis	glucose	5.4	34	1.4			62
Paecilomyces varioti[d]	sulfite waste liquor	9.1–10.11	57–63				75

[a]On a dry wt basis.
[b]Protein = % nitrogen × 6.25.
[c]Energy of approximately 12.6 kJ/g (3.0 kcal/g).
[d]Commonly known as Pekilo.

Table 7. Amino Acid Content of Nonphotosynthetic Microorganisms Grown on Various Substrates, g/16 g N[a]

Organism	Substrate	Ala	Arg	Asp	Cys	Glu	Gly	His	Ile	Leu	Lys	Met	Phe	Pro	Ser	Thr	Try	Tyr	Val	Reference
Bacteria																				
Lactobacillus bulgaricus 2217		9.0	4.5	10.5	0.4	9.1	3.5	2.2	4.5	6.1	9.3	2.2	3.2	3.6	2.6	4.3		3.3	5.8	81
Lactobacillus delbrueckii B443		7.5	4.6	14.6	0.5	12.4	4.3	2.0	5.5	7.6	9.6	1.9	4.2	3.3	2.8	4.4		3.3	5.9	81
Methylophilus (Pseudomonas) methylotrophus[b]	methanol	6.8	4.5	8.5	0.6	9.6	4.9	1.8	4.3	6.8	5.9	2.4	3.4	3.0	3.4	4.6	0.9	3.1	5.2	64
Yeasts																				
Candida (Saccharomycopsis) lipolytica[c]	n-alkanes	7.4	4.8	10.2	1.1	11.3	4.8	2.0	4.5	7.0	7.0	1.8	4.4	4.4	4.8	4.9	1.4	3.5	5.4	66
Candida utilis[d]	sulfite waste	5.8	5.4	9.2		15.6	3.6	1.2	3.8	7.6	4.8	1.1	8.6	6.0	5.0	5.4	2.4	6.2	3.8	5
	ethanol	5.5	5.4	8.8	0.4	14.6	4.5	2.1	4.5	7.1	6.6	1.4	4.1	3.4	4.7	5.5	1.2	3.3	5.7	80
Kluyveromyces marxianus var. *fragilis*	cheese whey							2.1	4.0	6.1	6.9	1.9	2.8			5.8	1.4	2.4	5.4	80
	cheese whey permeate	7.1	4.3	7.4	0.3	14.9	3.7	1.8	3.7	5.8	6.0	1.0	3.2	3.0	4.0	4.7	2.6	0.9	4.5	72
Saccharomyces cerevisiae	molasses		5.0		1.6		4.0		5.5	7.9	8.2	2.5	4.5			4.8	1.2	5.0	5.5	5
Molds																				
Fusarium graminearum[e]	glucose	6.3	5.4	8.4	0.5	11.9	4.3	2.1	3.5	4.6	6.1	1.6	4.5	4.5	4.5	4.3	1.8	3.1	4.9	74
Morchella hortensis		4.5	4.0	4.6	0.4	15.4	3.0	1.9	2.4	5.0	3.0	0.7	2.3	4.5	2.8	2.7	1.0	1.9	2.9	62
Paecilomyces varioti	spent sulfite liquor				1.1				4.3	6.9	6.4	1.5	3.7			4.6	1.2	3.4	5.1	75
FAO reference									4.2	4.8	4.2	2.2	2.8				1.4		4.2	82

[a]Dry wt basis. [b]Commonly called Pruteen. [c]Commonly called Toprina. [d]Commonly called Torula. [e]Commonly called Myco-protein.

Table 8. Protein Quality and Digestibility of Nonphotosynthetic Microorganisms

Organism[a]	Substrate	Animal	Microorganism in diet, %	Protein digestibility, %	PER[b]	BV,[c] %	Feed conversion ratio,[d] kg/kg wt gain	Reference
Bacteria								
Methylophilus methylotrophus	methanol	chicken	9.8				2.3 (2.33)	64
		pig	6.7				3.13 (3.34)	
Yeasts								
Candida lipolytica	n-alkanes	rat		96		61		78
		rat[e]		96		91		78
		chicken	10	88			2.58 (2.68)	78
		pig	7.5	92			3.04 (3.11)	78
		rat		85–88	0.9–1.4	32–48		83
Candida utilis	sulfite waste liquor	rat[f]		90	2.0–2.3	88		83
	ethanol	rat	8		2.10			84
Kluyveromyces marxianus var. *fragilis*	cheese whey	rat			2.26			71
Saccharomyces cerevisiae	molasses	rat			2.63			84
Molds								
Fusarium graminearum		rat			2.4			85
		rat[g]			3.4			85
		human				8.4		85
		pig	40				1.76	86

[a]Dried plus addition of indicated compound. [b]Protein efficiency ratio (PER). See Table 4. [c]Biological value (BV) = % of absorbed nitrogen retained in body tissue; complete retention = 100. [d]Data in parentheses for control group, with no single-cell protein in diet. [e]0.3% DL-methionine added to feed. [f]0.5% DL-methionine added to feed. [g]Methionine added to feed.

alkaloids that can be carried into the final product (91). In addition, from an economic standpoint, the entire plant must be utilizable for LPC.

LPC Processes. Process development for LPC production dates from the United Kingdom and Hungary from 1920–1940 (89,90). Table 9 presents some of the processing methods that are used or under development in the 1990s.

Various mechanical methods can be employed for rupturing leaf cells to prepare LPC (95). Leaf structural factors affecting protein release by mechanical processes include leaf weight, cell numbers, leaf thickness, intercellular space, and protein content as a function of leaf maturity. Dynamic compression is considered to be superior to shearing for commercial scale leaf rupturing processes. Other studies have shown that screw expellers should be modified to provide angled paddles for disintegrating leaves before they are passed into the pressing section (89). Heating and drying conditions used during the processing of LPC must be controlled carefully to minimize nonenzymatic browning, Maillard reactions, and reactions between the proteins and unsaturated fatty acids.

Solvent extraction removes chlorophyll and other pigments to give a light-colored product but increases processing costs. Furthermore, solvent extraction removes β-carotene and reduces vitamin A activity (89) (see TERPENOIDS; VITAMINS). Supercritical CO_2 extraction at 30 and 70 MPa (4,350 and 10,150 psi) and 40°C removed 90 and 70% carotene and lutein, respectively, from alfalfa LPC (96). This process avoids organic solvent residues and recovers valuable by-products.

Leaf materials also contain lipoxidases and highly unsaturated lipids. LPC process conditions should inactivate lipoxidases to obtain a stable product (92,97).

The USDA Western Regional Research Center has developed an improvement to the Pro-Xan process (Table 9) for preparing a bland, colorless LPC product from alfalfa suitable for human consumption (92,93,98). Aqueous sodium metabisulfite [7681-57-4], $Na_2S_2O_3$, is added to alfalfa prior to expressing the juice to lighten the color of the LPC and protect cystine and methionine from oxidation.

Table 9. Selected Processes for Leaf Protein Concentrate Production, 1993

Process	Description	Reference
Rothamsted	pulping in ribbed rollers; precipitating protein at 80°C or at pH 4.0; drying in air below 80°C	90
Pro-Xan	chopping; ammoniation to pH 8.5; roll or twin-screw pressing; sieve purification; coagulation with steam at 85°C; dewatering; drying	92,93
Vepex	mechanical disintegration; multistage pressing; add-back of liquor to press cake; coagulation at 82°C with addition of flocculents; centrifugation; evaporation; drying	94
Instituto di Industrie Agarie[a]	chopping and screw pressing; centrifugation; coagulation at pH 8.5; treatment with polyelectrolyte; centrifugation; precipitation at pH 4.0; centrifugation; drying	94

[a]Pisa, Italy.

The juice, after expressing, is heated by steam injection to 60-65°C for 10–20 s and cooled to 45°C. The proteins associated with the chloroplasts are then removed by centrifugation. The suspended solids are removed from the liquid phase in the centrifuge by a plate-and-frame filter press, and the proteins in the filtrate are precipitated at 80°C for 2–4 min. The protein precipitate is removed by centrifugation, and the protein is washed at pH 4–5 and spray dried to give a white LPC having potential food applications. The chloroplast protein fraction is adjusted to pH 8.5, and heated to 90–95°C to coagulate the protein. The protein is dewatered by centrifugation, granulated, and dried in a rotary dryer to give the feed-grade Pro-Xan II LPC (93,98).

The Vepex process developed in Hungary (Table 9) involves disintegration of plant materials followed by double screw pressing to maximize juice production. Green chloroplastic protein is removed by direct steam-injection heat treatment at 82°C with the addition of flocculents and centrifugation. The white protein fraction is separated from the chlorophyll-free process juice by direct steam injection at 80°C, followed by centrifugation and drying (94).

LPC Product Quality. Table 10 gives approximate analyses of several LPC products. Amino acid analyses of LPC products have been published including those from alfalfa, wheat leaf, barley, and lupin (101); soybean, sugar beet, and tobacco (102); Pro-Xan LPC products (100,103); and for a variety of other crop plants (104,105). The composition of LPCs varies widely depending on the raw materials and processes used. Amino acid profiles are generally satisfactory except for low sulfur amino acid contents, ie, cystine and methionine.

Enzyme degradation of leaf protein may occur during crushing and separation from the fiber. The amino acids produced by this enzyme action are soluble in the juice and may be lost unless all of the juice is recovered.

Table 11 presents data on the protein quality of a variety of LPC products obtained from rat-feeding studies. Typical protein efficiency ratio (PER) values for LPCs derived from alfalfa range from 1.41 without supplementation to 2.57 with 0.4% methionine added; casein can be adjusted to a PER of 2.50 (98,100). Biological values (BV) of mixtures of LPCs, such as barley and rye grass or soybean and alfalfa, may be higher than either LPC alone. The effect has been attributed to the enhanced biological availability of lysine in these mixtures (99).

Table 10. Analysis of Leaf Protein Products, wt %[a]

Leaf protein	Source	Protein[b]	Fat	Crude fiber	Ash	Reference
white LPC[c]	alfalfa	88.7	0.6	1.0	0.4	93
white LPC	quinoa	93.2	0.7	0.9	2.0	99
Pro-Xan[d]	alfalfa	61.9	8.9	1.7	11.1	93
Brassica napus[e]		58.5	15.6	2.0	10.0	100

[a]Dry wt basis.
[b]Protein = % nitrogen × 6.25.
[c]Nitrogen-free extract = 9.3%, and soluble solids = 0.3%.
[d]Whole leaf protein concentrate, 16.5% nitrogen-free extract and 7.9% soluble solids.
[e]Commonly known as Late Korean rape.

Table 11. Nutritive Value of Leaf Protein Concentrates and Other Protein Products[a]

Product	True digestibility,%	PER[b]	BV[c]	NPU[d]	Reference
	Leaf protein				
alfalfa Pro-Xan	77.9	1.41			100
alfalfa white LPC	91		59	54	99
Brassica carinata[e]	84	2.12			100
Brassica napus	72	1.80			100
barley	88		63	55	106
barley plus ryegrass LPC	81		73	59	106
soybean	90		55	49	99
soybean plus alfalfa white LPC	90		62	56	99
	Other protein				
fish meal	88		54	48	99
skim milk	96		52	50	99

[a]Rat-feeding studies. LPC = leaf protein concentrate.
[b]Protein efficiency ratio (PER) = weight gain (g) for a 10% protein level in the diet of rats as compared to the standard of 2.5 for casein.
[c]Biological value (BV) = % of absorbed nitrogen retained in body tissue; complete retention = 100.
[d]Net protein utilization (NPU) = % digestibility × BV; complete utilization = 100.
[e]Commonly known as Ethiopian mustard.

Human-feeding studies on LPC have been conducted in Jamaica. Diets for malnourished infants contained half of the nitrogen as LPC, and nitrogen retention was equivalent to that obtained with milk (107). An LPC derived from alfalfa has been shown to alleviate the symptoms of the protein deficiency disease Kwashiorkor (108). In India, LPC added to low protein diets, ie, 12.6 g/d protein, to provide an intake of 45–48.8 g/d protein resulted in improved nitrogen retention and digestibility in 10–12 yr-old children (107). Allergic reactions including facial edema were common in 1 out of 8 children. There are definite limits to the use of LPC products in human diets, and raw materials used for LPC production must be evaluated for possible allergenic problems. There is a lack of published information on the evaluation of the nutritional value of white LPCs in human diets.

Functional Properties of LPCs. LPC products prepared from *Brassica* sp., soybeans, sugar beets, and tobacco have been investigated for their functional properties including nitrogen solubility, fat- and water-binding capacity, emulsification, gelation, and foaming capacity and stability (102,109). The emulsification properties and foam stability of alfalfa LPC indicate potential applications in salad dressings and as a substitute for egg whites in baking (109).

Vegetable oil gels can be formed from a heated emulsion of alfalfa LPC and peanut oil (110). However, the flavor and texture of these gels are not generally acceptable. In general, whole green LPCs suffer from undesirable sensory properties. White LPCs, from which chlorophyll, phenolic compounds, and flavonoid pigments are removed, have desirable sensory properties and can be prepared by the process described previously. However, no economically viable processes have been developed for food-grade white LPC production.

The 1993 market for LPC-type products in the United States was for dried alfalfa meal for animal feed. This product is sold for both protein and carotenoid content. The USDA Pro-Xan product attempts to obtain improved xanthophyll contents for use in egg-laying rations in addition to protein contents. The limitations to commercial development of LPC products for human food use are high capital costs as compared with the low yields of protein, seasonal availability of raw materials, and the need in the United States for FDA approval of the products.

Seed-Meal Concentrates and Isolates.   Seed-meal protein products include flours, concentrates, and isolates, particularly soy protein products. These can be used as extenders for meat, seafood, poultry, eggs, or cheese (see SOYBEANS AND OTHER OILSEEDS). Detailed information on soybean and other seed-meal production processes is available (13,14,18).

Soybean concentrate production involves the removal of soluble carbohydrates, peptides, phytates, ash, and substances contributing undesirable flavors from defatted flakes after solvent extraction of the oil. Typical concentrate production processes include moist heat treatment to insolubilize proteins, followed by aqueous extraction of soluble constituents; aqueous alcohol extraction; and dilute aqueous acid extraction at pH 4.5.

Commercial soy protein concentrates typically contain 70 to 72% crude protein, ie, nitrogen $\times$ 6.25, dry wt basis. Soy protein isolates are prepared from desolventized, defatted flakes. A three-stage aqueous countercurrent extraction at pH 8.5 is used to disperse proteins and dissolve water-soluble constituents. Centrifugation then removes the extracted flakes, and the protein is precipitated from the aqueous phase by acidifying with HCl at pH 4.5. The protein precipitate is washed with water, redispersed at pH 7, and then spray dried. Typical commercial soy protein isolates contain greater than 90% crude protein, dry wt basis.

A modification of the conventional soy protein isolate process has been investigated on a small pilot-plant scale. It is based on the absorption of water from the aqueous protein after extraction at pH 8.5 using temperature-sensitive polyisopyropylacrylamide gels, followed by spray drying to give a 96% protein isolate (111).

Soy protein concentrates and isolates can be formed into fibrous structures by twin-screw extrusion texturization processes. The functional characteristics of these structures are influenced by pH adjustment during processing (112). Soybean protein also can be formed into fibers by forming a spinning dope from a slurry at pH 10–11, which is then aged at 40–50°C. This slurry is forced through a spinneret into an acid-coagulating bath. The fibers are heated to reduce the diameter to about 75 μm. The fibers can be formed with binders (eg, egg albumin), colored, and flavored to give the desired product characteristics (113).

Products prepared from soy protein products and resembling chicken, ham, frankfurters, and bacon are available commercially. Soy protein isolates are used in place of milk proteins or sodium caseinate in products such as coffee creamers, whipped toppings, yogurt, and infant formulas (see DAIRY SUBSTITUTES). Soy protein products also are used in snacks and in baked foods.

Hydration; water, fat, and flavor binding; gelation; emulsifying; foaming; and whipping characteristics vary among different soy protein products and complete substitution of animal proteins by these products is not always possible (114).

Soy protein products may impart a beany flavor to foods when used at levels greater than 20%. Undesirable components are present in the beans prior to processing and also may be generated during processing. Off-flavors and odors also may arise from oxidation of lipid components and from degradation of phenolic compounds during thermal processing of foods containing soy proteins. These undesirable flavor components can be diminished, but not completely eliminated, by extraction with alcohols such as methanol, ethanol, or isopropanol, or with an azeotropic mixture of hexane and an alcohol (115).

Vegetable proteins other than that from soy have potential applicability in food products. Functional characteristics of vegetable protein products are important factors in determining their uses in food products. Concentrates or isolates of proteins from cotton (qv) seed (116), peanuts (117), rape seed (canola) (118,119), sunflower (120), safflower (121), oats (122), lupin (123), okra (124), and corn germ (125,126) have been evaluated for functional characteristics, and for utility in protein components of baked products (127), meat products (128), and milk-type beverages (129) (see DAIRY SUBSTITUTES).

Functional properties of canola protein products can be improved by succinylation (130,131). Controlled acetylation can reduce undesirable phenolic constituents as well (132). However, antinutrients in canola and other vegetable protein products such as glucosinolates, phytic acid, and phenolic compounds have severely limited food applications of these products.

Hydrolyzed Vegetable Protein. To modify functional properties, vegetable proteins such as those derived from soybean and other oil seeds can be hydrolyzed by acids or enzymes to yield hydrolyzed vegetable proteins (HVP). Hydrolysis of peptide bonds by acids or proteoyltic enzymes yields lower molecular weight products useful as food flavorings. However, the protein functionalities of these hydrolysates may be reduced over those of untreated protein.

Deamidation of soy and other seed meal proteins by hydrolysis of the amide bond, and minimization of the hydrolysis of peptide bonds, improves functional properties of these products. For example, treatment of soy protein with dilute (0.05 N) HCl, with or without a cation-exchange resin (Dowex 50) as a catalyst (133), with anions such as bicarbonate, phosphate, or chloride at pH 8.0 (134), or with peptide glutaminase at pH 7.0 (135), improved solubility, whipability, water binding, and emulsifying properties.

HVP products prepared by hydrolysis with HCl contain varying amounts of glycerol chlorohydrins, such as 3-chloro-1,2-propanediol [96-24-2] and 1,3-dichloro-2-propanol [96-23-1], depending on reaction conditions and lipid contents of the starting material (135). As a result of their toxicities, regulating agencies in many countries have restricted the contents of these compounds in food.

Under FDA regulations, HVP products are permitted as optional ingredients in standardized canned foods such as pears, mushrooms, and tuna, and as a flavoring ingredient in nonstandardized foods (137). The U.S. Department of Agriculture has cleared HVP as a flavoring ingredient in various meat products (138).

Fish Protein Concentrates and Isolates. Fish protein concentrates (FPC) and isolates (FPI) are produced for human food use from whole edible species of fish using sanitary processing methods; fish meal and fish solubles are produced for animal feed. FPC raw materials include whole hake, hake-like fish, and herring of the genera *Clupea*, and menhaden and anchovy of the species *Engraulis mordax* without removal of heads, fins, tails, or intestinal contents. FPI raw materials include edible portions of fish body generally recognized as safe for human consumption after removal of heads, fins, tails, bones, scales, viscera, and intestinal contents (139). In the United States, FDA regulations describe the production processes for preparing FPC and FPI (139). The FDA regulations also specify that FPC and FPI contain minimum protein contents of 75 and 90%, respectively, a maximum fat content of 0.5%, and a maximum moisture content of 10% by weight. FPC must be free of *Escherichia coli*, *Salmonella*, and other food pathogens and have a total bacterial plate count of not more than 10,000 per gram.

Amino acid profiles of FPC are excellent and compare favorably with whole egg except for tryptophan and lysine (140). Hake and Atlantic FPCs prepared by isopropanol extraction have PERs of 3.29 and 3.05, respectively, as compared with 3.0 for casein (140). Numerous human feeding studies have been conducted with FPC. The results indicate that high quality, bland FPC products can be used as protein supplements but they are not suitable for use as a sole source of protein.

Fish protein concentrates vary widely in functional characteristics, ranging from those having high protein content and low water solubility, to those having lower protein contents but improved water solubility. Attempts have been made to improve functional properties of fish protein by enzyme hydrolysis (141), or by modification of the myofibrillar protein by succinylation (142).

Economic conditions in the United States have not favored the production of FPC and FPI having desirable functional and nutritional characteristics at prices competitive with those of conventional protein sources.

Textured and Structured Fishery Products. Numerous seafood analogue products, eg, crab, shrimp, and lobster analogues, have been prepared by modifying the structural and textural properties of fish proteins. Surimi, originally developed in Japan, is prepared from mechanically deboned fish muscle, such as Alaska pollock (*Theragara chalcogramma*), by freshwater leaching to yield a light-colored, bland, refined protein that can be used as a matrix for seafood analogues.

A typical process for manufacturing surimi-based seafood analogue products involves (143) mincing Alaska pollock; washing it at 10°C with water having pH 6.5–7.0 and low Ca^{2+}, Mg^{2+}, Fe^{2+}, and Mn^{2+} contents; rinsing, draining, and screw-press dewatering to 82% moisture; adding sucrose and sorbitol (91:1) to a final concentration of 9% as cryoprotectants; fabricating with starch, egg white, lactalbumin, and fat or oil to give the desired texture; incorporating flavoring ingredients; and cooking.

In addition to sucrose and sorbitol, polydextrose can be used as a cryoprotectant (144) (see SUGAR ALCOHOLS). Also, the type of starch used, ungelatinized or pregelatinized, affects the extent of water binding by surimi gels during mixing and cooking (145). More detailed information on the technologies for manufacturing seafood analogues is available (15,16).

A number of investigations have been directed toward improving the functional characteristics of fish proteins by enzymatic hydrolysis. Treating comminuted and defatted sardines (*Sardina pilchardus*) with subtilisin (Alcalase), to give a 5% degree of hydrolysis, solubilized the proteins and gave a product having improved emulsifying properties over those of sodium caseinate (146).

Eviscerated and ground mullet (*Mugil cephalus*) was hydrolyzed with bacterial alkaline proteases without adding water, followed by centrifugation to remove 80% of the liquid, and drying (147). A high degree of protein solubilization (70–80%) was achieved in the final product, which contained 83 to 86% protein. However, rat-feeding studies indicated that the hydrolysates had about 10 to 15% lower nutritional value than that measured by PER and feed efficiency values, ie, feed consumed/weight gain.

Synthetic Protein Products

Plastein Synthesis. Plasteins are mixtures of high molecular weight proteinaceous peptides. They are synthesized by enzyme-catalyzed growth of peptide chains from lower molecular weight peptides. The process by which plasteins are formed is called the plastein reaction and is the reverse of the proteolytic enzyme hydrolysis of peptide bonds of proteins (148). Japanese investigators have conducted extensive studies on the utility of the plastein reaction and of plasteins in food technology. The enzymatic modification of proteins from such products as soybeans, codfish, algae (*Chlorella*), wheat, milk, and baker's and hydrocarbon-grown yeasts followed by plastein synthesis have been investigated (149,150). The plasteins prepared were bland and did not have objectionable tastes, odors, or colors.

Plasteins are formed from soy protein hydrolysates with a variety of microbial proteases (149). Preferred conditions for hydrolysis and synthesis are obtained with an enzyme-to-substrate ratio of 1:100, and a temperature of 37°C for 24–72 h. A substrate concentration of 30 wt %, 80% hydrolyzed, gives an 80% net yield of plastein from the synthesis reaction. However, these results are based on a 1% protein solution used in the hydrolysis step; this would be too low for an economical process (see MICROBIAL TRANSFORMATIONS).

Fish protein concentrate and soy protein concentrate have been used to prepare a low phenylalanine, high tyrosine peptide for use with phenylketonuria patients (150). The process includes pepsin hydrolysis at pH 1.5; pronase hydrolysis at pH 6.5 to liberate aromatic amino acids; gel filtration on Sephadex G-15 to remove aromatic amino acids; incubation with papain and ethyl esters of L-tyrosine and L-tryptophan, ie, plastein synthesis; and ultrafiltration (qv). The plastein has a bland taste and odor and does not contain free amino acids. Yields of 69.3 and 60.9% from FPC and soy protein concentrate, respectively, have been attained.

A pepsin hydrolysate of flounder fish protein isolate has been used as the substrate (40% w/v) for plastein synthesis, using either pepsin at pH 5 or alpha chymotrypsin at pH 7, with an enzyme–substrate ratio of 1:100 w/v at 37°C for 24 h (151). The plastein yields for pepsin and alpha chymotrypsin after precipitation with ethanol were 46 and 40.5%, respectively.

Fish silage prepared by autolysis of rainbow trout viscera waste was investigated as a substrate for the plastein reaction using pepsin (pH 5.0), papain (pH 6–7), and chymotrypsin (pH 8.0) at 37°C for 24 h (152). Precipitation with ethanol was the preferred recovery method. Concentration of the protein hydrolysate by open-pan evaporation at 60°C gave equivalent yields and color of the final plastein to those of the freeze-dried hydrolysate.

The sulfur amino acid content of soy protein can be enhanced by preparing plasteins from soy protein hydrolysate and sources of methionine or cystine, such as ovalbumin hydrolysate (plastein AB), wool keratin hydrolysate (plastein AC), or L-methionine ethyl ester [3082-77-7] (alkali saponified plastein) (153). Typical PER values for a 1:2 mixture of plastein AC and soybean, and a 1:3 mixture of alkali-saponified plastein and soybean protein, were 2.86 and 3.38, respectively, as compared with 1.28 for the soy protein hydrolysate and 2.40 for casein.

Plasteins are still in the experimental stage of development. Further work is needed on the scale-up of processing conditions for plastein synthesis which would lead to commercially useful products and on the functional utility of plasteins as ingredients in foods.

Synthetic Proteins.   Protein-like polypeptides can be synthesized chemically from ammonia, water, and carbon dioxide, or from mixtures of amino acids which are now manufactured by chemical or microbiological synthesis (154). Polyamino acids can be produced in the laboratory by simultaneous polymerization of mixtures of amino acids at 180°C for 3–6 h under dry conditions. Protenoids, containing all of the common amino acids in peptide linkages, can be obtained if sufficient amounts of aspartic and glutamic acids are included in the reaction mixture. Apparently these protenoids are digestible by mammalian proteinases and can serve as sources of nutrients for the bacteria *Lactobacillus arabinosus* and *Proteus vulgaris*. The possibility exists for protenoids to be nutritionally imbalanced, have mammalian toxicity, and undesirable tastes, odors, and stability. These problems must be investigated further before any assessment of the utility of polyamino acids or protenoids can be made. No progress has been made in the early 1990s on developing useful food protein ingredients by purely chemical synthetic methods.

Product Quality and Safety

The Protein Advisory Group, ad hoc, is the working group of the WHO United Nations system involving WHO, FAO, and the United Nations International Children's Emergency Fund (UNICEF). It has developed guidelines for the evaluation of novel sources of protein, eg, single-cell protein; clinical testing of novel sources of protein; human testing of supplementary food mixtures; and nutritional and safety aspects of novel protein sources for animal feed (155).

In general, nonconventional protein foods must be competitive with conventional plant and animal protein sources on the bases of cost delivered to the consumer, nutritional value to humans or animals, functional value in foods, sensory quality, and social and cultural acceptability. Also, requirements of regulatory agencies in different countries for freedom from toxins or toxic residues in single-cell protein products, toxic glycosides in leaf protein products, pathogenic microorganisms, heavy metals and toxins in fish protein concentrates, or inhibitory or toxic peptide components in synthetic peptides must be met before new nonconventional food or feed protein products can be marketed.

In the United States, novel food ingredients or food ingredients produced by novel processes must be cleared by the FDA. In the case of meat and poultry, novel ingredients must also be cleared by the U.S. Department of Agriculture's Food Safety and Inspection Service (FSIS).

BIBLIOGRAPHY

"Proteins from Petroleum" in *ECT* 2nd ed., Suppl. Vol., pp. 836–854, by E. R. Elzinga and A. I. Laskin, Esso Research and Engineering Co.; "Foods, Nonconventional," in *ECT* 3rd ed., Vol. 11, pp. 184–207, by J. H. Litchfield, Battelle, Columbus Laboratories.

1. I. Goldberg, *Single Cell Protein*, Springer-Verlag, Berlin, 1985.
2. J. H. Litchfield in J. L. Marx, ed., *A Revolution in Biotechnology*, Cambridge University Press, Cambridge, U.K., 1989, pp. 71–81.
3. D. Sharp, *Bioprotein Manufacture: A Critical Assessment*, Ellis Horwood, Chichester, U.K., 1989, p. 140.
4. I. Goldberg and R. Williams, eds., *Biotechnology and Food Ingredients*, Van Nostrand Reinhold Co., Inc., New York, 1991.
5. G. Reed and T. W. Nagodawithana, *Yeast Technology*, 2nd ed., Van Nostrand Reinhold Co., Inc., New York, 1991.
6. A. Halasz and R. Lasztity, *Use of Yeast Biomass in Food Production*, CRC Press, Boca Raton, Fla., 1990.
7. M. Moo-Young and K. F. Gregory, eds., *Microbial Biomass Proteins*, Elsevier Applied Science Publishers, Ltd., London, 1986.
8. A. Richmond, *CRC Handbook of Microalgae Mass Culture*, CRC Press, Boca Raton, Fla., 1986.
9. G. Shelef and C. J. Soeder, eds., *Algae Biomass Production and Use*, Elsevier, Amsterdam, the Netherlands, 1980.
10. T. Stadler, J. Mollion, M.-C. Verdus, Y. Karamanos, H. Morvan, and D. Christiaen, eds., *Algal Biotechnology*, Elsevier Applied Science Publishers, Ltd., London, 1988.
11. M. A. Borowitzka and L. J. Borowitzka, *Micro-Algal Biotechnology*, Cambridge University Press, Cambridge, U.K., 1988.
12. L. Telek and H. D. Graham, eds., *Leaf Protein Concentrates*, Avi Publishing Co., Westport, Conn., 1983.
13. D. K. Salunkhe, J. V. Chavan, R. N. Adsule, and S. S. Kadam, eds., *World Oilseeds: Chemistry, Technology and Utilization*, Van Nostrand Reinhold Co., Inc., New York, 1992.
14. F. H. Steinke, D. H. Waggle, and M. N. Volgarev, eds., *New Protein Foods in Human Health*, CRC Press, Boca Raton, Fla., 1992.
15. R. E. Martin and R. L. Cullette, eds., *Engineered Seafood Including Surimi*, Noyes Data Corp., Park Ridge, N.J., 1990.
16. T. C. Lanier and C. M. Lee, eds., *Surimi Technology*, Marcell Dekker, Inc., New York, 1992.
17. A. M. Pearson and T. R. Dutson, eds., *Edible Meat By-Products: Advances in Meat Research*, Vol. 5, Elsevier Science Publishing Co., Inc., New York, 1989.
18. J. E. Kinsella and W. G. Soucie, eds., *Food Proteins*, The American Oil Chemists Society, Champaign, Ill., 1990.
19. F. V. Kosikowski, *Cheese and Fermented Milk Foods*, 2nd ed., F. V. Kosikowski and Associates, Brooktondale, N.Y., 1982.
20. J. H. Litchfield, *Science* **219**, 740 (1983).
21. J. H. Litchfield, *Chem. Tech.* **8**, 218 (1978).

22. S. C. Prescott and C. G. Dunn, *Industrial Microbiology*, 3rd ed., McGraw-Hill Book Co., Inc., New York, 1959, Chapt. 3.
23. C. J. Soeder in Ref. 9, pp. 9–19.
24. A. Vonshak and A. Richmond, *Biomass*, **15**, 233 (1988).
25. J. R. Benemann, *Develop. Ind. Microbiol.* **31**, 247 (1990).
26. R. A. Kay, *Crit. Rev. Food Sci. Nutr.* **30**, 555 (1991).
27. A. Klausner, *Bio/Technology* **4**, 947 (1986).
28. D. D. Duxbury, *Food Proc.* **50**(12), 50 (1989).
29. M. Javanmardian and B. O. Palsson, *Biotechnol. Bioeng.* **38**, 1182 (1991).
30. M. Oguchi, K. Otsubo, K. Nitta, A. Shimada, S. Fujil, T. Koyano, and K. Miki in R. D. MacElroy, B. G. Thompson, T. W. Tibbitts, and T. Volk, eds., *Controlled Ecological Life Support Systems, Natural and Artificial Ecosystems*, NASA Conf. Pub. 10040, National Aeronautics and Space Administration, Ames Research Center, Moffett Field, Calif., 1989, pp. 165–173.
31. C. J. Soeder and E. Hegewald in Ref. 10, p. 68.
32. F. Comacho Rubio, M. E. Martinez Sancho, S. Sanchez Villasclaras, and A. Delgado Perez, *Process Biochem.*, **24**(4), 133 (1989).
33. P. Soong in Ref. 9, pp. 97–113.
34. K. Kawaguchi in Ref. 9, pp. 25–33.
35. Y.-K. Lee and C.-S. Low, *Biotechnol. Bioeng.* **38**, 995 (1991).
36. L. V. Venkataraman, B. P. Nigam, and P. K. Ramanathan in Ref. 9, pp. 81–95.
37. H. Durand-Chastel in Ref. 9, pp. 51–64.
38. H. Durand-Chastel and G. Clement in A. Chavez, H. Bourges, and S. Basta, eds., *Proceedings of the 9th International Congress on Nutrition*, Vol. 3, S. Karger, Basel, Switzerland, 1975, pp. 84–90.
39. A. Vonshak, A. Abeliovich, S. Boussiba, S. Arad, and A. Richmond, *Biomass* **2**, 175 (1982).
40. M. Kobayashi and S. I. Kutara, *Process Biochem.* **13**(9), 27 (1978).
41. M. Kobayashi, M. Kubayashi, and H. Nakanishi, *J. Ferment. Technol.* **49**, 817 (1971).
42. K. Driessens, L. Liessens, S. Masduki, W. Verstraele, H. Nelis, and A. de Leeheer, *Process Biochem.* **22**(6), 160 (1987).
43. R. H. Shipman, I. C. Kao, and L. T. Fan, *Biotechnol. Bioeng.* **17**, 1561 (1975).
44. F. H. Mohn in Ref. 9, pp. 547–571.
45. J. Benemann, B. Koopman, J. Weissman, D. Eisenberg, and R. Goebel in Ref. 9, pp. 457–495.
46. J. de la Noue and N. de Pauw, *Biotech. Adv.* **6**, 725 (1988).
47. *Earthrise Newsletter No. 12*. Earthrise Company, San Rafael, Calif., 1988, 2 pp.
48. E. F. McFarren, M. L. Schafer, J. E. Campbell, K. H. Lewis, E. T. Jensen, and E. J. Schantz, *Adv. Food Res.* **10**, 135 (1960).
49. G. Clement in S. R. Tannenbaum and D. I. C. Wang, eds., *Single Cell Protein II*, MIT Press, Cambridge, Mass., 1975, pp. 467–474.
50. J. H. Litchfield, *Bioscience* **30**, 387 (1980).
51. P. Tapie and A. Bernard, *Biotechol. Bioeng.* **32**, 873 (1988).
52. O. P. Walz and H. Brune in Ref. 9, pp. 733–744.
53. M. Anusuya Devi and L. V. Venkataraman, *J. Food Sci.* **49**, 24 (1984).
54. *Food Nutrition Meeting Report*, Ser. No. 52, Food and Agricultural Organizational, World Health Organization, Rome, 1973.
55. J. A. Lubitz, *J. Food Sci.* **28**, 229 (1963).
56. B. Lipstein and S. Hurwitz in Ref. 9, pp. 667–685.
57. D. L. R. Narasimha, G. S. Venkataraman, S. K. Duggal, and B. O. Eggum, *J. Sci. Food Agri.* **33**, 456 (1982).
58. C. I. Waslien, *Crit. Rev. Food Sci. Nutr.* **6**, 77 (1975).

59. G. A. Leveille, H. E. Sauberlich, and J. Shockley, *J. Nutr.* **76**, 423 (1962).
60. S. K. Lee, H. M. Fox, C. Kies, and R. Dam, *J. Nutr.* **92**, 281 (1967).
61. H. D. Payer, W. Pabst, and K. H. Runkel in Ref. 9, pp. 785–797.
62. J. H. Litchfield in H. J. Peppler and D. Perlman, eds., *Microbial Technology*, 2nd ed., Vol. 2, Academic Press, Inc., New York, 1979, pp. 93–155.
63. J. H. Litchfield in Ref. 4, pp. 65–109.
64. J. S. Gow, J. D. Littlehailes, S. R. L. Smith, and R. B. Walter in S. R. Tannenbaum and D. I. C. Wang, eds., *Single Cell Protein II*, MIT Press, Cambridge, Mass., 1975, pp. 370–384.
65. D. G. MacLennan, J. S. Gow, and D. A. Stringer, *Process Biochem.* **8**(6), 22 (1973).
66. G. H. Evans in R. I. Mateles and S. R. Tannenbaum, *Single Cell Protein*, MIT Press, Cambridge, Mass., 1975, pp. 243–254.
67. U.S. Pat. 3,846,238 (Nov. 5, 1974), G. H. Evans and J. L. Shennan (to British Petroleum Co.).
68. U.S. Pat. 3,865,691 (Feb. 11, 1975), J. A. Ridgeway, T. A. Lappin, B. M. Benjamin, J. B. Corns, and C. Akin (to Standard Oil Co. of Indiana).
69. N. H. Mermelstein, *Food Technol.* **43**(7), 50 (1989).
70. D. O. Hitzman in Ref. 7, pp. 27–32.
71. S. Bernstein, G. H. Tzeng, and D. Sisson in A. E. Humphrey and E. L. Gaden, Jr., eds., *Single-Cell Protein from Renewable and Nonrenewable Resources*, John Wiley & Sons, Inc., New York, 1977, pp. 35–44.
72. S. Bernstein and P. E. Plantz, *Food Eng.* **48**(11), 74 (1977).
73. L. K. Shay and G. H. Wagner, *J. Dairy Sci.* **69**, 676 (1986).
74. G. L. Solomons in H. W. Blanch, S. Drew, and D. I. C. Wang, eds., *Comprehensive Biotechnology*, Vol. 3, Pergamon Press, Oxford, U.K., 1985, pp. 483–505.
75. H. Romantschuk in Ref. 64, pp. 344–355.
76. H. Romantschuk and M. Lehtomaki, *Process Biochem.* **13**(3), 16, 17, 23 (1978).
77. L. K. Shay and G. H. Wegner, *Food Technol.* **38**(10), 61 (1985).
78. C. A. Shacklady and E. Gatumel in H. Gounelle de Pontanel, ed., *Proteins from Hydrocarbons*, Academic Press, New York, 1973, pp. 27–52.
79. *Type B Torula Dried Yeast*, Lake States Yeast, Rinelander, Wis., 1986.
80. *Torutein Product Bulletin*, Amoco Foods Co., Chicago, Ill., 1974.
81. M. S. Erdman, W. G. Bergen, and C. A. Reddy. *Appl. Environ. Microbiol.* **33**, 901 (1977).
82. J. H. Litchfield in Ref. 74, pp. 463–481.
83. R. Bressani in Ref. 65, pp. 90–121.
84. G. Sarwar, R. W. Peace, and H. G. Botting in Ref. 7, pp. 107–116.
85. G. L. Solomons in Ref. 7, pp. 19–26.
86. I. Duthie in Ref. 64, pp. 505–544.
87. N. S. Scrimshaw in C. W. Robinson and J. A. Howell, eds., *Comprehensive Biotechnology*, Vol. 4. Pergamon Press, Oxford, U.K., 1985, pp. 673–684.
88. *CFR Title 21, Food and Drugs 172.325, 172.896*, U. S. Government Printing Office, Washington, D.C., 1992.
89. N. W. Pirie in Ref. 12, pp. 1–9.
90. N. W. Pirie, ed., *Leaf Protein: Its Agronomy, Preparation, Quality, and Use*, IBP Handbook No. 20, Blackwell Scientific Publications, Oxford, U.K., 1971, pp. 53–62, 86–91, 155–163.
91. L. Telek in Ref. 12, pp. 295–395.
92. G. O. Kohler and E. M. Bickoff, in Ref. 90, pp. 69–77.
93. G. O. Kohler, R. H. Edwards, and D. E. de Fremery in Ref. 12, pp. 508–524.
94. L. Koch in Ref. 12, pp. 601–632.
95. T. O. Addy, L. F. Whitney, and C. S. Chen in Ref. 12, pp. 490–507.

96. F. Favati, J. W. King, J. P. Friedrich, and K. Eskins, *J. Food Sci.* **53**, 1532 (1988).
97. S. Nagy and H. E. Nordby in Ref. 12, pp. 268–294.
98. G. O. Kohler, S. G. Wildman, N. A. Jorgenson, R. V. Enochian, and W. J. Bray in M. Milner, N. S. Scrimshaw, and D. I. C. Wang, eds., *Protein Resources and Technology: Status and Research Needs*, Avi Publishing Co., Westport, Conn., 1978, pp. 543–569.
99. R. Carlsson and P. Hanczakowski, *J. Sci. Food Agr.* **36**, 946 (1985).
100. C. K. Lyon, P. F. Knowles, and G. O. Kohler, *J. Sci. Food Agr.* **34**, 849 (1983).
101. M. Byers in Ref. 12, pp. 135–175.
102. S. J. Sheen, *J. Agric. Food Chem.* **39**, 681, (1991).
103. R. V. Enochian, G. O. Kohler, R. H. Edwards, D. D. Kuzmicky, and C. J. Vosloh, Jr. in Ref. 12, pp. 525–545.
104. C. Savangikar and M. Ohshima, *J. Agric. Food Chem.* **35**, 82 (1987).
105. K. Sunder Rao, R. Dominic, K. Singh, C. Kaluwin, D. E. Rivett, and G. P. Jones, *J. Agric. Food Chem.* **38**, 2137 (1990).
106. J. Maciejewicz-Rys and P. Hanczakowski, *J. Sci. Food Agr.* **50**, 99 (1990).
107. N. Singh, in Ref. 91, pp. 131–134.
108. R. P. Devedas and N. K. Murthy, *World Rev. Nutr. Diet* **31**, 159 (1978).
109. B. E. Knuckles and G. O. Kohler, *J. Agr. Food Chem.* **30**, 748 (1982).
110. W. E. Barbeau and J. E. Kinsella, *J. Food Sci.* **52**, 1030 (1987).
111. S. J. Trank, D. W. Johnson, and E. L. Cussler, *Food Technol.* **43**(6), 78 (1989).
112. L. R. Dahl and R. Villota, *J. Food Sci.* **56**, 1002 (1991).
113. A. M. Pearson, *Bioscience* **26**, 249 (1976).
114. J. E. Kinsella, *J. Am. Oil Chem. Soc.* **56**, 242 (1979).
115. G. MacLeod and J. Ames, *Crit. Rev. Food Sci. Nutri.* **27**, 219 (1988).
116. W. H. Martinez, *J. Am. Oil Chem. Soc.* **56**, 280 (1979).
117. J. J. Spadaro and H. K. Gardner, Jr., *J. Am. Oil Chem., Soc.* **56**, 422 (1979).
118. Y-M Tzeng, L-L Diosady, and L. J. Rubin, *J. Food Sci.* **55**, 1147 (1990).
119. K. D. Schwenke, J. Kroll, R. Lange, M. Kujawa, and W. Schnack, *J. Sci. Food Agr.* **51**, 391 (1990).
120. S. M. Claughton and R. J. Pearce, *J. Food Sci.* **54**, 357 (1989).
121. R. Tasneem and V. Prakash, *J. Sci. Food Agr.* **59**, 237 (1992).
122. C.-Y. Ma and G. Khanzada, *J. Food Sci.* **52**, 1583 (1987).
123. T. V. Hung, P. D. Hanson, V. C. Amenta, W. S. A. Kyle, and R. S. T. Yu, *J. Sci. Food Agr.* **41**, 131 (1987).
124. L. A. Bryant, J. Montecalvo, Jr., K. S. Murey, and B. Loy, *J. Food Sci.* **53**, 810 (1988).
125. C. S. Lin, and J. F. Zayas, *J. Food Sci.* **52**, 1308 (1987).
126. C-S Lin and J. F. Zayas, *J. Food Sci.* **52**, 1615 (1987).
127. P. Yue, N. Hettiarachchy, and B. L. D'Appolonia, *J. Food Sci.* **56**, 992 (1991).
128. R. Ohlson and K. Anjou, *J. Am. Oil Chem. Soc.* **56**, 431 (1979).
129. A. I. Ihekoronye, *J. Sci. Food Agr.* **54**, 89 (1991).
130. A. T. Paulson and M. A. Tung, *J. Food Sci.* **52**, 1557 (1987).
131. A. T. Paulson and M. A. Tung, *J. Food Sci.* **53**, 817 (1988).
132. R. Ponnampalam, M. A. Vijayalakshmi, L. Lemieux, and J. Amiot, *J. Food Sci.* **52**, 1552 (1987).
133. F. F. Shih, *J. Food Sci.* **52**, 1529 (1987).
134. F. F. Shih, *J. Food Sci.* **56**, 452 (1991).
135. J. S. Hamada, *J. Food Sci.* **56**, 1725 (1991).
136. J. Velisek, T. Davidek, V. Kubel, and I. Viden, *J. Food Sci.* **56**, 139 (1991).
137. *CFR Title 21, Food and Drugs 155.170, 155.201, 161.190*, U. S. Government Printing Office, Washington, D.C., 1992.
138. *The Food Chemical News Guide*, 228.1 (Aug. 17, 1992).

139. *CFR Title 21, Foods and Drugs, 172.340, 172.385*, U. S. Government Printing Office, Washington, D.C., 1992.

140. V. D. Sidwell, B. R. Stillings, and G. M. Knobl, Jr., *Food Technol.* **24**, 876 (1970).

141. C. Cheftel, M. Phern, D. I. C. Wang, and S. R. Tannenbaum, *J. Agr. Food Chem.* **19**, 155 (1971).

142. R. Miller and H. S. Groninger, Jr., *J. Food Sci.* **41**, 268 (1976).

143. C. M. Lee, *Food Technol.* **40**(3), 115 (1986).

144. G. A. MacDonald and T. Lanier, *Food Technol.* **45**(3), 150 (1991).

145. K. H. Chung and C. M. Lee, *J. Food Sci.* **56**, 263 (1991).

146. G. B. Quaglia and E. Orban, *J. Food Sci.* **55**, 1571 (1990).

147. B. D. Rebeca, M. T. Pena-Vera, and M. Diaz-Castaneda, *J. Food Sci.* **56**, 309 (1991).

148. S. Eriksen and I. S. Fagerson, *J. Food Sci.* **41**, 490 (1976).

149. M. Fujimaki, H. Kato, S. Arai, and M. Yamashita, *J. Appl. Bacteriol.* **34**, 119 (1971).

150. M. Yamashita, S. Arai, and M. Fujimaki, *J. Food Sci.* **41**, 1029 (1976).

151. J. Montecalvo, Jr., S. M. Constantinides, and C. S. T. Yang, *J. Food Sci.* **49**, 1305 (1984).

152. M. R. Raghunath and A. R. McCurdy, *J. Sci. Food Agric.* **54**, 655 (1991).

153. M. Yamashita, S. Ari, S. J. Tsai, and M. Fujimaki, *J. Agr. Food Chem.* **19**, 1151 (1971).

154. S. W. Fox, J. W. Frankenfeld, D. Romsos, B. M. Robinson, and S. A. Miller in Ref. 98, pp. 569–583.

155. *PAG Bulletin, Policy Statement No. 4, Single-Cell Protein*, 1970; *Guidelines No. 6, Preclinical Testing of Novel Sources of Protein*, 1970; *No. 7, Human Testing of Supplementary Food Mixtures*, 1970; *No. 8. Protein-Rich Mixtures for Use as Supplementary Foods*, 1971; *No. 12, Production of Single Cell Protein for Human Consumption*, 1972; *No. 15, Nutritional and Safety Aspects of Novel Protein Sources for Animal Feeding*, 1974; Protein Advisory Group of the United Nations, FAO/WHO/UNICEF, United Nations, New York.

JOHN H. LITCHFIELD
Battelle Memorial Institute

FOOD STANDARDS.   See FOOD PROCESSING; REGULATORY AGENCIES.

FOOD TOXICANTS, NATURALLY OCCURRING

Food products are fundamentally mixtures of chemical compounds. That certain of these compounds may produce human toxicities has been known since before recorded history. Because human diets are normally composed of large numbers of different foods, only minute quantities of any specific toxic material are consumed. Thus dilution exerts a significant protective effect against acute intoxication. Toxicants are substances which, upon ingestion, produce changes in homeostasis that are threatening to the normal function of the organism. There are substantial differences in the toxicity thresholds of individuals to specific agents. Factors affecting toxicity include body weight, sex, age, general state of health, and the presence of potentiating or inhibitory substances. Some naturally occurring food toxicants are listed in Table 1 (1–11). Structures are shown in Figure 1. See also References 1 and 2.

Toxic Proteins, Peptides, Amides, and Amino Acids

Nitrogenous compounds are the most frequently implicated natural toxicants in foods. These compounds may be grouped either according to gross manifestations or specific structural characteristics. Accordingly, vitamin-destroying enzymes, hemagglutenins, enzyme inhibitors, and many hepatotoxins, are of protein, peptide, or amino acid composition. Many of the hepatotoxins are also carcinogens.

Enzyme inhibitors (qv) of a protein nature are of significant concern because of widespread occurrence. The most common of these affect the pancreatic enzymes, trypsin [9002-07-7] and chymotrypsin [9004-07-3], and are found in legumes, eg, soybeans, peas, most common beans, blackeyed peas, lupine, and khesari, as well as in egg whites and potatoes (3–9). Dozens of protein molecules of molecular weights of ca 15,000–30,000 are included in this group. Although the mode of action is difficult to characterize, these compounds form strong proteolysis-resistant complexes with their enzyme substrates, and consequently block trypsin and chymotrypsin activity. These compounds also enhance the production of amylase [9000-92-4] by the pancreas or at least stimulate its release by as much as 300%. This may be explained by the action of identified glycoprotein amylase inhibitors present in many legumes (10). Pancreatic hypertrophy, which has been hypothesized to produce increased secretion of cystine-rich enzymes, occurs and leads to a net loss of sulfur-containing amino acids from the body. These inhibitors also seem to block the feedback inhibition of pancreatic enzyme production which is normally based on intestinal concentration. Thus a cyclic and accelerating effect is produced whereby trypsin and chymotrypsin increase almost without limit, but their specific activities are blocked (11,12).

Protein inhibitors are often active against a variety of enzymes, although each molecule may possess a separate and very distinct binding site for each enzyme. For example, many trypsin and chymotrypsin inhibitors are identical compounds (12).

Many protein inhibitors cause little nutritional difficulty because these compounds are heat labile under ordinary processing and cooking procedures including microwaving (13), and significant numbers are water soluble. Many are found in highest concentrations in the outer portions of plants, eg, wheat bran; thus normal peeling and milling operations also give some protection. It has been demonstrated that treatment with compounds such as sodium sulfite [7757-83-7], ascorbic acid [50-81-7], and cupric sulfate [7758-98-7] (14,15), as well as fermentation with *Rhizopus oligosporus*, such as in the production of tempeh, is also an effective means of reducing trypsin inhibitor activity in both fresh and hardened common beans (16).

Table 1. Naturally Occurring Food Toxicants

Compound	CAS Registry Number	Molecular formula	Toxin classification	Typical food sources	Structure number[a]
aflatoxin B$_1$	[1162-65-8]	$C_{17}H_{12}O_6$	mycotoxin	moldy grains, nut, oil seeds	(1)
amygdalin	[29883-15-6]	$C_{20}H_{27}NO_{11}$[b]	cyanogenic glycoside	apricot pits, peach pits	(2)
avidin (chicken)	[1405-69-2]		forms insoluble complex within biotin	raw egg white	
caffeic acid	[331-39-5]	$C_9H_8O_4$	destroys thiamine	bracken fern	(3)
caffeine	[58-08-2]	$C_8H_{10}N_4O_2$	alkaloid, stimulant	coffee, tea, cola, nuts	(4)
capsaicin	[404-86-4]	$C_{18}H_{27}NO_3$	amide	*Capsicum* peppers	(5)
goitrin	[500-12-9]	C_5H_7NOS	goitrogen	cabbage, kale, onions, cress, cauliflower, broccoli, turnips	(6)
myristicin	[607-91-0]	$C_{11}H_{12}O_3$	alkaloid, psychoactive	nutmeg, mace, carrots	(7)
oxalic acid	[144-62-7]	$C_2H_2O_4$	reacts with calcium to reduce availability	rhubarb	(8)
phytic acid	[83-86-3]	$C_6H_{18}O_{24}P_6$	reacts with calcium to reduce availability	oats	(9)
solanine	[20562-02-1]	$C_{45}H_{73}NO_{15}$	glycoalkaloid, antiacetylcholinesterase	potatoes, tomatoes, apples, eggplant	(10)
thiaminase	[9030-35-7]	[c]	enzyme, inactivates thiamine	raw fish	
tyramine	[51-67-2]	$C_8H_{11}NO$	vasoactive amine	cheeses, bananas, plantains, tomatoes, pineapple	(11)

[a]See Figure 1.
[b]Mol wt ca 60,000.
[c]Mol wt ca (75–100) $\times 10^3$.

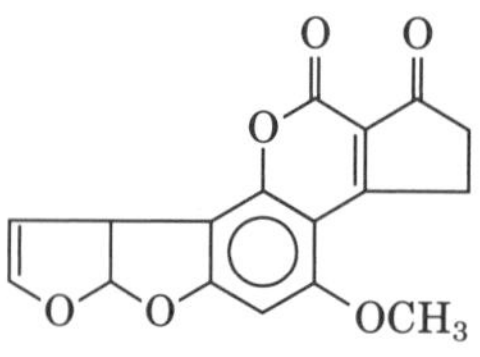

(1)

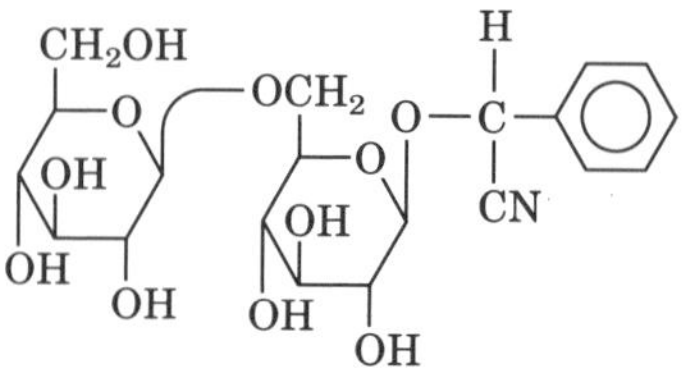

(2)

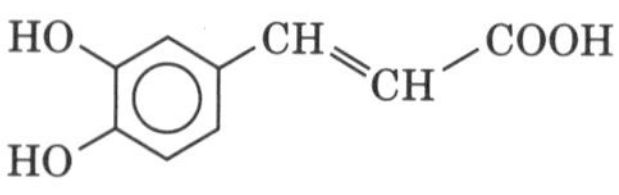

(3)

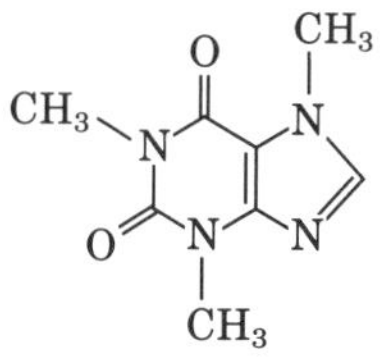

(4)

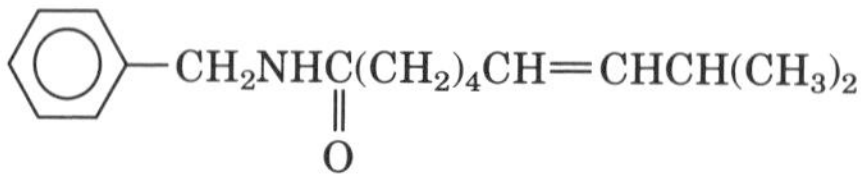

(5)

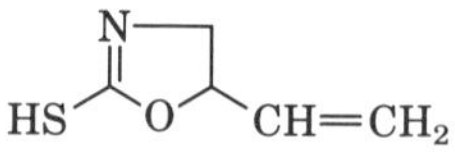

(6)

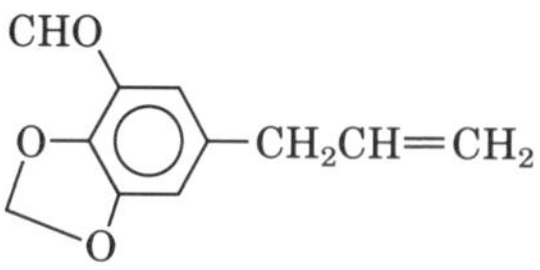

(7)

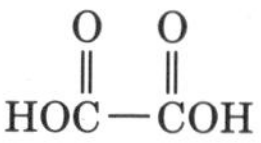

(8)

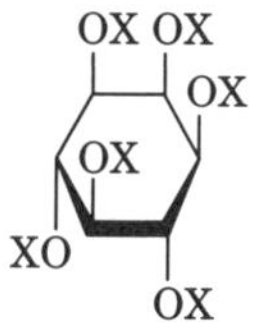

$X = OP(O)OH_2$

(9)

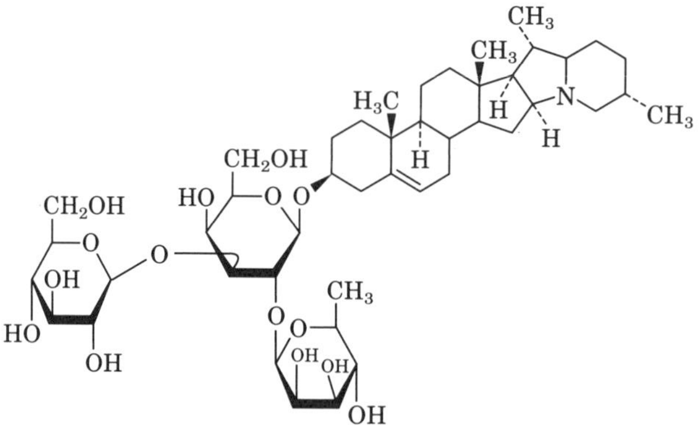

(10)

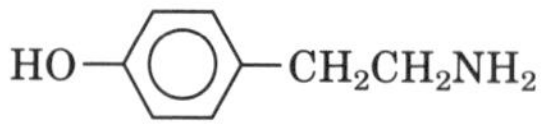

(11)

Fig. 1. Structures of naturally occurring food toxicants listed in Table 1.

Lupine seed, though used primarily in animal feeds (see FEEDS AND FEED ADDITIVES), does have potential for use in human applications as a replacement for soy flour, and is reported to contain both trypsin inhibitors and hemagglutenins (17). The former are heat labile at 90°C for 8 minutes; the latter seem much more stable to normal cooking temperatures. Various tropical root crops, including yam, cassava, and taro, are also known to contain both trypsin and chymotrypsin inhibitors, and certain varieties of sweet potatoes may also be implicated (18).

Reports of specific amino acid toxicities from normal eating patterns are rare (see AMINO ACIDS). Capsaicin (**5**), the amide responsible for the pungency of *Capsicum* peppers, is highly irritating at concentrations of 1:1,000,000 and highly toxic when taken in large doses (19). Of the essential amino acids, methionine [*63-68-3*] seems the least tolerated, especially on restricted protein intakes, where toxic doses have been reported in the range of 10–46 g/d (14). Tyrosine [*60-18-4*] behaves similarly, and both result in hepatic and neurological lesions when fed to rats at toxic levels. Leucine [*61-90-5*] has been responsible for depressed nicotinamide adenine dinucleotide (NAD) synthesis at levels of 14–20 g/d, whereas the toxicity of lysine [*56-87-1*], which causes abdominal cramps and diarrhea, occurs only with dose levels in excess of 64 g/d (20).

Of the nonessential amino acids, glycine [*56-40-6*] may be responsible for isolated toxicities at high (>30 g/d) dosages but is nonetheless generally considered nontoxic. Recognition of the Chinese Restaurant Syndrome in 1968 stimulated interest in the toxicity of glutamic acid [*617-65-2*], particularly as monosodium glutamate (MSG). Numbness in the back and neck, weakness, palpitation, and other evidence of neurological block have been observed in susceptible individuals with intakes of 3–25 g/d of MSG. Commercial and restaurant use of MSG has since been significantly reduced; however, these effects are closely related to L-glutamic acid [*56-86-0*] intolerance, and remain of interest because many native proteins contain 30% or more of this amino acid. Although all amino acids except alanine [*56-41-7*] have been shown to be toxic, the probability of intoxication is very remote. Humans seem able to tolerate all amino acids in excess of 10 times the recommended intake.

Lathyrus sativus (khesari), which constitutes a principal food crop for many in India, has been known for decades to cause neolathyrism. The causative agent, *N*-oxalyl-L-α-diaminopropionic acid (ODAP), produces partial or total loss of control of the lower limbs with associated neurological symptoms (21). Common methods of preparation, including soaking in lime water and boiling, are effective in destroying this amino acid.

The seeds of legumes may contain hemagglutenins and lectins that may cause destruction of the epithelia of the gastrointestinal tract; interfere with cell mitosis; cause hemorrhage; impede renal, cardiac, and hepatic function; and produce red blood cell agglutination (22). Many of these compounds are rendered inactive by moist heat, and the toxicity may be further reduced or neutralized by digestive enzymes, making them poorly absorbed. Because lectins reach the colon mostly in an inactive state, they appear to protect humans from colon cancer by causing hypersecretion of intestinal mucus, or by direct toxic effect on tumor cells (23).

Saponins disrupt red blood cells and may produce diarrhea and vomiting. They may also have a beneficial effect by complexing with cholesterol [*57-88-5*] and thus lowering serum cholesterol levels (24,25). In humans, intestinal microflora seem to either destroy saponins or inactivate them in small concentrations.

Acute toxicoses resulting from consumption of toxic mushrooms is infrequent, yet of increasing concern because of the practice of gathering fungi in the wild. The most serious of these toxicoses result from the Amanita family of mushrooms which contains several toxic peptides belonging to the amatoxin and phallotoxin groups. Of increased concern to 1990s toxicologists is the upsurge in reported cases of Amanita identification and poisonings in areas previously thought to be Amanita free. Whether these toxic species have always inhabited these environs or have been more recently introduced remains uncertain. Other toxic fungi also contain an array of hepatotoxic peptides and hallucinogens, some having been utilized for their drug effects by Native Americans for centuries. Most of these reactions are associated with progressive hepatic degeneration, and although only a few are fatal, their treatment is normally only symptomatic.

Both the common cultivated mushroom as well as the Shitake mushroom contain hydrazines that have been shown to be carcinogenic precursors in experimental animals. *Agaricus bisporus*, the primary commercial mushroom in the United States, contains the hydrazine analogue agaritine [2757-90-6] (**12**). Agaritine forms 4-hydroxymethylphenylhydrazine which can be transformed *in vitro* to 4-methylphenylhydrazine (4-MPH), the first hydrazine shown to be carcinogenic (26). Agaratine is destroyed by moist heat, and 4-MPH has not been isolated itself in mushrooms (27). Hence, toxic significance in the human diet remains unproven. In extreme cases, 100 g/d of fresh mushrooms approximates the carcinogenic dose observed in animals. Hydrazine levels, however, vary considerably as a function of variety, processing, storage, and preparation. One week of refrigerated storage reduces levels significantly, and all hydrazines are lost during canning and/or cooking (28,29) (see also HYDRAZINE AND ITS DERIVATIVES).

$$CH_2OH$$

$$HN-NH-\overset{\overset{\textstyle O}{\|}}{C}-CH_2CH_2\overset{\overset{\textstyle NH_3^+}{|}}{CH}-COO^-$$

(**12**)

Phytoalexins

Phytoalexins are low molecular weight compounds produced in plants as a defense mechanism against microorganisms. They do, however, exhibit toxicity to humans and other animals in addition to microbes (30). Coumarins, glycoalkaloids, isocoumarins, isoflavonoids, linear furanocoumarins, stilbenes, and terpenes all fall into the category of phytoalexins (31). Because phytoalexins are natural components of plants, and because their concentration may increase as a response to production and management stimuli, it is useful to recognize the possible effects of phytoalexins in the human diet.

Linear furanocoumarins are potent photosensitizing agents in celery, parsley, parsnips, limes, and figs. The most commonly reported symptoms include contact dermatitis and photodermatitis, particularly on the hands and forearms (32,33). Many linear furanocoumarins are present in these plant materials, all seem stable to cooking temperatures, and a few have even been reported to be mildly mutagenic (34). Coumarin (**13**) and its derivatives are also present in many species of citrus, occurring primarily in the peel, and hence find their way into human diets largely through the use of cold-pressed citrus-peel oils which are used as flavoring agents (35). Whereas both linear furanocoumarins and coumarins are widely distributed and possess toxic properties, reported cases of human toxicity have been limited largely to contact dermatitis in individuals handling large quantities of plants containing these compounds. The chronic low level consumption of these two classes of compounds, as would occur in a normal varied diet, has not been conclusively associated with human illness.

(13)   (14)

Other common phytoalexins in food materials are pisatin, cinnamylphenols, glyceolin, phaseolin [13401-40-6] (14), and 5-deoxykieritol in peas, beans, soybeans, and lima beans; viniferin in grapes; momilactones and oryzalexins in rice; α-tomatine in tomato; lubimen in eggplant; and capsidiol in green peppers.

Enumerable phytoalexins, including furanosesquiterpene, ipomeamarone, eudesmanes, and others, have been isolated from mold-infected sweet potatoes (31). The clinical symptoms seem to revolve around lung edema (36). Whereas high concentrations of these chemicals can occur in damaged sweet potatoes, the occurrence is much less (by as much as 20-fold) in nondamaged sweet potatoes (37). Of possible concern to human health is the fact that blemishes sufficient to result in large increases in concentration of lung-edema toxins are not always easily detected by the naked eye. Additionally, these compounds are heat stable (38).

Oligosaccharides

Oligosaccharides, specifically the α-galactosides raffinose [512-69-6], stachyose [470-55-3], and verbascose [546-62-3], are widely present in legumes and are indigestible by humans because of a lack of α-galactosidase. As a result, these compounds undergo fermentation in the colon with the concomitant production of CO_2, H_2, and CH_4, commonly referred to as flatulence. Reports have shown germination to be effective in reducing α-galactoside content of cowpeas and other legumes (39,40).

Goitrogens are compounds that produce goiter by interfering with thyroxine synthesis in the thyroid gland. Foodborne goitrogens are often characterized by the presence of sulfur and most are thiocyanates or closely related compounds. Because of their widespread occurrence in Cruciferae, eg, cabbage, kale, onions, cress, broccoli, cauliflower, rutabaga, turnip, and radish, goitrogens are among the most common and longest recognized substances of toxic nature in the human food supply (41). Thiocyanates actively compete with iodine for the same binding sites on the tyrosine molecule. Examples of goitrogenic compounds, other than thiocyanates, are derivatives of 2-3H-thioxazolidinone, found in watercress and mustard; 5-phenyl-2-thioxazolidinone in wintercress; and L-5-vinyl-2-thioxazolidinone in cabbage and kale. It has also been demonstrated that the quantity of the precursor found in plant materials seems related to available sulfur in the soil (42). However, studies on thioglucoside concentration in rutabaga have demonstrated no difference between roots grown on low S soils as compared to those grown over coal-fly ash (43). The enzyme responsible for the conversion of the precursor into the active molecule is destroyed by the heat associated with normal cooking operations (44).

Oxalates, Phytates, and Other Chelates

Of nutrient chelates in the human diet, oxalates and phytates are the most common. Oxalic acid (**8**), found principally in spinach, rhubarb leaves, beet leaves, some fruits, and mushrooms, is a primary chelator of calcium. Oxalate present in pineapple, kiwifruit, and possibly in other foods, occurs as calcium oxalate [563-72-4], CaC_2O_4. This compound is in the form of needle-like crystals, known as raphides, which can produce painful sensations in the mouth when eaten raw (45). The effects of oxalic acid in the diet may be twofold. First, it forms strong chelates with dietary calcium, rendering the calcium unavailable for absorption and assimilation. Secondly, absorbed oxalic acid causes assimilated Ca to be precipitated as insoluble salts that accumulate in the renal glomeruli and contribute to the formation of renal calculi (46).

Phytic acid (**9**), although restricted to a more narrow range of food products, mainly grains, complexes a broader spectrum of minerals than does oxalic acid. Decreased availability of P is probably the most widely recognized result of excessive intakes of phytic acid, yet Ca, Cu, Zn, Fe, and Mn are also complexed and rendered unavailable by this compound (47–49). Phytic acid has also been reported to reduce the activity of α-amylase and to decrease the activity of both proteolytic and lipolytic enzymes (50).

In addition to these purely nutritional effects, dietary chelates, especially oxalic acid, may produce more classic toxicity when ingested in excessive quantities. Although there is no doubt that intakes of 5 g or more of oxalic acid could be fatal to humans, the probability of such ingestion is remote. The consumption of some 4 kg of rhubarb leaves or roughly twice that quantity of spinach would be required. Therefore, concern regarding the possible hazardous effects of ingesting plant foods containing oxalate seems unwarranted. High intakes of both calcium and vitamin D help to offset the deleterious effects of oxalates (see VITAMINS).

Vasoactive and Psychoactive Amines and Alkaloids

Most compounds producing hypertensive episodes are classified as amines and are found in greatest concentration in banana, plantain, tomato, avocado, pineapple, broad beans, and various cheeses. Amines that are vasoactive include dopamine [51-61-6], $C_8H_{11}NO_2$; tyramine (**11**); histamine [51-45-6], $C_5H_9N_3$; tryptamine [61-54-1], $C_{10}H_{12}N_2$; noradrenaline [51-41-2], $C_8H_{11}NO_3$; and dihydroxyphenylalanine [59-92-7] (DOPA), $C_9H_{11}NO_4$ (51).

Patients receiving monoamine oxidase inhibitors (MAOI) as antidepressant therapy have been especially subject to the hypertensive effects of vasoactive amines (52). These dietary amines have also been implicated as causative agents in migraine. Other naturally occurring alkaloids (qv) have been recognized for centuries as possessing neurological stimulant and depressant properties.

Caffeine (**4**), a xanthine derivative, has been consumed for thousands of years and is present in over 60 plant species including coffee (qv) beans, tea (qv) leaves, cacao beans (see CHOCOLATE AND COCOA), and cola nuts. In addition to naturally occurring sources, caffeine has been used as a food ingredient (flavoring) for over 100 years. Caffeine is rapidly and completely absorbed, being distributed quickly throughout the body with no so-called barriers to limit absorption or distribution. Compounding this rapid absorption is the fact that caffeine is inefficiently excreted by the kidneys. Caffeine produces stimulatory effects by facilitating mental and muscular effort and diminishes drowsiness and fatigue. It has been suggested that these effects may be the result of caffeine acting as an antagonist of adenosine in respiratory, renal, and neural cardiovascular tissues (53). Individual thresholds of toxicity vary considerably, but symptoms such as restlessness, increased respiration, muscular tension and twitching, and tachycardia may imply acute toxicity. Although caffeine content in beverages is quite variable, 20 cups or more of coffee per day (85–100 mg/cup) would be necessary in most individuals to approach acute toxicity, and the human fatal dose has been estimated at about 10 g (150–200 cups of coffee) in a 24 h period. Considered a stimulant in usually encountered doses, caffeine's acute toxicity may actually be depressant in character (54,55).

Depressant symptoms, which include burning abdominal pain, decreased excitability, convulsions, nausea, and coma, become the general syndrome for all oral alkaloid poisoning. Discorine [*3329-91-7*] (**15**), a γ-unsaturated lactone found in yams, is an alkaloid having the empirical formula $C_{13}H_{21}O_2N$; it has been isolated from Nigerian yams, thus establishing the discorine-type alkaloids as representative of the *Dioscorea* genus of plants (56). Seeds of *Senecio* contain alkaloids belonging to the pyrrolizidine group and, in addition to producing characteristic signs of alkaloid toxicity, these are suspected of being hepatic carcinogens. Ergotism is the disease caused by intoxication by one of at least twelve alkaloid derivatives of lysergic acid produced by a parasitic fungus (ergot) on grains. Myristicin (**7**), found in both nutmeg and mace, is a psychoactive agent that may be fatal in infants who consume as little as two whole nutmegs. Its toxicity resembles alcohol intoxication.

(**15**)   (**16**)

Several glycoalkaloids present in food are of toxicological interest. Solanine (10), found in potatoes, tomatoes, apples, eggplant, and sugar beets, has been responsible for several cases of moderate to severe poisoning. Solanine is a cholinesterase inhibitor (see CHOLINE), and toxic doses are probably ca 200 mg. Market potatoes contain about 1–5 mg of solanine per 100 g fresh weight. The USDA establishes solanine levels of 20 mg/100 g as the limit for safe consumption. Because the greatest proportion of solanine occurs in the skin and eyes of potatoes, the practice of peeling diminishes the intake of this alkaloid. However, the increased consumption of potato skins as snack items has resulted in increased glycoalkaloid intake in some individuals. Toxic symptoms include headache, nausea, and diarrhea. Solanine is poorly absorbed, rapidly excreted, and readily hydrolyzed to the less toxic solanidine [80-78-4] (16) in the intestinal tract. Lethal concentrations are estimated at about 3 mg/kg body weight (57,58).

Many studies have reported a link between consumption of sunburned potatoes, ie, those exposed to the sun and having an accumulation of chlorophyll and solanine under the skin, with incidences of teratogenic effects and even death (59–61). Because sunburned potatoes in the commercial marketplace are relatively rare, and because the long-term effects of consumption of potatoes at the maximum established limits of solanine concentration are uncertain, there is equal uncertainty of the true incidence of human toxicity (62).

Solasodine [126-17-0], $C_{27}H_{43}NO_2$, a spirosalane alkaloid, is found in eggplant, and has been reported to be teratogenic in hamsters (63), but not in rats (64). Toxicity in humans has not been reported. Tomato products contain α-tomatine, a steroidal glycoalkaloid which has been reported to be toxic to a wide range of organisms, including microorganisms, cattle, and mice (65). Most of the glycoalkaloids occurring in plant materials exert a natural pesticidal function. It has been estimated that consumption of natural pesticides is some 10,000 times greater than consumption of manufactured ones (66).

Antinutrients

Any substance that destroys, inactivates, or in other ways renders unavailable an essential dietary constituent can be termed an antinutrient. The most widely studied are antivitamins. The presence of antivitamins in certain foods means that merely assuring an adequate intake of a vitamin is no guarantee that a deficiency state cannot exist physiologically. The enzyme thiaminase acts by either specific splitting of the thiamine molecule or nonspecific hydrolysis (67). Niacin inhibitors, acting through nicotinamide mononucleotide (NMN) depression in erythrocytes, have been studied in corn and millet, and a biotin antagonist, avidin, has long been recognized in raw egg white. Avidin forms a stable complex with biotin, thus rendering the vitamin unavailable for metabolic reaction (68). Linatine [10139-06-7], found in flaxseed, is the only pyridoxine antagonist known and seems to function by the formation of a stable complex (69). Yeast and pea seedlings contain specific pantothenic acid antagonists, although the structure and mode of action are unexplained (70). Riboflavin antagonism, found only in the Akee plum of Jamaica, is rather rare, but is of interest because it can be fatal (71).

Many plant substances possess antivitamin D activity but the mode of action and in most cases the identity remain unknown. Rachitogenic factors have been observed in yeast. Because of the metabolic interrelationships that exist between vitamin D, Ca, and P, it is sometimes difficult to differentiate between chelators of mineral elements and true antivitamins. One reported vitamin D antagonist in oats was later identified as phytic acid (72).

The antagonisms that exist between unsaturated fatty acids, and carotene and vitamin E are complicated and largely undefined. Linoleic acid acts as an antivitamin to *dl*-α-tocopherol [*59-02-9, 1406-18-9, 10191-41-0*] (vitamin E) by reducing availability through direct intestinal destruction. Various lipoxidases destroy carotenes and vitamin A (73). Dicoumarol [*66-76-2*] (3,3′-methylenebis(4-hydroxycoumarin)) is a true antimetabolite of vitamin K [*12001-79-5*] but seems to occur only in clover and related materials that are used primarily as animal feeds (74).

At various times, antivitamin factors specific to vitamin B_{12}, folic acid, and choline have been reported. However, it is uncertain whether these are true antimetabolites or if they may result from metabolic interrelationships with other dietary constituents.

Investigations have focused on the content of polyphenolics, tannins, and related compounds in various foods and the influence on nutrient availability and protein digestibility. It has been established that naturally occurring concentrations of polyphenoloxidase and polyphenols in products such as mushrooms can result in reduced iron bioavailability (75). Likewise, several studies have focused on decreased protein digestibility caused by the tannins of common beans and rapeseed (canola) (76–78).

Vitamin Toxicity

Reported cases of vitamin toxicity owing to overdose are usually associated with increased over-the-counter availability of supplemental vitamins and indiscriminate supplementation. The misconception that if a little is good a lot is better has compounded toxicological problems with the vitamins. Fat-soluble vitamins tend to accumulate in the body with relatively inactive mechanism for excretion and cause greater toxicological difficulties than do water-soluble vitamins.

Infants may be sensitive to doses of vitamin A [*11103-57-4*] in the range of 75,000–200,000 IU (22.5–60 mg), although the toxic dose in adults is probably 2–5 million IU (90.6–1.5 g). Intakes in this range from normal food supplies without oral supplements are simply beyond imagination (79). Vitamin D [*1406-16-2*] toxicity is much more difficult to substantiate clinically. Humans can synthesize active forms of the vitamin in the skin upon irradiation of 7-dehydrocholesterol. Toxic symptoms are relatively nonspecific, and dangerous doses seem to lie in the range of 1000–3000 IU/kg body wt (25–75 μg/kg body wt) (80). Cases of toxicity of both vitamins E and K have been reported, but under ordinary circumstances these vitamins are considered relatively innocuous (81).

Of the water-soluble vitamins, intakes of nicotinic acid [*59-67-6*] on the order of 10 to 30 times the recommended daily allowance (RDA) have been shown to cause flushing, headache, nausea, and moderate lowering of serum cholesterol with concurrent increases in serum glucose. Toxic levels of folic acid [59-30-3] are ca 20 mg/d in infants, and probably approach 400 mg/d in adults. The body seems able to tolerate very large intakes of ascorbic acid [50-81-7] (vitamin C) without ill effect, but levels in excess of 9 g/d have been reported to cause increases in urinary oxalic acid excretion. Urinary and blood uric acid also rise as a result of high intakes of ascorbic acid, and these factors may increase the tendency for formation of kidney or bladder stones. All other water-soluble vitamins possess an even wider margin of safety and present no practical problem (82).

Essential Minerals and Heavy Trace Elements

Ingestion of at least 10 times normal levels of essential minerals would be required to approach toxic proportions (see MINERAL NUTRIENTS). The only exceptions occur in cases of plant foods grown on soils unusually high in Mo [7439-98-7], Se [7782-49-2], and Cu [7740-50-8]. Levels can reach toxic quantities in these cases, but these are rare occurrences.

Cases involving human toxicity from heavy trace elements, such as Pb [7439-92-1], Hg [7439-97-6], As [7440-38-2], and Cd [7440-43-9], are much more common but are almost exclusively traced to accidental contamination rather than true natural occurrences. Consumption of seafoods taken from waters where industrial pollution raised mercury levels has been responsible for some reported cases of mercury poisoning. Methylmercury acetate [108-07-6] seems to be the toxic compound of most interest to humans and is formed primarily from microbial action. Infrequent problems with lead and cadmium have been traced to steels and equipment containing these elements coming in contact with foods during handling or processing (see FOOD PROCESSING). Naturally high levels of arsenic have been reported in some crustaceans and fishes, yet food problems have been restricted to meats taken from livestock treated with arsenic-containing antibiotics (83) or to accidental contamination (see MEAT AND MEAT PRODUCTS).

Heavy metals are of importance in human toxicity because the body possesses only inactive mechanisms for their excretion; thus chronic, low level intakes can accumulate to toxic proportions. Treatment has likewise been relatively unsuccessful, except for symptomatic relief. No effective means has been discovered to increase excretion.

Cyanogenic Glycosides

Complex glycosides, which upon hydrolysis yield hydrogen cyanide [74-90-8], are commonly found among plant materials. The toxicity of this class of compounds, found in the bitter almond, pits of stone fruits, sorghum, and lima beans, is directly related to HCN liberation upon digestive hydrolysis. Amygdalin (**2**), the cyanogenic glycoside of apricot pits, has been promoted as both a cancer cure and a vitamin. However, it has never been associated with a medically verified cancer cure, and no trained nutritionist recognizes the compound as a vitamin. Indeed, many cases of poisoning, and even death, have been ascribed to its use. Approximately 50 apricot pits contain a fatal dose of amygdalin (84).

The cyanogenic glycosides, phaseolunatin [554-35-8], $C_{10}H_{17}NO_6$, and vicianin [155-57-7], $C_{19}H_{25}NO_{10}$, have been isolated from lima beans and vetch, respectively. Several studies have reported that heating (cooking) acts to decrease the quantity of HCN liberated by these compounds upon enzymatic hydrolysis.

Nitrates, Nitrites, and Nitrosamines

The carcinogenicity of nitrosamines has created widespread concern over the safety of food products that are significant sources of nitrates and nitrites. Nitrosamines are readily formed by reaction of secondary amines with nitrites at acid pH, conditions which may occur in the gastrointestinal tract.

Nitrates are found in fairly high concentrations in beets, spinach, kale, collards, eggplant, celery, and lettuce. Additionally, nitrates and nitrites are commonly used in the curing solutions of bacon, ham, and other cured meats. In cured meats, nitrates and nitrites control the growth of microorganisms, particularly *Clostridium botulinum*, and also serve as color preservatives.

Although the potentially carcinogenic nitrosamines may be present in foods, particularly cured meats, occurrence is infrequent and at low levels. USDA regulations stipulate that ascorbic acid be added to cured meats at five times the level of nitrates and nitrites to prevent the formation of carcinogenic *N*-nitroso compounds (see FOOD ADDITIVES). Nitrosamines vary widely in molecular structures, and some 100 or more may be capable of producing pathological effects. Most are specific hepatotoxins, producing hepatic parenchymal cell necrosis, and may also act in a synergistic capacity with other carcinogens, notably the polycyclic hydrocarbons (85).

Sodium Chloride

Sodium chloride [7647-14-5] is an essential dietary component. It is necessary for proper acid–base balance and for electrolyte transfer between the intra- and extracellular spaces. The adult human requirement for NaCl probably ranges between 5–8 g/d. The normal diet provides something in excess of 10 g/d NaCl, and adding salt during cooking or at the table increases this intake.

Excessive intake of NaCl contributes to increased fluid retention and there may be a relationship between NaCl intake and hypertension. Few topics have caused such debate among nutritionists and physicians. There is no doubt that the relationship between hypertension and NaCl intake is of significance when compounded by obesity, or in about 30% of hypertensives. Edema resulting from excessive salt intake, especially in relatively inactive persons, is a principal contributing factor in elevated blood pressure. Both consumers and food processors have reduced use of NaCl.

Toxins

Mycotoxins. The condition produced by the consumption of moldy foods containing toxic material is referred to as mycotoxicosis. Molds and fungi fall into this category and several derive their toxicity from the production of oxalic acid, although the majority of mycotoxins are much more complex.

Mycotoxins find their way into the human diet by way of mold-contaminated cereal and legume crops, meat, and milk products. Corn and peanuts probably represent the most common sources of mycotoxins in the human diet. Many mycotoxins are acutely toxic as well as being potent carcinogens (86).

Many parasitic fungi have been shown to produce toxins; however, the toxins of *Aspergillus* and *Penicillium* have perhaps the greatest potency against humans. Polycyclic peptides are among the most dangerous toxic compounds produced by fungi. A number of mycotoxins contain the diketopiperazine nucleus and all are composed of amino acid units. Islanditoxin [*10089-09-5*], $C_{24}H_{31}Cl_2N_5O_7$ (**17**), produced by *Penicillium islandicum* on rice and other grains, is a Cl-containing bicyclic peptide which includes α-aminobutyric acid [*80-60-4*], $C_4H_9NO_2$, β-phenyl-β-aminopropionic acid [*13921-90-9*], $C_9H_{11}NO_2$, serine [*56-45-1*], $C_3H_7NO_3$, and dichloroproline [*60548-67-6*], $C_5H_7Cl_2NO_3$. Several anthraquinone derivatives, produced by members of the *Penicillium* group, are also of toxicological interest. Examples are skyrin [*602-06-2*], $C_{30}H_{18}O_{10}$, luteoskyrin [*21844-44-6*], and iridaskyrin [*568-42-3*], $C_{30}H_{18}O_{10}$. Various mycotoxins also include alkaloids and xanthones, coumarin, and terpene derivatives. Some are thermolabile, others thermostable. The latter are of greatest importance in food products because routine processing and cooking do little to reduce their potential toxicity. Toxic metabolites are usually produced only after mycelia have become established in the substrate. For this reason, foods are usually free of toxins until such mycelial development occurs (87). The mode of action of mycotoxins varies according to chemical nature. Some act by blocking cell-wall synthesis, by disrupting membrane integrity, or by decoupling oxidative phosphorylation and inhibiting respiratory pathways. Others are chelates that inhibit synthesis of proteins or DNA.

(**17**)

Aflatoxins, first noted as metabolites from *Aspergillus flavus*, are also produced by *A. parasiticus*, *A. niger*, *A. ruber*, *A. ostinaus* Wehmer, *A. wentii*, *A. vesicolor*, *Penicillium puberculum*, *P. citrium*, *P. variable*, and *P. frequentans*, as species of *Rhizopus*. *Aspergillus flavus* is a common contaminant of a multitude of foods, including soybeans, groundnuts, cassava, peas, pears, cacao pods, Brazil nuts, pecans, millet, corn, and wheat. Because of this widespread occurrence on plant materials the rare isolation of aflatoxins in food products is not surprising (88).

The term aflatoxin is not compound-specific. Different toxic compounds have been designated as B_1 and G_1 [1165-39-5], and B_2 [7220-81-7] and G_2 [7241-98-7] (dihydro derivatives of B_1 and G_1), as well as M_1 [6795-23-9], M_2 [6885-57-0], P_1 [32215-02-4], GM_1 [23532-00-5], and B_3 [23315-33-5]. Aflatoxin B_1 (**1**) is of greatest occurrence in nature, followed by G_1, B_2, and G_2. Water content of medium, temperature, pH, and light are among those environmental factors that affect aflatoxin production.

Aflatoxin B_1, produced by *Aspergillus flavus*, is probably the most potent hepatocarcinogen found in nature (89). Measurable levels of aflatoxin residues have been found in U.S. cereal grains and peanuts. These compounds seem resistant to cooking and processing (90,91); hence ordinary methods of preparation do not limit their intake. Because carcinogens may take many years to be recognized as causing illness, the real implication of aflatoxin ingestion in normal human diets is uncertain (92).

Acute toxicoses, as well as potential long-term effects of aflatoxin ingestion, have been extensively reported (93). Autopsy reports have noted a positive correlation between aflatoxin B_1 and victims of Reye's Syndrome in Thailand (94), but this evidence should only be considered as suggestive and preliminary (95).

Intoxication by aflatoxins is referred to as aflatoxicosis. Edema and necrosis of hepatic and renal tissues seem characteristic of aflatoxicosis, and hemorrhagic enteritis accompanied by nervous symptoms often appear in experimental animals. The mode of action of aflatoxins involve an interaction with DNA and inhibition of the polymerases responsible for DNA and RNA synthesis (96).

Cereals, bakery products, and oilseed products are those foods that are of highest risk in regard to aflatoxin contamination (see WHEAT AND OTHER CEREAL GRAINS). One study showed 74% of peanut butter samples to contain aflatoxins (97) (see NUTS). Another study (31) suggests a possible link of aflatoxin ingestion with mental retardation, but this connection remains largely uncorroborated. Specific links between liver carcinoma and aflatoxin ingestion have not been substantiated fully, and it is probable that only minimal dangers are to be encountered from commercially processed foods.

Many procedures have been studied for detoxification of aflatoxins, including heat and treatment with ammonia, methylamine, or sodium hydroxide coupled with extraction from an acetone–hexane–water solvent system. Because in detoxification it is important to free the toxin from cellular constituents to which it is bound, a stabilization of proteins using a tanning compound such as acetaldehyde (qv) or glutaraldehyde may be a solution to the problem (98).

Seafood Toxins. Virtually scores of fish and shellfish species have been reported to have toxic manifestations. Most of these toxicities have been shown to be microbiological in origin. There are a few, however, that are natural components of seafoods.

Several species of the moray eel (*Gymnothorax*) have caused toxic reactions, especially in Japan. The toxic principle appears to be proteinaceous and is found predominately in the blood but it may occur in the flesh as well. Its exact structure remains somewhat uncertain.

Amnesic shellfish poisoning resulted in four deaths in northeastern Canada in the early 1990s, and domoic acid, the causative agent, was first documented on the Washington and Oregon coasts in 1991. The toxin is produced by a single-celled phytoplankton that constitutes part of the food chain of some shellfish, including Dungeness crab. Domoic acid is concentrated in the tissues of these shellfish, and upon ingestion can result in symptoms including memory loss (hence the name of the condition), and in large doses, ultimately death. Domoic acid is not affected by common methods of food preparation and no medical treatment is presently available for acute poisoning (99).

Pufferfish toxin, isolated from a dozen or more species, has been identified as having the empirical formula $C_{11}H_{17}N_3O_8$, but the structure is not well-established, nor is it certain that the same structure is universally responsible for poisoning, although this is assumed to be the case. The so-called paralytic shellfish poisoning reported in many areas of the world has a microbiological etiology, and is thus more accurately a contamination rather than a natural toxicosis. The paralytic effects of the poisoning begin as a tingling sensation in the lips, tongue, and extremities, and gradually progress into nausea and convulsions. Japanese statistics indicate mortality rates approaching 65% (100).

The liver of sharks and other oily fishes sometimes accumulate toxic levels of vitamin A, and cases of acute poisoning have been reported both among Eskimos and the Japanese.

Other Toxins.   Some consumers have become concerned over the possible toxicity of free radicals in lipid oxidation products, including degradation products of cholesterol, in food fats and oils. Whereas complicated autoxidation and polymerization of fats does occur at temperatures of 200–300°C, such conditions are rarely encountered in normal cooking. Toxicity of these degradation products in experimental animals has been reported only using excessively large doses and presents no hazard to human health under ordinary circumstances. Free radicals resulting from lipid degradation are extremely short-lived (nanoseconds), and thus consumption is unlikely. Accordingly, free radicals from food possess no known toxicological significance (101).

Cyasin, a component of the nut of the cycad tree, a native of tropical environs, produces an acute toxicity in addition to drastically increasing the incidence of Lou Gerhig's disease (amyotropic lateral sclerosis). Cyasin is carcinogenic (102).

Estrogens are ingested by humans from both plant (cereal grains, legumes, potatoes, carrots, parsley, tree fruits, yams, vegetable oils) and animal (liver, egg yolk) sources. Whereas estrogens are known to promote tumor growth, the activity of food estrogen is one-tenth to one-thousandth the level of the most common human circulating estrogen. There has been no implication of foodborne estrogen in the etiology of human cancer, but the effects of long-term subphysiological intakes remain unclear (103). Some studies have concluded that because phytoestrogens are not mutagenic, they may be tumor promoters rather than tumor initiators (104).

A toxic component of braken fern, perhaps either quercetin (105) or ptaquiloside, a glucoside (106), has a mixed history of carcinogenicity. It is sometimes implicated in an increased incidence of bladder cancer in animals and esophageal cancer in humans. Multiple other dietary components seem to either promote or interfere with its action, and the significance of braken fern in human carcinogenesis remains unproven.

Halogenated compounds, found in high concentrations in seaweeds consumed in Japan and Hawaii, have been suspected of being carcinogenic, largely based on epidemiological extrapolation (high incidences of hepatic carcinoma). However, direct human causation has not been established (107).

Urethane [51-79-6] (ethyl carbamate) occurs as a natural by-product in fermented products such as wine, liquors, yogurt, beer, bread, olives, cheeses, and soy sauces. Whereas urethane has a known cancer etiology in experimental animals, no such relationship has yet been proven in humans (108,109). Alcohol may act by blocking the metabolism of urethane, and thus exert a protective effect in humans consuming alcoholic beverages (110).

Polycyclic aromatic hydrocarbons (PAHs) are carcinogens produced by the thermal breakdown of organic materials. These are widely distributed in both food and the environment, and are some of the principal carcinogens in cigarette tar and air pollution. Of over 20 PAHs isolated, benzopyrene and quinoline compounds are the most commonly encountered in foods, particularly those which are broiled or fried (111). Shellfish living in petroleum contaminated waters may also contain PAHs (112).

Mutagenic PAHs have been measured in a number of cooked meats, the concentration being dependent on temperature and cooking time. Well-done meats contain more mutagenic PAHs than those cooked rare or medium rare (113,114). Microwaving and moist-heat cooking methods produce few PAHs, presumably because of short cooking times and the lack of browning (115).

Epidemiologic studies in Japan indicate an increased risk of stomach cancer owing to consumption of broiled fish and meats (116). In the United States, stomach cancer incidence has steadily declined since the 1940s, whereas consumption of broiled food has increased (108). In addition, the average human intake of PAHs is only 0.002 of that required to produce cancer in half of animals fed. Test results are often contradictory (117) and many components of food, such as vitamin A, unsaturated fatty acids, thiols, nitrites, and even saliva itself, tend to inhibit the mutagenic activity of PAHs (118–120). Therefore, the significance of PAHs in the human diet remains unknown (121,109).

Legislation and Regulatory Considerations

There exists little specific legislation dealing with natural toxicants in foods. The *1958 Food Additives Amendment to the Federal Food, Drug, and Cosmetic Act* stipulates that no substance that has been shown to be carcinogenic to either humans or animals may be added to the food supply. Accordingly, those foods that contain added carcinogens are subject to the Delaney Clause. Maximum tolerances of heavy metals, such as Pb and Hg, have been established by FDA at 0.5 ppm in the food product. For aflatoxins, there is presently a zero tolerance in effect (based on the Delaney Clause), and screening is generally on a qualitative basis. With these exceptions, natural toxicants in food products are generally not treated by specific food legislation (122).

Naturally occurring toxicants have been reviewed in greater detail elsewhere (22,28,31,60,61).

BIBLIOGRAPHY

"Food Toxicants, Naturally Occurring" in *ECT* 3rd ed., Vol. 11, pp. 208–220, by F. H. Hoskins, Louisiana State University.

1. S. A. Miller, *Ann. N.Y. Acad.* **300**, 397 (1977).
2. R. L. Hall and S. L. Taylor, *Food Tech.* **43**(9), 270 (1989).
3. I. E. Leiner, *Toxic Constituents of Plant Foodstuffs*, Academic Press, Inc., New York, 1980.
4. J. J. Rackis and M. R. Gumbmann, *Antinutrients and Natural Toxicants in Foodstuffs*, Academic Press, Inc., New York, 1981.
5. A. E. Bender, *Natural Toxicants in Food*, VCH Publishers, New York, 1987, pp. 110–124.
6. B. Merz, *J. Am. Med. Assoc.* **249** 2746 (1983).
7. V. Frattali and R. F. Steiner, *Biochemistry* **7**, 521 (1968).
8. M. Friedman, *Nutritional and Toxicological Significance of Enzyme Inhibitors in Foods*, Plenum Press, New York, 1986.
9. S. S. Deshpande and S. S. Nielsen, *J. Food Sci.* **52**, 1130 (1987).
10. F. M. Lagolo, F. F. Filo, and E. W. Menezes, *Food Tech.* **45**(9), 119 (1991).
11. R. Huber and co-workers, *Naturwissenschaften* **57**, 389 (1970).
12. G. A. Silverstone, *Diet-Related Diseases*, AVI, Westport, Conn., 1985, pp. 44–60.
13. F. H. Smith and A. J. Clawson, *J. Am. Oil Chem. Soc.* **47**, 443 (1970).
14. H. Yoshida and G. Kajomoto, *J. Food Sci.* **52**, 1756 (1988).
15. D. J. Sessa, J. K. Haneg, and T. C. Nelson, *J. Agric. Food Chem.* **38**, 1469 (1990).
16. M. Friedman and M. R. Gumbmann, *J. Food Sci.* **51**, 1239 (1986).
17. O. Paredes-Lopez and G. I. Harry, *J. Food Sci.* **54**, 968 (1989).
18. J. H. Bradbury and B. C. Hammer, *J. Agric. Food Chem.* **38**, 1448 (1990).
19. V. Fratali, *J. Biol. Chem.* **244**, 274 (1969).
20. C. C. Toh, T. S. Lee, and A. K. Kiang, *Br. J. Pharmacol.* **10**, 175 (1955).
21. K. Jahan and K. Ahmad, *Food Nutr. Bull.* **6**(2), 52 (1984).
22. J. M. Jones, *Food Safety*, Eagan Press, St. Paul, Minn., 1992.
23. I. E. Leiner, in R. L. Ory, ed., *Antinutrients and Natural Toxicants in Foods*, F & N Press, Westport, Conn., 1981, pp. 143–158.
24. D. Okenfull and G. S. Sidhu, in P. R. Check, ed., *Toxicants of Plant Origin*, Vol. 2, CRC Press, Boca Raton, Fla., 1989, pp. 97–142.
25. Y. Birk and I. Peri, in I. E. Leiner, ed., *Toxic Constituents of Plant Foostuffs*, 2nd ed., Academic Press, Inc., New York 1980, p. 161.
26. B. Toth, *Cancer Res.* **35** 3693 (1975).
27. A. E. Ross, D. L. Nagel, and B. Toth, *Food Chem. Toxicol.* **20**, 903 (1982).
28. B. Toth, *Carcinogens and Mutagens in the Environment*, Vol. 3, CRC Press, Boca Raton, Fla., 1983, pp. 99–108.
29. W. J. Visek and S. K. Clinton, *Xenobiotic Metabolism*, ACS Symposium Series 277, Washington, D.C., 1985, pp. 293–307.
30. J. Paxton, *Plant Dis.* **64**, 734 (1980).
31. R. C. Beier, *Rev. Environ. Con. Toxicol.* **113**, 47 (1990).
32. R. C. Beier, G. W. Ivie, and E. H. Oertli, *ACS Symp. Series* **234**, 295 (1983).
33. R. C. Beier and co-workers, *Food Chem. Toxicol.* **2**, 163 (1983).
34. G. W. Ivie, J. T. MacGregor, and B. D. Hammock, *Mutat. Res.* **79**, 73 (1980).

35. W. L. Stanley and L. Jurd, *J. Agric. Food Chem.* **19**, 1106 (1971).
36. B. J. Wilson and co-workers, *Nature* **231**, 52 (1971).
37. D. T. Coxon, R. F. Curtis, and B. Howard, *Food Cosmet. Toxicol.* **13**, 87 (1971).
38. B. J. Wilson, D. T. C. Yang, and M. R. Boyd, *Nature* **227**, 521 (1970).
39. W. R. Bonorden and B. G. Swanson, *J. Sci. Food Agric.* **84**, 1 (1992).
40. I. A. Nnanna and R. D. Phillips, *J. Food Sci.* **55**, 151, (1990).
41. K. Chan Shiek and K. T. Madhusudhan, *J. Food Sci.* **53**, 1234 (1988).
42. E. Josefson, *Phytochemistry* **6**, 1617 (1967); R. Gmelen, *Präparative Pharmazie* **5**(2), 17 (1969); **5**(3), 33 (1969).
43. J. L. Anderson, D. J. Lisk, and G. S. Stoewsand, *J. Food Sci.* **55**, 556 (1990).
44. E. P. Lichtenstein and co-workers, *J. Agric. Food Chem.* **10**, 30 (1962).
45. P. Friis and A. Kjaer, *Acta Chem. Scand.* **20**, 698 (1966).
46. R. F. Crampton and F. A. Charlesworth, *Brit. Med. J.* **31**(3), 209 (1975).
47. C. O. Perera and co-workers, *J. Food Sci.* **55**, 1066 (1990).
48. A. Hodgkinson and P. M. Zarembski, *Calc. Tiss. Res.* **2**, 115 (1968).
49. R. E. Ferrel and co-workers, *Cereal Sci.* **14**, 110 (1969).
50. J. G. Reinhold and co-workers, *Lancet i*, 283 (1973).
51. N. Sapeika, *Trans. R. Soc. Afr.* **41**, 1 (1974).
52. M. E. Nyman and I. M. Bjorek, *J. Food Sci.* **54**, 1332 (1989).
53. A. S. Sandberg and U. Svanberg, *J. Food Sci.* **56**, 1330 (1991).
54. H. R. Roberts and J. J. Barone, *J. Food Sci.* **37**(9), 32 (1983).
55. R. W. Von Borstel, *J. Food Sci.* **37**(9), 40 (1983).
56. B. Blackwell and co-workers, *Br. J. Psychol.* **113**, 349 (1967).
57. A. M. C. Davies and P. J. Blindow, *J. Sci. Food Agric.* **35**, 553 (1984).
58. N. I. Mondy and B. Gasselin, *J. Food Sci.*, 53, 756 (1988).
59. K. T. H. Farrer, *Fancy Eating That!*, Melbourne University Press, Australia, 1983.
60. M. R. A. Morgan and D. T. Coxon, *Natural Toxicants in Food*, DCH Publishers, New York, 1987, pp. 221–230.
61. R. P. Sharma and D. K. Salunke, *Toxicants of Plant Origin*, CRC Press, Boca Raton, Fla., 1989, pp. 179–236.
62. T. R. Gormley, G. Downey, and D. O'Beirne, *Food, Health and the Consumer*, Elsevier, New York, 1987.
63. R. F. Keeler and co-workers, *Bull. Environ. Contam. Toxicol.* **15**, 522 (1976).
64. R. F. Keeler, *Lancet* **1**, 1187 (1973).
65. S. J. Jadhav, R. P. Sharma, and D. K. Salunke, *Crit. Rev. Toxicol.* **9**, 21 (1981).
66. B. N. Ames, *Science* **221**, 1256 (1983).
67. K. Nishie and co-workers, *Toxicol. Appl. Pharmacol.* **19**, 81 (1971).
68. R. G. Green, *Minn. Wildl. Dis Inv.* **3**, 83 (1937).
69. M. D. Melamad and N. M. Green, *Biochem. J.* **89**, 591 (1963).
70. N. D. Smashevskii, *Biol. Nauki (Moscow)*, (9), 80 (1968).
71. F. H. Kratzer, *Poult. Sci.* **26**, 90 (1947).
72. J. G. Reinhold and co-workers, *J. Nutr.* **106**, 493 (1976).
73. D. C. Harrison and E. Mellanby, *Biochem. J.* **33**, 1660 (1939).
74. H. Dam, *Pharmacol. Rev.* **9**, 1 (1957).
75. T. L. Aw and B. G. Swanson, *J. Food Sci.*, **50**, 67 (1985).
76. S. L. Balloun and E. L. Johnson, *Arch. Biochem. Biophys.* **42**, 355 (1953).
77. P. A. DaDamio and D. B. Thomson, *J. Food Sci.* **57**, 458 (1992).
78. F. Shahidi and M. Naczk, *J. Food Sci.* **54**, 1082 (1989).
79. G. Bartolozzi, *Rev. Clin. Pediat.* **80**, 231 (1967).
80. D. Rosenberg and co-workers, *Pediatrics* **80**, 231 (1967).
81. L. Piedrabuena, *Nutr. Abstr. Rev.* **40**, 48 (1970).
82. M. P. Lamdew, *N. Engl. J. Med.* **284**, 336 (1970).

83. G. K. Murphy and co-workers, *Envir. Sci. Technol.* **5**, 436 (1971).
84. V. Herbert, *Am. J. Clin. Nutr.* **32**, 1121 (1979).
85. T. N. Blumer and co-workers, eds., *Effect of Nitrates and Nitrites in Food as Related to Health*, Institute of Nutrition, University of North Carolina, Raleigh, 1973.
86. L. B. Bullerman, *J. Food Prot.* **42**, 65 (1979).
87. D. M. Hegsted, *Am. J. Clin. Nutr.* **31**, 1504 (1978).
88. C. Schlatter, *Bibl. Nutr. Dieta.* **40**, 55 (1988).
89. Y. Fu-Sun and S. Kong-Nien, *Adv. Cancer Res.* **47**, 297 (1986).
90. P. B. Hamilton, *Refu. Vet.* **39**, 17 (1982).
91. A. E. Rutledge, *Proceedings First International Symposium Feed Composition*, Utah State University, Logan, 1976, pp. 345–350.
92. C. W. Hesseltine, *Sixth International Symposium Mycotoxins and Phytotoxins*, Pretoria, South Africa, 1985, pp. 1–18.
93. C. A. Linsell, *Ann. Nutr. Aliment.* **31**, 997 (1977).
94. W. O. Caster and co-workers, *Int. J. Vitam. Nutr. Res.* **56**, 291 (1986).
95. D. W. Denning, *Acute Poison Rev.* **4**, 175 (1987).
96. J. J. Langone and H. Van Vunakis, *J. Natl. Cancer Inst.* **56**, 591 (1976).
97. C. S. Gagliardi and R. Mooney, *J. Food Prod.* **54**, 627 (1991).
98. M. Moorman, *Dairy Food Environ. Sanit.* **10**, 207 (1990).
99. *Northwest Parks and Wildlife* **2**(5), 8 (1992).
100. T. Kawabata, *Fish as Food*, Vol. 2, Academic Press, Inc., New York, 1962, p. 467.
101. W. L. Clark and G. W. Serbia, *Food Tech.* **45**(2), 84 (1991).
102. J. M. Concon, *Food Toxicology*, Marcel Dekker, Inc., New York, 1988.
103. G. A. Silverstone, *Diet Related Diseases*, AVI, Westport, Conn., 1985.
104. P. A. Murphy, *Food Tech.* **36**(1), 60 (1982).
105. G. T. Bryan and A. M. Pamukcu, *Carcinogens and Mutagens in the Environment*, CRC Press, Boca Raton, Fla., 1982.
106. I. Hirono, *Toxicants of Plant Origin*, CRC Press, Boca Raton, Fla., 1989.
107. H. F. Mower, *Carcinogens and Mutagens in the Environment*, CRC Press, Boca Raton, Fla., 1983.
108. R. J. Shamberger, *Nutrition and Cancer*, Plenum Press, New York, 1984.
109. NAS/NRC, *Diet and Health: Implications for Reducing Chronic Disease Risk*, NAS Press, Washington, D.C., 1989.
110. T. Monmaney, M. Hager, and S. Katz, *Newsweek on Health*, 12 (1987).
111. C. A. Krone, S. M. J. Yen, and W. T. Iwaoka, *Environ. Health Perspect.* **67**, 75–88 (1986).
112. B. P. Dunn, in Ref. 105, pp. 175–178.
113. J. McCann, *Diet, Nutrition and Cancer*, Alan R. Liss, New York, 1983, pp. 137–140.
114. H. V. Aeschbacher, *Genetic Toxicology of the Diet*, Alan R. Liss, New York, 1986, pp. 133–144.
115. L. F. Bjeldanes, *Nutr. Update* **1**, 105 (1985).
116. M. Kuratsune, M. Ikeda, and T. Hayashi, *Environ. Health Perspect.* **67**, 143 (1986).
117. B. N. Ames, R. Magaw, and L. S. Gold, *Science* **236**, 271 (1987).
118. M. Friedman, *Nutritional and Toxicological Aspects of Food Safety*, Plenum Press, New York, 1984.
119. M. Friedman, *Adv. Exp. Biol. Med.* **117**, 31 (1984).
120. A. J. Miller, *Food Tech.* **39**(2), 75 (1985).
121. T. Sugimura, *Environ. Health Perspect.* **67**, 5 (1986).
122. P. Handler, in *How Safe is Safe? The Design of Policy on Drugs and Food Additives*, NAS/NRC National Academy Press, Washington, D.C., 1974.

FRED H. HOSKINS
Washington State University

FORENSIC CHEMISTRY

Forensic chemistry can be defined as the application of chemistry to the law. In American jurisprudence, courts and judges are established to make factual determinations of matters brought before them. The fact-finding of the courts must often grapple with complex scientific issues and the legal system (1) has a particular way to deal with these technical and scientific matters. The court calls on subject matter experts (2), ie, expert witnesses, to explain and to interpret for the triers of fact the meaning of the scientific concept underlying a case and how it may be interpreted. To testify as an expert witness in a particular field or area of endeavor, the individual must qualify as an expert in a specific area, ie, have special knowledge, skill, training experience, or education. Courts have used subject matter experts for generations (3). The judge sitting on the particular case makes the determination whether an individual is qualified to give expert testimony.

Forensic science is an applied science having a focus on practical scientific issues that come up during criminal investigations or at trial. Some components are unique to the field because it is conducted within the legal arena. Forensic science issues in chemistry and biochemistry in criminal investigations are discussed herein. There are a host of other forensic science areas, eg, forensic medicine, forensic dentistry, forensic anthropology, forensic psychiatry, and forensic engineering, any of which may overlap with forensic chemistry.

Physical Evidence

Forensic scientists work with physical evidence, ie, "data presented to a court or jury in proof of the facts in issue and which may include the testimony of witnesses, records, documents or objects." Physical evidence is real or tangible and can literally include almost anything, eg, the transient scent of perfume on the clothing of an assault victim; the metabolite of a drug detected in the urine of an individual in a driving-under-the-influence-of-drugs case; the scene of an explosion; or bullets removed from a murder victim's body.

The courts are the ultimate consumer of the forensic scientist's information. Before expert testimony may be presented to a trier of fact (judge or jury), a legal determination must be made whether the information can be presented. Courts understand that lay juries may place undue reliance on experts and technology. To guard against this possibility, the court employs certain safeguards. Many jurisdictions use the Frye rule (4). In this landmark case, the United States Supreme Court considered whether or not the polygraph test was admissible and stated a general rule that courts go a long way in admitting expert testimony deduced from a well-recognized scientific principle or discovery. However, the theory from which the deduction is made must be sufficiently established to have gained general acceptance in the particular field to which it belongs. In 1993 the U.S. Supreme Court (5), liberalized the admissibility standard in federal cases, holding that scientific, technical, or other specialized evidence is admissible if it assists the trier of fact to understand the evidence. This results in a lower threshold for admissibility than general acceptance required by many state courts.

Examination of physical evidence provides two subtle and different types of conclusion as may be illustrated by the following examples. Consider a hit-and-run case involving an automobile (6) and a decedent. An examination of the victim's clothing turns up some blue paint. A suspect vehicle is located; the vehicle is blue. Infrared spectroscopy of the surface, solubility tests in various solvents, and microscopic examination of cross sections demonstrate that the composition of the paint from the vehicle and from paint samples recovered on the victim's clothing are identical (7). A laboratory report stating that the two specimens are identical is likely to prejudice a jury into concluding that the paints are identical, and therefore, it was the defendant's car that hit the pedestrian. A more carefully worded laboratory report would explain that the samples are identical and that the paint could have come from the car in question, or any other similarly painted car. Many automobile manufacturers use the same blend of paint on thousands, and likely hundreds of thousands of vehicles. No matter how much testing is done, the results are the same: the samples are indistinguishable. This concept is known as class or group characteristics. All members of a class or group have identical characteristics. Types of physical evidence which exhibit class characteristics are paint (qv), glass (qv), fibers (qv), fabric, building material, etc. This type of physical evidence is said to be identified. The best that chemical and physical examinations can ever do is to place items into groups of similarly manufactured items. It is not possible to differentiate one item of evidence as being uniquely distinguishable from another.

Some types of physical evidence, because of the manner in which the material is made, are unique; such evidence can be individualized. Examination can show an item of individualized evidence is unique and comes from one, and only one source. The classic example is fingerprints. No two individuals, even identical twins, have the same fingerprints. An examiner's report stating that a suspect's fingerprints are identical to latent fingerprints discovered on a weapon would prove, without doubt, that the suspect held the gun. Other categories of evidence exhibiting individualization are handwriting, markings on bullets fired from the same gun, broken pieces of glass or plastic which can be physically fit together again, and forensic deoxyribonucleic acid (DNA) evidence. In cases involving these types of evidence, the forensic examiner can state that the physical evidence came from a single source, to the exclusion of all others (8).

Historically, physical evidence has taken on increasing importance in criminal matters. Court decisions have consistently looked askance at a defendant's admissions of guilt and even question eyewitness testimony. Physical evidence has traditionally been viewed as impartial and unbiased, and not subject to the problems associated with confessions made by an accused or the testimony of witnesses.

Physical evidence serves two purposes. In some cases it is used to prove a component or element of a crime. For example, in a case involving trafficking in cocaine [50-36-2], the prosecutor must prove that the white powder found in the criminal's possession was cocaine (Table 1). The forensic chemist tests the substance and issues a report. If the powder is methamphetamine [537-46-2], the charge must be amended.

The other purpose for which physical evidence is used is to develop associative evidence in a case. Physical evidence may help to prove a victim or suspect was at a specific location, or that the two came in contact with one another. In one case, building material debris (wooden splinters, tar paper, insulation material) was found on a blanket used to wrapped a body that was found dumped at the side of a road. The evidence suggested an attic and eventually led detectives to the location where the murder occurred.

Most of the forensic science or crime laboratories located in North America are associated with law enforcement agencies, medical examiner–coroner departments, or prosecutors' offices. There are a large number of independent consultants, also. Laboratories exist at the municipal, county, state, and federal levels of government. There are approximately 300 government-operated forensic science laboratories in the United States.

Forensic science laboratories are generally divided into separate specialty areas. These typically include forensic toxicology, solid-dose drug testing, forensic serology, trace evidence analysis, firearms and tool mark examination, questioned documents examination, and latent fingerprint examination. Laboratories principally employ chemists, biochemists, and biologists at various degree levels. In some specialty areas, eg, firearms examination, questioned documents, and fingerprint examination, experts may not have an academic degree. Some laboratories employ examiners that specialize only in one area whereas other laboratories maintain a generalist philosophy, rotating examiners through several forensic science disciplines during a practitioner's career.

The bulk of the scientific testing in crime laboratories involves the analysis and characterization of either synthetic or biochemical organic substances or both. Additionally there are a number of evidence categories classified as inorganic. Most assays are simply qualitative and designed to answer the questions: what is it, where did it come from, and could it have come from a specific source? Quantitative analyses may also be carried out on samples involving drug evidence, toxicology evidence, and blood alcohol testing, wherever such information has probative value to the investigation or to the court in its deliberation.

Forensic Testing

Toxicology. Psychoactive substances, illicit and ethical (licit) drugs and alcohol (ethanol), are the greatest source of physical evidence analyzed in most crime laboratories (see PSYCHOPHARMACOLOGICAL AGENTS). Drug testing falls into two categories: solid dose samples and toxicology (qv) related cases, eg, blood, urine, or tissue specimens in post-mortem cases or cases involving driving under the influence of alcohol or drugs, as well as workplace or employee drug testing.

Alcohol. The number of driving under the influence of alcohol (DUI) cases reflects the enormity of the drunken driving problem in the United States (9). Tests to measure blood alcohol concentration are conducted on blood, urine, or breath (10). In the case of urine and breath, the alcohol concentration measured is reported in terms of the equivalent blood alcohol concentration. Most states in the United States presume that a person is under the influence of alcohol with respect to driving a motor vehicle at a blood alcohol concentration of 0.10%, ie, an ethanol concentration $\geq$ 10 g/100 mL of blood. Some states maintain a lower necessary concentration of 0.08%. In some European countries levels are as low

Table 1. Commonly Encountered Drugs of Abuse

Common name	CAS Registry Number	Chemical name	Molecular formula	Structure
amphetamine	[300-62-9]	α-methylbenzeneethanamine	$C_9H_{13}N$	
cocaine	[50-36-2]	{1R-(exo,exo)}-3-(benzoyloxy)-8-methyl-8-azabicyclo{3.2.1}octane-2-carboxylic acid methyl ester	$C_{17}H_{21}NO_4$	
heroin	[561-27-3]	7,8-didehydro-4,5α-epoxy-17-methylmorphinan-3,6,α-diol diacetate	$C_{21}H_{23}NO_5$	
lysergic acid diethylamide (LSD)	[50-37-3]	9,10-didehydro-N,N-diethyl-6-methylergoline-8β-carboxamide	$C_{20}H_{25}N_3O$	

Name	CAS Number	Molecular Formula	Structure
mescaline	[54-04-6]	$C_{11}H_{17}NO_3$	3,4,5-trimethoxybenzeneethanamine
methamphetamine	[537-46-2]	$C_{10}H_{15}N$	N,α-dimethylbenzeneethanamine
morphine	[57-27-2]	$C_{17}H_{19}NO_3$	7,8-didehydro-4,5-epoxy-17-methylmorphanin-3,6-diol
phencyclidine (PCP)	[77-10-1]	$C_{17}H_{25}N$	1-(1-phenylcyclohexyl)piperidine
Δ-9-tetrahydrocannabinol (Δ-9-THC)[a]		$C_{21}H_{30}O_2$	tetrahydro-6,6,9-trimethyl-3-pentyl-6H-debenzo[b,d]pyran-1-ol

[a]The active ingredient in marijuana and hashish.

as 0.05%. A blood alcohol concentration of 0.10% in a 68-kg (150-lb) person is the equivalent of about four drinks of 80 proof alcoholic beverage or four 340-g (12-oz) beers in the body at the time of the test (see BEER; BEVERAGE SPIRITS, DISTILLED; WINE). Ethanol is metabolized at the equivalent rate of about one drink per hour.

Blood and urine are most often analyzed for alcohol by headspace gas chromatography (qv) using an internal standard, eg, 1-propanol. Assays are straightforward and lend themselves to automation (see AUTOMATED INSTRUMENTATION). Urine samples are collected as a voided specimen, ie, subjects must void their bladders, wait about 20 minutes, and then provide the urine sample. Voided urine samples provide the most accurate determination of blood alcohol concentrations. Voided urine alcohol concentrations are divided by a factor of 1.3 to determine the equivalent blood alcohol concentration. The 1.3 value is used because urine has approximately one-third more water in it than blood and, at equilibrium, there is about one-third more alcohol in the urine as in the blood.

Breath alcohol testing is accomplished by a number of techniques. The oldest reliable procedure involves bubbling a measured volume of deep-lung air containing alcohol through an acidic solution of potassium dichromate, $K_2Cr_2O_7$. Deep-lung air is the last portion of expired breath. It is collected in breath alcohol testing to ensure that the alcohol concentration in the breath is in equilibrium with the alcohol in the alveolar blood supply. Products from a simple oxidation–reduction reaction forming Cr^{3+} are measured photometrically. The amount of Cr^{3+} formed is directly proportional to the alcohol concentration of the breath which is proportional to the blood alcohol concentration. Newer instruments rely on infrared spectroscopy to measure the blood alcohol concentration in breath. A fixed quantity of breath is captured and the alcohol concentration measured by determining the absorbance at the C—H stretch of ethanol (see INFRARED AND RAMAN SPECTROSCOPY, INFRARED TECHNOLOGY).

Other Substances. Driving under the influence of alcohol cases are complicated because people sometimes consume alcohol with other substances (11–13). The most common illicit substances taken with alcohol are marijuana and cocaine (see Table 1) (14). In combination with alcohol, some drugs have an additive effect. When a blood or urine alcohol sample is tested for alcohol and the result is well below the legal concentration threshold yet the test results are not consistent with the arresting officers observation that the subject was stuporous, further toxicological tests for the possible presence of drugs are indicated.

Forensic toxicology laboratories having large caseloads rely on immunoassay (qv) techniques to screen specimens. Immunoassay technology involves the manufacture of antibodies that are specific to particular drugs or to a class of drugs. For example, morphine (Table 1) can be chemically bound to a protein and injected into a host animal, such as a goat. After several weeks the animal is bled and antibodies for morphine and related drugs can be isolated and purified (15).

There are several immunological techniques in use. Enzyme multiplied immunological technique (EMIT) employs an enzymatic reaction to determine concentration, whereas radioimmunoassay (RIA) uses radioactively tagged reagents such as ^{125}I to measure concentration. Antibodies combine with the drug and drug metabolites present in blood or urine, in competition with a labeled drug which is in the reagent mixture. The tagged moiety may be radioactively labeled or attached to an enzyme that causes cells to lyse. A competing reaction between tagged and untagged portions provides a semiquantitative result by allowing the determination of the amount of radioactively labeled reagent remaining (RIA) or the extent of cell lysing (EMIT). These values are related to the concentration of the drug. These procedures yield results in a matter of minutes and are easily automated. However, one drawback is that there is cross-reactivity between similar drugs. For example, using the morphine antibody, most of the opiate alkaloids (qv) cross-react to some extent. Thus, using this test alone, it would be impossible to differentiate between codeine [76-57-3] and morphine. As a result, immunoassay procedures are best used as screening techniques and must be confirmed by other, more selective analytical procedures.

Thin-layer chromatography (tlc) (16) is frequently used. The procedure allows for rapid screening for most drugs of abuse using simple, inexpensive technology. A drawback to tlc, however, is that the technique is not especially sensitive and low levels of drugs may be missed.

Gas chromatography (gc) and gas chromatography–mass spectroscopy (gc/ms) are the most common analytical procedures used in modern forensic toxicology laboratories (17) (see ANALYTICAL METHODS, HYPHENATED INSTRUMENTS). Drugs are separated from their biological matrices, ie, blood, urine, liver, etc, by liquid–liquid or solid-phase extraction (qv) utilizing the solubility of the suspect drug in acid or alkaline aqueous solution relative to the organic solution containing the specimen. As quantitative analysis is often required, an internal standard is added to the assay at the beginning. For gas chromatography analysis, a chemically similar compound is used, whereas for gc/ms analysis, a deuterated version of the questioned drug is added to the specimen as an internal standard.

The interpretation of forensic toxicology (18) results is often challenging. Courts frequently ask if an amount of drug detected in a specimen could cause a specific type of behavior, ie, would someone be under the influence of a drug at a specific concentration, would a particular drug concentration cause diminished capacity, or was the drug the cause of death? In a random employee drug testing case, a worker screened positive for opiates by EMIT and gc/ms analysis of the urine specimen showed low levels of morphine. Although one possibility was that the individual was a heroin user, a review of foods eaten in the prior 24 hours suggested a more innocent cause: a poppy-seed bagel.

Solid-Dose Narcotics and Dangerous Drugs. Solid-dose drug testing (19) differs from forensic toxicology in that the solid form of the drug is tested, rather than a biological specimen containing the drug and its metabolite. The typical drugs of abuse (Table 1) in North America are heroin; cocaine, ie, free-base, crack, and the HCl salt; marijuana; hashish, a concentrated form of marijuana; amphetamine; methamphetamine; phencyclidine; and LSD. There are also many other illicit and legitimate pharmaceuticals (qv) (20) that find a way into the illegal drug market and thus must be analyzed in forensic science laboratories.

Forensic science laboratories may have different missions and therefore conduct different types of testing on samples (21,22). For example, the United States Department of Justice, Drug Enforcement Administration (DEA) forensic laboratories assist authorities in criminal intelligence-gathering efforts. As such, DEA chemists routinely analyze both the illicit drug and excipient, the material used in the cutting or diluting of the pure drug, in a given specimen. The excipient may provide information as to where the sample was produced.

Local and state forensic laboratories generally do not engage in excipient testing. Most provide qualitative and quantitative analysis of the evidence to determine if an illegal substance is present and if so, the amount of the drug present. The quantity of drug seized by the authorities may be important in jurisdictions which give enhanced sentences for larger amounts of the pure drug, or in some cases the total weight on the drug and diluent in possession of the defendant.

The large numbers of drug trafficking arrests made by police agencies and the resulting high volume of cases submitted to most crime laboratories make rapid analytical schemes the norm. Laboratories usually rely on a combination of screening tests followed by instrumental-based confirmatory analyses. Functional group color spot tests are followed by microcrystalline tests and then often a combination of gc, gc/ms, tlc, and Fourier transform ir spectroscopy for further identity confirmation. Microcrystalline tests are unique to drug testing in forensic science laboratories. The test involves adding a drop of a solution of a heavy-metal salt, eg, gold or platinum, to a few milligrams of the drug and observing the formation of characteristic crystals under the microscope. Microcrystalline tests are highly specific and very slight structural changes in a family of drugs can easily be detected.

The most common illicit drugs in the United States today are heroin, cocaine, marijuana, hashish, phenyclidine, LSD, and methamphetamine. These make up at least 90% of the total drugs seized.

Trace Evidence. Trace evidence (23) refers to minute, sometimes microscopic material found during the examination of a crime scene or a victim's or suspect's clothing (see TRACE AND RESIDUE ANALYSIS). Trace evidence often helps police investigators (24) develop connections between suspect and victim and the crime scene. The theory behind trace evidence was first articulated by a French forensic scientist: the Locard Exchange Principle notes that it is not possible to enter a location, such as a room, without changing the environment. An individual brings trace materials into the area and takes trace materials away. The challenge to the forensic scientist is to locate, collect, preserve, and characterize the trace evidence.

Searching a crime scene is a complex process (25), involving police, crime scene technicians, and forensic scientists. The procedure requires careful documentation, collection, and preservation of the evidence. Trace evidence (26) in criminal investigations typically consists of hairs (27,28); both natural and synthetic fibers (qv) (29,30), fabrics; glass (qv) (31,32); plastics (33); soil; plant material; building material such as cement (qv), paint (qv), stucco, wood (qv), etc (34), flammable fluid residues (35,36), eg, in arson investigations; explosive residues, eg, from bombings (37,38) (see EXPLOSIVES AND PROPELLENTS), and so on.

Perhaps the simplest examination done is the physical match. A small fragment of wood, plastic, or other material is recovered and fitted into a large piece found on the suspect or at the scene of the crime (39). Other examinations result only in demonstrating class characteristics (40). Such information may be used in a prosecution as circumstantial evidence in a trial. However, it is important that the forensic scientist neither inflate nor minimize (41,42) its importance.

Microscopy (qv) plays a key role in examining trace evidence owing to the small size of the evidence and a desire to use nondestructive testing (qv) techniques whenever possible. Polarizing light microscopy (43,44) is a method of choice for crystalline materials. Microscopy and microchemical analysis techniques (45,46) work well on small samples, are relatively nondestructive, and are fast. Evidence such as soil, minerals, synthetic fibers, explosive debris, foodstuff, cosmetics (qv), and the like, lend themselves to this technique as do comparison microscopy, refractive index, and density comparisons with known specimens. Other microscopic procedures involving infrared, visible, and ultraviolet spectroscopy (qv) also are used to examine many types of trace evidence.

More traditional analytical techniques (47) also are used. Capillary column gas chromatography is the method of choice for characterizing flammable fluid residues (48) in arson cases. Trace residues may be collected by heated headspace techniques or absorption–deabsorption of the residue from an appropriate solid matrix. The challenge in arson cases is interpreting the resulting chromatograms. Flammables subjected to high temperatures or weathering, eg, exposure to the elements over a period of time, appear significantly different from a sample of the original flammable fluid. It is also important to consider the effects of high heat and subsequent distillation on the arson scene's components such as carpeting, paints, wood products, various foams, etc. These can sometimes be confused with flammable residues (49).

Scanning electron microscopy (sem) and energy dispersive x-ray analysis (edx) are used frequently in gunshot residue examination (50,51) and to characterize evidence of an inorganic origin. When a gun is fired, gases from the explosion of the primer condense on the shooter's hands (52). These residues contain barium, antimony, and lead and are spherical. Collecting these residues and examining them by sem/edx is a straightforward way to determine whether someone recently fired a firearm.

Pattern recognition examinations are important in footwear (53) and tire impression cases. Often, tire impressions and footprints (54) are left at crime scenes and forensic scientists are asked to compare the impressions with shoes or tires. Interpretation of such evidence requires an understanding of the manufacturing process (55), a critical study of the large variety of different patterns, and experience in the way these items wear. One concern in footwear examination is whether a formation on a shoe sole is a wear mark, a mark unique to that shoe alone, or simply a defect caused during the manufacturing process. For example, if an injection molding process is used, tiny imperfections in a sole may be found on each and every shoe sole and not be a unique mark imperfection. Thus it is critical for the examiner to have a clear appreciation of the manufacturing processes involved.

Lasers (qv) and other high intensity or alternative light sources are useful in crime laboratories to visualize latent fingerprints, seminal fluid stains, obliterated writings, and erasures, and to aid in specialized photographic work. Infrared and ultraviolet light sources are also used to view items of evidence.

Forensic Serology.   Blood, often associated with crimes of violence, is powerful physical evidence. Its presence suggests association with the criminal act and blood can be used to associate suspects and locations with the bleeder. Blood is a complex mixture of cellular material, proteins, and enzymes and several tests are available for suspected bloody evidence. A typical test protocol involves (1) determining whether blood is present, (2) determining if it is human blood, (3) typing the blood, and (4) when applicable, performing DNA typing.

Detecting the presence of small, even invisible, amounts of blood is routine. Physical characteristics of dried stains give minimal information, however, as dried blood can take on many hues. Many of the chemical tests for the presence of blood rely on the catalytic peroxidase activity of heme (56,57). Minute quantities of blood catalyze oxidation reactions between colorless materials, eg, phenolphthalein, luco malachite green, luminol, etc, to colored or luminescent ones. The oxidant is typically hydrogen peroxide or sodium perborate (see AUTOMATED INSTRUMENTATION, HEMATOLOGY).

Species origin tests, used to determine whether the specimen is human or from another source, are immunological in nature. Host animals, usually rabbits, are injected with protein from another species. The animal creates antibodies to the unknown material. Serum from the host animal, containing species (human, bovine, equine, canine, etc) specific antibodies, is tested against a dilute solution of blood (antigens) collected as evidence. A positive reaction is determined by a visible band where the antibodies and antigens come into contact.

Blood collected as evidence in criminal acts is usually dried and deposited on a variety of substrates. Sample size is usually on the order of a 2 or 3 mm diameter stain. Traditional typing (58) involves ABO blood grouping, and characterizing stable polymorphic proteins or enzymes present in blood by means of electrophoresis (see BLOOD FRACTIONATION; ELECTROSEPARATIONS, ELECTROPHORESIS). The dried blood stain is extracted with saline then reconstituted onto a cotton thread. The thread is embedded into a gel and electrophoresed using appropriate control and known samples. Each genetic type has a known, independent population distribution and, using the product rule, the results of several typing tests on different enzymes and proteins can be multiplied together to determine the likelihood that the blood came from a certain donor.

More recently, the forensic application of DNA testing has dramatically enhanced the ability to determine the source of a blood sample (59,60) (see NUCLEIC ACIDS). Two procedures are in forensic use: restriction fragment length polymorphism (RFLP) and polymerase chain reaction (PCR). Using RFLP, DNA is extracted from a sample and is cut up into fragments at specific sequence sites using restriction enzymes. The fragments are separated by size by means of electrophoresis and transferred onto a nylon membrane. Radioactive, single-locus probes which recognize specific sequences on DNA are added to the membrane which is subsequently placed in contact with x-ray film to show the location of bands. The procedure is repeated four times using different probes and a probability of the sample coming from a specific donor is calculated. The resulting calculations often show that a sample is unique to one and only one source.

Utilizing PCR, the analysis works using much smaller samples. Samples of DNA are amplified, which is the biological equivalent of molecularly photocopying the DNA (see GENETICS ENGINEERING). Exceedingly small samples of DNA can be duplicated for subsequent testing. This procedure yields much smaller probabilities of a match.

BIBLIOGRAPHY

"Forensic Chemistry" in *ECT* 1st ed., Vol. 6, pp. 848–857, by P. L. Kirk, University of California; in *ECT* 3rd ed., Vol. 11, pp. 220–230, by T. P. Perros, George Washington University.

1. P. C. Giannelli, *J. Forensic Sci.* **34**(3), 730–748 (1989).
2. T. Hodgkinson, *Law Qly. Rev.* **104**(4), 198–202 (1988).
3. A. Moenssens, F. E. Imbau, and J. E. Starrs, *Scientific Evidence in Criminal Cases*, 3rd ed., Foundation Press, Mineola, N.Y., 1986.
4. *Frye v. United States*, 293 F. 1013, D.C. Cir. 1923.
5. Daubert v. Dow Pharmaceutical, Inc., 113 S. Ct. 2786 (1993).
6. D. R. Cousins and co-workers, *Forensic Sci. Int.* **43**(2), 183–197 (1989); G. Dabdoub and co-workers, *J. Forensic Sci.* **34**(6), 1395–1404 (1989).
7. S. G. Ryland and R. J. Kopec, *J. Forensic Sci.* **24**(1), 140–147 (1979).
8. F. C. Drummond and P. A. Pizzola, *J. Forensic Sci.* **35**(3), 746–752 (1990).
9. *Alcohol and the Impaired Driver. A Manual on the Medicolegal Aspects of Chemical Tests for Intoxication.* American Medical Association, Chicago, Ill., 1970.
10. M. F. Mason and K. M. Dubowski, *Clin. Chem.* **20**, 126 (1974).
11. J. C. Garriott and N. Latman, *J. Forensic Sci.* **21**(2), 398–415 (1976).
12. R. F. Turk, A. J. McBay, and P. Hudson. *J. Forensic Sci.* **19**(1), 90–97 (1974).
13. R. E. Willette, ed., *Drugs and Driving*, NIDA Research Monograph 11, U.S. Department of Health, Education and Welfare, Washington, D.C., 1977.
14. G. D. Lundberg, J. M. White, and K. I. Hoffman, *J. Forensic Sci.* **24**(1), 207–215 (1979).
15. A. C. Moffett, ed., *Clark's Idolation and Identification of Drugs*, The Pharmaceutical Press, London, 1986, pp. 148–159.
16. Ref. 15, pp. 160–177.
17. J. Chamberlain, *Analysis of Drugs in Biological Fluids*, CRC Press, Boca Raton, Fla., 1985.
18. R. L. Hawks and C. N. Chiang, eds, *Urine Testing for Drugs of Abuse*, NIDA Research Monograph 73, National Institute of Drug Abuse, Rockville, Md., 1986.
19. R. M. Baum, *C&EN*, 7–16 (Sept. 9, 1985).
20. R. D. Daigle, *J. Psychoactive Drugs*, **22**(1), 77–80 (1990).
21. R. D. James, *FBI Law Enf. Bull.* **58**(4), 16–21 (1989).
22. T. C. Kram, D. A. Cooper, and A. C. Allen, *Anal. Chem.* **53**(12), 1379A–1385A (1981).
23. N. Petraco, *J. Forensic Sci.* **31**(1), 321–328 (1986).
24. J. G. Collinson, *J. Forensic Sci. Soc.* **10**(4), 199 (1970).
25. B. A. J. Fisher, *Techniques of Crime Scene Investigation*, 5th ed., Elsevier Science Publishing Co., Inc., New York, 1992.
26. N. Petraco, *J. Forensic Sci.* **32**(5), 1422–1425 (1987).
27. *FBI Law Enf. Bull.* **45**(5), 9–15 (May 1976).
28. J. W. Hicks, *Microscopy of Hair: A Practical Guide and Manual.* Federal Bureau of Investigation, U.S. Government Printing Office, Washington, D.C., 1977.
29. R. R. Bresee, *J. Forensic Sci.* **32**(2), 510–521 (1987).
30. J. Robertson, *Forensic Examination of Fibers*, Ellis Harwood, a Division of Simon and Schuster, New York, 1992.

31. W. Fong, *J. Forensic Sci. Soc.* **11**(5), 267 (1971).

32. W. Fong, *J. Forensic Sci.* **18**(4), 398–404 (1973).

33. D. S. Pierce, *J. Forensic Identific.* **40**(2), 51–59, (1990).

34. S. Palenik, *The Microscope* **30**, 93–100 (1982).

35. P. J. Loscalzo, P. R. DeForest, and J. M. Chao. *J. Forensic Sci.* **25**(1), 162–167 (1980).

36. J. D. DeHaan, *Kirk's Fire Investigation.* 2nd ed., John Wiley and Sons, Inc., New York, 1983.

37. D. D. Garner, *J. Energetic Mater.* **4**(1-4), 133–148 (1986).

38. *Bomb Investigations.* National Bomb Data Center, Picatinny Arsenal, Dover, N.J., 1974.

39. U. G. von Bremen and L. K. R. Blunt, *J. Forensic Sci.* **28**(3), 644–654 (1983).

40. E. D. Hamm, *J. Forensic Identific.* **39**(5), 277–292 (1989).

41. H. Hollien, *J. Forensic Sci.* **35**(6), 1414–1423 (1990).

42. B. Knight, *J. Forensic Sci. Soc.* **29**(1), 53–60, (1989).

43. W. C. McCrone, L. C. McCrone, and J. G. Delly, *Polarized Light Microscopy*, Ann Arbor Science Publishers, Inc., Ann Arbor, Mich., 1979.

44. P. R. Deforest, in R. Saferstein, ed., *Forensic Science Handbook*, Prentice-Hall, Inc., Englewood Cliffs, N.J., 1982, Chapt. 9, pp. 416–528.

45. S. Palenik, in Ref. 44, Vol. II, 1988, Chapt. 4, pp. 161–208.

46. E. M. Chemot and C. W. Mason, *Handbook of Chemical Microscopy*, 3rd ed. Vol. I, and *Handbook of Chemical Microscopy*, 2nd ed. Vol. II, John Wiley and Sons, Inc., New York 1958 and 1940.

47. I. C. Stone, J. N. Lomonte, L. A. Fletcher, and W. T. Lowry, *J. Forensic Sci.* **23**(1), 78–83 (1978).

48. J. F. Boudreau and co-workers, *Arson and Arson Investigation Survey and Assessment*, National Institute of Law Enforcement and Criminal Justice, Law Enforcement Assistance Administration, U.S. Department of Justice, U.S. Government Printing Office, Washington, D.C., 1977.

49. J. J. O'Donnell, *Fire Arson Invest.* **39**(4), 25–27 (1989).

50. J. Andrasko and A. C. Maehly. *J. Forensic Sci.* **22**(2), 279–287 (1977).

51. D. G. Havekost, C. A. Peters, and R. D. Koons, *J. Forensic Sci.* **35**(5), 1096–1114 (1990).

52. S. Basu, *J. Forensic Sci.* **27**(1), 72–91 (1982).

53. E. E. Hueske, *J. Forensic Identific.* **41**(2), 92–95 (1991).

54. M. J. Cassidy, *Footwear Identification*, Royal Canadian Mounted Police, Ontario, Canada, 1980.

55. W. J. Bodziak, *J. Forensic Sci.* **31**(1), 153–176 (1986).

56. A. Fiori, in F. Lundquist, ed., *Methods of Forensic Science*, Vol. I, Wiley-Interscience, New York, 1962, pp. 243–290.

57. H. C. Lee, in Ref. 44, pp. 267–337

58. R. E. Gaensslen, *Law Enf. Communic.*, 23–30, (Feb. 1980).

59. *Genetic Witness - Forensic Uses of DNA Tests*, OTA-BA-438, U.S. Congress, Office of Technology Assessment, U.S. Government Printing Office, Washington, D.C., 1990.

60. *Committee on DNA Technology in Forensic Science Report* National Research Council, Committee on DNA Technology in Forensic Science, National Academy Press, Washington, D.C., 1992.

General References

R. Saferstein, ed., *Forensic Science Handbook.* Prentice-Hall, Englewood Cliffs, N.J., 1982 and *Forensic Science Handbook, Vol. II*, 1988.

P. L. Kirk, *Crime Investigation*, 2nd ed., John Wiley & Sons, Inc., New York, 1974.

R. H. Cravey and R. C. Baselt, *Introduction to Forensic Toxicology*, Biomedical Publications, Davis, Calif., 1981.

R. E. Gaensslen, *Sourcebook in Forensic Serology, Immunology and Biochemistry*, National Institute of Justice, U.S. Department of Justice, U.S. Government Printing Office, Washington, D.C., 1983.

BARRY A. J. FISHER
Scientific Services Bureau, Los Angeles County
Sheriff's Department

FORMALDEHYDE

Formaldehyde [*50-00-0*], $H_2C{=}O$, is the first of the series of aliphatic aldehydes. It was discovered by Butlerov in 1859 and has been manufactured since the beginning of the twentieth century. Annual worldwide production capacity now exceeds 15×10^6 t (calculated as 37% solution). Because of its relatively low cost, high purity, and variety of chemical reactions, formaldehyde has become one of the world's most important industrial and research chemicals (1).

Physical Properties

At ordinary temperatures, pure formaldehyde is a colorless gas with a pungent, suffocating odor. Physical properties are summarized in Table 1; thermodynamic values for temperatures ranging from 0–6000 K are given in the *JANAF Interim Thermochemical Tables* (11,12). Other properties are listed in Reference 9.

Formaldehyde is produced and sold as water solutions containing variable amounts of methanol. These solutions are complex equilibrium mixtures of methylene glycol, $CH_2(OH)_2$, poly(oxymethylene glycols), and hemiformals of these glycols. Ultraviolet spectroscopic studies (13–15) indicate that even in highly concentrated solutions the content of unhydrated HCHO is <0.04 wt %.

Density and refractive index are nearly linear functions of formaldehyde and methanol concentration. Based on available data (16–19), the density may be expressed in g/cm^3 by the following approximation:

$$\text{density} = [1.119 + 0.003\,(F - 45) - 0.0027\,M]\,[1.0 + 0.00055\,(55 - t)] \quad (1)$$

where F and M are the formaldehyde and methanol concentrations, wt %, respectively, and t is in °C.

The refractive index (20) may be expressed by a simple approximation for solutions containing 30–50 wt % HCHO and 0–15 wt % CH_3OH:

Table 1. Properties of Monomeric Formaldehyde

Property	Value	Reference
density, g/cm^3		
at $-80°C$	0.9151	2
at $-20°C$	0.8153	2
boiling point at 101.3 kPa,[a] °C	-19	3
melting point, °C	-118	3
vapor pressure, Antoine constants, Pa[b]		4
A	9.21876	
B	959.43	
C	243.392	
heat of vaporization,[c] ΔH_v at 19°C, kJ/mol[d]	23.3	3,5
heat of formation, $\Delta H_f°$ at 25°C, kJ/mol[d]	-115.9	6
std free energy, $\Delta G_f°$ at 25°C, kJ/mol[d]	-109.9	6
heat capacity, $C_p°$, J/(mol·K)[d]	35.4	6
entropy, $S°$, J/(mol·K)[d]	218.8	6
heat of combustion, kJ/mol[d]	563.5	6
heat of solution at 23°C kJ/mol[d]		7
in water	62	
in methanol	62.8	
in 1-propanol	59.5	
in 1-butanol	62.4	
critical constants		8,9
temperature, °C	137.2–141.2	
pressure, MPa[e]	6.784–6.637	
flammability in air		
lower/upper limits, mol %	7.0/73	10
ignition temperature, °C	430	10

[a]To convert kPa to mm Hg, multiply by 7.5.

[b]$\text{Log}_{10}P_{Pa} = A - B/(C+t)$; t = °C. To convert $\log_{10}P_{Pa}$ to $\log_{10 \text{ mm Hg}}$, subtract 2.1225 from A.

[c]At 164 to 251 K, $\Delta H_v = (27{,}384 + 14.56T - 0.1207\ T^2)$ J/mol[d] (3).

[d]To convert J to cal, divide by 4.184.

[e]To convert MPa to atm, divide by 0.101.

$$n_{\text{D}}^{18} = 1.3295 + 0.00125\ F + 0.0000113\ M \tag{2}$$

Viscosities have been measured for representative commercial formaldehyde solutions (21). Over the ranges of 30–50 wt % HCHO, 0–12 wt % CH$_3$OH, and 25–40°C, viscosity in mPa·s($=$cP) may be approximated by

$$\text{viscosity} = 1.28 + 0.039\ F + 0.05\ M - 0.024\ t \tag{3}$$

Significant discrepancies in formaldehyde partial pressures above aqueous solutions (22,23) can occur due to nonequilibrium conditions in the liquid phase. However, these problems have been overcome and consistent results obtained (8,18,22,24–26).

In methanol–formaldehyde–water solutions, increasing the concentration of either methanol or formaldehyde reduces the volatility of the other. Vapor-liquid-equilibrium data (8,27) for several methanolic formaldehyde solutions are given in Table 2. The flash point varies with composition, decreasing from 83 to 60°C as the formaldehyde and methanol concentrations increase (17,18).

Formaldehyde solutions exist as a mixture of oligomers, $HO(CH_2O)_nH$. Their distribution has been determined for 6–50 wt % HCHO solutions with low methanol using nmr and gas chromatographic techniques (28,29). Averages of the equilibrium constants for equation 4 are $K_2 = 7.1$, $K_3 = 4.7$, and $K_4 = 3.4$. The equilibrium constants appear to be nearly independent of temperature over the range of 30–65°C. Methanol stabilizes aqueous formaldehyde solutions by decreasing the average value of n (28,30). Hence, methanolic solutions can be stored at relatively low temperatures without precipitation of polymer.

$$HO(CH_2O)_nH + HOCH_2OH \xrightleftharpoons{K_n} HO(CH_2O)_{n+1}H + H_2O \qquad (4)$$

The following approximation was derived from data at 25–80°C (28,30,31) for solutions containing 7–55 wt % formaldehyde and 0–14 wt % methanol:

$$\text{monomer (mol \%)} = 100 - 12.3 \sqrt{F} + (1.44 - 0.0164\,F)\,M \qquad (5)$$

where the monomer is the mole percent of formaldehyde present as formaldehyde hemiformal and methylene glycol, F is the wt % HCHO, and M is the wt % CH_3OH.

Commercial formaldehyde–alcohol solutions are clear and remain stable above 16–21°C. They are readily obtained by dissolving highly concentrated formaldehyde in the desired alcohol.

Table 2. Vapor Pressure above Formaldehyde Solutions, kPa[a]

Wt %, liquid HCHO:CH₃OH:H₂O	40°C			60°C		
	HCHO	CH₃OH	H₂O	HCHO	CH₃OH	H₂O
37:7:56	0.039	0.053	0.308	0.143	0.122	0.851
37:1:62	0.031	0.007	0.322	0.130	0.016	0.891
50:7:43	0.048	0.041	0.265	0.176	0.096	0.749
50:1:49	0.040	0.005	0.283	0.162	0.012	0.795
55:35:10	0.109	0.286	0.095	0.321	0.693	0.285

[a]To convert kPa to mm Hg, multiply by 7.5.

Chemical Properties

Formaldehyde is noted for its reactivity and its versatility as a chemical intermediate. It is used in the form of anhydrous monomer solutions, polymers, and derivatives (see ACETAL RESINS).

Anhydrous, monomeric formaldehyde is not available commercially. The pure, dry gas is relatively stable at 80–100°C but slowly polymerizes at lower temperatures. Traces of polar impurities such as acids, alkalies, and water greatly accelerate the polymerization. When liquid formaldehyde is warmed to room temperature in a sealed ampul, it polymerizes rapidly with evolution of heat (63 kJ/mol or 15.05 kcal/mol). Uncatalyzed decomposition is very slow below 300°C; extrapolation of kinetic data (32) to 400°C indicates that the rate of decomposition is ca 0.44%/min at 101 kPa (1 atm). The main products are CO and H_2. Metals such as platinum (33), copper (34), and chromia and alumina (35) also catalyze the formation of methanol, methyl formate, formic acid, carbon dioxide, and methane. Trace levels of formaldehyde found in urban atmospheres are readily photooxidized to carbon dioxide; the half-life ranges from 35–50 minutes (36).

At ordinary temperatures, formaldehyde gas is readily soluble in water, alcohols, and other polar solvents. Its heat of solution in water and the lower aliphatic alcohols is approximately 63 kJ/mol (15 kcal/mol). The reaction of unhydrated formaldehyde with water is very fast; the first-order rate constant at 22°C is $9.8\ s^{-1}$ (37).

Formaldehyde is readily reduced to methanol by hydrogen over many metal and metal oxide catalysts. It is oxidized to formic acid or carbon dioxide and water. The Cannizzaro reaction gives formic acid and methanol. Similarly, a vapor-phase Tischenko reaction is catalyzed by copper (34) and boric acid (38) to produce methyl formate:

$$2\ HCHO \rightarrow HCOOCH_3 \qquad\qquad (6)$$

Formaldehyde condenses with itself in an aldol-type reaction to yield lower hydroxy aldehydes, hydroxy ketones, and other hydroxy compounds; the reaction is autocatalytic and is favored by alkaline conditions. Condensation with various compounds gives methylol ($-CH_2OH$) and methylene ($=CH_2$) derivatives. The former are usually produced under alkaline or neutral conditions, the latter under acidic conditions or in the vapor phase. In the presence of alkalies, aldehydes and ketones containing α-hydrogen atoms undergo aldol reactions with formaldehyde to form mono- and polymethylol derivatives. Acetaldehyde and 4 moles of formaldehyde give pentaerythritol (PE):

$$CH_3CHO\ +\ 3\ HCHO \rightarrow C(CH_2OH)_3CHO \qquad\qquad (7)$$

$$C(CH_2OH)_3CHO\ +\ HCHO\ +\ NaOH \rightarrow \underset{\text{pentaerythritol}}{C(CH_2OH)_4}\ +\ HCOONa \qquad\qquad (8)$$

In the vapor phase at 285°C, acrolein is formed from acetaldehyde (39):

$$HCHO\ +\ CH_3CHO \rightarrow HOCH_2CH_2CHO \rightarrow CH_2{=}CHCHO\ +\ H_2O \qquad\qquad (9)$$

Methylene derivatives are readily formed in the vapor phase with compounds having a hydrogen in the alpha position to an electron-withdrawing group. Acrylic (40,41) and methacrylic (42) acids (or esters) are produced at 300–425°C from acetic and propionic acids (or esters), respectively, using alkali and alkaline-earth catalysts. The same catalysts are effective for producing acrylonitrile from acetonitrile (42). With ketones, addition of two formaldehyde units occurs (43); for example, 2-butanone is converted to 2-methyl-1,4-pentadiene-3-one.

Chloromethylation of the aromatic nucleus occurs readily with alkyl and alkoxy substituents accelerating the reaction and halo, chloromethyl, carboxyl, and nitro groups retarding it.

$$\text{R}-\text{C}_6\text{H}_5 + \text{HCHO} + \text{HCl} \longrightarrow \text{R}-\text{C}_6\text{H}_4-\text{CH}_2\text{Cl} + \text{H}_2\text{O} \tag{10}$$

Reaction with phenol gives hydroxymethylphenols as the principal products (44). Through proper selection of reaction conditions and catalyst a 1:1 mixture of o- and p-isomers is obtained (45):

$$\text{HCHO} + \text{C}_6\text{H}_5\text{OH} \xrightarrow{\text{B}^-} \text{(HO)C}_6\text{H}_4\text{CH}_2\text{OH} \xrightarrow[\text{Pt/Pb}]{[\text{O}]} \text{(HO)C}_6\text{H}_4\text{CHO} \tag{11}$$

Both isomers can be oxidized to the corresponding benzaldehydes in high yield.

An important synthetic process for forming a new carbon–carbon bond is the acid-catalyzed condensation of formaldehyde with olefins (Prins reaction):

$$\text{C}{=}\text{C} + \text{HCHO} + \text{H}_2\text{O} \longrightarrow -\underset{\underset{\text{OH}}{|}}{\text{C}}-\underset{\underset{\text{CH}_2\text{OH}}{|}}{\text{C}}- \longrightarrow \text{C}{=}\text{C}-\text{CH}_2\text{OH} + \text{H}_2\text{O} \tag{12}$$

In acetic acid solvent, ethylene gives 1,3-propanediol acetates (46) and propylene gives 1,3-butanediol acetates (47). A similar reaction readily occurs with olefinic alcohols and ethers, diolefins, and mercaptans (48).

A commercial process based on the Prins reaction is the synthesis of isoprene from isobutylene and formaldehyde through the intermediacy of 4,4-dimethyl-1,3-dioxane (49–51):

$$2\,\text{HCHO} + \text{CH}_3-\text{C(CH}_3){=}\text{CH}_2 \xrightarrow{400^\circ\text{C}} \text{[4,4-dimethyl-1,3-dioxane]} \xrightarrow[\text{H}^+]{250^\circ\text{C}} \text{CH}_2{=}\text{C(CH}_3)-\text{CH}{=}\text{CH}_2 + \text{HCHO} + \text{H}_2\text{O} \tag{13}$$

Liquid-phase condensation of formaldehyde with propylene, catalyzed by BF_3 or H_2SO_4, gives butadiene (52,53).

With acidic catalysts in the liquid phase, formaldehyde and alcohols give formals, eg, dimethoxymethane from methanol:

$$\text{HCHO} + 2\,\text{CH}_3\text{OH} \rightarrow \text{CH}_3\text{OCH}_2\text{OCH}_3 + \text{H}_2\text{O} \tag{14}$$

Under neutral or slightly alkaline conditions, only the unstable hemiformal (CH_3O—CH_2OH, methoxymethanol) is produced. Alpha-chloromethyl ether is synthesized from aqueous formaldehyde, methanol, and hydrogen chloride (54). However, under anhydrous conditions, a carcinogenic by-product, bis(chloromethyl)ether is also formed (55).

Hydrogen cyanide reacts with aqueous formaldehyde in the presence of bases to produce glyconitrile [107-16-4] (56,57):

$$HCHO + HCN \rightarrow HOCH_2—C{\equiv}N \tag{15}$$

This extremely toxic material is an intermediate in the synthesis of nitrilotriacetic acid (NTA), EDTA, and glycine [56-40-6].

Monosubstituted acetylenes add formaldehyde in the presence of copper, silver, and mercury acetylide catalysts to give acetylenic alcohols (58) (Reppe reaction). Acetylene itself adds two molecules (see ACETYLENE-DERIVED CHEMICALS).

$$2\ HCHO + HC{\equiv}CH \rightarrow HOCH_2C{\equiv}CCH_2OH \tag{16}$$

Primary and secondary amines readily give alkylaminomethanols; the latter condense upon heating or under alkaline conditions to give substituted methyleneamines (59). With ammonia, the important industrial chemical, hexamine, is produced. Tertiary amines do not react.

$$HCHO + (CH_3)_2NH \rightarrow [(CH_3)_2N—CH_2OH] \xrightarrow{(CH_3)_2NH} (CH_3)_2N—CH_2—N(CH_3)_2 + H_2O \tag{17}$$

The well-known Mannich reaction is illustrated by the following example:

$$CH_3COCH_3 + (CH_3)_2NH{\cdot}HCl + HCHO \rightarrow CH_3COCH_2CH_2N(CH_3)_2{\cdot}HCl + H_2O \tag{18}$$

A detailed account of these reactions has been given (60).

Methylamines are formed by heating formaldehyde with primary or secondary amines or their salts under acid conditions (61):

$$R_2NH{\cdot}HCl + 2\ HCHO \rightarrow R_2N(CH_3){\cdot}HCl + HCOOH \tag{19}$$

Mono- and dimethylol derivatives are made by reaction of formaldehyde with unsubstituted amides. Dimethylolurea, an item of commercial importance and an intermediate in urea–formaldehyde resins, is formed in high yield under controlled conditions (62):

$$2\ HCHO + NH_2\overset{\overset{\displaystyle O}{\|}}{C}NH_2 \xrightarrow[\text{pH}=7\text{--}9]{15\text{--}25^{\circ}C} HOCH_2NH—\overset{\overset{\displaystyle O}{\|}}{C}—NHCH_2OH \tag{20}$$

Melamine reacts similarly to produce methylol derivatives, which form the familiar melamine–formaldehyde resins on heating (63) (see AMINO RESINS).

Reaction of formaldehyde, methanol, acetaldehyde, and ammonia over a silica alumina catalyst at 500°C gives pyridine [110-86-1] and 3-picoline [108-99-6] (64). This forms the basis of commercial processes for making pyridines (qv) from various aldehydes.

Formaldehyde reacts with syn gas (CO, H_2) to produce added value products. Ethylene glycol (EG) may be produced in a two-stage process or the intermediate, glycolaldehyde, isolated from the first stage (65):

$$HCHO + CO + H_2 \xrightarrow[\substack{30\ MPa \\ 150°C}]{Rh} HOCH_2-\overset{\overset{\displaystyle O}{\|}}{C}H \xrightarrow[\text{[H]}]{Rh,Pd} HOCH_2CH_2OH \qquad (21)$$

EG may also be produced via glycolic acid using catalysts containing strong acids (66), cobalt carbonyl (67–69), rhodium oxide (68), or HF solvent (70,71) (see GLYCOLS, ETHYLENE GLYCOL).

$$HCHO + CO + H_2O \xrightarrow{30\ MPa} HOCH_2-\overset{\overset{\displaystyle O}{\|}}{C}OH \xrightarrow{\text{[H]}} HOCH_2CH_2OH \qquad (22)$$

Manufacture

Historically, formaldehyde has been and continues to be manufactured from methanol. Following World War II, however, as much as 20% of the formaldehyde produced in the United States was made by the vapor-phase, noncatalytic oxidation of propane and butanes (72). This nonselective oxidation process produces a broad spectrum of coproducts (73) which requires a complex costly separation system (74). Hence, the methanol process is preferred. The methanol raw material is normally produced from synthesis gas that is produced from methane.

Most of the world's commercial formaldehyde is manufactured from methanol and air either by a process using a silver catalyst or one using a metal oxide catalyst. Reactor feed to the former is on the methanol-rich side of a flammable mixture and virtually complete reaction of oxygen is obtained; conversely, feed to the metal oxide catalyst is lean in methanol and almost complete conversion of methanol is achieved.

Silver Catalyst Process. In early formaldehyde plants methanol was oxidized over a copper catalyst, but this has been almost completely replaced with silver (75). The silver-catalyzed reactions occur at essentially atmospheric pressure and 600 to 650°C (76) and can be represented by two simultaneous reactions:

$$CH_3OH + 0.5\ O_2 \rightarrow HCHO + H_2O \qquad \Delta H = -156\ kJ\ (-37.28\ kcal) \qquad (23)$$

$$CH_3OH \rightarrow HCHO + H_2 \qquad \Delta H = +85\ kJ\ (20.31\ kcal) \qquad (24)$$

Between 50 and 60% of the formaldehyde is formed by the exothermic reaction (eq. 23) and the remainder by endothermic reaction (eq. 24) with the net result of a reaction exotherm. Carbon monoxide and dioxide, methyl formate, and formic acid are by-products. In addition, there are also physical losses, liquid-phase reactions, and small quantities of methanol in the product, resulting in an overall plant yield of 86–90% (based on methanol).

Figure 1 is a flow diagram of a typical formaldehyde plant (76–79) employing silver catalyst. A feed mixture is generated by sparging air into a pool of heated methanol and combining the vapor with steam. The mixture passes through a superheater to a catalyst bed of silver crystals or layers of silver gauze. The product is then rapidly cooled in a steam generator and then in a water-cooled heat exchanger and fed to the bottom of an absorption tower. The bulk of the methanol, water, and formaldehyde is condensed in the bottom water-cooled section of the tower, and almost complete removal of the remaining methanol and formaldehyde from the tail gas occurs in the top of the absorber by countercurrent contact with clean process water. Absorber bottoms go to a distillation tower where methanol is recovered for recycle to the reactor. The base stream from distillation, an aqueous solution of formaldehyde, is usually sent to an anion exchange unit which reduces the formic acid to specification level. The product contains up to 55% formaldehyde and less than 1.5% methanol.

A typical catalyst bed is very shallow (10 to 50 mm) (76,77). In some plants the catalyst is contained in numerous small parallel reactors; in others, catalyst-bed diameters up to 1.7 and 2.0 m (77,80) and capacities of up to 135,000 t/yr per reactor are reported (78). The silver catalyst has a useful life of three to eight months and can be recovered. It is easily poisoned by traces of transition group metals and by sulfur.

The reaction occurs at essentially adiabatic conditions with a large temperature rise at the inlet surface of the catalyst. The predominant temperature con-

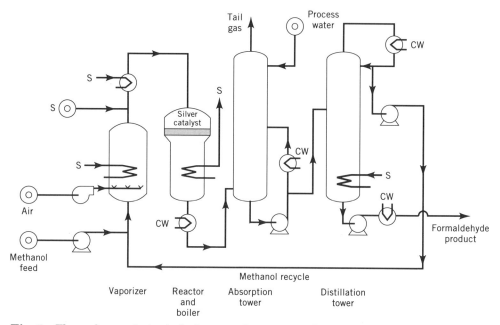

Fig. 1. Flow scheme of a typical silver catalyst process. S = steam; CW = cooling water.

trol is thermal ballast in the form of excess methanol or steam, or both, which is in the feed. If a plant is to produce a product containing 50 to 55% formaldehyde and no more than 1.5% methanol, the amount of steam that can be added is limited, and both excess methanol and steam are needed as ballast. Recycled methanol required for a 50–55% product is 0.25–0.50 parts per part of fresh methanol (76,77).

With increasing energy costs, maximum methanol conversion is desirable, eliminating the need for the energy-intensive distillation for methanol recovery. If a dilute product containing 40 to 44% formaldehyde and 1.0 to 1.5% methanol is acceptable, then the ballast steam can be increased to a level where recycled methanol is eliminated with a significant savings in capital cost and energy (81,82). In another process, tail gas from the absorber is recycled to the reactor. This additional gas plus steam provides the necessary thermal ballast without the need for excess methanol (82–84). This process can produce 50% formaldehyde with about 1.0% methanol without a distillation tower (79,84). Methanol recovery can be obviated in two-stage oxidation systems where, for example, part of the methanol is converted with a silver catalyst, the product is cooled, excess air is added, and the remaining methanol is converted over a metal oxide catalyst such as that described below (85). In another two-stage process, both first and second stages use silver catalysts (86–88). Formaldehyde–methanol solutions can be made directly from methanol oxidation product by absorption in methanol (89).

The absorber tail gas contains about 20 mol % hydrogen and has a higher heating value of ca 2420 kJ/m^3 (65 Btu/SCF). With increased fuel costs and increased attention to the environment, tail gas is burned for the twofold purpose of generating steam and eliminating organic and carbon monoxide emissions.

Aqueous formaldehyde is corrosive to carbon steel, but formaldehyde in the vapor phase is not. All parts of the manufacturing equipment exposed to hot formaldehyde solutions must be a corrosion-resistant alloy such as type-316 stainless steel. Theoretically, the reactor and upstream equipment can be carbon steel, but in practice alloys are required in this part of the plant to protect the sensitive silver catalyst from metal contamination.

Metal Oxide Catalyst Process. Oxidation of methanol to formaldehyde with vanadium pentoxide catalyst was first patented in 1921 (90), followed in 1933 by a patent for an iron oxide–molybdenum oxide catalyst (91), which is still the choice in the 1990s. Catalysts are improved by modification with small amounts of other metal oxides (92), support on inert carriers (93), and methods of preparation (94,95) and activation (96). In 1952, the first commercial plant using an iron–molybdenum oxide catalyst was put into operation (97). It is estimated that 70% of the new formaldehyde installed capacity is the metal oxide process (98).

In contrast to the silver process, all of the formaldehyde is made by the exothermic reaction (eq. 23) at essentially atmospheric pressure and at 300–400°C. By proper temperature control, a methanol conversion greater than 99% can be maintained. By-products are carbon monoxide and dimethyl ether, in addition to small amounts of carbon dioxide and formic acid. Overall plant yields are 88–92%.

Figure 2 is a flow scheme for a typical metal oxide catalyst formaldehyde plant (99–102). Vaporized methanol is mixed with air and optionally recycled tail gas and passed through catalyst-filled tubes in a heat-exchanger reactor. Heat

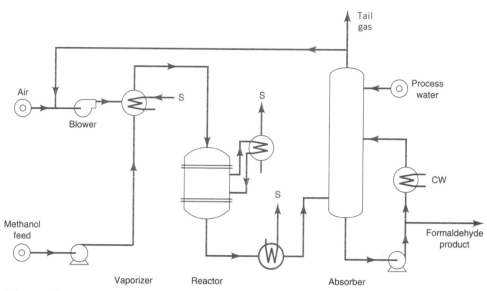

Fig. 2. Flow scheme of a typical metal oxide catalyst process. S = steam; CW = cooling water.

released by the exothermic reaction is removed by vaporization of a high boiling heat-transfer fluid on the outside of the tubes. Steam is normally produced by condensing the heat-transfer fluid. A typical reactor has short tubes of 1.0 to 1.5 m and a large shell diameter of 2.5 m or more. Product leaving the bottom of the reactor is cooled and passed to the base of an absorber. Formaldehyde concentration in the product is adjusted by controlling the amount of water added to the top of the absorber. A product up to 55% formaldehyde and less than 1% methanol can be made. Formic acid is removed by ion exchange.

The absence of a methanol recovery tower is an obvious advantage over the conventional silver process (Fig. 1). However, the equipment has to be large to accommodate the greater flow of gas. The air–methanol feed mixture must be on the methanol lean side of a flammable mixture. If the oxygen in the total reactor feed is reduced to about 10 mol % by partially replacing air with recycled tail gas, then the methanol in the feed can be increased somewhat without the danger of forming an explosive mixture (103), and for a given quantity of production, gas flow (air plus recycle) can be reduced by 17 to 37% (104). Even with gas recycle, the metal oxide process must handle a substantial volume of gas which is 3.0 to 3.5 times the gas flow of a conventional silver-catalyzed process.

A typical metal oxide catalyst has an effective life of 12 to 18 months (76,100,105). Compared to silver, it is much more tolerant to trace contamination. It requires less frequent change-outs, but a longer down time for each replacement. In contrast to a silver-catalyst plant, there is little economic justification to incinerate the metal oxide plant tail gas for the purpose of generating steam. The tail gas is essentially nitrogen and oxygen with combustible components (dimethyl ether, carbon monoxide, formaldehyde, methanol) representing only a few percent of the total. However, increasing environmental pressures make vent incineration more desirable. With the addition of auxiliary fuel, the vent can be oxidized by thermal incineration at temperatures of 700 to 900°C. As an alternative, the stream can be oxidized at 450 to 550°C in a catalytic incinerator which can be thermally self-sufficient with supplemental fuel required only for startup or abnormal operating conditions (104).

The requirements for the material of construction are the same as for the silver catalyst process except the use of alloys to protect the catalyst is not as important.

Development of New Processes.   There has been significant research activity to develop new processes for producing formaldehyde. Even though this work has been extensive, no commercial units are known to exist based on the technologies discussed in the following.

One possible route is to make formaldehyde directly from methane by partial oxidation. This process has been extensively studied (106–108). The incentive for such a process is reduction of raw material costs by avoiding the capital and expense of producing the methanol from methane.

Another possible route for producing formaldehyde is by the dehydrogenation of methanol (109–111) which would produce anhydrous or highly concentrated formaldehyde solutions. For some formaldehyde users, minimization of the water in the feed reduces energy costs, effluent generation, and losses while providing more desirable reaction conditions.

A third possible route is to produce formaldehyde from methylal that is produced from methanol and formaldehyde (112,113). The incentive for such a process is twofold. First, a higher concentrated formaldehyde product of 70% could be made by methylal oxidation as opposed to methanol oxidation, which makes a 55% product (112). This higher concentration is desirable for some formaldehyde users. Secondly, formaldehyde in aqueous recycle streams from other units could be recovered by reacting with methanol to produce methylal as opposed to recovery by other more costly means, eg, distillation and evaporation. Development of this process is complete (113).

Economic Aspects

Essentially all formaldehyde is produced as aqueous solutions containing 25–56 wt % HCHO and 0.5–15 wt % CH_3OH. All information on capacity, demand, and prices is reported on a 37 wt % formaldehyde basis.

Worldwide production capacity in 1989 was estimated to be over 15.5×10^6 t as 37 wt % formaldehyde (98). The United States, Canada, Europe, and Japan account for nearly 70% of the total capacity (98). Worldwide demand for formaldehyde in 1989 was estimated to be about 85–90% of capacity (98).

Plant capacities in the United States are shown in Table 3 (114). Capacity, production, and sales data between 1966 and 1989 are given in Table 4 (115,116). During the 1980s, production averaged 67% of capacity. During the same period, sales averaged about 32% of production, illustrating the large extent of captive use by formaldehyde producers.

Table 3. 1992 Capacity[a] of U.S. Formaldehyde Producers[b]

Producer	10^3 t/yr
Aqualon Co.	77
Borden, Inc.	1157
BTL Specialty Resins, Inc.	100
Capital Resins Corp.	45
D. B. Western	41
Degusssa Corp.	30
E. I. du Pont de Nemours & Co., Inc.	499
Dyno Polymers Inc.	136
Georgia-Pacific Corp.	748
Hoechst-Celanese Corp.	968
International Specialty Products, Inc.	91
Monsanto Co.	216
Neste Resins Corp.	272
Perstorp, Inc.	113
Spurlock Adhesives, Inc.	25
Wright Chemical Corp.	54
Total	*4572*

[a]37% basis.
[b]Ref. 114.

The growth rate of U.S. formaldehyde uses has declined since the 1960s as shown (116):

Period	Annual growth rates, %
1961–1970	10.8
1970–1980	2.3
1980–1990	1.4

U.S. formaldehyde prices for 1966–1989 are shown in Table 4 (115). Since the cost of methanol represents over 60% of formaldehyde's production costs, the formaldehyde price normally reflects the methanol price. Also, freight is a significant cost for formaldehyde since 1–3 kg of water may be shipped with every kg of formaldehyde. The significant price increase in the early 1970s was due to the sudden rise in hydrocarbon prices caused by the Organization of Petroleum Exporting Companies (OPEC) cartel increasing oil prices.

Specifications and Quality Control

Formaldehyde is sold in aqueous solutions with concentrations ranging from 25 to 56 wt % HCHO. Product specifications for typical grades (18,117–119) are summarized in Table 5. Formaldehyde is sold as low methanol (uninhibited) and high methanol (inhibited) grades. Methanol is used to retard paraformaldehyde formation.

Table 4. U.S. Formaldehyde Capacity, Production, and Sales[a]

Year	Capacity,[b] 10^3 t/yr	Production,[c] 10^3 t/yr	Sales,[c] 10^3 t/yr	Price,[b] $/kg
1966	1610	1680	617	0.083[d]
1969	2410	2000	729	0.074[d]
1972	3420	2560	825	0.077[d]
1981	3880	2590	838	0.177[e]
1984	3940	2640	825	0.197[d]
1989	4310	2890	na[f]	0.224[e]

[a]Based on 37 wt % formaldehyde (2 wt % methanol).
[b]Ref. 115.
[c]Ref. 116.
[d]Delivered bulk price for uninhibited 37% HCHO.
[e]Bulk price, fob, for uninhibited 37% HCHO.
[f]Not available.

Procedures for determining the quality of formaldehyde solutions are outlined by ASTM (120). Analytical methods relevant to Table 5 follow: formaldehyde by the sodium sulfite method (D2194); methanol by specific gravity (D2380); acidity as formic acid by titration with sodium hydroxide (D2379); iron by colorimetry (D2087); and color (APHA) by comparison to platinum–cobalt color standards (D1209).

Table 5. Formaldehyde Specifications[a]

Property	Methanol inhibited grades			Low methanol uninhibited grades		
formaldehyde, wt %	37	37	37	44	50	56
methanol, wt % (max)	6–8	12–15	1.0–1.8	1.5	1.5–2.0	2.0
acidity, wt % (max)[b]	0.02	0.02	0.02	0.03	0.05	0.04
iron, ppm (max)	0.5	0.05	0.5–1.0	0.5	0.5	0.75
color, APHA (max)	10	10	10	10	10	10

[a]The specification range is the high and low specifications of producers surveyed.
[b]As wt % formic acid.

Storage and Handling

As opposed to gaseous, pure formaldehyde, solutions of formaldehyde are unstable. Both formic acid (acidity) and paraformaldehyde (solids) concentrations increase with time and depend on temperature. Formic acid concentration builds at a rate of 1.5–3 ppm/d at 35°C and 10–20 ppm/d at 65°C (17,18). Trace metallic impurities such as iron can boost the rate of formation of formic acid (121). Although low storage temperature minimizes acidity, it also increases the tendency to precipitate paraformaldehyde.

Paraformaldehyde solids can be minimized by storing formaldehyde solutions above a minimum temperature for less than a given time period. The addition of methanol as an inhibitor or of another chemical as a stabilizer allows storage at lower temperatures and/or for longer times. Stabilizers for formaldehyde solutions (122–125) include hydroxypropylmethylcellulose, methyl- and ethylcelluloses, poly(vinyl alcohol)s, or isophthalobisguanamine at concentrations ranging from 10 to 1000 ppm. Inhibited formaldehyde typically contains 5–15 wt % methanol.

Most formaldehyde producers recommend a minimum storage temperature for both stabilized and unstabilized solutions. Figure 3 is a plot of data (17,18, 122,126) for uninhibited (<2.0 wt % methanol) formaldehyde. The minimum temperature to prevent paraformaldehyde formation in unstabilized 37% formaldehyde solutions stored for one to about three months is as follows: 35°C with less than 1% methanol; 21°C with 7% methanol; 7°C with 10% methanol; and 6°C with 12% methanol (127).

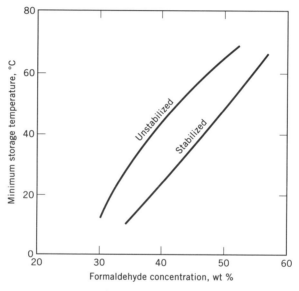

Fig. 3.   Storage temperature vs formaldehyde concentration (<2.0 wt % CH$_3$OH).

Materials of construction preferred for storage vessels (17–19,128) are 304-, 316-, and 347-type stainless steels or lined carbon steel.

Health and Safety Factors

Sources of human exposure to formaldehyde are engine exhaust, tobacco smoke, natural gas, fossil fuels, waste incineration, and oil refineries (129). It is found as a natural component in fruits, vegetables, meats, and fish and is a normal body metabolite (130,131). Facilities that manufacture or consume formaldehyde must control workers' exposure in accordance with the following workplace exposure limits in ppm: action level, 0.5; TWA, 0.75; STEL, 2 (132). In other environments such as residences, offices, and schools, levels may reach 0.1 ppm HCHO due to use of particle board and urea–formaldehyde foam insulation in construction.

Formaldehyde causes eye, upper respiratory tract, and skin irritation and is a skin sensitizer. Although sensory irritation, eg, eye irritation, has been reported at concentrations as low as 0.1 ppm in uncontrolled studies, significant eye/nose/throat irritation does not generally occur until concentrations of 1 ppm, based on controlled human chamber studies. Odor detection has commonly been reported to occur in the range of 0.06–0.5 ppm (133–135).

Formaldehyde is classified as a probable human carcinogen by the International Agency for Research on Cancer (IARC) and as a suspected human carcinogen by the American Conference of Governmental Industrial Hygienists (ACGIH). This is based on limited human evidence and on sufficient evidence in experimental animals (136). Lifetime inhalation studies with rodents have shown nasal cancer at formaldehyde concentrations that overwhelmed cellular defense mechanisms, ie, 6 to 15 ppm. No nasal cancer was seen at 2 ppm or lower levels (137).

Formaldehyde is not considered a teratogen and has not been reported to cause adverse reproductive effects (138,139). *In vitro* mutagenicity assays with formaldehyde have yielded positive responses, while *in vivo* assays have been largely negative (130).

Uses

Formaldehyde is a basic chemical building block for the production of a wide range of chemicals finding a wide variety of end uses such as wood products, plastics, and coatings. Table 6 shows the distribution of formaldehyde production in the United States from 1966 through 1989 (115). Production percentages reported in the following discussion are for the United States.

Table 6. U.S. Distribution of Formaldehyde Production According to Uses, 1966–1989[a]

	Percent of consumption[b]				
Product	1966	1972	1981	1984	1989
phenol–formaldehyde resins	23	25	20	21	22
urea–formaldehyde resins	30[c]	25	30	27	25
acetal resins	2	8	7	9	9
1,4-butanediol	na	2	7	9	11
melamine resins	na[c]	8	4	4	4
pentaerythritol	12	7	5	7	7
hexamethylenetetramine	6	6	4	7	6
urea–formaldehyde concentrates	5	5	4	6	6
methylene diisocyanate	na	na	3	4	5
ethylene glycol	14	na	na	na	na
others	8	14	16	6	5

[a]Ref. 115.
[b]na = not available.
[c]Melamine resins are included in urea–formaldehyde resins.

Amino and Phenolic Resins. The largest use of formaldehyde is in the manufacture of urea–formaldehyde, phenol–formaldehyde, and melamine–formaldehyde resins, accounting for over one-half (51%) of the total demand (115). These resins find use as adhesives for binding wood products that comprise particle board, fiber board, and plywood. Plywood is the largest market for phenol–formaldehyde resins; particle board is the largest for urea–formaldehyde resins. Under certain conditions, urea–formaldehyde resins may release formaldehyde that has been alleged to create health or environmental problems (see AMINO RESINS).

Phenol–formaldehyde resins are used as molding compounds (see PHENOLIC RESINS). Their thermal and electrical properties allow use in electrical, automotive, and kitchen parts. Other uses for phenol–formaldehyde resins include phenolic foam insulation, foundry mold binders, decorative and industrial laminates, and binders for insulating materials.

Urea–formaldehyde resins are also used as molding compounds and as wet strength additives for paper products. Melamine–formaldehyde resins find use in decorative laminates, thermoset surface coatings, and molding compounds such as dinnerware.

1,4-Butanediol. 1,4-Butanediol [110-63-4], made from formaldehyde and acetylene, is a significant market for formaldehyde representing 11% of its demand (115). It is used to produce tetrahydrofuran (THF), which is used for polyurethane elastomers; γ-butyrolactone, which is used to make various pyrrolidinone derivatives; poly(butylene terephthalate) (PBT), which is an engineering plastic; and polyurethanes. Formaldehyde growth in the acetylenic chemicals market is threatened by alternative processes to produce 1,4-butanediol not requiring formaldehyde as a raw material (140) (see ACETYLENE-DERIVED CHEMICALS).

Polyols. Several important polyhydric alcohols or polyols are made from formaldehyde. The principal ones include pentaerythritol, made from acetaldehyde and formaldehyde; trimethylolpropane, made from n-butyraldehyde and formaldehyde; and neopentyl glycol, made from isobutyraldehyde and formaldehyde. These polyols find use in the alkyd resin (qv) and synthetic lubricants markets. Pentaerythritol [115-77-5] is also used to produce rosin/tall oil esters and explosives (pentaerythritol tetranitrate). Trimethylolpropane [77-99-6] is also used in urethane coatings, polyurethane foams, and multifunctional monomers. Neopentyl glycol [126-30-7] finds use in plastics produced from unsaturated polyester resins and in coatings based on saturated polyesters.

The formaldehyde demands for pentaerythritol, trimethylolpropane, and neopentyl glycol are about 7, 2, and 1%, respectively, of production.

Acetal Resins. These are high performance plastics produced from formaldehyde that are used for automotive parts, in building products, and in consumer goods. Acetal resins (qv) are either homopolymers or copolymers of formaldehyde. Typically, the resin is produced from anhydrous formaldehyde or trioxane. The acetal resins formaldehyde demand are 9% of production (115).

Hexamethylenetetramine. Pure hexamethylenetetramine [100-97-0] (also called hexamine and HMTA) is a colorless, odorless, crystalline solid of adamantane-like structure (141). It sublimes with decomposition at >200°C but does not melt. Its solubility in water varies little with temperature, and at 25°C it is 46.5% in the saturated solution. It is a weak monobase; aqueous solutions are in the pH 8–8.5 range (142). Hexamethylenetetramine is readily prepared by treating aqueous formaldehyde with ammonia followed by evaporation and crystallization of the solid product. The reaction is fast and essentially quantitative (142).

The production of hexamethylenetetramine consumes about 6% of the U.S. formaldehyde supply (115). Its principal use is as a thermosetting catalyst for phenolic resins. Other significant uses are for the manufacture of RDX (cyclonite) high explosives, in molding compounds, and for rubber vulcanization accelerators. Some hexamethylenetetramine is made as an unisolated intermediate in the manufacture of nitrilotriacetic acid.

Slow-Release Fertilizers.   Products containing urea–formaldehyde are used to manufacture slow-release fertilizers. These products can be either solids, liquid concentrates, or liquid solutions. This market consumes almost 6% of the formaldehyde produced (115) (see CONTROLLED RELEASE TECHNOLOGY, AGRICULTURAL).

Methylenebis(4-phenyl isocyanate).   This compound is also known as methyl diisocyanate [101-68-8] (MDI) and is produced by the condensation of aniline and formaldehyde with subsequent phosgenation. Its principal end use is rigid urethane foams; other end uses include elastic fibers and elastomers. Total formaldehyde use is 5% of production (115).

Chelating Agents.   The chelating agents produced from formaldehyde include the aminopolycarboxylic acids, their salts, and organophosphonates. The largest demand for formaldehyde is for ethylenediaminetetraacetic acid [60-00-4] (EDTA); the next largest is for nitrilotriacetic acid [139-13-9] (NTA). Chelating agents find use in industrial and household cleaners and for water treatment. Overall, chelating agents represent a modest demand for formaldehyde of about 3%.

Formaldehyde–Alcohol Solutions.   These solutions are blends of concentrated aqueous formaldehyde, the alcohol, and the hemiacetal. Methanol decreases the average molecular weight of formaldehyde oligomers by formation of lower molecular weight hemiacetals. These solutions are used to produce urea and melamine resins; the alcohol can act as the resin solvent and as a reactant. The low water content can improve reactivity and reduce waste disposal and losses. Typical specifications for commercially available products are shown in Table 7 (117).

Paraformaldehyde.   Paraformaldehyde [9002-81-7], or paraform, is a solid mixture of linear poly(oxymethylene glycols) of fairly short chain length, $HO(CH_2O)_nH$ (143) (the range of n is 8–100). The average degree of polymerization is only roughly given by the formaldehyde content. The specifications of commercial paraformaldehyde are given in Table 8 (117). Gaseous formaldehyde can be generated from paraformaldehyde by heating (144). A current process for paraformaldehyde is based on continuous, staged vacuum evaporations, starting with 50% aqueous formaldehyde (145,146).

The rate at which paraformaldehyde dissolves (hydrolyzes) in water is at a minimum pH 3–5; it increases rapidly at lower or higher pH and at higher tem-

Table 7. Specifications and Physical Properties of Formaldehyde–Alcohol Solutions

Property	Methanol	Methanol	1-Butanol	Isobutyl alcohol
alcohol, wt %	48	55	53	52
HCHO, wt %	43	35	40	40
H_2O, wt %	9	10	7	8
bp, °C	88.0	91.6	104.5	103
specific gravity at 40°C	0.977	1.051	0.963	0.973
flash point, °C	33[a]	44[a]	57[b]	53.8[b]
U_f/L_f, mol %[c]		47.0/7.0	21.0/2.5	19.0/2.0

[a]Tag closed cup (ASTM D56).
[b]Tag open cup (ASTM D1310).
[c]Upper and lower flammable limits in air.

Table 8. Specifications of Commercial Paraformaldehyde[a]

Property	Prills	Powder[b]
formaldehyde, wt %	91.0–93.0	95.0–97.0
water, wt %, max	9.0	5.0
acidity as formic, wt %, max	0.03	0.03
iron, ppm, max	2	2
ash wt %, max	0.01	0.01

[a]Ref. 117.
[b]10 wt % max retained on 74 mm (200 mesh) sieve.

peratures. Once dissolved in water, paraformaldehyde behaves like methanol-free formaldehyde of the same concentration.

Paraformaldehyde is used by resin manufacturers seeking low water content or more favorable control of reaction rates. It is often used in making phenol–, urea–, resorcinol–, and melamine–formaldehyde resins.

Trioxane and Tetraoxane. The cyclic symmetrical trimer of formaldehyde, trioxane [110-88-3], is prepared by acid-catalyzed liquid- or vapor-phase processes (147–151). It is a colorless crystalline solid that boils at 114.5°C and melts at 61–62°C (17,152). The heats of formation are -176.9 kJ/mol (-42.28 kcal/mol) from monomeric formaldehyde and -88.7 kJ/mol (-21.19 kcal/mol) from 60% aqueous formaldehyde. It can be produced by continuous distillation of 60% aqueous formaldehyde containing 2–5% sulfuric acid. Trioxane is extracted from the distillate with benzene or methylene chloride and recovered by distillation (153) or crystallization (154). It is mainly used for the production of acetal resins (qv).

A practical synthesis has been claimed for the cyclic tetramer of formaldehyde, 1,3,5,7-tetraoxane [293-30-1], which has a boiling point of 175°C and a melting point of 112°C (155). It has found some use in textile treatment in Japan.

Other Applications. Formaldehyde is used for the manufacture of a great variety of other chemicals. Formaldehyde derivatives, such as dimethylol dihydroxyethylene, are used in textiles to produce permanent press fabrics. Other formaldehyde derivatives are used in this industry to produce fire-retardant fabrics. Pyridine chemicals, made from formaldehyde, acetaldehyde, and ammonia, are used for agricultural chemicals. Formaldehyde and paraformaldehyde have found use as a corrosion inhibitor, hydrogen sulfide scavenger, and biocide in oil production operations such as drilling, waterflood, and enhanced oil recovery. Other uses for formaldehyde and formaldehyde derivatives include fungicides, embalming fluids, silage preservatives, and disinfectants.

BIBLIOGRAPHY

"Formaldehyde" in *ECT* 1st ed., Vol. 6, pp. 857–875, by J. F. Walker, E. I. du Pont de Nemours & Co., Inc.; in *ECT* 2nd ed., Vol. 10, pp. 77–99, by J. F. Walker, Consultant; in *ECT* 3rd ed., Vol. 11, pp. 231–250, by H. R. Gerberich, A. L. Stautzenberger, and W. C. Hopkins, Celanese Chemical Co., Inc.

1. J. F. Walker, *Formaldehyde*, 3rd ed., Reinhold Publishing Corp., New York, 1974.
2. A. Kekule, *Chem. Ber.* **25**, 2435 (1892).
3. R. Spence and W. Wild, *J. Chem. Soc.*, 506 (1935).
4. S. Obe, *Computer Aided Data Book of Vapor Pressure*, Data Book Publishing Co., Tokyo, 1976, p. 109.
5. M. M. Brazhnikow and co-workers, *Zh. Prikl. Khim. (Leningrad)* **49**, 1041 (1976).
6. D. R. Stull and co-workers, *The Chemical Thermodynamics of Organic Compounds*, John Wiley & Sons, Inc., New York, 1969, pp. 62, 438.
7. G. Reuss and co-workers, in W. Gerhartz, ed., *Ullmann's Encyclopedia of Industrial Chemistry*, 5th ed., VCH Verlagsgesellschaft mbH, Weinheim, Germany, 1988, pp. 619–651.
8. Yu. M. Blazhin and co-workers, *Zh. Prikl. Khim. (Leningrad)* **49**(1), 174 (1976).
9. R. W. Gallant, *Hydrocarbon Process* **47**(5), 151 (1968).
10. *Fire Protection Guide on Hazardous Materials*, 9th ed., National Fire Protection Association, Quincy, Mass., 1986, pp. 49–51.
11. *JANAF Interim Thermochemical Tables*, Thermal Laboratory, The Dow Chemical Co., Midland, Mich., Mar. 31, 1961.
12. Ref. 1, pp. 40, 42.
13. R. Bieber and G. Trümpler, *Helv. Chim. Acta* **30**, 1860 (1947).
14. Ref. 1, pp. 60–61.
15. A. V. Rudnev and co-workers, *Zh. Fiz. Khim.* **51**, 2031 (1977).
16. Ref. 1, pp. 86–91, 500–501.
17. *Formaldehyde Product Bulletin*, Celanese Chemical Co., Inc., Dallas.
18. *Formaldehyde Solutions*, E. I. du Pont de Nemours & Co., Inc., Wilmington, Del.
19. *Formaldehyde Technical Bulletin*, Georgia-Pacific Corp., Portland, Oreg.
20. G. Natta and M. Baccarredda, *G. Chim. Ind. Appl.* **15**, 273 (1933).
21. Ref. 1, p. 93.
22. B. Olsson and S.-G. Svensson, *Trans. Inst. Chem. Eng.* **53**, 97 (1975).
23. Ref. 1, pp. 115, 129.
24. A. E. Obraztsov and co-workers, *Zh. Prikl. Khim. (Leningrad)* **42**, 2393 (1969).
25. U.S. Pat. 2,527,654 (Oct. 31, 1950), C. Pyle and J. A. Lane (to E. I. du Pont de Nemours & Co., Inc.).
26. Yu. M. Blazhin and co-workers, *Zh. Prikl. Khim. (Leningrad)* **50**(1), 39 (1977).
27. S. J. Green and R. E. Vener, *Ind. Eng. Chem.* **47**, 103 (1955).
28. W. Dankelman and J. M. H. Daemen, *Anal. Chem.* **48**, 401 (1976).
29. Ya. Slonim and co-workers, *Vysokomol. Soedin. Ser. B* **XVII**(12), 919 (1975).
30. Z. Fiala and M. Kavrátil, *Coll. Czech. Comm.* **39**, 2200 (1974).
31. D. A. Young, unpublished results, Celanese Research Co., Summit, N.J., 1978.
32. C. J. M. Fletcher, *Proc. Roy. Soc. (London)* **A146**, 357 (1934).
33. M. J. Marshall and D. F. Stedman, *Trans. Roy. Soc. Can.* **17** (Sect. III), 53 (1923).

34. Y. Miyazaki and I. Yasumori, *Bull. Chem. Soc. Jpn.* **40**, 2012 (1967).
35. H. Tropsch and O. Roehlen, *Abh. Kenntnis Kohle* **7**, 15 (1925).
36. *Formaldehyde*, Environmental Health Criteria 89, World Health Organization, Geneva, Switzerland, 1989, p. 17.
37. H. C. Sutton and T. M. Downes, *Chem. Comm.*, 1 (1972).
38. P. R. Stapp, *J. Org. Chem.* **88**, 1435 (1978).
39. U.S. Pat. 2,246,037 (June 17, 1941), M. Gallagher and R. L. Hasche (to Eastman Kodak).
40. Jpn. Kokai 45 21,928 (July 24, 1970), (to Air Reduction).
41. Jpn. Kokai 71 16,728 (May 8, 1971), K. Kimura and I. Hiroo (to Toa Synthetic Chemical Co.).
42. U.S. Pat. 3,933,888 (Jan. 20, 1976), F. W. Schlaefer (to Rohm and Haas).
43. H. Hanna and S. Malinowski, *Chim. Ind. (Milan)* **51**, 1078 (1969).
44. Ref. 1, p. 306.
45. H. Fiege and co-workers, *Proceedings 2nd International Haarmann-Reimer Symposium*, 1980, pp. 63–75.
46. Jpn. Kokai 76 91,203 (Aug. 10, 1976), F. Wataru and T. Urasaki (to Teijin, Ltd.).
47. B. N. Bobylev and co-workers, *Neftekhimiya* **9**(1), 71 (1969).
48. E. Arundale and L. A. Mikeska, *Chem. Rev.* **51**, 505 (1952).
49. M. Hellin and M. Davidson, *Inf. Chim.* **1980**(206), 163,181.
50. K. Naito and K. Ogino, *Pet. Petrochem. Int.* **13**(11), 44 (1973).
51. T. Yashima and co-workers, *Nippon Kagaku Kaishi*, 325 (1974).
52. U.S. Pat. 2,335,691 (Nov. 30, 1943), H. O. Mottern (to Jasco, Inc.).
53. U.S. Pat. 2,412,762 (Dec. 17, 1946), A. R. Workman (to Cities Service).
54. L. F. Fieser and M. Fieser, *Reagents for Organic Synthesis*, John Wiley & Sons, Inc., New York, 1967, p. 132.
55. Belg. Pat. 817,848 (July 19, 1973), L. A. Ens (to Dow Chemical Co.).
56. U.S. Pat. 2,890,238 (June 9, 1959), A. R. Sexton (to Dow Chemical Co.).
57. G. Reuss and co-workers, in Ref. 7, pp. 619–651.
58. U.S. Pat. 2,232,867 (Feb. 25, 1941), E. Keyssner and W. Reppe (to General Anilin and Film).
59. L. Henry, *Bull. Acad. Roy. Belg.* **28**, 359, 366 (1894).
60. R. Adams, *Organic Reactions*, Vol. I, John Wiley & Sons, Inc., New York, 1942, pp. 303–330.
61. Ref. 1, p. 365.
62. Ref. 1, p. 379.
63. Ref. 1, p. 394.
64. U.S. Pats. 2,744,904 (May 8, 1956), and 2,807,618 (Sept. 24, 1957), F. E. Cislak and W. R. Wheeler (to Reilly Tar and Chemical).
65. U.S. Pat. 4,200,765 (Apr. 29, 1980), R. W. Goetz (to National Distillers).
66. Ref. 1, p. 634.
67. Ger. Offen. 2,427,954 (Jan. 9, 1975), T. Yukawa, K. Kawasaki, and H. Wakamatsu (to Ajinomoto).
68. Belg. Pat. 858,628 (Jan. 2, 1978), R. G. Wall (to Chevron).
69. U.S. Pat. 4,016,209 (Apr. 5, 1977), S. Suzuki (to Chevron).
70. U.S. Pat. 3,911,003 (Oct. 7, 1975), S. Suzuki (to Chevron).
71. S. Suzuki and co-workers, *Proc. Symposium Catal. Conv. Syn. Gas Alcohols Chem.*, 1984, pp. 221–247.
72. Ref. 1, p. 7.
73. R. L. Mitchell, *Pet. Refiner* **35**, 179 (July 1956).
74. W. C. Hopkins and J. J. Fritsch, *Chem. Eng. Prog.* **51**, 361 (Aug. 1955).

75. H. R. Gerberich, *Precious Metals Catalysis Course*, International Precious Metals Institute, New Orleans, La., Mar. 6–9, 1988.
76. A. R. Chauvel and co-workers, *Hydrocarbon Process.* **52**, 179 (Sept. 1973).
77. D. G. Sleeman, *Chem. Eng.* **75**, 42 (Jan. 1, 1968).
78. *Hydrocarbon Process.* **52**, 135 (Nov. 1973).
79. J. H. Marten and M. T. Butler, *Oil Gas J.* **72**, 71 (Mar. 11, 1974).
80. U.S. Pat. 4,072,717 (Feb. 2, 1978), G. Halbritter (to BASF).
81. U. Gerloff, *Hydrocarbon Process.* **46**, 169 (June 1967).
82. H. Diem, *Chem. Eng.* **85**, 83 (Feb. 27, 1978).
83. Fr. Pat. 1,487,093 (June 30, 1967), (to E. I. du Pont de Nemours & Co., Inc.).
84. Y. Kuraishi and K. Yoshikawa, *Chem. Econ. Eng. Rev.* **14**, 31 (June 1982).
85. U.S. Pat. 2,519,788 (Aug. 22, 1950), W. A. Payne (to E. I. du Pont de Nemours & Co., Inc.).
86. U.S. Pat. 2,462,413 (Feb. 22, 1949), W. B. Meath (to Allied Chemical).
87. U.S. Pat. 4,450,301 (May 22, 1984), W. P. McMillan and co-workers (to Celanese).
88. U.S. Pat. 4,076,754 (Feb. 28, 1978), G. L. Kiser and B. G. Hendricks (to E. I. du Pont de Nemours & Co., Inc.).
89. U.S. Pat. 3,629,997 (Dec. 28, 1971), C. W. DeMuth (to Borden).
90. U.S. Pat. 1,383,059 (June 28, 1921), G. C. Bailey and A. E. Craver (to Barrett Co.).
91. U.S. Pat. 1,913,405 (June 13, 1933), V. E. Meharg and H. Adkins (to Bakelite Corp.).
92. U.S. Pat. 3,198,753 (Aug. 3, 1965), F. Traina (to Montecatini).
93. Brit. Pat. 1,080,508 (Aug. 23, 1967), (to Perstorp AB).
94. U.S. Pat. 3,408,309 (Oct. 29, 1968), A. W. Gessner (to Lummus).
95. U.S. Pat. 3,855,153 (Dec. 17, 1974), G. M. Chang (to Reichhold).
96. U.S. Pat. 2,813,309 (Nov. 5, 1957), C. L. Allyn and co-workers (to Reichhold).
97. *Chem. Eng.* **61**, 109 (Nov. 1954).
98. M. Henning, "Formaldehyde: A Global Perspective," presentation and paper from *1989 Asian Methanol Conference*, Singapore, Oct. 30–Nov. 1, 1989.
99. *Hydrocarbon Process.* **50**, 161 (Nov. 1971).
100. *Hydrocarbon Process.* **55**, 150 (Nov. 1975).
101. *Hydrocarbon Process.*, 141 (Nov. 1985).
102. *Hydrocarbon Process.*, 158 (Mar. 1991).
103. U.S. Pat. 2,436,287 (Feb. 17, 1948), W. F. Brondyke and J. A. Monier (to E. I. du Pont de Nemours & Co., Inc.); Brit. Pat. 589,292 (June 17, 1947), W. F. Brondyke and J. A. Monier (to E. I. du Pont de Nemours & Co., Inc.).
104. C. W. Horner, *Chem. Eng.* **84**, 108 (July 4, 1977).
105. *Chem. Week* **105**, 79 (Nov. 19, 1969).
106. J. Hargreaves, G. Hutchings, and R. Joyner, *Nature* **348**, 28 (Nov. 1990).
107. M. Brown and N. Parkyns, *Catal. Today* **8**(3), 305 (1991).
108. N. Spencer and C. Pereira, *J. Catal.* **116**(2), 399 (1989).
109. Eur. Pat. 405,348 (Jan. 2, 1991), H. Beck and co-workers (to Hoechst AG).
110. Eur. Pat. 261,867 (Mar. 30, 1988), M. Sagou and H. Fujii (to Sumitomo).
111. Jpn. Pat. 61,205,226 (Sept. 11, 1986) M. Misonoo and T. Yamamoto (to Polyplastics).
112. S. Stinson, *Chem. Eng. News*, 40–41 (Apr. 24, 1989).
113. J. Masamoto and co-workers, *J. Macromolecular Sci., Pure Appl. Chem.* **A29**(6), 441–456 (1992).
114. *Chem. Mark. Rep.*, (Aug. 24, 1992).
115. *Chem. Mark. Rep.*, (Sept. 18, 1989); *Chem. Mark. Rep.*, (Jan. 1, 1984); *Chem. Mark. Rep.*, (July 1, 1981); *Chem. Mark. Rep.*, (Oct. 1, 1972); *Chem. Mark. Rep.*, (July 1, 1969); *Chem. Mark. Rep.*, (Apr. 1, 1966).

116. *Synthetic Organic Chemicals: United States Production and Sales*, United States International Trade Commission, U.S. Government Printing Office, Washington, D.C., 1990.
117. *Hoechst-Celanese Chemical Group International Product Index*, Sept., 1991; *Material Safety Data Sheets*, Hoechst-Celanese, Corp., Dallas, Tex., 1991.
118. *Perstorp Formaldehyde Product Bulletin*, Feb. 1992.
119. *BASF Formaldehyde Product Bulletin*, Hoechst-Celanese Corp., Dallas, Tex., June 1988.
120. *Annual Book of ASTM Standards*, Vol. 6.03, American Society for Testing Materials, Philadelphia, Pa., 1991.
121. Ref. 1, p. 98.
122. *Formaldehyde Product Bulletin*, Badische Anilin- und Soda-Fabrik AG, Ludwigshafen, Germany, Sept. 1975.
123. U.S. Pat. 3,137,736 (June 16, 1964), R. H. Prinz and R. C. Kerr (to Celanese).
124. Ref. 1, pp. 93–100.
125. Eur. Pat. 319,613 (June 14, 1989), E. Roerdink and co-workers (to Methanol Chemie Nederland).
126. *Formaldehyde Product Bulletin*, Deutsche Gold- und Silber-Scheideanstalt AG, Frankfurt, 1978.
127. Ref. 1, p. 95.
128. Ref. 1, pp. 93–100.
129. *Formaldehyde*, Environmental Health Criteria 89, World Health Organization, Geneva, Switzerland, 1989, p. 15.
130. H. d'A. Heck, *Crit. Rev. Toxicol.* **20**, 397–426 (1990).
131. B. A. Owen and co-workers, *Reg. Toxicol. Pharmacol.* **11**, 220–236 (1990).
132. *Chem. Eng. News*, 16 (June 1, 1992).
133. T. J. Kulle and co-workers, *J. Air Pollution Control Assoc.* **37**, 919–924 (1987).
134. A. Weber-Tschopp, T. Fischer, and E. Grandjean, *Int. Arch. Occup. Environ. Health* **39**, 207–218 (1977).
135. Ref. 36, p. 139.
136. *IARC Monographs Suppl.* **7**, 211–216 (1987).
137. W. D. Kerns and co-workers, *Cancer Res.* **43**, 4382 (1983).
138. W. J. Martin, *Reproductive Toxicol.* **4**, 237–239 (1990).
139. L. J. Enders, *Contempory AB/GYN*, 1–4 (Feb. 1987).
140. N. Harris and M. Tuck, *Hydrocarbon Process.*, 79–82 (May 1990).
141. R. G. Dickinson and A. L. Redmond, *J. Am. Chem. Soc.* **45**, 28 (1943).
142. Ref. 1, pp. 511–551.
143. Ref. 1, pp. 140–205.
144. U.S. Pat. 3,883,309 (May 13, 1975), K. Ishigawa and co-workers (to Kanebo Ltd.).
145. U.S. Pats. 2,568,016 and 2,568,017 (Sept. 18, 1951), A. F. MacLean and W. E. Heinz (to Celanese).
146. U.S. Pat. 4,036,891 (July 19, 1977), J. C. T. Moller and O. E. Hansen.
147. J. Mahieux, *Hydrocarbon Process.* **48**(5), 163 (1969).
148. U.S. Pat. 3,310,572 (Mar. 21, 1967), H. Delle and H. Mann (to Degussa).
149. U.S. Pat. 3,637,751 (Jan. 25, 1972), H. Fuchs and H. Sperber (to BASF).
150. U.S. Pat. 4,110,298 (Aug. 29, 1978), W. J. Wells and A. L. Stautzenberger (to Celanese).
151. Neth. Appl. 6,415,198 (June 30, 1966), (to Skanska Attikfabriken Aktiebolag).
152. Ref. 1, pp. 191–199.
153. U.S. Pat. 4,043,873 (Aug. 23, 1977), J. Ackermann and co-workers (to Societa Italiana Resine).

154. U.S. Pat. 4,125,540 (Nov. 14, 1978), A. Sugio and co-workers (to Mitsubishi).
155. U.S. Pat. 3,426,041 (Feb. 4, 1969), Y. Miyake and co-workers (to Toyo Koatsu Industries).

H. ROBERT GERBERICH
GEORGE C. SEAMAN
Hoechst-Celanese Corporation

FORMALS. See FORMALDEHYDE.

FORMAMIDE. See FORMIC ACID AND DERIVATIVES.

FORMIC ACID AND DERIVATIVES

FORMIC ACID

Formic acid [*64-18-6*] (methanoic acid) is the first member of the homologous series of alkyl carboxylic acids. It occurs naturally in the defensive secretions of a number of insects, particularly of ants. Although the acid nature of the vapors above ants' nests had been known since at least 1488, the pure acid was not isolated until 1671, when the British chemist John Ray described the isolation of the pure acid by distillation of ants (1). This remained the main preparative method for more than a century until a convenient laboratory method was discovered by Gay-Lussac (2). The preparation of formates using carbon monoxide was described by Berthelot in 1856.

Formic acid was a product of modest industrial importance until the 1960s when it became available as a by-product of the production of acetic acid by liquid-phase oxidation of hydrocarbons. Since then, first-intent processes have appeared, and world capacity has climbed to around 330,000 t/yr, making this a medium-volume commodity chemical. Formic acid has a variety of industrial uses, including silage preservation, textile finishing, and as a chemical intermediate.

Physical Properties

Formic acid is a colorless, highly corrosive liquid with an intense and pungent smell. Its basic physical properties are given in Table 1. Formic acid is totally miscible with water, acetone, ether, ethyl acetate, methanol, and ethanol. It dissolves to the extent of about 10% in benzene, toluene, and xylenes and to a lesser extent in aliphatic hydrocarbons. Formic acid is unusual in forming a high boiling azeotrope (bp = 107°C at normal pressure) with water, containing 77.5% by weight of the acid. Azeotropic mixtures are also formed with many other substances (12). Formic acid and water form a eutectic mixture which freezes at −48.5°C and contains 70% by weight of formic acid.

In the vapor phase formic acid forms a hydrogen-bonded dimer:

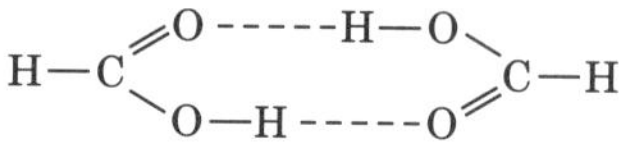

At room temperature and atmospheric pressure, 95% of the vapor consists of dimers (13). The properties of the vapor deviate considerably from ideal gas behavior because of the dimerization. In the solid state, formic acid forms infinite chains consisting of monomers linked by hydrogen bonds (14):

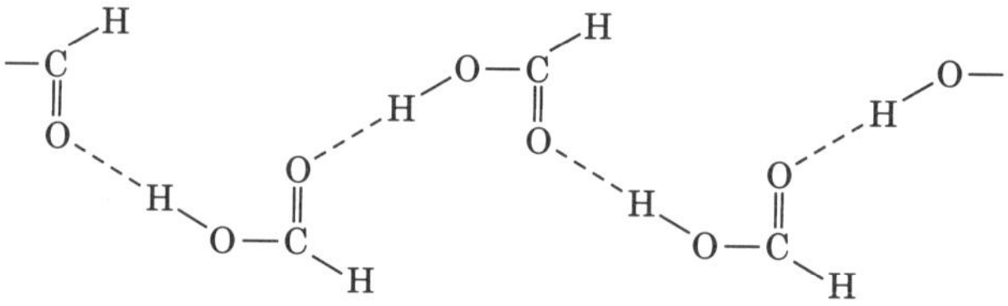

Table 1. Physical Properties of Pure Formic Acid

Property	Value	Reference
molecular weight	46.026	
melting point, °C	8.4	3
boiling point, °C	100.7	4
density, g/cm³ at 20°C	1.220	5
refractive index n_D^{20}	1.3714	5
surface tension at 20°C, mN/m(= dyn/cm)	37.67	6
viscosity at 20°C, mPa·s(= cP)	1.784	7
specific conductance, Ω/m	6.08×10^{-3}	8
dielectric constant at 20°C	57.9	9
latent heat of vaporization at boiling point, kJ/kg[a]	483	10
free energy of formation, ΔG_f^0(g), kJ/mol[a]	−351	11
enthalpy of formation, ΔH_f^0(g), kJ/mol[a]	−377	11

[a]To convert kJ to kcal, divide by 4.184.

The chain structure persists in liquid formic acid, but in solution in hydrocarbons the dimeric form predominates (15).

Chemical Properties

Formic acid exhibits many of the typical chemical properties of the aliphatic carboxylic acids, eg, esterification and amidation, but, as is common for the first member of an homologous series, there are distinctive differences in properties between formic acid and its higher homologues. The smaller inductive effect of hydrogen in comparison to an alkyl group leads, for example, to formic acid (pK_a = 3.74) being a considerably stronger acid than acetic acid (pK_a = 4.77) (16) and in fact the strongest of the simple, unsubstituted carboxylic acids. The formyl hydrogen can also be looked on as conferring some aldehydic character, and formic acid, for example, reduces aqueous silver nitrate to produce metallic silver.

Formic acid forms esters with primary, secondary, and tertiary alcohols. The high acidity of formic acid makes use of the usual mineral acid catalysts unnecessary in simple esterifications (17). Formic acid reacts with most amines to form formylamino compounds. With certain diamines imidazole formation occurs, a reaction that has synthetic utility (18):

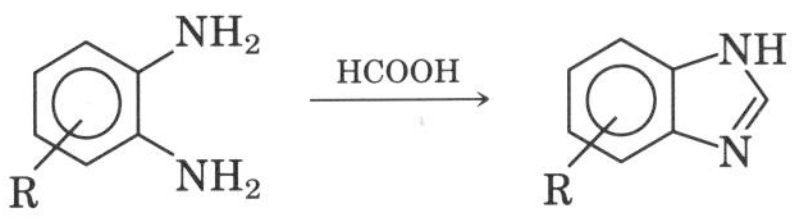

Formic acid can decompose either by dehydration, $HCOOH \rightarrow H_2O + CO$ (ΔG^0 = -30.1 kJ/mol; ΔH^0 = 10.5 kJ/mol) or by dehydrogenation, $HCOOH \rightarrow H_2 + CO_2$ (ΔG^0 = -58.6 kJ/mol; ΔH^0 = -31.0 kJ/mol). The kinetics of these reactions have been extensively studied (19). In the gas phase metallic catalysts favor dehydrogenation, whereas oxide catalysts favor dehydration. Dehydration is the predominant mode of decomposition in the liquid phase, and is catalyzed by strong acids. The mechanism is believed to be as follows (19):

$$HCOOH + H^+ \rightarrow HCOOH_2^+ \rightarrow HCO^+ + H_2O$$
$$HCO^+ \rightarrow H^+ + CO$$

The formyl cation, HCO^+, is also likely to be an intermediate in the modification of the Koch reaction whereby formic acid reacts with olefins to give carboxylic acids (20):

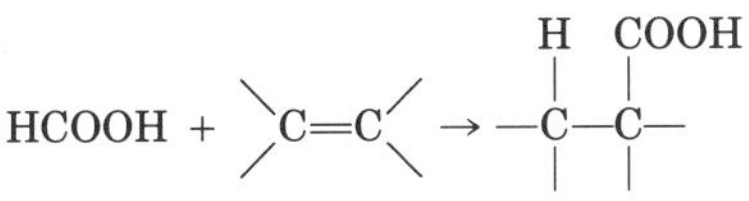

This reaction occurs readily in the presence of sulfuric or hydrofluoric acid. In the absence of such strong acids, formic acid reacts readily with olefins to give formate esters (21).

The anhydride of formic acid has not been isolated, but mixed anhydrides are known, and, with acetic acid, the latter have utility as formylating agents (22). The only known formyl halide is the fluoride, which has a boiling point of $-29°C$.

Manufacture

Formic acid is currently produced industrially by three main processes: (1) acidolysis of formate salts, which are in turn by-products of other processes; (2) as a coproduct with acetic acid in the liquid-phase oxidation of hydrocarbons; or (3) carbonylation of methanol to methyl formate, followed either by direct hydrolysis of the ester or by the intermediacy of formamide.

The reaction of formate salts with mineral acids such as sulfuric acid is the oldest industrial process for the production of formic acid, and it still has importance in the 1990s. Sodium formate [141-53-7] and calcium formate [544-17-2] are available industrially from the production of pentaerythritol and other polyhydric alcohols and of disodium dithionite (23). The acidolysis is technically straightforward, but the unavoidable production of sodium sulfate is a clear disadvantage of this route.

Liquid-phase oxidation of lower hydrocarbons has for many years been an important route to acetic acid [64-19-7]. In the United States, butane has been the preferred feedstock, whereas in Europe naphtha has been used. Formic acid is a coproduct of such processes. Between 0.05 and 0.25 tons of formic acid are produced for every ton of acetic acid. The reaction product is a highly complex mixture, and a number of distillation steps are required to isolate the products and to recycle the intermediates. The purification of the formic acid requires the use of azeotroping agents (24). Since the early 1980s hydrocarbon oxidation routes to acetic acid have declined somewhat in importance owing to the development of the rhodium-catalyzed route from CO and methanol (see ACETIC ACID).

The carbonylation of methanol [67-56-1] to methyl formate in the presence of basic catalysts has been practiced industrially for many years. In older processes for formic acid utilizing this reaction, the methyl formate [107-31-3] reacts with ammonia to give formamide [75-12-7], which is hydrolyzed to formic acid in the presence of sulfuric acid:

$$2\ CO + 2\ CH_3OH \longrightarrow 2\ HCOOCH_3 \xrightarrow{2\ NH_3} 2\ HCONH_2 + 2\ CH_3OH$$
$$2\ HCONH_2 + H_2SO_4 + 2\ H_2O \longrightarrow 2\ HCOOH + (NH_4)_2SO_4$$

Coproduction of ammonium sulfate is a disadvantage of the formamide route, and it has largely been supplanted by processes based on the direct hydrolysis of methyl formate. If the methanol is recycled to the carbonylation step the stoichiometry corresponds to the production of formic acid by hydration of carbon monoxide, a reaction which is too thermodynamically unfavorable to be carried out directly on an industrial scale.

$$CO + CH_3OH \rightarrow HCOOCH_3$$
$$HCOOCH_3 + H_2O \rightarrow HCOOH + CH_3OH$$
$$CO + H_2O \rightarrow HCOOH$$

A number of variants of this process have been patented (25). The conditions for the carbonylation step are broadly similar in each, but they differ in their approach to the hydrolysis stage.

The carbonylation of methanol to give formic acid is carried out in the liquid phase with the aid of a basic catalyst such as sodium methoxide. It is important to minimize the presence of water and carbon dioxide in the starting materials, as these cause deactivation of the catalyst. The reaction is an equilibrium, and elevated pressures are necessary to give good conversions. Typical reaction conditions appear to be 80°C, 4.5 MPa (44 atm) pressure and 2.5% w/w of catalyst. Under these conditions the methanol conversion is around 30% (25).

The hydrolysis of methyl formate is technologically demanding for a number of reasons. The hydrolysis equilibrium is relatively unfavorable (26), but is dependent on the water concentration in a way that favors the use of high stoichiometric excesses of water, with consequent problems of finding an energy-efficient method of removing the excess water. Furthermore, methyl formate is highly volatile (bp = 32°C) and formic acid is a sufficiently strong acid to catalyze the reesterification. It is therefore difficult to remove unreacted methyl formate without a significant amount of reesterification. In the Kemira-Leonard process (27) these problems are overcome by use of two hydrolysis reactors in series, followed by flash distillation to rapidly remove unconverted ester and distillation in a column of low residence time. The process operated by BASF (25) uses hydrolysis in the presence of a large excess of water followed by liquid–liquid extraction of formic acid with a nitrogen base and subsequent distillation. The process operated in the Ukraine (28) uses a fixed bed of cationic ion-exchange resin as hydrolysis catalyst. Other approaches to the hydrolysis problem have also been patented (29,30).

Other potential processes for production of formic acid that have been patented but not yet commercialized include liquid-phase oxidation (31) of methanol to methyl formate, and hydrogenation of carbon dioxide (32). The catalytic dehydrogenation of methanol to methyl formate (33) has not yet been adapted for formic acid production.

Shipping and Storage

Formic acid is commonly shipped in road or rail tankers or drums. For storage of the 85% acid at lower temperatures, containers of stainless steel (ASTM grades 304, 316, or 321), high density polyethylene, polypropylene, or rubber-lined carbon steels can be used (34). For higher concentrations, Austenitic stainless steels (ASTM 316) are recommended.

The DOT hazard classification of formic acid is "corrosive material." A DOT white label is mandatory for transportation. The EC classification is "corrosive."

Attention must be paid to the fact that formic acid, particularly at higher concentrations and temperatures, can, on storage, slowly decompose to liberate carbon monoxide. This can lead to safety hazards from the ensuing buildup in pressure or from the toxicity of carbon monoxide. It has been estimated that, in the absence of leakage, a full 2.5-L bottle of formic acid could develop a pressure of over 700 kPa (7 atm) over a period of one year at 25°C (35).

Economic Aspects

World installed capacity for formic acid is around 330,000 t/yr. Around 60% of the production is based on methyl formate. Of the remainder, about 60% comes from liquid-phase oxidation and 40% from formate salt-based processes. The largest single producer is BASF, which operates a 100,000 t/yr plant at Ludwigshafen in Germany. The only significant U.S. producer of formic acid is Hoechst-Celanese, which operates a butane oxidation process.

The principal use worldwide for formic acid is as a silage additive, an application that is not well developed in the United States; the U.S. market for formic acid is therefore relatively small (ca 30,000 t/yr) by world standards. Typical U.S. prices for formic acid (mid-1992) were around $0.90/kg.

Specifications and Analytical Methods

Formic acid is manufactured in concentrations of 85, 90, 95, 98, and > 99%, the diluent being water. A typical product sales specification is given in Table 2 (36).

Table 2. Specifications for Commercial Formic Acid

Property	Value
appearance	clear, free from matter in suspension
color	20 Hazen maximum
assay	98 or 85% mass minimum
nonvolatile matter	0.005% mass maximum
chloride as Cl^-	3 ppm maximum
sulfate as SO_4^{2-}	5 ppm maximum
iron as Fe	2 ppm maximum

The concentration of aqueous solutions of the acid can be determined by titration with sodium hydroxide, and the concentration of formate ion by oxidation with permanganate and back titration. Volatile impurities can be estimated by gas–liquid chromatography. Standard analytical methods are detailed in References 37 and 38.

Health and Safety Factors

The main hazard in normal handling of formic acid is likely to arise from its corrosive effect on the skin and mucous membranes. Suitable protective equipment should be worn when handling the acid, and rubber or PVC gloves, rubber boots, and goggles are needed during bulk handling operations. In the event of contact with the skin, the affected area should be washed thoroughly with water, and medical attention should be obtained if redness and blistering occur and persist. Medical attention should be sought in all cases of ingestion or contact with the eyes (39).

Exposure to formic acid vapor causes irritation of the eyes and respiratory tract. The TLV/TWA occupational exposure limit is 5 ppm (40). Self-contained breathing apparatus should be used when there is a risk of exposure to high vapor concentrations.

Formic acid is combustible (flash point = 69°C), but the explosive hazard is considered slight. The decomposition to CO requires appropriate precautions to be taken when entering tanks or other confined spaces that have contained the acid.

Formic acid is readily biodegraded, and therefore does not bioaccumulate. The theoretical biological demand (BOD) is 350 mg/g (25).

Uses

The largest single use of formic acid is as a silage additive. This market is well developed in Europe, but scarcely exists in the United States. If formic acid is applied to freshly cut grass prior to ensilation, the nutritional value of the ensuing silage is enhanced. Specifically, lactic fermentation is promoted, and the undesirable formation of butyric acid avoided. Milk yields of cattle fed with silage made in this way can be improved through the winter. Mixtures of formic acid with propionic acid or formate salts are also used for this purpose.

Formic acid can also be used as an antisalmonella additive in animal feeds, for decontamination of feed raw materials, and prevention of flock infection in the poultry industry by treatment of the finished feed.

Outside the agricultural area, formic acid finds a range of diverse applications either as an intermediate or as a cost-effective, relatively volatile strong acid. In the textile industry, formic acid is used as an acidulant in the dyeing of both natural and synthetic fibers (see DYES, APPLICATION AND EVALUATION). It is also used to neutralize alkaline solutions and facilitate rinsing during laundering. In the leather industry, formic acid finds a number of applications, particularly in tanning (see LEATHER).

In the rubber industry, formic acid is used for coagulating latex in the production of natural rubber and in the production of certain rubber chemicals (qv) (see RUBBER, NATURAL).

Formic acid is used as an intermediate in the production of a number of drugs, dyes, flavors, and perfume components. It is used, for example, in the synthesis of aspartame and in the manufacture of formate esters for flavor and fragrance applications.

BIBLIOGRAPHY

"Formic Acid" in *ECT* 1st ed., Vol. 6, pp. 875–883, by B. Toubes, Victor Chemical Works; "Formic Acid" under "Formic Acid and Derivatives" in *ECT* 2nd ed., Vol. 10, pp. 99–103, by J. F. Walker, Consultant; in *ECT* 3rd ed., Vol. 11, pp. 251–258, by F. S. Wagner, Jr., Celanese Chemical Co., Inc.

1. J. R. Partington, *Chem. Ind.*, 765 (1933).
2. F. D. Chattaway, *Chem. News* **CIX**, 109 (1914).
3. R. C. Wihot, J. Chao and K. R. Hall, *J. Phys. Chem. Ref. Data* **14**(1), 124 (1985).

4. W. R. Angus, G. T. W. Llewelyn, and G. Stott, *Trans. Faraday Soc.* **50**, 1311 (1954).
5. R. R. Dreisbach and R. A. Martin, *Ind. Eng. Chem.* **41**, 2875 (1949).
6. J. J. Jasper, *J. Phys. Chem. Ref. Data* **1**, 851 (1972).
7. S. E. Sheppard and R. C. Houck, *J. Rheol.* **1**, 349 (1929).
8. T. C. Wehman and A. I. Popov, *J. Phys. Chem.* **72**, 4031 (1968).
9. J. F. Johnson and R. H. Cole, *J. Am. Chem. Soc.* **73**, 4536 (1951).
10. A. S. Coolidge, *J. Am. Chem. Soc.* **52**, 1874 (1930).
11. D. R. Stull, E. F. Westrum, and G. C. Sinke, *The Chemical Thermodynamics of Organic Compounds*, John Wiley & Sons, Inc., New York, 1969, p. 448.
12. L. H. Horsley, *Azeotropic Data III, Adv. Chem. Ser.* **116**, 66 (1973).
13. J. Chao and B. J. Zwolinski, *J. Phys. Chem. Ref. Data* **7**(1), 363 (1978).
14. I. Nahringbauer, *Acta Cryst.* **B34**, 315 (1978).
15. R. J. Jacobsen, Y. Mikawa, and J. W. Brasch, *Spectrochim. Acta* **23A**, 2199 (1967).
16. J. E. Prue and A. J. Read, *Trans. Faraday Soc.* **62**, 1271 (1966).
17. H. W. Gibson, *Chem. Rev.* **69**, 673 (1969).
18. E. C. Wagner and W. H. Millett, *Org. Synth.* **Coll. Vol. II**, 65 (1943).
19. P. Mars, J. J. F. Scholten, and P. Zwietering, *Adv. Catal.* **14**, 35 (1963).
20. G. A. Olah and J. A. Olah, *Friedel Crafts and Related Reactions*, Vol. 3, Wiley-Interscience, New York, 1964, p. 1284.
21. H. B. Knight, R. E. Koos, and D. Swern, *J. Am. Chem. Soc.* **75**, 6212 (1953).
22. L. I. Krimen, *Org. Synth.* **50**, 1 (1970).
23. M. P. Czalkowski and A. R. Bayne, *Hydrocarbon Process.* **59**(11), 103 (1980).
24. W. Hunsmann and K. F. Simmrock, *Chem. Ing. Tech.* **36**(10), 1054 (1966).
25. W. Reutemann and H. Kiecka, *Ullmann's Encyclopaedia of Industrial Chemistry*, Vol. A12, VCH, New York, 1989, p. 13.
26. R. F. Schultz, *J. Am. Chem. Soc.* **61**, 1443 (1939).
27. Eur. Pat. 5,998 (Dec. 12, 1979), J. D. Leonard.
28. Belg. Pat. BE 893,357 (Nov. 29, 1982), I. I. Moiseev, O. A. Tagaev, and N. M. Zhavoronkov.
29. U.S. Pat. 3,907,884 (Sept. 23, 1975), J. B. Lynn, O. A. Homberg, and A. H. Singleton (to Bethlehem Steel Corp.).
30. Ger. Offen. DE 3,428,319 (Feb. 13, 1986), F. Praun, H. U. Hoeg, G. Bub, and M. Zoelffel (to Hüls AG).
31. Eur. Pat. Appl. 60,718 (Sept. 22, 1982), D. J. Drury and J. Pennington (to BP Chemicals Ltd.).
32. PCT Int. Pat. Appl. WO 86 02,066 (Apr. 10, 1986), J. J. Anderson, D. J. Drury, J. E. Hamlin, and A. G. Kent (to BP Chemicals Ltd.).
33. Ger. Offen. DE 2,753,634 (Oct. 27, 1977), M. Yoneoka and M. Osugi (to Mitsubishi Gas Chemical Co.).
34. *Bulk Storage of Formic Acid*, BP Chemicals, Ltd., London, 1990.
35. L. Bretherick, *Bretherick's Handbook of Reactive Chemical Hazards,* 4th ed., Butterworths, London, 1990, p. 150.
36. *Formic Acid*, Product Technigram, BP Chemicals, Ltd., London, 1991.
37. *British Standard BS4341*, British Standards Institute, Milton, Keynes, 1968.
38. R. M. Speights and A. J. Barnard *Encyclopaedia of Industrial Chemical Analysis*, Vol. 13, Wiley-Interscience, New York, 1971, p. 117.
39. *Formic Acid*, Material Safety Data Sheet, BP Chemicals, Ltd., London, 1990.
40. N. I. Sax and R. J. Lewis, *Dangerous Properties of Industrial Materials*, 7th ed., Van Nostrand Reinhold Co., Inc., New York, 1989, p. 1766.

DAVID J. DRURY
BP Chemicals Ltd.

FORMAMIDE

Formamide [75-12-7] (methanamide), $HCONH_2$, is the first member of the primary amide series and is the only one liquid at room temperature. It is hygroscopic and has a faint odor of ammonia. Formamide is a colorless to pale yellowish liquid, freely miscible with water, lower alcohols and glycols, and lower esters and acetone. It is virtually immiscible in almost all aliphatic and aromatic hydrocarbons, chlorinated hydrocarbons, and ethers. By virtue of its high dielectric constant, close to that of water and unusual for an organic compound, formamide has a high solvent capacity for many heavy-metal salts (1) and for salts of alkali and alkaline-earth metals (2). It is an important solvent, in particular for resins and plasticizers. As a chemical intermediate, formamide is especially useful in the synthesis of heterocyclic compounds, pharmaceuticals, crop protection agents, pesticides, and for the manufacture of hydrocyanic acid [74-90-8].

Even though formamide was synthesized as early as 1863 by W. A. Hoffmann from ethyl formate [109-94-4] and ammonia, it only became accessible on a large scale, and thus industrially important, after development of high pressure production technology. In the 1990s, formamide is mainly manufactured either by direct synthesis from carbon monoxide and ammonia, or more importantly in a two-stage process by reaction of methyl formate (from carbon monoxide and methanol) with ammonia.

Properties

Tables 1 and 2 list the important physical properties of formamide. Formamide is more highly hydrogen bonded than water at temperatures below 80°C but the degree of molecular association decreases rapidly with increasing temperature. Because of its high dielectric constant, formamide is an excellent ionizing solvent for many inorganic salts and also for peptides, proteins (eg, keratin), polysaccharides (eg, cellulose [9004-34-6], starch [9005-25-8]), and resins.

Formamide decomposes thermally either to ammonia and carbon monoxide or to hydrocyanic acid and water. Temperatures around 100°C are critical for formamide, in order to maintain the quality required. The lowest temperature range at which appreciable decomposition occurs is 180–190°C. Boiling formamide decomposes at atmospheric pressure at a rate of about 0.5%/min. In the absence of catalysts the reaction forming NH_3 and CO predominates, whereas hydrocyanic acid formation is favored in the presence of suitable catalysts, eg, aluminum oxides, with yields in excess of 90% at temperatures between 400 and 600°C.

Formamide is hydrolyzed very slowly at room temperature. The rate of hydrolysis increases rapidly in the presence of acids or bases and is further accelerated at elevated temperatures.

Reactions

As a result of its bifunctionality, formamide is a highly reactive intermediate that is useful in a wide variety of synthetic applications. The destructive distillation

Table 1. Physical Properties of Formamide

Property	Condition	Value	Reference
molecular formula		CH_3NO	
molecular weight		45.04	
melting point, °C		2.55	3
boiling point, °C	101.3 kPa[a]	210.5 (decomp.)	3
	2.4 kPa[a]	111–112	4
	1.5 kPa[a]	105–106	5
density, g/cm^3	20°C	1.1334	3
	25°C	1.1292	3
	50°C	1.1078	6
	80°C	1.0823	7
dielectric constant	20°C	109 ± 1.5	8
dipole moment, C·m[b]	30°C	1.12×10^{-29}	3
heat of combustion, kJ/mol[c]		−568.2	3
heat of vaporization, kJ/mol[c]		64.98	3
heat of fusion, kJ/mol[c]		6.694	7
standard heat of formation, kJ/mol[c]		−254.0	3
heat capacity, C_p, J/(K·mol)[c]	25°C	107.62	3
pK_a	20°C	−0.48	3
electrical conductivity, S($=\Omega^{-1}$)	25°C	2×10^{-7}	3
refractive index n_D^{15}	15°C	1.4491	6,9
n_D^{20}	20°C	1.44754	3
n_D^{25}	25°C	1.44682	3
coefficient of expansion, cm^3/K		0.000775	3
specific heat, kJ/(kg·K)[c]	19°C	2.30	10
surface tension, mN/m($=$dyn/cm)	20°C	58.35	3
	25°C	58.15	3
	50°C	55.72	10
dynamic viscosity, mPa·s($=$cP)	15°C	4.320	
	20°C	3.764	3
	25°C	3.302	3
	30°C	2.926	
flash point, °C[d]		175	4
ignition temperature, °C[e]		>500	4
explosive limits in air, vol %		2.7–19	4
speed of sound, m/s	23°C	1661	11

[a]To convert kPa to mm Hg, multiply by 7.5.
[b]To convert C·m to debye, divide by 3.336×10^{-30}.
[c]To convert J to cal, divide by 4.184.
[d]ASTM D92-85.
[e]ANSI/ASTM D2155-66.

of formamide at atmospheric pressure gives rise to the two principal decomposition products, ammonia and carbon monoxide, together with smaller amounts of hydrocyanic acid and water. In order to produce high yields of hydrocyanic acid, formamide can be dehydrated catalytically over aluminum oxides or silicates at temperatures exceeding 500°C (13–17). Although stable to hydrolysis at room

Table 2. Vapor Pressure of Formamide at
Various Temperatures[a]

Temperature, °C	Vapor pressure, kPa[b]
20	0.008
55.1	0.10
65.6	0.20
72.1	0.30
80.9	0.50
93.7	1.00
101.7	1.51
116.4	3.00
125.6	4.50
136.2	7.00
145.3	10.05

[a]Ref. 12.
[b]To convert kPa to mm Hg, multiply by 7.5.

temperature, formic acid and ammonia, as an ammonium salt, are formed at elevated temperatures. Treatment of formamide with acids and bases significantly accelerates the hydrolysis. In the past, the reaction of sulfuric acid with formamide served as the basis for the industrial-scale manufacture of formic acid, but this was replaced more recently by a new economic process based on methyl formate. In the presence of hydrogen halides, formamide reacts with alcohols to yield esters of formic acid. Thus ethyl formate is obtained from formamide, hydrochloric acid, and ethanol. In the absence of Lewis acids, which catalyze the esterification of formamide, reaction does not commence until the temperature exceeds 160°C (18). Fatty acid amides are formed in the photochemical addition reaction of formamide with olefins (19). Straight-chain acid amides have been obtained from formamide and α-olefins in the presence of peroxides (20). Formamide is useful as a formylating agent for primary and secondary amines.

Aqueous solutions of formamide react with formaldehyde [50-00-0] in neutral or alkaline media to yield the dimethylol derivative, N,N-bis(hydroxymethyl) formamide [6921-98-8] (21), which in solution is in equilibrium with the monomethylol derivative [13052-19-2] and formaldehyde. With benzaldehyde in the presence of pyridine, formamide condenses to yield benzylidene bisformamide [14328-12-2]. Similar reactions occur with ketones, which, however, require more drastic reaction conditions. Formamide is a valuable reagent in the synthesis of heterocyclic compounds. Synthetic routes to various types of compounds like imidazoles, oxazoles, pyrimidines, triazines, xanthines, and even complex purine alkaloids, eg, theophylline [58-55-9], theobromine [83-67-0], and caffeine [58-08-2], have been devised (22).

Manufacture and Processing

Early in the twentieth century, the first attempts to manufacture formamide directly from ammonia and carbon monoxide under high temperature and pressure encountered difficult technical problems and low yields (23). Only the introduction of alkali alkoxides in alcoholic solution, ie, the presence of alcoholate as a catalyst, led to the development of satisfactory large-scale formamide processes (24).

$$CO + NH_3 \xrightarrow{\text{ROM/ROH}} HCONH_2$$

However, BASF developed a two-step process (25). After methyl formate [107-31-3] became available in satisfactory yields at high pressure and low temperatures, its conversion to formamide by reaction with ammonia gave a product of improved quality and yield in comparison with the earlier direct synthesis.

$$CO + CH_3OH \rightarrow HCOOCH_3$$
$$HCOOCH_3 + NH_3 \rightarrow HCONH_2 + CH_3OH$$

Direct Synthesis. The direct synthesis of formamide from ammonia and carbon monoxide requires pressures between 0.8 and 1.7 MPa and temperatures between 75–80°C (26). Stainless steel is used as construction material, since both starting materials and products are corrosive and would lead to iron contamination if brought into contact with it. Various attempts to replace the alcoholate catalyst with improved systems have failed (27). It is essential that pure feed gases be used in the process. If the carbon monoxide source is low grade product (eg, off-gases from furnaces or reforming operations) the raw gas has to be purified, because impurities such as water, carbon dioxide, or hydrogen sulfide may lead to undesired side reactions or may cause catalyst poisoning. In particular, sodium formate [141-53-7] in the presence of water can cause severe problems by forming insoluble precipitates. If ammonia is employed in excess, it first has to be taken off overhead in a distillation step and recycled. In a second column, methanol is recovered at the top and formamide is taken off at the bottom. The additional product purification steps are required for catalyst recovery and salt removal, and also for overhead distillation of formamide.

Two-Step Process. The significant advantage of the two-step process is that it only requires commercial-grade methyl formate and ammonia. Thus the crude product leaving the reactor comprises, in addition to excess starting materials, only low boiling substances, which are easily separated off by distillation. The formamide obtained is of sufficient purity to meet all quality requirements without recourse to the costly overhead distillation that is necessary after the direct synthesis from carbon monoxide and ammonia.

In the BASF process (28), methyl formate is continuously fed into a reactor equipped with a circulation pump and an external heat exchanger. The ammonia, which is introduced through a nozzle, reacts to form formamide and methanol, releasing heat that is removed by an external heat exchanger. The typical reaction pressure is 0.2–0.3 MPa and the temperature is about 50–60°C. The processing is designed to serve the purposes of recovering methanol in high purity, so that it can be recycled into the methyl formate synthesis and, secondly, of producing formamide that meets the certified specifications. The off-gas leaving the reactor is scrubbed with water to remove excess ammonia, methanol, and methyl formate. The bulk of the released methanol in the liquid phase leaving the reactor is separated from formamide in a column overhead. The crude methanol is purified in a distillation system, whereby low boiling methyl formate and ammonia are separated. In order to complete the removal of low boiling by-products from the crude formamide, it has to undergo a vacuum distillation. The formamide so obtained is of >99.5% purity.

In addition to the processes mentioned above, there are also ongoing efforts to synthesize formamide directly from carbon dioxide [124-38-9], hydrogen [1333-74-0], and ammonia [7664-41-7] (29–32). Catalysts that have been proposed are Group VIII transition-metal coordination compounds. Under moderate reaction conditions, ie, 100–180°C, 1–10 MPa (10–100 bar), turnovers of up to 1000 mole formamide per mole catalyst have been achieved. However, since expensive noble metal catalysts are needed, further work is required prior to the technical realization of an industrial process for formamide synthesis based on carbon dioxide.

Shipment

Formamide is a registered substance, eg, in TSCA (75-12-7), EINECS (200-842-0), and MITI (2-681), and can, therefore, be produced in and imported into the United States, EEC, and Japan in compliance with the abovementioned acts.

In unalloyed steel containers formamide discolors slowly during shipment and storage. Both copper and brass are also subject to corrosion, particularly in the presence of water. Lead is less readily attacked. Aluminum and stainless steel are resistant to attack by formamide and should be used for shipping and storage containers where the color of the product is important or when metallic impurities must be minimized. Formamide attacks natural rubber but not neoprene. As a result of the solvent action of formamide, most protective paints and finishes are unsatisfactory when in contact with formamide. Therefore, formamide is best shipped in containers made of stainless steel or in drums made of, or coated with, polyethylene. Formamide supplied by BASF is packed in Lupolen drums (230 kg) or Lupolen canisters (60 kg) both in continental Europe and overseas.

Economic Aspects

The estimated capacity of formamide was approximately 100,000 t/yr worldwide in 1990. In 1993, there are only three significant producers; BASF in Germany is the leading manufacturer. Smaller quantities of formamide are produced in the former Czechoslovakia (Sokolov) and Japan (Nitto) by direct synthesis from carbon monoxide and ammonia. Most of the formamide produced is utilized directly by the manufacturers. The market price for formamide (ca 1993) is about $2.00/kg.

Specifications, Standards, and Quality Control

The quality of formamide supplied by BASF is certified as having a minimum assay of 99.5%. The principal impurities in the material are ammonium formate, methanol, water, and traces of iron. The quality of formamide supplied by BASF is certified to meet the specifications given in Table 3.

Direct determination of the formamide content by gc methods proves to be inaccurate because of its tendency to decompose at elevated temperatures. This also limits the accuracy of the classical Kjehldahl determination. The purity of formamide is, therefore, more reliably determined by analysis of its impurities and substraction of the combined contents from 100%.

For the purpose of quantitative analysis, formamide can be hydrolyzed under basic conditions to alkali formate and ammonia that can be determined by conventional methods.

Methanol can be converted to a dye after oxidation to formaldehyde and subsequent reaction with chromatropic acid [148-25-4]. The dye formed can be determined photometrically. However, gc methods are more convenient. Ammonium formate [540-69-2] is converted thermally to formic acid and ammonia. The latter is trapped by formaldehyde, which makes it possible to titrate the residual acid by conventional methods. The water content can be determined by standard Karl Fischer titration. In order to determine iron, it has to be reduced to the iron(II) form and converted to its bipyridyl complex. This compound is red and can be determined photometrically. Contamination with iron and impurities with polymeric hydrocyanic acid are mainly responsible for the color number of the merchandized formamide (< 20 APHA). Hydrocyanic acid is detected by converting it to a blue dye that is analyzed and determined photometrically.

For purposes of product identification and quality control it is useful not only to employ the abovementioned analytical methods but also to measure physical constants such as the density, refractive index, melting point, and pH value of the material.

Table 3. Specifications of Formamide

Quality	Value[a]	Method of determination
formamide assay	99.5%[b]	by difference
methanol content	0.1%	by photometry or gc
ammonium formate content	0.1%	by titration
water content	0.1%	DIN 51777, ASTM E203-75
iron content	0.0001%	by photometry

[a]Value is maximum unless otherwise stated.
[b]Value given is minimum.

Storage

The shelf life is unlimited in sealed containers. The product is neither explosive nor spontaneously flammable in air. Pure formamide is slightly hygroscopic, but stable to hydrolysis at room temperature. It is stable to the effects of light and air below ca 100°C. However, the product is combustible and should accordingly be stored with adequate precautions. The product is not included in Annex I (List of Dangerous Substances) of EEC Directive 76/907/EEC (including the latest amendment), but on the basis of the present data and in connection with the "EEC

Guide to Labelling" it is a dangerous substance and should be labeled as follows: "Hazard symbol: Xn. R-phrase: 47. S-phrase: 53" (EEC Directive 91/325/EEC). To prevent contact with formamide, an approved organic vapor respirator, a face shield, goggles, coveralls, and other protective clothing should be worn as necessary. Spilled material must be disposed of in accordance with local, state, and federal regulations.

Health and Safety

Formamide exhibits no particular acute toxicity with oral, dermal, and other applications in rats and other species. LD_{50} values are in the range of 2.7–17 g/kg. The acute inhalation hazard is also low. In the rabbit skin test it is not irritating, but it is mildly irritating to guinea pigs. In the rabbit eye the substance caused slight to strong irritation. Formamide is not a sensitizer in guinea pigs.

Oral applications of 1500 mg/kg/d for two weeks caused death of rats. Repeated oral applications (4 wks) of 30 μL/kg to rats caused no toxic symptoms, 100 μL/kg resulted in minor toxic effects, and 300 μL/kg induced impairment of several organs and lethality. Repeated oral applications (26 ×) of 0.1 mL/kg caused no toxic symptoms in cats or rabbits. Inhalation of 1500 ppm (2 wks) were without effects, whereas in a similar study with comparable concentrations, toxicity was observed. Exposure of 3 ppm (4 mo) to rats induced some organ toxicity. Repeated dermal applications of 3000 mg/kg in rats (3 mo) caused increased mortality, 300 and 1000 mg/kg caused only changes in blood, whereas 100 mg/kg were without effects.

In several mutagenicity tests formamide did not demonstrate a genotoxic potential. In a further study chromosome damage in rats was described but not clearly verified. Formamide produced adverse effects on the offspring after oral and dermal applications in different species. Formamide had no influence on the fertility of male rats after inhalation. TLV (1993–1994) is 10 ppm (18 mg/m^3) (33). Precautions that should be observed when handling formamide include avoidance of prolonged inhalation of vapors or contact of the liquid with skin or eyes. When handling the chemical its teratogenic property has to be taken into account (34).

Small quantities of spilled formamide can be washed away with plenty of water. Larger amounts should be absorbed appropriately or pumped into containers for proper disposal by incineration or biological degradation in a sewage water treatment plant.

Uses

Formamide has a wide variety of applications and has become an important intermediate in the chemical industry where it is used for producing heterocyclic compounds, pharmaceuticals, crop protection agents, fungicides, and pesticides and has gained importance as a solvent. It is hydrolyzed to formic acid by sulfuric acid. With twice the stoichiometric proportion of sulfuric acid, ammonium hydrogen sulfate is obtained (35), which as a saturated solution itself is able to hydrolyze formamide (36), thus allowing the process to be carried out in two consecutive steps (37). For environmental reasons the production of formic acid based on formamide is declining. In the formamide vacuum process, anhydrous hydrocyanic acid is produced by passing a mixture of formamide and a small proportion of air at reduced pressure through a heated tube or over a catalyst (38).

The good solvent properties of formamide have been exploited widely in the manufacture and processing of plastics. Formamide has been recommended for removing coatings of wire enamel from copper conductors (39). It can also be used in the spinning of acrylonitrile copolymers (40). It is employed in the antistatic finishing of plastics or formation of conductive coatings on plastic particles (41). If acrolein and α-substituted acroleins are polymerized in the presence of formamide, clear resin solutions are obtained that can be used for fabricating films. The resins can also be cured by storing (42).

BIBLIOGRAPHY

"Formamide" under "Formic Acid," in *ECT* 1st ed., Vol. 6, pp. 881–882, by B. Toubes, Victor Chemical Works; "Formamide" under "Formic Acid and Derivatives," in *ECT* 2nd ed., Vol. 10, pp. 103–108, by H. Louderback, E. I. du Pont de Nemours & Co., Inc.; in *ECT* 3rd ed., Vol. 11, pp. 258–263, by C. L. Eberling, BASF Aktiengesellschaft.

1. H. Röhler, *Z. Elektrochem.* **16**, 419 (1910).
2. E. Colton, R. E. Brooker, *J. Phys. Chem.* **62**, 1595 (1958).
3. J. A. Riddick, W. B. Bunger, and T. K. Sakano, *Organic Solvents—Physical Properties and Methods of Purification*, 4th ed., John Wiley & Sons, Inc., New York, 1986, p. 654.
4. BASF AG Ludwigshafen, unpublished results, 1989–1990.
5. C. Synowietz and K. Schäfer, eds., *Chemikerkalender*, 3rd ed., Springer-Verlag, Berlin, 1984, p. 288.
6. H. Bipp, in B. Elvers, S. Hawkins, M. Ravenscroft, J. Rounsaville, and G. Schuls, eds., *Ullmann's Encyclopedia of Industrial Chemistry*, 5th ed., Vol. A12, VCH Publishers, Weinheim, Germany, pp. 1–5.
7. G. Körösi and E. Sz. Kovats, *J. Chem. Eng. Data* **26**, 323 (1981).
8. J. Falbe, in M. Regitz, ed., *Römpps Chemie-Lexikon*, 9th ed., Vol. 2, Georg Thieme Verlag, 1990, pp. 1425-1426.
9. H. Boit, ed., *Beilsteins Handbuch der Organischen Chemie, Viertes Ergänzungswerk*, Vol. II/2, Sys. No. 156; H 26, E IV 45, Springer-Verlag, Berlin, 1975.
10. *Formamide*, Product Information Bulletin, Du Pont, Wilmington, Del., 1961.
11. M. A. Goodman and S. L. Whittenburg, *J. Chem. Eng. Data* **29**, 125 (1984).
12. BASF AG Ludwigshafen, unpublished results, 1984.
13. Ger. Pat. 476 662 (Dec. 20, 1924), R. Fick (to IG Farbenindustrie AG).
14. Ger. Pat. 498 733 (Jan. 1, 1929), T. Ewan (to ICI).
15. U.S. Pat. 2,042,451 (June 2, 1936), H. A. Bond and N. D. Scott (to Du Pont).
16. Ger. Pat. 561,816 (Oct. 24, 1930), H. Dohse (to IG Farbenindustrie AG).
17. Ger. Pat. 477,437 (Apr. 11, 1926), E. Münch and F. Nicolai (to IG Farbenindustrie AG).
18. H. Bredereck, R. Gompper, and G. Theilig, *Chem. Ber.* **87**, 537 (1954).
19. D. Elad, *Chem. Ind.* 362 (1962).
20. A. Rieche, E. Schmitz, and E. Gründemann, *Angew. Chem.* **73**, 621 (1961).
21. S. L. Vail, C. M. Moran, and H. B. Moore, *J. Org. Chem.* **27**, 2067 (1962).
22. H. Bredereck, R. Gompper, H. G. v. Schuh, and G. Theilig, *Angew. Chem.* **71**, 753 (1959).
23. Ger. Pat. 390,798 (July 10, 1921), K. H. Meyer (to IG Farbenindustrie/BASF AG).

24. Ger. Pat. 550,749 (Aug. 26, 1924), R. Witzel (to IG Farbenindustrie/BASF AG).
25. Ger. Pat. 1,215,130 (July 3, 1962), E. Germann and H. Fritz (to BASF AG).
26. Ger. Pat. 624,508 (Jan. 2, 1936), (to Du Pont).
27. U.S. Pat. 3,099,689 (July 30, 1963), H. J. Craigg (to Ethyl Corp.).
28. Ger. Pat. 2,623,173 (May 22, 1976), H. Hohenschutz, M. Strohmeyer, M. Herr, and H. Kiefer (to BASF AG).
29. U.S. Pat. 3,530,182 (Dec. 26, 1967), P. Haynes and co-workers (to Shell Co.).
30. P. Haynes, L. H. Slaugh, and J. F. Kohnle, *Tet. Lett.* **5**, 365 (1970).
31. Jpn. Pat. 52,036,617 (Sept. 17, 1975), K. Yoshihisa and S. Kenji (to Mitsui Petrochemicals).
32. Jpn. Pat. 52,087,112 (Jan. 14, 1976), K. Yoshihisa and S. Kenji (to Mitsui Petrochemicals).
33. *TLVs: Threshold Limit Values and Biological Exposure Indices*, American Conference of Governmental Industrial Hygienists (ACGIH), Cincinnati, Ohio, 1990–1991.
34. *BIBRA Toxicity Profile*, The British Industrial Biological Research Association, Carshalton, Surrey, U.K., 1990.
35. Ger. Pat. 710,170 (July 31, 1941), W. Klempt (to Bergwerksverband zur Verwertung von Schutzrechten der Kohlentechnik GmbH).
36. Ger. Pat. 714,970 (Nov. 20, 1941), W. Klempt (to Bergwerksverband zur Verwertung von Schutzrechten der Kohlentechnik GmbH).
37. Ger. Pat. 720,013 (Mar. 26, 1942), W. Klempt (to Bergwerksverband zur Verwertung von Schutzrechten der Kohlentechnik GmbH).
38. Eur. Pat. 209,039 (July 7, 1986) U. von Oehsen, K. Stecher, W. Köhler, B. Müller, F. Brunnmüller, and R. Schneider (to BASF AG).
39. U.S. Pat. 2,737,465 (Mar. 6, 1956), L. Pessel (to Radio Corporation of America).
40. U.S. Pat. 2,776,945 (Jan. 8, 1957), F. J. Rahl and H. H. Weinstock, Jr. (to Allied Chemical & Dye Corp.).
41. U.S. Pat. 2,879,234 (Mar. 24, 1959), M. A. Coler.
42. Ger. Pat. 1,059,661 (June 18, 1959), K. -H. Rink (to Degussa).

General References

L. F. Fieser and M. Fieser, *Reagents for Organic Synthesis*, Vol. 3, John Wiley & Sons, Inc., New York, 1972, p. 147.
Formamide, BASF Technical Leaflet 1983; BASF data sheet, BASF AG, Operating Division Zwischenprodukte, Ludwigshafen, Germany, 1989.

A. HÖHN
BASF AG

DIMETHYLFORMAMIDE

N,N-Dimethylformamide [*68-12-2*], DMF, is a clear, colorless, hygroscopic liquid with a slight amine odor. The odor is due to trace amounts of dimethylamine, which, along with formic acid, is formed as a by-product from hydrolysis by trace amounts of water. The rate of hydrolysis in the absence of catalysts and at ambient temperature is quite slow, but the odor of dimethylamine is detectable at less than 1 ppm (1). DMF was first synthesized in 1893 by heating a mixture of sodium formate and dimethylammonium chloride, and later prepared by direct reaction of dimethylamine with formic acid (2). Although it does not occur widely in nature, traces of DMF have been detected in sausage, cooked mushrooms, grapes, and wine. The pure liquid is kinetically stable to its boiling point, but its decomposition is catalyzed by strong bases and some transition metals. It is also susceptible to photochemical degradation. The solvent properties of DMF are particularly attractive. Because of its high dielectric constant, aprotic nature, wide liquid range, and low volatility, it is frequently used for chemical reactions and other applications requiring high dissolving power. It is often referred to as "the universal solvent."

Physical and Chemical Properties

Tables 1 and 2 list some physical properties and thermodynamic information for DMF (3,4).

Table 1. Physical Properties of DMF, $HCON(CH_3)_2$

Property	Value
molecular formula	C_3H_7NO
molecular weight	73.09
boiling point, °C	153.0
freezing point, °C	−60.4
density, g/cm^3	
20°C	0.949
40°C	0.931
vapor pressure, kPa^a	
20°C	0.38
40°C	1.3
60°C	3.9
vapor density	2.5
critical values	
temperature, °C	374
pressure, MPa^b	4.42
volume, cm^3/g	3.65
refractive index, n_D^{25}	1.4269
dielectric constant, ϵ, 10 kHz, 25°C	36.7
surface tension, 25°C, mN/m(= dyn/cm)	36.42
dipole moment, C·m,^{c} 20°C	1.27×10^{-29}
viscosity, 25°C, mPa·s(= cP)	0.802
solubility parameter, δ, MPa$^{1/2 d}$	24.8
donor number^e	26.6
hydrogen-bonding index^f	6.4

^{a}To convert kPa to mm Hg, multiply by 7.5.
^{b}To convert MPa to psi, multiply by 145.
^{c}To convert C·m to debye, divide by 3.336×10^{-30}.
^{d}To convert MPa$^{1/2}$ to (cal/cm^3)$^{1/2}$, divide by 2.05.
^{e}Ref. 5.
^{f}On a Du Pont scale.

Table 2. Thermodynamic Properties of DMF[a]

Property	Value
flash point, °C	
TCC	58
TOC	67
autoignition temperature, °C	445
flammability limits in air, vol %	
lower (100°C)	2.2
upper (100°C)	15.2
free energy of formation, ΔG_g^0, 25°C, kJ/mol	−88.4
enthalpy of formation, 25°C, kJ/mol	
ΔH_g^0,	−191.7
ΔH_l^0,	−239.3
heat of combustion, liquid, 25°C, kJ/mol	1921
heat of vaporization, kJ/mol	
25°C	47.6
140°C	38.3
heat capacity, kJ/(kg·K)	
liquid at 25°C	2.03
gas at 127°C, 101.3 kPa[b]	1.61
entropy, S, gas at 25°C, 101.3 kPa,[b] kJ/(mol·K)	0.33

[a]To convert kJ to kcal, divide by 4.184.
[b]101.3 kPa = 1 atm.

Manufacture and Shipment

There are two processes used commercially for DMF manufacture. A two-step process involves carbonylation of methanol [67-56-1] to methyl formate [107-31-3], and reaction of the formate with dimethylamine.

$$CH_3OH + CO \rightarrow HCOOCH_3 \tag{1}$$

$$HCOOCH_3 + (CH_3)_2NH \rightarrow HCON(CH_3)_2 + CH_3OH \tag{2}$$

The methanol carbonylation is performed in the presence of a basic catalyst such as sodium methoxide and the product isolated by distillation. In one continuous commercial process (6), the methyl formate and dimethylamine react at 350 kPa (3.46 atm) and from 110 to 120°C to effect a conversion of about 90%. The reaction mixture is then fed to a reactor–stripper operating at about 275 kPa (2.7 atm), where the reaction is completed and DMF and methanol are separated from the lighter by-products. The crude material is then purified in a separate distillation column operating at atmospheric pressure.

A second process is the direct carbonylation of dimethylamine [124-40-3] in the presence of a basic catalyst or a transition metal. This carbonylation is often run in the presence of methanol in order to help solubilize the catalyst (7), and presumably proceeds through methyl formate as an intermediate.

$$(CH_3)_2NH + CO \rightarrow HC\overset{\overset{\textstyle O}{\|}}{-}N(CH_3)_2 \qquad (3)$$

Again, the basic catalyst is typically sodium methoxide, although other bases such as phenoxides (8) and basic anion-exchange resins (9) have also been used. The reaction using sodium methoxide is performed at 4.9 MPa (48 atm) and 120°C (10).

A modification of the direct process has recently been reported using a circulating reactor of the Buss Loop design (11). In addition to employing lower temperatures, this process is claimed to have lower steam and electricity utility requirements than a more traditional reactor (12) for the direct carbonylation, although cooling water requirements are higher. The reaction can also be performed in the presence of an amidine catalyst (13). Related processes have been reported that utilize a mixture of methylamines as the feed, but require transition-metal catalysts (14).

Another method of preparation involving methyl formate has been reported wherein the formate reacts with ammonia and methanol in the presence of ammonium chloride at 255°C and 2.9 MPa (28.6 atm) (15). In this case, monomethylformamide is present in considerable quantities as a by-product.

In another DMF process, hydrocyanic acid reacts with methanol in the presence of water and a titanium catalyst (16), or in the presence of dimethylamine and a catalyst (17).

DMF can also be manufactured from carbon dioxide, hydrogen, and dimethylamine in the presence of halogen-containing transition-metal compounds (18). The reaction has also been performed with metal oxides and salts of alkali metals as promoters (19).

The direct, one-step production of DMF from carbon monoxide, hydrogen, and ammonia has also been reported. A ruthenium carbonyl catalyst is used, either in a polar organic solvent (20) or in a phosphonium molten salt medium (21).

Formamide has been alkylated with methanol in the presence of a metal catalyst to give DMF (22). The alkylation reaction can also be catalyzed by tetralkylammonium salts (23).

Oxidative processes can also be used to prepare DMF. For example, it can be produced from tetraoxane (a source of formaldehyde (qv)), oxygen, and dimethylamine over Pd–Al$_2$O$_3$ (24) or from trimethylamine and oxygen in the presence of a metal halide catalyst (25).

DMF can be purchased in steel drums (DOT 17E, UN1A1, 410 lbs net = 186 kg), tank trucks, and railcars. On Oct. 1, 1993, new regulations in the United States were established for DMF under HM-181; the official shipping name is N,N-dimethylformamide (shipping designation UN 2265, Packing Group III, Flammable Liquid). Formerly, it was classified as a Combustible Liquid in bulk quantities, but as "Not Regulated" in drums (49 CFR). International overseas shipments have an IMCO classification of 3.3.

Economic Aspects

World capacity of DMF is about 267,000 t, with plans for an additional 50,000 t announced for India and Indonesia. Table 3 lists capacities of principal producers. World production is considerably less than capacity and is probably around 125,000 t. In the United States, 1991 production was about 25,000 t, and a significant portion of that was exported, mostly to Taiwan. DMF production and export are closely tied to economic conditions and exchange rates, as demonstrated by the dramatic turnaround from 1983 when a substantial portion of U.S. demand was met by imports (26). Demand in the United States has slowed since the early 1980s because of decreased acrylic fiber production (see FIBERS, ACRYLIC). As with all organic solvents, growing concerns about emissions have

Table 3. DMF Manufacturers and Capacities

Manufacturer	Location	Capacity, t	World capacity, %
Air Products and Chemicals	Pensacola, Fla., United States	7,000	2.6
BASF Aktiengesellschaft	Ludwigshafen, Germany	60,000	22.5
BASF Quimica da Bahia SA	Camaçari, Brazil	6,000	2.2
Celanese Mexicana	Cosoleacaque, Veracruz, Mexico	6,000	2.2
Chinook	Toronto, Canada	10,000	3.8
E. I. du Pont de Nemours & Co., Inc.	Belle, W. Va., United States	41,000	15.4
Ertisa	Huelva, Spain	5,000	1.9
ICI-Petrochemicals Division	Billingham, U.K.	15,000	5.6
Korea Fertilizer Co.	Ulsan, Republic of Korea	8,000	3.0
Lee Chang Yung Chemical Industry	Hsinchu City, Taiwan	10,000	3.8
Leuna Werke	Leuna, Germany	19,000	7.1
Mitsubishi Gas Chemicals	Matsuhama, Niigata, Japan	20,000	7.5
Nitto Chemical Industries	Tsurami, Yokohama, Japan	35,000	13.1
PT Langgeng Kimindo Pratama	Karawang, West Java, Indonesia	40,000[a]	
Rashtriya, Chemicals and Fertilizer	Thane, India	2,500[a]	
UCB	Gent, Belgium	25,000	9.4
Total		*267,000*	

[a]Planned; not included in total.

caused many users to make more efficient use of DMF. Increasingly, it is being recovered and purified for reuse. Because many applications require pure DMF, this recovery is not trivial. However, this trend toward recycling will certainly continue and impact future demand for this solvent. The 1992 selling price for DMF was $1.39/kg.

Specifications, Analytical Methods, and Storage and Handling

Table 4 lists the specifications set by Du Pont, the largest U.S. producer of DMF (4). Water in DMF is determined either by Karl Fischer titration or by gas chromatography. The chromatographic method is more reliable at lower levels of water (< 500 ppm) (4). DMF purity is determined by gc. For specialized laboratory applications, conductivity measurements have been used as an indication of purity (27). DMF in water can be measured by refractive index, hydrolysis to DMA followed by titration of the liberated amine, or, most conveniently, by infrared analysis. A band at 1087 cm^{-1} is used for the ir analysis.

A number of methods are available for determination of DMF in air. Air can be drawn through an absorbent to concentrate the DMF vapors. The absorbent is then heated and the released DMF is analyzed by gc. Infrared analysis of an air sample can also be used, although enhanced sensitivity can be attained by scrubbing an air sample with perchloroethylene and then analyzing the resultant solution by ir. The most convenient procedure is to use indicating absorbent tubes that change color when air contaminated with DMF is passed through them. In the United States, these sampling tubes are available from Lab Safety Supply, Inc. (Janesville, Wisconsin) or Mine Safety Appliances (Pittsburgh, Pennsylvania).

Pure DMF is essentially noncorrosive to metals; however, copper, tin, and their alloys should be avoided. Ideal materials for its handling and storage are nonalloy (carbon) steels, stainless steels, and aluminum. Seals and other soft materials should be made of polytetrafluoroethylene, polyethylene, or high molecular weight polypropylene. Ethylene–propylene rubber O-rings and diaphragms can also be used with DMF. Oils and grease are not suitable as lubricants; however, graphite can be used to lubricate moving parts in contact with DMF. Since DMF is hygroscopic, it should be kept under a blanket of dry nitrogen. High purity DMF required for acrylic fiber production is best stored in aluminum tanks.

Table 4. Specifications for DMF[a]

Quality	Value
water, %	0.05
basicity (as dimethylamine, ppm)	15
formic acid, ppm	20
iron, ppm	0.05
color (APHA)	15
pH at 25°C, 20% aqueous solution	6.5–9.0

[a]Ref. 4.

Uses

The two largest uses for DMF in the United States had been in pharmaceutical processing and acrylic fiber production. Until the late 1980s, acrylic fiber spinning was the largest single use for DMF and accounted for around 25% of total U.S. consumption. However, in 1990, Du Pont ceased U.S. production of acrylic fibers (28). The remaining U.S. producers use dimethylacetamide [127-19-5], DMAC, in their fiber-spinning processes and acrylic fiber production no longer consumes significant quantities of DMF in the United States.

Worldwide demand for DMF in acrylic fiber production has held up better than in the United States. The high solubility of polyacrylonitrile in DMF, coupled with DMF's high water miscibility, makes it an attractive solvent for this application. Its principal competition in this area comes from DMAC.

The combined pharmaceutical applications account for an estimated 25% of DMF consumption. In the pharmaceutical industry, DMF is used in many processes as a reaction and crystallizing solvent because of its remarkable solvent properties. For example, hydrocortisone acetate [50-03-3], dihydrostreptomycin sulfate [5490-27-7], and amphotericin A [1405-32-9] are pharmaceutical products whose crystallization is facilitated by the use of DMF. It is also a good solvent for the fungicide griseofulvin [126-07-8] and is used in its production.

Another significant application for DMF is as a solvent for depositing polyurethane coatings on leather and artificial leather fabrics. This use, too, has fallen somewhat since the late 1970s, as changing fashions have decreased the demand for "wet look" fabrics, and it now accounts for about 5% of U.S. demand for DMF. Again, this use is more significant in the rest of the world than it is in the U.S. Other polymers that dissolve in DMF include poly(vinyl chloride), vinyl chloride–vinyl acetate copolymers, and some polyamides. DMF is also used as a solvent in epoxy formulations. Dicyandiamide [461-58-5], an important epoxy curing agent, is soluble in DMF, and solutions are used in preparing formulations for use in laminated printed circuit boards.

Another use is in various extraction and absorption processes for the purification of acetylene or butadiene and for separation of aliphatic hydrocarbons, which have limited solubility in DMF, from aromatic hydrocarbons. DMF has also been used to recover CO_2 from flue gases. Because of the high solubility of SO_2 in DMF, this method can even be used for exhaust streams from processes using high sulfur fuels. The CO_2 is not contaminated with sulfur-containing impurities, which are recovered from the DMF in a separate step (29).

Because of its ability to dissolve both inorganic- and organic-based residual fluxes, DMF is used as a quench and cleaner combination for hot-dip tinned parts. The high solubility of inorganic salts in DMF has led to the use of such solutions in high voltage capacitors. DMF is also used in some industrial paint stripping applications. However, toxicity concerns limit its use in consumer products of this type. The high solvent strength of DMF also contributes to its use as a carrier for inks and dyes in various printing and fiber-dyeing applications.

DMF is used extensively as a solvent, reagent, and catalyst in synthetic organic chemistry. Several comprehensive reviews describe its uses in this area (30–32).

Health and Safety Factors

The acute toxicity of DMF is relatively low. The LD_{50} by oral ingestion in rats is 2800 mg/kg and the LC_{50} for mice is 9400 mg/m^3 (2 h). Skin absorption is also an important route by which DMF can be introduced into the body, and an LD_{50} of 4720 mg/kg has been observed in rabbit-skin exposure studies.

The chemical has been linked to alcohol intolerance among workers, which is manifested by a reddening of the skin upon ingesting alcoholic beverages shortly after exposure to DMF. The skin reddening is temporary and is usually limited to the face and neck. Chronic exposure to high levels of DMF causes liver damage. In some studies, blood tests on exposed workers showed evidence of liver damage even when no symptoms were reported by the workers (33).

Earlier reports of a link between testicular cancer and DMF exposure have not been corroborated in a study of 4000 Du Pont employees (34). Very recently, inhalation studies in mice and rats have shown no oncogenic effect from DMF (35). The International Agency for Research on Cancer (IARC) has concluded that evidence associating DMF with cancer in animals is "inadequate," but has classified DMF as "possibly carcinogenic to humans" (Group 2B) (36).

Although DMF has led to increased embryo mortality in pregnant animals at doses close to the lethal level for the pregnant animal, DMF exposures below the OSHA limits should not represent a hazard to pregnant workers as long as prudent work practices are followed (4).

The American Conference of Governmental Industrial Hygienists (ACGIH) has recommended that time weighted-average exposures for DMF not exceed 10 ppm or 30 mg/m^3 (skin designation, 1989 standard) for an eight-hour work day. In the United States, OSHA has accepted the ACGIH limits in setting regulations for worker exposures. As with other industrial chemicals, regulations and expert opinion evolve over time, and DMF exposure limits may be tightened in the future. A Biological Exposure Index (BEI) of 40 mg DMF metabolites/g of creatinine in urine has also been adopted by ACGIH and applies in cases where there is significant potential for absorption of DMF liquid or vapor through the skin.

Although it is a versatile and generally stable solvent, DMF must be used with care in some applications. It can react violently with very strong reducing and oxidizing agents. Runaway reactions have been reported under certain conditions with nitric acid and its salts, permanganates, dichromates, bromine, chlorine gas, sodium hydride, sodium borohydride, and alkyl aluminum compounds (37). DMF is widely used as a solvent for many of these reagents, and in some instances, the conditions leading to the violent reactions are not well understood. Incidents have occurred after several successful experiments (38). Any scale-up involving DMF should proceed with caution and workers should consult the literature carefully. DMF can also react violently with some organic halides, particularly in the presence of iron salts and other metal contaminants; violent reactions have been reported with ethylene dibromide (39), carbon tetrachloride, and hexachlorobenzene. Additional caution is called for when using inorganic halide reagents such as thionyl chloride and phosphorus oxychloride, since dimethylcarbamoyl chloride, a suspect carcinogen, can be formed from

DMF and these reagents (40). The reaction with thionyl chloride can also be accelerated by iron or zinc contaminants, resulting in a sudden pressure rise (41).

BIBLIOGRAPHY

"Dimethylformamide" under "Formic Acids and Derivatives," in *ECT* 2nd ed., Vol. 10, pp. 109–114, by H. Louderback, E. I. du Pont de Nemours & Co., Inc.; in *ECT* 3rd ed., Vol. 11, pp. 263–268, by C. L. Eberling, BASF Aktiengesellschaft.

1. K. Verschueren, *Handbook of Environmental Data on Organic Chemicals*, Van Nostrand Reinhold Co., Inc., New York, p. 260.
2. J. A. Mitchell and E. E. Reid, *J. Am. Chem. Soc.* **53**, 1879–1883 (1931).
3. Technical data, Design Institute for Physical Property Data (DIPPR) of the American Institute of Chemical Engineers (AIChE), through STN International, Columbus, Ohio, 1992.
4. *Dimethylformamide. Properties, Uses, Storage, and Handling*, Du Pont product brochure H-10902, Oct. 1988.
5. S. Hahn, W. M. Miller, R. N. Lichtenthaler, and J. M. Prausnitz, *J. Sol'n. Chem.* **14**, 129–137 (1985).
6. U.S. Pat. 3,072,725 (Jan. 8, 1963), R. C. Surman (to E. I. du Pont de Nemours & Co., Inc.).
7. U.S. Pat. 4,098,820 (July 4, 1978), W. Couteau and J. Ramioulle (to UCB, SA).
8. U.S. Pat. 4,565,866 (Jan. 21, 1986), R. G. Duranieau, J. F. Knifton, and G. P. Speranza (to Texaco Inc.).
9. U.S. Pat. 4,101,577 (July 18, 1978), D. L. Smathers (to E. I. du Pont de Nemours & Co., Inc.).
10. Brit. Pat. 1,213,173 (Nov. 18, 1970), (to Nitto Chemical Industry Co., Ltd.).
11. *Hydrocarbon Process. Int. Ed.* **64**(11), 128 (Nov. 1985).
12. *Hydrocarbon Process. Int. Ed.* **62**(11), 90 (Nov. 1983).
13. U.S. Pat. 4,539,427 (Sept. 3, 1985), M. J. Green (to BP Chemicals, Ltd.).
14. U.S. Pat. 3,446,842 (May 27, 1969), K. Nozaki (to Shell Oil Co.); U.S. Pat. 4,229,373 (Oct. 21, 1980), K. Sano, H. Kiga, and T. Ikarashi (to Mitsubishi Gas Chemical Co., Inc.).
15. Jpn. Kokai Tokkyo Koho JP 82 80,349 (May 19, 1982) (to Mitsubishi Gas Chemical Co., Inc.).
16. U.S. Pat. 3,882,175 (May 6, 1975), N. Kominami, Y. Fukuoka, and K. Sasaki (to Asahi Chemical Industry Co., Ltd.).
17. Jpn. Kokai 72 08,051 (Nov. 27, 1968), K. Takahashi and Y. Fukuoka (to Asahi Chemical Industry Co., Ltd.).
18. U.S. 3,530,182 (Sept. 22, 1970), P. Haynes, J. F. Kohnle, and L. H. Slaugh (to Shell Oil Co.).
19. U.S. 4,269,998 (May 26, 1981), T. Imai (to UOP, Inc.).
20. U.S. 4,558,157 (Dec. 10, 1985), J. A. Marsella and G. P. Pez (to Air Products and Chemicals, Inc.).
21. U.S. 4,556,734 (Dec. 3, 1985), J. F. Knifton (to Texaco Inc.).
22. Jpn Pat. 73 28,417 (Sept. 1, 1973), S. Senoo, Y. Fukuoka, and K. Sasaki (to Asahi Chemical Industry Co., Ltd.).
23. U.S. 4,853,485 (Aug. 1, 1989), H. E. Bellis (to E. I. du Pont de Nemours & Co., Inc.).
24. U.S. 4,304,937 (Dec. 8, 1981), N. Joji, T. Nobuhiro, and F. Yohei (to Asahi Kasei Kogyo Kabushiki Kaisha).

25. U.S. Pat. 4,281,193 (July 28, 1981), H. E. Bellis (to E. I. du Pont de Nemours & Co., Inc.).

26. *Chem. Mark. Rep.* 11 (July 4, 1983).

27. J. Juillard, in J. F. Coetzee, ed., *Recommended Methods for Purification of Solvents and Tests for Impurities*, Pergamon Press, Oxford, U.K., 1982, pp. 32–37.

28. M. Reisch, *Chem. Eng. News* **68**(40), 9–10 (Oct. 1, 1990).

29. U.S. Pat. 4,528,002 (July 9, 1985), G. Linde (to Linde Aktiengesellschaft).

30. R. S. Kittila, *Dimethylformamide Chemical Uses*, E. I. du Pont de Nemours & Co., Inc., Wilmington, Del., 1967.

31. R. S. Kittila, *Supplement to Dimethylformamide Chemical Uses*, E. I. du Pont de Nemours & Co., Inc., Wilmington, Del., 1973.

32. S. S. Pizey, *Synthetic Reagents*, Vol. 1, Ellis Horwood, Ltd., Chichester, 1974, Chapt. 1.

33. A. Gescher, *Chem. Brit.* **26**, 435–438, (1990).

34. J. L. Chen, W. E. Fayerweather, and S. Pell, *J. Occup. Med.* **30**, 819–821 (1988), cited in Ref. 33.

35. G. L. Kennedy, personal communication, presented at the *Society of Toxicology Meeting*, New Orleans, La., Mar. 14–19, 1993.

36. *NIOSH Alert, Request for Assistance in Preventing Adverse Health Effects from Exposure to Dimethylformamide (DMF)*, DHHS (NIOSH) Publication No. 90-105, Center for Disease Control, U.S. Dept. of Health and Human Services, Washington, D.C., Sept. 1990.

37. L. Bretherick, *Bretherick's Handbook of Reactive Chemical Hazards*, 4th ed., Butterworths, London, 1990, pp 386–387.

38. G. DeWall, *Chem. Eng. News* **60**(37), 5 (Sept. 13, 1982).

39. *Ethylene Dibromide, Laboratory Hazards Data Sheet No. 112*, Royal Society of Chemistry, London, 1991.

40. Ref. 4, pp. 15–17.

41. M. S. Joshi, *Chem. Eng. News* **64**(14), 2 (Apr. 7, 1986).

General References

DMF-Dimethylformamide, product brochure, Air Products and Chemicals, Inc., Allentown, Pa., 1984.

G. L. Kennedy, *CRC Crit. Rev. Toxicol.* **17**, 129–182 (1986). Extensive review of toxicology.

J. A. Riddick, W. B. Burger, and T. K. Sakano, *Organic Solvents, Physical Properties and Methods of Purification*, 4th ed., John Wiley & Sons, Inc., New York, 1986.

H. Bipp and H. Kieczka in *Ullmann's Encyclopedia of Industrial Chemistry*, 5th ed., Vol. A12, VCH Publishers, New York, 1989, pp. 1–12.

M. Fieser, *Fieser and Fieser's Reagents for Organic Synthesis*, Vol. 14, John Wiley & Sons, Inc., New York, 1989, pp. 148–149. See also Vol. 12, 1986, pp. 203–204, which contains references to earlier DMF citations in this series.

R. Keller and co-workers, *Properties of Nonaqueous Electrolytes*, NASA CR-1425, 1969, and CR-72407, 1968, Washington, D.C.

JOHN A. MARSELLA
Air Products and Chemicals, Inc.

FRACTALS. See SUPPLEMENT.

FRACTIONATION, BLOOD

CELL SEPARATION

Whole blood is seldom used in modern blood transfusion. Blood is separated into its components. Transfusion therapy optimizes the use of the blood components, using each for a specific need. Red cell concentrates are used for patients needing oxygen transport, platelets are used for hemostasis, and plasma is used as a volume expander or a source of proteins needed for clotting of the blood.

The discovery in 1900 of the existence of blood groups, together with improved understanding of the importance of sterile conditions, paved the way to modern blood transfusion therapy. In 1915, the feasibility of storage of whole blood was demonstrated. During World War I, the optimal concentration of citrate for use as an anticoagulant was determined. This anticoagulant was used until 1942, when the acid–citrate–dextrose (ACD) solution was developed.

A method for the fractionation of plasma, allowing albumin, γ-globulin, and fibrinogen to become available for clinical use, was developed during World War II (see also FRACTIONATION, BLOOD–PLASMA FRACTIONATION). A stainless steel blood cell separation bowl, developed in the early 1950s, was the earliest blood cell separator. A disposable polycarbonate version of the separation device, now known as the Haemonetics Latham bowl for its inventor, was first used to collect platelets from a blood donor in 1971. Another cell separation rotor was developed to facilitate white cell collections. This donut-shaped rotor has evolved to the advanced separation chamber of the COBE Spectra apheresis machine.

Blood Component Therapy

Blood is composed of a cellular portion, the formed elements, suspended in plasma. The formed elements constitute approximately 40–45% of the blood volume, ie, the hematocrit. When a test tube with blood is centrifuged, the formed elements are packed onto the bottom of the tube, leaving plasma on top (Fig. 1).

The formed elements consist primarily of red blood cells, ie, erythrocytes. Less than 1/600 of the total volume of the formed elements is composed of white blood cells, ie, leukocytes, and less than 1/800 are platelets, ie, thrombocytes. Table 1 gives the typical constitution of human blood.

Each component of blood has a function in the body. Red cells transport oxygen and carbon dioxide between the lungs and cells in the tissues. White cells function as defense of the body. Platelets are important for hemostasis, ie, the maintenance of vascular integrity. Plasma, an aqueous solution containing various proteins and fatty acids, transports cells, food, and hormones throughout the body. Some proteins in plasma play a role in clotting, others are messengers between cells.

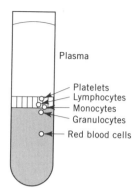

Fig. 1. Distribution of component cells by density in a centrifuged sample of human blood.

Table 1. Hematological Values of Human Blood Components

Parameter	Value range
Physical properties	
blood volume by body weight, mL/kg	80–85
blood osmolarity, mOsm	285–295
blood pH	7.35–7.45
hematocrit, vol %	35–48
Plasma protein fraction	
fraction of plasma, wt %	7–9
concentration, g/100 mL plasma	6.0–8.4
albumin, mass % of protein	50–80
globulins, mass % of protein	15–45
fibrinogen, mass % of protein	3–8
Cell components	
erythrocytes[a]	
10^9 cells/mL	4.8–5.4
fraction of cells, vol %	91
leukocytes[b]	
10^6 cells/mL	4–11
fraction of cells, vol %	5
granulocytes, vol % WBC	55–65
monocytes, vol % WBC	3–9
lymphocytes, vol % WBC	25–33
thrombocytes[c]	
10^6 cells/mL	200–400
fraction of cells, vol %	4

[a]Red blood cells (RBC).   [b]White blood cells (WBC).   [c]Platelets.

A new field of transfusion medicine, cell therapy, has developed with the better understanding of the function of different cell types in the body. In cell therapy, various malignancies are treated by transfusion of specific cell types from blood. Therefore, more and more specialized methods for separating blood into the various components are required.

Collection of Blood and Blood Components. A number of developments since the 1960s have fueled the need for advanced blood cell separation technology. Advancements in medicine have resulted in increasing requirements for blood and its various cellular and plasma-derived components. National programs strive for self-sufficiency in meeting the demand for blood-derived products. Increased awareness of transfusion transmitted disease has resulted in a quest for a 100% safe blood supply even though chances of transmitting blood-borne disease are very slim. In the United States chances of contracting hepatitis A or B through blood transfusion are less than 1 in 20,000, and transmission of the HIV virus through blood transfusion is estimated to be less than 1 in 300,000 in 1993.

Blood can be collected in the form of whole blood donations. In the United States, one unit, ie, 450 mL, of blood is collected from a healthy volunteer blood donor who is allowed to donate blood once every 10 weeks. A unit of blood is typically separated into a red cell fraction, ie, red cell concentrate; a platelet fraction, ie, random donor platelets (RDP); and plasma.

Blood components are also collected through apheresis. In apheresis, advanced blood cell separators are used to collect one or more specific blood components from a donor. The cell separators collect blood into a separation chamber, isolate the desired blood components, and return the blood components not needed to the donor. This procedure is performed on-line within one sterile disposable tubing set. The two principal components collected through apheresis are plasma and single-donor platelets (SDP).

The objectives of collection, separation, preparation, and storage of blood components are (*1*) to provide a safe blood product through careful screening and testing of the donor and collected product; (*2*) to maintain sterility of the product by adequate cleansing of the venepuncture site(s) and sterile processing methods which are essential to avoid bacterial contamination of the products; (*3*) to maintain viability and function of the components, ie, separation and storage methods need to be optimal for the specific transfused product; and (*4*) to make optimal use of blood components by transfusion of only those components indicated for the malignancy of the patients.

Function and Use of Blood Components

Primary blood components include plasma, red blood cells (erythrocytes), white blood cells (leukocytes), platelets (thrombocytes), and stem cells. Plasma consists of water; dissolved proteins, ie, fibrinogen, albumins, and globulins; coagulation factors; and nutrients. The principal plasma-derived blood products are single-donor plasma (SDP), produced by sedimentation from whole blood donations; fresh frozen plasma (FFP), collected both by apheresis and from whole blood collections; cryoprecipitate, produced by cryoprecipitation of FFP; albumin, collected through apheresis; and coagulation factors, produced by fractionation from FFP and by apheresis (see FRACTIONATION, BLOOD–PLASMA FRACTIONATION).

Red Blood Cells.   Red blood cells (RBC) transport and deliver oxygen and carbon dioxide between the tissues and lungs. Red blood cell transfusions increase the oxygen carrying capacity in anemic patients.

Packed red cells are prepared from whole blood. These are collected in blood collection units having integrally attached transfer packs. The red cells are sedimented by centrifugation, and the plasma and buffy coat are expressed from the bag. Further processing of the packed red cells may be needed for a number of clinical indications. To reduce the white blood cell (WBC) contamination in a red cell product, two separation techniques are used.

The red cells may be washed with physiologic saline by using a number of centrifugation and dilution steps. Granulocytes, which form aggregates with platelets in the red cell product, are removed. The total leukocyte reduction varies from 70 to 95%. Approximately 15% of the red cells are lost in this process, and 95% of the plasma from the packed red cells is removed; this may be advantageous in patients having an immunoglobulin deficiency.

The red cells also may be filtered to reduce the white cell content. This technique is needed if there is a chance of the patient developing graft versus host disease (GvHD), ie, transfused white cells attack the cells of the patient.

Red cells may be salvaged for autologous transfusion from blood shed during a procedure in the operating room. This process is called intraoperative blood cell salvage. Shed blood is collected from the operative wound and then filtered, separated, washed, and centrifuged. The red cells, the only component of the shed blood not affected by the operation, are separated by centrifugation in a special chamber. While centrifuging these cells, physiologic saline is circulated through the bed of red cells to wash any debris out of the red cell layer, eg, free hemoglobin, cell stroma, and bone chips.

White Blood Cells.   White blood cells, or leukocytes, have varying function and morphology. Mononuclear leukocytes include lymphocyte B and T-cells, monocytes, and progenitor cells. Polynuclear granulocytes include neutrophils, basophils, and eosinophils. The most important groups in cell separation are lymphocytes, monocytes, and granulocytes.

Contamination of blood products with lymphocytes can lead to transfusion-induced reactions ranging from a mild fever to severe reactions such as alloimmunization and graft versus host disease (GvHD), in which the transfused lymphocytes (graft) survive the defensive immune reaction of the patient (host) and start a reaction which destroys the cells of the host. The patient also may develop an immune response to the human leukocyte antigen (HLA) type of the graft's cells and reject all platelet transfusions that do not match their own HLA system. The HLA system, found on blood platelets and lymphocytes, is more complicated than, but similar to, the ABO blood group system of red cells.

Although the significant efforts in blood cell separation (ca 1993) are aimed at producing virtually leukocyte-free blood products and improving both cell separation and filtration, leukocyte collection is required for some use in blood transfusion therapy. Lymphocytes produce cytokines, which play a role in the communication between blood cells and can be used for the production of interferons for diabetes and interleukins for medical research (see INSULIN AND OTHER ANTIDIABETIC DRUGS). Natural killer cells (NK-cells), a lymphocyte subgroup, may be applied in the treatment of immune disease. Typically the strategy is to collect the patient's own lymphocytes by apheresis, separate out NK-cells, culture and treat them *ex vivo*, and reinfuse them. Transfusion of leukocytes also may be beneficial in fighting sepsis in neonates and granulocytes may be transfused to cleanse the blood from bacteria. White cells also are transfused in the collection of blood-forming cells, or stem cells; this is discussed separately.

Platelets. Blood platelets play a key role in the prevention of blood loss from intact vessels, and the arrest of bleeding from injured vessels. They release adenosine diphosphate [58-64-0] (ADP) and thromboxane A_2 [57576-52-0] which results in vascular contraction and, indirectly, in the formation of fibrin clot. Platelet transfusions are indicated for patients with thrombocytopenia, ie, a shortage of healthy platelets; or thrombocytopathy, ie, platelet malignancy associated with spontaneous hemorrhages.

Transfusion-induced autoimmune disease has been a significant complication in the treatment of patients who require multiple platelet transfusions. Platelets and lymphocytes carry their own blood group system, ie, the human leukocyte antigen (HLA) system, and it can be difficult to find an HLA matched donor. A mismatched platelet transfusion does not induce immediate adverse reactions, but may cause the patient to become refractory to the HLA type of the transfused platelets. The next time platelets with an HLA type similar to that of the transfused platelets are transfused, they are rejected by the patient and thus have no clinical efficacy. Exposure to platelets originating from different donors is minimized by the use of apheresis platelets. One transfusable dose (unit) of apheresis platelets contains 3–5×10^{11} platelets. An equal dose of platelets from whole blood donation requires platelets from six to eight units of whole blood. Furthermore, platelets can be donated every 10 days, versus 10 weeks for whole blood donations.

White cell contamination of a platelet product can induce GvHD. It is believed that GvHD can be minimized by a contamination of less than 5×10^6 white cells per therapeutic dose of platelets, ie, 3–5×10^{11}. Blood cell separation technology is directed toward consistently achieving this goal. Combinations of centrifugation, countercurrent washing, filtration, and uv-irradiation are being investigated to avoid alloimmunization or GvHD.

Stem Cells. Stem cells, or hematopoietic progenitor cells, have the capability of endlessly reproducing themselves and forming new blood cells. High dose chemoradiotherapy, used in the treatment of a number of malignant diseases, destroys all cells in the body that have reproducing capabilities, including stem cells. Patients are treated after each round of chemoradiation therapy by transfusion of stem cells to reconstitute their own blood supply. Stem cells may be collected either from the bone marrow or from the peripheral blood stream, ie, administering growth factors to patients cause the release of stem cells into the blood stream. Stem cells have separation properties similar to lymphocytes and monocytes, and therefore are found in the lower layers of the buffy coat.

Centrifugation Methods

Each type of blood cell has its own distribution of mass densities (Fig. 2). Most blood cell separators are based on the formation of blood components into layers

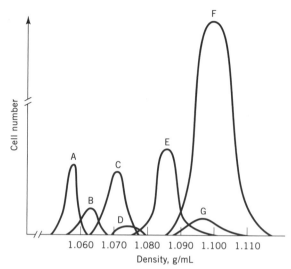

Fig. 2. Mass density distribution of blood components: A, platelets; B, monocytes; C, lymphocytes; D, basophils; E, neutrophils; F, erythrocytes; and G, eosinophils.

by density gradient only. Some cell separators, ie, Haemonetics MCS, apply methods based on a combination of mass density and cell size.

Density Gradient Separation. Based on specific density, each cell in a test tube finds its own position (see Fig. 1), ie, red cells at the bottom, then granulocytes, monocytes, lymphocytes, platelets, and plasma on top. Table 2 lists average mass density of the cellular components of blood. The actual numbers vary slightly from person to person.

Many cell separation methods are based on the formation of layers by mass-density gradient. The simplest method is based on spinning down a bag of blood and expressing off the different layers. The more complex apheresis machines, eg, Baxter Fenwall CS3000 and COBE Spectra, are based on continuous-flow principles. These machines have complicated centrifugal separation chambers which can collect one specific blood component with high purity from a donor and return the other components to the donor. Other blood cell separators utilize a batch processing method and discontinuous flow, eg, the Haemonetics V50plus and Mo-

Table 2. Properties of Blood Components

Component	Average density, g/mL	Diameter, μm	Shape
plasma	1.025–1.029		
platelets	1.040	3	disk, 1-μm thick
white blood cells			
lymphocytes	1.070	8–15	spherical
monocytes	1.075–1.080	15–22	spherical
granulocytes	1.087–1.092	10–14	spherical
red blood cells	1.093–1.096	7.2–7.9	donut, 2-μm thick

bile collection system. These machines fill the separation chamber, collect the desired product, and then return the processed blood to the donor.

In some cases, density gradient solutions are used to separate a specific layer of cells. A solution, like Ficoll or Percoll, with a mass density between the density of the cells that are to be collected and the other cells, is added to the blood product.

Countercurrent Separation and Elutriation. The process known as elutriation in cell separation is a refined method for separation of cells having close mass densities. Cells can be separated by making use of differences in the critical velocity of cells. If the mass densities of two cells are identical, but the sizes are different, then the larger particle has a higher critical velocity than the smaller one.

According to Newton's second law, the sum of the forces on a particle, ie, one spherical cell in plasma, should equal its mass, m_p, times the acceleration of the cell or particle, a_p:

$$F_c - F_D - F_B = m_p a_p \tag{1}$$
$$-F_g + F_D + F_B = m_p a_p$$

where F_g (gravity) and F_c (centripetal) are forces on the cell, F_B (buoyant) and F_D (drag) are forces by the fluid on the surface of the cell. After a brief transient period, the cell falls with a constant settling velocity. For a fast rotating system, equation 1 reduces to

$$F_c - F_D - F_B = 0 \tag{2}$$

If the particle is at a given radius from the center of rotation, the plasma now has to apply an inward drag force, F_D, on the cell to maintain the position of the particle:

$$F_D = F_c + F_B \tag{3}$$

When this inward drag force, F_D, is exceeded by the plasma, the particle moves inward with the plasma. The inward velocity the plasma needs to exceed in order to drag the particle inward is called the critical velocity, U_{cr}, of the particle:

$$U_{cr} = \frac{2r_p^2}{9} \frac{(\rho_p - \rho_f)\,\omega^2 R}{\mu_f \phi(C)} \tag{4}$$

U_{cr} in m/s is critical velocity; r_p, radius (size) of the particle; ρ_p, mass density of the particle; ρ_f, fluid mass density; ω, rotational velocity; R, position of the particle relative to the center of rotation; μ_f, viscosity of the fluid; and $\phi(C)$, a function of the concentration of particles in the fluid.

Cell separation techniques that use an inward flow component are referred to as countercurrent separation techniques. The concept of countercurrent separation is complicated by biological variations of all parameters in equation 4.

The two principal applications of countercurrent flow are found in the Beckman elutriators and the Haemonetics apheresis equipment. The Beckman elutriators are capable of very specific cell separation of small batches of cells. The Haemonetics surge technique can separate platelets and lymphocytes from four liters of donor blood in one hour and forty minutes.

Beckman Elutriation Method. The Beckman elutriation method uses a chamber designed so that the centrifugal effect of the radial inward fluid flow is constant (Fig. 3). The separation chambers are made of transparent epoxy resin which facilitates observation of the movements of the cell boundary in strobe light illumination. This enables detection of the radius at which the cells are separating. When a mixture of cells, eg, mononuclear white cells, enters the chamber, separation can be achieved by fine tuning centrifuge speed and inward fluid flow to the specific cell group. This is a laboratory method suitable for relatively small numbers of cells. Chambers are available in sizes to handle $2–3 \times 10^8$, $1–2 \times 10^9$, and 1×10^{10} cells. The Beckman chambers can be applied to collect mononuclear cells from bone marrow aspirates.

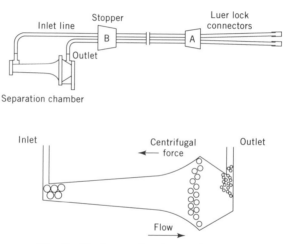

Fig. 3.   Beckman elutriation chamber.

Haemonetics Bowl Technology.   Haemonetics disposable bowl technology has evolved from the original plasma separation chamber. The two principal shapes of bowls are the Latham bowl and the blow-molded bowl (Fig. 4).

Haemonetics' apheresis technology utilizes a discontinuous flow method, eg, countercurrent separation, to collect blood components such as plasma, single-donor platelets, lymphocytes, and stem cells. Anticoagulated blood from a donor enters the bowl via the inlet port and feed tube, as shown in Figure 5. At the end of the feed tube, the blood meets the base of the bowl and fluid is accelerated to the angular velocity of the bowl, ie, $\omega = 4800$ rpm (540 rad/s). As a result of the centripetal force, blood migrates into the separation chamber, ie, the space between the body and outer core. In this separation chamber, blood separates into layers according to the mass density of the components. When the separation chamber is full, plasma is forced out of the separation chamber into the upper

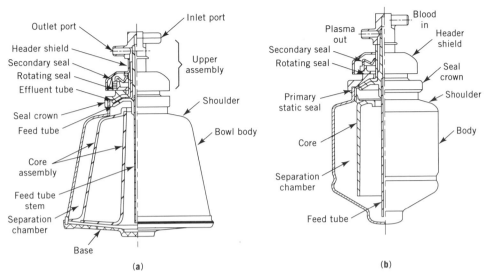

Fig. 4. Haemonetics disposable bowls: (**a**) Latham bowl, and (**b**) blow-molded bowl.

assembly where it contacts the effluent tube that is not rotating. The bowl has a unique rotating seal which maintains sterility of the bowl contents while allowing for rotation of the bowl. Plasma leaves the bowl through the effluent tube and outlet port into the effluent line, which leads to the collection bags.

A small (25-kg), portable apheresis system, available in 1993, is designed to meet a wide variety of blood cell separation needs. The role of the apheresis system is to control the behavior, separation, and collection of blood components from the bowl while maintaining maximum donor safety. The system controls the flow rates of blood and components through variable pump speeds. It directs the flow of components out of the bowl, by fully automatic opening and closing of valves based on the output of the system sensors. The system monitors the separation of blood components in the bowl by an optics system that aims at the shoulder of the bowl. A sensor on the effluent line monitors the flow of components out of the bowl.

The special design of the Latham bowl allows for a specific blood cell separation known as SURGE. This technique makes use of the principle of critical velocity. The Latham bowl is filled until the buffy coat, ie, layer of platelets and white cells, moves in front of the bowl optics. At this point the machine starts to recirculate plasma through the bowl at increasing rates. The smallest particles, ie, platelets, are the first to leave the bowl. Their high number causes the effluent line to turn foggy. The optical density of the fluid in the effluent line is monitored by the line sensor. A special algorithm then determines when to open and close the appropriate valves, as well as the optimum recirculation rate.

Filtration

Filtration (qv) is applied in blood cell separation to remove leukocytes from red blood cell (RBC) and platelet concentrates. Centrifugational blood cell separators

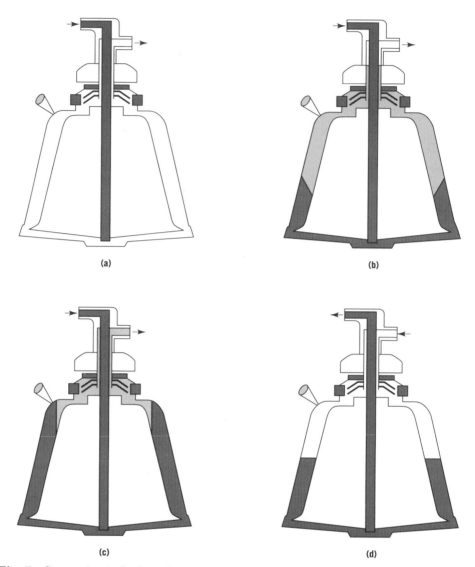

Fig. 5. Separation in Latham bowl: (**a**) whole blood is pumped down the feed tube and enters bowl at bottom; (**b**) centrifugal force spins denser cellular components outside, leaving plasma or platelet-rich plasma (PRP) in inner band; (**c**) when bowl is full, plasma flows out effluent tube, followed by platelets and then leukocytes, until bowl is almost completely full of red cells; (**d**) after draw is completed, bowl stops spinning and uncollected components are pumped through the feed tube and returned to donor.

do not reduce white blood cells (WBC) in red cell and platelet products sufficiently to avoid clinical complications such as GvHD and alloimmunization. A post-apheresis filtration step is needed to further reduce the WBC load. Modern filters are capable of a 3-log reduction in white cell contamination of the blood product, eg, apheresis single-donor platelet units having a typical white cell contamination of 5×10^8 white cells in 4×10^{11} platelets can be reduced to a 5×10^5 white cell contamination, a sufficiently low number to avoid severe transfusion reactions.

Filter Design.   Modern leukocyte-reduction filters have become highly efficient as a result of careful filter design and advanced biomaterials. The binding of leukocytes to the filter media is weak, and hence the flow of blood components through the filter must be well controlled. Effects, such as channeling and bypass, are detrimental to the quality of the product. General design considerations include high flow rate through the filter, low retention volume, and hydrophilic filter media that does not require priming prior to filtration. The filter should also give a high yield of the source product, and loss of platelets or red cells in the filter as a result of either adhesion or retention volume should be minimal.

Two types of leuko-reduction filters are applied in blood cell separation: those for filtering red cells and those for platelet filtration. Filters designed for use with RBCs consist of two filtration layers: an upstream screen filter for trapping large particles and microaggregates, and a downstream adsorption filter for leukocyte reduction. The filters for platelet concentrates have only the adsorption filter. The two principal filter designs commercially available (ca 1993) are a relatively flat, large diameter disk-shaped filter, ie, the RC and PL filters of Pall Corp., and filters having a relatively small cross-sectional area but greater depth, eg, the SepaCell filters of Asahi Corp.

The white cell adsorption filter layer is typically of a nonwoven fiber design. The biomaterials of the fiber media are surface modified to obtain an optimal avidity and selectivity for the different blood cells. Materials used include polyesters, eg, poly(ethylene terephthalate) and poly(butylene terephthalate), cellulose acetate, methacrylate, polyamides, and polyacrylonitrile. Filter materials are not cell specific and do not provide for specific filtration of lymphocytes out of the blood product rather than all leukocytes.

Mechanisms of binding leukocytes to filter fibers are not yet fully understood. The longer a cell is in contact with the material, the more likely it is to attach to the filter. Therefore, too high a flow rate of blood product through the filter can impair performance. Cells that have attached to the filter can be detached if their critical shear force is exceeded. This can occur owing to increased local flow rates as a result of channeling; channeling may be induced by trapped air bubbles in the filter as a result of ineffective priming of the filter. The need for high flow rates also can impair peak performance of the filters, eg, in urgent bedside filtration.

Leuko-reduction can be performed at the time of collection by apheresis in the blood lab or at the patient's bedside. Economic, quality assurance, logistic, clinical, and liability considerations play a role in this. Leukocyte reduction filters are quite expensive, and leukocyte reduction may not be essential in all cases of transfusion.

Mechanisms of Leukocyte Adsorption.   The exact mechanism of leukocyte adhesion to filter media is not yet fully understood. Multiple mechanisms simultaneously contribute to the adhesion of cells to biomaterials, however, physical and biological mechanisms have been distinguished. Physical mechanisms include barrier phenomenon, surface tension, and electrostatic charge; biological mechanisms include cell activation and cell to cell binding.

Barrier Phenomenon.   In red cell filtration, the blood first comes into contact with a screen filter. This screen filter, generally a 7–10-μm filter, does not allow microaggregate debris through. As the blood product passes through the deeper layer of the filter, the barrier phenomenon continues as the fiber density increases. As the path becomes more and more tortuous the cells are more likely to be trapped in the filter.

Surface Tension.   Interfacial surface tension between fluid and filter media is considered to play a role in the adhesion of blood cells to synthetic fibers. Interfacial tension is a result of the interaction between the surface tension of the fluid and the filter media. Direct experimental evidence has shown that varying this interfacial tension influences the adhesion of blood cells to biomaterials. The viscosity of the blood product is important in the shear forces of the fluid to the attached cells; viscosity of a red cell concentrate is at least 500 times that of a platelet concentrate. This has a considerable effect on the shear and flow rates through the filter. The surface stickiness plays a role in the critical shear force for detachment of adhered blood cells.

Electrostatic Charge Density.   Red blood cells, platelets, and white blood cells have a net negative surface charge at physiologic pH. The charge density per unit of cell surface area is different for these three cell lines. It is suspected that white cells carry the highest net negative charge density; therefore, filter media having the correct net positive charge is expected to selectively remove white cells from the filtered blood products.

Cell Activation.   Several studies have shown that platelets and white cells undergo shape changes when adhering to filter media. The cells are activated by contact with the filter media and form pseudopods which attach to the filter media. The cells' membranes may need a certain degree of viability to be able to actively attach to the filter media. When white cells are treated with metabolic inhibitors, the capability of leukocyte reduction by the filter is reduced.

Cell Adhesion.   The membranes of leukocytes and platelets contain a variety of components that promote cell-surface contact. Although numerous cell-surface molecules are likely to play a role in cell-surface adhesion, the group of selectins are of particular interest to research on this subject. Selectins are molecules that are known to promote leukocyte–platelet adhesion. However, selectin-based models have not been able to account for the fact that platelets are allowed to pass through the filter and leukocytes are not.

Cell–Cell Interactions.   Older generations of leukocyte filters depended partly on the formation of platelet–leukocyte–thrombin formations. It is not clear whether this mechanism plays a role in third-generation filters.

Emerging Cell Specific Technologies

A number of cell specific technologies for cell separation are emerging. Cell specific typing through surface antigen marking is possible. The feasibility of separating two subgroups from white blood cell concentrates is being investigated; ie, stem cells, which are marked by the CD34 antigen, and lymphokine activated killer cells (LAK) (CD8). Purified stem cells can be used for cancer therapy (see CHEMOTHERAPEUTICS, ANTICANCER), but also for curing various hematologic genetic diseases through DNA engineering of the purified stem cells (see GENETIC ENGINEERING). The LAK cells are used to treat immune diseases.

Technologies to purify cells from white cell concentrates are in the research stage. Principles used include antibodies covalently bound to a surface, antibody-coated microbeads in a column, magnetic microparticles that have been coated with antibodies, and hollow fibers that have been coated with antibodies.

Regulations, Storage, and Shipment

Blood transfusion is highly regulated worldwide by government institutions, such as the USFDA, and through associations of blood banks, such as the American Association of Blood Banks (AABB). Strict regulations on good manufacturing practices (GMP) have been established to ensure maximum safety of the transfused products.

Each blood component has specific storage requirements in terms of optimal temperature, additives, expiration, and storage containers. Red blood cells (RBC) from whole blood, provided in 200 mL units, have an expiration of 42 days. Frozen, deglycerolized RBC, in 170 mL containers, and washed red cells, in 200 mL containers, both expire 24 hours after thawing and washing, respectively; leukocyte-reduced RBC, in 200 mL containers, are viable for 24 hours.

For optimal functionality, platelets require a stable and well-balanced pH, gas exchange, ambient temperature, and gentle agitation. Special plastics have been developed for optimal storage of platelets.

BIBLIOGRAPHY

"Blood Fractionation" in *ECT* 1st ed., Vol. 2, pp. 556–584, by L. E. Strong, Kalamazoo College; in *ECT* 2nd ed., Vol. 3, pp. 576–602, L. E. Strong, Earlham College; in *ECT* 3rd ed., Vol. 4, pp. 25–61, by M. H. Stryker and A. A. Waldman, The New York Blood Center.

1. M. Bock and co-workers, *Transfusion*, 333–334 (1991).
2. A. Boyum and co-workers, "Density Dependent Separation of White Blood Cells," in J. R. Harris, ed., *Blood Separation and Plasma Fractionation*, Wiley-Liss, New York, 1991.
3. A. Brand, *Transfusion*, 377 (1989).
4. J. P. Crowley, "Transfusion of Plasma," in E. C. Rossi, T. L. Simon, and G. S. Moss, eds., *Principles of Transfusion Medicine*, Williams & Wilkins, Baltimore, Md., 1991.
5. J. P. Dutcher, "Platelet Transfusion Therapy in Patients with Malignancy," in J. P. Dutcher, ed., *Modern Transfusion Therapy*, CRC Press, Boca Raton, Fla., 1990.
6. J. P. Dutcher, "Granulocyte Transfusions in Patients with Malignancy," in Ref. 5.
7. W. H. Dzik, *Transfusion*, 334–339 (1992).
8. W. H. Dzik, "Characteristics and Controversies of Blood Collected by Intra Operative Salvage."
9. M. K. Elias and co-workers, *Annal. Hematol.*, 302–306 (1991).
10. W. F. Ganong, *Review of Medical Physiology*, 13th ed., 1985.
11. L. N. Halpern, *Plasma Ther. Trans. Tech.*, 405–408 (1986).
12. C. A. Kasparin and M. Rzasa, eds., *Transfusion Therapy: A Practical Approach*, AABB, Arlington, Va., 1991.
13. T. S. Kickler and W. Bell, *Transfusion*, 411–414 (1989).
14. A. Latham, *Vox Sanguinis* **51**, 249–252 (1986).
15. J. E. Menitove and L. J. McCarthy, eds., *Hemostatic Disorders and the Blood Bank*, AABB, Arlington, Va., 1984.

16. M. F. Murphy and A. H. Waters, "Leukocyte Depletion of Red Cell and Platelet Concentrates," in Ref. 2.
17. G. Myllyla, "Whole Blood and Plasma Procurement and the Impact of Plasmapheresis," in Ref. 2.
18. S. T. Nance, ed., *Blood Safety: Current Challenges*, AABB, Bethesda, Md., 1992.
19. J. A. F. Napier, "Separation of Whole Blood: an Overview," in Ref. 2.
20. P. M. Ness, "Red Cell Products—Uses and Abuses," in Ref. 5.
21. D. H. Pamphillon and co-workers, *Transfusion*, 379–383 (1989).
22. T. H. Price, "Platelet Pheresis and Leukapheresis," in Ref. 4.
23. E. I. Rossi, and co-workers, "Transfusion in Transition," in Ref. 4.
24. S. G. Sandler, "Management of the Donation Process," in Ref. 4.
25. D. W. Schoendorfer, *Transfusion*, 182–189 (1983).
26. T. L. Simon, "Platelet Transfusion Therapy," in Ref. 4.
27. T. L. Simon, "Apheresis Principles and Practices," in Ref. 4.
28. G. Sirchia, *Transfusion*, 402–405 (1987).
29. C. Th. Smit Sibinga, "Platelet Apheresis and Leukapheresis," in Ref. 2.
30. M. Spivacek, "Autologous Transfusion," in Ref. 5.
31. D. Surgenor and co-workers, *New Engl. J. Med.*, 1646–1651 (1990).
32. I. I. Sussman, "Indications and Use of Fresh Frozen Plasma, Cryoprecipitate and Individual Coagulation Factors," in Ref. 5.
33. H. F. Taswell and A. A. Pineda, eds., *Autologous Transfusion and Hemotherapy*, Blackwell Scientific, Boston, Mass., 1991.
34. J. L. Tullis and D. M. Surgenor, *Science* **26**, 792–797 (1956).

THEODORE HEIN SMIT SIBINGA
Haemonetics

PLASMA FRACTIONATION

Human blood plasma contains over 700 different proteins (qv) (1). Some of these are used in the treatment of illness and injury and form a set of pharmaceutical products that have become essential to modern medicine (Table 1). Preparation of these products is commonly referred to as blood plasma fractionation, an activity often regarded as a branch of medical technology, but which is actually a process industry engaged in the manufacture of specialist biopharmaceutical products derived from a natural biological feedstock (see PHARMACEUTICALS).

History. Methods for the fractionation of plasma were developed as a contribution to the U.S. war effort in the 1940s (2). Following publication of a seminal treatise on the physical chemistry of proteins (3), a research group was established which was subsequently commissioned to develop a blood volume expander for the treatment of military casualties. Process methods were developed for the preparation of a stable, physiologically acceptable solution of albumin [103218-45-7], the principal osmotic protein in blood. Early preparations, derived from equine and bovine plasma, caused allergic reactions when tested in humans and were replaced by products obtained from human plasma (4). Process studies were still being carried out in the pilot-plant laboratory at Harvard in December 1941 when the small supply of experimental product was rushed to Hawaii to treat casualties at the U.S. naval base at Pearl Harbor. On January 5, 1942 the decision was made

Table 1. Pharmaceutical Plasma Derivatives[a]

Product	Clinical application	Molecular weight $\times 10^3$	Normal plasma concentration, g/L
	Albumin[b]		
human serum albumin	protein and volume replacement	68	31–33
plasma protein fraction	volume replacement	68	36–40
	Coagulation proteins[c]		
Factor VIII	hemophilia A treatment	300	3×10^{-4}
Factor IX complex	treatment of hemophilia B and other coagulation disorders	57	5×10^{-3}
antiinhibitor coagulant complex[d]	hemophilia A treatment where Factor VIII antibodies are present		
	Inhibitors[c]		
α-1-proteinase inhibitor	emphysema treatment	52	1.5
antithrombin III	antithrombin III deficiencies treatment	58	0.1
	Immunoglobulins[e]		
immune globulin intravenous (normal)	immunoglobulin (IgG) replacement; treatment of immune disorders	150	12.5
immune globulin intravenous	treatment of cytomegalovirus (CMV) infection in immune-suppressed individuals	150	
immune serum globulin (normal)	prevention of hepatitis A and rubella infections	150	12.5
hepatitis B immune globulin	prevention of hepatitis B infection	150	
pertussis immune globulin	prevention of whooping cough infection	150	
rabies immune globulin	prevention of rabies infection	150	
rho(D) immune globulin	prevention of hemolytic disease of the newborn	150	
tetanus immune globulin	treatment or prevention of tetanus infection	150	
vaccinia immune globulin	prevention of small-pox infection	150	
varicella immune globulin	prevention of chicken-pox infection	150	

[a]U.S. Licensed.   [b]Active component is albumin.   [c]Active component is indicated product.   [d]Active component is not known.   [e]Active component is IgG.

Table 2. 1993 Plasma Fractionators

Organization	Capacity, m³/yr	Location
United States, commercial		
Alpha Therapeutics	1800	Los Angeles
Cutter Biologicals	1800	West Haven, Conn.
Hyland Therapeutics	1800	Glendale, Calif.
Armour Pharmaceuticals	1300	Collegeville, Pa.
Immuno	450	Rochester, Mich.
Ortho Diagnostic	50	Raritan, N.J.
United States, not-for-profit		
Melville Biologics/NYBC	400	Melville, N.Y.
Michigan Department of Public Health	85	Lansing, Mich.
Massachusetts Public Health Labs	50	Boston
Worldwide, commercial		
Institute Merieux	1200	Lyon, France
Immuno AG	1000	Vienna
Nuovi Lab-Forman Biugini	700	Lucca, Italy
Sclavo SpA	600	Siena, Italy
Behringwerke AG	500	Marburg, Germany
Green Cross Corp.	350	Osaka, Japan
Green Cross Corp.	300	Seoul
Institue Grifols	300	Barcelona
Kabi Pharmacia	300	Stockholm
Worldwide, not-for-profit		
Bio-Products Laboratory	600	Elstree, U.K.
National Center for Blood Transfusion	600	Paris
Centre Regionale de Transfusion	550	Lille, France
DRK-Blutspend. Niedersachsen	400	Springe, Germany
SRC Central Laboratory	400	Bern, Switzerland
Central Laboratory N .C	250	Amsterdam
Chemo-Sero-Ther. Research Institute	250	Kumamoto, Japan
Commonwealth Serum Laboratory	250	Melbourne, Australia

to embark on large-scale manufacture at a number of U.S. pharmaceutical plants (4,5).

Full details of this work were published (6) and the processes, or variants of them, were introduced in a number of other countries. In the United States, the pharmaceutical industry continued to provide manufacturing sites, treating plasma fractionation as a normal commercial activity. In many other countries processing was undertaken by the Red Cross or blood transfusion services that emerged following World War II. In these organizations plasma fractionation was part of a larger operation to provide whole blood, blood components, and specialist medical services on a national basis. These different approaches resulted in the development of two distinct sectors in the plasma fractionation industry; ie, a commercial or for-profit sector based on paid donors and a noncommercial or not-for-profit sector based on unpaid donors.

In 1993 there were over 100 organizations undertaking plasma fractionation worldwide, having plant capacities ranging from 4 to 1800 m³/yr. Virtually all of these plants use methods based on those originally devised. Table 2 lists the six commercial manufacturers in the United States and the largest plasma fractionators worldwide.

Manufacturing and Processing

Plasma fractionation is unusual in pharmaceutical manufacturing because it involves the processing of proteins and the preparation of multiple products from a single feedstock. A wide range of unit operations are utilized to accomplish these tasks. They are listed in Table 3; some are common to a number of products and all must be closely integrated. The overall manufacturing operation can be represented as a set of individual product streams, each based on the processing of an intermediate product derived from a mainstream fractionation process (Fig. 1).

Principal Unit Operations. Figure 2 shows the principal unit operations involved in a typical plasma fractionation operation.

Protein Precipitation. The separation of proteins according to differences in solubility plays a significant role in plasma fractionation; a number of precipitation steps are used in the processes for albumin, immunoglobulin (immune globulin) and Factor VIII [9001-27-8] manufacture. Solubility behavior, a unique property of a protein, is determined by size, composition, and conformation, as well as by the environment in which the molecules are located. The protein surface can be regarded as mostly hydrophilic and the protein interior as largely hydrophobic. These properties are determined by the nature and distribution of the amino acid residues that make up the protein (7) (see AMINO ACIDS). The ionizable and polar amino acids are involved in charge repulsion, which plays an important role in preventing the aggregation of protein molecules. It is for this reason that proteins normally display a solubility minimum at their iso-ionic pH.

The solubility of a protein also is determined by the physical and chemical nature of its environment. Properties of the solution that influence protein solvation and protein–protein interactions are particularly important. These interactions, believed to be predominantly electrostatic in nature (8), are therefore influenced by the temperature, dielectric constant, and ionic strength of the solution. The presence of substances that compete preferentially for water molecules reduces protein solubility; a number of substances of this type, such as neutral salts, organic solvents, and nonionic polymers, have been used to precipitate proteins (9). Other available precipitation reagents function by interacting directly with the protein, ie, either changing the surface charge or linking protein molecules together to form aggregates that exceed the solubility limit. Reagents in this category include metal ions, organic dyes, and polyelectrolytes (9). A large number of parameters are potentially available for the manipulation of protein solubility and many of these have been applied to the separation of plasma proteins (10,11).

Table 3. Plasma Fractionation Unit Operations[a]

Unit operation	Method/technology	FVIII	FIX	IgG (im)	IgG (iv)	PPF	Albumin
Protein separation[b]							
fractional precipitation[c]	cold-ethanol precipitation		++	+	+	+	+
fractional extraction[d]	from cold-ethanol ppt			+	+	+	++
solid–liquid separation	centrifugation	+	++	++	++	++	++
	depth filtration			++	++	++	++
selective adsorption	depth filtration	+	+	+	+	+	+
selective adsorption/desorption[e]	ion-exchange chromatography	+	+				
	immuno-affinity chromatography	++	++		++		++
Virus inactivation, in-process							
heat treatment	carbohydrate stabilized	++	++		++		
	fatty acid stabilized					++	++
chemical treatment[f]	solvent–detergent treated	++	++		++	++	
Formulation and finishing[g]							
selective adsorption	depth filtration			+	+	+	+
membrane filtration	cross-flow filtration	++	++	++	++	+	+
	dead-end filtration	+	+	+	+	+	+
stabilization	chemical additives	+	+		+	+	+
dispensing	aseptic-dispensing	+	+	+	+	+	+
drying	freeze drying	+	+	++	++		
Virus inactivation, terminal							
heat treatment	pasteurization					+	+
	dry heating	++	++				

[a] +, method in common use; ++, optional method, depending on procedures used by different manufacturers; im, intramuscular; iv, intravenous. [b] Size exclusion by gel filtration is an optional method for FVIII. [c] Charge reduction (pH, temperature) and other precipitation are common methods of FVIII fractionation. [d] Extraction from other precipitates is an optional method for FVIII fractionation. [e] Affinity chromatography is an optional method for FIX. [f] Optional chemical treatments include potassium thiocyanate [333-20-0] for FIX and acid/enzyme treatment for IgG (iv). [g] Selective proteolysis by acid/enzyme treatment is an optional method for IgG (iv).

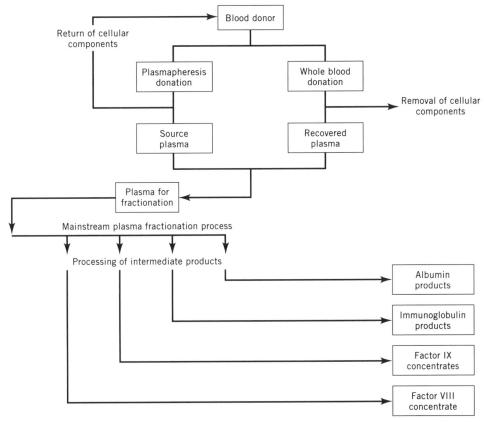

Fig. 1. Organization of the supply and fractionation of human plasma.

Precipitation was the principal protein separation technology initially chosen for the development of an industrially suitable fractionation process. Ethanol [64-17-5] (qv) was selected as the precipitation reagent because of its volatility. It can be subsequently removed using available drying technology (12).

In designing this fractionation scheme (6), known properties of the most significant plasma proteins were utilized to establish a set of process parameters. The solution was adjusted to a pH close to the iso-ionic point of the proteins to be precipitated. Small increases in ionic strength were used to specifically increase the solubility of the more soluble proteins (12,13), thereby increasing the degree of discrimination between different proteins. A progressive increase in ethanol concentration was used to generally decrease protein solubility. Concentrations ranged from 8 to 40%, with the more soluble proteins, eg, albumin, being precipitated at the final stage. The temperature of the solution was held close to the freezing point to avoid protein denaturation, and the use of very dilute solutions was avoided, as proteins are known to be more stable at higher concentrations.

The ability of ethanol to precipitate proteins was believed to result from a reduction in the solution dielectric constant (14). However, the difference between

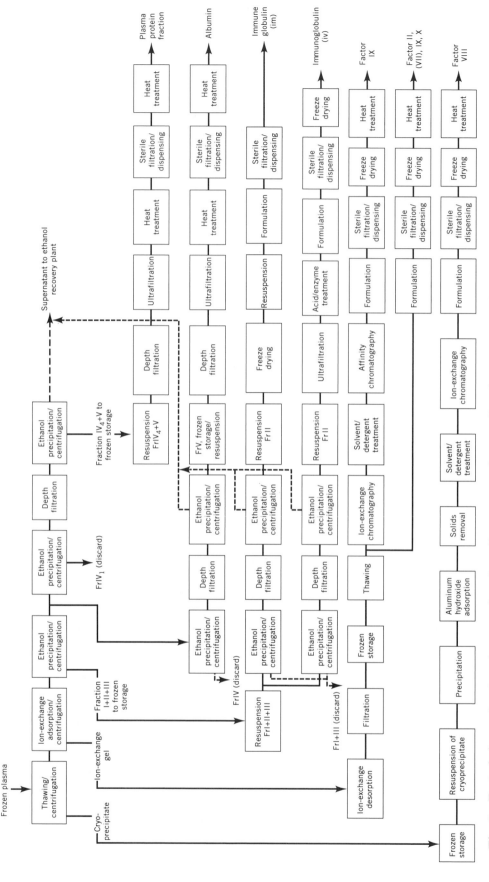

Fig. 2. Typical plasma fractionation operation where Fr represents fraction. Based on processes at the Protein Fractionation Centre, Scottish National Blood Transfusion Service, Edinburgh.

water and ethanol–water mixtures is relatively small at $-5°C$, compared to values at $20°C$ (15), leading to questions about the original interpretation (16,17). From a surface-thermodynamic analysis it has been proposed that the relative surface tension values of the different components favor the binding of water to ethanol rather than the plasma proteins, enabling ethanol to have a dehydrating effect (17). This alternative mechanism of cold-ethanol precipitation, which has yet to be confirmed, appears to be similar to theories developed in the 1970s concerning the precipitation of proteins by salting-out (18–20).

As of this writing cold-ethanol precipitation remains the dominant separations technology used in plasma fractionation (6,21–24). The methods have been reviewed in detail elsewhere (10,11,25,26). The process of cold ethanol fractionation was originally established as a batch operation. However, the use of continuous-flow processing under automatic control was subsequently suggested; details of these proposals are available (27–29). With the benefit of a mainframe computer for process control (qv), continuous-flow processing was introduced successfully into routine use in 1976 (30,31). Studies have been promising (32) but, despite proposals for continuous-flow processing (33,34), the stirred batch tank has remained the dominant mode of operation in the industry (Fig. 3).

Solid–Liquid Separation. The separation of proteins by precipitation technology is accomplished when the solid and liquid phases have been separated from one another. Centrifugation, using either tubular bowl or multichamber centrifuges, is used for this purpose (9,35) (see SEPARATION, CENTRIFUGAL). The machines must be refrigerated to ensure that the correct temperature for precipita-

Fig 3. Example of stirred batch vessels for ethanol fractionation. Courtesy of Swiss Red Cross, Bern.

tion is obtained at the point of separation. Protein precipitates consist of large numbers of small particles, typically 0.1 to 1.0 μm in diameter, which aggregate together to form a larger particle or floc. This aggregate can be broken down by the shear forces experienced in some types of pump and in the entry region to centrifuges. Consequently, performance of the centrifugation operation is influenced by the shear characteristics of the process equipment as well as by the size, density, and strength of the particles (36). Both of these types of centrifuge function by retaining the solids within the rotating bowl, while the feed suspension and resultant supernatant flow continuously. One consequence of this mode of operation is a relatively small solids holding capacity per machine and the need to use a large number of centrifuges to process the contents of a large fractionation vessel. Some manufacturers have introduced depth filtration (qv) for the removal of the solids at some of the precipitation stages. However, protein precipitates consist of highly compressible particles, and filter aids (37) must be used if blinding of the filter surface is to be avoided.

Protein Adsorption. The selective adsorption (qv) of a protein or group of proteins to a solid-phase reagent, followed in most cases by some form of selective desorption, also constitutes a principal form of protein separations technology in plasma fractionation. Solid-phase reagents can be categorized according to the forces responsible for adsorption; eg, charge interactions (ion-exchange (qv)), biochemical interactions (affinity), and highly bioselective interactions (immunoaffinity). Some adsorbents have been less well characterized, eg, kaolin–silica composites [1335-30-4] used in depth filters for the adsorption of lipoproteins from albumin and immunoglobulin solutions, and aluminum hydroxide [21645-51-2] used for the removal of coagulation factor contaminants from Factor VIII solutions.

The forces involved in the separation of proteins by ion-exchange adsorption are believed to be similar to those associated with protein precipitation (18). However, separation is usually achieved by manipulation of pH and ionic strength alone. Ion-exchange chromatography (qv) has been used in the preparation of Factor IX [9001-28-9] concentrates since the early 1970s (38–40) and is also used in the preparation of Factor VIII concentrates (41–43). A number of manufacturers have utilized ion-exchange adsorption in immunoglobulin (44–46) and albumin (47–49) manufacture, usually as an adjunct to cold-ethanol fractionation. Some procedures have been developed to replace cold-ethanol precipitation by ion-exchange chromatography, but this approach has been confined to relatively small-scale plants (50); one larger scale operation (51) was closed before substantial experience was gained. Ion-exchange chromatography also has a role in the preparation of newer plasma products such as α-1-proteinase inhibitor [9041-92-3] (52).

Affinity chromatography is used in the preparation of more highly purified Factor IX concentrates (53–55) as well as in the preparation of products such as antithrombin III [9000-94-6] (56,57). Heparin [9005-49-6], a sulfated polysaccharide (58), is the ligand used most commonly in these applications because it possesses specific binding sites for a number of plasma proteins (59,60).

Immunoaffinity chromatography utilizes the high specificity of antigen–antibody interactions to achieve a separation. The procedure typically involves the binding, to a solid phase, of a mouse monoclonal antibody which reacts either directly with the protein to be purified or with a closely associated protein which itself binds the product protein. The former approach has been applied in the preparation of Factor VIII (43) and Factor IX (61) concentrates. The latter method has been used in the preparation of Factor VIII (42) by immobilization of a monoclonal antibody to von Willebrand factor [109319-16-6] (62), a protein to which Factor VIII binds noncovalently. Further purification is necessary downstream of the immunoaffinity step to remove traces of mouse protein which may leach from the solid phase (63), as well as the toxic chemicals required to disrupt the strong antigen–antibody bond for product desorption (43,61). Ion-exchange chromatography, affinity chromatography, and ultrafiltration (qv) have all been used for this purpose.

Membrane Separations. The availability and development of microporous synthetic membranes and associated process technology has had a significant impact on the manufacture of plasma derivatives (see MEMBRANE TECHNOLOGY). Two very different areas of application exist: (1) the concentration of protein solutions and the removal of low molecular weight solutes use membrane systems that retain macromolecular substances, ie, ultrafiltration; (2) the removal of bacteria utilizes membranes sized to retain particles and larger microorganisms while allowing the macromolecular proteins to pass through, ie, sterile filtration.

Ultrafiltration/Diafiltration. This application of membrane technology is normally carried out in a cross-flow mode using membranes of a nominal pore size in the molecular weight range 10 to 100, depending on the size of the protein being processed. Membrane systems are available in a number of formats, including hollow fiber (see HOLLOW-FIBER MEMBRANES), spiral cartridge, and thin-channel (64,65). The performance of the operation depends on characteristics of the membrane, eg, pore size, porosity, strength, and hydrophobicity; properties of the process solution, eg, the solubility and concentration of protein; fluid mechanics (qv) at the membrane surface (66–68); and pressure difference across the membrane, which acts as the driving force (64,65). Ultrafiltration technology was introduced into plasma fractionation for the removal of residual quantities of ethanol from albumin solutions (69–72), as an alternative to methods such as freeze drying and vacuum distillation. The ability to remove solutes according to molecular size also enables ultrafiltration to be used in the diafiltration mode for the exchange of salts (73,74) and the removal of metal ions, such as aluminum (75,76). Similar technology has been introduced into processes for the manufacture of immunoglobulins (77,78) and coagulation factor concentrates (79–81). Ultrafiltration is generally regarded as the method of choice for the formulation of intermediate and final products. More recent attention has been directed toward further optimization of conditions (82), and the automation (83) and validation (84) of the operation.

Sterile Filtration. Plasma derivatives are prepared for infusion into patients and must therefore be sterile at the point of use. This is achieved by filtering the finally formulated solutions, usually through a series of filters, down to a final 0.2 μm membrane filter (85). Significant advances in the technology include the replacement of filter disks by cartridges of much greater capacity (86) and lower degree of protein adsorption (87,88). The critical nature of this step makes validation of the procedure on a batch-by-batch basis an essential requirement (89–91). Process operations beyond this point, such as dispensing, freeze drying, and closure of the final container, must all be carried out under sterile conditions (see STERILIZATION TECHNIQUES).

Freeze-Drying. Plasma derivatives must have a defined shelf-life, usually for a period of two years. The final dosage form must ensure that a product retains its biological activity throughout the period specified. Products that are not stable in solution for this length of time are normally freeze-dried and reconstituted using sterile water for injection at the time of use. Freeze-drying involves the separation of water from nonvolatile constituents by sublimation from a frozen state. To carry out the process effectively it is necessary to specify the operating conditions carefully at a number of stages (92,93).

The product must be formulated and frozen in a manner which ensures that there is no fluid phase remaining. To achieve this, it is necessary to cool the product to a temperature below which no significant liquid–solid phase transitions exist. This temperature can be determined by differential scanning calorimetry or by measuring changes in resistivity (94,95).

Most of the water is sublimated from the frozen mass by heating the product under reduced pressure. The operating conditions must be such that the product remains in a solid state while sublimation is taking place. The completion of sublimation can be observed by an increase in product temperature. This increase occurs when the energy being introduced is no longer consumed by the latent heat of sublimation, but is absorbed by the product instead.

When sublimation is complete, further heat is applied and the pressure further reduced in a controlled manner to drive-off residual adsorbed moisture. A final moisture content of ≤2% is normally specified.

The development of freeze-drying for the production of blood derivatives was pioneered during World War II (96,97). It is used for the stabilization of coagulation factor (98,99) and intravenous immunoglobulin (IgG iv) products, and also for the removal of ethanol from intramuscular immunoglobulin (IgG im) solutions prior to their final formulation (Fig. 2).

Inactivation and Removal of Viruses. In developing methods of plasma fractionation, the possibility of transmitting infection from human viruses present in the starting plasma pool has been recognized (4,5). Consequently, studies of product stability encompass investigation of heat treatment of products in both solution (100) and dried (101) states to establish virucidal procedures that could be applied to the final product. Salts of fatty acid anions, such as sodium caprylate [*1984-06-1*], and the acetyl derivative of the amino acid tryptophan, sodium acetyl-tryptophanate [*87-32-1*], are capable of stabilizing albumin solutions to 60°C for 10 hours (100); this procedure prevents the transmission of viral hepatitis (102,103). The degree of protein stabilization obtained (104) and the safety of the product in clinical practice have been confirmed (105,106). The procedure has also been shown to inactivate the human immunodeficiency virus (HIV) (107).

The early immunoglobulin products prepared by cold-ethanol fractionation were found to be free from transmitting hepatitis infection (106,108); this was not the case with products prepared by alternative methods (109). Subsequently, some batches of intravenous immunoglobulin transmitted hepatitis infection (110), emphasizing the importance of establishing validated procedures for dealing with potential viral contaminants (111).

A number of coagulation factor products were introduced for the treatment of coagulation disorders during the 1970s (see BLOOD, COAGULANTS AND ANTI-COAGULANTS). The risk of infection with hepatitis led to the withdrawal of fibrinogen [9001-32-5] concentrates from clinical use (112) and to the restriction of other products to the treatment of life threatening conditions such as hemophilia (113). Although some research into the removal (114) or inactivation (115,116) of the hepatitis B virus was underway, the emergence of acquired immune deficiency syndrome (AIDS) in the early 1980s placed the hemophilia population at risk of infection from these products. The impact of HIV cannot be overstated (117–120). Intensive research has led to the development of viral inactivation technologies capable of destroying not only HIV but also viruses responsible for hepatitis B and hepatitis non-A, non-B, eg, hepatitis C. Further methods have been developed for the testing of blood and plasma donations (121). The organization, validation, and control of plasma fractionation operations has been revised to increase the security and reliability of manufacturing procedures (122). These points have been reviewed in detail (122–125).

Viruses that have been transmitted by plasma derivatives are listed in Table 4 together with some of their characteristics. A large number of methods have been studied for the inactivation of viruses (126,127), but not all of these have been found to be fully effective (128). The principal methods, categorized as either in-process or terminal according to position in the process, are listed in Table 5. This distinction is important in the design and organization of manufacturing to ensure that recontamination cannot occur downstream of a virus inactivation step (122,136). The effectiveness of these methods has resulted in plasma products having a high degree of safety with regard to viral infection (126,128,137,138). However, further measures may be required to deal with viruses that are resistant to solvent–detergent treatments, eg, hepatitis A virus (139) and parvovirus B19 (140), both of which are small, nonlipid-enveloped viruses. The use of separation technologies such as adsorption chromatography (141,142) or membrane filtration (143) has been considered in this context. However, the objective of removing some virus particles at a concentration of 10^5/mL (144,145) requires that a stage efficiency of >99.9999% be guaranteed on every occasion. This level of performance is not normally associated with bioseparations technology. The

Table 4. Viruses Transmitted by Human Plasma Derivatives

Virus	Diameter, nm	Nucleic acid	Strandedness
HIV-1	100	RNA	single
hepatitis A[a]	27	RNA[b]	
hepatitis B	42	DNA	partially double
hepatitis C	30–60	RNA	single
hepatitis delta	36	RNA	single
parvovirus, B19[a]	24	DNA	single

[a]No lipid envelope.
[b]This RNA is symmetrical.

Table 5. Virus Inactivation Methods

Method	Stabilizers	Temperature, °C	Time, h	Application	References
In-process solutions					
heating	sodium *N*-acetyl-DL-tryptophanate and/or sodium caprylate	60	10	albumin	102,105
	carbohydrate with or without an amino acid[a]	60	10	coagulation factors	129
	carbohydrate and a neutral salt[b]	60	10	coagulation factors	61,130
acid treatment	neutral salts[c]	60	10	antithrombin III	115
	carbohydrate	35	20	immunoglobulin	131
solvent–detergent treatment		25	6	coagulation factors, other protein	132
moist heating[d]	lyophilization	60	10	coagulation factors	133
Terminal heating					
in solution	sodium *N*-acetyl-DL-tryptophanate and/or sodium caprylate	60	10	albumin	102,105
in freeze-dried state	lyophilization with excipients[e]	80	72	coagulation factors	134,135

[a]For example, sucrose [57-50-1] with or without glycine [56-40-6].
[b]For example, sucrose and sodium sulfate [7757-82-6].
[c]For example, sodium citrate [994-36-5].
[d]Of the dried preparation.
[e]For example, sucrose.

difficulty of validating such operations on a batch-by-batch basis (122) also suggests that this approach may be inferior to well-designed inactivation technologies such as heat treatment.

Process Rationale. The products of plasma fractionation must be both safe and efficaceous, having an active component, protein composition, formulation, stability, and dose form appropriate to the intended clinical application. Processing must address a number of specific issues for each product. Different manufacturers may choose a different set or combination of unit operations for this purpose.

Plasma Collection. Human plasma is collected from donors either as a plasma donation, from which the red cells and other cellular components have been removed and returned to the donor by a process known as plasmapheresis, or in the form of a whole blood donation. These are referred to as source plasma and recovered plasma, respectively (Fig. 1). In both instances the donation is collected into a solution of anticoagulant (146) to prevent the donation from clotting and to maintain the stability of the various constituents. Regulations in place to safeguard the donor specify both the frequency of donation and the volume that can be taken on each occasion (147).

Procedures for the collection of whole blood are similar throughout the world. An interval from at least 8 weeks (United States) to 12 weeks (United Kingdom) is required between a donation of 450 mL blood, which yields about 250 mL plasma. In some countries a smaller volume of blood is collected, eg, 350–400 mL in Italy, Greece, and Turkey and as little as 250 mL in some Asian countries (147). Regulations concerning plasmapheresis donations vary more widely across the world; eg, up to 300 mL of plasma can be taken in Europe in contrast to 1000 mL in the United States, both on a weekly basis. Consequently, both the mode of donation and the country in which it is given can have a profound effect on plasma collection (Table 6).

Table 6. Donation Method and Plasma Collection[a]

Requirements	Source plasma		Recovered plasma
	United States	United Kingdom	United Kingdom
volume of plasma per donor, L/yr	50–60	10–15	0.5
minimum number of donors for 1 t/yr plasma	17–20	65–98	1944
donations required for 1 t/yr plasma	1700	3333	3887

[a]Figures are based on typical practice, described in a number of reports (147–151).

Initial Plasma Processing. Following donation, the separated plasma is frozen (152) and transported to a fractionation plant, where it is held in frozen storage before being released for processing. On entering processing, plasma is vulnerable to bacterial contamination and proteolytic degradation. The more labile constituents are particularly at risk. The early process steps aim for a degree of purification, the creation of a stable environment free from bacterial growth, and, where possible, a significant reduction in process volume. These objectives can be met by precipitation processes (153). Ideally, a range of intermediate products are produced at this stage that are held in storage pending release for further purification. The subdivision of processes in this manner carries a number of advantages including flexibility in scheduling and batch sizing, as well as in maximizing the utilization of limiting or expensive resources.

Factor VIII, immunoglobulin, and albumin are all held as protein precipitates, the first as cryoprecipitate and the others as the Cohn fractions $FI + II + III$ (or $FII + III$) and $FIV_4 + V$ (or FV), respectively (Table 7, Fig. 2). Similarly, Fractions $FIV_1 + FIV_4$ can provide an intermediate product for the preparation of antithrombin III and α-1-proteinase inhibitor. This ability to reduce plasma to a number of compact, stable, intermediate products, together with the bacteriacidal properties of cold-ethanol, are the principal reasons these methods are still used industrially.

Factor VIII Process. The Factor VIII molecule consists of multiple polypeptides having molecular weights of approximately $80-210 \times 10^3$. The purified form consists of a light (mol wt $= 80 \times 10^3$) and a heavy (mol wt $= 90-210 \times 10^3$) chain (154), associated via a calcium linkage (155). The molecule is also bound noncovalenty to von Willebrand factor (mol wt $= 220 \times 10^3$) forming complexes in the molecular weight range $1-10 \times 10^6$. Factor VIII is a particularly labile molecule vulnerable to degradation both by proteolysis (156,157) and by depletion of calcium ions (157,158). Factor VIII is contained in cryoprecipitate (159), the precipitate which forms when frozen plasma is thawed. This has enabled the molecule to be removed from the nonideal environment of the plasma stream at the very beginning of the manufacturing process (Fig. 2). Factor VIII is also vul-

Table 7. Composition of Protein Precipitates[a]

Cohn Fraction	Principal components	Initial plasma protein, %
I	fibrinogen; fibronectin; Factor VIII	5–10
II + III[b]	IgG; IgA; IgM; Factors II, V,[c] VII, IX, X; α- and β-globulins; fibrinogen; plasminogen; plasmin; β-lipoprotein	20–25
(III)	IgA; IgM; Factors II, V; β-lipoprotein; plasminogen; plasmin	6–7
(II-3)	IgG; fibrinogen; β-lipoprotein	4–5
(II-1,2)	IgG	3–4
IV$_1$	α-1-proteinase inhibitor; antithrombin III; IgM; ceruloplasmin[d]; α- and β-globulins; α-lipoprotein; albumin	5–10
IV$_4$	transferrin; haptoglobin; ceruloplasmin; α- and β-globulins; albumin	5–10
V	albumin; α-globulins	50–60

[a]Refs. 10 and 25.
[b]Subfractions given in parentheses.
[c]Factor V [*9001-24-5*].
[d]Ceruloplasmin [*9031-37-2*].

nerable during cryoprecipitation because it is resolubilized as the temperature of the thawed suspension rises. Consequently, processing must be carried out both rapidly and with a high degree of temperature control if loss of Factor VIII is to be minimized (160,161).

Fibrinogen and fibronectin are the other principal proteins in cryoprecipitate. Both are poorly soluble, adherent proteins that can limit the capacity of subsequent chromatographic and filtration operations. As for all protein precipitates, some supernatant remains trapped within the amorphous cryoprecipitate particles and is carried over with the mass of solids; other coagulation factors, which in their activated form can degrade Factor VIII, are of particular concern in this regard. The concentration of fibrinogen and fibronectin is normally reduced using further precipitation steps (129,134,162–168); residual coagulation factors of the prothrombin complex can be removed by adsorption to aluminium hydroxide (163). It is necessary to maintain a sufficient concentration of ionized calcium throughout the process to prevent the dissociation of Factor VIII (169,170).

At this point, most manufacturers include a virus inactivation step, eg, either incubation in the presence of a solvent–detergent mixture or heat treatment (Table 5). Further purification is subsequently required to remove chemicals, eg, tri-*n*-butyl phosphate [126-73-8] and polysorbate-80 [9005-65-6] used in the solvent–detergent treatment (132), or stabilizers used during pasteurization (129,130). This is achieved chromatographically using either ion-exchange adsorption (41), immunoadsorption (42,43), or size exclusion chromatography. Further protein purification is also obtained at this point, enabling the potency and solubility of the final product to be increased and constituents which may be responsible for immune disturbances in hemophiliacs (171,172) to be removed. Factor VIII is a trace substance in plasma and highly purified preparations contain very little protein. It is necessary to carefully formulate the solution to be compatible both with the final stages of manufacture, eg, sterile filtration, dispensing, and freeze drying, and with the product container. This can be achieved by increasing the solution ionic strength and by adding excipients, such as amino acids or human albumin, to prevent the adsorption of Factor VIII to surfaces and to provide a bulking agent for freeze-drying (81).

Factor IX Processes. Direct ion-exchange adsorption is used to recover Factor IX from the supernatant that remains following cryoprecipitation (38–40). Alternatively, Factor IX can be recovered from Cohn Fraction III (173,174). A number of coagulation factors tend to copurify with Factor IX, including Factor II [9001-26-7], Factor VII [9001-25-6], and Factor X [9001-29-0]. These form part of the coagulation cascade, which leads to blood clotting when activated (175,176). It is essential to avoid creating these activated states during processing, as they can lead to coagulation occurring within the process solution. To avoid these difficulties, careful attention must be paid to process materials and reagents, and an adequate degree of anticoagulation must be available at all times.

Products prepared in this manner typically contain significant quantities of coagulation Factors II, VII, and X and other proteins as well as Factor IX (177). Although these preparations have been relatively well tolerated since their introduction in the early 1970s, some patients have experienced thrombotic reactions (178,179), primarily when relatively high doses have been used. The constituents responsible for these reactions have not been fully identified but candidate substances include phospholipid with activated coagulation factors (106,180); Factor IX degradation products (181); or the infusion of Factors II, VII, and X in patients deficient only in Factor IX (53). Further purification, introduced in an attempt to eliminate or reduce this problem (182), involves the use of a second anion-exchange adsorption/desorption step, followed by affinity chromatography using dextran sulfate [9011-18-1] (53,55) or heparin (54) bound to a solid-phase matrix. Immunoaffinity chromatography has also been used for this purpose (61). Methods for preparing both the more highly purified Factor IX product and the standard Factor II, (VII), IX, and X product are illustrated in Figure 2. Fraction VII is not present in all standards.

Immunoglobulin Processes. Virtually all immunoglobulin products consist of IgG, a globular protein composed of two heavy polypeptide chains that run the length of the molecule and two light-chains which are about half the length; together these form a Y-like topology. During processing, IgG is vulnerable to both proteolytic degradation, which results in fragmentation (106,183), and to self-association leading to aggregation of the molecules (184). Most IgG products are prepared using cold-ethanol fractionation, with the precipitate Cohn Fraction I + II + III, or Fraction II + III, being separated from the plasma stream and then processed to obtain Fraction II, which is predominantly IgG (Table 7). Specific conditions under which precipitates are formed for the four principal fractionation schemes in use may be found in the literature (6,21–24). These conditions must be selected and controlled carefully to avoid copurification of proteolytic enzymes such as plasmin [9001-90-5], which can lead to fragmentation of the IgG molecule, with a concomitant loss of biological activity. Aggregation can be minimized by avoiding denaturing conditions and by ensuring that solutions are appropriately formulated at every step (185).

The resuspended and formulated Fraction II precipitate normally contains some aggregated IgG and trace substances that can cause hypotensive reactions in patients, such as the enzyme prekallikrein activator (186). These features restrict this type of product to intramuscular administration. Further processing is required if products suitable for intravenous administration are required. Processes used for this purpose include treatment at pH 4 with the enzyme pepsin [9001-75-6] being added if necessary (131,184), or further purification by ion-exchange chromatography (44). These and other methods have been fully reviewed (45,185,187,188). Intravenous immunoglobulin products are usually supplied in the freeze-dried state but a product stable in the solution state is also available (189).

Albumin Processes. The manufacture of albumin products, like the immunoglobulins, is usually based on cold-ethanol fractionation with the precipitate Fraction V being composed predominantly of albumin (Table 7). All albumin products are pasteurized at 60°C for 10 hours to inactivate potential viral contaminants (Table 5); the purification process must remove components that would denature or aggregate on heating. This can be achieved with as little as 80% of the protein in the form of albumin, if the remaining proteins are α- and β-globulins, and not immunoglobulin or fibrinogen (190). This observation has led to the development of two types of product (106), ie, human albumin containing >96% albumin, and plasma protein fraction (PPF) containing >83% albumin. The first of these is prepared from Fraction V (6,22,23) and the latter from Fraction $IV_4 + V$ (24). During the 1970s, hypotensive reactions with plasma protein fraction were associated with the presence of prekallikrein activator (PKA) (191), a potent vasodilator generated from Factor XII [9001-30-3] (Hageman Factor) by surface activation during processing. Operating conditions can be controlled to avoid PKA contamination, usually by ensuring that the Fraction IV cut, which precedes Fraction $IV_4 + V$ precipitation, is sufficiently large to remove PKA; rapid processing can also be used to minimize the degree of PKA contamination (31). The careful selection and control of precipitation parameters to avoid contamination with PKA applies to the more purified human albumin product as well as to the less purified plasma protein fraction. The presence of sodium acetate [127-09-3] in PPF has also been associated with adverse reactions (192). Acetate buffers have been used for pH control (6), and residual quantities may remain if the albumin precipitate is not washed and ethanol is removed by sublimation or evaporation. The use of diafiltration for the formulation of products (ca 1980s) means that salts, such as sodium acetate, and metal ions, such as aluminum, can be removed easily. Metal ions may have been introduced with reagents such as sodium hydroxide [1310-73-2] or by leaching from depth filters (75,76).

Lipoproteins may denature on heating and if present during pasteurization can result in the formation of haze or turbidity in the final product. This material was removed traditionally by filtration through asbestos (qv) sheets (6); however, health hazards associated with asbestos have led to its replacement by alternative filter materials (23,37,193). These media have been less effective than asbestos and further measures have been required to ensure the visual clarity of albumin products, eg, further filtration developments for lipid removal (194), preferential denaturation of contaminants using in-process heat treatment, and anion-exchange chromatography (49).

Other Processes. A number of other plasma products are entering into clinical use; growth is expected in at least some of these areas. Fibrinogen, previously withdrawn because of the hepatitis risk (112), can now be supplied in a virally inactivated form suitable either for infusion or as part of a fibrin [9001-31-4] sealant kit used for wound healing (195). Fibrinogen can be recovered from cryoprecipitate, Cohn Fraction I (Table 7), or from side fractions of Factor VIII processing. Low solubility, characteristic of fibrinogen, means that precipitation technology is particularly suitable for its preparation. However, ion-exchange chromatography may be required to remove chemicals used in virucidal treatments.

Another by-product of Factor VIII processing having clinical value is von Willebrand factor. It has been recovered from side fractions using ion-exchange and affinity chromatography (196).

Alpha-1-proteinase inhibitor and antithrombin III are used to treat people with hereditary deficiencies of these proteins. Both can be recovered from Cohn Fraction IV (Table 7) using ion-exchange chromatography (52) and affinity chromatography (197), respectively. Some manufacturers recover antithrombin III directly from the plasma stream by affinity adsorption (56,198,199).

Economic Aspects

Estimates for a number of economic aspects of plasma fractionation can be made (200–206). The world capacity for plasma fractionation exceeded 20,000 t of plasma in 1990 and has increased by about 75% since 1980, with strong growth in the not-for-profit sector (Fig. 4). The quantity of plasma processed in 1993 was about 17,000 t/yr; the commercial sector accounts for about 70% of this, with over 8000 t/yr in the form of source plasma from paid donors (Fig. 5). Plant capacities and throughput are usually quoted in terms of principal products, such as albumin and Factor VIII. These figures may not encompass manufacture of other products.

The clinical use of plasma products varies widely between countries (Table 8) with commercial products being imported in some instances to meet demand.

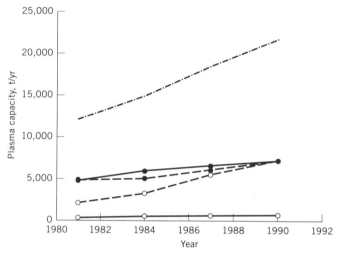

Fig. 4. Capacity of plasma fractionation plants: ○, not-for-profit capacities; ●, for-profit capacities in (——), the United States and (— — —), Europe; (—·—·—) world total.

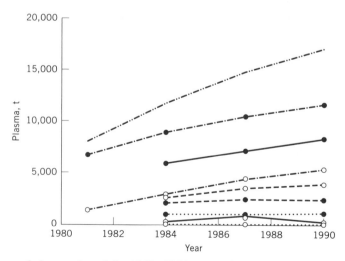

Fig. 5. Processed plasma, by origin, 1980–1990: ○, not-for-profit, ●, for-profit. (——) Source plasma; (— — —), recovered plasma; (····) placental; (—·—·—), totals by distributor; (—··—··—), world totals.

Table 8. Plasma Derivatives Distributed per 10^6 Population[a]

Country	Plasma derivative		
	Albumin, kg	IgG iv, kg	Factor VIII, IU × 10^6
Belgium	601	9.4	2.21
Denmark	320	3.8	3.54
England and Wales	160	4.1	1.99
France	508	17.1	1.74
Germany[b]	596	17.4	3.63
Italy	497	19.0	1.36
Japan	614	79.0	2.67
the Netherlands	392	6.8	2.58
Norway	297	1.8	1.76
Portugal	99	2.9	1.31
Scotland	245	5.7	2.02
Spain	158	4.2	1.51
Sweden	243	5.4	4.23
Switzerland	536	16.1	2.11
United States	521	35.3	2.54
Average	*386*	*15.2*	*2.35*

[a]Estimated values for 1990–1991.
[b]The former West Germany.

In the United States, the market for plasma products increased from $250,000,000 in 1980 to over $850,000,000 in 1991. This expansion resulted from a 60% increase in the use of albumin, a 70% increase in the use of Factor X concentrate, and the introduction of intravenous immunoglobulin (IgG iv). The 1987 quantity of IgG iv was 10.7 kg per million population; by 1991 this usage had increased approximately 220%. The introduction of highly purified preparations of Factor VIII resulted in a marked increase in price; ie, in the period 1980–1987 the average price was $0.10/IU, but by 1989 the price had increased approximately 500% (207,208). The value of these highly purified products has been questioned (207), and some initial difficulties with supply as well as price have been experienced (208,209). Similar price increases emerged in 1991 for Factor IX products; the 1992 price for high purity Factor IX ($0.70/IU) was significantly greater than the average 1980–1990 price of Factor IX complex ($0.09/IU). These significant changes in price have had a profound effect on the share of the U.S. market occupied by individual products, eg, albumin products represented 70% of the total market in 1980 but only 33% in 1991 (Fig. 6), although the demand for albumin increased by about 50% over this period.

Many countries aspire to supply their requirement for plasma products from their own plasma resource, a position supported by the World Health Organization (WHO) (210). Where multiple products are prepared from a common feedstock, the product in shortest supply dictates the scale of the manufacturing operation. Factor VIII and albumin have provided the driving force for plasma fractionation since the 1970s; however, in the United States since the late 1980s,

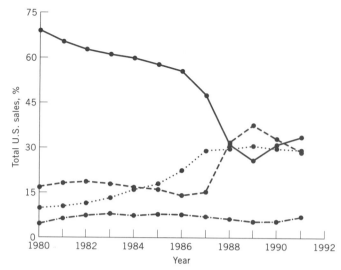

Fig. 6. Share of U.S. market occupied by human albumin/plasma protein fraction, (——); Factor VIII concentrate, (— — —); intravenous immunoglobulin (····); and other products, (—·—).

intravenous immunoglobulin has become more significant. The quantity of plasma required is determined by clinical demand and by the yield of the product from the manufacturing process. For Factor VIII the quantity of plasma required increases sharply at lower yields, a trend that increases with increasing demand. The yield of high purity Factor VIII was about 150 IU/L plasma (206) in 1993. This emphasizes the critical importance of Factor VIII yield for those countries concerned with national self-sufficiency. The extent to which the demand for plasma products can be met from national supplies of plasma (Fig. 7) is determined by the volume of plasma collected as well as by product yields. The relatively large volume of plasma taken from each paid U.S. donor (Table 6) has been particularly important in meeting U.S. demand, as well as much of the importation required by other countries.

Regulation and Control

The preparation of clinical products from human plasma is regulated as a pharmaceutical manufacturing operation by national authorities who are responsible for giving authorization to distribute a product in their country. To obtain authorization, manufacturers must obtain approval for their manufacturing operation and for each of their products. In some countries this is granted by awarding a manufacturing license for the overall operation, and individual product licenses for each of the products involved. This is done by the Food and Drug Administration (FDA) in the United States and by the Medicines Control Agency (MCA) in the United Kingdom.

To obtain a manufacturing license it is necessary to comply with current good manufacturing practice (cGMP). In the United States the cGMP guidelines,

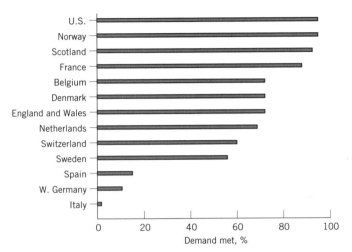

Fig. 7. Plasma product demand met by national plasma supplies. Based on total cost for all products, 1990–1991.

published in the *Code of Federal Regulations* (211), encompass organization and personnel; buildings and facilities; equipment; control of components, containers, and closures; production and process controls; packaging and labeling; holding and distribution; laboratory controls; records and reports; and return of products. Manufacturing guidelines have also been published by WHO (149), as well as by other national authorities (151,212). There is interest in reducing differences in standards that exist between countries (213), eg, European Community member states must comply with relevant EC directives under the aegis of the EC Committee for Proprietary Medicinal Products (CPMP) (214).

Products licensed for use in the United States in 1993 are listed in Table 1. Product licenses cover method of preparation and product characterization, including safety and toxicity, as well as the intended clinical applications. Where monoclonal antibodies are used as process reagents, these must comply with regulatory guidelines for the preparation of biotechnology products (215–217). National control authorities are responsible for ensuring that products continue to meet the relevant standards; in some countries independent testing is carried out on a batch-by-batch basis. This is done in the United States by the FDA, and in the United Kingdom by the National Institute for Biological Standards and Control.

Specifications and Analytical Methods

Fundamental product specifications are published, with those for other drug substances, in national or regional pharmacopoeias. Plasma products described in the *U.S. Pharmacopeia* (USP) (218) include human albumin, plasma protein fraction, antihemophilic factor, Factor IX complex, normal immune globulin, and a variety of specific immune globulins such as hepatitis B, tetanus, and varicella-zoster. The *European Pharmacopoeia* (EP) (219) lists human albumin solution, plasma protein solution, freeze-dried coagulation Factors VIII and IX, normal immunoglobulin, and a variety of other immunoglobulins. Principal characteristics of human albumin, plasma protein fraction, normal immune globulin, intravenous immunoglobulin (IgG iv), and Factor VIII and Factor IX concentrates are summarized in Tables 9–11. Draft versions of new monographs are published for comment in the journals *Pharmacopoeial Forum* and *Pharmeuropa* for the United States, and in European *Pharmacopoeias*, respectively. Proposals from the latter have been summarized for albumin (Table 9) and IgG iv (Table 10).

Pharmacopoeial monographs describe products and methods that are well established and may not necessarily reflect state-of-the-art manufacturing or clinical practices. Issues concerning current practice may be dealt with by learned societies, eg, the International Society of Thrombosis and Haemostasis (ISTH) provides international coordination in the treatment of coagulation disorders. Specialist ISTH subcommittees hold open meetings biannually and their recommendations are published in the journal *Thrombosis and Haemostasis*. Examples of topics covered include the assessment of the safety (222) and pharmacokinetics (223,224) of Factor VIII concentrates. Other organizations having this type of role are the World Federation of Hemophilia and the International Society of Blood Transfusion. Recommendations may also be made by national groups of clinicians (222). For products other than coagulation factors, international consensus meetings may be held under the auspices of bodies such as the National Institute for

Table 9. Properties of Human Albumin and Plasma Protein Fraction[a]

Characteristic	Human albumin		Plasma protein fraction	
	USP XXII	EP[b,c] 1990	USP XXII	EP[d] 1987
form	aqueous solution	slightly viscous liquid	solution	aqueous solution
color		colorless to amber		clear to pale yellow
protein concentration, g/L	50	40–50	50	40–50
	250	150–250		
albumin content, %	≥96	≥95	≥83	≥85
α- and β-globulins, %			≤17	≤15
γ-globulins, %			≤1	
haem, OD at A_{403}[e]	≤0.25	≤0.15		≤0.15
potassium, mmol/g		≤0.05	≤2.0[f]	≤0.05
sodium, mmol/L	130–160	≤160	130–160	≤160
pH		6.7–7.3	6.7–7.3	6.7–7.3

[a]Plasma protein solution.
[b]EP draft revision lists color as colorless, yellow, or green; protein concentration, 35 to 50 g/L; and prekallikrein activator of ≤35 IU/mL. Ref. 220.
[c]Aggregates, ≤5%; alkaline phosphatase [9001-79-0], ≤0.1 units/g; aluminum, ≤200 μg/L.
[d]Aggregates, ≤10%; alkaline phosphatase, ≤0.1 μg.
[e]OD at A_{403} = optical density at absorbance at 403 nm.
[f]mEq/L.

Table 10. Properties of Normal Immunoglobulin and Intravenous Immunoglobulin

| Characteristic | Immunoglobulin | | Immunoglobulin intravenous |
	USP[a] XXII	EP[b] 1992	EP[c] draft
form	solution	solution or freeze-dried powder	liquid or freeze-dried powder
color		clear to light brown solution or white to slightly yellow powder	colorless to pale yellow liquid or white to slightly yellow friable mass
number of donors for plasma pool	≥1000	≥1000	≥1000
protein concentration, g/L	150–180	100–180	30–120
immunoglobulin G, %	≥90	≥90	≥95[d]
aggregate, %		≤10	≤3

[a]Antidiphtheria activity, ≥2 IU/mL; positive antimeasles and antipolio activity.
[b]≤5% fragments after 4 weeks at 37°C; antihepatitis A activity, ≥100 IU/mL.
[c]≥80% Fc function (portion of fully functional immunoglobulin molecule); prekallikrein activator, ≤35 IU/mL; HBsAs antibody, ≥1 IU/g; normal subclass distribution osmolality, ≤280 mosmol; 4 to 7.4 pH. Ref. 221.
[d]Excluding albumin stabilizer.

Table 11. Properties of Factor VIII (Antihemophilic Factor) and Factor IX Concentrates

| Characteristic | Factor VIII (AHF) | | Factor IX complex | |
	USP XXII	EP[a] 1986	USP XXII	EP[b] 1987
form	freeze-dried powder	powder or friable solid	freeze-dried powder	powder or friable solid
color		white or pale yellow		white; may be blue, yellow, or green on reconstitution
potency				
IU/mL		≥3		≥20
% of label	80–120	80–125	80–120	80–125
purity IU/g	≥100	≥100		≥600

[a]≤80% Fibrinogen; freedom from coagulation at 20–25°C ≥3 h; ≤2% residual moisture; 6.8–7.4 pH; reconstitution time at 20–25°C is ≤30 min.
[b]Heparin, ≤5 IU/IU FIX; nonactivated partial thromboplastin time, ≥150 s; thrombin-fibrinogen time test, ≥24 h at RT, >6 h at 37°C; ≤2% residual moisture; 6.5–7.5 pH; reconstitution time at 20–25°C is ≤ 10 min.

Health (226,227) and the American Association of Blood Banks (228) in the United States.

A wide range of analytical methods are employed in characterizing plasma products, and details of reference methods can be found in *European Pharmacopoeia* monographs and in WHO (149) and FDA (212) guidelines. Reference methods for the measurement of total protein involve determination of total protein nitrogen. Some highly purified coagulation factor products have a total protein content below the sensitivity of the standard methods. These may also contain amino acids as stabilizers which provide nitrogen over and above the protein content of the product. In these circumstances alternative methods are required (229), such as the dye-binding method (230).

Determination of the potency of Factor VIII is also difficult. This is normally measured by the ability of the sample to correct the clotting time of plasma deficient in Factor VIII. A number of methods and practices have evolved for this purpose (231), but these give very different results, particularly when activation of products may also occur (232). International standards have been used, but further standardization of the analytical method and harmonization of working standards is underway (233,234) under the auspices of the ISTH and the EC.

Health, Safety, and Environmental Factors

The possibility that infectious donations of plasma may enter the fractionation process places staff at risk; the transmission of hepatitis B to fractionation workers had been reported (235) before the screening of plasma for hepatitis B infection was introduced. The extensive testing of donations that takes place (ca 1993) (121) reduces this risk substantially. Nevertheless it is assumed that plasma for fractionation may be contaminated with viruses such as hepatitis B, hepatitis C, and HIV, and appropriate precautions should be taken. For example, automation should be used where possible, all equipment should be decontaminated for maintenance and repair, and all staff should wear appropriate protective clothing. These precautions apply to laboratory facilities as well as to the manufacturing plant. Materials that are potentially contaminated, such as used filter pads and waste fractions, should be sterilized prior to disposal. Fractionation staff now receive hepatitis B vaccination routinely and normal immunoglobulin may be administered following accidental exposure to potentially infectious materials (236). The need to validate methods of virus inactivation has led some manufacturers to establish containment laboratories for handling HIV and other viruses; these facilities must comply with national regulations concerning the handling of dangerous pathogens.

Ethanol (qv), the principal bulk reagent in plasma fractionation, is categorized as a highly flammable material with vapor concentration of 3–19% ethanol being explosive (237) at temperatures above the flash point of 13°C (238). These properties must be considered in the design and specification of equipment and facilities involved in ethanol fractionation. Once the fractionation process has been completed, there are waste solutions containing up to 40% ethanol which require disposal. Some manufacturers recycle this material using distillation (qv), a procedure which must be regulated and controlled to the satisfaction of local customs and excise authorities.

BIBLIOGRAPHY

"Blood Fractionation" in *ECT* 1st ed., Vol. 2, pp. 556–584, by L. E. Strong, Kalamazoo College; in *ECT* 2nd ed., Vol. 3, pp. 576–602, by L. E. Strong, CBA Project, Earlham College; in *ECT* 3rd ed., Vol. 4, pp. 25–61, by M. H. Stryker and A. A. Waldman, The New York Blood Center.

1. N. L. Anderson and N. G. Anderson, *Electrophorisis* **12**, 883–906 (1991).
2. J. T. Edsall, *Protein Sci.* **1**, 1526–1530 (1992).
3. E. J. Cohn, *Physiol. Rev.* **5**, 349–437 (1925).
4. C. A. Janeway, in J. T. Sgouris and A. René, eds., *Proceedings of the Workshop on Albumin*, DHEW Publ. No. (NIH) 76-925, U.S. Government Printing Office, Washington, D.C., 1976, pp. 3–21.
5. J. W. Palmer, in Ref. 4, pp. 255–269.
6. E. J. Cohn and co-workers, *J. Am. Chem. Soc.* **68**, 459–475 (1946).
7. S. D. Black and D. R. Mould, *Anal. Biochem.* **193**, 72–82 (1991).
8. T. Arakawa and S. N. Timasheff, *Methods Enzymol.* **114**, 49–77 (1985).
9. D. J. Bell, M. Hoare, and P. Dunnill, *Adv. Biochem. Eng.* **26**, 1–72 (1983).
10. J. M. Curling, ed., *Methods of Plasma Protein Fractionation*, Academic Press, Inc., New York, 1980.
11. M. H. Stryker, M. J. Bertolini, and Y. L. Hao, *Adv. Biotechnol. Processes* **4**, 275–336 (1985).
12. J. T. Edsall, *Adv. Protein Chem.* **3**, 383–479 (1947).
13. J. F. Taylor, in H. Neurath and K. Bailey, eds., *The Proteins*, Vol. 1, part A, Academic Press, Inc., New York, 1953, pp. 1–58.
14. E. J. Cohn, in E. J. Cohn and J. T. Edsall, eds., *Proteins, Amino Acids and Peptides*, Reinhold Publishing Co., New York, 1943, pp. 569–585.
15. J. Wyman, *J. Am. Chem. Soc.* **53**, 3292–3301 (1931).
16. F. Rothstein, in R. Harrison, ed., *Protein and Peptide Purification: Process Development and Scale-Up*, Marcel Dekker, Inc., New York, in press.
17. C. J. Van Oss, *J. Protein Chem.* **8**, 661–668 (1989).
18. W. Melander and C. Horvath, *Arch. Biochem. Biophys.* **183**, 200–215 (1977).
19. C. J. Van Oss and co-workers, *J. Protein Chem.* **4**, 245–263 (1985).
20. T. Arakawa, R. Bhat, and S. N. Timasheff, *Biochemistry* **29**, 1914–1923 (1990).
21. J. L. Oncley and co-workers, *J. Am. Chem. Soc.* **71**, 541–550 (1949).
22. P. Kistler and H. S. Nitschmann, *Vox Sang.* **7**, 414–424 (1962).
23. S. Holst, M. Martinez, and B. Zarth, *J. Parenteral Drug Assoc.* **32**, 15–21 (1978).
24. J. H. Hink and co-workers, *Vox Sang.* **2**, 174–186 (1957).
25. R. B. Pennell, in F. W. Putnam, ed., *The Plasma Proteins*, Vol. 1., Academic Press, Inc., New York, 1960, pp. 9–49.
26. H. E. Schultze and J. F. Heremans, *Molecular Biology of Human Proteins*, Vol. 1, Elsevier Publishing Co., Amsterdam, 1966, pp. 236–317.
27. E. J. Cohn, in *Papers on the Separation of the Formed Elements of Plasma*, Laboratory of Physical Chemistry, Harvard University, Boston, 1950.
28. J. G. Watt, *Vox Sang.* **18**, 42–61 (1970).
29. J. G. Watt, *Vox Sang.* **23**, 126–134 (1972).
30. P. R. Foster and J. G. Watt, in Ref. 10, pp. 17–31.
31. P. R. Foster and co-workers, *J. Chem. Tech. Biotechnol.* **36**, 461–466 (1986).

32. G. Mitra and J. Lundblad, *Biotechnol. Bioeng.* **20**, 1037–1044 (1978).

33. C. E. Chang, *Biotechnol. Bioeng.* **31**, 841–846 (1988).

34. F. Rothstein, in *Proceedings of the Biopharma Conference '91*, Aster Publishing Corp., Eugene, Oreg., 1991, pp. 95–118.

35. H. Hemfort, in H. E. Sandberg, ed., *Proceedings of the International Workshop on Technology for Protein Separation and Improvement of Blood Plasma Fractionation*, DHEW Publ. No. (NIH) 78-1422, U.S. Government Printing Office, Washington, D.C., 1978, pp. 85–96.

36. N. J. Titchner-Hooker and R. V. McIntosh, *Bioproc. Eng.* **8**, 215–222 (1993).

37. J. V. Fiore, W. P. Olson, and S. L. Holst, in Ref. 10, pp. 239–268.

38. G. W. R. Dike, E. Bidwell, and C. R. Rizza, *Br. J. Haematol.* **22**, 469–489 (1972).

39. S. H. Middleton, I. H. Bennett, and J. K. Smith, *Vox Sang.* **24**, 441–456 (1973).

40. J. H. Heystek, H. Brummelhuis, and H. Krijnen, *Vox Sang.* **25**, 113–123 (1973).

41. T. Burnouf, M. Burnouf-Radesovich, and J. J. Huart, *Vox Sang.* **60**, 8–15 (1991).

42. M. E. Hrindra, F. Feldman, and A. B. Schreiber, *Semin. Hematol.* **27**(2, suppl. 2), 19–24 (1990).

43. M. Griffin, *Ann. Hematol.* **63**, 131–137 (1991).

44. A. D. Friesen, J. M. Bowman, and W. C. H. Bees, *Vox Sang.* **48**, 201–212 (1985).

45. J. A. Hooper, M. Alpern, and S. Mankarious, in H. W. Krijnen, P. F. W. Strengers, and W. G. van Aken, eds., *Immunoglobulins*, Netherlands Red Cross, Amsterdam, 1988, pp. 361–380.

46. H. Suomela, in Ref. 10, pp. 107–116.

47. J. M. Curling, in Ref. 10, pp. 78–91.

48. J. F. Stoltz and co-workers, *Colloque INSERM* **175**, 191–200 (1989).

49. J. E. More and M. J. Harvey, in J. R. Harris, ed., *Blood Separation and Plasma Fractionation*, John Wiley & Sons, Inc., New York, 1991, pp. 261–306.

50. F. Hasko and co-workers, *Develop. Biol. Stand.* **67**, 39–48 (1987).

51. A. D. Friesen, *Develop. Biol. Stand.* **67**, 3–13 (1987).

52. M. H. Coan and co-workers, *Vox Sang.* **48**, 333–342 (1985).

53. D. Menaché and co-workers, *Blood* **64**, 1220–1227 (1984).

54. T. Burnouf and co-workers, *Vox Sang.* **57**, 225–232 (1989).

55. S. W. Herring and co-workers, *J. Lab. Clin. Med.* **121**, 394–405 (1993).

56. M. Miller-Anderson, H. Borg, and L. O. Anderson, *Thromb. Res.* **5**, 439–452 (1974).

57. H. Nunez and W. N. Drohan, *Semin. Hematol.* **28**, 24–30 (1991).

58. J. Choay, *Semin. Thromb. Hemostas.* **15**, 359–364 (1989).

59. D. Josic, F. Bal, and H. Schwinn, *J. Chromatog.* **632**, 1–10 (1993).

60. C. W. Pratt and F. C. Church, *Blood Coagulation and Fibrinolysis* **4**, 479–490 (1993).

61. M. E. Hrindra and co-workers, *Semin. Hematol.* **28**(3, suppl. 6), 6–14 (1991).

62. U.S. Pat. 4,361,509 (Nov. 30, 1982), T. S. Zimmerman and C. A. Fulcher (to Scripps Clinic and Research Foundation).

63. H. Bessos and co-workers, *Prep. Chromatog.* **1**, 207–220 (1991).

64. M. Cheryan, *Ultrafiltration Handbook*, Technomic Publishing Co., Inc., Lancaster, Pa., 1986.

65. R. Rautenbach and R. Albrecht, *Membrane Processes*, John Wiley & Sons, Inc., New York, 1989.

66. J. Murkes and C. G. Carlsson, *Crossflow Filtration*, John Wiley & Sons, Inc., New York, 1988.

67. E. Iritani and co-workers, *J. Chem. Eng. Japan* **24**, 177–183 (1991).

68. W. S. Opong and A. L. Zydney, *J. Colloid Int. Sci.* **142**, 41–60 (1991).

69. H. Friedli and co-workers, *Vox Sang.* **31**, 283–288 (1976).

70. P. K. Ng, J. Lundblad, and G. Mitra, *Separation Sci.* **11**, 499–502 (1976).

71. H. Friedli, E. Fournier, and T. Volk, *Vox Sang.* **33**, 93–96 (1977).

72. R. Schmitthauesler, *Proc. Biochem.* **12**, 13–15, 46 (Oct. 1977).
73. P. K. Ng, G. Mitra, and J. L. Lundblad, *J. Pharm. Sci.* **67**, 431–433 (1978).
74. S. L. Holst, L. Sarno, and M. Martinez, in A. R. Cooper, ed., *Ultrafiltration Membranes and Applications*, Plenum Press, New York, 1980, pp. 575–589.
75. D. S. Milliner and co-workers, *N. Engl. J. Med.* **312**, 1390 (1985).
76. B. Cuthbertson and co-workers, *Br. Med. J.* **295**, 1062 (1987).
77. H. Friedli and P. Kistler, in Ref. 74, pp. 565–574.
78. L. Martinache and M. P. Henon, in Ref. 10, pp. 223–235.
79. G. Mitra and J. L. Lundblad, *Vox Sang.* **40**, 109–114 (1981).
80. G. Mitra and P. Ng, in W. C. McGregor, ed., *Membrane Separations in Biotechnology*, Marcel Dekker, Inc., New York, 1986, pp. 115–134.
81. R. V. McIntosh and P. R. Foster, *Transf. Sci.* **11**, 55–66 (1990).
82. M. Meireles, P. Aimar, and V. Sanchez. *Biotechnol. Bioeng.* **38**, 528–534 (1991).
83. I. Peine, J. Swenson, and J. Bendedictus, *J. Parenteral Sci. Technol.* **36**, 79–85 (1982).
84. S. L. Michaels, *J. Parenteral Sci. Technol.* **45**, 218–223 (1991).
85. R. Duberstein, in Ref. 10, pp. 269–290; A. R. Reti, in Ref. 10, pp. 291–304.
86. T. H. Meltzer, ed., *Filtration in the Pharmaceutical Industry*, Marcel Dekker, Inc., New York, 1987.
87. P. J. M. Van den Oetelaar, I. M. Mentink, and G. J. Brinks, *Drug Dev. Ind. Pharm.* **15**, 97–106 (1989).
88. D. Pendlebury, *Lab. Practice* **40**, 47 (1991).
89. S. H. Goldsmith and G. P. Grundelman, *Pharm. Manufact.*, 31–37 (Nov. 1985).
90. R. V. Levy, K. S. Souza, and C. B. Neville, *Pharm. Technol.* **14**(9), 160–173 (1990).
91. R. V. Levy and co-workers, *Pharm. Technol.* **15**(5), 58–68 (1991).
92. N. A. Williams and G. P. Polli, *J. Parenteral Sci. Technol.* **38**, 48–59 (1984).
93. J. C. May and F. Brown, eds., "Biological Product Freeze-Drying and Formulation," *Develop. Biol. Stand.* **74** (1992).
94. J. Ford and P. Timmins, eds., *Pharmaceutical Thermal Analysis*, Interpharma Press, Buffalo, N.Y., 1991, pp. 279–298.
95. L. R. Rey, *An. N.Y. Acad. Sci.* **35**, 510–534 (1960).
96. H. T. Meryman, *Cryobiology* **28**, 307–313 (1991).
97. R. I. N. Greaves, *J. Am. Med. Assoc.* **124**, 76–79 (1944).
98. A. E. Nicholson, *Develop. Biol. Stand.* **36**, 69–75 (1977).
99. D. T. H. Liu and co-workers, *Vox Sang.* **38**, 216–221 (1980).
100. P. D. Boyer and co-workers, *J. Biol. Chem.* **162**, 181–198 (1946).
101. J. D. Ferry and co-workers, *J. Am. Chem. Soc.* **69**, 409–416 (1947).
102. S. S. Gellis and co-workers, *J. Clin. Invest.* **27**, 239–244 (1948).
103. R. Murray and W. C. L. Diefenbach, *Proc. Soc. Exper. Biol. Med.* **84**, 230–231 (1953).
104. A. Shrake, J. S. Finlayson, and P. D. Ross, *Vox Sang.* **47**, 7–18 (1984).
105. J. T. Edsall, *Vox Sang.* **46**, 338–340 (1984).
106. J. S. Finlayson and D. L. Aronson, *Semin. Thromb. Hemost.* 6, 1–74 (1979); **6**, 85–139 (1980).
107. B. Cuthbertson and co-workers, *Lancet* **11**, 41 (1987).
108. J. P. Gregerson, J. Hilfenhaus, and J. F. Lemp, *J. Biol. Stand.* **17**, 377–379 (1989).
109. L. F. Barker and R. Murray, *Am. J. Med. Sci.* **263**, 27–33 (1971).
110. P. L. Yap and P. E. Williams, in P. L. Yap, ed., *Applications of Intravenous Immunoglobulin Therapy*, Churchill Livingston, Edinburgh, 1992, pp. 43–62.
111. B. Cuthbertson and co-workers, *J. Inf.* **15**, 125–133 (1987).
112. J. P. Hile, *Fed. Register* **43**, 1131–1132 (1978).
113. D. L. Aronson, *Am. J. Hematol.* **27**, 7–12 (1988).
114. A. J. Johnson and co-workers, *J. Lab. Clin. Med.* **88**, 91–101 (1976).
115. W. H. Holleman and co-workers, *Thromb. Haemostas.* **38**, 201 (1977).

116. Jpn. Pat. 7,859,018 (May 27, 1978), T. Fukushima and co-workers (to Green Cross Corp.).
117. W. Fricke and co-workers, *Transfusion* **32**, 707–709 (1992).
118. Centers For Disease Control, *Morbid. Mortal. Weekly Rep.* **36**, 593–595 (1987).
119. J. M. Bader, *Lancet* **340**, 1087 (1992).
120. *Council Directive 89/381/EEC*, Official Journal of the European Communities, No. L181/44, Luxembourg, 1989.
121. R. Y. Dodd and W. Kline, in G. Rock and M. J. Seghatchian, eds., *Quality Assurance in Transfusion Medicine*, Vol. 1, CRC Press, Boca Raton, Fla., 1992, pp. 445–484.
122. P. R. Foster and B. Cuthbertson, in R. Madhok, C. D. Forbes, and B. L. Evatt, eds., *Blood, Blood Products and HIV*, 2nd ed., Chapman and Hall, London, 1994, pp. 211–251.
123. J. J. Morgenthaler, ed., *Virus Inactivation in Blood Products*, Karger, Basel, 1989.
124. B. Cuthbertson, K. G. Reid, and P. R. Foster, in Ref. 49, pp. 385–435.
125. B. Horowitz, *Develop. Biol. Stand.* **75**, 43–82 (1991).
126. T. Burnouf, *Biologicals* **20**, 91–100 (1992).
127. C. K. Kasper and co-workers, *Transfusion* **33**, 422–434 (1993).
128. W. A. Fricke and M. A. Lamb, *Semin. Thromb. Haemostas.* **19**, 54–61 (1993).
129. U.S. Pat. 4,297,344 (Oct. 27, 1981), H. Schwinn and co-workers (to Behringwerke AG); U.S. Pat. 4,440,679 (Apr. 3, 1984), P. M. Fernandes and J. L. Lundblad (to Cutter Laboratories, Inc.).
130. U.S. Pat. 4,876,241 (Oct. 24, 1989), F. Feldman and co-workers (to Armour Pharm. Co.).
131. K. G. Reid and co-workers, *Vox Sang.* **55**, 75–80 (1988).
132. B. Horowitz and co-workers, *Transfusion* **25**, 516–522 (1985).
133. P. N. Barret and co-workers, *Thromb. Haemost.* **58**, 371 (1987).
134. L. Winkelman and co-workers, *Vox Sang.* **57**, 97–103 (1989).
135. R. V. McIntosh and co-workers, *Thromb. Haemost.* **58**, 306 (1987).
136. P. R. Foster and co-workers, *Lancet* **2**, 43 (1988).
137. J. S. Epstein and W. A. Fricke, *Arch. Pathol. Lab. Med.* **114**, 335–340 (1990).
138. P. M. Mannucci, *Vox Sang.* **64**, 197–203 (1993).
139. P. M. Mannucci, *Lancet* **339**, 819 (1992); A. Gerritzen and co-workers, *Lancet* **340**, 1231–1232 (1992); I. J. Temperley and co-workers, *Lancet* **340**, 1465 (1992); K. Peerlinck and J. Vermylen, *Lancet* **341**, 179 (1993).
140. O. B. Corsi and co-workers, *J. Med. Virol.* **25**, 165–170 (1988).
141. T. F. Schwarz and co-workers, *J. Med. Virol.* **35**, 28–31 (1991).
142. D. Piszkiewicz, C. S. Sun, and S. C. Tondreau, *Thromb. Res.* **55**, 627–634 (1989).
143. Y. Hamamoto, *Vox Sang.* **56**, 230–236 (1989).
144. D. P. Thomas, *Br. J. Haematol.* **70**, 393–395 (1988).
145. B. Horowitz, *Yale J. Biol. Med.* **63**, 361–369 (1990).
146. *U.S. Pharmacopoeia XXII*, USP Convention Inc., Rockville, Md., 1989, pp. 99–103.
147. P. L. Mollison, C. P. Engelfriet, and M. Contreras, *Blood Transfusion in Clinical Medicine*, 9th ed., Blackwell Scientific Publications, Oxford, 1993, pp. 1–47.
148. *Technical Manual*, 10th ed., American Association of Blood Banks, Arlington, Va., 1990.
149. World Health Organization, *Technical Report Series, no. 786*, WHO, Geneva, 1989, pp. 94–176.
150. M. S. Blackistone, *Plasma Quarterly* **7**, 167–169 (1985).
151. *Guidelines for the Blood Transfusion Services in the United Kingdom*, National Institute For Biological Standards and Control, Potters Bar, 1992.

152. R. V. McIntosh and co-workers, in C. T. Smit Sibinga, P. C. Das, and H. T. Meryman, eds., *Cryopreservation and Low Temperature Biology in Blood Transfusion*, Kluwer Academic Publishers, Norwell, Mass., 1990, pp. 11–24.
153. P. R. Foster, in L. R. Weatherley, ed., *Engineering Processes For Bioseparations*, Butterworth-Heineman, Oxford, in press.
154. P. A. Foster and T. S. Zimmerman, *Blood Reviews* **3**, 180–191 (1989).
155. M. E. Mikaelsson, N. Forsman, and U. M. Oswaldsson, *Blood* **62**, 1006–1015 (1983).
156. V. Atichartakarn and co-workers, *Blood* **51**, 281–297 (1978).
157. G. A. Rock and co-workers, *Thromb. Res.* **29**, 521–535 (1983).
158. P. R. Foster, I. H. Dickson, and T. A. McQuillan, *Br. J. Haematol.* **53**, 343 (1983).
159. J. G. Pool, E. J. Hershgold, and A. R. Pappenhagen, *Nature* **203**, 312 (1964).
160. P. R. Foster and B. J. White, *Lancet* **2**, 574 (1978).
161. P. R. Foster and co-workers, *Vox Sang.* **42**, 180–189 (1982).
162. R. H. Wagner and co-workers, *Thromb. Diath. Haemostas.* **11**, 64–74 (1964).
163. J. Newman and co-workers, *Br. J. Haematol.* **21**, 1–20 (1971).
164. U.S. Pat. 3,682,881 (Aug. 8, 1972), L. F. Fekete and E. Shanbrom (to Baxter Laboratories Inc.); U.S. Pat. 3,973,002 (Aug. 3, 1976), J. J. Hagen and C. Glaser (to E. R. Squibb & Sons, Inc.).
165. U.S. Pat. 4,406,886 (Sept. 27, 1983), M. Bier and P. R. Foster (to University Patents Inc.).
166. U.S. Pat. 4,478,825 (Oct. 23, 1984), J. W. Bloom (to Armour Pharm. Co.).
167. L. Thorell and B. Blomback, *Thromb. Res.* **35**, 431–450 (1984).
168. P. K. Ng, H. C. Eguizabel, and G. Mitra, *Thromb. Res.* **42**, 825–834 (1986).
169. P. R. Foster and co-workers, *Thromb. Haemostas.* **50**, 117 (1983).
170. P. R. Foster and co-workers, *Vox Sang.* **55**, 81–89 (1988).
171. H. G. Watson and C. A. Ludlam, *Blood Reviews* **6**, 26–33 (1992).
172. C. V. Prowse, *Blood Coag. Fibrinolysis* **3**, 597–604 (1992).
173. M. Wickerhauser and J. T. Sgouris, *Vox Sang.* **22**, 137–160 (1972).
174. G. S. Gilchrist and co-workers, *N. Engl. J. Med.* **280**, 291–295 (1969).
175. R. G. Macfarlane, in R. Biggs, ed., *Human Blood Coagulation, Haemostasis and Thrombosis*, 2nd ed., Blackwell Scientific Publications, Oxford, 1976, pp. 1–31.
176. S. D. Carson and J. P. Brozna, *Blood Coagulation and Fibrinolysis* **4**, 281–292 (1993).
177. L. Pejaudier and co-workers, *Vox Sang.* **52**, 1–9 (1987).
178. D. C. Triantaphyllopoulous, *Am. J. Clin. Path.* **57**, 603–610 (1972).
179. C. K. Kasper, *N. Engl. J. Med.* **289**, 592 (1973).
180. A. R. Giles and co-workers, *Blood* **59**, 401–407 (1982).
181. F. Feldman and R. Kleszynski, *Presentation at European Workshop on FVIII and FIX Concentrates*, National Institute For Biological Standards and Control, London, May 1992.
182. I. R. MacGregor and co-workers, *Thromb. Haemostas.* **66**, 609–613 (1991).
183. H. Suomela, R. Hekali, and E. Vahvaselka, *Develop. Biol. Stand.* **44**, 99–105 (1979).
184. S. Barandun and co-workers, *Vox Sang.* **7**, 157–174 (1962).
185. R. H. Rousell and J. P. McCue, in Ref. 49, pp. 307–340.
186. B. M. Alving and co-workers, *J. Lab. Clin. Med.* **96**, 334–346 (1980).
187. R. van Furth, P. C. J. Leijh, and F. Klein, *J. Infect. Dis.* **149**, 511–517 (1984).
188. J. L. Lundblad and D. D. Schroeder, in Ref. 110, pp. 17–41.
189. U.S. Pat. 4,499,073 (Feb. 12, 1985), R. A. Tenold (to Cutter Labs, Inc.).
190. D. J. Mulford, E. H. Mealey, and L. W. Welton, *J. Clin. Inv.* **34**, 983–986 (1955).
191. B. M. Alving and co-workers, *N. Engl. J. Med.* **299**, 66–70 (1978).
192. G. N. Olinger and co-workers, *Ann. Surg.* **190**, 305–311 (1979).
193. J. Rossitto, *Pharm. Technol. Int.*, 39–55 (Apr. 1979).

194. K. C. Hou and T. Campbell, *Fluid/Particle Sep. J.* **1**, 108–110 (Dec. 1988).
195. J. W. Gibble and P. M. Ness, *Transfusion* **30**, 741–747 (1990).
196. M. Burnouf-Radosevich and T. Burnouf, *Vox Sang.* **62**, 1–11 (1992).
197. D. L. Hoffman, *Am. J. Med.* **87**(Suppl. 3B), 3–23 (1989).
198. J. K. Smith and co-workers, *Vox Sang.* **48**, 325–332 (1985).
199. P. Feldman and L. Winkelman, in Ref. 49, pp. 341–383.
200. American Blood Resources Association, *Plasma Quarterly* **6**(2), 36–38 (1984).
201. *Worldwide Directory of Plasma Fractionators*, Marketing Research Bureau, Laguna Beach, Calif., 1990.
202. *World Analysis of the Blood Banking and Blood Plasma Products Market*, Market Intelligence Research Corp., Mountain View, Calif., 1991.
203. *The Plasma Fractions Market in Europe*, Marketing Research Bureau, Laguna Beach, Calif., 1990.
204. *The Plasma Fractions Market in the United States*, Marketing Research Bureau, Laguna Beach, Calif., 1991.
205. G. Myllya, in Ref. 49, pp. 15–42.
206. T. Stagnaro, *J. Am. Blood. Res. Assoc.*, 46–49 (Summer 1992).
207. J. D. Cash, *Lancet* **I**, 1270 (1988).
208. G. F. Pierce and co-workers, *J. Am. Med. Assoc.* **261**, 3434–3438 (1989).
209. R. Lipton, L. M. Aledort, and M. Hilgartner, *Transfusion* **30**, 573 (1990).
210. *Bull. WHO* **69**, 17–26 (1991).
211. *Code of Federal Regulations 21*, part 211, U.S. Government Printing Office, Washington, D.C., 1992.
212. Ref. 211, part 640.
213. J. R. Sharp, *Good Manufacturing Practice*, Interpharm Press, Buffalo Grove, Ill., 1991.
214. *The Rules Governing Medicinal Products in the EEC*, Vols. I–III, Office for Official Publications of the EEC, Luxembourg, 1989; Vol. III, addendum I, 1990; Vol. III, addendum 2, 1992; Vol. IV, 1992.
215. Committee For Proprietary Medicinal Products, *Notes to Applicants for Marketing Authorisations on the Production and Quality Control of Monoclonal Antibodies of Murine Origin Intended for use in Man*, Commission of the European Communities, Brussels, 1987.
216. W. E. Kinas and co-workers, *Pharmacopeial Forum* **16**, 136–152 (1990).
217. F. Feldman and co-workers, in G. Rock and M. J. Segatchian, eds., *Quality Assurance in Transfusion Medicine*, Vol. II, CRC Press, Boca Raton, Fla., 1993, pp. 259–284.
218. *The United States Pharmacopeia, Twenty-Second Revision*, The United States Pharmacopeial Convention Inc., Rockville, Md., 1989.
219. *European Pharmacopoeia*, 2nd ed., Maisonneuve SA, Sainte-Ruffine, France, 1980–1992.
220. *Pharmeuropa* **4**, 236–239 (1992).
221. *Pharmeuropa* **3**, 259–268 (1991).
222. P. M. Mannucci and M. Colombo, *Thromb. Haemostas.* **61**, 532–534 (1989).
223. M. Morfini, M. Lee, and A. Messori, *Thromb. Haemostas.* **66**, 384–386 (1991).
224. M. L. Lee, E. D. Gomperts, and H. S. Kingdon, *Thromb. Haemostas.* **69**, 87 (1993).
225. E. Mayne and co-workers, *Blood Coag. Fibrinolysis* **3**, 205–214 (1992).
226. J. T. Sgouris and A. René, eds., *Proceedings of the Workshop on Albumin, DHEW, Publ. No. (NIH) 76-925*, U.S. Government Printing Office, Washington, D.C., 1976.
227. NIH Consensus Conference, *J. Am. Med. Assoc.* **264**, 3189–3193 (1990).
228. A. I. Chernoff, H. G. Klein, and L. A. Sherman, eds., *Transfusion* **29**, 711–742 (1989).
229. A. L. Löf and co-workers, *Vox Sang.* **63**, 172–177 (1992).
230. M. Bradford, *Anal. Biochem.* **72**, 248–254 (1976).

231. I. M. Nilsson, T. W. Barrowcliffe, and K. Schimpf, eds., *Scand. J. Haematol.* **33**, Suppl. 41 (1984).
232. T. W. Barrowcliffe and co-workers, *Lancet* **336**, 124 (1990).
233. T. W. Barrowcliffe, in C. T. Smit Sibinga, P. C. Das, and P. M. Mannucci, eds., *Coagulation and Blood Transfusion*, Kluwer Academic Publications, Dordrecht, the Netherlands, 1991, pp. 175–185.
234. T. W. Barrowcliffe, *Semin. Thromb. Hemostas.* **19**, 73–79 (1993).
235. J. S. Taylor, E. Schmunes, and W. A. Holmes, *J. Am. Med. Assoc.* **230**, 850–853 (1974).
236. L. B. Seef and co-workers, *Gastroenterology* **72**, 111–121 (1977).
237. R. C. Weast, ed., *Handbook of Chemistry and Physics*, 55th ed., CRC Press, Cleveland, Ohio, 1974, p. D-85.
238. S. Budavari, ed., *The Merck Index*, 11th ed., Merck & Co. Inc, Rathway, N.J., 1989, p. 594.

PETER R. FOSTER
Scottish National Blood Transfusion Service

FRACTURE MECHANICS

Fracture mechanics is a methodology which characterizes the resistance of a material to crack propagation. Provided specific requirements are met, a material property can be measured which describes the performance of the material when a sharp or natural crack is present. This property is called the plane strain fracture toughness and is independent of the specimen geometry used to make the measurement. The measured fracture toughness can then be used in the design of a component or structure to avoid fracture. The concepts of fracture mechanics for brittle crack growth were originally proposed in 1920 (1–3). However, it was not until World War II that the technology was substantially developed. Crack propagation problems were experienced with the Liberty ships constructed to carry supplies across the Atlantic ocean. Over a hundred ships fractured in half and many others had serious cracks, leading to extensive research into brittle fracture in metals. Later, the development of nuclear power plants and other critical structures made accurate predictions of the fracture behavior of thick-walled metal parts essential.

Fracture mechanics is now quite well established for metals, and a number of ASTM standards have been defined (4–6). For other materials, standardization efforts are underway (7,8). The techniques and procedures are being adapted from the metals literature. The concepts are applicable to any material, provided the structure of the material can be treated as a continuum relative to the size-scale of the primary crack. There are many textbooks on the subject covering the application of fracture mechanics to metals, polymers, and composites (9–15) (see COMPOSITE MATERIALS).

Inherent Flaws

Perhaps the single most important concept in fracture mechanics is the existence of inherent flaws. Any material contains imperfections or defects. Examples are voids at grain boundaries in metals or a dirt particle in a polymer. It is from these defects that cracks eventually begin to grow, because defects cause local stress concentrations. A structure having no defects would not fracture. A structure having a large defect fractures at a lower load than an otherwise identical structure containing a smaller defect. Therefore, fracture is controlled by both the defect size, effectively the length of the crack already present, and the magnitude of the load which can be applied before failure occurs. The most severe defect is a sharp crack because this causes the most severe stress concentration. Fracture mechanics describes the different combinations of applied load and defect size which can lead to failure in a body containing a sharp crack.

Linear Elastic Fracture Mechanics

A crack in a body may grow as a result of loads applied in any of the three coordinate directions, leading to different possible modes of failure. The most common is an in-plane opening mode (Mode I). The other two are shear loading in the crack plane (Mode II) and antiplane shear (Mode III), as defined in Figure 1. Only Mode I loading is considered herein.

Crack Tip Stresses. The simplest case for fracture mechanics analysis is a linear elastic material where stress, σ, is proportional to strain, ϵ, giving

$$\sigma = E\epsilon \tag{1}$$

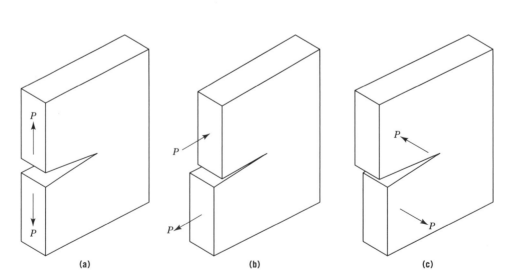

Fig. 1. Three modes of fracture where P is load: (**a**) Mode I, (**b**) Mode II, and (**c**) Mode III.

where E is the elastic modulus. In this case the stresses and strains around the tip of a sharp crack, for a given applied remote load, can be shown to vary inversely with the square root of the distance r from the crack tip (16) and go to infinity precisely at the crack tip. This is most easily visualized by considering the distribution of stress around the end of the primary axis of an elliptical hole in an infinite plate, as shown in Figure 2. The Y direction stress, σ_Y, along the X axis can be written as (11):

$$\sigma_Y = \sigma[1 + a^{0.5}(2\rho + 2r)/(\rho + 2r)^{1.5}] \tag{2}$$

for $\rho << a$, where a is the primary axis length, b is the minor axis, $\rho = b^2/a$, and σ is the remotely applied stress on the system. As b tends to zero, equivalent to the sharp crack case, σ_Y goes to infinity exactly at the end of the crack and varies inversely with the square root of the distance r from the end of the crack:

$$\sigma_Y = \sigma[1 + (a/2r)^{0.5}] \tag{3}$$

Therefore, the magnitude of the stress at small distances from the crack tip is a function of the crack length, a, and the remotely applied stress, σ. Close to the crack tip ($r << a$) the stress can be scaled using a parameter called the stress intensity factor, K (9–11):

$$\sigma_Y = K/(2\pi r)^{0.5} \tag{4}$$

where K is a function of the nominal remotely applied stress in the uncracked body and the crack length:

$$K = \sigma Y a^{0.5} \tag{5}$$

where Y is a factor which corrects for different geometries and is a function of the size of the crack relative to the dimensions of the test specimen or cracked struc-

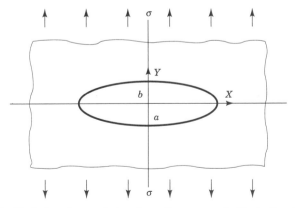

Fig. 2. An elliptical hole in a large plate where the arrows indicate the direction of stress. Terms are defined in text.

ture. For a crack in an infinite plate loaded in tension (the case considered here) $Y = \pi^{0.5}$. All of the stresses around the crack tip can be expressed in terms of K, and a more general form of equation 4 is

$$\sigma_\theta = (K/(2\pi r)^{0.5})[f(\theta)] \tag{6}$$

Because the material is assumed to be linear elastic, the local displacements around the crack tip can also be expressed in terms of K.

Clearly the stresses and strains are not actually infinite in real materials, but it is found that in relatively brittle materials, where the amount of plastic yielding is small, failure is controlled by K and fracture occurs when a critical value of K is applied. This critical value is the fracture toughness of the material, denoted K_{IC} where the subscript I indicates Mode I loading as shown in Figure 1. A zone of plastic yielding close to the crack tip prevents the stresses and strains from becoming infinite, but as long as this zone is small the overall toughness of the material can be characterized using K_{IC}.

Energy Release Rates.   The analysis can equivalently be conducted in terms of the energy dissipated in growing the crack, rather than the stresses. The energy release rate, G, is the rate at which energy is released from the overall system as the crack grows. This is the energy released per unit area of crack surface formed, rather than a time-dependent rate, and so describes the energy available to drive the crack through the material. G can be thought of as the energy which would be available to drive the crack if the crack were to grow at the current applied remote stress level. In fact, the crack does not begin to grow until this available energy is sufficient to overcome the resistance of the material to crack growth. As for K_{IC}, there is a critical energy release rate, G_{IC}, at which the crack begins to grow. The two approaches are complementary and, by considering the rates of change of the stresses and displacements around the crack tip per unit of crack growth, it can be shown that for plane strain conditions (9–11):

$$K^2 = EG/(1 - \nu^2) \tag{7}$$

in general and

$$K_{IC}^2 = EG_{IC}/(1 - \nu^2) \tag{8}$$

in particular at fracture, where ν is Poisson's ratio. Therefore, G can be expressed in terms of the applied remote stress and the crack length using equations 5 and 7 giving

$$G = \sigma^2 Y^2 a(1 - \nu^2)/E \tag{9}$$

G can be calculated from the change in compliance, ie, the reciprocal of stiffness, of the structure or test specimen. For a cracked body of arbitrary shape loaded to a load P and displacement u, as shown in Figure 3a, if some increment of crack growth, δa, occurs this leads to an increase in the compliance of the specimen, causing a reduction in the load, an increase in the displacement, or both. Referring to Figure 3b, the change in stored energy is the area ABC and

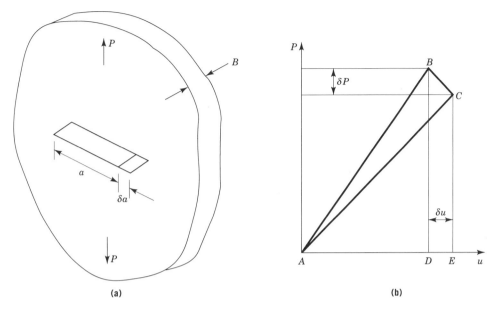

Fig. 3. The effect of crack growth on potential energy in a loaded body where (**a**) is a cracked body of arbitrary shape with a load P applied, and (**b**) is the change in potential energy in the body owing to incremental crack growth, δa. Other terms are defined in text.

the work done by the applied load is $BCDE$. Hence, an expression for the change in total potential energy of the system, U, for an incremental change in load and displacement can be written (9–11):

$$\delta U = 0.5\ Pu - 0.5(P + \delta P)(u + \delta u) + \delta u(P + 0.5\ \delta P) = 0.5(P\delta u - u\delta P) \qquad (10)$$

Because G is defined as the energy released per unit area of crack surface formed, or more correctly the energy which would be released if the crack were to grow at the present applied load, then:

$$G = (1/B)(dU/da) = (P^2/2B)(dC/da) = 0.5B(u/C)^2(dC/da) \qquad (11)$$

where B is the thickness of the specimen, as shown in Figure 3, and C is the compliance of the specimen. This relationship is useful for experimental determination of the energy release rate.

Crack Tip Yielding and Constraint. In any real material there is some plastic yielding at the crack tip, even though the overall fracture behavior of the material can be characterized using stress intensity concepts which consider the stresses in the elastic region. From the elastic solutions for the stresses close to the crack tip, for example equation 4 for the Y direction stresses, and an appropriate yield criterion, such as that of von Mises (9–11), it is possible to calculate the size and shape of the plastically yielded zone at the crack tip. Figure 4a shows this zone schematically. The plastic zone is larger near the free surfaces of the

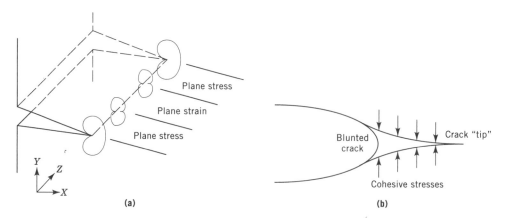

Fig. 4. (a) The crack tip plastic zone and (b) the Dugdale plastic zone model. Terms are defined in text.

specimen than at the center. This is because the high stresses and strains in the plane of the specimen (the XY plane) lead to a tendency to contract laterally (in the Z direction) at the crack tip. This can occur at the free surfaces, but near the center the unyielded material surrounding the crack tip tends to resist this contraction. A positive stress in the Z direction is generated. Because yield criteria such as those of von Mises are based on the difference between the principal stresses in the three coordinate directions, this Z direction stress acts to restrict the total amount of plastic yielding in the center of the specimen. The degree of constraint and the size of the plastic zone is approximately constant above some minimum distance into the specimen.

This constraint effect is the reason why there is a minimum specimen thickness requirement for valid fracture toughness testing. The plastic work done in the crack tip plastic zone as the crack grows is a significant part of the total work which must be done to grow the crack. For thin specimens, the size of this zone, and the total amount of work done, is a function of the thickness of the specimen. Therefore, the measured toughness is also a function of the thickness of the specimen. However, the true fracture toughness is a material property independent of the test specimen geometry (including thickness) used to measure the value. This is only the case if a relatively thick test specimen is used. When the thickness is sufficient to generate Z direction constraint, the toughness measured is referred to as the plane strain fracture toughness of the material, because deformation is largely confined to the XY plane.

The distance from the crack tip, along the X-axis, at which the von Mises equivalent stress falls below the yield stress, defines the size of the plastic zone, r_p. For the plane stress case of unconstrained yielding, which corresponds to the free surface of the specimen in Figure 4, this gives

$$r_p = (1/2\pi)(K/\sigma_0)^2 \tag{12}$$

where σ_0 is the yield stress of the material. This equation assumes a perfectly plastic material. In the center of the specimen the plastic zone is substantially smaller by an amount dependent on the Poisson's ratio of the material. An alternative estimate of the length of the plastic zone has been obtained (17) by treating the plastic zone as a cohesive zone, partially drawing the crack faces back together behind the effective crack front, as shown in Figure 4**b**. This leads to an expression of similar form to equation 12, although the constant is larger. This model is useful as it introduces the concept of crack tip blunting. Behind the cohesive zone the crack has opened out, and is no longer sharp, as a result of the plastic yielding that is occurring at the crack tip. The amount of blunting before crack growth occurs can be related to the energy needed to grow the crack, G_{IC}, and the yield stress of the material (9–11).

Elastic–Plastic Fracture Mechanics

In more ductile materials the assumptions of linear elastic fracture mechanics (LEFM) are not valid because the material yields more at the crack tip, so that the stresses and strains around the tip are no longer dominated by the linear elastic analysis. In order to characterize the fracture behavior of the material it is necessary to look inside the plastic zone at the distribution of stress and strain. If it is assumed that the material deforms in tension according to a power law hardening stress–strain curve,

$$\sigma = \sigma_0 \epsilon^n \tag{13}$$

then it is possible to calculate the variation of stress and strain within the yielded crack tip zone. This stress and strain distribution is known as an HRR field (18–20). The magnitude of the product of stress and strain is characterized by a different parameter, the J-integral, and this product varies inversely with the distance from the crack tip, the same as in the linear elastic case. However, the individual variation of stress or strain alone is no longer the simple inverse square root form of the linear elastic case, but a more complex expression dependent on the shape of the stress–strain curve.

    The J-integral derives its name from the fact that it is a contour integral which circles the crack tip. It describes the rate per unit area of crack surface at which energy would have to cross the integration path, or be released from strained material inside the integration path, in order to grow the crack. The value of J obtained is independent of the path used to evaluate it in all regions that are characterized by the HRR stress field (18). The HRR field is only valid as long as the stresses and strains are proportional at the crack tip, ie, when strain increases stress also increases. Therefore, the J-integral only characterizes the crack tip stresses and strains in the region where this condition is met. The HRR analysis leads to a prediction of an infinite product of stress and strain precisely at the tip of a sharp crack, as with the elastic analysis. In practice, very close to the crack tip there is a region of reducing stress (unloading) owing to the blunting of the crack tip.

The parameters K, G, and J are complementary and the relationship between them can be seen by considering the boundary between the beginning of the plastic zone and the surrounding elastic (unyielded) material, as shown in Figure 5. J can also be used as a measure of the toughness of a material, like K_{IC}, by defining a critical value J_{IC} (5), or by plotting the applied J to produce various amounts of crack growth (6). The magnitude of the blunting at the crack tip resulting from local yielding can be quantified using the crack-opening displacement (COD), which is defined as the separation of the crack faces at the points where two lines drawn back from the crack tip at 45° to the crack path intersect the crack faces. The COD is proportional to the applied J-integral and inversely proportional to the yield stress of the material. The proportionality constant is dependent on the shape of the stress–strain curve of the material (18). The COD has been proposed as a fracture criterion on the basis that crack growth begins when a critical COD value is reached (9).

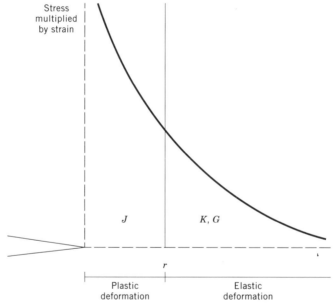

Fig. 5. The boundary between elastic and plastic zones at the crack tip. Terms are defined in text.

Micromechanisms

For a perfectly brittle material the fracture process involves the breaking of bonds at an atomic level to form new surfaces. This was the original concept (1) and was based on studies of the fracture of glass (qv). The energy needed per unit area of new surface formed, γ, was considered. Because the growth of a crack forms two new surfaces, the energy release rate needed to drive the crack in a brittle material is 2γ. However, in most engineering materials there is also a significant amount of plastic energy which must be dissipated in order to grow the crack, which gives a condition for crack growth of the form:

$$G \text{ or } J = (2\gamma + W_p) \tag{14}$$

where W_p is the plastic work that must be done to grow the crack. Fracture mechanics attempts to quantify the amount of energy needed to grow the crack, by treating the behavior on a macroscale continuum basis. The objective is to subtract all nonessential work done in deforming the test specimen to determine the energy specific to the crack growth process alone. Locally at the crack tip various microscale processes are occurring to actually fail the material (21). There is some process zone, inside the plastic zone adjacent to the blunted crack tip, where material has been deformed so much that it begins to rupture. Examples of the mechanisms occurring in the process zone are cleavage along grain boundaries or void growth and coalescence in metals (9,10). In some polymers crazing may occur owing to the triaxial stresses at the crack tip (11,22). It is often possible to get some information about the fracture behavior from the appearance of the fracture surface after the crack has grown.

Practical Testing Procedures

Specimen Requirements. Fracture toughness tests divide into two distinct groups, K_{IC} (or G_{IC}) measurements for relatively brittle materials and J-integral tests for more ductile materials. The reason for this becomes apparent when the K_{IC} test is considered. Fracture mechanics testing is required to follow certain procedures so that a valid measurement is obtained. Specifically, the initial crack in the test specimen must be sharp and the thickness and depth of the specimen must be sufficient to generate the plane strain constraint discussed earlier. A poorly conducted fracture mechanics test leads to an overestimate, rather than an underestimate, of the true toughness of the material. Hence, care is needed to ensure a valid measurement. Basically a thick specimen is always preferable to a thin specimen, but if the available material thickness is limited, then a test can be conducted to determine a provisional toughness K_Q. Using the yield stress of the material, σ_0, an assessment can then be made of the validity of K_Q using the expression

$$B_{\min}, (D - a)_{\min}, a_{\min} = 2.5(K_Q/\sigma_0)^2 \tag{15}$$

where $B_{\min}$ is the minimum specimen thickness to obtain a valid measurement for a material with that toughness and yield stress, $(D - a)_{\min}$ is the minimum uncracked ligament in the direction of crack growth, and $a_{\min}$ is the minimum initial crack length. For example, if $B_{\min}$ is greater than the specimen thickness that was actually used to measure K_Q, then K_Q is, unfortunately, not a valid measurement and a thicker specimen is needed to generate the plane strain constraint. Sidegrooves can sometimes be used to generate additional constraint so that a valid measurement can be obtained with a restricted thickness. However, this complicates the calculation of the toughness (23).

In tougher materials the minimum thickness required by equation 15 can become excessive. In such cases the J-integral test is an attractive alternative. Because this test considers the stress distribution around the crack inside the plastic zone, it can be used to obtain a valid toughness measurement in a thinner specimen, because more extensive plastic yielding does not invalidate the analysis. The equivalent of equation 15 for the J-integral test is

$$B_{\min}, (D - a)_{\min}, a_{\min} = 25(J_Q/\sigma_0) \tag{16}$$

where J_Q is a provisional value of J_{IC}.

The other important point is to achieve a sharp initial crack or notch. The initial notch is typically about half the depth of the specimen and a pre-notch is usually sawn or machined into the specimen. This notch must be sharpened. In metals this is often done by cycling the specimen in fatigue loading at a low load to grow a natural crack some distance beyond the machined notch (9). In other materials, such as polymers, a sharp crack can be achieved by skillfully sharpening the crack tip using a new razor blade (11). Although fracture toughness tests can be conducted at any loading rate, most are done at low rates such as a cross-head rate of 10 mm/min, using a suitable screw driven or servohydraulic load frame.

Fracture Toughness Testing. Some typical fracture toughness test geometries are shown in Figure 6. The single-edge notch (SEN) or center notch (CN) specimens are loaded in tension. The compact tension (CT) specimen applies an opening load to the crack, using pins loading through holes in the specimen. The three-point bend (TPB) specimen is supported at each end and loaded downward at the center opposite the notch. Different calibration factors (Y in eq. 5) are tabulated for the different geometries so that the applied stress intensity can be calculated from the load at fracture and the crack length. To obtain valid plane strain fracture toughness values the ASTM standards (4,7) set certain requirements for the dimensions of the specimen.

A fracture toughness test, where unstable brittle fracture occurs at a critical applied load, leads to a load–displacement curve like that shown schematically in Figure 7**a**, having an essentially linear loading curve and a catastrophic drop in load at fracture. Gross nonlinearity in the loading curve generally indicates that excessive yielding is occurring, although it may also indicate slow stable crack growth prior to unstable fracture. Because it is the fracture toughness at the initiation of crack growth that is desired, the load at this point must be determined. The ASTM standards suggest drawing a line through the straight part of the loading curve and then drawing a second line AB with a 5% lower slope, as shown in Figure 7**b**. If the maximum load falls between these lines, then this load is used to calculate the toughness of the material. If the maximum load is outside these lines, then a reduced load at the point where the 95% slope line crosses the loading curve should be used. Other conditions must also be met, as described in the standards. If the reduced load is less than 90% of the maximum load, or if brittle fracture does not occur, then the test is invalid and a J-integral test must be considered. Some typical K_{IC} values in units of MPa·m$^{1/2}$ are steel, 40–90; titanium, 38; aluminum, 30; 30% glass-reinforced nylon-6,6, 6.0; nylon-6,6, 3.5; ABS resins, 2.0; poly(methyl methacrylate) (PMMA), 1.2; and epoxies, 0.6.

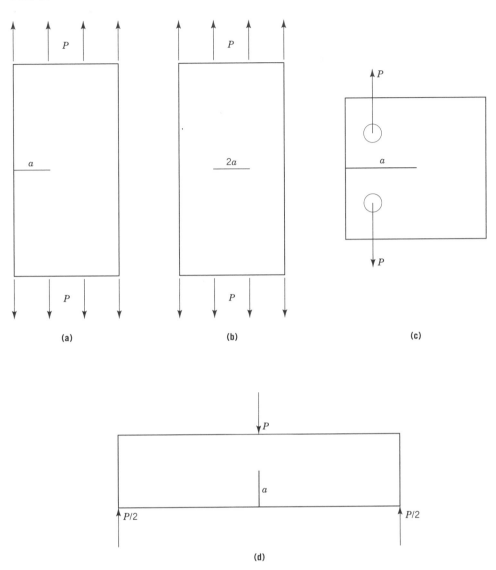

Fig. 6. Fracture toughness test specimens: (**a**) single-edge notch; (**b**) center notch; (**c**) compact tension; and (**d**) three-point bend. Terms are defined in text.

J-Integral Testing. The J-integral test is rather different from the K_{IC} test because a curve of crack growth resistance against crack growth is generated (24). An example of such a curve is shown schematically in Figure 8. This requires more than one specimen, with each specimen loaded to different levels to generate different amounts of crack growth (5). After the test each specimen is broken open, for example by cooling in liquid nitrogen and impacting at a high rate to generate brittle failure, and the amount of crack growth is measured from the appearance of the fracture surface. The crack front typically has some curvature. More crack growth occurs in the center of the specimen, so that either an average crack

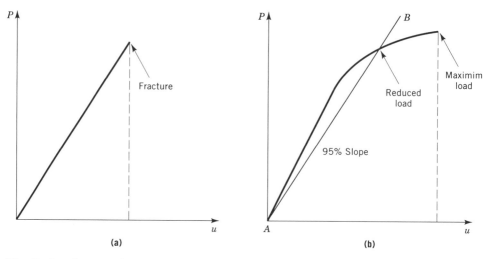

Fig. 7. Load versus deflection for (**a**) perfectly brittle fracture and (**b**) slight nonlinearity. Terms are defined in text.

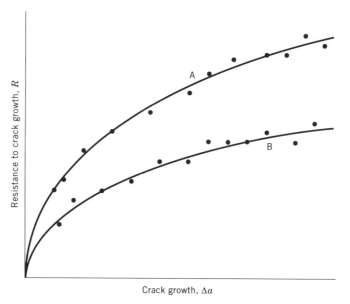

Fig. 8. J-integral test resistance curves for materials A and B.

growth can be obtained by taking a number of measurements across the thickness (5), or the maximum crack growth can be used, which leads to a conservative estimate of toughness. This curvature arises from the plane strain conditions at the center of the specimen and the higher apparent toughness plane stress state at the surface. The corresponding applied J is calculated from the input energy, the area under the load–deflection curve, using an equation of the form

$$J = 2U/[B(D - a)] \qquad (17)$$

where B, D, and a are the specimen thickness, depth, and crack length, respectively. This equation is only valid when certain constraints on specimen dimensions are met (6,10). The resistance curve can be then be constructed by plotting the applied J values against the resulting crack growth. The standards (5,6) set requirements for the number and spacing of the data points used to construct the curve. An alternative single-specimen method uses just one specimen, which is progressively loaded to higher and higher loads (higher applied J values), partially unloading the specimen between each increment of load increase. From the slope of the partial unloading curves the compliance of the specimen is obtained and the crack length is calculated from this compliance (6). The resistance curve can then be constructed from the area under the curve up to each partial unloading point and the corresponding crack growth.

An attempt has been made to define a single critical J_{IC} value, like K_{IC} for brittle fracture, from these curves (5). However, the amount of crack growth which is used to define this critical value is inevitably rather arbitrary. A more recent approach (6) is to fit a power law curve of the form

$$J = A(\Delta a)^n \qquad (18)$$

to the data, where Δa is the measured crack growth and the coefficients A and n can be used for relative comparison of materials. A material having a higher resistance curve, ie, more energy per unit of crack growth, is nominally tougher, but full interpretation of these curves is rather complex because the slope of the curve must also be considered, relative to the system applying the load, to assess the stability of the crack. Because the resistance curve of material A in Figure 8 is always above that of material B, then material A is the tougher of the two.

Fatigue

Fatigue cracking is slow crack growth under repeated cyclic loading or constant static load. For cyclic loading unnotched fatigue data are often presented in the form of $S–N$ curves, where applied stress, S, is plotted against cycles to failure, N. This method leads to widely scattered data because the total lifetime is the sum of the time to initiate a crack from an existing inherent flaw, the size of which is variable between specimens, and the time to grow the crack to final failure of the specimen. This variability is overcome by using fracture mechanics because a deliberate initial crack is introduced into the specimen, which is then cyclically loaded. Because the load applied to each cycle is generally quite low, relative to the single-cycle failure load, the amount of plastic yielding at the crack tip should not be excessive and the problem is well suited to characterization using linear elastic fracture mechanics (LEFM) (25). For cyclic loading between some maximum and minimum load, the corresponding maximum and minimum applied K can be calculated, provided the length of the crack is continually monitored during the test. Data can then be plotted in the form of the rate of change of crack length per cycle, da/dN, against the range of applied K per cycle, $\Delta K = (K_{max} - K_{min})$.

A da/dN curve such as that shown schematically in Figure 9 is generally obtained, consisting of three regions. The first region is a slower growth rate at small applied ΔK levels, with the possibility of a threshold ΔK below which no crack growth at all occurs. The main crack growth is described by the Paris law (26):

$$da/dN = D(\Delta K)^m \tag{19}$$

and the final region is more rapid crack growth at higher applied ΔK levels. A log–log plot of the main crack growth data allows the constants D and m in equation 19 to be estimated. It is then possible to integrate equation 19 to predict the number of cycles for a crack to grow from some initial length to a final length. However, some caution is needed in making such a calculation because there are many variables in fatigue behavior that must be considered. In particular, the ratio of the maximum applied stress to the minimum stress during each cycle is an important parameter. This ratio is called the cycle or R ratio. A unique curve of the form described by equation 19 is only obtained for a constant R ratio, because the same ΔK can be applied for any R value. Therefore, it is more complete to write

$$da/dN = f(\Delta K, R) \tag{20}$$

The conditions at the crack tip during fatigue loading can be quite different, depending on the R ratio. For example, if the R ratio is negative then the crack is

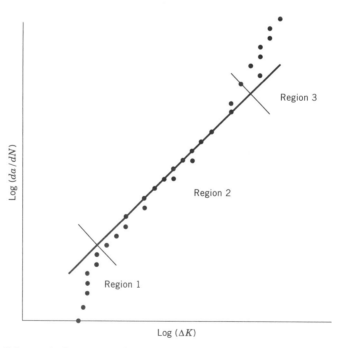

Fig. 9. Schematic fatigue crack growth data showing the regions of growth rate.

forced closed during part of the cycle. The stress intensity calculation has no meaning when the applied loading is compressive, and it could be argued that the crack growth rate is then controlled by the tensile loading part of the cycle alone. However, there is a plastic zone at the crack tip, so that the crack does not close perfectly as soon as the applied loading becomes negative, which means that the compressive part of the loading also makes some contribution to the generation of damage at the crack tip. Provided the experimental data have been obtained at the appropriate ΔK and R for the application where fatigue life is to be predicted, then an allowance for these factors can be incorporated in the data. Further complications can arise if the component or structure being analyzed experiences random loadings of different severity during its service life. The prediction of total fatigue life is then dependent to some extent on the sequence in which these different amplitude loadings occur. An overload cycle, above previous applied stress intensity levels, generates compressive residual stresses at the crack tip which inhibit crack growth when the applied stress intensity reduces to its original range in subsequent cycles (27). Therefore, a simple analysis based on integrating equation 19 to determine the lifetime gives a conservative prediction (an overestimate of the growth rate) by ignoring the interaction between cycles of different amplitude. This is a topic for more specialized texts (10,28). Equation 19 provides the powerful capability to estimate the number of cycles needed for a crack to grow some amount. By periodically monitoring the length of fatigue cracks in structures such as pressure vessels and aircraft, and considering *da/dN* data, a decision can be made about whether it is safe for the structure to remain in service until the next inspection.

Fracture mechanics concepts can also be applied to fatigue crack growth under a constant static load, but in this case the material behavior is nonlinear and time-dependent (29,30). Slow, stable crack growth data can be presented in terms of the crack growth rate per unit of time against the applied K or J, if the nonlinearity is not too great. For extensive nonlinearity a viscoelastic analysis can become very complex (11) and a number of schemes based on the time rate of change of J have been proposed (31,32).

The measurement of crack length in a laboratory specimen can be achieved using various methods. The surface of the specimen can be observed with a traveling microscope to measure the apparent growth. In metals a method based on the changing electrical resistance of the specimen, resulting from the growth of the crack, can be used (9,33). Ultrasonic techniques (9) can also be used (see ULTRASONICS), or the crack length can be deduced by calculation from measurements of the compliance of the specimen and the applied load (34). An ASTM standard for fracture mechanics fatigue testing has been prepared (35). Restrictions on the specimen dimensions for a valid test apply, of similar form to equation 15. However, for fatigue testing there is a maximum specimen thickness, rather than a minimum thickness, because of concerns about the curvature of the crack front and also the possiblity of heat generation inside the specimen during cycling. Heat generation is of particular concern in testing polymers where the thermal conductivity is low and where viscoelastic effects may cause significant heat generation (11,36,37). Heating can be reduced by lowering the frequency of the cyclic loading, which is typically in the range of 1–5 Hz.

Impact Loading

The fracture toughness of a material can vary with the strain rate at which the test is conducted. This is why materials such as polymers tend to be more brittle when loaded rapidly. For this reason, fracture mechanics tests are also conducted under impact loading conditions. The three-point bend specimen is probably the most commonly used geometry for impact tests. A falling weight with a suitable loading head is released from a measured height so that it impacts the back of the specimen at a known speed. In principle the fracture toughness can be calculated as before, using an equation of the form of equation 5 and the load at fracture. In practice, as the loading rate increases, dynamic vibration of the specimen occurs during loading, after the initial impact, and the load–displacement signal recorded by a transducer attached to the impacting head tends to show substantial oscillations. This makes it difficult, if not impossible, to determine the load at fracture accurately (38). Alternative approaches, such as measuring the displacement to failure rather than the load, may be necessary (39).

The commonly used Izod test and Charpy test, for example, as described in the ASTM standards for metals and polymers (40,41) (see also PLASTICS TESTING), are not fracture mechanics tests for a number of reasons. First, a blunt notch is used, whereas in a fracture mechanics test a sharp notch is of vital importance. Second, the total energy to fail the specimen, rather than the energy specific to the crack growth process, is recorded. Third, no account is taken of the plane stress–plane strain effects discussed earlier. Therefore, for a constant specimen thickness, two different materials having similar fracture toughnesses but different yield stresses may record different performances in an Izod or Charpy test, simply because one material happens to be in plane stress and the other in plane strain at that specimen thickness. However, a fracture mechanics variation on the basic Charpy test is possible for materials where LEFM is valid. If a series of impact tests are performed on Charpy-type three-point bend specimens with sharp notches and a range of initial crack lengths, then the energy to break can be plotted against a function of the compliance of the specimen. It can be shown (11) that:

$$G_{IC} = U/BD\phi \tag{21}$$

where G_{IC} is the critical plane strain energy release rate, U is the energy to break the specimen, B and D are the thickness and depth of the specimen, respectively, and ϕ is a rather complex function of the compliance of the specimen, which depends on its dimensions, including the crack length. Therefore, by plotting U against $BD\phi$ for a range of initial notch lengths, a straight line should be obtained having a slope equal to the critical energy release rate of the material. As for other fracture toughness tests, care must be taken to check that the thickness of the specimen is sufficient for a valid measurement.

Dynamic Crack Propagation

In the preceding sections the measurement of fracture toughness for a stationary crack was considered. The load up to fracture may be applied slowly or rapidly, as in an impact test, but the crack itself is not moving until fracture occurs. The more rapidly the load is applied, the higher the strain rate around the crack tip. Hence, some dependence of fracture toughness on loading rate is generally observed. Therefore, it is to be expected that the fracture toughness of a material where the crack is propagating at some speed also varies with the speed at which the crack is growing. Because the toughness of a material tends to fall with increasing rate, up to some limit, once a crack begins to propagate in an unstable manner it accelerates to a high (over 100 m/s) speed. Therefore, the calculation of the fracture toughness becomes an extremely complex problem, as the available energy to drive the crack is controlled by dynamic effects. Specialized techniques are needed to calculate the energy release rate during rapid crack propagation (42,43) and to measure the fracture toughness experimentally (44,45). It may appear that such calculations are of rather academic interest, because the structure has apparently failed if the crack is running at such speeds. However, there is considerable interest in understanding dynamic fracture problems to determine the conditions for a rapidly propagating crack to arrest before total failure occurs (46). For example, in continuous steel or polyethylene pipelines (qv) there is the worrisome possibility that a crack which started to propagate could run along the axis of the pipe for a large distance, perhaps driven by gas pressure inside the pipe. Dynamic fracture has, therefore, received considerable attention in terms of both theoretical and experimental investigations.

Environment and Temperature

All of the tests discussed herein could be conducted at different temperatures and in different environments. The effect of a particular environment or temperature can be significant and the experimenter should take care to ensure that the data is collected under the appropriate conditions for the problem being considered. When testing at elevated temperatures it is important to check the validity of the measurement using equation 15, based on a yield stress measured at the temperature of the fracture toughness test. Stress corrosion cracking is the term used for the combined effect of an aggressive environment and an applied stress on the propagation of a crack (47). In the case of metals the effect may be either erosion of material at the crack tip in an anodic process, or the evolution of hydrogen in a cathodic process which then diffuses into the metal and causes hydrogen embrittlement (9,48). For polymers, solvents that diffuse very slowly into the polymer when it is not loaded may enter the crack tip region quite rapidly when a stress is applied, probably because of the formation of voids or crazes under load. The solvent then plasticizes the crack tip, leading to a deterioration of properties and more rapid crack growth (11). The variety of combinations of materials, environments, and potential interactions is so great that any fracture toughness measurement must be performed in the environment in which the component being designed is to be used.

Applications and Design Issues

Fracture toughness is often used purely as a relative comparison of the cracking resistance of different materials. However, it has the potential to also be used for quantitative prediction of actual component or structure performance. If the size of the crack or inherent flaw present in a component can be determined, then by performing a stress analysis of the structure, probably by finite element analysis (49), or referring to tabulated stress intensity factor solutions for different geometries (50,51), it is possible to calculate the applied stress intensity loading on the crack for a particular load on the structure. In a complex three-dimensional structure, substantial computational effort may be needed to accurately determine the applied stress intensity. A crack growing from a defect generally has a curved crack front profile, eg, a penny-shaped crack, unlike the straight crack, two-dimensional laboratory specimens considered herein. This complicates the calculation of applied stress intensity. However, once the applied stress intensity at the design load has been determined, an assessment of whether fracture will occur can be made. A number of rather elaborate design procedures have been developed, one of which is the R6 design code (52,53). This design procedure essentially considers the two competing failure processes of gross plastic yielding across the uncracked cross section, leading to failure by plastic collapse of the structure, and failure by crack growth. It attempts to define combinations of these two modes which would lead to failure when neither alone would be sufficient.

For a crack growing under J-integral controlled conditions in a structure, the crack may grow some distance and then stop, or it may propagate unstably. To determine the stability of a crack in a structure it is necessary to know the form of the resistance curve of the material and also the energy which could be released from the structure to drive the crack. A construction of the form shown

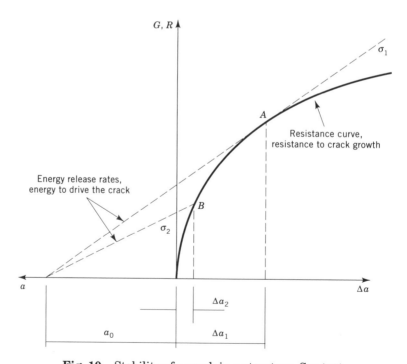

Fig. 10. Stability of a crack in a structure. See text.

schematically in Figure 10 can be used to assess stability. The energy release rate of the structure is proportional to the crack length, for a given applied stress (from eq. 9), whereas the energy needed for an amount of crack growth Δa follows the resistance curve. The tangent point A, for a particular initial crack length a_0 and applied stress σ_1, defines the transition from stable crack growth to unstable fracture. For a lower applied stress σ_2 the crack grows some amount, up to point B, and then stops without instability occurring. Note that a_0 is much greater than Δa and Figure 10 is not drawn to scale. For a brittle material characterized with G_{IC}, the resistance curve would be a horizontal line at G_{IC} on the Y axis. As soon as the applied energy release rate was equal to G_{IC}, unstable fracture would occur. Brittle materials have a flat resistance curve.

The use of fatigue data and crack length measurements to predict the remaining service life of a structure under cyclic loading is possibly the most common application of fracture mechanics for performance prediction. In complex structures the growth of cracks is routinely monitored at intervals, and from data about crack growth rates and the applied loadings at that point in the structure, a decision is made about whether the structure can continue to operate safely until the next scheduled inspection.

NOMENCLATURE

a	crack length
B	specimen thickness
C	specimen compliance
COD	crack opening displacement
D	specimen depth in the direction of crack growth
E	elastic modulus
G	energy release rate
G_{IC}	plane strain critical value of G
HRR	Hutchinson, Rice, and Rosengren
J	J-integral value
J_{IC}	plane strain critical value of J
K	stress intensity factor
K_{IC}	plane strain fracture toughness
LEFM	linear elastic fracture mechanics
N	number of cycles
P	load
r	distance from the crack tip
r_p	plastic zone radius
R	ratio of maximum to minimum stress in fatigue, or resistance to crack growth in J-integral testing
u	displacement
U	energy
W_p	plastic work
Y	geometric correction factor in the K equation
γ	surface energy
δ	crack opening displacement
Δa	crack growth
ϵ	strain

ν Poisson's ratio
σ stress
σ_Y stress in Y direction
σ_0 yield stress
ϕ energy calibration factor

BIBLIOGRAPHY

1. A. A. Griffith, *Phil. Trans. Roy. Soc.* **A221**, 163 (1920).
2. K. B. Broberg, *Eng. Fract. Mechanics* **16**, 497 (1982).
3. J. M. Barsom, *Fracture Mechanics Retrospective—Early Classic Papers (1913–1965)*, ASTM Publications, Philadelphia, 1987.
4. ASTM E399-90, "Plane Strain Fracture Toughness of Metallic Materials," *Annual Book of ASTM Standards*, ASTM Publications, Philadelphia, 1993.
5. ASTM E813-89, J_{IC}, "A Measure of Fracture Toughness," *Annual Book of ASTM Standards*, ASTM Publications, Philadelphia, 1993.
6. ASTM E1152-87, "Determining J–R Curves," *Annual Book of ASTM Standards*, ASTM Publications, Philadelphia, 1993.
7. ASTM D5045-91, "Plane Strain Fracture Toughness and Strain Energy Release Rate of Plastic Materials," *Annual Book of ASTM Standards*, ASTM Publications, Philadelphia, 1993.
8. J. G. Williams and M. J. Cawood, *Polym. Testing* **9**, 15 (1990).
9. J. F. Knott, *Fundamentals of Fracture Mechanics*, Butterworths, London, 1979.
10. D. Broek, *Elementary Engineering Fracture Mechanics*, Martinus Nijhoff, Dordrecht, the Netherlands, 1987.
11. J. G. Williams, *Fracture Mechanics of Polymers*, Ellis Horwood, Chichester, UK, 1984.
12. D. Broek, *The Practical Use of Fracture Mechanics*, Kluwer Academic Publishers, Dordrecht, the Netherlands, 1989.
13. K. Friedrich, *Application of Fracture Mechanics to Composite Materials*, Elsevier Science Publishers, Amsterdam, the Netherlands, 1989.
14. M. F. Kanninen and C. H. Popelar, *Advanced Fracture Mechanics*, Oxford University Press, New York, 1985.
15. R. W. Hertzberg, *Deformation and Fracture Mechanics of Engineering Materials*, John Wiley & Sons, Inc., New York, 1983.
16. M. L. Williams, *J. Appl. Mechanics* **28**, 109 (1957).
17. D. S. Dugdale, *J. Mechanics Phys. Solids* **8**, 100 (1960).
18. J. R. Rice and G. F. Rosengren, *J. Mechanics Phys. Solids* **16**, 1 (1968).
19. J. W. Hutchinson, *J. Mechanics Phys. Solids* **16**, 13 (1968).
20. *Ibid.*, p. 337.
21. S. Murakami, *JSME Int. J.* **30**, 701 (1987).
22. E. Passaglia, *J. Phys. Chem. Solids* **48**, 1075 (1987).
23. C. N. Freed and J. M. Krafft, *J. Mater.* **1**, 770 (1966).
24. W. Schmitt and R. Kienzler, *Eng. Fract. Mechanics* **32**, 409 (1989).
25. R. J. Allen, G. S. Booth, and T. Jutla, *Fatigue Fract. Eng. Mater. Struct.* **11**, 45 (1988).
26. P. C. Paris, M. P. Gomez, and W. E. Anderson, *Trend Eng.* **13**, 9 (1961).
27. D. J. Smith and S. J. Garwood, *Int. J. Pressure Vessels Piping* **41**, 255 (1990).
28. ASTM STP 462, *Effects of Environment and Complex Load History on Fatigue Life*, ASTM Publications, Philadelphia, 1968.
29. H. P. Van Leeuwen, *Eng. Fract. Mechanics* **9**, 951 (1977).

30. D. W. Rees, *Progr. Nucl. Energy* **19**, 211 (1987).

31. *Fracture Mechanics*, ASTM STP514, ASTM STP514, ASTM Publications, Philadelphia, 1972.

32. A. G. Atkins and Y-W. Mai, *Elastic and Plastic Fracture*, Ellis Horwood, Chichester, U.K., 1985.

33. K. J. Miller, *Proc. Inst. Mech. Eng., Part C: Mech. Eng. Sci.* **205**, 291 (1991).

34. A. Saxena and E. J. Hudak, *Int. J. Fract.* **14**, 453 (1978).

35. ASTM E647-93, "Measurement of Fatigue Crack Growth Rates," *Annual Book of ASTM Standards*, ASTM Publications, Philadelphia, 1993.

36. R. W. Hertzberg and J. A. Manson, *Fatigue of Engineering Plastics*, Academic Press, New York, 1980.

37. J. A. Sauer and G. C. Richardson, *Int. J. Fract.* **16**, 499 (1980).

38. R. A. Mines, *Int. J. Impact Eng.* **9**, 441 (1990).

39. J. G. Williams and G. C. Adams, *Int. J. Fract.* **33**, 209 (1987).

40. ASTM E23-93, "Notched Bar Impact Testing of Metallic Materials," *Annual Book of ASTM Standards*, ASTM Publications, Philadelphia, 1993.

41. ASTM D256-92, "Impact Resistance of Plastics and Electrical Materials," *Annual Book of ASTM Standards*, ASTM Publications, Philadelphia, 1993.

42. A. N. Atluri and T. Nishioka, *Int. J. Fract.* **27**, 245 (1985).

43. T. Nakamura, C. F. Shih, and L. B. Freund, *Int. J. Fract.* **27**, 229 (1985).

44. J. F. Kalthoff, *Int. J. Fract.* **27**, 277 (1985).

45. J. Duffy and C. F. Shih in K. Salama and co-workers, eds., *Proceedings of the Seventh International Conference on Fracture (ICF7)*, Pergamon Press, Oxford, U.K., Vol. 5, (1989), p. 633.

46. E. J. Ripling and P. B. Crosley, *Eng. Fract. Mechanics* **23**, 21 (1986).

47. K. Sieradzki and R. C. Newman, *J. Phys. Chem. Solids* **48**, 1101 (1987).

48. R. P. Gangloff, *Mater. Sci. Eng.* **A103**, 157 (1988).

49. S. N. Atluri, *Computational Methods in the Mechanics of Fracture*, Elsevier Science Publishers, Amsterdam, the Netherlands, 1986.

50. D. P. Rooke and D. J. Cartwright, *Compendium of Stress Intensity Factors*, Her Majesty's Stationery Office, London, 1974.

51. Y. Murakami, ed., *Stress Intensity Factors Handbook*, Pergamon Press, Oxford, U.K., 1987.

52. I. Milne, R. A. Ainsworth, A. R. Dowling, and A. T. Stewart, *Int. J. Pressure Vessels Piping* **32**, 3 (1988).

53. S. T. Rolfe and J. M. Barsom, *Fracture and Fatigue Control in Structures*, Prentice-Hall, Inc., Englewood Cliffs, N.J., 1977.

BARRY A. CROUCH
E. I. du Pont de Nemours & Co., Inc.

FREONS.   See REFRIGERATION AND REFRIGERANTS.

FRICTION MATERIAL.   See BRAKE LININGS AND CLUTCH FACINGS.

FRIEDEL-CRAFTS REACTIONS

In 1877, at the Sorbonne in Paris, Charles Friedel and his American associate, James Mason Crafts, showed that anhydrous aluminum chloride could be used as a condensing agent in a general synthetic method for furnishing an infinite number of hydrocarbons. In work stretching over 14 years, they extended their discoveries of the catalytic effect of aluminum chloride in a variety of organic reactions: (1) reactions of alkyl and acyl halides and unsaturated compounds with aromatic and aliphatic hydrocarbons; (2) reactions of acid anhydrides with aromatic hydrocarbons; (3) reactions of oxygen, sulfur, sulfur dioxide, carbon dioxide, and phosgene with aromatic hydrocarbons; (4) cracking of aliphatic and aromatic hydrocarbons; and (5) polymerization of unsaturated hydrocarbons. The diversity of reactions is astounding.

Many important industrial processes such as the production of high octane gasoline, ethylbenzene (eventually leading to polystyrene), synthetic rubber, plastics, and detergent alkylates are based on Friedel-Crafts chemistry. The scope of the reactions is extremely wide as they form a large part of the more general field of electrophilic reactions, the class of reactions involving electron deficient carbocationic intermediates. The published literature is very extensive, and for more detailed information, monographs (1,2) and other comprehensive reviews (3–5) should be consulted (see also ALKYLATION; ALUMINUM COMPOUNDS; BORON COMPOUNDS; CATALYSIS).

The observation that aluminum chloride [7446-70-0] is not the only specific catalyst in the reaction was included in the first papers of Friedel and Crafts. Ferric chloride [7705-08-0] and zinc chloride [7646-85-7], as well as the double salt sodium aluminum chloride [7784-16-9], were employed *inter alia*, but were generally found less reactive. In many Friedel-Crafts reactions substantial amounts of $AlCl_3$ (or related halides) must be used, greatly exceeding what would be needed as catalyst. The reason for this is that the Lewis or Brønsted acids are tied up, either forming molar complexes with the reagents and products (such as carbonyl compounds, etc) or participating as counter ions in so-called "red oil" formation (systems composed of carbocationic complexes resulting from protonation or alkylation of aromatics or aliphatic unsaturated systems). This was a significant drawback of Friedel-Crafts chemistry. The emergence of highly acidic solid catalysts, such as H-ZSM-5, an acidic form of Zeolite (aluminosilicate), or Nafion-H (perfluoroalkanesulfonic acid polymer developed by Du Pont), now allows Friedel-Crafts reactions without such complex formation. Reagents are passed through heterogeneous solid acid catalysts, representing a significant advance in the practical application of Friedel-Crafts reactions.

To define Friedel-Crafts reactions, it was necessary to come to a clear understanding that not one but a number of electrophilic reactions are classified as Friedel-Crafts type (2). Friedel-Crafts-type reactions are generally considered to be any substitution, isomerization, elimination, cracking, polymerization, or addition reaction that takes place under the catalytic effect of Lewis acid-type acidic halides (with or without cocatalysts) or Brønsted acids. Friedel-Crafts reactions are not limited to the formation of carbon–carbon bonds but also lead to formation or cleavage of carbon–oxygen, carbon–nitrogen, carbon–sulfur, carbon–halogen, carbon–metals, and many other types of bonds. Friedel-Crafts reactions can be divided into two general categories: alkylations and acylations. Within these two broad areas there is considerable diversity.

Alkylation

ALKYLATION OF AROMATIC COMPOUNDS

In the usual Friedel-Crafts alkylation of aromatics, a hydrogen atom of the aromatic nucleus is replaced by an alkyl group through the interaction of an alkylating agent in the presence of a catalyst. The most frequently used alkylating agents are alkyl halides, alkenes, alkynes, alcohols, esters (of carboxylic and inorganic acids), ethers, aldehydes and ketones, cycloalkanes and cycloalkenes, thiols (mercaptans), sulfides, amines (via diazotization), thiocyanates, and the like. The overall reactions using alkyl halides and alcohols as alkylating agents in the presence of aluminum chloride can be written as follows:

$$C_6H_6 + RX \xrightarrow{AlCl_3} C_6H_6R + HX$$

$$C_6H_6 + ROH \xrightarrow{AlCl_3} C_6H_6R + H_2O$$

Friedel-Crafts alkylation using alkenes has important industrial applications. The ethylation of benzene with ethylene to ethylbenzene used in the manufacture of styrene, is one of the largest scale industrial processes. The reaction is done under the catalysis of $AlCl_3$ in the presence of a proton source, ie, H_2O, HCl, etc, although other catalysts have also gained significance.

Propylation of benzene with propylene, catalyzed by supported phosphoric acid (or related catalysts such as $AlCl_3$), gives cumene [98-82-8] in another important industrial process. Cumene (qv), through the intermediacy of cumene hydroperoxide, is used in the manufacture of phenol (qv). Resorcinol similarly can be made from m-diisopropylbenzene (6).

Tertiary, benzyl, and allylic nitro compounds can also be used as Friedel-Crafts alkylating agents; eg, reaction of $(CH_3)_3CNO_2$ (2-nitro-2-methyl propane [594-70-7]) with anisole in the presence of $SnCl_4$ gives 4-t-butylanisole [5396-38-3] (7). Solid acids, such as perfluorodecanesulfonic acid [335-77-3], and perfluorooctanesulfonic acid [1763-23-1] have been used as catalysts for regioselective alkylations (8).

Mechanism.   The mechanism of alkylation and of other related Friedel-Crafts reactions is best explained by the carbocation concept. The alkylation of benzene with isopropyl chloride may be used as a general example:

$$(CH_3)_2CHCl + AlCl_3 \rightleftharpoons [CH_3\overset{+}{C}HCH_3]\,[AlCl_4]^-$$

Lewis acid catalysts, such as $AlCl_3$ or BF_3, coordinate strongly with non-bonded electron pairs but they interact only weakly with bonded electron pairs. Therefore, n-donor reagents, such as alkyl halides, can react with Lewis acid catalysts even under complete exclusion of moisture or any other proton source:

$$RX + MX_3 \rightleftharpoons R^{\delta+}\cdots X\cdots MX_3^{\delta-} \rightleftharpoons [R^+]MX_4^-$$

However, strong protic acid catalysts are needed when π- or σ- donor alkylating agents are used to produce carbocationic or highly polarized donor-acceptor-complexes as the reactive alkylating intermediates:

$$RCH{=}CH_2 + HMX_{4\ or\ 6} \rightleftharpoons R^+CHCH_3\ MX_{4\ or\ 6}^-$$
$$R_2CHCH_3 + HMX_{4\ or\ 6} \rightleftharpoons R_2{}^+CHMX_{4\ or\ 6}^- + CH_4$$

In superacidic media, the carbocationic intermediates, which were long postulated to exist during Friedel-Crafts type reactions (9–11) can be observed, and even isolated as salts. The structures of these carbocations have been studied in high acidity–low nucleophilicity solvent systems using spectroscopic methods such as nmr, ir, Raman, esr, and x-ray crystallography.

Dealkylation, Transalkylation, and Disproportionation.   The action of aluminum chloride also removes alkyl groups from alkylbenzenes (dealkylation, disproportionation) (12). Alkylbenzenes, when heated with $AlCl_3$, form mixtures of benzene and polyalkylated benzenes:

$$2\ C_6H_5R \xrightarrow{AlCl_3,\ heat} C_6H_6 + C_6H_4R_2,\ etc$$

The tendency of alkyl groups to disproportionate is as follows: $CH_3 < C_2H_5 < n\text{-}C_3H_7 < i\text{-}C_3H_7 < tert\text{-}C_4H_9$ (13).

For example, in the industrially important alkylation of benzene with ethylene to ethylbenzene, polyethylbenzenes are also produced. The overall formation of polysubstituted products is minimized by recycling the higher ethylation products for the ethylation of fresh benzene (14). By adding the calculated equilibrium amount of polyethylbenzene to the benzene feed, a high conversion of ethylene to monoethylbenzene can be achieved (15) (see also XYLENES AND ETHYLBENZENE).

Polyalkylations. It had been assumed (16) that the tendency toward poly-substitution during Friedel-Crafts alkylation is due to the greater reactivity of the initially produced alkylbenzenes toward further substitution. This kinetic effect, however, has been shown to be limited; the alkyl groups on a benzene nucleus have only a small activating effect on the reaction rate. The actual alkylation occurs in a heterogeneous reaction system, specifically in the acidic catalyst layer, and the reason for polysubstitution is the preferential extraction of the initial alkylates into this catalyst layer allowing ready further alkylation (17). The tendency toward polysubstitution may be minimized by the use of a mutual solvent for both the hydrocarbon and catalyst, by high speed stirring, by operating in the vapor phase or at a temperature sufficiently high for the aluminum chloride to be soluble in the hydrocarbon layer.

Orientation. If more than one alkyl group is introduced into benzene the question of orientation arises. Alkylation in the presence of aluminum chloride and other Friedel-Crafts catalysts usually yields a considerable proportion of meta dialkylbenzenes, as well as the expected ortho and para isomers (16). In general, the more drastic the reaction conditions (longer reaction time and high temperature, large amount of catalyst, absence of solvent, etc), the greater is the tendency toward the formation of meta derivatives. Although electron releasing groups such as alkyl, amino, etc, groups on the aromatic ring direct electrophiles preferentially to the ortho and para positions, the meta isomers are usually the thermodynamically favored products, and hence are formed via secondary isomerization particularly under more forcing conditions.

Shape selective catalysts, such as Zeolites of the H-ZSM-5 type, are capable of directing alkyl groups preferentially to the para position (18). The ratio of the catalyst to the substrate also plays a role in controlling the regiochemistry of the alkylations. For example, selective alkylation of anilines at the para position is achieved using alkylating agents and $AlCl_3$ in equimolar ratio (19).

Isomerization. In the course of alkylation two different types of isomerizations can take place. The first involves the alkylating agent which may undergo rearrangement and is generally converted to a more highly branched isomer. Thus the alkylation of benzene with n-propyl chloride in the presence of aluminum chloride yields n-propylbenzene and isopropylbenzene (cumene), and with n-butyl chloride sec-butylbenzene is formed. Under the usual conditions of a Friedel-Crafts alkylation, the alkyl group rearrangement takes place via carbocations favoring the formation of more stable highly branched species, ie, primary → secondary → tertiary.

Alternatively the alkylated aromatic products may rearrange. n-Butylbenzene [104-57-8] is readily isomerized to isobutylbenzene [538-93-2] and sec-butylbenzene [135-98-8] under the catalytic effect of Friedel-Crafts catalysts. The tendency toward rearrangement depends on the alkylating agent and the reaction conditions (catalyst, solvent, temperature, etc).

Because of isomerization, alkylation of benzene with tertiary alkyl halides can also yield secondary alkylbenzenes rather than only tertiary alkylbenzenes (20). For example, the tert-hexylbenzene, which is first formed by the reaction of benzene with 2-chloro-2,3-dimethylbutane and $AlCl_3$ isomerizes largely to 2,2-dimethyl-3-phenylbutane by a 1,2-CH_3 shift. With ferric chloride as the catalyst, tert-hexylbenzene does not undergo isomerization and is isolated as such.

The isomerization of dialkylbenzenes generally results in a considerable increase of the meta isomer, because as thermodynamically controlled equilibrium is approached the relative ratios of the ortho, meta, and para isomers are predicted by their relative thermodynamic stabilities (21). Similar thermodynamically controlled equilibria are obtained in the treatment of dihalo- (22) and polyhalobenzenes, as well as haloalkylbenzenes (23) with Friedel-Crafts catalyst systems. Aryl groups migrate in a way similar to the migration of alkyl and cycloalkyl groups.

Preformed Carbocationic Intermediates. Propargyl cations stabilized by hexacarbonyl dicobalt have been used to effect Friedel-Crafts alkylation of electron-rich aromatics, such as anisole, *N,N*-dimethylaniline and 1,2,4,-trimethoxybenzene (24). Intramolecular reactions have been found to be regio and stereoselective, and have been used in the preparation of derivatives of 9*H*-fluorenes and dibenzofurans (25).

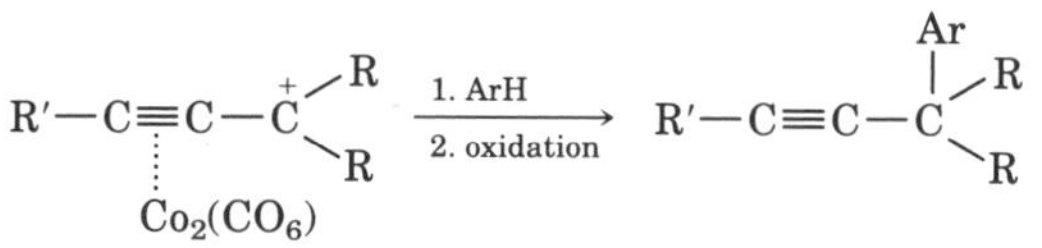

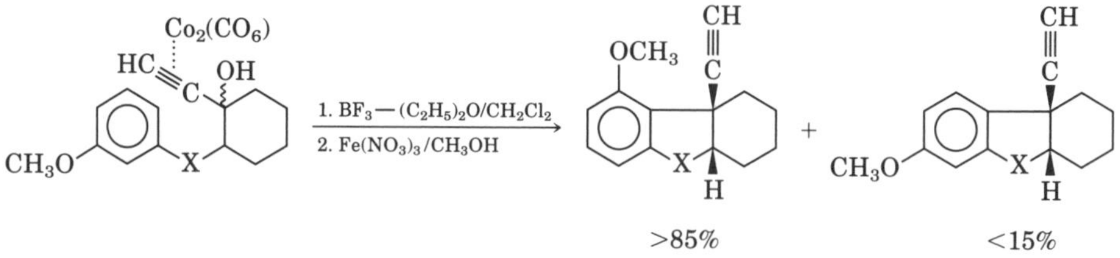

R′—C≡C—C
R
R
Co2(CO6)
1. ArH
2. oxidation
Ar
R′—C≡C—C
R
R

Co2(CO6)
HC
OH
C
CH3O
X
1. BF3—(C2H5)2O/CH2Cl2
2. Fe(NO3)3/CH3OH
CH
C
OCH3
X
H
>85%
+
CH
C
CH3O
X
H
<15%

The Friedel-Crafts alkylation of aromatics with the resonance-stabilized trichlorocyclopropenium triflate offers a synthetic pathway to triaryl cyclopropenium salts (26). The trichlorocyclopropenium ion has also been shown to undergo Friedel-Crafts reaction with alkenes and alkynes to give trivinyl and tri(halovinyl) cyclopropenium ions.

Cl
CF3SO3
Cl
Cl
C6H5X
(X = F, H)
Ar
CF3SO3
Ar
Ar

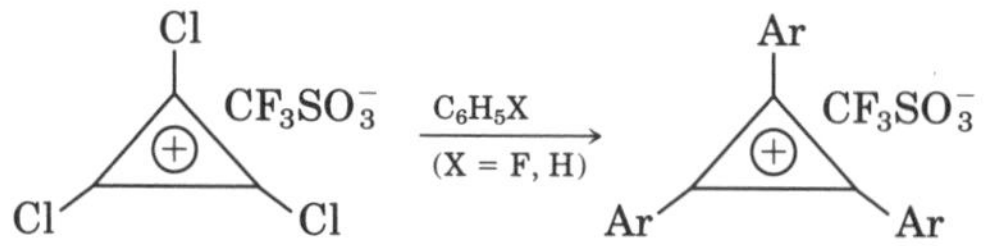

Some destabilized carbocations, although not observed directly, could be generated *in situ* under Friedel-Crafts reaction conditions, for example, the destabilized diethyl malonyl cation (27).

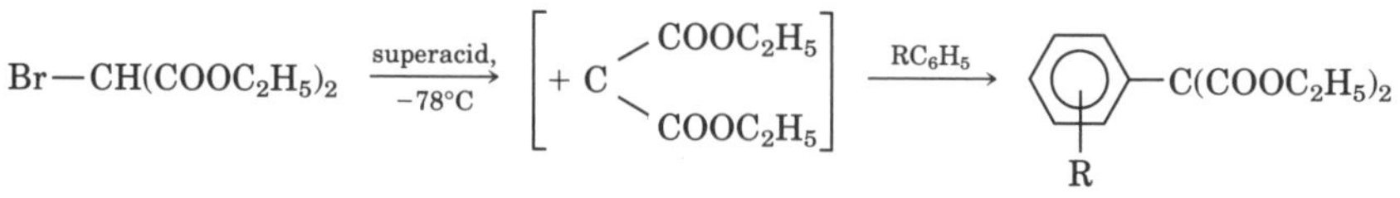

Br—CH(COOC2H5)2
superacid,
−78°C
+ C
COOC2H5
COOC2H5
RC6H5
C(COOC2H5)2
R

Highly unstable vinyl cations, generated *in situ* from vinyl triflates have also been arylated (the triflate group is replaced by the aromatics) to give vinyl aromatics under Friedel-Crafts conditions (28).

Alkylation with Alkanes. Superacids such as $HF-SbF_5$ abstract hydride ion from alkanes to form carbocations at low temperatures. The carbocations generated from C1–C5 alkanes, mixtures of alkanes and alkenes, as well as alkylbenzenes, alkylate benzene to give alkylated aromatics (29). Acid-catalyzed alkylation of benzene with cyclohexane has been brought about by using isobutane as the promoter. The principal products are cycloalkylbenzenes, substituted indanes, or tetralins, and C1–C6 alkylbenzenes (30). Ultrasound induced reactions of 2,2,4-trimethylpentane with benzene in the presence of catalysts such as CF_3SO_3H/SbF_5, $H[B(SO_3CF_3)_4]$, $AlBr_3$, HF/TaF_5, $HF-NbF_5$, and $HF-TiF_4$ proceed to give mainly *t*-butyl and di-*t*-butyl benzenes (31).

Stereoselective Alkylations. Benzene is stereoselectively alkylated with chiral 4-valerolactone in the presence of aluminum chloride with 50% net inversion of configuration (32). The stereoselectivity is explained by the coordination of the Lewis acid with the carbonyl oxygen of the lactone, resulting in the S_{N_2} type displacement at the C—O bond. Partial racemization of the substrate (incomplete inversion of configuration) results by internal return of the ion-pair complex with concomitant rotation of the C—C bond.

Both 2-methyltetrahydrofuran [*96-47-9*] (32) and 3-chlorobutanoic acid [*1951-12-8*] (33) alkylate benzene with inversion of configuration. *N*-benzoyl-4-mesyloxy-L-proline reacts with benzene to give the corresponding 4-phenyl derivative with inversion of configuration (34).

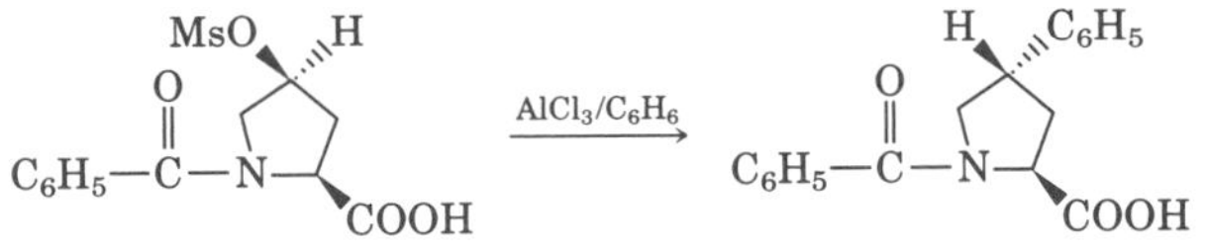

2-Methyloxetane [*2167-39-7*] alkylates benzene to give 3-phenyl-1-butanol:

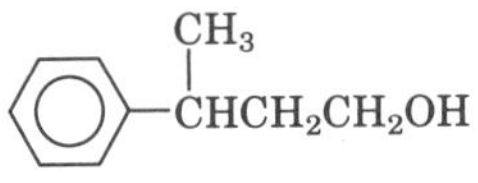

with 20% inversion of configuration, using $AlCl_3$ or $TiCl_4$ as catalyst, whereas partial retention was observed with $SnCl_4$ (35). $AlCl_3$ catalyzed alkylation of benzene with enantiomerically pure 2-chloro-1-phenylpropane or 1-chloro-2-phenylpropane results in the formation of 1,2-diphenylpropane and 1,1-diphenylpropane, the former being formed with 45–100% retention of configuration. A π-phenyl assisted cation intermediate, with unsymmetrical bridging was proposed to account for the lack of racemization.

Chiral glyoxylates have been used to effect *ortho*-hydroxyalkylation of phenols via coordinative complexes. In this way, optically active 2-hydroxymandelic esters have been obtained with up to 94% diastereoselectivity (36).

R* = menthyl, etc; R = 3-t-butyl, 4-methoxy, etc

Stereoselectivity was also observed in the Friedel-Crafts reaction of optically active phenyloxirane with toluene and anisole. The product diarylethanol had an enantiomeric ratio of 60:40 (37).

Synthetic utility of stereoselective alkylations in natural product chemistry is exemplified by the preparation of optically active 2-arylglycine esters (38). Chirally specific α-amino acids with methoxyaryl groups attached to the α-carbon were prepared by reaction of the dimethyl ether of a chiral bis-lactam derivative with methoxy arenes. Using SnCl$_4$ as the Lewis acid, enantioselectivities ranging from 65 to 95% were obtained.

A combination of $C_{p2}ZrCl_2$–AgClO$_4$ (where C_p = cyclopentadienyl) effectively promotes the Friedel-Crafts coupling of glycosyl fluorides with aromatic compounds, such as trimethoxybenzene or methoxynaphthalenes. The derived C-aryl glycosides are potent antitumor agents (39).

Trimethylsilyl triflate–silver perchlorate (1:1) also catalyzes the highly stereoselective glycosylation of β-naphthol to provide predominantly the β-anomer.

Optically active 2-arylalkanoic acid esters have been prepared by the Friedel-Crafts alkylation of arenes with optically active α-sulfonyloxy esters (40). Friedel-Crafts alkylation of benzene with (S)-methyl 2-(chlorosulfonyloxy)- or 2-(mesyloxy)propionate proceeded with predominant inversion of configuration (>97%) to give (S)-methyl 2-phenylpropionate.

where X = Cl or CH$_3$

In view of the ready availability of optically pure lactic acid derivatives this reaction offers an attractive general method for the preparation of optically pure aromatic ester derivatives (41). Stereoselective alkylation (15–60% inversion) of benzene with optically active 1,2- 1,3- and 1,5-dihaloalkanes was also reported (42).

Haloalkylation. Haloalkyl groups can be introduced directly by processes similar to Friedel-Crafts alkylation into aromatic and, to some extent, aliphatic compounds. Because haloalkylations involve bi- or polyfunctional alkylating agents, they must be performed under conditions that promote the initial haloalkylation but not, to any substantial degree, subsequent further alkylations with the initially formed haloalkylated products.

The most frequently used haloalkylating agents are: aldehydes and hydrogen halides, haloalkyl ethers, haloalkyl sulfides, acetals and hydrogen halides, di- and polyhaloalkanes, haloalkenes, haloalcohols, haloalkyl sulfates, haloalkyl *p*-tosylates, and miscellaneous further haloalkyl esters. Haloalkylations include halomethylation, haloethylation, and miscellaneous higher haloalkylations. Under specific conditions, bis- and polyhaloalkylation can also be achieved.

Haloalkylations are accompanied by further alkylation by the initially formed haloalkylated product, yielding diarylalkanes or cyclialkylated products, eg, benzene reacts with CCl_4 in the presence of $AlCl_3$ to give $C_6H_5CCl_3$ and $(C_6H_5)_2CCl_2$ (43). With dichloromethane, the initially formed benzyl chloride is further alkylated to diphenylmethane.

Acidic halide catalysts used in chloromethylation include (44) $ZnCl_2$, $ZnCl_2 + AlCl_3$, $SnCl_4$, $SnCl_2$, $AlCl_3$ + ketones, $AlCl_3$ + pyridine, $AlCl_3$ + tertiary amines, $AsCl_3$, $FeCl_3$, BF_3, $TiCl_4$, TiF_4, $BiCl_3$, $SbCl_3$, and $SbCl_5$. The most widely used protic acid catalysts are HCl, H_2SO_4, H_3PO_4, $H_2SnCl_6 \cdot 6H_2O$, $ClSO_3H$ + CH_3OH, CH_3COOH, as well as benzene-, naphthalene-, and *p*-toluenesulfonic acids (44). Zinc chloride is probably the most frequently used catalyst. Its activity is sometimes increased by fusion with a small amount of aluminum chloride. In other instances, however, sufficient catalytic effect is obtained with a mineral acid alone.

Reactions have been brought about under heterogeneous conditions with an excess of aromatic compound serving as the solvent, and under homogeneous conditions, as in glacial acetic acid solution (45). Solvents used are diethyl ether, dioxane, dimethoxymethane, carbon tetrachloride, chloroform, ethylene chloride, perchloroethylene, nitrobenzene, ligroin (light petroleum distillate), and carbon disulfide.

Chloromethylation can be efficiently effected using chloromethyl methyl ether, in the presence of Lewis acids. However, due to its carcinogenic nature, extreme care should be exercised in its use (46). Alternatively, paraformaldehyde–HCl/$ZnCl_2$ can be used as the chloromethylating agent.

Selective monohaloalkylations were achieved when chlorobromoalkanes such as 1-chloro-3-bromo-3-methylbutane were used as haloalkylating agents (47). The bromine is replaced preferentially:

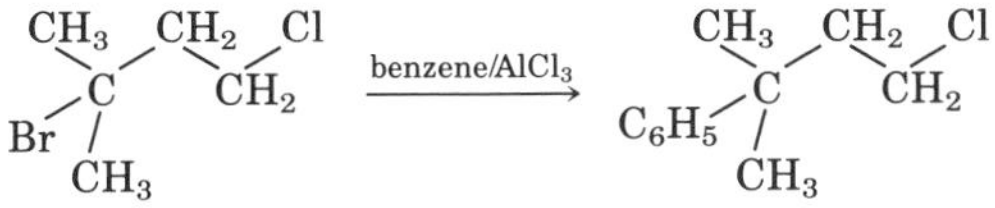

Fluorochloro, fluorobromo, and fluoroiodoalkanes react selectively with aromatics under boron trifluoride catalysis to provide chloro-, bromo- and iodoalkylated products (48). The higher reactivity of the C—F bond over C—Cl, C—Br, and C—I bonds under Lewis acid catalysis results in the observed products.

Alkenyl halides have also been used for haloalkylation reactions in the presence of protic acids (49).

The antifungal agent clotrimazole [23593-75-1] has been prepared by the Friedel-Crafts reaction of o-chlorobenzotrichloride [2136-89-2] and benzene in the presence of AlCl$_3$, followed by treatment with imidazole (50).

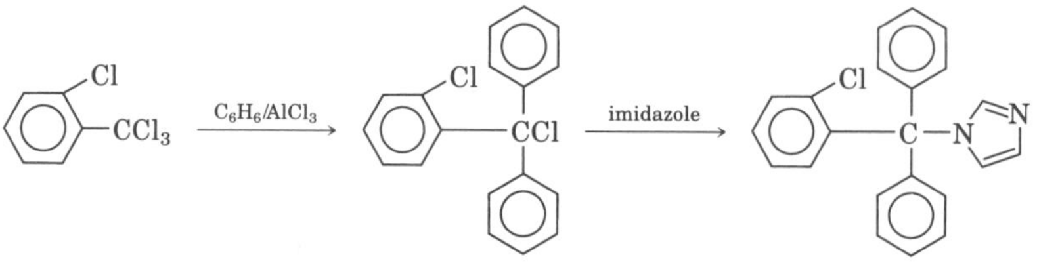

Fluoroalkylations are frequently performed indirectly using tandem reactions. Arenes react with sodium borohydride in trifluoroacetic acid to afford otherwise difficult to obtain 1,1,1-trifluoro-2,2-diarylethanes. Presumably sodium borohydride reacts initially with the trifluoroacetic acid to produce the trifluoroacetaldehyde or its equivalent, which rapidly undergoes Friedel-Crafts-type condensation to give an intermediate carbinol. The carbinol further alkylates benzene under the reaction conditions giving the observed product. The reaction with sterically crowded arenes such as mesitylene and durene proceeds only up to the carbinol stage (51). Although trichloroacetaldehyde is known to give similar products, the protic acid catalyzed Friedel-Crafts reactions of trifluoroacetaldehyde (fluoral) with arenes have not yet been reported. Trifluoroalkylated aromatics are also prepared using tris-[trifluoroacetoxy]borane in trifluoroacetic acid (52).

$$ArH \xrightarrow[\text{2. NaOH/H}_2\text{O, 0°C}]{\text{1. CF}_3\text{COOH/NaBH}_4} ArCH(CF_3)OH \xrightarrow{ArH} ArCH(CF_3)Ar$$

Cycloalkylation. Friedel-Crafts alkylation of aromatics with difunctional compounds leads partly to open-chain akylates but, if conditions are suitable, mainly to cyclic products (53). These compounds react at each end of the difunctional open chain, and thus attach a new ring to the aromatic nucleus. The difunctional alkylating agents first form monofunctional alkyl derivative with the aromatic compound which undergoes subsequent intramoleuclar alkylation forming the new fused ring. Extensive intermolecular reactions between the bifunctional alkylating agent and the aromatics can, however, also take place.

Ring formation readily occurs in the alkylation of aromatics with di- and polyhalides, eg, the reaction of di- and trihalomethanes with aromatics in the presence of aluminum chloride. In the reaction of dichloromethane and benzene, besides diarylmethanes, anthracene derivatives are also formed (54).

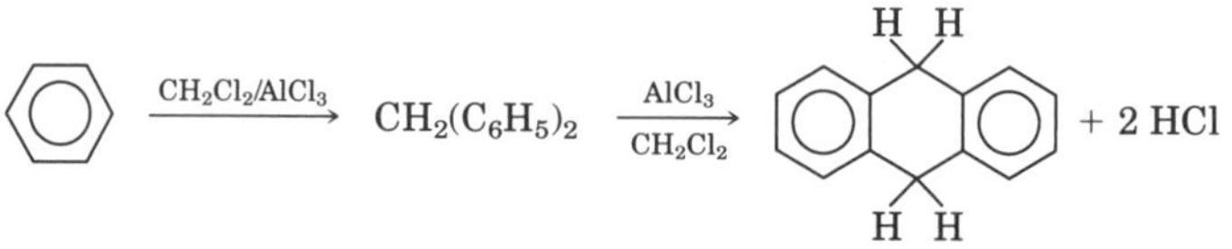

9,10-dihydroanthracene [613-31-0]

Cycloalkylation of aromatic compounds with 1,4- or 1,3-glycols in the presence of aluminum chloride leads to the formation of derivatives of tetrahydronaphthalene or indane, respectively (55).

1-methyl-1,2,3,4-tetrahydronaphthalene

Dienes can also be used in Friedel-Crafts cycloalkylations. For example, treatment of phenol with 2,5-dimethyl-2,4-hexadiene gives 5,5,8,8-tetramethyl,6,7-dihydro-2-naphthol.

Intramolecular alkylation of haloalkylbenzenes is perhaps the simplest intramolecular counterpart of the classical Friedel-Crafts alkylation reaction. Primary, secondary, or tertiary phenylalkyl chlorides undergo ring closure. Compounds of the general formula $C_6H_5(CH_2)_nCl$ cyclize preferentially to give six-membered ring compounds. When a tertiary carbocation intermediate is involved, five-membered rings can be formed:

1,1-dimethylindane [4912-92-9]

Aromatic hydrocarbons with an unsaturated side chain undergo ring closure when heated with Lewis acids (56).

Alkylbenzenes containing an α-tertiary hydrogen, such as in isopropyl and *sec*-butyl side chains, rapidly transfer this hydrogen as a hydride ion producing an α-carbocation analogous to that formed in the protonation of styrene compounds. Subsequent alkylation of an olefin followed by ring closure produces indanes by much the same process. *p*-Cymene [99-87-6], on treatment with a branched olefin, such as methylcyclohexene, or trimethylethylene in the presence of an acid catalyst, yields 1,3,3,6-tetramethyl-1-*p*-tolylindane:

Synthetic and mechanistic studies of suitably substituted aryl and alkyl carbinols to give octahydrophenanthrenes have been reported (57).

Superacids such as HF-SbF$_5$ effect cycloalkylation of aryl alkyl ketones to give tetralone derivatives (58). Tandem intramolecular cycloalkylations can be achieved when functional groups are located in close proximity (59).

Friedel-Crafts metallocyclization of (halogenomethyl(aryl)phosphine) platinum(II) complexes in the presence of triphenylphosphine gives a cationic metallacyclic species (60).

Synthetic and mechanistic aspects of intramolecular cyclization in the tricyclic diterpenoid area have been studied in detail. In general, the presence of electron withdrawing groups such as carbonyl in the side chain retard the rates of cyclization (61).

Epoxide-opening intramolecular cyclizations, in light of their importance in natural product chemistry, have received much attention (62).

Double Friedel-Crafts alkylation of configurationally pure pyrocene (a substituted lactone) with aromatics results in the formation of cycloalkylation product with retention of configuration at the chiral center (63).

Tertiary 3- and 4-phenylalkanols undergo cycloalkylation readily with H$_2$SO$_4$. Primary and most secondary 3-phenylalkanols, however, do not, even at higher temperatures with H$_3$PO$_4$ (64). Cyclization of phenylalkyl alcohols to tetrahydronaphthalenes and indanes has been extensively investigated (65).

Arylation of Aromatic Compounds.   In contrast to Friedel-Crafts alkylations, arylations of aromatics are not as well known, and usually require drastic conditions. They include: (1) dehydrogenating condensation (Scholl reaction); (2) arylation with aryl halides; and (3) arylation with diazonium halides.

The elimination of two ring hydrogens accompanied by the formation of an aryl–aryl bond under Friedel-Crafts conditions is known as the Scholl reaction (1). The dehydrogenating condensations can take place by either inter- or intramolecular pathways. Intermolecular Scholl reactions are numerous and include such reactions as formation of biphenyl from benzene, of perylene from naphthalene (through binaphthyl), of 2,2'-dipyridyl from pyridine, and the formation of high molecular weight polycondensed aromatics.

Aryl halides are rarely used as electrophilic arylating agents in Friedel-Crafts-type reactions. However, under suitable conditions, simple halobenzenes may be used as Friedel-Crafts arylating agents (66). Their decreasing order of reactivity is F >> Cl > Br. Aromatic hydrocarbons are arylated with fluorobenzene in the presence of aluminum halides to the corresponding biphenyls. The directive effects are in agreement with electrophilic aromatic substitution.

In the investigation of the decomposition reaction of aryldiazonium tetrafluoroborates in nitrobenzene, it was found that in addition to fluorobenzene, 3,3'-dinitrobiphenyl was formed (67). An ionic type of arylation reaction seems to take place. Decomposition of aryldiazonium tetrafluoro-, tetrachloro-, and tetrabromoborates in aromatic solvents leads to electrophilic ring arylation (68).

ALKYLATION OF ALIPHATIC COMPOUNDS

The first reported alkylation of branched-chain alkanes by ethylene, over aluminum chloride (69), made it possible to alkylate alkanes (except methane and ethane) with straight chain or branched alkenes.

$$\text{isobutane} \xrightarrow{\text{ethylene}} \text{hexanes} \xrightarrow{\text{ethylene}} \text{octanes} \xrightarrow{\text{ethylene}} \text{decanes}$$

The first industrial aliphatic alkylation process, the sulfuric acid catalyzed alkylation of branched alkanes with alkenes, was developed in 1938 (70). Because of its simplicity, low cost, the wide variety of raw materials, giving high octane alkylate, this process is still widely used. In the presence of 98.5–99.5% sulfuric acid, isobutane $(CH_3)_2CHCH_3$, isopentane $(CH_3)_2CHCH_2CH_3$, and isohexane $(CH_3)_2CHCH_2CH_2CH_3$ are easily alkylated by alkenes, producing alkylates with high octane numbers (see ALKYLATION; CATALYSIS). Anhydrous HF has replaced H_2SO_4 as the catalyst to a certain extent. The suggested reaction pathway is as follows:

$$R_2C{=}CH_2 \xrightarrow{\text{H}^+} R_2\overset{+}{C}{-}CH_3 \xrightarrow{\text{R}_3\text{CH}} R_2CHCH_3 + R_3C^+$$

$$R_3C^+ + R_2C{=}CH_2 \longrightarrow R_2\overset{+}{C}{-}CH_2CR_3 \xrightarrow{\text{H}^-} R_2CHCH_2CR_3$$

Products do not contain 2,2,3-trimethylbutane or 2,2,3,3-tetramethylbutane, which would be expected as the primary alkylation products of direct alkylation of isobutane with propylene and isobutylene, respectively. In fact, the process involves alkylation of the alkenes by the carbocations produced from the isoalkanes via intermolecular hydride abstraction.

On the other hand, under superacidic conditions, alkanes are readily alkylated via front-side σ-insertion by carbocationic alkylating agents. The direct alkylation of the tertiary C—H σ-bond of isobutylene with isobutane has been demonstrated (71). The sterically unfavorable reaction of *tert*-butyl fluoroantimonate with isobutane gave a C_8 fraction, 2% of which was 2,2,3,3-tetramethylbutane:

$$(CH_3)_3C^+ \ + \ HC(CH_3)_3 \ \rightleftarrows \ (CH_3)_3CC(CH_3)_3 \ + \ H^+$$

When using $HF:TaF_5$ in a flow system for alkylation of excess ethane with ethylene (in a 9:1 molar ratio), only *n*-butane was obtained; isobutane was not detectable even by gas chromatography (72). Only direct σ-alkylation can account for these results. If the ethyl cation alkylated ethylene, the reaction would proceed through butyl cations, inevitably leading also to the formation of isobutane (through *t*-butyl cation).

Lower alkanes such as methane and ethane have been polycondensed in superacid solutions at 50°C, yielding higher liquid alkanes (73). The proposed mechanism for the oligocondensation of methane requires the involvement of protonated alkanes (pentacoordinated carbonium ions) and oxidative removal of hydrogen by the superacid system.

Natural gas, or methane itself can be converted into gasoline range hydrocarbons (C_4–C_6) by oxidative condensation over solid superacid catalysts (TaF_5/ fluoridated Al_2O_3, SbF_5/fluorinated graphite, or TaF_5/AlF_3). Gasolines can be further converted into lower boiling (bp <196°C) hydrocarbons (diesel fuel-range hydrocarbons) by contact with a Friedel-Crafts catalyst, composed of high sodium ZSM-5 zeolite-bentonite at high temperatures (74).

The condensation of alkyl halides with olefinic hydrocarbons and haloolefins is catalyzed by Friedel-Crafts catalysts (75). The primary reaction consists of the addition of the alkyl group and the halogen atom to the double bond. In general, the reaction takes place more readily with tertiary than primary and secondary halides. Depending largely upon the particular reactants and the catalyst, the products are mainly alkyl halides containing quaternary carbon atoms, and are obtained in yields up to 75%. These compounds may further be dehydrohalogenated to the corresponding alkylated olefins and haloolefins.

ISOMERIZATION OF SATURATED HYDROCARBONS

The ability of Friedel-Crafts-type superacids to protonate not only π-bonds but also the weaker σ-bonds of hydrocarbons at relatively low temperatures allows the isomerization of saturated hydrocarbons, which is of substantial practical importance (1). Straight-chain alkanes of the gasoline range (C_5–C_8 alkanes), have considerably lower octane numbers than their branched isomers. The isomerization of alkanes can be effected by catalysts such as $AlCl_3$, $AlBr_3$, HF–SbF_5, HF–TaF_5, HSO_3F–SbF_5, and the like. The stronger superacids allow the reaction at relatively low temperatures, which favor the branched alkanes (76). Processes for the isomerization of straight-chain alkanes in the gas phase over acid or zeolite (molecular-sieve) catalysts have been developed (see CATALYSIS; MOLECULAR SIEVES).

Isomerization of straight-chain hydrocarbons is of particular importance for lead-free gasoline. Addition of high octane aromatic hydrocarbons or olefins is

questionable based on environmental considerations (77). An efficient octane en-
hancing additive is methyl *tert*-butyl ether (MTBE).

Friedel-Crafts acids such as $AlCl_3$, $AlBr_3$ (molten), or $AlBr_3$ (in CS_2 or low
boiling hydrocarbon solvents) were found to be useful in the isomerization of cyclic
hydrocarbons to cage hydrocarbons. For example, *endo*- or *exo*-trimethylenenor-
bornane, tetrahydrobinor-*S*, and heptacyclooctadecanes have been converted to
adamantane, diamantane, and triamantane, respectively, in 10–60% yield (78).
Use of superacidic catalysts, such as $B(OSO_2CF_3)_3$ and $CF_3SO_3H–B(OSO_2CF_3)_3$
lead to higher yields ranging from 70 to 90% (79).

Acylation

ACYLATION OF AROMATIC COMPOUNDS

Ketone Synthesis.   In the Friedel-Crafts ketone synthesis, an acyl group is
introduced into the aromatic nucleus by an acylating agent such as an acyl halide,
acid anhydride, ester, or the acid itself. Ketenes, amides, and nitriles also may be
used; aluminum chloride and boron trifluoride are the most common catalysts
(see KETONES).

Reaction of the aromatic (eg, C_6H_6), acyl halide (RCOX), and aluminum halide (AlX_3) liberates hydrogen halide and produces a complex of aromatic ketone
and aluminum halide from which the ketone is liberated by hydrolysis:

$$C_6H_6 + RCOX + AlX_3 \longrightarrow HX + \left[\begin{array}{c} C_6H_5 \\ \diagdown \\ \diagup \\ R \end{array} C{=}OAlX_3 \right] \xrightarrow{H_2O} C_6H_5{-}\overset{\overset{\displaystyle O}{\|}}{C}{-}R$$

Apparently a molar equivalent of catalyst ($AlCl_3$) combines with the acyl
halide, giving a 1:1 addition compound, which then acts as the active acylating
agent. Reaction with aromatics gives the $AlCl_3$ complex of the product ketone
liberating HX:

$$RCOX + AlX_3 \rightarrow RCOX{\cdot}AlX_3 \rightleftharpoons RCO^+AlX_4^-$$

$$ArH + RCO^+ AlX_4^- \rightarrow ArCOR{\cdot}AlX_3 + HX$$

Evidence supporting the formation of 1:1 addition compounds is substantiated by
the actual isolation of stable acyl halide–Lewis acid complexes.

If the catalyst gets firmly attached to the resulting ketone, it is no longer
able to activate more acyl halide (80). However, halogen exchange between acyl
and aluminum halides continues in the presence of excess ketone. Apparently,
the presence of ketone does not prevent formation of acyl cation but lowers its
efficiency as an acylating agent. Similarly, the ability of certain solvents such as
nitrobenzene, nitromethane, and carbon disulfide, to modify Friedel-Crafts acylation reactions is related to their ability to complex acyl cations, thus reducing
their electron deficiency.

A further consequence of association of acylating agents with basic compounds is an increase in the bulk of the reagent, and greater resistance to attack at the more sterically hindered positions of aromatic compounds. Thus acylation of chrysene and phenanthrene in nitrobenzene or in carbon disulfide occurs to a considerable extent in an outer ring, whereas acylation of naphthalene leads to extensive reaction at the less reactive but sterically less hindered 2-position.

Friedel-Crafts acylation of aromatics is of considerable practical value owing to the importance of aryl ketones and aldehydes. Acylating agents in general are more reactive than alkylating agents, and the reaction of acyl halides with aromatic compounds in the presence of a Friedel-Crafts catalyst proceed readily. These reactions are also successful with aromatic amino derivatives, although as a general rule, electronegative groups have an inhibiting effect. Usually, it is difficult to introduce more than one acyl group into an aromatic ring, although some diacylation products are reported in the literature. A high yield of diacetyl-mesitylene was obtained when mesitylene was acetylated with a sixfold excess of catalyst and using the Elbs method (81). In another study (82), 63% conversion to the diacylated product was obtained in the acetylation of fluoroanthrene. Reaction of benzene with 2,6-dimethoxybenzoyl chloride gives products indicating acylation of the ring of the acyl chloride as well as acylation of benzene, yielding 2,6-dimethoxybenzophenone [*25855-75-8*] (83).

In the Friedel-Crafts acylation of ortho–para directing aromatic compounds para derivatives are formed, in most cases, preferentially over the ortho isomers. The reason for this behavior was long attributed to steric factors, although exceptions were known (84). In a systematic study of the effect of substituents in acylating agents on the selectivity of Friedel-Crafts acylation (85), the AlCl$_3$-catalyzed benzoylation of toluene with benzoyl chloride gave 8% 2- and 91% 4-methylbenzophenone. In contrast, acylation with 2,6-dinitrobenzoyl chloride gave 42% ortho isomer and 55% para isomer. These results show that the relative reactivity of the electrophilic acylating agents affects the isomer distribution (86).

1-Phenylnorbornane is benzoylated faster than isopropylbenzene or toluene despite the bulkiness of the norbornyl group probably because of hyperconjugation (87). Hyperconjugation of the C—C bond is at least as or more important as that of the C—H bond since 1-phenylnorbornane has no α-hydrogen atom.

Industrial Applications. Perfluoroacylbenzene sulfonates, used as additives in fire-extinguishing compounds and galvanizing baths have been prepared (88). Perfluoroacylbenzenesulfonate salts prepared by Friedel-Crafts reaction of perfluoroacyl halides and benzene, and subsequent sulfonation have been used as surfactants (89).

1-(5-Chloro-6-methoxy-2-naphthyl)-1-propanone has been prepared by Friedel-Crafts reaction of propanoyl chloride with 1-chloro-2-methoxy naphthalene, and is used as an intermediate for the antiinflammatory/analgesic naproxen [22204-53-1] (90) (see ANALGESICS, ANTIPYRETICS, AND ANTIINFLAMMATORY AGENTS).

Trihaloacyl aromatics have been prepared by Friedel-Crafts acylation of aromatics with CX_3COCl (X = Cl, Br) in the presence of $AlCl_3$. They are used as monomers in the preparation of polycarbonates, polyesters, polyamides, polyketones, and polyurethanes (91).

Friedel-Crafts acylation of phenyl methyl thioether with 3-methylbutanoyl chloride gives 4-acetylated thioether, which is used as an intermediate for antihypertensive drugs (92).

Dihydroxytetrahydronapthacenedione derivatives, used as intermediates for the anthracycline antibiotics have been prepared by Friedel-Crafts reaction of tetralin derivatives with orthophthaloyl chloride [88-95-9] in high yields (93).

Friedel-Crafts acylation of 3,3-dimethyl-2-indolinone by succinic anhydride gives 3,3-dimethyl-5-(3-carboxypropionyl)-2-indoline, which is used as an intermediate in the preparation of inotropic agents for treatment of heart failure (94). Antibacterial phlorophenone derivatives have been prepared by Friedel-Crafts acylation with propanoyl chloride (95).

Friedel-Crafts reaction of p-methoxybenzene with γ-butyrolactone gives the dimethoxytetralone, which serves as an intermediate for anthracyclinones, such as daunomycinone [21794-55-8] (96).

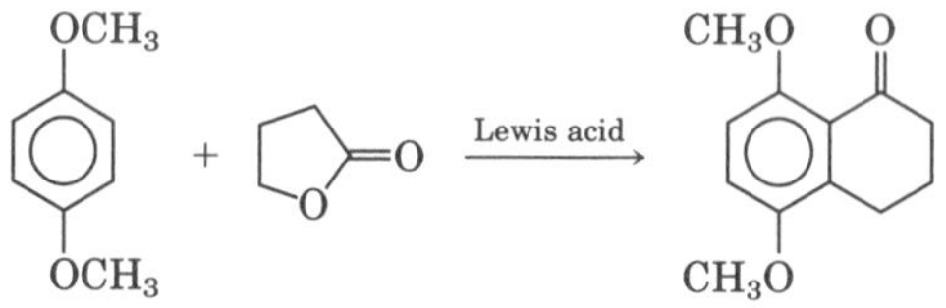

1-Naphthylacetic acid derivatives, showing antiinflammatory, analgesic, and antipyretic activities are prepared by Friedel-Crafts acylation of methyl 1-naphthyl acetate at the 4 position with $(CH_3)_2CHCOCl$ followed by Clemmensen reduction (97).

Friedel-Crafts acetylation of cyclohexylbenzene gave 4-cyclohexylacetophenone which was used as an intermediate for the preparation of 2,4-dioxo-4-substituted-1-butanoic acid derivatives useful in treating urinary tract calcium oxalate lithiasis (98).

Acyl fluoride−Lewis acid complexes are stable acylium (oxocarbonium) salts that can be used advantageously in a modification of the Perrier synthesis (99). The reaction of acyl chlorides or bromides with anhydrous complex fluoro-silver salts also leads to the preparation of acylium salts that are highly reactive acylating agents. Because these metathetic reactions are brought about in acid-free media (the only acid present is by-product conjugate acid formed in the substitution reaction with aromatics), they are applicable to systems otherwise sensitive to Lewis or Brønsted acid catalysts.

$$RCOCl + AgSbF_6 \rightarrow RCO^+SbF_6^- + AgCl$$

Stereoselective Acylations.   Intramolecular Friedel-Crafts acylation reaction of N-aralkyl α-amino acid derivatives gives cyclic ketones with high enantioselectivity (100). This methodology has been used for the enantiospecific syntheses of tylophorine [482-20-2] and cryptopleurine [87302-53-2], the principal representatives of phenanthroindolizidine and phenanthroquinolizidine alkaloids (qv) (101).

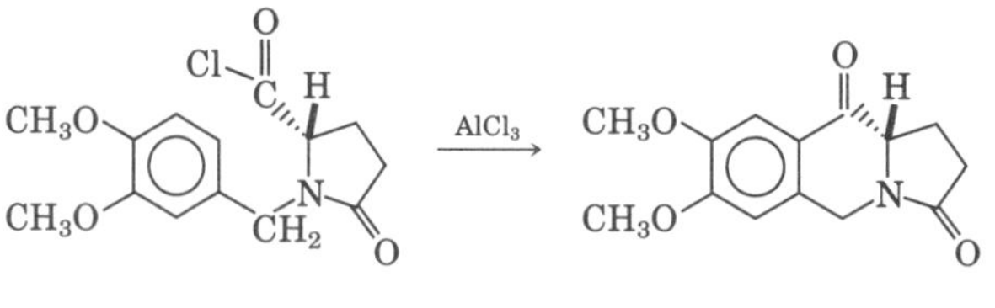

The reaction can be used in the large-scale production of the optically active amino acid derivatives. The chirality of the α- carbon is substantially retained and resolution of the product is avoided.

N-Trifluoroacetylamino acid chlorides also undergo intermolecular Friedel-Crafts acylation reaction with complete preservation of chirality to provide similar natural products (102,103).

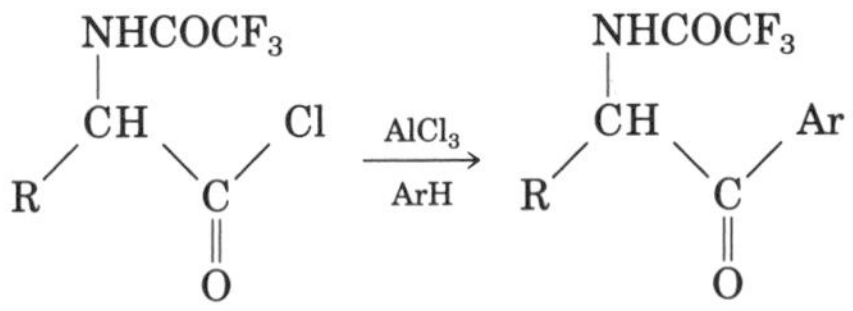

Similarly, N-carboxy-α-amino acid anhydrides react with aromatics such as toluene, xylenes, and mesitylene to give α-amino acylated products in moderate yields with almost complete retention of configuration of the α-amino acid.

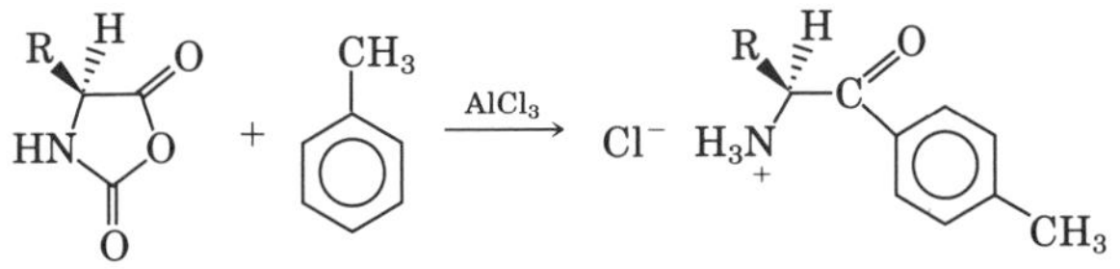

The N-carboxyl group is lost during the reaction, and no additional deprotection step is required (104). Benzene reacts with N-carboxyglycine anhydride to give aminomethyl phenyl ketone; however, it does not react with other N-carboxy-α-amino acid anhydrides (105).

Cycloacylation. Cyclic ketones can be prepared by intramolecular Friedel-Crafts acylation of an aromatic ring that has an acyl halide group in an attached side chain (1). The method is used in the preparation of hydrindones, tetralones, chromones, xanthones, etc. Cyclization of ketoacyl chlorides leading to the formation of coumarandiones, chromandiones, indandiones, and anthraquinones are further examples of intramolecular cycloacylations; five- and six-membered ring ketones are generally obtained in good yield by this method. Polyphosphoric acid (PPA) is often employed as a catalyst for cycloacylations.

In an alternative procedure the acids themselves undergo cyclization catalyzed by hydrogen fluoride. γ-Phenylbutyric acid [*36541-31-8*] can be cyclized to α-tetralone [*29059-07-2*] by this method:

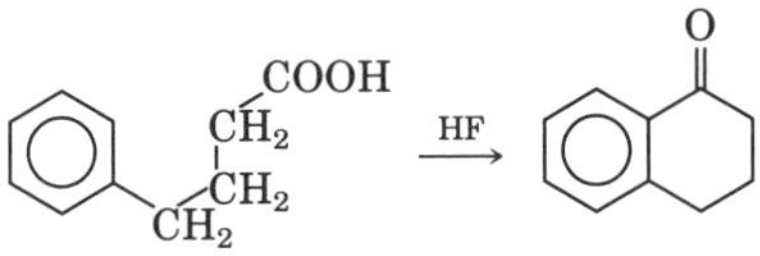

Cycloacylations readily take place in intermolecular acylations involving bifunctional acylating agents. Both functional groups may be acyl (as in the case of α,ω-diacyl halides) or one may be an alkylating group (as in unsaturated acyl halides or certain haloacyl halides) (18).

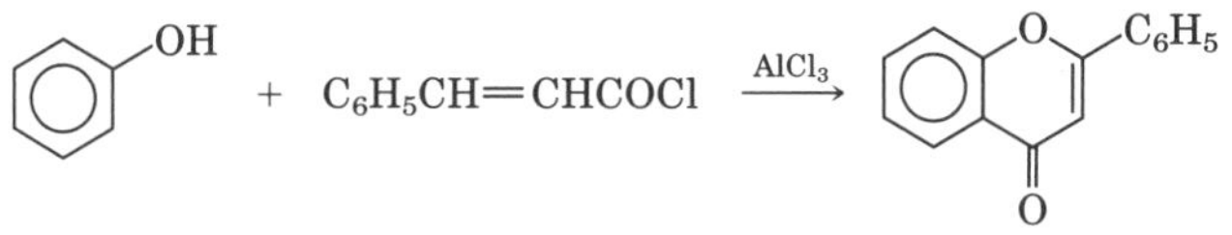

Polystyrene can be cross-linked by its acylation with bifunctional acylating agents such as adipoyl, sebacoyl, or malonyl chlorides in the presence of $AlCl_3$ in CS_2 solution at 0°C (106).

Aldehyde Synthesis.   Formylation would be expected to take place when formyl chloride or formic anhydride reacts with an aromatic compound in the presence of aluminum chloride or other Friedel-Crafts catalysts. However, the acid chloride and anhydride of formic acid are both too unstable to be of preparative interest.

In the presence of aluminum chloride and a small amount of cuprous halide, a mixture of hydrogen chloride and carbon monoxide serves as a formylating agent of aromatics (Gattermann-Koch reaction) (107):

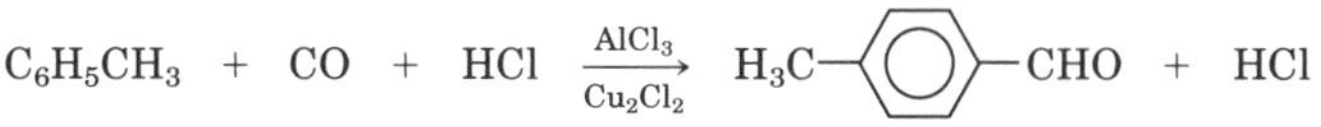

Intermediate formation of formyl chloride is not necessary since the actual alkylating agent, HCO^+, can be produced by protonation of carbon monoxide or its complexes. However, it is difficult to obtain an equimolar mixture of anhydrous hydrogen chloride and carbon monoxide. Suitable laboratory preparations involve the reaction of chlorosulfonic acid with formic acid or the reaction of benzoyl chloride with formic acid:

$$HCOOH + ClSO_3H \rightarrow H_2SO_4 + CO + HCl$$
$$HCOOH + C_6H_5COCl \rightarrow C_6H_5COOH + CO + HCl$$

The Gattermann-Koch synthesis is suitable for the preparation of simple aromatic aldehydes from benzene and its substituted derivatives, as well as from polycyclic aromatics. The para isomers are produced preferentially. Aromatics with meta-directing substituents cannot be formylated (108).

Formylation of aromatics using superacid catalyst systems such as $HF-BF_3$ (109), $HF-SbF_5$ (110), $HF-CF_3SO_3H-BF_3$ (111), and CF_3SO_3H (112), and $FSO_3H–SbF_5$ (113) has been achieved in good yields. Formylation of alkylbenzenes using $SbF_5–FSO_3H$ resulted in the formation of alkylbenzaldehydes and formylalkylbenzenesulfonyl fluorides. The composition of the products was dependent on the acidity of the medium. At high acidities, the formyl derivative was the primary product, whereas lowering of the acidity by decreasing the amount of SbF_5 resulted in sulfonyl compounds.

Aromatics containing electron releasing groups such as phenols, dimethyl-aminobenzene and indole are formylated by 2-ethoxy-1,3-dithiolane in the presence of boron trifluoroetherate, followed by hydrolysis (114). The preformed dithiolanium tetrafluoroborate also undergoes Friedel-Crafts reaction with aromatics such as dimethylaminobenzene and indole (115), and was used to generate dithiolanium derivatives (formyl precursors) from the enoltrimethylsilyl ether derivatives (116).

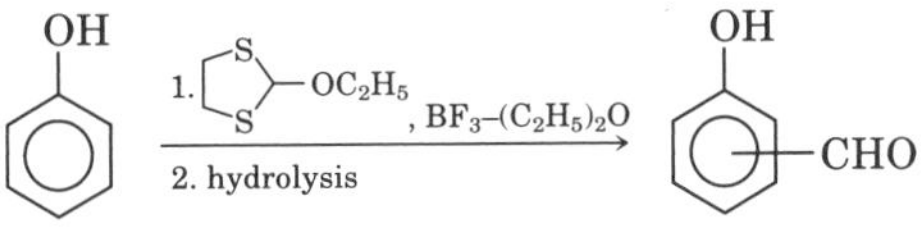

Whereas the above reactions are applicable to activated aromatics, deactivated aromatics can be formylated by reaction with hexamethylenetetramine in strong acids such as 75% polyphosphoric acid, methanesulfonic acid, or trifluoroacetic acid to give salicylaldehyde derivatives (117). Formyl fluoride (HCOF) has also been used as formylating agent in the Friedel-Crafts reaction of aromatics (118). Formyl fluoride [*1493-02-3*] in the presence of BF_3 was found to be an efficient electrophilic formylating agent, giving 53% para-, 43% ortho- and 3.5% meta-toluic aldehydes upon formylation of toluene (110).

Attempts to use acetic-formic anhydride with Friedel-Crafts catalysts resulted only in acetylation. However, using anhydrous HF as a catalyst, a small amount of aldehyde is also formed in accordance with the fact that acetic–formic anhydride gives both acetyl and formyl fluoride with HF. By continuous removal of the low boiling HCOF, the reaction can be shifted to the formation of this compound (118).

In the presence of strongly acidic media, such as triflic acid, hydrogen cyanide or trimethylsilyl cyanide formylates aromatics such as benzene. Diprotonotated nitriles were proposed as the active electrophilic species in these reactions (119).

Friedel-Crafts acylation using nitriles (other than HCN) and HCl is an extension of the Gattermann reaction, and is called the Houben-Hoesch reaction (120–122). These reactions give ketones and are usually applicable to only activated aromatics, such as phenols and phenolic ethers. The protonated nitrile, ie, the nitrilium ion, acts as the electrophilic species in these reactions. Nonactivated benzene can also be acylated with the nitriles under superacidic conditions 95% trifluoromethanesulfonic acid containing 5% SbF_5 ($H_0 > -18$) (119). A dicationic diprotonated nitrile intermediate was suggested for these reactions, based on the fact that the reactions do not proceed under less acidic conditions. The significance of dicationic superelectrophiles in Friedel-Crafts reactions has been discussed (123,124).

Aromatic and heterocyclic compounds are formylated by reaction with dialkyl- or alkylarylformamides in the presence of phosphorus oxychloride or phosgene (Vilsmeier aldehyde synthesis) (125). The Vilsmeier reaction is a Friedel-Crafts type formylation (126), since the intermediate cation formed by the interaction of phosphorus oxychloride with formamide is a typical electrophilic reagent. Ionic addition compounds of formamide with phosgene or phosphorus oxychloride are also known (127).

$$ArH + R_2NCHO \xrightarrow{POCl_3} ArCHO + R_2NH$$

Aromatic compounds are formylated also by dichloromethyl methyl ether or trialkyl orthoformates (128).

Nitrile Synthesis. Cyanogen bromide [506-68-3] condenses with toluene in the presence of aluminum chloride to give *p*-tolunitrile (129).

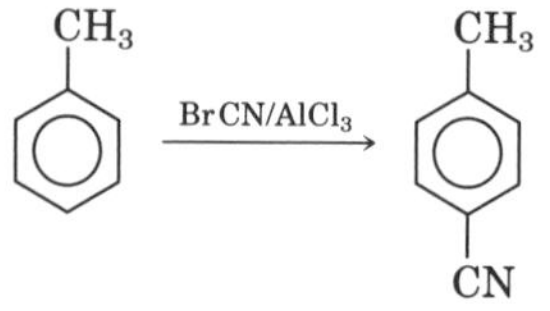

An alternative method consists of the reaction of trichloroacetonitrile [545-06-2] with a hydrocarbon and AlCl$_3$. A ketimine is formed which is hydrolyzed by treatment with potassium hydroxide into the nitrile and chloroform. The reaction proceeds with aromatics such as toluene and phenols (130).

Tertiary alkyl chlorides have been converted to the tertiary nitriles with trimethylsilyl nitrile in dichloromethane in the presence of SnCl$_4$ (131). The reaction was applied to the synthesis of several bridgehead nitriles, such as 1-adamantyl and 1-diamantyl nitriles from the corresponding chloro or bromo derivatives using SnCl$_4$ or AlBr$_3$ catalysts (132).

Preparation of Arylcarboxylic Acids and Derivatives. The general Friedel-Crafts acylation principle can be successfully applied to the preparation of aromatic carboxylic acids. Carbonyl halides (phosgene, carbonyl chloride fluoride, or carbonyl fluoride) [353-50-4] are diacyl halides of carbonic acid. Phosgene [75-44-5] or oxalyl chloride [79-37-8] react with aromatic hydrocarbons to give aroyl chlorides that yield acids on hydrolysis (133):

$$C_6H_6 + COCl_2 \xrightarrow[-HCl]{AlCl_3} C_6H_5COCl \xrightarrow{H_2O} C_6H_5COOH$$

However, excess aromatics can give the corresponding ketones:

$$C_6H_5COCl \xrightarrow[-HCl]{C_6H_6/AlCl_3} C_6H_5COC_6H_5$$

Carbon disulfide as solvent favors the formation of the acid since the intermediate complex formed, C$_6$H$_5$COCl·AlCl$_3$, is insoluble in it and by precipitation avoids secondary ketone formation.

Ketone formation can also be avoided if one of the functional acyl halogens in phosgene is blocked. Carbamyl chlorides, readily obtained by the reaction of phosgene with ammonia or amines, are suitable reagents for the preparation of amides in direct Friedel-Crafts acylation of aromatics. The resulting amides can be hydrolyzed to the corresponding acids (134):

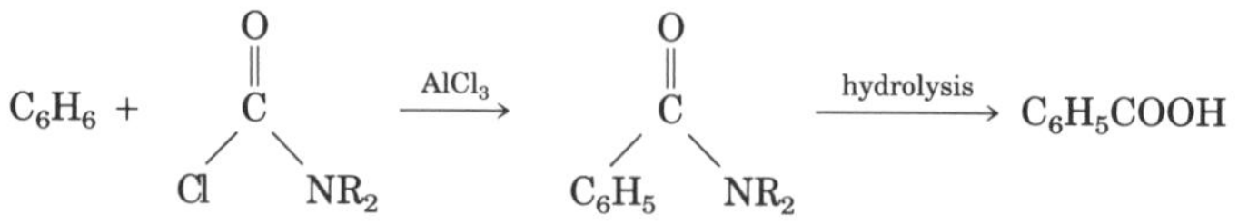

The practical application of this reaction has been demonstrated in the preparation of terephthalic acid from toluene, in which case oxidation follows hydrolysis (135). The reaction also proceeds well with substituted carbamyl chlorides such as $C_6H_5(C_2H_5)NCOCl$.

Amides result from the reaction of aromatic hydrocarbons with isocyanates, such as phenyl isocyanate [103-71-9], in the presence of aluminum chloride. Phenyl isothiocyanate [103-72-0] similarly gives thioanilides (136).

$$C_6H_5NCO + C_6H_6 \xrightarrow{\text{AlCl}_3} C_6H_6NHCOC_6H_5$$

In these reactions the active acylating agent is the carbamyl chloride, formed by the reaction of the isocyanate with hydrogen chloride (137):

$$C_6H_5NCO + HCl \rightarrow C_6H_5NHCOCl$$

Carbon dioxide can be considered the acid anhydride of carbonic acid. Accordingly, it reacts with benzene, albeit in low yield, to give benzoic acid and benzophenone in the presence of aluminum chloride.

With more reactive substances, zinc or ferric chlorides may be substituted as catalysts (138). More elevated temperatures and high pressure are, however, generally needed, and only very reactive substrates (such as phenols) react readily.

ANALOGUES OF ACYLATION REACTIONS

Sulfonylation. Under Friedel-Crafts reaction conditions, sulfonyl halides and sulfonic acid anhydrides sulfonylate aromatics (139), a reaction that can be considered the analogue of the related acylation with acyl halides and anhydrides. The products are sulfones. Sulfonyl chlorides are the most frequently used reagents, although the bromides and fluorides also react:

$$C_6H_6 + RSO_2Cl \xrightarrow{\text{AlCl}_3} C_6H_5SO_2R + HCl$$

Methanesulfonic and benzenesulfonic anhydrides are the most frequently used anhydrides in Friedel-Crafts sulfonylation reactions:

$$C_6H_6 + (CH_3SO_2)_2O \rightarrow C_6H_5SO_2CH_3 + CH_3SO_3H$$

Benzenesulfonic anhydride has been claimed to be superior to benzenesulfonyl chloride (140). Catalysts used besides aluminum chloride are ferric chloride, antimony pentachloride, aluminum bromide, and boron trifluoride (141).

Sulfonation. The general Friedel-Crafts acylation principle can be applied to sulfonation with halides and anhydrides of sulfuric acid (halosulfuric acids, sulfur trioxide). Aluminum chloride and boron trifluoride are effective catalysts in certain sulfonations with halosulfuric acids. In general, sulfonations by sulfuric acid or oleum should also be considered within the scope of Friedel-Crafts acylations, providing both sulfonating agent and catalyst. However, strong Lewis acid catalysts such as $GaCl_3$ are required for the sulfonation of deactivated aromatics, such as nitrobenzene (142).

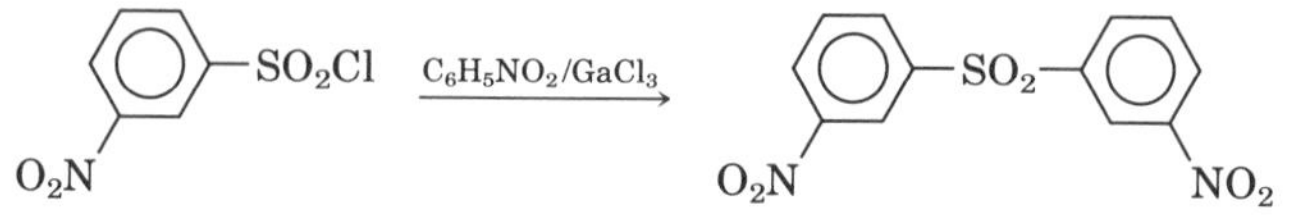

2,4,5,4′-tetrachlorodiphenyl sulfone [*116-29-0*], used as a pesticide has been prepared by the Friedel-Crafts sulfonation of 2,4,5-trichlorobenzenesulfonyl chloride with chlorobenzene (143).

Sulfonation of aromatic hydrocarbons with sulfuric acid is catalyzed by hydrogen fluoride or, at lower temperatures, by boron trifluoride (144). The products obtained are more uniform and considerably less sulfuric acid is needed, probably because BF_3 forms complexes with the water formed in the reaction, and thus prevents dilution of the sulfuric acid.

The migration of alkyl or halogen atoms during the sulfonation of polyalkyl (polyhalo) benzenes containing at least four substituents is known as the Jacobsen rearrangement (145). Thus 2,3,5,6-tetramethylbenzenesulfonic acid is rearranged chiefly to 2,3,4,5-tetramethylbenzenesulfonic acid (see SULFUR COMPOUNDS).

Sulfination. Benzene and its homologues react with SO_2 in the presence of $AlCl_3$ and HCl to form sulfinic acids (146)

$$C_6H_6 + SO_2 \xrightarrow[\text{HCl}]{\text{AlCl}_3} C_6H_5SO_2H$$

Sulfuration. Hydrocarbons, such as cyclohexane and *n*-pentane have been converted to dicyclohexyl and dipentyl sulfides by monoclinic sulfur in the presence of $AlCl_3$, $GaCl_3$, or $FeCl_3$. These reactions are accelerated by the addition of HCl or HBr and give a mixture of products in many instances (147). Alkylbenzenes also react with sulfur in the presence of $AlCl_3$ to give a mixture of sulfurated products consisting mainly of diaryl sulfides, 1,2-dithiol-3-thiones, and substituted thiophenes (148).

Elemental sulfur reacts with alkanes such as cyclopentane in the presence of superacidic trifluoromethanesulfonic acid to give symmetrical dialkyl sulfides in moderate yields.

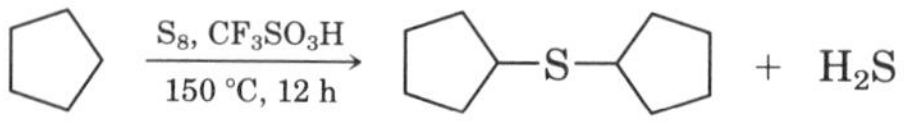

These reactions involve the intermediate formation of thiols, followed by condensation to the sulfides. The observation of isomerized products in suitable cases indicates the intermediate formation of carbocations, either by protolysis of alkanes by the superacid or reversible ionization of the thiol products (149).

Nitration. The general Friedel-Crafts acylation principle can also be applied to nitration involving nitryl halides and dinitrogen pentoxide (the halides and anhydride of nitric acid). In a more general sense, nitration with nitric acid catalyzed by proton acids (H_2SO_4, $HClO_4$, HF, etc) or by Lewis halides (BF_3, $AlCl_3$, etc) should also be considered as Friedel-Crafts-type reactions in analogy to the ketone syntheses where carboxylic acids are used as acylating agents. Friedel-Crafts nitration using nitryl chloride, NO_2Cl, with aluminum chloride as catalyst has been reported (150). Anhydrous silver tetrafluoroborate is also a suitable methatetic cation-forming agent in these reactions (151).

Dinitrogen tetroxide is an effective Friedel-Crafts nitrating agent (152) for aromatics in the presence of aluminum chloride, ferric chloride, or sulfuric acid (153). Dinitrogen pentoxide is a powerful nitrating agent, even in the absence of catalysts, preferably in sulfuric acid solution (154). Solid dinitrogen pentoxide is known to be the nitronium nitrate, $(NO_2)^+(NO_3)^-$. The use of BF_3 as catalyst has been reported (155).

Stable nitronium salts such as $(NO_2)^+(BF_4)^-$, $(NO_2)^+(AsF_6)^-$, $(NO_2^+)_2(SiF_6^{2-})^-$, $(NO_2)^+(HS_2O_7)^-$, etc, are extremely powerful Friedel-Crafts nitrating agents for aromatics (156). Nitronium tetrafluoroborate is easily obtained from N_2O_5, HF, and BF_3, or from nitric acid, HF, and BF_3 (157).

Nitrations can be performed in homogeneous media, using tetramethylene sulfone or nitromethane (nitroethane) as solvent. A large variety of aromatic compounds have been nitrated with nitronium salts in excellent yields in nonaqueous media. Sensitive compounds, otherwise easily hydrolyzed or oxidized by nitric acid, can be nitrated without secondary effects. Nitration of aromatic compounds is considered an irreversible reaction. However, the reversibility of the reaction has been demonstrated in some cases, eg, 9-nitroanthracene, as well as pentamethylnitrobenzene transnitrate benzene, toluene, and mesitylene in the presence of superacids (158) (see NITRATION).

Perchlorylation. The stability of perchloryl fluoride [7616-94-6], $FClO_3$, makes possible perchlorylation of aromatics, a reaction closely related to the Friedel-Crafts acylation (159):

$$C_6H_6 + FClO_3 \xrightarrow{AlCl_3} C_6H_5ClO_3 + HF$$

Aluminum bromide and chloride are equally active catalysts, whereas boron trifluoride is considerably less active probably because of its limited solubility in aromatic hydrocarbons. The perchloryl aromatics are interesting compounds but must be handled with care because of their explosive nature and sensitivity to mechanical shock and local overheating.

Halogenation. The halogenation of a wide variety of aromatic compounds proceeds readily in the presence of ferric chloride, aluminum chloride, and related Friedel-Crafts catalysts. Halogenating agents used are elementary halogens (chlorine, bromine, or iodine), and interhalogens (such as iodine monochloride and bromine monochloride). Although iodination can also be achieved, oxidative conditions must be provided in order to remove the HI formed, which otherwise tends to reduce the ring-iodinated compounds (160). The extent of halogenation is regulated by the amount of halogenating agent used and by reaction conditions. Alkylated benzenes, phenols, phenol ethers, and polynuclear hydrocarbons undergo ring halogenation with ease.

Bromination can be conveniently effected by transfer of bromine from one nucleus to another. As the Friedel-Crafts isomerization of bromoaromatic compounds generally takes place through an intermolecular mechanism, the migrating bromine atom serves as a source of positive bromine, thus effecting ring brominations (161,162). 2,4,6-Tribromophenol, for example, has been prepared by bromination of phenol with dibromobenzene.

Hydroxylation.   The direct synthesis of phenol from benzene is of interest since phenol is widely used in industry and benzene is a relatively cheap starting material for phenol. Electrophilic hydroxylation of benzene with H_2O_2 or peracids, catalyzed by $AlCl_3$, BF_3, etc, usually results in mixtures containing also polyhydroxybenzenes and oxidized products. Using superacidic systems selective monohydroxylation is, however, possible. The phenyloxonium ion formed by protonation of the hydroxy group during the reaction is unreactive towards electrophiles, resulting in no further hydroxylation.

H_2O_2 in the presence of HF/BF_3 acts as an effective and economical reagent for aromatic hydroxylation (163). Hydroxylations of phenols and amines in similar high acidity media are very effective (163). Xylenes were hydroxylated by bis(trimethylsilyl) peroxide and $AlCl_3$ in poor yields (164). Bis(trimethylsilyl) peroxide $(CH_3)_3SiOOSi(CH_3)_3$ can be used with triflic acid (CF_3SO_3H) and acts as an effective hydroxylating agent of aromatics such as toluene, mesitylene and naphthalene (165). Sodium perborate (a safe and inexpensive commercial chemical) can be used in conjunction with the triflic acid to hydroxylate aromatics (166).

Amination.   Direct amination of aromatics can be achieved through the use of $NH_2OH \cdot HCl$, or NH_2OSO_3H in the presence of aluminum chloride or chloramine (NH_2Cl) (167). Usually more than two equivalents of Friedel-Crafts catalysts are employed. The catalysts coordinate with the reagent assisting in the increased polarization of the N—O bond. Although these reactions are electrophilic in nature, it is unlikely that free NH_2^+ is involved. Aromatics have been aminated with *in situ* prepared amino diazonium ion through the reaction of sodium azide and aluminum chloride (168).

Subsequently it was found that trimethylsilyl azide in triflic acid is a more efficient and improved reagent for aminations (169). Amination of toluene in the presence of trichloramine–$AlCl_3$ proceeds predominantly at the *m*-position.

ACYLATION OF ALIPHATIC COMPOUNDS

Similar to alkylation, not only aromatic but also aliphatic and cycloaliphatic compounds undergo Friedel-Crafts acylation reactions.

Olefins and cycloolefins give unsaturated ketones with acyl halides (the Nenitzescu reaction) (170). Saturated chloroketones are formed as intermediates followed by elimination of HCl. Similar products may be obtained by using acid

anhydrides. 1-Methylcyclopentene on acylation with acetyl bromide in the presence of $AlCl_3$ gives 1,3-diacetyl-2-methylcyclopentene and 1-acetyl-2-methylcyclopentene in a ratio of 40:60 (171).

The β-halo ketone intermediates formed in the foregoing reactions arise from the capture of carbocationic intermediates by halide of the gegenions. In some cases, solvents such as acetonitrile can act as the competing nucleophilic species. For example, β-amido ketones could be obtained by the acylation of alkenes in acetonitrile (172).

Using α,β-unsaturated acyl halides, alkenes are acylated to give α,β,α′,β′-unsaturated ketones, which undergo spontaneous intramolecular Nazarov cyclizations to cyclopentenones, important precursors of natural products (173).

Cyclododecene (cis and trans) was similarly transformed to the cyclic enone, which is an important intermediate in the preparation of perfumary products, such as muscone [541-91-3].

Intramolecular Friedel-Crafts acylations of olefins also give cyclic α,β-unsaturated cyclic ketones. Cyclopropane fused bicyclo[5.3.0]octenones, thus obtained, were used in the preparation of the marine sesquiterpenes, africanol [53823-07-7] and dactylol [58542-75-9] (174).

The rather harsh conditions that are used in the acylation of olefins can be avoided in suitable cases by the incorporation of silyl groups. Thus, acylations of trialkylsilylolefins under Friedel-Crafts conditions give α,β-unsaturated ketones by replacement of the trialkylsilyl group by the acyl group. These reactions proceed under mild conditions, and are highly regioselective. The regioselectivity and the high rates of these reactions are due to the hyperconjugative stabilization of the carbocationic intermediates by the β-silicon substituent (175).

2-Methylcyclobutenyltrimethylsilane undergoes Friedel-Crafts acylation affording the 2-methyl-1-acetylcyclobutene [67223-99-8], which was an intermediate in the synthesis of grandisol [26532-22-9] (176).

Allylsilanes undergo highly regioselective acylation to give β,γ-unsaturated ketones (177). Acylation of γ,γ-dialkylallyltrialkylsilane provides a route to the construction of difficultly accessible quaternary carbon centers.

90%

Allylsilanes are more reactive than vinylsilanes in Friedel-Crafts reactions, as shown in the selective acylation of 2,3-disilylalkenes. The allylsilanes, α-silyloxyallyltrialkylsilanes, have been used as enolate equivalents in the preparation of 1,4-diketones (178). The mild reaction conditions required for these reactions tolerate many other functional groups, providing valuable synthetic routes.

Conjugated dienes, upon complexation with metal carbonyl complexes, are activated for Friedel-Crafts acylation reaction at the allylic position. Such reactions are increasingly being used in the stereoselective synthesis of acylated dienes. Friedel-Crafts acetylation of dicarbonyl(η⁴-cyclohexadienetriphenylphosphine iron) proceeded under mild conditions in near quantitative yield to give dicarbonyl(η^4-5-*endo*-acetyl) cyclohexa-1,3-dienetriphenylphosphine iron (179).

$$CH_3-\overset{O}{\underset{\|}{C}}-Cl \; + \quad \overset{Fe(CO)_2P(C_6H_5)_3}{\text{⬡}} \quad \xrightarrow{AlCl_3} \quad \overset{Fe(CO)_2P(C_6H_5)_3}{\underset{\overset{|}{COCH_3}}{\text{⬡}}}$$

Chiral diene–iron tricarbonyl complexes were acylated using aluminum chloride to give acylated diene–iron complexes with high enantiomeric purity (>96% ee). For example, *trans*-piperylene–iron tricarbonyl reacted with acyl halides under Friedel-Crafts conditions to give 1-acyl-1,3-pentadiene–iron tricarbonyl complex without any racemization. These complexes can be converted to a variety of enantiomerically pure tertiary alcohols (180).

Acylation of acetylenic compounds provides *trans*-β-chlorovinyl ketones (181). Vinyl cations were proposed to be the intermediates in these reactions.

$$R-C{\equiv}C-R \quad \xrightarrow[-70°C]{R'COCl/AlCl_3} \quad \underset{R}{\overset{\overset{\displaystyle O}{\underset{\|}{R'-C}}}{}} \underset{Cl}{\overset{R}{C{=}C}}$$

When aromatics are present, they can capture the intermediate vinyl cation to give β-aryl-α,β-unsaturated ketones (182). Thus acylation of alkyl or aryl acetylenes with acylium salts in the presence of aromatics gives α,β-unsaturated ketones with a trisubstituted double bond. The mild reaction conditions employed do not cause direct acylation of aromatics.

$$ArH \; + \; R-C{\equiv}C-H \; + \; R'-\overset{+}{C}{\equiv}O \; BF_4^- \quad \xrightarrow[-50°C]{CH_2Cl_2} \quad \underset{Ar}{\overset{R}{C}}{=}CH-\overset{\overset{\displaystyle O}{\underset{\|}{}}}{C}-R'$$

Introduction of a silyl groups at the termini of acetylenes provides a highly selective and mild method for the acylation reactions, the silicon substituent again being replaced by the acyl group. Acylation of bis(trimethylsilyl)ethyne provides terminal alkynyl ketones (183). Macrocyclization of ω-trimethylsilylethynyl alkanoyl chlorides under high dilution conditions provides 11–15 membered cyclic ynones, which are intermediates in the preparation of muscone. The regioselectivity of the reaction is in accordance with the β-carbocation stabilizing influence of the silicon substituent (184).

Even saturated hydrocarbons give ketones with acyl chlorides (20). For example, cyclohexane and acetyl chloride react in the presence of aluminum chloride to give 1-acetyl-2-methylcyclopentane.

Aliphatic Aldehyde Syntheses. Friedel-Crafts-type aliphatic aldehyde syntheses are considerably rarer than those of aromatic aldehydes. However, the hydroformylation reaction of olefins (185) and the related oxo synthesis are effected by strong acid catalysts, eg, tetracarbonylhydrocobalt, $HCo(CO)_4$ (see OXO PROCESS).

The Gattermann-Koch reaction when applied to alkenes or alkanes gives ketones or acids but not aldehydes. However, the Vilsmeier aldehyde synthesis can be applied to aliphatic compounds. For example, 1,2-dialkoxyethylenes react with N-methylformanilide and $POCl_3$ to give alkoxymalondialdehydes:

$$ROCH{=}CHOR \; + \; \underset{\text{(N-methylformanilide)}}{\text{[structure]}} \xrightarrow{POCl_3} \begin{array}{c} CHO \\ | \\ HC{-}OR \\ | \\ CHO \end{array}$$

Syntheses of Aliphatic Carboxylic Acids and Derivatives.

Alkenes are carbonylated in the presence of acid catalysts at 75–100°C and under pressures of 60–90 MPa (600–900 atm) to give carboxylic acids (186).

$$(CH_3)_2C{=}CH_2 + CO + H_2O \xrightarrow[BF_3{\cdot}2H_2O]{H_2SO_4} (CH_3)_3CCOOH$$

Olefins are carbonylated in concentrated sulfuric acid at moderate temperatures (0–40°C) and low pressures with formic acid, which serves as the source of carbon monoxide (Koch-Haaf reaction) (187). Liquid hydrogen fluoride, preferably in the presence of boron trifluoride, is an equally good catalyst and solvent system (see CARBOXYLIC ACIDS).

Carbocations generated from alkanes using superacids react with carbon monoxide under mild conditions to form carboxylic acid (188). In this process isomeric carboxylic acids are produced as a mixture. However, when the reaction is run with catalytic amounts of bromine (0.3 mmol eq) in HF-SbF_5 solution, regioselective carboxylation is obtained. n-Propane was converted almost exclusively to isobutyric acid under these conditions.

Aliphatic Nitration.

Alkanes undergo electrophilic nitration with nitronium salts such as $(NO_2)^+(PF_6)^-$ in a protic solvent such as CH_2Cl_2 and sulfolane (189). Adamantane and diamantane have also been nitrated with nitronium tetrafluoroborate in nitromethane solvents (190).

Polymerization.

Considerable interest has been focused on olefin polymerizations catalyzed by $AlCl_3$ and BF_3 and a variety of other Lewis-acid halides and protic acids. Three main types are recognized (191): (1) conversion of low molecular weight olefins to gasoline-range olefins (192); (2) conversion to intermediate molecular weight polymers for use as synthetic lubricant oils; and (3) conversion to high molecular weight polymers such as synthetic rubber. Under suitable conditions, the degree of polymerization may be controlled to produce any of the above types. Typical monomers that can readily be converted to high molecular weight polymers are isobutylene, styrene, α-methylstyrene, butadiene, isoprene, and vinyl alkyl ethers.

Even alkanes, when treated with superacids, can undergo oligocondensation. For example, highly branched polyalkanes, of molecular weight up to 700, were obtained by treating gaseous alkanes (C_1–C_4) with liquid superacids at room temperature (81).

Cross-linked macromolecular gels have been prepared by Friedel-Crafts cross-linking of polystyrene with a dihaloaromatic compound, or Friedel-Crafts cross-linking of styrene–chloroalkyl styrene copolymers. These polymers in their sulfonated form have found use as thermal stabilizers, especially for use in drilling fluids (193). Cross-linking polymers with good heat resistance were also prepared by Friedel-Crafts reaction of diacid halides with haloaryl ethers (194).

Friedel-Crafts reaction of naphthalene or tetrahydronaphthalene derivatives with those of styrene or alkylbenzenes has been used in the preparation of high viscous fluids for traction drive (195). Similarly, Friedel-Crafts reaction of tetraline and α-methylstyrene followed by catalytic hydrogenation provided 1-(1-decalyl)-2-cyclohexyl propane, which is used as a highly heat resistant fluid (196).

Hydrocarbon resins (qv) are prepared by copolymerization of vinyltoluene, styrene, and α-methylstyrene in the presence of a Friedel-Crafts catalyst ($AlCl_3$). These resins are compatible with wax and ethylene–vinyl acetate copolymer (197).

Polymer-type antioxidants have been prepared by Friedel-Crafts reaction of *p*-cresol and *p*- and/or *m*-chloromethylstyrene in the presence of boron trifluoride-etherate (198). The oligomeric product resulting from the alkylation of phenyl-α-naphthylamine using C12–15 propylene oligomer in the presence of $AlCl_3$ or activated white clays is used as an antioxidant additive for lubricating oils (199).

Catalysts

Friedel-Crafts catalysts are electron acceptors, ie, Lewis acids. The alkylating ability of benzyl chloride was selected to evaluate the relative catalytic activity of a large number of Lewis acid halides. The results of this study suggest four categories of catalyst activity (200) (Table 1).

Acid Halides (Lewis Acids). All metal halide-type Lewis catalysts, generally known as Friedel-Crafts catalysts, have an electron-deficient central metal

Table 1. Friedel-Crafts Catalyst Activities

Group	Characteristic	Examples
A	very active, high yields but extensive intra- and intermolecular isomerization	$AlCl_3$, $AlBr_3$, AlI_3, $GaCl_3$, $GaCl_2$, $GaBr_3$, GaI_3, $ZrCl_4$, $HfCl_4$, $HfBr_4$, HfI_4, SbF_5, NbF_5, $NbCl_5$, TaF_5, $TaCl_5$, $TaBr_5$, MoF_6, and $MoCl_5$
B	moderately active, high yields without significant side reactions	$InCl_3$, $InBr_3$, $SbCl_5$, WCl_6, $ReCl_5$, $FeCl_3$, $AlCl_3$–RNO_2, $AlBr_3$–RNO_2, $GaCl_3$–RNO_2, SbF_5–RNO_2, and $ZnCl_2$
C	weak, low yields without side reactions	BCl_3, BBr_3, BI_3, $SnCl_4$, $TiCl_4$, $TiBr_4$, $ReCl_3$, $FeCl_2$, and $PtCl_4$
D	very weak or inactive	many metal, alkaline-earth, and rare-earth element halides

atom capable of electron acceptance from the basic reagents. The most frequently used are aluminum chloride and bromide, followed by $BeCl_2$, $CdCl_2$, $ZnCl_2$, BF_3, BCl_3, BBr_3, $GaCl_3$, $GaBr_3$, $TiCl_4$, $ZrCl_4$, $SnCl_4$, $SnBr_4$, $SbCl_5$, $SbCl_3$, $BiCl_3$, $FeCl_3$, and UCl_4.

In addition, boron, aluminum, and gallium tris(trifluoromethanesulfonates) (triflates), $M(OTf)_3$ and related perfluoroalkanesulfonates were found effective for Friedel-Crafts alkylations under mild conditions (200). These Lewis acids behave as pseudo halides. Boron tris(triflate) shows the highest catalytic activity among these catalysts. A systematic study of these catalysts in the alkylation of aromatics such as benzene and toluene has been reported (201).

Easy availability and the low cost of aluminum chloride are partially responsible for its wide use in industry. Although aluminum chloride is frequently thought of as $AlCl_3$, at ordinary temperatures it is, in fact, the dimer Al_2Cl_6, which prevails up to 440°C; between 440 and 880°C there is an equilibrium mixture of monomer and dimer. At higher temperatures, only the monomer exists, although above 1000°C some ionic dissociation takes place. Under the usual Friedel-Crafts reaction conditions, the catalytically active species is always the monomeric Lewis acid $AlCl_3$ (see ALUMINUM COMPOUNDS).

Although Friedel and Crafts in their original work described investigations with anhydrous aluminum chloride, it is very difficult to obtain a Lewis acid-type metal halide in an absolutely anhydrous state and exclude moisture or other impurities during the course of reaction. In view of these facts, it is clear that neither the original inventors, Friedel and Crafts, nor the thousands of subsequent researchers who did most successful work with aluminum chloride and related catalyst systems, worked under truly anhydrous conditions. Impurities such as water, oxygen, hydrogen halides, organic halides, etc, were present in almost all cases. The presence of traces of moisture has been found to accelerate rather than hinder the reactions. In many cases, the presence of these so-called initiators or cocatalysts is indeed essential (202). The beneficial action of traces of moisture has been observed especially in reactions involving the olefinic double bond (alkylation with olefins, polymerization, etc).

The inactivity of pure anhydrous Lewis acid halides in Friedel-Crafts polymerization of olefins was first demonstrated in 1936 (203); it was found that pure, dry aluminum chloride does not react with ethylene. Subsequently it was shown (204) that boron trifluoride alone does not catalyze the polymerization of isobutylene when kept absolutely dry in a vacuum system. However, polymers form upon admission of traces of water. The active catalyst is boron trifluoride hydrate, $BF_3 \cdot H_2O$, ie, a conjugate protic acid $H^+(BF_3OH)^-$.

Cocatalysts of two types occur: (1) proton-donor substances, such as hydroxy compounds and proton acids, and (2) cation-forming substances (other than proton), including alkyl and acyl halides which form carbocations and other donor substances leading to oxonium, sulfonium, halonium, etc, complexes.

Metal Alkyls and Alkoxides.   Metal alkyls (eg, aluminum boron, zinc alkyls) are fairly active catalysts. Hyperconjugation with the electron-deficient metal atom, however, tends to decrease the electron deficiency. The effect is even stronger in alkoxides which are, therefore, fairly weak Lewis acids. The present discussion does not encompass catalyst systems of the Ziegler-Natta type (such as AlR_3 + $TiCl_4$), although certain similarities with Friedel-Crafts systems are apparent.

The most important application of metal alkoxides in reactions of the Friedel-Crafts type is that of aluminum phenoxide as a catalyst in phenol alkylation (205). Phenol is sufficiently acidic to react with aluminum with the formation of $(C_6H_5O)_3Al$. Aluminum phenoxide, when dissolved in phenol, greatly increases the acidic strength. It is believed that, similar to alkoxoacids (206) an aluminum phenoxoacid is formed, which is a strong conjugate acid of the type $HAl(OC_6H_5)_4$. This acid is then the catalytically active species (see ALKOXIDES, METAL).

Protic Acids (Brønsted Acids). Sulfuric acid is among the most used Brønsted acids for the Friedel-Crafts reactions, especially in hydrocarbon conversions, and in alkylation for the preparation of high octane gasoline. Anhydrous HF has replaced in part sulfuric acid, because of its convenience, although the toxic hazardous nature of HF is causing environmental concerns in its industrial use. Trifluoromethanesulfonic acid [1493-13-6] (and related superacids) are also gaining significance. Triflic acid does not react with aromatics (whereas sulfuric acid causes sulfonation) and thus offers substantial advantages with aromatic systems.

Acidic Oxides and Sulfides (Acidic Chalcogenides). Chalcogenide catalysts include a great variety of solid oxides and sulfides; the most widely used are alumina or silica (either natural or synthetic), in which other oxides such as chromia, magnesia, molybdena, thoria, tungsten oxide, and zirconia may also be present, as well as certain sulfides such as sulfides of molybdenum. The composition and structure of different types of bauxites, floridin, Georgia clay, and other natural aluminosilicates are still not well known. Some synthetic catalysts, other than silica–alumina compositions, representative of the acidic chalcogenides are BeO, Cr_2O_3, P_2O_5, TiO_2, and $Al_2(SO_4)_3$ which may be regarded as Al_2O_3, $3SO_3$, $Al_2O_3 \cdot xCr_2O_3$, $Al_2O_3 \cdot Fe_2O_3$, $Al_2O_3 \cdot MnO$, $Al_2O_3 \cdot CoO$, $Al_2O_3 \cdot Mo_2O_3$, $Al_2O_3 \cdot V_2O_3$, $Cr_2O_3 \cdot Fe_2O_3$, MoS_2, and MoS_3. In contrast to sulfuric acid which may be regarded as a fully hydrated chalcogenide, the members of this group are seldom very highly hydrated under conditions of use.

Silica–alumina has been studied most extensively. Dehydrated silica–alumina is inactive as isomerization catalyst but addition of water increases activity until a maximum is reached; additional water then decreases activity. The effect of water suggests that Brønsted acidity is responsible for catalyst activity (207). Silica–alumina is quantitatively at least as acidic as 90% sulfuric acid (208).

Acidic Cation-Exchange Resins. Brønsted acid catalytic activity is responsible for the successful use of acidic cation-exchange resins, which are also solid acids. Cation-exchange catalysts are used in esterification, acetal synthesis, ester alcoholysis, acetal alcoholysis, alcohol dehydration, ester hydrolysis, and sucrose inversion. The solid acid type permits simplified procedures when high boiling and viscous compounds are involved because the catalyst can be separated from the products by simple filtration. Unsaturated acids and alcohols that can polymerize in the presence of proton acids can thus be esterified directly and without polymerization.

Sulfonated styrene–divinylbenzene cross-linked polymers have been applied in many of the previously mentioned reactions and also in the acylation of thiophene with acetic anhydride and acetyl chloride (209). Resins of this type (Dowex 50, Amberlite IR-112, and Permutit Q) are particularly effective catalysts in the alkylation of phenols with olefins (such as propylene, isobutylene, diisobutylene), alkyl halides, and alcohols (210) (see ION EXCHANGE).

Superacids. *Brønsted Superacids.* In the 1960s a class of acids hundreds of millions times stronger than mineral acids was discovered; acids stronger than 100% sulfuric acid are called superacids (211). The determination of acidity by pH measurement does not hold for very concentrated acid solution. Hammett's logarithmic acidity function is generally used (212).

$$H_o = pK_{BH}{}^+ - \log \frac{BH^+}{B}$$

where $pK_{BH}{}^+$ is the dissociation constant of the conjugate acid and $\frac{BH^+}{B}$ is the ionization ratio. Typical H_o values are -12.6 for 100% H_2SO_4, and -11.0 for anhydrous HF. Although more recent H_o measurements on completely anhydrous HF have shown acidities comparable to that of FSO_3H (-15.1) (213).

Fluorosulfuric acid [7789-21-1] (HSO_3F) is one of the strongest Brønsted acids known with $H_o = -15.1$. This acidity is somewhat lower than that of $H_2SO_4 - SO_3$, ie, $H_2S_2O_7$. However, because of its stability, ease of purification, its wide liquid range (mp = $-89°C$, bp = $162°C$) and relatively low viscosity (1.56 mPa·s($=$cP) at $28°C$), it is more convenient to use.

Perfluoroalkanesulfonic acids also show high acidity. The parent trifluoromethanesulfonic acid (triflic acid), CF_3SO_3H, is commercially prepared by electrochemical fluorination of methanesulfonic acid (214). It has an H_o value of -14.1 (215,216). The higher homologues show slightly decreasing acidities.

Super Lewis Acids. Acid systems stronger than anhydrous $AlCl_3$ are classified as super Lewis acids (211). By this definition, Lewis acids such as SbF_5, NbF_5, AsF_5, and TaF_5 are so categorized.

Brønsted-Lewis Superacids. Conjugate Friedel-Crafts acids prepared from protic and Lewis acids, such as HCl–$AlCl_3$ and HCl–$GaCl_3$ are, indeed, superacids with an estimated H_o value of -15 to -16 and are effective catalysts in hydrocarbon transformation (217).

In the early 1960s acid systems were prepared comprising a pentafluoride of group V elements, particularly SbF_5 and a strong Brønsted acid such as HF, FSO_3H, CF_3SO_3H, etc (218). Magic Acid [33843-68-4] ($HSO_3F–SbF_5$) is one of the strongest members of the system; fluoroantimonic acid [16950-06-4], HF–SbF_5, even surpasses Magic Acid in its acidity. The acidity of HF or HSO_3F is increased sharply by adding SbF_5 (219,220). These very highly acidic systems are being utilized in transformations such as isomerization of straight-chain alkanes (221), alkane–alkene alkylations (222), and the like (223). $CF_3SO_3H–SbF_5$ and $CF_3SO_3H–B(OTf)_3$ have been shown to be highly effective catalysts for Friedel-Crafts alkylation and isomerization reactions (224).

Solid Superacids. Most large-scale petrochemical and chemical industrial processes are preferably done, whenever possible, over solid catalysts. Solid acid systems have been developed with considerably higher acidity than those of acidic oxides. Graphite-intercalated $AlCl_3$ is an effective solid Friedel-Crafts catalyst but loses catalytic activity because of partial hydrolysis and leaching of the Lewis acid halide from the graphite. Aluminum chloride can also be complexed to sulfonate polystyrene resins but again the stability of the catalyst is limited.

More stable catalysts are obtained by using fluorinated graphite or fluorinated alumina as backbones, and Lewis acid halides, such as SbF_5, TaF_5, and NbF_5, which have a relatively low vapor pressure. These Lewis acids are attached to the fluorinated solid supports through fluorine bridging. They show high reactivity in Friedel-Crafts type reactions including the isomerization of straight-chain alkanes such as n-hexane.

Another type of solid superacid is based on perfluorinated resin sulfonic acid such as the acid form of Du Pont's Nafion resin, a copolymer of a perfluorinated epoxide and vinylsulfonic acid or solid, high molecular weight perfluoroalkane-sulfonic acids such as perfluorodecanesulfonic acid, $CF_3(CF_2)_9SO_3H$. Such solid catalysts have been found efficient in many alkylations of aromatic hydrocarbons (225) and other Friedel-Crafts reactions (226).

Superacidic Zeolites. The well-defined crystal structures of both natural and synthetic zeolites permit selective hydrocarbon transformations. The selectivity of the zeolites can be improved by deactivations of external acid sites with amines, replacement of the cationic sites by transition metal ions by ion-exchange, or by modification of the silica–alumina ratio. Some zeolites such as H-ZSM-5 and the like display superacidic character at high temperatures (see MOLECULAR SIEVES). The have found wide utility in electrophilic aromatic alkylations, transalkylations, disproportionation, hydrocarbon synthesis, and more importantly, methanol conversion to hydrocarbons, including fuel gas and gasoline (227). H-ZSM-5 was also used as an efficient catalyst for the thermal degradation of polypropylene into gasoline range hydrocarbons (228). Various pillared clays obtained by reaction of metal trihalides with the hydroxyl groups on clays act as selective catalysts, especially in transalkylations (229).

BIBLIOGRAPHY

"Friedel-Crafts Reactions" in *ECT* 1st ed., Vol. 6, pp. 883–892, by G. W. Pedlow and P. E. Hoch, General Aniline & Film Corp.; in *ECT* 2nd ed., Vol. 10, pp. 135–166, by G. A. Olah and C. A. Cupas, Western Reserve University; in *ECT* 3rd ed., Vol. 11, pp. 269–300, by G. A. Olah and D. Meidar, University of Southern California.

1. G. A. Olah, ed., *Friedel-Crafts and Related Reactions*, Vols. 1–4, Wiley-Interscience, New York, 1963–1965.
2. G. A. Olah, *Friedel-Crafts Chemistry*, John Wiley & Sons, Inc., New York, 1973.
3. G. Kranzlein, *Aluminum Chlorid in der Organischen Chemie*, 3rd ed., Verlag Chemie, Berlin, 1939.
4. C. A. Thomas, *Anhydrous Aluminum Chloride in Organic Chemistry*, Reinhold Publishing Corp., New York, 1961.
5. R. M. Roberts and A. A. Khalaf, *Friedel-Crafts Alkylation Chemistry*, Mercel Dekker, Inc., New York, 1984.
6. USSR Pat. 1,479,449 (May 1989), V. V. Lobkina and co-workers.
7. N. Ono and co-workers, *J. Chem. Soc. Chem. Commun.*, 1285 (1986).
8. X. Fu, and co-workers, *Synth. Commun.* **21**, 1273 (1991).

9. G. A. Olah, ed., *Reactive Intermediates in Organic Chemistry*, John Wiley & Sons, Inc., New York, 1968–1972.

10. G. A. Olah and P. v. R. Schleyer, eds., *Carbonium Ions*, John Wiley & Sons, Inc., New York, 1968–1972.

11. D. Bethell and V. Gold, *Carbonium Ions: An Introduction*, Academic Press, Inc., New York, 1967.

12. O. Jacobsen, *Chem. Ber.* **18**, 338 (1885).

13. N. G. Bun-Hoi and P. Cagneunt, *Bull. Soc. Chim. Fr.* **9**, 887 (1942).

14. A. W. Francis and E. E. Reid, *Ind. Eng. Chem.* **38**, 1194 (1946).

15. L. F. Albright and A. R. Goldsly, eds., *Industrial and Laboratory Alkylations*, American Chemical Society, Washington, D.C., 1977; K. Weisgermel and H. J. Arpe *Industrial Organic Chemistry*, Verlag Chemie, Berlin, 1978.

16. C. C. Price, in R. Adams, ed., *Organic Reactions*, Vol. 3, John Wiley & Sons, Inc., New York, 1946, p. 1.

17. A. W. Francis, *Chem. Rev.* **43**, 257 (1948).

18. W. W. Kaeding, *J. Catalysis* **95**, 512 (1985).

19. Eur. Pat. Appl. EP 79093 A1 18 (May 1983), F. Baardman and co-workers (to Shell Internationale Research).

20. L. Schmerling and J. P. West *J. Am. Chem. Soc.* **76**, 1917 (1959).

21. W. J. Taylor and co-workers, *J. Res. Natl. Bur. Stand.* **37**, 95 (1946).

22. G. A. Olah, W. S. Tolgysei, and R. E. A. Dear, *J. Org. Chem.* **27**, 3445, 3449 (1962).

23. G. A. Olah and M. Meyer, *J. Org. Chem.* **27**, 3464 (1962).

24. K. M. Nicholas, *Acc. Chem. Res.* **207**, 20 (1987).

25. D. D. Grove and co-workers, *Tetrahedron Lett.* **31**, 6277 (1990).

26. R. Weiss, H. Kolbl, and C. Schlierf, *J. Org. Chem.* **41**, 2258 (1976).

27. D. Fletcher and co-workers, *Tetrahedron Lett.* **27**, 4853 (1986).

28. M. A. Garcia and co-workers, *Chem. Ber.* **120**, 1255 (1987).

29. G. A. Olah and co-workers, *J. Am. Chem. Soc.* **97**, 6807 (1975).

30. R. Miethchen, S. Steege, and C. -F. Kroger, *J. Prakt. Chem.* **325**, 823 (1983) and references therein; R. Miethchen, C. Rohse, and C.-F. Kroger, *Z. Chem.* **24**, 145 (1986).

31. R. Miethchen, K. Kohlheim, and A. Muller, *Z. Chem.* **26**, 168 (1986) and references therein.

32. J. I. Brauman and A. J. Pandell, *J. Am. Chem. Soc.* **88**, 5421 (1967); J. I. Brauman and A. Solladie-Cavalo, *Chem. Commun.*, 1124 (1968).

33. S. Suga and co-workers, *Tetrahedron Lett.*, 3283 (1969).

34. D. R. Kronenthal and co-workers, *Tetrahedron Lett.* **31**, 1241 (1990).

35. M. Segi and co-workers, *Bull. Chem. Soc. Jpn.* **55**, 167 (1982).

36. F. Bigi and co-workers, *Tetrahedron: Asymmetry* **1**, 861 (1990).

37. S. K. Taylor and co-workers, *J. Heterocycl. Chem.* **20**, 1745 (1983).

38. U. Schollkopf and S. Gruttner, *Angew. Chem. Int. Ed. Engl.* **26**, 683 (1987).

39. T. Matsumoto, M. Katsuki, and K. Suzuki, *Tetrahedron Lett.* **30**, 833 (1989).

40. Ger. Pat. 3,404,336 (1985) G. Fritz, M. Eggersdorfer, and H. Siegel (to BASF).

41. O. Piccolo and co-workers, *J. Org. Chem.* **50**, 3945 (1985).

42. S. Masuda and co-workers, *J. Chem. Soc. Chem. Commun.*, 86 (1980).

43. D. Raabe and H. H. Hoerhold, *J. Prakt. Chem.* **329**, 1131 (1987).

44. Ref. 1, Vol. 2, Part 2, Chapt. 21.

45. H. F. Fieser and M. D. Gates, Jr., *J. Am. Chem. Soc.* **62**, 2335 (1940).

46. S. Kajigaeshi and co-workers, *Synthesis*, 335 (1984).

47. L. V. Bugrova and I. P. Tsukervanik, *Zh. Org. Khim.* **1**, 714 (1965).

48. G. A. Olah and S. J. Kuhn, *J. Org. Chem.* **29**, 2317 (1964).

49. C. D. Nenitzescu and D. A. Isacescu, *Chem. Ber.* **66**, 1100 (1933).

50. *Chem. Abstr.* **114**, p228920h (1991).

51. C. F. Nutaitis and G. W. Gribble, *Synthesis*, 756 (1985).
52. J. M. Briody and G. L. Marshall, *Synthesis*, 939 (1982).
53. H. A. Bruson and J. W. Kroeger, *J. Am. Chem. Soc.* **62**, 36 (1940).
54. Y. Lavaux and M. Lombard, *Bull. Soc. Chim. Fr.* **7**, 913 (1910).
55. I. P. Tsukervanik and N. I. Bogdanova, *J. Gen. Chem. USSR* **22**, 410 (1953).
56. Bergmann, *Chem. Rev.* **29**, 529 (1941).
57. B. R. Davis, S. J. Johnson, and P. D. Woodgate, *Aust. J. Chem.* **40**, 1283 (1987).
58. N. Yoneda, Y. Takahashi, and A. Suzuki, *Chem. Lett.*, 231 (1978).
59. M. F. Ansell and M. E. Selleck, *J. Chem. Soc.*, 1238 (1956).
60. Z.-U. Yang, and G. B. Young, *J. Chem. Soc., Dalton Trans.* **9**, 2019 (1984).
61. A. M. El-Khawaga and co-workers, *Rev. Roum. Chim.* **30**, 599 (1985).
62. S. K. Taylor and co-workers, *J. Org. Chem.* **48**, 2449 (1983); S. K. Taylor and co-workers, *J. Org. Chem.* **52**, 425 (1987); S. K. Taylor and co-workers, *J. Org. Chem.* **53**, 3309 (1988).
63. Eur. Pat. Appl. EP 71,006 (1983), G. Suzukamo and Y. Sakito (to Sumitomo Chemical Co., Ltd.).
64. A. A. Khalaf and R. M. Roberts, *J. Org. Chem.* **34**, 3571 (1968).
65. R. O. Roblin, D. Davidson, and M. T. Bogert, *J. Am. Chem. Soc.* **57**, 151 (1935).
66. D. V. Nightingale, *Chem. Rev.* **25**, 329 (1939); G. A. Olah, W. S. Tolgyesi, and R. E. A. Dear, *J. Org. Chem.* **27**, 3441 (1962).
67. G. Makarova, M. K. Matveeva, and E. A. Gribehenko, *Bull. Acad. Sci. USSR Div. Chem. Sci.*, 1399 (1958); A. N. Nesmeyanov and L. G. Makrova, *Bull. Acad. Sci. USSR Div. Chem. Sci.*, 213 (1947).
68. G. A. Olah and W. S. Tolgysesi, *J. Org. Chem.* **26**, 2053 (1961).
69. V. N. Ipatieff and co-workers, *J. Am. Chem. Soc.* **58**, 913 (1936).
70. S. F. Birch and co-workers, *J. Inst. Pet. Technol.* **24**, 303 (1938).
71. G. A. Olah and J. A. Olah, *J. Am. Chem. Soc.* **93**, 1256 (1970).
72. M. Siskin, R. H. Schlosberg, and M. P. Kocsi, *Symposium on New Hydrocarbon Chemistry, ACS San Francisco Meeting*, Aug. 29–Sept. 3, 1976; R. H. Schlosberg and co-workers, *J. Am. Chem. Soc.* **98**, 7723 (1976).
73. G. A. Olah and R. H. Schlosberg, *J. Am. Chem. Soc.* **90**, 2726 (1968).
74. Brit. Pat. Appl. 2,136,013 (Sept. 1984), D. Seddon and S. Bessell (to Broken Hill Pty. Co. Ltd., Commonwealth Scientific and Industrial Research Organization.
75. L. Schmerling, *J. Am. Chem. Soc.* **67**, 1152 (1945); **68**, 1650 (1946).
76. U.S. Pat. 4,116,880 (1978), G. A. Olah (to University of Southern California).
77. *Occupation Safety and Health Administration Standards and Regulations of U.S. Government*, U.S. Government Printing Office, Washington, D.C., 1990.
78. R. C. Fort and P. v. R. Schleyer, *Chem. Rev.* **64**, 277 (1964); T. M. Gund and co-workers, *Tetrahedron Lett.*, 3877 (1970); F. S. Hollowood and M. A. McKervey, *J. Org. Chem.* **45**, 4954 (1980); T. M. Gund, W. Thielecke, and P. v. R. Schleyer, *Org. Synth.* **53**, 30 (1973); T. Courtney and co-workers, *J. Chem. Soc. Perkin Trans.* **1**, 2691 (1972); G. N. Schrauzer, B. N. Bastian, and G. A. Fosselian, *J. Am. Chem. Soc.* **45**, 4954 (1980).
79. O. Farooq and co-workers, *J. Org. Chem.* **53**, 2840 (1988).
80. H. Ulich and G. Heyne, *Z. Elektrochem.* **41**, 509 (1935).
81. D. T. Roberts, Jr., and L. E. Calihan, *J. Macromol. Sci. Chem.*, 1629, 1641 (1973); P. H. Goore and J. A. Hiskins, *J. Chem. Soc. C.*, 517 (1970).
82. N. P. Bun-Hoi, P. Mabille, and Do-Cao Thang, *Bull. Soc. Chim. Fr.*, 981 (1968).
83. P. Fletcher and W. Marlow, *J. Chem. Soc. C.*, 937 (1970).
84. D. Papa, E. Schwenk, and A. Klingsberg, *J. Am. Chem. Soc.* **68**, 2133 (1946).
85. G. A. Olah and S. Kobayashi, *J. Am. Chem. Soc.* **93**, 6964 (1971).
86. Ref. 1, Vol. 1, p. 581.

87. F. R. Jensen and B. E. Swart, *J. Am. Chem. Soc.* **91**, 5686 (1969).

88. Ger. Pat. DD 239,788 (Oct. 1986), B. Dreher and co-workers (to Akademie der Wissenschaften der DDR).

89. Pol. Pat. 146,900 (July 1989), D. Prescher and co-workers (to Polska Akademia Nauk, Instytut Chemii Organicznet).

90. Eur. Pat. 301,311 (Feb. 1989), C. Giordano (to Zambon Group, SpA).

91. S. African Pat. 86 04,124 (Feb. 1988), G. M. St. George and M. E. Walters. (to Dow Chemical Co.); Eur. Pat. Appl. EP 189,266 (July 30, 1986), M. E. Walters (to Dow Chemical Co.)

92. U.S. Pat. 4,632,930 (Dec. 30, 1986), D. J. Carini and R. R. Wexler (to E. I. du Pont de Nemours & Co., Inc.).

93. Jpn. Pat. 62, 132, 838 (June 1987), N. Tanno, H. Sato, and K. Ishizumi (to Sumitomo Pharmaceuticals Co., Ltd.).

94. Eur. Pat. Appl. EP 161,918 (Nov. 21, 1985), D. W. Robertson (to Eli Lilly and Co.).

95. Eur. Pat. Appl. EP 69,536 (Jan. 12, 1983), T. G. Fourie (to Noristan Ltd.).

96. Eur. Pat. Appl. EP 63,945 (Nov. 3, 1982), L. A. Mitscher (to University of Kansas, Endowment Assoc.).

97. H. Pacheco, M. A. Descours-Saint-Martino, D. Yavordios, J. Koeberle, U.S. Pat. 4,356,188 (Oct. 26, 1982), H. Pachero and co-workers (to Institut de Recherche Scientifique).

98. U.S. Pat. 4,336,397 (June 22, 1982), E. J. Cragoe and co-workers (to Merck and Co., Inc.).

99. G. A. Olah and S. J. Kuhn, *J. Org. Chem.* **26**, 237 (1961).

100. T. F. Buckley III and H. Rapoport, *J. Am. Chem. Soc.* **103**, 6157 (1981); U.S. Pat. 4,618,710 A (Oct. 21, 1986) H. B. Rapoport and T. F. Buckley III (to University of California, Berkeley).

101. T. F. Buckley III and H. Rapoport, *J. Org. Chem.* **48**, 4222 (1983).

102. J. E. Nordlander and co-workers, *J. Org. Chem.* **49**, 4107 (1984).

103. J. E. Nordlander and co-workers, *J. Org. Chem.* **50**, 3481 (1985).

104. I. Osamu and co-workers, *J. Org. Chem.* **57**, 7334 (1992).

105. F. S. Statham, *J. Chem. Soc.*, 213 (1951).

106. F. Wolf, K. Fredrich, and G. Schwachula, *Plaste. Kaut.* **16**, 727 (1969).

107. L. Gattermann and J. A. Koch, *Chem. Ber.* **30**, 1622 (1897).

108. L. Bert, *Compt. Rend.* **221**, 77 (1945).

109. S. Fujiyama and T. Kasahara, *Hydrocarbon Process.*, 147 (1978).

110. G. A. Olah and co-workers, *J. Am. Chem. Soc.* **98**, 296 (1976).

111. G. A. Olah, K. Laali, and O. Farooq, *J. Org. Chem.* **50**, 1483 (1985); G. A. Olah, Ohanessian, and M. Arvanaghi, *Chem. Rev.* **87**, 671 (1987).

112. B. L. Booth, T. A. Ell-Fekky, and G. M. F. Noori, *J. Chem. Soc. Perkin Trans.* **1**, 181 (1980).

113. M. Tanaka, M. T. Iyoda, and Y. Souma, *J. Org. Chem.* **57**, 2677 (1992).

114. S. Jo and co-workers, *Bull. Chem. Soc. Jpn.* **54**, 2120 (1981).

115. G. A. Olah and V. P. Reddy, University of Southern California, 1991, unpublished results.

116. V. P. Reddy, D. Bellew, and G. K. S. Prakash, *Synthesis*, 1209 (1992).

117. Y. Suzuki and H. Takahashi, *Chem Pharm. Bull.* **31**, 1751 (1983).

118. G. A. Olah and S. J. Kuhn *J. Org. Chem.* **26**, 237 (1961).

119. M. Yato, T. Ohwada, and K. Shudo, *J. Am. Chem. Soc.* **113**, 691 (1991).

120. W. Ruske, in Ref. 1, Vol. 3, Part I, Chapt. 32.

121. K. Hoesch, *Ber. Dtsch. Chem. Ges.* **48**, 1122 (1915).

122. J. Houben, *Ber. Dtsch. Chem. Ges.* **59**, 2878 (1926).

123. G. A. Olah, G. K. S. Prakash, and K. Lammertsma, *Res. Chem. Intermed.* **12**, 141 (1989).

124. G. A. Olah, *Angew. Chem. Int. Ed. Engl.* **32**, 767 (1993).

125. A. Vilsmeier, *Chem. Ztg.* **75**, 133 (1951).

126. H. Lorenz and R. Witzinger, *Helv. Chim. Acta* **28**, 600 (1945).

127. Z. Arnold and F. Sorm, *Chem. Listy* **51**, 1082 (1957); C. Jutz, *Chem. Ber.* **91**, 850 (1958).

128. A. Rieche, H. Gross, and E. Hoft, *Chem. Ber.* **95**, 88 (1960).

129. R. Scholl and F. Kacer, *Chem. Ber.* **36**, 322 (1903).

130. J. Houben and W. Fischer, *Chem. Ber.* **66**, 339 (1933).

131. M. Reetz and I. Chatziiosifidis, *Angew. Chem.* **93**, 1075 (1981).

132. G. A. Olah, O. Farooq, and G. K. S. Prakash, *Synthesis*, 1140 (1985).

133. C. Friedel, J. M. Crafts, and E. Ador, *Compt. Rend.* **85**, 673 (1877).

134. H. Hopff, *Angew. Chem.* **60**, 245 (1948).

135. F. Runge, H. Reinhardt, and G. Kuhnhanss, *Chem. Tech (Berlin)*, 644 (1956).

136. U.S. Pat. 1,892,990 (Jan. 3, 1933), F. Linner (to Beck, Koller & Co.).

137. M. T. Bogert and M. Meyer, *J. Am. Chem. Soc.* **44**, 1568 (1922).

138. J. F. Norris and J. E. Wood, *J. Am. Chem. Soc.* **62**, 1428 (1940).

139. S. C. J. Olivier, *Rec. Trav. Chim.* **33**, 91, 263, 244 (1914).

140. L. S. Field, *J. Am. Chem. Soc.* **74**, 394 (1952).

141. W. E. Truce and C. W. Vriesen, *J. Am. Chem. Soc.* **75**, 5032 (1953).

142. A. Amer, A. M. El-Massry, and C. U. Pittman, Jr., *Chem. Scr.* **351**, 29 (1989).

143. E. Skotnicki, *Przem. Chem.* **67**, 177 (1988).

144. R. J. Thomas, W. F. Anzilotti, and G. F. Hennion, *Ind. Eng. Chem.* **32**, 408 (1940).

145. L. I. Smith, in R. Adams, ed., *Organic Reactions*, Vol. 1, John Wiley & Sons, Inc., New York, 1942, p. 370.

146. E. Knoevenagel and J. Kenner, *Chem. Ber.* **41**, 3315 (1908).

147. H. Hopff, R. Roggero, and G. Valkanas, *Rev. Chim. (Bucharest)* **7**, 921 (1962); Brit. Pat. 783,037 (1957), J. R. Geigi (to A.G).

148. M. G. Voronkov and co-workers, in J. S. Pizey, ed., *Reactions of Sulfur with Organic Compounds*, Consultants Bureau, New York, 1987, Chapt. 3.

149. G. A. Olah, Q. Wang, and G. K. S. Prakash, *J. Am. Chem. Soc.* **112**, 3698 (1990).

150. C. C. Price and C. A. Sears, *J. Am. Chem. Soc.* **75**, 3276 (1953).

151. G. A. Olah, A. E. Pavlath, and S. Kuhn, *Chem. Ind. (London)*, 50 (1957).

152. A. Schaarschmidt, *Chem. Ber.* **57**, 2065 (1924); *Angew. Chem.* **39**, 1457 (1926).

153. L. A. Pinek, *J. Am. Chem. Soc.* **49**, 2536 (1927).

154. A. Klemenz and K. Scholler, *Z. Anorg. Allgem. Chem.* **141**, 231 (1927).

155. G. B. Bachman and J. L. Denver, *J. Am. Chem. Soc.* **80**, 5871 (1958).

156. G. A. Olah, S. J. Kuhn, and A. Mlinko, *J. Chem. Soc.*, 4257 (1956).

157. S. J. Kuhn and G. A. Olah, *J. Am. Chem. Soc.* **83**, 4564 (1961).

158. G. A. Olah and co-workers, *J. Am. Chem. Soc.* **101**, 1805 (1979).

159. C. E. Inmann, R. E. Oesterling, and E. A. Tyezkowsky, *J. Am. Chem. Soc.* **88**, 3819 (1966).

160. E. Wertyporoch, *Liebigs Ann. Chem.* **493**, 153 (1932).

161. G. A. Olah and M. Meyer, *J. Org. Chem.* **27**, 3464 (1962).

162. G. A. Olah, W. S. Tolgyesi, and R. E. A. Dear, *J. Org. Chem.* **27**, 3455 (1962).

163. G. A. Olah, A. P. Fung, and T. Keumi, *J. Org. Chem.* **46**, 4305 (1981).

164. J.-C. Jacquesy and co-workers, *Bull. Soc. Chim. Fr.* **4**, 625 (1986); *Tetrahedron Lett.* **25**, 1479 (1984). J.-C. Jacquesy, M.-P. Jounannetaud, and G. Morellet, *Tetrahedron Lett.* **24**, 3099 (1983).

165. G. A. Olah and T. D. Ernst, *J. Org. Chem.* **54**, 1204 (1989).

166. G. K. S. Prakash and co-workers, *Synlett.* **39** (1991); *Revue Roumaine de Chemie* **36**, 4567 (1991).
167. C. Graebe, *Chem. Ber.* **34**, 1778 (1901); J. S. Turski; P. Kovacic and R. P. Bennet, *J. Am. Chem. Soc.* **83**, 221 (1961); P. Kovacic and J. L. Foote, *J. Am. Chem. Soc.* **83**, 743 (1961).
168. A. Mertenz and co-workers, *J. Am. Chem. Soc.* **105**, 5657 (1983).
169. G. A. Olah and T. D. Ernst, *J. Org. Chem.* **54**, 1203 (1989).
170. H. Wieland and L. Bettag, *Chem. Ber.* **55**, 2246 (1922).
171. R. Pardo and M. Santelli, *Tetrahedron Lett.* **22**, 3843 (1981).
172. I. D. Gridnev, A. V. Shastin, and E. S. Balenkova, *Zh. Org. Chem.* **23**, 1546 (1987).
173. S. Hacini, R. Pardo, and M. Santelli, *Tetrahedron Lett.*, 4553 (1979).
174. J. Tsuji, K. Kasuga, and T. Takahashi, *Bull. Chem. Soc. Jpn.* **52**, 216 (1979); L. A. Pacquette and W. H. Ham, *J. Am. Chem. Soc.* **109**, 3025 (1987).
175. I. Fleming, J. Dunogues, and R. H. Smithers, *Org. React. (NY)* **37**, 57 (1989).
176. E. Negishi and co-workers, *J. Am. Chem. Soc.* **105**, 6344 (1983).
177. J.-P. Pillot, J. Dunogues, and R. Calas, *Tetrahderon Lett.*, 1871, (1976).
178. M. Laguerre, J. Dunogues, and R. Calas, *Tetrahedron Lett.*, 57 (1978); A. Hosomi, H. Hashimoto, and H. Sakurai, *J. Org. Chem.* **43**, 2551 (1978).
179. A. J. Birch and co-workers, *J. Organomet. Chem.* **260**, C59 (1984).
180. M. Franck-Neumann, P. Chemla, and D. Martina, *Synlett.*, 641 (1990).
181. A. E. Pohland and W. R. Benson, *Chem. Rev.* **66**, 161 (1966); H. Martens, F. Janssens, and G. Hoormaert, *Tetrahedron 31*, 177 (1975).
182. A. A. Schegolev and co-workers, *Synthesis*, 324 (1977).
183. L. Birkofer, A. Ritter, and H. Uhlenbrauck, *Chem. Ber.* **96**, 3280 (1963).
184. K. Utimoto and co-workers, *Tetrahedron Lett.*, 2301 (1978).
185. G. Salamon, *Rec. Trav. Chim.* **68**, 903 (1949).
186. T. A. Ford, H. W. W. Jacobsen, and F. C. McGrew, *J. Am. Chem. Soc.* **70**, 3743 (1948).
187. H. Koch and W. Haaf, *Angew. Chem.* **70**, 311 (1958); *Liebigs Ann. Chem.* **618**, 251 (1958).
188. S. Delavarenne and co-workers, *J. Am. Chem. Soc.* **111**, 383 (1989); S. Delavarene and co-workers, *J. Chem. Soc. Chem. Commun.*, 1049 (1989).
189. G. A. Olah and H. C. Lin, *J. Am. Chem. Soc.* **93**, 1259 (1971).
190. G. A. Olah and co-workers, *J. Am. Chem. Soc.* **115**, 7246 (1993).
191. A. G. Oblad, G. A. Millis, and H. Heinemann, in P. H. Emmet, ed., *Catalysis*, Vol. 6, Reinhold Publishing Corp., New York, 1958, p. 341.
192. V. N. Ipatieff, B. B. Corson, and G. Egloff, *Ind. Eng. Chem.* **27**, 1077 (1935).
193. Fr. Pat. 2,587,708 (Mar. 1987), R. Audebert and co-workers (to CECA SA).
194. Eur. Pat. 277,476 (Aug. 1988), R. H. Lubowitz and C. H. Sheppard (to Boeing Co.).
195. Eur. Pat. 240,814 (Oct. 1984), T. Minokami, T. Tsubouchi, and H. Hata (to Idemitsu Kosan Co., Ltd.).
196. Jpn. Pat. 63 00 388 (1988), K. Abe, T. Minogami, and K. Hata (to Idemitsu Kosan Co., Ltd.).
197. Eur. Pat. 190,868 (Aug. 13, 1986), G. Wouters (to Exxon Chemical Patents, Inc.).
198. Jpn. Pat. 61 221,284 (Oct. 1986), G. Arimatsu and co-workers (to Toyobo Co., Ltd.).
199. Jpn. Pat. 62 181,396 (1987), N. Ishida, N. Yokoyama, and H. Takashima (to Nippon Oil Co., Ltd.).
200. G. A. Olah, S. Kobayashi, and M. Tashiro, *J. Am. Chem. Soc.* **96**, 7448 (1972).
201. G. A. Olah and co-workers, *J. Am. Chem. Soc.* **110**, 2560 (1988).
202. C. D. Nenitzescu and E. I. P. Cantuiari, *Chem. Ber.* **66**, 1097 (1933).
203. V. N. Ipatieff and A. V. Grosse, *J. Am. Chem. Soc.* **58**, 915 (1936).
204. A. G. Evans and co-workers, *Nature* **157**, 102 (1946).
205. A. J. Kolka, J. P. Napolitano, and G. G. Ecke, *J. Org. Chem.* **21**, 712 (1956).

206. H. Meerwein and T. Bersin, *Liebigs Ann. Chem.* **476**, 113 (1920).
207. F. E. Condon, in P. H. Emmett, ed., *Catalysis*, Vol. 6, Reinhold Publishing Corp., New York, 1949, p. 200.
208. A. Roe, in R. Adams, ed., *Organic Reactions*, Vol. 5, John Wiley & Sons, Inc., New York, 1949, p. 200.
209. U.S. Pat. 2,711,414 (June 21, 1955), T. R. Norton (to the Dow Chemical Co.).
210. B. Loev and J. T. Massengele, *J. Org. Chem.* **22**, 988 (1957).
211. G. A. Olah, G. K. Prakash, and J. Sommer, *Superacids*, John Wiley & Sons, Inc., New York, 1985, pp. 24–27 and references therein.
212. L. P. Hammett and A. J. Deyrup, *J. Am. Chem. Soc.* **54**, 2721 (1932).
213. R. J. Gillespie and J. Liang, *J. Am. Chem. Soc.* **110**, 6053 (1980); T. A. O'Donnell, *J. Fluor. Chem.* **25**, 75 (1984).
214. R. N. Hazeldine and J. M. Kidd, *J. Chem. Soc.*, 4228 (1954).
215. R. D. Howells and J. D. McCowen, *Chem. Rev.*, 69 (1977).
216. J. Groudin, R. Sagnes, and A. Commeyras, *Bull. Soc. Chim. Fr.*, 1779 (1976).
217. D. A. McCowley and A. P. Line, *J. Am. Chem. Soc.* **74**, 6246 (1956).
218. G. A. Olah, G. K. S. Prakash, and J. Sommer, *Science* **206**, 13 (1979) and references therein.
219. M. Kilpatgrick and P. E. Lubersky, *J. Am. Chem. Soc.* **76**, 5863 (1954).
220. R. J. Gillespie and T. E. Peel, *J. Am. Chem. Soc.* **95**, 5173 (1973).
221. U.S. Pat. 3,838,489 (Oct. 1, 1974), J. E. Mahan and J. R. Norell (to Phillips Petroleum Co.)
222. U.S. Pat. 3,708,553 (Jan. 2, 1973), G. A. Olah (to Esso Research and Engineering Co.).
223. G. A. Olah and J. Sommer, *Recherche* **10**, 625 (1979).
224. G. A. Olah and co-workers, *J. Org. Chem.* **55**, 1516 (1990); G. A. Olah and co-workers, *J. Am. Chem. Soc.* **110**, 2560 (1988); G. A. Olah, K. Laali, and O. Farooq, *J. Org. Chem.* **50**, 1483 (1985).
225. G. A. Olah and D. Meidar, *Nouv. J. Chim.* **3**, 269 (1979); G. A. Olah and co-workers, *J. Catalysis* **61**, 96 (1980).
226. G. A. Olah, D. Meidar, and J. A. Olah, *Nouv. J. Chim.* **3**, 275 (1978); G. A. Olah and J. Kaspi, *Nouv. J. Chem.* **2**, 585 (1979).
227. Eur. Pat. 193,282 (1986), E. W. Valyocsik, D. H. Olson, and P. G. Rodewald (to Mobil Oil Corp.).
228. R. C. Mordi, R. Fields, and J. Dwyer, *J. Chem. Soc. Chem. Commun.* **4**, 374 (1992).
229. Eur. Pat. 130,005 (1985), M. P. Atkins (British Petroleum Co. PLC).

GEORGE A. OLAH
V. PRAKASH REDDY
G. K. SURYA PRAKASH
University of Southern California

FRUIT JUICES

The fruit juice industry was started in the United States in 1869. Concord grapes were juiced from their vines, and the juice was filtered, bottled, then pasteurized in a hot water bath (1). The European fruit juice industry began in 1896 with the publication of detailed instructions for commercial production of unfermented fruit juices. That same year apple and grape juices were first produced commercially at Berne, Switzerland (2). In the United States, nationwide expansion of commercial juice production developed about 1929, when economic necessity forced American consumers into an unprecedented era of home canning, including fruit and vegetable juices. Consumers' acquired taste for these products led to the expansion of the commercial industry in the 1930s, mainly processing overripe products not suitable for whole fruit canning (3). During the 1930s, development of flash pasteurization and enamel-lined cans afforded superior juice quality and unprecedented growth for the industry.

During the 1940s frozen single-strength fruit juices were introduced. Frozen concentrated fruit juices were first sold in the U.S. market during 1945–1946. This freezing technology had a dramatic effect on the industry and was responsible for most of the growth of the Florida citrus industry. In 1993, frozen concentrated fruit juice for manufacturing was the principal juice product of international commerce. Consumers have demanded more ready-to-serve juice products, especially chilled single-strength juices. Commercial aseptic packaging of single-strength juices and juice drinks worldwide permits packaging in soft plastic packages that can be stored at higher temperatures for convenience and economy for the consumer. Aseptic packaging is also used for more economical storage and transport at higher temperatures of bulk single-strength and concentrated juices (see FOOD PACKAGING).

Raw Materials

The early juice industry was largely a salvage operation. The principal source of raw material was misshapen, poorly colored, or skin blemished fruit unsuitable for the fresh, canned, or frozen fruit market. In the 1990s, raw materials are selected for suitability for juice production, except for apple juice production which still uses much cull fruit (4).

Variety and maturity are important factors affecting suitability for juice production (5). For some juices, such as citrus (5) or grape (6), only a few varieties are used to provide a distinctive flavor or to ensure freedom from bitterness. When a range of varieties is available, eg, apples, blends are used to achieve uniformity of flavor (4). In general, a blend is perceived as having a better flavor than a single varietal juice. Fruit grown in a warmer climate and having a high sugar (qv) content may be mixed with fruit from a cooler region to achieve a desired sugar–acid balance. This sugar–acid ratio is based on °Brix readings, ie, wt % sugar, obtained with a refractometer, and on total acidity obtained by titration (7).

Juice factories frequently employ field persons to advise growers on the application of sprays to the growing crops so that residues on harvested fruit are within prescribed limits. They also may sample the crop before harvest for analysis, and coordinate harvesting with factory production schedules. Payment for raw materials is frequently based on specifications that are either official government grades or stated market standards. Official graders may be employed to test each load.

Manufacturing

The composition of a number of fruit juices and juice drinks is given in Table 1. Figure 1 describes many of the steps in the production of fruit juices. The processing of citrus and deciduous fruit into juice is discussed separately. Citrus have a thick, relatively tough peel that must be kept separate from the juice during extraction. Deciduous fruit are crushed whole and the juice is then separated from the pulp, peel, and seeds, usually in a pressing operation.

Citrus Processing. In Florida, citrus fruit generally are transported by truck for processing. The truck is unloaded by gravity feed to a conveyor belt that conveys the fruit to a storage bin (see CONVEYING). An official government inspector checks an 18-kg sample of the fruit for minimum standards of °Brix and acidity before it is certified as meeting maturity requirements for processing. The fruit is washed with a detergent as it passes over roller brushes, then rinsed and dried. Washing removes debris and dirt, and reduces the number of microorganisms on the fruit thereby making subsequent juice sterilization processes more effective. Graders remove unwholesome fruit as the fruit passes over roller conveyors and is segregated automatically into several sizes prior to juice extraction by one of several types of extractors (9). Use of computer controlled sizing and grading of each load of fruit, based on fruit color, size, shape, and weight, is increasing. There is potential for total electronic grading and thus total automated quality control.

In one extractor (FMC Inc.), the fruit is located between two cups having sharp-edged metal tubes at their base. The upper cup descends and the many fingers on each cup mesh to express the juice as the tubes cut holes in the top and bottom of the fruit. On further compression, the rag, seeds, and juice sacs are compressed into the bottom tube between the two plugs of peel. A piston moves up inside the bottom tube forcing the juice through perforations in the tube wall. A simultaneous water spray washes the peel oil expressed during extraction away from the peel as an oil–water emulsion; the peel oil is recovered separately from the emulsion.

In another extractor (Automatic Machinery and Electonics Inc. (AMC)) the individual fruits are cut in half as they pass a stationary knife. The halves are oriented in a vertical plane, picked up by synthetic rubber cups, and positioned across plastic serrated reamers revolving in a synchronized carrier in a vertical plane. As the fruit halves progress around the extractor turntable, the rotating reamers exert increasing pressure and express the juice. The oil and pulp contents in the juice increase with greater reaming pressure. The recoverable oil is removed in a separate step prior to juice extraction. Needle-sharp spikes prick the peel of the whole fruit, releasing oil that is washed away with water and recovered from the oil–water emulsion.

Table 1. Composition of Fruit Juices[a,b]

Juice	Water, %	Food energy, kJ[c]	Protein, g	Fat, g	Carbo-hydrate, g	Fiber, g	Ash, g	Ca, mg	Fe, mg	P, mg	K, mg	Na, mg	Vitamin A, IU[d]	Thiamin, mg	Ribo-flavin, mg	Niacin, mg	Ascorbic acid, mg
acerola	94.3	88	0.4	0.3	4.8	0.3	0.2	10	0.5	9	97	3	509	0.02	0.06	0.4	1600
apple	87.9	196	0.1	0.1	11.7	0.2	0.2	7	0.4	7	119	3	1	0.02	0.02	0.1	1
apricot nectar	84.9	236	0.4	0.1	14.4	0.2	0.3	7	0.4	9	114	3	1316	0.01	0.01	0.3	1
cranberry, cocktail	85.0	243	0.1	0.1	14.9		0.1	3	0.2	1	24	4		0.01	0.02	0.1	43
grapefruit																	
white	90.1	158	0.5	0.1	9.0		0.3	7	0.2	11	153	1	7	0.04	0.02	0.2	29
sweetened	87.4	194	0.6	0.1	11.1		0.8	8	0.4	11	162	2	0	0.04	0.02	0.3	27
grape	84.1	256	0.6	0.1	15.0		0.3	9	0.2	11	132	3	8	0.03	0.04	0.3	e
lemon	92.5	89	0.4	0.3	6.5		0.4	11	0.1	9	102	21	15	0.04	0.01	0.2	25
lime	92.5	87	0.3	0.2	6.7		0.3	12	0.2	10	75	16	16	0.03	0.01	0.2	6
orange																	
chilled	88.4	186	0.8	0.3	10.1		0.5	10	0.2	11	190	1	78	0.11	0.02	0.3	33
conc, diluted	88.1	188	0.7	0.1	10.8	0.1	0.4	9	0.1	16	190	1	78	0.08	0.02	0.2	39
orange–grapefruit	88.6	180	0.6	0.1	10.3		0.4	8	0.5	14	158	3	119	0.06	0.03	0.3	29
papaya nectar	85.0	238	0.2	0.2	14.5		0.2	10	0.3	0	31	5	111	0.01	0.01	0.2	3
passion fruit																	
purple	85.6	213	0.4	0.1	13.6	0.1	0.3	4	0.2	13	190		717		0.13	1.5	30
yellow	84.2	253	0.7	0.2	14.5	0.2	0.5	4	0.4	25	278	6	2410		0.10	2.2	18
peach nectar	85.6	225	0.3	e	13.9	0.1	0.2	5	0.2	6	40	7	258	0.01	0.01	0.3	5
pear nectar	84.0	249	0.1	e	15.8	0.3	0.1	5	0.3	3	13	4	1	0.01	0.01	0.1	1
pineapple	85.5	233	0.3	0.1	13.8	0.1	0.3	17	0.3	8	134	1	5	0.05	0.02	0.3	11
prune	81.2	296	0.6	e	17.5	e	0.7	12	1.2	25	276	4	3	0.02	0.07	0.8	4
tangerine	88.9	180	0.5	0.2	10.1	0.1	0.3	18	0.2	14	178	1	420	0.06	0.02	0.1	31
sweetened	87.0	209	0.5	0.2	12.0	0.1	0.3	18	0.2	14	178	1	420	0.06	0.02	0.1	22

[a]Ref. 8. [b]Per 100 g, edible portion. [c]To convert kJ to kcal, divide by 4.184. [d]IU = international units. [e]Trace.

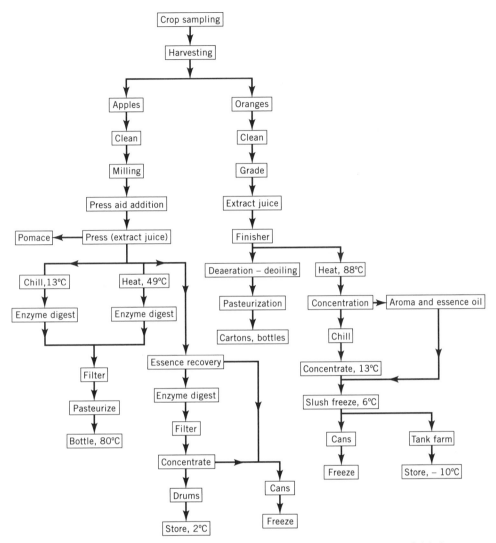

Fig. 1. Manufacturing process for citrus (orange) and deciduous (apple) juices.

In the extraction of citrus juices it is desirable to have as gentle an extraction pressure as possible. There should be minimal contact time between juice and pulp to reduce the amount of bitter substances expressed from the peel into the juice. The amount of suspended solids in citrus juice is controlled in a subsequent separation in a finisher. A screw action is used to force the juice through a perforated screen and separate the larger pulp particles from the juice. The oil level in the juice is adjusted by vaporizing under a vacuum (10). The separated pulp is washed and finished several times to produce a solution which is then either added back to the juice to increase juice yield, or concentrated to produce pulp wash solids, also called water extract of orange solids, which can be used as a cloudy beverage base.

Navel orange juice, grapefruit juice, and pulp wash solids can be excessively bitter. A new (ca 1992) commercial process for debittering these liquids involves separating the pulp by centrifugation or ultrafiltration (qv), passing the clarified juice through a column packed with a neutral resin to selectively remove the bitter juice components, and recombining the pulp with the debittered juice prior to concentration in an evaporator. The debittered juice or pulp wash solids are used in blended orange or grapefruit juice products.

Concentration and Aroma Recovery. Most of the citrus juice sold and transported internationally is as frozen concentrate, usually 60–65 °Brix. Because single-strength citrus juices generally are 7–12 °Brix solutions, a fivefold or greater concentration has occurred, making the concentrate a more economical product to freeze, store, and transport.

Citrus juices are almost exclusively concentrated in a thermally accelerated short-time evaporator (TASTE) which has optimized efficiency for the amount of water removed per kg of steam used; eg, in a seven-effect evaporator, 2.77 kg (6.1 lbs) of water are removed for each kg of steam used. Each effect consists of a bundle of tubes, with falling films of juice flowing inside, mounted inside a cylinder so that steam flowing into the first-effect cylinder can condense on the outside walls of the tube bundle as it transfers heat to vaporize a portion of the water contained in the juice. The hot water vapor thus produced is used to heat juice in the next effect. This results in successively lower temperatures with each successive effect, starting at 98°C in the hottest effect and reaching 45°C in the last effect. Concentrated juice is discharged from the last effect to a vacuum flash cooler producing a product temperature of about 13°C. The system is under vacuum with a progressively higher vacuum in each successive effect sufficient to cause water evaporation.

During water removal, much of the desirable flavor characteristic of the citrus juice is carried off with the vapor, especially in the early effects. Essence recovery units are used routinely to condense the volatile flavor substances as a water-phase essence, industrially termed aroma, and an oil-phase essence oil. These are valuable commercial flavor fractions saved separately to be added back later to concentrated juice. The concentrate is cooled quickly and stored at −10°C in up to 378,541 L (100,000 gallon) refrigerated stainless steel tanks, ie, bulk storage or tank farms. At the time a processor is ready to prepare a commercial juice pack for retail sale, concentrate from several batches in bulk storage is blended to achieve the desired sugar-to-acid ratio and color. Water-phase essence and oil fractions are added to restore flavor lost during concentration. Most of the single-strength orange juice marketed worldwide is prepared by reconstitution of such blended concentrates to 11.8 °Brix juice, followed by pasteurization and chilling prior to retail packaging.

A newer juice concentration process, requiring minimal heat treatment, has been applied commercially in Japan to citrus juice concentration. The pulp is separated from the juice by ultrafiltration and pasteurized. The clarified juice containing the volatile flavorings is concentrated at 10°C by reverse osmosis (qv) and the concentrate and pulp are recombined to produce a 42–51 °Brix citrus juice concentrate. The flavor of this concentrate has been judged superior to that of commercially available concentrate, and close to that of fresh juice (11).

Pasteurized single-strength orange juice (POJ) not from concentrate, often called premium orange juice, is increasing in sales in the United States; the market share of frozen concentrated orange juice is decreasing (12). POJ requires special extraction and storage conditions to ensure high quality. Juices are expressed with extractors designed to produce juice having low oil content (<0.035%). Juice must be chilled and stored aseptically or in frozen blocks to provide a year-round supply.

Unpasteurized orange and grapefruit juices also are increasingly popular products in the United States. Fruit must be extracted under strict sanitary conditions and the juice kept just above 0°C during its approximately 17 day shelf life. Year round juice supply is difficult because long-term storage of fruit or unpasteurized juice is not feasible.

Lemon and Lime Juice. Lemons and Persian limes can be extracted using the same FMC and AMC extractors described above. The juice can be concentrated in a TASTE evaporator, an APV Crepaco, Inc. evaporator, or other types of evaporators (13). Although lime juice, and especially lemon juice, are widely used as condiments on food, the bulk of concentrated juice is used to make frozen concentrated lemonade and limeade. Frozen concentrated limeade, first sold in 1951, is made by the addition of lime oil to lime juice and then the addition of sugar until the °Brix reading reaches 50. The product is frozen and stored at −18°C or lower (13).

Lemon and lime peel oils are the most valuable essential oils recovered from citrus. They are widely used in the soft drink industry in lemon–lime flavored beverages, and lime oil is an ingredient in cola flavor as well (14) (see CARBONATED BEVERAGES; OILS, ESSENTIAL). Lemon oil is used extensively in household products for its pleasant clean aroma. The principal lime oil of commerce is collected from the Key, or Mexican, lime which is too small to be extracted in conventional juice extractors. The whole fruit is crushed in a screw press and the oil is obtained by vacuum deoiling, ie, partial distillation. Because the acidic juice and peel oil are in contact prior to distillation for oil recovery, chemical changes to the lime oil occur which give the oil a strong aroma and flavor not normally present (13). This stronger flavored oil has become the accepted oil of commerce for most lime flavoring uses (see FLAVORS AND SPICES).

Clarified lime juice, made by mixing juice with filter aid prior to passing through a filter press, is the one clarified citrus juice that is a significant article of commerce. The pasteurized bottled juice is popular for drink mixes, punch bases, and fountain drinks (13).

Deciduous Fruit Processing. Fruit used for juicing must be sound fruit, free from gross damage or contamination, especially mold or rot which can lead to tainted juices. Mechanical harvesting can cause bruising and skin penetration which may cause off-flavors or the growth of pathogenic organisms (15). Mechanical harvesting is used extensively on apples and black currants, whereas berries are mainly picked by hand. Most deciduous fruits, including apples, pears, grapes, berries, black currants, and red currants, are processed by milling, ie, comminuting the whole fruit and then pressing the juice from the resulting mash.

Milling. All fruit used for juicing must be thoroughly washed. A variety of methods are used to break the fruit to release juice. These methods depend on the structure of raw material, the clarity desired in the final juice, enzymatic discoloration, and destruction of pectin. The most common disintegrator used in North America is the hammermill, or a variation of it, where fixed or free-swinging hammers force the fruit particles through a screen. The degree of disintegration is varied by type of hammer used, ie, blunt hammers for impact disintegration or sharp hammers for chopping; diameters of the screen holes beneath the hammers; and hammer speed. Some processors believe this type of breakdown gives higher juice yields than other methods (15).

Another common system is the fixed knife mill. This consists of a chamber containing fixed knives and a rotating three-armed spider that shreds the fruit by forcing them against the fixed knives. Fruit are gravity fed into the chamber from an opening in the top. Small holes under the knives permit the mash to fall through to a conveyor.

Milling is essential for relatively hard fruit such as apples, whereas grapes and berries need only light crushing. Some soft fruit, such as raspberries and strawberries, are received frozen and must be defrosted prior to processing. Grapes and other fruit harvested in clusters are mechanically destemmed prior to crushing. For crushing, a series of intermeshed arms mounted on two cylinders shred the grapes as they pass through to produce a mash.

The pulp from stone fruit, such as apricots and peaches, must be removed in a way that prevents damage to the stone; the stone often contains components which can affect juice flavor and storage stability. A stoned fruit mill contains hard rubber-lobed wheels that rotate together to force the fruit down and strip most of the pulp from the intact stone. An alternative process for peaches involves halving and pitting the fruit in a Filper twist pitter, pulping the fruit halves, and heating the pulp to 82°C in a steam cooler. A fruit disintegrator is then used to produce a puree which is put through a finisher. Such pulpy fruit purees are treated with sugar syrup to produce nectar drinks containing 25–50% juice (16).

Enzyme Treatment of Pulp.   Because of their cell structure, some fruits are not easily juiced by milling and pressing. Commercial pectolytic enzymes are widely used to break down cell structure and dissolve pectins in the juice to improve extraction efficiency. The principal disadvantage to enzyme use is the cost relative to the additional juice yield obtained. For fresh apples and pears, efficient juicing can occur without the use of enzymes, but after a period of storage the cell structure changes and enzyme treatment is of greater benefit. Many other fruits benefit from enzyme treatment prior to juicing as well, including strawberries, cherries, plums, black currants, and raspberries (14). Fruit for processing is received either frozen or at ambient temperature. It usually must be mixed with the pectolytic enzyme and heated for 1–2 h for optimum pressing efficiency, eg, 1 h at 15–30°C for apples (see ENZYME APPLICATIONS).

Press Aids.   Fibrous materials are added to the fruit pulp to facilitate juice liberation, or to provide a stable structure to withstand the application of pressure (14). Several types of press aid are used, eg, diatomaceous earth (see DIATOMITE), coarse wood flour, rice hulls, and bleached or unbleached wood pulp. Diatomaceous earth and wood flour may be sprinkled over press cloths, or added to the pulp before the cloths are filled for hydraulic pressing, thereby making discharge of the spent cakes easier and also increasing the amount of juice. When continuous presses are used, more fibrous press aids are required to develop a cake structure that permits compression without slippage against a revolving screw. The most common materials for this are bleached and unbleached wood pulp which liberates a minimum of soluble materials that may affect the flavor or color of the juice adversely. Unbleached wood pulp is available in bale form and must be shredded, whereas the sheet form of bleached pulp is frequently defibrated in a hammermill.

The press aid must be thoroughly dispersed throughout the fruit pulp before it enters the press. For this purpose, a mixing trough equipped with open-loop paddles made from square barstock has been used. The composition of the fruit entering the press by gravity is controlled by feeding the pulp to the mixer with a variable speed screw pump. The addition of unbleached pulp, when it is used, is controlled by the rate of feed to the shredder; when bleached pulp is used, by the rate of feeding to the hammermill. The amount of bleached or unbleached pulp added to the fruit may vary 0.5–5 wt %, depending on type, variety, and maturity of the fruit, degree of defibration of the press aid, and thoroughness of dispersion in the fruit pulp. If too little fiber is added, the juice yield is low and there is an exceptional amount of suspended solids in the juice. Too much press aid gives a decreased juice yield owing to absorption by the press aid; the cake may be so dry that it cannot be discharged from the press. Rice hulls are frequently added at the rate of 0.25–1% to prevent blinding of the screen holes in the press cage by fruit particles.

Pressing.   There are a variety of fruit presses. Some presses are more suitable for one type of fruit than for others, but most can be used for any fruit with varying degrees of success. Some presses require the use of a press aid such as diatomaceous earth or wood chips. Total juice yield from the original fruit defines press efficiency. Additional yield can be obtained by adding a small amount of water to the press cake and pressing again.

A rack and frame press uses heavy nylon cloth positioned in a wooden frame inside a rack. A measured amount of apple or other fruit mash is added from a hopper above the frame. The mash is leveled with a hand trowel and the edges of the nylon cloth are folded over the mash to encase it and create a cheese. The frame is removed, and a second rack is placed on top of the first cheese; the process is repeated until a stack of cheeses is prepared. A hydraulic ram then applies gradually increasing pressure on the stack and expresses the juice. A high yield of juice (80%) is obtained and no press aid is required. Because this process is labor intensive (17), it is mostly used for small farm and pilot-plant operations.

The Bucher-Guyer horizontal rotary press is a highly automated batch process machine that requires no press aid. The press consists of a horizontal hydraulic ram inside a rotating cylinder containing many flexible rods covered with a knitted synthetic fabric. The rods have serrated surfaces to allow juice which passes through the fabric to flow to the discharge ends. Hydraulic pressure is applied for a preset time, the ram is retracted, and the cylinder is rotated to break up the press cake. This cycle is repeated several times before the press cake is removed from the cylinder and the press is cleaned (16). Juice yield for this horizontal rotary press is 84%; with secondary water addition it is increased to 92% (15).

The Stoll press uses a horizontal hydraulic ram and a chamber with screens on the sides that holds the mixture of fruit mash and press aid. Hydraulic pressure is applied to the mash first at 2750–3440 kPa (400–500 psi) and then at 17,200–20,600 kPa (2500–3000 psi) for preset times; the pomace is removed after each cycle. One person can operate several presses at once, each with a 4–5 t/h throughput.

A belt press commonly uses two belts of woven material. Fruit mash is deposited on, and spread out evenly between, the two belts which converge and press the mash between them. The belts wrap around a series of successively smaller cylinders or pass between rollers (less efficient) for the pressing action. Juice yield for the wraparound belt press is 72%, and increases to 92% with secondary water addition (15).

Screw presses have been used extensively in North America, but are not popular in Europe. The mash is added at the top of a vertical rotating screw. As the mash moves down it is compressed in the taper and juice flows out through slatted conical walls. Compression of the pomace blocks channels for juice flow and inhibits juice extraction. Extensive use of press aids is necessary for efficient operation.

Concentration and Aroma Recovery. Concentration of juice from deciduous fruit is best carried out using an evaporator that causes as little thermal degradation as possible and that permits recovery of volatile materials important to the aroma of the fresh fruit, ie, essence. Evaporators that use a high temperature for a short time and operate under a vacuum, such as the APV Crepaco falling film plate evaporator or the Alfa Laval centrifugal evaporator, are most commonly used. The TASTE evaporator, almost universally used for citrus, is rarely if ever used to concentrate deciduous fruit juices.

Multiple effect falling film vacuum plate evaporators use a plate-and-frame heat exchanger to heat the juice to the required temperature for boiling under partial vacuum. The plate evaporator consists of multiple gasketed plates arranged in series within a compact frame. As feed juice flows in a single pass as a falling film, the water is vaporized on contact with steam heated plates and is discharged as a liquid–vapor mixture to a large volume separator. The concentrated juice flows out the bottom and the water vapors are used to heat the next effect or are condensed. For maximum energy efficiency the warm concentrate stream is used to preheat feed juice (18). Falling film plate evaporators are especially useful for heat-sensitive fruit juice because of short heat contact time, rapid start-up and shutdown times, and minimum juice loss from a small liquid hold up volume (15). They also can be used by a single processor for multiple fruit crops during the year.

Centrifugal evaporators are also suitable for concentrating fruit juices because these have very low heat contact times. The expanding flow centrifugal evaporator contains a stack of rotating, steam-filled hollow cones. As juice is sprayed onto the inside of the cones near the apex, centrifugal force spreads juice across the inner surfaces of the cones. The resulting concentrate is forced to the stationary walls of the evaporator and collected. Steam condensate on the outer surface of each cone is similarly forced out by centrifugal force so that the heating surface in contact with the product is constantly heated with hot dry steam. Thus juice concentrate and steam condensate are continuously removed. Juice can be concentrated in one pass with a few seconds contact time (15).

There are two methods available for aroma recovery. In one method, a portion of the water is stripped from the juice prior to concentration and fractionally distilled to recover a concentrated aqueous essence solution. Apple juice requires

10% water removal, peach 40%, and Concord grape 25–30% to remove volatile flavor as an essence. Fractional distillation affords an aqueous essence flavor solution of 100–200-fold strength, which means the essence is 100 to 200 times more concentrated in flavor than the starting juice. A second method of essence recovery is to condense the volatiles from the last effect of the evaporator; they are enriched in volatile flavor components (18).

Essences generally are stored separately from the bulk concentrates for stability, and their addition prior to retail packaging is essential to restoring much of the natural fresh flavor of the starting juice otherwise lost during processing. Unlike citrus, which affords both an aqueous and an oil-phase essence, only an aqueous-phase essence is obtained for deciduous fruit. Virtually no essential oil is present in the peel or juice in the latter.

Clarification. The clarities of fruit juices vary widely. Citrus and pineapple juices have high insoluble solids contents, whereas cranberry, grape, and many apple juices have little or no suspended solids and are nearly translucent. In some juices, such as citrus, many of the flavoring constituents adhere to the cloud, and the clarified juices lack flavor (19). For this reason, and the unsightly appearance of sediment in bottled juice, considerable research has been devoted in the citrus industry to the causes of solids flocculation. Preservation of the initial pectin content by rapid heating and cooling to inactivate the natural enzymes has been found to be important. Size of the suspended particles is another factor influencing flocculation; this may be adjusted by screening, homogenization, colloid milling, or selective filtration.

In the production of opalescent or natural-type apple juice, ascorbic acid is added to the fruit pulp before pressing, or to the juice as it comes from the press, to retain more of the apple flavor (4). Ascorbic acid addition and pasteurization of the juice as soon as possible after pressing prevent polyphenol oxidation, which causes browning and contributes to pulp flocculation.

In the production of clear juices, the natural pectin is degraded with pectolytic enzymes to facilitate filtration and prevent the subsequent precipitation of the pectin. Other sources of colloidal haze in filtered juices are polymerized tannins, or compounds formed by the combination of tannins and proteins. The haze also may result from the denaturation of proteins. To overcome tannin precipitates, gelatin is sometimes added to the juice during enzyme digestion. The amount and type of gelatin added to form a floc must be carefully controlled because an excess of gelatin causes a haze. Laboratory tests are frequently conducted by adding graduated amounts of gelatin and tannin to samples of juice and noting the amounts required for greatest clarity after standing overnight. Such tests are frequently necessary because of the variation in the natural tannin and protein contents of the juice. The use of kieselsol [7631-86-9], a synthetic silica sol, in combination with gelatin (4), or honey and pectinase enzyme mixtures (20), increase the speed of the clarification process so that in-line clarification is possible.

Digestion of juice with commercial pectolytic enzyme preparations may be carried out at 13°C for 15–18 h or at 49°C for 1 or 2 h. Intermediate digestion temperatures are not used because of the possibility of alcoholic fermentation. If a higher digestion temperature is used, the juice may be clarified by rotary vacuum precoat filtration followed by a small-cartridge or leaf-type filter to remove any filter aid. If juices are digested at low temperatures, sufficient time usually elapses for flocculation of the solids, and the clear juice is decanted and passed through a pressure-leaf filter. The filter aids used are diatomaceous earth of fine to medium grades depending on the clarity desired and the nature of the solids. Completeness of pectin destruction is of importance in maintaining good filtration rates and filtrate clarity.

Use of ultrafiltration (UF) membranes is becoming increasingly popular for clarification of apple juice. All particulate matter and cloud is removed, but enzymes pass through the membrane as part of the clarified juice. Thus pasteurization before UF treatment to inactivate enzymes prevents haze formation from enzymatic activity. Retention of flavor volatiles is lower than that using a rack-and-frame press, but higher than that using rotary vacuum precoat-filtration (21).

Natural style juices that contain the cloud are increasing in popularity, especially apple juices, because these retain more fresh flavor if processed carefully. Optimum processing conditions chill the fruit to 4°C before milling, add 500 ppm ascorbic acid to retard browning, press under nitrogen, and flash pasteurize the juice as quickly as possible (4).

To avoid formation of tartrate crystals in bottled grape juice, the filtered juice is stored at -2°C, or just above the freezing point, for several weeks to permit the deposition of potassium acid tartrate crystals (6). Prior to storage, the juice is pasteurized at 88°C and then cooled in storage at -2°C before it is pumped into clean storage tanks. Unless care is taken in the sanitation of the pasteurization and cooling equipment, piping, and tanks, yeast growth may cause alcohol production or generation of off-flavors.

Pasteurization. Fruit juices generally have pH ≤ 4 and spoilage is usually limited to bacteria and yeast and mold growth (14). Flash pasteurization minimizes flavor changes owing to heat treatment and is recommended for premium quality and natural style juices. For flash pasteurization several pasteurization methods are used, eg, a plate heat exchanger with a separate holding tube. The plate heat exchanger is a bank of stainless steel plates separated from each other so that heating medium and juice can transfer heat through the plates without direct contact with the product.

Regeneration-type plate pasteurizers consist of a preheating section in which hot pasteurized juice is used as the heating medium to preheat incoming juice and simultaneously cool the pasteurized juice, thus saving energy (9). The preheated juice is further heated with steam or hot water to pasteurization temperature. A holding section within the plate pasteurizer maintains the temperature long enough to assure adequate pasteurization. A cooling section using chilled water may be needed, but the initial preheating section using pasteurized juice may be so efficient that the exit temperature for pasteurized juice approaches ambient temperature and no further cooling is needed (15).

Pectic enzymes are inactivated by pasteurization. Citrus juices require higher temperatures for enzyme deactivation than for pasteurization. Heat treatment at 85–94°C for 30 s inactivates pectic enzymes (9) and is more than adequate for pasteurization.

Hot filling of the container with juice pasteurizes the package as well as the product, thus ensuring product sterility. In this process, bottles or cans are filled with juice preheated to a temperature sufficiently hot to pasteurize the container and closure (lid) without further heating. Immediate inversion of the sealed container is necessary to assure sterilization of the lid. After the pasteurization period the containers are cooled as quickly as possible. An alternative procedure is to heat the pack with hot water sprays in a tunnel, usually in three stages. Initially the spray preheats the pack in the first one-third of the tunnel; sprays in the next third heat the pack to pasteurization temperature; and the final third is a cooling section with progressively cooler sprays so that a cooled pack emerges (15).

Aseptic packaging in flexible laminated multilayer plastic containers has become popular worldwide because the packaging materials are economical and the product can be stored at ambient temperature (22). In such packaging the juice must be heat pasteurized, cooled to prevent heat damage to the package, and introduced into the package in a sterile atmosphere. The package is sterilized with hydrogen peroxide (qv) or other approved sterile solution prior to juice addition. These aseptically packaged juice drinks have a shelf life of 6–8 months at room temperature.

Chemical Preservatives. Many governments require a listing of ingredients, including preservatives, on the container label, and only permit use of those materials considered to be noninjurious to health. The most common preservatives used in fruit juices are the potassium and sodium salts of sorbic, benzoic, and sulfurous acids. The maximum amount of sodium benzoate [532-32-1] permitted is 0.1 wt %; a common rate of addition to apple juice is 0.05–0.075 wt % (23). In an acidic juice, the benzoic acid [65-85-0] may impart an objectionable flavor at concentrations greater than 0.08 wt %. The storage life of juices treated with chemical preservatives is seriously curtailed if they are heavily contaminated during preparation, or not refrigerated as much as possible. The amount of sulfur dioxide [7446-09-5], SO_2, added as a gas or as a sulfite salt varies widely with pH and solids content of the juice or concentrate. At low pH, free SO_2 is the most effective form. At high pH, dissociated SO_2 is bound by sugars, aldehydes, etc, and is not available as an antiseptic. Effective amounts of SO_2 cannot be used in canned beverages because of accelerated corrosion, with thiamin, or with certain natural coloring agents (23).

Freezing Preservation. Fruit juices can be preserved by freezing. The juice is filled into waxed paper or plastic containers approved for food use so that a 10 vol % headspace is left for expansion during freezing. The juice should be as cold as possible before filling and in some cases the juice or concentrate is chilled to a slush in a wiped-film heat exchanger. The containers are placed on refrigerated shelves with provision for forced-air circulation around them, or in the case of round containers, conveyed with rotation through a liquid refrigerant. The frozen juice should be held at a uniform temperature of $-18°C$ to reduce ice crystal growth and damage to the colloidal solids. Frozen orange juice that has not been pasteurized or otherwise processed is sold in limited quantity in the United States. The convenience of having a ready-to-serve single-strength juice is partly lost by the consumer having to thaw the product before use.

Dehydration. Drying pure fruit juices is difficult because of the low melting points of the solids and their hygroscopic nature owing to the presence of fructose (20). Under conditions of low water vapor pressure, as in freeze-drying, the juice must be frozen in such a manner that the ice crystals are thoroughly distributed throughout the mass, or a concentrated syrup results rather than dry solids. However, by the addition of various ingredients such as dextrose, sucrose, lactose, or maltodextrin, the hygroscopicity of the solids may be reduced. Such additions have a dilution effect so that there are less fruit solids in the final powdered form. Vitamin and flavor losses during drying can be restored when formulating the finished product (23). In general the flavor and appearance of dried juice powders are poorer than the equivalent liquid concentrated fruit beverages and so these have not become popular with the consumer.

Tropical Fruit

Pineapple juice has been the most popular tropical fruit juice consumed in the United States for many years. It has long been a by-product of the pineapple canning industry. Juice pressed from the core, and from trimmings of fruit cut into cylinders for canning, is combined with juice collected from the cutting table to provide most of the pineapple juice sold commercially (24). This juice contains up to 40% pulp. The pulp content is reduced, usually by finishing with a screw-type finisher followed by centrifugation, before pasteurization and concentration. Pineapple juice can be concentrated in a falling film plate-type evaporator with an essence recovery step prior to concentration (24,25). As with other fruit juices, concentrated pineapple juice with added aqueous essence of either 60 °Brix with high pulp content, or 72 °Brix with low pulp content, is the principal product traded internationally. A higher quality pineapple juice made using whole ripe pineapples is being sold at a premium price in the European Economic Community (EEC), accounting for nearly 20% of the pineapple juice consumed.

Pineapple juice has been available commercially since 1932, but the production and sale of other tropical fruit juices has more recently received significant attention in Europe and, especially, in North America (24). Many tropical fruit juices are too pulpy or have harsh or exotic flavors which make 100% juice products unacceptable to most U.S. consumers. They are more acceptable as nectars containing 25–50% juice or as blended fruit drinks where their strong flavors are diluted or modified.

Because of high pulp content, most tropical fruit are difficult to pasteurize or concentrate in a conventional manner without severe flavor change from excess heat treatment. Some are stored and sold as frozen purees, but others, such as yellow passion fruit, can be concentrated effectively and are sold as 50 °Brix concentrates. The APV Crepaco Paravap, similar to the falling film plate evaporator, has plates especially designed for high vapor velocities and turbulences which afford a high rate of heat transfer and are amenable to concentration of fruit purees (18). As with other fruit juices, recovery of aqueous essence is usually necessary for production of high quality concentrates.

Mango and papaya are tropical fruits available in limited supply as concentrated juices or purees. Available single-strength purees include guava, banana, kiwifruit, lulo, soursop, and umbu (24). The international market for tropical fruit drinks is in its infancy, but is expanding. Blends with more traditional juices such as orange and apple are some of the more successful drinks marketed.

Fruit Juice Drinks

In most markets fruit juice must be 100% juice and contain no additives. Fruit nectars consist of pulp, juice, sugar, and water, and contain from 25 to 50% juice, depending on the fruit used. In the EEC the minimum juice content is 50% for orange and apple, 40% for apricot, and 25% for passion fruit and guava (26). Other fruit juice drinks include cocktails, which usually contain at least 25% juice; and a variety of juice drinks which can contain from 1.5–70% juice (Table 2).

　　Cranberry juice, too acidic to be consumed as a 100% juice drink, has been sold since 1929 as cranberry juice cocktail. Juice extraction usually involves pressing the juice from thawed cranberries in a tapered screw press, which affords a 60–64% juice yield. The juice is diluted with two volumes of water and sugar is added to raise the °Brix to 15 to produce a juice cocktail. Under the Federal Food, Drug and Cosmetic Act, cranberry juice cocktail must contain not less than 25% single-strength cranberry juice with soluble solids content of 14–16 °Brix, vitamin C content of 30–60 mg/177 mL (6 oz), and citric acid content not less than 0.55 g/100 mL (16).

　　Commonly, a juice drink contains 10% fruit juice, which usually is a blend of several fruits. The 1990 Federal Nutrition and Labeling Act requires declaration of juice content so that the consumer can make a more informed choice (3). With cocktails and juice drinks, added sugars, acids, flavorings, colorings, and

Table 2. Comparison of Fruit Juice and Beverage[a]

Characteristics	Unsweetened fruit juice	Fruit beverage, including nectars
composition,%	100 (fruit)	1.5–70 (fruit or juice)
total sugars, %	8–17	0.1–20
energy, kJ/100 mL[b]	146.4–271.9	2.1–313.8
acid,[c] %	0.6–1.5	0.1–0.8
pH	2.5–4.0	2.5–4.0
nutrients	vitamins, mainly C, some A; potassium	as added; vitamin C common, a minority is multivitamin
other ingredients	acid, vitamins, preservatives	sugar, acid, intense sweeteners, flavorings, colorings, vitamins, minerals, preservatives
flavor	distinctive full flavor and body, blends of flavors not common	juice enhanced by added extract or flavorings, range of mouthfeel, blended flavors common
nutrition	no low calorie juices	usually lower calorie; can be fortified as desired
health and safety	distinctive of fruit	sugar level and acidity selected by manufacturer

[a]Ref. 23.
[b]To convert kJ to kcal, divide by 4.184.
[c]As citric acid. High acid fruit juice may have up to 6% acid.

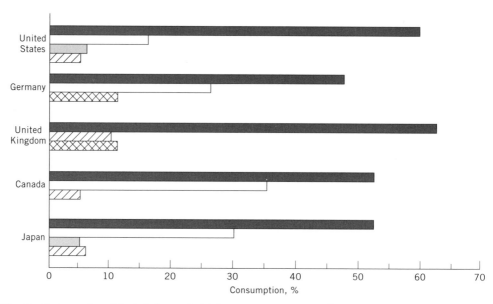

Fig. 2. International fruit juice and drink consumption (12): ■ orange, □ apple, ▨ grape-fruit, ▨ grape, and ▨ multivitamin or blends. Values are in percent of total juice/drink consumption.

nutrients can be used to provide a wide variety of stable products of uniform quality. Because drinks require less juice than 100% juice products, the drinks can be sold at a lower price.

Consumption in the fruit drink/nectar market varies widely from country to country. In the United States juice consumption is 73% vs only 27% for drink/nectar consumption. In Japan the reverse is true with drink/nectar accounting for 73% of the market. In the United Kingdom, 80% of the market is derived from fruit juice, but in Europe as a whole, drink/nectar outsells pure juice by 30% (12).

World Supply and Consumption

World trade in fruit juices for 1989 totaled $3.9 billion, of which 58% was orange, 13.8% apple, 5.4% grape, 4.6% pineapple, 4% grapefruit, 3.1% mixed, 2.9% other citrus, and an estimated 5% tropical fruit other than pineapple, mostly banana, passion fruit, and mango (27). However, these figures do not include the large amount of fruit grown, processed, and consumed within a single country.

The principal fruit juice and beverage markets are the United States, where 12.1×10^6 L is consumed per year; followed by Canada, 5.6×10^6 L; Western Europe, 1.2×10^6 L; and Japan, 0.76×10^6 L. In all significant markets orange juice predominates (Fig. 2). In the United States apple juice is second, followed by grape and then grapefruit juice (12).

BIBLIOGRAPHY

"Fruit Juices" in *ECT* 2nd ed., Vol. 10, pp. 167–178, by M. K. Veldhuis, U.S. Fruit and Vegetable Products, USDA; in *ECT* 3rd ed., Vol. 11, pp. 300–316, by J. C. Moyer, New York State Agricultural Experiment Station.

1. C. S. Pederson, in P. E. Nelson and D. K. Tressler, eds., *Fruit and Vegetable Juice Processing Technology*, 3rd ed., AVI Publishing Co., Westport, Conn., 1980, pp. 268–309.
2. S. W. Shear, in D. K. Tressler and M. A. Joslyn, eds., *Fruit and Vegetable Juice Processing Technology*, AVI Publishing Co., Westport, Conn., 1961, pp. 1–63.
3. M. Brown, R. L. K. Kilmer, and K. Bedigian, in S. Nagy, C. S. Chen, and P. E. Shaw, eds., *Fruit Juice Processing Technology*, AgScience, Auburndale, Fla., 1993, pp. 1–22.
4. A. G. H. Lea, in D. Hicks, ed., *Production and Packaging of Non-carbonated Fruit Juices and Fruit Beverages*, Van Nostrand Reinhold, New York, 1990, pp. 182–225.
5. H. E. Nordby and S. Nagy, in Ref. 1, pp. 35–96.
6. M. R. McLellan and E. J. Race, in Ref. 4, pp. 226–242.
7. K. Helrich, ed., *Official Methods of Analysis*, Vol. 2, 15th ed., Association of Official Analytical Chemists, Arlington, Va., pp. 910–928.
8. S. E. Gebhardt, R. Cutrufelli, and R. H. Matthews, "Composition of Foods," in *Agriculture Handbook No. 8-9*, U.S. Dept. of Agriculture, Washington, D.C., Aug. 1992.
9. H. M. Rebeck, in Ref. 4, pp. 1–32.
10. D. A. Kimball, *Citrus Processing Quality Control and Technology*, Van Nostrand Reinhold, New York, 1991.
11. S. Cross, *Proc. Fla. State Hortic. Soc.* **102**, 146–152 (1989).
12. P. Mittal, in R. F. Matthews, ed., *Proceedings of the 1990 Food Industry Short Course*, Institute of Food and Agricultural Sciences, University of Florida, Gainesville.
13. H. E. Swisher and L. H. Swisher, in Ref. 1, pp. 144–179.
14. A. H. Johnson, in A. H. Johnson and M. S. Peterson, eds., *Encyclopedia of Food Technology*, AVI Publishing Co., Westport, Conn., 1974, pp. 159–162.
15. J. W. Downes, in Ref. 4, pp. 158–181.
16. B. S. Luh, in Ref. 1, pp. 436–505.
17. V. L. Bump, in D. L. Downing, ed., *Processed Apple Products*, Van Nostrand Reinhold, New York, 1989, pp. 53–82.
18. *APV Plate Evaporators*, APV technical bulletin T-1-100, section 13, APV Crepaco, Inc., Tonawanda, N.Y., 1992.
19. S. Nagy and P. E. Shaw, in I. D. Morton and A. J. MacLeod, eds., *Food Flavours Part C. The Flavour of Fruits*, Elsevier, New York, 1990, pp. 93–124.
20. M. R. McLellan, R. W. Kime, and L. R. Lind, *J. Food Sci.* **50**, 206–208 (1985).
21. M. A. Rao and co-workers, *J. Food Sci.* **52**, 375–377 (1987).
22. P. E. Shaw, in I. D. Morton and A. J. MacLeod, eds., *Food Flavours Part B. The Flavour of Beverages*, Elsevier, New York, 1986, pp. 337–368.
23. D. Hicks, in Ref. 4, pp. 264–306.
24. J. Hooper, in Ref. 4, pp. 114–136.
25. F. P. Mehrlich and G. E. Felton, in Ref. 1, pp. \$180–211.
26. R. Korbech-Olsen, *Int. Trade Forum* **26**(4), 12–17 (1990).
27. *The World Market for Fruit Juice: Citrus and Tropical*, CCP:CI 91/4, U.N. Food and Agriculture Organization, Rome, May 1991.

PHILIP E. SHAW
Citrus and Subtropical Products Laboratory, USDA

FUEL CELLS

Fuel cells are electrochemical devices that convert the chemical energy of a fuel directly into electrical and thermal energy. In a typical fuel cell, gaseous fuels are fed continuously to the anode (negative electrode) compartment, and an oxidant, eg, oxygen or air, is fed continuously to the cathode (positive electrode) compartment. The electrochemical reactions take place at the electrodes to produce an electric (direct) current. The fuel cell theoretically has the capability of producing electrical energy for as long as the fuel and oxidant are fed to the electrodes. In reality, degradation or malfunction of components limits the practical operating life of fuel cells.

Besides the direct production of electricity, heat is also produced in fuel cells. This heat can be effectively utilized for the generation of additional electricity or for other purposes, depending on the temperature. A practical consideration for fuel cells is compatibility with the available fuels and oxidants. As of this writing the electrochemical reactions involving hydrogen and oxygen (or air) are the only practical ones. The oxygen (qv) is usually derived from air. Hydrogen (qv) is available from several fuel sources, eg, steam-reformed fossil fuels (see COAL; PETRO-LEUM), coal gasification (see COAL CONVERSION PROCESSES), steam-reformed methanol (qv), etc (see FUELS, SYNTHETIC–GASEOUS FUELS). Very little practical success has been achieved using the direct electrochemical oxidation of hydrocarbon-based liquids or gases in fuel cells operating at 200°C or lower.

General Characteristics

One of the main attractive features of fuel cell systems is the expected high fuel-to-electricity efficiency. This efficiency, which runs from 40–60% based on the lower heating value (LHV) of the fuel, is higher than that of almost all other energy conversion systems (see POWER GENERATION). In addition, high temperature fuel cells produce high grade heat which is available for cogeneration applications. Because fuel cells operate at near constant efficiency, independent of size, small fuel cells operate nearly as efficiently as large ones. Thus fuel cell power plants can be configured in a wide range of electrical levels from watts to megawatts. Fuel cells are quiet and operate with virtually no noxious emissions, but they are sensitive to certain fuel contaminants, eg, CO, H_2S, NH_3, and halides, depending on the type of fuel cell. These contaminants must be minimized in the fuel gas. The two primary impediments to the widespread use of fuel cells are high initial cost and short operational lifetime. These two aspects are the focus of research.

Types of Fuel Cells

A variety of fuel cells has been developed for terrestrial and space applications. Fuel cells are usually classified according to the type of electrolyte used in the cells as polymer electrolyte fuel cell (PEFC), alkaline fuel cell (AFC), phosphoric acid fuel cell (PAFC), molten carbonate fuel cell (MCFC), and solid oxide fuel cell (SOFC). These fuel cells are listed in Table 1 in the approximate order of increasing operating temperature, ranging from $\sim 80°C$ for PEFCs to $\sim 1000°C$ for SOFCs. The physicochemical and thermomechanical properties of materials used for the cell components, ie, electrodes, electrolyte, bipolar separator, current collector, etc, determine the practical operating temperature and useful life of the cells. The properties of the electrolyte are especially important. Solid polymer and aqueous electrolytes are limited to temperatures of ca 200°C or lower because of high water-vapor pressure and/or rapid degradation at higher temperatures. The operating temperature of high temperature fuel cells is determined by the melting point (MCFC) or ionic-conductivity requirements (SOFC) of the electrolyte. The operating temperature dictates the type of fuel that can be utilized.

The low temperature fuel cells utilizing aqueous electrolytes are, in most practical applications, restricted to hydrogen as the fuel. The presence of carbon monoxide (qv) and sulfur-containing gases are detrimental to fuel cell performance because these poison the anode in low temperature fuel cells. In high temperature fuel cells, the list of usable fuels is more extensive for two reasons: the inherently rapid electrode kinetics and the lessened need for high electrocatalytic activity at high temperature. In addition, there are options for hydrocarbon fuels which can be utilized either directly or indirectly (see HYDROCARBONS).

In low temperature fuel cells, ie, AFC, PAFC, PEFC, protons or hydroxyl ions are the principal charge carriers in the electrolyte, whereas in the high temperature fuel cells, ie, MCFC, SOFC, carbonate and oxide ions are the charge carriers in the molten carbonate and solid oxide electrolytes, respectively. Fuel cells that use zirconia-based solid oxide electrolytes must operate at about 1000°C because the transport rate of oxygen ions in the solid oxide is adequate for practical applications only at such high temperatures. Another option is to use extremely thin solid oxide electrolytes to minimize the ohmic losses.

Polymer Electrolyte Fuel Cell. The electrolyte in a PEFC is an ion-exchange (qv) membrane, a fluorinated sulfonic acid polymer, which is a proton conductor (see MEMBRANE TECHNOLOGY). The only liquid present in this fuel cell is the product water; thus corrosion problems are minimal. Water management in the membrane is critical for efficient performance. The fuel cell must operate under conditions where the by-product water does not evaporate faster than it is produced because the membrane must be hydrated to maintain acceptable proton conductivity. Because of the limitation on the operating temperature, usually less than 120°C, H_2-rich gas having little or no ($<$ a few ppm) CO is used, and higher Pt loadings than those used in PAFCs are usually required in both the anode and cathode.

The advantages of PEFCs are no free corrosive liquid in the cell, simple fabrication of the cell, ability to withstand large pressure differentials, materials corrosion problems are minimal, and demonstrated long life, ie, membrane life exceeds 100,000 h. On the other hand, the disadvantages of PEFCs are that the fluorinated polymer electrolyte is traditionally expensive, water management in the membrane is critical for efficient operation, and long-term high performance with low catalyst loadings in the electrodes needs to be demonstrated.

The use of organic cation-exchange membrane polymers in fuel cells was conceived in the 1950s. The early membranes tested in PEFCs include the hydro-

Table 1. Fuel Cell Components and Operating Conditions[a,b]

Characteristic	PEFC	AFC[c]	AFC[d]	PAFC	MCFC	SOFC[e]
anode	Pt black or Pt/C	80% Pt–20% Pd	Ni	Pt/C	Ni–10% Cr	Ni–ZrO_2 cermet
cathode	Pt black or Pt/C	90% Au–10% Pt	Li-doped NiO	Pt/C	Li-doped NiO	Sr-doped $LaMnO_3$
pressure, MPa[f]	0.1–0.5	0.4	~0.4	0.1–1	0.1–1	0.1
temperature, °C	80	80–90	260	200	650	1000
electrolyte, wt %	Nafion[g]	35–45% KOH	85% KOH	100% H_3PO_4	62% Li_2CO_3–38% K_2CO_3[h]	yttria-stabilized ZrO_2

[a]AFC = alkaline fuel cell; MCFC = molten carbonate fuel cell; PAFC = phosphoric acid fuel cell; PEFC = polymer electrolyte fuel cell; and SOFC = solid oxide fuel cell.

[b]All cells are bipolar having a filter-press or flat-plate construction, except where otherwise indicated.

[c]Used in the space shuttle *Orbiter*.

[d]Used in the *Apollo* program.

[e]Tubular cells.

[f]To convert MPa to psi, multiply by 145.

[g]Fluorinated sulfonic acid, registered trademark of E. I. du Pont de Nemours & Co., Inc.

[h]In mol %.

carbon-type polymers such as cross-linked polystyrene–divinylbenzene–sulfonic acids and sulfonated phenol–formaldehyde. The U.S. Gemini Space Program in the 1960s used a fuel cell module which had the dimensions: 31.7-cm diameter, 63.5-cm high, 30 kg, 1 kW at 23.3–26.5 V; the cell operated at 37 mA/cm^2 at 0.78 V on pure H_2 and O_2 at 138–207 kPa (20–30 psi) and ~35°C. Membranes were polystyrene–divinylbenzene–sulfonic acid cross-linked within an inert fluorocarbon film. The life of PEFCs was limited by oxidative degradation of the polymer electrolyte. When these polystyrenes were replaced with fluorine-substituted polystyrenes, eg, polytrifluorostyrene sulfonic acid, the life of PEFCs was extended by four to five times. However, the operating temperature of PEFCs using fluorinated polystyrenes was limited to less than 75°C. The development of Nafion (Du Pont), ie, perfluorocarbon sulfonate sulfonic acid, membranes yielded electrochemical stability in PEFCs at temperatures up to about 100°C. This polymer consists of the ionomer units shown where $n = 6$–10 and $m \geq 1$.

$$+(CF_2-CF_2)_n-CF_2-CF+$$
$$[OCF_2-CF(CF_3)]_m-OCF_2-CF_2-SO_3H$$

Nafion and its derivatives all have two features in common. The polymer chains consist mainly of a poly(tetrafluoroethylene) (PTFE) backbone, which statistically forms segments several units in length, and perfluorinated vinyl polyether, a few ether links long. The latter joins the PTFE segments to form a flexible branch pendent to the main perfluoro-chain and carries a terminal acidic group to provide the cation-exchange capacity. These perfluorinated ionomer membranes with sulfonic acid groups meet all the required characteristics of ion-exchange membranes for use in fuel cells, as well as for use in H_2O and alkali hydroxide electrolysis cells. Nafion, first used in fuel cells in 1966, is the most widely used ion-exchange membrane in PEFCs (see also ALKALI AND CHLORINE PRODUCTS; ELECTROCHEMICAL PROCESSING).

The Nafion membranes, fully fluorinated polymers, exhibit exceptionally high chemical stability in strong bases in strong oxidizing and reducing acids, H_2O_2, Cl_2, H_2, and O_2 at temperatures up to 125°C. A high degree of dissociation and a high (>4 molal) concentration of mobile H^+ ions ensure good ionic conductivity in Nafion. A conductivity of >0.05 (ohm·cm)$^{-1}$ at 25°C is considered to be acceptable for use in fuel cells. A review on the conductivity properties of Nafion is available (1). The range of equivalent weights for Nafion that is of greatest interest in PEFCs is 1100 to 1350. This provides a highly acidic environment, ie, comparable to a 10 wt % H_2SO_4 solution, in a hydrated membrane.

The porous electrodes in PEFCs are bonded to the surface of the ion-exchange membranes which are 0.12- to 0.25-mm thick by pressure and at a temperature usually between the glass-transition temperature and the thermal degradation temperature of the membrane. These conditions provide the necessary environment to produce an intimate contact between the electrocatalyst and the membrane surface. The early PEFCs contained Nafion membranes and about 4 mg/cm^2 of Pt black in both the cathode and anode. Such electrode/membrane combinations, using the appropriate current collectors and supporting structure in PEFCs and water electrolysis cells, are capable of operating at pressures up to

20.7 MPa (3000 psi), differential pressures up to 3.5 MPa (500 psi), and current densities of 2000 mA/cm².

Very substantial advances have been made in terms of improvements in electrode structures and increases in the Pt utilization as illustrated in Figure 1. It appears that Pt loadings of less than 0.2 mg Pt/cm² are adequate to obtain acceptable performance in PEFCs using pure H_2 as the fuel (see THIN FILMS). Whereas early electrodes contained 4 mg Pt/cm², the most recent developments in electrode fabrication have permitted Pt loadings to be reduced to 0.13 mg Pt/cm² in a thin-film structure, while maintaining high performance.

In a H_2/O_2 PEFC, H_2 is oxidized at the anode to H^+ ions, which are transported through the membrane to the cathode. During operation of the fuel cell, water is transported through the Nafion membrane with the H^+ ions because of the electroosmotic effect. At 100°C, 3.5 to 4 H_2O molecules are transported with each H^+ ion (3). The transport of water through the membrane presents a water-management problem. There is a tendency for the anode side to dehydrate, resulting in a reduction in ionic conductivity and a decrease in the power output from the cell. One solution to this problem is to humidify the fuel gas entering the cell, but the addition of too much H_2O can be detrimental because it may form a liquid film on the electrocatalyst which interferes with the transport of H_2 to the electrocatalyst. At the cathode H_2O is produced which must be rapidly removed to avoid flooding of the electrode.

As of this writing, the primary focus of research and development in PEFC technology is a fuel-cell system for terrestrial transportation applications requiring the development of low cost cell components. Reformed methanol is expected to be a principal fuel source for PEFCs in transportation applications. Because the operating temperature of PEFCs is much lower than that of PAFCs (see Table

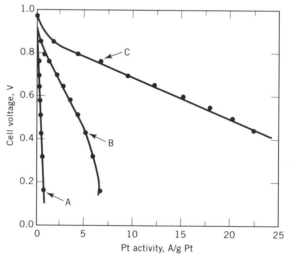

Fig. 1. Increase in Pt utilization in PEFCs, where A represents the GE space technology fuel cell, 4 mg Pt/cm²; B represents Prototech, 0.45 mg Pt/cm²; and C represents thin film, 0.13 mg Pt/cm² (2).

1), poisoning of the anode electrocatalyst by CO from steam-reformed methanol is a problem.

A newer series of perfluorinated ionomers (qv), developed by Dow Chemical Co., provides an attractive alternative to Nafion in PEFCs (4,5). This newer polymer has a PTFE-like backbone similar to those of Nafion, but the pendent side chain containing the sulfonic acid group is shorter. Instead of the long side chain of Nafion, the side chain of the Dow polymer consists of $OCF_2-CF_2-SO_3H$. This polymer possesses ion-exchange properties similar to that of Nafion, and it is also available with higher acid strength and lower equivalent weights, ie, 600–950. Even at these low equivalent weights the Dow membrane has good mechanical strength and does not hydrate excessively, whereas Nafion of comparable equivalent weight would form a highly gelled polymer, having poor or no mechanical integrity. The physical and transport properties of ion-exchange membranes are largely determined by the amount of absorbed H_2O. For a given equivalent weight, the Dow polymer absorbs less water ($\sim$50%) than Nafion, but it has comparable ionic conductivity and lower permeability. This polymer has a higher glass transition temperature (165 vs 110°C for Nafion); thus PEFCs containing this material should be capable of operating at temperatures above 100°C where poisoning of the electrocatalyst by CO is less problematic and electrode kinetics are more rapid.

Alkaline Fuel Cell. The electrolyte in the alkaline fuel cell is concentrated (85 wt %) KOH in fuel cells that operate at high ($\sim$250°C) temperature, or less concentrated (35–50 wt %) KOH for lower (<120°C) temperature operation. The electrolyte is retained in a matrix of asbestos (qv) or other metal oxide, and a wide range of electrocatalysts can be used, eg, Ni, Ag, metal oxides, spinels, and noble metals. Oxygen reduction kinetics are more rapid in alkaline electrolytes than in acid electrolytes, and the use of non-noble metal electrocatalysts in AFCs is feasible. However, a significant disadvantage of AFCs is that alkaline electrolytes, ie, NaOH, KOH, do not reject CO_2. Consequently, as of this writing, AFCs are restricted to specialized applications where CO_2-free H_2 and O_2 are utilized.

The most successful application of AFC technology was in the U.S. space programs, ie, *Apollo* and the Space Shuttle. The AFC used in the U.S. *Apollo* space program was based on technology originally developed in the 1930s (6). This original fuel cell operated at 200 to 240°C, 45 wt % KOH, and pressure maintained at 4 to 5.6 MPa (580–810 psi) to prevent the electrolyte from boiling. The anode consisted of a dual-porosity Ni electrode, ie, a two-layer structure having porous Ni of 16-μm maximum pore diameter on the electrolyte side and 30-μm pore diameter on the gas side. The cathode consisted of a similar dual-layer porous structure of lithiated NiO. The three-phase boundary in the porous electrodes was maintained by a differential gas pressure across the electrode because at that time a wetproofing agent was not available.

The AFC fuel-cell module for the U.S. *Apollo* space program (57-cm diameter, 112-cm high, $\sim$110 kg, 1.42 kW at 27–31 V, 0.6-kW average power) utilized pure H_2 and O_2 and concentrated electrolyte (85 wt % KOH) to permit cell operation at lower (ca 400 kPa (58 psi) reactant gas pressure) pressure without electrolyte boiling. Using this concentrated electrolyte, cell performance is not as high as in the less concentrated electrolyte; thus the operating temperature was increased to 260°C. The typical performance of this AFC was 0.85 V at 150 mA/cm², which compared favorably to the performance of the original cell operating at about 10 times higher pressure.

The alkaline fuel cells in the space shuttle *Orbiter* (fuel-cell module: 35-cm high, 38-cm wide, 101-cm long, 91 kg, 12 kW at 27.5 V, 7-kW average power) operate in the same pressure range as for the *Apollo* program but at a lower (80 to 90°C) temperature and a higher (470 mA/cm^2 at 0.86 V) current density. The electrodes contain high loadings of noble metals (anode: 10 mg of 80% Pt–20% Pd/cm^2 on a Ag-plated Ni screen, cathode: 20 mg of 90% Au–10% Pt/cm^2 on a Ag-plated Ni screen) that are bonded with PTFE to achieve high performance at lower temperatures. A wide variety of materials (eg, potassium titanate, ceria, asbestos, zirconium phosphate gel) have been used to retain the alkaline solution in the microporous electrolyte separators for AFCs. A brief survey of the advanced-technology components in AFCs for space applications is available (7).

Phosphoric Acid Fuel Cell.   Concentrated phosphoric acid is used for the electrolyte in PAFC, which operates at 150 to 220°C. At lower temperatures, phosphoric acid is a poor ionic conductor (see PHOSPHORIC ACID AND THE PHOSPHATES), and CO poisoning of the Pt electrocatalyst in the anode becomes more severe when steam-reformed hydrocarbons (qv) are used as the hydrogen-rich fuel. The relative stability of concentrated phosphoric acid is high compared to other common inorganic acids; consequently, the PAFC is capable of operating at elevated temperatures. In addition, the use of concentrated (~100%) acid minimizes the water-vapor pressure so water management in the cell is not difficult. The porous matrix used to retain the acid is usually silicon carbide SiC, and the electrocatalyst in both the anode and cathode is mainly Pt.

In the mid-1960s, the conventional porous electrodes were PTFE-bonded Pt black, and the loadings were about 9 mg Pt/cm^2. As of the early 1990s, Pt supported on carbon black has replaced Pt black in porous PTFE-bonded electrode structures as the electrocatalyst. A dramatic reduction in Pt loading has also occurred. Platinum loadings are about 0.25 mg/cm^2 in the anode and about 0.50 mg/cm^2 in the cathode, although at both electrodes, Pt may be combined with other metals. The operating temperature, and correspondingly the acid concentration, of PAFCs has increased to achieve higher cell performance. Temperatures of about 200°C and acid concentrations of 100% H$_3$PO$_4$ are commonly used. In addition, the operating pressure of PAFCs has surpassed 0.5 MPa (70 psi) in many tests, and commercial electric utility systems are planned to operate at ~0.8 MPa (115 psi).

Molten Carbonate Fuel Cell.   The electrolyte in the MCFC is usually a combination of alkali (Li, Na, K) carbonates retained in a ceramic matrix of LiAlO$_2$ particles. The fuel cell operates at 600 to 700°C where the alkali carbonates form a highly conductive molten salt and carbonate ions provide ionic conduction. At the operating temperatures in MCFCs, Ni-based materials containing chromium (anode) and nickel oxide (cathode) can function as electrode materials, and noble metals are not required.

In MCFCs there are no materials that can serve to wet-proof a porous structure against ingress by molten carbonates. Consequently, the technology to obtain a stable three-phase interface in MCFC porous electrodes is different from that of PAFCs. In the MCFC, the stable interface in the electrodes is achieved by carefully tailoring the pore structures of the electrodes and the electrolyte matrix

(LiAlO$_2$) so that the capillary forces establish a dynamic equilibrium in the different porous structures to prevent the electrodes from flooding with electrolyte, particularly the cathode. On the other hand, the anode is relatively insensitive to the degree of filling by the electrolyte.

In a conventional fuel cell system, a carbonaceous fuel is fed to a fuel processor where it is steam-reformed to produce H$_2$ as well as such other products as CO, CO$_2$, and H$_2$O. The hydrogen (qv) is then introduced into the fuel cell and electrochemically oxidized. Efforts are underway to develop a fuel-cell system that eliminates the need for a separate fuel processor by providing for the reforming of carbonaceous fuels in the fuel cell, ie, internal reforming, and near the electrochemically active sites. This technique appears practical in high temperature fuel cells such as MCFCs where stream-reforming of hydrocarbons to produce hydrogen can be sustained using catalysts at the cell operating temperatures. By integrating the reforming reaction and the electrochemical oxidation reaction of hydrogen in the fuel cell, the internal-reforming MCFC (IRMCFC), as illustrated schematically in Figure 2, is realized. The IRMCFC eliminates the need for the external fuel processor and its ancillary equipment, and provides a highly efficient, simple, reliable, and potentially cost-effective alternative to the conventional MCFC system.

Steam reforming of CH$_4$ is commonly carried out at 750 to 900°C, thus at the lower operating temperature of MCFCs a high activity catalyst is required. The internal reforming of methane in IRMCFCs, where the steam-reforming reaction

$$CH_4 + H_2O \rightarrow CO + 3\,H_2 \qquad (1)$$

occurs simultaneously with the electrochemical oxidation of hydrogen in the anode compartment, has been demonstrated. The steam-reforming reaction is endothermic, whereas the overall fuel cell reaction is exothermic. In an IRMCFC, the heat required for reaction 1 is supplied by the waste heat from the fuel cell reaction eliminating the need for an external heat exchanger, required by a con-

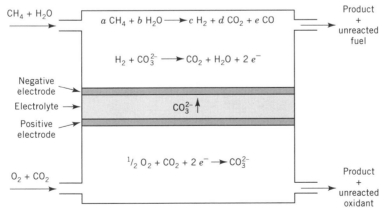

Fig. 2. Internal reforming in MCFCs.

ventional fuel processor. In addition, the product steam produced during the oxidation of hydrogen can be used to drive the reforming reaction and the water gas shift reaction to produce additional H_2. The forward direction of reaction 1 is favored by high temperature and low pressure, thus an IRMCFC is best suited to operate near atmospheric pressure. A supported Ni catalyst, eg, Ni supported on MgO or $LiAlO_2$, provides sufficient catalytic activity to sustain the steam-reforming reaction at 650°C to produce H_2 at the necessary rate.

Two manifold concepts have been developed to direct the reactant fuel and oxidant gases to the MCFC stack and to remove the exhaust gases. These are illustrated schematically in Figure 3, and are referred to as internally and externally manifolded stacks. The external manifold (Fig. 3**b**) is commonly used in fuel cell stacks formed by repeated stacking of thin flat-plate cells. The reactant fuel and oxidant gases are directed into and out of the stack by using a manifold that is connected to the sides of the thin cells, and cross flow of the two gases is the common flow pattern. On the other hand, the gases used in the internally manifolded stack (Fig. 3**a**) are directed into and out of the stack through openings that are formed in the stack components, and not through an external manifold. The gas flow patterns with the internally manifolded stack are more complex, and examples involving both cross flow and parallel flow have been described (8). The advantage of an internal manifold is that electrolyte migration along the length of the stack does not occur because the stack has no porous manifold gaskets in contact with the electrolyte. Furthermore this design permits easy stack assembly and scale-up without the need for a heavy external manifold. On the other hand, stacks having internal manifolds have a more complicated design, and fabrication of components is more complex.

Solid Oxide Fuel Cell. In a SOFC, there is no liquid electrolyte present that is susceptible to movement in the porous electrode structure, and electrode flood-

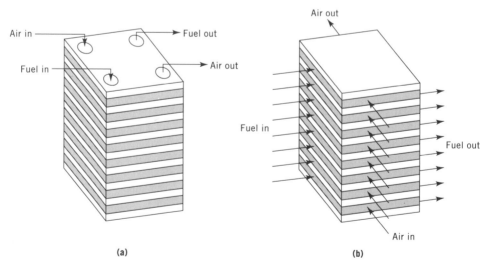

Fig. 3. Schematics of gas manifolds for MCFC stacks: (**a**) internally manifolded fuel cell stack; (**b**) externally manifolded fuel cell stack.

ing is not a problem. Consequently, the three-phase interface that is necessary for efficient electrochemical reaction involves two solid phases (solid electrolyte/electrode) and a gas phase. A critical requirement of the porous electrodes for the SOFC is that they are sufficiently thin and porous to provide an extensive electrode/electrolyte interfacial region for electrochemical reaction. Because of the high (typically 1000°C) operating temperatures of SOFCs, the materials used in the cell components are limited by chemical stability in oxidizing and/or reducing environments, chemical stability of contacting materials, acceptable conductivity, and thermomechanical compatibility. The principal effort to date has been on demonstrating the viability of a tubular structure (Fig. 4) in SOFCs. The tubular design uses a porous ceramic support tube of about 100-cm length and 1.27-cm diameter. However, more recently the development of planar electrolyte structures has received considerable attention.

An electrochemical vapor deposition (EVD) technique has been developed that produces thin layers of refractory oxides that are suitable for the electrolyte and cell interconnection in SOFCs (9). In this technique, the appropriate metal chloride ($MeCl_x$) vapor is introduced on one side of a porous support tube, and H_2/H_2O gas is introduced on the other side. The gas environments on both sides of the support tube act to form two galvanic couples, ie,

$$MeCl_x + y\ O^{2-} \rightarrow MeO_y + \frac{1}{2}x\ Cl_2 + 2y\ e^- \qquad (2)$$

and

$$H_2O + 2\ e^- \rightarrow H_2 + O^{2-} \qquad (3)$$

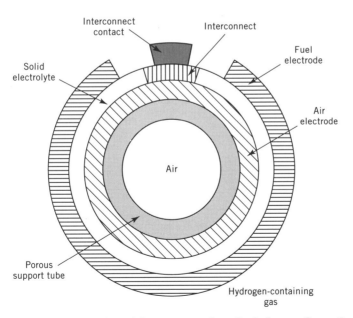

Fig. 4. Schematic representation of the cross section of tubular configuration for SOFC.

The result is the formation of a dense and uniform metal oxide layer in which the deposition rate is controlled by the diffusion rate of ionic species and the concentration of electronic charge carriers. This procedure is used to fabricate the thin layer of solid electrolyte (yttria-stabilized zirconia) and the interconnection (Mg-doped lanthanum chromite).

The anode consists of metallic Ni with a Y_2O_3-stabilized ZrO_2 skeleton, which serves to inhibit sintering of the metal particles and to provide a thermal expansion coefficient comparable to those of the other cell materials. The anode structure is fabricated, having a porosity of 20 to 40% to facilitate mass transport of reactant and product gases. The Sr-doped lanthanum manganite, $La_{1-x}Sr_xMnO_3$, where $x = 0.10$–0.15, that is most commonly used for the cathode material is a p-type semiconductor. Similar to the anode, the cathode is a porous structure that must permit rapid mass transport of reactant and product gases. The cell interconnection material (Mg-doped lanthanum chromite, $LaCr_{1-x}Mg_xO_3$, $x = 0.02$–0.10), on the other hand, must be impervious to fuel and oxidant gases and must possess good electronic conductivity. In addition, the cell interconnection is exposed to both the cathode and anode environments, thus it must be chemically stable under O_2 partial pressures of about 10^5 Pa (14.5 psi) down to 10^{-13} Pa at 1000°C. The solid electrolyte, commonly Y_2O_3-stabilized ZrO_2, must be free of porosity that permits gas to permeate from one side of the electrolyte layer to the other, and it should be thin to minimize ohmic loss. In addition, the electrolyte must have a transport number for O^{2-} as near to unity as possible, and a transport number for electronic conduction as near to zero as possible. Zirconia-based electrolytes are suitable for SOFCs because these exhibit pure anionic conductivity over a wide range of O_2 partial pressures (10^5–10^{-15} Pa (14.5–10^{-13} psi)). The solid electrolyte in SOFCs must be only about 25–50 μm thick if its ohmic loss at 1000°C is to be comparable to that of the electrolyte in PAFCs. Fortunately, thin electrolyte structures of about 40-μm thickness can be fabricated by EVD, as well as by tape casting and other ceramic processing techniques (see CERAMICS). The other cell components should permit electronic conduction, and interdiffusion of ionic species in these components at 1000°C should not have a significant effect on their electronic conductivity. Other severe restrictions placed on the cell components are that they must be stable to the gaseous environments in the cell, they must be capable of withstanding thermal cycling, and the thermal expansion coefficients of the cell components must be compatible.

Applications

Fuel cells operating on pure H_2 and O_2 provide a useful power source in remote areas such as in space or under the sea where system weight and volume are important parameters. On the other hand, fuel cell power plants operating on fossil fuels and air offer the potential for environmentally acceptable, highly efficient, and low cost power generation. Thus fuel cells can be considered for terrestrial applications where environmental pollution or noise would be objectionable, and they can be located near the point of use of the electricity such as on an urban site, rather than at a remote location. An analysis (10), summarized in Table 2, describes the minimum technical requirements for fuel cells for four dif-

Table 2. Minimum Technical Requirements for Fuel Cell Applications[a]

Characteristic	Buildings	Industry	Transportation	Utility
technology	PEFC/PAFC	SOFC	PEFC	MCFC
efficiency, %	35	40	35	45
system life, yr	2/15	15	2+	20
capacity, kW	5–200	200–2000	5–200	2000+
operating temperature, °C	90/194	1000	90	700
heat recovery	important	important		important

[a]Ref. 10.

ferent types of applications, ie, for buildings, industry, transportation, and utilities. The fuel cell technologies which are most likely to be used in the various applications, and their anticipated capacities are given. The life of the power plants listed in Table 2 are longer than the projected life for the fuel cell stacks; thus periodic stack replacement is required. Larger fuel cell units utilized by the utility and industrial sectors are likely to be operated at high temperatures where the excess heat can be utilized to generate additional electrical energy. On the other hand, fuel cells for transportation applications are more convenient if they operate near ambient temperatures where the time to reach operating conditions is shorter.

An excellent summary of the development of fuel cells from the work carried out in 1839 (11) to the pioneering efforts in the 1930s (6) is available (12) as are other notable reviews (13–15) that document the historical development of fuel cells. Development of fuel-cell stacks and integration to other supporting equipment increased with the advent of the U.S. Space Program. At the same time, efforts were underway to demonstrate the feasibility of fuel cells for terrestrial applications (13,14). For example, in the mid-1950s Union Carbide Corp. demonstrated several terrestrial-type systems. Then in the 1960s a serious attempt to demonstrate fuel-cell technology for utility-related applications was initiated with the Team to Advance Research on Gas Energy Transformation (TARGET) Program, which resulted in the test of 12.5 kW PAFCs at 35 sites in the United States, Japan, and Canada. Further advances toward commercialization of PAFCs have been relatively slow. The demonstration of multimegawatt PAFC power plants occurred in the 1980s and early 1990s at Tokyo Electric Power Company (Goi, Japan).

In 1981 a national program, called the Moonlight Project, was initiated for energy conservation in Japan. A primary part of this strategy has been to develop fuel cell technologies, and as of this writing Japan is believed to have the largest commercialization program for fuel cells in the world. The Japanese government started a project to demonstrate a 1-MW unpressurized PAFC (Tokyo Gas Co.) and a 5-MW pressurized PAFC (Kansai Electric Power Co.) at utility sites. In addition, the installation of PAFC power plants totaling 20 MW at various sites in Japan should be completed by 1994. More modest efforts are underway to demonstrate MCFC and SOFC technologies in Japan. A survey of fuel-cell development and demonstration activities can be found in Reference 16.

Phosphoric Acid Fuel Cell. Tests of 4.8 MW and 11 MW power plants using PAFCs have been successfully completed in Japan. The fuel cell stacks (PC23) of the latter, built by International Fuel Cells, Inc. (IFC), began operation in March 1991. Eighteen 700-kW stacks were used. IFC designed the power conversion section and Toshiba Corp. provided the rest of the power plant system, ie, fuel reformer, thermal management, balance-of-plant equipment. This, the largest PAFC built and operated as of this writing, achieved full-rated output in April 1991. At its rated power, a gross a-c power efficiency of 43.6% (41.8% net) was obtained, and the measured NO_x exhaust was <3 ppm. By the end of August 1992, the total electricity generated exceeded 23,435 MW·h, having a power generation time of 4041 h. Another demonstration of a smaller (1 MW) power plant using two PAFC stacks manufactured by IFC is planned for Milan, Italy. The manufacture of IFC's 200-kW PAFC power plants in Europe is to be carried out by CLC srl, Genoa, Italy and Ansaldo SpA (Italy).

A subsidiary of IFC and Toshiba Corp. called ONSI Corp. was formed for the commercial development, production, and marketing of packaged PAFC power plants of up to 1-MW capacities. ONSI is commercially manufacturing 200-kW PAFC systems for use in a PC25 power plant. The power plants are manufactured in a highly automated facility, using robotic techniques to assemble the repeating electrode, bipolar separator, etc, units into the fuel cell stack.

The PC25 power plant, 2.5-m wide × 7.6-m long × 3.2-m high, is capable of remote unattended operation using pipeline gas. This power plant has a rated power output of 200 kW, operates at ambient pressure, and has achieved an electrical efficiency of over 40% (LHV). It is capable of operating in the cogeneration mode, producing 222 kW (760,000 Btu/h) of hot (74°C) water, and an overall energy efficiency of 85%. The first PC25 power plant, tested at IFC, accumulated over 12,000 h of operation, starting in 1991. The initial stack voltage was ~200 V, whereas the long-term voltage remained steady at ~190 V. As of this writing, the longest continuously operating PC25 (5400 h) is located in Japan, and has a total operating time of 7900 h.

The availability of both electric power and useful thermal energy makes these power plants attractive for dispersed locations for small commercial and industrial buildings. The applications of 200-kW PAFC power plants are summarized in Table 3. By July 1993, 29 out of a total of 52 planned power plants were installed. The first of these power plants was operated by the South Coast Air Quality Management District (SCAQMD) in Diamond Bar, California. The measured emissions from the PAFC were low. The NO_x was 0.5 ppmv, CO 2 ppmv, and total hydrocarbons (THC) 4 ppmv. There was no SO_2, smoke, or particulates. The SCQAMD was able to drastically curtail the institutional issues of siting, licensing, and permitting for installing PAFC power plants in their jurisdiction encompassing the Los Angeles basin area.

The performance characteristics of several versions of PAFC stacks developed by IFC for utility (Configuration A) and on-site (Configuration B) applications are summarized in Table 4. These stacks are water-cooled. Cooling plates are located at five-cell intervals in the stack for utility applications and at seven-cell intervals in stacks for on-site applications. The cell voltage and efficiency of the PAFC power plant are clearly enhanced by operating at elevated pressures. The PAFC (Configuration B) for on-site applications such as hospitals, hotels, and

Table 3. Applications and Locations of 200-kW PAFCS[a,b]

| | Location | | | | | Total |
| | North America | | Asia | | | number of |
Application/type	United States	Canada	Japan	South Korea	Europe	plants
hospital	6					6
nursing home	1					1
hotel	2					2
office	2	1	4	1		8
education	2					2
public safety	1					1
district heating system			3		4	7
light industrial	4		6		3	13
R&D[c] facility			7		2	9
special application	2		1			3
Total	20	1	21	1	9	52

[a]PC25 power plants.
[b]Ref. 17.
[c]R&D = research and development.

Table 4. Performance Characteristics of IFC PAFC Powerplants[a]

Stack configuration[b]	Electrode area, m^2	Pressure, kPa^c	Current density, mA/cm^2	Cell voltage, V	Power density, mW/cm^2	Remarks[d]
B	0.45	100	215	0.650	140	SNG/air
B	0.45	820	215	0.740	159	SNG/air
B	0.93	820	431	0.715	307	fuel NS
A	0.93	820	215	0.760	164	fuel NS

[a]Ref. 18.
[b]A = utility application; B = on-site application.
[c]To convert kPa to psi, multiply by 0.145.
[d]SNG = simulated natural gas; NS = not specified.

the light industry, operates at atmospheric pressure. Increasing the operating pressure to 0.82 MPa (120 psi) and electrode area to 0.93 m^2 of the PAFC stack designated as Configuration B results in enhanced power density. Configuration A, which is used in the 11-MW demonstration program in Japan, operates at 0.82 MPa (120 psi).

Westinghouse Electric Corp. initiated a program to develop air-cooled PAFC stacks, containing cooling plates at six-cell intervals. Full size 100-kW stacks (468 cells, 0.12-m^2 electrode area) were built, and a module containing four of these stacks was tested. An air-cooled stack operated at 0.480 MPa yielded a cell voltage of 0.7 V at 267 mA/cm^2 (187 mW/cm^2). Demonstration of this technology is planned for a site in Norway.

There are several organizations in Japan that are actively involved in PAFC demonstration programs. Fuji Electric Co. is taking the lead in the commercialization of PAFCs in Japan, with units ranging from 50 kW to 5 MW. Over 65 PAFC power plants having total capacity of greater than 10 MW are on order. 50-kW units having a capability for cogeneration, electrical efficiency of 40% and total efficiency of 80% (LHV), using town gas are being developed. The larger fuel cell power plants are being sited at gas and electric utilities, as well as industrial companies in Japan. In addition, Fuji is involved in the United States Department of Energy (DOE) program to demonstrate the viability of PAFCs for powering city buses. Other Japanese companies involved in the development of PAFCs are Mitsubishi Electric Co. (MELCO) and Toshiba Corp. MELCO was involved with a demonstration test of a 200-kW PAFC stack that operated for over 13,000 h, and a short stack that operated for 16,000 h with no acid addition. The company is planning additional tests of 200-kW PAFC stacks, starting in 1994 at Osaka Gas (Japan). Toshiba Corp. is collaborating with IFC to market the 200-kW PAFC (PC25), as well as supporting the development of improved PAFC stacks (200 kW, 50 kW) based on the PC25, and much larger units (670 kW, 1 MW).

The Office of Transportation Technologies of the U.S. DOE is supporting programs to develop fuel cells for transportation applications. As of this writing, the viability of a PAFC-powered bus (9-m long) operating on an urban route is being investigated. This program is coordinated by H-Power (Belleville, N.J.) in collaboration with subcontractors Bus Manufacturing USA, Inc. (bus design and fabrication), Booz Allen & Hamilton (bus system integration), Fuji Electric Co. Ltd. (50-kW PAFC hardware), Soleq Corp. (power train and controls), and Transportation Manufacturing Corp. (transit bus industry guidance and 12-m bus conceptual design). The PAFC (0.2 m^2 electrode area, 175 cells) is designed to operate at 240 mA/cm^2 (0.66 V/cell) and 190°C on H$_2$ from steam-reformed CH$_3$OH. The overall design efficiency is 38% (based on LHV CH$_3$OH). Delivery of the buses for dynamometer testing began in 1993. Demonstration runs are planned for Washington, D.C. and Los Angeles.

Molten Carbonate Fuel Cell.   The MCFC is well-suited to utilize fuels that are produced in coal gasifiers or from other sources. In one approach, Energy Research Corp. (ERC) is steam-reforming natural gas to form H$_2$ and CO$_2$ in the fuel cell (internal reforming), eliminating the need for a dedicated fuel processor. The reaction is carried out using a catalyst that is located in the fuel manifold near the anode inlet and in the fuel cell stack itself. Because the steam reforming reaction is endothermic, heat that is generated during the production of electrical energy is utilized. The MCFC stacks developed by ERC operate at atmospheric pressure and utilize external gas manifolds.

The MCFC appears to be in the final stages of prototype testing, and a 234-cell MCFC stack (0.37 m^2 cell area, 70 kW) tested at ERC operates with a cross-flow gas distribution pattern (see Fig. 3). A 70-kW stack from ERC was also tested, starting in the fall of 1991, at a site of Pacific Gas and Electric Co. in San Ramon, California. In addition, ERC has tested a 120-kW, 244-cell MCFC stack (0.56 m^2 electrode area). The purpose of these tests was to demonstrate the feasibility of the technology for scaling up to large size MCFC stacks. A large demonstration plant utilizing the ERC technology in large fuel cell stacks (125 kW stacks, 0.56

m^2 electrode area) to produce a 2-MW power plant in Santa Clara, California is expected to be operational by late 1994 or early 1995.

In 1987 M-C Power (Burr Ridge, Illinois) was formed to commercialize the MCFC technology developed by the Institute of Gas Technology (IGT), Chicago, in partnership with Ishikawajima-Harima Heavy Industries Co., Ltd. (IHI) of Japan. In 1991 M-C Power demonstrated the largest stack to date (70 cells) using an internally manifolded heat exchanger (IMHEX) fuel-cell-stack (19,20) in a 1580-h test. The gas flow field in this design is illustrated in Figure 5. This configuration has large manifolds for the gas inlets and outlets and an octagonal flow field. Both cross and parallel flow occur in different areas of the electrode. This design permits heat to flow more efficiently in the stack and minimizes the tendency of the cells to shrink over operating time, which occurs because of structural changes in the cell components, eg, sintering of porous electrodes. In 1992, M-C Power began a test of the first full-area commercial IMHEX 20-cell, 20-kW stack (0.93 m^2 cell area), and exceeded 2500 h of operation. A demonstration of MCFC technology using a 250-kW power plant in 1994 at the UNOCAL Science and Technology Center in Brea, California and one at a hospital in San Diego are planned.

A summary of the MCFC technologies is presented in Table 5. Both ERC and MCP have developed pilot continuous manufacturing facilities to produce MCFC cell components. Based on the results of an analysis to define the optimum operating pressure, ERC selected 0.1 MPa (14.5 psi) to obtain the lowest cost of electricity and longest cell life in MCFC stacks operating with internal fuel reforming. On the other hand, the goal of the MCP program is to operate at pressures above ambient, ie, 0.3 MPa (43.5 psi), using externally reformed fuels. The demonstrated lifetimes of MCFC stacks are short of the goal of 40,000 h. Efforts to extend the life include minimizing voltage-induced electrolyte migration in the stack, which appears to be a principal reason for performance degradation.

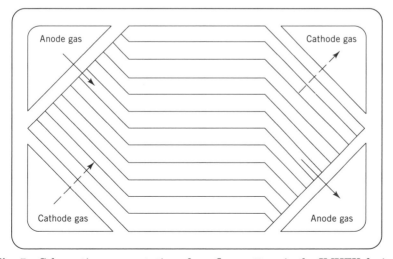

Fig. 5.   Schematic representation of gas flow pattern in the IMHEX design.

Table 5. Performance Status of Continuous MCFC Technologies.[a,b]

Parameter	Energy Research Corp.	M-C Power Corp.
module size, m^2	0.37	0.93
cells, number	234	20
current density, mA/cm^2	172	172
power density, mA/cm^2	113 (151)	129 (215)
operating pressure, MPa^c	0.1 (0.1)	0.1 (0.3)
electrical efficiency, $\%^d$	(60–65)	(60–65)
stack lifetime, h	10,000 (40,000)	2,500 (40,000)

[a]Ref. 18.
[b]Data in parentheses are the goal.
[c]To convert MPa to psi, multiply by 145.
[d]2-MW natural gas cell.

The Japanese Government is actively involved with Hitachi, Ltd., IHI, and MELCO to develop 100-kW MCFC power plants and systems. A target for this program is to develop a 1-MW class MCFC power plant leading to a test at a utility site by 1997. Hitachi is exploring a concept that revolves around the use of 25-kW blocks of cells (22 cells, 1.2 m^2 electrode area), which are integrated to form large stacks, ie, four 25-kW blocks to form a 100-kW stack. Performance of these stacks at both 0.1 MPa (14.5 psi) and 0.59 MPa (87 psi) has been demonstrated for over 5000 h. Large (1.4 m^2) cells have been tested by IHI for 9700 h, and tests of a 100-kW stack are under way as of this writing. MELCO is involved in the development of both direct internal-reformed (DIR) and indirect internal-reformed (IIR) MCFC technologies. The distinction between these involves the location of the steam-reforming catalyst such as Ni supported on MgO. In the DIR MCFC, the catalyst is located in the anode gas channels, and simultaneous reforming and electrochemical oxidation of hydrogen occur in close proximity. In the IIR MCFC, the reforming catalyst is located in separate and independent reaction chambers in the cell stack. Because the catalyst is located in the anode compartment of the DIR MCFC, it is susceptible to contamination and deactivation by the molten carbonate electrolyte. A successful demonstration of 4500 h showing 99% CH_4 conversion with a DIR 5-kW MCFC (ten 5000 cm^2 cells) was completed. MELCO also demonstrated a 100-kW IIR MCFC stack (192 cells) in a test lasting more than 2000 h.

Programs to develop MCFC technology are also under way in Europe. Ansaldo SpA (Italy) is setting up facilities to produce 1-m^2 cells in an automated process, and their goal is to test 100-kW stacks in 1994. The 100-kW stack is also to be tested by IBERDROLA in Spain as part of a complete power plant system. Two Dutch companies, Stork and Royal Schelde, have joined with the Dutch government to form Brandstofcel Nederland (BCN), which plans to test a 50-kW MCFC and two 250-kW MCFC stacks in 1994.

Solid Oxide Fuel Cell. A comprehensive review of the SOFC technology can be found in the literature (21). A tubular or cylindrical configuration was adopted for SOFCs to alleviate problems with gas seals experienced in the initial, planar design. A tubular design having a porous support tube of about 30-cm length and

1.27-cm diameter (active area of about 110 cm^2) developed by Westinghouse Electric Corp. (WEC) produced about 18 W. The manifolding of the oxidant and fuel gases for these types of tubular cells is illustrated in Figure 6. The oxidant gas is introduced via a central Al_2O_3 tube, and the fuel gas is supplied to the exterior of the closed-end tube. In this arrangement, the Al_2O_3 tube extends to the proximity of the closed end of the support tube, and the oxidant flows back past the cathode surface to the open end. The fuel gas flows past the anode on the exterior of the cell, and in a parallel direction (co-flow) to the oxidant gas. The spent gases are exhausted into a common plenum where the remaining active gases react, and the generated heat serves to preheat the incoming oxidant stream. One attractive feature of this arrangement is that it eliminates the need for leak-free gas manifolding of the fuel and oxidant streams.

The tubular design is the most advanced SOFC technology. Tests of a nominal WEC 25-kW SOFC unit were started in 1992 at Rokko Island near Osaka,

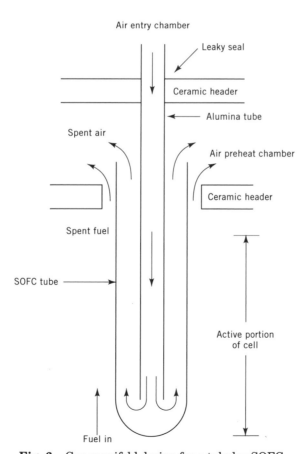

Fig. 6. Gas-manifold design for a tubular SOFC.

Japan in a joint program by Kansai Electric Co., Osaka Gas, and Tokyo Gas Co. This unit consists of 1152 cells, 50-cm length, which are contained in two independently controlled and operated sections. The SOFC has surpassed 2500 h of operation, and achieved d-c power of 36 kW of steady output and a peak of 44 kW. The WEC technology was originally based on the use of a porous support tube (see Fig. 4). Tubular cells 100-cm long are now standard, but commercial cells are expected to be 200 cm in length. A cell has been operated at 875°C for more than 28,000 h at 250 mA/cm^2 and with a performance degradation of about 1%/1000 h of operation. The average performance from a test matrix of 12 SOFC cells was an average cell voltage of 0.516 V at 377 mA/cm^2. An effort is under way to eliminate the support tube, and instead utilize a self-supported air electrode (SSAE), which should decrease the materials and fabrication costs, and improve cell performance. Tubular cells employing SSAE (36-, 50-, and 77-cm long) have been tested for up to 6800 h.

Efforts are under way to develop SOFCs that have a bipolar, planar design. The planar design permits a more efficient utilization of the weight and volume in the fuel cell stack. The effort at AlliedSignal Aerospace Co. (Torrance, California), for example, is to build a monolithic SOFC capable of achieving high power densities because of its compact design (Fig. 7). This planar design resembles the corrugated structure found in cardboard. The typical layer thicknesses are 100 μm, and cell-to-cell distances are 1 to 2 mm. Both co-flow and cross-flow of the

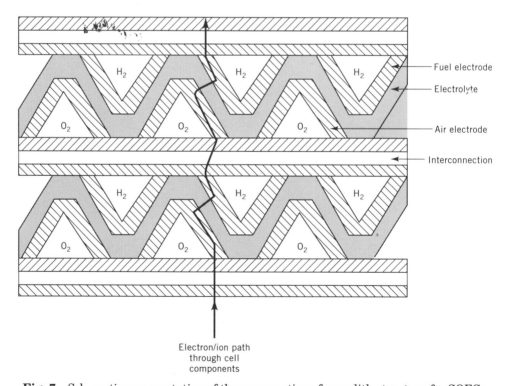

Fig. 7. Schematic representation of the cross section of monolith structure for SOFCs.

reac̭ant gases have been considered. The monolithic structure was fabricated in the green state and then the complete structure was co-fired to obtain small (100 and 50 W) SOFC stacks. Other options for fabricating monolithic structures have also been considered (21). Using a single cell (5 cm^2), a performance of 0.75 V at 500 mA/cm^2 was obtained at 1000°C on 97% H_2/3% H_2O and air. More recently, AlliedSignal has shifted effort to evaluate the flat-plate planar design. Cell components of up to 23 cm × 23 cm have been produced, and thin electrolyte layers (1–10 μm) on NiO/ZrO$_2$ have been fabricated by tape calendering. Because of the thin electrolyte layers, a lower operating temperature can be used. A power density of 0.3 W/cm^2 was obtained using a small (<10 cm^2) planar cell at 800°C using 97% H_2/3% H_2O and air, and no significant performance degradation was observed for more than 1000 h.

Ceramatec, Inc. (Salt Lake City, Utah) and Ztek Corp. (Waltham, Massachusetts) are also pursuing the development of planar SOFCs. Ceramatec has obtained performance of ~300 mA/cm^2 at 0.5 V/cell in 5-cell stacks and 200 mA/cm^2 at 0.5 V/cell in 40-cell stacks. Ceramatec is collaborating with Sulzer-Innotec (Switzerland) to develop planar SOFC power plants for cogeneration applications by 1996. Ceramatec is integrating cells of 12-cm diameter into an internally manifold stack designed by Sulzer. Ztek has fabricated compact (16 cells/ 2.54 cm) SOFC stacks of the planar design. Single cells achieved power densities of 250 mW/cm^2, and 25 W was obtained using a 10-cell stack, which was tested for 1000 h.

Research into SOFC programs in Japan was started in 1981 under the Moonlight Project. Japanese companies such as Fuji Electric Corp., Mitsubishi Heavy Industries, Ltd., Tonen Corp., and Sanyo Electric Co. are participating in the program. This effort is in its early stages and is concentrating on cell components and small cell development.

Several research and development (R&D) programs on SOFCs are under way in Europe. Siemens AG (Germany) is developing a planar SOFC having metallic bipolar plates using, eg, Cr-based alloys, and internal gas manifolding. Several sizes of stacks were manufactured, and a 10-cell stack delivered 103 W at 235 mA/cm^2. Soft-glass seals developed by Dornier GmbH (Germany) demonstrated over 3000 h of operation without suffering gas leaks. They also obtained over 3000-h operation using a 5-cell stack (25 cm^2 cell area) without experiencing gas leaks through the seals in planar SOFCs, and observed a performance of 350 mA/cm^2 at 0.6 V on H_2 and air. Research on cell components, such as the solid electrolyte and anode materials, is being pursued at the Riso National Laboratory (Denmark) and Imperial College (United Kingdom).

Polymer Electrolyte Fuel Cell. The PEFC is being developed in the United States primarily for transportation applications. The cost of the Nafion membrane, traditionally very high, may present a barrier to the acceptance of PEFCs in the consumer sector. Ion-exchange membranes developed by Dow Chemical and Chlorine Engineering (Japan) in the latter 1980s have, however, shown encouraging results in tests of PEFCs. These developments and the potential market for PEFCs in the transportation sector should help to lower the cost of the polymer electrolyte.

A team consisting of Allison Gas Turbine Division of General Motors Corp. (prime contractor) and Los Alamos National Laboratory, Dow Chemical Co., Ballard Power Systems (Vancouver, Canada), and GM Research Laboratories (sub-contractors), are developing PEFCs as a midterm option for transportation applications. These fuel cells are expected to offer significant advantages, ie, reduced weight and size, over PAFCs. State-of-the-art 5-kW PEFCs are under test, and technical issues on component development, fuel processing, water and thermal management, electronic controls, start-up, and transient operations are being addressed. This program, which is sponsored by the U.S. DOE, has a goal of developing a 50-kW PEFC power plant utilizing steam-reformed CH_3OH for vehicular applications.

Ballard Power Systems, the leader in the manufacture of PEFC stacks, has sold at least fifty 3- to 5-kW units worldwide. Ballard is involved in a program in Canada to demonstrate a 120-kW PEFC stack to power a transit 20-passenger, 9752-kg bus. For this demonstration, on-board compressed hydrogen, sufficient for 150-km range, is the fuel.

Energy Partners, Inc. (West Palm Beach, Florida), acquired fuel cell technology from Treadwell Corp. (Thomaston, Connecticut), which supplied electrochemical equipment to the U.S. Navy. Energy Partners, Inc. are involved in developing PEFCs for propulsion applications in transportation and submersible vehicles. A 20-kW PEFC stack was designed for demonstration tests.

A 15-kW PEFC power plant and power system for the supply of primary power in an unmanned underwater vehicle (UUV) is being developed by IFC to replace the Zn–AgO battery stacks that are used as of this writing. The 112-cm diameter UUV is expected to operate with high reliability for more than 5000 h in an enclosed environment. A 20-cell stack was built and tested for 2100 h, and achieved an initial performance level of 0.82 V at 280 mA/cm^2 using pure H_2/O_2. Based on the successful results of the 20-cell stack, 80-cell PEFC stacks were built and tested. Plans are also under way at IFC to develop PEFC technology to provide auxiliary power in a nuclear submarine, with a baseline design consisting of a 1-MW power plant.

The Japanese Government initiated a program in 1992 to promote the development of PEFCs for both portable and stationary applications. The goal is to demonstrate a 1-kW module having a power density of 0.3 W/cm^2 at a cell voltage greater than 0.75 V by 1995. A few research projects are under way in Japan.

Siemens AG has been involved in R&D on PEFCs, and Vickers Shipbuilding & Engineering Ltd. (United Kingdom) is evaluating PEFCs from Ballard Power Systems for power generation. A 35-cell stack was successfully tested for more than 300 h. Plans are under way to test a 20-kW PEFC.

Because of the low operating temperature and ease of fabrication for low power units, PEFCs are the most likely fuel cell to be introduced in portable power packs. PEFCs in sizes of 300–500 W are being considered as a power source, eg, 4-h duration, 300 W, 1.2 kW, for the modern soldier operating in the enclosed environment of a self-contained protective suit, which has facilities for air conditioning, radio communication, etc. Analytic Power Corp. (Boston) is assessing the use of PEFCs for this application.

Alkaline Fuel Cell. Commercial development of AFCs outside of the aerospace and military sector has not been vigorously pursued in the United States. Regenerative fuel cells, ie, those capable of producing electricity by consumption of oxygen and hydrogen, as well as generating these gases by electrolysis, utilizing both AFC and PEFC technologies, have been proposed for space applications. High performance AFC cells based on lightweight cell hardware were developed by IFC. A 4-cell stack (~470 cm^2 cell area) operated at power densities of up to 3.4 W/cm^2. The specific mass was 0.56 kg/kW compared to ~6.5 kg/kW for the AFC in the Space Shuttle. This dramatic improvement was achieved by changing the cell operating conditions to ~150°C and 1.4 MPa (203 psi), as well as by developing improved materials and structures for cell components.

There is an effort in Belgium, spearheaded by Elenco NV, to demonstrate the viability of AFCs in transportation applications. The focus is to develop low cost components for AFCs. A 78-kW AFC stack which operates at about 70°C and uses pure H_2 fuel and recirculating KOH for the electrolyte has been developed. Elenco is collaborating with Ansaldo Richerche (Italy) to supply the electric drive train, Air Products Nederland (the Netherlands) to supply the hydrogen storage capability, and SAFT (France) to supply the Cd/NiOOH batteries for a power source. Demonstrations (Eureka Bus Program) of this technology in city buses (18-m long) in Amsterdam and Brussels are planned in 1994.

The development of AFC technology is also being pursued in Sweden for potential application in utility power generation and transportation. A strong research effort on gas diffusion electrodes and electrocatalysis for AFCs has been undertaken. An analysis performed by the Royal Institute of Technology concluded that CO_2 can be economically scrubbed from fuel gases obtained by coal gasification, thus making large, low cost AFC power plants attractive for utility power generation.

Economic Aspects

From the standpoint of commercialization of fuel cell technologies, there are two challenges: initial cost and reliable life. The initial selling price of the 200-kW PAFC power plant from IFC was about $3500/kW. A competitive price is projected to be about $1500/kW or less for the utility and commercial on-site markets. For transportation applications, cost is also a critical issue. The fuel cell must compete with conventional mass-produced propulsion systems. Furthermore, it is not clear if the manufacturing cost per kilowatt of small fuel cell systems can be lower than the cost of much larger units. The life of a fuel cell stack must be five years minimum for utility applications, and reliable, maintenance-free operation must be achieved over this time period. The projection for the PAFC stack is a five year life, but reliable operation has yet to be demonstrated for this period.

Several activities, if successful, would strongly boost the prospects for fuel cell technology. These include the development of (1) an active electrocatalyst for the direct electrochemical oxidation of methanol; (2) improved electrocatalysts for oxygen reduction; and (3) a more CO-tolerant electrocatalyst for hydrogen. A comprehensive assessment of the research needs for advancing fuel cell technologies, conducted in the 1980s, is available (22).

A viable electrocatalyst operating with minimal polarization for the direct electrochemical oxidation of methanol at low temperature would strongly enhance the competitive position of fuel cell systems for transportation applications. Fuel cells that directly oxidize CH_3OH would eliminate the need for an external reformer in fuel cell systems resulting in a less complex, more lightweight system occupying less volume and having lower cost. Improvement in the performance of PEFCs for transportation applications, which operate close to ambient temperatures and utilize steam-reformed CH_3OH, would be a more CO-tolerant anode electrocatalyst. Such an electrocatalyst would reduce the need to pretreat the steam-reformed CH_3OH to lower the CO content in the anode fuel gas. Platinum–ruthenium alloys show encouraging performance for the direct oxidation of methanol.

For high temperature fuel cells, there is still a strong need to develop lower cost materials for cell components. In the case of SOFCs, improved fabrication processes and materials that permit acceptable performance in fuel cells at lower operating temperatures are also highly desirable.

One factor contributing to the inefficiency of a fuel cell is poor performance of the positive electrode. This accounts for overpotentials of 300–400 mV in low temperature fuel cells. An electrocatalyst that is capable of oxygen reduction at lower overpotentials would benefit the overall efficiency of the fuel cell. Despite extensive efforts expended on electrocatalysis studies of oxygen reduction in fuel cell electrolytes, platinum-based metals are still the best electrocatalysts for low temperature fuel cells.

BIBLIOGRAPHY

"Fuel Cells" in *ECT* 1st ed., Suppl. 2, pp. 355–376, by H. L. Recht, Atomics International; "Fuel Cells" under "Batteries, Primary" in *ECT* 2nd ed., Vol. 3, pp. 139–159, by E. B. Yeager, Western Reserve University; in *ECT* 3rd ed., Vol. 3, pp. 545–568, by E. J. Cairns and R. R. Witherspoon, General Motors Research Laboratories.

1. R. S. Yeo, in R. S. Yeo, T. Katan, and D. T. Chin, eds., *Proceedings of the Symposium on Transport Processes in Electrochemical Systems*, The Electrochemical Society, Inc., Pennington, N.J., 1982, p. 178.
2. F. A. Uribe and co-workers, in D. Scherson and co-workers, eds., *Proceedings of the Workshop on Structural Effects in Electrocatalysis and Oxygen Electrochemistry*, The Electrochemical Society, Inc., Pennington, N.J., 1992, p. 494.
3. A. B. LaConti, A. R. Fragala, and J. R. Boyack, in J. D. E. McIntyre, S. Srinivasan, and F. G. Will, eds., *Proceedings of the Symposium on Electrode Materials and Processes for Energy Conversion and Storage*, The Electrochemical Society, Inc., Pennington, N.J., 1977, p. 354.
4. G. A. Eisman, *J. Power Sources*, **29**, 389 (1990); J. W. Van Zee and co-workers, eds., *Proceedings of the Symposium on Diaphragms, Separators, and Ion-Exchange Membranes*, The Electrochemical Society, Inc., Pennington, N.J., 1986, p. 156.
5. B. R. Ezzell, W. P. Carl, and W. A. Mod, *Industrial Membrane Processes*, AIChE Symposium Series, Vol. 82, no. 248, American Institute of Chemical Engineers, New York, 1986, p. 45.
6. F. T. Bacon, *Electrochim. Acta* **14**, 569 (1969); F. T. Bacon and T. M. Fry, *Proc. Royal Soc. London* **334A**, 427 (1973).
7. D. W. Sheibley and R. A. Martin, *Prog. Batteries Solar Cells* **6**, 155 (1987).
8. S. Sato, in J. R. Selman and co-workers, eds., *Proceedings of the Second Symposium on Molten Carbonate Fuel Cell Technology*, The Electrochemical Society, Inc., Pennington, N.J., 1990, p. 137.

9. A. O. Isenberg, in J. D. E. McIntyre, S. Srinivasan, and F. G. Will, eds., *Proceedings of the Symposium on Electrode Materials and Processes for Energy Conversion and Storage*, The Electrochemical Society, Inc., Pennington, N.J., 1977, p. 682.

10. R. Fiskum, *Proceedings of the 28th Intersociety Energy Conversion Engineering Conference*, Vol. 1, American Chemical Society, Washington, D.C., 1993, p. 1.1171.

11. W. R. Grove, *Phil. Mag.* **14**, 127 (1839).

12. H. A. Liebhafsky and E. J. Cairns, *Fuel Cells and Fuel Batteries*, John Wiley & Sons, Inc., New York, 1968, p. 18.

13. M. Barak, *Adv. Energy Conv.* **6**, 29 (1966).

14. K. V. Kordesch, *J. Electrochem. Soc.* **125**, 77C (1978).

15. A. J. Appleby, *J. Power Sources* **29**, 3 (1990).

16. J. H. Hirschenhoffer, in Ref. 10, p. 1.1163.

17. H. Healy, paper presented at *28th Intersociety Energy Conversion Engineering Conference*, Atlanta, Ga., Aug. 8–13, 1993.

18. *Fuel Cells Technology Status Report*, DOE/METC-92/0276, U.S. Dept. of Energy, Washington, D.C., July 1992.

19. U.S. Pat. 5,045,413 (Sept. 3, 1992), L. G. Marianowski, R. J. Petri, and M. G. Lawson (to Institute of Gas Technology).

20. U.S. Pat. 5,077,148 (Dec. 31, 1991), F. C. Schora and co-workers (to Institute of Gas Technology).

21. N. Q. Minh, *J. Am. Ceram. Soc.* **76**, 563 (1993).

22. *Energy* **11**(1/2), (1986).

KIMIO KINOSHITA
ELTON J. CAIRNS
University of California, Berkeley